Student Solutions Manual

Cindy Trimble & Associates

Beginning & Intermediate Algebra

FOURTH EDITION

Elayn Martin-Gay

Upper Saddle River, NJ 07458

Editorial Director, Mathematics: Christine Hoag
Editor-in-Chief: Paul Murphy
Sponsoring Editor: Mary Beckwith
Assistant Editor: Georgina Brown
Associate Managing Editor: Bayani Mendoza de Leon
Project Manager: Kristy S. Mosch
Art Director: Heather Scott
Supplement Cover Manager: Paul Gourhan
Supplement Cover Designer: Victoria Colotta
Operations Specialist: Ilene Kahn

Pearson Prentice Hall
Pearson Education, Inc.
Upper Saddle River, NJ 07458

Printed in the United States of America
10 9 8 7 6 5 4 3 2 1

ISBN-13: 978-0-13-603081-2 Standalone
ISBN-10: 0-13-603081-5 Standalone
ISBN-13: 978-0-13-603082-9 Value Pack
ISBN-10: 0-13-603082-3 Value Pack

Pearson Education Ltd., London
Pearson Education Singapore, Pte. Ltd.
Pearson Education Canada, Inc.
Pearson Education—Japan
Pearson Education Australia PTY, Limited
Pearson Education North Asia, Ltd., Hong Kong
Pearson Educación de Mexico, S.A. de C.V.
Pearson Education Malaysia, Pte. Ltd.
Pearson Education Upper Saddle River, New Jersey

Contents

Chapter 1

Exercise Set 1.1

Answers will vary on Exercises 1–21.

Section 1.2

Practice Exercises

1. **a.** $5 < 8$ since 5 is to the left of 8 on the number line.

   b. $6 > 4$ since 6 is to the right of 4 on the number line.

   c. $16 < 82$ since 16 is to the left of 82 on the number line.

2. **a.** $9 \geq 3$ is true, since $9 > 3$ is true.

   b. $3 \geq 8$ is false, since neither $3 > 8$ nor $3 = 8$ is true.

   c. $25 \leq 25$ is true, since $25 = 25$ is true.

   d. $4 \leq 14$ is true, since $4 < 14$ is true.

3. **a.** $3 < 8$

   b. $15 \geq 9$

   c. $6 \neq 7$

4. The integer −52 represents owing the bank 52 dollars.

5. **a.** The natural number is 25.

   b. The whole number is 25.

   c. The integers are 25, −15, −99.

   d. The rational numbers are 25, $\frac{7}{3}$, −15, $\frac{-3}{4}$, −3.7, 8.8, −99.

   e. The irrational number is $\sqrt{5}$.

   f. The real numbers are 25, $\frac{7}{3}$, −15, $\frac{-3}{4}$, $\sqrt{5}$, −3.7, 8.8, −99.

6. **a.** $0 < 3$ since 0 is to the left of 3 on the number line.

   b. $15 > -5$ since 15 is to the right of −5 on the number line.

   c. $3 = \frac{12}{4}$ since $\frac{12}{4}$ simplifies to 3.

7. **a.** $|-8| = 8$ since −8 is 8 units from 0 on a number line.

   b. $|9| = 9$ since 9 is 9 units from 0 on a number line.

   c. $|-2.5| = 2.5$ since −2.5 is 2.5 units from 0 on a number line.

   d. $\left|\frac{5}{11}\right| = \frac{5}{11}$ since $\frac{5}{11}$ is $\frac{5}{11}$ unit from 0 on a number line.

   e. $\left|\sqrt{3}\right| = \sqrt{3}$ since $\sqrt{3}$ is $\sqrt{3}$ units from 0 on a number line.

8. **a.** $|8| = |-8|$ since $8 = 8$.

   b. $|-3| > 0$ since $3 > 0$.

   c. $|-7| < |-11|$ since $7 < 11$.

   d. $|3| > |2|$ since $3 > 2$.

   e. $|0| < |-4|$ since $0 < 4$.

Vocabulary and Readiness Check

1. The <u>whole</u> numbers are {0, 1, 2, 3, 4, ...}.

2. The <u>natural</u> numbers are {1, 2, 3, 4, 5, ...}.

3. The symbols ≠, ≤, and > are called <u>inequality</u> symbols.

4. The <u>integers</u> are {..., −3, −2, −1, 0, 1, 2, 3, ...}.

5. The <u>real</u> numbers are {all numbers that correspond to points on the number line}.

6. The <u>rational</u> numbers are $\left\{\frac{a}{b} \middle| a \text{ and } b \text{ are integers}, b \neq 0\right\}$.

7. The <u>irrational</u> numbers are {nonrational numbers that correspond to points on the number line}.

8. The distance between a number b and 0 on a number line is $|b|$.

Exercise Set 1.2

1. $7 > 3$ since 7 is to the right of 3 on the number line.

3. $6.26 = 6.26$

5. $0 < 7$ since 0 is to the left of 7 on the number line.

7. $-2 < 2$ since -2 is to the left of 2 on the number line.

9. $32 < 212$ since 32 is to the left of 212 on the number line.

11. $2631 > 2456$ since 2631 is to the right of 2456 on the number line.

13. $11 \le 11$ is true, since $11 = 11$.

15. $10 > 11$ is false, since 10 is to the left of 11 on the number line.

17. $3 + 8 \ge 3(8)$ is false, since 11 is to the left of 24 on the number line.

19. $7 > 0$ is true, since 7 is to the right of 0 on the number line.

21. $30 \le 45$

23. Eight is less than twelve is written as $8 < 12$.

25. Five is greater than or equal to four is written as $5 \ge 4$.

27. Fifteen is not equal to negative two is written as $15 \ne -2$.

29. 535 represents an altitude of 535 feet.
-8 represents 8 feet below sea level.

31. $-21{,}350$ represents a population decrease of 21,350.

33. 350 represents a deposit of \$350. -126 represents a withdrawal of \$126.

35. The tallest bars represent the greatest number of visitors; 1998, 1999

37. Look for the bars that have heights greater than 280; 1998, 1999, 2000

39. In 2001, there were 279 million visitors. In 2006, there were 273 million visitors.
279 million > 273 million

41. The number 0 belongs to the sets of: whole numbers, integers, rational numbers, and real numbers.

43. The number -2 belongs to the sets of: integers, rational numbers, and real numbers.

45. The number 6 belongs to the sets of: natural numbers, whole numbers, integers, rational numbers, and real numbers.

47. The number $\frac{2}{3}$ belongs to the sets of: rational numbers and real numbers.

49. The number $-\sqrt{5}$ belongs to the sets of: irrational numbers and real numbers.

51. False; rational numbers may be non-integers.

53. True

55. True

57. True

59. False; an irrational number may not be written as a fraction.

61. $-10 > -100$ since -10 is to the right of -100 on the number line.

63. $32 > 5.2$ since 32 is to the right of 5.2 on the number line.

65. $\frac{18}{3} < \frac{24}{3}$ since $6 < 8$.

67. $-51 < -50$ since -51 is to the left of -50 on the number line.

69. $|-5| > -4$ since $5 > -4$.

71. $|-1| = |1|$ since $1 = 1$.

73. $|-2| < |-3|$ since $2 < 3$.

75. $|0| < |-8|$ since $0 < 8$.

77. $-0.04 > -26.7$ since -0.04 is to the right of -26.7 on the number line.

79. The sun is brighter since $-26.7 < -0.04$.

81. The sun is the brightest since -26.7 is to the left of all other numbers listed.

83. $20 \le 25$ has the same meaning as $25 \ge 20$.

85. $6 > 0$ has the same meaning as $0 < 6$.

87. $-12 < -10$ has the same meaning as $-10 > -12$.

89. Answers may vary

Section 1.3

Practice Problems

1. a. $36 = 4 \cdot 9 = 2 \cdot 2 \cdot 3 \cdot 3$

b. $75 = 3 \cdot 25 = 3 \cdot 5 \cdot 5$

2. a. $\frac{63}{72} = \frac{3 \cdot 3 \cdot 7}{2 \cdot 2 \cdot 2 \cdot 3 \cdot 3} = \frac{7}{2 \cdot 2 \cdot 2} = \frac{7}{8}$

b. $\frac{64}{12} = \frac{2 \cdot 2 \cdot 2 \cdot 2 \cdot 2 \cdot 2}{2 \cdot 2 \cdot 3} = \frac{2 \cdot 2 \cdot 2 \cdot 2}{3} = \frac{16}{3}$

c. $\frac{7}{25} = \frac{7}{5 \cdot 5}$
There are no common factors other than 1, so $\frac{7}{25}$ is already in lowest terms.

3. $\frac{3}{8} \cdot \frac{7}{9} = \frac{3 \cdot 7}{8 \cdot 9} = \frac{3 \cdot 7}{2 \cdot 2 \cdot 2 \cdot 3 \cdot 3} = \frac{7}{2 \cdot 2 \cdot 2 \cdot 3} = \frac{7}{24}$

4. a. $\frac{3}{4} \div \frac{4}{9} = \frac{3}{4} \cdot \frac{9}{4} = \frac{3 \cdot 9}{4 \cdot 4} = \frac{27}{16}$

b. $\frac{5}{12} \div 15 = \frac{5}{12} \cdot \frac{1}{15} = \frac{5 \cdot 1}{12 \cdot 15} = \frac{5}{12 \cdot 3 \cdot 5} = \frac{1}{36}$

c. $\frac{7}{6} \div \frac{7}{15} = \frac{7}{6} \cdot \frac{15}{7} = \frac{7 \cdot 15}{6 \cdot 7} = \frac{15}{6} = \frac{3 \cdot 5}{2 \cdot 3} = \frac{5}{2}$

5. a. $\frac{8}{5} - \frac{3}{5} = \frac{8-3}{5} = \frac{5}{5} = 1$

b. $\frac{8}{5} - \frac{2}{5} = \frac{8-2}{5} = \frac{6}{5}$

c. $\frac{3}{5} + \frac{1}{5} = \frac{3+1}{5} = \frac{4}{5}$

d. $\frac{5}{12} + \frac{1}{12} = \frac{5+1}{12} = \frac{6}{12} = \frac{1}{2}$

6. $\frac{2}{3} = \frac{2}{3} \cdot \frac{7}{7} = \frac{2 \cdot 7}{3 \cdot 7} = \frac{14}{21}$

7. a.
$$\begin{aligned} \frac{5}{11} + \frac{1}{7} &= \frac{5 \cdot 7}{11 \cdot 7} + \frac{1 \cdot 11}{7 \cdot 11} \\ &= \frac{35}{77} + \frac{11}{77} \\ &= \frac{35+11}{77} \\ &= \frac{46}{77} \end{aligned}$$

b.
$$\begin{array}{rcrcr} 9\frac{1}{13} & = & 9\frac{2}{26} & = & 8\frac{28}{26} \\ -5\frac{1}{2} & = & -5\frac{13}{26} & = & -5\frac{13}{26} \\ \hline & & & & 3\frac{15}{26} \end{array}$$

c.
$$\begin{aligned} \frac{1}{3} + \frac{29}{30} - \frac{4}{5} &= \frac{10}{30} + \frac{29}{30} - \frac{4 \cdot 6}{5 \cdot 6} \\ &= \frac{10+29}{30} - \frac{24}{30} \\ &= \frac{39-24}{30} \\ &= \frac{15}{30} \\ &= \frac{1}{2} \end{aligned}$$

Vocabulary and Readiness Check

1. A quotient of two numbers, such as $\frac{5}{8}$, is called a <u>fraction</u>.

2. In the fraction $\frac{3}{11}$, the number 3 is called the <u>numerator</u> and the number 11 is called the <u>denominator</u>.

3. To factor a number means to write it as a <u>product</u>.

4. A fraction is said to be <u>simplified</u> when the numerator and the denominator have no common factors other than 1.

5. In $7 \cdot 3 = 21$, the numbers 7 and 3 are called <u>factors</u> and the number 21 is called the <u>product</u>.

6. The fractions $\frac{2}{9}$ and $\frac{9}{2}$ are called <u>reciprocals</u>.

7. Fractions that represent the same quantity are called <u>equivalent</u> fractions.

8. 3 of the 8 equal parts are shaded; $\frac{3}{8}$

9. 1 of the 4 equal parts are shaded; $\frac{1}{4}$

10. 5 of the 7 equal parts are shaded; $\frac{5}{7}$

11. 2 of the 5 equal parts are shaded; $\frac{2}{5}$

Exercise Set 1.3

1. $33 = 3 \cdot 11$

3. $98 = 2 \cdot 49 = 2 \cdot 7 \cdot 7$

5. $20 = 4 \cdot 5 = 2 \cdot 2 \cdot 5$

7. $75 = 3 \cdot 25 = 3 \cdot 5 \cdot 5$

9. $45 = 9 \cdot 5 = 3 \cdot 3 \cdot 5$

11. $\frac{2}{4} = \frac{2}{2 \cdot 2} = \frac{1}{2}$

13. $\frac{10}{15} = \frac{2 \cdot 5}{3 \cdot 5} = \frac{2}{3}$

15. $\frac{3}{7} = \frac{3}{7}$

17. $\frac{18}{30} = \frac{2 \cdot 3 \cdot 3}{2 \cdot 3 \cdot 5} = \frac{3}{5}$

19. $\frac{1}{2} \cdot \frac{3}{4} = \frac{1 \cdot 3}{2 \cdot 4} = \frac{1 \cdot 3}{2 \cdot 2 \cdot 2} = \frac{3}{8}$

21. $\frac{2}{3} \cdot \frac{3}{4} = \frac{2 \cdot 3}{3 \cdot 4} = \frac{2 \cdot 3}{3 \cdot 2 \cdot 2} = \frac{1}{2}$

23. $\frac{1}{2} \div \frac{7}{12} = \frac{1}{2} \cdot \frac{12}{7} = \frac{1 \cdot 12}{2 \cdot 7} = \frac{1 \cdot 2 \cdot 2 \cdot 3}{2 \cdot 7} = \frac{2 \cdot 3}{7} = \frac{6}{7}$

25. $\frac{3}{4} \div \frac{1}{20} = \frac{3}{4} \cdot \frac{20}{1} = \frac{3 \cdot 20}{4 \cdot 1} = \frac{3 \cdot 2 \cdot 2 \cdot 5}{2 \cdot 2} = \frac{3 \cdot 5}{1} = 15$

27. $\frac{7}{10} \cdot \frac{5}{21} = \frac{7 \cdot 5}{10 \cdot 21} = \frac{7 \cdot 5}{2 \cdot 5 \cdot 3 \cdot 7} = \frac{1}{2 \cdot 3} = \frac{1}{6}$

29. $2\frac{7}{9} \cdot \frac{1}{3} = \frac{25}{9} \cdot \frac{1}{3} = \frac{25 \cdot 1}{9 \cdot 3} = \frac{5 \cdot 5 \cdot 1}{3 \cdot 3 \cdot 3} = \frac{25}{27}$

31.
$$\begin{aligned} \text{Area} &= \frac{11}{12} \cdot \frac{3}{5} \\ &= \frac{11 \cdot 3}{12 \cdot 5} \\ &= \frac{11 \cdot 3}{2 \cdot 2 \cdot 3 \cdot 5} \\ &= \frac{11}{2 \cdot 2 \cdot 5} \\ &= \frac{11}{20} \text{ sq mi} \end{aligned}$$

33. $\frac{4}{5} - \frac{1}{5} = \frac{4-1}{5} = \frac{3}{5}$

35. $\frac{4}{5} + \frac{1}{5} = \frac{4+1}{5} = \frac{5}{5} = 1$

37. $\frac{17}{21} - \frac{10}{21} = \frac{17-10}{21} = \frac{7}{21} = \frac{7}{3 \cdot 7} = \frac{1}{3}$

39. $\frac{23}{105} + \frac{4}{105} = \frac{23+4}{105} = \frac{27}{105} = \frac{3 \cdot 3 \cdot 3}{3 \cdot 5 \cdot 7} = \frac{3 \cdot 3}{5 \cdot 7} = \frac{9}{35}$

41. $\frac{7}{10} = \frac{7 \cdot 3}{10 \cdot 3} = \frac{21}{30}$

43. $\frac{2}{9} = \frac{2 \cdot 2}{9 \cdot 2} = \frac{4}{18}$

45. $\frac{4}{5} = \frac{4 \cdot 4}{5 \cdot 4} = \frac{16}{20}$

47. $\frac{2}{3}+\frac{3}{7}=\frac{2\cdot 7}{3\cdot 7}+\frac{3\cdot 3}{7\cdot 3}=\frac{14}{21}+\frac{9}{21}=\frac{23}{21}$

49. $$\begin{aligned}2\frac{13}{15}-1\frac{1}{5}&=\frac{43}{15}-\frac{6}{5}\\&=\frac{43}{15}-\frac{6\cdot 3}{5\cdot 3}\\&=\frac{43-18}{15}\\&=\frac{25}{15}\\&=1\frac{2}{3}\end{aligned}$$

51. $$\begin{aligned}\frac{5}{22}-\frac{5}{33}&=\frac{5\cdot 3}{22\cdot 3}-\frac{5\cdot 2}{33\cdot 2}\\&=\frac{15}{66}-\frac{10}{66}\\&=\frac{15-10}{66}\\&=\frac{5}{66}\end{aligned}$$

53. $\frac{12}{5}-1=\frac{12}{5}-\frac{5}{5}=\frac{12-5}{5}=\frac{7}{5}$

55. $$\begin{aligned}1-\frac{3}{10}-\frac{5}{10}&=\frac{10}{10}-\frac{3}{10}-\frac{5}{10}\\&=\frac{10-3-5}{10}\\&=\frac{2}{10}\\&=\frac{2}{2\cdot 5}\\&=\frac{1}{5}\end{aligned}$$

The unknown part is $\frac{1}{5}$.

57. $1-\frac{1}{4}-\frac{3}{8}=\frac{8}{8}-\frac{1\cdot 2}{4\cdot 2}-\frac{3}{8}=\frac{8-2-3}{8}=\frac{3}{8}$

The unknown part is $\frac{3}{8}$.

59. $$\begin{aligned}1-\frac{1}{2}-\frac{1}{6}-\frac{2}{9}&=\frac{18}{18}-\frac{1\cdot 9}{2\cdot 9}-\frac{1\cdot 3}{6\cdot 3}-\frac{2\cdot 2}{9\cdot 2}\\&=\frac{18-9-3-4}{18}\\&=\frac{2}{18}\\&=\frac{1}{9}\end{aligned}$$

The unknown part is $\frac{1}{9}$.

61. $\frac{10}{21}+\frac{5}{21}=\frac{10+5}{21}=\frac{15}{21}=\frac{3\cdot 5}{3\cdot 7}=\frac{5}{7}$

63. $\frac{10}{3}-\frac{5}{21}=\frac{10\cdot 7}{3\cdot 7}-\frac{5}{21}=\frac{70}{21}-\frac{5}{21}=\frac{65}{21}$

65. $\frac{2}{3}\cdot\frac{3}{5}=\frac{2\cdot 3}{3\cdot 5}=\frac{2}{5}$

67. $\frac{3}{4}\div\frac{7}{12}=\frac{3}{4}\cdot\frac{12}{7}=\frac{3\cdot 12}{4\cdot 7}=\frac{3\cdot 3\cdot 4}{4\cdot 7}=\frac{9}{7}$

69. $\frac{5}{12}+\frac{4}{12}=\frac{5+4}{12}=\frac{9}{12}=\frac{3\cdot 3}{3\cdot 4}=\frac{3}{4}$

71. $5+\frac{2}{3}=\frac{15}{3}+\frac{2}{3}=\frac{15+2}{3}=\frac{17}{3}$

73. $\frac{7}{8}\div 3\frac{1}{4}=\frac{7}{8}\div\frac{13}{4}=\frac{7}{8}\cdot\frac{4}{13}=\frac{7\cdot 4}{8\cdot 13}=\frac{7\cdot 4}{2\cdot 4\cdot 13}=\frac{7}{26}$

75. $\frac{7}{18}\div\frac{14}{36}=\frac{7}{18}\cdot\frac{36}{14}=\frac{7\cdot 36}{18\cdot 14}=\frac{7\cdot 2\cdot 18}{18\cdot 2\cdot 7}=1$

77. $\frac{23}{105}-\frac{2}{105}=\frac{23-2}{105}=\frac{21}{105}=\frac{21}{21\cdot 5}=\frac{1}{5}$

79. $1\frac{1}{2}+3\frac{2}{3}=\frac{3}{2}+\frac{11}{3}$
$=\frac{3\cdot 3}{2\cdot 3}+\frac{11\cdot 2}{3\cdot 2}$
$=\frac{9}{6}+\frac{22}{6}$
$=\frac{9+22}{6}$
$=\frac{31}{6}$
$=5\frac{1}{6}$

81. $\frac{2}{3}-\frac{5}{9}+\frac{5}{6}=\frac{2\cdot 2\cdot 3}{3\cdot 2\cdot 3}-\frac{5\cdot 2}{9\cdot 2}+\frac{5\cdot 3}{6\cdot 3}$
$=\frac{12}{18}-\frac{10}{18}+\frac{15}{18}$
$=\frac{12-10+15}{18}$
$=\frac{17}{18}$

83. $5+4\frac{1}{8}+4\frac{1}{8}+15\frac{3}{4}+15\frac{3}{4}+10\frac{1}{2}$
$=\frac{40}{8}+\frac{33}{8}+\frac{33}{8}+\frac{126}{8}+\frac{126}{8}+\frac{84}{8}$
$=\frac{40+33+33+126+126+84}{8}$
$=\frac{442}{8}$
$=55\frac{1}{4}$ feet

85. $5\frac{1}{50}+1\frac{3}{25}=5\frac{1}{50}+1\frac{6}{50}=6\frac{7}{50}$ meters

87. Answers may vary

89. $5\frac{1}{2}-2\frac{1}{8}=\frac{11}{2}-\frac{17}{8}$
$=\frac{11\cdot 4}{2\cdot 4}-\frac{17}{8}$
$=\frac{44}{8}-\frac{17}{8}$
$=\frac{44-17}{8}$
$=\frac{27}{8}$
$=3\frac{3}{8}$ miles

91. $\frac{7}{50}$ are in the physical sciences.

93. $1-\frac{4}{25}-\frac{7}{50}-\frac{7}{50}-\frac{7}{100}-\frac{21}{100}-\frac{3}{100}$
$=\frac{100}{100}-\frac{4\cdot 4}{25\cdot 4}-\frac{7\cdot 2}{50\cdot 2}-\frac{7\cdot 2}{50\cdot 2}-\frac{7}{100}-\frac{21}{100}-\frac{3}{100}$
$=\frac{100-16-14-14-7-21-3}{100}$
$=\frac{25}{100}$
$=\frac{1}{4}$

$\frac{1}{4}$ are in the biological and agricultural sciences.

95. $\frac{960}{3054}=\frac{6\cdot 160}{6\cdot 509}=\frac{160}{509}$ were Old Navy stores.

97. Area $=\frac{1}{2}bh=\frac{1}{2}\cdot\frac{7}{8}\cdot\frac{4}{9}=\frac{7\cdot 4}{2\cdot 2\cdot 4\cdot 9}=\frac{7}{36}$ sq ft

Section 1.4

Practice Exercises

1. a. $1^3=1\cdot 1\cdot 1=1$

b. $5^2=5\cdot 5=25$

c. $\left(\frac{1}{10}\right)^2=\left(\frac{1}{10}\right)\left(\frac{1}{10}\right)=\frac{1}{100}$

d. $9^1=9$

e. $\left(\frac{2}{5}\right)^3=\left(\frac{2}{5}\right)\left(\frac{2}{5}\right)\left(\frac{2}{5}\right)=\frac{8}{125}$

2. a. $6+3\cdot 9=6+27=33$

b. $4^3\div 8+3=64\div 8+3=8+3=11$

c. $\left(\frac{2}{3}\right)^2\cdot|-8|=\frac{4}{9}\cdot 8=\frac{32}{9}$ or $3\frac{5}{9}$

d. $\frac{9(14-6)}{|-2|}=\frac{9(8)}{2}=\frac{72}{2}=36$

e. $\frac{7}{4}\cdot\frac{1}{4}-\frac{1}{4}=\frac{7}{16}-\frac{4}{16}=\frac{3}{16}$

3. $\frac{6^2-5}{3+|6-5|\cdot 8}=\frac{36-5}{3+|1|\cdot 8}=\frac{31}{3+8}=\frac{31}{11}$

4. $$\begin{aligned}4[25-3(5+3)] &= 4[25-3(8)]\\ &= 4[25-24]\\ &= 4[1]\\ &= 4\end{aligned}$$

5. $\frac{36\div 9+5}{5^2-3}=\frac{4+5}{25-3}=\frac{9}{22}$

6. a. $2x+y=2(2)+5=4+5=9$

b. $\frac{4x}{3y}=\frac{4(2)}{3(5)}=\frac{8}{15}$

c. $\frac{3}{x}+\frac{x}{y}=\frac{3}{2}+\frac{2}{5}=\frac{15}{10}+\frac{4}{10}=\frac{19}{10}$

d. $x^3+y^2=2^3+5^2=8+25=33$

7. $$\begin{aligned}9x-6&=7x\\ 9(4)-6&\stackrel{?}{=}7(4)\\ 36-6&\stackrel{?}{=}28\\ 30&=28 \quad \text{False}\end{aligned}$$
4 is not a solution of $9x-6=7x$.

8. a. Six times a number is $6x$, since $6x$ denotes the product of 6 and x.

b. A number decreased by 8 is $x-8$ because "decreased by" means subtract.

c. The product of a number and 9 is $x\cdot 9$ or $9x$.

d. Two times a number is $2x$, plus 3 is $2x+3$.

e. The sum of 7 and a number x is $7+x$.

9. a. A number x increased by 7 is $x+7$, so $x+7=13$.

b. Two less than a number x is $x-2$, so $x-2=11$.

c. Double a number x is $2x$, added to 9 is $2x+9$, so $2x+9\neq 25$.

d. Five times 11 is 5(11), so $5(11)\geq x$, where x is an unknown number.

Calculator Explorations

1. $5^4=625$
2. $7^4=2401$
3. $9^5=59{,}049$
4. $8^6=262{,}144$
5. $2(20-5)=30$
6. $3(14-7)+21=3(7)+21=21+21=42$
7. $24(862-455)+89=9857$
8. $99+(401+962)=1462$
9. $\frac{4623+129}{36-34}=2376$
10. $\frac{956-452}{89-86}=168$

Vocabulary and Readiness Check

1. In the expression 5^2, the 5 is called the base and the 2 is called the exponent.
2. The symbols ( ), [ ], and { } are examples of grouping symbols.
3. A symbol that is used to represent a number is called a variable.
4. A collection of numbers, variables, operation symbols, and grouping symbols is called an expression.
5. A mathematical statement that two expressions are equal is called an equation.
6. A value for the variable that makes an equation a true statement is called a solution.
7. Deciding what values of a variable make an equation a true statement is called solving the equation.
8. To simplify the expression $1+3\cdot 6$, first multiply.

9. To simplify the expression (1 + 3) · 6, first add.

10. To simplify the expression (20 – 4) · 2, first subtract.

11. To simplify the expression 20 – 4 ÷ 2, first divide.

Exercise Set 1.4

1. $3^5 = 3 \cdot 3 \cdot 3 \cdot 3 \cdot 3 = 243$

3. $3^3 = 3 \cdot 3 \cdot 3 = 27$

5. $1^5 = 1 \cdot 1 \cdot 1 \cdot 1 \cdot 1 = 1$

7. $5^1 = 5$

9. $\left(\frac{1}{5}\right)^3 = \left(\frac{1}{5}\right)\left(\frac{1}{5}\right)\left(\frac{1}{5}\right) = \frac{1 \cdot 1 \cdot 1}{5 \cdot 5 \cdot 5} = \frac{1}{125}$

11. $\left(\frac{2}{3}\right)^4 = \left(\frac{2}{3}\right)\left(\frac{2}{3}\right)\left(\frac{2}{3}\right)\left(\frac{2}{3}\right) = \frac{2 \cdot 2 \cdot 2 \cdot 2}{3 \cdot 3 \cdot 3 \cdot 3} = \frac{16}{81}$

13. $7^2 = 7 \cdot 7 = 49$

15. $4^2 = 4 \cdot 4 = 16$

17. $(1.2)^2 = (1.2) \cdot (1.2) = 1.44$

19. $5 + 6 \cdot 2 = 5 + 12 = 17$

21. $4 \cdot 8 - 6 \cdot 2 = 32 - 12 = 20$

23. $2(8 - 3) = 2(5) = 10$

25. $2 + (5 - 2) + 4^2 = 2 + 3 + 4^2 = 2 + 3 + 16 = 21$

27. $5 \cdot 3^2 = 5 \cdot 9 = 45$

29. $\frac{1}{4} \cdot \frac{2}{3} - \frac{1}{6} = \frac{2}{12} - \frac{1}{6} = \frac{1}{6} - \frac{1}{6} = 0$

31. $\frac{6-4}{9-2} = \frac{2}{7}$

33.
$$\begin{aligned} 2[5 + 2(8 - 3)] &= 2[5 + 2(5)] \\ &= 2[5 + 10] \\ &= 2[15] \\ &= 30 \end{aligned}$$

35. $\frac{19 - 3 \cdot 5}{6 - 4} = \frac{19 - 15}{6 - 4} = \frac{4}{2} = 2$

37. $\frac{|6-2|+3}{8+2 \cdot 5} = \frac{|4|+3}{8+2 \cdot 5} = \frac{4+3}{8+2 \cdot 5} = \frac{4+3}{8+10} = \frac{7}{18}$

39. $\frac{3+3(5+3)}{3^2+1} = \frac{3+3(8)}{3^2+1} = \frac{3+3(8)}{9+1} = \frac{3+24}{9+1} = \frac{27}{10}$

41.
$$\begin{aligned} \frac{6+|8-2|+3^2}{18-3} &= \frac{6+|6|+3^2}{18-3} \\ &= \frac{6+6+3^2}{18-3} \\ &= \frac{6+6+9}{18-3} \\ &= \frac{21}{15} \\ &= \frac{3 \cdot 7}{3 \cdot 5} \\ &= \frac{7}{5} \end{aligned}$$

43. No; since in the absence of grouping symbols we always perform multiplications or divisions before additions or subtractions in any expression.

45. **a.** $(6 + 2) \cdot (5 + 3) = 8 \cdot 8 = 64$

b. $(6 + 2) \cdot 5 + 3 = 8 \cdot 5 + 3 = 40 + 3 = 43$

c. $6 + 2 \cdot 5 + 3 = 6 + 10 + 3 = 19$

d. $6 + 2 \cdot (5 + 3) = 6 + 2 \cdot 8 = 6 + 16 = 22$

47. Let $y = 3$.
$3y = 3(3) = 9$

49. Let $x = 1$ and $z = 5$.
$\frac{z}{5x} = \frac{5}{5(1)} = \frac{5}{5} = 1$

51. Let $x = 1$.
$3x - 2 = 3(1) - 2 = 3 - 2 = 1$

53. Let $x = 1$ and $y = 3$.
$|2x + 3y| = |2(1) + 3(3)| = |2 + 9| = |11| = 11$

55. Let $y = 3$.
$5y^2 = 5(3)^2 = 5(9) = 45$

57. Let $x = 12$, $y = 8$ and $z = 4$.

$$\frac{x}{z} + 3y = \frac{12}{4} + 3(8) = 3 + 24 = 27$$

59. Let $x = 12$ and $y = 8$.

$$\begin{aligned} x^2 - 3y + x &= (12)^2 - 3(8) + 12 \\ &= 144 - 24 + 12 \\ &= 132 \end{aligned}$$

61. Let $x = 12$, $y = 8$ and $z = 4$.

$$\frac{x^2 + z}{y^2 + 2z} = \frac{(12)^2 + 4}{(8)^2 + 2(4)} = \frac{144 + 4}{64 + 8} = \frac{148}{72} = \frac{37}{18}$$

63. Evaluate $16t^2$ for each value of t.

$t = 1$: $16(1)^2 = 16(1) = 16$

$t = 2$: $16(2)^2 = 16(4) = 64$

$t = 3$: $16(3)^2 = 16(9) = 144$

$t = 4$: $16(4)^2 = 16(16) = 256$

Time t (in seconds)	Distance $16t^2$ (in feet)
1	16
2	64
3	144
4	256

65. Let $x = 5$.

$$\begin{aligned} 3x + 30 &= 9x \\ 3(5) + 30 &\stackrel{?}{=} 9(5) \\ 15 + 30 &\stackrel{?}{=} 45 \\ 45 &= 45, \text{ true} \end{aligned}$$

5 is a solution of the equation.

67. Let $x = 0$.

$$\begin{aligned} 2x + 6 &= 5x - 1 \\ 2(0) + 6 &\stackrel{?}{=} 5(0) - 1 \\ 0 + 6 &\stackrel{?}{=} 0 - 1 \\ 6 &= -1, \text{ false} \end{aligned}$$

0 is not a solution of the equation.

69. Let $x = 8$.

$$\begin{aligned} 2x - 5 &= 5 \\ 2(8) - 5 &\stackrel{?}{=} 5 \\ 16 - 5 &\stackrel{?}{=} 5 \\ 9 &= 5, \text{ false} \end{aligned}$$

8 is not a solution of the equation.

71. Let $x = 2$.

$$\begin{aligned} x + 6 &= x + 6 \\ 2 + 6 &\stackrel{?}{=} 2 + 6 \\ 8 &= 8, \text{ true} \end{aligned}$$

2 is a solution of the equation.

73. Let $x = 0$.

$$\begin{aligned} x &= 5x + 15 \\ (0) &\stackrel{?}{=} 5(0) + 15 \\ 0 &\stackrel{?}{=} 0 + 15 \\ 0 &= 15, \text{ false} \end{aligned}$$

0 is not a solution of the equation.

75. More than means to add; $x + 15$

77. Five subtracted from a number is represented as $x - 5$.

79. Three times a number increased by 22 is represented by $3x + 22$.

81. One increased by two equals the quotient of nine and three is $1 + 2 = 9 \div 3$.

83. Three is not equal to four divided by two is represented by $3 \neq 4 \div 2$.

85. The sum of 5 and a number is 20 is represented by $5 + x = 20$.

87. Thirteen minus three times a number is 13 is represented by $13 - 3x = 13$.

89. The quotient of 12 and a number is $\frac{1}{2}$ is represented by $\frac{12}{x} = \frac{1}{2}$.

91. Answers may vary.

93. $(20 - 4) \cdot 4 \div 2 = (16) \cdot 4 \div 2 = 64 \div 2 = 32$

95. Let $l = 8$ and $w = 6$.
$2l + 2w = 2(8) + 2(6) = 16 + 12 = 28$ m

97. Let $l = 120$ and $w = 100$.
$lw = (120)(100) = 12{,}000$ sq ft

99. Let $P = 650$, $T = 3$, and $I = 126.75$.

$$\frac{I}{PT} = \frac{126.75}{(650)(3)} = \frac{126.75}{1950} = 0.065 = 6.5\%$$

101. Let $m = 84$
$3.00 + 0.12(84) = 3.00 + 10.08 = \$13.08$

Section 1.5

Practice Exercises

1. 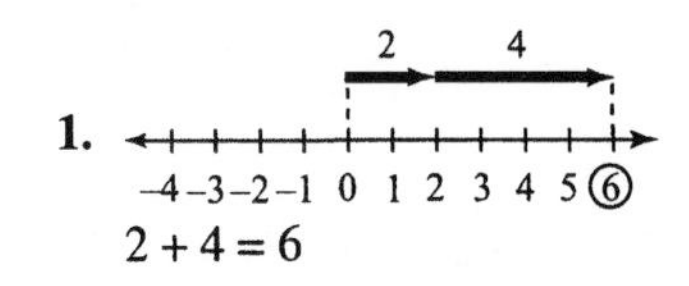

$2+4=6$

2. −3 −2
−5 −4 −3 −2 −1 0 1 2 3 4 5
$-2+(-3)=-5$

3. **a.** $-5+(-8)$
Add the absolute values.
$5+8=13$
The common sign is negative, so
$-5+(-8)=-13$.

b. $-31+(-1)$
Add the absolute values.
$31+1=32$
The common sign is negative, so
$-31+(-1)=-32$.

4. 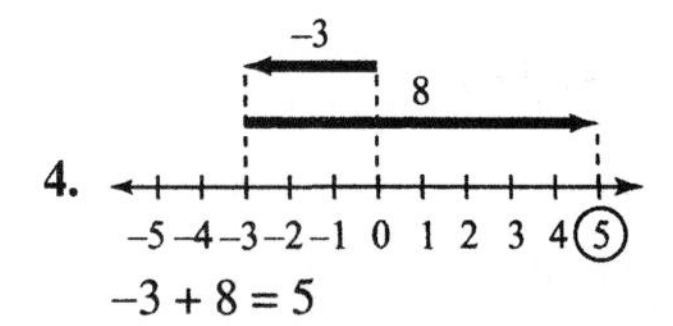

$-3+8=5$

5. **a.** $15+(-18)$
Subtract the absolute values.
$18-15=3$
Use the sign of the number with the largest absolute value.
$15+(-18)=-3$

b. $-19+20=20-19=1$

c. $-0.6+0.4=-(0.6-0.4)=-(0.2)=-0.2$

6. **a.** $-\frac{3}{5}+\left(-\frac{2}{5}\right)=-\frac{5}{5}=-1$

b. $3+(-9)=-6$

c. $2.2+(-1.7)=0.5$

d. $-\frac{2}{7}+\frac{3}{10}=-\frac{20}{70}+\frac{21}{70}=\frac{1}{70}$

7. **a.** $8+(-5)+(-9)=3+(-9)=-6$

b. $[-8+5]+[-5+|-2|]=[-3]+[-5+2]$
$=-3+[-3]$
$=-6$

8. $-5+8+(-2)=3+(-2)=1$
The overall gain is \$1.

9. **a.** The opposite of $-\frac{5}{9}$ is $\frac{5}{9}$.

b. The opposite of 8 is −8.

c. The opposite of 6.2 is −6.2.

d. The opposite of −3 is 3.

10. **a.** Since $|-15|=15$, $-|-15|=-15$.

b. $-\left(-\frac{3}{5}\right)=\frac{3}{5}$

c. $-(-5y)=5y$

d. $-(-8)=8$

Vocabulary and Readiness Check

1. Two numbers that are the same distance from 0 but lie on opposite sides of 0 are called <u>opposites</u>.

2. The sum of a number and its opposite is always <u>0</u>.

3. If n is a number, then $-(-n)=\underline{n}$.

4. $-80+(-127)=$ <u>negative number</u>.

5. $-162+164=$ <u>positive number</u>.

6. $-162+162=\underline{0}$.

7. $-1.26+(-8.3)=$ <u>negative number</u>.

8. $-3.68+0.27=$ <u>negative number</u>.

9. $-\frac{2}{3}+\frac{2}{3}=\underline{0}$.

Exercise Set 1.5

1. $6+3=9$

3. $-6+(-8)=-14$

5. $8 + (-7) = 1$

7. $-14 + 2 = -12$

9. $-2 + (-3) = -5$

11. $-9 + (-3) = -12$

13. $-7 + 3 = -4$

15. $10 + (-3) = 7$

17. $5 + (-7) = -2$

19. $-16 + 16 = 0$

21. $27 + (-46) = -19$

23. $-18 + 49 = 31$

25. $-33 + (-14) = -47$

27. $6.3 + (-8.4) = -2.1$

29. $|-8| + (-16) = 8 + (-16) = -8$

31. $117 + (-79) = 38$

33. $-9.6 + (-3.5) = -13.1$

35. $-\frac{3}{8} + \frac{5}{8} = \frac{2}{8} = \frac{1}{4}$

37. $-\frac{7}{16} + \frac{1}{4} = -\frac{7}{16} + \frac{1 \cdot 4}{4 \cdot 4} = -\frac{7}{16} + \frac{4}{16} = -\frac{3}{16}$

39.
$$\begin{aligned} -\frac{7}{10} + \left(-\frac{3}{5}\right) &= -\frac{7}{10} + \left(-\frac{3 \cdot 2}{5 \cdot 2}\right) \\ &= -\frac{7}{10} + \left(-\frac{6}{10}\right) \\ &= -\frac{13}{10} \end{aligned}$$

41. $-15 + 9 + (-2) = -6 + (-2) = -8$

43. $-21 + (-16) + (-22) = -37 + (-22) = -59$

45. $-23 + 16 + (-2) = -7 + (-2) = -9$

47. $|5 + (-10)| = |-5| = 5$

49. $6 + (-4) + 9 = 2 + 9 = 11$

51. $[-17 + (-4)] + [-12 + 15] = [-21] + [3] = -18$

53. $|9 + (-12)| + |-16| = |-3| + 16 = 3 + 16 = 19$

55.
$$\begin{aligned} -1.3 + [0.5 + (-0.3) + 0.4] &= -1.3 + [0.2 + 0.4] \\ &= -1.3 + [0.6] \\ &= -0.7 \end{aligned}$$

57. $-15 + 9 = -6$
The high temperature in Anoka was $-6°$.

59. $-17{,}657 + 1230 = -16{,}427$
You are 16,427 feet below the rim.

61.
$$\begin{aligned} (-155) + (-3895) + (-5200) &= (-4050) + (-5200) \\ &= -9250 \end{aligned}$$
The total net income was $-\$9250$ million.

63. $(-5) + (-2) + (-2) = (-7) + (-2) = -9$
Her score was 9 under par.

65. The opposite of 6 is -6.

67. The opposite of -2 is 2.

69. The opposite of 0 is 0.

71. Since $|-6|$ is 6, the opposite of $|-6|$ is -6.

73. Answers may vary

75. $-|-2| = -2$

77. $-|0| = -0 = 0$

79. $-\left|-\frac{2}{3}\right| = -\frac{2}{3}$

81. Answers may vary

83. Let $x = -4$.
$$\begin{aligned} x + 9 &= 5 \\ (-4) + 9 &\stackrel{?}{=} 5 \\ 5 &= 5, \text{ true} \end{aligned}$$
-4 is a solution of the equation.

85. Let $y = -1$.
$$\begin{aligned} y + (-3) &= -7 \\ (-1) + (-3) &\stackrel{?}{=} -7 \\ -4 &= -7, \text{ false} \end{aligned}$$
-1 is not a solution of the equation.

87. Look for the tallest bar. The temperature is the highest in July.

89. Look for the bar whose length has a positive value closest to 0; October

91. $[(-9.1)+14.4+8.8]\div 3=[5.3+8.8]\div 3$
$=[14.1]\div 3$
$=4.7$

The average was 4.7°F.

93. Since a is a positive number, $-a$ is a <u>negative</u> number.

95. Since a is a positive number, $a + a$ is a <u>positive</u> number.

Section 1.6

Practice Exercises

1. a. $-7-6=-7+(-6)=-13$

b. $-8-(-1)=-8+1=-7$

c. $9-(-3)=9+3=12$

d. $5-7=5+(-7)=-2$

2. a. $8.4-(-2.5)=8.4+2.5=10.9$

b. $-\frac{5}{8}-\left(-\frac{1}{8}\right)=-\frac{5}{8}+\frac{1}{8}=-\frac{4}{8}=-\frac{1}{2}$

c. $-\frac{3}{4}-\frac{1}{5}=-\frac{3}{4}+\left(-\frac{1}{5}\right)$
$=-\frac{15}{20}+\left(-\frac{4}{20}\right)$
$=-\frac{19}{20}$

3. $-2-5=-2+(-5)=-7$

4. a. $-15-2-(-4)+7=-15+(-2)+4+7=-6$

b. $3.5+(-4.1)-(-6.7)=3.5+(-4.1)+6.7$
$=6.1$

5. a. $-4+[(-8-3)-5]=-4+[(-8+(-3))-5]$
$=-4+[(-11)-5]$
$=-4+[-11+(-5)]$
$=-4+[-16]$
$=-20$

b. $|-13|-3^2+[2-(-7)]=13-9+[2+7]$
$=13-9+9$
$=13$

6. a. $\frac{7-x}{2y+x}=\frac{7-(-3)}{2(4)+(-3)}=\frac{7+3}{8+(-3)}=\frac{10}{5}=2$

b. $y^2+x=(4)^2+(-3)=16+(-3)=13$

7. $282-(-75)=282+75=\$357$

8. a. $x=90°-62°=28°$

b. $y=180°-43°=137°$

Vocabulary and Readiness Check

1. 7 minus a number <u>$7-x$</u>

2. 7 subtracted from a number <u>$x-7$</u>.

3. A number decreased by 7 <u>$x-7$</u>

4. 7 less a number <u>$7-x$</u>

5. A number less than 7 <u>$7-x$</u>

6. A number subtracted from 7 <u>$7-x$</u>

Exercise Set 1.6

1. $-6-4=-6+(-4)=-10$

3. $4-9=4+(-9)=-5$

5. $16-(-3)=16+3=19$

7. $\frac{1}{2}-\frac{1}{3}=\frac{1}{2}+\left(-\frac{1}{3}\right)$
$=\frac{1\cdot 3}{2\cdot 3}+\left(-\frac{1\cdot 2}{3\cdot 2}\right)$
$=\frac{3}{6}+\left(-\frac{2}{6}\right)$
$=\frac{1}{6}$

9. $-16-(-18)=-16+18=2$

11. $-6-5=-6+(-5)=-11$

13. $7-(-4)=7+4=11$

15. $-6-(-11)=-6+11=5$

17. $16-(-21)=16+21=37$

19. $9.7-16.1=9.7+(-16.1)=-6.4$

21. $-44-27=-44+(-27)=-71$

23. $-21-(-21)=-21+21=0$

25. $-2.6-(-6.7)=-2.6+6.7=4.1$

27. $-\frac{3}{11}-\left(-\frac{5}{11}\right)=-\frac{3}{11}+\frac{5}{11}=\frac{2}{11}$

29. $$\begin{aligned}-\frac{1}{6}-\frac{3}{4}&=-\frac{1}{6}+\left(-\frac{3}{4}\right)\\&=-\frac{1\cdot 2}{6\cdot 2}+\left(-\frac{3\cdot 3}{4\cdot 3}\right)\\&=-\frac{2}{12}+\left(-\frac{9}{12}\right)\\&=-\frac{11}{12}\end{aligned}$$

31. $8.3-(-0.62)=8.3+0.62=8.92$

33. $8-(-5)=8+5=13$

35. $-6-(-1)=-6+1=-5$

37. $7-8=7+(-8)=-1$

39. $-8-15=-8+(-15)=-23$

41. Answers may vary

43. $$\begin{aligned}-10-(-8)+(-4)-20&=-10+8+(-4)+(-20)\\&=-2+(-4)+(-20)\\&=-6+(-20)\\&=-26\end{aligned}$$

45. $$\begin{aligned}5-9+(-4)-8-8&=5+(-9)+(-4)+(-8)+(-8)\\&=-4+(-4)+(-8)+(-8)\\&=-8+(-8)+(-8)\\&=-16+(-8)\\&=-24\end{aligned}$$

47. $-6-(2-11)=-6-(-9)=-6+9=3$

49. $3^3-8\cdot 9=27-8\cdot 9=27-72=27+(-72)=-45$

51. $2-3(8-6)=2-3(2)=2-6=2+(-6)=-4$

53. $$\begin{aligned}(3-6)+4^2&=[3+(-6)]+4^2\\&=[-3]+4^2\\&=[-3]+16\\&=13\end{aligned}$$

55. $$\begin{aligned}&-2+[(8-11)-(-2-9)]\\&=-2+[(8+(-11))-(-2+(-9))]\\&=-2+[(-3)-(-11)]\\&=-2+[(-3)+11]\\&=-2+[8]\\&=6\end{aligned}$$

57. $$\begin{aligned}|-3|+2^2+[-4-(-6)]&=3+2^2+[-4+6]\\&=3+2^2+[2]\\&=3+4+[2]\\&=7+[2]\\&=9\end{aligned}$$

59. Let $x=-5$ and $y=4$.
$x-y=-5-4=-5+(-4)=-9$

61. Let $x=-5$, $y=4$, and $t=10$.
$$\begin{aligned}|x|+2t-8y&=|-5|+2(10)-8(4)\\&=5+2(10)-8(4)\\&=5+20-32\\&=25-32\\&=25+(-32)\\&=-7\end{aligned}$$

63. Let $x=-5$ and $y=4$.
$$\frac{9-x}{y+6}=\frac{9-(-5)}{4+6}=\frac{9+5}{4+6}=\frac{14}{10}=\frac{2\cdot 7}{2\cdot 5}=\frac{7}{5}$$

65. Let $x=-5$ and $y=4$.
$y^2-x=4^2-(-5)=16+5=21$

67. Let $x=-5$ and $t=10$.
$$\begin{aligned}\frac{|x-(-10)|}{2t}&=\frac{|-5-(-10)|}{2(10)}\\&=\frac{|-5+10|}{2(10)}\\&=\frac{|5|}{2(10)}\\&=\frac{5}{20}\\&=\frac{5}{4\cdot 5}\\&=\frac{1}{4}\end{aligned}$$

69. The change in temperature is the difference between the last temperature and the first temperature.
$-56 - 44 = -56 + (-44) = -100$
The temperature dropped 100°.

71. Gains: +2
Losses: −5, −20
$2 + (-5) + (-20) = -3 + (-20) = -23$
Total loss of 23 yards

73. $-475 - 94 = -475 + (-94) = -569$
He was born in 569 B.C.

75. Rises: +120
Drops: −250, −178
$120 + (-250) + (-178) = -130 + (-178) = -308$
The overall vertical change was a drop of 308 feet.

77. $19{,}340 - (-512) = 19{,}340 + 512 = 19{,}852$
19,852 feet higher

79. $y = 180 - 50 = 180 + (-50) = 130$
The supplementary angle is 130°.

81. $x = 90 - 60 = 90 + (-60) = 30$
The complementary angle is 30°.

83. Let $x = -4$.
$$\begin{aligned} x - 9 &= 5 \\ -4 - 9 &\stackrel{?}{=} 5 \\ -4 + (-9) &\stackrel{?}{=} 5 \\ -13 &= 5, \text{ false} \end{aligned}$$
−4 is not a solution of the equation.

85. Let $x = -2$.
$$\begin{aligned} -x + 6 &= -x - 1 \\ -(-2) + 6 &\stackrel{?}{=} -(-2) - 1 \\ 2 + 6 &\stackrel{?}{=} 2 + (-1) \\ 8 &= 1, \text{ false} \end{aligned}$$
−2 is not a solution of the equation.

87. Let $x = 2$.
$$\begin{aligned} -x - 13 &= -15 \\ -2 - 13 &\stackrel{?}{=} -15 \\ -2 + (-13) &\stackrel{?}{=} -15 \\ -15 &= -15, \text{ true} \end{aligned}$$
2 is a solution of the equation.

89. The change in temperature is the difference between the given month's temperature and the previous month's.
F: $-23.7 - (-19.3) = -23.7 + 19.3 = -4.4°$
Mr: $-21.1 - (-23.7) = -21.1 + 23.7 = 2.6°$
Ap: $-9.1 - (-21.1) = -9.1 + 21.1 = 12°$
Ma: $14.4 - (-9.1) = 14.4 + 9.1 = 23.5°$
Jn: $29.7 - 14.4 = 29.7 + (-14.4) = 15.3°$
Jy: $33.6 - 29.7 = 33.6 + (-29.7) = 3.9°$
Au: $33.3 - 33.6 = 33.3 + (-33.6) = -0.3°$
S: $27.0 - 33.3 = 27.0 + (-33.3) = -6.3°$
O: $8.8 - 27.0 = 8.8 + (-27.0) = -18.2°$
N: $-6.9 - 8.8 = -6.9 + (-8.8) = -15.7°$
D: $-17.2 - (-6.9) = -17.2 + 6.9 = -10.3°$

91. Look for the negative number whose absolute value is the greatest; October

93. True; answers may vary

95. True; answers may vary

97. Negative; $4.362 - 7.0086 = -2.6466$

Integrated Review

1. The opposite of a positive number is a negative number.

2. The sum of two negative numbers is a negative number.

3. The absolute value of a negative number is a positive number.

4. The absolute value of zero is 0.

5. The reciprocal of a positive number is a positive number.

6. The sum of a number and its opposite is 0.

7. The absolute value of a positive number is a positive number.

8. The opposite of a negative number is a positive number.

	Number	Opposite	Absolute Value
9.	$\frac{1}{7}$	$-\frac{1}{7}$	$\frac{1}{7}$
10.	$-\frac{12}{5}$	$\frac{12}{5}$	$\frac{12}{5}$
11.	3	−3	3
12.	$-\frac{9}{11}$	$\frac{9}{11}$	$\frac{9}{11}$

13. $-19+(-23)=-42$

14. $7-(-3)=7+3=10$

15. $-15+17=2$

16. $-8-10=-8+(-10)=-18$

17. $18+(-25)=-7$

18. $-2+(-37)=-39$

19. $-14-(-12)=-14+12=-2$

20. $5-14=5+(-14)=-9$

21. $4.5-7.9=4.5+(-7.9)=-3.4$

22. $-8.6-1.2=-8.6+(-1.2)=-9.8$

23. $-\frac{3}{4}-\frac{1}{7}=-\frac{21}{28}-\frac{4}{28}=-\frac{21}{28}+\left(-\frac{4}{28}\right)=-\frac{25}{28}$

24. $\frac{2}{3}-\frac{7}{8}=\frac{16}{24}-\frac{21}{24}=\frac{16}{24}+\left(-\frac{21}{24}\right)=-\frac{5}{24}$

25. $$\begin{aligned}-9-(-7)+4-6&=-9+7+4-6\\&=-9+7+4+(-6)\\&=-4\end{aligned}$$

26. $$\begin{aligned}11-20+(-3)-12&=11+(-20)+(-3)+(-12)\\&=-9+(-3)+(-12)\\&=-12+(-12)\\&=-24\end{aligned}$$

27. $$\begin{aligned}24-6(14-11)&=24-6[14+(-11)]\\&=24-6(3)\\&=24-18\\&=24+(-18)\\&=6\end{aligned}$$

28. $$\begin{aligned}30-5(10-8)&=30-5[10+(-8)]\\&=30-5(2)\\&=30-10\\&=30+(-10)\\&=20\end{aligned}$$

29. $(7-17)+4^2=[7+(-17)]+4^2=(-10)+16=6$

30. $$\begin{aligned}9^2+(10-30)&=9^2+[10+(-30)]\\&=81+(-20)\\&=61\end{aligned}$$

31. $$\begin{aligned}|-9|+3^2+(-4-20)&=9+9+[-4+(-20)]\\&=9+9+(-24)\\&=18+(-24)\\&=-6\end{aligned}$$

32. $$\begin{aligned}|-4-5|+5^2+(-50)&=|-4+(-5)|+5^2+(-50)\\&=|-9|+25+(-50)\\&=9+25+(-50)\\&=34+(-50)\\&=-16\end{aligned}$$

33. $$\begin{aligned}-7+[(1-2)+(-2-9)]&=-7+[(-1)+(-11)]\\&=-7+[-12]\\&=-19\end{aligned}$$

34. $$\begin{aligned}-6+[(-3+7)+(4-15)]&=-6+[(4)+(-11)]\\&=-6+(-7)\\&=-13\end{aligned}$$

35. $1-5=1+(-5)=-4$

36. $-3-(-2)=-3+2=-1$

37. $\frac{1}{4}-\left(-\frac{2}{5}\right)=\frac{1}{4}+\frac{2}{5}=\frac{5}{20}+\frac{8}{20}=\frac{13}{20}$

38. $-\frac{5}{8}-\left(\frac{1}{10}\right)=-\frac{25}{40}-\frac{4}{40}=-\frac{25}{40}+\left(-\frac{4}{40}\right)=-\frac{29}{40}$

39. $$\begin{aligned}&2(19-17)^3-3(-7+9)^2\\&=2[19+(-17)]^3-3(-7+9)^2\\&=2(2)^3-3(2)^2\\&=2(8)-3(4)\\&=16-12\\&=16+(-12)\\&=4\end{aligned}$$

40. $$\begin{aligned}&3(10-9)^2+6(20-19)^3\\&=3[10+(-9)]^2+6[20+(-19)]^3\\&=3(1)^2+6(1)^3\\&=3+6\\&=9\end{aligned}$$

41. $x-y=-2-(-1)=-2+1=-1$

42. $x+y=-2+(-1)=-3$

43. $y+z=-1+9=8$

44. $z-y=9-(-1)=9+1=10$

45. $\dfrac{|5z-x|}{y-x}=\dfrac{|5(9)-(-2)|}{-1-(-2)}=\dfrac{|45+2|}{-1+2}=\dfrac{|47|}{1}=47$

46. $\dfrac{|-x-y+z|}{2z}=\dfrac{|-(-2)-(-1)+9|}{2(9)}$
$=\dfrac{|2+1+9|}{18}$
$=\dfrac{|12|}{18}$
$=\dfrac{12}{18}$
$=\dfrac{2}{3}$

Section 1.7

Practice Exercises

1. a. $8(-5)=-40$

b. $(-3)(-4)=12$

c. $(-6)(9)=-54$

2. a. $(-1)(-5)(-6)=5(-6)=-30$

b. $(-3)(-2)(4)=6(4)=24$

c. $(-4)(0)(5)=0(5)=0$

d. $(-2)(-3)-(-4)(5)=6-(-20)$
$=6+20$
$=26$

3. a. $(0.23)(-0.2)=-[(0.23)(0.2)]=-0.046$

b. $\left(-\dfrac{3}{5}\right)\cdot\left(\dfrac{4}{9}\right)=-\dfrac{3\cdot 4}{5\cdot 9}=-\dfrac{12}{45}=-\dfrac{4}{15}$

c. $\left(-\dfrac{7}{12}\right)(-24)=\dfrac{7\cdot 24}{12\cdot 1}=7\cdot 2=14$

4. a. $(-6)^2=(-6)(-6)=36$

b. $-6^2=-(6\cdot 6)=-(36)=-36$

c. $(-4)^3=(-4)(-4)(-4)=16(-4)=-64$

d. $-4^3=-(4\cdot 4\cdot 4)=-[16(4)]=-64$

5. a. The reciprocal of $\dfrac{8}{3}$ is $\dfrac{3}{8}$ since $\dfrac{8}{3}\cdot\dfrac{3}{8}=1$.

b. The reciprocal of 15 is $\dfrac{1}{15}$ since $15\cdot\dfrac{1}{15}=1$.

c. The reciprocal of $-\dfrac{2}{7}$ is $-\dfrac{7}{2}$ since $\left(-\dfrac{2}{7}\right)\left(-\dfrac{7}{2}\right)=1$.

d. The reciprocal of -5 is $-\dfrac{1}{5}$ since $(-5)\left(-\dfrac{1}{5}\right)=1$.

6. a. $\dfrac{16}{-2}=16\left(-\dfrac{1}{2}\right)=-8$

b. $24\div(-6)=24\left(-\dfrac{1}{6}\right)=-4$

c. $\dfrac{-35}{-7}=\dfrac{35}{7}=\dfrac{5\cdot 7}{7}=5$

7. a. $\dfrac{-18}{-6}=\dfrac{18}{6}=\dfrac{3\cdot 6}{6}=3$

b. $\dfrac{-48}{3}=-\dfrac{48}{3}=-\dfrac{3\cdot 16}{3}=-16$

c. $\dfrac{3}{5}\div\left(-\dfrac{1}{2}\right)=\dfrac{3}{5}\cdot(-2)=-\dfrac{6}{5}$

d. $-\dfrac{4}{9}\div 8=-\dfrac{4}{9}\cdot\dfrac{1}{8}=-\dfrac{4}{9\cdot 4\cdot 2}=-\dfrac{1}{9\cdot 2}=-\dfrac{1}{18}$

8. a. $\dfrac{0}{-2}=0$

b. $\dfrac{-4}{0}$ is undefined.

c. $\dfrac{-5}{6(0)}=\dfrac{-5}{0}$ is undefined.

9. a. $\dfrac{(-8)(-11)-4}{-9-(-4)}=\dfrac{88-4}{-9+4}=\dfrac{84}{-5}=-\dfrac{84}{5}$

b. $\frac{3(-2)^3-9}{-6+3}=\frac{3(-8)-9}{-3}$
$=\frac{-24-9}{-3}$
$=\frac{-33}{-3}$
$=11$

10. a. $7y-x=7(-2)-(-5)=-14+5=-9$

b. $x^2-y^3=(-5)^2-(-2)^3$
$=25-(-8)$
$=25+8$
$=33$

c. $\frac{2x}{3y}=\frac{2(-5)}{3(-2)}=\frac{-10}{-6}=\frac{5}{3}$

Calculator Explorations

1. $-38(26-27)=38$

2. $-59(-8)+1726=2198$

3. $134+25(68-91)=-441$

4. $45(32)-8(218)=-304$

5. $\frac{-50(294)}{175-265}=163.\overline{3}$

6. $\frac{-444-444.8}{-181-324}=1.76$

7. $9^5-4550=54,499$

8. $5^8-6259=384,366$

9. $(-125)^2=15,625$

10. $-125^2=-15,625$

Vocabulary and Readiness Check

1. If n is a real number, then $n\cdot 0=\underline{0}$ and $0\cdot n=\underline{0}$.

2. If n is a real number, but not 0, then $\frac{0}{n}=\underline{0}$ and we say $\frac{n}{0}$ is undefined.

3. The product of two negative numbers is a positive number.

4. The quotient of two negative numbers is a positive number.

5. The quotient of a positive number and a negative number is a negative number.

6. The product of a positive number and a negative number is a negative number.

7. The reciprocal of a positive number is a positive number.

8. The opposite of a positive number is a negative number.

Exercise Set 1.7

1. $-6(4)=-24$

3. $2(-1)=-2$

5. $-5(-10)=50$

7. $-3\cdot 4=-12$

9. $-7\cdot 0=0$

11. $2(-9)=-18$

13. $-\frac{1}{2}\left(-\frac{3}{5}\right)=\frac{1\cdot 3}{2\cdot 5}=\frac{3}{10}$

15. $-\frac{3}{4}\left(-\frac{8}{9}\right)=\frac{3\cdot 8}{4\cdot 9}=\frac{24}{36}=\frac{2\cdot 12}{3\cdot 12}=\frac{2}{3}$

17. $5(-1.4)=-7$

19. $-0.2(-0.7)=0.14$

21. $-10(80)=-800$

23. $4(-7)=-28$

25. $(-5)(-5)=25$

27. $\frac{2}{3}\left(-\frac{4}{9}\right)=-\frac{2\cdot 4}{3\cdot 9}=-\frac{8}{27}$

29. $-11(11)=-121$

31. $-\frac{20}{25}\left(\frac{5}{16}\right)=-\frac{20\cdot 5}{25\cdot 16}=-\frac{100}{400}=-\frac{1}{4}$

33. $(-1)(2)(-3)(-5)=-2(-3)(-5)=6(-5)=-30$

35. $(-2)(5)-(-11)(3)=-10-(-33)=-10+33=23$

37. $(-6)(-1)(-2)-(-5)=-12+5=-7$

39. True; example: $(-2)(-2)(-2)=-8$
False; example: $(-2)(-2)(-2)(-2)=16$

41. False

43. $$\begin{aligned}(-2)^4 &= (-2)(-2)(-2)(-2)\\ &= 4(-2)(-2)\\ &= -8(-2)\\ &= 16\end{aligned}$$

45. $-1^5=-(1)(1)(1)(1)(1)=-1$

47. $(-5)^2=(-5)(-5)=25$

49. $-7^2=-(7)(7)=-49$

51. Reciprocal of 9 is $\frac{1}{9}$ since $9\cdot\frac{1}{9}=1$.

53. Reciprocal of $\frac{2}{3}$ is $\frac{3}{2}$ since $\frac{2}{3}\cdot\frac{3}{2}=1$.

55. Reciprocal of -14 is $-\frac{1}{14}$ since $-14\cdot-\frac{1}{14}=1$.

57. Reciprocal of $-\frac{3}{11}$ is $-\frac{11}{3}$ since $-\frac{3}{11}\cdot-\frac{11}{3}=1$.

59. Reciprocal of 0.2 is $\frac{1}{0.2}$ since $0.2\cdot\frac{1}{0.2}=1$.

61. Reciprocal of $\frac{1}{-6.3}$ is -6.3 since

$\frac{1}{-6.3}\cdot-6.3=1$.

63. $\frac{18}{-2}=18\cdot-\frac{1}{2}=-9$

65. $\frac{-16}{-4}=-16\cdot-\frac{1}{4}=4$

67. $\frac{-48}{12}=-48\cdot\frac{1}{12}=-4$

69. $\frac{0}{-4}=0\cdot-\frac{1}{4}=0$

71. $-\frac{15}{3}=-15\cdot\frac{1}{3}=-5$

73. $\frac{5}{0}$ is undefined.

75. $\frac{-12}{-4}=-12\cdot-\frac{1}{4}=3$

77. $\frac{30}{-2}=30\cdot-\frac{1}{2}=-15$

79. $\frac{6}{7}\div-\frac{1}{3}=\frac{6}{7}\cdot\left(-\frac{3}{1}\right)=-\frac{6\cdot 3}{7\cdot 1}=-\frac{18}{7}$

81. $-\frac{5}{9}\div\left(-\frac{3}{4}\right)=-\frac{5}{9}\cdot\left(-\frac{4}{3}\right)=\frac{5\cdot 4}{9\cdot 3}=\frac{20}{27}$

83. $-\frac{4}{9}\div\frac{4}{9}=-\frac{4}{9}\cdot\frac{9}{4}=-1$

85. $\frac{-9(-3)}{-6}=\frac{27}{-6}=-\frac{9}{2}$

87. $\frac{12}{9-12}=\frac{12}{-3}=-4$

89. $\frac{-6^2+4}{-2}=\frac{-36+4}{-2}=\frac{-32}{-2}=16$

91. $\frac{8+(-4)^2}{4-12}=\frac{8+16}{4-12}=\frac{24}{-8}=-3$

93. $\frac{22+(3)(-2)}{-5-2}=\frac{22+(-6)}{-5-2}=\frac{16}{-7}=-\frac{16}{7}$

95. $\frac{-3-5^2}{2(-7)}=\frac{-3-25}{2(-7)}=\frac{-3+(-25)}{-14}=\frac{-28}{-14}=2$

97. $\frac{6-2(-3)}{4-3(-2)}=\frac{6-(-6)}{4-(-6)}=\frac{6+6}{4+6}=\frac{12}{10}=\frac{6}{5}$

99. $\frac{-3-2(-9)}{-15-3(-4)} = \frac{-3-(-18)}{-15-(-12)} = \frac{-3+18}{-15+12} = \frac{15}{-3} = -5$

101. $\frac{|5-9|+|10-15|}{|2(-3)|} = \frac{|-4|+|-5|}{|-6|} = \frac{4+5}{6} = \frac{9}{6} = \frac{3}{2}$

103. Let $x = -5$ and $y = -3$.
$3x + 2y = 3(-5) + 2(-3) = -15 + (-6) = -21$

105. Let $x = -5$ and $y = -3$.

$$\begin{aligned} 2x^2 - y^2 &= 2(-5)^2 - (-3)^2 \\ &= 2(25) - 9 \\ &= 50 + (-9) \\ &= 41 \end{aligned}$$

107. Let $x = -5$ and $y = -3$.
$x^3 + 3y = (-5)^3 + 3(-3) = -125 + (-9) = -134$

109. Let $x = -5$ and $y = -3$.
$\frac{2x-5}{y-2} = \frac{2(-5)-5}{-3-2} = \frac{-10-5}{-3-2} = \frac{-15}{-5} = 3$

111. Let $x = -5$ and $y = -3$.
$\frac{-3-y}{x-4} = \frac{-3-(-3)}{-5-4} = \frac{-3+3}{-5-4} = \frac{0}{-9} = 0$

113. $4(-6203) = -24,812$
The net income will be –$24,812 million.

115. Let $x = 7$.

$$\begin{aligned} -5x &= -35 \\ -5(7) &\stackrel{?}{=} -35 \\ -35 &= -35, \text{ true} \end{aligned}$$

7 is a solution of the equation.

117. Let $x = -20$.

$$\begin{aligned} \frac{x}{10} &= 2 \\ \frac{-20}{10} &\stackrel{?}{=} 2 \\ -2 &= 2, \text{ false} \end{aligned}$$

–20 is not a solution of the equation.

119. Let $x = 5$.

$$\begin{aligned} -3x - 5 &= -20 \\ -3(5) - 5 &\stackrel{?}{=} -20 \\ -15 - 5 &\stackrel{?}{=} -20 \\ -20 &= -20, \text{ true} \end{aligned}$$

5 is a solution of the equation.

121. Answers may vary

123. –1 and 1 are their own reciprocals.

125. Since q is negative, r is negative, and t is positive, then $\frac{q}{r \cdot t}$ is positive.

127. It is not possible to determine whether $q + t$ is positive or negative.

129. Since q is negative, r is negative, and t is positive, then $t(q + r)$ is negative.

131.

$$\begin{aligned} -2 + \frac{-15}{3} &= \frac{-2 \cdot 3}{1 \cdot 3} + \frac{-15}{3} \\ &= \frac{-6 + (-15)}{3} \\ &= \frac{-21}{3} \\ &= -7 \end{aligned}$$

133. $2[-5 + (-3)] = 2(-8) = -16$

The Bigger Picture

1. $-0.2(25) - 5$
2. $86 - 100 = -14$
3. $-\frac{1}{7} + \left(-\frac{3}{5}\right) = -\frac{5}{35} - \frac{21}{35} = -\frac{26}{35}$
4. $\frac{-40}{-5} = 8$
5. $(-7)^2 = (-7)(-7) = 49$
6. $-7^2 = -(7 \cdot 7) = -49$
7. $\frac{|-42|}{-|-2|} = \frac{42}{-2} = -21$
8. $\frac{8.6}{0}$ is undefined.
9. $\frac{0}{8.6} = 0$
10. $-25 - (-13) = -25 + 13 = -12$
11. $-8.3 - 8.3 = -16.6$

12. $-\frac{8}{9}\left(-\frac{3}{16}\right)=\frac{3\cdot 8}{9\cdot 16}=\frac{1\cdot 1}{3\cdot 2}=\frac{1}{6}$

13. $2+3(8-11)^3=2+3(-3)^3$
$=2+3(-27)$
$=2+(-81)$
$=-79$

14. $-2\frac{1}{2}\div\left(-3\frac{1}{4}\right)=-\frac{5}{2}\div\left(-\frac{13}{4}\right)$
$=-\frac{5}{2}\left(-\frac{4}{13}\right)$
$=\frac{4\cdot 5}{2\cdot 13}$
$=\frac{2\cdot 5}{13}$
$=\frac{10}{13}$

15. $20\div 2\cdot 5=10\cdot 5=50$

16. $-2[(1-5)-(7-17)]=-2[(-4)-(-10)]$
$=-2[-4+10]$
$=-2[6]$
$=-12$

Section 1.8

Practice Exercises

1. a. $x\cdot 8=\underline{8\cdot x}$

b. $x+17=\underline{17+x}$

2. a. $(2+9)+7=\underline{2+(9+7)}$

b. $-4\cdot(2\cdot 7)=\underline{(-4\cdot 2)\cdot 7}$

3. a. $(5+x)+9=(x+5)+9=x+(5+9)=x+14$

b. $5(-6x)=[5\cdot(-6)]x=-30x$

4. a. $5(x-y)=5(x)-5(y)=5x-5y$

b. $-6(4+2t)=-6(4)+(-6)(2t)=-24-12t$

c. $2(3x-4y-z)=2(3x)+2(-4y)+2(-z)$
$=6x-8y-2z$

d. $(3-y)\cdot(-1)=3(-1)+(-y)(-1)=-3+y$

e. $-(x-7+2s)=(-1)(x-7+2s)$
$=(-1)x+(-1)(-7)+(-1)(2s)$
$=-x+7-2s$

f. $2(7x+4)+6=2(7x)+2(4)+6$
$=14x+8+6$
$=14x+14$

5. a. $5\cdot w+5\cdot 3=5(w+3)$

b. $9w+9z=9\cdot w+9\cdot z=9(w+z)$

6. a. $(7\cdot 3x)\cdot 4=(3x\cdot 7)\cdot 4$; commutative property of multiplication

b. $6+(3+y)=(6+3)+y$; associative property of addition

c. $8+(t+0)=8+t$; identity element for addition

d. $-\frac{3}{4}\cdot\left(-\frac{4}{3}\right)=1$; multiplicative inverse property

e. $(2+x)+5=5+(2+x)$; commutative property of addition

f. $3+(-3)=0$; additive inverse property

g. $(-3b)\cdot 7=(-3\cdot 7)\cdot b$; commutative and associative properties of multiplication

Vocabulary and Readiness Check

1. $x+5=5+x$ is a true statement by the <u>commutative property of addition</u>.

2. $x\cdot 5=5\cdot x$ is a true statement by the <u>commutative property of multiplication</u>.

3. $3(y+6)=3\cdot y+3\cdot 6$ is true by the <u>distributive property</u>.

4. $2\cdot(x\cdot y)=(2\cdot x)\cdot y$ is a true statement by the <u>associative property of multiplication</u>.

5. $x+(7+y)=(x+7)+y$ is a true statement by the <u>associative property of addition</u>.

6. The numbers $-\frac{2}{3}$ and $-\frac{3}{2}$ are called <u>reciprocals or multiplicative inverses</u>.

7. The numbers $-\frac{2}{3}$ and $\frac{2}{3}$ are called <u>opposites or additive inverses</u>.

Exercise Set 1.8

1. $x + 16 = 16 + x$

3. $-4 \cdot y = y \cdot (-4)$

5. $xy = yx$

7. $2x + 13 = 13 + 2x$

9. $(xy) \cdot z = x \cdot (yz)$

11. $2 + (a + b) = (2 + a) + b$

13. $4 \cdot (ab) = 4a \cdot (b)$

15. $(a + b) + c = a + (b + c)$

17. $8 + (9 + b) = (8 + 9) + b = 17 + b$

19. $4(6y) = (4 \cdot 6)y = 24y$

21. $\frac{1}{5}(5y) = \left(\frac{1}{5} \cdot 5\right)y = 1 \cdot y = y$

23.
$$\begin{aligned}(13 + a) + 13 &= (a + 13) + 13 \\ &= a + (13 + 13) \\ &= a + 26\end{aligned}$$

25. $-9(8x) = (-9 \cdot 8)x = -72x$

27. $\frac{3}{4}\left(\frac{4}{3}s\right) = \left(\frac{3}{4} \cdot \frac{4}{3}\right)s = 1s = s$

29. Answers may vary

31. $4(x + y) = 4x + 4y$

33. $9(x - 6) = 9x - 9 \cdot 6 = 9x - 54$

35. $2(3x + 5) = 2(3x) + 2(5) = 6x + 10$

37. $7(4x - 3) = 7(4x) - 7(3) = 28x - 21$

39. $3(6 + x) = 3(6) + 3x = 18 + 3x$

41. $-2(y - z) = -2y - (-2)z = -2y + 2z$

43. $-7(3y + 5) = -7(3y) + (-7)(5) = -21y - 35$

45.
$$\begin{aligned}5(x + 4m + 2) &= 5x + 5(4m) + 5(2) \\ &= 5x + 20m + 10\end{aligned}$$

47.
$$\begin{aligned}-4(1 - 2m + n) &= -4(1) - (-4)(2m) + (-4)n \\ &= -4 + 8m - 4n\end{aligned}$$

49.
$$\begin{aligned}-(5x + 2) &= -1(5x + 2) \\ &= -1(5x) + (-1)(2) \\ &= -5x - 2\end{aligned}$$

51.
$$\begin{aligned}-(r - 3 - 7p) &= -1(r - 3 - 7p) \\ &= -1r - (-1)(3) - (-1)(7p) \\ &= -r + 3 + 7p\end{aligned}$$

53.
$$\begin{aligned}\frac{1}{2}(6x + 8) &= \frac{1}{2}(6x) + \frac{1}{2}(8) \\ &= \left(\frac{1}{2} \cdot 6\right)x + \left(\frac{1}{2} \cdot 8\right) \\ &= 3x + 4\end{aligned}$$

55.
$$\begin{aligned}-\frac{1}{3}(3x - 9y) &= -\frac{1}{3}(3x) - \left(-\frac{1}{3}\right)(9y) \\ &= \left(-\frac{1}{3} \cdot 3\right)x - \left(-\frac{1}{3} \cdot 9\right)y \\ &= -1 \cdot x + 3 \cdot y \\ &= -x + 3y\end{aligned}$$

57.
$$\begin{aligned}3(2r + 5) - 7 &= 3(2r) + 3(5) - 7 \\ &= 6r + 15 + (-7) \\ &= 6r + 8\end{aligned}$$

59.
$$\begin{aligned}-9(4x + 8) + 2 &= -9(4x) + (-9)(8) + 2 \\ &= -36x - 72 + 2 \\ &= -36x - 70\end{aligned}$$

61.
$$\begin{aligned}-4(4x + 5) - 5 &= -4(4x) + (-4)(5) - 5 \\ &= -16x + (-20) + (-5) \\ &= -16x - 25\end{aligned}$$

63. $4 \cdot 1 + 4 \cdot y = 4(1 + y)$

65. $11x + 11y = 11(x + y)$

67. $(-1) \cdot 5 + (-1) \cdot x = -1(5 + x) = -(5 + x)$

69. $30a + 30b = 30(a + b)$

71. $3 \cdot 5 = 5 \cdot 3$; commutative property of multiplication

73. $2 + (x + 5) = (2 + x) + 5$; associative property of addition

75. $9(3 + 7) = 9 \cdot 3 + 9 \cdot 7$; distributive property

77. $(4 \cdot y) \cdot 9 = 4 \cdot (y \cdot 9)$; associative property of multiplication

79. $0 + 6 = 6$; identity element of addition

81. $-4(y + 7) = -4 \cdot y + (-4) \cdot 7$; distributive property

83. $-4 \cdot (8 \cdot 3) = (8 \cdot -4) \cdot 3$; associative and commutative properties of multiplication

85.

Expression	Opposite	Reciprocal
8	-8	$\frac{1}{8}$

87.

Expression	Opposite	Reciprocal
x	$-x$	$\frac{1}{x}$

89.

Expression	Opposite	Reciprocal
$2x$	$-2x$	$\frac{1}{2x}$

91. No

93. Yes

95. Answers may vary

Chapter 1 Vocabulary Check

1. The symbols $\neq$, <, and > are called inequality symbols.

2. A mathematical statement that two expressions are equal is called an equation.

3. The absolute value of a number is the distance between that number and 0 on the number line.

4. A symbol used to represent a number is called a variable.

5. Two numbers that are the same distance from 0 but lie on opposite sides of 0 are called opposites.

6. The number in a fraction above the fraction bar is called the numerator.

7. A solution of an equation is a value for the variable that makes the equation a true statement.

8. Two numbers whose product is 1 are called reciprocals.

9. In 2^3, the 2 is called the base and the 3 is called the exponent.

10. The number in a fraction below the fraction bar is called the denominator.

11. Parentheses and brackets are examples of grouping symbols.

12. A set is a collection of objects.

Chapter 1 Review

1. $8 < 10$ since 8 is to the left of 10 on the number line.

2. $7 > 2$ since 7 is to the right of 2 on the number line.

3. $-4 > -5$ since -4 is to the right of -5 on the number line.

4. $\frac{12}{2} > -8$ since $6 > -8$.

5. $|-7| < |-8|$ since $7 < 8$.

6. $|-9| > -9$ since $9 > -9$.

7. $-|-1| = -1$ since $-1 = -1$.

8. $|-14| = -(-14)$ since $14 = 14$.

9. $1.2 > 1.02$ since 1.2 is to the right of 1.02 on the number line.

10. $-\frac{3}{2} < -\frac{3}{4}$ since $-\frac{3}{2}$ is to the left of $-\frac{3}{4}$ on the number line.

11. Four is greater than or equal to negative three is written as $4 \geq -3$.

12. Six is not equal to five is written as $6 \neq 5$.

13. 0.03 is less than 0.3 is written as $0.03 < 0.3$.

14. $400 > 155$ or $155 < 400$

15. **a.** The natural numbers are 1 and 3.

b. The whole numbers are 0, 1, and 3.

c. The integers are –6, 0, 1, and 3.

d. The rational numbers are –6, 0, 1, $1\frac{1}{2}$, 3, and 9.62.

e. The irrational number is π.

f. The real numbers are all numbers in the given set.

16. **a.** The natural numbers are 2 and 5.

b. The whole numbers are 2 and 5.

c. The integers are –3, 2, and 5.

d. The rational numbers are –3, –1.6, 2, 5, $\frac{11}{2}$, and 15.1.

e. The irrational numbers are $\sqrt{5}$ and 2π.

f. The real numbers are all numbers in the given set.

17. Look for the negative number with the greatest absolute value. The greatest loss was on Friday.

18. Look for the largest positive number. The greatest gain was on Wednesday.

19. $36 = 4 \cdot 9 = 2 \cdot 2 \cdot 3 \cdot 3$

20. $120 = 8 \cdot 15 = 2 \cdot 2 \cdot 2 \cdot 3 \cdot 5$

21. $\frac{8}{15} \cdot \frac{27}{30} = \frac{8 \cdot 27}{15 \cdot 30} = \frac{2 \cdot 4 \cdot 3 \cdot 3 \cdot 3}{3 \cdot 5 \cdot 2 \cdot 3 \cdot 5} = \frac{12}{25}$

22. $\frac{7}{8} \div \frac{21}{32} = \frac{7}{8} \cdot \frac{32}{21} = \frac{7 \cdot 32}{8 \cdot 21} = \frac{7 \cdot 8 \cdot 4}{8 \cdot 3 \cdot 7} = \frac{4}{3}$

23.
$$\begin{aligned}\frac{7}{15} + \frac{5}{6} &= \frac{7 \cdot 2}{15 \cdot 2} + \frac{5 \cdot 5}{6 \cdot 5} \\ &= \frac{14}{30} + \frac{25}{30} \\ &= \frac{14 + 25}{30} \\ &= \frac{39}{30} \\ &= \frac{3 \cdot 13}{3 \cdot 10} \\ &= \frac{13}{10}\end{aligned}$$

24.
$$\begin{aligned}\frac{3}{4} - \frac{3}{20} &= \frac{3 \cdot 5}{4 \cdot 5} - \frac{3}{20} \\ &= \frac{15}{20} - \frac{3}{20} \\ &= \frac{15 - 3}{20} \\ &= \frac{12}{20} \\ &= \frac{3 \cdot 4}{5 \cdot 4} \\ &= \frac{3}{5}\end{aligned}$$

25.
$$\begin{aligned}2\frac{3}{4} + 6\frac{5}{8} &= \frac{11}{4} + \frac{53}{8} \\ &= \frac{11 \cdot 2}{4 \cdot 2} + \frac{53}{8} \\ &= \frac{22}{8} + \frac{53}{8} \\ &= \frac{22 + 53}{8} \\ &= \frac{75}{8} \\ &= 9\frac{3}{8}\end{aligned}$$

26. $7\frac{1}{6}-2\frac{2}{3}=\frac{43}{6}-\frac{8}{3}$
$=\frac{43}{6}-\frac{8\cdot 2}{3\cdot 2}$
$=\frac{43}{6}-\frac{16}{6}$
$=\frac{43-16}{6}$
$=\frac{27}{6}$
$=\frac{9\cdot 3}{2\cdot 3}$
$=\frac{9}{2}$
$=4\frac{1}{2}$

27. $5\div\frac{1}{3}=5\cdot\frac{3}{1}=15$

28. $2\cdot 8\frac{3}{4}=2\cdot\frac{35}{4}=\frac{2\cdot 35}{2\cdot 2}=\frac{35}{2}=17\frac{1}{2}$

29. $1-\frac{1}{6}-\frac{1}{4}=\frac{12}{12}-\frac{1\cdot 2}{6\cdot 2}-\frac{1\cdot 3}{4\cdot 3}$
$=\frac{12}{12}-\frac{2}{12}-\frac{3}{12}$
$=\frac{12-2-3}{12}$
$=\frac{7}{12}$

The unknown part is $\frac{7}{12}$.

30. $P=2l+2w$
$P=2\left(1\frac{1}{3}\right)+2\left(\frac{7}{8}\right)$
$=\frac{2}{1}\cdot\frac{4}{3}+\frac{2}{1}\cdot\frac{7}{8}$
$=\frac{8}{3}+\frac{14}{8}$
$=\frac{8\cdot 8}{3\cdot 8}+\frac{14\cdot 3}{8\cdot 3}$
$=\frac{64}{24}+\frac{42}{24}$
$=\frac{64+42}{24}$
$=\frac{106}{24}$
$=4\frac{10}{24}$
$=4\frac{5}{12}$ meters
$A=lw$
$A=1\frac{1}{3}\cdot\frac{7}{8}$
$=\frac{4}{3}\cdot\frac{7}{8}$
$=\frac{4\cdot 7}{3\cdot 2\cdot 4}$
$=\frac{7}{6}$
$=1\frac{1}{6}$ sq meters

31. P = the sum of the lengths of the sides
$P=\frac{5}{11}+\frac{8}{11}+\frac{3}{11}+\frac{3}{11}+\frac{2}{11}+\frac{5}{11}=\frac{26}{11}=2\frac{4}{11}$ in.
A = the sum of the two areas, each given by lw
$A=\frac{5}{11}\cdot\frac{5}{11}+\frac{3}{11}\cdot\frac{3}{11}=\frac{25}{121}+\frac{9}{121}=\frac{34}{121}$ sq in.

32. $7\frac{1}{2}-6\frac{1}{8}=\frac{15}{2}-\frac{49}{8}$
$=\frac{15\cdot 4}{2\cdot 4}-\frac{49}{8}$
$=\frac{60}{8}-\frac{49}{8}$
$=\frac{60-49}{8}$
$=\frac{11}{8}$
$=1\frac{3}{8}$ ft

33. $1\frac{1}{8}+1\frac{13}{16}=\frac{9}{8}+\frac{29}{16}$
$=\frac{9\cdot 2}{8\cdot 2}+\frac{29}{16}$
$=\frac{18}{16}+\frac{29}{16}$
$=\frac{18+29}{16}$
$=\frac{47}{16}$
$=2\frac{15}{16}$ lb

34. $1\frac{1}{2}+1\frac{11}{16}+1\frac{3}{4}+1\frac{5}{8}+\frac{11}{16}+1\frac{1}{8}$
$=\frac{3}{2}+\frac{27}{16}+\frac{7}{4}+\frac{13}{8}+\frac{11}{16}+\frac{9}{8}$
$=\frac{3\cdot 8}{2\cdot 8}+\frac{27}{16}+\frac{7\cdot 4}{4\cdot 4}+\frac{13\cdot 2}{8\cdot 2}+\frac{11}{16}+\frac{9\cdot 2}{8\cdot 2}$
$=\frac{24+27+28+26+11+18}{16}$
$=\frac{134}{16}$
$=8\frac{3}{8}$ lb

35. Total weight = weight of girls + weight of boys
$8\frac{3}{8}+2\frac{15}{16}=\frac{67}{8}+\frac{47}{16}$
$=\frac{67\cdot 2}{8\cdot 2}+\frac{47}{16}$
$=\frac{134+47}{16}$
$=\frac{181}{16}$
$=11\frac{5}{16}$ lb

36. Look for the largest number. Jioke weighed the most.

37. Look for the smallest number. Odera weighed the least.

38. $1\frac{13}{16}-\frac{11}{16}=\frac{29}{16}-\frac{11}{16}$
$=\frac{29-11}{16}$
$=\frac{18}{16}$
$=1\frac{2}{16}$
$=1\frac{1}{8}$ lb

39. $5\frac{1}{2}-1\frac{5}{8}=\frac{11}{2}-\frac{13}{8}$
$=\frac{11\cdot 4}{2\cdot 4}-\frac{13}{8}$
$=\frac{44-13}{8}$
$=\frac{31}{8}$
$=3\frac{7}{8}$ lb

40. $4\frac{5}{32}-1\frac{1}{8}=\frac{133}{32}-\frac{9}{8}$
$=\frac{133}{32}-\frac{9\cdot 4}{8\cdot 4}$
$=\frac{133-36}{32}$
$=\frac{97}{32}$
$=3\frac{1}{32}$ lb

41. $2^4=2\cdot 2\cdot 2\cdot 2=16$

42. $5^2=5\cdot 5=25$

43. $\left(\frac{2}{7}\right)^2=\frac{2}{7}\cdot\frac{2}{7}=\frac{4}{49}$

44. $\left(\frac{3}{4}\right)^3=\frac{3}{4}\cdot\frac{3}{4}\cdot\frac{3}{4}=\frac{27}{64}$

45. $6\cdot 3^2+2\cdot 8=6\cdot 9+2\cdot 8=54+16=70$

46. $68-5\cdot 2^3=68-5\cdot 8=68-40=28$

47. $3(1+2\cdot 5)+4=3(1+10)+4$
$=3(11)+4$
$=33+4$
$=37$

48. $8+3(2\cdot 6-1)=8+3(12-1)$
$=8+3(11)$
$=8+33$
$=41$

49. $\dfrac{4+|6-2|+8^2}{4+6\cdot 4}=\dfrac{4+|4|+64}{4+24}$
$=\dfrac{4+4+64}{4+24}$
$=\dfrac{72}{28}$
$=\dfrac{4\cdot 18}{4\cdot 7}$
$=\dfrac{18}{7}$

50. $5[3(2+5)-5]=5[3(7)-5]$
$=5[21-5]$
$=5[16]$
$=80$

51. The difference of twenty and twelve is equal to the product of two and four is written as $20-12=2\cdot 4$.

52. The quotient of nine and two is greater than negative five is written as $\frac{9}{2}>-5$.

53. Let $x = 6$ and $y = 2$.
$2x+3y=2(6)+3(2)=12+6=18$

54. Let $x = 6$, $y = 2$, and $z = 8$.
$x(y+2z)=6[2+2(8)]=6[2+16]=6[18]=108$

55. Let $x = 6$, $y = 2$, and $z = 8$.
$\frac{x}{y}+\frac{z}{2y}=\frac{6}{2}+\frac{8}{2(2)}=\frac{6}{2}+\frac{8}{4}=3+2=5$

56. Let $x = 6$ and $y = 2$.
$x^2-3y^2=(6)^2-3(2)^2$
$=36-3(4)$
$=36-12$
$=36+(-12)$
$=24$

57. Let $a = 37$ and $b = 80$.
$180-a-b=180-37-80$
$=180+(-37)+(-80)$
$=143+(-80)$
$=63°$

58. Let $x = 3$.
$7x-3=18$
$7(3)-3\stackrel{?}{=}18$
$21-3\stackrel{?}{=}18$
$18=18$, true
3 is a solution to the equation.

59. Let $x = 1$.
$3x^2+4=x-1$
$3(1)^2+4\stackrel{?}{=}1-1$
$3+4\stackrel{?}{=}0$
$7=0$, false
1 is not a solution to the equation.

60. The additive inverse of -9 is 9.

61. The additive inverse of $\frac{2}{3}$ is $-\frac{2}{3}$.

62. The additive inverse of $|-2|$ is -2 since $|-2|=2$.

63. The additive inverse of $-|-7|$ is 7 since $-|-7|=-7$.

64. $-15+4=-11$

65. $-6+(-11)=-17$

66. $\frac{1}{16}+\left(-\frac{1}{4}\right)=\frac{1}{16}+\left(-\frac{1\cdot 4}{4\cdot 4}\right)$
$=\frac{1}{16}+\left(-\frac{4}{16}\right)$
$=-\frac{3}{16}$

67. $-8+|-3|=-8+3=-5$

68. $-4.6+(-9.3)=-13.9$

69. $-2.8+6.7=3.9$

70. $-282+728=446$ feet

71. $6-20=6+(-20)=-14$

72. $-3.1-8.4=-3.1+(-8.4)=-11.5$

73. $-6-(-11)=-6+11=5$

74. $4-15=4+(-15)=-11$

75.
$$\begin{aligned}-21-16+3(8-2)&=-21+(-16)+3[8+(-2)]\\&=-21+(-16)+3[6]\\&=-21+(-16)+18\\&=-37+18\\&=-19\end{aligned}$$

76.
$$\begin{aligned}\frac{11-(-9)+6(8-2)}{2+3\cdot4}&=\frac{11+9+6[8+(-2)]}{2+3\cdot4}\\&=\frac{11+9+6[6]}{2+3\cdot4}\\&=\frac{11+9+36}{2+12}\\&=\frac{56}{14}\\&=4\end{aligned}$$

77. Let $x=3$, $y=-6$, and $z=-9$.
$$\begin{aligned}2x^2-y+z&=2(3)^2-(-6)+(-9)\\&=2(9)+6+(-9)\\&=18+6+(-9)\\&=24+(-9)\\&=15\end{aligned}$$

78. Let $x=3$ and $y=-6$.
$$\begin{aligned}\frac{y-x+5x}{2x}&=\frac{y+4x}{2x}\\&=\frac{-6+4(3)}{2(3)}\\&=\frac{-6+12}{6}\\&=\frac{6}{6}\\&=1\end{aligned}$$

79. The multiplicative inverse of -6 is $-\frac{1}{6}$ since $-6\cdot-\frac{1}{6}=1$.

80. The multiplicative inverse of $\frac{3}{5}$ is $\frac{5}{3}$ since $\frac{3}{5}\cdot\frac{5}{3}=1$.

81. $6(-8)=-48$

82. $(-2)(-14)=28$

83. $\frac{-18}{-6}=3$

84. $\frac{42}{-3}=-14$

85. $\frac{4\cdot(-3)+(-8)}{2+(-2)}=\frac{-12+(-8)}{2+(-2)}=\frac{-20}{0}$
The expression is undefined.

86. $\frac{3(-2)^2-5}{-14}=\frac{3(4)-5}{-14}=\frac{12-5}{-14}=\frac{7}{-14}=-\frac{1}{2}$

87. $\frac{-6}{0}$ is undefined.

88. $\frac{0}{-2}=0$

89.
$$\begin{aligned}-4^2-(-3+5)\div(-1)\cdot2&=-16-(2)\div(-1)\cdot2\\&=-16+2\cdot2\\&=-16+4\\&=-12\end{aligned}$$

90.
$$\begin{aligned}-5^2-(2-20)\div(-3)\cdot3&=-25-(-18)\div(-3)\cdot3\\&=-25-6\cdot3\\&=-25-18\\&=-43\end{aligned}$$

91. Let $x=-5$ and $y=-2$.
$x^2-y^4=(-5)^2-(-2)^4=25-16=9$

92. Let $x=-5$ and $y=-2$.
$x^2-y^3=(-5)^2-(-2)^3=25-(-8)=25+8=33$

93. $\frac{-9+(-7)+1}{3}=\frac{-15}{3}=-5$
Her average score per round was 5 under par.

94. $\frac{-1+0+(-3)+0}{4}=\frac{-4}{4}=-1$
His average score per round was 1 under par.

95. $-6+5=5+(-6)$; commutative property of addition

96. $6\cdot1=6$; multiplicative identity property

97. $3(8 - 5) = 3 \cdot 8 + 3 \cdot (-5)$; distributive property

98. $4 + (-4) = 0$; additive inverse property

99. $2 + (3 + 9) = (2 + 3) + 9$; associative property of addition

100. $2 \cdot 8 = 8 \cdot 2$; commutative property of multiplication

101. $6(8 + 5) = 6 \cdot 8 + 6 \cdot 5$; distributive property

102. $(3 \cdot 8) \cdot 4 = 3 \cdot (8 \cdot 4)$; associative property of multiplication

103. $4 \cdot \frac{1}{4} = 1$; multiplicative inverse property

104. $8 + 0 = 8$; additive identity property

105. $5(y - 2) = 5(y) + 5(-2) = 5y - 10$

106. $-3(z + y) = -3(z) + (-3)(y) = -3z - 3y$

107. $\begin{aligned}-(7-x+4z) &= (-1)(7)+(-1)(-x)+(-1)(4z)\\ &= -7+x-4z\end{aligned}$

108. $\frac{1}{2}(6z-10) = \frac{1}{2}(6z)+\frac{1}{2}(-10) = 3z-5$

109. $\begin{aligned}-4(3x+5)-7 &= -4(3x)+(-4)(5)-7\\ &= -12x-20-7\\ &= -12x-27\end{aligned}$

110. $\begin{aligned}-8(2y+9)-1 &= -8(2y)+(-8)(9)-1\\ &= -16y-72-1\\ &= -16y-73\end{aligned}$

111. $-|-11| < |11.4|$ since $-|-11| = -11$ and $|11.4| = 11.4$.

112. $-1\frac{1}{2} > -2\frac{1}{2}$ since $-1\frac{1}{2}$ is to the right of $-2\frac{1}{2}$ on the number line.

113. $-7.2 + (-8.1) = -15.3$

114. $14 - 20 = 14 + (-20) = -6$

115. $4(-20) = -80$

116. $\frac{-20}{4} = -5$

117. $-\frac{4}{5}\left(\frac{5}{16}\right) = -\frac{4}{16} = -\frac{1}{4}$

118. $-0.5(-0.3) = 0.15$

119. $8 \div 2 \cdot 4 = 4 \cdot 4 = 16$

120. $(-2)^4 = (-2)(-2)(-2)(-2) = 16$

121. $\frac{-3-2(-9)}{-15-3(-4)} = \frac{-3+18}{-15+12} = \frac{15}{-3} = -5$

122. $\begin{aligned}5+2[(7-5)^2+(1-3)] &= 5+2[2^2+(-2)]\\ &= 5+2[4+(-2)]\\ &= 5+2[2]\\ &= 5+4\\ &= 9\end{aligned}$

123. $-\frac{5}{8} \div \frac{3}{4} = -\frac{5}{8} \cdot \frac{4}{3} = -\frac{20}{24} = -\frac{5}{6}$

124. $\frac{-15+(-4)^2+|-9|}{10-2\cdot 5} = \frac{-15+16+9}{10-10} = \frac{1+9}{0}$ is undefined.

Chapter 1 Test

1. The absolute value of negative seven is greater than five is written as $|-7| > 5$.

2. The sum of nine and five is greater than or equal to four is written as $(9 + 5) \geq 4$.

3. $-13 + 8 = -5$

4. $-13 - (-2) = -13 + 2 = -11$

5. $12 \div 4 \cdot 3 - 6 \cdot 2 = 3 \cdot 3 - 6 \cdot 2 = 9 - 12 = -3$

6. $(13)(-3) = -39$

7. $(-6)(-2) = 12$

8. $\frac{|-16|}{-8} = \frac{16}{-8} = -2$

9. $\frac{-8}{0}$ is undefined.

10. $\frac{|-6|+2}{5-6} = \frac{6+2}{5+(-6)} = \frac{8}{-1} = -8$

11. $\frac{1}{2}-\frac{5}{6}=\frac{1\cdot 3}{2\cdot 3}-\frac{5}{6}=\frac{3}{6}-\frac{5}{6}=\frac{3-5}{6}=\frac{-2}{6}=-\frac{1}{3}$

12. $$\begin{aligned}-1\frac{1}{8}+5\frac{3}{4}&=-\frac{9}{8}+\frac{23}{4}\\&=-\frac{9}{8}+\frac{2\cdot 23}{2\cdot 4}\\&=-\frac{9}{8}+\frac{46}{8}\\&=\frac{-9+46}{8}\\&=\frac{37}{8}\\&=4\frac{5}{8}\end{aligned}$$

13. $$\begin{aligned}(2-6)\div\frac{-2-6}{-3-1}-\frac{1}{2}&=(2-6)\div\frac{-8}{-4}-\frac{1}{2}\\&=-4\div 2-\frac{1}{2}\\&=-2-\frac{1}{2}\\&=-2\frac{1}{2}\end{aligned}$$

14. $3(-4)^2-80=3(16)-80=48+(-80)=-32$

15. $$\begin{aligned}6[5+2(3-8)-3]&=6\{5+2[3+(-8)]+(-3)\}\\&=6\{5+2[-5]+(-3)\}\\&=6\{5+(-10)+(-3)\}\\&=6\{-5+(-3)\}\\&=6\{-8\}\\&=-48\end{aligned}$$

16. $\frac{-12+3\cdot 8}{4}=\frac{-12+24}{4}=\frac{12}{4}=3$

17. $\frac{(-2)(0)(-3)}{-6}=\frac{0(-3)}{-6}=\frac{0}{-6}=0$

18. $-3>-7$ since -3 is to the right of -7 on the number line.

19. $4>-8$ since 4 is to the right of -8 on the number line.

20. $2<|-3|$ since $2<3$.

21. $|-2|=-1-(-3)$ since $|-2|=2$ and $-1-(-3)=-1+3=2$.

22. $2221<10{,}993$ or $10{,}993>2221$

23. **a.** The natural numbers are 1 and 7.

b. The whole numbers are 0, 1 and 7.

c. The integers are $-5, -1, 0, 1$, and 7.

d. The rational numbers are $-5, -1, \frac{1}{4}, 0, 1, 7$, and 11.6.

e. The irrational numbers are $\sqrt{7}$ and 3π.

f. The real numbers are all numbers in the given set.

24. Let $x=6$ and $y=-2$.
$x^2+y^2=(6)^2+(-2)^2=36+4=40$

25. Let $x=6$, $y=-2$ and $z=-3$.
$x+yz=6+(-2)(-3)=6+6=12$

26. Let $x=6$ and $y=-2$.
$$\begin{aligned}2+3x-y&=2+3(6)-(-2)\\&=2+18+2\\&=20+2\\&=22\end{aligned}$$

27. Let $x=6$, $y=-2$ and $z=-3$.
$\frac{y+z-1}{x}=\frac{-2+(-3)-1}{6}=\frac{-5+(-1)}{6}=\frac{-6}{6}=-1$

28. $8+(9+3)=(8+9)+3$; associative property of addition

29. $6\cdot 8=8\cdot 6$; commutative property of multiplication

30. $-6(2+4)=-6\cdot 2+(-6)\cdot 4$; distributive property

31. $\frac{1}{6}(6)=1$; multiplicative inverse property

32. The opposite of -9 is 9.

33. The reciprocal of $-\frac{1}{3}$ is -3.

34. Look for the negative number that has the greatest absolute value. The second down had the greatest loss of yardage.

35. Gains: 5, 29
Losses: −10, −2

$$\begin{aligned}\text{Total gain or loss} &= 5+(-10)+(-2)+29\\ &= (-5)+(-2)+29\\ &= -7+29\\ &= 22 \text{ yards gained}\end{aligned}$$

Yes, they scored a touchdown.

36. Since −14 + 31 = 17, the temperature at noon was 17°.

37. 356 + 460 + (−166) = 650
The net income was $650 million.

38. Change in value per share = −1.50
Change in total value = 280(−1.50) = −420
Total loss of $420

Chapter 2

Section 2.1

Practice Exercises

1. a. The numerical coefficient of t is 1, since t is $1t$.

b. The numerical coefficient of $-7x$ is -7.

c. The numerical coefficient of $-\frac{w}{5}$ is $-\frac{1}{5}$, since $-\frac{w}{5}$ means $-\frac{1}{5}\cdot w$.

d. The numerical coefficient of $43x^4$ is 43.

e. The numerical coefficient of $-b$ is -1, since $-b$ is $-1b$.

2. a. $-4xy$ and $5yx$ are like terms, since $xy = yx$ by the commutative property.

b. $5q$ and $-3q^2$ are unlike terms, since the exponents on q are not the same.

c. $3ab^2$, $-2ab^2$, and $43ab^2$ are like terms, since each variable and its exponent match.

d. y^5 and $\frac{y^5}{2}$ are like terms, since the exponents on y are the same.

3. a. $4x^2+3x^2=(4+3)x^2=7x^2$

b. $-3y + y = -3y + 1y = (-3 + 1)y = -2y$

c. $5x-3x^2+8x^2=5x+(-3+8)x^2=5x+5x^2$

4. a. $3y+8y-7+2=(3+8)y+(-7+2)=11y-5$

b.
$$\begin{aligned}6x-3-x-3&=6x-1x+(-3-3)\\&=(6-1)x+(-3-3)\\&=5x-6\end{aligned}$$

c. $\frac{3}{4}t-t=\frac{3}{4}t-1t=\left(\frac{3}{4}-1\right)t=-\frac{1}{4}t$

d.
$$\begin{aligned}9y+3.2y+10+3&=(9+3.2)y+(10+3)\\&=12.2y+13\end{aligned}$$

e. $5z-3z^4$
These two terms cannot be combined because they are unlike terms.

5. a. $3(2x - 7) = 3(2x) + 3(-7) = 6x - 21$

b.
$$\begin{aligned}&-5(3x-4z-5)\\&=-5(3x)+(-5)(-4z)+(-5)(-5)\\&=-15x+20z+25\end{aligned}$$

c.
$$\begin{aligned}&-(2x-y+z-2)\\&=-1(2x-y+z-2)\\&=-1(2x)-1(-y)-1(z)-1(-2)\\&=-2x+y-z+2\end{aligned}$$

6. a. $4(9x + 1) + 6 = 36x + 4 + 6 = 36x + 10$

b.
$$\begin{aligned}-7(2x-1)-(6-3x)&=-14x+7-6+3x\\&=-11x+1\end{aligned}$$

c. $8 - 5(6x + 5) = 8 - 30x - 25 = -30x - 17$

7. "Subtract $7x - 1$ from $2x + 3$" translates to $(2x+3)-(7x-1)=2x+3-7x+1=-5x+4$

8. a.

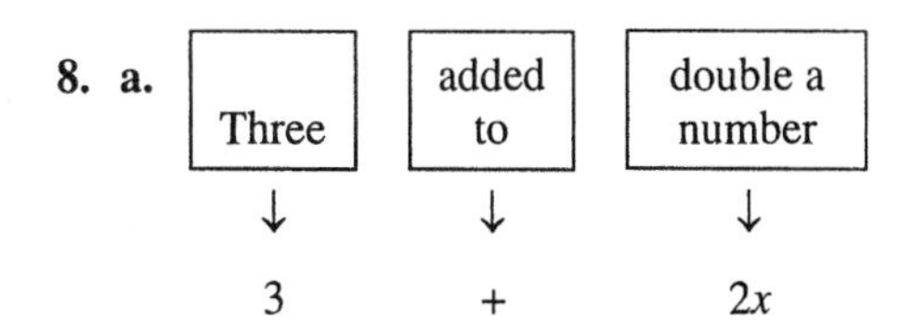

b.

the sum of 5 and a number	subtracted from	six	
↓	↓	↓	
$(5 + x)$	$-$	6	$= 5 + x - 6$

$(5 + x) - 6 = 5 + x - 6 = x - 1$

c.

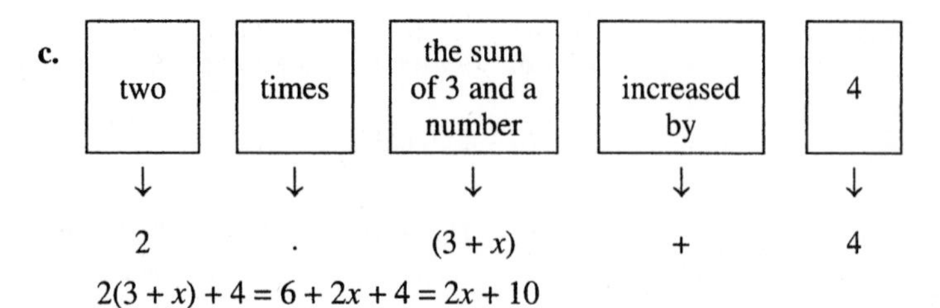

$2(3 + x) + 4 = 6 + 2x + 4 = 2x + 10$

d.

a number	added to	half the number	added to	5 times the number
↓	↓	↓	↓	↓
x	$+$	$\frac{1}{2}x$	$+$	$5x$

$$x + \frac{1}{2}x + 5x = \frac{13}{2}x$$

Vocabulary and Readiness Check

1. $23y^2 + 10y - 6$ is called an <u>expression</u> while $23y^2$, $10y$, and -6 are each called a <u>term</u>.

2. To simplify $x + 4x$, we <u>combine like terms</u>.

3. The term y has an understood <u>numerical coefficient</u> of 1.

4. The terms $7z$ and $7y$ are <u>unlike</u> terms and the terms $7z$ and $-z$ are <u>like</u> terms.

5. For the term $-\frac{1}{2}xy^2$, the number $-\frac{1}{2}$ is the <u>numerical coefficient</u>.

6. $5(3x - y)$ equals $15x - 5y$ by the <u>distributive</u> property.

7. The numerical coefficient of $-7y$ is -7.

8. The numerical coefficient of $3x$ is 3.

9. The numerical coefficient of x is 1.

10. The numerical coefficient of $-y$ is -1.

11. The numerical coefficient of $-\frac{5y}{3}$ is $-\frac{5}{3}$.

12. The numerical coefficient of $-\frac{2}{3}z$ is $-\frac{2}{3}$.

13. $5y$ and y are like terms.

14. $-2x^2y$ and $6xy$ are unlike terms.

15. $2z$ and $3z^2$ are unlike terms.

16. b^2a and $-\frac{7}{8}ab^2$ are like terms.

Exercise Set 2.1

1. $7y + 8y = (7 + 8)y = 15y$

3. $8w - w + 6w = (8 - 1 + 6)w = 13w$

5. $$\begin{aligned} 3b-5-10b-4 &= 3b-10b-5-4 \\ &= (3-10)b-9 \\ &= -7b-9 \end{aligned}$$

7. $m-4m+2m-6=(1-4+2)m-6=-m-6$

9. $$\begin{aligned} 5g-3-5-5g &= (5g-5g)+(-3-5) \\ &= (5-5)g+(-8) \\ &= 0g-8 \\ &= -8 \end{aligned}$$

11. $$\begin{aligned} 6.2x-4+x-1.2 &= 6.2x+x-4-1.2 \\ &= (6.2+1)x-5.2 \\ &= 7.2x-5.2 \end{aligned}$$

13. $$\begin{aligned} 6x-5x+x-3+2x &= 6x-5x+x+2x-3 \\ &= (6-5+1+2)x-3 \\ &= 4x-3 \end{aligned}$$

15. $7x^2+8x^2-10x^2=(7+8-10)x^2=5x^2$

17. $$\begin{aligned} &6x+0.5-4.3x-0.4x+3 \\ &= 6x-4.3x-0.4x+0.5+3 \\ &= (6-4.3-0.4)x+(0.5+3) \\ &= 1.3x+3.5 \end{aligned}$$

19. Answers may vary

21. $5(y - 4) = 5(y) - 5(4) = 5y - 20$

23. $-2(x + 2) = -2(x) + (-2)(2) = -2x - 4$

25. $7(d - 3) + 10 = 7d - 21 + 10 = 7d - 11$

27. $$\begin{aligned} -5(2x-3y+6) &= -5(2x)-(-5)(3y)+(-5)(6) \\ &= -10x+15y-30 \end{aligned}$$

29. $-(3x - 2y + 1) = -3x + 2y - 1$

31. $$\begin{aligned} 5(x+2)-(3x-4) &= 5x+10-3x+4 \\ &= 2x+14 \end{aligned}$$

33. $$\begin{array}{ccc} 6x+7 & \text{added to} & 4x-10 \\ \downarrow & \downarrow & \downarrow \\ (6x+7) & + & (4x-10) = 6x+4x+7-10 \\ & & = 10x-3 \end{array}$$

35. $$\begin{array}{ccc} 3x-8 & \text{minus} & 7x+1 \\ \downarrow & \downarrow & \downarrow \\ (3x-8) & - & (7x+1) = 3x-8-7x-1 \\ & & = 3x-7x-8-1 \\ & & = -4x-9 \end{array}$$

37. $$\begin{array}{ccc} m-9 & \text{minus} & 5m-6 \\ \downarrow & \downarrow & \downarrow \\ (m-9) & - & (5m-6) = m-9-5m+6 \\ & & = m-5m-9+6 \\ & & = -4m-3 \end{array}$$

39. $2k - k - 6 = (2 - 1)k - 6 = k - 6$

41. $$\begin{aligned} -9x+4x+18-10x &= -9x+4x-10x+18 \\ &= (-9+4-10)x+18 \\ &= -15x+18 \end{aligned}$$

43. $$\begin{aligned} -4(3y-4)+12y &= -4(3y)-(-4)(4)+12y \\ &= -12y+16+12y \\ &= -12y+12y+16 \\ &= 16 \end{aligned}$$

45. $3(2x-5)-5(x-4)=6x-15-5x+20=x+5$

47. $-2(3x-4)+7x-6=-6x+8+7x-6=x+2$

49. $5k - (3k - 10) = 5k - 3k + 10 = 2k + 10$

51. $(3x+4)-(6x-1)=3x+4-6x+1=-3x+5$

53. $3.4m-4-3.4m-7=3.4m-3.4m-4-7=-11$

55. $\frac{1}{3}(7y-1)+\frac{1}{6}(4y+7)=\frac{7}{3}y-\frac{1}{3}+\frac{4}{6}y+\frac{7}{6}$
$=\frac{7}{3}y+\frac{2}{3}y-\frac{1}{3}+\frac{7}{6}$
$=\frac{9}{3}y-\frac{2}{6}+\frac{7}{6}$
$=3y+\frac{5}{6}$

57. $2+4(6x-6)=2+24x-24=-22+24x$

59. $0.5(m+2)+0.4m=0.5m+1+0.4m=0.9m+1$

61. $10-3(2x+3y)=10-6x-9y$

63. $6(3x-6)-2(x+1)-17x=18x-36-2x-2-17x$
$=18x-2x-17x-36-2$
$=-x-38$

65. $\frac{1}{2}(12x-4)-(x+5)=6x-2-x-5=5x-7$

67.

twice a number	decreased by	4
↓	↓	↓
$2x$	$-$	4

69.

seven	added to	double a number
↓	↓	↓
7	+	$2x$

71.

three-fourths of a number	increased by	12
↓	↓	↓
$\frac{3}{4}x$	+	12

73.

5 times a number	added to	−2	added to	7 times a number
↓	↓	↓	↓	↓
$5x$	+	−2	+	$7x$

$5x+(-2)+7x=12x-2$

75.

8	times	the sum of a number and 6
↓	↓	↓
8	·	$(x+6)$

$8(x+6)=8x+48$

77.

double a number	minus	the sum of the number and 10
↓	↓	↓
$2x$	$-$	$(x+10)$

$2x-(x+10)=2x-x-10=x-10$

79.

7	multiplied by	the quotient of a number and 6
↓	↓	↓
7	$\cdot$	$\frac{x}{6}$

$7\left(\frac{x}{6}\right)=\frac{7x}{6}$

81.

2	added to	3 times a number	added to	−9	added to	4 times a number
↓	↓	↓	↓	↓	↓	↓
2	+	$3x$	+	−9	+	$4x$

$2+3x+(-9)+4x=7x-7$

83. $y-x^2=3-(-1)^2=3-1=2$

85. $a-b^2=2-(-5)^2=2-25=-23$

87. $yz-y^2=(-5)(0)-(-5)^2=0-25=-25$

89. $5x+(4x-1)+5x+(4x-1)=5x+4x-1+5x+4x-1$
$=18x-2$
The perimeter is $(18x-2)$ feet.

91. 1 cone + 1 cylinder $\stackrel{?}{=}$ 3 cubes
1 cube + 2 cubes $\stackrel{?}{=}$ 3 cubes
3 cubes = 3 cubes: Balanced

93. 2 cylinders + 1 cube $\stackrel{?}{=}$ 3 cones + 2 cubes
$2\cdot 2$ cubes + 1 cube $\stackrel{?}{=}$ 3 cubes + 2 cubes
4 cubes + 1 cube $\stackrel{?}{=}$ 3 cubes + 2 cubes
5 cubes = 5 cubes: Balanced

95. Answers may vary

97. $12(x+2)+(3x-1)=12x+24+3x-1=15x+23$
The total length is $(15x+23)$ inches.

99. $5b^2c^3+8b^3c^2-7b^3c^2=5b^2c^3+b^3c^2$

101. $3x-(2x^2-6x)+7x^2 = 3x-2x^2+6x+7x^2$
$= 5x^2+9x$

103. $-(2x^2y+3z)+3z-5x^2y$
$= -2x^2y-3z+3z-5x^2y$
$= -7x^2y$

Section 2.2

Practice Exercises

1. $x+3=-5$
$x+3-3=-5-3$
$x=-8$
Check: $x+3=-5$
$-8+3 \stackrel{?}{=} -5$
$-5=-5$
The solution is −8.

2. $y-0.3=-2.1$
$y-0.3+0.3=-2.1+0.3$
$y=-1.8$
Check: $y-0.3=-2.1$
$-1.8-0.3 \stackrel{?}{=} -2.1$
$-2.1=-2.1$
The solution is −1.8.

3. $8x-5x-3+9=x+x+3-7$
$3x+6=2x-4$
$3x+6-2x=2x-4-2x$
$x+6=-4$
$x+6-6=-4-6$
$x=-10$
Check:
$8x-5x-3+9=x+x+3-7$
$8(-10)-5(-10)-3+9 \stackrel{?}{=} -10+(-10)+3-7$
$-80+50-3+9 \stackrel{?}{=} -10+(-10)+3-7$
$-24=-24$
The solution is −10.

4. $2=4(2a-3)-(7a+4)$
$2=4(2a)+4(-3)-7a-4$
$2=8a-12-7a-4$
$2=a-16$
$2+16=a-16+16$
$18=a$
Check by replacing a with 18 in the original equation.

5. $\frac{4}{5}x=16$
$\frac{5}{4}\cdot\frac{4}{5}x=\frac{5}{4}\cdot 16$
$\left(\frac{5}{4}\cdot\frac{4}{5}\right)x=\frac{5}{4}\cdot 16$
$1x=20$
$x=20$
Check: $\frac{4}{5}x=16$
$\frac{4}{5}\cdot 20 \stackrel{?}{=} 16$
$16=16$
The solution is 20.

6. $8x=-96$
$\frac{8x}{8}=\frac{-96}{8}$
$x=-12$
Check: $8x=-96$
$8(-12) \stackrel{?}{=} -96$
$-96=-96$
The solution is −12.

7. $\frac{x}{5}=13$
$5\cdot\frac{x}{5}=5\cdot 13$
$x=65$
Check: $\frac{x}{5}=13$
$\frac{65}{5} \stackrel{?}{=} 13$
$13=13$
The solution is 65.

8. $6b-11b=18+2b-6+9$
$-5b=21+2b$
$-5b-2b=21+2b-2b$
$-7b=21$
$\frac{-7b}{-7}=\frac{21}{-7}$
$b=-3$
Check by replacing b with −3 in the original equation. The solution is −3.

9. a. The other number is $9-2=7$.

b. The other number is $9-x$.

c. The other piece has length $(9 - x)$ feet.

10. Let x = first integer.
$x + 2$ = second even integer.
$x + 4$ = third even integer.
$x + (x + 2) + (x + 4) = 3x + 6$

Vocabulary and Readiness Check

1. The difference between an equation and an expression is that an equation contains an equal sign, whereas an expression does not.

2. Equivalent equations are equations that have the same solution.

3. A value of the variable that makes the equation a true statement is called a solution of the equation.

4. The process of finding the solution of an equation is called solving the equation for the variable.

5. By the addition property of equality, $x = -2$ and $x + 10 = -2 + 10$ are equivalent equations.

6. The equations $x = \frac{1}{2}$ and $\frac{1}{2} = x$ are equivalent equations. The statement is true.

7. By the multiplication property of equality, $y = \frac{1}{2}$ and $5 \cdot y = 5 \cdot \frac{1}{2}$ are equivalent equations.

8. The equations $\frac{z}{4} = 10$ and $4 \cdot \frac{z}{4} = 10$ are not equivalent equations. The statement is false.

9. The equations $-7x = 30$ and $\frac{-7x}{-7} = \frac{30}{7}$ are not equivalent equations. The statement is false.

10. By the multiplication property of equality, $9x = -63$ and $\frac{9x}{9} = \frac{-63}{9}$ are equivalent equations.

11. $3a = 27$
$a = \frac{27}{3} = 9$

12. $9c = 54$
$c = \frac{54}{9} = 6$

13. $5b = 10$
$b = \frac{10}{5} = 2$

14. $7t = 14$
$t = \frac{14}{7} = 2$

Exercise Set 2.2

1.
$$x + 7 = 10$$
$$x + 7 - 7 = 10 - 7$$
$$x = 3$$
Check: $x + 7 = 10$
$3 + 7 \stackrel{?}{=} 10$
$10 = 10$
The solution is 3.

3.
$$x - 2 = -4$$
$$x - 2 + 2 = -4 + 2$$
$$x = -2$$
Check: $x - 2 = -4$
$-2 - 2 \stackrel{?}{=} -4$
$-4 = -4$
The solution is -2.

5.
$$3 + x = -11$$
$$3 + x - 3 = -11 - 3$$
$$x = -14$$
Check: $3 + x = -11$
$3 + (-14) \stackrel{?}{=} -11$
$-11 = -11$
The solution is -14.

7.
$$r - 8.6 = -8.1$$
$$r - 8.6 + 8.6 = -8.1 + 8.6$$
$$r = 0.5$$
Check: $x - 8.6 = -8.1$
$0.5 - 8.6 \stackrel{?}{=} -8.1$
$-8.1 = -8.1$
The solution is 0.5.

9. $8x = 7x - 3$
$8x - 7x = 7x - 7x - 3$
$x = -3$
Check: $8x = 7x - 3$
$8(-3) \stackrel{?}{=} 7(-3) - 3$
$-24 \stackrel{?}{=} -21 - 3$
$-24 = -24$
The solution is −3.

11. $5b - 0.7 = 6b$
$5b - 5b - 0.7 = 6b - 5b$
$-0.7 = b$
Check: $5b - 0.7 = 6b$
$5(-0.7) - 0.7 \stackrel{?}{=} 6(-0.7)$
$-3.5 - 0.7 \stackrel{?}{=} -4.2$
$-4.2 = -4.2$
The solution is −0.7.

13. $7x - 3 = 6x$
$7x - 6x - 3 = 6x - 6x$
$x - 3 = 0$
$x - 3 + 3 = 0 + 3$
$x = 3$
Check: $7x - 3 = 6x$
$7(3) - 3 \stackrel{?}{=} 6(3)$
$21 - 3 \stackrel{?}{=} 18$
$18 = 18$
The solution is 3.

15. $3x - 6 = 2x + 5$
$3x - 2x - 6 = 2x - 2x + 5$
$x - 6 = 5$
$x - 6 + 6 = 5 + 6$
$x = 11$
The solution is 11.

17. $3t - t - 7 = t - 7$
$2t - 7 = t - 7$
$2t - t - 7 = t - t - 7$
$t - 7 = -7$
$t - 7 + 7 = -7 + 7$
$t = 0$
The solution is 0.

19. $7x + 2x = 8x - 3$
$9x = 8x - 3$
$9x - 8x = 8x - 8x - 3$
$x = -3$
The solution is −3.

21. $-2(x+1) + 3x = 14$
$-2x - 2 + 3x = 14$
$-2 + x = 14$
$2 - 2 + x = 14 + 2$
$x = 16$
The solution is 16.

23. $-5x = -20$
$\frac{-5x}{-5} = \frac{-20}{-5}$
$x = 4$
The solution is 4.

25. $3x = 0$
$\frac{3x}{3} = \frac{0}{3}$
$x = 0$
The solution is 0.

27. $-x = -12$
$\frac{-x}{-1} = \frac{-12}{-1}$
$x = 12$
The solution is 12.

29. $3x + 2x = 50$
$5x = 50$
$\frac{5x}{5} = \frac{50}{5}$
$x = 10$
The solution is 10.

31. $\frac{2}{3}x = -8$
$\frac{3}{2}\left(\frac{2}{3}x\right) = \frac{3}{2}(-8)$
$x = -12$
The solution is −12.

33. $\frac{1}{6}d = \frac{1}{2}$
$6\left(\frac{1}{6}d\right) = 6\left(\frac{1}{2}\right)$
$d = 3$
The solution is 3.

35. $\frac{a}{-2} = 1$
$-2\left(\frac{a}{-2}\right) = -2(1)$
$a = -2$
The solution is −2.

37. $\frac{k}{7} = 0$

$7\left(\frac{k}{7}\right) = 7(0)$

$k = 0$

The solution is 0.

39. Answers may vary

41. $2x - 4 = 16$

$2x - 4 + 4 = 16 + 4$

$2x = 20$

$\frac{2x}{2} = \frac{20}{2}$

$x = 10$

Check: $2x - 4 = 16$

$2(10) - 4 \stackrel{?}{=} 16$

$20 - 4 \stackrel{?}{=} 16$

$16 = 16$

The solution is 10.

43. $-x + 2 = 22$

$-x + 2 - 2 = 22 - 2$

$-x = 20$

$x = -20$

Check: $-x + 2 = 22$

$-(-20) + 2 \stackrel{?}{=} 22$

$20 + 2 \stackrel{?}{=} 22$

$22 = 22$

The solution is −20.

45. $6a + 3 = 3$

$6a + 3 - 3 = 3 - 3$

$6a = 0$

$\frac{6a}{6} = \frac{0}{6}$

$a = 0$

Check: $6a + 3 = 3$

$6(0) + 3 \stackrel{?}{=} 3$

$0 + 3 \stackrel{?}{=} 3$

$3 = 3$

The solution is 0.

47. $6x + 10 = -20$

$6x + 10 - 10 = -20 - 10$

$6x = -30$

$\frac{6x}{6} = \frac{-30}{6}$

$x = -5$

Check: $6x + 10 = -20$

$6(-5) + 10 \stackrel{?}{=} -20$

$-30 + 10 \stackrel{?}{=} -20$

$-20 = -20$

The solution is −5.

49. $5 - 0.3k = 5$

$5 - 5 - 0.3k = 5 - 5$

$-0.3k = 0$

$\frac{-0.3k}{-0.3} = \frac{0}{-0.3}$

$k = 0$

Check: $5 - 0.3k = 5$

$5 - 0.3(0) \stackrel{?}{=} 5$

$5 - 0 \stackrel{?}{=} 5$

$5 = 5$

The solution is 0.

51. $-2x + \frac{1}{2} = \frac{7}{2}$

$-2x + \frac{1}{2} - \frac{1}{2} = \frac{7}{2} - \frac{1}{2}$

$-2x = \frac{6}{2}$

$-2x = 3$

$\frac{-2x}{-2} = \frac{3}{-2}$

$x = -\frac{3}{2}$

Check: $-2x + \frac{1}{2} = \frac{7}{2}$

$-2\left(-\frac{3}{2}\right) + \frac{1}{2} \stackrel{?}{=} \frac{7}{2}$

$\frac{6}{2} + \frac{1}{2} \stackrel{?}{=} \frac{7}{2}$

$\frac{7}{2} = \frac{7}{2}$

The solution is $-\frac{3}{2}$.

53. $\frac{x}{3} + 2 = -5$

$\frac{x}{3} + 2 - 2 = -5 - 2$

$\frac{x}{3} = -7$

$3 \cdot \frac{x}{3} = 3 \cdot -7$

$x = -21$

Check: $\frac{x}{3}+2=-5$

$\frac{-21}{3}+2 \stackrel{?}{=} -5$

$-7+2 \stackrel{?}{=} -5$

$-5=-5$

The solution is –21.

55. $10=2x-1$

$10+1=2x-1+1$

$11=2x$

$\frac{11}{2}=\frac{2x}{2}$

$\frac{11}{2}=x$

Check: $10=2x-1$

$10 \stackrel{?}{=} 2\left(\frac{11}{2}\right)-1$

$10 \stackrel{?}{=} 11-1$

$10=10$

The solution is $\frac{11}{2}$.

57. $6z-8-z+3=0$

$5z-5=0$

$5z-5+5=0+5$

$5z=5$

$\frac{5z}{5}=\frac{5}{5}$

$z=1$

Check: $6z-8-z+3=0$

$6\cdot 1-8-1+3 \stackrel{?}{=} 0$

$6-8-1+3 \stackrel{?}{=} 0$

$0=0$

The solution is 1.

59. $10-3x-6-9x=7$

$4-12x=7$

$4-4-12x=7-4$

$-12x=3$

$\frac{-12x}{-12}=\frac{3}{-12}$

$x=-\frac{1}{4}$

Check: $10-3x-6-9x=7$

$10-3\left(-\frac{1}{4}\right)-6-9\left(-\frac{1}{4}\right) \stackrel{?}{=} 7$

$10+\frac{3}{4}-6+\frac{9}{4} \stackrel{?}{=} 7$

$4+\frac{12}{4} \stackrel{?}{=} 7$

$4+3 \stackrel{?}{=} 7$

$7=7$

The solution is $-\frac{1}{4}$.

61. $\frac{5}{6}x=10$

$\frac{6}{5}\cdot\frac{5}{6}x=\frac{6}{5}\cdot 10$

$x=12$

Check: $\frac{5}{6}x=10$

$\frac{5}{6}\cdot 12 \stackrel{?}{=} 10$

$10=10$

The solution is 12.

63. $1=0.4x-0.6x-5$

$1=-0.2x-5$

$1+5=-0.2x-5+5$

$6=-0.2x$

$\frac{6}{-0.2}=\frac{-0.2x}{-0.2}$

$-30=x$

Check: $1=0.4x-0.6x-5$

$1 \stackrel{?}{=} 0.4(-30)-0.6(-30)-5$

$1 \stackrel{?}{=} -12+18-5$

$1=1$

The solution is –30.

65. $z-5z=7z-9-z$

$-4z=6z-9$

$-4z-6z=6z-6z-9$

$-10z=-9$

$\frac{-10z}{-10}=\frac{-9}{-10}$

$z=\frac{9}{10}$

Check: $z-5z=7z-9-z$

$$\frac{9}{10}-5\left(\frac{9}{10}\right)\stackrel{?}{=}7\left(\frac{9}{10}\right)-9-\frac{9}{10}$$
$$\frac{9}{10}-\frac{45}{10}\stackrel{?}{=}\frac{63}{10}-\frac{90}{10}-\frac{9}{10}$$
$$-\frac{36}{10}=-\frac{36}{10}$$

The solution is $\frac{9}{10}$.

67. $0.4x-0.6x-5=1$
$$-0.2x-5=1$$
$$-0.2x-5+5=1+5$$
$$-0.2x=6$$
$$\frac{-0.2x}{-0.2}=\frac{6}{-0.2}$$
$$x=-30$$

Check: $0.4x-0.6x-5=1$
$$0.4(-30)-0.6(-30)-5\stackrel{?}{=}1$$
$$-12+18-5\stackrel{?}{=}1$$
$$1=1$$

The solution is −30.

69. $6-2x+8=10$
$$14-2x=10$$
$$14-14-2x=10-14$$
$$-2x=-4$$
$$\frac{-2x}{-2}=\frac{-4}{-2}$$
$$x=2$$

Check: $6-2x+8=10$
$$6-2\cdot 2+8\stackrel{?}{=}10$$
$$6-4+8\stackrel{?}{=}10$$
$$10=10$$

The solution is 2.

71. $-3a+6+5a=7a-8a$
$$6+2a=-a$$
$$6+2a-2a=-a-2a$$
$$6=-3a$$
$$\frac{6}{-3}=\frac{-3a}{-3}$$
$$-2=a$$

Check: $-3a+6+5a=7a-8a$
$$-3(-2)+6+5(-2)\stackrel{?}{=}7(-2)-8(-2)$$
$$6+6-10\stackrel{?}{=}-14+16$$
$$2=2$$

The solution is −2.

73. $20=-3(2x+1)+7x$
$$20=-6x-3+7x$$
$$20=x-3$$
$$20+3=x-3+3$$
$$23=x$$

Check: $20=-3(2x+1)+7x$
$$20\stackrel{?}{=}-3(2\cdot 23+1)+7\cdot 23$$
$$20\stackrel{?}{=}-3(46+1)+161$$
$$20\stackrel{?}{=}-3(47)+161$$
$$20\stackrel{?}{=}-141+161$$
$$20=20$$

The solution is 23.

75. The other number is $20-p$.

77. The length of the other piece is $(10-x)$ feet.

79. The supplement of the angle $x°$ is $(180-x)°$.

81. April received $(n+284)$ votes.

83. The length of the Golden Gate Bridge is $(m-60)$ feet.

85. Ortiz received $(n+47{,}628)$ votes.

87. The area of the Sahara Desert is $7x$ square miles.

89. Sum = first integer + second integer
Sum $= x+(x+2)=x+x+2=2x+2$

91. Sum = first integer + third integer
Sum $= x+(x+2)=x+x+2=2x+2$

93. Let x = first room number.
$x+2$ = second room number
$x+4$ = third room number
$x+6$ = fourth room number
$x+8$ = fifth room number
$x+(x+2)+(x+4)+(x+6)+(x+8)=5x+20$

95. $5x+2(x-6)=5x+2x-12=7x-12$

97. $-(x-1)+x=-x+1+x=1$

99. $(-3)^2=(-3)(-3)=9$
$-3^2=-3\cdot 3=-9$
$(-3)^2>-3^2$

101. $(-2)^3=(-2)(-2)(-2)=-8$
$-2^3=-2\cdot 2\cdot 2=-8$
$(-2)^3=-2^3$

103. $$\begin{aligned} 180-[x+(2x+7)] &= 180-[x+2x+7] \\ &= 180-[3x+7] \\ &= 180-3x-7 \\ &= 173-3x \end{aligned}$$
The third angle is $(173 - 3x)°$.

105. Answers may vary

107. $$\begin{aligned} x-4 &= -9 \\ x-4+(4) &= -9+(4) \\ x &= -5 \end{aligned}$$
The answer is 4.

109. Answers may vary

111. Answers may vary

113. $$\begin{aligned} 6x &= \underline{\quad\quad} \\ 6(-8) &= \underline{\quad\quad} \\ -48 &= \underline{\quad\quad} \end{aligned}$$

115. $$\begin{aligned} 9x &= 2100 \\ \frac{9x}{9} &= \frac{2100}{9} \\ x &= \frac{700}{3} \end{aligned}$$
Each dose should be $\frac{700}{3}$ milligrams.

117. Check $y = 1.2$: $$\begin{aligned} 8.13 + 5.85y &= 20.05y - 8.91 \\ 8.13+5.85(1.2) &\stackrel{?}{=} 20.05(1.2)-8.91 \\ 8.13+7.02 &\stackrel{?}{=} 24.06-8.91 \\ 15.15 &= 15.15 \end{aligned}$$
Solution

119. $$\begin{aligned} -3.6x &= 10.62 \\ \frac{-3.6x}{-3.6} &= \frac{10.62}{-3.6} \\ x &= -2.95 \end{aligned}$$

121. $$\begin{aligned} 7x-5.06 &= -4.92 \\ 7x-5.06+5.06 &= -4.92+5.06 \\ 7x &= 0.14 \\ \frac{7x}{7} &= \frac{0.14}{7} \\ x &= 0.02 \end{aligned}$$

Section 2.3

Practice Exercises

1. $$\begin{aligned} 2(4a-9)+3 &= 5a-6 \\ 8a-18+3 &= 5a-6 \\ 8a-15 &= 5a-6 \\ 8a-15-5a &= 5a-6-5a \\ 3a-15 &= -6 \\ 3a-15+15 &= -6+15 \\ 3a &= 9 \\ \frac{3a}{3} &= \frac{9}{3} \\ a &= 3 \end{aligned}$$
Check: $$\begin{aligned} 2(4a-9)+3 &= 5a-6 \\ 2[4(3)-9]+3 &\stackrel{?}{=} 5(3)-6 \\ 2(12-9)+3 &\stackrel{?}{=} 15-6 \\ 2(3)+3 &\stackrel{?}{=} 9 \\ 6+3 &\stackrel{?}{=} 9 \\ 9 &= 1 \end{aligned}$$
The solution is 3 or the solution set is $\{3\}$.

2. $$\begin{aligned} 7(x-3) &= -6x \\ 7x-21 &= -6x \\ 7x-21-7x &= -6x-7x \\ -21 &= -13x \\ \frac{-21}{-13} &= \frac{-13x}{-13} \\ \frac{21}{13} &= x \end{aligned}$$
Check: $$\begin{aligned} 7(x-3) &= -6x \\ 7\left(\frac{21}{13}-3\right) &\stackrel{?}{=} -6\left(\frac{21}{13}\right) \\ 7\left(\frac{21}{13}-\frac{39}{13}\right) &\stackrel{?}{=} -\frac{126}{13} \\ 7\left(-\frac{18}{13}\right) &\stackrel{?}{=} -\frac{126}{13} \\ -\frac{126}{13} &= -\frac{126}{13} \end{aligned}$$
The solution is $\frac{21}{13}$.

3.
$$\frac{3}{5}x-2=\frac{2}{3}x-1$$
$$15\left(\frac{3}{5}x-2\right)=15\left(\frac{2}{3}x-1\right)$$
$$15\left(\frac{3}{5}x\right)-15(2)=15\left(\frac{2}{3}x\right)-15(1)$$
$$9x-30=10x-15$$
$$9x-30-9x=10x-15-9x$$
$$-30=x-15$$
$$-30+15=x-15+15$$
$$-15=x$$

Check: $\frac{3}{5}x-2=\frac{2}{3}x-1$
$$\frac{3}{5}\cdot-15-2\stackrel{?}{=}\frac{2}{3}\cdot-15-1$$
$$-9-2\stackrel{?}{=}-10-1$$
$$-11=-11$$

The solution is −15.

4.
$$\frac{4(y+3)}{3}=5y-7$$
$$3\cdot\frac{4(y+3)}{3}=3\cdot(5y-7)$$
$$4(y+3)=3(5y-7)$$
$$4y+12=15y-21$$
$$4y+12-4y=15y-21-4y$$
$$12=11y-21$$
$$12+21=11y-21+21$$
$$33=11y$$
$$\frac{33}{11}=\frac{11y}{11}$$
$$3=y$$

To check, replace y with 3 in the original equation. The solution is 3.

5.
$$0.35x+0.09(x+4)=0.30(12)$$
$$100[0.35x+0.09(x+4)]=100[0.03(12)]$$
$$35x+9(x+4)=3(12)$$
$$35x+9x+36=36$$
$$44x+36=36$$
$$44x+36-36=36-36$$
$$44x=0$$
$$\frac{44x}{44}=\frac{0}{44}$$
$$x=0$$

To check, replace x with 0 in the original equation. The solution is 0.

6.
$$4(x+4)-x=2(x+11)+x$$
$$4x+16-x=2x+22+x$$
$$3x+16=3x+22$$
$$3x+16-3x=3x+22-3x$$
$$16=22$$

There is no solution.

7.
$$12x-18=9(x-2)+3x$$
$$12x-18=9x-18+3x$$
$$12x-18=12x-18$$
$$12x-18+18=12x-18+18$$
$$12x=12x$$
$$12x-12x=12x-12x$$
$$0=0$$

The solution is all real numbers.

Calculator Explorations

1. Solution $(-24=-24)$
2. Solution $(-4=-4)$
3. Not a solution $(19.4\neq 10.4)$
4. Not a solution $(-11.9\neq -60.1)$
5. Solution $(17{,}061=17{,}061)$
6. Solution $(-316=-316)$

Vocabulary and Readiness Check

1. $x=-7$ is an equation.
2. $x-7$ is an expression.
3. $4y-6+9y+1$ is an expression.
4. $4y-6=9y+1$ is an equation.
5. $\frac{1}{x}-\frac{x-1}{8}$ is an expression.
6. $\frac{1}{x}-\frac{x-1}{8}=6$ is an equation.
7. $0.1x+9=0.2x$ is an equation.
8. $0.1x^2+9y-0.2x^2$ is an expression.

Exercise Set 2.3

1.
$$\begin{aligned}-4y+10&=-2(3y+1)\\-4y+10&=-6y-2\\-4y+10-10&=-6y-2-10\\-4y&=-6y-12\\-4y+6y&=-6y-12+6y\\2y&=-12\\\frac{2y}{2}&=\frac{-12}{2}\\y&=-6\end{aligned}$$

3.
$$\begin{aligned}15x-8&=10+9x\\15x-8+8&=10+9x+8\\15x&=18+9x\\15x-9x&=18+9x-9x\\6x&=18\\\frac{6x}{6}&=\frac{18}{6}\\x&=3\end{aligned}$$

5.
$$\begin{aligned}-2(3x-4)&=2x\\-6x+8&=2x\\-6x+6x+8&=2x+6x\\8&=8x\\\frac{8}{8}&=\frac{8x}{8}\\1&=x\end{aligned}$$

7.
$$\begin{aligned}5(2x-1)-2(3x)&=1\\10x-5-6x&=1\\4x-5&=1\\4x-5+5&=1+5\\4x&=6\\\frac{4x}{4}&=\frac{6}{4}\\x&=\frac{3}{2}\end{aligned}$$

9.
$$\begin{aligned}-6(x-3)-26&=-8\\-6x+18-26&=-8\\-6x-8&=-8\\-6x-8+8&=-8+8\\-6x&=0\\\frac{-6x}{-6}&=\frac{0}{-6}\\x&=0\end{aligned}$$

11.
$$\begin{aligned}8-2(a+1)&=9+a\\8-2a-2&=9+a\\-2a+6&=9+a\\-2a+6-6&=9+a-6\\-2a&=3+a\\-2a-a&=3+a-a\\-3a&=3\\\frac{-3a}{-3}&=\frac{3}{-3}\\a&=-1\end{aligned}$$

13.
$$\begin{aligned}4x+3&=-3+2x+14\\4x+3&=2x+11\\4x-2x+3&=2x-2x+11\\2x+3&=11\\2x+3-3&=11-3\\2x&=8\\\frac{2x}{2}&=\frac{8}{2}\\x&=4\end{aligned}$$

15.
$$\begin{aligned}-2y-10&=5y+18\\-2y-5y-10&=5y-5y+18\\-7y-10&=18\\-7y-10+10&=18+10\\-7y&=28\\\frac{-7y}{-7}&=\frac{28}{-7}\\y&=-4\end{aligned}$$

17.
$$\begin{aligned}\frac{2}{3}x+\frac{4}{3}&=-\frac{2}{3}\\3\left(\frac{2}{3}x+\frac{4}{3}\right)&=3\left(-\frac{2}{3}\right)\\2x+4&=-2\\2x+4-4&=-2-4\\2x&=-6\\\frac{2x}{2}&=\frac{-6}{2}\\x&=-3\end{aligned}$$

19.
$$\frac{3}{4}x-\frac{1}{2}=1$$
$$4\left(\frac{3}{4}x-\frac{1}{2}\right)=4(1)$$
$$3x-2=4$$
$$3x-2+2=4+2$$
$$3x=6$$
$$\frac{3x}{3}=\frac{6}{3}$$
$$x=2$$

21.
$$0.50x+0.15(70)=35.5$$
$$100[0.50x+0.15(70)]=100(35.5)$$
$$50x+15(70)=3550$$
$$50x+1050=3550$$
$$50x+1050-1050=3550-1050$$
$$50x=2500$$
$$\frac{50x}{50}=\frac{2500}{50}$$
$$x=50$$

23.
$$\frac{2(x+1)}{4}=3x-2$$
$$4\left[\frac{2(x+1)}{4}\right]=4(3x-2)$$
$$2(x+1)=12x-8$$
$$2x+2=12x-8$$
$$2x-12x+2=12x-12x-8$$
$$-10x+2=-8$$
$$-10x+2-2=-8-2$$
$$-10x=-10$$
$$\frac{-10x}{-10}=\frac{-10}{-10}$$
$$x=1$$

25.
$$x+\frac{7}{6}=2x-\frac{7}{6}$$
$$6\left(x+\frac{7}{6}\right)=6\left(2x-\frac{7}{6}\right)$$
$$6x+7=12x-7$$
$$6x-12x+7=12x-12x-7$$
$$-6x+7=-7$$
$$-6x+7-7=-7-7$$
$$-6x=-14$$
$$\frac{-6x}{-6}=\frac{-14}{-6}$$
$$x=\frac{7}{3}$$

27.
$$0.12(y-6)+0.06y=0.08y-0.7$$
$$100[0.12(y-6)+0.06y]=100[0.08y-0.7]$$
$$12(y-6)+6y=8y-70$$
$$12y-72+6y=8y-70$$
$$18y-72=8y-70$$
$$18y-8y-72=8y-8y-70$$
$$10y-72=-70$$
$$10y-72+72=-70+72$$
$$10y=2$$
$$\frac{10y}{10}=\frac{2}{10}$$
$$y=\frac{1}{5}=0.2$$

29.
$$4(3x+2)=12x+8$$
$$12x+8=12x+8$$
$$12x+8-12x=12x+8-12x$$
$$8=8$$
All real numbers are solutions.

31.
$$\frac{x}{4}+1=\frac{x}{4}$$
$$4\left(\frac{x}{4}+1\right)=4\left(\frac{x}{4}\right)$$
$$x+4=x$$
$$x-x+4=x-x$$
$$4=0$$
There is no solution.

33.
$$3x-7=3(x+1)$$
$$3x-7=3x+3$$
$$3x-3x-7=3x-3x+3$$
$$-7=3$$
There is no solution.

35.
$$-2(6x-5)+4=-12x+14$$
$$-12x+10+4=-12x+14$$
$$-12x+14=-12x+14$$
$$-12x+14+12x=-12x+14+12x$$
$$14=14$$
All real numbers are solutions.

37.
$$\frac{6(3-z)}{5}=-z$$
$$5\left[\frac{6(3-z)}{5}\right]=5(-z)$$
$$6(3-z)=5(-z)$$
$$18-6z=-5z$$
$$18-6z+6z=-5z+6z$$
$$18=z$$

39.
$$\begin{aligned}-3(2t-5)+2t&=5t-4\\-6t+15+2t&=5t-4\\-4t+15&=5t-4\\-4t+15+4&=5t-4+4\\-4t+19&=5t\\-4t+19+4t&=5t+4t\\19&=9t\\\frac{19}{9}&=\frac{9t}{9}\\\frac{19}{9}&=t\end{aligned}$$

41.
$$\begin{aligned}5y+2(y-6)&=4(y+1)-2\\5y+2y-12&=4y+4-2\\7y-12&=4y+2\\7y-12+12&=4y+2+12\\7y&=4y+14\\7y-4y&=4y+14-4y\\3y&=14\\\frac{3y}{3}&=\frac{14}{3}\\y&=\frac{14}{3}\end{aligned}$$

43.
$$\begin{aligned}\frac{3(x-5)}{2}&=\frac{2(x+5)}{3}\\6\left[\frac{3(x-5)}{2}\right]&=6\left[\frac{2(x+5)}{3}\right]\\9(x-5)&=4(x+5)\\9x-45&=4x+20\\9x-4x-45&=4x-4x+20\\5x-45&=20\\5x-45+45&=20+45\\5x&=65\\\frac{5x}{5}&=\frac{65}{5}\\x&=13\end{aligned}$$

45.
$$\begin{aligned}0.7x-2.3&=0.5\\10(0.7x-2.3)&=10(0.5)\\7x-23&=5\\7x-23+23&=5+23\\7x&=28\\\frac{7x}{7}&=\frac{28}{7}\\x&=4\end{aligned}$$

47.
$$\begin{aligned}5x-5&=2(x+1)+3x-7\\5x-5&=2x+2+3x-7\\5x-5&=5x-5\\5x-5x-5&=5x-5x-5\\-5&=-5\end{aligned}$$
All real numbers are solutions.

49.
$$\begin{aligned}4(2n+1)&=3(6n+3)+1\\8n+4&=18n+9+1\\8n+4&=18n+10\\8n+4-4&=18n+10-4\\8n&=18n+6\\8n-18n&=18n+6-18n\\-10n&=6\\\frac{-10n}{-10}&=\frac{6}{-10}\\n&=-\frac{3}{5}\end{aligned}$$

51.
$$\begin{aligned}x+\frac{5}{4}&=\frac{3}{4}x\\4\left(x+\frac{5}{4}\right)&=4\left(\frac{3}{4}x\right)\\4x+5&=3x\\4x+5-4x&=3x-4x\\5&=-x\\\frac{5}{-1}&=\frac{-x}{-1}\\-5&=x\end{aligned}$$

53.
$$\begin{aligned}\frac{x}{2}-1&=\frac{x}{5}+2\\10\left(\frac{x}{2}-1\right)&=10\left(\frac{x}{5}+2\right)\\5x-10&=2x+20\\5x-10+10&=2x+20+10\\5x&=2x+30\\5x-2x&=2x+30-2x\\3x&=30\\\frac{3x}{3}&=\frac{30}{3}\\x&=10\end{aligned}$$

55.
$$\begin{aligned}2(x+3)-5&=5x-3(1+x)\\2x+6-5&=5x-3-3x\\2x+1&=2x-3\\2x-2x+1&=2x-2x-3\\1&=-3\end{aligned}$$
There is no solution.

57.
$$\begin{aligned}0.06-0.01(x+1)&=-0.02(2-x)\\100[0.06-0.01(x+1)]&=100[-0.02(2-x)]\\6-(x+1)&=-2(2-x)\\6-x-1&=-4+2x\\5-x&=-4+2x\\5-x-2x&=-4+2x-2x\\5-3x&=-4\\5-5-3x&=-4-5\\-3x&=-9\\\frac{-3x}{-3}&=\frac{-9}{-3}\\x&=3\end{aligned}$$

59.
$$\begin{aligned}\frac{9}{2}+\frac{5}{2}y&=2y-4\\2\left(\frac{9}{2}+\frac{5}{2}y\right)&=2(2y-4)\\9+5y&=4y-8\\9+5y-4y&=4y-8-4y\\9+y&=-8\\9+y-9&=-8-9\\y&=-17\end{aligned}$$

61.
$$\begin{aligned}-2y-10&=5y+18\\-2y-10-18&=5y+18-18\\-2y-28&=5y\\-2y-28+2y&=5y+2y\\-28&=7y\\\frac{-28}{7}&=\frac{7y}{7}\\-4&=y\end{aligned}$$

63.
$$\begin{aligned}0.6x-0.1&=0.5x+0.2\\10(0.6x-0.1)&=10(0.5x+0.2)\\6x-1&=5x+2\\6x-5x-1&=5x-5x+2\\x-1&=2\\x-1+1&=2+1\\x&=3\end{aligned}$$

65.
$$\begin{aligned}0.02(6t-3)&=0.12(t-2)+0.18\\100[0.02(6t-3)]&=100[0.12(t-2)+0.18]\\2(6t-3)&=12(t-2)+18\\12t-6&=12t-24+18\\12t-6&=12t-6\\12t-12t-6&=12t-12t-6\\-6&=-6\end{aligned}$$
All real numbers are solutions.

67.

−8	minus	a number
↓	↓	↓
−8	−	x

69.

−3	plus	twice a number
↓	↓	↓
−3	+	$2x$

71.

9	times	a number	plus	20
↓	↓	↓	↓	↓
9	·	$(x$	+	$20) = 9(x+20)$

73.
$$\begin{aligned}x+(2x-3)+(3x-5)&=x+2x-3+3x-5\\&=6x-8\end{aligned}$$
The perimeter is $(6x-8)$ meters.

75. a.
$$\begin{aligned}x+3&=x+3\\x+3-x&=x+3-x\\3&=3\\3-3&=3-3\\0&=0\end{aligned}$$
All real numbers are solutions.

b. Answers may vary

c. Answers may vary

77.
$$\begin{aligned}5x+1&=5x+1\\5x+1-5x&=5x+1-5x\\1&=1\end{aligned}$$
All real numbers are solutions. The answer is a.

79.
$$\begin{aligned}2x-6x-10&=-4x+3-10\\-4x-10&=-4x-7\\-4x-10+4x&=-4x-7+4x\\-10&=-7\end{aligned}$$
There is no solution. The answer is b.

81.
$$\begin{aligned}9x-20&=8x-20\\9x-20-8x&=8x-20-8x\\x-20&=-20\\x-20+20&=-20+20\\x&=0\end{aligned}$$
The answer is c.

83. Answers may vary

85. a. Since the perimeter is the sum of the lengths of the sides, $x+x+x+2x+2x=28$.

b. $7x = 28$

$$\frac{7x}{7} = \frac{28}{7}$$

$$x = 4$$

c. $2x = 2(4) = 8$

The lengths are $x = 4$ centimeters and $2x = 8$ centimeters.

87. Answers may vary

89.

$$\begin{aligned}
1000(7x-10) &= 50(412+100x) \\
7000x - 10{,}000 &= 20{,}600 + 5000x \\
7000x - 5000x - 10{,}000 &= 20{,}600 + 5000x - 5000x \\
2000x - 10{,}000 &= 20{,}600 \\
2000x - 10{,}000 + 10{,}000 &= 20{,}600 + 10{,}000 \\
2000x &= 30{,}600 \\
\frac{2000x}{2000} &= \frac{30{,}600}{2000} \\
x &= 15.3
\end{aligned}$$

91.

$$\begin{aligned}
0.035x + 5.112 &= 0.010x + 5.107 \\
1000(0.035x + 5.112) &= 1000(0.010x + 5.107) \\
35x + 5112 &= 10x + 5107 \\
35x - 10x + 5112 &= 10x - 10x + 5107 \\
25x + 5112 &= 5107 \\
25x + 5112 - 5112 &= 5107 - 5112 \\
25x &= -5 \\
\frac{25x}{25} &= \frac{-5}{25} \\
x &= -\frac{1}{5} = -0.2
\end{aligned}$$

93.

$$\begin{aligned}
x(x-3) &= x^2 + 5x + 7 \\
x^2 - 3x &= x^2 + 5x + 7 \\
x^2 - x^2 - 3x &= x^2 - x^2 + 5x + 7 \\
-3x &= 5x + 7 \\
-3x - 5x &= 5x - 5x + 7 \\
-8x &= 7 \\
\frac{-8x}{-8} &= \frac{7}{-8} \\
x &= -\frac{7}{8}
\end{aligned}$$

95.
$$\begin{aligned} 2z(z+6) &= 2z^2+12z-8 \\ 2z^2+12z &= 2z^2+12z-8 \\ 2z^2-2z^2+12z &= 2z^2-2z^2+12z-8 \\ 12z &= 12z-8 \\ 12z-12z &= 12z-12z-8 \\ 0 &= -8 \end{aligned}$$
There is no solution.

The Bigger Picture

1.
$$\begin{aligned} 3x-4 &= 3(2x-1)+7 \\ 3x-4 &= 6x-3+7 \\ 3x-4 &= 6x+4 \\ 3x-4-6x &= 6x+4-6x \\ -3x-4 &= 4 \\ -3x-4+4 &= 4+4 \\ -3x &= 8 \\ \frac{-3x}{-3} &= \frac{8}{-3} \\ x &= -\frac{8}{3} \end{aligned}$$

2.
$$\begin{aligned} 5+2x &= 5(x+1) \\ 5+2x &= 5x+5 \\ 5+2x-5x &= 5x+5-5x \\ 5-3x &= 5 \\ 5-3x-5 &= 5-5 \\ -3x &= 0 \\ \frac{-3x}{-3} &= \frac{0}{-3} \\ x &= 0 \end{aligned}$$

3.
$$\begin{aligned} \frac{x+3}{2} &= 1 \\ 2\left(\frac{x+3}{2}\right) &= 2(1) \\ x+3 &= 2 \\ x+3-3 &= 2-3 \\ x &= -1 \end{aligned}$$

4.
$$\begin{aligned} \frac{x-2}{2}-\frac{x-4}{3} &= \frac{5}{6} \\ 6\left(\frac{x-2}{2}-\frac{x-4}{3}\right) &= 6\left(\frac{5}{6}\right) \\ 3(x-2)-2(x-4) &= 5 \\ 3x-6-2x+8 &= 5 \\ x+2 &= 5 \\ x+2-2 &= 5-2 \\ x &= 3 \end{aligned}$$

5.
$$\begin{aligned} \frac{7}{5}+\frac{y}{10} &= 2 \\ 10\left(\frac{7}{5}+\frac{y}{10}\right) &= 10(2) \\ 2(7)+y &= 20 \\ 14+y &= 20 \\ 14+y-14 &= 20-14 \\ y &= 6 \end{aligned}$$

6.
$$\begin{aligned} 5+2x &= 2(x+1) \\ 5+2x &= 2x+2 \\ 5+2x-2x &= 2x+2-2x \\ 5 &= 2 \quad \text{False} \end{aligned}$$
This false statement indicates that there is no solution.

7.
$$\begin{aligned} 4(x-2)+3x &= 9(x-1)-2 \\ 4x-8+3x &= 9x-9-2 \\ 7x-8 &= 9x-11 \\ 7x-8-9x &= 9x-11-9x \\ -2x-8 &= -11 \\ -2x-8+8 &= -11+8 \\ -2x &= -3 \\ \frac{-2x}{-2} &= \frac{-3}{-2} \\ x &= \frac{3}{2} \end{aligned}$$

8.
$$\begin{aligned} 6(x+1)-2 &= 6x+4 \\ 6x+6-2 &= 6x+4 \\ 6x+4 &= 6x+4 \\ 6x+4-6x &= 6x+4-6x \\ 4 &= 4 \quad \text{True} \end{aligned}$$
This true statement indicates that all real numbers are solutions of the equation.

Integrated Review

1.
$$\begin{aligned} x-10 &= -4 \\ x-10+10 &= -4+10 \\ x &= 6 \end{aligned}$$

2.
$$\begin{aligned} y+14 &= -3 \\ y+14-14 &= -3-14 \\ y &= -17 \end{aligned}$$

3.
$$\begin{aligned} 9y &= 108 \\ \frac{9y}{9} &= \frac{108}{9} \\ y &= 12 \end{aligned}$$

4. $-3x = 78$
$$\frac{-3x}{-3} = \frac{78}{-3}$$
$$x = -26$$

5. $-6x + 7 = 25$
$$-6x + 7 - 7 = 25 - 7$$
$$-6x = 18$$
$$\frac{-6x}{-6} = \frac{18}{-6}$$
$$x = -3$$

6. $5y - 42 = -47$
$$5y - 42 + 42 = -47 + 42$$
$$5y = -5$$
$$\frac{5y}{5} = \frac{-5}{5}$$
$$y = -1$$

7. $\frac{2}{3}x = 9$
$$\frac{3}{2}\left(\frac{2}{3}x\right) = \frac{3}{2}(9)$$
$$x = \frac{27}{2}$$

8. $\frac{4}{5}z = 10$
$$\frac{5}{4}\left(\frac{4}{5}z\right) = \frac{5}{4}(10)$$
$$z = \frac{25}{2}$$

9. $\frac{r}{-4} = -2$
$$-4\left(\frac{r}{-4}\right) = -4(-2)$$
$$r = 8$$

10. $\frac{y}{-8} = 8$
$$-8\left(\frac{y}{-8}\right) = -8(8)$$
$$y = -64$$

11. $6 - 2x + 8 = 10$
$$-2x + 14 = 10$$
$$-2x + 14 - 14 = 10 - 14$$
$$-2x = -4$$
$$\frac{-2x}{-2} = \frac{-4}{-2}$$
$$x = 2$$

12. $-5 - 6y + 6 = 19$
$$-6y + 1 = 19$$
$$-6y + 1 - 1 = 19 - 1$$
$$-6y = 18$$
$$\frac{-6y}{-6} = \frac{18}{-6}$$
$$y = -3$$

13. $2x - 7 = 2x - 27$
$$2x - 2x - 7 = 2x - 2x - 27$$
$$-7 = -27$$
There is no solution.

14. $3 + 8y = 8y - 2$
$$3 + 8y - 8y = 8y - 8y - 2$$
$$3 = -2$$
There is no solution.

15. $-3a + 6 + 5a = 7a - 8a$
$$2a + 6 = -a$$
$$2a - 2a + 6 = -a - 2a$$
$$6 = -3a$$
$$\frac{6}{-3} = \frac{-3a}{-3}$$
$$-2 = a$$

16. $4b - 8 - b = 10b - 3b$
$$3b - 8 = 7b$$
$$3b - 3b - 8 = 7b - 3b$$
$$-8 = 4b$$
$$\frac{-8}{4} = \frac{4b}{4}$$
$$-2 = b$$

17. $-\frac{2}{3}x = \frac{5}{9}$
$$-\frac{3}{2}\left(-\frac{2}{3}x\right) = -\frac{3}{2}\left(\frac{5}{9}\right)$$
$$x = -\frac{5}{6}$$

18.
$$-\frac{3}{8}y = -\frac{1}{16}$$
$$-\frac{8}{3}\left(-\frac{3}{8}y\right) = -\frac{8}{3}\left(-\frac{1}{16}\right)$$
$$y = \frac{1}{6}$$

19.
$$10 = -6n + 16$$
$$10 - 16 = -6n + 16 - 16$$
$$-6 = -6n$$
$$\frac{-6}{-6} = \frac{-6n}{-6}$$
$$1 = n$$

20.
$$-5 = -2m + 7$$
$$-5 - 7 = -2m + 7 - 7$$
$$-12 = -2m$$
$$\frac{-12}{-2} = \frac{-2m}{-2}$$
$$6 = m$$

21.
$$3(5c - 1) - 2 = 13c + 3$$
$$15c - 3 - 2 = 13c + 3$$
$$15c - 5 = 13c + 3$$
$$15c - 13c - 5 = 13c - 13c + 3$$
$$2c - 5 = 3$$
$$2c - 5 + 5 = 3 + 5$$
$$2c = 8$$
$$\frac{2c}{2} = \frac{8}{2}$$
$$c = 4$$

22.
$$4(3t + 4) - 20 = 3 + 5t$$
$$12t + 16 - 20 = 3 + 5t$$
$$12t - 4 = 3 + 5t$$
$$12t - 5t - 4 = 3 + 5t - 5t$$
$$7t - 4 = 3$$
$$7t - 4 + 4 = 3 + 4$$
$$7t = 7$$
$$\frac{7t}{7} = \frac{7}{7}$$
$$t = 1$$

23.
$$\frac{2(z+3)}{3} = 5 - z$$
$$3\left[\frac{2(z+3)}{3}\right] = 3(5 - z)$$
$$2z + 6 = 15 - 3z$$
$$2z + 3z + 6 = 15 - 3z + 3z$$
$$5z + 6 = 15$$
$$5z + 6 - 6 = 15 - 6$$
$$5z = 9$$
$$\frac{5z}{5} = \frac{9}{5}$$
$$z = \frac{9}{5}$$

24.
$$\frac{3(w+2)}{4} = 2w + 3$$
$$4\left[\frac{3(w+2)}{4}\right] = 4(2w + 3)$$
$$3w + 6 = 8w + 12$$
$$3w - 8w + 6 = 8w - 8w + 12$$
$$-5w + 6 = 12$$
$$-5w + 6 - 6 = 12 - 6$$
$$-5w = 6$$
$$\frac{-5w}{-5} = \frac{6}{-5}$$
$$w = -\frac{6}{5}$$

25.
$$-2(2x - 5) = -3x + 7 - x + 3$$
$$-4x + 10 = -4x + 10$$
$$-4x + 4x + 10 = -4x + 4x + 10$$
$$10 = 10$$
All real numbers are solutions.

26.
$$-4(5x - 2) = -12x + 4 - 8x + 4$$
$$-20x + 8 = -20x + 8$$
$$-20x + 20x + 8 = -20x + 20x + 8$$
$$8 = 8$$
All real numbers are solutions.

27.
$$\begin{aligned}0.02(6t-3) &= 0.04(t-2)+0.02\\ 100[0.02(6t-3)] &= 100[0.04(t-2)+0.02]\\ 2(6t-3) &= 4(t-2)+2\\ 12t-6 &= 4t-8+2\\ 12t-6 &= 4t-6\\ 12t-4t-6 &= 4t-4t-6\\ 8t-6 &= -6\\ 8t-6+6 &= -6+6\\ 8t &= 0\\ \frac{8t}{8} &= \frac{0}{8}\\ t &= 0\end{aligned}$$

28.
$$\begin{aligned}0.03(m+7) &= 0.02(5-m)+0.03\\ 100[0.03(m+7)] &= 100[0.02(5-m)+0.03]\\ 3(m+7) &= 2(5-m)+3\\ 3m+21 &= 10-2m+3\\ 3m+21 &= 13-2m\\ 3m+2m+21 &= 13-2m+2m\\ 5m+21 &= 13\\ 5m+21-21 &= 13-21\\ 5m &= -8\\ \frac{5m}{5} &= \frac{-8}{5}\\ m &= -\frac{8}{5} = -1.6\end{aligned}$$

29.
$$\begin{aligned}-3y &= \frac{4(y-1)}{5}\\ 5(-3y) &= 5\left[\frac{4(y-1)}{5}\right]\\ -15y &= 4y-4\\ -15y-4y &= 4y-4y-4\\ -19y &= -4\\ \frac{-19y}{-19} &= \frac{-4}{-19}\\ y &= \frac{4}{19}\end{aligned}$$

30.
$$\begin{aligned}-4x &= \frac{5(1-x)}{6}\\ 6(-4x) &= 6\left[\frac{5(1-x)}{6}\right]\\ -24x &= 5-5x\\ -24x+5x &= 5-5x+5x\\ -19x &= 5\\ \frac{-19x}{-19} &= \frac{5}{-19}\\ x &= -\frac{5}{19}\end{aligned}$$

31.
$$\begin{aligned}\frac{5}{3}x-\frac{7}{3} &= x\\ 3\left(\frac{5}{3}x-\frac{7}{3}\right) &= 3(x)\\ 5x-7 &= 3x\\ 5x-5x-7 &= 3x-5x\\ -7 &= -2x\\ \frac{-7}{-2} &= \frac{-2x}{-2}\\ \frac{7}{2} &= x\end{aligned}$$

32.
$$\begin{aligned}\frac{7}{5}n+\frac{3}{5} &= -n\\ 5\left(\frac{7}{5}n+\frac{3}{5}\right) &= 5(-n)\\ 7n+3 &= -5n\\ 7n-7n+3 &= -5n-7n\\ 3 &= -12n\\ \frac{3}{-12} &= \frac{-12n}{-12}\\ -\frac{1}{4} &= n\end{aligned}$$

33.
$$\begin{aligned}\frac{1}{10}(3x-7) &= \frac{3}{10}x+5\\ 10\left[\frac{1}{10}(3x-7)\right] &= 10\left(\frac{3}{10}x+5\right)\\ 3x-7 &= 3x+50\\ 3x-7-3x &= 3x+50-3x\\ -7 &= 50\end{aligned}$$

There is no solution.

34.
$$\begin{aligned}\frac{1}{7}(2x-5) &= \frac{2}{7}x+1\\ 7\left[\frac{1}{7}(2x-5)\right] &= 7\left(\frac{2}{7}x+1\right)\\ 2x-5 &= 2x+7\\ 2x-5-2x &= 2x+7-2x\\ -5 &= 7\end{aligned}$$

There is no solution.

35. $5+2(3x-6)=-4(6x-7)$
$$5+6x-12=-24x+28$$
$$6x-7=-24x+28$$
$$6x-7+24x=-24x+28+24x$$
$$30x-7=28$$
$$30x-7+7=28+7$$
$$30x=35$$
$$\frac{30x}{30}=\frac{35}{30}$$
$$x=\frac{7}{6}$$

36. $3+5(2x-4)=-7(5x+2)$
$$3+10x-20=-35x-14$$
$$10x-17=-35x-14$$
$$10x-17+35x=-35x-14+35x$$
$$45x-17=-14$$
$$45x-17+17=-14+17$$
$$45x=3$$
$$\frac{45x}{45}=\frac{3}{45}$$
$$x=\frac{1}{15}$$

Section 2.4

Practice Exercises

1. Let x = the number.
$$3x-6=2x+3$$
$$3x-6-2x=2x+3-2x$$
$$x-6=3$$
$$x-6+6=3+6$$
$$x=9$$
The number is 9.

2. Let x = the number.
$$3x-4=2(x-1)$$
$$3x-4=2x-2$$
$$3x-4-2x=2x-2-2x$$
$$x-4=-2$$
$$x-4+4=-2+4$$
$$x=2$$
The number is 2.

3. Let x = the length of short piece, then $4x$ = the length of long piece.
$$x+4x=45$$
$$5x=45$$
$$\frac{5x}{5}=\frac{45}{5}$$
$$x=9$$
$4x=4(9)=36$
The short piece is 9 inches and the long piece is 36 inches.

4. Let x = number of Republicans, then $x+6$ = number of Democrats.
$$x+x+6=50$$
$$2x+6=50$$
$$2x+6-6=50-6$$
$$2x=44$$
$$\frac{2x}{2}=\frac{44}{2}$$
$$x=22$$
$x+6=22+6=28$
There were 22 Republican and 28 Democratic Governors.

5. x = degree measure of first angle
$3x$ = degree measure of second angle
$x+55$ = degree measure of third angle
$$x+3x+(x+55)=180$$
$$5x+55=180$$
$$5x+55-55=180-55$$
$$5x=125$$
$$\frac{5x}{5}=\frac{125}{5}$$
$$x=25$$
$3x=3(25)=75$
$x+55=25+55=80$
The measures of the angles are 25°, 75°, and 80°.

6. Let x = the first even integer, then
$x+2$ = the second even integer, and
$x+4$ = the third even integer.
$$x+(x+2)+(x+4)=144$$
$$3x+6=144$$
$$3x+6-6=144-6$$
$$3x=138$$
$$\frac{3x}{3}=\frac{138}{3}$$
$$x=46$$
$x+2=46+2=48$
$x+4=46+4=50$
The integers are 46, 48, and 50.

Vocabulary and Readiness Check

1. $2x$; $2x - 31$

2. $3x$; $3x + 17$

3. $x + 5$; $2(x + 5)$

4. $x - 11$; $7(x - 11)$

5. $20 - y$; $\frac{20-y}{3}$ or $(20 - y) \div 3$

6. $-10 + y$; $\frac{-10+y}{9}$ or $(-10 + y) \div 9$

Exercise Set 2.4

1. Let x = the number.
$$\begin{aligned} 2x+7 &= x+6 \\ 2x+7-x &= x+6-x \\ x+7 &= 6 \\ x+6-7 &= 6-7 \\ x &= -1 \end{aligned}$$
The number is -1.

3. Let x = the number.
$$\begin{aligned} 3x-6 &= 2x+8 \\ 3x-6-2x &= 2x+8-2x \\ x-6 &= 8 \\ x-6+6 &= 8+6 \\ x &= 14 \end{aligned}$$
The number is 14.

5. Let x = the number.
$$\begin{aligned} 2(x-8) &= 3(x+3) \\ 2x-16 &= 3x+9 \\ 2x-2x-16 &= 3x-2x+9 \\ -16 &= x+9 \\ -16-9 &= x+9-9 \\ -25 &= x \end{aligned}$$
The number is -25.

7. Let x = the number.
$$\begin{aligned} 4(-2+x) &= 5x+\frac{1}{2} \\ -8+4x &= 5x+\frac{1}{2} \\ -8+4x-4x &= 5x+\frac{1}{2}-4x \\ -8 &= x+\frac{1}{2} \\ -8-\frac{1}{2} &= x+\frac{1}{2}-\frac{1}{2} \\ -8\frac{1}{2} &= x \\ -\frac{17}{2} &= x \end{aligned}$$
The number is $-\frac{17}{2}$.

9. Let x = length of the shorter piece and
$2x + 2$ = length of the longer piece.
$$\begin{aligned} x+2x+2 &= 17 \\ 3x+2 &= 17 \\ 3x+2-2 &= 17-2 \\ 3x &= 15 \\ \frac{3x}{3} &= \frac{15}{3} \\ x &= 5 \end{aligned}$$
$2x + 2 = 2(5) + 2 = 12$
The shorter piece is 5 feet and the longer piece is 12 feet.

11. Let x = weight of Armanty meteorite, then
$3x$ = weight of Hoba West meteorite.
$$\begin{aligned} x+3x &= 88 \\ 4x &= 88 \\ \frac{4x}{4} &= \frac{88}{4} \\ x &= 22 \end{aligned}$$
$3x = 3(22) = 66$
The Armanty meteorite weighs 22 tons and the Hoba West meteorite weighs 66 tons.

13. Let x = number of cinema screens in the U.S., then $x + 5806$ = number of cinema screens in China.

$$x + x + 5806 = 78,994$$
$$2x + 5806 = 78,994$$
$$2x + 5806 - 5806 = 78,994 - 5806$$
$$2x = 73,188$$
$$\frac{2x}{2} = \frac{73,188}{2}$$
$$x = 36,594$$

$x + 5806 = 36,594 + 5806 = 42,400$

The U.S. has 36,594 cinema screens, and China has 42,400.

15. Let x = the measure of each of the two equal angles, and $2x + 30$ = the measure of the third.

$$x + x + 2x + 30 = 180$$
$$4x + 30 = 180$$
$$4x + 30 - 30 = 180 - 30$$
$$4x = 150$$
$$\frac{4x}{4} = \frac{150}{4}$$
$$x = 37.5$$

$2x + 30 = 2(37.5) + 30 = 105$

The angles are 37.5°, 37.5°, and 105°.

	First Integer	Next Integers			Indicated Sum
17. Three consecutive integers:	Integer: x	$x + 1$	$x + 2$		Sum of the three consecutive integers simplified: $(x + 1) + (x + 2) = 2x + 3$
19. Three consecutive even integers:	Even integer: x	$x + 2$	$x + 4$		Sum of the first and third even consecutive integers, simplified: $x + (x + 4) = 2x + 4$
21. Four consecutive integers:	Integer: x	$x + 1$	$x + 2$	$x + 3$	Sum of the four consecutive integers, simplified: $x + (x + 1) + (x + 2) + (x + 3) = 4x + 6$
23. Three consecutive odd integers:	Odd integer: x	$x + 2$	$x + 4$		Sum of the second and third consecutive odd integers, simplified: $(x + 2) + (x + 4) = 2x + 6$

25. Let x = the number of the left page and $x + 1$ = the number of the right page.

$$x + x + 1 = 469$$
$$2x + 1 = 469$$
$$2x + 1 - 1 = 469 - 1$$
$$2x = 468$$
$$\frac{2x}{2} = \frac{468}{2}$$
$$x = 234$$

$x + 1 = 234 + 1 = 235$

The page numbers are 234 and 235.

27. Let x = the code for Belgium,
$x + 1$ = the code for France,
$x + 2$ = the code for Spain.

$$x+x+1+x+2=99$$
$$3x+3=99$$
$$3x+3-3=99-3$$
$$3x=96$$
$$\frac{3x}{3}=\frac{96}{3}$$
$$x=32$$

$x+1=32+1=33$
$x+2=32+2=34$
The codes are Belgium: 32; France: 33; Spain: 34.

29.
$$x+2x+(1+5x)=25$$
$$8x+1=25$$
$$8x+1-1=25-1$$
$$8x=24$$
$$\frac{8x}{8}=\frac{24}{8}$$
$$x=3$$

$2x=2(3)=6$
$1+5x=1+5(3)=16$
The lengths of the pieces are 3 inches, 6 inches, and 16 inches.

31. Let x = the number.

$$10-5x=3x$$
$$10-5x+5x=3x+5x$$
$$10=8x$$
$$\frac{10}{8}=\frac{8x}{8}$$
$$\frac{5}{4}=x$$

The number is $\frac{5}{4}$.

33. Let x = carats in Angola, then
$4x$ = carats in Botswana.

$$x+4x=40{,}000{,}000$$
$$5x=40{,}000{,}000$$
$$\frac{5x}{5}=\frac{40{,}000{,}000}{5}$$
$$x=8{,}000{,}000$$

$4x=4(8{,}000{,}000)=32{,}000{,}000$
Botswana produces 32,000,000 carats and Angola produces 8,000,000 carats.

35. Let x = the measure of the smallest angle,
$x + 2$ = the measure of the second, and
$x + 4$ = the measure of the third.

$$x+x+2+x+4=180$$
$$3x+6=180$$
$$3x+6-6=180-6$$
$$3x=174$$
$$\frac{3x}{3}=\frac{174}{3}$$
$$x=58$$

$x+2=58+2=60$
$x+4=58+4=62$
The angles are 58°, 60°, and 62°.

37. Let x = first integer (Russia),
$x + 1$ = second integer (Austria),
$x + 2$ = third integer (Canada),
$x + 3$ = fourth integer (United States).

$$x+(x+1)+(x+2)+(x+3)=94$$
$$4x+6=94$$
$$4x+6-6=94-6$$
$$4x=88$$
$$\frac{4x}{4}=\frac{88}{4}$$
$$x=22$$

$x+1=22+1=23$
$x+2=22+2=24$
$x+3=22+3=25$
The number of medals for each country is Russia: 22; Austria: 23; Canada: 24; United States: 25.

39. Let x = the number.

$$3(x+5)=2x-1$$
$$3x+15=2x-1$$
$$3x+15-2x=2x-1-2x$$
$$x+15=-1$$
$$x+15-15=-1-15$$
$$x=-16$$

The number is −16.

41. Let x = votes for Pavich, then
$x + 20{,}196$ = votes for Weller.

$$x+x+20{,}196=196{,}554$$
$$2x+20{,}196=196{,}554$$
$$2x+20{,}196-20{,}196=196{,}554-20{,}196$$
$$2x=176{,}358$$
$$\frac{2x}{2}=\frac{176{,}358}{2}$$
$$x=88{,}179$$

$x+20{,}196=88{,}179+20{,}196=108{,}375$
Pavich received 88,179 votes and Weller received 108,375 votes.

43. Let x = smaller angle, then
$3x + 8$ = larger angle.

$$x+(3x+8)=180$$
$$4x+8=180$$
$$4x+8-8=180-8$$
$$4x=172$$
$$\frac{4x}{4}=\frac{172}{4}$$
$$x=43$$

$3x + 8 = 3(43) + 8 = 137$
The angles measure 43° and 137°.

45. Let x = the number.

$$\frac{x}{4}+\frac{1}{2}=\frac{3}{4}$$
$$4\left(\frac{x}{4}+\frac{1}{2}\right)=4\left(\frac{3}{4}\right)$$
$$x+2=3$$
$$x+2-2=3-2$$

The number is 1.

47. Let x = the measure of each of the two smaller angles, and $2x - 15$ = the measure of each of the larger angles.

$$2x+2(2x-15)=360$$
$$2x+4x-30=360$$
$$6x-30=360$$
$$6x-30+30=360+30$$
$$6x=390$$
$$\frac{6x}{6}=\frac{390}{6}$$
$$x=65$$

$2x - 15 = 2(65) - 15 = 115$
The smaller angles are each 65° and the larger angles are each 115°.

49. Let x = speed of TGV, then
$x + 3.8$ = speed of Maglev.

$$x+x+3.8=718.2$$
$$2x+3.8=718.2$$
$$2x+3.8-3.8=718.2-3.8$$
$$2x=714.4$$
$$\frac{2x}{2}=\frac{714.4}{2}$$
$$x=357.2$$

$x + 3.8 = 357.2 + 3.8 = 361$
The speed of the TGV is 357.2 mph and the speed of the Maglev is 361 mph.

51. Let x = the number.

$$\frac{1}{3}\cdot x=\frac{5}{6}$$
$$3\cdot\frac{1}{3}x=3\cdot\frac{5}{6}$$
$$x=\frac{5}{2}$$

The number is $\frac{5}{2}$.

53. Let x = number of counties in Montana and
$x + 2$ = number in California.

$$x+x+2=114$$
$$2x+2=114$$
$$2x+2-2=114-2$$
$$2x=112$$
$$\frac{2x}{2}=\frac{112}{2}$$
$$x=56$$

$x + 2 = 56 + 2 = 58$
There are 56 counties in Montana and 58 counties in California.

55. Let x = points for Bears, then
$x + 12$ = points for Colts.

$$x+(x+12)=46$$
$$2x+12=46$$
$$2x+12-12=46-12$$
$$2x=34$$
$$\frac{2x}{2}=\frac{34}{2}$$
$$x=17$$

$x + 12 = 17 + 12 = 29$
The Bears scored 17 points and the Colts scored 29 points.

57. Let x = smaller angles, then
$x + 76.5$ = third angle.

$$x+x+(x+76.5)=180$$
$$3x+76.5=180$$
$$3x+76.5-76.5=180-76.5$$
$$3x=103.5$$
$$\frac{3x}{3}=\frac{103.5}{3}$$
$$x=34.5$$

$x + 76.5 = 34.5 + 76.5 = 111$
The angles measure 34.5°, 34.5°, and 111°.

59. Let x = length of the first piece,
$2x$ = length of the second piece, and
$5x$ = length of the third piece.

$$x+2x+5x=40$$
$$8x=40$$
$$\frac{8x}{8}=\frac{40}{8}$$
$$x=5$$

$2x = 2(5) = 10$
$5x = 5(5) = 25$
The lengths are 5, 10, and 25 inches.

61. Select the tallest bar. Hawaii spends the most money on tourism.

63. Let x = amount spent by Florida, then
$x + 1.7$ = amount spent by Texas.

$$x+x+1.7=60.5$$
$$2x+1.7=60.5$$
$$2x+1.7-1.7=60.5-1.7$$
$$2x=58.8$$
$$\frac{2x}{2}=\frac{58.8}{2}$$
$$x=29.4$$

$x + 1.7 = 29.4 + 1.7 = 31.1$
Florida spends \$29.4 million and Texas spends \$31.1 million.

65. Answers may vary

67. $2W + 2L = 2(7) + 2(10) = 14 + 20 = 34$

69. $\pi r^2 = \pi\cdot(15)^2 = \pi\cdot 225 = 225\pi$

71. Answers may vary

Section 2.5

Practice Exercises

1. Let $d = 580$ and $r = 5$.

$$d=r\cdot t$$
$$580=5t$$
$$\frac{580}{5}=\frac{5t}{5}$$
$$116=t$$

It takes 116 seconds or 1 minute 56 seconds.

2. Let $l = 40$ and $P = 98$.

$$P=2l+2w$$
$$98=2\cdot 40+2w$$
$$98=80+2w$$
$$98-80=80+2w-80$$
$$18=2w$$
$$\frac{18}{2}=\frac{2w}{2}$$
$$9=w$$

The dog run is 9 feet wide.

3. Let $C = 8$.

$$F=\frac{9}{5}C+32$$
$$F=\frac{9}{5}\cdot 8+32$$
$$F=\frac{72}{5}+\frac{160}{5}$$
$$F=\frac{232}{5}=46.4$$

The equivalent temperature is 46.4°F.

4. Let w = width of sign, then
$5w + 3$ = length of sign.

$$P=2l+2w$$
$$66=2(5w+3)+2w$$
$$66=10w+6+2w$$
$$66=12w+6$$
$$66-6=12w+6-6$$
$$60=12w$$
$$\frac{60}{12}=\frac{12w}{12}$$
$$5=w$$

$5w + 3 = 5(5) + 3 = 28$
The sign has length 28 inches and width 5 inches.

5.

$$I=Prt$$
$$\frac{I}{Pt}=\frac{Prt}{Pt}$$
$$\frac{I}{Pt}=r \text{ or } r=\frac{I}{Pt}$$

6.

$$H=5as+10a$$
$$H-10a=5as+10a-10a$$
$$H-10a=5as$$
$$\frac{H-10a}{5a}=\frac{5as}{5a}$$
$$\frac{H-10a}{5a}=s \text{ or } s=\frac{H-10a}{5a}$$

7.
$$\begin{aligned} N &= F + d(n-1) \\ N - F &= F + d(n-1) - F \\ N - F &= d(n-1) \\ \frac{N-F}{n-1} &= \frac{d(n-1)}{n-1} \\ \frac{N-F}{n-1} &= d \text{ or } d = \frac{N-F}{n-1} \end{aligned}$$

8.
$$\begin{aligned} A &= \frac{1}{2}a(b+B) \\ 2 \cdot A &= 2 \cdot \frac{1}{2}a(b+B) \\ 2A &= a(b+B) \\ 2A &= ab + aB \\ 2A - ab &= ab + aB - ab \\ 2A - ab &= aB \\ \frac{2A-ab}{a} &= \frac{aB}{a} \\ \frac{2A-ab}{a} &= B \text{ or } B = \frac{2A-ab}{a} \end{aligned}$$

Exercise Set 2.5

1. Let $A = 45$ and $b = 15$.
$$\begin{aligned} A &= bh \\ 45 &= 15h \\ \frac{45}{15} &= \frac{15h}{15} \\ 3 &= h \end{aligned}$$

3. Let $S = 102$, $l = 7$, and $w = 3$.
$$\begin{aligned} S &= 4lw + 2wh \\ 102 &= 4(7)(3) + 2(3)h \\ 102 &= 84 + 6h \\ 102 - 84 &= 84 - 84 + 6h \\ 18 &= 6h \\ \frac{18}{6} &= \frac{6h}{6} \\ 3 &= h \end{aligned}$$

5. Let $A = 180$, $B = 11$, and $b = 7$.
$$\begin{aligned} A &= \frac{1}{2}h(B+b) \\ 180 &= \frac{1}{2}h(11+7) \\ 2(180) &= 2\left[\frac{1}{2}h(18)\right] \\ 360 &= 18h \\ \frac{360}{18} &= \frac{18h}{18} \\ 20 &= h \end{aligned}$$

7. Let $P = 30$, $a = 8$, and $b = 10$.
$$\begin{aligned} P &= a + b + c \\ 30 &= 8 + 10 + c \\ 30 &= 18 + c \\ 30 - 18 &= 18 - 18 + c \\ 12 &= c \end{aligned}$$

9. Let $C = 15.7$, and $\pi \approx 3.14$.
$$\begin{aligned} C &= 2\pi r \\ 15.7 &\approx 2(3.14)r \\ 15.7 &\approx 6.28r \\ \frac{15.7}{6.28} &\approx \frac{6.28r}{6.28} \\ 2.5 &\approx r \end{aligned}$$

11. Let $I = 3750$, $P = 25{,}000$, and $R = 0.05$.
$$\begin{aligned} I &= PRT \\ 3750 &= 25{,}000(0.05)T \\ 3750 &= 1250T \\ \frac{3750}{1250} &= \frac{1250T}{1250} \\ 3 &= T \end{aligned}$$

13. Let $V = 565.2$, $r = 6$, and $\pi \approx 3.14$.
$$\begin{aligned} V &= \frac{1}{3}\pi r^2 h \\ 565.2 &\approx \frac{1}{3}(3.14)(6)^2 h \\ 565.2 &\approx 37.68h \\ \frac{565.2}{37.68} &\approx \frac{37.68h}{37.68} \\ 15 &\approx h \end{aligned}$$

15.
$$\begin{aligned} f &= 5gh \\ \frac{f}{5g} &= \frac{5gh}{5g} \\ \frac{f}{5g} &= h \end{aligned}$$

17.
$$\begin{aligned} V &= lwh \\ \frac{V}{lh} &= \frac{lwh}{lh} \\ \frac{V}{lh} &= w \end{aligned}$$

19.
$$\begin{aligned} 3x + y &= 7 \\ 3x - 3x + y &= 7 - 3x \\ y &= 7 - 3x \end{aligned}$$

21.
$$A = P + PRT$$
$$A - P = P - P + PRT$$
$$A - P = PRT$$
$$\frac{A-P}{PT} = \frac{PRT}{PT}$$
$$\frac{A-P}{PT} = R$$

23.
$$V = \frac{1}{3}Ah$$
$$3V = 3\left(\frac{1}{3}Ah\right)$$
$$3V = Ah$$
$$\frac{3V}{h} = \frac{Ah}{h}$$
$$\frac{3V}{h} = A$$

25.
$$P = a + b + c$$
$$P - (b + c) = a + b + c - (b + c)$$
$$P - b - c = a + b + c - b - c$$
$$P - b - c = a$$

27.
$$S = 2\pi rh + 2\pi r^2$$
$$S - 2\pi r^2 = 2\pi rh + 2\pi r^2 - 2\pi r^2$$
$$S - 2\pi r^2 = 2\pi rh$$
$$\frac{S - 2\pi r^2}{2\pi r} = \frac{2\pi rh}{2\pi r}$$
$$\frac{S - 2\pi r^2}{2\pi r} = h$$

29. a.
$$A = lw$$
$$A = 11.5(9)$$
$$A = 103.5$$
$$P = 2l + 2w$$
$$P = 2(11.5) + 2(9)$$
$$P = 23 + 18$$
$$P = 41$$
The area is 103.5 square feet and the perimeter is 41 feet.

b. Baseboards have to do with perimeter because they are installed around the edges. Carpet has to do with area because it is installed in the middle of the room.

31. a.
$$A = \frac{1}{2}h(b_1 + b_2)$$
$$A = \frac{1}{2}(12)(56 + 24)$$
$$A = 6(80)$$
$$A = 480$$
$$P = l_1 + l_2 + l_3 + l_4$$
$$P = 24 + 20 + 56 + 20$$
$$P = 120$$
The area is 480 square inches and the perimeter is 120 inches.

b. The frame has to do with perimeter because it surrounds the edge of the picture. The glass has to do with area because it covers the entire picture.

33. $A = 3990$ and $w = 57$.
$$A = lw$$
$$3990 = l \cdot 57$$
$$\frac{3990}{57} = \frac{57l}{57}$$
$$70 = l$$
The length is 70 feet.

35. Let $F = 14$.
$$14 = \frac{9}{5}C + 32$$
$$5(14) = 5\left(\frac{9}{5}\right)C + 5(32)$$
$$70 = 9C + 160$$
$$70 - 160 = 9C + 160 - 160$$
$$-90 = 9C$$
$$\frac{-90}{9} = \frac{9C}{9}$$
$$-10 = C$$
The equivalent temperature is $-10°C$.

37. Let $d = 25{,}000$ and $r = 4000$.
$$d = rt$$
$$25{,}000 = 4000t$$
$$\frac{25{,}000}{4000} = \frac{4000t}{4000}$$
$$6.25 = t$$
It will take 6.25 hours.

39. Let $P = 260$ and $w = \frac{2}{3}l$.

$$P = 2l + 2w$$
$$260 = 2l + 2\left(\frac{2}{3}l\right)$$
$$260 = \frac{10}{3}l$$
$$3(260) = 3\left(\frac{10}{3}l\right)$$
$$780 = 10l$$
$$\frac{780}{10} = \frac{10l}{10}$$
$$78 = l$$
$$w = \frac{2}{3}l = \frac{2}{3}(78) = 52$$

The width is 52 feet and the length is 78 feet.

41. Let $P = 102$, a = the length of the shortest side, $b = 2a$, and $c = a + 30$.

$$P = a + b + c$$
$$102 = a + 2a + a + 30$$
$$102 = 4a + 30$$
$$102 - 30 = 4a + 30 - 30$$
$$72 = 4a$$
$$\frac{72}{4} = \frac{4a}{4}$$
$$18 = a$$
$$b = 2a = 2(18) = 36$$
$$c = a + 30 = 18 + 30 = 48$$

The lengths are 18 feet, 36 feet, and 48 feet.

43. Let $d = 138$ and $t = 2.5$.

$$d = rt$$
$$138 = r \cdot 2.5$$
$$\frac{138}{2.5} = \frac{r \cdot 2.5}{2.5}$$
$$55.2 = r$$

The speed is 55.2 mph.

45. Let $l = 8$, $w = 6$, and $h = 3$.

$$V = lwh$$
$$V = 8(6)(3) = 144$$

Let x = number of piranha and volume per fish = 1.5.

$$144 = 1.5x$$
$$\frac{144}{1.5} = \frac{1.5x}{1.5}$$
$$96 = x$$

96 piranhas can be placed in the tank.

47. Let $h = 60$, $B = 130$, and $b = 70$.

$$A = \frac{1}{2}(B + b)h$$
$$A = \frac{1}{2}(130 + 70)60 = \frac{1}{2}(200)(60) = 6000$$

Let x = number of bags of fertilizer and the area per bag = 4000.

$$4000x = 6000$$
$$\frac{4000x}{4000} = \frac{6000}{4000}$$
$$x = 1.5$$

Two bags must be purchased.

49. Let $d = 16$, so $r = 8$.

$$A = \pi r^2 = \pi(8)^2 = 64\pi$$

Let $d = 10$, so $r = 5$.

$$A = 2\pi r^2 = 2\pi(5)^2 = 50\pi$$

One 16-inch pizza has more area and therefore gives more pizza for the price.

51.
$$x + x + x + 2.5x + 2.5x = 48$$
$$8x = 48$$
$$\frac{8x}{8} = \frac{48}{8}$$
$$x = 6$$
$$2.5x = 2.5(6) = 15$$

Three sides measure 6 meters and two sides measure 15 meters.

53. $r = 361$ and $d = 72.2$.

$$d = rt$$
$$72.2 = 361t$$
$$\frac{72.2}{361} = \frac{361t}{361}$$
$$0.2 = t$$

It will take 0.2 hour or 0.2(60) = 12 minutes.

55. Let x = the length of a side of the square and $x + 5$ = the length of a side of the triangle.

$$P(\text{triangle}) = P(\text{square}) + 7$$
$$3(x + 5) = 4x + 7$$
$$3x + 15 = 4x + 7$$
$$3x - 3x + 15 = 4x - 3x + 7$$
$$15 = x + 7$$
$$15 - 7 = x + 7 - 7$$
$$8 = x$$
$$x + 5 = 8 + 5 = 13$$

The side of the triangle is 13 inches.

57. Let $d = 135$ and $r = 60$.

$$\begin{aligned} d &= rt \\ 135 &= 60t \\ \frac{135}{60} &= \frac{60t}{60} \\ 2.25 &= t \end{aligned}$$

It would take 2.25 hours.

59. Let $A = 1,813,500$ and $w = 150$.

$$\begin{aligned} A &= lw \\ 1,813,500 &= l(150) \\ \frac{1,813,500}{150} &= \frac{150l}{150} \\ 12,090 &= l \end{aligned}$$

The length is 12,090 feet.

61. Let $F = 122$.

$$\begin{aligned} 122 &= \frac{9}{5}C + 32 \\ 5(122) &= 5\left(\frac{9}{5}\right)C + 5(32) \\ 610 &= 9C + 160 \\ 610 - 160 &= 9C + 160 - 160 \\ 450 &= 9C \\ \frac{450}{9} &= \frac{9C}{9} \\ 50 &= C \end{aligned}$$

The equivalent temperature is 50°C.

63. Let $l = 199$, $w = 78.5$, and $h = 33$.

$V = lwh$

$V = 199(78.5)(33) = 515,509.5$

The volume must be 515,509.5 cubic inches.

65. Let $\pi \approx 3.14$ and $d = 9.5$ so $r = 4.75$.

$$V = \frac{4}{3}\pi r^3 \approx \frac{4}{3}(3.14)(4.75)^3 \approx 449$$

The volume is 449 cubic inches.

67. Let $C = 167$.

$$\begin{aligned} F &= \frac{9}{5}C + 32 \\ &= \frac{9}{5}(167) + 32 \\ &= 300.6 + 32 \\ &= 332.6 \end{aligned}$$

The equivalent temperature is 332.6°F.

69. Nine divided by the sum of a number and five is $\frac{9}{x+5}$.

71. Three times the sum of a number and four is $3(x + 4)$.

73. Triple the difference of a number and twelve is $3(x - 12)$.

75.

$$\begin{aligned} \square - \bigcirc \cdot \mathbb{I} &= \triangle \\ -\bigcirc \cdot \mathbb{I} &= \triangle - \square \\ \frac{-\bigcirc \mathbb{I}}{-\mathbb{I}} &= \frac{\triangle - \square}{-\mathbb{I}} \\ \bigcirc &= \frac{\square - \triangle}{\mathbb{I}} \end{aligned}$$

77. Let $C = -78.5$.

$$\begin{aligned} F &= \frac{9}{5}C + 32 \\ &= \frac{9}{5}(-78.5) + 32 \\ &= -141.3 + 32 \\ &= -109.3 \end{aligned}$$

The equivalent temperature is −109.3°F.

79. Let $d = 93,000,000$ and $r = 186,000$.

$$\begin{aligned} d &= rt \\ 93,000,000 &= 186,000t \\ \frac{93,000,000}{186,000} &= \frac{186,000t}{186,000} \\ 500 &= t \end{aligned}$$

It will take 500 seconds or $8\frac{1}{3}$ minutes.

81. Let $t = 365$ and $r = 20$.

$d = rt = 20(365) = 7300$ inches

$$\frac{7300 \text{ inches}}{1} \cdot \frac{1 \text{ foot}}{12 \text{ inch}} \approx 608.33 \text{ feet}$$

It moves about 608.33 feet.

83. Let $d = 2$ then $r = 1$.

$$15 \text{ feet} = \frac{15 \text{ feet}}{1} \cdot \frac{12 \text{ inches}}{1 \text{ foot}} = 180 \text{ inches, so}$$

$h = 180$.

$V = \pi r^2 h$

$V = (\pi)(1)^2(180) = 180\pi \approx 565.5$

The volume of the column is 565.5 cubic inches.

85. The original parallelogram has an area $V = bh$. The altered box has a base $2b$, a height $2h$, and a new area.

$A = 2b(2h) = 4bh$

The area is multiplied by 4.

Section 2.6

Practice Exercises

1. Let x = the unknown percent.

$$35 = x \cdot 56$$
$$\frac{35}{56} = \frac{56x}{56}$$
$$0.625 = x$$

The number 35 is 62.5% of 56.

2. Let x = the unknown number.

$$198 = 55\% \cdot x$$
$$198 = 0.55x$$
$$\frac{198}{0.55} = \frac{0.55x}{0.55}$$
$$360 = x$$

The number 198 is 55% of 360.

3. **a.** From the circle graph, 4% of trips made by American travelers are for combined business/pleasure.

b. From the circle graph,
17% + 66% + 4% = 87% of trips are for business, pleasure, or combined business/pleasure.

c. Since 4% are trips for business/pleasure, find 4% of 325.
$0.04 \cdot 325 = 13$
We can expect 13 of the Americans to be traveling for business/pleasure.

4. Let x = discount.

$$x = 85\% \cdot 480$$
$$x = 0.85 \cdot 480$$
$$x = 408$$

The discount is $408.
New price = $480 – $408 = $72

5. Increase = 299,800 – 198,900 = 100,900
Let x = percent increase.

$$100,900 = x \cdot 198,900$$
$$\frac{100,900}{198,900} = \frac{198,900x}{198,900}$$
$$0.507 \approx x$$

The percent increase is 50.7%.

6. Let x = number of new films in 2004.

$$x + 0.028x = 535$$
$$1.028x = 535$$
$$\frac{1.028x}{1.028} = \frac{535}{1.028}$$
$$x \approx 520$$

There were 520 new feature films released in 2004.

7. Let x = number of liters of 2% solution.

Eyewash	No. of gallons	· Acid Strength	= Amt. of Acid
2%	x	2%	$0.02x$
5%	$6-x$	5%	$0.05(6-x)$
Mix: 3%	6	3%	$0.03(6)$

$$\begin{aligned} 0.02x+0.05(6-x) &= 0.03(6) \\ 0.02x+0.3-0.05x &= 0.18 \\ -0.03x+0.3 &= 0.18 \\ -0.03x+0.3-0.3 &= 0.18-0.3 \\ -0.03x &= -0.12 \\ \frac{-0.03x}{-0.03} &= \frac{-0.12}{-0.03} \\ x &= 4 \end{aligned}$$

$6-x=6-4=2$

She should mix 4 liters of 2% eyewash with 2 liters of 5% eyewash.

Vocabulary and Readiness Check

1. No, 25% + 25% + 40% = 90% ≠ 100%.

2. No, 30% + 30% + 30% = 90% ≠ 100%.

3. Yes, 25% + 25% + 25% + 25% = 100%.

4. Yes, 40% + 50% + 10% = 100%.

Exercise Set 2.6

1. Let x = the number.

$x = 16\% \cdot 70$

$x = 0.16 \cdot 70$

$x = 11.2$

11.2 is 16% of 70.

3. Let x = the percent.

$28.6 = x \cdot 52$

$\frac{28.6}{52} = \frac{52x}{52}$

$0.55 = x$

28.6 is 55% of 52.

5. Let x = the number.

$45 = 25\% \cdot x$

$45 = 0.25x$

$\frac{45}{0.25} = \frac{0.25x}{0.25}$

$180 = x$

45 is 25% of 180.

7. Animal Feed = 51%
Ethanol = 18%
51% + 18% = 69%
69% of corn production is used for animal feed or ethanol.

9. 18% of 10,535 = 0.18(10,535) = 1896.3
1896.3 million bushels or 1,896,300,000 bushels were used to make ethanol.

11. Let x = amount of discount.

$$x = 8\% \cdot 18{,}500$$
$$x = 0.08 \cdot 18{,}500$$
$$x = 1480$$

New price $= 18{,}500 - 1480 = 17{,}020$

The discount was \$1480 and the new price is \$17,020.

13. Let x = tip.

$$x = 15\% \cdot 40.50$$
$$x = 0.15 \cdot 40.5$$
$$x = 6.075 \approx 6.08$$

Total = 40.50 + 6.08 = 46.58
The total cost is \$46.58.

15. Increase = 280 − 208 = 72
Let x = percent.

$$72 = x \cdot 208$$
$$\frac{72}{208} = \frac{208x}{208}$$
$$35 \approx x$$

The percent increase is 35%.

17. Decrease = 40 − 28 = 12
Let x = percent.

$$12 = x \cdot 40$$
$$\frac{12}{40} = \frac{40x}{40}$$
$$0.3 = x$$

The percent decrease is 30%.

19. Let x = the original price and $0.25x$ = the discount.

$$x - 0.25x = 78$$
$$0.75x = 78$$
$$\frac{0.75x}{0.75} = \frac{78}{0.75}$$
$$x = 104$$

The original price was \$104.

21. Let x = last year's salary, and $0.04x$ = pay raise.

$$x + 0.04x = 44{,}200$$
$$1.04x = 44{,}200$$
$$\frac{1.04x}{1.04} = \frac{44{,}200}{1.04}$$
$$x = 42{,}500$$

Last year's salary was \$42,500.

23. Let x = the amount of pure acid.

	No. of gallons	· Strength	= Amt. of Acid
100%	x	1.00	x
40%	2	0.4	2(0.4)
70%	$x + 2$	0.7	$0.7(x + 2)$

$$x + 2(0.4) = 0.7(x + 2)$$
$$x + 0.8 = 0.7x + 1.4$$
$$x - 0.7x + 0.8 = 0.7x - 0.7x + 1.4$$
$$0.3x + 0.8 = 1.4$$
$$0.3x + 0.8 - 0.8 = 1.4 - 0.8$$
$$0.3x = 0.6$$
$$\frac{0.3x}{0.3} = \frac{0.6}{0.3}$$
$$x = 2$$

Mix 2 gallons of pure acid.

25. Let x = the number of pounds at \$7/lb.

	No. of lb	· Cost/lb	= Value
\$7/lb	x	7	$7x$
\$4/lb	14	4	4(14)
\$5/lb	$x + 14$	5	$5(x + 14)$

$$7x + 4(14) = 5(x + 14)$$
$$7x + 56 = 5x + 70$$
$$7x - 5x + 56 = 5x - 5x + 70$$
$$2x + 56 = 70$$
$$2x + 56 - 56 = 70 - 56$$
$$2x = 14$$
$$\frac{2x}{2} = \frac{14}{2}$$
$$x = 7$$

Add 7 pounds of \$7/pound coffee.

27. Let x = the number.
$x = 23\% \cdot 20$
$x = 0.23 \cdot 20$
$x = 4.6$
23% of 20 is 4.6.

29. Let x = the number.
$40 = 80\% \cdot x$
$40 = 0.8x$
$\frac{40}{0.8} = \frac{0.8x}{0.8}$
$50 = x$
40 is 80% of 50.

31. Let x = the percent.
$144 = x \cdot 480$
$\frac{144}{480} = \frac{480x}{480}$
$0.3 = x$
144 is 30% of 480.

33. From the graph, the height of the bar is 71. Therefore, 71% of the population in Fairbanks, Alaska, shop by catalog.

35. 65% of 275,043 = $0.65 \cdot 275{,}043 \approx 178{,}778$
We predict 178,778 catalog shoppers live in Anchorage.

37.

Ford Motor Company Model Year 2006 Vehicle Sales Worldwide		
	Thousands of Vehicles	Percent of Total (Rounded to Nearest Percent)
North America	3051	$\frac{3051}{6597} \approx 46\%$
Europe	1846	$\frac{1846}{6597} \approx 28\%$
Asia-Pacific-Africa	589	$\frac{589}{6597} \approx 9\%$
South America	381	$\frac{381}{6597} \approx 6\%$
Rest of the World	730	Example: $\frac{730}{6597} \approx 11\%$
Total	6597	100%
Source: Ford Motor Company		

39. Let x = the decrease in price.
$x = 0.25(256) = 64$
The decrease in price is \$64.
The sale price is $256 - 64 = \$192$.

41. Increase = $86 - 40 = 46$
Let x = the percent.
$46 = x \cdot 40$
$\frac{46}{40} = \frac{40x}{40}$
$1.15 = x$
The percent increase is 115%.

43. Let x = the number of cards (in millions) issued in 2001.
Increase = $7.26x$
No. in 2001 + increase = no. in 2006
$x + 7.26 = 1900$
$8.26x = 1900$
$\frac{8.26x}{8.26} = \frac{1900}{8.26}$
$x = 230$
230 million cards were issued in 2006.

45. Let x = the amount of 20% alloy.

	No. of Oz	· Strength	= Amt. of Copper
50%	200	0.5	200(0.5)
20%	x	0.2	$0.2x$
Mix	$x + 200$	0.3	$0.3(x + 200)$

$$200(0.5) + 0.2x = 0.3(x + 200)$$
$$100 + 0.2x = 0.3x + 60$$
$$100 + 0.2x - 0.2x = 0.3x - 0.2x + 60$$
$$100 = 0.1x + 60$$
$$100 - 60 = 0.1x + 60 - 60$$
$$40 = 0.1x$$
$$\frac{40}{0.1} = \frac{0.1x}{0.1}$$
$$400 = x$$

Mix with 400 ounces of 20% alloy.

47. Let x = mark-up.
$x = 70\% \cdot 27$
$x = 0.7 \cdot 27$
$x = 18.9$
Adult price = 27 + 18.9 = 45.9
The mark-up is \$18.90 and the adult ticket price is \$45.90.

49. Increase = 144 − 36 = 108
Let x = percent.
$$108 = x \cdot 36$$
$$\frac{108}{36} = \frac{36x}{36}$$
$$3 = x$$
The percent increase is 300%.

51. Let x = the number of employees prior to layoff.
Decrease = $0.35x$
$$x - 0.35x = 78$$
$$0.65x = 78$$
$$\frac{0.65x}{0.65} = \frac{78}{0.65}$$
$$x = 120$$
There were 120 employees prior to the layoffs.

53. Let x = pounds of peanuts.

	pounds	price/lb	price
peanuts	x	5	$5x$
bites	10	2	2(10)
trail mix	$10 + x$	3	$3(10 + x)$

$$5x + 2(10) = 3(10 + x)$$
$$5x + 20 = 30 + 3x$$
$$5x + 20 - 3x = 30 + 3x - 3x$$
$$20 + 2x = 30$$
$$20 + 2x - 20 = 30 - 20$$
$$2x = 10$$
$$\frac{2x}{2} = \frac{10}{2}$$
$$x = 5$$
Therefore, 5 pounds of chocolate-covered peanuts should be mixed.

55. Decrease = 2.19 − 2.09 = 0.10
Let x = percent.
$$0.10 = x \cdot 2.19$$
$$\frac{0.10}{2.19} = \frac{2.19x}{2.19}$$
$$0.046 \approx x$$
The percent decrease is 4.6%.

57. Let x = number of decisions in 1982–1983, then $0.457x$ = decrease.
$$x - 0.457x = 182$$
$$0.543x = 182$$
$$\frac{0.543x}{0.543} = \frac{182}{0.543}$$
$$x \approx 335$$
There were 335 decisions in 1982–1983.

59. Let x = increase.
$x = 48\% \cdot 577$
$x = 0.48 \cdot 577$
$x \approx 277$
Naga pepper = 577 + 277 = 854
The Naga Jolokia pepper measures 854 thousand Scoville units.

61. 42% of 860 = $0.42 \cdot 860 \approx 361$
You would expect 361 college students to rank flexible hours as their top priority.

63. $-5 > -7$

65. $|-5| = -(-5)$

67. $(-3)^2 = 9;\ -3^2 = -9$
$(-3)^2 > -3^2$

69. No; answers may vary

71. No; answers may vary

73. 230 is x percent of 2400.

$x(2400) = 230$

$$x = \frac{230}{2400} \approx 0.096 = 9.6\%$$

This is about 9.6% of the daily value.

75. Let x = percent of calories from fat.

$x(130) = 35$

$$x = \frac{35}{130} \approx 0.269 = 26.9\%$$

This is less than 30% so it satisfies the recommendation.

77. Let x = percent of calories from protein.

$x(280) = 4(12)$

$$x = \frac{48}{280} \approx 0.171 = 17.1\%$$

About 17.1% of the calories in one serving come from protein.

Section 2.7

Practice Exercises

1. Let x = time down, then $x + 1$ = time up.

	Rate ·	Time =	Distance
Up	1.5	$x+1$	$1.5(x+1)$
Down	4	x	$4x$

$$\begin{aligned} d &= d \\ 1.5(x+1) &= 4x \\ 1.5x + 1.5 &= 4x \\ 1.5 &= 2.5x \\ \frac{1.5}{2.5} &= \frac{2.5x}{2.5} \\ 0.6 &= x \end{aligned}$$

Total Time $= x + 1 + x = 0.6 + 1 + 0.6 = 2.2$

The entire hike took 2.2 hours.

2. Let x = speed of eastbound train, then $x - 10$ = speed of westbound train.

	r ·	t =	d
East	x	1.5	$1.5x$
West	$x-10$	1.5	$1.5(x-10)$

$$\begin{aligned} 1.5x + 1.5(x-10) &= 171 \\ 1.5x + 1.5x - 15 &= 171 \\ 3x - 15 &= 171 \\ 3x &= 186 \\ \frac{3x}{3} &= \frac{186}{3} \\ x &= 62 \end{aligned}$$

$x - 10 = 62 - 10 = 52$

The eastbound train is traveling at 62 mph and the westbound train is traveling at 52 mph.

3. Let x = the number of \$20 bills, then $x + 47$ = number of \$5 bills.

Denomination	Number	Value
\$5 bills	$x+47$	$5(x+47)$
\$20 bills	x	$20x$

$$\begin{aligned} 5(x+47) + 20x &= 1710 \\ 5x + 235 + 20x &= 1710 \\ 235 + 25x &= 1710 \\ 25x &= 1475 \\ x &= 59 \end{aligned}$$

$x + 47 = 59 + 47 = 106$

There are 106 \$5 bills and 59 \$20 bills.

4. Let x = amount invested at 11.5%, then
$30{,}000 - x$ = amount invested at 6%.

	Principal ·	Rate ·	Time =	Interest
11.5%	x	0.115	1	$x(0.115)(1)$
6%	$30{,}000 - x$	0.06	1	$0.06(30{,}000 - x)(1)$
Total	30,000			2790

$$\begin{aligned}0.115x + 0.06(30{,}000 - x) &= 2790\\ 0.115x + 1800 - 0.06x &= 2790\\ 1800 + 0.055x &= 2790\\ 0.055x &= 990\\ \frac{0.055x}{0.055} &= \frac{990}{0.055}\\ x &= 18{,}000\end{aligned}$$

$30{,}000 - x = 30{,}000 - 18{,}000 = 12{,}000$
She invested \$18,000 at 11.5% and \$12,000 at 6%.

Exercise Set 2.7

1. Let x = the time traveled by the jet plane.

	Rate ·	Time =	Distance
Jet	500	x	$500x$
Prop	200	$x + 2$	$200(x + 2)$

$$\begin{aligned}d &= d\\ 500x &= 200(x+2)\\ 500x &= 200x + 400\\ 300x &= 400\\ \frac{300x}{300} &= \frac{400}{300}\\ x &= \frac{4}{3}\end{aligned}$$

The jet traveled for $\frac{4}{3}$ hours.

$d = rt$

$d = 500\left(\frac{4}{3}\right) = 666\frac{2}{3}$

The planes are $666\frac{2}{3}$ miles from the starting point.

3. Let x = the average speed on the winding road and $x + 20$ on the level.

	Rate	· Time	= Distance
Winding	x	4	$4x$
Level	$x + 20$	3	$3(x + 20)$
Total			305

$$4x+3(x+20)=305$$
$$4x+3x+60=305$$
$$7x+60=305$$
$$7x=245$$
$$\frac{7x}{7}=\frac{245}{7}$$
$$x=35$$

$x + 20 = 35 + 20 = 55$
The average speed on level road was 55 mph.

5. The value of y dimes is $0.10y$.

7. The value of $x + 7$ nickels is $0.05(x + 7)$.

9. The value of $4y$ \$20 bills is $20(4y)$ or $80y$.

11. The value of $35 - x$ \$50 bills is $50(35 - x)$.

13. Let x = number of \$10 bills, then $20 + x$ number of \$5 bills.

	Number of Bills	Value of Bills
\$5 bills	$20 + x$	$5(20 + x)$
\$10 bills	x	$10x$
Total		280

$$5(20+x)+10x=280$$
$$100+5x+10x=280$$
$$100+15x=280$$
$$15x=180$$
$$x=12$$

$20 + x = 32$
There are 12 \$10 bills and 32 \$5 bills.

15. Let x = the amount invested at 9% for one year.

	Principal	· Rate =	Interest
9%	x	0.09	$0.09x$
8%	$25{,}000 - x$	0.08	$0.08(25{,}000 - x)$
Total	25,000		2135

$$0.09x+0.08(25{,}000-x)=2135$$
$$0.09x+2000-0.08x=2135$$
$$0.01x+2000=2135$$
$$0.01x=135$$
$$\frac{0.01x}{0.01}=\frac{135}{0.01}$$
$$x=13{,}500$$

$25{,}000 - x = 25{,}000 - 13{,}500 = 11{,}500$
She invested \$11,500 at 8% and \$13,500 at 9%.

17. Let x = the amount invested at 11% for one year.

	Principal	· Rate =	Interest
11%	x	0.11	$0.11x$
4%	$10{,}000 - x$	-0.04	$-0.04(10{,}000 - x)$
Total	10,000		650

$$0.11x-0.04(10{,}000-x)=650$$
$$0.11x-400+0.04x=650$$
$$0.15x-400=650$$
$$0.15x=1050$$
$$\frac{0.15x}{0.15}=\frac{1050}{0.15}$$
$$x=7000$$

$10{,}000 - x = 10{,}000 - 7000 = 3000$
He invested \$7000 at 11% and \$3000 at 4%.

19. Let x = the number of adult tickets, then $500 - x$ = the number of child tickets.

	Number ·	Rate =	Cost
Adult	x	43	$43x$
Child	$500 - x$	28	$28(500 - x)$
Total	500		16,805

$$43x + 28(500 - x) = 16{,}805$$
$$43x + 14{,}000 - 28x = 16{,}805$$
$$14{,}000 + 15x = 16{,}805$$
$$15x = 2805$$
$$x = 187$$

$500 - x = 500 - 187 = 313$

Sales included 187 adult tickets and 313 child tickets.

21. Let x = the amount invested at 10% for one year.

	Principal ·	Rate =	Interest
10%	x	0.10	$0.10x$
8%	$54{,}000 - x$	0.08	$0.08(54{,}000 - x)$

$$0.10x = 0.08(54{,}000 - x)$$
$$0.10x = 4320 - 0.08x$$
$$0.18x = 4320$$
$$\frac{0.18x}{0.18} = \frac{4320}{0.18}$$
$$x = 24{,}000$$

$54{,}000 - x = 54{,}000 - 24{,}000 = 30{,}000$

Invest \$30,000 at 8% and \$24,000 at 10%.

23. Let x = the time they are able to talk.

	Rate ·	Time =	Distance
Alan	55	x	$55x$
Dave	65	$x - 1$	$65(x - 1)$
Total			250

$$55x + 65(x - 1) = 250$$
$$55x + 65x - 65 = 250$$
$$120x - 65 = 250$$
$$120x = 315$$
$$\frac{120x}{120} = \frac{315}{120}$$
$$x = 2\frac{5}{8}$$

They can talk for $2\frac{5}{8}$ hours or

2 hours $37\frac{1}{2}$ minutes.

25. Let x = number of nickels, then
$3x$ = number of dimes.

	Number	Value
Nickels	x	$0.05x$
Dimes	$3x$	$0.10(3x)$
Total		56.35

$$0.05x + 0.10(3x) = 56.35$$
$$0.05x + 0.3x = 56.35$$
$$0.35x = 56.35$$
$$x = 161$$

$3x = 3(161) = 483$

They collected 161 nickels and 483 dimes.

27. Let x = the amount invested at 9% for one year.

	Principal ·	Rate =	Interest
9%	x	0.09	$0.09x$
6%	3000	0.06	0.06(3000)
Total			585

$$0.09x + 0.06(3000) = 585$$
$$0.09x + 180 = 585$$
$$0.09x = 405$$
$$\frac{0.09x}{0.09} = \frac{405}{0.09}$$
$$x = 4500$$

Should invest \$4500 at 9%.

29. Let x = the rate of hiker 1.

	Rate ·	Time =	Distance
Hiker 1	x	2	$2x$
Hiker 2	$x + 1.1$	2	$2(x + 1.1)$
Total			11

$$2x + 2(x + 1.1) = 11$$
$$2x + 2x + 2.2 = 11$$
$$4x + 2.2 = 11$$
$$4x = 8.8$$
$$\frac{4x}{4} = \frac{8.8}{4}$$
$$x = 2.2$$

$x + 1.1 = 2.2 + 1.1 = 3.3$

Hiker 1: 2.2 mph; Hiker 2: 3.3 mph

31. Let x = the time spent rowing upstream.

	Rate ·	Time =	Distance
Upstream	5	x	$5x$
Downstream	11	$4-x$	$11(4-x)$

$$5x = 11(4-x)$$
$$5x = 44 - 11x$$
$$16x = 44$$
$$\frac{16x}{16} = \frac{44}{16}$$
$$x = 2.75$$

He rowed upstream for 2.75 hours.

$d = rt$

$d = 5(2.75) = 13.75$

He rowed 13.75 miles each way for a total of 27.5 miles.

33. $3 + (-7) = -4$

35. $\frac{3}{4} - \frac{3}{16} = \frac{4}{4}\cdot\frac{3}{4} - \frac{3}{16} = \frac{12}{16} - \frac{3}{16} = \frac{12-3}{16} = \frac{9}{16}$

37. $-5 - (-1) = -5 + 1 = -4$

39. Let x = number of \$100 bills, then $x + 46$ = number of \$50 bills, and $7x$ = number of \$20 bills.

	Number	Value
\$100 bills	x	$100x$
\$50 bills	$x+46$	$50(x+46)$
\$20 bills	$7x$	$20(7x)$
Total		9550

$$100x + 50(x+46) + 20(7x) = 9550$$
$$100x + 50x + 2300 + 140x = 9550$$
$$290x + 2300 = 9550$$
$$290x = 7250$$
$$x = 25$$

$x + 46 = 71$

$7x = 7(25) = 175$

There were 25 \$100 bills, 71 \$50 bills, and 175 \$20 bills.

41.
$$R = C$$
$$24x = 100 + 20x$$
$$4x = 100$$
$$\frac{4x}{x} = \frac{100}{4}$$
$$x = 25$$

Should sell 25 skateboards to break even.

43.
$$R = C$$
$$7.50x = 4.50x + 2400$$
$$3x = 2400$$
$$\frac{3x}{3} = \frac{2400}{3}$$
$$x = 800$$

Should sell 800 books to break even.

45. Answers may vary

Section 2.8

Practice Exercises

1. $x < 5$

Place a parenthesis at 5 since the inequality symbol is <. Shade to the left of 5. The solution set is $(-\infty, 5)$.

5

2.
$$x + 11 \ge 6$$
$$x + 11 - 11 \ge 6 - 11$$
$$x \ge -5$$

The solution set is $[-5, \infty)$.

–5

3.
$$-5x \ge -15$$
$$\frac{-5x}{-5} \le \frac{-15}{-5}$$
$$x \le 3$$

The solution set is $(-\infty, 3]$.

3

4.
$$3x > -9$$
$$\frac{3x}{3} > \frac{-9}{3}$$
$$x > -3$$

The solution set is $(-3, \infty)$.

–3

5.
$$45 - 7x \le -4$$
$$45 - 7x - 45 \le -4 - 45$$
$$-7x \le -49$$
$$\frac{-7x}{-7} \ge \frac{-49}{-7}$$
$$x \ge 7$$
The solution set is $[7, \infty)$.

6.
$$3x + 20 \le 2x + 13$$
$$3x + 20 - 2x \le 2x + 13 - 2x$$
$$x + 20 \le 13$$
$$x + 20 - 20 \le 13 - 20$$
$$x \le -7$$
The solution set is $(-\infty, -7]$.

7.
$$6 - 5x > 3(x - 4)$$
$$6 - 5x > 3x - 12$$
$$6 - 5x - 3x > 3x - 12 - 3x$$
$$6 - 8x > -12$$
$$6 - 8x - 6 > -12 - 6$$
$$-8x > -18$$
$$\frac{-8x}{-8} < \frac{-18}{-8}$$
$$x < \frac{9}{4}$$
The solution set is $\left(-\infty, \frac{9}{4}\right)$.

8.
$$3(x - 4) - 5 \le 5(x - 1) - 12$$
$$3x - 12 - 5 \le 5x - 5 - 12$$
$$3x - 17 \le 5x - 17$$
$$3x - 17 - 5x \le 5x - 17 - 5x$$
$$-2x - 17 \le -17$$
$$-2x - 17 + 17 \le -17 + 17$$
$$-2x \le 0$$
$$\frac{-2x}{-2} \ge \frac{0}{-2}$$
$$x \ge 0$$
The solution set is $[0, \infty)$.

9. $-3 \le x < 1$
Graph all numbers greater than or equal to -3 and less than 1. Place a bracket at -3 and a parenthesis at 1.
The solution set is $[-3, 1)$.

10.
$$-4 < 3x + 2 \le 8$$
$$-4 - 2 < 3x + 2 - 2 \le 8 - 2$$
$$-6 < 3x \le 6$$
$$\frac{-6}{3} < \frac{3x}{3} \le \frac{6}{3}$$
$$-2 < x \le 2$$
The solution set is $(-2, 2]$.

11.
$$1 < \frac{3}{4}x + 5 < 6$$
$$4(1) < 4\left(\frac{3}{4}x + 5\right) < 4(6)$$
$$4 < 3x + 20 < 24$$
$$4 - 20 < 3x + 20 - 20 < 24 - 20$$
$$-16 < 3x < 4$$
$$\frac{-16}{3} < \frac{3x}{3} < \frac{4}{3}$$
$$-\frac{16}{3} < x < \frac{4}{3}$$
The solution set is $\left(-\frac{16}{3}, \frac{4}{3}\right)$.

12. Let x = number of classes.
$$300 + 375x \le 1500$$
$$300 + 375x - 300 \le 1500 - 300$$
$$375x \le 1200$$
$$\frac{375x}{375} \le \frac{1200}{375}$$
$$x \le 3.2$$
Kasonga can afford at most 3 community college classes this semester.

Vocabulary and Readiness Check

1. $6x - 7(x + 9)$ is an expression.
2. $6x = 7(x + 9)$ is an equation.
3. $6x < 7(x + 9)$ is an inequality.

4. $5y - 2 \geq -38$ is an inequality.

5. $\frac{9}{7} = \frac{x+2}{14}$ is an equation.

6. $\frac{9}{7} - \frac{x+2}{14}$ is an expression.

7. -5 is not a solution to $x \geq -3$.

8. $|-6| = 6$ is not a solution to $x < 6$.

9. 4.1 is not a solution to $x < 4.01$.

10. -4 is not a solution to $x \geq -3$.

Exercise Set 2.8

1. $[2, \infty), x \geq 2$

-2

3. $(-\infty, -5), x < -5$

-5

5. $x \leq -1, (-\infty, -1]$

-1

7. $x < \frac{1}{2}, \left(-\infty, \frac{1}{2}\right)$

$\frac{1}{2}$

9. $y \geq 5, [5, \infty)$

5

11. $2x < -6$

$x < -3, (-\infty, -3)$

-3

13. $x - 2 \geq -7$

$x \geq -5, [-5, \infty)$

-5

15. $-8x \leq 16$

$\frac{-8x}{-8} \geq \frac{16}{-8}$

$x \geq -2, [-2, \infty)$

-2

17. $3x - 5 > 2x - 8$

$x - 5 > -8$

$x > -3, (-3, \infty)$

-3

19. $4x - 1 \leq 5x - 2x$

$4x - 1 \leq 3x$

$x - 1 \leq 0$

$x \leq 1, (-\infty, 1]$

1

21. $x - 7 < 3(x + 1)$

$x - 7 < 3x + 3$

$-2x - 7 < 3$

$-2x < 10$

$\frac{-2x}{-2} > \frac{10}{-2}$

$x > -5, (-5, \infty)$

-5

23. $-6x + 2 \geq 2(5 - x)$

$-6x + 2 \geq 10 - 2x$

$-4x + 2 \geq 10$

$-4x \geq 8$

$\frac{-4x}{-4} \leq \frac{8}{-4}$

$x \leq -2, (\infty, -2]$

-2

25. $4(3x - 1) \leq 5(2x - 4)$

$12x - 4 \leq 10x - 20$

$2x - 4 \leq -20$

$2x \leq -16$

$x \leq -8, (-\infty, -8]$

-8

27. $3(x + 2) - 6 > -2(x - 3) + 14$

$3x + 6 - 6 > -2x + 6 + 14$

$3x > -2x + 20$

$5x > 20$

$x > 4, (4, \infty)$

4

29. $-2x \le -40$
$\frac{-2x}{-2} \ge \frac{-40}{-2}$
$x \ge 20$, $[20, \infty)$
20

31. $-9 + x > 7$
$x > 16$, $(16, \infty)$
16

33. $3x - 7 < 6x + 2$
$-3x - 7 < 2$
$-3x < 9$
$\frac{-3x}{-3} > \frac{9}{-3}$
$x > -3$, $(-3, \infty)$
−3

35. $5x - 7x \ge x + 2$
$-2x \ge x + 2$
$-3x \ge 2$
$\frac{-3x}{-3} \le \frac{2}{-3}$
$x \le -\frac{2}{3}$, $\left(-\infty, -\frac{2}{3}\right]$
$-\frac{2}{3}$

37. $\frac{3}{4}x > 2$
$x > \frac{8}{3}$, $\left(\frac{8}{3}, \infty\right)$
$\frac{8}{3}$

39. $3(x - 5) < 2(2x - 1)$
$3x - 15 < 4x - 2$
$-x - 15 < -2$
$-x < 13$
$\frac{-x}{-1} > \frac{13}{-1}$
$x > -13$, $(-13, \infty)$
−13

41. $4(2x + 1) < 4$
$8x + 4 < 4$
$8x < 0$
$x < 0$, $(-\infty, 0)$
0

43. $-5x + 4 \ge -4(x - 1)$
$-5x + 4 \ge -4x + 4$
$-x + 4 \ge 4$
$-x \ge 0$
$\frac{-x}{-1} \le \frac{0}{-1}$
$x \le 0$, $(-\infty, 0]$
0

45. $-2(x - 4) - 3x < -(4x + 1) + 2x$
$-2x + 8 - 3x < -4x - 1 + 2x$
$-5x + 8 < -2x - 1$
$-3x + 8 < -1$
$-3x < -9$
$\frac{-3x}{-3} > \frac{-9}{-3}$
$x > 3$, $(3, \infty)$
3

47. $-3x + 6 \ge 2x + 6$
$-5x + 6 \ge 6$
$-5x \ge 0$
$\frac{-5x}{-5} \le \frac{0}{-5}$
$x \le 0$, $(-\infty, 0]$
0

49. Answers may vary

51. $-1 < x < 3$, $(-1, 3)$
−1   3

53. $0 \le y < 2$, $[0, 2)$
0   2

55. $-3 < 3x < 6$
$-1 < x < 2$, $(-1, 2)$
−1   2

57. $2 \le 3x - 10 \le 5$
$12 \le 3x \le 15$
$4 \le x \le 5$, [4, 5]

59. $-4 < 2(x-3) \le 4$
$-4 < 2x - 6 \le 4$
$2 < 2x \le 10$
$1 < x \le 5$, (1, 5]

61. $-2 < 3x - 5 < 7$
$3 < 3x < 12$
$1 < x < 4$, (1, 4)

63. $-6 < 3(x-2) \le 8$
$-6 < 3x - 6 \le 8$
$0 < 3x \le 14$
$0 < x \le \frac{14}{3}$, $\left(0, \frac{14}{3}\right]$

65. Answers may vary

67. $2x + 6 > -14$
$2x > -20$
$x > -10$

69. Let x = the number of people invited.
$34x + 50 \le 3000$
$34x \le 2950$
$x \le 86.8$
They may invite 86 people.

71. Let x = the length.
$2l + 2w = P$
$2x + 2(15) \le 100$
$2x + 30 \le 100$
$2x \le 70$
$x \le 35$
The length can be no greater than 35 cm.

73. Let x = the rate for \$5000 for one year.

	Principal ·	Rate =	Interest
11%	10,000	0.11	0.11(10,000)
?	5000	x	$5000x$
Total			1600

$0.11(10,000) + 5000x \ge 1600$
$1100 + 5000x \ge 1600$
$5000x \ge 500$
$x \ge 0.1$
Should invest the \$5000 at 10% or more.

75. Let x = his score on the third game.
$$\frac{146 + 201 + x}{3} \ge 180$$
$$3\left(\frac{146 + 201 + x}{3}\right) \ge 3(180)$$
$347 + x \ge 540$
$x \ge 193$
He must bowl at least 193.

77. $x < 200$ recommended
$200 \le x \le 240$ borderline
$x > 240$ high

79. Let x = the unknown number.
$-5 < 2x + 1 < 7$
$-6 < 2x < 6$
$-3 < x < 3$
All numbers between −3 and 3

81. $-39 \le \frac{5}{9}(F - 32) \le 45$
$-351 \le 5(F - 32) \le 405$
$-351 \le 5F - 160 \le 405$
$-191 \le 5F \le 565$
$-38.2° \le F \le 113°$

83. $(2)^3 = (2)(2)(2) = 8$

85. $(1)^{12} = (1)(1)(1)(1)(1)(1)(1)(1)(1)(1)(1)(1) = 1$

87. $\left(\frac{4}{7}\right) = \left(\frac{4}{7}\right)\left(\frac{4}{7}\right) = \frac{16}{49}$

89. Read the value on the vertical axis corresponding to 2003; \$52.70

91. The greatest drop was in 2005.

93. $C = 3.14d$

$2.9 \le 3.14d \le 3.1$

$0.924 \le d \le 0.987$

The diameter must be between 0.924 cm and 0.987 cm.

95. $x(x+4) > x^2 - 2x + 6$

$x^2 + 4x > x^2 - 2x + 6$

$4x > -2x + 6$

$6x > 6$

$x > 1, (1, \infty)$

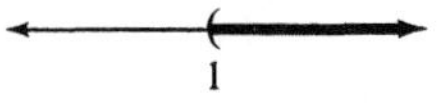

97. $x^2 + 6x - 10 < x(x-10)$

$x^2 + 6x - 10 < x^2 - 10x$

$6x - 10 < -10x$

$16x - 10 < 0$

$16x < 10$

$x < \frac{10}{6}$

$x < \frac{5}{8}, \left(-\infty, \frac{5}{8}\right)$

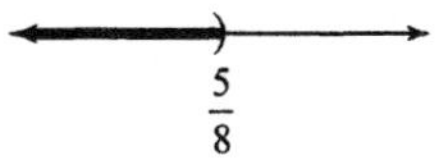

The Bigger Picture

1. $-5x = 15$

$\frac{-5x}{-5} = \frac{15}{-5}$

$x = -3$

The solution is −3.

2. $-5x > 15$

$\frac{-5x}{-5} < \frac{15}{-5}$

$x < -3$

The solution set is (−∞, −3).

3. $9y - 14 = -12$

$9y - 14 + 14 = -12 + 14$

$9y = 2$

$\frac{9y}{9} = \frac{2}{9}$

$y = \frac{2}{9}$

The solution is $\frac{2}{9}$.

4. $9x - 3 = 5x - 4$

$9x - 3 - 5x = 5x - 4 - 5x$

$4x - 3 = -4$

$4x - 3 + 3 = -4 + 3$

$4x = -1$

$\frac{4x}{4} = \frac{-1}{4}$

$x = -\frac{1}{4}$

The solution is $-\frac{1}{4}$.

5. $4(x-2) \le 5x + 7$

$4x - 8 \le 5x + 7$

$4x - 8 - 5x \le 5x + 7 - 5x$

$-x - 8 \le 7$

$-x - 8 + 8 \le 7 + 8$

$-x \le 15$

$\frac{-x}{-1} \ge \frac{15}{-1}$

$x \ge -15$

The solution set is [−15, ∞).

6. $5(4x-1) = 2(10x-1)$

$20x - 5 = 20x - 2$

$20x - 5 - 20x = 20x - 2 - 20x$

$-5 = -2$

Since this is a false statement, there is no solution.

7. $-5.4 = 0.6x - 9.6$

$-5.4 + 9.6 = 0.6x - 9.6 + 9.6$

$4.2 = 0.6x$

$\frac{4.2}{0.6} = \frac{0.6x}{0.6}$

$7 = x$

The solution is 7.

8. $\frac{1}{3}(x-4) < \frac{1}{4}(x+7)$

$12\left[\frac{1}{3}(x-4)\right] < 12\left[\frac{1}{4}(x+7)\right]$

$4(x-4) < 3(x+7)$

$4x - 16 < 3x + 21$

$4x - 16 - 3x < 3x + 21 - 3x$

$x - 16 < 21$

$x - 16 + 16 < 21 + 16$

$x < 37$

The solution set is (−∞, 37).

9. $$\begin{aligned} 3y-5(y-4) &= -2(y-10) \\ 3y-5y+20 &= -2y+20 \\ -2y+20 &= -2y+20 \\ -2y+20+2y &= -2y+20+2y \\ 20 &= 20 \end{aligned}$$
All real numbers are solutions.

10. $$\begin{aligned} \frac{7(x-1)}{3} &= \frac{2(x+1)}{5} \\ 15\left[\frac{7(x-1)}{3}\right] &= 15\left[\frac{2(x+1)}{5}\right] \\ 35(x-1) &= 6(x+1) \\ 35x-35 &= 6x+6 \\ 35x-35-6x &= 6x+6-6x \\ 29x-35 &= 6 \\ 29x-35+35 &= 6+35 \\ 29x &= 41 \\ \frac{29x}{29} &= \frac{41}{29} \\ x &= \frac{41}{29} \end{aligned}$$
The solution is $\frac{41}{29}$.

Chapter 2 Vocabulary Check

1. Terms with the same variables raised to exactly the same powers are called like terms.

2. A linear equation in one variable can be written in the form $ax + b = c$.

3. Equations that have the same solution are called equivalent equations.

4. Inequalities containing two inequality symbols are called compound inequalities.

5. An equation that describes a known relationship among quantities is called a formula.

6. A linear inequality in one variable can be written in the form $ax + b < c$, (or $>$, $\leq$, $\geq$).

7. The numerical coefficient of a term is its numerical factor.

Chapter 2 Review

1. $5x - x + 2x = 6x$

2. $0.2z - 4.6x - 7.4z = -4.6x - 7.2z$

3. $\frac{1}{2}x+3+\frac{7}{2}x-5=\frac{8}{2}x-2=4x-2$

4. $\frac{4}{5}y+1+\frac{6}{5}y+2=\frac{10}{5}y+3=2y+3$

5. $2(n-4)+n-10=2n-8+n-10=3n-18$

6. $3(w+2)-(12-w)=3w+6-12+w=4w-6$

7. $(x+5)-(7x-2)=x+5-7x+2=-6x+7$

8. $$\begin{aligned} (y-0.7)-(1.4y-3) &= y-0.7-1.4y+3 \\ &= -0.4y+2.3 \end{aligned}$$

9. Three times a number decreased by 7 is $3x - 7$.

10. Twice the sum of a number and 2.8 added to 3 times a number is $2(x + 2.8) + 3x$.

11. $$\begin{aligned} 8x+4 &= 9x \\ 8x+4-8x &= 9x-8x \\ 4 &= x \end{aligned}$$

12. $$\begin{aligned} 5y-3 &= 6y \\ 5y-3-5y &= 6y-5y \\ -3 &= y \end{aligned}$$

13. $$\begin{aligned} \frac{2}{7}x+\frac{5}{7}x &= 6 \\ \frac{7}{7}x &= 6 \\ x &= 6 \end{aligned}$$

14. $$\begin{aligned} 3x-5 &= 4x+1 \\ -5 &= x+1 \\ -6 &= x \end{aligned}$$

15. $$\begin{aligned} 2x-6 &= x-6 \\ x-6 &= -6 \\ x &= 0 \end{aligned}$$

16. $$\begin{aligned} 4(x+3) &= 3(1+x) \\ 4x+12 &= 3+3x \\ x+12 &= 3 \\ x &= -9 \end{aligned}$$

17. $$\begin{aligned} 6(3+n) &= 5(n-1) \\ 18+6n &= 5n-5 \\ 18+n &= -5 \\ n &= -23 \end{aligned}$$

18. $5(2+x)-3(3x+2)=-5(x-6)+2$
$10+5x-9x-6=-5x+30+2$
$-4x+4=-5x+32$
$x+4=32$
$x=28$

19. $x-5=3$
$x-5+\underline{5}=3+\underline{5}$
$x=8$

20. $x+9=-2$
$x+9-\underline{9}=-2-\underline{9}$
$x=-11$

21. $10-x$; choice b.

22. $x-5$; choice a.

23. Complementary angles sum to 90°.
$(90-x)°$; choice b.

24. Supplementary angles sum to 180°.
$180-(x+5)=180-x-5=175-x$
$(175-x)°$; choice c.

25. $\frac{3}{4}x=-9$
$\frac{4}{3}\left(\frac{3}{4}x\right)=\frac{4}{3}(-9)$
$x=-12$

26. $\frac{x}{6}=\frac{2}{3}$
$6\cdot\frac{x}{6}=6\cdot\frac{2}{3}$
$x=4$

27. $-5x=0$
$\frac{-5x}{-5}=\frac{0}{-5}$
$x=0$

28. $-y=7$
$\frac{-y}{-1}=\frac{7}{-1}$
$y=-7$

29. $0.2x=0.15$
$\frac{0.2x}{0.2}=\frac{0.15}{0.2}$
$x=0.75$

30. $\frac{-x}{3}=1$
$-3\cdot\frac{-x}{3}=-3\cdot1$
$x=-3$

31. $-3x+1=19$
$-3x=18$
$\frac{-3x}{-3}=\frac{18}{-3}$
$x=-6$

32. $5x+25=20$
$5x=-5$
$\frac{5x}{5}=\frac{-5}{5}$
$x=-1$

33. $7(x-1)+9=5x$
$7x-7+9=5x$
$7x+2=5x$
$2=-2x$
$\frac{2}{-2}=\frac{-2x}{-2}$
$-1=x$

34. $7x-6=5x-3$
$2x-6=-3$
$2x=3$
$\frac{2x}{2}=\frac{3}{2}$
$x=\frac{3}{2}$

35. $-5x+\frac{3}{7}=\frac{10}{7}$
$7\left(-5x+\frac{3}{7}\right)=7\cdot\frac{10}{7}$
$-35x+3=10$
$-35x=7$
$x=-\frac{7}{35}$
$x=-\frac{1}{5}$

36. $5x+x=9+4x-1+6$
$6x=4x+14$
$2x=14$
$x=7$

37. Let x = the first integer, then
$x + 1$ = the second integer, and
$x + 2$ = the third integer.
sum $= x + (x + 1) + (x + 2) = 3x + 3$

38. Let x = the first integer, then
$x + 2$ = the second integer
$x + 4$ = the third integer
$x + 6$ = the fourth integer.
sum $= x + (x + 6) = 2x + 6$

39.
$$\begin{aligned} \frac{5}{3}x + 4 &= \frac{2}{3}x \\ 3\left(\frac{5}{3}x + 4\right) &= 3\left(\frac{2}{3}x\right) \\ 5x + 12 &= 2x \\ 12 &= -3x \\ -4 &= x \end{aligned}$$

40.
$$\begin{aligned} \frac{7}{8}x + 1 &= \frac{5}{8}x \\ 8\left(\frac{7}{8}x + 1\right) &= 8\left(\frac{5}{8}x\right) \\ 7x + 8 &= 5x \\ 8 &= -2x \\ -4 &= x \end{aligned}$$

41.
$$\begin{aligned} -(5x + 1) &= -7x + 3 \\ -5x - 1 &= -7x + 3 \\ 2x - 1 &= 3 \\ 2x &= 4 \\ x &= 2 \end{aligned}$$

42.
$$\begin{aligned} -4(2x + 1) &= -5x + 5 \\ -8x - 4 &= -5x + 5 \\ -3x - 4 &= 5 \\ -3x &= 9 \\ x &= -3 \end{aligned}$$

43.
$$\begin{aligned} -6(2x - 5) &= -3(9 + 4x) \\ -12x + 30 &= -27 - 12x \\ 30 &= -27 \end{aligned}$$
There is no solution.

44.
$$\begin{aligned} 3(8y - 1) &= 6(5 + 4y) \\ 24y - 3 &= 30 + 24y \\ -3 &= 30 \end{aligned}$$
There is no solution.

45.
$$\begin{aligned} \frac{3(2 - z)}{5} &= z \\ 3(2 - z) &= 5z \\ 6 - 3z &= 5z \\ 6 &= 8z \\ \frac{6}{8} &= z \\ \frac{3}{4} &= z \end{aligned}$$

46.
$$\begin{aligned} \frac{4(n + 2)}{5} &= -n \\ 4(n + 2) &= -5n \\ 4n + 8 &= -5n \\ 8 &= -9n \\ -\frac{8}{9} &= n \end{aligned}$$

47.
$$\begin{aligned} 0.5(2n - 3) - 0.1 &= 0.4(6 + 2n) \\ 10[0.5(2n - 3) - 0.1] &= 10[0.4(6 + 2n)] \\ 5(2n - 3) - 1 &= 4(6 + 2n) \\ 10n - 15 - 1 &= 24 + 8n \\ 10n - 16 &= 24 + 8n \\ 2n - 16 &= 24 \\ 2n &= 40 \\ n &= 20 \end{aligned}$$

48.
$$\begin{aligned} -9 - 5a &= 3(6a - 1) \\ -9 - 5a &= 18a - 3 \\ -9 &= 23a - 3 \\ -6 &= 23a \\ -\frac{6}{23} &= a \end{aligned}$$

49.
$$\begin{aligned} \frac{5(c + 1)}{6} &= 2c - 3 \\ 5(c + 1) &= 6(2c - 3) \\ 5c + 5 &= 12c - 18 \\ -7c + 5 &= -18 \\ -7c &= -23 \\ c &= \frac{23}{7} \end{aligned}$$

50. $\frac{2(8-a)}{3}=4-4a$
$2(8-a)=3(4-4a)$
$16-2a=12-12a$
$10a+16=12$
$10a=-4$
$a=\frac{-4}{10}$
$a=-\frac{2}{5}$

51. $200(70x-3560)=-179(150x-19,300)$
$14,000x-712,000=-26,850x+3,454,700$
$40,850x-712,000=3,454,700$
$40,850x=4,166,700$
$x=102$

52. $1.72y-0.04y=0.42$
$1.68y=0.42$
$y=0.25$

53. Let x = length of a side of the square, then
$50.5 + 10x$ = the height.
$x+(50.5+10x)=7327$
$11x+50.5=7327$
$11x=7276.5$
$x=661.5$
$50.5 + 10x = 50.5 + 10(661.5) = 6665.5$
The height is 6665.5 inches.

54. Let x = the length of the shorter piece and
$2x$ = the length of the other.
$x+2x=12$
$3x=12$
$x=4$
$2x = 2(4) = 8$
The lengths are 4 feet and 8 feet.

55. Let x = number of Keebler plants, then
$2x - 1$ = number of Kellogg plants.
$x+(2x-1)=53$
$3x-1=53$
$3x=54$
$x=18$
$2x - 1 = 2(18) - 1 = 35$
There were 18 Keebler plants and 35 Kellogg plants.

56. Let x = first integer, then
$x + 1$ = second integer, and
$x + 2$ = third integer.
$x+(x+1)+(x+2)=-114$
$3x+3=-114$
$3x=-117$
$x=-39$
$x + 1 = -39 + 1 = -38$
$x + 2 = -39 + 2 = -37$
The integers are –39, –38, –37.

57. Let x = the unknown number.
$\frac{x}{3}=x-2$
$3\cdot\frac{x}{3}=3(x-2)$
$x=3x-6$
$-2x=-6$
$x=3$
The number is 3.

58. Let x = the unknown number.
$2(x+6)=-x$
$2x+12=-x$
$12=-3x$
$-4=x$
The number is –4.

59. Let $P = 46$ and $l = 14$.
$P=2l+2w$
$46=2(14)+2w$
$46=28+2w$
$18=2w$
$9=w$

60. Let $V = 192$, $l = 8$, and $w = 6$.
$V=lwh$
$192=8(6)h$
$192=48h$
$4=h$

61. $y=mx+b$
$y-b=mx$
$\frac{y-b}{x}=m$

62. $r=vst-5$
$r+5=vst$
$\frac{r+5}{vt}=s$

63. $2y-5x=7$
$$-5x=-2y+7$$
$$x=\frac{-2y+7}{-5}$$
$$x=\frac{2y-7}{5}$$

64. $3x-6y=-2$
$$-6y=-3x-2$$
$$y=\frac{-3x-2}{-6}$$
$$y=\frac{3x+2}{6}$$

65. $C=\pi D$
$$\frac{C}{D}=\pi$$

66. $C=2\pi r$
$$\frac{C}{2r}=\pi$$

67. Let $V=900$, $l=20$, and $h=3$.
$$V=lwh$$
$$900=20w(3)$$
$$900=60w$$
$$15=w$$
The width is 15 meters.

68. Let x = width, then $x+6$ = length.
$$60=2x+2(x+6)$$
$$60=2x+2x+12$$
$$60=4x+12$$
$$48=4x$$
$$12=x$$
$$x+6=12+6=18$$
The dimensions are 18 feet by 12 feet.

69. Let $d=10{,}000$ and $r=125$.
$$d=rt$$
$$10{,}000=125t$$
$$80=t$$
It will take 80 minutes or 1 hour and 20 minutes.

70. Let $F=104$.
$$C=\frac{5}{9}(F-32)$$
$$=\frac{5}{9}(104-32)$$
$$=\frac{5}{9}(72)$$
$$=40$$
The temperature was 40°C.

71. Let x = the percent.
$$9=x\cdot 45$$
$$\frac{9}{45}=\frac{45x}{45}$$
$$0.2=x$$
9 is 20% of 45.

72. Let x = the percent.
$$59.5=x\cdot 85$$
$$\frac{59.5}{85}=\frac{85x}{85}$$
$$0.7=x$$
59.5 is 70% of 85.

73. Let x = the number.
$$137.5=125\%\cdot x$$
$$137.5=1.25x$$
$$\frac{137.5}{1.25}=\frac{1.25x}{1.25}$$
$$110=x$$
137.5 is 125% of 110.

74. Let x = the number.
$$768=60\%\cdot x$$
$$768=0.6x$$
$$\frac{768}{0.6}=\frac{0.6x}{0.6}$$
$$1280=x$$
768 is 60% of 1280.

75. Let x = mark-up.
$$x=11\%\cdot 1900$$
$$x=0.11\cdot 1900$$
$$x=209$$
New price = 1900 + 209 = 2109
The mark-up is \$209 and the new price is \$2109.

76. Find 66.9% of 76,000.
$$0.669\cdot 76{,}000=50{,}844$$
We would expect 50,844 people to use the Internet.

77. Let x = gallons of 40% solution.

Strength	gallons	Concentration	
40%	x	0.4	$0.4x$
10%	$30-x$	0.1	$0.1(30-x)$
20%	30	0.2	0.2(30)

$$0.4x+0.1(30-x)=0.2(30)$$
$$0.4x+3-0.1x=6$$
$$0.3x+3=6$$
$$0.3x=3$$
$$x=10$$

$30-x=30-10=20$

Mix 10 gallons of 40% acid solution with 20 gallons of 10% acid solution.

78. Increase = 21.0 − 20.7 = 0.3

Let x = percent.

$$0.3=x\cdot 20.7$$
$$\frac{0.3}{20.7}=\frac{20.7x}{20.7}$$
$$0.0145\approx x$$

The percent increase is 1.45%.

79. From the graph, the height of 'Almost hit a car' is 18%.

80. Choose the tallest graph. The most common effect is swerving into another lane.

81. Find 21% of 4600.

$0.21\cdot 4600=966$

We would expect 966 customers to have cut someone off.

82. 46% + 41% + 21% + 18% = 126%

No; answers may vary

83. Let x = time up, then $3-x$ = time down.

	Rate	Time	Distance
Up	10	x	$10x$
Down	50	$3-x$	$50(3-x)$

$$d=d$$
$$10x=50(3-x)$$
$$10x=150-50x$$
$$60x=150$$
$$x=2.5$$

$$\begin{aligned}\text{Total distance} &= 10x + 50(3-x)\\ &= 10(2.5)+50(3-2.5)\\ &= 25+50(0.5)\\ &= 25+25\\ &= 50\end{aligned}$$

The distance traveled was 50 km.

84. Let x = the amount invested at 10.5% for one year.

	Principal ·	Rate =	Interest
10.5%	x	0.105	0.105
8.5%	$50{,}000 - x$	0.085	$0.085(50{,}000 - x)$
Total	50,000		4550

$$\begin{aligned}0.105x + 0.085(50{,}000 - x) &= 4550\\ 0.105x + 4250 - 0.085x &= 4550\\ 0.02x + 4250 &= 4550\\ 0.02x &= 300\\ x &= 15{,}000\end{aligned}$$

$50{,}000 - x = 50{,}000 - 15{,}000 = 35{,}000$

Invest \$35,000 at 8.5% and \$15,000 at 10.5%.

85. Let x = the number of dimes,
$2x$ = the number of quarters, and
$500 - x - 2x$ the number of nickels.

	No. of Coins ·	Value =	Amt. of Money
Dimes	x	0.1	$0.1x$
Quarters	$2x$	0.25	$0.25(2x)$
Nickels	$500 - 3x$	0.05	$0.05(500 - 3x)$
Total	500		88

$$\begin{aligned}0.1x + 0.25(2x) + 0.05(500 - 3x) &= 88\\ 0.1x + 0.5x + 25 - 0.15x &= 88\\ 0.45x + 25 &= 88\\ 0.45x &= 63\\ x &= 140\end{aligned}$$

$500 - 3x = 500 - 3(140) = 500 - 420 = 80$

There were 80 nickels in the pay phone.

114. $-5x < 20$

$$\frac{-5x}{-5} > \frac{20}{-5}$$

$x > -4$, $(-4, \infty)$

−4

115. $-3(1+2x)+x \geq -(3-x)$

$-3-6x+x \geq -3+x$

$-3-5x \geq -3+x$

$-5x \geq x$

$-6x \geq 0$

$$\frac{-6x}{-6} \leq \frac{0}{-6}$$

$x \leq 0$, $(-\infty, 0]$

0

Chapter 2 Test

1. $2y - 6 - y - 4 = y - 10$

2. $2.7x + 6.1 + 3.2x - 4.9 = 5.9x + 1.2$

3. $4(x-2)-3(2x-6) = 4x-8-6x+18$
$= -2x+10$

4. $7 + 2(5y - 3) = 7 + 10y - 6 = 10y + 1$

5. $-\frac{4}{5}x = 4$

$$-\frac{5}{4}\cdot\left(-\frac{4}{5}x\right) = -\frac{5}{4}\cdot 4$$

$x = -5$

6. $4(n-5) = -(4-2n)$

$4n-20 = -4+2n$

$2n-20 = -4$

$2n = 16$

$n = 8$

7. $5y-7+y = -(y+3y)$

$6y-7 = -4y$

$-7 = -10y$

$\frac{7}{10} = y$

8. $4z+1-z = 1+z$

$3z+1 = 1+z$

$2z+1 = 1$

$2z = 0$

$z = 0$

9. $\frac{2(x+6)}{3} = x-5$

$2(x+6) = 3(x-5)$

$2x+12 = 3x-15$

$12 = x-15$

$27 = x$

10. $\frac{1}{2}-x+\frac{3}{2} = x-4$

$$2\left(\frac{1}{2}-x+\frac{3}{2}\right) = 2(x-4)$$

$1-2x+3 = 2x-8$

$-2x+4 = 2x-8$

$-4x+4 = -8$

$-4x = -12$

$x = 3$

11. $-0.3(x-4)+x = 0.5(3-x)$

$10[-0.3(x-4)+x] = 10[0.5(3-x)]$

$-3(x-4)+10x = 5(3-x)$

$-3x+12+10x = 15-5x$

$7x+12 = 15-5x$

$12x+12 = 15$

$12x = 3$

$x = \frac{3}{12} = \frac{1}{4} = 0.25$

12. $-4(a+1)-3a = -7(2a-3)$

$-4a-4-3a = -14a+21$

$-7a-4 = -14a+21$

$7a-4 = 21$

$7a = 25$

$a = \frac{25}{7}$

13. $-2(x-3) = x+5-3x$

$-2x+6 = -2x+5$

$6 = 5$

There is no solution.

14. Let x = the number.

$x+\frac{2}{3}x = 35$

$$3\left(x+\frac{2}{3}x\right) = 3(35)$$

$3x+2x = 105$

$5x = 105$

$x = 21$

The number is 21.

15. Let $l = 35$, and $w = 20$.
$2A = 2lw = 2(35)(20) = 1400$
Let x = the number of gallons needed at 200 square feet per gallon.
$1400 = 200x$
$7 = x$
7 gallons are needed.

16. Let x = one area code, then
$2x$ = other area code.
$x + 2x = 1203$
$3x = 1203$
$\frac{3x}{3} = \frac{1203}{3}$
$x = 401$
$2x = 2(401) = 802$
The area codes are 401 and 802.

17. Let x = the amount invested at 10% for one year.

Principal · Rate = Interest

	Principal	Rate	Interest
10%	x	0.10	$0.1x$
12%	$2x$	0.12	$0.12(2x)$
Total			2890

$0.1x + 0.12(2x) = 2890$
$0.1x + 0.24x = 2890$
$0.34x = 2890$
$x = 8500$
$2x = 2(8500) = 17{,}000$
He invested \$8500 at 10% and \$17,000 at 12%.

18. Let x = the time they travel.

Rate · Time = Distance

	Rate	Time	Distance
Train 1	50	x	$50x$
Train 2	64	x	$64x$
Total			285

$50x + 64x = 285$
$114x = 285$
$x = 2\frac{1}{2}$

They must travel for $2\frac{1}{2}$ hours.

19. Let $y = -14$, $m = -2$, and $b = -2$.
$y = mx + b$
$-14 = -2x - 2$
$-12 = -2x$
$6 = x$

20. $V = \pi r^2 h$
$\frac{V}{\pi r^2} = \frac{\pi r^2 h}{\pi r^2}$
$\frac{V}{\pi r^2} = h$

21. $3x - 4y = 10$
$-4y = -3x + 10$
$y = \frac{-3x + 10}{-4}$
$y = \frac{3x - 10}{4}$

22. $3x - 5 \ge 7x + 3$
$-4x - 5 \ge 3$
$-4x \ge 8$
$\frac{-4x}{-4} \le \frac{8}{-4}$
$x \le -2$, $(-\infty, -2]$

−2

23. $x + 6 > 4x - 6$
$-3x + 6 > -6$
$-3x > -12$
$\frac{-3x}{-3} < \frac{-12}{-3}$
$x < 4$, $(-\infty, 4)$

4

24. $-2 < 3x + 1 < 8$
$-3 < 3x < 7$
$-1 < x < \frac{7}{3}$, $\left(-1, \frac{7}{3}\right)$

−1 $\frac{7}{3}$

25. $\frac{2(5x+1)}{3} > 2$

$2(5x+1) > 6$

$10x + 2 > 6$

$10x > 4$

$x > \frac{4}{10} = \frac{2}{5}, \left(\frac{2}{5}, \infty\right)$

$\frac{2}{5}$

Chapter 2 Cumulative Review

1. a. the natural numbers are 11 and 112.

b. The whole numbers are 0, 11, and 112.

c. The integers are −3, −2, 0, 11, and 112.

d. The rational numbers are −3, −2, −1.5, 0, $\frac{1}{4}$, 11, and 112.

e. The irrational number is $\sqrt{2}$.

f. All the numbers in the given set are real numbers.

2. a. The natural numbers are 2, 7, and 8.

b. The whole numbers are 0, 2, 7, and 8.

c. The integers are −185, 0, 2, 7, and 8.

d. The rational numbers are −185, $-\frac{1}{5}$, 0, 2, 7, and 8.

e. The irrational number is $\sqrt{3}$.

f. All the numbers in the given set are real numbers.

3. a. $|4| = 4$

b. $|-5| = 5$

c. $|0| = 0$

d. $\left|-\frac{1}{2}\right| = \frac{1}{2}$

e. $|5.6| = 5.6$

4. a. $|5| = 5$

b. $|-8| = 8$

c. $\left|-\frac{2}{3}\right| = \frac{2}{3}$

5. a. $40 = 2 \cdot 2 \cdot 2 \cdot 5$

b. $63 = 3 \cdot 3 \cdot 7$

6. a. $44 = 2 \cdot 2 \cdot 11$

b. $90 = 2 \cdot 3 \cdot 3 \cdot 5$

7. $\frac{2}{5} = \frac{2}{5} \cdot \frac{4}{4} = \frac{8}{20}$

8. $\frac{2}{3} = \frac{2}{3} \cdot \frac{8}{8} = \frac{16}{24}$

9. $3[4+2(10-1)] = 3[4+2(9)]$
$= 3[4+18]$
$= 3[22]$
$= 66$

10. $5[16-4(2+1)] = 5[16-4(3)]$
$= 5[16-12]$
$= 5[4]$
$= 20$

11. Let $x = 2$.
$3x + 10 = 8x$
$3(2) + 10 \stackrel{?}{=} 8(2)$
$6 + 10 \stackrel{?}{=} 16$
$16 = 16$
2 is a solution of the equation.

12. Let $x = 3$.
$5x - 2 = 4x$
$5(3) - 2 \stackrel{?}{=} 4(3)$
$15 - 2 \stackrel{?}{=} 12$
$13 \neq 12$
3 is not a solution of the equation.

13. $-1 + (-2) = -3$

14. $(-2) + (-8) = -10$

15. $-4 + 6 = 2$

16. $-3 + 10 = 7$

17. a. $-(-10) = 10$

b. $-\left(-\frac{1}{2}\right) = \frac{1}{2}$

c. $-(-2x) = 2x$

d. $-|-6| = -(6) = -6$

18. a. $-(-5) = 5$

b. $-\left(-\frac{2}{3}\right) = \frac{2}{3}$

c. $-(-a) = a$

d. $-|-3| = -(3) = -3$

19. a. $5.3 - (-4.6) = 5.3 + 4.6 = 9.9$

b. $-\frac{3}{10} - \frac{5}{10} = -\frac{3}{10} + \left(-\frac{5}{10}\right)$
$= \frac{-3-5}{10}$
$= -\frac{8}{10}$
$= -\frac{4}{5}$

c. $-\frac{2}{3} - \left(-\frac{4}{5}\right) = -\frac{2}{3}\cdot\frac{5}{5} + \frac{4}{5}\cdot\frac{3}{3}$
$= -\frac{10}{15} + \frac{12}{15}$
$= \frac{2}{15}$

20. a. $-2.7 - 8.4 = -2.7 + (-8.4) = -11.1$

b. $-\frac{4}{5} - \left(-\frac{3}{5}\right) = -\frac{4}{5} + \frac{3}{5} = \frac{-4+3}{5} = -\frac{1}{5}$

c. $\frac{1}{4} - \left(-\frac{1}{2}\right) = \frac{1}{4} + \frac{1}{2}\cdot\frac{2}{2} = \frac{1}{4} + \frac{2}{4} = \frac{3}{4}$

21. a. $x = 90 - 38 = 90 + (-38) = 52$
The complementary angle is 52°.

b. $y = 180 - 62 = 180 + (-62) = 118$
The supplementary angle is 118°.

22. a. $x = 90 - 72 = 90 + (-72) = 18$
The complementary angle is 18°.

b. $y = 180 - 47 = 180 + (-47) = 133$
The supplementary angle is 133°.

23. a. $(-1.2)(0.05) = -0.06$

b. $\frac{2}{3}\cdot\left(-\frac{7}{10}\right) = -\frac{2\cdot 7}{3\cdot 10} = -\frac{14}{30} = -\frac{7}{15}$

c. $\left(-\frac{4}{5}\right)(-20) = \frac{4\cdot 20}{5} = \frac{80}{5} = 16$

24. a. $(4.5)(-0.08) = -0.36$

b. $-\frac{3}{4}\cdot -\frac{8}{17} = \frac{3\cdot 8}{4\cdot 17} = \frac{24}{68} = \frac{6}{17}$

25. a. $\frac{-24}{-4} = 6$

b. $\frac{-36}{3} = -12$

c. $\frac{2}{3} \div \left(-\frac{5}{4}\right) = \frac{2}{3}\left(-\frac{4}{5}\right) = -\frac{8}{15}$

d. $-\frac{3}{2} \div 9 = -\frac{3}{2} \div \frac{9}{1} = -\frac{3}{2}\cdot\frac{1}{9} = -\frac{3}{18} = -\frac{1}{6}$

26. a. $\frac{-32}{8} = -4$

b. $\frac{-108}{-12} = 9$

c. $-\frac{5}{7} \div \left(\frac{-9}{2}\right) = -\frac{5}{7}\left(-\frac{2}{9}\right) = \frac{10}{63}$

27. a. $x + 5 = 5 + x$

b. $3 \cdot x = x \cdot 3$

28. a. $y + 1 = 1 + y$

b. $y \cdot 4 = 4 \cdot y$

29. a. $8 \cdot 2 + 8 \cdot x = 8(2 + x)$

b. $7s + 7t = 7(s + t)$

30. a. $4 \cdot y + 4 \cdot \frac{1}{3} = 4\left(y + \frac{1}{3}\right)$

b. $0.10x + 0.10y = 0.10(x + y)$

31. $(2x-3)-(4x-2) = 2x-3-4x+2 = -2x-1$

32. $(-5x+1)-(10x+3) = -5x+1-10x-3$
$= -15x-2$

33.
$$\begin{aligned} y+0.6 &= -1.0 \\ y+0.6-0.6 &= -1.0-0.6 \\ y &= -1.6 \end{aligned}$$

34.
$$\begin{aligned} \frac{5}{6}+x &= \frac{2}{3} \\ 6\left(\frac{5}{6}\right)+6(x) &= 6\left(\frac{2}{3}\right) \\ 5+6x &= 4 \\ 6x &= -1 \\ x &= -\frac{1}{6} \end{aligned}$$

35.
$$\begin{aligned} 7 &= -5(2a-1)-(-11a+6) \\ 7 &= -10a+5+11a-6 \\ 7 &= a-1 \\ 7+1 &= a-1+1 \\ 8 &= a \end{aligned}$$

36.
$$\begin{aligned} -3x+1-(-4x-6) &= 10 \\ -3x+1+4x+6 &= 10 \\ x+7 &= 10 \\ x &= 3 \end{aligned}$$

37. $\frac{y}{7} = 20$
$y = 140$

38. $\frac{x}{4} = 18$
$x = 72$

39.
$$\begin{aligned} 4(2x-3)+7 &= 3x+5 \\ 8x-12+7 &= 3x+5 \\ 8x-5 &= 3x+5 \\ 5x-5 &= 5 \\ 5x &= 10 \\ x &= 2 \end{aligned}$$

40.
$$\begin{aligned} 6x+5 &= 4(x+4)-1 \\ 6x+5 &= 4x+16-1 \\ 6x+5 &= 4x+15 \\ 2x+5 &= 15 \\ 2x &= 10 \\ x &= 5 \end{aligned}$$

41. Let x = a number.
$$\begin{aligned} 2(x+4) &= 4x-12 \\ 2x+8 &= 4x-12 \\ 8 &= 2x-12 \\ 20 &= 2x \\ 10 &= x \end{aligned}$$
The number is 10.

42. Let x = a number.
$$\begin{aligned} x+4 &= 3x-8 \\ 4 &= 2x-8 \\ 12 &= 2x \\ 6 &= x \end{aligned}$$
The number is 6.

43.
$$\begin{aligned} V &= lwh \\ \frac{V}{wh} &= \frac{lwh}{wh} \\ \frac{V}{wh} &= l \end{aligned}$$

44.
$$\begin{aligned} C &= 2\pi r \\ \frac{C}{2\pi} &= \frac{2\pi r}{2\pi} \\ \frac{C}{2\pi} &= r \end{aligned}$$

45. $x+4 \le -6$
$x \le -10$, $(-\infty, -10]$

−10

46. $x-3 > 2$
$x > 5$, $(5, \infty)$

5

Chapter 3

Section 3.1

Practice Exercises

1. a. We look for the shortest bar, which is the bar representing Germany. We move from the right edge of this bar vertically downward to the Internet user axis. Germany has approximately 45 million Internet users.

   b. India has approximately 50 million Internet users. Germany has approximately 45 million Internet users. We subtract $50 - 45 = 5$ or 5 million. India has 5 million more Internet users than Germany.

2. a. We locate the number 40 along the time axis and move vertically upward until the line is reached. From this point on the line, we move horizontally to the left until the pulse rate axis is reached. Reading the number of beats per minute, we find that the pulse rate is 70 beats per minute 40 minutes after a cigarette is lit.

   b. The number 0 on the time axis corresponds to the time when the cigarette is being lit. We move vertically upward to the point on the line and then horizontally to the left to the pulse rate axis. The pulse rate is 60 beats per minute when the cigarette is being lit.

   c. We find the highest point of the line graph, which represents the highest pulse rate. From this point, we move vertically downward to the time axis. We find the pulse rate is the highest at 5 minutes, which means 5 minutes after lighting a cigarette.

3. a. Point (4, –3) lies in quadrant IV.

   b. Point (–3, 5) lies in quadrant II.

   c. Point (0, 4) lies on an axis, so it is not in any quadrant.

   d. Point (–6, 1) lies in quadrant II.

   e. Point (–2, 0) lies on an axis, so it is not in any quadrant.

   f. Point (5, 5) lies in quadrant I.

   g. Point $\left(3\frac{1}{2}, 1\frac{1}{2}\right)$ lies in quadrant I.

   h. Point (–4, –5) lies in quadrant III.

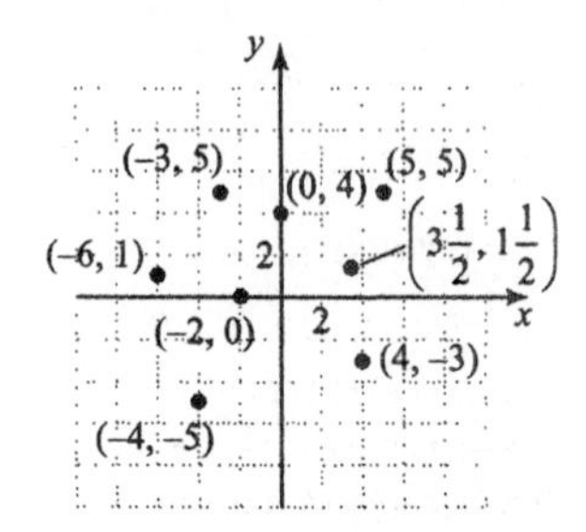

4. a. The ordered pairs are (2000, 92), (2001, 84), (2002, 73), (2003, 64), (2004, 65), (2005, 67), and (2006, 96).

   b. We plot the ordered pairs. We label the horizontal axis "Year" and the vertical axis "Wildfires (in thousands)."

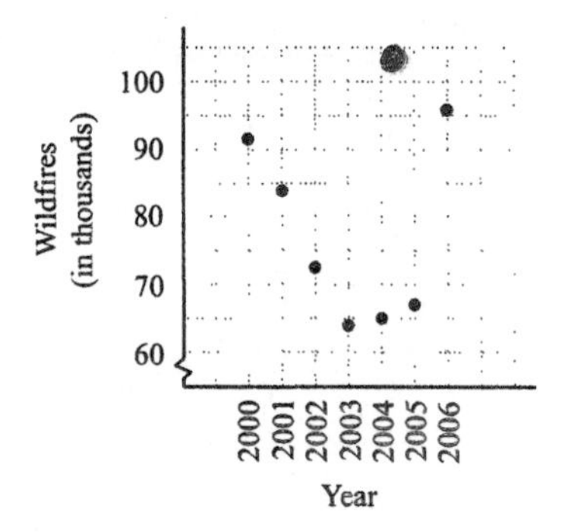

5. a. Let $x = 3$ and $y = 1$.
$$\begin{aligned} x + 3y &= 6 \\ 3 + 3(1) &= 6 \\ 3 + 3 &= 6 \\ 6 &= 6 \quad \text{true} \end{aligned}$$
(3, 1) is a solution.

   b. Let $x = 6$ and $y = 0$.
$$\begin{aligned} x + 3y &= 6 \\ 6 + 3(0) &= 6 \\ 6 + 0 &= 6 \\ 6 &= 6 \quad \text{true} \end{aligned}$$
(6, 0) is a solution.

c. Let $x = -2$ and $y = \frac{2}{3}$.

$$x + 3y = 6$$
$$-2 + 3\left(\frac{2}{3}\right) = 6$$
$$-2 + 2 = 6$$
$$0 = 6 \quad \text{false}$$

$\left(-2, \frac{2}{3}\right)$ is not a solution.

6. a. Let $x = 0$ and solve for y.

$$2x - y = 8$$
$$2(0) - y = 8$$
$$0 - y = 8$$
$$-y = 8$$
$$y = -8$$

The ordered pair is (0, −8).

b. Let $y = 4$ and solve for x.

$$2x - y = 8$$
$$2x - 4 = 8$$
$$2x = 12$$
$$x = 6$$

The ordered pair is (6, 4).

c. Let $x = -3$ and solve for y.

$$2x - y = 8$$
$$2(-3) - y = 8$$
$$-6 - y = 8$$
$$-y = 14$$
$$y = -14$$

The ordered pair is (−3, −14).

7. a. Replace x with −2 in the equation and solve for y.

$$y = -4x$$
$$y = -4(-2)$$
$$y = 8$$

The ordered pair is (−2, 8).

b. Replace y with −12 in the equation and solve for x.

$$y = -4x$$
$$-12 = -4x$$
$$3 = x$$

The ordered pair is (3, −12).

c. Replace x with 0 in the equation and solve for y.

$$y = -4x$$
$$y = -4(0)$$
$$y = 0$$

The ordered pair is (0, 0).

The completed table is shown below.

x	y
−2	8
3	−12
0	0

8. a. Let $x = -10$.

$$y = \frac{1}{5}x - 2$$
$$y = \frac{1}{5}(-10) - 2$$
$$y = -2 - 2$$
$$y = -4$$

Ordered pair: (−10, −4)

b. Let $x = 0$.

$$y = \frac{1}{5}x - 2$$
$$y = \frac{1}{5}(0) - 2$$
$$y = 0 - 2$$
$$y = -2$$

Ordered pair: (0, −2)

c. Let $y = 0$.

$$y = \frac{1}{5}x - 2$$
$$0 = \frac{1}{5}x - 2$$
$$2 = \frac{1}{5}x$$
$$10 = x$$

Ordered pair: (10, 0)

The completed table is shown below.

x	y
−10	−4
0	−2
10	0

9. When $x = 0$,
$y = -1800x + 12{,}000$
$y = -1800 \cdot 0 + 12{,}000$
$y = 0 + 12{,}000$
$y = 12{,}000$

When $x = 1$,
$y = -1800x + 12{,}000$
$y = -1800 \cdot 1 + 12{,}000$
$y = -1800 + 12{,}000$
$y = 10{,}200$

When $x = 2$,
$y = -1800x + 12{,}000$
$y = -1800 \cdot 2 + 12{,}000$
$y = -3600 + 12{,}000$
$y = 8400$

When $x = 3$,
$y = -1800x + 12{,}000$
$y = -1800 \cdot 3 + 12{,}000$
$y = -5400 + 12{,}000$
$y = 6600$

When $x = 4$,
$y = -1800x + 12{,}000$
$y = -1800 \cdot 4 + 12{,}000$
$y = -7200 + 12{,}000$
$y = 4800$

The completed table is shown below.

x	0	1	2	3	4
y	12,000	10,200	8400	6600	4800

Vocabulary and Readiness Check

1. The horizontal axis is called the <u>x-axis</u>.

2. The vertical axis is called the <u>y-axis</u>.

3. The intersection of the horizontal axis and the vertical axis is a point called the <u>origin</u>.

4. The axes divide the plane into regions, called <u>quadrants</u>. There are <u>four</u> of these regions.

5. In the ordered pair of numbers (−2, 5), the number −2 is called the <u>x-coordinate</u> and the number 5 is called the <u>y-coordinate</u>.

6. Each ordered pair of numbers corresponds to <u>one</u> point in the plane.

7. An ordered pair is a <u>solution</u> of an equation in two variables if replacing the variables by the coordinates of the ordered pair results in a true statement.

Exercise Set 3.1

1. We look for the tallest bar, which is the bar representing France. France is the most popular tourist destination.

3. We look for bars extending above the horizontal line at 40 on the vertical axis, which are the bars representing France, U.S., Spain, and China. These countries have more than 40 million tourists per year.

5. The bar for the United Kingdom (U.K.) extends to the horizontal line at 30 on the vertical axis. The United Kingdom has approximately 30 million tourists per year.

7. From 2000 on the year axis, we move vertically up to the point on the line graph. This point is on the attendance axis at approximately 72,600. The Super Bowl attendance in 2000 was approximately 72,600.

9. We find the highest point on the line graph and move vertically downward to the year axis. The year with the greatest Super Bowl attendance was 2007. From the highest point on the graph, we move horizontally to the attendance axis. The highest attendance was approximately 104,000.

11. From 2002 on the year axis, we move vertically up to the point on the line graph. When we move horizontally to the vertical axis. The number of students per teacher was approximately 15.9 in 2002.

13. The number of students per teacher shows the greatest decrease between 1996 and 1998. Notice that the line graph is steepest between 1996 and 1998.

15. The points on the line graph for 1994, 1996, and 1998 lie above the horizontal line at 16 on the vertical axis. The point on the line graph for 2000 appears to lie on the horizontal line at 16 on the vertical axis. The point for 2002 is the first that lies below this horizontal line. The first year shown that the number of students per teacher fell below 16 was 2002.

17. a. Point (1, 5) lies in quadrant I.

b. Point (–5, –2) lies in quadrant III.

c. Point (–3, 0) lies on the *x*-axis, so it is not in any quadrant.

d. Point (0, –1) lies on the *y*-axis, so it is not in any quadrant.

e. Point (2, –4) lies in quadrant IV.

f. Point $\left(-1, 4\frac{1}{2}\right)$ lies in quadrant II.

g. (3.7, 2.2) lies in quadrant I.

h. Point $\left(\frac{1}{2}, -3\right)$ lies in quadrant IV.

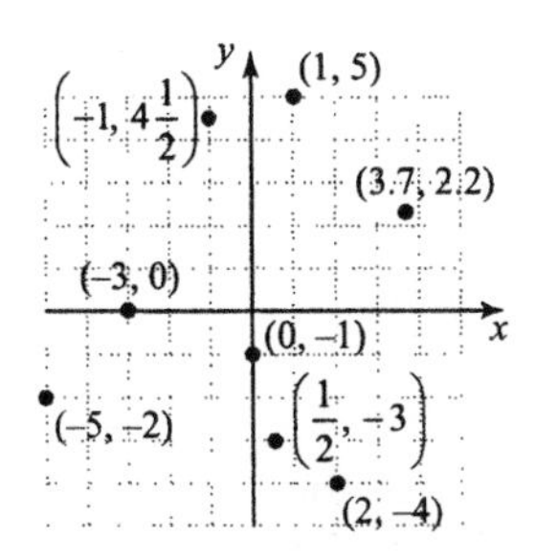

19. Point *A* lies at the origin. Its coordinates are given by the ordered pair (0, 0).

21. Point *C* lies three units to the right and two units above the origin. Its coordinates are given by the ordered pair (3, 2).

23. Point *E* lies two units to the left and two units below the origin. Its coordinates are given by the ordered pair (–2, –2).

25. Point *G* lies two units to the right and one unit below the origin. Its coordinates are given by the ordered pair (2, –1).

27. Point *B* lies on the *y*-axis three units below the origin. Its coordinates are given by the ordered pair (0, –3).

29. Point *D* lies one unit to the right and three units above the origin. Its coordinates are given by the ordered pair (1, 3).

31. Point *F* lies three units to the left and one unit below the origin. Its coordinates are given by the ordered pair (–3, –1).

33. a. The ordered pairs are (2002, 12), (2003, 14), (2004, 14), (2005, 11), and (2006, 12).

b. We plot the ordered pairs. We label the horizontal axis "Year" and the vertical axis "Regular-Season Games Won by Super Bowl Winner."

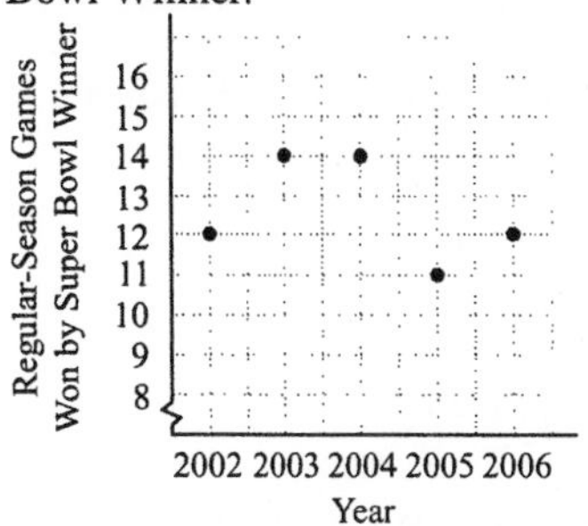

35. a. The ordered pairs are (2001, 1770), (2003, 2800), (2005, 3904), (2007, 7500), and (2009, 10,800).

b. We plot the ordered pairs. We label the horizontal axis "Year" and the vertical axis "Ethanol Fuel Production (in millions of gallons)."

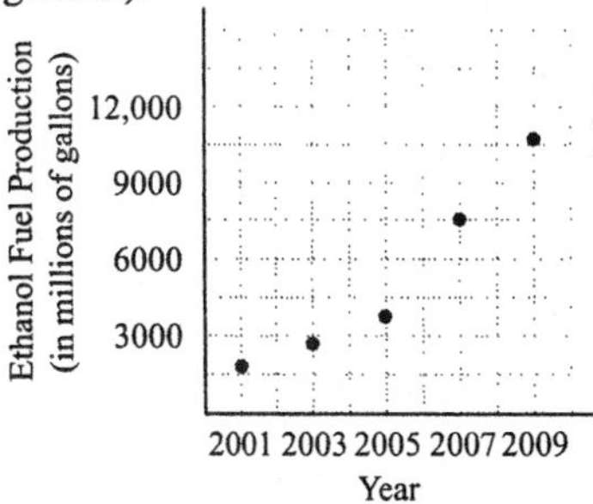

c. The ethanol production is increasing as the years increase.

37. a. The ordered pairs are (2313, 2), (2085, 1), (2711, 21), (2869, 39), (2920, 42), (4038, 99), (1783, 0), and (2493, 9).

b. We plot the ordered pairs. We label the horizontal axis "Distance from Equator (in miles)" and the vertical axis "Average Annual Snowfall (in inches)."

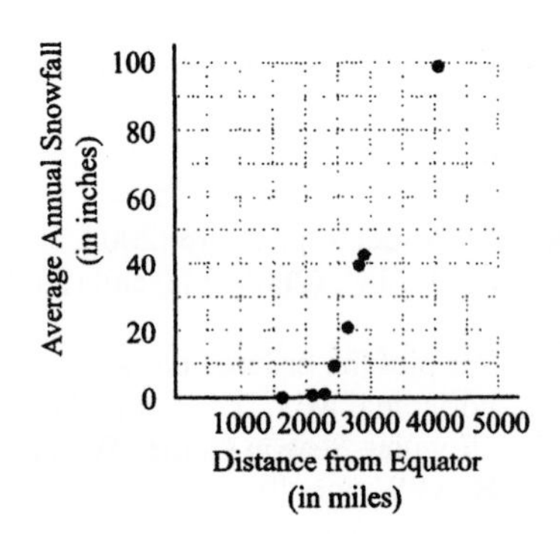

c. The farther from the equator, the more snowfall.

39. For (3, 1), let $x = 3$ and $y = 1$.

$2x + y = 7$
$2(3) + 1 = 7$
$6 + 1 = 7$
$7 = 7$ true

Yes, (3, 1) is a solution.

For (7, 0), let $x = 7$ and $y = 0$.

$2x + y = 7$
$2(7) + 0 = 7$
$14 + 0 = 7$
$14 = 7$ false

No, (7, 0) is not a solution.

For (0, 7), let $x = 0$ and $y = 7$.

$2x + y = 7$
$2(0) + 7 = 7$
$0 + 7 = 7$
$7 = 7$ true

Yes, (0, 7) is a solution.

41. For (0, 0), let $x = 0$ and $y = 0$.

$x = -\frac{1}{3}y$
$0 = -\frac{1}{3}(0)$
$0 = 0$ true

Yes, (0, 0) is a solution.

For (3, –9), let $x = 3$ and $y = -9$.

$x = -\frac{1}{3}y$
$3 = -\frac{1}{3}(-9)$
$3 = 3$ true

Yes, (3, –9) is a solution.

43. For (4, 5), let $x = 4$ and $y = 5$.

$x = 5$
$4 = 5$ false

No, (4, 5) is not a solution.

For (5, 4), let $x = 5$ and $y = 4$.

$x = 5$
$5 = 5$ true

Yes, (5, 4) is a solution.

For (5, 0), let $x = 5$ and $y = 0$.

$x = 5$
$5 = 5$ true

Yes, (5, 0) is a solution.

45. Replace y with –2 and solve for x.

$x - 4y = 4$
$x - 4(-2) = 4$
$x + 8 = 4$
$x = -4$

The ordered pair is (–4, –2).

Replace x with 4 and solve for y.

$x - 4y = 4$
$4 - 4y = 4$
$-4y = 0$
$y = 0$

The ordered pair is (4, 0).

47. Replace x with –8 and solve for y.

$y = \frac{1}{4}x - 3$
$y = \frac{1}{4}(-8) - 3$
$y = -2 - 3$
$y = -5$

The ordered pair is (–8, –5).

Replace y with 1 and solve for x.

$y = \frac{1}{4}x - 3$
$1 = \frac{1}{4}x - 3$
$4 = \frac{1}{4}x$
$16 = x$

The ordered pair is (16, 1).

49. Replace x with 0 and solve for y.

$y = -7x$
$y = -7(0)$
$y = 0$
The ordered pair is (0, 0).

Replace x with -1 and solve for y.

$y = -7x$
$y = -7(-1)$
$y = 7$
The ordered pair is $(-1, 7)$.

Replace y with 2 and solve for x.

$y = -7x$
$2 = -7x$
$-\frac{2}{7} = x$

The ordered pair is $\left(-\frac{2}{7}, 2\right)$.

The completed table is shown below.

x	y
0	0
-1	7
$-\frac{2}{7}$	2

51. Replace x with 0 and solve for y.

$y = -x + 2$
$y = -0 + 2$
$y = 2$
The ordered pair is (0, 2).

Replace y with 0 and solve for x.

$y = -x + 2$
$0 = -x + 2$
$x = 2$
The ordered pair is (2, 0).

Replace x with -3 and solve for y.

$y = -x + 2$
$y = -(-3) + 2$
$y = 3 + 2$
$y = 5$
The ordered pair is $(-3, 5)$.
The completed table is shown below.

x	y
0	2
2	0
-3	5

53. Replace x with 0 and solve for y.

$y = \frac{1}{2}x$
$y = \frac{1}{2}(0)$
$y = 0$
The ordered pair is (0, 0).

Replace x with -6 and solve for y.

$y = \frac{1}{2}x$
$y = \frac{1}{2}(-6)$
$y = -3$
The ordered pair is $(-6, -3)$.

Replace y with 1 and solve for x.

$y = \frac{1}{2}x$
$1 = \frac{1}{2}x$
$2 = x$
The ordered pair is (2, 1).
The completed table is shown below.

x	y
0	0
-6	-3
2	1

55. Replace x with 0 and solve for y.

$x + 3y = 6$
$0 + 3y = 6$
$3y = 6$
$y = 2$
The ordered pair is (0, 2).

Replace y with 0 and solve for x.

$x + 3y = 6$
$x + 3(0) = 6$
$x + 0 = 6$
$x = 6$
The ordered pair is (6, 0).

Replace y with 1 and solve for x.

$x + 3y = 6$
$x + 3(1) = 6$
$x + 3 = 6$
$x = 3$

The ordered pair is (3, 1).
The completed table is shown below.

x	y
0	2
6	0
3	1

57. Replace x with 0 and solve for y.

$$y = 2x - 12$$
$$y = 2(0) - 12$$
$$y = 0 - 12$$
$$y = -12$$

The ordered pair is (0, –12).

Replace y with –2 and solve for x.

$$y = 2x - 12$$
$$-2 = 2x - 12$$
$$10 = 2x$$
$$5 = x$$

The ordered pair is (5, –2).

Replace x with 3 and solve for y.

$$y = 2x - 12$$
$$y = 2(3) - 12$$
$$y = 6 - 12$$
$$y = -6$$

The ordered pair is (3, –6).
The completed table is shown below.

x	y
0	–12
5	–2
3	–6

59. Replace x with 0 and solve for y.

$$2x + 7y = 5$$
$$2(0) + 7y = 5$$
$$7y = 5$$
$$y = \frac{5}{7}$$

The ordered pair is $\left(0, \frac{5}{7}\right)$.

Replace y with 0 and solve for x.

$$2x + 7y = 5$$
$$2x + 7(0) = 5$$
$$2x = 5$$
$$x = \frac{5}{2}$$

The ordered pair is $\left(\frac{5}{2}, 0\right)$.

Replace y with 1 and solve for x.

$$2x + 7y = 5$$
$$2x + 7(1) = 5$$
$$2x + 7 = 5$$
$$2x = -2$$
$$x = -1$$

The ordered pair is (–1, 1).
The completed table is shown below.

x	y
0	$\frac{5}{7}$
$\frac{5}{2}$	0
–1	1

61. Replace y with 0 and solve for x.

$$x = -5y$$
$$x = -5(0)$$
$$x = 0$$

The ordered pair is (0, 0).

Replace y with 1 and solve for x.

$$x = -5y$$
$$x = -5(1)$$
$$x = -5$$

The ordered pair is (–5, 1).

Replace x with 10 and solve for y.

$$x = -5y$$
$$10 = -5y$$
$$-2 = y$$

The ordered pair is (10, –2).
The completed table is shown below.

x	y
0	0
–5	1
10	–2

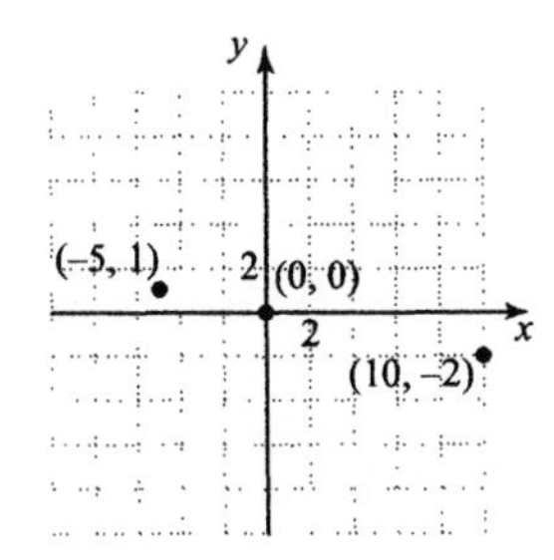

63. Replace x with 0 and solve for y.

$y = \frac{1}{3}x + 2$

$y = \frac{1}{3}(0) + 2$

$y = 2$

The ordered pair is (0, 2).

Replace x with −3 and solve for y.

$y = \frac{1}{3}x + 2$

$y = \frac{1}{3}(-3) + 2$

$y = -1 + 2$

$y = 1$

The ordered pair is (−3, 1).

Replace y with 0 and solve for x.

$y = \frac{1}{3}x + 2$

$0 = \frac{1}{3}x + 2$

$-\frac{1}{3}x = 2$

$x = -6$

The ordered pair is (−6, 0).
The completed table is shown below.

x	y
0	2
−3	1
−6	0

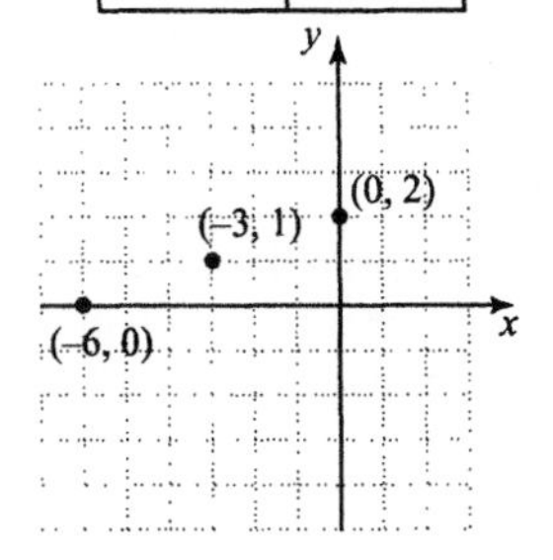

65. a. When $x = 100$,

$y = 80x + 5000$

$y = 80(100) + 5000$

$y = 8000 + 5000$

$y = 13,000$

When $x = 200$,

$y = 80x + 5000$

$y = 80(200) + 5000$

$y = 16,000 + 5000$

$y = 21,000$

When $x = 300$,

$y = 80x + 5000$

$y = 80(300) + 5000$

$y = 24,000 + 5000$

$y = 29,000$

The completed table is shown below.

x	100	200	300
y	13,000	21,000	29,000

b. Replace y with 8600 and solve for x.

$y = 80x + 5000$

$8600 = 80x + 5000$

$3600 = 80x$

$45 = x$

Thus, 45 computer desks can be produced for \$8600.

67. a. When $x = 1$,

$y = -2.35x + 55.92$

$y = -2.35(1) + 55.92$

$y = -2.35 + 55.92$

$y = 53.57$

When $x = 3$,

$y = -2.35x + 55.92$

$y = -2.35(3) + 55.92$

$y = -7.05 + 55.92$

$y = 48.87$

When $x = 5$,

$y = -2.35x + 55.92$

$y = -2.35(5) + 55.92$

$y = -11.75 + 55.92$

$y = 44.17$

The completed table is shown below.

x	1	3	5
y	53.57	48.87	44.17

b. Replace y with 46 and solve for x.

$$y = -2.35x + 55.92$$
$$46 = -2.35x + 55.92$$
$$-9.92 = -2.35x$$
$$4 \approx x$$

The yearly average amount spent on recorded music was approximately \$46 in year 4 or 2005.

69. Four years after 2000, or in 2004, there were 1308 Target stores.

71. In year 0, there appear to be approximately 975 Target stores. In year 1, there appear to be approximately 1050 Target stores. The increase for year 1 is approximately $1050 - 975 = 75$ stores.

In year 1, there appear to be approximately 1050 Target stores. In year 2, there appear to be approximately 1150 Target stores. The increase for year 2 is approximately $1150 - 1050 = 100$ stores.

In year 2, there appear to be approximately 1150 Target stores. In year 3, there appear to be approximately 1225 Target stores. The increase for year 3 is approximately $1225 - 1150 = 75$ stores.

73. The graph of the ordered pair (a, b) is the same as the graph of the ordered pair (b, a) when $a = b$.

75. Subtract x from each side.

$$x + y = 5$$
$$y = 5 - x$$

77. Subtract $2x$ from each side. Then divide each side by 4.

$$2x + 4y = 5$$
$$4y = -2x + 5$$
$$y = -\frac{1}{2}x + \frac{5}{4}$$

79. Divide each side by -5.

$$10x = -5y$$
$$-2x = y$$
$$y = -2x$$

81. Subtract x from each side. Then divide each side by -3.

$$x - 3y = 6$$
$$-3y = -x + 6$$
$$y = \frac{1}{3}x - 2$$

83. False; the point $(-1, 5)$ lies in quadrant II.

85. True

87. In quadrant III, both coordinates are negative: (negative, negative).

89. In quadrant IV, the x-coordinate is positive and the y-coordinate is negative: (positive, negative).

91. At the origin, both coordinates are zero: (0, 0).

93. A point of the form (0, number) is located on the y-axis.

95. No; answers may vary.

97. Answers may vary

99. The point four units to the right of the y-axis and seven units below the x-axis has ordered pair $(4, -7)$.

101.

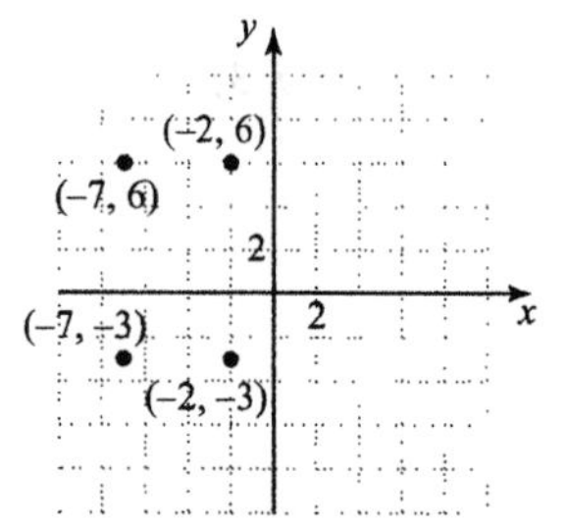

a. The fourth vertex is $(-2, 6)$. The rectangle is 9 units by 5 units.

b. The perimeter is $9 + 5 + 9 + 5 = 28$ units.

c. The area is $9 \times 5 = 45$ square units.

Section 3.2

Practice Exercises

1. a. $3x + 2.7y = -5.3$ is a linear equation in two variables because it is written in the form $Ax + By = C$ with $A = 3$, $B = 2.7$, and $C = -5.3$.

b. $x^2 + y = 8$ is not a linear equation in two variables because x is squared.

c. $y = 12$ is a linear equation in two variables because it can be written in the form $Ax + By = C$: $0x + y = 12$.

d. $5x = -3y$ is a linear equation in two variables because it can be written in the form $Ax + By = C$: $5x + 3y = 0$.

2. Find three ordered pair solutions.
Let $x = 0$.
$$x + 3y = 9$$
$$0 + 3y = 9$$
$$3y = 9$$
$$y = 3$$

Let $x = 3$.
$$x + 3y = 9$$
$$3 + 3y = 9$$
$$3y = 6$$
$$y = 2$$

Let $y = 1$.
$$x + 3y = 9$$
$$x + 3(1) = 9$$
$$x + 3 = 9$$
$$x = 6$$
The ordered pairs are (0, 3), (3, 2), and (6, 1).

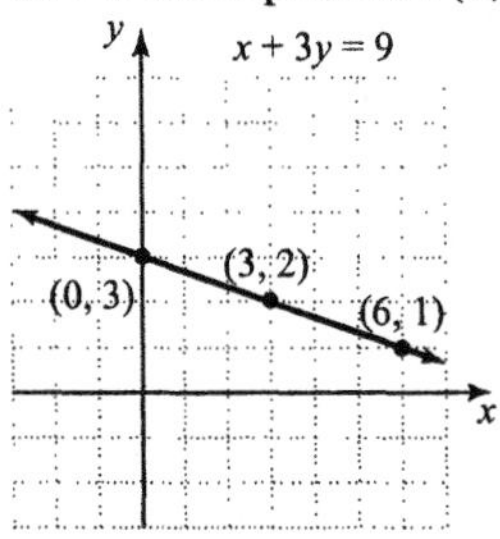

3. Find three ordered pair solutions.
Let $x = 0$.
$$3x - 4y = 12$$
$$3(0) - 4y = 12$$
$$-4y = 12$$
$$y = -3$$

Let $y = 0$.
$$3x - 4y = 12$$
$$3x - 4(0) = 12$$
$$3x = 12$$
$$x = 4$$

Let $x = 2$.
$$3x - 4y = 12$$
$$3(2) - 4y = 12$$
$$6 - 4y = 12$$
$$-4y = 6$$
$$y = -\frac{6}{4} = -\frac{3}{2}$$
The ordered pairs are (0, −3), (4, 0), and $\left(2, -\frac{3}{2}\right)$.

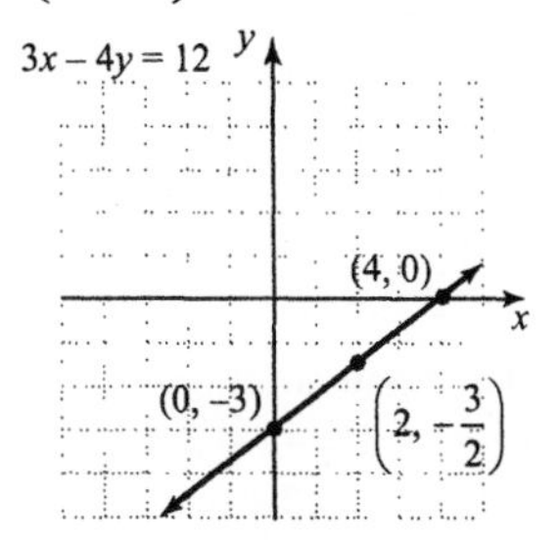

4. Find three ordered pair solutions.
If $x = 1$, $y = -2(1) = -2$.
If $x = 0$, $y = -2(0) = 0$.
If $x = -1$, $y = -2(-1) = 2$.

x	y
1	−2
0	0
−1	2

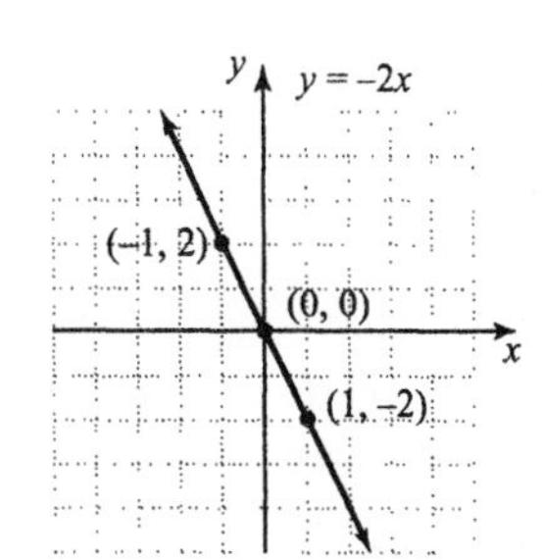

5. Find three ordered pair solutions.

If $x = 2$, $y = \frac{1}{2}(2) + 3 = 1 + 3 = 4$.

If $x = 0$, $y = \frac{1}{2}(0) + 3 = 0 + 3 = 3$.

If $x = -4$, $y = \frac{1}{2}(-4) + 3 = -2 + 3 = 1$.

x	y
2	4
0	3
–4	1

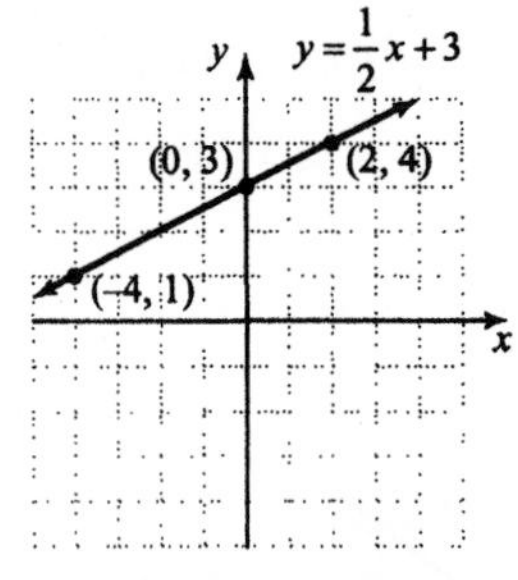

6. Find three ordered pair solutions.
   If $x = 1$, $y = -2(1) + 3 = -2 + 3 = 1$.
   If $x = 0$, $y = -2(0) + 3 = 0 + 3 = 3$.
   If $x = 3$, $y = -2(3) + 3 = -6 + 3 = -3$.

x	y
1	1
0	0
3	–3

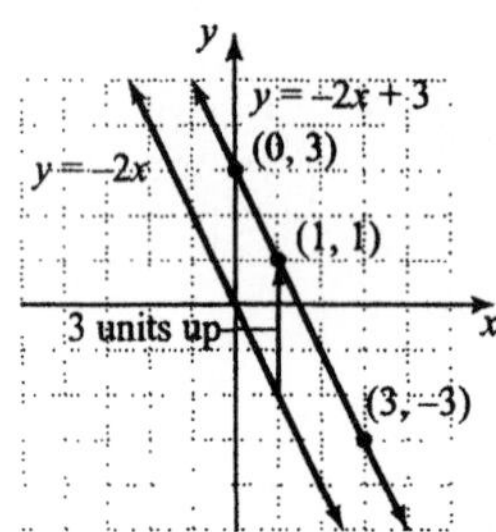

The graph of $y = -2x + 3$ is the same as the graph of $y = -2x$ except that the graph of $y = -2x + 3$ is moved three units upward.

7. **a.** Find three ordered pair solutions.
   If $x = 0$, $y = 22.2(0) + 371 = 0 + 371 = 371$.
   If $x = 6$,
   $y = 22.2(6) + 371 = 133.2 + 371 = 504.2$.
   If $x = 9$,
   $y = 22.2(9) + 371 = 199.8 + 371 = 570.8$.

x	y
0	371
6	504.2
9	570.8

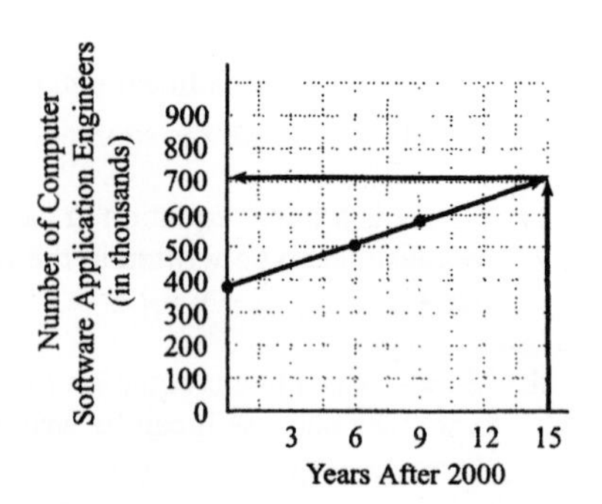

b. The graph shows that we predict approximately 700 thousand computer software application engineers in the year 2015.

Calculator Explorations

1. $y = -3x + 7$

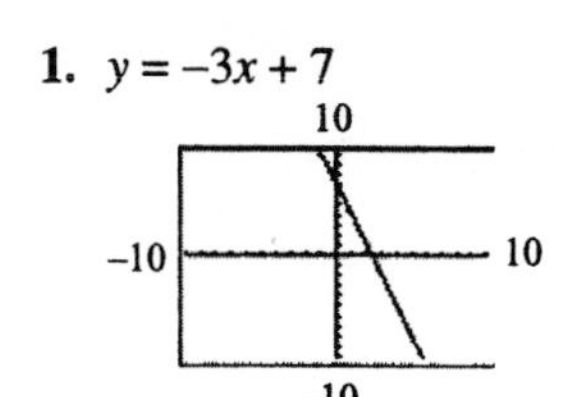

2. $y = -x + 5$

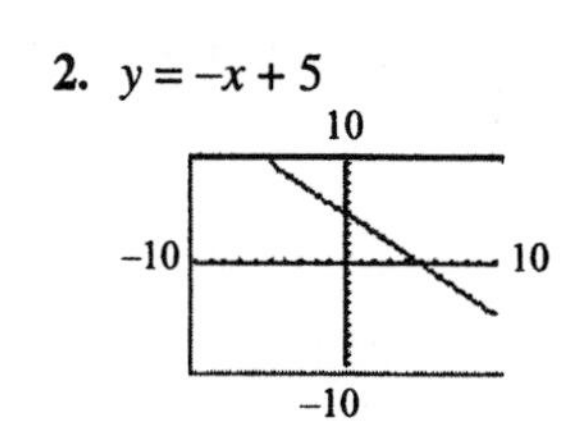

3. $y = 2.5x - 7.9$

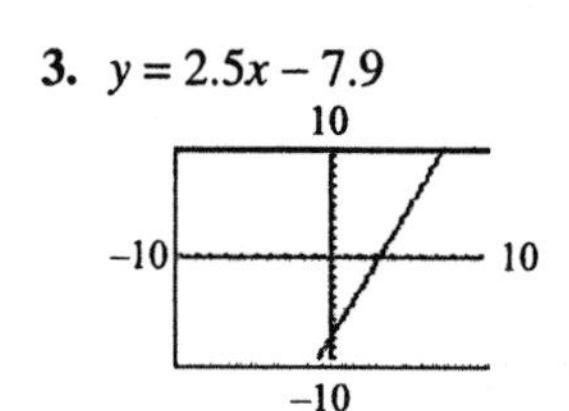

4. $y = -1.3x + 5.2$

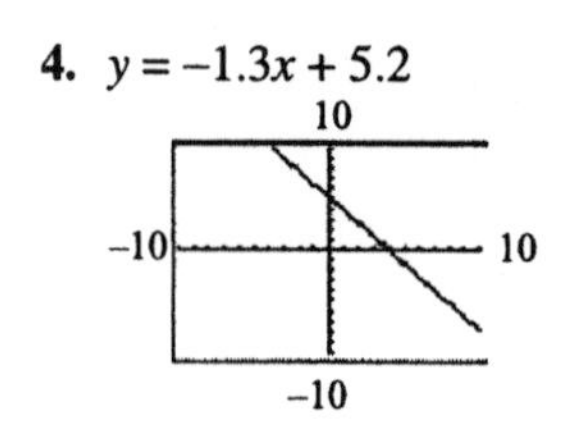

5. $y = -\frac{3}{10}x + \frac{32}{5}$

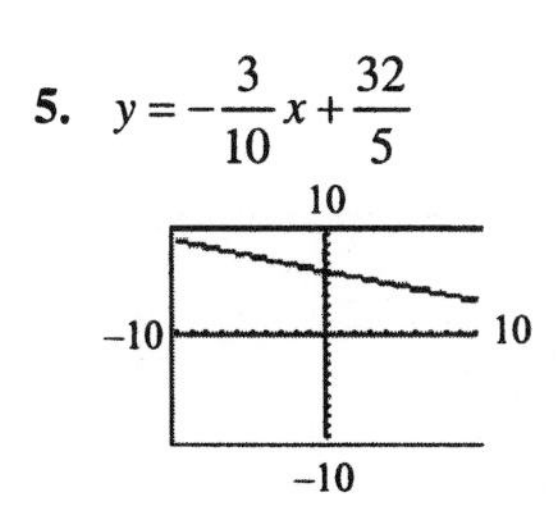

6. $y = \frac{2}{9}x - \frac{22}{3}$

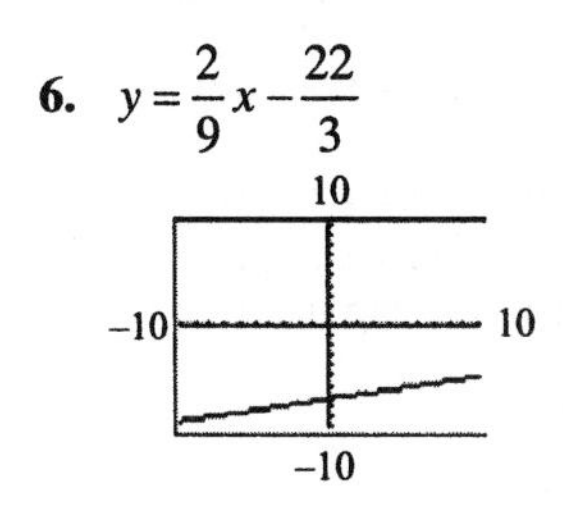

Exercise Set 3.2

1. Yes; it can be written in the form $Ax + By = C$.

3. Yes; it can be written in the form $Ax + By = C$.

5. No; x is squared.

7. Yes; it can be written in the form $Ax + By = C$.

9. Let $y = 0$.
$x - y = 6$
$x - 0 = 6$
$x = 6$

Let $x = 4$.
$x - y = 6$
$4 - y = 6$
$-y = 2$
$y = -2$

Let $y = -1$
$x - y = 6$
$x - (-1) = 6$
$x + 1 = 6$
$x = 5$

x	y
6	0
4	−2
5	−1

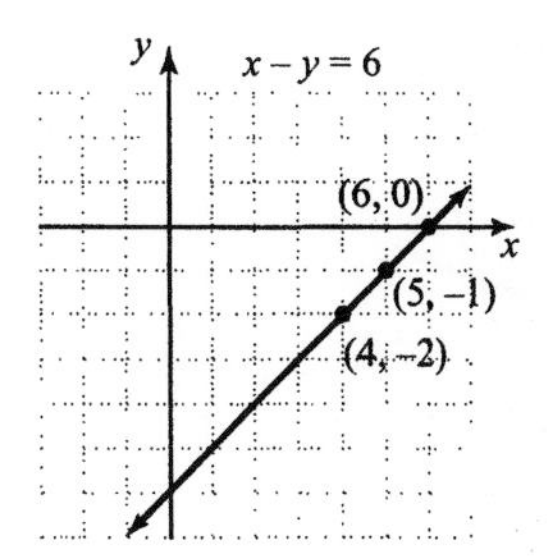

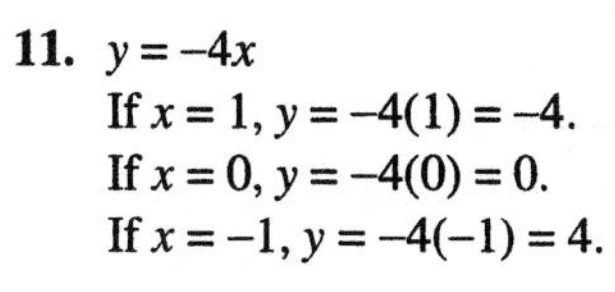

11. $y = -4x$
If $x = 1$, $y = -4(1) = -4$.
If $x = 0$, $y = -4(0) = 0$.
If $x = -1$, $y = -4(-1) = 4$.

x	y
1	−4
0	0
−1	4

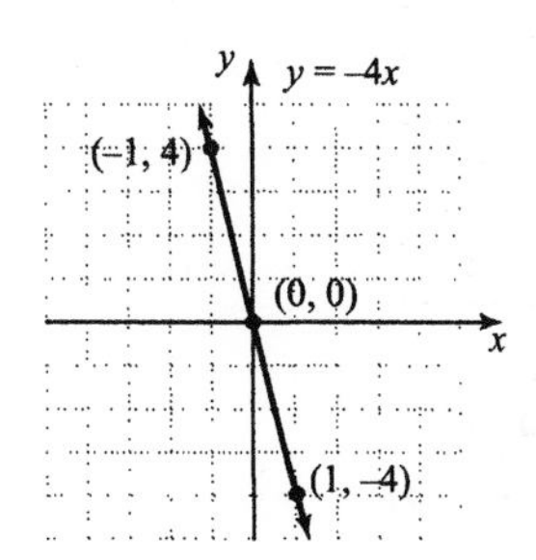

13. $y = \frac{1}{3}x$

If $x = 0$, $y = \frac{1}{3}(0) = 0$.

If $x = 6$, $y = \frac{1}{3}(6) = 2$.

If $x = -3$, $y = \frac{1}{3}(-3) = -1$.

x	y
0	0
6	2
−3	−1

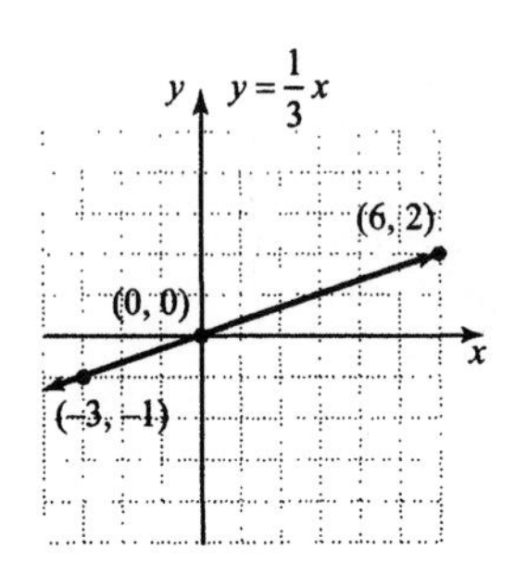

15. $y = -4x + 3$
If $x = 0$, $y = -4(0) + 3 = 0 + 3 = 3$.
If $x = 1$, $y = -4(1) + 3 = -4 + 3 = -1$.
If $x = 2$, $y = -4(2) + 3 = -8 + 3 = -5$.

x	y
0	3
1	−1
2	−5

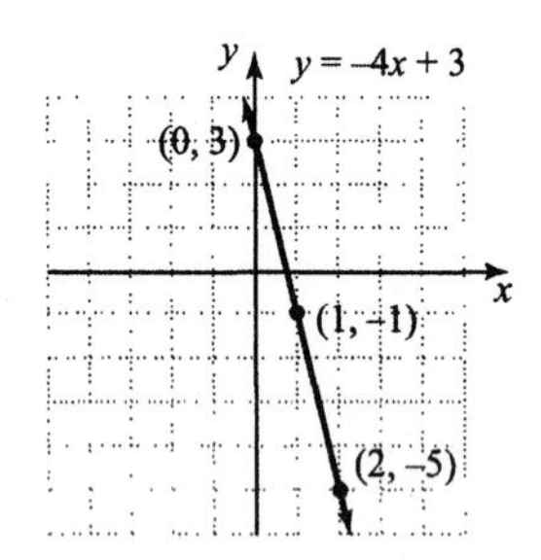

17. $x + y = 1$

x	y
0	1
1	0
2	−1

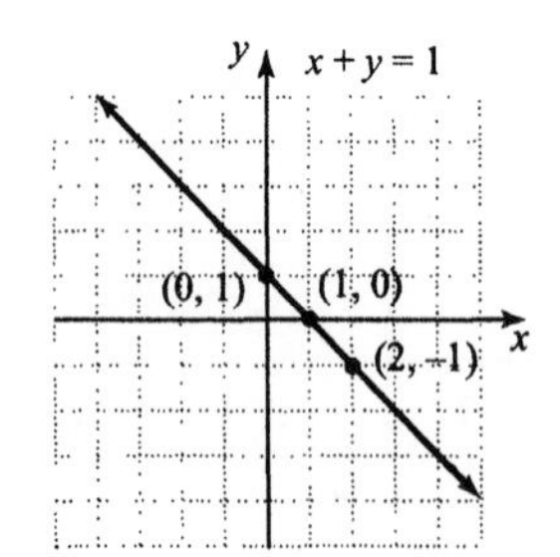

19. $x - y = -2$

x	y
−2	0
0	2
2	4

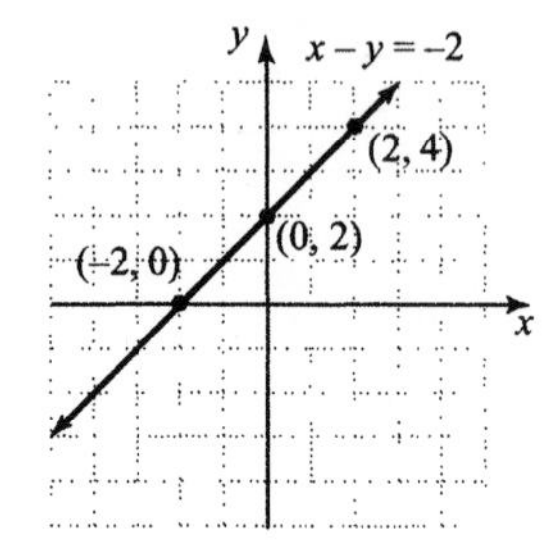

21. $x - 2y = 6$

x	y
−4	−5
0	−3
4	−1

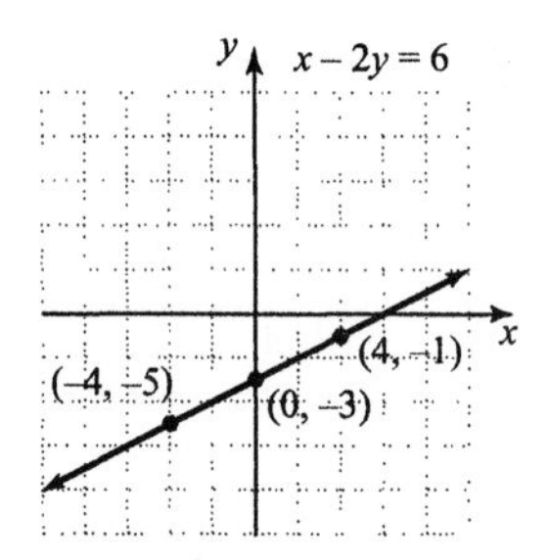

23. $y = 6x + 3$

x	y
−1	−3
0	3
1	9

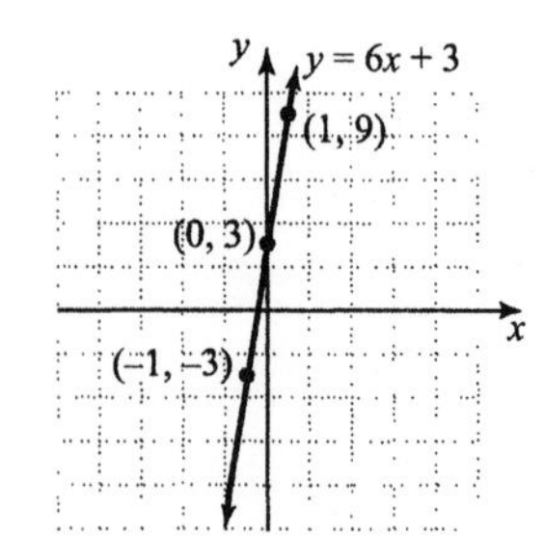

25. $x = -4$

x	y
–4	–1
–4	0
–4	2

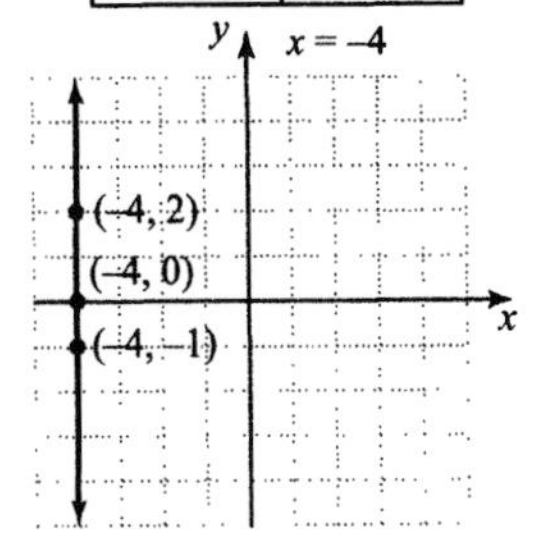

27. $y = 3$

x	y
–1	3
0	3
2	3

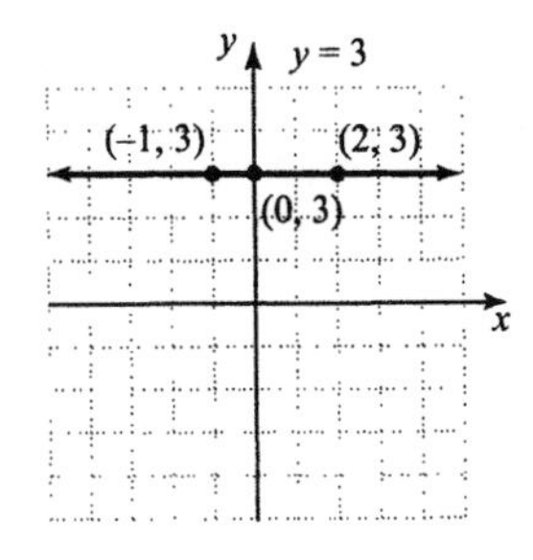

29. $y = x$

x	y
–1	–1
0	0
2	2

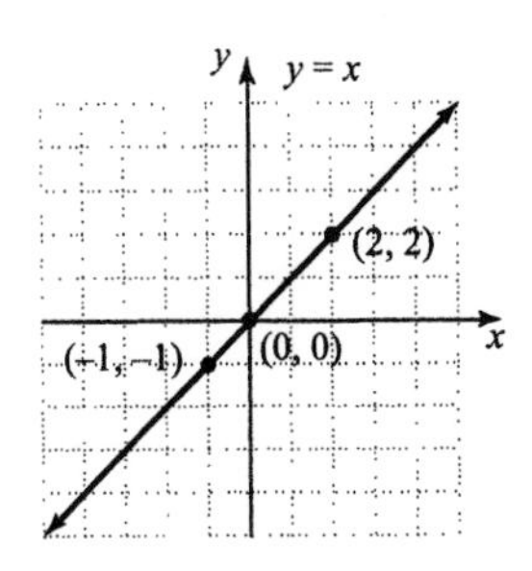

31. $x = -3y$

x	y
–6	2
0	0
6	–2

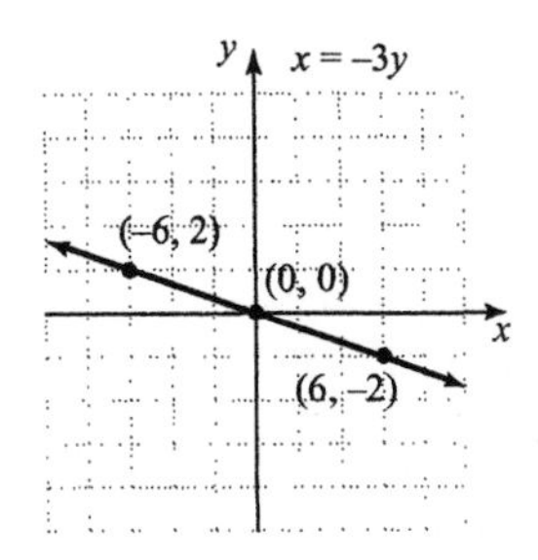

33. $x + 3y = 9$

x	y
–9	6
0	3
3	2

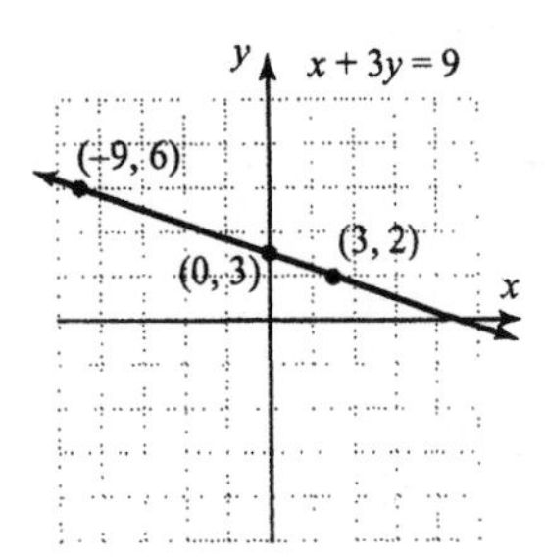

35. $y = \frac{1}{2}x + 2$

x	y
–4	0
0	2
4	4

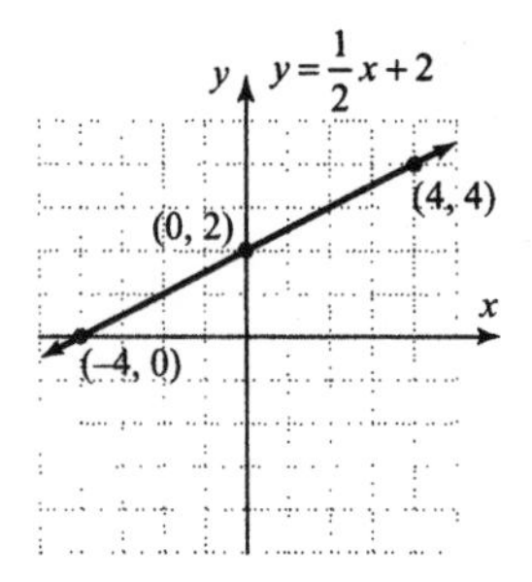

37. $3x - 2y = 12$

x	y
0	−6
2	−3
4	0

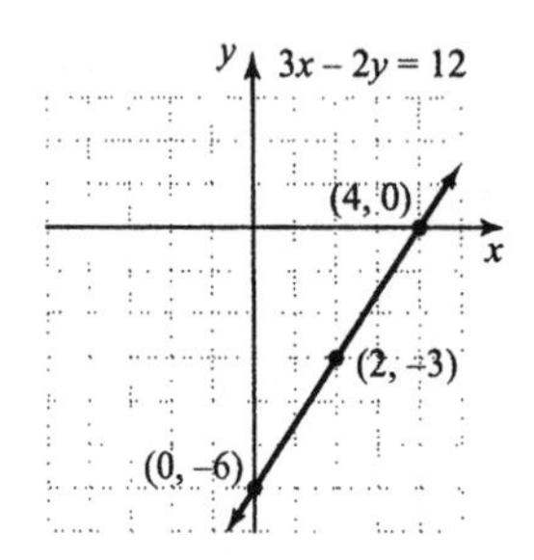

39. $y = -3.5x + 4$

x	y
0	4
1	0.5
2	−3

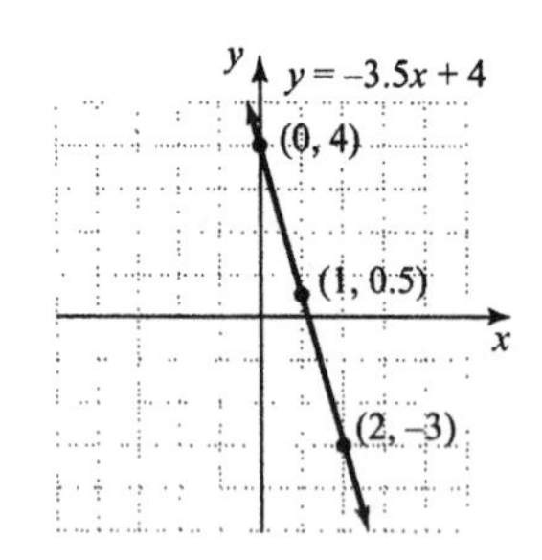

41. $y = 5x$

x	y
−1	−5
0	0
1	5

$y = 5x + 4$

x	y
−1	−1
0	4
1	9

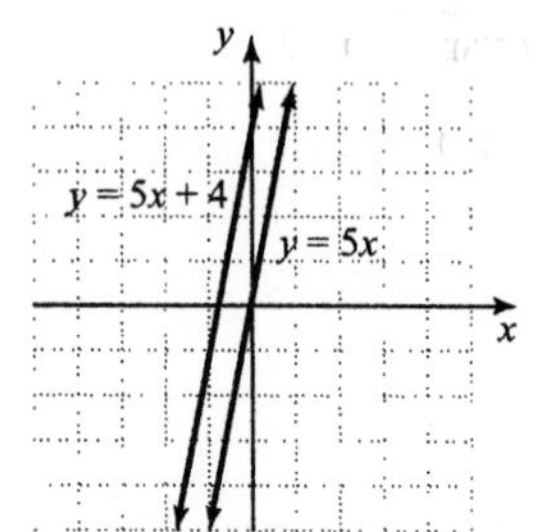

Answers may vary; possible answer: The graph of $y = 5x + 4$ is the same as the graph of $y = 5x$ except it is moved 4 units upward.

43. $y = -2x$

x	y
−2	4
0	0
2	−4

$y = -2x - 3$

x	y
−2	1
0	−3
2	−7

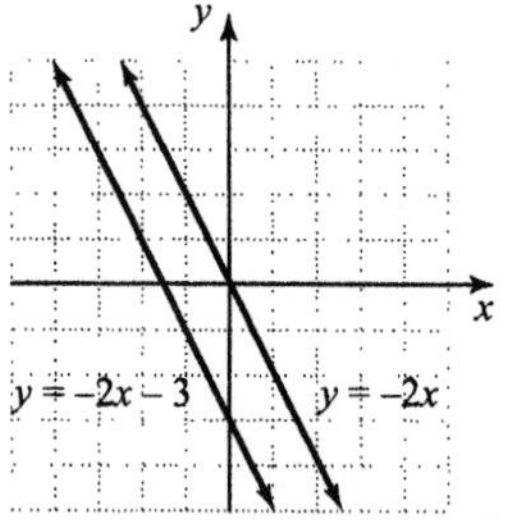

Answers may vary; possible answer: The graph of $y = -2x - 3$ is the same as the graph of $y = -2x$ except it is moved 3 units downward.

45. $y = \frac{1}{2}x$

x	y
−4	−2
0	0
4	2

$y = \frac{1}{2}x + 2$

x	y
−4	0
0	2
4	4

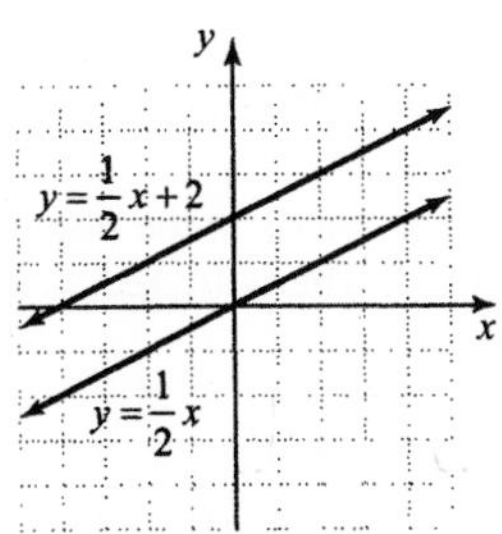

Answers may vary; possible answer: The graph of $y = \frac{1}{2}x + 2$ is the same as the graph of $y = \frac{1}{2}x$ except it is moved 2 units upward.

47. Comparing $y = 5x + 5$ to $y = mx + b$, we see that $b = 5$. We see that graph c crosses the *y*-axis at (0, 5).

49. Comparing $y = 5x - 1$ to $y = mx + b$, we see that $b = -1$. We see that graph d crosses the *y*-axis at (0, −1).

51. **a.** Using the equation, let $x = 8$.
$y = 0.5x + 3$
$y = 0.5(8) + 3 = 4 + 3 = 7$
The ordered pair is (8, 7).

b. Eight years after 1997, in 2005, there were 7 million snowboarders.

c. The year 2012 is 15 years after 1997, so let $x = 15$.
$y = 0.5x + 3$
$y = 0.5(15) + 3 = 7.5 + 3 = 10.5$
If the trend continues, there will be 10.5 million snowboarders in 2012.

53. Let $x = 5$.
$y = 54x + 275$
$y = 54(5) + 275 = 270 + 275 = 545$
The expected minimum salary after 5 years' experience is $545 thousand.

55.

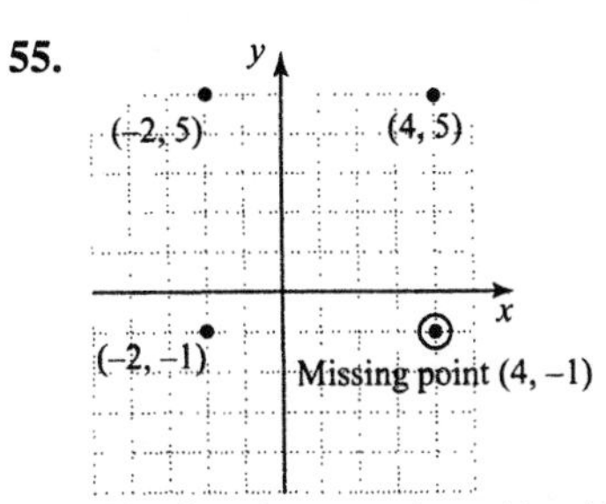

The fourth vertex is at (4, −1).

57.
$$\begin{aligned} 3(x-2)+5x &= 6x-16 \\ 3x-6+5x &= 6x-16 \\ 8x-6 &= 6x-16 \\ 2x-6 &= -16 \\ 2x &= -10 \\ x &= -5 \end{aligned}$$

59.
$$\begin{aligned} 3x+\frac{2}{5} &= \frac{1}{10} \\ 10(3x)+10\left(\frac{2}{5}\right) &= 10\left(\frac{1}{10}\right) \\ 30x+4 &= 1 \\ 30x &= -3 \\ x &= -\frac{1}{10} \end{aligned}$$

61. The equation is $y = x + 5$.

x	y
−2	3
0	5
2	7

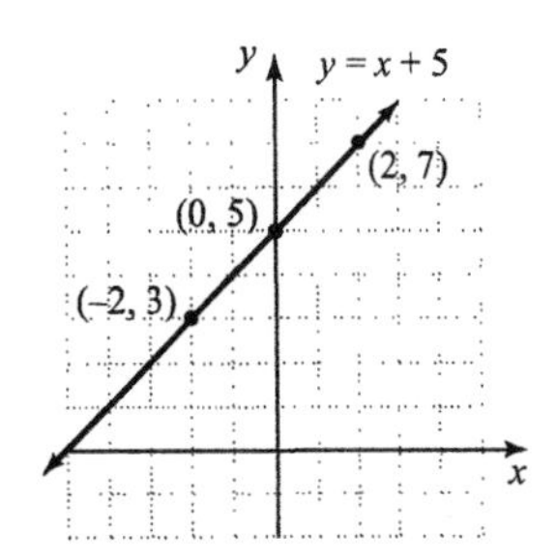

63. The equation is $2x + 3y = 6$.

x	y
3	0
0	2
−3	4

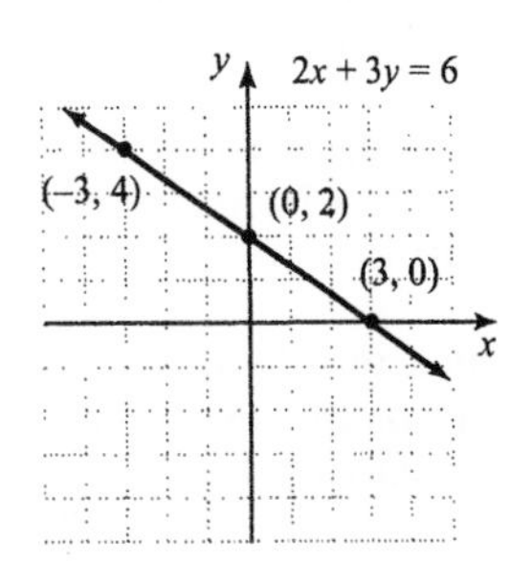

65. $x+y+5+5=22$
$x+y+10=22$
$x+y=12$
Let $x=3$.
$3+y=12$
$y=9$ centimeters

67. Answers may vary

69. $y=x^2$

x	y
0	0
1	1
–1	1
2	4
–2	4

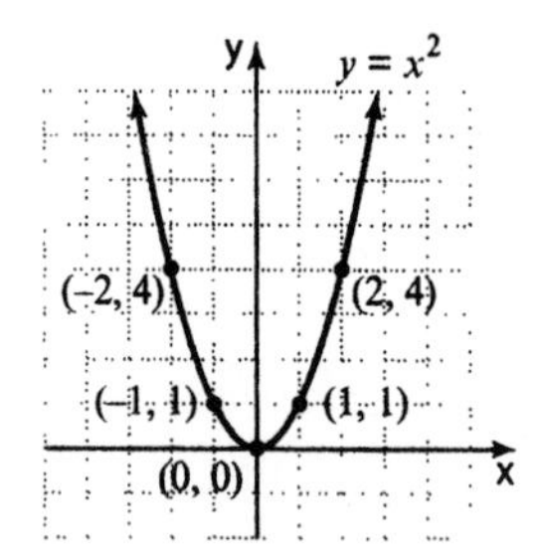

Section 3.3

Practice Exercises

1. The graph crosses the x-axis at the point (–4, 0). The x-intercept is (–4, 0).
The graph crosses the y-axis at the point (0, –6). The y-intercept is (0, –6).

2. The graph crosses the x-axis at the point (–1, 0) and at the point (–0.5, 0). The x-intercepts are (–1, 0) and (–0.5, 0).
The graph crosses the y-axis at the point (0, 1). The y-intercept is (0, 1).

3. The graph crosses both the x-axis and the y-axis at the point (0, 0). The x-intercept is (0, 0), and the y-intercept is (0, 0).

4. The graph does not cross the x-axis. There is no x-intercept. The graph crosses the y-axis at the point (0, 3). The y-intercept is (0, 3).

5. The graph crosses the x-axis at the point (–1, 0) and at the point (5, 0). The x-intercepts are (–1, 0) and (5, 0).
The graph crosses the y-axis at the point (0, –2) and at the point (0, 2). The y-intercepts are (0, –2) and (0, 2).

6. Let $y=0$.
$x+2y=-4$
$x+2(0)=-4$
$x+0=-4$
$x=-4$

Let $x=0$.
$x+2y=-4$
$0+2y=-4$
$2y=-4$
$y=-2$

The x-intercept is (–4, 0), and the y-intercept is (0, –2).
Let $x=2$.
$x+2y=-4$
$2+2y=-4$
$2y=-6$
$y=-3$

x	y
–4	0
0	–2
2	–3

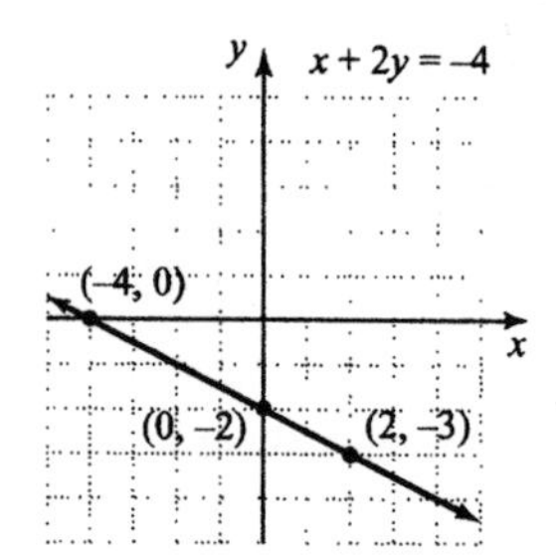

7. Let $y=0$.
$x=3y$
$x=3(0)$
$x=0$

Let $x=0$.
$x=3y$
$0=3y$
$0=y$

Both the x-intercept and the y-intercept are (0, 0).
Let $y=-1$
$x=3(-1)$
$x=-3$

Let $y=1$.
$x=3(1)$
$x=3$

x	y
0	0
3	1
–3	–1

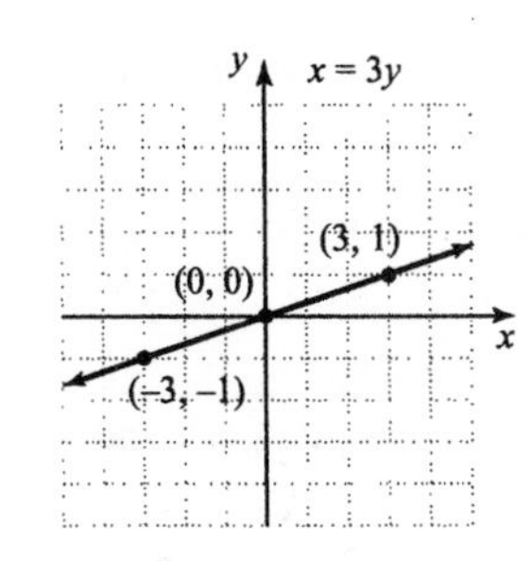

8. Let $y = 0$.

$$3x = 2y + 4$$
$$3x = 2(0) + 4$$
$$3x = 4$$
$$x = \frac{4}{3}$$

Let $x = 0$.

$$3x = 2y + 4$$
$$3(0) = 2y + 4$$
$$-4 = 2y$$
$$-2 = y$$

Let $x = 2$.

$$3x = 2y + 4$$
$$3(2) = 2y + 4$$
$$6 = 2y + 4$$
$$2 = 2y$$
$$1 = y$$

x	y
0	−2
$\frac{4}{3}$	0
2	1

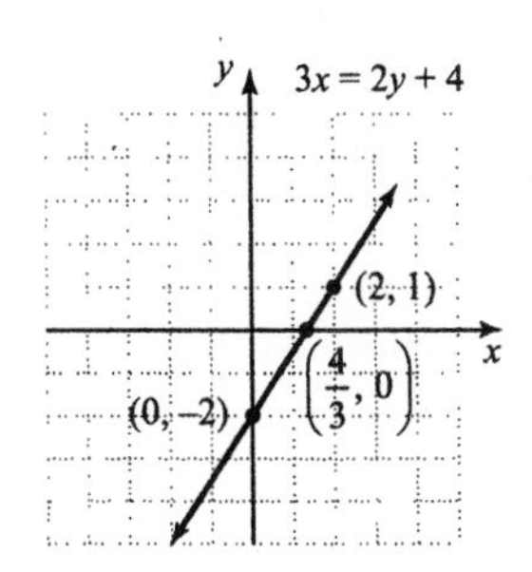

9. For any x-value chosen, notice that y is 2.

x	y
−5	2
0	2
5	2

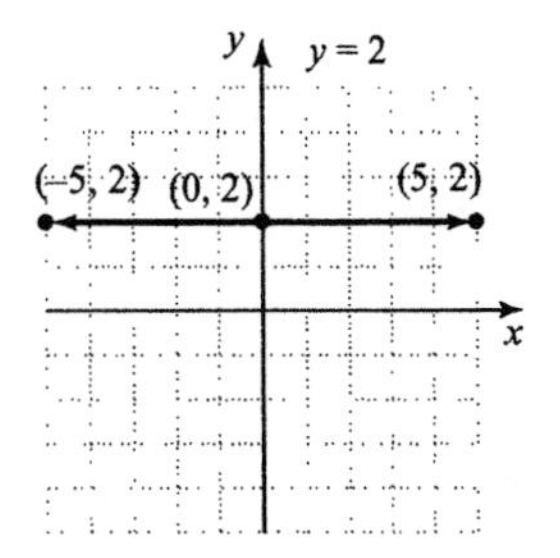

10. For any y-value chosen, notice that x is −2.

x	y
−2	−4
−2	0
−2	4

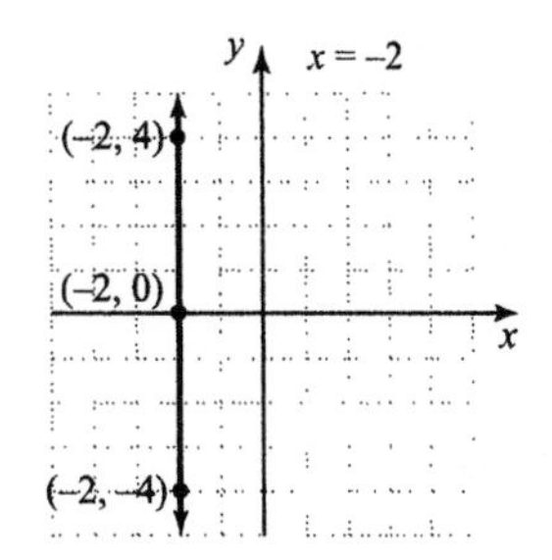

Calculator Explorations

1. $x = 3.78y$

$$y = \frac{x}{3.78}$$

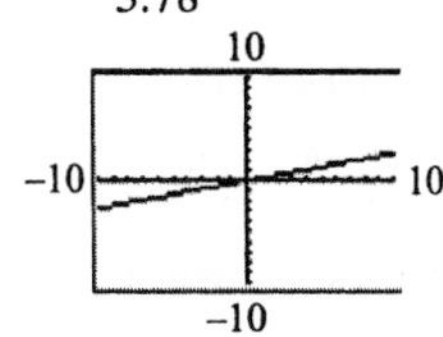

2. $-2.61y = x$

$$y = \frac{x}{-2.61}$$

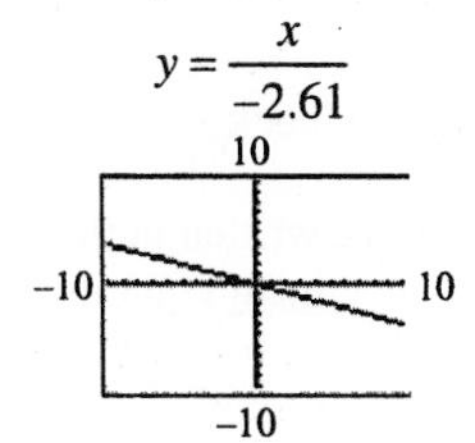

3. $3x+7y=21$
$$7y=-3x+21$$
$$y=-\frac{3}{7}x+3$$

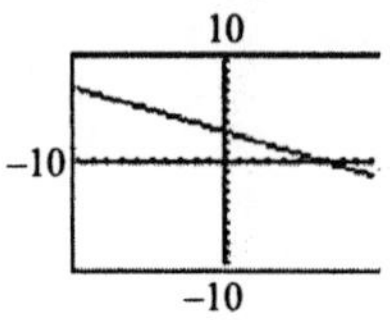

10

−10 10

−10

4. $-4x+6y=12$
$$6y=4x+12$$
$$y=\frac{2}{3}x+2$$

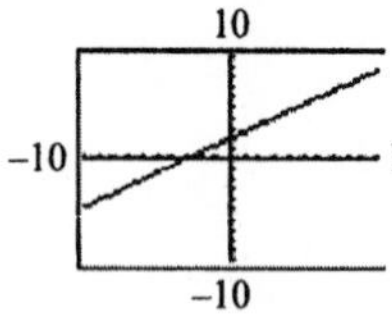

10

−10 10

−10

5. $-2.2x+6.8y=15.5$
$$6.8y=2.2x+15.5$$
$$y=\frac{2.2}{6.8}x+\frac{15.5}{6.8}$$

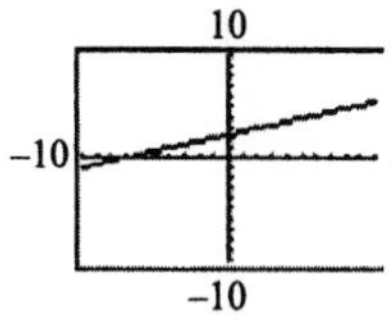

10

−10 10

−10

6. $5.9x-0.8y=-10.4$
$$-0.8y=-5.9x-10.4$$
$$y=\frac{5.9}{0.8}x+\frac{10.4}{0.8}$$

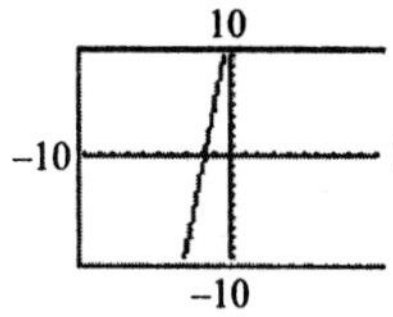

10

−10 10

−10

Vocabulary and Readiness Check

1. An equation that can be written in the form $Ax + By = C$ is called a linear equation in two variables.

2. The form $Ax + By = C$ is called standard form.

3. The graph of the equation $y = -1$ is a horizontal line.

4. The graph of the equation $x = 5$ is a vertical line.

5. A point where a graph crosses the y-axis is called a y-intercept.

6. A point where a graph crosses the x-axis is called a x-intercept.

7. Given an equation of a line, to find the x-intercept (if there is one), let $y = 0$ and solve for x.

8. Given an equation of a line, to find the y-intercept (if there is one), let $x = 0$ and solve for y.

9. False; for example, the horizontal line $y = 2$ does not have an x-intercept.

10. True

11. True

12. False; the graph of $y = 5x$ contains the point (1, 5) but not the point (5, 1).

Exercise Set 3.3

1. x-intercept: (−1, 0); y-intercept: (0, 1)

3. x-intercept: (−2, 0), (2, 0)

5. x-intercepts: (−2, 0), (1, 0), (3, 0)
y-intercept: (0, 3)

7. x-intercepts: (−1, 0), (1, 0)
y-intercepts: (0, 1), (0, −2)

9. Infinite; because the line could be vertical ($x = 0$) or horizontal ($y = 0$).

11. 0; because the circle could completely reside within one quadrant.

13. $x - y = 3$
$y = 0$: $x - 0 = 3$, $x = 3$
$x = 0$: $0 - y = 3$, $y = -3$
x-intercept: (3, 0); y-intercept: (0, −3)

x	y
3	0
0	−3

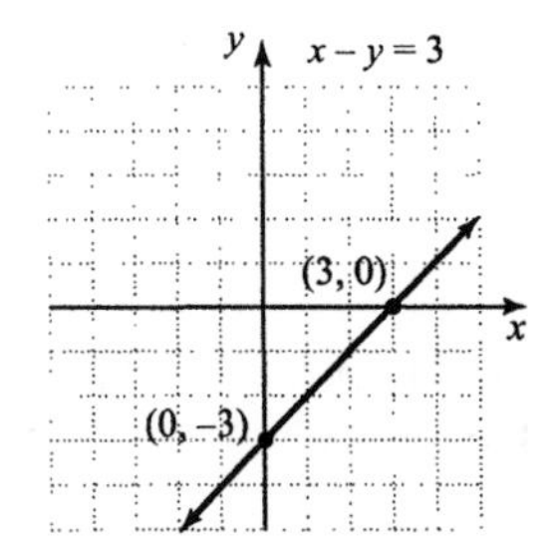

15. $x = 5y$
$y = 0$: $x = 5(0) = 0$
$x = 0$: $0 = 5y$, $y = 0$
x-intercept: (0, 0); y-intercept: (0, 0)
$y = 1$: $x = 5(1) = 5$

x	y
0	0
5	1

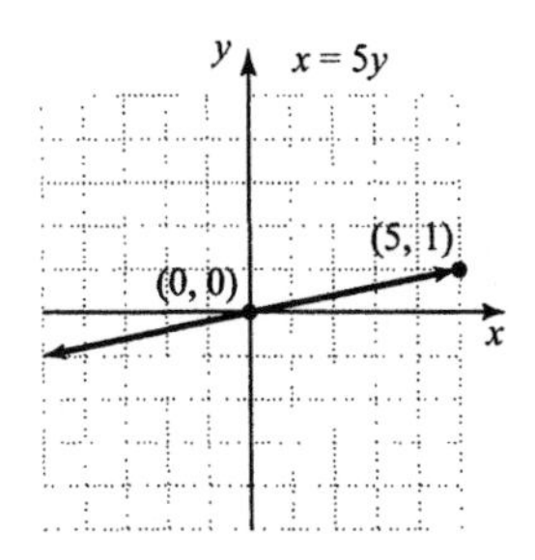

17. $-x + 2y = 6$
$y = 0$: $-x + 2(0) = 6$, $x = -6$
$x = 0$: $-0 + 2y = 6$, $y = 3$
x-intercept: (−6, 0); y-intercept: (0, 3)

x	y
−6	0
0	3

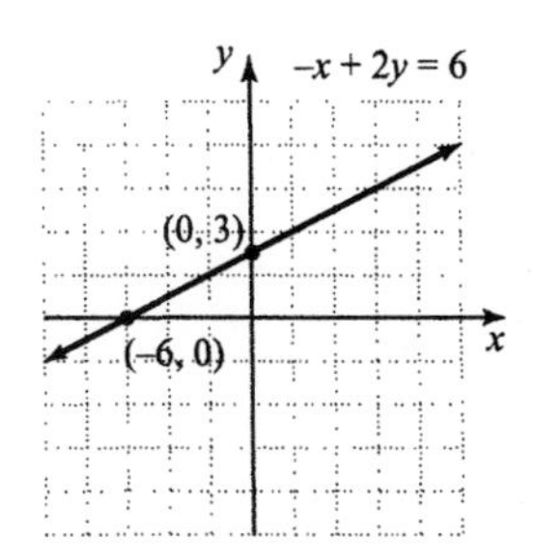

19. $2x - 4y = 8$
$y = 0$: $2x - 4(0) = 8$, $x = 4$
$x = 0$: $2(0) - 4y = 8$, $y = -2$
x-intercept: (4, 0); y-intercept: (0, −2)

x	y
4	0
0	−2

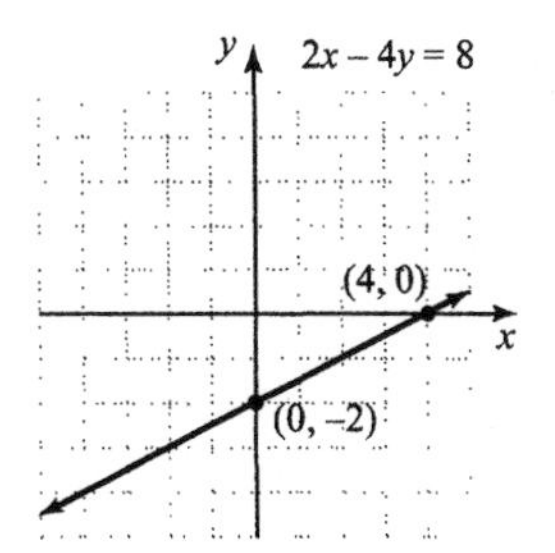

21. $y = 2x$
$y = 0$: $0 = 2x$, $0 = x$
$x = 0$: $y = 2(0)$, $y = 0$
x-intercept: (0, 0); y-intercept: (0, 0)
$x = 1$: $y = 2(1)$, $y = 2$

x	y
0	0
1	2

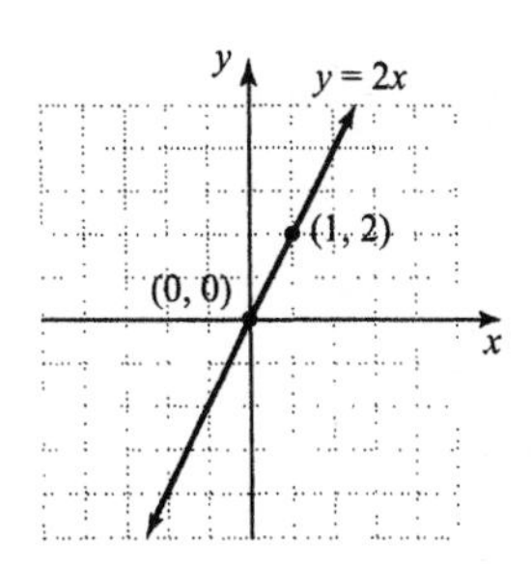

23. $y = 3x + 6$
$y = 0$: $0 = 3x + 6$, $-6 = 3x$, $-2 = x$
$x = 0$: $y = 3(0) + 6$, $y = 6$
x-intercept: (−2, 0); y-intercept: (0, 6)

x	y
−2	0
0	6

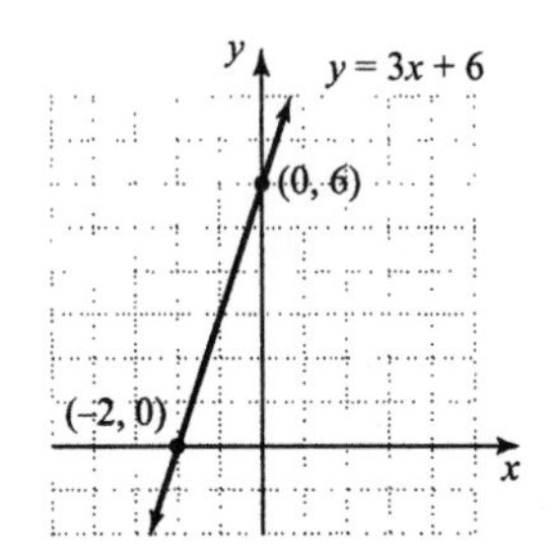

25. $x = -1$ for all values of y.

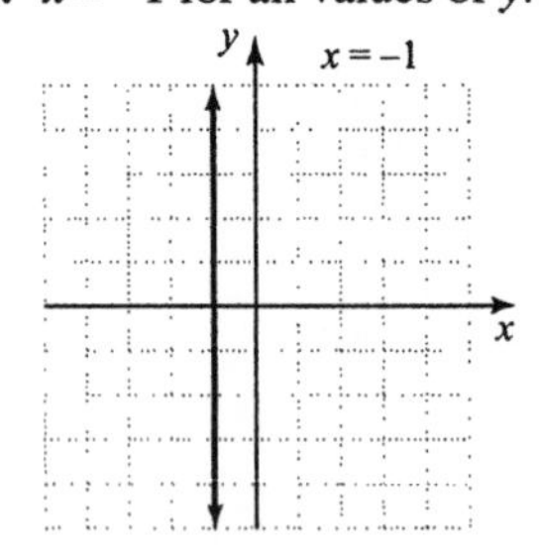

27. $y = 0$ for all values of x.

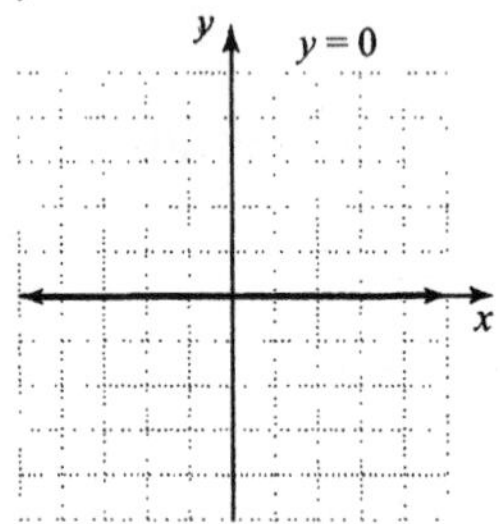

29. $y + 7 = 0$

$y = -7$ for all values of x.

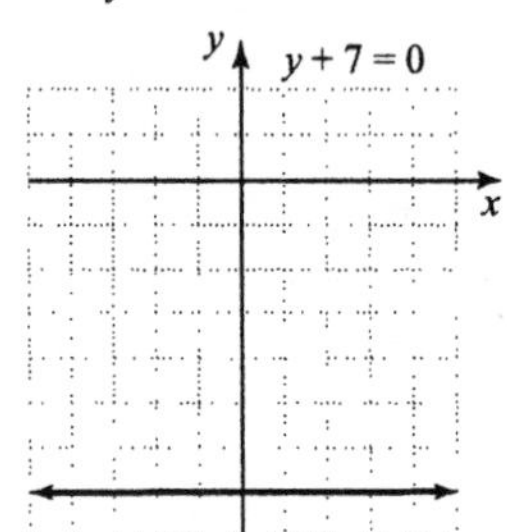

31. $x + 3 = 0$; $x = -3$ for all values of y.

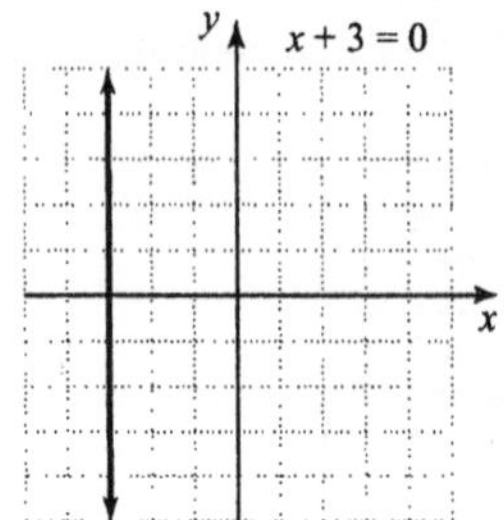

33. $x = y$

x-intercept: (0, 0); y-intercept: (0, 0)

Second point: (4, 4)

x	y
4	4
0	0

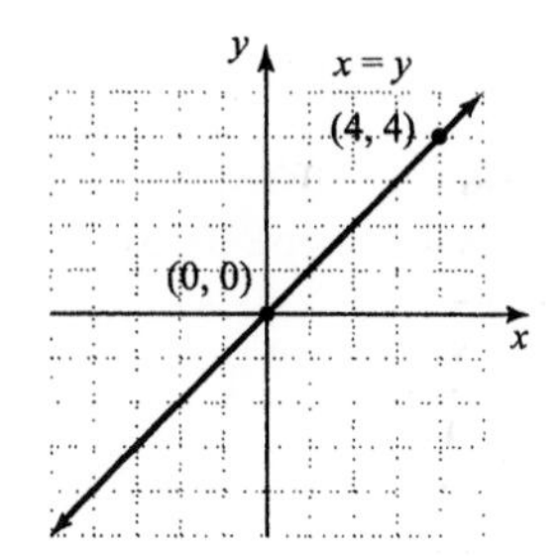

35. $x + 8y = 8$

x-intercept: (8, 0); y-intercept: (0, 1)

x	y
8	0
0	1

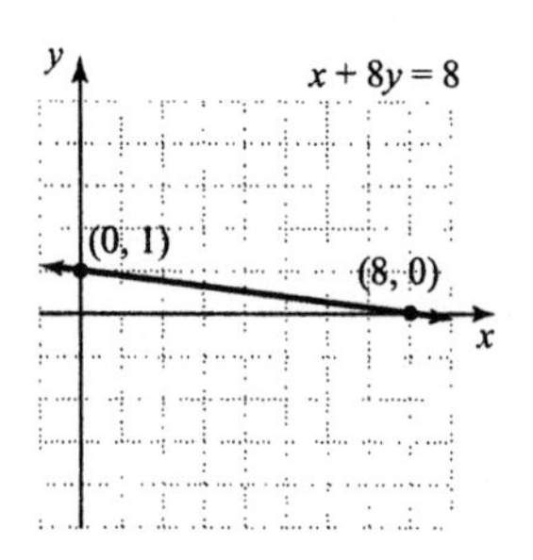

37. $5 = 6x - y$

x-intercept: $\left(\frac{5}{6}, 0\right)$; y-intercept: (0, −5)

x	y
$\frac{5}{6}$	0
0	−5

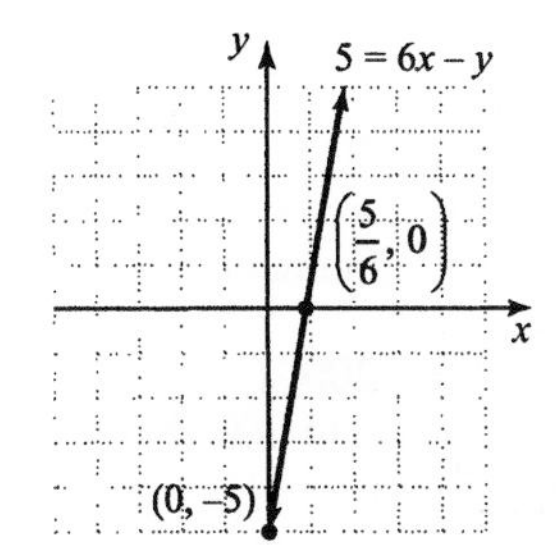

39. $-x + 10y = 11$

x-intercept: $(-11, 0)$; y-intercept: $\left(0, \frac{11}{10}\right)$

x	y
-11	0
0	$\frac{11}{10}$

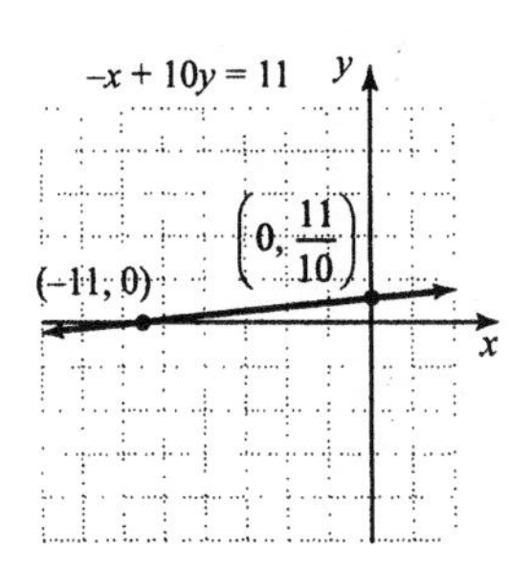

41. $x = -4\frac{1}{2}$ for all values of y.

x	y
$-4\frac{1}{2}$	0
$-4\frac{1}{2}$	3

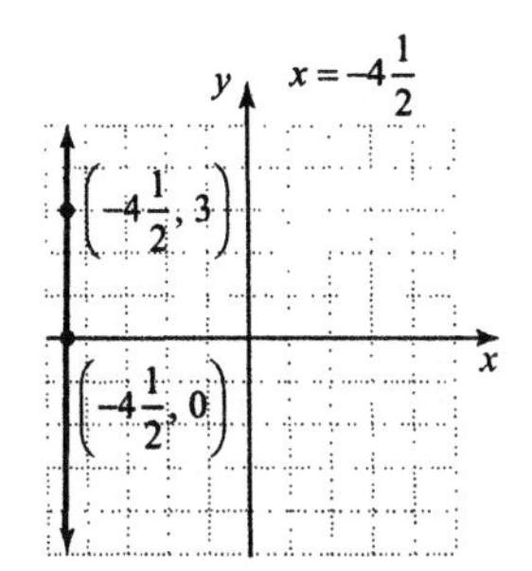

43. $y = 3\frac{1}{4}$ for all values of x.

x	y
0	$3\frac{1}{4}$
2	$3\frac{1}{4}$

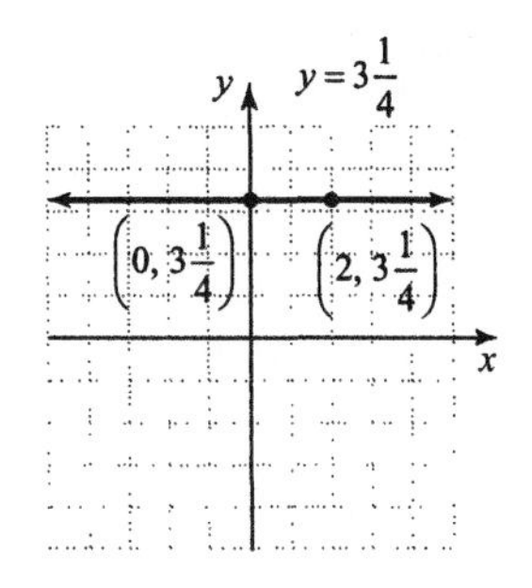

45. $y = -\frac{2}{3}x + 1$

x-intercept: $\left(\frac{3}{2}, 0\right)$; y-intercept: $(0, 1)$

x	y
$\frac{3}{2}$	0
0	1

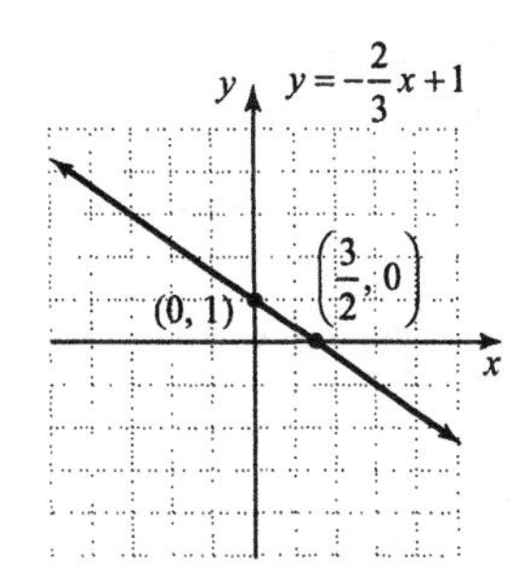

47. $4x - 6y + 2 = 0$

x-intercept: $\left(-\frac{1}{2}, 0\right)$; y-intercept: $\left(0, \frac{1}{3}\right)$

x	y
$-\frac{1}{2}$	0
0	$\frac{1}{3}$

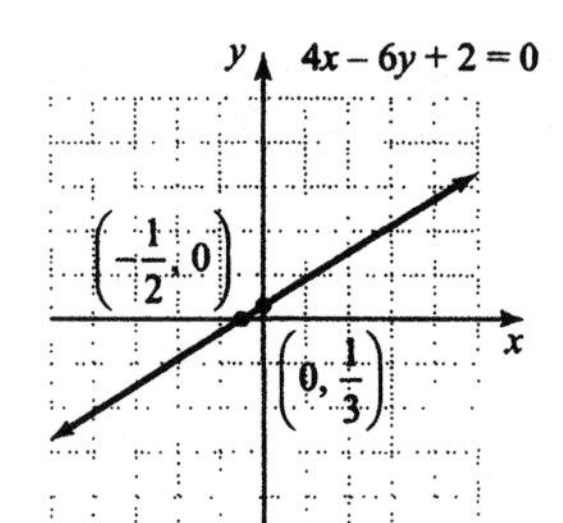

49. $y = 3$
The graph is a horizontal line with y-intercept (0, 3).
C

51. $x = -1$
The graph is a vertical line with x-intercept (−1, 0).
E

53. $y = 2x + 3$
The y-intercept is (0, 3) and the x-intercept is $\left(-\frac{3}{2}, 0\right)$.
B

55. $\frac{-6-3}{2-8} = \frac{-9}{-6} = \frac{3}{2}$

57. $\frac{-8-(-2)}{-3-(-2)} = \frac{-6}{-1} = 6$

59. $\frac{0-6}{5-0} = \frac{-6}{5} = -\frac{6}{5}$

61. $y = 1181x + 6505$

a. Let $x = 0$ and solve for y. The y-intercept is (0, 6505).

b. In 2003, the revenue for Disney Parks and Resorts was about \$6505 million.

63. $y = -0.075x + 1.65$

a. $y = 0$:
$$0 = -0.075x + 1.65$$
$$0.075x = 1.65$$
$$x = 22$$
(22, 0)

b. 22 years after 2002 (2024); 0 people will attend movies at the theatre.

c. Answers may vary

65. $3x + 6y = 1200$

a. $x = 0$: $3(0) + 6y = 1200$, $y = 200$
(0, 200) corresponds to no chairs and 200 desks being manufactured.

b. $y = 0$: $3x + 6(0) = 1200$, $x = 400$
(400, 0) corresponds to 400 chairs and no desks being manufactured.

c.

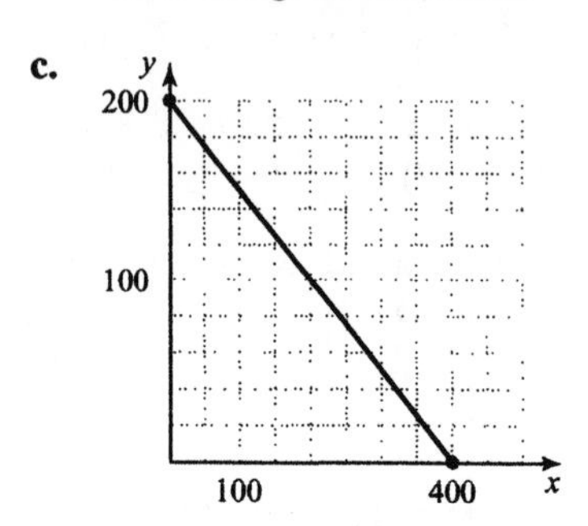

d. $y = 50$:
$$3x + 6(50) = 1200$$
$$3x + 300 = 1200$$
$$3x = 900$$
$$x = 300$$
300 chairs can be made.

67. Parallel to $y = -1$ is horizontal.
y-intercept is (0, −4), so $y = -4$ for all values of x. $y = -4$

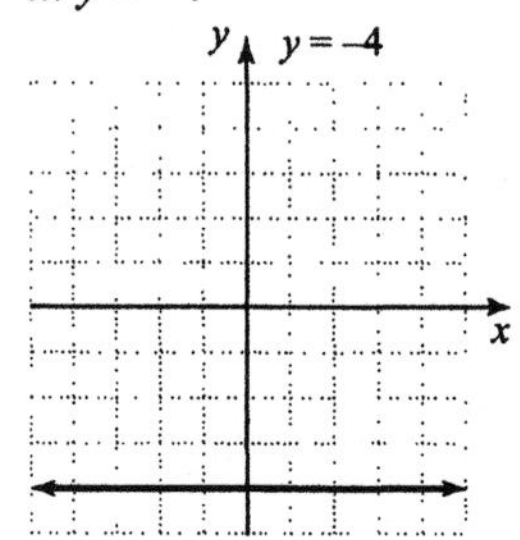

69. Answers may vary

71. Answers may vary

Section 3.4

Practice Exercises

1. If we let (x_1, y_1) be (−4, 11), then $x_1 = -4$ and $y_1 = 11$. Also, let (x_2, y_2) be (2, 5) so that $x_2 = 2$ and $y_2 = 5$.

$$m = \frac{y_2 - y_1}{x_2 - x_1} = \frac{5-11}{2-(-4)} = \frac{-6}{6} = -1$$

The slope of the line is −1.

2. Let (x_1, y_1) be (3, 1) and (x_2, y_2) be (–3, –1).

$$m = \frac{y_2 - y_1}{x_2 - x_1} = \frac{-1-1}{-3-3} = \frac{-2}{-6} = \frac{1}{3}$$

3. $y = \frac{2}{3}x - 2$

The equation is in slope-intercept form, $y = mx + b$. The coefficient of x, $\frac{2}{3}$, is the slope.

The constant term, –2, is the y-value of the y-intercept, (0, –2).

4. Write the equation in slope-intercept form by solving the equation for y.

$$6x - y = 5$$
$$-y = -6x + 5$$
$$y = 6x - 5$$

The coefficient of x, 6, is the slope. The constant term, –5, is the y-value of the y-intercept, (0, –5).

5. Write the equation in slope-intercept form by solving the equation for y.

$$5x + 2y = 8$$
$$2y = -5x + 8$$
$$\frac{2y}{2} = \frac{-5x}{2} + \frac{8}{2}$$
$$y = -\frac{5}{2}x + 4$$

The coefficient of x, $-\frac{5}{2}$, is the slope, and the y-intercept is (0, 4).

6. Recall that $y = 3$ is a horizontal line. Two ordered pair solutions of $y = 3$ and (1, 3) and (3, 3).

$$m = \frac{y_2 - y_1}{x_2 - x_1} = \frac{3-3}{3-1} = \frac{0}{2} = 0$$

The slope of the line $y = 3$ is 0.

7. Recall that the graph of $x = -4$ is a vertical line. Two ordered pair solutions of $x = -4$ and (–4, 1) and (–4, 3).

$$m = \frac{y_2 - y_1}{x_2 - x_1} = \frac{3-1}{-4-(-4)} = \frac{2}{0}$$

The slope of the vertical line $x = -4$ is undefined.

8. a. The slope of the line $y = -5x + 1$ is –5. We solve the second equation for y.

$$x - 5y = 10$$
$$-5y = -x + 10$$
$$\frac{-5y}{-5} = \frac{-x}{-5} + \frac{10}{-5}$$
$$y = \frac{1}{5}x - 2$$

The slope of the second line is $\frac{1}{5}$. Since the product of the slopes is $\frac{1}{5}(-5) = -1$, the lines are perpendicular.

b. Solve each equation for y.

$$x + y = 11 \qquad 2x + y = 11$$
$$y = -x + 11 \qquad y = -2x + 11$$

The slopes are –1 and –2. The slopes are not the same, and their product is not –1. Thus, the lines are neither parallel nor perpendicular.

c. Solve each equation for y.

$$2x + 3y = 21 \qquad 6y = -4x - 2$$
$$3y = -2x + 21 \qquad \frac{6y}{6} = \frac{-4x}{6} - \frac{2}{6}$$
$$\frac{3y}{3} = \frac{-2x}{3} + \frac{21}{3} \qquad y = -\frac{2}{3}x - \frac{1}{3}$$
$$y = -\frac{2}{3}x + 7$$

The slopes are $-\frac{2}{3}$ and $-\frac{2}{3}$. Since the lines have the same slope and different y-intercepts, they are parallel.

9. $\text{grade} = \frac{\text{rise}}{\text{run}} = \frac{1794}{7176} = 0.25 = 25\%$

The grade is 25%.

10. Use (2, 2) and (6, 5) to calculate slope.

$$m = \frac{5-2}{6-2} = \frac{3}{4} = \frac{0.75 \text{ dollar}}{1 \text{ pound}}$$

The Wash-n-Fold charges \$0.75 per pound of laundry.

Calculator Explorations

1. $y_1 = 3.8x$
$y_2 = 3.8x - 3$
$y_3 = 3.8x + 9$

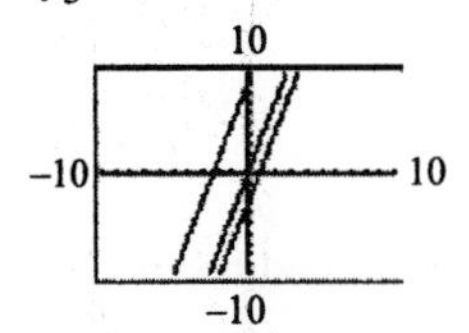

2. $y_1 = -4.9x$
$y_2 = -4.9x + 1$
$y_3 = -4.9x + 8$

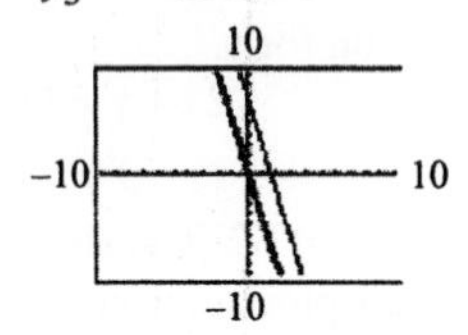

3. $y_1 = \frac{1}{4}x$
$y_2 = \frac{1}{4}x + 5$
$y_3 = \frac{1}{4}x - 8$

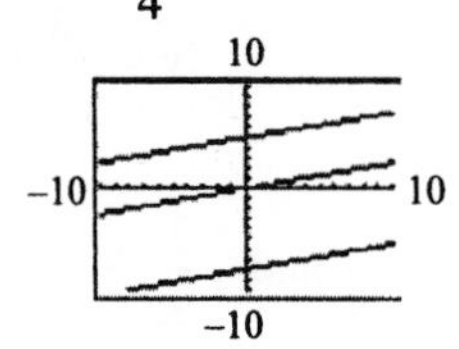

4. $y_1 = -\frac{3}{4}x$
$y_2 = -\frac{3}{4}x - 5$
$y_3 = -\frac{3}{4}x + 6$

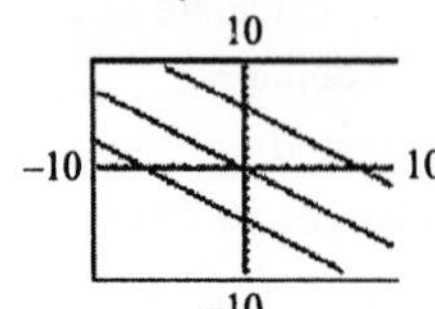

Vocabulary and Readiness Check

1. The measure of the steepness or tilt of a line is called slope.

2. If an equation is written in the form $y = mx + b$, the value of the letter m is the value of the slope of the graph.

3. The slope of a horizontal line is 0.

4. The slope of a vertical line is undefined.

5. If the graph of a line moves upward from left to right, the line has positive slope.

6. If the graph of a line moves downward from left to right, the line has negative slope.

7. Given two points of a line, slope $= \frac{\text{change in } y}{\text{change in } x}$.

8. The line goes down. The slope is negative.

9. The line goes up. The slope is positive.

10. The line is vertical. The slope is undefined.

11. The line is horizontal. The slope is 0.

12. The slope is positive. The line is "upward."

13. The slope is negative. The line is "downward."

14. The slope is 0. The line is horizontal.

15. The slope is undefined. The line is vertical.

Exercise Set 3.4

1. $(x_1, y_1) = (-1, 5)$ and $(x_2, y_2) = (6, -2)$
$$m = \frac{y_2 - y_1}{x_2 - x_1} = \frac{-2-5}{6-(-1)} = \frac{-7}{7} = -1$$

3. $(x_1, y_1) = (-4, 3)$ and $(x_2, y_2) = (-4, 5)$
$$m = \frac{y_2 - y_1}{x_2 - x_1} = \frac{5-3}{-4-(-4)} = \frac{2}{0}$$
The slope is undefined.

5. $(x_1, y_1) = (-2, 8)$ and $(x_2, y_2) = (1, 6)$
$$m = \frac{y_2 - y_1}{x_2 - x_1} = \frac{6-8}{1-(-2)} = \frac{-2}{3} = -\frac{2}{3}$$

7. $(x_1, y_1) = (5, 1)$ and $(x_2, y_2) = (-2, 1)$
$$m = \frac{y_2 - y_1}{x_2 - x_1} = \frac{1-1}{-2-5} = \frac{0}{-7} = 0$$

9. $(x_1, y_1) = (-1, 2)$ and $(x_2, y_2) = (2, -2)$
$m = \frac{y_2 - y_1}{x_2 - x_1} = \frac{-2-2}{2-(-1)} = \frac{-4}{3} = -\frac{4}{3}$

11. $(x_1, y_1) = (2, 3)$ and $(x_2, y_2) = (2, -1)$
$m = \frac{y_2 - y_1}{x_2 - x_1} = \frac{-1-3}{2-2} = \frac{-4}{0}$
The slope is undefined.

13. $(x_1, y_1) = (-3, -2)$ and $(x_2, y_2) = (-1, 3)$
$m = \frac{y_2 - y_1}{x_2 - x_1} = \frac{3-(-2)}{-1-(-3)} = \frac{5}{2}$

15. The slope of line 1 is positive, and the slope of line 2 is negative. Thus, line 1 has the greater slope.

17. Both line 1 and line 2 have positive slopes, but line 2 is steeper than line 1. Thus, line 2 has the greater slope.

19. (0, 0) and (2, 2)
$m = \frac{y_2 - y_1}{x_2 - x_1} = \frac{2-0}{2-0} = \frac{2}{2} = 1$
D

21. A vertical line has undefined slope.
B

23. (2, 0) and (4, −1)
$m = \frac{y_2 - y_1}{x_2 - x_1} = \frac{-1-0}{4-2} = -\frac{1}{2}$
E

25. $x = 6$ is a vertical line, so it has an undefined slope.

27. $y = -4$ is a horizontal line, so it has a slope $m = 0$.

29. $x = -3$ is a vertical line, so it has an undefined slope.

31. $y = 0$ is a horizontal line, so it has a slope $m = 0$.

33. $y = 5x - 2$
The equation is in slope-intercept form. The coefficient of x, 5, is the slope.

35. $y = -0.3x + 2.5$
The equation is in slope-intercept form. The coefficient of x, −0.3, is the slope.

37. Solve for y.
$$2x + y = 7$$
$$y = -2x + 7$$
The coefficient of x, −2, is the slope.

39. Solve for y.
$$2x - 3y = 10$$
$$-3y = -2x + 10$$
$$\frac{-3y}{-3} = \frac{-2x}{-3} + \frac{10}{-3}$$
$$y = \frac{2}{3}x - \frac{10}{3}$$
The coefficient of x, $\frac{2}{3}$, is the slope.

41. The graph of $x = 1$ is a vertical line. The slope is undefined.

43. Solve for y.
$$x = 2y$$
$$\frac{1}{2}x = y \text{ or } y = \frac{1}{2}x$$
The coefficient of x, $\frac{1}{2}$, is the slope.

45. The graph of $y = -3$ is a horizontal line. The slope is 0.

47. Solve for y.
$$-3x - 4y = 6$$
$$-4y = 3x + 6$$
$$\frac{-4y}{-4} = \frac{3x}{-4} + \frac{6}{-4}$$
$$y = -\frac{3}{4}x - \frac{3}{2}$$
The coefficient of x, $-\frac{3}{4}$, is the slope.

49. Solve for y.
$$20x - 5y = 1.2$$
$$-5y = -20x + 1.2$$
$$\frac{-5y}{-5} = \frac{-20x}{-5} + \frac{1.2}{-5}$$
$$y = 4x - 0.24$$
The coefficient of x, 4, is the slope.

51. $y = \frac{2}{9}x + 3,\ y = -\frac{2}{9}x$

The slopes are $\frac{2}{9}$ and $-\frac{2}{9}$. The slopes are not the same, and their product is not −1. The lines are neither parallel nor perpendicular.

53. The slope of $y = 3x - 9$ is 3. Solve the other equation for y.

$x - 3y = -6$
$-3y = -x - 6$
$\frac{-3y}{-3} = -\frac{x}{-3} - \frac{6}{-3}$
$y = \frac{1}{3}x + 2$

The slope is $\frac{1}{3}$. The slopes are not the same, and their product is not -1. The lines are neither parallel nor perpendicular.

55. Solve the equations for y.

$6x = 5y + 1$
$6x - 1 = 5y$
$\frac{6x}{5} - \frac{1}{5} = \frac{5y}{5}$
$y = \frac{6}{5}x - \frac{1}{5}$

$-12x + 10y = 1$
$10y = 12x + 1$
$\frac{10y}{10} = \frac{12x}{10} + \frac{1}{10}$
$y = \frac{6}{5}x + \frac{1}{10}$

The lines have the same slope, $\frac{6}{5}$, but different y-intercepts. The lines are parallel.

57. Solve the equations for y.

$6 + 4x = 3y$
$\frac{6}{3} + \frac{4x}{3} = \frac{3y}{3}$
$y = \frac{4}{3}x + 2$

$3x + 4y = 8$
$4y = -3x + 8$
$\frac{4y}{4} = -\frac{3x}{4} + \frac{8}{4}$
$y = -\frac{3}{4}x + 2$

The slopes are $\frac{4}{3}$ and $-\frac{3}{4}$. Their product is -1, so the lines are perpendicular.

59. $\text{pitch} = \frac{6}{10} = \frac{3}{5}$

61. $\text{grade} = \frac{\text{rise}}{\text{run}} = \frac{2}{16} = 0.125 = 12.5\%$

63. $\text{grade} = \frac{\text{rise}}{\text{run}} = \frac{2580}{6450} = 0.40 = 40\%$

65. $\text{grade} = \frac{\text{rise}}{\text{run}} = \frac{10}{12.66} = 0.79 = 79\%$

67. Use (2002, 74) and (2007, 89) to calculate slope.

$m = \frac{89 - 74}{2007 - 2002} = \frac{15}{5} = \frac{3 \text{ million households}}{1 \text{ year}}$

Every 1 year, there are/should be 3 million more U.S. households with personal computers.

69. Use (5000, 2100) and (20,000, 8400) to calculate slope.

$m = \frac{8400 - 2100}{20{,}000 - 5000} = \frac{6300}{15{,}000} = \frac{0.42 \text{ dollar}}{1 \text{ mile}}$

It costs \$0.42 per 1 mile to own and operate a compact car.

71. $y - (-6) = 2(x - 4)$
$y + 6 = 2x - 8$
$y = 2x - 14$

73. $y - 1 = -6(x - (-2))$
$y - 1 = -6(x + 2)$
$y - 1 = -6x - 12$
$y = -6x - 11$

75. $(-3, -3)$ and $(0, 0)$

$m = \frac{y_2 - y_1}{x_2 - x_1} = \frac{0 - (-3)}{0 - (-3)} = \frac{3}{3} = 1$

a. $m = 1$

b. $m = -1$

77. $(-8, -4)$ and $(3, 5)$

$m = \frac{y_2 - y_1}{x_2 - x_1} = \frac{5 - (-4)}{3 - (-8)} = \frac{9}{11}$

a. $m = \frac{9}{11}$

b. $m = -\frac{11}{9}$

79. $(2, 1)$ and $(0, 0)$: $m = \frac{0 - 1}{0 - 2} = \frac{-1}{-2} = \frac{1}{2}$

$(2, 1)$ and $(-2, -1)$: $m = \frac{-1 - 1}{-2 - 2} = \frac{-2}{-4} = \frac{1}{2}$

$(2, 1)$ and $(-4, -2)$: $m = \frac{-2 - 1}{-4 - 2} = \frac{-3}{-6} = \frac{1}{2}$

$(0, 0)$ and $(-2, -1)$: $m = \frac{-1 - 0}{-2 - 0} = \frac{-1}{-2} = \frac{1}{2}$

(0, 0) and (−4, −2): $m = \frac{-2-0}{-4-0} = \frac{-2}{-4} = \frac{1}{2}$

(−2, −1) and (−4, −2): $m = \frac{-2-(-1)}{-4-(-2)} = \frac{-1}{-2} = \frac{1}{2}$

Since the slope of the line between each pair of points is the same, the points lie on the same line.

81. Answers may vary

83. In 2001, the average fuel economy was approximately 28.5 miles per gallon.

85. The lowest point on the graph corresponds to the year 2000. The average fuel economy was approximately 28.1 miles per gallon.

87. The line segment from 2000 to 2001 is the steepest, so it has the greatest slope.

89. $\text{pitch} = \frac{\text{rise}}{\text{run}}$

$\frac{1}{3} = \frac{x}{18}$

$3x = 18$

$x = 6$

91. a. (2006, 1657) and (2001, 1132)

b. $m = \frac{y_2 - y_1}{x_2 - x_1} = \frac{1132-1657}{2001-2006} = \frac{-525}{-5} = 105$

c. For the years 2001 through 2006, the price per acre of U.S. farmland rose approximately $105 per year.

93. (1, 1), (−4, 4) and (−3, 0)

$m_1 = \frac{0-1}{-3-1} = \frac{1}{4},\ m_2 = \frac{0-4}{-3-(-4)} = -4$

$m_1 m_2 = -1$, so the sides are perpendicular.

95. (2.1, 6.7) and (−8.3, 9.3)

$m = \frac{y_2 - y_1}{x_2 - x_1} = \frac{9.3-6.7}{-8.3-2.1} = \frac{2.6}{-10.4} = -0.25$

97. (2.3, 0.2) and (7.9, 5.1)

$m = \frac{y_2 - y_1}{x_2 - x_1} = \frac{5.1-0.2}{7.9-2.3} = \frac{4.9}{5.6} = 0.875$

99. $y = -\frac{1}{3}x + 2$

$y = -2x + 2$

$y = -4x + 2$

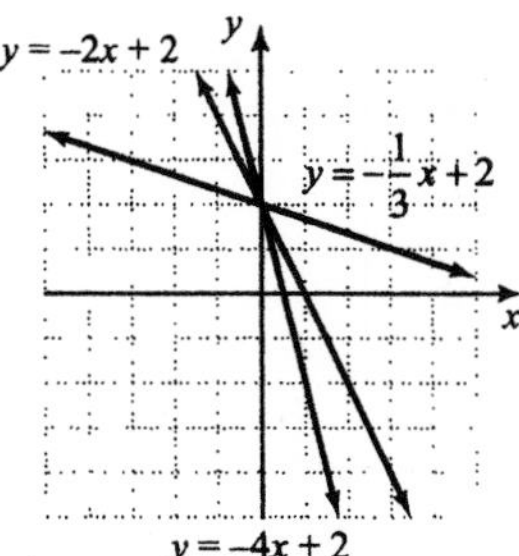

The line becomes steeper.

Integrated Review

1. (0, 0) and (2, 4)

$m = \frac{y_2 - y_1}{x_2 - x_1} = \frac{4-0}{2-0} = \frac{4}{2} = 2$

2. Horizontal line, $m = 0$

3. (0, 1) and (3, −1)

$m = \frac{y_2 - y_1}{x_2 - x_1} = \frac{-1-1}{3-0} = -\frac{2}{3}$

4. Vertical line, slope is undefined.

5. $y = -2x$

$m = -2, b = 0$

x	y
0	0
1	−2
−1	2

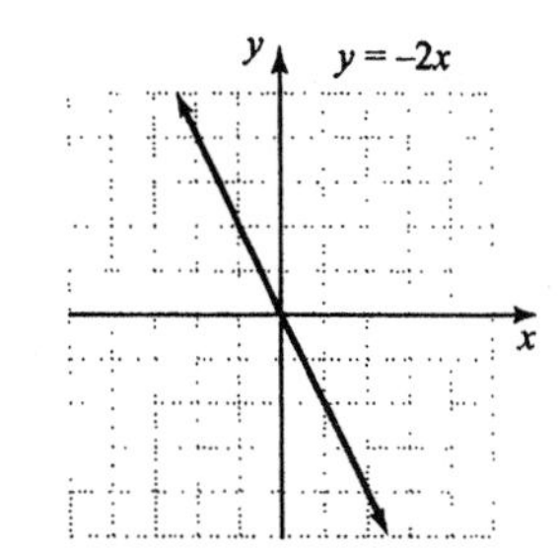

6. $x + y = 3$
$y = -x + 3$
$m = -1, b = 3$

x	y
0	3
3	0
1	2

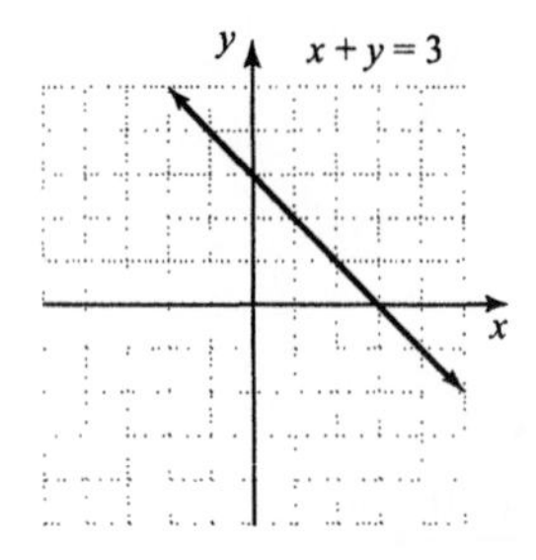

7. $x = -1$ for all values of y.
Vertical line; slope is undefined.

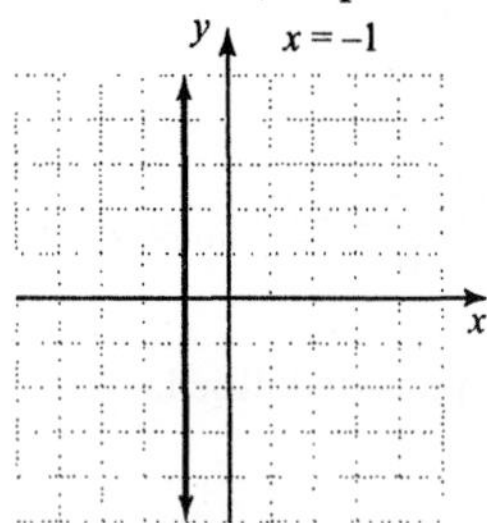

8. $y = 4$ for all values of x.
Horizontal line; $m = 0$

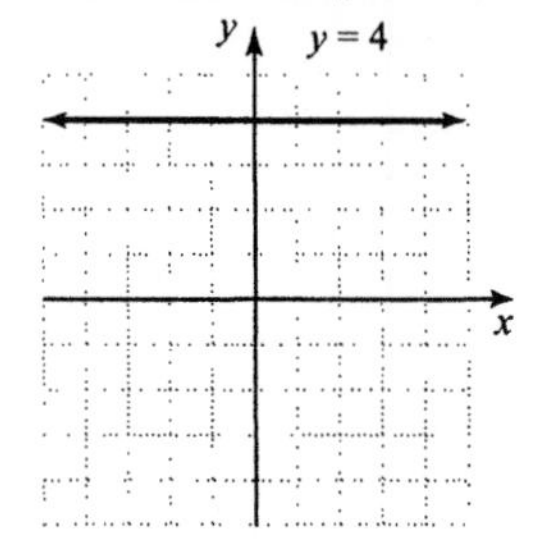

9. $x - 2y = 6$
$-2y = -x + 6$
$y = \frac{1}{2}x - 3$

$m = \frac{1}{2}, b = -3$

x	y
0	−3
2	−2
4	−1

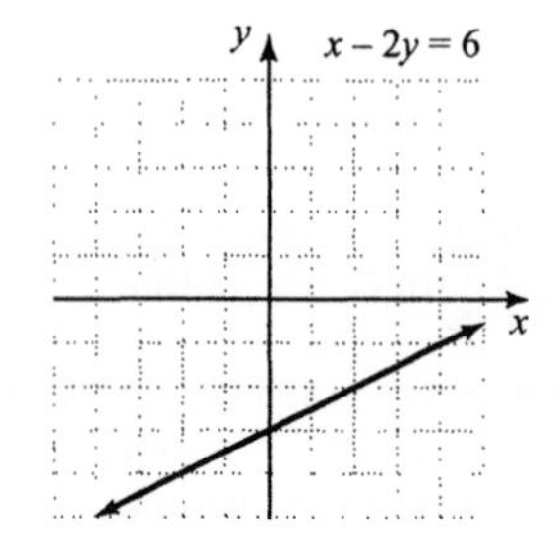

10. $y = 3x + 2$
$m = 3, b = 2$

x	y
0	2
−1	−1
−2	−4

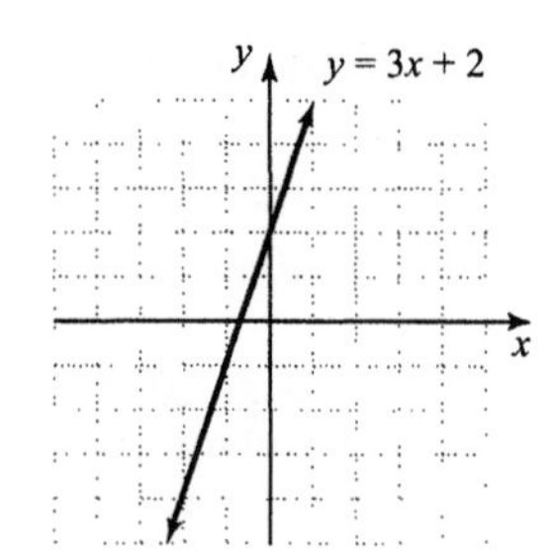

11. $5x + 3y = 15$

x	y
0	5
3	0

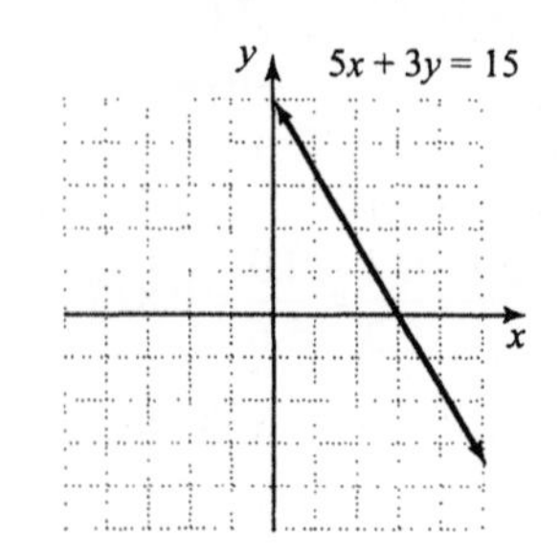

12. $2x - 4y = 8$

x	y
0	−2
4	0

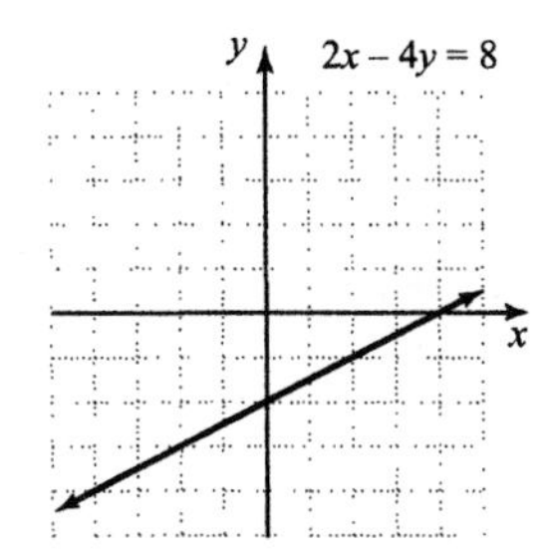

13. The slope of the first line is $-\frac{1}{5}$. Solve the second equation for y.

$$3x = -15y$$
$$\frac{3x}{-15} = \frac{-15y}{-15}$$
$$y = -\frac{1}{5}x$$

The slope of the second line is also $-\frac{1}{5}$. Since the lines have the same slope but different y-intercepts, the lines are parallel.

14. Solve the equations for y.

$$x - y = \frac{1}{2} \qquad 3x - y = \frac{1}{2}$$
$$-y = -x + \frac{1}{2} \qquad -y = -3x + \frac{1}{2}$$
$$y = x - \frac{1}{2} \qquad y = 3x - \frac{1}{2}$$

The slopes are 1 and 3. Since the slopes are not equal and their product is not −1, the lines are neither parallel nor perpendicular.

15. a. Let $x = 0$.
$y = -75(0) + 1650 = 1650$
The y-intercept is (0, 1650).

b. In 2002, there were 1650 million admissions to movie theaters in the United States.

c. The equation is in slope-intercept form. The coefficient of x, −75, is the slope.

d. For the years 2002 through 2005, the number of movie theater admissions decreased at a rate of 75 million per year.

16. a. Let $x = 9$.
$y = 3.3(9) - 3.1 = 29.7 - 3.1 = 26.6$
The ordered pair is (9, 26.6).

b. In 2009, the predicted revenue for online advertising is $26.6 billion.

Section 3.5

Practice Exercises

1. y-intercept: (0, 7); slope: $\frac{1}{2}$

Let $m = \frac{1}{2}$ and $b = 7$.

$$y = mx + b$$
$$y = \frac{1}{2}x + 7$$

2. $y = \frac{2}{3}x - 5$

The slope is $\frac{2}{3}$, and the y-intercept is (0, −5).

We plot (0, −5). From this point, we move up 2 units and then right 3 units. We stop at the point (3, −3).

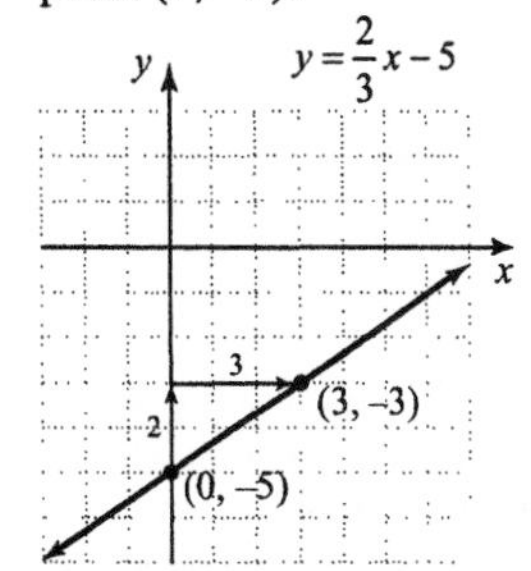

3. Solve the equation for y.

$$3x - y = 2$$
$$-y = -3x + 2$$
$$y = 3x - 2$$

The slope is 3, and the y-intercept is (0, −2). We plot (0, −2). From this point, we move up 3 units and then right 1 unit. We stop at the point (1, 1).

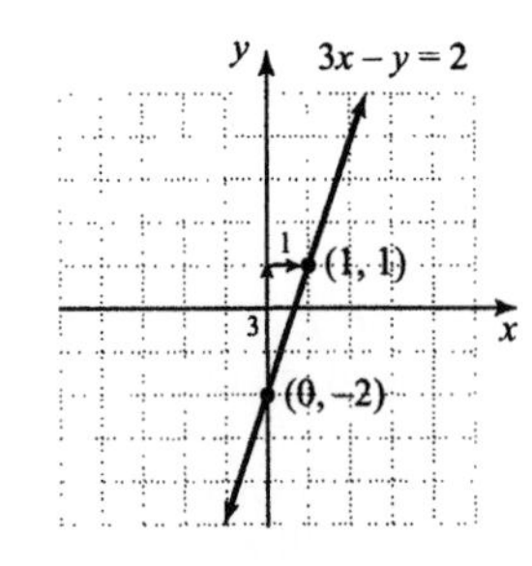

4. Line passing through (2, 3) with slope 4

$$\begin{aligned} y - y_1 &= m(x - x_1) \\ y - 3 &= 4(x - 2) \\ y - 3 &= 4x - 8 \\ -4x + y &= -5 \\ 4x - y &= 5 \end{aligned}$$

5. Line through (−1, 6) and (3, 1)

$$m = \frac{1-6}{3-(-1)} = \frac{-5}{4} = -\frac{5}{4}$$

Use the slope $-\frac{5}{4}$ and the point (3, 1).

$$\begin{aligned} y - y_1 &= m(x - x_1) \\ y - 1 &= -\frac{5}{4}(x - 3) \\ 4(y - 1) &= 4\left(-\frac{5}{4}\right)(x - 3) \\ 4y - 4 &= -5(x - 3) \\ 4y - 4 &= -5x + 15 \\ 5x + 4y &= 19 \end{aligned}$$

6. The equation of a vertical line can be written in the form $x = c$, so an equation for a vertical line passing through (3, −2) is $x = 3$.

7. Since the graph of $y = -2$ is a horizontal line, any line parallel to it is also vertical. The equation of a horizontal line can be written in the form $y = c$. An equation for the horizontal line passing through (4, 3) is $y = 3$.

8. a. Write two ordered pairs, (30, 150,000) and (50, 120,000).

$$\begin{aligned} m &= \frac{120{,}000 - 150{,}000}{50 - 30} \\ &= \frac{-30{,}000}{20} \\ &= -1500 \end{aligned}$$

Use the slope −1500 and the point (30, 150,000).

$$\begin{aligned} y - y_1 &= m(x - x_1) \\ y - 150{,}000 &= -1500(x - 30) \\ y - 150{,}000 &= -1500x + 45{,}000 \\ y &= -1500x + 195{,}000 \end{aligned}$$

b. Find y when $x = 60$.

$y = -1500x + 195{,}000$
$y = -1500(60) + 195{,}000$
$y = -90{,}000 + 195{,}000$
$y = 105{,}000$

To sell 60 condos per month, the price should be $105,000.

Calculator Explorations

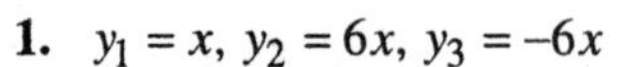

1. $y_1 = x,\ y_2 = 6x,\ y_3 = -6x$

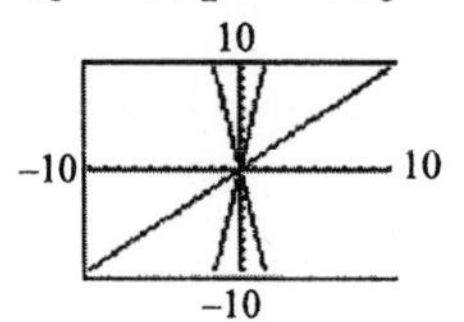

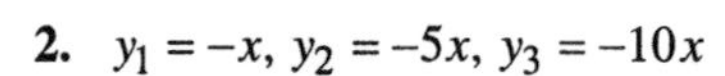

2. $y_1 = -x,\ y_2 = -5x,\ y_3 = -10x$

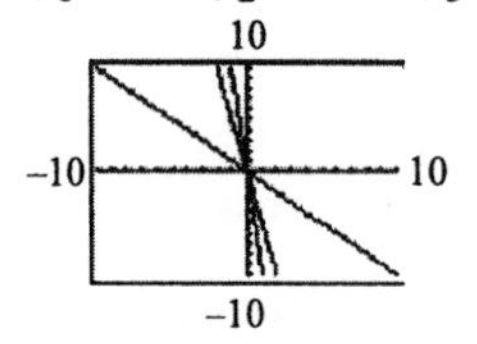

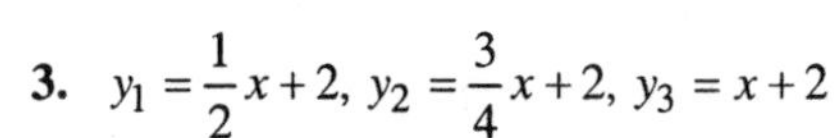

3. $y_1 = \frac{1}{2}x + 2,\ y_2 = \frac{3}{4}x + 2,\ y_3 = x + 2$

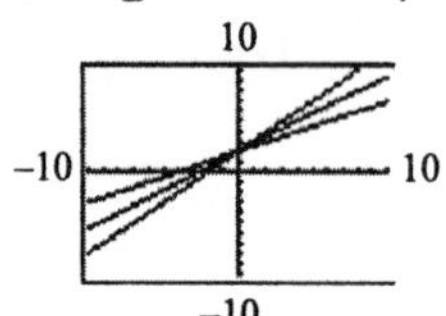

4. $y_1 = x + 1,\ y_2 = \frac{5}{4}x + 1,\ y_3 = \frac{5}{2}x + 1$

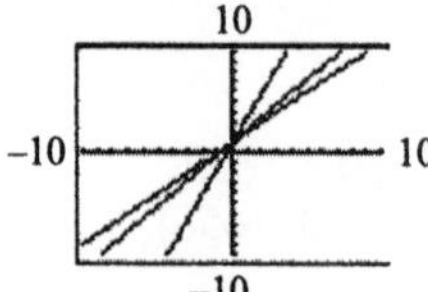

5. $y_1 = -7x + 5,\ y_2 = 7x + 5$

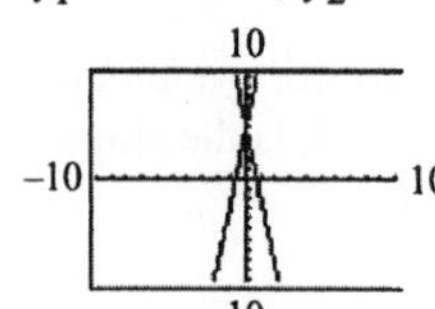

6. $y_1 = 3x - 1,\ y_2 = -3x - 1$

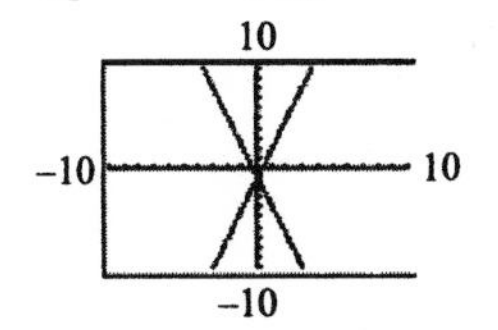

Vocabulary and Readiness Check

1. The form $y = mx + b$ is called slope-intercept form. When a linear equation in two variables is written in this form, m is the slope of its graph and $(0, b)$ is its y-intercept.

2. The form $y - y_1 = m(x - x_1)$ is called point-slope form. When a linear equation in two variables is written in this form, m is the slope of its graph and (x_1, y_1) is a point on the graph.

3. $y - 7 = 4(x + 3)$; point-slope form

4. $5x - 9y = 11$; standard form

5. $y = \frac{1}{2}$; horizontal line

6. $x = -17$; vertical line

7. $y = \frac{3}{4}x - \frac{1}{3}$; slope-intercept form

Exercise Set 3.5

1. $m = 5, b = 3$
$y = mx + b$
$y = 5x + 3$

3. $m = -4,\ b = -\frac{1}{6}$
$y = mx + b$
$y = -4x + \left(-\frac{1}{6}\right)$
$y = -4x - \frac{1}{6}$

5. $m = \frac{2}{3},\ b = 0$
$y = mx + b$
$y = \frac{2}{3}x + 0$
$y = \frac{2}{3}x$

7. $m = 0, b = -8$
$y = mx + b$
$y = 0x + (-8)$
$y = -8$

9. $m = -\frac{1}{5},\ b = \frac{1}{9}$
$y = mx + b$
$y = -\frac{1}{5}x + \frac{1}{9}$

11. $y = 2x + 1$

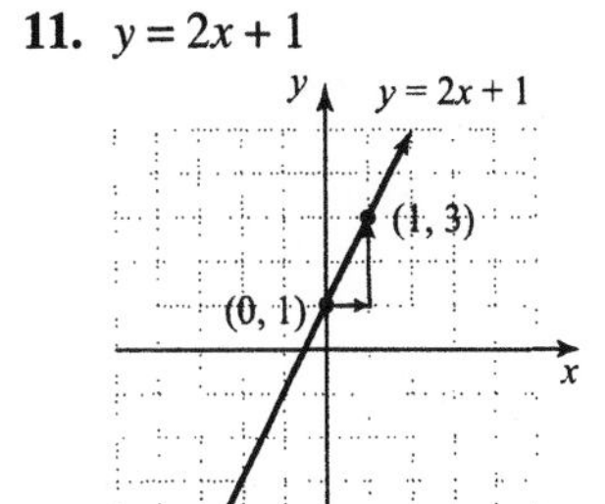

13. $y = \frac{2}{3}x + 5$

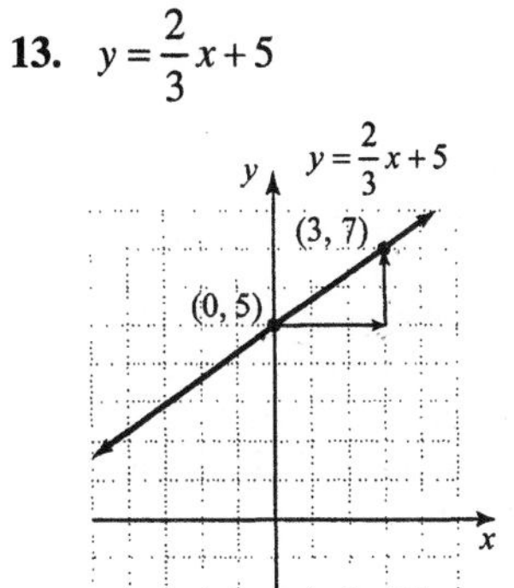

15. $y = -5x$

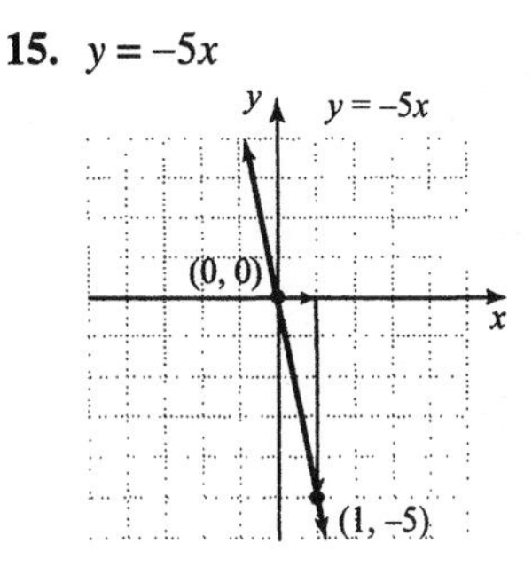

17. $4x + y = 6$
$y = -4x + 6$

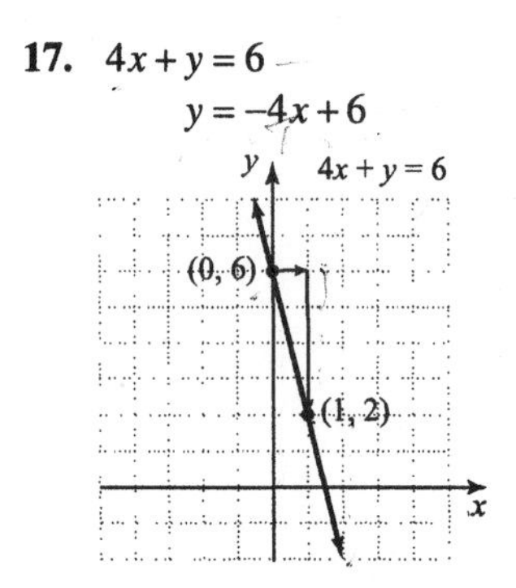

19. $4x - 7y = -14$
$-7y = -4x - 14$
$y = \frac{4}{7}x + 2$

y
4x − 7y = −14
(7, 6)
(0, 2)
x

21. $x = \frac{5}{4}y$
$\frac{4}{5}x = y$
$y = \frac{4}{5}x$

y
$y = \frac{5}{4}y$
(5, 4)
(0, 0)
x

23. $m = 6$; (2, 2)
$y - y_1 = m(x - x_1)$
$y - 2 = 6(x - 2)$
$y - 2 = 6x - 12$
$-6x + y = -10$ or $6x - y = 10$

25. $m = -8$; (−1, −5)
$y - y_1 = m(x - x_1)$
$y - (-5) = -8(x - (-1))$
$y + 5 = -8x - 8$
$8x + y = -13$

27. $m = \frac{3}{2}$; (5, −6)
$y - y_1 = m(x - x_1)$
$y - (-6) = \frac{3}{2}(x - 5)$
$2(y + 6) = 3(x - 5)$
$2y + 12 = 3x - 15$
$-3x + 2y = -27$
$3x - 2y = 27$

29. $m = -\frac{1}{2}$; (−3, 0)
$y - y_1 = m(x - x_1)$
$y - 0 = -\frac{1}{2}(x - (-3))$
$y = -\frac{1}{2}(x + 3)$
$-2y = x + 3$
$-x - 2y = 3$
$x + 2y = -3$

31. (3, 2) and (5, 6)
$m = \frac{y_2 - y_1}{x_2 - x_1} = \frac{6 - 2}{5 - 3} = \frac{4}{2} = 2$
$m = 2$; (3, 2)
$y - y_1 = m(x - x_1)$
$y - 2 = 2(x - 3)$
$y - 2 = 2x - 6$
$-2x + y = -4$
$2x - y = 4$

33. (−1, 3) and (−2, −5)
$m = \frac{y_2 - y_1}{x_2 - x_1} = \frac{-5 - 3}{-2 - (-1)} = \frac{-8}{-1} = 8$
$m = 8$; (−1, 3)
$y - y_1 = m(x - x_1)$
$y - 3 = 8(x - (-1))$
$y - 3 = 8x + 8$
$-8x + y = 11$
$8x - y = -11$

35. (2, 3) and (−1, −1)

$m = \frac{y_2 - y_1}{x_2 - x_1} = \frac{-1-3}{-1-2} = \frac{-4}{-3} = \frac{4}{3}$

$m = \frac{4}{3}$; (2, 3)

$y - y_1 = m(x - x_1)$

$y - 3 = \frac{4}{3}(x-2)$

$3(y-3) = 4(x-2)$

$3y - 9 = 4x - 8$

$-4x + 3y = 1$

$4x - 3y = -1$

37. (0, 0) and $\left(-\frac{1}{8}, \frac{1}{13}\right)$

$m = \frac{\frac{1}{13} - 0}{-\frac{1}{8} - 0} = \frac{\frac{1}{13}}{-\frac{1}{8}} = \frac{1}{13}\left(-\frac{8}{1}\right) = -\frac{8}{13}$

$m = -\frac{8}{13}$; (0, 0)

$y - y_1 = m(x - x_1)$

$y - 0 = -\frac{8}{13}(x - 0)$

$y = -\frac{8}{13}x$

$13y = -8x$

$8x + 13y = 0$

39. Vertical line, point (0, 2)

$x = c$

$x = 0$

41. Horizontal line, point (−1, 3)

$y = c$

$y = 3$

43. Vertical line, point $\left(-\frac{7}{3}, -\frac{2}{5}\right)$

$x = c$

$x = -\frac{7}{3}$

45. $y = 5$ is horizontal.
Parallel to $y = 5$ is horizontal; $y = c$.
Point (1, 2)
$y = 2$

47. $x = -3$ is vertical.
Perpendicular to $x = -3$ is horizontal; $y = c$.
Point (−2, 5)
$y = 5$

49. $x = 0$ is vertical.
Parallel to $x = 0$ is vertical; $x = c$.
Point (6, −8)
$x = 6$

51. $m = -\frac{1}{2}$; $\left(0, \frac{5}{3}\right)$

$y = mx + b$

$y = -\frac{1}{2}x + \frac{5}{3}$

53. (10, 7) and (7, 10)

$m = \frac{y_2 - y_1}{x_2 - x_1} = \frac{10-7}{7-10} = \frac{3}{-3} = -1$

$m = -1$; (10, 7)

$y - y_1 = m(x - x_1)$

$y - 7 = -1(x - 10)$

$y - 7 = -x + 10$

$y = -x + 17$

55. Undefined slope, through $\left(-\frac{3}{4}, 1\right)$

A line with undefined slope is vertical. A vertical line has an equation of the form $x = c$.

$x = -\frac{3}{4}$

57. $m = 1$; (−7, 9)

$y - y_1 = m(x - x_1)$

$y - 9 = 1[x - (-7)]$

$y - 9 = x + 7$

$y = x + 16$

59. $m = -5, b = 7$

$y = mx + b$

$y = -5x + 7$

61. *x*-axis is horizontal.
Parallel to *x*-axis is horizontal; $y = c$.
Point (6, 7)
$y = 7$

63. (2, 3) and (0, 0)

$m = \frac{y_2 - y_1}{x_2 - x_1} = \frac{3-0}{2-0} = \frac{3}{2}$; $b = 0$

$y = mx + b$

$y = \frac{3}{2}x + 0$

$y = \frac{3}{2}x$

65. y-axis is vertical.
Perpendicular to y-axis is horizontal; $y = c$.
Point (−2, −3)
$y = -3$

67. $m = -\frac{4}{7}$; (−1, −2)

$$y - y_1 = m(x - x_1)$$
$$y - (-2) = -\frac{4}{7}[x - (-1)]$$
$$y + 2 = -\frac{4}{7}x - \frac{4}{7}$$
$$y = -\frac{4}{7}x - \frac{4}{7} - 2$$
$$y = -\frac{4}{7}x - \frac{18}{7}$$

69. a. (1, 32) and (3, 96)

$$m = \frac{y_2 - y_1}{x_2 - x_1} = \frac{96 - 32}{3 - 1} = \frac{64}{2} = 32$$

$m = 32$; (1, 32)

$$s - s_1 = m(t - t_1)$$
$$s - 32 = 32(t - 1)$$
$$s - 32 = 32t - 32$$
$$s = 32t$$

b. If $t = 4$, then $s = 32(4) = 128$ ft/sec.

71. a. Use (0, 29,000) and (3, 71,000).

$$m = \frac{71{,}000 - 29{,}000}{3 - 0} = \frac{42{,}000}{3} = 14{,}000$$

$b = 29{,}000$
$y = mx + b$
$y = 14{,}000x + 29{,}000$

b. Let $x = 2010 - 2004 = 6$.

$$y = 14{,}000(6) + 29{,}000$$
$$= 84{,}000 + 29{,}000$$
$$= 113{,}000$$

We predict there will be 113,000 hybrids in 2010.

73. a. Use (0, 79.6) and (6, 85).

$$m = \frac{85 - 79.6}{6 - 0} = \frac{5.4}{6} = 0.9$$

$b = 79.6$
$y = mx + b$
$y = 0.9x + 79.6$

b. Let $x = 2010 - 2000 = 10$.
$y = 0.9(10) + 79.6 = 9 + 79.6 = 88.6$
We predict there will be 88.6 persons per square mile in 2010.

75. a. The ordered pairs are (0, 14.7) and (10, 14.14).

b.
$$m = \frac{14.14 - 14.7}{10 - 0} = \frac{-0.56}{10} = -0.056$$

$b = 14.7$
$y = mx + b$
$y = -0.056x + 14.7$

c. Let $x = 2016 - 1996 = 20$.

$$y = -0.056(20) + 14.7$$
$$= -1.12 + 14.7$$
$$= 13.58$$

We predict that there will be 13.58 births per thousand population in 2016.

77. a. The ordered pairs are (0, 5) and (3, 20).

b.
$$m = \frac{20 - 5}{3 - 0} = \frac{15}{3} = 5$$

$b = 5$
$y = mx + b$
$y = 5x + 5$

c. Let $x = 2012 - 2003 = 9$.
$y = 5(9) + 5 = 45 + 5 = 50$
We predict that the membership will be 50 thousand, or 50,000, in 2012.

79. If $x = 2$, then

$$x^2 - 3x + 1 = (2)^2 - 3(2) + 1 = 4 - 6 + 1 = -1$$

81. If $x = -1$, then

$$x^2 - 3x + 1 = (-1)^2 - 3(-1) + 1 = 1 + 3 + 1 = 5$$

83. No

85. Yes

87. Answers may vary

89. $y = 3x - 1$, $m_1 = 3$

a. Parallel: $m_2 = m_1 = 3$; (−1, 2)

$$y - y_1 = m_2(x - x_1)$$
$$y - 2 = 3(x - (-1))$$
$$y - 2 = 3x + 3$$
$$-3x + y = 5$$
$$3x - y = -5$$

b. Perpendicular: $m_2 = -\frac{1}{m_1} = -\frac{1}{3}$; $(-1, 2)$

$$y - y_1 = m_2(x - x_1)$$
$$y - 2 = -\frac{1}{3}(x - (-1))$$
$$3(y - 2) = -1(x + 1)$$
$$3y - 6 = -x - 1$$
$$x + 3y = 5$$

91. $3x + 2y = 7,\ y = -\frac{3}{2}x + \frac{7}{2},\ m_1 = -\frac{3}{2}$

a. Parallel: $m_2 = m_1 = -\frac{3}{2}$; $(3, -5)$

$$y - y_1 = m_2(x - x_1)$$
$$y - (-5) = -\frac{3}{2}(x - 3)$$
$$2(y + 5) = -3(x - 3)$$
$$2y + 10 = -3x + 9$$
$$3x + 2y = -1$$

b. Perpendicular: $m_2 = -\frac{1}{m_1} = \frac{2}{3}$; $(3, -5)$

$$y - y_1 = m_2(x - x_1)$$
$$y - (-5) = \frac{2}{3}(x - 3)$$
$$3(y + 5) = 2(x - 3)$$
$$3y + 15 = 2x - 6$$
$$2x - 3y = 21$$

Section 3.6

Practice Exercises

1. The domain is the set of all x-values $\{0, 1, 5\}$. The range is the set of all y-values: $\{-2, 0, 3, 4\}$.

2. a. $\{(4, 1), (3, -2), (8, 5), (-5, 3)\}$
Each x-value is assigned to only one y-value, so this set of ordered pairs is a function.

b. $\{(1, 2), (-4, 3), (0, 8), (1, 4)\}$
The x-value 1 is assigned to two y-values, 2 and 4, so this set of ordered pairs is not a function.

3. a. This is the graph of the relation $\{(-2, 1), (3, -3), (3, 2)\}$. The x-coordinate 3 is paired with two y-coordinates, -3 and 2, so this is not the graph of a function.

b. This is the graph of the relation $\{(-2, 1), (0, 1), (1, -3), (3, 2)\}$. Each x-coordinate has exactly one y-coordinate, so this is the graph of a function.

4. a. This is the graph of a function since no vertical line will intersect this graph more than once.

b. This is the graph of a function since no vertical line will intersect this graph more than once.

c. This is the graph of a function since no vertical line will intersect this graph more than once.

d. This is not the graph of a function. Vertical lines can be drawn that intersect the graph in two points. An example of one is shown.

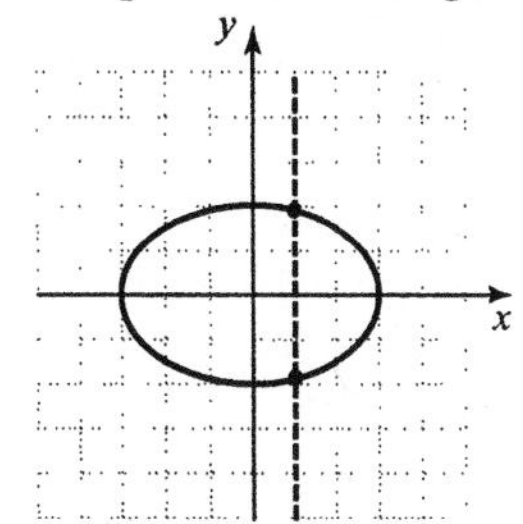

5. a. $y = 2x$ is a function because its graph is a nonvertical line.

b. $y = -3x - 1$ is a function because its graph is a nonvertical line.

c. $y = 8$ is a function because its graph is a nonvertical line.

d. $x = 2$ is not a function because its graph is a vertical line.

6. a. Since June is the sixth month, we look for 6 on the horizontal axis. From this point, we move vertically upward until the graph is reached. From the point on the graph, we move horizontally to the left to the vertical axis. The vertical axis there reads about 69°F.

b. We find 40°F on the temperature axis and move horizontally to the right. We eventually reach the point corresponding to 11, or November.

c. Yes, this is the graph of a function. It passes the vertical line test.

7. $h(x) = x^2 + 5$

a. $h(2) = 2^2 + 5 = 4 + 5 = 9$
(2, 9)

b. $h(-5) = (-5)^2 + 5 = 25 + 5 = 30$
(−5, 30)

c. $h(0) = 0^2 + 5 = 0 + 5 = 5$
(0, 5)

8. a. $h(x) = 6x + 3$
In this function, x can be any real number. The domain of $h(x)$ is the set of all real numbers, or $(-\infty, \infty)$ in interval notation.

b. $f(x) = \frac{1}{x^2}$
Recall that we cannot divide by 0 so that the domain of $f(x)$ is the set of all real numbers except 0. In interval notation, we write $(-\infty, 0) \cup (0, \infty)$.

9. a.

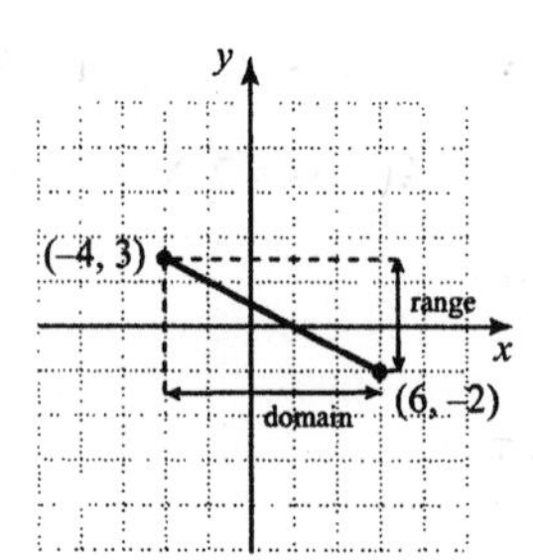

The domain is [−4, 6].
The range is [−2, 3].

b.

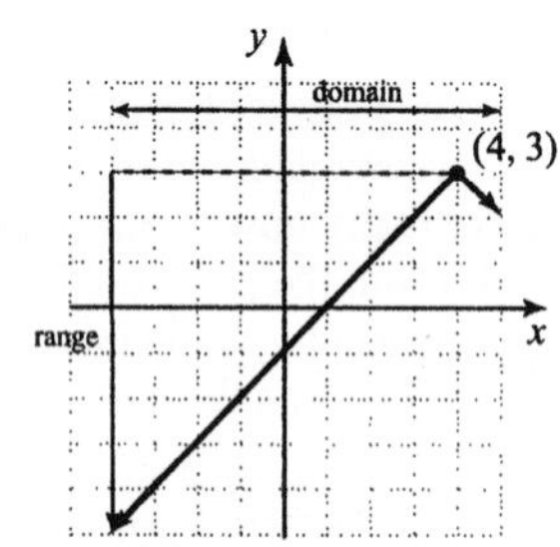

The domain is $(-\infty, \infty)$.
The range is $(-\infty, 3]$.

Vocabulary and Readiness Check

1. A set of ordered pairs is called a <u>relation</u>.

2. A set of ordered pairs that assigns to each x-value exactly one y-value is called a <u>function</u>.

3. The set of all y-coordinates of a relation is called the <u>range</u>.

4. The set of all x-coordinates of a relation is called the <u>domain</u>.

5. All linear equations are functions except those whose graphs are <u>vertical</u> lines.

6. All linear equations are functions except those whose equations are of the form <u>$x = c$</u>.

7. If $f(3) = 7$, the corresponding ordered pair is <u>(3, 7)</u>.

8. The domain of $f(x) = x + 5$ is <u>$(-\infty, \infty)$</u>.

Exercise Set 3.6

1. {(2, 4), (0, 0), (−7, 10), (10, −7)}
Domain: {−7, 0, 2, 10}
Range: {−7, 0, 4, 10}

3. {(0, −2), (1, −2), (5, −2)}
Domain: {0, 1, 5}
Range: {−2}

5. Every point has a unique x-value: it is a function.

7. Two or more points have the same x-value: it is not a function.

9. No; two points have x-coordinate 1.

11. Yes; no two points have the same x-coordinate.

13. Yes; no vertical line can be drawn that intersects the graph more than once.

15. No; there are many vertical lines that intersect the graph twice, $x = 1$, for example.

17. Yes; $y = x + 1$ is a non-vertical line.

19. Yes; $y - x = 7$ is a non-vertical line.

21. Yes; $y = 6$ is a non-vertical line.

23. No; $x = -2$ is a vertical line.

25. No; does not pass the vertical line test.

27. The point on the graph above June corresponds to approximately 9:30 P.M. on the time axis.

29. The sunset is at approximately 3 P.M. twice, on January 1 and on December 1.

31. Yes; it passes the vertical line test.

33. \$4.25 per hour; the segment representing dates before October 1996 corresponds to 4.25 on the vertical axis.

35. 2009; the first line segment above 7.00 on the vertical axis represents dates beginning July 24, 2009.

37. Yes; answers may vary

39. $f(x) = 2x - 5$
$f(-2) = 2(-2) - 5 = -4 - 5 = -9$
$f(0) = 2(0) - 5 = -5$
$f(3) = 2(3) - 5 = 6 - 5 = 1$

41. $f(x) = x^2 + 2$
$f(-2) = (-2)^2 + 2 = 4 + 2 = 6$
$f(0) = (0)^2 + 2 = 2$
$f(3) = (3)^2 + 2 = 9 + 2 = 11$

43. $f(x) = 3x$
$f(-2) = 3(-2) = -6$
$f(0) = 3(0) = 0$
$f(3) = 3(3) = 9$

45. $f(x) = |x|$
$f(-2) = |-2| = 2$
$f(0) = |0| = 0$
$f(3) = |3| = 3$

47. $h(x) = -5x$
$h(-1) = -5(-1) = 5$
$h(0) = -5(0) = 0$
$h(4) = -5(4) = -20$

49. $h(x) = 2x^2 + 3$
$h(-1) = 2(-1)^2 + 3 = 2 + 3 = 5$
$h(0) = 2(0)^2 + 3 = 3$
$h(4) = 2(4)^2 + 3 = 2 \cdot 16 + 3 = 32 + 3 = 35$

51. $f(3) = 6$ corresponds to the ordered pair (3, 6).

53. $g(0) = -\frac{1}{2}$ corresponds to the ordered pair $\left(0, -\frac{1}{2}\right)$.

55. $h(-2) = 9$ corresponds to the ordered pair (−2, 9).

57. $(-\infty, \infty)$

59. $x + 5 \neq 0 \Rightarrow x \neq -5$, therefore $(-\infty, -5) \cup (-5, \infty)$ or all real numbers except −5.

61. $(-\infty, \infty)$

63. D: $(-\infty, \infty)$, R: $x \geq -4$, $[-4, \infty)$

65. D: $(-\infty, \infty)$, R: $(-\infty, \infty)$

67. D: $(-\infty, \infty)$, R: $\{2\}$

69. (−2, 1)

71. (−3, −1)

73. $f(-5) = 12$

75. (3, −4)

77. $f(5) = 0$

79. $H(x) = 2.59x + 47.24$

a. $H(46) = 2.59(46) + 47.24 = 166.38$ cm

b. $H(39) = 2.59(39) + 47.24 = 148.25$ cm

81. Answers may vary

83. $y = x + 7$
$f(x) = x + 7$

85. $g(x) = -3x + 12$

a. $g(s) = -3(s) + 12 = -3s + 12$

b. $g(r) = -3(r) + 12 = -3r + 12$

87. $f(x) = x^2 - 12$

a. $f(12) = (12)^2 - 12 = 132$

b. $f(a) = (a)^2 - 12 = a^2 - 12$

Chapter 3 Vocabulary Check

1. An ordered pair is a solution of an equation in two variables if replacing the variables by the coordinates of the ordered pair results in a true statement.

2. The vertical number line in the rectangle coordinate system is called the y-axis.

3. A linear equation can be written in the form $Ax + By = C$.

4. An x-intercept is a point of the graph where the graph crosses the x-axis.

5. The form $Ax + By = C$ is called standard form.

6. A y-intercept is a point of the graph where the graph crosses the y-axis.

7. The equation $y = 7x - 5$ is written in slope-intercept form.

8. The equation $y + 1 = 7(x - 2)$ is written in point-slope form.

9. To find an x-intercept of a graph, let $y = 0$.

10. The horizontal number line in the rectangular coordinate system is called the x-axis.

11. To find a y-intercept of a graph, let $x = 0$.

12. The slope of a line measures the steepness or tilt of a line.

13. A set of ordered pairs that assigns to each x-value exactly one y-value is called a function.

14. The set of all x-coordinates of a relation is called the domain of the relation.

15. The set of all y-coordinates of a relation is called the range of the relation.

16. A set of ordered pairs is called a relation.

Chapter 3 Review

1–6.

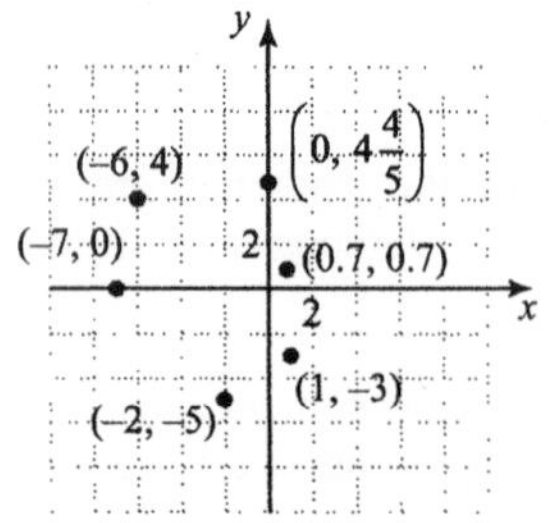

7. a. (8.00, 1), (7.50, 10), (6.50, 25), (5.00, 50), (2.00, 100)

b.

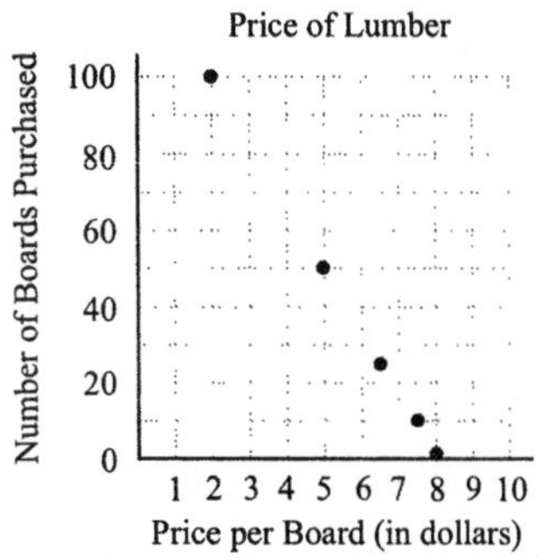

8. a. (2001, 9.8), (2002, 15.1), (2003, 14.6), (2004, 14.0), (2005, 13.8), (2006, 13.6)

b.

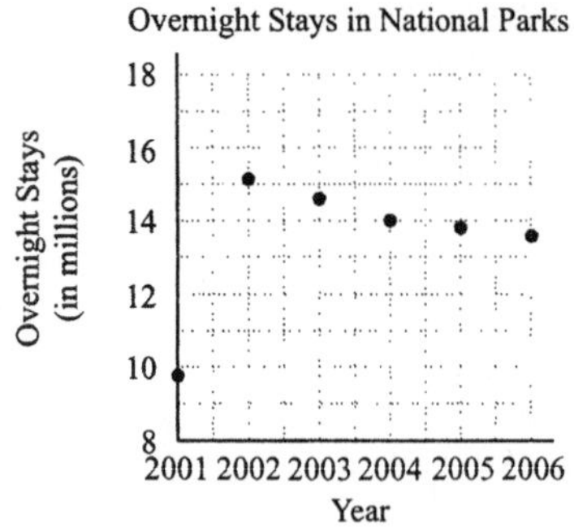

9. $7x - 8y = 56$

(0, 56)

$7(0) - 8(56) \stackrel{?}{=} 56$

$-448 \neq 0$ No

(8, 0)

$7(8) - 8(0) \stackrel{?}{=} 56$

$56 = 56$ Yes

10. $-2x+5y=10$

$(-5, 0)$

$-2(-5)+5(0) \stackrel{?}{=} 10$

$10=10$ Yes

$(1, 1)$

$-2(1)+5(1) \stackrel{?}{=} 10$

$3 \neq 10$ No

11. $x=13$

$(13, 5)$

$(13) \stackrel{?}{=} 13$

$13=13$ Yes

$(13, 13)$

$(13) \stackrel{?}{=} 13$

$13=13$ Yes

12. $y=2$

$(7, 2)$

$(2) \stackrel{?}{=} 2$

$2=2$ Yes

$(2, 7)$

$(7) \stackrel{?}{=} 2$

$7 \neq 2$ No

13. $-2+y=6x$, $x=7$

$-2+y=6(7)$

$-2+y=42$

$y=44$

$(7, 44)$

14. $y=3x+5$, $y=-8$

$-8=3x+5$

$-13=3x$

$-\frac{13}{3}=x$

$\left(-\frac{13}{3}, -8\right)$

15. $9=-3x+4y$

$y=0$: $9=-3x+4(0)$, $9=-3x$, $-3=x$

$y=3$: $9=-3x+4(3)$, $9=-3x+12$, $-3=-3x$, $1=x$

$x=9$: $9=-3(9)+4y$, $9=-27+4y$, $36=4y$, $9=y$

x	y
−3	0
1	3
9	9

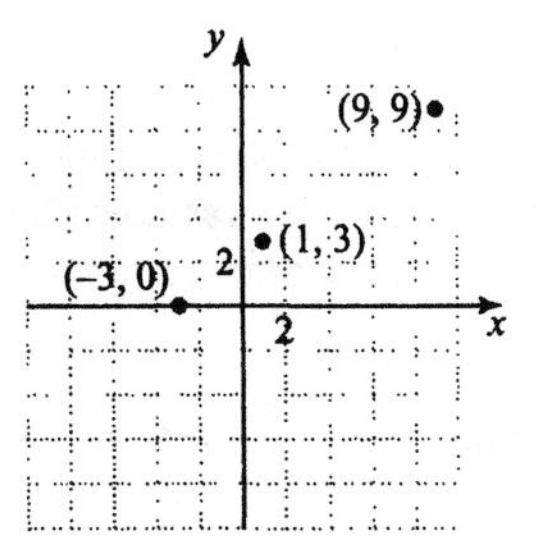

16. $y=5$ for all values of x.

x	y
7	5
−7	5
0	5

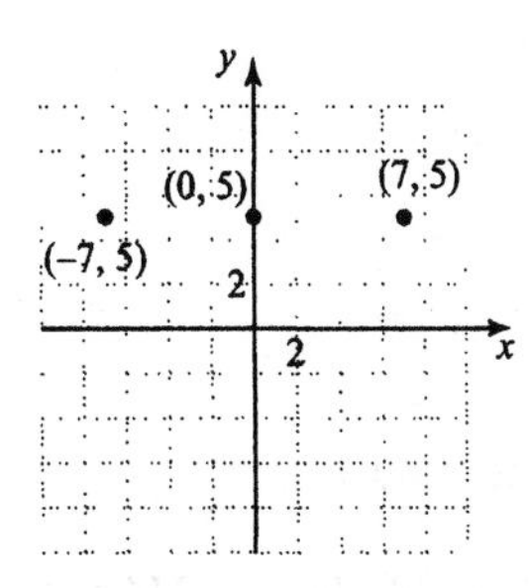

17. $x=2y$

$y=0$: $x=2(0)=0$

$y=5$: $x=2(5)=10$

$y=-5$: $x=2(-5)=-10$

x	y
0	0
10	5
−10	−5

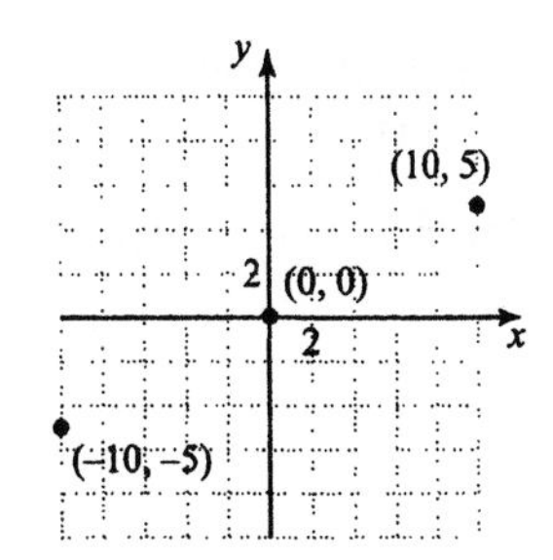

18. a. $y = 5x + 2000$
$x = 1$: $y = 5(1) + 2000 = 2005$
$x = 100$: $y = 5(100) + 2000 = 2500$
$x = 1000$: $y = 5(1000) + 2000 = 7000$

x	1	100	1000
y	2005	2500	7000

b. Let $y = 6430$.
$6430 = 5x + 2000$
$4430 = 5x$
$886 = x$
886 CD holders can be produced.

19. $x - y = 1$

x	y
1	0
0	−1

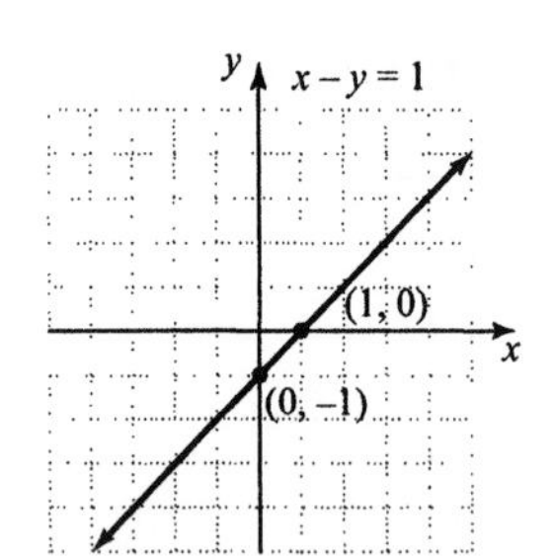

20. $x + y = 6$

x	y
6	0
0	6

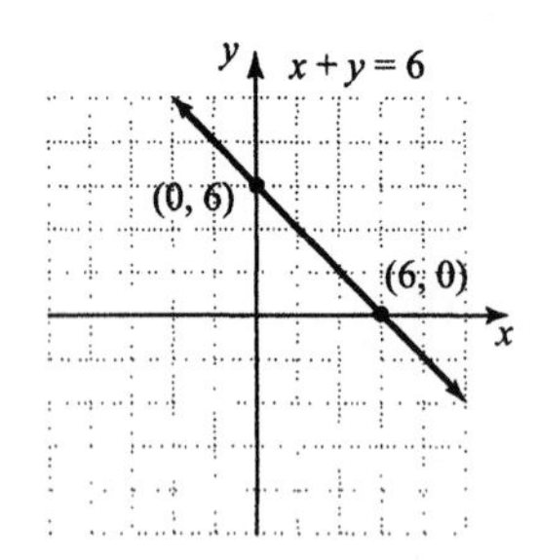

21. $x - 3y = 12$

x	y
12	0
0	−4

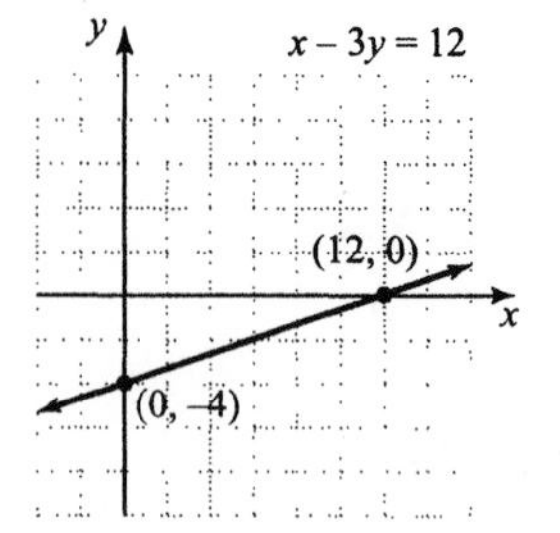

22. $5x - y = -8$

x	y
−2	−2
0	8

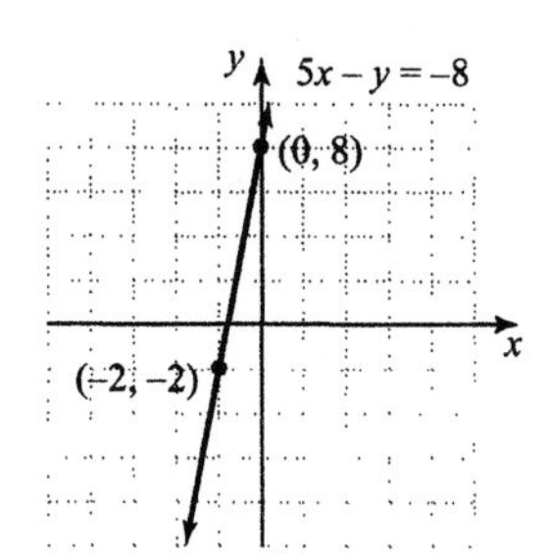

23. $x = 3y$

x	y
0	0
6	2

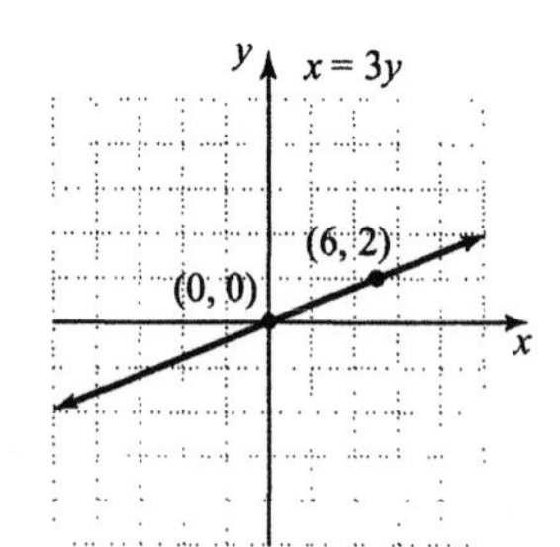

24. $y = -2x$

x	y
0	0
4	−8

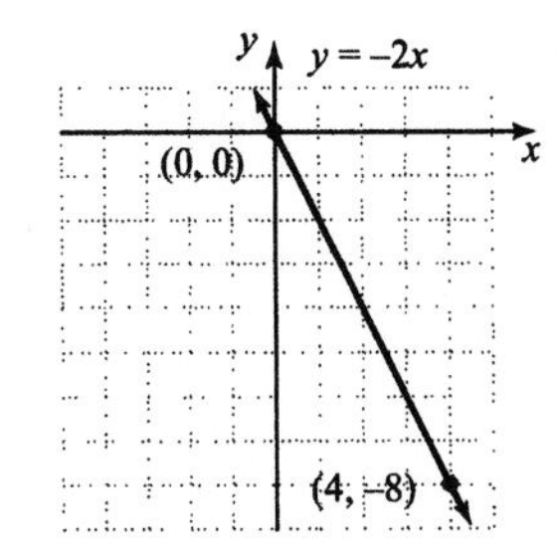

25. $2x - 3y = 6$

x	y
0	−2
3	0

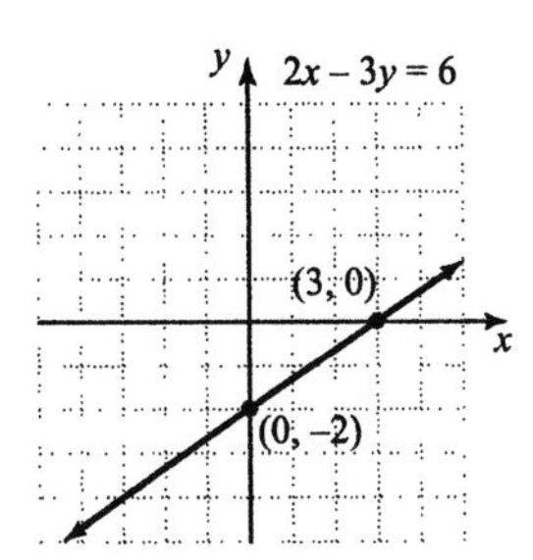

26. $4x - 3y = 12$

x	y
0	−4
3	0

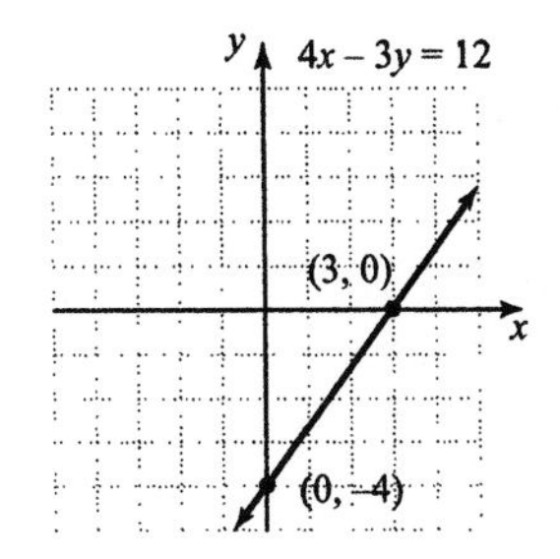

27. $y = 3x + 111$

U.S. Long-Distance Revenue

Revenue (in billions of dollars)

140
130
120
110

0 1 2 3 4 5 6 7 8

Years after 1999

Expect a revenue of $135 billion in 2007.

28. x-intercept: (4, 0)
y-intercept: (0, −2)

29. y-intercept: (0, −3)

30. x-intercepts: (−2, 0), (2, 0)
y-intercepts: (0, 2), (0, −2)

31. x-intercepts: (−1, 0), (2, 0), (3, 0)
y-intercept: (0, −2)

32. $x - 3y = 12$

x	y
0	−4
12	0

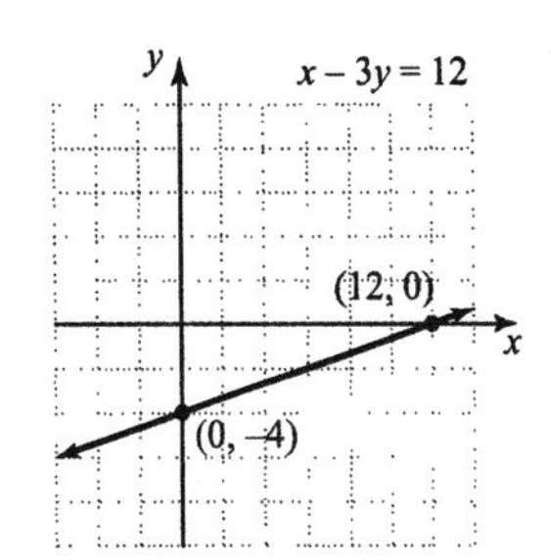

33. $-4x + y = 8$

x	y
0	8
−2	0

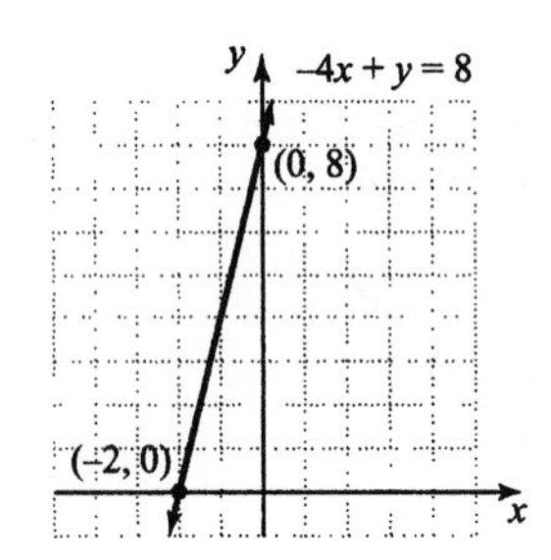

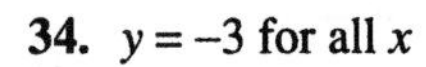

34. $y = -3$ for all x

x	y
0	−3

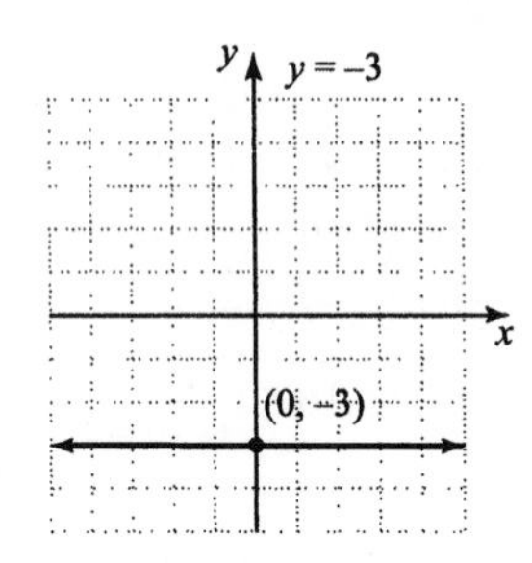

35. $x = 5$ for all y

x	y
5	0

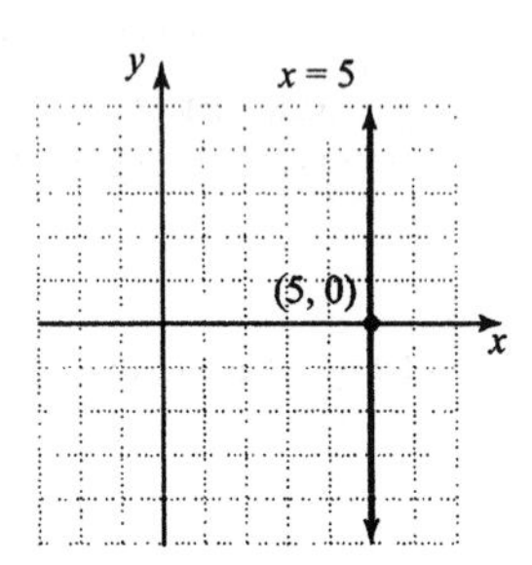

36. $y = -3x$
Find a second point.

x	y
0	0
3	−9

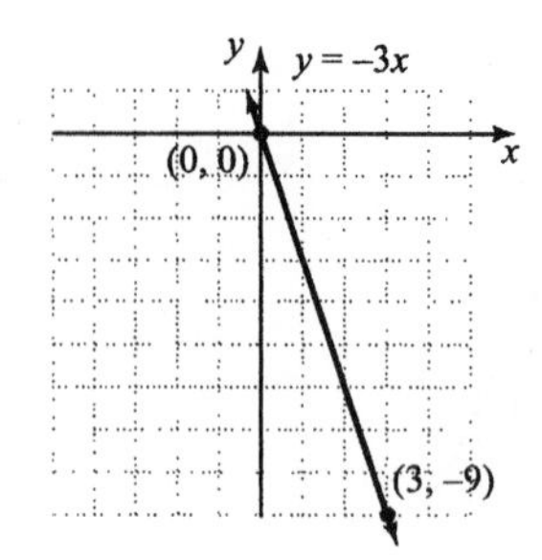

37. $x = 5y$
Find a second point.

x	y
0	0
5	1

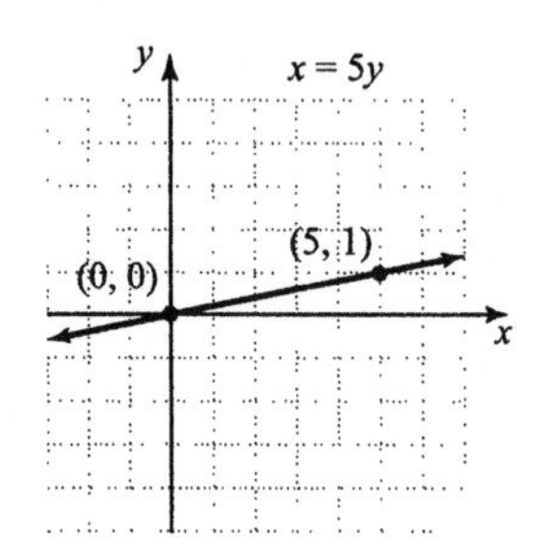

38. $x - 2 = 0$
$x = 2$ for all y

x	y
2	0

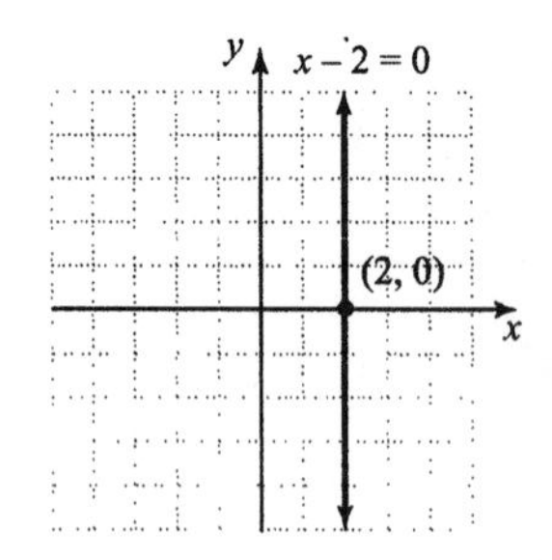

39. $y + 6 = 0$
$y = -6$ for all x

x	y
0	−6

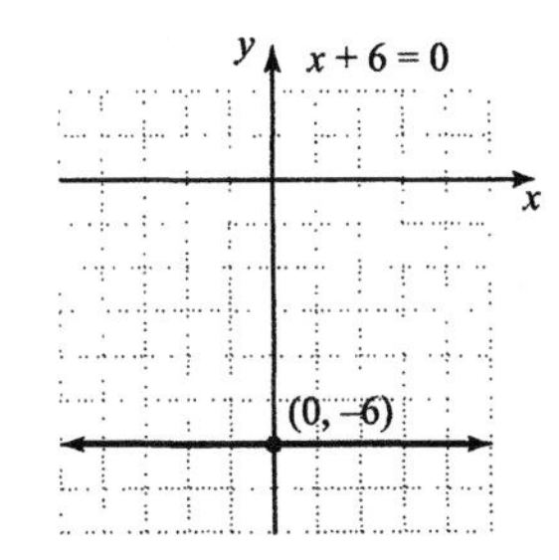

40. (−1, 2), and (3, −1)

$$m = \frac{y_2 - y_1}{x_2 - x_1} = \frac{-1-2}{3-(-1)} = -\frac{3}{4}$$

41. (−2, −2) and (3, −1)

$$m = \frac{y_2 - y_1}{x_2 - x_1} = \frac{-1-(-2)}{3-(-2)} = \frac{1}{5}$$

42. $m = 0$
d

43. $m = -1$
b

44. Slope is undefined.
c

45. $m = 3$
a

46. $m = \frac{2}{3}$
e

47. (2, 5) and (6, 8)

$$m = \frac{y_2 - y_1}{x_2 - x_1} = \frac{8-5}{6-2} = \frac{3}{4}$$

48. (4, 7) and (1, 2)

$$m = \frac{y_2 - y_1}{x_2 - x_1} = \frac{2-7}{1-4} = \frac{-5}{-3} = \frac{5}{3}$$

49. (1, 3) and (−2, −9)

$$m = \frac{y_2 - y_1}{x_2 - x_1} = \frac{-9-3}{-2-1} = \frac{-12}{-3} = 4$$

50. (−4, 1), and (3, −6)

$$m = \frac{y_2 - y_1}{x_2 - x_1} = \frac{-6-1}{3-(-4)} = \frac{-7}{7} = -1$$

51. $y = 3x + 7$
The equation is in slope-intercept form. The slope is the coefficient of x, or 3.

52. Solve for y.

$$x - 2y = 4$$
$$-2y = -x + 4$$
$$y = \frac{1}{2}x - 2$$

The slope is $\frac{1}{2}$.

53. $y = -2$
This is the equation of a horizontal line. The slope is 0.

54. $x = 0$
This is the equation of a vertical line. The slope is undefined.

55. Solve the equations for y.

$$x - y = 6 \qquad\qquad x + y = 3$$
$$-y = -x + 6 \qquad\qquad y = -x + 3$$
$$y = x - 6$$

The slopes are 1 and −1. Since their product is −1, the lines are perpendicular.

56. Solve the equations for y.

$$3x + y = 7 \qquad\qquad -3x - y = 10$$
$$y = -3x + 7 \qquad\qquad -y = 3x + 10$$
$$y = -3x - 10$$

The slopes are both −3. Since the lines have the same slope but different y-intercepts, they are parallel.

57. The first line, $y = 4x + \frac{1}{2}$, has slope 4. Solve the second equation for y.

$$4x + 2y = 1$$
$$2y = -4x + 1$$
$$y = -2x + \frac{1}{2}$$

The second line has slope −2. Since the slopes are not the same and their product is not −1, the lines are neither parallel nor perpendicular.

58. $x = 4, y = -2$
The first equation's graph is a vertical line, and the second equation's graph is a horizontal line. These lines are perpendicular.

59. Use the points (1985, 232) and (2006, 608).

$m = \dfrac{608-232}{2006-1985} = \dfrac{376}{21} \approx \dfrac{17.90 \text{ dollars}}{1 \text{ year}}$

Every 1 year, monthly daycare costs increase by \$17.90.

60. Use the points (2004, 46) and (2009, 56.5).

$m = \dfrac{56.5-46}{2009-2004} = \dfrac{10.5}{5} \approx \dfrac{2.1 \text{ billion dollars}}{1 \text{ year}}$

Every 1 year, \$2.1 billion more dollars are spent on technology.

61. $3x + y = 7$

$y = -3x + 7$

$y = mx + b$

$m = -3$, y-intercept = (0, 7)

62. $x - 6y = -1$

$-6y = -x - 1$

$y = \dfrac{1}{6}x + \dfrac{1}{6}$

$y = mx + b$

$m = \dfrac{1}{6}$, y-intercept = $\left(0, \dfrac{1}{6}\right)$

63. $y = 2$

$y = mx + b$

$m = 0$, y-intercept = (0, 2)

64. $x = -5$

$y = mx + b$

m is undefined.

There is no y-intercept.

65. $m = -5$, $b = \dfrac{1}{2}$

$y = mx + b$

$y = -5x + \dfrac{1}{2}$

66. $m = \dfrac{2}{3}$, $b = 6$

$y = mx + b$

$y = \dfrac{2}{3}x + 6$

67. $y = 3x - 1$

$y = mx + b$

$m = 3$, $b = -1$

68. $y = -3x$

$y = mx + b$

$m = -3$, $b = 0$

69. $5x - 3y = 15$

$-3y = -5x + 15$

$y = \dfrac{5}{3}x - 5$

$y = mx + b$

$m = \dfrac{5}{3}$, $b = -5$

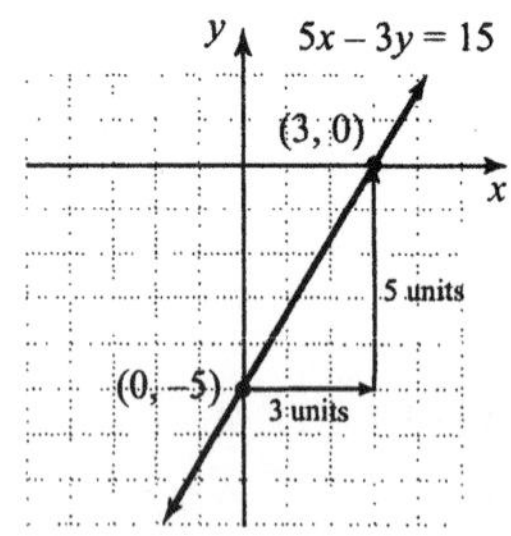

70. $-x + 2y = 8$

$2y = x + 8$

$y = \dfrac{1}{2}x + 4$

$y = mx + b$

$m = \dfrac{1}{2}$, $b = 4$

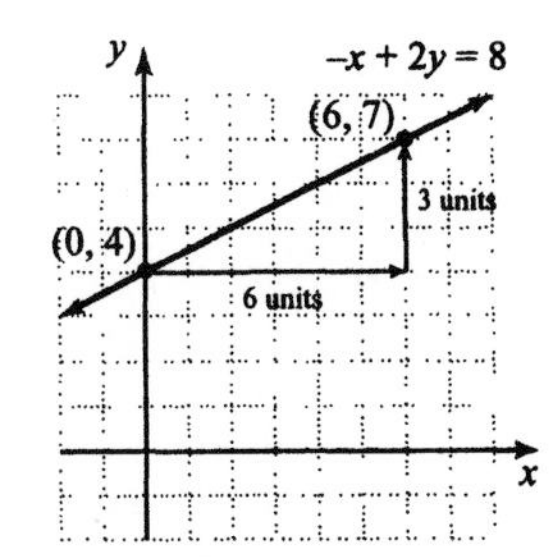

71. $y=-4x$
$m=-4, b=0$
c

72. $y=-2x+1$
$m=-2, b=1$
d

73. $y=2x-1$
$m=2, b=-1$
b

74. $y=2x$
$m=2, b=0$
a

75. $m=-3$; $(0, -5)$

$$y=mx+b$$
$$y=-3x-5$$
$$3x+y=-5$$

76. $m=\frac{1}{2}; \left(0, -\frac{7}{2}\right)$

$$y=mx+b$$
$$y=\frac{1}{2}x-\frac{7}{2}$$
$$2y=x-7$$
$$x-2y=7$$

77. Horizontal line, point $(-2, -3)$
$y=c$
$y=-3$

78. Horizontal line, point $(0, 0)$
$y=c$
$y=0$

79. $m=-6$; $(2, -1)$

$$y-y_1=m(x-x_1)$$
$$y-(-1)=-6(x-2)$$
$$y+1=-6x+12$$
$$6x+y=11$$

80. $m=12; \left(\frac{1}{2}, 5\right)$

$$y-y_1=m(x-x_1)$$
$$y-5=12\left(x-\frac{1}{2}\right)$$
$$y-5=12x-6$$
$$12x-y=1$$

81. $(0, 6)$ and $(6, 0)$

$$m=\frac{y_2-y_1}{x_2-x_1}=\frac{0-6}{6-0}=\frac{-6}{6}=-1$$
$m=-1$; $(0, 6)$
$$y-y_1=m(x-x_1)$$
$$y-6=-1(x-0)$$
$$y-6=-x$$
$$x+y=6$$

82. $(0, -4)$ and $(-8, 0)$

$$m=\frac{y_2-y_1}{x_2-x_1}=\frac{0-(-4)}{-8-0}=\frac{4}{-8}=-\frac{1}{2}$$
$m=-\frac{1}{2}$; $(0, -4)$
$$y-y_1=m(x-x_1)$$
$$y-(-4)=-\frac{1}{2}(x-0)$$
$$y+4=-\frac{1}{2}x$$
$$2y+8=-x$$
$$x+2y=-8$$

83. Vertical line, point $(5, 7)$
$x=c$
$x=5$

84. Horizontal line, point $(-6, 8)$
$y=c$
$y=8$

85. $y=8$ is horizontal.
Perpendicular to $y=8$ is vertical; $x=c$.
Point $(6, 0)$
$x=6$

86. $x=-2$ is vertical.
Perpendicular to $x=-2$ is horizontal; $y=c$,
point $(10, 12)$
$y=12$

87. Two points have the same *x*-value: it is not a function.

88. Every point has a unique x-value: it is a function.

89. Yes; $7x - 6y = 1$ is a non-vertical line.

90. Yes; $y = 7$ is a non-vertical line.

91. No; $x = 2$ is a vertical line.

92. Yes; for each value of x there is only one value of y.

93. No; the graph does not pass the vertical line test.

94. Yes; the graph passes the vertical line test.

95. $f(x) = -2x + 6$

a. $f(0) = -2(0) + 6 = 6$

b. $f(-2) = -2(-2) + 6 = 4 + 6 = 10$

c. $f\left(\frac{1}{2}\right) = -2\left(\frac{1}{2}\right) + 6 = -1 + 6 = 5$

96. $h(x) = -5 - 3x$

a. $h(2) = -5 - 3(2) = -11$

b. $h(-3) = -5 - 3(-3) = 4$

c. $h(0) = -5 - 3(0) = -5$

97. $g(x) = x^2 + 12x$

a. $g(3) = (3)^2 + 12(3) = 45$

b. $g(-5) = (-5)^2 + 12(-5) = -35$

c. $g(0) = (0)^2 + 12(0) = 0$

98. $h(x) = 6 - |x|$

a. $h(-1) = 6 - |-1| = 6 - 1 = 5$

b. $h(1) = 6 - |1| = 6 - 1 = 5$

c. $h(-4) = 6 - |-4| = 6 - 4 = 2$

99. $(-\infty, \infty)$

100. $x - 2 \neq 0 \Rightarrow x \neq 2$, therefore $(-\infty, 2) \cup (2, \infty)$ or all real numbers except 2.

101. D: $[-3, 5]$, R: $[-4, 2]$

102. D: $(-\infty, \infty)$, R: $x \geq 0$, $[0, \infty)$

103. D: $\{3\}$, R: $(-\infty, \infty)$

104. D: $(-\infty, \infty)$, R: $x \leq 2$, $(-\infty, 2]$

105. $2x - 5y = 9$

Let $y = 1$.
$2x - 5(1) = 9$
$2x - 5 = 9$
$2x = 14$
$x = 7$

Let $x = 2$.
$2(2) - 5y = 9$
$4 - 5y = 9$
$-5y = 5$
$y = -1$

Let $y = -3$.
$2x - 5(-3) = 9$
$2x + 15 = 9$
$2x = -6$
$x = -3$

x	y
7	1
2	−1
−3	−3

106. $x = -3y$

Let $x = 0$.
$0 = -3y$
$0 = y$

Let $y = 1$.
$x = -3(1)$
$x = -3$

Let $x = 6$.
$6 = -3y$
$-2 = y$

x	y
0	0
−3	1
6	−2

107. $2x - 3y = 6$

Let $y = 0$.
$2x - 3(0) = 6$
$2x = 6$
$x = 3$

Let $x = 0$.
$2(0) - 3y = 6$
$-3y = 6$
$y = -2$

x-intercept: $(3, 0)$
y-intercept: $(0, -2)$

108. $-5x + y = 10$

Let $y = 0$.
$$-5x + 0 = 10$$
$$-5x = 10$$
$$x = -2$$

Let $x = 0$.
$$-5(0) + y = 10$$
$$y = 10$$

x-intercept: $(-2, 0)$
y-intercept: $(0, 10)$

109. $x - 5y = 10$

x	y
10	0
0	−2

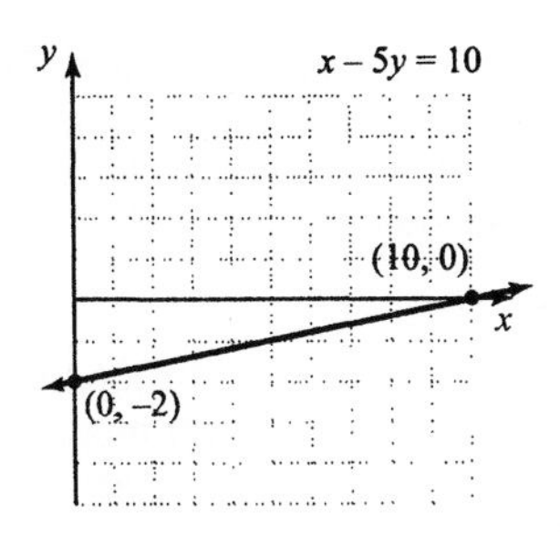

110. $x + y = 4$

x	y
4	0
0	4

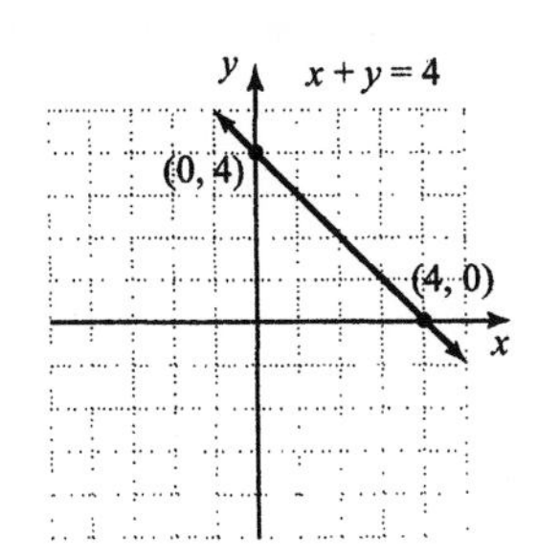

111. $y = -4x$

x	y
0	0
1	−4

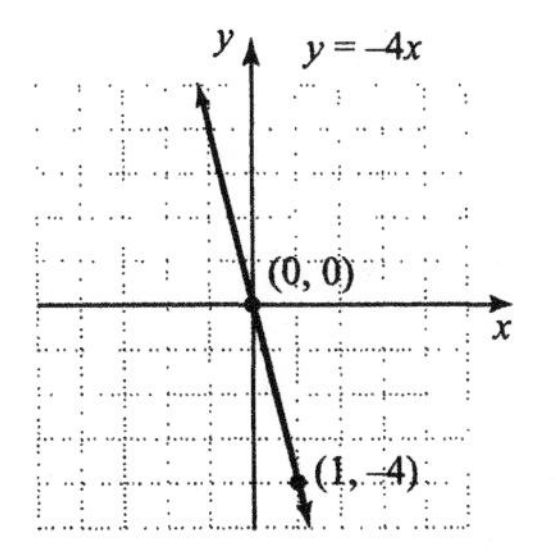

112. $2x + 3y = -6$

x	y
−3	0
0	−2

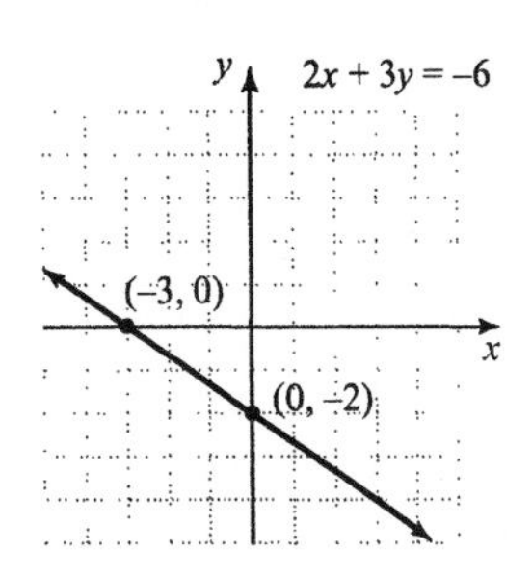

113. $x = 3$
This is the equation of a vertical line with x-intercept $(3, 0)$.

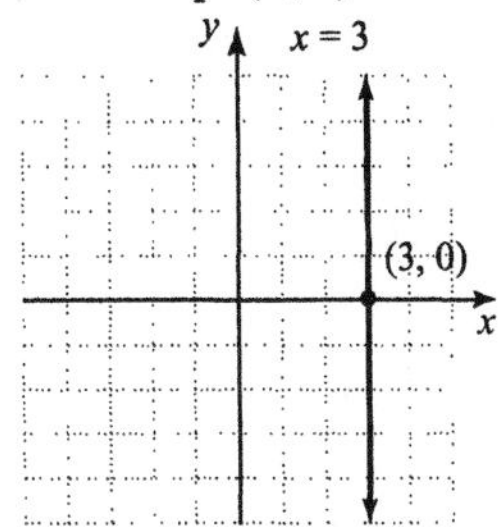

114. $y = -2$
This is the equation of a horizontal line with y-intercept $(0, -2)$.

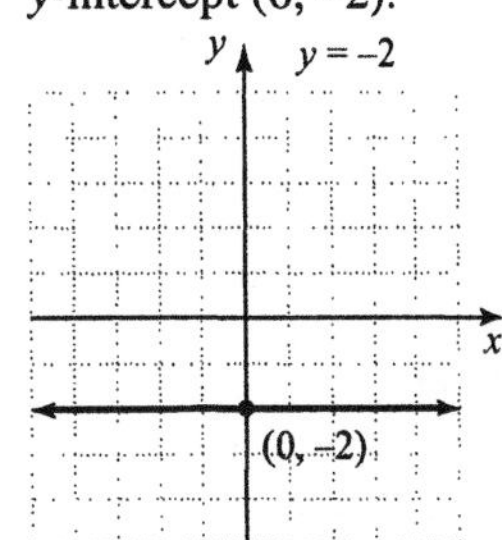

115. (3, –5) and (–4, 2)

$$m = \frac{y_2 - y_1}{x_2 - x_1} = \frac{2-(-5)}{-4-3} = \frac{7}{-7} = -1$$

116. (1, 3) and (–6, –8)

$$m = \frac{y_2 - y_1}{x_2 - x_1} = \frac{-8-3}{-6-1} = \frac{-11}{-7} = \frac{11}{7}$$

117. (0, –4) and (2, 0)

$$m = \frac{y_2 - y_1}{x_2 - x_1} = \frac{0-(-4)}{2-0} = \frac{4}{2} = 2$$

118. (0, 2) and (6, 0)

$$m = \frac{y_2 - y_1}{x_2 - x_1} = \frac{0-2}{6-0} = \frac{-2}{6} = -\frac{1}{3}$$

119. Solve for *y*.

$$-2x + 3y = -15$$
$$3y = 2x - 15$$
$$y = \frac{2}{3}x - 5$$

The slope is $\frac{2}{3}$. The *y*-intercept is (0, –5).

120. Solve for *y*.

$$6x + y - 2 = 0$$
$$y = -6x + 2$$

The slope is –6. The *y*-intercept is (0, 2).

121. $m = -5$; (3, –7)

$$y - y_1 = m(x - x_1)$$
$$y - (-7) = -5(x - 3)$$
$$y + 7 = -5x + 15$$
$$5x + y = 8$$

122. $m = 3$; (0, 6)

$$y = mx + b$$
$$y = 3x + 6$$
$$3x - y = -6$$

123. (–3, 9) and (–2, 5)

$$m = \frac{y_2 - y_1}{x_2 - x_1} = \frac{5-9}{-2-(-3)} = \frac{-4}{1} = -4$$

$m = -4$; (–2, 5)

$$y - y_1 = m(x - x_1)$$
$$y - 5 = -4(x - (-2))$$
$$y - 5 = -4(x + 2)$$
$$y - 5 = -4x - 8$$
$$4x + y = -3$$

124. (3, 1) and (5, –9)

$$m = \frac{y_2 - y_1}{x_2 - x_1} = \frac{-9-1}{5-3} = \frac{-10}{2} = -5$$

$m = -5$; (3, 1)

$$y - y_1 = m(x - x_1)$$
$$y - 1 = -5(x - 3)$$
$$y - 1 = -5x + 15$$
$$5x + y = 16$$

125. The highest point on the graph is above 2002 on the horizontal axis and corresponds to approximately 27.1 on the vertical axis, so the greatest beef production was 27.1 billion pounds in 2002.

126. The lowest point on the graph is above 2004 on the horizontal axis and corresponds to approximately 24.6 on the vertical axis, so the least beef production was 24.6 billion pounds in 2004.

127. The points for 2002, 2003, and 2006 lie above 25.0 on the vertical axis, so beef production was greater than 25 billion pounds in these years.

128. In 2003 and in 2004, there were decreases in beef production from the preceding years. In 2005 and in 2006, there were increases over the preceding years. The point for 2005 is only slightly higher than the point for 2004, denoting a small increase. However, the point for 2006 is much higher than the point for 2005, denoting a greater increase. The greatest increase occurred in 2006.

Chapter 3 Test

1. $y = \frac{1}{2}x$

$m = \frac{1}{2};\ b = 0$

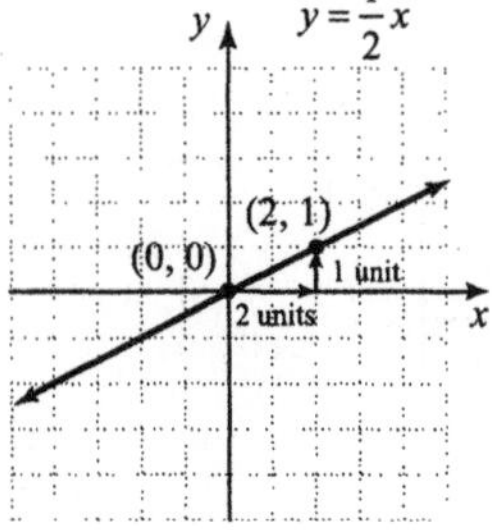

y
$y = \frac{1}{2}x$
(2, 1)
(0, 0)
1 unit
2 units
x

2. $2x + y = 8$

x	y
4	0
0	8

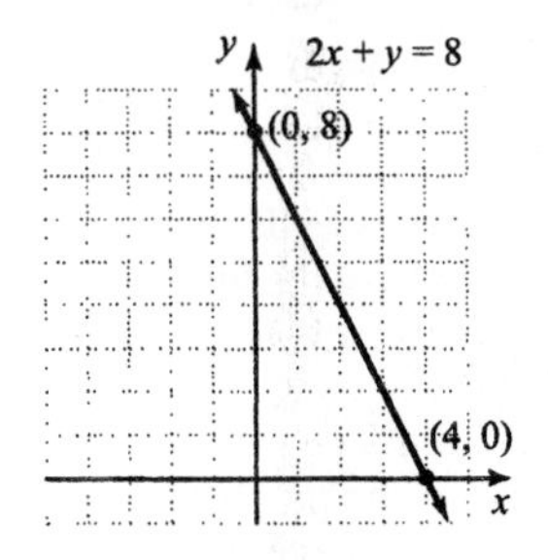

3. $5x - 7y = 10$

x	y
2	0
–5	–5

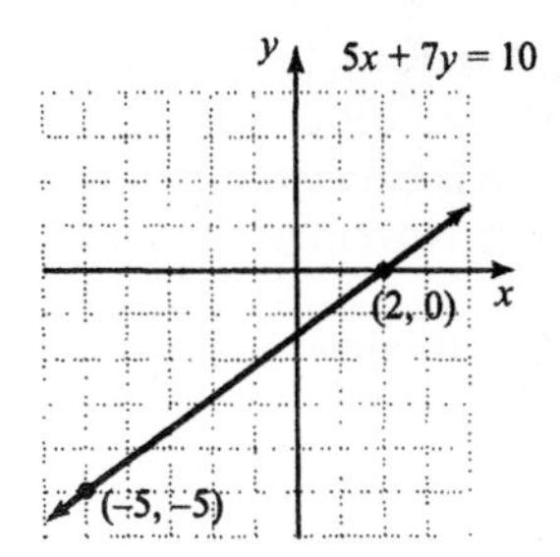

4. $y = -1$ for all values of x.

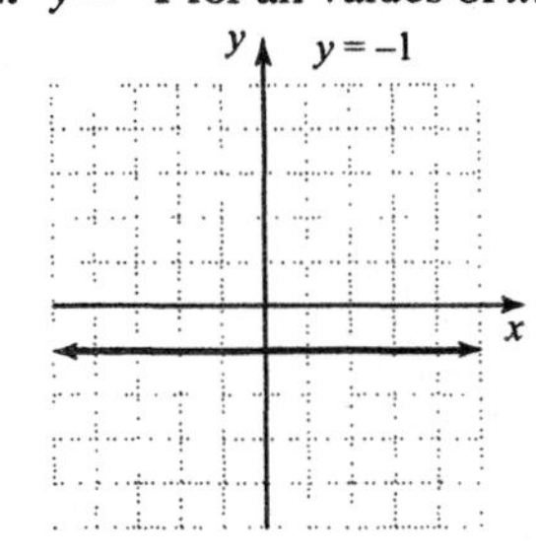

5. $x - 3 = 0$
$x = 3$ for all values of y.

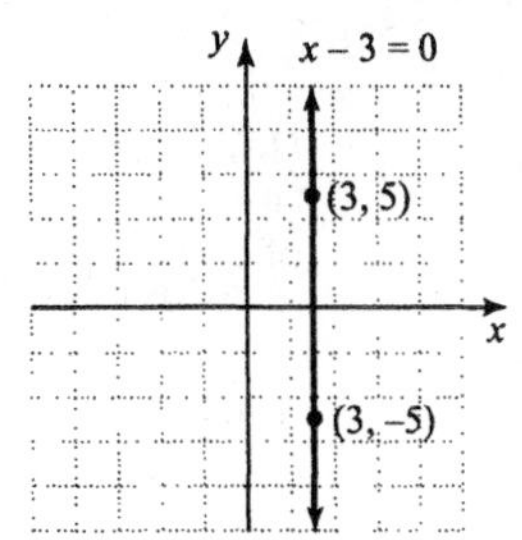

6. $(-1, -1)$ and $(4, 1)$
$$m = \frac{y_2 - y_1}{x_2 - x_1} = \frac{1-(-1)}{4-(-1)} = \frac{2}{5}$$

7. Horizontal line: $m = 0$

8. $(6, -5)$ and $(-1, 2)$
$$m = \frac{y_2 - y_1}{x_2 - x_1} = \frac{2-(-5)}{-1-6} = \frac{7}{-7} = -1$$

9. $-3x + y = 5$
$$y = 3x + 5$$
$$y = mx + b$$
$$m = 3$$

10. $x = 6$ is a vertical line. The slope is undefined.

11. $7x - 3y = 2$
$$-3y = -7x + 2$$
$$y = \frac{7}{3}x - \frac{2}{3}$$
$$y = mx + b$$
$$m = \frac{7}{3},\ b = -\frac{2}{3},\ \left(0, -\frac{2}{3}\right)$$

12. $y = 2x - 6,\ m_1 = 2$
$$-4x = 2y,\ -2x = y$$
$$y = -2x,\ m_2 = -2$$
$m_1 \neq m_2$ and $m_1 m_2 \neq -1$, neither

13. $m = -\frac{1}{4};\ (2, 2)$
$$y - y_1 = m(x - x_1)$$
$$y - 2 = -\frac{1}{4}(x - 2)$$
$$4(y-2) = -(x-2)$$
$$4y - 8 = -x + 2$$
$$x + 4y = 10$$

14. (0, 0) and (6, −7)

$$m = \frac{y_2 - y_1}{x_2 - x_1} = \frac{-7-0}{6-0} = -\frac{7}{6}$$

$m = -\frac{7}{6}$; (0, 0)

$$y - y_1 = m(x - x_1)$$
$$y - 0 = -\frac{7}{6}(x - 0)$$
$$6y = -7x$$
$$7x + 6y = 0$$

15. (2, −5) and (1, 3)

$$m = \frac{y_2 - y_1}{x_2 - x_1} = \frac{3-(-5)}{1-2} = \frac{8}{-1} = -8$$

$m = -8$; (1, 3)

$$y - y_1 = m(x - x_1)$$
$$y - 3 = -8(x - 1)$$
$$y - 3 = -8x + 8$$
$$8x + y = 11$$

16. $x = 7$ is vertical.
Parallel to $x = 7$ is vertical;
$x = c$, point (−5, −1)
$x = -5$

17. $m = \frac{1}{8}$, $b = 12$

$$y = mx + b$$
$$y = \frac{1}{8}x + 12$$
$$8y = x + 96$$
$$x - 8y = -96$$

18. Yes; it passes the vertical line test.

19. No; it does not pass the vertical line test.

20. $h(x) = x^3 - x$

a. $h(-1) = (-1)^3 - (-1) = -1 + 1 = 0$

b. $h(0) = (0)^3 - (0) = 0$

c. $h(4) = (4)^3 - (4) = 64 - 4 = 60$

21. $x + 1 \neq 0 \Rightarrow x \neq -1$, therefore
$(-\infty, -1) \cup (-1, \infty)$ or all real numbers except −1.

22. D: $(-\infty, \infty)$, R: $x \leq 4$, $(-\infty, 4]$

23. D: $(-\infty, \infty)$, R: $(-\infty, \infty)$

24. $f(7) = 20$ corresponds to the ordered pair (7, 20).

25. The bar for Denmark extends to about 210 on the horizontal axis. The average water use per person per day in Denmark is approximately 210 liters.

26. The bar for Australia extends to about 490 on the horizontal axis. The average water use per person per day in Australia is approximately 490 liters.

27. The highest point on the graph corresponds to 7 on the horizontal axis, denoting July. The average high temperature is the greatest in July.

28. April corresponds to 4 on the horizontal axis. Moving horizontally to the left from the point on the graph above 4, we reach approximately 63 on the vertical axis. The average high temperature for April is approximately 63°F.

29. The points for months 1, 2, 3, 11, and 12 lie below 60 on the vertical axis. Thus, the average high temperature is below 60°F in January, February, March, November, and December.

Chapter 3 Cumulative Review

1. a. $2 < 3$

b. $7 > 4$

c. $72 > 27$

2. $\frac{56}{64} = \frac{7 \cdot 8}{8 \cdot 8} = \frac{7}{8}$

3. $\frac{2}{15} \cdot \frac{5}{13} = \frac{2 \cdot 5}{3 \cdot 5 \cdot 13} = \frac{2}{39}$

4. $\frac{10}{3}+\frac{5}{21}=\frac{10\cdot 7}{3\cdot 7}+\frac{5}{21}$
$=\frac{70+5}{21}$
$=\frac{75}{21}$
$=\frac{3\cdot 25}{3\cdot 7}$
$=\frac{25}{7}$
$=3\frac{4}{7}$

5. $\frac{3+|4-3|+2^2}{6-3}=\frac{3+|1|+2^2}{6-3}=\frac{3+1+4}{6-3}=\frac{8}{3}$

6. $16-3\cdot 3+2^4=16-3\cdot 3+16$
$=16-9+16$
$=23$

7. a. $-8+(-11)=-19$

b. $-5+35=30$

c. $0.6+(-1.1)=-0.5$

d. $-\frac{7}{10}+\left(-\frac{1}{10}\right)=-\frac{8}{10}=-\frac{4}{5}$

e. $11.4+(-4.7)=6.7$

f. $-\frac{3}{8}+\frac{2}{5}=-\frac{3\cdot 5}{8\cdot 5}+\frac{2\cdot 8}{5\cdot 8}=\frac{-15+16}{40}=\frac{1}{40}$

8. $|9+(-20)|+|-10|=|-11|+|-10|=11+10=21$

9. a. $-14-8+10-(-6)=-14+(-8)+10+6$
$=-6$

b. $1.6-(-10.3)+(-5.6)=1.6+10.3+(-5.6)$
$=6.3$

10. $-9-(3-8)=-9-(-5)=-9+5=-4$

11. Let $x=-2$ and $y=-4$.

a. $5x-y=5(-2)-(-4)=-10+4=-6$

b. $x^4-y^2=(-2)^4-(-4)^2=16-16=0$

c. $\frac{3x}{2y}=\frac{3(-2)}{2(-4)}=\frac{-6}{-8}=\frac{3}{4}$

12. $\frac{x}{-10}=2$
Let $x=-20$.
$\frac{-20}{-10}\stackrel{?}{=}2$
$2=2$ True
-20 is a solution to the equation.

13. a. $10+(x+12)=10+x+12=x+22$

b. $-3(7x)=-21x$

14. $(12+x)-(4x-7)=12+x-4x+7=19-3x$

15. a. $-3y$: -3

b. $22z^4$: 22

c. $y=1y$: 1

d. $-x=-1x$: -1

e. $\frac{x}{7}=\frac{1}{7}x$: $\frac{1}{7}$

16. $-5(x-7)=-5x-(-5)(7)=-5x+35$

17. $x-7=10$
$x-7+7=10+7$
$x=17$

18. $5(3+z)-(8z+9)=-4$
$15+5z-8z-9=-4$
$-3z+6=-4$
$-3z=-10$
$z=\frac{10}{3}$

19. $12a-8a=10+2a-13-7$
$4a=2a-10$
$4a-2a=2a-2a-10$
$2a=-10$
$\frac{2a}{2}=\frac{-10}{2}$
$a=-5$

20.
$$\frac{x}{4}-1=-7$$
$$4\left(\frac{x}{4}\right)-4(1)=4(-7)$$
$$x-4=-28$$
$$x=-24$$

21. Sum
= first integer + second integer + third integer
$$\text{Sum}=x+(x+1)+(x+2)$$
$$=x+x+1+x+2$$
$$=3x+3$$

22.
$$\frac{x}{3}-2=\frac{x}{3}$$
$$3\left(\frac{x}{3}\right)-3(2)=3\left(\frac{x}{3}\right)$$
$$x-6=x$$
$$-6=0$$
This is false. There is no solution.

23.
$$\frac{2(a+3)}{3}=6a+2$$
$$2(a+3)=3(6a+2)$$
$$2a+6=18a+6$$
$$-16a+6=6$$
$$-16a=0$$
$$a=0$$

24.
$$x+2y=6$$
$$x-x+2y=6-x$$
$$2y=6-x$$
$$\frac{2y}{2}=\frac{6-x}{2}$$
$$y=\frac{6-x}{2}$$

25. Let x = the number of Republican representatives and $x + 31$ = the number of Democratic representatives.
$$x+x+31=435$$
$$2x+31=435$$
$$2x=404$$
$$x=202$$
$$x+31=233$$
There were 202 Republican representatives and 233 Democratic.

26.
$$5(x+4)\geq 4(2x+3)$$
$$5x+20\geq 8x+12$$
$$-3x+20\geq 12$$
$$-3x\geq -8$$
$$\frac{-3x}{-3}\leq\frac{-8}{-3}$$
$$x\leq\frac{8}{3}, \left(-\infty, \frac{8}{3}\right]$$

27. The perimeter of a rectangle is given by the formula $P = 2l + 2w$. Let l = the length of the garden.
$$P=2l+2w$$
$$140=2l+2w$$
$$140=2l+2(30)$$
$$140=2l+60$$
$$80=2l$$
$$40=l$$
The length of the garden is 40 feet.

28.
$$-3<4x-1\leq 2$$
$$-2<4x\leq 3$$
$$-\frac{1}{2}<x\leq\frac{3}{4}, \left(-\frac{1}{2}, \frac{3}{4}\right]$$

29.
$$y=mx+b$$
$$y-b=mx+b-b$$
$$y-b=mx$$
$$\frac{y-b}{m}=\frac{mx}{m}$$
$$\frac{y-b}{m}=x$$

30. $y=-5x$

x	y
0	0
−1	5
2	−10

31. Let x = the amount of 70% acid.
No. of liters · Strength = Amt of Acid

70%	x	0.7	$0.7x$
40%	$12-x$	0.4	$0.4(12-x)$
50%	12	0.5	$0.5(12)$

$$0.7x + 0.4(12 - x) = 0.5(12)$$
$$0.7x + 4.8 - 0.4x = 6$$
$$0.3x + 4.8 = 6$$
$$0.3x = 1.2$$
$$x = 4$$

$12 - x = 12 - 4 = 8$

Mix 4 liters of 70% acid with 8 liters of 40% acid.

32. $y = -3x + 5$

x	y
−1	8
0	5
1	2

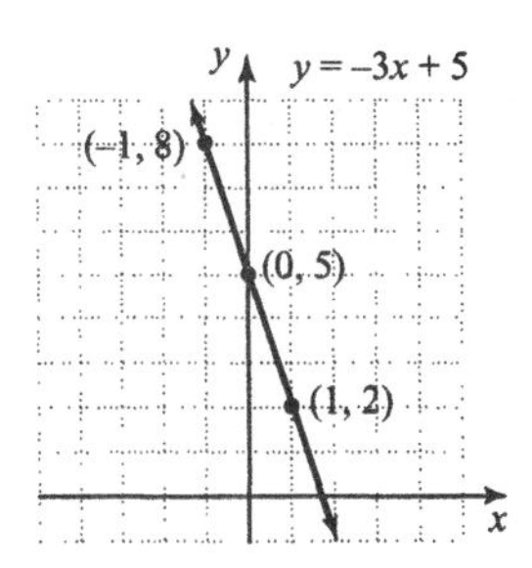

33. $x \geq -1$, $[-1, \infty)$

−1

34. $2x + 4y = -8$

x-intercept, $y = 0$

$2x + 4(0) = -8 \Rightarrow x = -4$: $(-4, 0)$

y-intercept, $x = 0$

$2(0) + 4y = -8 \Rightarrow y = -2$: $(0, -2)$

35.
$$-1 \leq 2x - 3 < 5$$
$$2 \leq 2x < 8$$
$$1 \leq x < 4,\ [1, 4)$$

1 4

36. $x = 2$

$x = 2$ for all values of y.

37. **a.** $x - 2y = 6$

$(6, 0)$

$(6) - 2(0) \stackrel{?}{=} 6$

$6 = 6$ Yes

b. $x - 2y = 6$

$(0, 3)$

$(0) - 2(3) \stackrel{?}{=} 6$

$-6 \neq 6$ No

c. $x - 2y = 6$

$\left(1, -\frac{5}{2}\right)$

$(1) - 2\left(-\frac{5}{2}\right) \stackrel{?}{=} 6$

$1 + 5 \stackrel{?}{=} 6$

$6 = 6$ Yes

38. $(0, 5)$ and $(-5, 4)$

$$m = \frac{y_2 - y_1}{x_2 - x_1} = \frac{4 - 5}{-5 - 0} = \frac{-1}{-5} = \frac{1}{5}$$

39. **a.** linear; because it can be written in the form $Ax + By = C$.

b. linear; because it can be written in the form $Ax + By = C$.

c. not linear; because y is squared.

d. linear; because it can be written in the form $Ax + By = C$.

40. $x = -10$ is a vertical line. The slope is undefined.

41. $y = -1$ is horizontal, slope is 0.

42.
$$2x - 5y = 10$$
$$-5y = -2x + 10$$
$$y = \frac{2}{5}x - 2$$

$y = mx + b$

$m = \frac{2}{5}$, $b = -2$

The slope is $\frac{2}{5}$.

The y-intercept is $(0, -2)$.

43. $m = \frac{1}{4};\ b = -3$

$y = mx + b$

$y = \frac{1}{4}x + (-3)$

$y = \frac{1}{4}x - 3$

44. (2, 3) and (0, 0)

$m = \frac{y_2 - y_1}{x_2 - x_1} = \frac{0-3}{0-2} = \frac{-3}{-2} = \frac{3}{2}$

Point: (0, 0)

$y - y_1 = m(x - x_1)$

$y - 0 = \frac{3}{2}(x - 0)$

$2y = 3x$

$3x - 2y = 0$

Chapter 4

Section 4.1

Practice Exercises

1. $\begin{cases} 4x - y = 2 \\ y = 3x \end{cases}$

   (4, 12)
   $4(4) - 12 \stackrel{?}{=} 2$
   $16 - 12 \stackrel{?}{=} 2$
   $4 = 2$ False
   (4, 12) is not a solution of the system.

2. $\begin{cases} x - 3y = -7 \\ 2x + 9y = 1 \end{cases}$

   (−4, 1)
   $-4 - 3(1) \stackrel{?}{=} -7$ $\quad$ $2(-4) + 9(1) \stackrel{?}{=} 1$
   $-4 - 3 \stackrel{?}{=} -7$ $\quad$ $-8 + 9 \stackrel{?}{=} 1$
   $-7 = -7$ True $\quad$ $1 = 1$ True
   (−4, 1) is a solution of the system.

3. $\begin{cases} x - y = 3 \\ x + 2y = 18 \end{cases}$

$x - y = 3$

x	y
−4	−7
0	−3
4	1

$x + 2y = 18$

x	y
−4	11
0	9
4	7

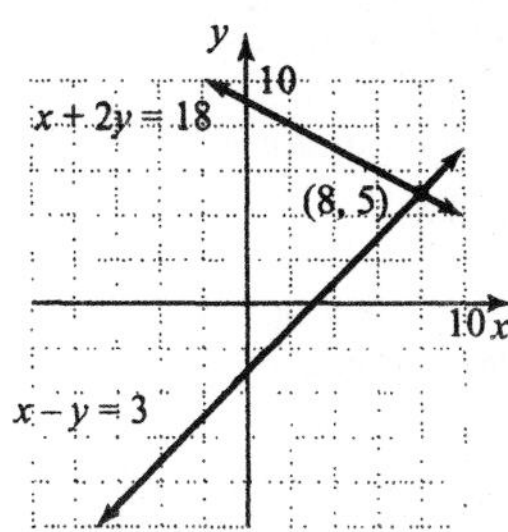

The two lines appear to intersect at (8, 5).

$x - y = 3$ $\quad$ $x + 2y = 18$
$8 - 5 \stackrel{?}{=} 3$ $\quad$ $8 + 2(5) \stackrel{?}{=} 18$
$3 = 3$ True $\quad$ $8 + 10 \stackrel{?}{=} 18$
$18 = 18$ True
(8, 5) is the solution of the system.

4. $\begin{cases} -4x + 3y = -3 \\ y = -5 \end{cases}$

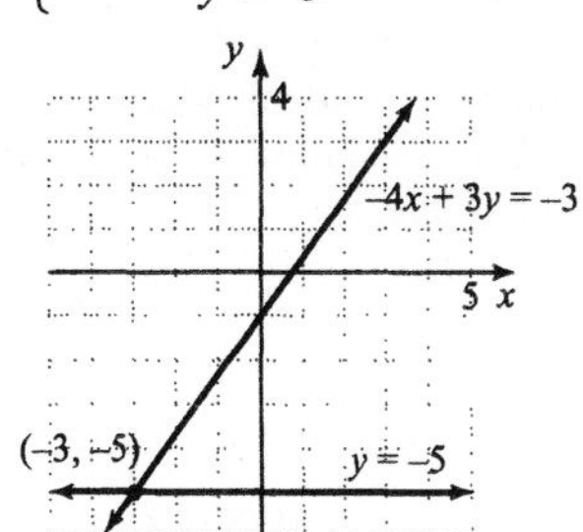

The two lines appear to intersect at (−3, −5).
Check.

$-4x + 3y = -3$ $\quad$ $y = -5$
$-4(-3) + 3(-5) \stackrel{?}{=} -3$ $\quad$ $-5 = -5$ True
$12 - 15 \stackrel{?}{=} -3$
$-3 = -3$ True
(−3, −5) is the solution of the system.

5. $\begin{cases} 3y = 9x \\ 6x - 2y = 12 \end{cases}$

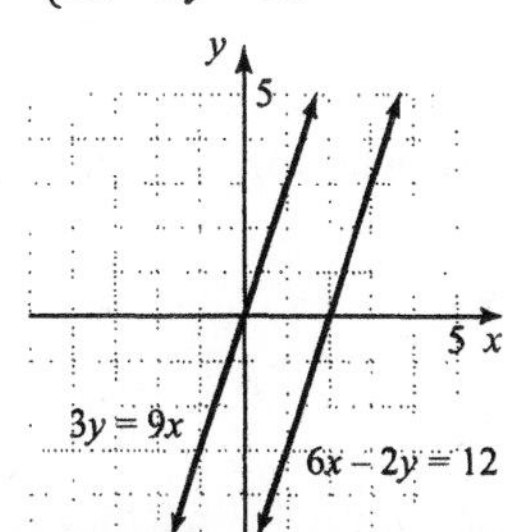

The lines appear to be parallel. To confirm this, write both equations in slope-intercept form.

$3y = 9x$ $\quad$ $6x - 2y = 12$
$y = 3x$ $\quad$ $-2y = -6x + 12$
$y = 3x - 6$

The slopes are the same, so the lines are parallel. Thus, there is no solution of the system and the system is inconsistent.

6. $\begin{cases} x - y = 4 \\ -2x + 2y = -8 \end{cases}$

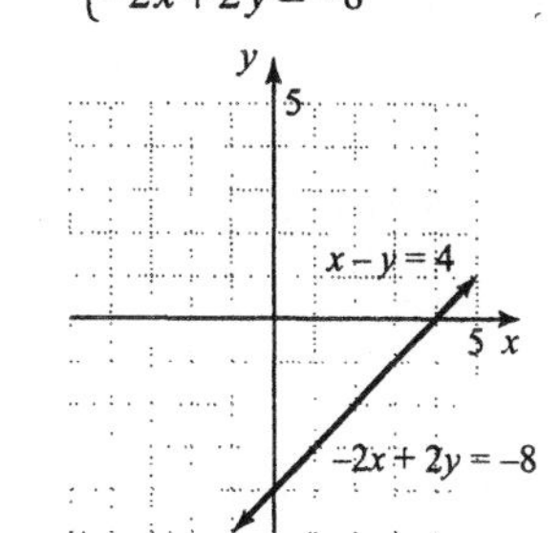

The graphs appear to be identical. To confirm this, write both equations in slope-intercept form.

$$\begin{aligned} x-y&=4 \\ -y&=-x+4 \\ y&=x-4 \end{aligned} \qquad \begin{aligned} -2x+2y&=-8 \\ -x+y&=-4 \\ y&=x-4 \end{aligned}$$

The equations are identical. Thus, there is an infinite number of solutions of the system; the system is consistent; the equations are dependent.

7. $\begin{cases} 5x+4y=6 \\ x-y=3 \end{cases}$

Write each equation in slope-intercept form.

$$\begin{aligned} 5x+4y&=6 \\ 4y&=-5x+6 \\ y&=-\frac{5}{4}x+\frac{3}{2} \end{aligned} \qquad \begin{aligned} x-y&=3 \\ -y&=-x+3 \\ y&=x-3 \end{aligned}$$

The slopes are not equal, so the two lines are neither parallel nor identical and must intersect. Therefore, this system has one solution and is consistent.

8. $\begin{cases} -\frac{2}{3}x+y=6 \\ 3y=2x+5 \end{cases}$

Write each equation in slope-intercept form.

$$\begin{aligned} -\frac{2}{3}x+y&=6 \\ y&=\frac{2}{3}x+6 \end{aligned} \qquad \begin{aligned} 3y&=2x+5 \\ y&=\frac{2}{3}x+\frac{5}{3} \end{aligned}$$

The slope of each line is $\frac{2}{3}$, but they have different y-intercepts. Therefore, the lines are parallel. The system has no solution and is inconsistent.

Calculator Explorations

1. $\begin{cases} y=-2.68x+1.21 \\ y=5.22x-1.68 \end{cases}$

The approximate point of intersection is (0.37, 0.23).

2. $\begin{cases} y=4.25x+3.89 \\ y=-1.88x+3.21 \end{cases}$

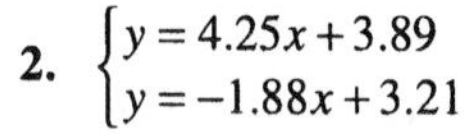

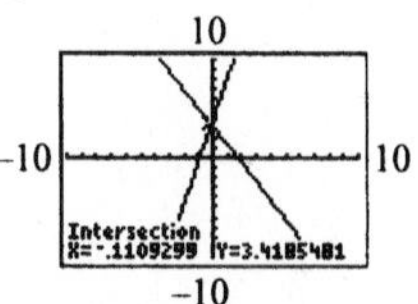

The approximate point of intersection is (−0.11, 3.42).

3. $\begin{cases} 4.3x-2.9y=5.6 \\ 8.1x+7.6y=-14.1 \end{cases}$

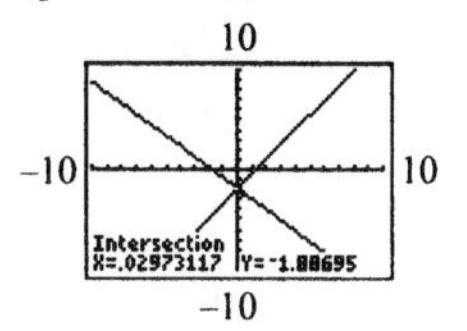

The approximate point of intersection is (0.03, −1.89).

4. $\begin{cases} -3.6x-8.6y=10 \\ -4.5x+9.6y=-7.7 \end{cases}$

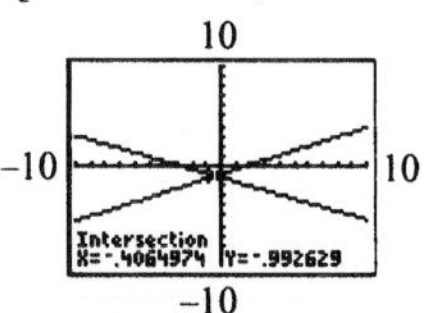

The approximate point of intersection is (−0.41, −0.99).

Vocabulary and Readiness Check

1. In a system of linear equations in two variables, if the graphs of the equations are the same, the equations are <u>dependent</u> equations.

2. Two or more linear equations are called a <u>system of linear equations</u>.

3. A system of equations that has at least one solution is called a <u>consistent</u> system.

4. A <u>solution</u> of a system of two equations in two variables is an ordered pair of numbers that is a solution of both equations in the system.

5. A system of equations that has no solution is called an <u>inconsistent</u> system.

6. In a system of linear equations in two variables, if the graphs of the equations are different, the equations are <u>independent</u> equations.

7. One solution, (–1, 3)

8. No solution

9. Infinite number of solutions

10. One solution, (3, 4)

Exercise Set 4.1

1. a. Let $x = 2$ and $y = 4$.

$$\begin{array}{ll} x+y=8 & 3x+2y=21 \\ 2+4 \stackrel{?}{=} 8 & 3(2)+2(4) \stackrel{?}{=} 21 \\ 6=8 \text{ False} & 6+8 \stackrel{?}{=} 21 \\ & 14=21 \text{ False} \end{array}$$

(2, 4) is not a solution of the system.

b. Let $x = 5$ and $y = 3$.

$$\begin{array}{ll} x+y=8 & 3x+2y=21 \\ 5+3 \stackrel{?}{=} 8 & 3(5)+2(3) \stackrel{?}{=} 21 \\ 8=8 \text{ True} & 15+6 \stackrel{?}{=} 21 \\ & 21=21 \text{ True} \end{array}$$

(5, 3) is a solution of the system.

3. a. Let $x = 3$ and $y = 4$.

$$\begin{array}{ll} 3x-y=5 & x+2y=11 \\ 3(3)-4 \stackrel{?}{=} 5 & 3+2(4) \stackrel{?}{=} 11 \\ 9-4 \stackrel{?}{=} 5 & 3+8 \stackrel{?}{=} 11 \\ 5=5 \text{ True} & 11=11 \text{ True} \end{array}$$

(3, 4) is a solution of the system.

b. Let $x = 0$ and $y = -5$.

$$\begin{array}{ll} 3x-y=5 & x+2y=11 \\ 3(0)-(-5) \stackrel{?}{=} 5 & 0+2(-5) \stackrel{?}{=} 11 \\ 0+5 \stackrel{?}{=} 5 & 0-10 \stackrel{?}{=} 11 \\ 5=5 \text{ True} & -10=11 \text{ False} \end{array}$$

(0, –5) is not a solution of the system.

5. a. Let $x = -3$ and $y = -3$.

$$\begin{array}{ll} 2y=4x+6 & 2x-y=-3 \\ 2(-3) \stackrel{?}{=} 4(-3)+6 & 2(-3)-(-3) \stackrel{?}{=} -3 \\ -6 \stackrel{?}{=} -12+6 & -6+3 \stackrel{?}{=} -3 \\ -6=-6 \text{ True} & -3=-3 \text{ True} \end{array}$$

(–3, –3) is a solution of the system.

b. Let $x = 0$ and $y = 3$.

$$\begin{array}{ll} 2y=4x+6 & 2x-y=-3 \\ 2(3) \stackrel{?}{=} 4(0)+6 & 2(0)-3 \stackrel{?}{=} -3 \\ 6 \stackrel{?}{=} 0+6 & 0-3 \stackrel{?}{=} -3 \\ 6=6 \text{ True} & -3=-3 \text{ True} \end{array}$$

(0, 3) is a solution of the system.

7. a. Let $x = -2$ and $y = 0$.

$$\begin{array}{ll} -2=x-7y & 6x-y=13 \\ -2 \stackrel{?}{=} -2-7(0) & 6(-2)-0 \stackrel{?}{=} 13 \\ -2=-2 \text{ True} & -12=13 \text{ False} \end{array}$$

(–2, 0) is not a solution of the system.

b. Let $x = \frac{1}{2}$ and $y = \frac{5}{14}$.

$$\begin{array}{ll} -2=x-7y & 6x-y=13 \\ -2 \stackrel{?}{=} \frac{1}{2}-7\left(\frac{5}{14}\right) & 6\left(\frac{1}{2}\right)-\left(\frac{5}{14}\right) \stackrel{?}{=} 13 \\ 2 \stackrel{?}{=} \frac{1}{2}-\frac{5}{2}=-\frac{4}{2} & 3-\frac{5}{14} \stackrel{?}{=} 13 \\ -2=-2 \text{ True} & \frac{37}{14}=13 \text{ False} \end{array}$$

$\left(\frac{1}{2}, \frac{5}{14}\right)$ is not a solution of the system.

9. $\begin{cases} x+y=4 \\ x-y=2 \end{cases}$

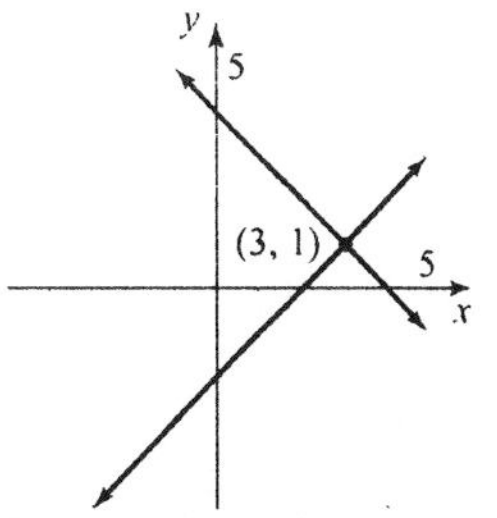

The solution of the system is (3, 1), consistent and independent.

11. $\begin{cases} x+y=6 \\ -x+y=-6 \end{cases}$

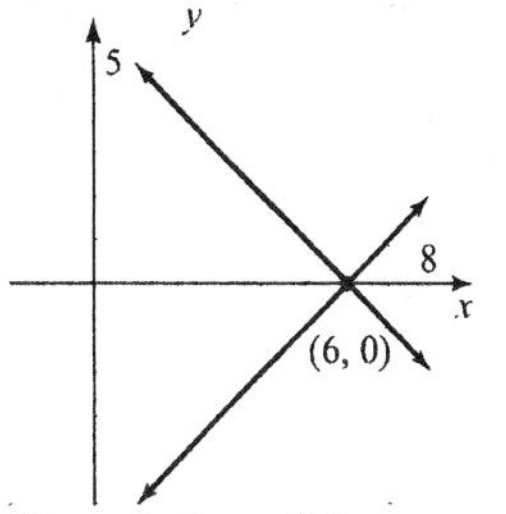

The solution of the system is (6, 0), consistent and independent.

13. $\begin{cases} y = 2x \\ 3x - y = -2 \end{cases}$

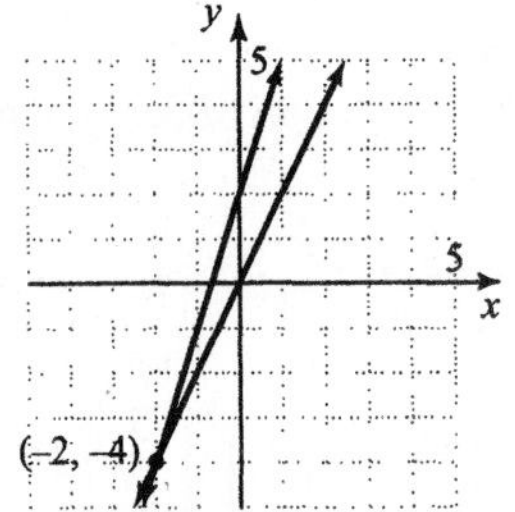

The solution of the system is (−2, −4), consistent and independent.

15. $\begin{cases} y = x + 1 \\ y = 2x - 1 \end{cases}$

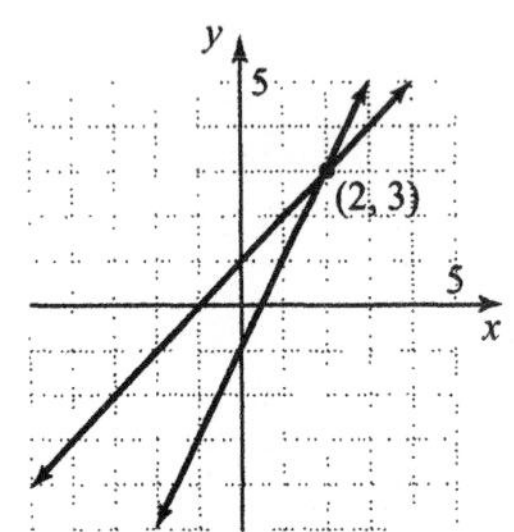

The solution of the system is (2, 3), consistent and independent.

17. $\begin{cases} 2x + y = 0 \\ 3x + y = 1 \end{cases}$

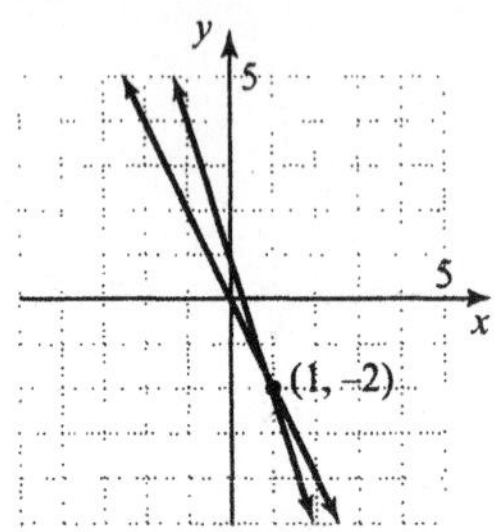

The solution of the system is (1, −2), consistent and independent.

19. $\begin{cases} y = -x - 1 \\ y = 2x + 5 \end{cases}$

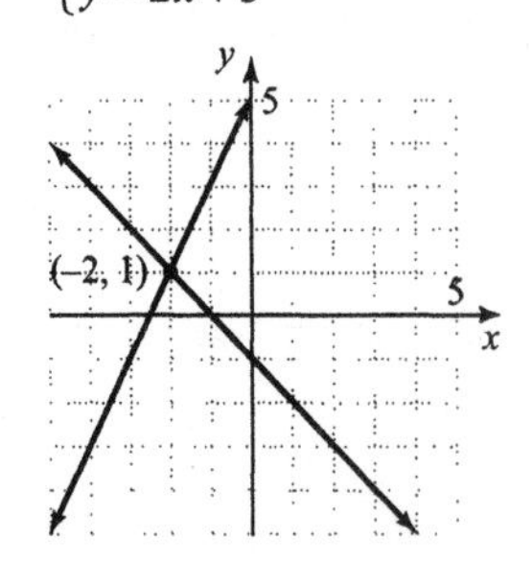

The solution of the system is (−2, 1), consistent and independent.

21. $\begin{cases} x + y = 5 \\ x + y = 6 \end{cases}$

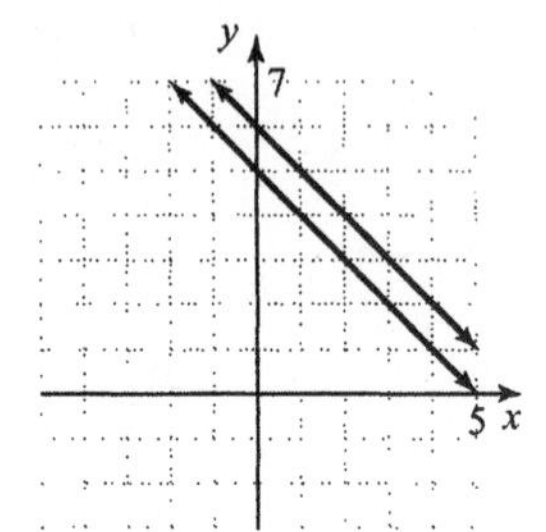

There is no solution, inconsistent and independent.

23. $\begin{cases} 2x - y = 6 \\ y = 2 \end{cases}$

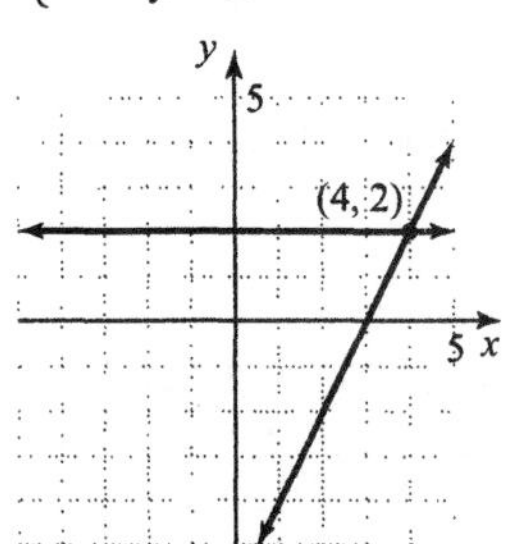

The solution of the system is (4, 2), consistent and independent.

25. $\begin{cases} x - 2y = 2 \\ 3x + 2y = -2 \end{cases}$

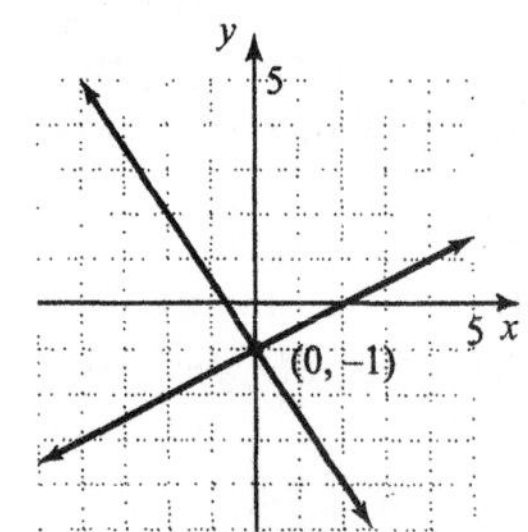

The solution of the system is (0, −1), consistent and independent.

27. $\begin{cases} 2x + y = 4 \\ 6x = -3y + 6 \end{cases}$

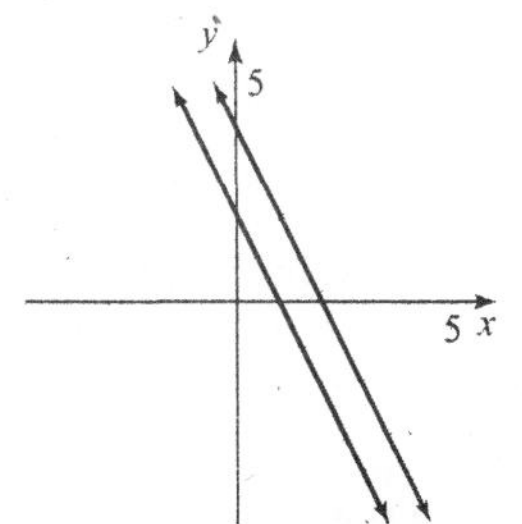

There is no solution, inconsistent and independent.

29. $\begin{cases} y - 3x = -2 \\ 6x - 2y = 4 \end{cases}$

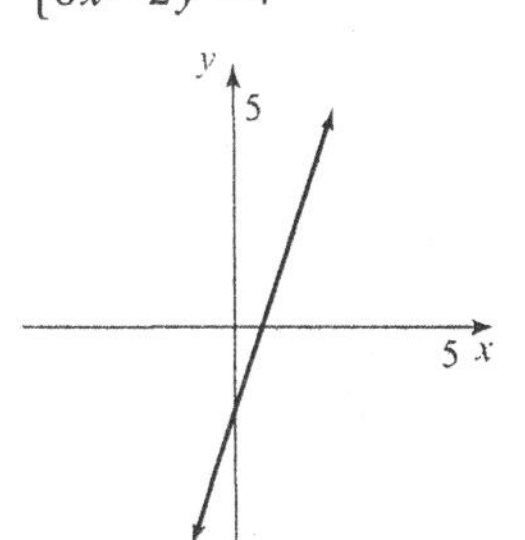

There is an infinite number of solutions, consistent and dependent.

31. $\begin{cases} x = 3 \\ y = -1 \end{cases}$

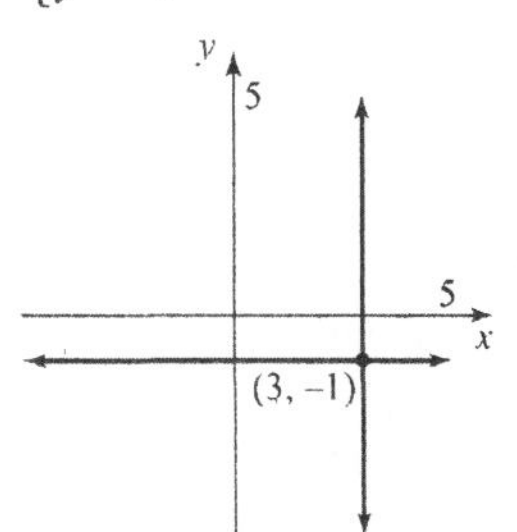

The solution of the system is (3, –1), consistent and independent.

33. $\begin{cases} y = x - 2 \\ y = 2x + 3 \end{cases}$

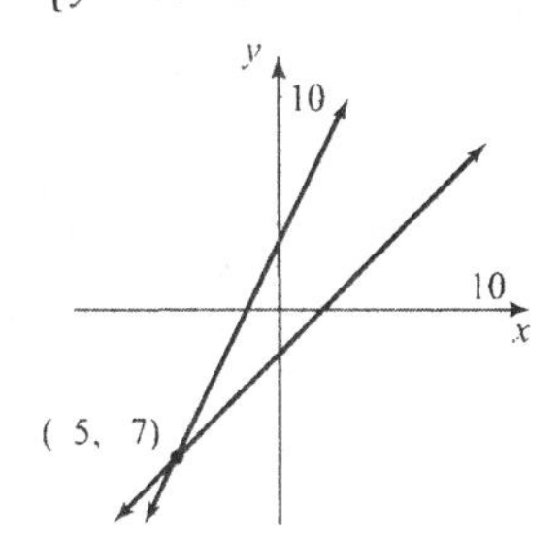

The solution of the system is (–5, –7), consistent and independent.

35. $\begin{cases} 2x - 3y = -2 \\ -3x + 5y = 5 \end{cases}$

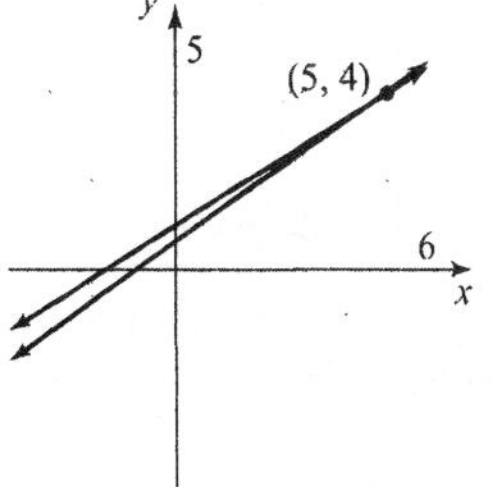

The solution of the system is (5, 4), consistent and independent.

37. $\begin{cases} 6x - y = 4 \\ \frac{1}{2}y = -2 + 3x \end{cases}$

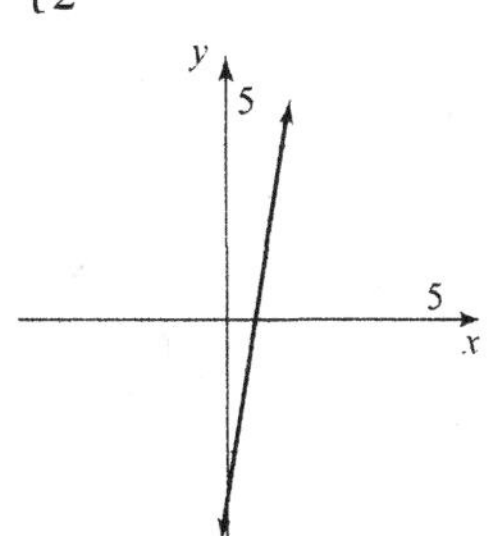

There is an infinite number of solutions, consistent and dependent.

39. $\begin{cases} 4x + y = 24 \\ x + 2y = 2 \end{cases} \rightarrow \begin{cases} y = -4x + 24 \\ y = -\frac{1}{2}x + 1 \end{cases}$

The lines are intersecting; there is one solution.

41. $\begin{cases} 2x + y = 0 \\ 2y = 6 - 4x \end{cases} \rightarrow \begin{cases} y = -2x \\ y = -2x + 3 \end{cases}$

The lines are parallel; there is no solution.

43. $\begin{cases} 6x - y = 4 \\ \frac{1}{2}y = -2 + 3x \end{cases} \rightarrow \begin{cases} y = 6x - 4 \\ y = 6x - 4 \end{cases}$

The lines are identical; there is an infinite number of solutions.

45. $\begin{cases} x = 5 \\ y = -2 \end{cases}$

The lines are intersecting; there is one solution.

47. $\begin{cases} 3y-2x=3 \\ x+2y=9 \end{cases} \rightarrow \begin{cases} y=\frac{2}{3}x+1 \\ y=-\frac{1}{2}x+\frac{9}{2} \end{cases}$

The lines are intersecting; there is one solution.

49. $\begin{cases} 6y+4x=6 \\ 3y-3=-2x \end{cases} \rightarrow \begin{cases} y=-\frac{2}{3}x+1 \\ y=-\frac{2}{3}x+1 \end{cases}$

The lines are identical; there is an infinite number of solutions.

51. $\begin{cases} x+y=4 \\ x+y=3 \end{cases} \rightarrow \begin{cases} y=-x+4 \\ y=-x+3 \end{cases}$

The lines are parallel; there is no solution.

53. $5(x-3)+3x=1$
$5x-15+3x=1$
$8x-15=1$
$8x=16$
$x=2$

The solution is 2.

55. $4\left(\frac{y+1}{2}\right)+3y=0$
$2(y+1)+3y=0$
$2y+2+3y=0$
$5y+2=0$
$5y=-2$
$y=-\frac{2}{5}$

The solution is $-\frac{2}{5}$.

57. $8a-2(3a-1)=6$
$8a-6a+2=6$
$2a+2=6$
$2a=4$
$a=2$

The solution is 2.

59. Answers may vary

61. Answers may vary

63. Answers may vary

65. The graph for fish is above the graph for shellfish for the years 2000, 2001, and 2002.

67. The graph for Toyota is lower than the graph for GM for the years 2001, 2002, and 2003.

69. Answers may vary

71. Answers may vary

73. a. Each table includes the point (4, 9). Therefore (4, 9) is a solution of the system.

b.

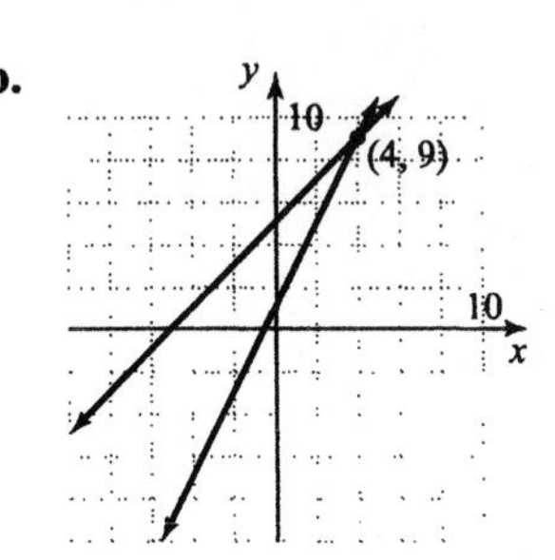

c. Yes

75. Answers may vary

Section 4.2

Practice Exercises

1. $\begin{cases} 2x-y=9 \\ x=y+1 \end{cases}$

Substitute $y + 1$ for x in the first equation.
$2x-y=9$
$2(y+1)-y=9$
$2y+2-y=9$
$y+2=9$
$y=7$

Let $y = 7$ in the second equation.
$x = y + 1 = 7 + 1 = 8$
The solution of the system is (8, 7).
Check.

$2x-y=9$
$2(8)-7 \stackrel{?}{=} 9$
$16-7 \stackrel{?}{=} 9$
$9=9$ True

$x=y+1$
$8 \stackrel{?}{=} 7+1$
$8=8$ True

The solution of the system is (8, 7).

2. $\begin{cases} 7x-y=-15 \\ y=2x \end{cases}$

Substitute $2x$ for y in the first equation.
$7x-y=-15$
$7x-2x=-15$
$5x=-15$
$x=-3$

Let $x = -3$ in the second equation.
$y = 2x = 2(-3) = -6$
The solution of the system is $(-3, -6)$.

3. $\begin{cases} x+3y=6 \\ 2x+3y=10 \end{cases}$

Solve the first equation for x.

$$\begin{aligned} x+3y&=6 \\ x&=-3y+6 \end{aligned}$$

Substitute $-3y + 6$ for x in the second equation.

$$\begin{aligned} 2x+3y&=10 \\ 2(-3y+6)+3y&=10 \\ -6y+12+3y&=10 \\ -3y+12&=10 \\ -3y&=-2 \\ y&=\frac{2}{3} \end{aligned}$$

Let $y=\frac{2}{3}$ in the equation for x.

$$x=3y+6=-3\left(\frac{2}{3}\right)+6=-2+6=4$$

The solution of the system is $\left(4, \frac{2}{3}\right)$.

4. $\begin{cases} 5x+3y=-9 \\ -2x+y=8 \end{cases}$

Solve the second equation for y.

$$\begin{aligned} -2x+y&=8 \\ y&=2x+8 \end{aligned}$$

Substitute $2x + 8$ for y in the first equation.

$$\begin{aligned} 5x+3y&=-9 \\ 5x+3(2x+8)&=-9 \\ 5x+6x+24&=-9 \\ 11x+24&=-9 \\ 11x&=-33 \\ x&=-3 \end{aligned}$$

Let $x = -3$ in the equation for y.
$y = 2x + 8 = 2(-3) + 8 = -6 + 8 = 2$
The solution of the system is $(-3, 2)$.

5. $\begin{cases} \frac{1}{4}x-y=2 \\ x=4y+8 \end{cases}$

Substitute $4y + 8$ for x in the first equation.

$$\begin{aligned} \frac{1}{4}x-y&=2 \\ \frac{1}{4}(4y+8)-y&=2 \\ y+2-y&=2 \\ 2&=2 \end{aligned}$$

The two linear equations are equivalent. Thus, the system has an infinite number of solutions.

6. $\begin{cases} 4x-3y=12 \\ -8x+6y=-30 \end{cases}$

Solve the first equation for x.

$$\begin{aligned} 4x-3y&=12 \\ 4x&=3y+12 \\ x&=\frac{3}{4}y+3 \end{aligned}$$

Substitute $\frac{3}{4}y+3$ for x in the second equation.

$$\begin{aligned} -8x+6y&=-30 \\ -8\left(\frac{3}{4}y+3\right)+6y&=-30 \\ -6y-24+6y&=-30 \\ -24&=-30 \end{aligned}$$

The false statement $-24 = -30$ indicates that the system has no solution and is inconsistent.

Vocabulary and Readiness Check

1. Since $x = 1$, $y = 4x = 4(1) = 4$ and the solution is $(1, 4)$.

2. There is no solution, since $0 = 34$ is a false statement.

3. There is an infinite number of solutions, since the statement $0 = 0$ is true for all values of the variables.

4. Since $y = 0$, $x = y + 5 = 0 + 5 = 5$ and the solution is $(5, 0)$.

5. Since $x = 0$ and $x + y = 0$, $y = -x = -0 = 0$ and the solution is $(0, 0)$.

6. There is an infinite number of solutions, since the statement $0 = 0$ is true for all values of the variables.

Exercise Set 4.2

1. $\begin{cases} x+y=3 \\ x=2y \end{cases}$

Substitute 2y for x in the first equation.

$2y+y=3$
$3y=3$
$y=1$

Let y = 1 in the second equation.

$x=2(1)=2$

The solution is (2, 1).

3. $\begin{cases} x+y=6 \\ y=-3x \end{cases}$

Substitute $-3x$ for y in the first equation.

$x+(-3x)=6$
$-2x=6$
$x=-3$

Let $x=-3$ in the second equation.

$y=-3(-3)=9$

The solution is (−3, 9).

5. $\begin{cases} y=3x+1 \\ 4y-8x=12 \end{cases}$

Substitute $3x+1$ for y in the second equation.

$4(3x+1)-8x=12$
$12x+4-8x=12$
$4x+4=12$
$4x=8$
$x=2$

Let $x=2$ in the first equation.

$y=3(2)+1=7$

The solution is (2, 7).

7. $\begin{cases} y=2x+9 \\ y=7x+10 \end{cases}$

Substitute $2x+9$ for y in the second equation.

$2x+9=7x+10$
$-5x+9=10$
$-5x=1$
$x=-\frac{1}{5}$

Let $x=-\frac{1}{5}$ in the first equation.

$y=2\left(-\frac{1}{5}\right)+9=-\frac{2}{5}+\frac{45}{5}=\frac{43}{5}$

The solution is $\left(-\frac{1}{5}, \frac{43}{5}\right)$.

9. $\begin{cases} 3x-4y=10 \\ y=x-3 \end{cases}$

Substitute $x-3$ for y in the first equation.

$3x-4(x-3)=10$
$3x-4x+12=10$
$-x=-2$
$x=2$

Let $x=2$ in the second equation

$y=2-3=-1$

The solution is (2, −1).

11. $\begin{cases} x+2y=6 \\ 2x+3y=8 \end{cases}$

Solve the first equation for x.

$x=6-2y$

Substitute $6-2y$ for x in the second equation.

$2(6-2y)+3y=8$
$12-4y+3y=8$
$-y=-4$
$y=4$

Let $y=4$ in $x=6-2y$.

$x=6-2(4)=-2$

The solution is (−2, 4).

13. $\begin{cases} 3x+2y=16 \\ x=3y-2 \end{cases}$

Substitute $3y-2$ for x in the first equation.

$3(3y-2)+2y=16$
$9y-6+2y=16$
$11y=22$
$y=2$

Let $y=2$ in the second equation.

$x=3(2)-2=4$

The solution is (4, 2).

15. $\begin{cases} 2x-5y=1 \\ 3x+y=-7 \end{cases}$

Solve the second equation for y.

$y=-7-3x$

Substitute $-7-3x$ for y in the first equation.

$2x-5(-7-3x)=1$
$2x+35+15x=1$
$17x=-34$
$x=-2$

Let $x=-2$ in $y=-7-3x$.

$y=-7-3(-2)=-1$

The solution is (−2, −1).

17. $\begin{cases} 4x+2y=5 \\ -2x=y+4 \end{cases}$

Solve the second equation for y.

$y=-2x-4$

Substitute $-2x-4$ for y in the first equation.

$$4x+2(-2x-4)=5$$
$$4x-4x-8=5$$
$$-8=5 \text{ False}$$

The system has no solution.

19. $\begin{cases} 4x+y=11 \\ 2x+5y=1 \end{cases}$

Solve the first equation for y.

$y=11-4x$

Substitute $11-4x$ for y in the second equation.

$$2x+5(11-4x)=1$$
$$2x+55-20x=1$$
$$-18x=-54$$
$$x=3$$

Let $x=3$ in $y=11-4x$.

$y=11-4(3)=-1$

The solution is $(3, -1)$.

21. $\begin{cases} x+2y+5=-4+5y-x \\ 2x+9=3y \\ \\ 2x+x=y+4 \\ -4+3x=y \end{cases}$

Substitute $3x-4$ for y in $2x+9=3y$.

$$2x+9=3(3x-4)$$
$$2x+9=9x-12$$
$$21=7x$$
$$3=x$$

Let $x=3$ in $3x-4=y$.

$y=3(3)-4=5$

The solution of the system is $(3, 5)$.

23. $\begin{cases} 6x-3y=5 \\ x+2y=0 \end{cases}$

Solve the second equation for x.

$x=-2y$

Substitute $-2y$ for x in the first equation.

$$6(-2y)-3y=5$$
$$-12y-3y=5$$
$$-15y=5$$
$$y=-\frac{1}{3}$$

Let $y=-\frac{1}{3}$ in $x=-2y$.

$$x=-2\left(-\frac{1}{3}\right)=\frac{2}{3}$$

The solution is $\left(\frac{2}{3}, -\frac{1}{3}\right)$.

25. $\begin{cases} 3x-y=1 \\ 2x-3y=10 \end{cases}$

Solve the first equation for y.

$y=3x-1$

Substitute $3x-1$ for y in the second equation.

$$2x-3(3x-1)=10$$
$$2x-9x+3=10$$
$$-7x=7$$
$$x=-1$$

Let $x=-1$ in $y=3x-1$.

$y=3(-1)-1=-4$

The solution is $(-1, -4)$.

27. $\begin{cases} -x+2y=10 \\ -2x+3y=18 \end{cases}$

Solve the first equation for x.

$x=2y-10$

Substitute $2y-10$ for x in the second equation.

$$-2(2y-10)+3y=18$$
$$-4y+20+3y=18$$
$$-y=-2$$
$$y=2$$

Let $y=2$ in $x=2y-10$.

$x=2(2)-10=-6$

The solution is $(-6, 2)$.

29. $\begin{cases} 5x+10y=20 \\ 2x+6y=10 \end{cases}$

Solve the first equation for x.

$$x+2y=4$$
$$x=4-2y$$

Substitute $4-2y$ for x in the second equation.

$$2(4-2y)+6y=10$$
$$8-4y+6y=10$$
$$2y=2$$
$$y=1$$

Let $y=1$ in $x=4-2y$.

$x=4-2(1)=2$

The solution is $(2, 1)$.

31. $\begin{cases} 3x+6y=9 \\ 4x+8y=16 \end{cases}$

Solve the first equation for x.

$x+2y=3$

$x=3-2y$

Substitute $3-2y$ for x in the second equation.

$4(3-2y)+8y=16$

$12-8y+8y=16$

$12=16$ False

The system has no solution.

33. $\begin{cases} \frac{1}{3}x-y=2 \\ x-3y=6 \end{cases}$

Solve the second equation for x.

$x=6+3y$

Substitute $6+3y$ for x in the first equation.

$\frac{1}{3}(6+3y)-y=2$

$2+y-y=2$

$2=2$

The equations in the original system are equivalent and there are an infinite number of solutions.

35. $\begin{cases} x=\frac{3}{4}y-1 \\ 8x-5y=-6 \end{cases}$

Substitute $\frac{3}{4}y-1$ for x in the second equation.

$8\left(\frac{3}{4}y-1\right)-5y=-6$

$6y-8-5y=-6$

$y=2$

Let $y=2$ in the first equation.

$x=\frac{3}{4}(2)-1=\frac{1}{2}$

The solution of the system is $\left(\frac{1}{2}, 2\right)$.

37. $\begin{cases} -5y+6y=3x+2(x-5)-3x+5 \\ y=3x+2x-10-3x+5 \\ y=2x-5 \\ \\ 4(x+y)-x+y=-12 \\ 4x+4y-x+y=-12 \\ 3x+5y=-12 \end{cases}$

Substitute $2x-5$ for y in the second equation.

$3x+5(2x-5)=-12$

$3x+10x-25=-12$

$13x=13$

$x=1$

Let $x=1$ in $y=2x-5$.

$y=2(1)-5=-3$

The solution is $(1, -3)$.

39. $3x+2y=6$

$-2(3x+2y)=-2(6)$

$-6x-4y=-12$

41. $-4x+y=3$

$3(-4x+y)=3(3)$

$-12x+3y=9$

43. $\begin{array}{r} 3n+6m \\ 2n-6m \\ \hline 5n \end{array}$

45. $\begin{array}{r} -5a-7b \\ 5a-8b \\ \hline -15b \end{array}$

47. Answers may vary

49. No; answers may vary.

51. c; answers may vary.

53. a. $\begin{cases} y=3.9x+443 \\ y=14.2x+314 \end{cases}$

Substitute $14.2x+314$ for y in the first equation.

$14.2x+314=3.9x+443$

$10.3x=129$

$x\approx 12.52$

Let $x=12.52$ in $y=3.9x+443$.

$y\approx 3.9(12.52)+443\approx 491.828$

The solution is $(13, 492)$.

b. In $1970+13=1983$, the number of men and the number of women receiving bachelor's degrees was the same.

c.

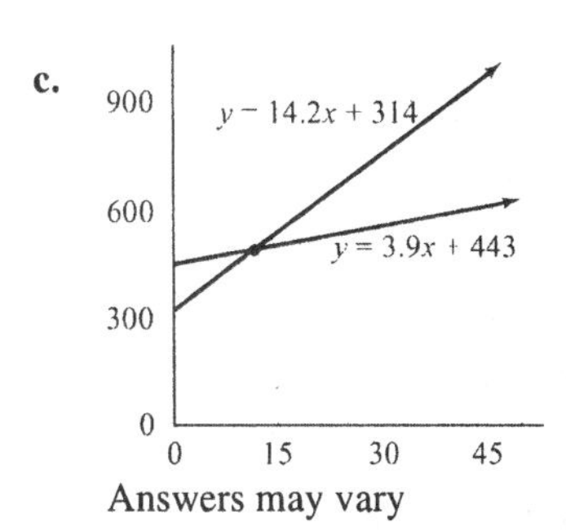

Answers may vary

55. $\begin{cases} y = 5.1x + 14.56 \\ y = -2x - 3.9 \end{cases}$

Substitute $-2x - 3.9$ for y in the first equation.

$-2x - 3.9 = 5.1x + 14.56$
$-7.1x = 18.46$
$x = -2.6$

Let $x = -2.6$ in $y = -2x - 3.9$.

$y = -2(-2.6) - 3.9 = 1.3$

The solution is $(-2.6, 1.3)$.

57. $\begin{cases} 3x + 2y = 14.05 \\ 5x + y = 18.5 \end{cases}$

Solve the second equation for y.

$y = -5x + 18.5$

Substitute $-5x + 18.5$ for y in the first equation.

$3x + 2(-5x + 18.5) = 14.05$
$3x - 10x + 37 = 14.05$
$-7x = -22.95$
$x \approx 3.279$

Let $x = 3.279$ in $y = -5x + 18.5$.

$y \approx -5(3.279) + 18.5 \approx 2.105$

The solution is approximately $(3.28, 2.11)$.

Section 4.3

Practice Exercises

1. $\begin{cases} x - y = 2 \\ x + y = 8 \end{cases}$

Add the left sides of the equations together and the right sides of the equations together.

$x - y = 2$
$\underline{x + y = 8}$
$2x \quad = 10$
$x = 5$

Let $x = 5$ in the first equation.

$x - y = 2$
$5 - y = 2$
$3 = y$

The solution is $(5, 3)$.

Check.

$x - y = 2$ $\qquad$ $x + y = 8$
$5 - 3 \stackrel{?}{=} 2$ $\qquad$ $5 + 3 \stackrel{?}{=} 8$
$2 = 2$ True $\qquad$ $8 = 8$ True

The solution of the system is $(5, 3)$.

2. $\begin{cases} x - 2y = 11 \\ 3x - y = 13 \end{cases}$

Multiply both sides of the first equation by -3 and add to the second equation.

$-3x + 6y = -33$
$\underline{3x - y = 13}$
$5y = -20$
$y = -4$

Let $y = -4$ in the first equation.

$x - 2y = 11$
$x - 2(-4) = 11$
$x + 8 = 11$
$x = 3$

The solution of the system is $(3, -4)$.

3. $\begin{cases} x - 3y = 5 \\ 2x - 6y = -3 \end{cases}$

Multiply both sides of the first equation by -2 and add to the second equation.

$-2x + 6y = -10$
$\underline{2x - 6y = -3}$
$0 = -13$ False

The system has no solution.

4. $\begin{cases} 4x - 3y = 5 \\ -8x + 6y = -10 \end{cases}$

Multiply the first equation by 2 and add to the second equation.

$8x - 6y = 10$
$\underline{-8x + 6y = -10}$
$0 = 0$ True

The equations are equivalent, so the system has an infinite number of solutions.

5. $\begin{cases} 4x + 3y = 14 \\ 3x - 2y = 2 \end{cases}$

Multiply the first equation by 2 and the second equation by 3 and add.

$8x + 6y = 28$
$\underline{9x - 6y = 6}$
$17x \quad = 34$
$x = 2$

Let $x = 2$ in the second equation.

$$\begin{aligned} 3x-2y&=2\\ 3(2)-2y&=2\\ 6-2y&=2\\ -2y&=-4\\ y&=2 \end{aligned}$$

The solution of the system is (2, 2).

6. $\begin{cases} -2x+\dfrac{3y}{2}=5 \\ -\dfrac{x}{2}-\dfrac{y}{4}=\dfrac{1}{2} \end{cases}$

Clear fractions by multiplying the first equation by 2 and the second by 4.

$\begin{cases} -4x+3y=10 \\ -2x-y=2 \end{cases}$

Multiply the second simplified equation by 3 and add.

$$\begin{aligned} -4x+3y&=10\\ -6x-3y&=6\\ \hline -10x\quad&=16 \end{aligned}$$

$$x=-\frac{16}{10}=-\frac{8}{5}$$

Now multiply the second simplified equation by −2 and add.

$$\begin{aligned} -4x+3y&=10\\ 4x+2y&=-4\\ \hline 5y&=6\\ y&=\frac{6}{5} \end{aligned}$$

The solution of the system is $\left(-\frac{8}{5}, \frac{6}{5}\right)$.

Exercise Set 4.3

1. $\begin{cases} 3x+y=5 \\ 6x-y=4 \end{cases}$

$$\begin{aligned} 3x+y&=5\\ 6x-y&=4\\ \hline 9x\quad&=9\\ x&=1 \end{aligned}$$

Let $x = 1$ in the first equation.

$$\begin{aligned} 3(1)+y&=5\\ 3+y&=5\\ y&=2 \end{aligned}$$

The solution of the system is (1, 2).

3. $\begin{cases} x-2y=8 \\ -x+5y=-17 \end{cases}$

$$\begin{aligned} x-2y&=8\\ -x+5y&=-17\\ \hline 3y&=-9\\ y&=-3 \end{aligned}$$

Let $y = -3$ in the first equation.

$$\begin{aligned} x-2(-3)&=8\\ x+6&=8\\ x&=2 \end{aligned}$$

The solution of the system is (2, −3).

5. $\begin{cases} 3x+y=-11 \\ 6x-2y=-2 \end{cases}$

Multiply the first equation by 2.

$$\begin{aligned} 6x+2y&=-22\\ 6x-2y&=-2\\ \hline 12x\quad&=-24\\ x&=-2 \end{aligned}$$

Let $x = -2$ in the first equation.

$$\begin{aligned} 3(-2)+y&=-11\\ -6+y&=-11\\ y&=-5 \end{aligned}$$

The solution of the system is (−2, −5).

7. $\begin{cases} 3x+2y=11 \\ 5x-2y=29 \end{cases}$

$$\begin{aligned} 3x+2y&=11\\ 5x-2y&=29\\ \hline 8x\quad&=40\\ x&=5 \end{aligned}$$

Let $x = 5$ in the first equation.

$$\begin{aligned} 3(5)+2y&=11\\ 15+2y&=11\\ 2y&=-4\\ y&=-2 \end{aligned}$$

The solution of the system is (5, −2).

9. $\begin{cases} x+5y=18 \\ 3x+2y=-11 \end{cases}$

Multiply the first equation by −3.

$$\begin{aligned} -3x-15y&=-54\\ 3x+\ 2y&=-11\\ \hline -13y&=-65\\ y&=5 \end{aligned}$$

Let $y = 5$ in the first equation.

$x+5(5)=18$
$x+25=18$
$x=-7$
The solution of the system is (−7, 5).

11. $\begin{cases} x+y=6 \\ x-y=6 \end{cases}$

$$\begin{aligned} x+y&=6 \\ x-y&=6 \\ \hline 2x \quad &=12 \\ x&=6 \end{aligned}$$

Let $x = 6$ in the first equation.
$6+y=6$
$y=0$
The solution of the system is (6, 0).

13. $\begin{cases} 2x+3y=0 \\ 4x+6y=3 \end{cases}$

Multiply the first equation by −2.

$$\begin{aligned} -4x-6y&=0 \\ 4x+6y&=3 \\ \hline 0&=3 \quad \text{False} \end{aligned}$$

The system has no solution.

15. $\begin{cases} -x+5y=-1 \\ 3x-15y=3 \end{cases}$

Multiply the first equation by 3.

$$\begin{aligned} -3x+15y&=-3 \\ 3x-15y&=3 \\ \hline 0&=0 \end{aligned}$$

There are an infinite number of solutions.

17. $\begin{cases} 3x-2y=7 \\ 5x+4y=8 \end{cases}$

Multiply the first equation by 2.

$$\begin{aligned} 6x-4y&=14 \\ 5x+4y&=8 \\ \hline 11x \quad &=22 \\ x&=2 \end{aligned}$$

Let $x = 2$ in the first equation.
$3(2)-2y=7$
$6-2y=7$
$-2y=1$
$y=-\frac{1}{2}$

The solution of the system is $\left(2, -\frac{1}{2}\right)$.

19. $\begin{cases} 8x=-11y-16 \\ 2x+3y=-4 \end{cases}$

Add $11y$ to both sides of the first equation and multiply the second equation by −4, then add.

$$\begin{aligned} 8x+11y&=-16 \\ -8x-12y&=16 \\ \hline -y&=0 \\ y&=0 \end{aligned}$$

Let $y = 0$ in the first equation.
$8x=-11(0)-16$
$8x=-16$
$x=-2$
The solution of the system is (−2, 0).

21. $\begin{cases} 4x-3y=7 \\ 7x+5y=2 \end{cases}$

Multiply the first equation by 5 and the second equation by 3.

$$\begin{aligned} 20x-15y&=35 \\ 21x+15y&=6 \\ \hline 41x \quad &=41 \\ x&=1 \end{aligned}$$

Let $x = 1$ in the first equation.
$4x-3y=7$
$4(1)-3y=7$
$4-3y=7$
$-3y=3$
$y=-1$
The solution of the system is (1, −1).

23. $\begin{cases} 4x-6y=8 \\ 6x-9y=12 \end{cases}$

Multiply the first equation by 3 and the second equation by −2.

$$\begin{aligned} 12x-18y&=24 \\ -12x+18y&=-24 \\ \hline 0&=0 \end{aligned}$$

The equations in the original system are equivalent and there is an infinite number of solutions.

25. $\begin{cases} 2x-5y=4 \\ 3x-2y=4 \end{cases}$

Multiply the first equation by −3 and the second equation by 2.

$$\begin{array}{r}-6x+15y=-12\\ 6x-4y=8\\ \hline 11y=-4\\ y=-\frac{4}{11}\end{array}$$

Multiply the first equation by −2 and the second equation by 5.

$$\begin{array}{r}-4x+10y=-8\\ 15x-10y=20\\ \hline 11x\quad=12\\ x=\frac{12}{11}\end{array}$$

The solution of the system is $\left(\frac{12}{11}, -\frac{4}{11}\right)$.

27. $\begin{cases}\frac{x}{3}+\frac{y}{6}=1\\ \frac{x}{2}-\frac{y}{4}=0\end{cases}$

Multiply the first equation by 6 and the second equation by 4.

$$\begin{array}{r}2x+y=6\\ 2x-y=0\\ \hline 4x\quad=6\\ x=\frac{3}{2}\end{array}$$

Multiply the second equation of the simplified system by −1.

$$\begin{array}{r}2x+y=6\\ -2x+y=0\\ \hline 2y=6\\ y=3\end{array}$$

The solution of the system is $\left(\frac{3}{2}, 3\right)$.

29. $\begin{cases}\frac{10}{3}x+4y=-4\\ 5x+6y=-6\end{cases}$

Multiply the first equation by 3 and the second equation by −2.

$$\begin{array}{r}10x+12y=-12\\ -10x-12y=12\\ \hline 0=0\end{array}$$

The system has an infinite number of solutions.

31. $\begin{cases}x-\frac{y}{3}=-1\\ -\frac{x}{2}+\frac{y}{8}=\frac{1}{4}\end{cases}$

Multiply the first equation by 3 and the second equation by 8.

$$\begin{array}{r}3x-y=-3\\ -4x+y=2\\ \hline -x\quad=-1\\ x=1\end{array}$$

Multiply the first equation of the simplified system by 4 and the second equation by 3.

$$\begin{array}{r}12x-4y=-12\\ -12x+3y=6\\ \hline -y=-6\\ y=6\end{array}$$

The solution of the system is (1, 6).

33. $\begin{cases}-4(x+2)=3y\\ 2x-2y=3\end{cases}\rightarrow\begin{cases}-4x-8=3y\\ 2x-2y=3\end{cases}$

$\rightarrow\begin{cases}-4x-3y=8\\ 2x-2y=3\end{cases}$

Multiply the second equation by 2.

$$\begin{array}{r}-4x-3y=8\\ 4x-4y=6\\ \hline -7y=14\\ y=-2\end{array}$$

Let $y = -2$ in the second equation.

$$\begin{aligned}2x-2(-2)&=3\\ 2x+4&=3\\ 2x&=-1\\ x&=-\frac{1}{2}\end{aligned}$$

The solution of the system is $\left(-\frac{1}{2}, -2\right)$.

35. $\begin{cases}\frac{x}{3}-y=2\\ -\frac{x}{2}+\frac{3y}{2}=-3\end{cases}$

Multiply the first equation by 3 and the second equation by 2.

$$\begin{array}{r}x-3y=6\\ -2x+3y=-6\\ \hline 0=0\end{array}$$

The equations of the original system are equivalent and there is an infinite number of solutions.

37. $\begin{cases} \frac{3}{5}x - y = -\frac{4}{5} \\ 3x + \frac{y}{2} = -\frac{9}{5} \end{cases}$

Multiply the first equation by 5 and the second equation by 10.

$$\begin{array}{l} 3x - 5y = -4 \\ \underline{30x + 5y = -18} \\ 33x \quad = -22 \\ x = -\frac{2}{3} \end{array}$$

Let $x = -\frac{2}{3}$ in $30x + 5y = -18$.

$$\begin{aligned} 30\left(-\frac{2}{3}\right) + 5y &= -18 \\ -20 + 5y &= -18 \\ 5y &= 2 \\ y &= \frac{2}{5} \end{aligned}$$

The solution of the system is $\left(-\frac{2}{3}, \frac{2}{5}\right)$.

39. $\begin{cases} 3.5x + 2.5y = 17 \\ -1.5x - 7.5y = -33 \end{cases}$

Multiply the first equation by 6 and the second equation by 2.

$$\begin{array}{l} 21x + 15y = 102 \\ \underline{-3x - 15y = -66} \\ 18x \quad = 36 \\ x = 2 \end{array}$$

Let $x = 2$ in $-3x - 15y = -66$.

$$\begin{aligned} -3(2) - 15y &= -66 \\ -6 - 15y &= -66 \\ -15y &= -60 \\ y &= 4 \end{aligned}$$

The solution of the system is (2, 4).

41. $\begin{cases} 0.02x + 0.04y = 0.09 \\ -0.1x + 0.3y = 0.8 \end{cases}$

Multiply the first equation by 100 and the second equation by 20.

$$\begin{array}{l} 2x + 4y = 9 \\ \underline{-2x + 6y = 16} \\ 10y = 25 \\ y = \frac{5}{2} = 2.5 \end{array}$$

Let $y = 2.5$ in $2x + 4y = 9$.

$$\begin{aligned} 2x + 4(2.5) &= 9 \\ 2x + 10 &= 9 \\ 2x &= -1 \\ x &= -\frac{1}{2} = -0.5 \end{aligned}$$

The solution of the system is (–0.5, 2.5).

43. $\begin{cases} 2x - 3y = -11 \\ y = 4x - 3 \end{cases}$

Substitute $4x - 3$ for y in the first equation.

$$\begin{aligned} 2x - 3(4x - 3) &= -11 \\ 2x - 12x + 9 &= -11 \\ -10x &= -20 \\ x &= 2 \end{aligned}$$

Let $x = 2$ in the second equation.

$y = 4(2) - 3 = 5$

The solution of the system is (2, 5).

45. $\begin{cases} x + 2y = 1 \\ 3x + 4y = -1 \end{cases}$

Multiply the first equation by –2.

$$\begin{array}{l} -2x - 4y = -2 \\ \underline{3x + 4y = -1} \\ x \quad = -3 \end{array}$$

Let $x = -3$ in the first equation.

$$\begin{aligned} -3 + 2y &= 1 \\ 2y &= 4 \\ y &= 2 \end{aligned}$$

The solution is (–3, 2).

47. $\begin{cases} 2y = x + 6 \\ 3x - 2y = -6 \end{cases}$

Subtract x from both sides of the first equation.

$$\begin{array}{l} -x + 2y = 6 \\ \underline{3x - 2y = -6} \\ 2x \quad = 0 \\ x = 0 \end{array}$$

Let $x = 0$ in the first equation.

$$\begin{aligned} 2y &= 0 + 6 \\ 2y &= 6 \\ y &= 3 \end{aligned}$$

The solution of the system is (0, 3).

49. $\begin{cases} y = 2x - 3 \\ y = 5x - 18 \end{cases}$

Substitute $5x - 18$ for y in the first equation.

$$\begin{aligned} 5x - 18 &= 2x - 3 \\ 3x &= 15 \\ x &= 5 \end{aligned}$$

Let $x = 5$ in the second equation.
$y = 5(5) - 18 = 7$
The solution of the system is (5, 7).

51. $\begin{cases} x+\frac{1}{6}y=\frac{1}{2} \\ 3x+2y=3 \end{cases}$

Multiply the first equation by −12.

$$\begin{array}{r} -12x-2y=-6 \\ 3x+2y=3 \\ \hline -9x \quad =-3 \\ x=\frac{1}{3} \end{array}$$

Substitute $\frac{1}{3}$ for x in the second equation.

$$3\left(\frac{1}{3}\right)+2y=3$$
$$1+2y=3$$
$$2y=2$$
$$y=1$$

The solution of the system is $\left(\frac{1}{3}, 1\right)$.

53. $\begin{cases} \frac{x+2}{2}=\frac{y+11}{3} \\ \frac{x}{2}=\frac{2y+16}{6} \end{cases}$

Multiply the first equation by 6 and the second equation by −6.

$$\begin{cases} 3(x+2)=2(y+11) \\ 3x+6=2y+22 \\ 3x-2y=16 \\ \\ -3x=-2y-16 \\ -3x+2y=-16 \end{cases}$$

Add the two equations.

$$\begin{array}{r} 3x-2y=16 \\ -3x+2y=-16 \\ \hline 0=0 \end{array}$$

There is an infinite number of solutions.

55. $\begin{cases} 2x+3y=14 \\ 3x-4y=-69.1 \end{cases}$

Multiply the first equation by 3 and the second equation by −2.

$$\begin{array}{r} 6x+9y=42 \\ -6x+8y=138.2 \\ \hline 17y=180.2 \\ y=10.6 \end{array}$$

Let $y = 10.6$ in the first equation.

$$2x+3(10.6)=14$$
$$2x+31.8=14$$
$$2x=-17.8$$
$$x=-8.9$$

The solution of the system is (−8.9, 10.6).

57. Let x = a number.
$2x + 6 = x - 3$

59. Let x = a number.
$20 - 3x = 2$

61. Let n = a number.
$4(n + 6) = 2n$

63. $\begin{cases} 4x+2y=-7 \\ 3x-y=-12 \end{cases}$

To eliminate y, multiply the second equation by 2.
$6x - 2y = -24$

65. $\begin{cases} 3x+8y=-5 \\ 2x-4y=3 \end{cases} = \begin{cases} 3x+8y=-5 \\ 4x-8y=6 \end{cases}$

The correct answer is **b**; answers may vary

67. Answers may vary

69. $\begin{cases} x+y=5 \\ 3x+3y=b \end{cases}$

Multiply the first equation by −3.

$$\begin{array}{r} -3x-3y=-15 \\ 3x+3y=b \\ \hline 0=b-15 \end{array}$$

a. The system has an infinite number of solutions if this statement is true.
$b = 15$

b. The system has no solution if this statement is false. b = any real number except 15.

71. $\begin{cases} 1.2x+3.4y=27.6 \\ 7.2x-1.7y=-46.56 \end{cases}$

Multiply the second equation by 2.

$$\begin{array}{r} 1.2x+3.4y=27.6 \\ \underline{14.4x-3.4y=-93.12} \\ 15.6x \quad\quad =-65.52 \\ x=-4.2 \end{array}$$

Let $x = -4.2$ in the first equation.

$$\begin{aligned} 1.2(-4.2)+3.4y&=27.6 \\ -5.04+3.4y&=27.6 \\ 3.4y&=32.64 \\ y&=9.6 \end{aligned}$$

The solution of the system is (–4.2, 9.6).

73. a. $\begin{cases} 7.4x-y=-258 \\ 12.6x-y=-231 \end{cases}$

Multiply the first equation by –1.

$$\begin{array}{r} -7.4x+y=258 \\ \underline{12.6x-y=-231} \\ 5.2x \quad =27 \\ x\approx 5 \end{array}$$

Let $x = 5$ in the first equation.

$$\begin{aligned} 7.4(5)-y&=-258 \\ 37-y&=-258 \\ -y&=-295 \\ y&=295 \end{aligned}$$

The solution of the system is approximately (5, 295) [or (5, 294) or (5, 296); answers may vary].

b. In 2009 (2004 + 5), the number of pharmacy technician jobs equals the number of network and data analyst jobs.

c. There will be approximately 294–296 thousand jobs.

Integrated Review

1. $\begin{cases} 2x-3y=-11 \\ y=4x-3 \end{cases}$

Substitute $4x - 3$ for y in the first equation.

$$\begin{aligned} 2x-3(4x-3)&=-11 \\ 2x-12x+9&=-11 \\ -10x&=-20 \\ x&=2 \end{aligned}$$

Let $x = 2$ in the second equation.

$y = 4(2) - 3 = 5$

The solution of the system is (2, 5).

2. $\begin{cases} 4x-5y=6 \\ y=3x-10 \end{cases}$

Substitute $3x - 10$ for y in the first equation.

$$\begin{aligned} 4x-5(3x-10)&=6 \\ 4x-15x+50&=6 \\ -11x&=-44 \\ x&=4 \end{aligned}$$

Let $x = 4$ in the second equation.

$y = 3(4) - 10 = 2$

The solution of the system is (4, 2).

3. $\begin{cases} x+y=3 \\ x-y=7 \end{cases}$

$$\begin{array}{r} x+y=3 \\ \underline{x-y=7} \\ 2x \quad =10 \\ x=5 \end{array}$$

Let $x = 5$ in the first equation.

$$\begin{aligned} 5+y&=3 \\ y&=-2 \end{aligned}$$

The solution of the system is (5, –2).

4. $\begin{cases} x-y=20 \\ x+y=-8 \end{cases}$

$$\begin{array}{r} x-y=20 \\ \underline{x+y=-8} \\ 2x \quad =12 \\ x=6 \end{array}$$

Let $x = 6$ in the second equation.

$$\begin{aligned} 6+y&=-8 \\ y&=-14 \end{aligned}$$

The solution of the system is (6, –14).

5. $\begin{cases} x+2y=1 \\ 3x+4y=-1 \end{cases}$

Solve the first equation for x.

$x = 1 - 2y$

Substitute $1 - 2y$ for x in the second equation.

$$\begin{aligned} 3(1-2y)+4y&=-1 \\ 3-6y+4y&=-1 \\ -2y&=-4 \\ y&=2 \end{aligned}$$

Let $y = 2$ in $x = 1 - 2y$.

$x = 1 - 2(2) = -3$

The solution is (–3, 2).

6. $\begin{cases} x+3y=5 \\ 5x+6y=-2 \end{cases}$

Solve the first equation for x.

$x = 5 - 3y$

Substitute $5 - 3y$ for x in the second equation.

$$5(5-3y)+6y=-2$$
$$25-15y+6y=-2$$
$$-9y=-27$$
$$y=3$$

Let $y = 3$ in $x = 5 - 3y$.

$x = 5 - 3(3) = -4$

The solution is $(-4, 3)$.

7. $\begin{cases} y = x+3 \\ 3x-2y=-6 \end{cases}$

Substitute $x + 3$ for y in the second equation.

$$3x-2(x+3)=-6$$
$$3x-2x-6=-6$$
$$x=0$$

Let $x = 0$ in the first equation.

$y = 0 + 3 = 3$

The solution is $(0, 3)$.

8. $\begin{cases} y=-2x \\ 2x-3y=-16 \end{cases}$

Substitute $-2x$ for y in the second equation.

$$2x-3(-2x)=-16$$
$$2x+6x=-16$$
$$8x=-16$$
$$x=-2$$

Let $x = -2$ in the first equation.

$y = -2(-2) = 4$

The solution is $(-2, 4)$.

9. $\begin{cases} y=2x-3 \\ y=5x-18 \end{cases}$

Substitute $5x - 18$ for y in the first equation.

$$5x-18=2x-3$$
$$3x=15$$
$$x=5$$

Let $x = 5$ in the second equation.

$y = 5(5) - 18 = 7$

The solution is $(5, 7)$.

10. $\begin{cases} y=6x-5 \\ y=4x-11 \end{cases}$

Substitute $6x - 5$ for y in the second equation.

$$6x-5=4x-11$$
$$2x=-6$$
$$x=-3$$

Let $x = -3$ in the first equation.

$y = 6(-3) - 5 = -23$

The solution is $(-3, -23)$.

11. $\begin{cases} x+\frac{1}{6}y=\frac{1}{2} \\ 3x+2y=3 \end{cases}$

Multiply the first equation by 6.

$\begin{cases} 6x+y=3 \\ 3x+2y=3 \end{cases}$

Multiply the first equation of the simplified system by -2.

$$\begin{array}{r} -12x-2y=-6 \\ 3x+2y=3 \\ \hline -9x \quad =-3 \end{array}$$
$$x=\frac{1}{3}$$

Multiply the second equation of the simplified system by -2.

$$\begin{array}{r} 6x \quad +y=3 \\ -6x-4y=-6 \\ \hline -3y=-3 \end{array}$$
$$y=1$$

The solution of the system is $\left(\frac{1}{3}, 1\right)$.

12. $\begin{cases} x+\frac{1}{3}y=\frac{5}{12} \\ 8x+3y=4 \end{cases}$

Multiply the first equation by 12.

$\begin{cases} 12x+4y=5 \\ 8x+3y=4 \end{cases}$

Multiply the first equation of the simplified system by 2 and the second equation by -3.

$$\begin{array}{r} 24x+8y=10 \\ -24x-9y=-12 \\ \hline -y=-2 \end{array}$$
$$y=2$$

Multiply the first equation of the simplified system by 3 and the second equation by -4.

$$\begin{array}{r} 36x+12y=15 \\ -32x-12y=-16 \\ \hline 4x \quad =-1 \end{array}$$
$$x=-\frac{1}{4}$$

The solution of the system is $\left(-\frac{1}{4}, 2\right)$.

13. $\begin{cases} x-5y=1 \\ -2x+10y=3 \end{cases}$

Multiply the first equation by 2.

$$\begin{aligned} 2x-10y&=2 \\ -2x+10y&=3 \\ \hline 0&=5 \quad \text{False} \end{aligned}$$

The system has no solution.

14. $\begin{cases} -x+2y=3 \\ 3x-6y=-9 \end{cases}$

Multiply the first equation by 3.

$$\begin{aligned} -3x+6y&=9 \\ 3x-6y&=-9 \\ \hline 0&=0 \end{aligned}$$

The equations in the original system are equivalent and there is an infinite number of solutions.

15. $\begin{cases} 0.2x-0.3y=-0.95 \\ 0.4x+0.1y=0.55 \end{cases}$

Multiply both equations by 10.

$\begin{cases} 2x-3y=-9.5 \\ 4x+y=5.5 \end{cases}$

Multiply the first equation of the simplified system by −2.

$$\begin{aligned} -4x+6y&=19 \\ 4x+y&=5.5 \\ \hline 7y&=24.5 \\ y&=3.5 \end{aligned}$$

Multiply the second equation of the simplified system by 3.

$$\begin{aligned} 2x-3y&=-9.5 \\ 12x+3y&=16.5 \\ \hline 14x&=7 \\ x&=0.5 \end{aligned}$$

The solution of the system is (0.5, 3.5).

16. $\begin{cases} 0.08x-0.04y=-0.11 \\ 0.02x-0.06y=-0.09 \end{cases}$

Multiply both equations by 100.

$\begin{cases} 8x-4y=-11 \\ 2x-6y=-9 \end{cases}$

Multiply the second equation of the simplified system by −4.

$$\begin{aligned} 8x-4y&=-11 \\ -8x+24y&=36 \\ \hline 20y&=25 \\ y&=1.25 \end{aligned}$$

Multiply the first equation of the simplified system by −3 and the second equation by 2.

$$\begin{aligned} -24x+12y&=33 \\ 4x-12y&=-18 \\ \hline -20x&=15 \\ x&=-0.75 \end{aligned}$$

The solution of the system is (−0.75, 1.25).

17. $\begin{cases} x=3y-7 \\ 2x-6y=-14 \end{cases}$

Substitute 3*y* − 7 for *x* in the second equation.

$$\begin{aligned} 2(3y-7)-6y&=-14 \\ 6y-14-6y&=-14 \\ -14&=-14 \end{aligned}$$

The equations in the original system are equivalent and there is an infinite number of solutions.

18. $\begin{cases} y=\frac{x}{2}-3 \\ 2x-4y=0 \end{cases}$

Substitute $\frac{x}{2}-3$ for *y* in the second equation.

$$\begin{aligned} 2x-4\left(\frac{x}{2}-3\right)&=0 \\ 2x-2x+12&=0 \\ 12&=0 \quad \text{False} \end{aligned}$$

There is no solution.

19. $\begin{cases} 2x+5y=-1 \\ 3x-4y=33 \end{cases}$

Multiply the first equation by 4 and the second equation by 5.

$$\begin{aligned} 8x+20y&=-4 \\ 15x-20y&=165 \\ \hline 23x&=161 \\ x&=7 \end{aligned}$$

Let *x* = 7 in the first equation.

$$\begin{aligned} 2(7)+5y&=-1 \\ 14+5y&=-1 \\ 5y&=-15 \\ y&=-3 \end{aligned}$$

The solution of the system is (7, −3).

20. $\begin{cases} 7x-3y=2 \\ 6x+5y=-21 \end{cases}$

Multiply the first equation by 5 and the second equation by 3.

$$\begin{array}{l} 35x-15y=10 \\ \underline{18x+15y=-63} \\ 53x \quad\quad = -53 \\ \quad\quad x=-1 \end{array}$$

Let $x = -1$ in the first equation.

$$\begin{aligned} 7(-1)-3y &= 2 \\ -7-3y &= 2 \\ -3y &= 9 \\ y &= -3 \end{aligned}$$

The solution of the system is (−1, −3).

21. Answers may vary

22. Answers may vary

Section 4.4

Practice Exercises

1. $\begin{cases} 3x+2y-z=0 & (1) \\ x-y+5z=2 & (2) \\ 2x+3y+3z=7 & (3) \end{cases}$

Multiply equation (2) by 2 and add to equation (1) to eliminate y.

$\begin{cases} 3x+2y-z=0 \\ 2(x-y+5z)=2(2) \end{cases}$

$\begin{cases} 3x-2y \quad -z=0 \\ 2x-2y+10z=4 \end{cases}$

$5x \quad +9z=4 \quad (4)$

Multiply equation (2) by 3 and add to equation (3) to eliminate y again.

$\begin{cases} 3(x-y+5z)=3(2) \\ 2x+3y+3z=7 \end{cases}$

$\begin{cases} 3x-3y+15z=6 \\ 2x+3y \quad +3z=7 \end{cases}$

$5x+ \quad 18z=13 \quad (5)$

Multiply equation (4) by −1 and add to equation (5) to eliminate x.

$\begin{cases} -1(5x+9z)=-1(4) \\ 5x+18z=13 \end{cases}$

$\begin{cases} -5x \quad -9z=-4 \\ 5x+18z=13 \end{cases}$

$$\begin{aligned} 9z &= 9 \\ z &= 1 \end{aligned}$$

Replace z with 1 in equation (4) or (5).

$$\begin{aligned} 5x+9z &= 4 \\ 5x+9(1) &= 4 \\ 5x &= -5 \\ x &= -1 \end{aligned}$$

Replace x with −1 and z with 1 in equation (1), (2), or (3).

$$\begin{aligned} x-y+5z &= 2 \\ -1-y+5(1) &= 2 \\ -y+4 &= 2 \\ -y &= -2 \\ y &= 2 \end{aligned}$$

The solution is (−1, 2, 1). To check, let $x = -1$, $y = 2$, and $z = 1$ in all three original equations of the system.

2. $\begin{cases} 6x-3y+12z=4 & (1) \\ -6x+4y-2z=7 & (2) \\ -2x+y-4z=3 & (3) \end{cases}$

Multiply equation (3) by 3 and add to equation (1) to eliminate x.

$\begin{cases} 6x-3y+12z=4 \\ 3(-2x+y-4z)=3(3) \end{cases}$

$\begin{cases} 6x-3y+12z=4 \\ -6x+3y-12z=9 \end{cases}$

$0=13$ False

Since the statement is false, this system is inconsistent and has no solution. The solution set is { } or ∅.

3. $\begin{cases} 3x+4y \quad\quad =0 & (1) \\ 9x \quad\quad -4z=6 & (2) \\ \quad -2y+7z=1 & (3) \end{cases}$

Equation (2) has no term containing the variable y. Eliminate y using equations (1) and (3). Multiply equation (3) by 2 and add to equation (1).

$\begin{cases} 3x+4y=0 \\ 2(-2y+7z)=2(1) \end{cases}$

$\begin{cases} 3x+4y \quad\quad =0 \\ \quad -4y+14z=2 \end{cases}$

$3x \quad +14z=2 \quad (4)$

Multiply equation (4) by −3 and add to equation (2) to eliminate x.

$\begin{cases} 9x-4z=6 \\ -3(3x+14z)=-3(2) \end{cases}$

$$\begin{cases} 9x - 4z = 6 \\ -9x - 52z = -6 \end{cases}$$

$$-56z = 0$$
$$z = 0$$

Replace z with 0 in equation (2) and solve for x.

$$9x - 4z = 6$$
$$9x - 4(0) = 6$$
$$9x = 6$$
$$x = \frac{6}{9} = \frac{2}{3}$$

Replace z with 0 in equation (3) and solve for y.

$$-2y + 7z = 1$$
$$-2y + 7(0) = 1$$
$$-2y = 1$$
$$y = -\frac{1}{2}$$

The solution is $\left(\frac{2}{3}, -\frac{1}{2}, 0\right)$.

4. $$\begin{cases} 2x + y - 3z = 6 & (1) \\ x + \frac{1}{2}y - \frac{3}{2}z = 3 & (2) \\ -4x - 2y + 6z = -12 & (3) \end{cases}$$

Multiply both sides of equation (2) by 2 to eliminate fractions, and multiply both sides of equation (3) by $-\frac{1}{2}$ since all coefficients in equation (3) are divisible by 2 and the coefficient of x is negative. The resulting system is

$$\begin{cases} 2x + y - 3z = 6 \\ 2x + y - 3z = 6 \\ 2x + y - 3z = 6 \end{cases}$$

Since the three equations are identical, there are infinitely many solutions of the system. The equations are dependent. The solution set can be written as $\{(x, y, z)|2x + y - 3z = 6\}$.

5. $$\begin{cases} x + 2y + 4z = 16 & (1) \\ x \quad\quad + 2z = -4 & (2) \\ \quad\quad y - 3z = 30 & (3) \end{cases}$$

Solve equation (2) for x and equation (3) for y.

$$x + 2z = -4 \qquad\qquad y - 3z = 30$$
$$x = -2z - 4 \qquad\qquad y = 3z + 30$$

Substitute $-2z - 4$ for x and $3z + 30$ for y in equation (1) and solve for z.

$$x + 2y + 4z = 16$$
$$(-2z - 4) + 2(3z + 30) + 4z = 16$$
$$-2z - 4 + 6z + 60 + 4z = 16$$
$$8z + 56 = 16$$
$$8z = -40$$
$$z = -5$$

Use $x = -2z - 4$ to find x:
$x = -2(-5) - 4 = 10 - 4 = 6$.
Use $y = 3z + 30$ to find y:
$y = 3(-5) + 30 = -15 + 30 = 15$.
The solution is (6, 15, −5).

Exercise Set 4.4

1. $$x + y + z = 3 \qquad\qquad -x + y + z = 5$$
$$(-1) + 3 + 1 \stackrel{?}{=} 3 \qquad\qquad -(-1) + 3 + 1 \stackrel{?}{=} 5$$
$$3 = 3 \qquad\qquad 5 = 5$$

a is true. b is true.

$$-x + y + 2z = 0 \qquad\qquad x + 2y - 3z = 2$$
$$-(-1) + 3 + 2(1) \stackrel{?}{=} 0 \qquad\qquad (-1) + 2(3) - 3(1) \stackrel{?}{=} 2$$
$$6 = 0 \qquad\qquad 2 = 2$$

c is false. d is true.

Therefore, equations a, b, and d.

3. Yes; answers may vary

5. $$\begin{cases} x - y + z = -4 & (1) \\ 3x + 2y - z = 5 & (2) \\ -2x + 3y - z = 15 & (3) \end{cases}$$

Add E1 and E2.

$$4x + y = 1$$

Add E1 and E3.

$$-x + 2y = 11$$

Solve the new system:

$$\begin{cases} 4x + y = 1 \\ -x + 2y = 11 \end{cases}$$

Multiply the second equation by 4.

$$\begin{cases} 4x + y = 1 \\ -4x + 8y = 44 \end{cases}$$

Add the equations.

$$4x + y = 1$$
$$-4x + 8y = 44$$
$$9y = 45$$
$$y = 5$$

Replace y with 5 in the equation $4x + y = 1$.

$$4x + 5 = 1$$
$$4x = -4$$
$$x = -1$$

Replace x with −1 and y with 5 in E1.

$(-1)-(5)+z=-4$
$-6+z=-4$
$z=2$
The solution is (–1, 5, 2).

7. $\begin{cases} x+y \quad\quad = 3 & (1) \\ \quad 2y \quad\quad = 10 & (2) \\ 3x+2y-3z=1 & (3) \end{cases}$

Solve E2 for y: $y=5$

Replace y with 5 in E1.
$x+5=3$
$x=-2$
Replace x with –2 and y with 5 in E3.
$3(-2)+2(5)-3z=1$
$-6+10-3z=1$
$4-3z=1$
$-3z=-3$
$z=1$
The solution is (–2, 5, 1).

9. $\begin{cases} 2x+2y+z=1 & (1) \\ -x+y+2z=3 & (2) \\ x+2y+4z=0 & (3) \end{cases}$

Add E2 and E3.
$3y+6z=3$ or $y+2z=1$
Multiply E2 by 2 and add to E1.
$-2x+2y+4z=6$
$\underline{2x+2y+z=1}$
$4y+5z=7$
Solve the new system:
$\begin{cases} y+2z=1 \\ 4y+5z=7 \end{cases}$
Multiply the first equation by –4.
$\begin{cases} -4y-8z=-4 \\ 4y+5z=7 \end{cases}$
Add the equations.
$-4y-8z=-4$
$\underline{4y+5z=7}$
$-3z=3$
$z=-1$
Replace z with –1 in the equation $y+2z=1$.
$y+2(-1)=1$
$y-2=1$
$y=3$
Replace y with 3 and z with –1 in E3.
$x+2(3)+4(-1)=0$
$x+6-4=0$
$x+2=0$
$x=-2$
The solution is (–2, 3, –1).

11. $\begin{cases} x-2y+z=-5 & (1) \\ -3x+6y-3z=15 & (2) \\ 2x-4y+2z=-10 & (3) \end{cases}$

Multiply E2 by $-\frac{1}{3}$ and E3 by $\frac{1}{2}$.

$\begin{cases} x-2y+z=-5 \\ x-2y+z=-5 \\ x-2y+z=-5 \end{cases}$

All three equations are identical. There are infinitely many solutions. The solution set is $\{(x, y, z)|x-2y+z=-5\}$.

13. $\begin{cases} 4x-y+2z=5 & (1) \\ \quad 2y+z=4 & (2) \\ 4x+y+3z=10 & (3) \end{cases}$

Multiply E1 by –1 and add to E3.
$-4x+y-2z=-5$
$\underline{4x+y+3z=10}$
$2y+z=5$ (4)
Multiply E4 by –1 and add to E2.
$-2y-z=-5$
$\underline{2y+z=4}$
$0=-1$ False
Inconsistent system; the solution set is ∅.

15. $\begin{cases} x+ \quad 5z=0 & (1) \\ 5x+y \quad\quad =0 & (2) \\ \quad y-3z=0 & (3) \end{cases}$

Multiply E3 by –1 and add to E2.
$-y+3z=0$
$\underline{5x+y \quad =0}$
$5x+ \quad 3z=0$ (4)
Multiply E1 by –5 and add to E4.
$-5x-25z=0$
$\underline{5x+3z=0}$
$-22z=0$
$z=0$
Replace z with 0 in E4.
$5x+3(0)=0$
$5x=0$
$x=0$
Replace x with 0 in E2.
$5(0)+y=0$
$y=0$
The solution is (0, 0, 0).

17. $\begin{cases} 6x - \quad 5z = 17 & (1) \\ 5x - y + 3z = -1 & (2) \\ 2x + y \quad = -41 & (3) \end{cases}$

Add E2 and E3.

$7x + 3z = -42 \quad (4)$

Multiply E4 by 5, multiply E1 by 3, and add.

$$\begin{aligned} 35x + 15z &= -210 \\ 18x - 15z &= 51 \\ \hline 53x \quad &= -159 \\ x &= -3 \end{aligned}$$

Replace x with –3 in E1.

$$\begin{aligned} 6(-3) - 5z &= 17 \\ -18 - 5z &= 17 \\ -5z &= 35 \\ z &= -7 \end{aligned}$$

Replace x with –3 in E3.

$$\begin{aligned} 2(-3) + y &= -41 \\ -6 + y &= -41 \\ y &= -35 \end{aligned}$$

The solution is (–3, –35, –7).

19. $\begin{cases} x + y + z = 8 & (1) \\ 2x - y - z = 10 & (2) \\ x - 2y - 3z = 22 & (3) \end{cases}$

Add E1 and E2.

$3x = 18$ or $x = 6$

Add twice E1 to E3.

$$\begin{aligned} 2x + 2y + 2z &= 16 \\ x - 2y - 3z &= 22 \\ \hline 3x - \quad z &= 38 \end{aligned}$$

Replace x with 6 in this equation.

$$\begin{aligned} 3(6) - z &= 38 \\ 18 - z &= 38 \\ -z &= 20 \\ z &= -20 \end{aligned}$$

Replace x with 6 and z with –20 in E1.

$$\begin{aligned} 6 + y + (-20) &= 8 \\ y - 14 &= 8 \\ y &= 22 \end{aligned}$$

The solution is (6, 22, –20).

21. $\begin{cases} x + 2y - z = 5 & (1) \\ 6x + y + z = 7 & (2) \\ 2x + 4y - 2z = 5 & (3) \end{cases}$

Add E1 and E2.

$7x + 3y = 12 \quad (4)$

Add twice E2 to E3.

$$\begin{aligned} 12x + 2y + 2z &= 14 \\ 2x + 4y - 2z &= 5 \\ \hline 14x + 6y \quad &= 19 \quad (5) \end{aligned}$$

Multiply E4 by –2 and add to E5.

$$\begin{aligned} -14x - 6y &= -24 \\ 14x + 6y &= 19 \\ \hline 0 &= -5 \quad \text{False} \end{aligned}$$

Inconsistent system; the solution set is ∅.

23. $\begin{cases} 2x - 3y + z = 2 & (1) \\ x - 5y + 5z = 3 & (2) \\ 3x + y - 3z = 5 & (3) \end{cases}$

Add –2 times E2 to E1.

$$\begin{aligned} 2x - 3y + z &= 2 \\ -2x + 10y - 10z &= -6 \\ \hline 7y - 9z &= -4 \quad (4) \end{aligned}$$

Add –3 times E2 to E3.

$$\begin{aligned} -3x + 15y - 15z &= -9 \\ 3x + y - 3z &= 5 \\ \hline 16y - 18z &= -4 \end{aligned}$$

Solve the new system:

$\begin{cases} 7y - 9z = -4 & (4) \\ 16y - 18z = -4 & (5) \end{cases}$

Multiply E4 by –2 and add to E5.

$$\begin{aligned} -14y + 18z &= 8 \\ 16y - 18z &= -4 \\ \hline 2y \quad &= 4 \\ y &= 2 \end{aligned}$$

Replace y with 2 in E4.

$$\begin{aligned} 7(2) - 9z &= -4 \\ -9z &= -18 \\ z &= 2 \end{aligned}$$

Replace y with 2 and z with 2 in E1.

$$\begin{aligned} 2x - 3(2) + 2 &= 2 \\ 2x &= 6 \\ x &= 3 \end{aligned}$$

The solution is (3, 2, 2).

25. $\begin{cases} -2x - 4y + 6z = -8 & (1) \\ x + 2y - 3z = 4 & (2) \\ 4x + 8y - 12z = 16 & (3) \end{cases}$

Add 2 times E2 to E1.

$$\begin{aligned} 2x + 4y - 6z &= 8 \\ -2x - 4y + 6z &= -8 \\ \hline 0 &= 0 \end{aligned}$$

Add –4 times E2 to E3.

$$\begin{aligned} -4x - 8y + 12z &= -16 \\ 4x + 8y - 12z &= 16 \\ \hline 0 &= 0 \end{aligned}$$

The system is dependent.
The solution set is $\{(x, y, z) | x + 2y - 3z = 4\}$.

27. $\begin{cases} 2x+2y-3z=1 & (1) \\ y+2z=-14 & (2) \\ 3x-2y=-1 & (3) \end{cases}$

Add E1 to E3.
$5x-3z=0$ (4)
Add twice E2 to E3.

$$\begin{array}{l} 2y+4z=-28 \\ 3x-2y=-1 \\ \hline 3x+4z=-29 \ (5) \end{array}$$

Multiply E4 by 4, multiply E5 by 3, and add.

$$\begin{array}{l} 20x-12z=0 \\ 9x+12z=-87 \\ \hline 29x=-87 \\ x=-3 \end{array}$$

Replace x with –3 in E4.

$$5(-3)-3z=0$$
$$3z=-15$$
$$z=-5$$

Replace z with –5 in E2.

$$y+2(-5)=-14$$
$$y-10=-14$$
$$y=-4$$

The solution is (–3, –4, –5).

29. $\begin{cases} x+2y-z=5 & (1) \\ -3x-2y-3z=11 & (2) \\ 4x+4y+5z=-18 & (3) \end{cases}$

Add E1 and E2.
$-2x-4z=16$ or $x+2z=-8$ (4)
Add twice E2 to E3.

$$\begin{array}{l} -6x-4y-6z=22 \\ 4x+4y+5z=-18 \\ \hline -2x-z=4 \ (5) \end{array}$$

Solve the new system:
$\begin{cases} x+2z=-8 & (4) \\ -2x-z=4 & (5) \end{cases}$

Add twice E4 to E5.

$$\begin{array}{l} 2x+4z=-16 \\ -2x-z=4 \\ \hline 3z=-12 \\ z=-4 \end{array}$$

Replace z with –4 in E4.

$$x+2(-4)=-8$$
$$x-8=-8$$
$$x=0$$

Replace x with 0 and z with –4 in E1.

$$0+2y-(-4)=5$$
$$2y=1$$
$$y=\frac{1}{2}$$

The solution is $\left(0, \frac{1}{2}, -4\right)$.

31. $\begin{cases} \frac{3}{4}x-\frac{1}{3}y+\frac{1}{2}z=9 & (1) \\ \frac{1}{6}x+\frac{1}{3}y-\frac{1}{2}z=2 & (2) \\ \frac{1}{2}x-y+\frac{1}{2}z=2 & (3) \end{cases}$

Multiply E1 by 12, multiply E2 by 6, and multiply E3 by 2.

$\begin{cases} 9x-4y+6z=108 & (4) \\ x+2y-3z=12 & (5) \\ x-2y+z=4 & (6) \end{cases}$

Add twice E5 to E4.

$$\begin{array}{l} 2x+4y-6z=24 \\ 9x-4y+6z=108 \\ \hline 11x=132 \\ x=12 \end{array}$$

Add E5 and E6.
$2x-2z=16$ or $x-z=8$
Replace x with 12 in this equation.

$$12-z=8$$
$$z=4$$

Replace x with 12 and z with 4 in E6.

$$12-2y+4=4$$
$$12-2y=0$$
$$-2y=-12$$
$$y=6$$

The solution is (12, 6, 4).

33. Let x = the first number, then $2x$ = the second number.

$$x+2x=45$$
$$3x=45$$
$$x=15$$
$$2x=2(15)=30$$

The numbers are 15 and 30.

35. $2(x-1)-3x=x-12$

$$2x-2-3x=x-12$$
$$-x-2=x-12$$
$$-2x=-10$$
$$x=5$$

37. $-y-5(y+5)=3y-10$
$-y-5y-25=3y-10$
$-6y-25=3y-10$
$-9y=15$
$y=-\frac{15}{9}=-\frac{5}{3}$

39. Answers may vary

41. Answers may vary

43. $\begin{cases} x+y+z=1 & (1) \\ 2x-y+z=0 & (2) \\ -x+2y+2z=-1 & (3) \end{cases}$

Add E1 and E3.
$3y+3z=0$ or $y+z=0$ (4)
Add −2 times E1 to E2.
$-2x-2y-2z=-2$
$2x-y+z=0$
$-3y-z=-2$ (5)
Add E4 and E5.
$-2y=-2$
$y=1$
Replace y with 1 in E4.
$1+z=0$
$z=-1$
Replace y with 1 and z with −1 in E1.
$x+1+(-1)=1$
$x=1$
The solution is (1, 1, −1), and
$\frac{x}{8}+\frac{y}{4}+\frac{z}{3}=\frac{1}{8}+\frac{1}{4}-\frac{1}{3}$
$=\frac{3}{24}+\frac{6}{24}-\frac{8}{24}$
$=\frac{1}{24}.$

45. $\begin{cases} x+y-w=0 & (1) \\ y+2z+w=3 & (2) \\ x-z=1 & (3) \\ 2x-y-w=-1 & (4) \end{cases}$

Add E1 and E2.
$x+2y+2z=3$ (5)
Add E2 and E4.
$2x+2z=2$ or $x+z=1$ (6)
Add E3 and E6.
$x-z=1$
$x+z=1$
$2x=2$
Replace x with 1 in E3.
$1-z=1$
$z=0$
Replace x with 1 and z with 0 in E5.
$1+2y+2(0)=3$
$1+2y=3$
$2y=2$
$y=1$
Replace y with 1, and z with 0 in E2.
$1+2(0)+w=3$
$1+w=3$
$w=2$
The solution is (1, 1, 0, 2).

47. $\begin{cases} x+y+z+w=5 & (1) \\ 2x+y+z+w=6 & (2) \\ x+y+z=2 & (3) \\ x+y=0 & (4) \end{cases}$

Add −1 times E4 to E3.
$-x-y=0$
$x+y+z=2$
$z=2$
Replace z with 2 in E1 and E2.
$\begin{cases} x+y+w=3 & (5) \\ 2x+y+w=4 & (6) \end{cases}$
Add −1 times E5 to E6.
$-x-y-w=-3$
$2x+y+w=4$
$x=1$
Replace x with 1 in E4.
$1+y=0$
$y=-1$
Replace x with 1, y with −1, and z with 2 in E1.
$1+(-1)+2+w=5$
$2+w=5$
$w=3$
The solution is (1, −1, 2, 3).

49. Answers may vary

Section 4.5

Practice Exercises

1. a. We are given a system of equations.
$\begin{cases} y=-0.16x+113.9 \\ y=1.06x+62.3 \end{cases}$
We want to know the year x in which the pounds y are the same. Since both equations are solved for y, we use the substitution method. Substitute $-0.16x+113.9$ for y in the second equation.

$$-0.16x + 113.9 = 1.06x + 62.3$$
$$-1.22x = -51.6$$
$$x = \frac{-51.6}{-1.22} \approx 42.30$$

Since we are only asked to give the year, we need only solve for x. The consumption of red meat and poultry will be the same about 42.30 years after 1995, or in about 2037.

b. Yes; answers may vary.

2. Let x = first number
y = second number
"A first number is five more than a second number" is translated as $x = y + 5$. "Twice the first number is 2 less than 3 times the second number" is translated as $2x = 3y - 2$.
We solve the following system.
$$\begin{cases} x = y + 5 \\ 2x = 3y - 2 \end{cases}$$
Since the first equation is solved for x, we use substitution. Substitute $y + 5$ for x in the second equation.
$$2(y+5) = 3y - 2$$
$$2y + 10 = 3y - 2$$
$$12 = y$$
Replace y with 12 in the equation $x = y + 5$ and solve for x.
$$x = 12 + 5 = 17$$
The numbers are 12 and 17.

3. Let x = price for adult admission and y = price per child admission.
$$\begin{cases} 3x + 3y = 75 \\ 2x + 4y = 62 \end{cases}$$
Multiply the first equation by 2 and the second equation by −3.
$$6x + 6y = 150$$
$$-6x - 12y = -186$$
$$-6y = -36$$
$$y = 6$$
Let $y = 6$ in the second equation.
$$2x + 4y = 62$$
$$2x + 4(6) = 62$$
$$2x + 24 = 62$$
$$2x = 38$$
$$x = 19$$

a. $x = 19$, so the adult price is $19.

b. $y = 6$, so the child price is $6.

c. $5(19) + 15(6) = 95 + 90 = 185 < 200$
No, the regular rates are less than the group rate.

4. Let x = speed of the V150
y = speed of the Atlantique
We summarize the information in a chart. Both trains have traveled two hours.

	Rate	•	Time	=	Distance
V150	x		2		$2x$
Atlantique	y		2		$2y$

The trains are 2150 kilometers apart, so the sum of the distances is 2150: $2x + 2y = 2150$.
The V150 is 75 kph faster than the Atlantique: $x = y + 75$.
We solve the following system.
$$\begin{cases} 2x + 2y = 2150 \\ x = y + 75 \end{cases}$$
Since the second equation is solved for x, we use substitution. Substitute $y + 75$ for x in the first equation.
$$2(y+75) + 2y = 2150$$
$$2y + 150 + 2y = 2150$$
$$4y + 150 = 2150$$
$$4y = 2000$$
$$y = 500$$
To find x, we replace y with 500 in the second equation.
$$x = 500 + 75 = 575$$
The speed of the V150 is 575 kph, and the speed of the Atlantique is 500 kph.

5. Let x = amount of 99% acid
y = amount of water (0%)
Both x and y are measured in liters. We use a table to organize the given data.

	Amount	Acid Strength	Amount of Pure Acid
99% acid	x	99%	$0.99x$
Water	y	0%	$0y$

The amount of 99% acid and water combined must equal 1 liter, so $x + y = 1$.
The amount of pure acid in the mixture must equal the sum of the amounts of pure acid in the 99% acid and in the water, so $0.99x + 0y = 0.05(1)$, which simplifies to $0.99x = 0.05$.

We solve the following system.

$$\begin{cases} x+y=1 \\ 0.99x=0.05 \end{cases}$$

Since the second equation does not contain y, we solve it for x.

$$0.99x=0.05$$
$$x=\frac{0.05}{0.99}\approx 0.05$$

To find y, we replace x with 0.05 in the first equation.

$$x+y=1$$
$$0.05+y=1$$
$$y=0.95$$

The teacher should use 0.05 liter of the 99% HCL solution and 0.95 liter of water.

6. Let x = the number of packages.
The firm charges the customer \$4.50 for each package, so the revenue equation is $R(x)=4.5x$.
Each package costs \$2.50 to produce and the equipment costs \$3000, so the cost equation is $C(x)=2.5x+3000$.
Since the break-even point is when $R(x) = C(x)$, we solve the equation $4.5x = 2.5x + 3000$.

$$4.5x=2.5x+3000$$
$$2x=3000$$
$$x=1500$$

The company must sell 1500 packages to break even.

7. Let x = measure of smallest angle
y = measure of largest angle
z = measure of third angle
The sum of the measures is 180°:
$x+y+z=180$.
The measure of the largest angle is 40° more than the measure of the smallest angle:
$y=x+40$.
The measure of the remaining angle is 20° more than the measure of the smallest angle:
$y=x+20$.
We solve the following system.

$$\begin{cases} x+y+z=1180 \\ y=x+40 \\ z=x+20 \end{cases}$$

We substitute $x + 40$ for y and $x + 20$ for z in the first equation.

$$x+(x+40)+(x+20)=180$$
$$3x+60=180$$
$$3x=120$$
$$x=40$$

Then $y = x + 40 = 40 + 40 = 80$ and $z = x + 20 = 40 + 20 = 60$.
The angle measures are 40°, 60°, and 80°.

Exercise Set 4.5

1. a. $l-w=8-5=3$

$$P=2l+2w$$
$$=2(8)+2(5)$$
$$=13+10$$
$$=23\neq 30$$

b. $l-w=8-7=1\neq 3$

c. $l-w=9-6=3$
$P=2l+2w=2(9)+2(6)=18+12=30$

Choice **c** is correct.

3. a. $2d+3n=2(3)+3(4)=6+12=18\neq 17$

b. $2d+3n=2(4)+3(3)=8+9=17$
$5d+4n=5(4)+4(3)=20+12=32$

c. $2d+3n=2(2)+3(5)=4+15=19\neq 17$

Choice **b** is correct.

5. a. $80+20=100$

$$80d+20q=80(0.10)+20(0.25)$$
$$=8+5$$
$$=13$$

b. $20+44=64\neq 100$

c. $60+40=100$

$$60d+40q=60(0.10)+40(0.25)$$
$$=6+10$$
$$=16\neq 13$$

Choice **a** is correct.

7. Let x = the larger number and y = the smaller number.

$$\begin{cases} x+y=15 \\ x-y=7 \end{cases}$$

9. Let x = the amount invested in the larger account and y = the amount invested in the smaller account.

$$\begin{cases} x+y=6500 \\ x=y+800 \end{cases}$$

11. Let x = the first number and y = the second number.

$$\begin{cases} x+y=83 \\ x-y=17 \end{cases}$$

$$\begin{aligned} x+y&=83 \\ x-y&=17 \\ \hline 2x \quad &=100 \\ x&=50 \end{aligned}$$

Let $x = 50$ in the first equation.

$$\begin{aligned} 50+y&=83 \\ y&=33 \end{aligned}$$

The numbers are 50 and 33.

13. Let x = the first number and y = the second number.

$$\begin{cases} x+2y=8 \\ 2x+y=25 \end{cases}$$

Multiply the first equation by −2.

$$\begin{aligned} -2x-4y&=-16 \\ 2x\ +y&=25 \\ \hline -3y&=9 \\ y&=-3 \end{aligned}$$

Let $y = -3$ in the first equation.

$$\begin{aligned} x+2(-3)&=8 \\ x-6&=8 \\ x&=14 \end{aligned}$$

The numbers are 14 and −3.

15. Let x = Taurasi's points and y = Augustus's points.

$$\begin{cases} x-y=116 \\ x+y=1604 \end{cases}$$

$$\begin{aligned} x-y&=116 \\ x+y&=1604 \\ \hline 2x \quad &=1720 \\ x&=860 \end{aligned}$$

Let $x = 860$ in the second equation.

$$\begin{aligned} x+y&=1604 \\ 860+y&=1604 \\ y&=744 \end{aligned}$$

Taurasi scored 860 points and Augustus scored 744 points.

17. Let x = the price of an adult's ticket and y = the price of a child's ticket.

$$\begin{cases} 3x+4y=159 \\ 2x+3y=112 \end{cases}$$

Multiply the first equation by −2 and the second equation by 3.

$$\begin{aligned} -6x-8y&=-318 \\ 6x+9y&=336 \\ \hline y&=18 \end{aligned}$$

Let $y = 18$ in the first equation.

$$\begin{aligned} 3x+4(18)&=159 \\ 3x+72&=159 \\ 3x&=87 \\ x&=29 \end{aligned}$$

An adult's ticket is $29 and a child's ticket is $18.

19. Let x = the number of quarters and y = the number of nickels.

$$\begin{cases} x+y=80 \\ 0.25x+0.05y=14.6 \end{cases}$$

Solve the first equation for y.

$y = 80 - x$

Substitute $80 - x$ for y in the second equation.

$$\begin{aligned} 0.25x+0.05(80-x)&=14.6 \\ 0.25x+4-0.05x&=14.6 \\ 0.20x&=10.6 \\ x&=53 \end{aligned}$$

Let $x = 53$ in $y = 80 - x$.

$y = 80 - 53$

$y = 27$

There are 53 quarters and 27 nickels.

21. Let x = price of Apple stock and y = price of Microsoft stock.

$$\begin{cases} 50x+60y=6035.90 \\ x-y=60.68 \end{cases}$$

Multiply the second equation by 60.

$$\begin{aligned} 50x+60y&=6035.90 \\ 60x-60y&=3640.80 \\ \hline 110x \quad &=9676.70 \\ x&=87.97 \end{aligned}$$

Let $x = 87.97$ in the second equation.

$$\begin{aligned} x-y&=60.68 \\ 87.97-y&=60.68 \\ -y&=-27.29 \\ y&=27.29 \end{aligned}$$

The price of Apple's stock was $87.97 and the price of Microsoft stock was $27.29.

23. Let x = the daily fee and y = the mileage charge.

$$\begin{cases} 4x+450y=240.50 \\ 3x+200y=146.00 \end{cases}$$

Multiply the first equation by 3 and the second equation by −4.

$$\begin{aligned} 12x+1350y &= 721.50 \\ -12x-800y &= -584.00 \\ \hline 550y &= 137.5 \\ y &= 0.25 \end{aligned}$$

Let $y = 0.25$ in the second equation.

$$\begin{aligned} 3x+200(0.25) &= 146.00 \\ 3x+50 &= 146.00 \\ 3x &= 96.00 \\ x &= 32.00 \end{aligned}$$

The daily fee is $32 and the mileage charge is $0.25 per mile.

25.

	d	$= r$	$\cdot\ t$
Downstream	18	$x+y$	2
Upstream	18	$x-y$	$4\frac{1}{2}$

$$\begin{cases} 2(x+y) = 18 \\ \frac{9}{2}(x-y) = 18 \end{cases}$$

Multiply the first equation by $\frac{1}{2}$ and the second equation by $\frac{2}{9}$.

$$\begin{aligned} x+y &= 9 \\ x-y &= 4 \\ \hline 2x \quad &= 13 \\ x &= 6.5 \end{aligned}$$

Let $x = 6.5$ in $x + y = 9$.

$$\begin{aligned} 6.5+y &= 9 \\ y &= 2.5 \end{aligned}$$

Pratap can row 6.5 miles per hour in still water. The rate of the current is 2.5 miles per hour.

27.

	d	$= r$	$\cdot\ t$
With the wind	780	$x+y$	$1\frac{1}{2}$
Into the wind	780	$x-y$	2

$$\begin{cases} \frac{3}{2}(x+y) = 780 \\ 2(x-y) = 780 \end{cases}$$

Multiply the first equation by $\frac{2}{3}$ and the second equation by $\frac{1}{2}$.

$$\begin{aligned} x+y &= 520 \\ x-y &= 390 \\ \hline 2x \quad &= 910 \\ x &= 455 \end{aligned}$$

Let $x = 455$ in $x + y = 520$.

$$\begin{aligned} 455+y &= 520 \\ y &= 65 \end{aligned}$$

The plane can fly 455 miles per hour in still air. The speed of the wind is 65 miles per hour.

29. Let x = the time spent walking and y = the time spent on the bicycle.

	r	$\cdot\ t$	$= d$
Walking	4	x	$4x$
Biking	20	y	$20y$

$$\begin{cases} x+y = 6 \\ 4x+20y = 96 \end{cases}$$

Multiply the first equation by -4.

$$\begin{aligned} -4x-4y &= -24 \\ 4x+20y &= 96 \\ \hline 16y &= 72 \\ y &= 4.5 \end{aligned}$$

He spent $4\frac{1}{2}$ hours on the bicycle.

31. Let x = ounces of 4% solution and y = ounces of 12% solution.

Concentration Rate	Ounces of Solution	Ounces of Pure Acid
0.04	x	$0.04x$
0.12	y	$0.12y$
0.09	12	0.09(12)

$$\begin{cases} x+y = 12 \\ 0.04x+0.12y = 0.09(12) \end{cases}$$

Multiply the first equation by -4 and the second equation by 100.

$$\begin{aligned} -4x-4y &= -48 \\ 4x+12y &= 108 \\ \hline 8y &= 60 \\ y &= 7.5 \end{aligned}$$

Let $y = 7.5$ in the first equation.

$x + 7.5 = 12$
$x = 4.5$

$4\frac{1}{2}$ ounces of 4% solution and $7\frac{1}{2}$ ounces of 12% solution should be mixed.

33. Let x = pounds of \$4.95 per pound beans and y = pounds of \$2.65 per pound beans.

	Cost Rate	Pounds of Beans	Dollars Cost
High Quality	4.95	x	$4.95x$
Low Quality	2.65	y	$2.65y$
Mixture	3.95	200	3.95(200)

$$\begin{cases} x + y = 200 \\ 4.95x + 2.65y = 3.95(200) \end{cases}$$

Solve the first equation for y.
$y = 200 - x$
Substitute $200 - x$ for y in the second equation.

$$4.95x + 2.65(200 - x) = 3.95(200)$$
$$4.95x + 530 - 2.65x = 790$$
$$2.30x = 260$$
$$x \approx 113.04$$

Let $x = 113.04$ in the first equation.
$113.04 + y = 200$
$y \approx 86.96$
He needs 113 pounds of \$4.95 per pound beans and 87 pounds of \$2.65 per pound beans.

35. Let x = the first angle and y = the second angle.

$$\begin{cases} x + y = 90 \\ x = 2y \end{cases}$$

Substitute $2y$ for x in the first equation.
$2y + y = 90$
$3y = 90$
$y = 30$
Let $y = 30$ in the second equation.
$x = 2(30) = 60$
The angles are 60° and 30°.

37. Let x = the first angle and y = the second angle.

$$\begin{cases} x + y = 90 \\ x = 3y + 10 \end{cases}$$

Substitute $3y + 10$ for x in the first equation.
$3y + 10 + y = 90$
$4y = 80$
$y = 20$
Let $y = 20$ in the second equation.
$x = 3(20) + 10 = 70$
The angles are 70° and 20°.

39. Let x = the number sold at \$9.50 and y = the number sold at \$7.50.

$$\begin{cases} x + y = 90 \\ 9.5x + 7.5y = 721 \end{cases}$$

Solve the first equation for y.
$y = 90 - x$
Substitute $90 - x$ for y in the second equation.

$$9.5x + 7.5(90 - x) = 721$$
$$9.5x + 675 - 7.5x = 721$$
$$2x = 46$$
$$x = 23$$

Let $x = 23$ in $y = 90 - x$.
$y = 90 - 23 = 67$
They sold 23 at \$9.50 and 67 at \$7.50.

41. Let x = the rate of the faster group and y = the rate of the slower group.

	r	$\cdot\ t$	$= d$
Faster group	x	240	$240x$
Slower group	y	240	$240y$

$$\begin{cases} x = y + \frac{1}{2} \\ 240x + 240y = 1200 \end{cases}$$

Substitute $y + \frac{1}{2}$ for x in the second equation.

$$240\left(y + \frac{1}{2}\right) + 240y = 1200$$
$$240y + 120 + 240y = 1200$$
$$480y = 1080$$
$$y = \frac{1080}{480} = 2\frac{1}{4}$$

Let $y = 2\frac{1}{4}$ in the first equation.

$$x = 2\frac{1}{4} + \frac{1}{2} = 2\frac{3}{4}$$

The rate of the faster group is $2\frac{3}{4}$ miles per

hour. The rate of the slower group is $2\frac{1}{4}$ miles per hour.

43. Let x = gallons of 30% solution and y = gallons of 60% solution.

Concentration Rate	Gallons of Solution	Gallons of Pure Fertilizer
0.30	x	$0.30x$
0.60	y	$0.60y$
0.50	150	0.50(150)

$$\begin{cases} x+y=150 \\ 0.30x+0.60y=0.50(150) \end{cases}$$

Multiply the first equation by −3 and the second equation by 10.

$$\begin{aligned} -3x-3y&=-450 \\ 3x+6y&=750 \\ \hline 3y&=300 \\ y&=100 \end{aligned}$$

Let y = 100 in the first equation.

$$\begin{aligned} x+100&=150 \\ x&=50 \end{aligned}$$

50 gallons of 30% solution and 100 gallons of 60% solution.

45. Let x = the width and y = the length.

$$\begin{cases} 2x+2y=144 \\ y=x+12 \end{cases}$$

Substitute x + 12 for y in the first equation.

$$\begin{aligned} 2x+2(x+12)&=144 \\ 2x+2x+24&=144 \\ 4x&=120 \\ x&=30 \end{aligned}$$

Let x = 30 in the second equation.

$y = 30 + 12 = 42$

The width is 30 inches and the length is 42 inches.

47. $\begin{cases} y=5.3x+39.5 & (1) \\ y=4.5x+45.5 & (2) \end{cases}$

Substitute $5.3x + 39.5$ for y in E2.

$$\begin{aligned} 5.3x+39.5&=4.5x+45.5 \\ 0.8x&=6 \\ x&=7.5 \end{aligned}$$

2000 + 7 = 2007

The year was 2007.

49. a. Answers may vary, but depend on the slope of each function.

b. $\begin{cases} y=-1379.4x+150{,}604 & (1) \\ y=478.4x+157{,}838 & (2) \end{cases}$

Substitute $478.4x + 157{,}838$ for y in E1.

$$\begin{aligned} 478.4x+157{,}838&=-1379.4x+150{,}604 \\ 1857.8x&=-7234 \\ x&=\frac{-7234}{1857.8}\approx-3.89 \end{aligned}$$

1995 + (−3.89) = 1991.11

They were the same in 1991.

51. $\begin{cases} x+y=180 \\ x=y-30 \end{cases}$

Replace x with $y-30$ in E1.

$$\begin{aligned} (y-30)+y&=180 \\ 2y-30&=180 \\ 2y&=210 \\ y&=105 \end{aligned}$$

Replace y with 105 in E2.

$x=105-30=75$

The value of x is 75° and the value of y is 105°.

53. $C(x)=30x+10{,}000$

$R(x)=46x$

$$\begin{aligned} 46x&=30x+10{,}000 \\ 16x&=10{,}000 \\ x&=625 \end{aligned}$$

625 units

55. $C(x)=1.2x+1500$

$R(x)=1.7x$

$$\begin{aligned} 1.7x&=1.2x+1500 \\ 0.5x&=1500 \\ x&=3000 \end{aligned}$$

3000 units

57. $C(x)=75x+160{,}000$

$R(x)=200x$

$$\begin{aligned} 200x&=75x+160{,}000 \\ 125x&=160{,}000 \\ x&=1280 \end{aligned}$$

1280 units

59. a. $R(x)=450x$

b. $C(x)=200x+6000$

c. $R(x) = C(x)$
$450x = 200x + 6000$
$250x = 6000$
$x = 24$ desks

61. Let x = units of Mix A, y = units of Mix B, and z = units of Mix C.

$$\begin{cases} 4x + 6y + 4z = 30 & (1) \\ 6x + y + z = 16 & (2) \\ 3x + 2y + 12z = 24 & (3) \end{cases}$$

Multiply E2 by −6 and add to E1.

$$\begin{array}{l} -36x - 6y - 6z = -96 \\ \underline{4x + 6y + 4z = 30} \\ -32x \quad - 2z = -66 \text{ or } 16x + z = 33 \quad (4) \end{array}$$

Multiply E2 by −2 and add to E3.

$$\begin{array}{l} -12x - 2y - 2z = -32 \\ \underline{3x + 2y + 12z = 24} \\ -9x + \quad 10z = -8 \quad (5) \end{array}$$

Multiply E4 by −10 and add to E5.

$$\begin{array}{l} -160x - 10z = -330 \\ \underline{-9x + 10z = -8} \\ -169x \quad = -338 \\ x = 2 \end{array}$$

Replace x with 2 in E4.
$16(2) + z = 33$
$32 + z = 33$
$z = 1$

Replace x with 2 and z with 1 in E2.
$6(2) + y + 1 = 16$
$y = 3$

You need 2 units of Mix A, 3 units of Mix B, and 1 unit of Mix C.

63. Let x = length of shortest side,
y = length of longest side, and
z = length of the other two sides

$$\begin{cases} x + y + 2z = 29 & (1) \\ y = 2x & (2) \\ z = x + 2 & (3) \end{cases}$$

Substitute $y = 2x$ and $z = x + 2$ in E1.
$x + (2x) + 2(x + 2) = 29$
$x + 2x + 2x + 4 = 29$
$5x = 25$
$x = 5$

Replace x with 5 in E2 and E3.
$y = 2(5)$ $\quad z = 5 + 2$
$y = 10$ $\quad z = 7$

The sides are 5 in., 7 in., 7 in., and 10 in.

65. Let x = the first number
y = the second number, and
z = the third number.

$$\begin{cases} x + y + z = 40 \\ x = y + 5 \\ x = 2z \end{cases}$$

$$\begin{cases} x + y + z = 40 & (1) \\ x - y = 5 & (2) \\ x - 2z = 0 & (3) \end{cases}$$

Add E1 and E2.
$2x + z = 45$ (4)

Multiply E3 by −2 and add to E4.

$$\begin{array}{l} -2x + 4z = 0 \\ \underline{2x + z = 45} \\ 5z = 45 \\ z = 9 \end{array}$$

Replace z with 9 in E3.
$x - 2(9) = 0$
$x = 18$

Replace x with 18 in E2.
$18 - y = 5$
$y = 13$

The numbers are 18, 13, and 9.

67. Let x = number of free throws,
y = number of two-point field goals, and
z = number of three-point field goals

$$\begin{cases} x + 2y + 3z = 860 \\ y = 2z - 65 \\ x = y - 34 \end{cases}$$

$$\begin{cases} x + 2y + 3z = 860 & (1) \\ y - 2z = -65 & (2) \\ x - y = -34 & (3) \end{cases}$$

Multiply E3 by −1 and add to E1.

$$\begin{array}{l} -x + y = 34 \\ \underline{x + 2y + 3z = 860} \\ 3y + 3z = 894 \text{ or } y + z = 298 \quad (4) \end{array}$$

Multiply E4 by −1 and add to E2.

$$\begin{array}{l} -y - z = -298 \\ \underline{y - 2z = -65} \\ -3z = -363 \\ z = 121 \end{array}$$

Replace z with 121 in E2.
$y - 2(121) = -65$
$y - 242 = -65$
$y = 177$

Replace y with 177 in E3.
$x - 177 = -34$
$x = 143$

She made 143 free throws, 177 two-point field goals, and 121 three-point field goals.

69. $\begin{cases} x+y+z=180 \\ y+2x+5=180 \\ z+2x-5=180 \end{cases}$

$\begin{cases} x+y+z=180 & (1) \\ 2x+y \quad =175 & (2) \\ 2x+ \quad z=185 & (3) \end{cases}$

Multiply E1 by −1 and add to E2.

$-x-y-z=-180$
$2x+y \quad =175$
$x \quad -z=-5 \quad (4)$

Add E3 and E4.

$3x=180$
$x=60$

Replace x with 60 in E4.

$60-z=-5$
$z=65$

Replace x with 60 in E2.

$2(60)+y=175$
$120+y=175$
$y=55$

$x = 60$, $y = 55$, and $z = 65$

71. $-3x<-9$

$\frac{-3x}{-3}>\frac{-9}{-3}$

$x>3$, $(3, \infty)$

73. $4(2x-1)\geq 0$

$8x-4\geq 0$

$8x\geq 4$

$x\geq\frac{1}{2}$, $\left[\frac{1}{2}, \infty\right)$

75. The minimum price is $0.49.
The maximum price is $0.65.
0.72 > 0.65 Impossible
0.29 < 0.49 Impossible
0.49 < 0.58 < 0.65 Possible
The answer is **a**.

77. Let x = the width and y = the length.

$\begin{cases} 2x+y=33 \\ y=2x-3 \end{cases}$

Substitute $2x - 3$ for y in the first equation.

$2x+2x-3=33$
$4x=36$
$x=9$

Let $x = 9$ in the second equation.

$y = 2(9) - 3 = 15$

The width is 9 feet and the length is 15 feet.

79. a. $\begin{cases} y=0.82x+17.2 \\ y=0.33x+30.5 \end{cases}$

Substitute $0.82x + 17.2$ for y in the second equation.

$0.82x+17.2=0.33x+30.5$
$0.49x=13.3$
$x\approx 27.14$

Let $x = 27.14$ in $y = 0.82x + 17.2$.

$y = 0.82(27.14) + 17.2$
$y\approx 39.4548$

The solution is approximately (27.1, 39.5).

b. For viewers who are 27.1 years over 18 (or 45.1 years of age) the percent who watch cable news and network news is the same, 39.5%.

c. Answers may vary

81. a. Replace $f(x)$ with y in each equation.

$\begin{cases} y=0.85x+41.75 & (1) \\ y=1.13x+10.49 & (2) \end{cases}$

Substitute $0.85x + 41.75$ for y in E2.

$0.85x+41.75=1.13x+10.49$
$31.26=0.28x$
$\frac{31.26}{0.28}=x$
$112\approx x$

Replace x with 112 in E1.

$y = 0.85(112) + 41.75 \approx 137$

The solution is (112, 137).

b. 112 months is 9 years, 4 months.
Since the first month is February 2007, the 112th month is 9 years and 4 months later, or June 2016.

Chapter 4 Vocabulary Check

1. In a system of linear equations in two variables, if the graphs of the equations are the same, the equations are dependent equations.
2. Two or more linear equations are called a system of linear equations.
3. A system of equations that has at least one solution is called a consistent system.
4. A solution of a system of two equations in two variables is an ordered pair of numbers that is a solution of both equations in the system.

5. Two algebraic methods for solving systems of equations are addition and substitution.

6. A system of equations that has no solution is called an inconsistent system.

7. In a system of linear equations in two variables, if the graphs of the equations are different, the equations are independent equations.

Chapter 4 Review

1. a. Let $x = 12$ and $y = 4$.

$$2x-3y=12$$
$$2(12)-3(4) \stackrel{?}{=} 12$$
$$24-12 \stackrel{?}{=} 12$$
$$12=12 \quad \text{True}$$

$$3x+4y=1$$
$$3(12)+4(4) \stackrel{?}{=} 1$$
$$36+16 \stackrel{?}{=} 1$$
$$52=1 \quad \text{False}$$

(12, 4) is not a solution of the system.

b. Let $x = 3$ and $y = -2$.

$$2x-3y=12$$
$$2(3)-3(-2) \stackrel{?}{=} 12$$
$$6+6 \stackrel{?}{=} 12$$
$$2=12 \quad \text{True}$$

$$3x+4y=1$$
$$3(3)+4(-2) \stackrel{?}{=} 1$$
$$9-8 \stackrel{?}{=} 1$$
$$1=1 \quad \text{True}$$

(3, −2) is a solution of the system.

c. Let $x = -3$ and $y = 6$.

$$2x-3y=12$$
$$2(-3)-3(6) \stackrel{?}{=} 12$$
$$-6-18 \stackrel{?}{=} 12$$
$$-24=12 \quad \text{False}$$

$$3x+4y=1$$
$$3(-3)+4(6) \stackrel{?}{=} 1$$
$$-9+24 \stackrel{?}{=} 1$$
$$15=1 \quad \text{False}$$

(−3, 6) is not a solution of the system.

2. a. Let $x = \frac{3}{4}$ and $y = -3$.

$$4x+y=0$$
$$4\left(\frac{3}{4}\right)-3 \stackrel{?}{=} 0$$
$$3-3 \stackrel{?}{=} 0$$
$$0=0 \quad \text{True}$$

$$-8x-5y=9$$
$$-8\left(\frac{3}{4}\right)-5(-3) \stackrel{?}{=} 9$$
$$-6+15 \stackrel{?}{=} 9$$
$$9=9 \quad \text{True}$$

$\left(\frac{3}{4}, -3\right)$ is a solution of the system.

b. Let $x = -2$ and $y = 8$.

$$4x+y=0$$
$$4(-2)+8 \stackrel{?}{=} 0$$
$$-8+8 \stackrel{?}{=} 0$$
$$0=0 \quad \text{True}$$

$$-8x-5y=9$$
$$-8(-2)-5(8) \stackrel{?}{=} 9$$
$$16-40 \stackrel{?}{=} 9$$
$$-24=9 \quad \text{False}$$

(−2, 8) is not a solution of the system.

c. Let $x = \frac{1}{2}$ and $y = -2$.

$$4x+y=0$$
$$4\left(\frac{1}{2}\right)-2 \stackrel{?}{=} 0$$
$$2-2 \stackrel{?}{=} 0$$
$$0=0 \quad \text{True}$$

$$-8x-5y=9$$
$$-8\left(\frac{1}{2}\right)-5(-2) \stackrel{?}{=} 9$$
$$-4+10 \stackrel{?}{=} 9$$
$$6=9 \quad \text{False}$$

$\left(\frac{1}{2}, -2\right)$ is not a solution of the system.

3. a. Let $x = -6$ and $y = -8$.

$$5x-6y=18$$
$$5(-6)-6(-8) \stackrel{?}{=} 18$$
$$-30+48 \stackrel{?}{=} 18$$
$$18=18 \quad \text{True}$$

$$\begin{aligned} 2y - x &= -4 \\ 2(-8) - (-6) &\stackrel{?}{=} -4 \\ -16 + 6 &\stackrel{?}{=} -4 \\ -10 &= -4 \quad \text{False} \end{aligned}$$

$(-6, -8)$ is not a solution of the system.

b. Let $x = 3$ and $y = \frac{5}{2}$.

$$\begin{aligned} 5x - 6y &= 18 \\ 5(3) - 6\left(\frac{5}{2}\right) &\stackrel{?}{=} 18 \\ 15 - 15 &\stackrel{?}{=} 18 \\ 0 &= 18 \quad \text{False} \end{aligned}$$

$$\begin{aligned} 2y - x &= -4 \\ 2\left(\frac{5}{2}\right) - 3 &\stackrel{?}{=} -4 \\ 5 - 3 &\stackrel{?}{=} -4 \\ 2 &= -4 \quad \text{False} \end{aligned}$$

$\left(3, \frac{5}{2}\right)$ is not a solution of the system.

c. Let $x = 3$ and $y = -\frac{1}{2}$.

$$\begin{aligned} 5x - 6y &= 18 \\ 5(3) - 6\left(-\frac{1}{2}\right) &\stackrel{?}{=} 18 \\ 15 + 3 &\stackrel{?}{=} 18 \\ 18 &= 18 \quad \text{True} \end{aligned}$$

$$\begin{aligned} 2y - x &= -4 \\ 2\left(-\frac{1}{2}\right) - 3 &\stackrel{?}{=} -4 \\ -1 - 3 &\stackrel{?}{=} -4 \\ -4 &= -4 \quad \text{True} \end{aligned}$$

$\left(3, -\frac{1}{2}\right)$ is a solution of the system.

4. a. Let $x = 2$ and $y = 2$.

$$\begin{aligned} 2x + 3y &= 1 \\ 2(2) + 3(2) &\stackrel{?}{=} 1 \\ 4 + 6 &\stackrel{?}{=} 1 \\ 10 &= 1 \quad \text{False} \end{aligned} \qquad \begin{aligned} 3y - x &= 4 \\ 3(2) - 2 &\stackrel{?}{=} 4 \\ 6 - 2 &\stackrel{?}{=} 4 \\ 4 &= 4 \quad \text{True} \end{aligned}$$

$(2, 2)$ is not a solution of the system.

b. Let $x = -1$ and $y = 1$.

$$\begin{aligned} 2x + 3y &= 1 \\ 2(-1) + 3(1) &\stackrel{?}{=} 1 \\ -2 + 3 &\stackrel{?}{=} 1 \\ 1 &= 1 \quad \text{True} \end{aligned} \qquad \begin{aligned} 3y - x &= 4 \\ 3(1) - (-1) &\stackrel{?}{=} 4 \\ 3 + 1 &\stackrel{?}{=} 4 \\ 4 &= 4 \quad \text{True} \end{aligned}$$

$(-1, 1)$ is a solution of the system.

c. Let $x = 2$ and $y = -1$.

$$\begin{aligned} 2x + 3y &= 1 \\ 2(2) + 3(-1) &\stackrel{?}{=} 1 \\ 4 - 3 &\stackrel{?}{=} 1 \\ 1 &= 1 \quad \text{True} \end{aligned}$$

$$\begin{aligned} 3y - x &= 4 \\ 3(-1) - 2 &\stackrel{?}{=} 4 \\ -3 - 2 &\stackrel{?}{=} 4 \\ -5 &= 4 \quad \text{False} \end{aligned}$$

$(2, -1)$ is not a solution of the system.

5. $\begin{cases} x + y = 5 \\ x - 1 = y \end{cases}$

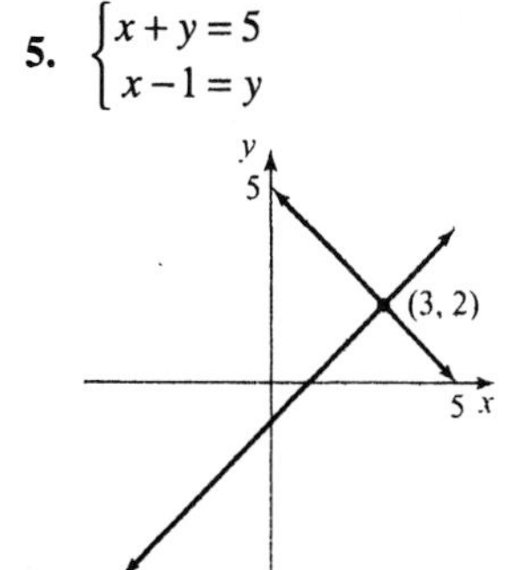

The solution of the system is $(3, 2)$.

6. $\begin{cases} x + y = 3 \\ x - y = -1 \end{cases}$

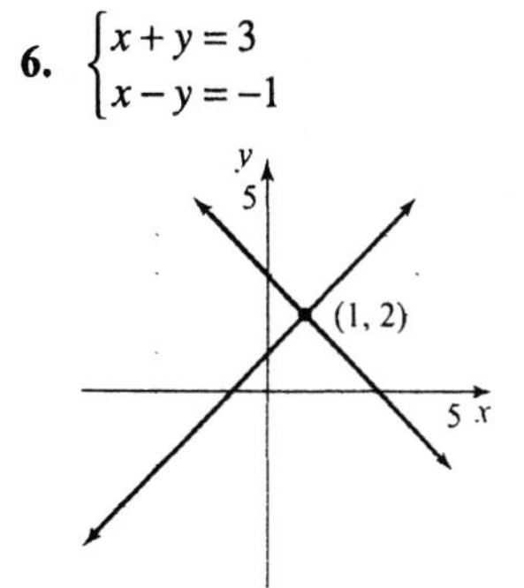

The solution of the system is $(1, 2)$.

7. $\begin{cases} x = 5 \\ y = -1 \end{cases}$

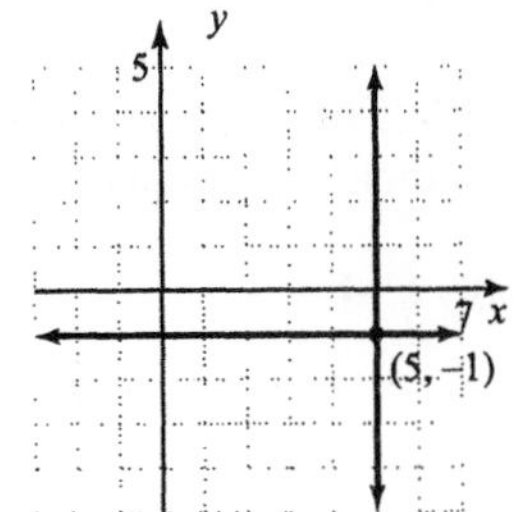

The solution of the system is (5, −1).

8. $\begin{cases} x = -3 \\ y = 2 \end{cases}$

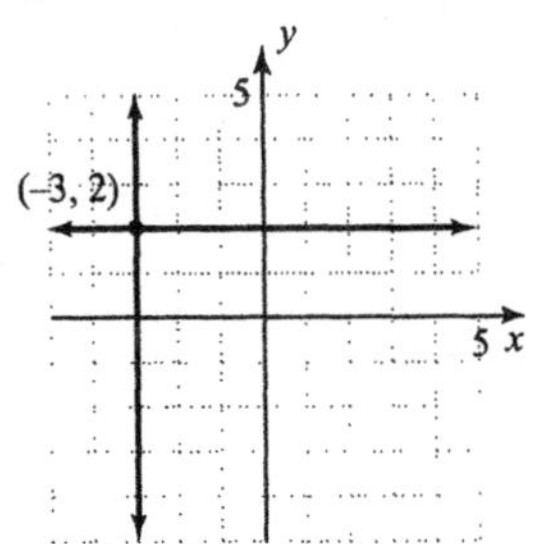

The solution of the system is (−3, 2).

9. $\begin{cases} 2x + y = 5 \\ x = -3y \end{cases}$

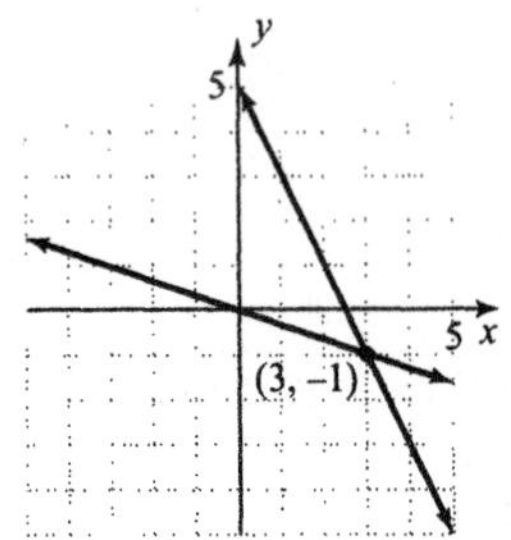

The solution of the system is (3, −1).

10. $\begin{cases} 3x + y = -2 \\ y = -5x \end{cases}$

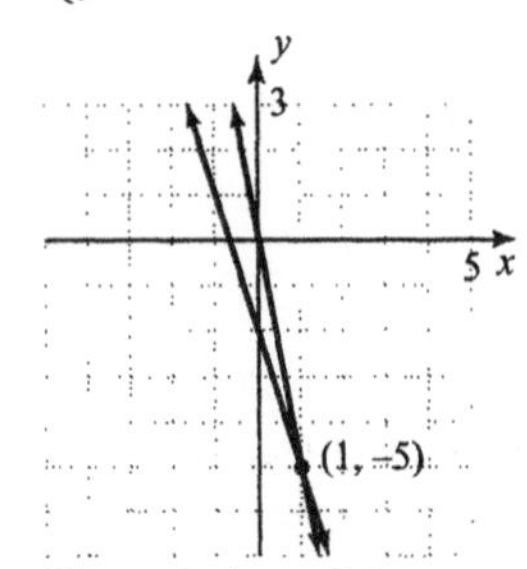

The solution of the system is (1, −5).

11. $\begin{cases} y = 3x \\ -6x + 2y = 6 \end{cases}$

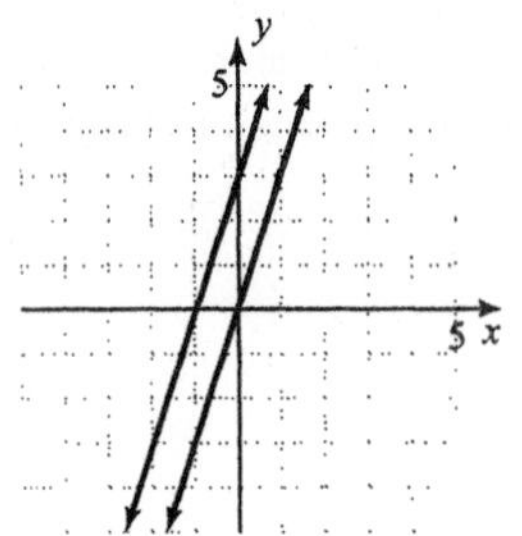

There is no solution.

12. $\begin{cases} x - 2y = 2 \\ -2x + 4y = -4 \end{cases}$

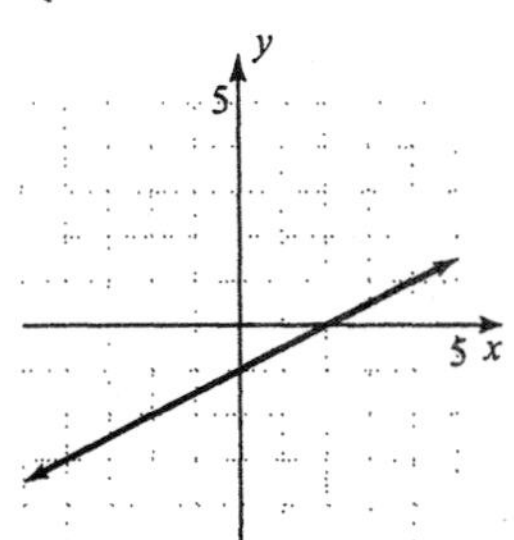

There is an infinite number of solutions.

13. $\begin{cases} y = 2x + 6 \\ 3x - 2y = -11 \end{cases}$

Substitute $2x + 6$ for y in the second equation.

$$\begin{aligned} 3x - 2(2x + 6) &= -11 \\ 3x - 4x - 12 &= -11 \\ -x &= 1 \\ x &= -1 \end{aligned}$$

Let $x = -1$ in the first equation.

$y = 2(-1) + 6 = 4$

The solution is (−1, 4).

14. $\begin{cases} y = 3x - 7 \\ 2x - 3y = 7 \end{cases}$

Substitute $3x - 7$ for y in the second equation.

$$\begin{aligned} 2x - 3(3x - 7) &= 7 \\ 2x - 9x + 21 &= 7 \\ -7x &= -14 \\ x &= 2 \end{aligned}$$

Let $x = 2$ in the first equation.

$y = 3(2) - 7 = -1$

The solution is (2, −1).

15. $\begin{cases} x+3y=-3 \\ 2x+y=4 \end{cases}$

Solve the first equation for x.

$x=-3y-3$

Substitute $-3y-3$ for x in the second equation.

$$\begin{aligned} 2(-3y-3)+y&=4 \\ -6y-6+y&=4 \\ -5y&=10 \\ y&=-2 \end{aligned}$$

Let $y=-2$ in $x=-3y-3$.

$x=-3(-2)-3=3$

The solution is (3, −2).

16. $\begin{cases} 3x+y=11 \\ x+2y=12 \end{cases}$

Solve the first equation for y.

$y=11-3x$

Substitute $11-3x$ for y in the second equation.

$$\begin{aligned} x+2(11-3x)&=12 \\ x+22-6x&=12 \\ -5x&=-10 \\ x&=2 \end{aligned}$$

Let $x=2$ in $y=11-3x$.

$y=11-3(2)=5$

The solution is (2, 5).

17. $\begin{cases} 4y=2x+6 \\ x-2y=-3 \end{cases}$

Solve the second equation for x.

$x=2y-3$

Substitute $2y-3$ for x in the first equation.

$$\begin{aligned} 4y&=2(2y-3)+6 \\ 4y&=4y-6+6 \\ 0&=0 \end{aligned}$$

The system has an infinite number of solutions.

18. $\begin{cases} 9x=6y+3 \\ 6x-4y=2 \end{cases}$

Solve the first equation for y.

$$\begin{aligned} 9x&=6y+3 \\ 9x-3&=6y \\ \frac{3}{2}x-\frac{1}{2}&=y \end{aligned}$$

Substitute $\frac{3}{2}x-\frac{1}{2}$ for y in the second equation.

$$\begin{aligned} 6x-4\left(\frac{3}{2}x-\frac{1}{2}\right)&=2 \\ 6x-6x+2&=2 \\ 2&=2 \end{aligned}$$

The system has an infinite number of solutions.

19. $\begin{cases} x+y=6 \\ y=-x-4 \end{cases}$

Substitute $-x-4$ for y in the first equation.

$$\begin{aligned} x+(-x-4)&=6 \\ x-x-4&=6 \\ -4&=6 \quad \text{False} \end{aligned}$$

There is no solution.

20. $\begin{cases} -3x+y=6 \\ y=3x+2 \end{cases}$

Substitute $3x+2$ for y in the first equation.

$$\begin{aligned} -3x+(3x+2)&=6 \\ -3x+3x+2&=6 \\ 2&=6 \quad \text{False} \end{aligned}$$

There is no solution.

21. $\begin{cases} 2x+3y=-6 \\ x-3y=-12 \end{cases}$

$$\begin{aligned} 2x+3y&=-6 \\ x-3y&=-12 \\ \hline 3x&=-18 \\ x&=-6 \end{aligned}$$

Let $x=-6$ in the first equation.

$$\begin{aligned} 2(-6)+3y&=-6 \\ -12+3y&=-6 \\ 3y&=6 \\ y&=2 \end{aligned}$$

The solution of the system is (−6, 2).

22. $\begin{cases} 4x+y=15 \\ -4x+3y=-19 \end{cases}$

$$\begin{aligned} 4x+y&=15 \\ -4x+3y&=-19 \\ \hline 4y&=-4 \\ y&=-1 \end{aligned}$$

Let $y=-1$ in the first equation.

$$\begin{aligned} 4x+(-1)&=15 \\ 4x-1&=15 \\ 4x&=16 \\ x&=4 \end{aligned}$$

The solution of the system is (4, −1).

23. $\begin{cases} 2x-3y=-15 \\ x+4y=31 \end{cases}$

Multiply the second equation by −2.

$$\begin{aligned} 2x-3y&=-15 \\ -2x-8y&=-62 \\ \hline -11y&=-77 \\ y&=7 \end{aligned}$$

Let $y = 7$ in the second equation.

$$\begin{aligned} x+4(7)&=31 \\ x+28&=31 \\ x&=3 \end{aligned}$$

The solution of the system is (3, 7).

24. $\begin{cases} x-5y=-22 \\ 4x+3y=4 \end{cases}$

Multiply the first equation by −4.

$$\begin{aligned} -4x+20y&=88 \\ 4x\ \ +3y&=4 \\ \hline 23y&=92 \\ y&=4 \end{aligned}$$

Let $y = 4$ in the first equation.

$$\begin{aligned} x-5(4)&=-22 \\ x-20&=-22 \\ x&=-2 \end{aligned}$$

The solution of the system is (−2, 4).

25. $\begin{cases} 2x-6y=-1 \\ -x+3y=\frac{1}{2} \end{cases}$

Multiply the second equation by 2.

$$\begin{aligned} 2x-6y&=-1 \\ -2x+6y&=1 \\ \hline 0&=0 \end{aligned}$$

There is an infinite number of solutions.

26. $\begin{cases} 0.6x-0.3y=-1.5 \\ 0.04x-0.02y=-0.1 \end{cases}$

Multiply the first equation by 20 and the second equation by −300.

$$\begin{aligned} 12x-6y&=-30 \\ -12x+6y&=30 \\ \hline 0&=0 \end{aligned}$$

There are an infinite number of solutions.

27. $\begin{cases} \frac{3}{4}x+\frac{2}{3}y=2 \\ x+\frac{y}{3}=6 \end{cases}$

Multiply the first equation by 12 and the second equation by 3.

$\begin{cases} 9x+8y=24 \\ 3x+y=18 \end{cases}$

Multiply the second equation in the simplified system by −3.

$$\begin{aligned} 9x+8y&=24 \\ -9x-3y&=-54 \\ \hline 5y&=-30 \\ y&=-6 \end{aligned}$$

Let $y = -6$ in $3x + y = 18$.

$$\begin{aligned} 3x+(-6)&=18 \\ 3x&=24 \\ x&=8 \end{aligned}$$

The solution of the system is (8, −6).

28. $\begin{cases} 10x+2y=0 \\ 3x+5y=33 \end{cases}$

Multiply the first equation by −5 and the second equation by 2.

$$\begin{aligned} -50x-10y&=0 \\ 6x+10y&=66 \\ \hline -44x\qquad&=66 \\ x&=-\frac{3}{2} \end{aligned}$$

Let $x=-\frac{3}{2}$ in the first equation.

$$\begin{aligned} 10\left(-\frac{3}{2}\right)+2y&=0 \\ -15+2y&=0 \\ 2y&=15 \\ y&=\frac{15}{2} \end{aligned}$$

The solution is $\left(-\frac{3}{2}, \frac{15}{2}\right)$.

29. $\begin{cases} x+\qquad z=4 & (1) \\ 2x-y\qquad=4 & (2) \\ x+y-z=0 & (3) \end{cases}$

Adding E2 and E3 gives $3x-z=4$ (4)

Adding E1 and E4 gives $4x=8$ or $x=2$

Replace x with 2 in E1.

$$\begin{aligned} 2+z&=4 \\ z&=2 \end{aligned}$$

Replace x with 2 and z with 2 in E3.

$$\begin{aligned} 2+y-2&=0 \\ y&=0 \end{aligned}$$

The solution is (2, 0, 2).

30. $\begin{cases} 2x+5y \quad\quad = 4 & (1) \\ x-5y+z=-1 & (2) \\ 4x \quad\quad -z=11 & (3) \end{cases}$

Add E2 and E3.

$5x-5y=10 \quad (4)$

Add E1 and E4.

$7x=14$

$x=2$

Replace x with 2 in E1.

$2(2)+5y=4$

$4+5y=4$

$5y=0$

$y=0$

Replace x with 2 in E3.

$4(2)-z=11$

$8-z=11$

$z=-3$

The solution is (2, 0, –3).

31. $\begin{cases} \quad 4y+2z=5 & (1) \\ 2x+8y \quad\quad =5 & (2) \\ 6x \quad\quad +4z=1 & (3) \end{cases}$

Multiply E1 by –2 and add to E2.

$-8y-4z=-10$

$\underline{2x+8y \quad\quad =5}$

$2x \quad\quad -4z=-5 \quad (4)$

Add E3 and E4.

$8x=-4$

$x=-\frac{1}{2}$

Replace x with $-\frac{1}{2}$ in E2.

$2\left(-\frac{1}{2}\right)+8y=5$

$-1+8y=5$

$8y=6$

$y=\frac{3}{4}$

Replace x with $-\frac{1}{2}$ in E3.

$6\left(-\frac{1}{2}\right)+4z=1$

$-3+4z=1$

$4z=4$

$z=1$

The solution is $\left(-\frac{1}{2}, \frac{3}{4}, 1\right)$.

32. $\begin{cases} 5x+7y \quad\quad =9 & (1) \\ \quad 14y-z=28 & (2) \\ 4x \quad\quad +2z=-4 & (3) \end{cases}$

Dividing E3 by 2 gives $2x + z = -2$.

Add this equation to E2.

$2x \quad\quad +z=-2$

$\underline{\quad 14y-z=28}$

$2x+14y \quad =26$ or $x+7y=13 \quad (4)$

Multiply E4 by –1 and add to E1.

$-x-7y=-13$

$\underline{5x+7y=9}$

$4x \quad\quad =-4$

$x=-1$

Replace x with –1 in E4.

$-1+7y=13$

$7y=14$

$y=2$

Replace x with –1 in E3.

$4(-1)+2z=-4$

$-4+2z=-4$

$2z=0$

$z=0$

The solution is (–1, 2, 0).

33. $\begin{cases} 3x-2y+2z=5 & (1) \\ -x+6y+z=4 & (2) \\ 3x+14y+7z=20 & (3) \end{cases}$

Multiply E2 by 3 and add to E1.

$3x-2y+2z=5$

$\underline{-3x+18y+3z=12}$

$16y+5z=17 \quad (4)$

Multiply E3 by –1 and add to E1.

$3x-2y+2z=5$

$\underline{-3x-14y-7z=-20}$

$-16y-5z=-15 \quad (5)$

Add E4 and E5.

$16y+5z=17$

$\underline{-16y-5z=-15}$

$0=2$ False

The system is inconsistent. The solution is ∅.

34. $\begin{cases} x+2y+3z=11 & (1) \\ \quad y+2z=3 & (2) \\ 2x \quad\quad +2z=10 & (3) \end{cases}$

Multiply E2 by –2 and add to E1.

$x+2y+3z=11$

$\underline{-2y-4z=-6}$

$x \quad\quad -z=5 \quad (4)$

Multiply E4 by 2 and add to E3.

$$\begin{array}{l} 2x+2z=10 \\ \underline{2x-2z=10} \\ 4x \quad\quad = 20 \\ \quad x=5 \end{array}$$

Replace x with 5 in E3.

$$\begin{aligned} 2(5)+2z&=10 \\ 10+2z&=10 \\ 2z&=0 \\ z&=0 \end{aligned}$$

Replace z with 0 in E2.

$$\begin{aligned} y+2(0)&=3 \\ y+0&=3 \\ y&=3 \end{aligned}$$

The solution is (5, 3, 0).

35. $\begin{cases} 7x-3y+2z=0 & (1) \\ 4x-4y-\ z=2 & (2) \\ 5x+2y+3z=1 & (3) \end{cases}$

Multiply E2 by 2 and add to E1.

$$\begin{array}{l} 7x-3y+2z=0 \\ \underline{8x-8y-2z=4} \\ 15x-11y \quad = 4 \ \ (4) \end{array}$$

Multiply E2 by 3 and add to E3.

$$\begin{array}{l} 12x-12y-3z=6 \\ \underline{5x\ +2y+3z=1} \\ 17x-10y \quad =7 \ \ (5) \end{array}$$

Solve the new system.

$\begin{cases} 15x-11y=4 & (4) \\ 17x-10y=7 & (5) \end{cases}$

Multiply E4 by −10, multiply E5 by 11, and add.

$$\begin{array}{l} -150x+110y=-40 \\ \underline{187x-110y=77} \\ 37x \quad\quad =37 \\ \quad x=1 \end{array}$$

Replace x with 1 in E4.

$$\begin{aligned} 15(1)-11y&=4 \\ 15-11y&=4 \\ -11y&=-11 \\ y&=1 \end{aligned}$$

Replace x with 1 and y with 1 in E1.

$$\begin{aligned} 7(1)-3(1)+2z&=0 \\ 4+2z&=0 \\ 2z&=-4 \\ z&=-2 \end{aligned}$$

The solution is (1, 1, −2).

36. $\begin{cases} x-3y-5z=-5 & (1) \\ 4x-2y+3z=13 & (2) \\ 5x+3y+4z=22 & (3) \end{cases}$

Multiply E1 by −4 and add to E2.

$$\begin{array}{l} -4x+12y+20z=20 \\ \underline{4x-\ 2y\ +\ 3z=13} \\ \quad 10y+23z=33 \ \ (4) \end{array}$$

Multiply E1 by −5 and add to E3.

$$\begin{array}{l} -5x+15y+25z=25 \\ \underline{5x+\ 3y\ +\ 4z=22} \\ \quad 18y+29z=47 \ \ (5) \end{array}$$

Solve the new system.

$\begin{cases} 10y+23z=33 & (4) \\ 18y+29z=47 & (5) \end{cases}$

Multiply E4 by 9, multiply E5 by −5 and add.

$$\begin{array}{l} 90y+207z=297 \\ \underline{-90y-145z=-235} \\ \quad 62z=62 \\ \quad z=1 \end{array}$$

Replace z with 1 in E4.

$$\begin{aligned} 10y+23(1)&=33 \\ 10y&=10 \\ y&=1 \end{aligned}$$

Replace y with 1 and z with 1 in E1.

$$\begin{aligned} x-3(1)-5(1)&=-5 \\ x-8&=-5 \\ x&=3 \end{aligned}$$

The solution is (3, 1, 1).

37. Let x = the larger number and y = the smaller number.

$\begin{cases} x+y=16 \\ 3x-y=72 \end{cases}$

$$\begin{array}{l} x+y=16 \\ \underline{3x-y=72} \\ 4x \quad =88 \\ \quad x=22 \end{array}$$

Let $x = 22$ in the first equation.

$$\begin{aligned} 22+y&=16 \\ y&=-6 \end{aligned}$$

The numbers are −6 and 22.

38. Let x = the number of orchestra seats and y = the number of balcony seats.

$\begin{cases} x+y=360 \\ 45x+35y=15{,}150 \end{cases}$

Solve the first equation for x.

$x = 360 - y$

Substitute $360 - y$ for x in the second equation.

$$\begin{aligned} 45(360-y)+35y&=15{,}150 \\ 16{,}200-45y+35y&=15{,}150 \\ -10y&=-1050 \\ y&=105 \end{aligned}$$

Let $y = 105$ in $x = 360 - y$.
$x = 360 - 105 = 255$
There are 255 orchestra seats and 105 balcony seats.

39. Let x = the riverboat's speed in still water and y = the rate of the current.

	d =	r ·	t
Downriver	340	$x+y$	14
Upriver	340	$x-y$	19

$$\begin{cases} 14(x+y) = 340 \\ 19(x-y) = 340 \end{cases}$$

Multiply the first equation by $\frac{1}{14}$ and the second equation by $\frac{1}{19}$.

$$x + y = \frac{340}{14} \approx 24.29$$
$$x - y = \frac{340}{19} \approx 17.89$$
$$2x \approx 42.18$$
$$x \approx 21.09$$

Multiply the second equation of the simplified system by -1.

$$x + y \approx 24.29$$
$$-x + y \approx -17.89$$
$$2y \approx 6.4$$
$$y \approx 3.2$$

The riverboat's speed in still water is 21.1 miles per hour. The rate of the current is 3.2 miles per hour.

40. Let x = amount of 6% solution and y = amount of 14% solution.

Concentration Rate	Amount of Solution	Amount of Pure Acid
0.06	x	$0.06x$
0.14	y	$0.14y$
0.12	50	$0.12(50)$

$$\begin{cases} x + y = 50 \\ 0.06x + 0.14y = 0.12(50) \end{cases}$$

Multiply the first equation by -6 and the second equation by 100.

$$-6x - 6y = -300$$
$$6x + 14y = 600$$
$$8y = 300$$
$$y = 37.5$$

Let $y = 37.5$ in the first equation.

$$x + 37.5 = 50$$
$$x = 12.5$$

$12\frac{1}{2}$ cc of 6% solution and $37\frac{1}{2}$ cc of 14% solution.

41. Let x = the cost of an egg and y = the cost of a strip of bacon.

$$\begin{cases} 3x + 4y = 3.80 \\ 2x + 3y = 2.75 \end{cases}$$

Multiply the first equation by -2 and the second equation by 3.

$$-6x - 8y = -7.60$$
$$6x + 9y = 8.25$$
$$y = 0.65$$

Let $y = 0.65$ in the first equation.

$$3x + 4(0.65) = 3.80$$
$$3x + 2.60 = 3.80$$
$$3x = 1.20$$
$$x = 0.40$$

An egg costs 40¢ and a strip of bacon costs 65¢.

42. Let x = the time spent walking and y = the time spent jogging.

	r ·	t =	d
Walking	4	x	$4x$
Jogging	7.5	y	$7.5y$

$$\begin{cases} x + y = 3 \\ 4x + 7.5y = 15 \end{cases}$$

Multiply the first equation by -4.

$$-4x - 4y = -12$$
$$4x + 7.5y = 15$$
$$3.5y = 3$$
$$y \approx 0.857$$

Let $y = 0.857$ in the first equation.

$$x + 0.857 = 3$$
$$x \approx 2.143$$

He spent 2.14 hours walking and 0.86 hours jogging.

43. Let x = the number of pennies,
y = the number of nickels, and
z = the number dimes.

$$\begin{cases} x+y+z=53 & (1) \\ 0.01x+0.05y+0.10z=2.77 & (2) \\ y=z+4 & (3) \end{cases}$$

Clear the decimals from E2 by multiplying by 100.

$x+5y+10z=277$ (4)

Replace y with $z+4$ in E1.

$x+z+4+z=53$
$x+2z=49$ (5)

Replace y with $z+4$ in E4.

$x+5(z+4)+10z=277$
$x+15z=257$ (6)

Solve the new system.

$$\begin{cases} x+\ 2z=49 & (5) \\ x+15z=257 & (6) \end{cases}$$

Multiply E5 by −1 and add to E6.

$-x-\ 2z=-49$
$x+15z=257$
$13z=208$
$z=16$

Replace z with 16 in E3.

$x+2(16)=49$
$x+32=49$
$x=17$

Replace z with 16 in E3.

$y=16+4=20$

He has 17 pennies, 20 nickels, and 16 dimes in his jar.

44. Let c = pounds of chocolate used,
n = pounds of nuts used, and
r = pounds of raisins used.

$$\begin{cases} r=2n & (1) \\ c+n+r=45 & (2) \\ 3.00c+2.70n+2.25r=2.80(45) & (3) \end{cases}$$

Replace r with $2n$ in E2.

$c+n+2n=45$
$c+3n=45$
$c=-3n+45$

Replace r with $2n$ and c with $-3n+45$ in E3.

$3.00(-3n+45)+2.70n+2.25(2n)=126$
$-9n+135+2.7n+4.5n=126$
$-1.8n+135=126$
$-1.8n=-9$
$n=5$

Replace n with 5 in E1.

$r=2(5)=10$

Replace n with 5 and r with 10 in E2.

$c+5+10=45$
$c+15=45$
$c=30$

She should use 30 pounds of creme-filled chocolates, 5 pounds of chocolate-covered nuts, and 10 pounds of chocolate-covered raisins.

45. Let x = length of the equal side and
y = length of the third side.

$$\begin{cases} 2x+y=73 & (1) \\ y=x+7 & (2) \end{cases}$$

Replace y with $x+7$ in E1.

$2x+x+7=73$
$3x=66$
$x=22$

Replace x with 22 in E2.

$y=22+7=29$

Two sides of the triangle have length 22 cm and the third side has length 29 cm.

46. Let f = the first number, s = the second number, and t = the third number.

$$\begin{cases} f+s+t=295 & (1) \\ f=s+5 & (2) \\ f=2t & (3) \end{cases}$$

Solve E2 for s and E3 for t.

$s=f-5$
$t=\dfrac{f}{2}$

Replace s with $f-5$ and t with $\dfrac{f}{2}$ in E1.

$f+f-5+\dfrac{f}{2}=295$
$\dfrac{5}{2}f=300$
$f=120$

Replace f with 300 in the equation $s=f-5$.

$s=120-5=115$

Replace f with 120 the equation $\dfrac{f}{2}$.

$t=\dfrac{120}{2}=60$

The first number is 120, the second number is 115, and the third number is 60.

47. $\begin{cases} x-2y=1 \\ 2x+3y=-12 \end{cases}$

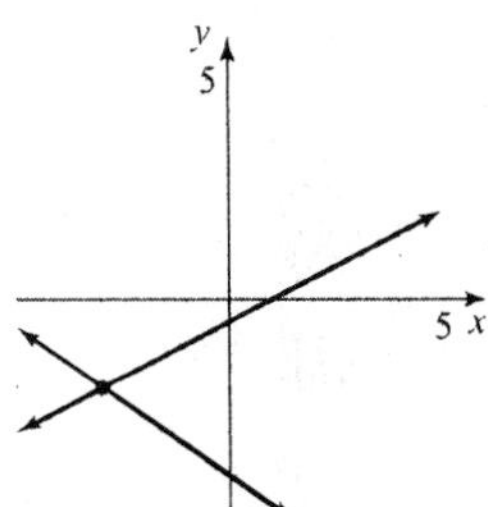

The solution is (−3, −2).

48. $\begin{cases} 3x-y=-4 \\ 6x-2y=-8 \end{cases}$

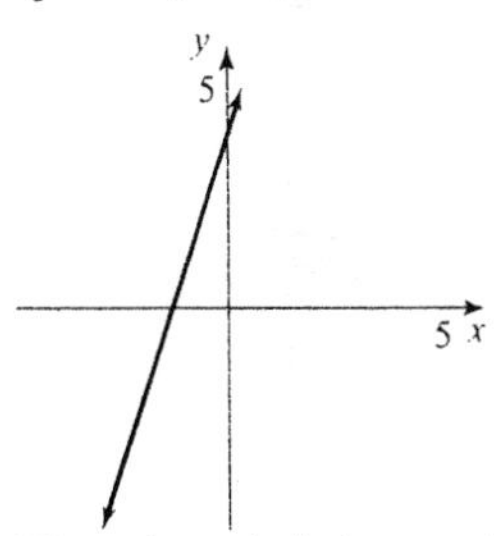

There is an infinite number of solutions.

49. $\begin{cases} x+4y=11 \\ 5x-9y=-3 \end{cases}$

Solve the first equation for x.

$x = 11 - 4y$

Substitute $11 - 4y$ for x in the second equation.

$5(11-4y)-9y=-3$

$55-20y-9y=-3$

$-29y=-58$

$y=2$

Let $y = 2$ in the first equation.

$x+4(2)=11$

$x+8=11$

$x=3$

The solution is (3, 2).

50. $\begin{cases} x+9y=16 \\ 3x-8y=13 \end{cases}$

Solve the first equation for x.

$x = 16 - 9y$

Substitute $16 - 9y$ for x in the second equation.

$3(16-9y)-8y=13$

$48-27y-8y=13$

$-35y=-35$

$y=1$

Let $y = 1$ in the first equation.

$x+9(1)=16$

$x+9=16$

$x=7$

The solution is (7, 1).

51. $\begin{cases} y=-2x \\ 4x+7y=-15 \end{cases}$

Substitute $-2x$ for y in the second equation.

$4x+7(-2x)=-15$

$4x-14x=-15$

$-10x=-15$

$x=\frac{3}{2}=1\frac{1}{2}$

Let $x=\frac{3}{2}$ in the first equation.

$y=-2\left(\frac{3}{2}\right)=-3$

The solution is $\left(1\frac{1}{2}, -3\right)$.

52. $\begin{cases} 3y=2x+15 \\ -2x+3y=21 \end{cases}$

Solve the first equation for x.

$3y=2x+15$

$3y-15=2x$

$\frac{3}{2}y-\frac{15}{2}=x$

Substitute $\frac{3}{2}y-\frac{15}{2}$ for x in the second equation.

$-2\left(\frac{3}{2}y-\frac{15}{2}\right)+3y=21$

$-3y+15+3y=21$

$15=21$ False

The system has no solution.

53. $\begin{cases} 3x-y=4 \\ 4y=12x-16 \end{cases}$

Solve the first equation for y.

$3x - 4 = y$

Substitute $3x - 4$ for y in the second equation.

$4(3x-4)=12x-16$

$12x-16=12x-16$

$0=0$

There is an infinite number of solutions.

54. $\begin{cases} x+y=19 \\ x-y=-3 \end{cases}$

$$\begin{aligned} x+y&=19 \\ x-y&=-3 \\ \hline 2x \quad &=16 \\ x&=8 \end{aligned}$$

Let $x = 8$ in the first equation.

$$\begin{aligned} 8+y&=19 \\ y&=11 \end{aligned}$$

The solution is (8, 11).

55. $\begin{cases} x-3y=-11 \\ 4x+5y=-10 \end{cases}$

Solve the first equation for x.

$x = 3y - 11$

Substitute $3y - 11$ for x in the second equation.

$$\begin{aligned} 4(3y-11)+5y&=-10 \\ 12y-44+5y&=-10 \\ 17y&=34 \\ y&=2 \end{aligned}$$

Let $y = 2$ in the first equation.

$$\begin{aligned} x-3(2)&=-11 \\ x-6&=-11 \\ x&=-5 \end{aligned}$$

The solution is (−5, 2).

56. $\begin{cases} -x-15y=44 \\ 2x+3y=20 \end{cases}$

Solve the first equation for x.

$$\begin{aligned} -x-15y&=44 \\ -x&=15y+44 \\ x&=-15y-44 \end{aligned}$$

Substitute $-15y - 44$ for x in the second equation.

$$\begin{aligned} 2(-15y-44)+3y&=20 \\ -30y-88+3y&=20 \\ -27y&=108 \\ y&=-4 \end{aligned}$$

Let $y = -4$ in $x = -15y - 44$.

$x = -15(-4) - 44 = 60 - 44 = 16$

The solution is (16, −4).

57. $\begin{cases} x-3y+2z=0 & (1) \\ 9y-z=22 & (2) \\ 5x+3z=10 & (3) \end{cases}$

Multiply E1 by 3 and add to E2.

$$\begin{aligned} 3x-9y+6z&=0 \\ 9y-z&=22 \\ \hline 3x \quad +5z&=22 \quad (4) \end{aligned}$$

Multiply E3 by −5 and E4 by 3 and add the results.

$$\begin{aligned} -25x-15z&=-50 \\ 9x+15z&=66 \\ \hline -16x \quad &=16 \\ x&=-1 \end{aligned}$$

Replace x with −1 in E3.

$$\begin{aligned} 5(-1)+3z&=10 \\ -5+3z&=10 \\ 3z&=15 \\ z&=5 \end{aligned}$$

Replace z with 5 in E2.

$$\begin{aligned} 9y-5&=22 \\ 9y&=27 \\ y&=3 \end{aligned}$$

The solution is (−1, 3, 5).

58. $\begin{cases} x-4y=4 \\ \frac{1}{8}x-\frac{1}{2}y=3 \end{cases}$

Multiply the second by −8 and add to the first equation to eliminate x.

$\begin{cases} x-4y=4 \\ -x+4y=-24 \end{cases}$

The equation 0 = −20 is false. The system has no solution. The solution set is { } or ∅.

59. Let x = the larger number and y = the smaller number.

$\begin{cases} x+y=12 \\ x+3y=20 \end{cases}$

Multiply the first equation by −1.

$$\begin{aligned} -x-y&=-12 \\ x+3y&=20 \\ \hline 2y&=8 \\ y&=4 \end{aligned}$$

Let $y = 4$ in the first equation.

$$\begin{aligned} x+4&=12 \\ x&=8 \end{aligned}$$

The numbers are 4 and 8.

60. Let x = the smaller number and y = the larger number.

$\begin{cases} x-y=-18 \\ 2x-y=-23 \end{cases}$

Multiply the first equation by −1.

$$\begin{aligned} -x+y&=18 \\ 2x-y&=-23 \\ \hline x \quad &=-5 \end{aligned}$$

Let $x = -5$ in the first equation.

$$\begin{aligned} -5 - y &= -18 \\ -y &= -13 \\ y &= 13 \end{aligned}$$

The numbers are −5 and 13.

61. Let x = the number of nickels and y = the number of dimes.

$$\begin{cases} x + y = 65 \\ 0.05x + 0.10y = 5.30 \end{cases}$$

Multiply the first equation by −5 and the second equation by 100.

$$\begin{aligned} -5x - 5y &= -325 \\ 5x + 10y &= 530 \\ \hline 5y &= 205 \\ y &= 41 \end{aligned}$$

Let $y = 41$ in the first equation.

$$\begin{aligned} x + 41 &= 65 \\ x &= 24 \end{aligned}$$

There are 24 nickels and 41 dimes.

62. Let x = the number of 13¢ stamps and y = the number of 22¢ stamps.

$$\begin{cases} x + y = 26 \\ 0.13x + 0.22y = 4.19 \end{cases}$$

Multiply the first equation by −13 and the second equation by 100.

$$\begin{aligned} -13x - 13y &= -338 \\ 13x + 22y &= 419 \\ \hline 9y &= 81 \\ y &= 9 \end{aligned}$$

Let $y = 9$ in the first equation.

$$\begin{aligned} x + 9 &= 26 \\ x &= 17 \end{aligned}$$

They purchased 17 13¢ stamps and 9 22¢ stamps.

63. Let x = length of the shortest side
y = length of the second side
z = length of the third side
We solve the system

$$\begin{cases} x + y + z = 126 \\ y = 2x \\ z = x + 14 \end{cases}$$

We substitute $2x$ for y and $x + 14$ for z in the first equation.

$$\begin{aligned} x + 2x + (x + 14) &= 126 \\ 4x + 14 &= 126 \\ 4x &= 112 \\ x &= 28 \end{aligned}$$

Now we find y and z.
$y = 2x = 2(28) = 56$
$z = x + 14 = 28 + 14 = 42$
The lengths are 28 units, 42 units, and 56 units.

Chapter 4 Test

1. False; one solution, infinitely many solutions, or no solutions are the only possibilities.

2. False; a solution of a system of equations must be a solution of each equation in the system.

3. True

4. False; $x = 0$ is part of the solution.

5. Let $x = 1$ and $y = -1$.

$$\begin{aligned} 2x - 3y &= 5 \\ 2(1) - 3(-1) &\stackrel{?}{=} 5 \\ 2 + 3 &\stackrel{?}{=} 5 \\ 5 &= 5 \quad \text{True} \end{aligned} \qquad \begin{aligned} 6x + y &= 1 \\ 6(1) + (-1) &\stackrel{?}{=} 1 \\ 6 - 1 &\stackrel{?}{=} 1 \\ 5 &= 1 \quad \text{False} \end{aligned}$$

$(1, -1)$ is not a solution of the system.

6. Let $x = 3$ and $y = -4$.

$$\begin{aligned} 4x - 3y &= 24 \\ 4(3) - 3(-4) &\stackrel{?}{=} 24 \\ 12 + 12 &\stackrel{?}{=} 24 \\ 24 &= 24 \quad \text{True} \end{aligned}$$

$$\begin{aligned} 4x + 5y &= -8 \\ 4(3) + 5(-4) &\stackrel{?}{=} -8 \\ 12 - 20 &\stackrel{?}{=} -8 \\ -8 &= -8 \quad \text{True} \end{aligned}$$

$(3, -4)$ is a solution of the system.

7. $$\begin{cases} y - x = 6 \\ y + 2x = -6 \end{cases}$$

y
5
(−4, 2)
3
x

The solution is $(-4, 2)$.

8. $$\begin{cases} 3x - 2y = -14 \\ x + 3y = -1 \end{cases}$$

Solve the second equation for x.
$x = -3y - 1$

Substitute $-3y - 1$ for x in the first equation.

$$\begin{aligned} 3(-3y-1)-2y &= -14 \\ -9y-3-2y &= -14 \\ -11y &= -11 \\ y &= 1 \end{aligned}$$

Let $y = 1$ in $x = -3y - 1$.

$x = -3(1) - 1 = -4$

The solution is $(-4, 1)$.

9. $\begin{cases} \frac{1}{2}x + 2y = -\frac{15}{4} \\ 4x = -y \end{cases}$

Solve the second equation for y.

$y = -4x$

Substitute $-4x$ for y in the first equation.

$$\begin{aligned} \frac{1}{2}x + 2(-4x) &= -\frac{15}{4} \\ \frac{1}{2}x - 8x &= -\frac{15}{4} \\ -\frac{15}{2}x &= -\frac{15}{4} \\ x &= \frac{1}{2} \end{aligned}$$

Let $x = \frac{1}{2}$ in the equation $y = -4x$.

$y = -4\left(\frac{1}{2}\right) = -2$

The solution is $\left(\frac{1}{2}, -2\right)$.

10. $\begin{cases} 3x + 5y = 2 \\ 2x - 3y = 14 \end{cases}$

Multiply the first equation by 2 and the second equation by -3.

$$\begin{aligned} 6x + 10y &= 4 \\ -6x + 9y &= -42 \\ \hline 19y &= -38 \\ y &= -2 \end{aligned}$$

Let $y = -2$ in the first equation.

$$\begin{aligned} 3x + 5(-2) &= 2 \\ 3x - 10 &= 2 \\ 3x &= 12 \\ x &= 4 \end{aligned}$$

The solution is $(4, -2)$.

11. $\begin{cases} 4x - 6y = 7 \\ -2x + 3y = 0 \end{cases}$

Multiply the second equation by 2.

$$\begin{aligned} 4x - 6y &= 7 \\ -4x + 6y &= 0 \\ \hline 0 &= 7 \end{aligned}$$

The system is inconsistent. There is no solution.

12. $\begin{cases} 3x + y = 7 \\ 4x + 3y = 1 \end{cases}$

Solve the first equation for y.

$y = 7 - 3x$

Substitute $7 - 3x$ for y in the second equation.

$$\begin{aligned} 4x + 3(7 - 3x) &= 1 \\ 4x + 21 - 9x &= 1 \\ -5x &= -20 \\ x &= 4 \end{aligned}$$

Let $x = 4$ in $y = 7 - 3x$.

$y = 7 - 3(4) = -5$

The solution is $(4, -5)$.

13. $\begin{cases} 3(2x + y) = 4x + 20 \\ \quad 6x + 3y = 4x + 20 \\ \quad 2x + 3y = 20 \\ \\ \quad x - 2y = 3 \end{cases}$

Multiply the second equation by -2.

$$\begin{aligned} 2x + 3y &= 20 \\ -2x + 4y &= -6 \\ \hline 7y &= 14 \\ y &= 2 \end{aligned}$$

Let $y = 2$ in the second equation.

$$\begin{aligned} x - 2(2) &= 3 \\ x - 4 &= 3 \\ x &= 7 \end{aligned}$$

The solution of the system is $(7, 2)$.

14. $\begin{cases} \frac{x-3}{2} = \frac{2-y}{4} \\ \frac{7-2x}{3} = \frac{y}{2} \end{cases}$

Multiply the first equation by 4 and the second equation by 6.

$$\begin{cases} 2(x-3) = 2-y \\ 2x-6 = 2-y \\ 2x+y = 8 \\ \\ 2(7-2x) = 3y \\ 14-4x = 3y \\ 4x+3y = 14 \end{cases}$$

Multiply the first equation by −3.

$$\begin{array}{r} -6x-3y = -24 \\ \underline{4x+3y = 14} \\ -2x \quad = -10 \\ x = 5 \end{array}$$

Let $x = 5$ in the first equation.

$$\begin{aligned} 2(5)+y &= 8 \\ 10+y &= 8 \\ y &= -2 \end{aligned}$$

The solution of the system is (5, −2).

15. Let x = the larger number and y = the smaller number.

$$\begin{cases} x+y = 124 \\ x-y = 32 \end{cases}$$

$$\begin{array}{r} x+y = 124 \\ \underline{x-y = 32} \\ 2x \quad = 156 \\ x = 78 \end{array}$$

Let $x = 78$ in the first equation.

$$\begin{aligned} 78+y &= 124 \\ y &= 46 \end{aligned}$$

The numbers are 78 and 46.

16. Let x = cc's of 12% solution and y = cc's of 16% solution.

Concentration Rate	cc's of Solution	cc's of salt
12%	x	$0.12x$
22%	80	0.22(80)
16%	y	$0.16y$

$$\begin{cases} x+80 = y \\ 0.12x+0.22(80) = 0.16y \end{cases}$$

Multiply the first equation by −16 and the second equation by 100.

$$\begin{array}{r} -16x-1280 = -16y \\ \underline{12x+1760 = 16y} \\ -4x+480 = 0 \\ -4x = -480 \\ x = 120 \end{array}$$

Should add 120 cc's of 12% solution

17. Let x = the number of thousands of farms in Texas and y = the number of thousands of farms in Missouri.

$$\begin{cases} x+y = 336 \\ x-y = 116 \end{cases}$$

$$\begin{array}{r} x+y = 336 \\ \underline{x-y = 116} \\ 2x \quad = 452 \\ x = 226 \end{array}$$

Let $x = 226$ in the first equation.

$$\begin{aligned} 226+y &= 336 \\ y &= 110 \end{aligned}$$

There are 226,000 farms in Texas and 110,000 farms in Missouri.

18. $$\begin{cases} 2x-3y \quad = 4 & (1) \\ 3y+2z = 2 & (2) \\ x \quad -z = -5 & (3) \end{cases}$$

Add E1 and E2.

$2x+2z = 6$ or $x+z = 3$ (4)

Add E3 and E4.

$$\begin{array}{r} x+z = 3 \\ \underline{x-z = -5} \\ 2x \quad = -2 \\ x = -1 \end{array}$$

Replace x with −1 in E3.

$-1-z = -5$

$-z = -4$ so $z = 4$

Replace x with −1 in E1.

$$\begin{aligned} 2(-1)-3y &= 4 \\ -2-3y &= 4 \\ -3y &= 6 \\ y &= -2 \end{aligned}$$

The solution is (−1, −2, 4).

19. $$\begin{cases} 3x-2y-z = -1 & (1) \\ 2x-2y \quad = 4 & (2) \\ 2x \quad -2z = -12 & (3) \end{cases}$$

Multiply E2 by −1 and add to E1.

$$\begin{array}{r} 3x-2y-z = -1 \\ \underline{-2x+2y \quad = -4} \\ x \quad -z = -5 \quad (4) \end{array}$$

Multiply E4 by −2 and add to E3.

$$\begin{array}{r} 2x-2z=-12 \\ -2x+2z=10 \\ \hline 0=-2 \end{array} \text{ False}$$

The system is inconsistent. The solution set is $\varnothing$.

20. Let x = measure of the smallest angle. Then the largest angle has a measure of $5x - 3$, and the remaining angle has a measure of $2x - 1$.The sum of the three angles must add to 180°:

$$\begin{aligned} a+b+c &= 180 \\ x+(5x-3)+(2x-1) &= 180 \\ x+5x-3+2x-1 &= 180 \\ 8x-4 &= 180 \\ 8x &= 184 \\ x &= 23 \end{aligned}$$

$5x-3=5(23)-3=115-3=112$

$2x-1=2(23)-1=46-1=45$

The angle measures are 23°, 45°, and 112°.

Chapter 4 Cumulative Review

1. a. $-1<0$

b. $7=\frac{14}{2}$

c. $-5>-6$

2. a. $5^2=5\cdot 5=25$

b. $2^5=2\cdot 2\cdot 2\cdot 2\cdot 2=32$

3. a. commutative property of multiplication

b. associative property of addition

c. identity element for addition

d. commutative property of multiplication

e. multiplicative inverse property

f. additive inverse property

g. commutative and associative properties of multiplication

4. Let $x = 8$, $y = 5$.

$y^2-3x=5^2-3(8)=25-24=1$

5. $(2x-3)-(4x-2)=2x-3-4x+2=-2x-1$

6.
$$\begin{aligned} &7-12+(-5)-2+(-2) \\ &=7+(-12)+(-5)+(-2)+(-2) \\ &=7+(-21) \\ &=-14 \end{aligned}$$

7.
$$\begin{aligned} 7 &= -5(2a-1)-(-11a+6) \\ 7 &= -10a+5+11a-6 \\ 7 &= a-1 \\ 7+1 &= a-1+1 \\ 8 &= a \end{aligned}$$

8. Let $x=-7$, $y=-3$.

$$\begin{aligned} 2y^2-x^2 &= 2(-3)^2-(-7)^2 \\ &= 2(9)-49 \\ &= 18-49 \\ &= -31 \end{aligned}$$

9.
$$\begin{aligned} \frac{5}{2}x &= 15 \\ \frac{2}{5}\cdot\frac{5}{2}x &= \frac{2}{5}\cdot 15 \\ x &= 6 \end{aligned}$$

10.
$$\begin{aligned} &0.4y-6.7+y-0.3-2.6y \\ &=0.4y+y+(-2.6y)+(-6.7)+(-0.3) \\ &=-1.2y-7 \end{aligned}$$

11.
$$\begin{aligned} \frac{x}{2}-1 &= \frac{2}{3}x-3 \\ 6\left(\frac{x}{2}-1\right) &= 6\left(\frac{2}{3}x-3\right) \\ 3x-6 &= 4x-18 \\ -x-6 &= -18 \\ -x &= -12 \\ x &= 12 \end{aligned}$$

12.
$$\begin{aligned} 7(x-2)-6(x+1) &= 20 \\ 7x-14-6x-6 &= 20 \\ x-20 &= 20 \\ x &= 40 \end{aligned}$$

13. Let x = the number.

$$\begin{aligned} 2(x+4) &= 4x-12 \\ 2x+8 &= 4x-12 \\ -2x+8 &= -12 \\ -2x &= -20 \\ x &= 10 \end{aligned}$$

The number is 10.

14. $5(y-5)=5y+10$
$5y-25=5y+10$
$-25=10$
False statement; there is no solution.

15. $y=mx+b$
$y-b=mx+b-b$
$y-b=mx$
$\frac{y-b}{m}=\frac{mx}{m}$
$\frac{y-b}{m}=x$

16. Let x = the number.
$5(x-1)=6x$
$5x-5=6x$
$-x-5=0$
$-x=5$
$x=-5$
The number is -5.

17. $-2x\le -4$
$\frac{-2x}{-2}\ge\frac{-4}{-2}$
$x\ge 2$, $[2, \infty)$

18. $P=a+b+c$
$P-a-c=a+b+c-a-c$
$P-a-c=b$

19. $x=-2y$

x	y
0	0
−4	2

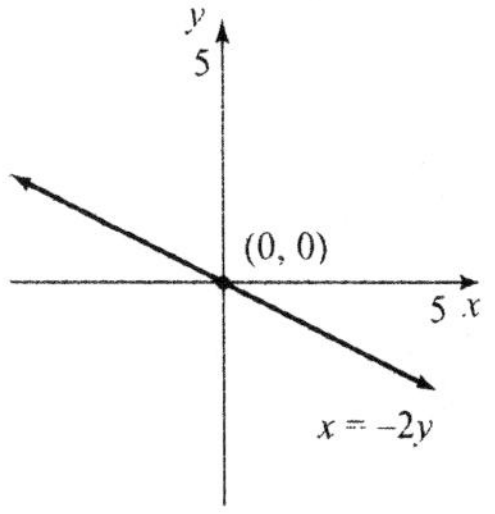

20. $3x+7\ge x-9$
$2x+7\ge -9$
$2x\ge -16$
$x\ge -8$, $[-8, \infty)$

21. $(-1, 5)$ and $(2, -3)$
$$m=\frac{y_2-y_1}{x_2-x_1}=\frac{-3-5}{2-(-1)}=\frac{-8}{3}=-\frac{8}{3}$$

22. $x-3y=3$

x	y
0	−1
3	0
9	2

23. $y=\frac{3}{4}x+6$
$y=mx+b$
$m=\frac{3}{4}$

24. $(-1, 3)$ and $(2, -8)$
$$m=\frac{y_2-y_1}{x_2-x_1}=\frac{-8-3}{2-(-1)}=-\frac{11}{3}$$
A parallel line has the same slope.
Slope is $-\frac{11}{3}$.

25. $3x-4y=4$
$-4y=-3x+4$
$y=\frac{-3x}{-4}+\frac{4}{-4}$
$y=\frac{3}{4}x-1$
$y=mx+b$
$m=\frac{3}{4}$, $b=-1$

Slope is $\frac{3}{4}$, y-intercept is $(0, -1)$.

26. $y=7x+0$
$y=mx+b$
$m=7$, $b=0$
Slope is 7, y-intercept is $(0, 0)$.

27. $m=-2$, with point $(-1, 5)$
$y-y_1=m(x-x_1)$
$y-5=-2[x-(-1)]$
$y-5=-2x-2$
$2x+y=3$

28. Line: $y = 4x - 5$ $m_1 = 4$
Line 2: $-4x + y = 7$ $y = 4x + 7$ $m_2 = 4$
$m_2 = m_1$
The lines are parallel.

29. A vertical line has an equation $x = c$.
Point, $(-1, 5)$
$x = -1$

30. $m = -5$, with point $(-2, 3)$

$$\begin{aligned} y - y_1 &= m(x - x_1) \\ y - 3 &= -5[x - (-2)] \\ y - 3 &= -5x - 10 \\ y &= -5x - 7 \end{aligned}$$

31. Domain is $\{-1, 0, 3\}$
Range is $\{-2, 0, 2, 3\}$

32. $f(x) = 5x^2 - 6$
$f(0) = 5(0)^2 - 6 = -6$
$f(-2) = 5(-2)^2 - 6 = 5(4) - 6 = 14$

33. a. function

b. not a function

34. a. not a function

b. function

c. not a function

35. $$\begin{cases} 3x - y = 4 \\ y = 3x - 4,\ m = 3 \\ \\ x + 2y = 8 \\ y = -\frac{1}{2}x + 4,\ m = -\frac{1}{2} \end{cases}$$

Because they have different slopes, there is only one solution.

36. a. Let $x = 1$ and $y = -4$.

$$\begin{aligned} 2x - y &= 6 \\ 2(1) - (-4) &\stackrel{?}{=} 6 \\ 2 + 4 &\stackrel{?}{=} 6 \\ 6 &= 6 \quad \text{True} \end{aligned}$$

$$\begin{aligned} 3x + 2y &= -5 \\ 3(1) + 2(-4) &\stackrel{?}{=} -5 \\ 3 - 8 &\stackrel{?}{=} -5 \\ -5 &= -5 \quad \text{True} \end{aligned}$$

$(1, -4)$ is a solution of the system.

b. Let $x = 0$ and $y = 6$.

$$\begin{aligned} 2x - y &= 6 \\ 2(0) - (6) &\stackrel{?}{=} 6 \\ 0 - 6 &\stackrel{?}{=} 6 \\ -6 &= 6 \quad \text{False} \end{aligned}$$

$3x + 2y = -5$
Test not needed

$(0, 6)$ is not a solution of the system.

c. Let $x = 3$ and $y = 0$.

$$\begin{aligned} 2x - y &= 6 \\ 2(3) - (0) &\stackrel{?}{=} 6 \\ 6 - 0 &\stackrel{?}{=} 6 \\ 6 &= 6 \quad \text{True} \end{aligned}$$

$$\begin{aligned} 3x + 2y &= -5 \\ 3(3) + 2(0) &\stackrel{?}{=} -5 \\ 9 + 0 &\stackrel{?}{=} -5 \\ 9 &= -5 \quad \text{False} \end{aligned}$$

$(3, 0)$ is not a solution of the system.

37. $$\begin{cases} x + 2y = 7 \\ 2x + 2y = 13 \end{cases}$$

Solve the first equation for x.
$x = 7 - 2y$
Substitute $7 - 2y$ for x in the second equation.

$$\begin{aligned} 2(7 - 2y) + 2y &= 13 \\ 14 - 4y + 2y &= 13 \\ -2y &= -1 \\ y &= \frac{1}{2} \end{aligned}$$

Let $y = \frac{1}{2}$ in $x = 7 - 2y$.

$$x = 7 - 2\left(\frac{1}{2}\right) = 6$$

The solution is $\left(6, \frac{1}{2}\right)$.

38. $$\begin{cases} 3x - 4y = 10 \\ y = 2x \end{cases}$$

Substitute $2x$ for y in the first equation.

$$\begin{aligned} 3x - 4(2x) &= 10 \\ 3x - 8x &= 10 \\ -5x &= 10 \\ x &= -2 \end{aligned}$$

Let $x = -2$ in the second equation.

$y = 2(-2) = -4$
The solution is (−2, −4).

39. $\begin{cases} x+y=7 \\ x-y=5 \end{cases}$

$$\begin{array}{r} x+y=7 \\ \underline{x-y=5} \\ 2x \quad =12 \\ x=6 \end{array}$$

Let $x = 6$ in the first equation.
$6+y=7$
$y=1$
The solution to the system is (6, 1).

40. $\begin{cases} x=5y-3 \\ x=8y+4 \end{cases}$

Substitute $8y + 4$ for x in the first equation.
$8y+4=5y-3$
$3y+4=-3$
$3y=-7$
$y=-\frac{7}{3}$
Let $y=-\frac{7}{3}$ in the second equation.
$x=8\left(-\frac{7}{3}\right)+4$
$x=-\frac{56}{3}+\frac{12}{3}$
$x=-\frac{44}{3}$
The solution is $\left(-\frac{44}{3}, -\frac{7}{3}\right)$.

41. $\begin{cases} 3x-y+z=-15 & (1) \\ x+2y-z=1 & (2) \\ 2x+3y-2z=0 & (3) \end{cases}$

Add E1 and E2.
$4x+y=-14$ (4)
Multiply E1 by 2 and add to E3.

$$\begin{array}{r} 6x-2y+2z=-30 \\ \underline{2x+3y-2z=0} \\ 8x+y \quad =-30 \ (5) \end{array}$$

Solve the new system:
$\begin{cases} 4x+y=-14 & (4) \\ 8x+y=-30 & (5) \end{cases}$
Multiply E4 by −1 and add to E5.

$$\begin{array}{r} -4x-y=14 \\ \underline{8x+y=-30} \\ 4x \quad =-16 \\ x=-4 \end{array}$$

Replace x with −4 in E4.
$4(-4)+y=-14$
$-16+y=-14$
$y=2$
Replace x with −4 and y with 2 in E1.
$3(-4)-(2)+z=-15$
$-12-2+z=-15$
$-14+z=-15$
$z=-1$
The solution is (−4, 2, −1).

42. $\begin{cases} x-2y+z=0 & (1) \\ 3x-y-2z=-15 & (2) \\ 2x-3y+3z=7 & (3) \end{cases}$

Multiply E1 by 2 and add to E2.

$$\begin{array}{r} 2x-4y+2z=0 \\ \underline{3x-y-2z=-15} \\ 5x-5y \quad =-15 \text{ or } x-y=-3 \ (4) \end{array}$$

Multiply E1 by −3 and add to E3.

$$\begin{array}{r} -3x+6y-3z=0 \\ \underline{2x-3y+3z=7} \\ -x+3y \quad =7 \ (5) \end{array}$$

Add E4 and E5.
$2y=4$
$y=2$
Replace y with 2 in E4.
$x-2=-3$
$x=-1$
Replace x with −1 and y with 2 in E1.
$-1-2(2)+z=0$
$-5+z=0$
$z=5$
The solution is (−1, 2, 5).

43. Let x = the first number and y = the second number.
$\begin{cases} x=y-4 \\ 4x=2y+6 \end{cases}$
Substitute $y - 4$ for x in the second equation.
$4(y-4)=2y+6$
$4y-16=2y+6$
$2y=22$
$y=11$
Let $y = 11$ in $x = y - 4$.
$x = 11 - 4 = 7$
The numbers are 7 and 11.

44. Let x = the first number and y = the second number.

$$\begin{cases} x+y=37 \\ x-y=21 \end{cases}$$

$$\begin{array}{r} x+y=37 \\ \underline{x-y=21} \\ 2x \qquad =58 \\ x=29 \end{array}$$

Let $x = 29$ in the first equation.

$$29+y=37$$
$$y=8$$

The numbers are 29 and 8.

Chapter 5

Section 5.1

Practice Exercises

1. a. $3^3 = 3 \cdot 3 \cdot 3 = 27$

b. Use 4 as a factor once, $4^1 = 4$

c. $(-8)^2 = (-8)(-8) = 64$

d. $-8^2 = -(8 \cdot 8) = -64$

e. $\left(\frac{3}{4}\right)^3 = \frac{3}{4} \cdot \frac{3}{4} \cdot \frac{3}{4} = \frac{27}{64}$

f. $(0.3)^4 = (0.3)(0.3)(0.3)(0.3) = 0.0081$

g. $3 \cdot 5^2 = 3 \cdot 25 = 75$

2. a. If x is 3, $3x^4 = 3 \cdot (3)^4$
$= 3 \cdot (3 \cdot 3 \cdot 3 \cdot 3)$
$= 3 \cdot 81$
$= 243$

b. If x is -4, $\frac{6}{x^2} = \frac{6}{(-4)^2} = \frac{6}{(-4)(-4)} = \frac{6}{16} = \frac{3}{8}$

3. a. $3^4 \cdot 3^6 = 3^{4+6} = 3^{10}$

b. $y^3 \cdot y^2 = y^{3+2} = y^5$

c. $z \cdot z^4 = z^1 \cdot z^4 = z^{1+4} = z^5$

d. $x^3 \cdot x^2 \cdot x^6 = x^{3+2+6} = x^{11}$

e. $(-2)^5 \cdot (-2)^3 = (-2)^{5+3} = (-2)^8$

f. $b^3 \cdot t^5$, cannot be simplified because b and t are different bases.

4. $(-5y^3)(-3y^4) = -5 \cdot y^3 \cdot -3 \cdot y^4$
$= -5 \cdot -3 \cdot y^3 \cdot y^4$
$= 15y^7$

5. a. $(y^7z^3)(y^5z) = (y^7 \cdot y^5) \cdot (z^3 \cdot z^1)$
$= y^{12} \cdot z^4$ or $y^{12}z^4$

b. $(-m^4n^4)(7mn^{10})$
$= (-1 \cdot 7) \cdot (m^4 \cdot m^1) \cdot (n^4 \cdot n^{10})$
$= (-7) \cdot (m^5) \cdot (n^{14})$ or $-7m^5n^{14}$

6. a. $(x^4)^3 = x^{4 \cdot 3} = x^{12}$

b. $(z^3)^7 = z^{3 \cdot 7} = z^{21}$

c. $[(-2)^3]^5 = (-2)^{3 \cdot 5} = (-2)^{15}$

7. a. $(pr)^5 = p^5 \cdot r^5 = p^5r^5$

b. $(6b)^2 = 6^2 \cdot b^2 = 36b^2$

c. $\left(\frac{1}{4}x^2y\right)^3 = \left(\frac{1}{4}\right)^3 \cdot (x^2)^3 \cdot y^3$
$= \frac{1}{64} \cdot x^6 \cdot y^3$
$= \frac{1}{64}x^6y^3$

d. $(-3a^3b^4c)^4 = (-3)^4 \cdot (a^3)^4 \cdot (b^4)^4 \cdot c^4$
$= 81a^{12}b^{16}c^4$

8. a. $\left(\frac{x}{y^2}\right)^5 = \frac{x^5}{(y^2)^5} = \frac{x^5}{y^{10}}$, $y \neq 0$

b. $\left(\frac{2a^4}{b^3}\right)^5 = \frac{2^5 \cdot (a^4)^5}{(b^3)^5} = \frac{32a^{20}}{b^{15}}$, $b \neq 0$

9. a. $\frac{z^8}{z^4} = z^{8-4} = z^4$

b. $\frac{(-5)^5}{(-5)^3} = (-5)^{5-3} = (-5)^2 = 25$

c. $\frac{8^8}{8^6} = 8^{8-6} = 8^2 = 64$

d. $\frac{q^5}{t^2}$ cannot be simplified because q and t are different bases.

e. Begin by grouping common bases.
$$\frac{6x^3y^7}{xy^5} = 6 \cdot \frac{x^3}{x} \cdot \frac{y^7}{y^5} = 6 \cdot x^{3-1} \cdot y^{7-5} = 6x^2y^2$$

10. a. $-3^0 = -1 \cdot 3^0 = -1 \cdot 1 = -1$

b. $(-3)^0 = 1$

c. $8^0 = 1$

d. $(0.2)^0 = 1$

e. $(xz)^0 = x^0 \cdot z^0 = 1 \cdot 1 = 1$

11. a. This is a quotient raised to a power, so we use the power of a quotient rule.
$$\left(\frac{5}{xz}\right)^3 = \frac{5^3}{x^3z^3} = \frac{125}{x^3z^3}$$

b. This is a product raised to a power, so we use the power of a product rule.
$$(2z^8x^5)^4 = 2^4(z^8)^4(x^5)^4 = 16z^{32}x^{20}$$

c. Use the power of a product or quotient rule; then use the power rule for exponents.
$$\left(\frac{-3x^3}{y^4}\right)^3 = \frac{(-3)^3(x^3)^3}{(y^4)^3} = -\frac{27x^9}{y^{12}}$$

Vocabulary and Readiness Check

1. Repeated multiplication of the same factor can be written using an exponent.

2. In 5^2, the 2 is called the exponent and the 5 is called the base.

3. To simplify $x^2 \cdot x^7$, keep the base and add the exponents.

4. To simplify $(x^3)^6$, keep the base and multiply the exponents.

5. The understood exponent on the term y is 1.

6. If $x^{\square} = 1$, the exponent is 0.

7. In 3^2, the base is 3 and the exponent is 2.

8. In $(-3)^6$, the base is -3 and the exponent is 6.

9. In -4^2, the base is 4 and the exponent is 2.

10. In $5 \cdot 3^4$, the base 5 has exponent 1 and the base 3 has exponent 4.

11. In $5x^2$, the base 5 has exponent 1 and the base x has exponent 2.

12. In $(5x)^2$, the base is $5x$ and the exponent is 2.

Exercise Set 5.1

1. $7^2 = 7 \cdot 7 = 49$

3. $(-5)^1 = -5$

5. $-2^4 = -2 \cdot 2 \cdot 2 \cdot 2 = -16$

7. $(-2)^4 = (-2)(-2)(-2)(-2) = 16$

9. $(0.1)^5 = (0.1)(0.1)(0.1)(0.1)(0.1) = 0.00001$

11. $\left(\frac{1}{3}\right)^4 = \left(\frac{1}{3}\right)\left(\frac{1}{3}\right)\left(\frac{1}{3}\right)\left(\frac{1}{3}\right) = \frac{1}{81}$

13. $7 \cdot 2^5 = 7 \cdot 2 \cdot 2 \cdot 2 \cdot 2 \cdot 2 = 224$

15. $-2 \cdot 5^3 = -2 \cdot 5 \cdot 5 \cdot 5 = -250$

17. Answers may vary

19. $x^2 = (-2)^2 = (-2)(-2) = 4$

21. $5x^3 = 5(3)^3 = 5 \cdot 3 \cdot 3 \cdot 3 = 135$

23. $2xy^2 = 2(3)(5)^2 = 2(3)(5)(5) = 150$

25. $\frac{2z^4}{5} = \frac{2(-2)^4}{5} = \frac{2(-2)(-2)(-2)(-2)}{5} = \frac{32}{5}$

27. $x^2 \cdot x^5 = x^{2+5} = x^7$

29. $(-3)^3 \cdot (-3)^9 = (-3)^{3+9} = (-3)^{12}$

31. $(5y^4)(3y) = 5(3)y^{4+1} = 15y^5$

33. $(x^9y)(x^{10}y^5) = x^{9+10}y^{1+5} = x^{19}y^6$

35. $(-8mn^6)(9m^2n^2) = --8(9)m^{1+2}n^{6+2}$
$= -72m^3n^8$

37. $(4z^{10})(-6z^7)(z^3) = 4(-6)z^{10+7+3} = -24z^{20}$

39. $A = (4x^2) \cdot (5x^3)$
$= (4 \cdot 5) \cdot (x^2 \cdot x^3)$
$= 20x^{2+3}$
$= 20x^5$

The area is $20x^5$ square feet.

41. $(x^9)^4 = x^{9 \cdot 4} = x^{36}$

43. $(pq)^8 = p^8q^8$

45. $(2a^5)^3 = 2^3 \cdot (a^5)^3 = 8 \cdot a^{5 \cdot 3} = 8a^{15}$

47. $(x^2y^3)^5 = (x^2)^5 \cdot (y^3)^5 = x^{2 \cdot 5} \cdot y^{3 \cdot 5} = x^{10}y^{15}$

49. $(-7a^2b^5c)^2 = (-7)^2 \cdot (a^2)^2 \cdot (b^5)^2 \cdot c^2$
$= 49a^{2 \cdot 2}b^{5 \cdot 2}c^2$
$= 49a^4b^{10}c^2$

51. $\left(\frac{r}{s}\right)^9 = \frac{r^9}{s^9}$

53. $\left(\frac{mp}{n}\right)^5 = \frac{(mp)^5}{n^5} = \frac{m^5 \cdot p^5}{n^5} = \frac{m^5p^5}{n^5}$

55. $\left(\frac{-2xz}{y^5}\right)^2 = \frac{(-2)^2x^2z^2}{y^{5 \cdot 2}} = \frac{4x^2z^2}{y^{10}}$

57. $A = (8z^5)^2 = 8^2 \cdot (z^5)^2 = 64 \cdot z^{5 \cdot 2} = 64z^{10}$

The area is $64z^{10}$ square decimeters.

59. $V = (3y^4)^3 = 3^3y^{4 \cdot 3} = 27y^{12}$

The volume is $27y^{12}$ cubic feet.

61. $\frac{x^3}{x} = \frac{x^3}{x^1} = x^{3-1} = x^2$

63. $\frac{(-4)^6}{(-4)^3} = (-4)^{6-3} = (-4)^3 = -64$

65. $\frac{p^7q^{20}}{pq^{15}} = p^{7-1}q^{20-15} = p^6q^5$

67. $\frac{7x^2y^6}{14x^2y^3} = \frac{7}{14}x^{2-2}y^{6-3} = \frac{1}{2}x^0y^3 = \frac{y^3}{2}$

69. $7^0 = 1$

71. $(2x)^0 = 1$

73. $-7x^0 = -7(1) = -7$

75. $5^0 + y^0 = 1 + 1 = 2$

77. $-9^2 = -9 \cdot 9 = -81$

79. $\left(\frac{1}{4}\right)^3 = \frac{1}{4} \cdot \frac{1}{4} \cdot \frac{1}{4} = \frac{1}{64}$

81. $\left(\frac{9}{qr}\right)^2 = \frac{9^2}{(qr)^2} = \frac{81}{q^2r^2}$

83. $a^2a^3a = a^{2+3+1} = a^6$

85. $(2x^3)(-8x^4) = 2(-8)x^{3+4} = -16x^7$

87. $(a^7b^{12})(a^4b^8) = a^{7+4}b^{12+8} = a^{11}b^{20}$

89. $(-2mn^6)(-13m^8n) = -2(-13)m^{1+8}n^{6+1}$
$= 26m^9n^7$

91. $(z^4)^{10} = z^{4 \cdot 10} = z^{40}$

93. $(-6xyz^3)^2 = (-6)^2x^2y^2z^{3 \cdot 2} = 36x^2y^2z^6$

95. $\frac{3x^5}{x^4} = 3x^{5-4} = 3x$

97. $(9xy)^2 = 9^2x^2y^2 = 81x^2y^2$

99. $2^0 + 2^5 = 1 + 32 = 33$

101. $\left(\frac{3y^5}{6x^4}\right)^3 = \frac{3^3y^{5\cdot3}}{6^3x^{4\cdot3}} = \frac{27y^{15}}{216x^{12}} = \frac{y^{15}}{8x^{12}}$

103. $\frac{2x^3y^2z}{xyz} = 2x^{3-1}y^{2-1}z^{1-1} = 2x^2y^1z^0 = 2x^2y$

105. $y - 10 + y = y + y - 10 = 2y - 10$

107. $7x + 2 - 8x - 6 = 7x - 8x + 2 - 6 = -x - 4$

109. $2(x-5)+3(5-x) = 2x-10+15-3x = -x+5$

111. $(x^{14})^{23} = x^{14\cdot23} = x^{322}$
Multiply the exponents; choice c.

113. $x^{14} + x^{23}$ cannot be simplified further; choice e.

115. Answers may vary

117. Answers may vary

119. $V = x^3 = 7^3 = 7\cdot7\cdot7 = 343$
The volume is 343 cubic meters.

121. Volume; volume measures capacity.

123. Answers may vary

125. $x^{5a}x^{4a} = x^{5a+4a} = x^{9a}$

127. $(a^b)^5 = a^{b\cdot5} = a^{5b}$

129. $\frac{x^{9a}}{x^{4a}} = x^{9a-4a} = x^{5a}$

131. $A = P\left(1+\frac{r}{12}\right)^6$

$A = 1000\left(1+\frac{0.09}{12}\right)^6$

$= 1000(1.0075)^6$

≈ 1045.85

You need \$1045.85 to pay off the loan.

Section 5.2

Practice Exercises

1. a. The exponent on y is 3, so the degree of $5y^3$ is 3.

b. $10xy$ can be written as $10x^1y^1$. The degree of the term is the sum of the exponents, so the degree is 1 + 1 = 2.

c. The degree of $z = z^1$ is 1.

d. $-3a^2b^5c$ can be written as $-3a^2b^5c^1$. The degree of the term is the sum of the exponents, so the degree is 2 + 5 + 1 or 8.

e. The constant, 8, can be written as $8x^0$ (since $x^0 = 1$). The degree of 8 or $8x^0$ is 0.

2. a. The degree of the trinomial $5b^2 - 3b + 7$ is 2, the greatest degree of any of its terms.

b. Rewrite the binomial as $7t^1 + 3$, the degree is 1.

c. The degree of the polynomial $5x^2 + 3x - 6x^3 + 4$ is 3.

3.

Term	numerical coefficient	degree of term
$-3x^3y^2$	−3	5
$4xy^2$	4	3
$-y^2$	−1	2
$3x$	3	1
−2	−2	0

4. a. $P(x) = -2x^2 - x + 7$

$P(1) = -2(1)^2 - 1 + 7 = 4$

b. $P(x) = -2x^2 - x + 7$

$P(-4) = -2(-4)^2 - (-4) + 7 = -21$

5. To find each height, we evaluate $P(t)$ when $t = 1$ and when $t = 2$.

$P(t) = -16t^2 + 130$

$$\begin{aligned} P(1) &= -16(1)^2 + 130 \\ &= -16 + 130 \\ &= 114 \end{aligned}$$

The height of the camera at 1 second is 114 feet.

$P(t) = -16t^2 + 130$

$$\begin{aligned} P(2) &= -16(2)^2 + 130 \\ &= -16(4) + 130 \\ &= -64 + 130 \\ &= 66 \end{aligned}$$

The height of the camera at 2 seconds is 66 feet.

6. a. $-4y + 2y = (-4 + 2)y = -2y$

b. These terms cannot be combined because z and $5z^3$ are not like terms.

c. $7a^2 - 5 - 3a^2 - 7 = 7a^2 - 3a^2 - 5 - 7$
$= 4a^2 - 12$

d.
$$\begin{aligned} &\frac{3}{8}x^3 - x^2 + \frac{5}{6}x^4 + \frac{1}{12}x^3 - \frac{1}{2}x^4 \\ &= \left(\frac{5}{6} - \frac{1}{2}\right)x^4 + \left(\frac{3}{8} + \frac{1}{12}\right)x^3 - x^2 \\ &= \left(\frac{5}{6} - \frac{3}{6}\right)x^4 + \left(\frac{9}{24} + \frac{2}{24}\right)x^3 - x^2 \\ &= \frac{2}{6}x^4 + \frac{11}{24}x^3 - x^2 \\ &= \frac{1}{3}x^4 + \frac{11}{24}x^3 - x^2 \end{aligned}$$

7. $9xy - 3x^2 - 4yx + 5y^2 = -3x^2 + (9 - 4)xy + 5y^2$
$= -3x^2 + 5xy + 5y^2$

8.
$$\begin{aligned} &x \cdot x + 2 \cdot x + 2 \cdot 2 + 5 \cdot x + x \cdot 3x \\ &= x^2 + 2x + 4 + 5x + 3x^2 \\ &= 4x^2 + 7x + 4 \end{aligned}$$

9. a.
$$\begin{aligned} &(4y^2 + x - 3y - 7) + (x + y^2 - 2) \\ &= 4y^2 + x - 3y - 7 + x + y^2 - 2 \\ &= 4y^2 + y^2 - 3y + x + x - 7 - 2 \\ &= 5^2 - 3y + 2x - 9 \end{aligned}$$

b.
$$\begin{aligned} &(-8a^2b - ab^2 + 10) + (-2ab^2 - 10) \\ &= -8a^2b - ab^2 + 10 - 2ab^2 - 10 \\ &= -8a^2b - ab^2 - 2ab^2 + 10 - 10 \\ &= -8a^2b - 3ab^2 \end{aligned}$$

10.
$$\begin{aligned} &(3x^2 - 9x + 11) + (-3x^2 + 7x^3 + 3x - 4) \\ &= 7x^3 + 3x^2 - 3x^2 - 9x + 3x + 11 - 4 \\ &= 7x^3 - 6x + 7 \end{aligned}$$

11. First, change the sign of each term of the second polynomial and then add.
$$\begin{aligned} &(3x^3 - 5x^2 + 4x) - (x^3 - x^2 + 6) \\ &= (3x^3 - 5x^2 + 4x) + (-x^3 + x^2 - 6) \\ &= 3x^3 - x^3 - 5x^2 + x^2 + 4x - 6 \\ &= 2x^3 - 4x^2 + 4x - 6 \end{aligned}$$

12.
$$\begin{aligned} &[(8x - 11) + (2x + 5)] - (3x + 5) \\ &= 8x - 11 + 2x + 5 - 3x - 5 \\ &= 8x + 2x - 3x - 11 + 5 - 5 \\ &= 7x - 11 \end{aligned}$$

13. a.
$$\begin{aligned} &(3a^2 - 4ab + 7b^2) + (-8a^2 + 3ab - b^2) \\ &= 3a^2 - 4ab + 7b^2 - 8a^2 + 3ab - b^2 \\ &= -5a^2 - ab + 6b^2 \end{aligned}$$

b.
$$\begin{aligned} &(5x^2y^2 - 6xy - 4xy^2) \\ &\qquad - (2x^2y^2 + 4xy - 5 + 6y^2) \\ &= 5x^2y^2 - 6xy - 4xy^2 - 2x^2y^2 \\ &\qquad - 4xy + 5 - 6y^2 \\ &= 3x^2y^2 - 10xy - 4xy^2 - 6y^2 + 5 \end{aligned}$$

Graphing Calculator Explorations

1.
$$\begin{aligned} &(2x^2 + 7x + 6) + (x^3 - 6x^2 - 14) \\ &= x^3 - 4x^2 + 7x - 8 \end{aligned}$$

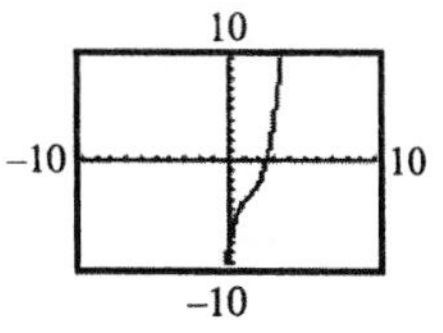

2. $(-14x^3 - x + 2) + (-x^3 + 3x^2 + 4x)$
$= -15x^3 + 3x^2 + 3x + 2$

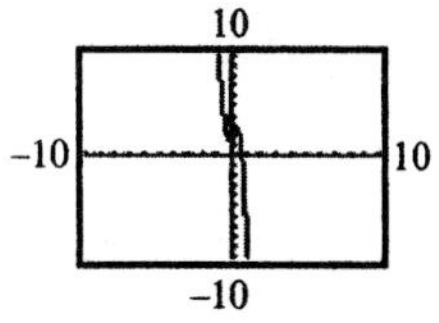

3. $(1.8x^2 - 6.8x - 1.7) - (3.9x^2 - 3.6x)$
$= -2.1x^2 - 3.2x - 1.7$

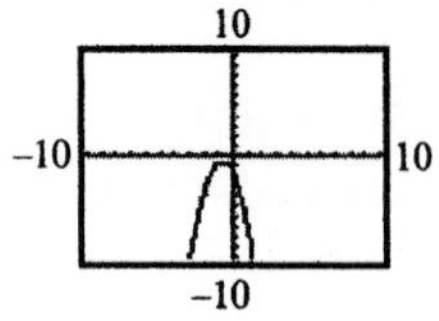

4. $(-4.8x^2 + 12.5x - 7.8) - (3.1x^2 - 7.8x)$
$= -7.9x^2 + 20.3x - 7.8$

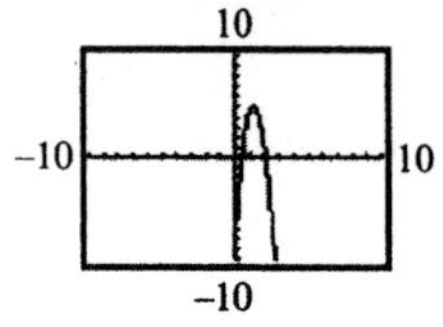

5. $(1.29x - 5.68) + (7.69x^2 - 2.55x + 10.98)$
$= 7.69x^2 - 1.26x + 5.3$

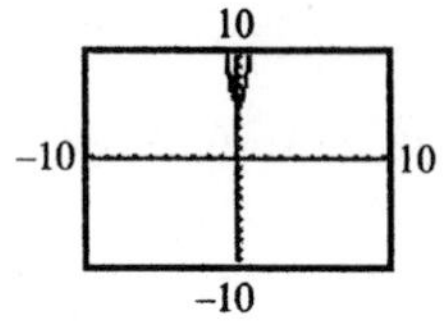

6. $(-0.98x^2 - 1.56x + 5.57) + (4.36x - 3.71)$
$= -0.98x^2 + 2.8x + 1.86$

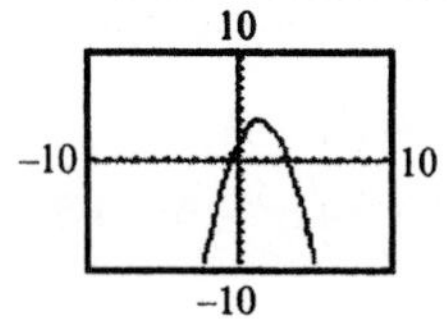

Vocabulary and Readiness Check

1. A binomial is a polynomial with exactly 2 terms.
2. A monomial is a polynomial with exactly one term.
3. A trinomial is a polynomial with exactly three terms.
4. The numerical factor of a term is called the coefficient.
5. A number term is also called a constant.
6. The degree of a polynomial is the greatest degree of any term of the polynomial.
7. $-9y - 5y = (-9 - 5)y = -14y$
8. $6m^5 + 7m^5 = (6 + 7)m^5 = 13m^5$
9. $x + 6x = (1 + 6)x = 7x$
10. $7z - z = (7 - 1)z = 6z$
11. $5m^2 + 2m$ Not like terms.
12. $8p^3 + 3p^2$ Not like terms.

Exercise Set 5.2

1. $x + 2$ is a binomial because it has two terms. The degree is 1 since x is x^1.

3. $9m^3 - 5m^2 + 4m - 8$ is neither a monomial, a binomial, nor a trinomial because it has more than three terms. The degree is 3, the greatest degree of any of its terms.

5. $12x^4y - x^2y^2 - 12x^2y^4$ is a trinomial because it has three terms. The degree is 6, the greatest degree of any of its terms.

7. $3zx - 5x^2$ is a binomial because it has two terms. The degree is 2 because 2 is the degree of the term with the highest degree.

	Polynomial	*Degree*
9.	$3xy^2 - 4$	3
11.	$5a^2 - 2a + 1$	2

13. $P(x) = x^2 + x + 1$
$P(7) = 7^2 + 7 + 1 = 49 + 7 + 1 = 57$

15. $Q(x) = 5x^2 - 1$

$Q(-10) = 5(-10)^2 - 1$
$= 5(100) - 1$
$= 500 - 1$
$= 499$

17. $P(x) = x^2 + x + 1$

$P(0) = 0^2 + 0 + 1 = 0 + 1 = 1$

19. $Q(x) = 5x^2 - 1$

$Q\left(\frac{1}{4}\right) = 5\left(\frac{1}{4}\right)^2 - 1 = 5\left(\frac{1}{16}\right) - 1 = \frac{5}{16} - \frac{16}{16} = -\frac{11}{16}$

21. $P(t) = -16t^2 + 1150$

$P(1) = -16(1)^2 + 1150 = -16 + 1150 = 1134$

After 1 second, the height is 1134 feet.

23. $P(t) = -16t^2 + 1150$

$P(3) = -16(3)^2 + 1150 = -144 + 1150 = 1006$

After 3 seconds, the height is 1006 feet.

25. $14x^2 + 9x^2 = (14 + 9)x^2 = 23x^2$

27. $15x^2 - 3x^2 - y = (15 - 3)x^2 - y = 12x^2 - y$

29. $8s - 5s + 4s = (8 - 5 + 4)s = 7s$

31. $0.1y^2 - 1.2y^2 + 6.7 - 1.9$
$= (0.1 - 1.2)y^2 + (6.7 - 1.9)$
$= -1.1y^2 + 4.8$

33. $\frac{2}{5}x^2 - \frac{1}{3}x^3 + x^2 - \frac{1}{4}x^3 + 6$
$= \left(-\frac{1}{3} - \frac{1}{4}\right)x^3 + \left(\frac{2}{5} + 1\right)x^2 + 6$
$= \left(-\frac{4}{12} - \frac{3}{12}\right)x^3 + \left(\frac{2}{5} + \frac{5}{5}\right)x^2 + 6$
$= -\frac{7}{12}x^3 + \frac{7}{5}x^2 + 6$

35. $6a^2 - 4ab + 7b^2 - a^2 - 5ab + 9b^2$
$= (6 - 1)a^2 + (-4 - 5)ab + (7 + 9)b^2$
$= 5a^2 - 9ab + 16b^2$

37. $(-7x + 5) + (-3x^2 + 7x + 5)$
$= -7x + 5 - 3x^2 + 7x + 5$
$= -3x^2 + (-7x + 7x) + (5 + 5)$
$= -3x^2 + 10$

39. $(2x^2 + 5) - (3x^2 - 9) = 2x^2 + 5 - 3x^2 + 9$
$= (2x^2 - 3x^2) + (5 + 9)$
$= -x^2 + 14$

41. $3x - (5x - 9) = 3x - 5x + 9$
$= (3x - 5x) + 9$
$= -2x + 9$

43. $(2x^2 + 3x - 9) - (-4x + 7)$
$= 2x^2 + 3x - 9 + 4x - 7$
$= 2x^2 + (3x + 4x) + (-9 - 7)$
$= 2x^2 + 7x - 16$

45.
$$\begin{array}{r} 3t^2 + 4 \\ + \ 5t^2 - 8 \\ \hline 8t^2 - 4 \end{array}$$

47.
$$\begin{array}{r} 4z^2 - 8z + 3 \\ - \ (6z^2 + 8z - 3) \\ \hline \end{array} \qquad \begin{array}{r} 4z^2 - 8z + 3 \\ + \ (-6z^2 - 8z + 3) \\ \hline -2z^2 - 16z + 6 \end{array}$$

49.
$$\begin{array}{r} 5x^3 - 4x^2 + 6x - 2 \\ - \ (3x^3 - 2x^2 \ \ - x - 4) \\ \hline \end{array} \qquad \begin{array}{r} 5x^3 - 4x^2 + 6x - 2 \\ + \ (-3x^3 + 2x^2 \ \ + x + 4) \\ \hline 2x^3 - 2x^2 + 7x + 2 \end{array}$$

51. $(81x^2 + 10) - (19x^2 + 5) = 81x^2 + 10 - 19x^2 - 5$
$= 62x^2 + 5$

53. $[(8x + 1) + (6x + 3)] - (2x + 2)$
$= 8x + 1 + 6x + 3 - 2x - 2$
$= 8x + 6x - 2x + 1 + 3 - 2$
$= 12x + 2$

55. $(-3y^2 - 4y) + (2y^2 + y - 1)$
$= -3y^2 - 4y + 2y^2 + y - 1$
$= -y^2 - 3y - 1$

57. $(5x+8)-(-2x^2-6x+8)$
$=5x+8+2x^2+6x-8$
$=2x^2+11x$

59. $(-8x^4+7x)+(-8x^4+x+9)$
$=-8x^4+7x-8x^4+x+9$
$=-16x^4+8x+9$

61. $(3x^2+5x-8)+(5x^2+9x+12)-(x^2-14)$
$=3x^2+5x-8+5x^2+9x+12-x^2+14$
$=7x^2+14x+18$

63. $(7x-3)-4x=7x-3-4x=3x-3$

65. $(7x^2+3x+9)-(5x+7)=7x^2+3x+9-5x-7$
$=7x^2-2x+2$

67. $[(8y^2+7)+(6y+9)]-(4y^2-6y-3)$
$=8y^2+7+6y+9-4y^2+6y+3$
$=4y^2+12y+19$

69. $[(-x^2-2x)+(5x^2+x+9)]-(-2x^2+4x-12)$
$=-x^2-2x+5x^2+x+9+2x^2-4x+12$
$=6x^2-5x+21$

71. $2x\cdot 2x+x\cdot 7+x\cdot x+x\cdot 5=4x^2+7x+x^2+5x$
$=5x^2+12x$

73. $9x+10+3x+12+4x+15+2x+7$
$=(9x+3x+4x+2x)+(10+12+15+7)$
$=18x+44$

75. $(-x^2+3x)+(2x^2+5)+(4x-1)$
$=-x^2+3x+2x^2+5+4x-1$
$=x^2+7x+4$
The perimeter is (x^2+7x+4) feet.

77. $(4y^2+4y+1)-(y^2-10)$
$=4y^2+4y+1-y^2+10$
$=3y^2+4y+11$
The length of the remaining piece is $(3y^2+4y+11)$ meters.

79. $(9a+6b-5)+(-11a-7b+6)$
$=9a+6b-5-11a-7b+6$
$=-2a-b+1$

81. $(4x^2+y^2+3)-(x^2+y^2-2)$
$=4x^2+y^2+3-x^2-y^2+2$
$=3x^2+5$

83. $(x^2+2xy-y^2)+(5x^2-4xy+20y^2)$
$=x^2+2xy-y^2+5x^2-4xy+20y^2$
$=6x^2-2xy+19y^2$

85. $(11r^2s+16rs-3-2r^2s^2)-(3sr^2+5-9r^2s^2)$
$=11r^2s+16rs-3-2r^2s^2-3sr^2-5+9r^2s^2$
$=8r^2s+16rs-8+7r^2s^2$

87. $7.75x+9.16x^2-1.27-14.58x^2-18.34$
$=(9.16-14.58)x^2+7.75x+(-1.27-18.34)$
$=-5.42x^2+7.75x-19.61$

89. $[(7.9y^4-6.8y^3+3.3y)+(6.1y^3-5)]$
$-(4.2y^4+1.1y-1)$
$=7.9y^4-6.8y^3+3.3y+6.1y^3-5-4.2y^4$
$-1.1y+1$
$=3.7y^4-0.7y^3+2.2y-4$

91. $3x(2x)=3\cdot 2\cdot x\cdot x=6x^2$

93. $(12x^3)(-x^5)=(12x^3)(-1x^5)$
$=(12)(-1)(x^3)(x^5)$
$=-12x^8$

95. $10x^2(20xy^2)=10\cdot 20x^2\cdot x\cdot y^2=200x^3y^2$

97. Answers may vary

99. Answers may vary

101. $10y-6y^2-y=(10-1)y-6y^2=9y-6y^2$
choice b

103. $(5x-3)+(5x-3)=(5x+5x)+(-3-3)$
$=(5+5)x-6$
$=10x-6$
choice e

105. **a.** $z + 3z = (1 + 3)z = 4z$

b. $z \cdot 3z = 3 \cdot z \cdot z = 3 \cdot z^{1+1} = 3z^2$

c. $-z - 3z = (-1 - 3)z = -4z$

d. $(-z)(-3z) = (-1 \cdot -3) \cdot z \cdot z = 3 \cdot z^{1+1} = 3z^2$

107. $(4x^{2a} - 3x^a + 0.5) - (x^{2a} - 5x^a - 0.2)$
$= 4x^{2a} - 3x^a + 0.5 - x^{2a} + 5x^a + 0.2$
$= 4x^{2a} - x^{2a} - 3x^a + 5x^a + 0.5 + 0.2$
$= 3x^{2a} + 2x^a + 0.7$

109. $(8x^{2y} - 7x^y + 3) + (-4x^{2y} + 9x^y - 14)$
$= 8x^{2y} - 7x^y + 3 - 4x^{2y} + 9x^y - 14$
$= 8x^{2y} - 4x^{2y} - 7x^y + 9x^y + 3 - 14$
$= 4x^{2y} + 2x^y - 11$

111. $P(x) + Q(x) = (3x + 3) + (4x^2 - 6x + 3)$
$= 4x^2 + 3x - 6x + 3 + 3$
$= 4x^2 - 3x + 6$

113. $Q(x) - R(x) = (4x^2 - 6x + 3) - (5x^2 - 7)$
$= 4x^2 - 5x^2 - 6x + 3 + 7$
$= -x^2 - 6x + 10$

115. $2[Q(x)] - R(x) = 2(4x^2 - 6x + 3) - (5x^2 - 7)$
$= 2(4x^2) - 2(6x) + 2(3) - 5x^2 + 7$
$= 8x^2 - 12x + 6 - 5x^2 + 7$
$= 3x^2 - 12x + 13$

117. $P(x) = 2x - 3$

a. $P(a) = 2a - 3$

b. $P(-x) = 2(-x) - 3 = -2x - 3$

c. $P(x + h) = 2(x + h) - 3 = 2x + 2h - 3$

119. $P(x) = 4x$

a. $P(a) = 4a$

b. $P(-x) = 4(-x) = -4x$

c. $P(x + h) = 4(x + h) = 4x + 4h$

121. $6.4x^2 + 37.9x + 2856.8$
$= 6.4(26)^2 + 37.9(26) + 2856.8$
$= 4326.4 + 985.4 + 2856.8$
$= 8168.6$
Costs are predicted to be \$8169 in 2010.

123. $(2.13x^2 + 21.89x + 1190) + (8.71x^2 - 1.46x + 2095)$
$= (2.13 + 8.71)x^2 + (21.89 - 1.46)x + (1190 + 2095)$
$= 10.84x^2 + 20.43x + 3285$

Section 5.3

Practice Exercises

1. $5y \cdot 2y = (5 \cdot 2)(y \cdot y) = 10y^2$

2. $(5z^3) \cdot (-0.4z^5) = (5 \cdot -0.4)(z^3 \cdot z^5) = -2z^8$

3. $\left(-\frac{1}{9}b^6\right)\left(-\frac{7}{8}b^3\right) = \left(-\frac{1}{9} \cdot -\frac{7}{8}\right)(b^6 \cdot b^3) = \frac{7}{72}b^9$

4. **a.** $3x(5x^5 + 5) = 3x(5x^5) + 3x(5) = 15x^6 + 15x$

b. $-5x^3(2x^2 - 9x + 2)$
$= -5x^3(2x^2) + (-5x^3)(-9x) + (-5x^3)(2)$
$= -10x^5 + 45x^4 - 10x^3$

5. Multiply each term of the first binomial by each term of the second.
$(5x - 2)(2x + 3)$
$= 5x(2x) + 5x(3) + (-2)(2x) + (-2)(3)$
$= 10x^2 + 15x - 4x - 6$
$= 10x^2 + 11x - 6$

6. Recall that $a^2 = a \cdot a$, so $(5x - 3y)^2 = (5x - 3y)(5x - 3y)$. Multiply each term of the first binomial by each term of the second.
$(5x - 3y)(5x - 3y)$
$= 5x(5x) + 5x(-3y) + (-3y)(5x) + (-3y)(-3y)$
$= 25x^2 - 15xy - 15xy + 9y^2$
$= 25x^2 - 30xy + 9y^2$

7. Multiply each term of the first polynomial by each term of the second.
$$(y+4)(2y^2-3y+5)$$
$$= y(2y^2)+y(-3y)+y(5)+4(2y^2)+4(-3y)+4(5)$$
$$= 2y^3-3y^2+5y+8y^2-12y+20$$
$$= 2y^3+5y^2-7y+20$$

8. Write $(s+2t)^3$ as $(s + 2t)(s + 2t)(s + 2t)$.
$$(s+2t)(s+2t)(s+2t)$$
$$= (s^2+2st+2st+4t^2)(s+2t)$$
$$= (s^2+4st+4t^2)(s+2t)$$
$$= (s^2+4st+4t^2)s+(s^2+4st+4t^2)(2t)$$
$$= s^3+4s^2t+4st^2+2s^2t+8st^2+8t^3$$
$$= s^3+6s^2t+12st^2+8t^3$$

9.
$$\begin{array}{r} 5x^2-3x+5 \\ \times \qquad\qquad x-4 \\ \hline -20x^2+12x-20 \\ 5x^3\ -3x^2\ +5x \qquad \\ \hline 5x^3-23x^2+17x-20 \end{array}$$

10.
$$\begin{array}{r} x^3-2x^2+1 \\ \times \qquad\qquad x^2+2 \\ \hline 2x^3-4x^2+2 \\ x^5-2x^4 \qquad\quad +x^2 \qquad \\ \hline x^5-2x^4+2x^3-3x^2+2 \end{array}$$

11.
$$\begin{array}{r} 5x^2+2x-2 \\ x^2-x+3 \\ \hline 15x^2+6x-6 \\ -5x^3-2x^2+2x \qquad \\ 5x^4+2x^3-2x^2 \qquad\qquad \\ \hline 5x^4-3x^3+11x^2+8x-6 \end{array}$$

Vocabulary and Readiness Check

1. The expression $5x(3x + 2)$ equals $5x \cdot 3x + 5x \cdot 2$ by the distributive property.

2. The expression $(x + 4)(7x - 1)$ equals $x(7x - 1) + 4(7x - 1)$ by the distributive property.

3. The expression $(5y-1)^2$ equals $(5y - 1)(5y - 1)$.

4. The expression $9x \cdot 3x$ equals $27x^2$.

5. $x^3 \cdot x^5 = x^{3+5} = x^8$

6. $x^2 \cdot x^6 = x^{2+6} = x^8$

7. $x^3 + x^5$ cannot be simplified.

8. $x^2 + x^6$ cannot be simplified.

9. $x^7 \cdot x^7 = x^{7+7} = x^{14}$

10. $x^{11} \cdot x^{11} = x^{11+11} = x^{22}$

11. $x^7 + x^7 = (1+1)x^7 = 2x^7$

12. $x^{11} + x^{11} = (1+1)x^{11} = 2x^{11}$

Exercise Set 5.3

1. $-4n^3 \cdot 7n^7 = (-4 \cdot 7)(n^3 \cdot n^7) = -28n^{10}$

3. $(-3.1x^3)(4x^9) = (-3.1 \cdot 4)(x^3 \cdot x^9) = -12.4x^{12}$

5. $\left(-\frac{1}{3}y^2\right)\left(\frac{2}{5}y\right) = \left(-\frac{1}{3} \cdot \frac{2}{5}\right)(y^2 \cdot y) = -\frac{2}{15}y^3$

7. $(2x)(-3x^2)(4x^5) = (2 \cdot -3 \cdot 4)(x \cdot x^2 \cdot x^5) = -24x^8$

9. $3x(2x+5) = 3x(2x)+3x(5) = 6x^2+15x$

11. $-2a(a+4) = -2a(a)+(-2a)(4) = -2a^2-8a$

13. $3x(2x^2-3x+4) = 3x(2x^2)+3x(-3x)+3x(4)$
$$= 6x^3-9x^2+12x$$

15. $-2a^2(3a^2-2a+3)$
$$= -2a^2(3a^2)+(-2a^2)(-2a)+(-2a^2)(3)$$
$$= -6a^4+4a^3-6a^2$$

17. $-y(4x^3-7x^2y+xy^2+3y^3)$
$$= -y(4x^3)+(-y)(-7x^2y)+(-y)(xy^2)+(-y)(3y^3)$$
$$= -4x^3y+7x^2y^2-xy^3-3y^4$$

19. $\frac{1}{2}x^2(8x^2-6x+1)$
$=\frac{1}{2}x^2(8x^2)+\frac{1}{2}x^2(-6x)+\frac{1}{2}x^2(1)$
$=4x^4-3x^3+\frac{1}{2}x^2$

21. $(x+4)(x+3)=x(x)+x(3)+4(x)+4(3)$
$=x^2+3x+4x+12$
$=x^2+7x+12$

23. $(a+7)(a-2)=a(a)+a(-2)+7(a)+7(-2)$
$=a^2-2a+7a-14$
$=a^2+5a-14$

25. $\left(x+\frac{2}{3}\right)\left(x-\frac{1}{3}\right)$
$=x(x)+x\left(-\frac{1}{3}\right)+\frac{2}{3}(x)+\frac{2}{3}\left(-\frac{1}{3}\right)$
$=x^2-\frac{1}{3}x+\frac{2}{3}x-\frac{2}{9}$
$=x^2+\frac{1}{3}x-\frac{2}{9}$

27. $(3x^2+1)(4x^2+7)$
$=3x^2(4x^2)+3x^2(7)+1(4x^2)+1(7)$
$=12x^4+21x^2+4x^2+7$
$=12x^4+25x^2+7$

29. $(2y-4)^2$
$=(2y-4)(2y-4)$
$=2y(2y)+2y(-4)+(-4)(2y)+(-4)(-4)$
$=4y^2-8y-8y+16$
$=4y^2-16y+16$

31. $(4x-3)(3x-5)$
$=4x(3x)+4x(-5)+(-3)(3x)+(-3)(-5)$
$=12x^2-20x-9x+15$
$=12x^2-29x+15$

33. $(3x^2+1)^2=(3x^2+1)(3x^2+1)$
$=3x^2(3x^2)+3x^2(1)+1(3x^2)+1(1)$
$=9x^4+3x^2+3x^2+1$
$=9x^4+6x^2+1$

35. a. $(3x+5)+(3x+7)=(3x+3x)+(5+7)$
$=6x+12$

b. $(3x+5)(3x+7)$
$=3x(3x)+3x(7)+5(3x)+5(7)$
$=9x^2+21x+15x+35$
$=9x^2+36x+35$

c. Answers may vary

37. $(x-2)(x^2-3x+7)$
$=x(x^2)+x(-3x)+x(7)+(-2)(x^2)$
$\quad+(-2)(-3x)+(-2)(7)$
$=x^3-3x^2+7x-2x^2+6x-14$
$=x^3-5x^2+13x-14$

39. $(x+5)(x^3-3x+4)$
$=x(x^3)+x(-3x)+x(4)+5(x^3)+5(-3x)+5(4)$
$=x^4-3x^2+4x+5x^3-15x+20$
$=x^4+5x^3-3x^2-11x+20$

41. $(2a-3)(5a^2-6a+4)$
$=2a(5a^2)+2a(-6a)+2a(4)+(-3)(5a^2)$
$\quad+(-3)(-6a)+(-3)(4)$
$=10a^3-12a^2+8a-15a^2+18a-12$
$=10a^3-27a^2+26a-12$

43. $(x+2)^3=(x+2)(x+2)(x+2)$
$=(x^2+2x+2x+4)(x+2)$
$=(x^2+4x+4)(x+2)$
$=(x^2+4x+4)x+(x^2+4x+4)2$
$=x^3+4x^2+4x+2x^2+8x+8$
$=x^3+6x^2+12x+8$

45. $(2y-3)^3$
$=(2y-3)(2y-3)(2y-3)$
$=(4y^2-6y-6y+9)(2y-3)$
$=(4y^2-12y+9)(2y-3)$
$=(4y^2-12y+9)2y+(4y^2-12y+9)(-3)$
$=8y^3-24y^2+18y-12y^2+36y-27$
$=8y^3-36y^2+54y-27$

47.
$$\begin{array}{r} 2x-11 \\ \times \quad 6x+1 \\ \hline 2x-11 \\ 12x^2-66x \quad \\ \hline 12x^2-64x-11 \end{array}$$

49.
$$\begin{array}{r} 2x^2+4x-1 \\ \times \quad 5x+1 \\ \hline 2x^2+4x-1 \\ 10x^3+20x^2-5x \quad \\ \hline 10x^3+22x^2-x-1 \end{array}$$

51.
$$\begin{array}{r} 2x^2-7x-9 \\ \times \quad x^2+5x-7 \\ \hline -14x^2+49x+63 \\ 10x^3-35x^2-45x \quad \\ 2x^4-7x^3-9x^2 \quad\quad \\ \hline 2x^4+3x^3-58x^2+4x+63 \end{array}$$

53. $-1.2y(-7y^6)=-1.2(-7)(y\cdot y^6)=8.4y^7$

55. $-3x(x^2+2x-8)$
$=-3x(x^2)+(-3x)(2x)+(-3x)(-8)$
$=-3x^3-6x^2+24x$

57. $(x+19)(2x+1)=x(2x)+x(1)+19(2x)+19(1)$
$=2x^2+x+38x+19$
$=2x^2+39x+19$

59. $\left(x+\frac{1}{7}\right)\left(x-\frac{3}{7}\right)$
$=x(x)+x\left(-\frac{3}{7}\right)+\frac{1}{7}(x)+\frac{1}{7}\left(-\frac{3}{7}\right)$
$=x^2-\frac{3}{7}x+\frac{1}{7}(x)-\frac{3}{49}$
$=x^2-\frac{2}{7}x-\frac{3}{49}$

61. $(3y+5)^2=(3y+5)(3y+5)$
$=3y(3y)+3y(5)+5(3y)+5(5)$
$=9y^2+15y+15y+25$
$=9y^2+30y+25$

63. $(a+4)(a^2-6a+6)$
$=a(a^2)+a(-6a)+a(6)+4(a^2)+4(-6a)+4(6)$
$=a^3-6a^2+6a+4a^2-24a+24$
$=a^3-2a^2-18a+24$

65. $(2x-5)^3$
$=(2x-5)(2x-5)(2x-5)$
$=(4x^2-10x-10x+25)(2x-5)$
$=(4x^2-20x+25)(2x-5)$
$=(4x^2-20x+25)2x+(4x^2-20x+25)(-5)$
$=8x^3-40x^2+50x-20x^2+100x-125$
$=8x^3-60x^2+150x-125$

67. $(4x+5)(8x^2+2x-4)$
$=4x(8x^2)+4x(2x)+4x(-4)+5(8x^2)$
$\quad +5(2x)+5(-4)$
$=32x^3+8x^2-16x+40x^2+10x-20$
$=32x^3+48x^2-6x-20$

69.
$$\begin{array}{r} 3x^2+2x-4 \\ \times \quad 2x^2-4x+3 \\ \hline 9x^2+6x-12 \\ -12x^3-8x^2+16x \quad \\ 6x^4+4x^3-8x^2 \quad\quad \\ \hline 6x^4-8x^3-7x^2+22x-12 \end{array}$$

71. $(2x-5)(2x+5)$
$=2x(2x)+2x(5)+(-5)(2x)+(-5)(5)$
$=4x^2+10x-10x-25$
$=4x^2-25$
The area is $(4x^2-25)$ square yards.

73. $\frac{1}{2}(3x-2)(4x)=2x(3x-2)$
$=2x(3x)+2x(-2)$
$=6x^2-4x$
The area is $(6x^2-4x)$ square inches.

75. $(5x)^2=(5x)(5x)=(5\cdot 5)(x\cdot x)=25x^2$

77. $(-3y^3)^2=(-3y^3)(-3y^3)$
$=(-3\cdot -3)(y^3\cdot y^3)$
$=9y^6$

79. left rectangle: $x\cdot x=x^2$
right rectangle: $x\cdot 3=3x$
left rectangle + right rectangle: x^2+3x

81. top left rectangle: $x \cdot x = x^2$
top right rectangle: $x \cdot 3 = 3x$
bottom left rectangle: $2 \cdot x = 2x$
bottom right rectangle: $2 \cdot 3 = 6$
entire figure: $x^2 + 3x + 2x + 6 = x^2 + 5x + 6$

83. $5a + 6a = (5 + 6)a = 11a$

85. $(5x)^2 + (2y)^2 = (5x)(5x) + (2y)(2y) = 25x^2 + 4y^2$

87. $(3x-1) + (10x-6) = (3x+10x) + (-1-6) = 13x - 7$

89. $(3x-1)(10x-6)$
$= 3x(10x) + 3x(-6) + (-1)(10x) + (-1)(-6)$
$= 30x^2 - 18x - 10x + 6$
$= 30x^2 - 28x + 6$

91. $(3x-1) - (10x-6) = 3x - 1 - 10x + 6$
$= (3x - 10x) + (-1 + 6)$
$= -7x + 5$

93. a. $(a+b)(a-b) = a(a) + a(-b) + b(a) + b(-b)$
$= a^2 - ab + ab - b^2$
$= a^2 - b^2$

b. $(2x+3y)(2x-3y)$
$= 2x(2x) + 2x(-3y) + 3y(2x) + 3y(-3y)$
$= 4x^2 - 6xy + 6xy - 9y^2$
$= 4x^2 - 9y^2$

c. $(4x+7)(4x-7)$
$= 4x(4x) + 4x(-7) + 7(4x) + 7(-7)$
$= 16x^2 - 28x + 28x - 49$
$= 16x^2 - 49$

d. Answers may vary

95. larger square: $(x+3)^2 = (x+3)(x+3)$
$= x(x) + x(3) + 3(x) + 3(3)$
$= x^2 + 3x + 3x + 9$
$= x^2 + 6x + 9$
smaller square: $2^2 = 2 \cdot 2 = 4$
shaded region: $x^2 + 6x + 9 - 4 = x^2 + 6x + 5$
The area of the shaded region is $(x^2 + 6x + 5)$ square units.

Section 5.4

Practice Exercises

1. $(x+2)(x-5)$
$= (x)(x) + (x)(-5) + (2)(x) + (2)(-5)$
$= x^2 - 5x + 2x - 10$
$= x^2 - 3x - 10$

2. $(4x-9)(x-1)$
$= 4x(x) + 4x(-1) + (-9)(x) + (-9)(-1)$
$= 4x^2 - 4x - 9x + 9$
$= 4x^2 - 13x + 9$

3. $3(x+5)(3x-1) = 3(3x^2 - x + 15x - 5)$
$= 3(3x^2 + 14x - 5)$
$= 9x^2 + 42x - 15$

4. $(4x-1)^2$
$= (4x-1)(4x-1)$
$= (4x)(4x) + (4x)(-1) + (-1)(4x) + (-1)(-1)$
$= 16x^2 - 4x - 4x + 1$
$= 16x^2 - 8x + 1$

5. a. $(b+3)^2 = b^2 + 2(b)(3) + 3^2 = b^2 + 6b + 9$

b. $(x-y)^2 = x^2 - 2(x)(y) + y^2 = x^2 - 2xy + y^2$

c. $(3y+2)^2 = (3y)^2 + 2(3y)(2) + 2^2 = 9y^2 + 12y + 4$

d. $(a^2 - 5b)^2 = (a^2)^2 - 2(a^2)(5b) + (5b)^2 = a^4 - 10a^2b + 25b^2$

6. a. $3(x+5)(x-5) = 3(x^2 - 5^2)$
$= 3x(x^2 - 25)$
$= 3x^2 - 75$

b. $(4b-3)(4b+3) = (4b)^2 - 3^2 = 16b^2 - 9$

c. $\left(x + \frac{2}{3}\right)\left(x - \frac{2}{3}\right) = x^2 - \left(\frac{2}{3}\right)^2 = x^2 - \frac{4}{9}$

d. $(5s+t)(5s-t) = (5s)^2 - t^2 = 25s^2 - t^2$

e. $(2y-3z^2)(2y+3z^2)=(2y)^2-(3z^2)^2$
$=4y^2-9z^4$

7. a. $(4x+3)(x-6)=4x^2-24x+3x-18$
$=4x^2-21x-18$

b. $(7b-2)^2=(7b)^2-2(7b)(2)+2^2$
$=49b^2-28b+4$

c. $(x+0.4)(x-0.4)=x^2-(0.4)^2=x^2-0.16$

d. $(x^2-3)(3x^4+2)=3x^6+2x^2-9x^4-6$

e. $(x+1)(x^2+5x-2)$
$=x(x^2+5x-2)+1(x^2+5x-2)$
$=x^3+5x^2-2x+x^2+5x-2$
$=x^3+6x^2+3x-2$

Vocabulary and Readiness Check

1. $(x+4)^2=x^2+2(x)(4)+4^2$
$=x^2+8x+16\neq x^2+16$
The statement is false.

2. $(x+6)(2x-1)=2x^2-x+12x-6$
$=2x^2+11x-6$
The statement is true.

3. $(x+4)(x-4)=x^2-4^2=x^2-16\neq x^2+16$
The statement is false.

4. $(x-1)(x^3+3x-1)$
$=x(x^3+3x-1)-1(x^3+3x-1)$
$=x^4+3x^2-x-x^3-3x+1$
$=x^4-x^3+3x^2-4x+1$
This is a polynomial of degree 4; the statement is false.

Exercise Set 5.4

1. $(x+3)(x+4)=x^2+4x+3x+12=x^2+7x+12$

3. $(x-5)(x+10)=x^2+10x-5x-50$
$=x^2+5x-50$

5. $(5x-6)(x+2)=5x^2+10x-6x-12$
$=5x^2+4x-12$

7. $(y-6)(4y-1)=4y^2-1y-24y+6$
$=4y^2-25y+6$

9. $(2x+5)(3x-1)=6x^2-2x+15x-5$
$=6x^2+13x-5$

11. $(x-2)^2=x^2-2(x)(2)+2^2=x^2-4x+4$

13. $(2x-1)^2=(2x)^2-2(2x)(1)+(1)^2$
$=4x^2-4x+1$

15. $(3a-5)^2=(3a)^2-2(3a)(5)+5^2$
$=9a^2-30a+25$

17. $(5x+9)^2=(5x)^2+2(5x)(9)+9^2$
$=25x^2+90x+81$

19. Answers may vary

21. $(a-7)(a+7)=a^2-7^2=a^2-49$

23. $(3x-1)(3x+1)=(3x)^2-1^2=9x^2-1$

25. $\left(3x-\frac{1}{2}\right)\left(3x+\frac{1}{2}\right)=(3x)^2-\left(\frac{1}{2}\right)^2=9x^2-\frac{1}{4}$

27. $(9x+y)(9x-y)=(9x)^2-y^2=81x^2-y^2$

29. $(2x+0.1)(2x-0.1)=(2x)^2-(0.1)^2=4x^2-0.01$

31. $(a+5)(a+4)=a^2+4a+5a+20=a^2+9a+20$

33. $(a+7)^2=a^2+2(a)(7)+7^2=a^2+14a+49$

35. $(4a+1)(3a-1)=12a^2-4a+3a-1$
$=12a^2-a-1$

37. $(x+2)(x-2)=x^2-2^2=x^2-4$

39. $(3a+1)^2=(3a)^2+2(3a)(1)+1^2=9a^2+6a+1$

41. $(x^2+y)(4x-y^4)=4x^3-x^2y^4+4xy-y^5$

43. $(x+3)(x^2-6x+1)$
$= x(x^2-6x+1)+3(x^2-6x+1)$
$= x^3-6x^2+x+3x^2-18x+3$
$= x^3-3x^2-17x+3$

45. $(2a-3)^2 = (2a)^2-2(2a)(3)+(3)^2$
$= 4a^2-12a+9$

47. $(5x-6z)(5x+6z) = (5x)^2-(6z)^2 = 25x^2-36z^2$

49. $(x^5-3)(x^5-5) = x^{10}-5x^5-3x^5+15$
$= x^{10}-8x^5+15$

51. $\left(x-\frac{1}{3}\right)\left(x+\frac{1}{3}\right) = x^2-\left(\frac{1}{3}\right)^2 = x^2-\frac{1}{9}$

53. $(a^3+11)(a^4-3) = a^7-3a^3+11a^4-33$

55. $3(x-2)^2 = 3[x^2-2(x)(2)+2^2]$
$= 3(x^2-4x+4)$
$= 3x^2-12x+12$

57. $(3b+7)(2b-5) = 6b^2-15b+14b-35$
$= 6b^2-b-35$

59. $(7p-8)(7p+8) = (7p)^2-(8)^2 = 49p^2-64$

61. $\left(\frac{1}{3}a^2-7\right)\left(\frac{1}{3}a^2+7\right) = \left(\frac{1}{3}a^2\right)^2-(7)^2$
$= \frac{1}{9}a^4-49$

63. $5x^2(3x^2-x+2) = 5x^2(3x^2)+5x^2(-x)+5x^2(2)$
$= 15x^4-5x^3+10x^2$

65. $(2r-3s)(2r+3s) = (2r)^2-(3s)^2 = 4r^2-9s^2$

67. $(3x-7y)^2 = (3x)^2-2(3x)(7y)+(7y)^2$
$= 9x^2-42xy+49y^2$

69. $(4x+5)(4x-5) = (4x)^2-5^2 = 16x^2-25$

71. $(8x+4)^2 = (8x)^2+2(8x)(4)+(4)^2$
$= 64x^2+64x+16$

73. $\left(a-\frac{1}{2}y\right)\left(a+\frac{1}{2}y\right) = a^2-\left(\frac{1}{2}y\right)^2 = a^2-\frac{1}{4}y^2$

75. $\left(\frac{1}{5}x-y\right)\left(\frac{1}{5}x+y\right) = \left(\frac{1}{5}x\right)^2-y^2 = \frac{1}{25}x^2-y^2$

77. $(a+1)(3a^2-a+1)$
$= a(3a^2-a+1)+1(3a^2-a+1)$
$= 3a^3-a^2+a+3a^2-a+1$
$= 3a^3+2a^2+1$

79. $(2x+1)^2 = (2x)^2+2(2x)(1)+1^2 = 4x^2+4x+1$
The area is $(4x^2+4x+1)$ square feet.

81. $\frac{50b^{10}}{70b^5} = \frac{50}{70}b^{10-5} = \frac{5b^5}{7}$

83. $\frac{8a^{17}b^{15}}{-4a^7b^{10}} = \frac{8}{-4}a^{17-7}b^{15-10} = -2a^{10}b^5$

85. $\frac{2x^4y^{12}}{3x^4y^4} = \frac{2}{3}x^{4-4}y^{12-4} = \frac{2y^8}{3}$

87. $(-1, 1)$ and $(2, 2)$
$m = \frac{y_2-y_1}{x_2-x_1} = \frac{2-1}{2-(-1)} = \frac{1}{3}$

89. $(-1, -2)$ and $(1, 0)$
$m = \frac{y_2-y_1}{x_2-x_1} = \frac{0-(-2)}{1-(-1)} = \frac{2}{2} = 1$

91. $(a-b)^2 = a^2-2ab+b^2$
Choice c.

93. $(a+b)^2 = a^2+2ab+b^2$
Choice d.

95. From FOIL, the first term in the result is $(x^{\square})^2 = x^{2\square}$. Thus, $2\square = 4$ so $\square = 2$.

97. $\frac{1}{2}(5a+b)(5a-b) = \frac{1}{2}(25a^2-b^2) = \frac{25a^2}{2}-\frac{b^2}{2}$
The area is $\left(\frac{25a^2}{2}-\frac{b^2}{2}\right)$ square units.

99. $(5x-3)^2-(x+1)^2$
$=(25x^2-30x+9)-(x^2+2x+1)$
$=25x^2-30x+9-x^2-2x-1$
$=(24x^2-32x+8)$
The shaded area is
$(24x^2-32x+8)$ square meters.

101. $(x+5)(x+5)=(x+5)^2$
$=x^2+2(x)(5)+5^2$
$=x^2+10x+25$
The area is $(x^2+10x+25)$ square units.

103. Answers may vary

105. $[(x+y)-3][(x+y)+3]=(x+y)^2-3^2$
$=x^2+2xy+y^2-9$

107. $[(a-3)+b][(a-3)-b]=(a-3)^2-b^2$
$=a^2-6a+9-b^2$

Integrated Review

1. $(5x^2)(7x^3)=(5\cdot 7)(x^2\cdot x^3)=35x^5$

2. $(4y^2)(8y^7)=(4\cdot 8)(y^2\cdot y^7)=32y^9$

3. $-4^2=-(4\cdot 4)=-16$

4. $(-4)^2=(-4)(-4)=16$

5. $(x-5)(2x+1)=2x^2+x-10x-5$
$=2x^2-9x-5$

6. $(3x-2)(x+5)=3x^2+15x-2x-10$
$=3x^2+13x-10$

7. $(x-5)+(2x+1)=x-5+2x+1=3x-4$

8. $(3x-2)+(x+5)=3x-2+x+5=4x+3$

9. $\frac{7x^9y^{12}}{x^3y^{10}}=7x^{9-3}y^{12-10}=7x^6y^2$

10. $\frac{20a^2b^8}{14a^2b^2}=\frac{20}{14}a^{2-2}b^{8-2}=\frac{10b^6}{7}$

11. $(12m^7n^6)^2=12^2m^{7\cdot 2}n^{6\cdot 2}=144m^{14}n^{12}$

12. $(4y^9z^{10})^3=4^3y^{9\cdot 3}z^{10\cdot 3}=64y^{27}z^{30}$

13. $3(4y-3)(4y+3)=3[(4y)^2-3^2]$
$=3(16y^2-9)$
$=48y^2-27$

14. $2(7x-1)(7x+1)=2[(7x)^2-1^2]$
$=2(49x^2-1)$
$=98x^2-2$

15. $(x^7y^5)^9=x^{7\cdot 9}y^{5\cdot 9}=x^{63}y^{45}$

16. $(3^1x^9)^3=3^{1\cdot 3}x^{9\cdot 3}=3^3x^{27}=27x^{27}$

17. $(7x^2-2x+3)-(5x^2+9)$
$=7x^2-2x+3-5x^2-9$
$=2x^2-2x-6$

18. $(10x^2+7x-9)-(4x^2-6x+2)$
$=10x^2+7x-9-4x^2+6x-2$
$=6x^2+13x-11$

19. $0.7y^2-1.2+1.8y^2-6y+1=2.5y^2-6y-0.2$

20. $7.8x^2-6.8x+3.3+0.6x^2-9$
$=8.4x^2-6.8x-5.7$

21. $(x+4y)^2=x^2+2(x)(4y)+(4y)^2$
$=x^2+8xy+16y^2$

22. $(y-9z)^2=y^2-2(y)(9z)+(9z)^2$
$=y^2-18yz+81z^2$

23. $(x+4y)+(x+4y)=x+4y+x+4y=2x+8y$

24. $(y-9z)+(y-9z)=y-9z+y-9z=2y-18z$

25. $7x^2-6xy+4(y^2-xy)=7x^2-6xy+4y^2-4xy$
$=7x^2-10xy+4y^2$

26. $5a^2-3ab+6(b^2-a^2)=5a^2-3ab+6b^2-6a^2$
$=-a^2-3ab+6b^2$

27. $(x-3)(x^2+5x-1)$
$= x(x^2+5x-1)-3(x^2+5x-1)$
$= x^3+5x^2-x-3x^2-15x+3$
$= x^3+2x^2-16x+3$

28. $(x+1)(x^2-3x-2)$
$= x(x^2-3x-2)+1(x^2-3x-2)$
$= x^3-3x^2-2x+x^2-3x-2$
$= x^3-2x^2-5x-2$

29. $(2x^3-7)(3x^2+10)$
$= 2x^3(3x^2)+2x^3(10)-7(3x^2)-7(10)$
$= 6x^5+20x^3-21x^2-70$

30. $(5x^3-1)(4x^4+5)$
$= 5x^3(4x^4)+5x^3(5)-1(4x^4)-1(5)$
$= 20x^7+25x^3-4x^4-5$

31. $(2x-7)(x^2-6x+1)$
$= 2x(x^2-6x+1)-7(x^2-6x+1)$
$= 2x^3-12x^2+2x-7x^2+42x-7$
$= 2x^3-19x^2+44x-7$

32. $(5x-1)(x^2+2x-3)$
$= 5x(x^2+2x-3)-1(x^2+2x-3)$
$= 5x^3+10x^2-15x-x^2-2x+3$
$= 5x^3+9^2-17x+3$

33. $5x^3+5y^3$ cannot be simplified.

34. $(5x^3)(5y^3)=5\cdot 5x^3y^3=25x^3y^3$

35. $(5x^3)^3=5^3x^{3\cdot 3}=125x^9$

36. $\dfrac{5x^3}{5y^3}=\dfrac{x^3}{y^3}$

37. $x+x=2x$

38. $x\cdot x=x^2$

Section 5.5

Practice Exercises

1. a. $5^{-3}=\dfrac{1}{5^3}=\dfrac{1}{125}$

b. $3y^{-4}=3\cdot\dfrac{1}{y^4}=\dfrac{3}{y^4}$

c. $3^{-1}+2^{-1}=\dfrac{1}{3}+\dfrac{1}{2}=\dfrac{2}{6}+\dfrac{3}{6}=\dfrac{5}{6}$

d. $(-5)^{-2}=\dfrac{1}{(-5)^2}=\dfrac{1}{(-5)(-5)}=\dfrac{1}{25}$

e. $\dfrac{1}{x^{-5}}=\dfrac{1}{\frac{1}{x^5}}=x^5$

f. $\dfrac{1}{4^{-3}}=\dfrac{1}{\frac{1}{4^3}}=\dfrac{4^3}{1}=64$

2. a. $\dfrac{1}{s^{-5}}=\dfrac{s^5}{1}=s^5$

b. $\dfrac{1}{2^{-3}}=\dfrac{2^3}{1}=8$

c. $\dfrac{x^{-7}}{y^{-5}}=\dfrac{y^5}{x^7}$

d. $\dfrac{4^{-3}}{3^{-2}}=\dfrac{3^2}{4^3}=\dfrac{9}{64}$

3. a. $\dfrac{x^{-3}}{x^2}=x^{-3-2}=x^{-5}=\dfrac{1}{x^5}$

b. $\dfrac{5}{y^{-7}}=5\cdot\dfrac{1}{y^{-7}}=5\cdot y^7=5y^7$

c. $\dfrac{z}{z^{-4}}=\dfrac{z^1}{z^{-4}}=z^{1-(-4)}=z^5$

4. a. $\left(\dfrac{3}{4}\right)^{-2}=\dfrac{3^{-2}}{4^{-2}}=\dfrac{4^2}{3^2}=\dfrac{16}{9}$

b. $\frac{x^2(x^5)^3}{x^7} = \frac{x^2 \cdot x^{15}}{x^7}$
$= \frac{x^{2+15}}{x^7}$
$= \frac{x^{17}}{x^7}$
$= x^{17-7}$
$= x^{10}$

c. $\left(\frac{5p^8}{q}\right)^{-2} = \frac{5^{-2}(p^8)^{-2}}{q^{-2}}$
$= \frac{5^{-2}p^{-16}}{q^{-2}}$
$= \frac{q^2}{5^2 p^{16}}$
$= \frac{q^2}{25p^{16}}$

d. $\frac{6^{-2}x^{-4}y^{-7}}{6^{-3}x^3y^{-9}} = 6^{-2-(-3)}x^{-4-3}y^{-7-(-9)}$
$= 6^1 x^{-7} y^2$
$= \frac{6y^2}{x^7}$

e. $(a^4b^{-3})^{-5} = a^{-20}b^{15} = \frac{b^{15}}{a^{20}}$

f. $\left(\frac{-3x^4y}{x^2y^{-2}}\right)^3 = \frac{(-3)^3x^{12}y^3}{x^2y^{-2}}$
$= \frac{-27x^{12}y^3}{x^6y^{-6}}$
$= -27x^{12-6}y^{3-(-6)}$
$= -27x^6y^9$

5. a. $0.000007 = 7 \times 10^{-6}$
The decimal point is moved 6 places, and the original number is less than 1, so the count is –6.

b. $20,700,000 = 2.07 \times 10^7$
The decimal point is moved 7 places, and the original number is 10 or greater, so the count is 7.

c. $0.0043 = 4.3 \times 10^{-3}$
The decimal point is moved 3 places, and the original number is less than 1, so the count is –3.

d. $812,000,000 = 8.12 \times 10^8$
The decimal point is moved 8 places, and the original number is 10 or greater, so the count is 8.

6. a. Move the decimal point 4 places to the left.
$3.67 \times 10^{-4} = 0.000367$

b. Move the decimal point 6 places to the right.
$8.954 \times 10^6 = 8,954,000$

c. Move the decimal point 5 places to the left.
$2.009 \times 10^{-5} = 0.00002009$

d. Move the decimal point 3 places to the right.
$4.054 \times 10^3 = 4054$

7. a. $(5 \times 10^{-4})(8 \times 10^6) = (5 \cdot 8) \times (10^{-4} \cdot 10^6)$
$= 40 \times 10^2$
$= 4000$

b. $\frac{64 \times 10^3}{32 \times 10^{-7}} = \frac{64}{32} \times 10^{3-(-7)}$
$= 2 \times 10^{10}$
$= 20,000,000,000$

Calculator Explorations

1. $5.31 \times 10^3 = 5.31$ EE 3
2. $-4.8 \times 10^{14} = -4.8$ EE 14
3. $6.6 \times 10^{-9} = 6.6$ EE -9
4. $-9.9811 \times 10^{-2} = -9.9811$ EE -2
5. $3,000,000 \times 5,000,000 = 1.5 \times 10^{13}$
6. $230,000 \times 1000 = 2.3 \times 10^8$
7. $(3.26 \times 10^6)(2.5 \times 10^{13}) = 8.15 \times 10^{19}$
8. $(8.76 \times 10^{-4})(1.237 \times 10^9) = 1.083612 \times 10^6$

Vocabulary and Readiness Check

1. The expression x^{-3} equals $\underline{\frac{1}{x^3}}$.

2. The expression 5^{-4} equals $\underline{\frac{1}{625}}$.

3. The number 3.021×10^{-3} is written in <u>scientific notation</u>.

4. The number 0.0261 is written in <u>standard form</u>.

5. $5x^{-2} = 5 \cdot \frac{1}{x^2} = \frac{5}{x^2}$

6. $3x^{-3} = 3 \cdot \frac{1}{x^3} = \frac{3}{x^3}$

7. $\frac{1}{y^{-6}} = \frac{1}{\frac{1}{y^6}} = \frac{y^6}{1} = y^6$

8. $\frac{1}{x^{-3}} = \frac{1}{\frac{1}{x^3}} = \frac{x^3}{1} = x^3$

9. $\frac{4}{y^{-3}} = \frac{4}{\frac{1}{y^3}} = 4 \cdot \frac{y^3}{1} = 4y^3$

10. $\frac{16}{y^7} = \frac{16}{\frac{1}{y^7}} = 16 \cdot \frac{y^7}{1} = 16y^7$

Exercise Set 5.5

1. $4^{-3} = \frac{1}{4^3} = \frac{1}{64}$

3. $(-2)^{-4} = \frac{1}{(-2)^4} = \frac{1}{16}$

5. $7x^{-3} = 7 \cdot \frac{1}{x^3} = \frac{7}{x^3}$

7. $\left(\frac{1}{2}\right)^{-5} = \frac{1^{-5}}{2^{-5}} = \frac{2^5}{1^5} = 32$

9. $\left(-\frac{1}{4}\right)^{-3} = \frac{(-1)^{-3}}{(4)^{-3}} = \frac{4^3}{(-1)^3} = \frac{64}{-1} = -64$

11. $3^{-1} + 2^{-1} = \frac{1}{3} + \frac{1}{2} = \frac{2}{6} + \frac{3}{6} = \frac{5}{6}$

13. $\frac{1}{p^{-3}} = p^3$

15. $\frac{p^{-5}}{q^{-4}} = \frac{q^4}{p^5}$

17. $\frac{x^{-2}}{x} = x^{-2-1} = x^{-3} = \frac{1}{x^3}$

19. $\frac{z^{-4}}{z^{-7}} = z^{-4-(-7)} = z^3$

21. $3^{-2} + 3^{-1} = \frac{1}{3^2} + \frac{1}{3} = \frac{1}{9} + \frac{1}{3} = \frac{1}{9} + \frac{3}{9} = \frac{4}{9}$

23. $\frac{-1}{p^{-4}} = -1(p^4) = -p^4$

25. $-2^0 - 3^0 = -1(1) - 1 = -1 - 1 = -2$

27. $\frac{x^2 x^5}{x^3} = x^{2+5-3} = x^4$

29. $\frac{p^2 p}{p^{-1}} = p^{2+1-(-1)} = p^4$

31. $\frac{(m^5)^4 m}{m^{10}} = \frac{m^{20} m}{m^{10}} = m^{20+1-10} = m^{11}$

33. $\frac{r}{r^{-3} r^{-2}} = r^{1-(-3)-(-2)} = r^6$

35. $(x^5 y^3)^{-3} = x^{5(-3)} y^{3(-3)} = x^{-15} y^{-9} = \frac{1}{x^{15} y^9}$

37. $\frac{(x^2)^3}{x^{10}} = \frac{x^6}{x^{10}} = x^{6-10} = x^{-4} = \frac{1}{x^4}$

39. $\dfrac{(a^5)^2}{(a^3)^4} = \dfrac{a^{10}}{a^{12}} = a^{10-12} = a^{-2} = \dfrac{1}{a^2}$

41. $\dfrac{8k^4}{2k} = \dfrac{8}{2} \cdot k^{4-1} = 4k^3$

43. $\dfrac{-6m^4}{-2m^3} = \dfrac{-6}{-2} \cdot m^{4-3} = 3m$

45. $\dfrac{-24a^6b}{6ab^2} = \dfrac{-24}{6} \cdot a^{6-1}b^{1-2} = -4a^5b^{-1} = -\dfrac{4a^5}{b}$

47. $(-2x^3y^{-4})(3x^{-1}y) = -2(3)x^{3+(-1)}y^{-4+1}$

$= -6x^2y^{-3}$

$= -\dfrac{6x^2}{y^3}$

49. $(a^{-5}b^2)^{-6} = a^{-5(-6)}b^{2(-6)} = a^{30}b^{-12} = \dfrac{a^{30}}{b^{12}}$

51. $\left(\dfrac{x^{-2}y^4}{x^3y^7}\right)^2 = \dfrac{x^{-2(2)}y^{4(2)}}{x^{3(2)}y^{7(2)}}$

$= \dfrac{x^{-4}y^8}{x^6y^{14}}$

$= x^{-4-6}y^{8-14}$

$= x^{-10}y^{-6}$

$= \dfrac{1}{x^{10}y^6}$

53. $\dfrac{4^2z^{-3}}{4^3z^{-5}} = 4^{2-3}z^{-3-(-5)} = 4^{-1}z^2 = \dfrac{z^2}{4}$

55. $\dfrac{2^{-3}x^{-4}}{2^2x} = 2^{-3-2}x^{-4-1}$

$= 2^{-5}x^{-5}$

$= \dfrac{1}{2^5x^5}$

$= \dfrac{1}{32x^5}$

57. $\dfrac{7ab^{-4}}{7^{-1}a^{-3}b^2} = 7^{1-(-1)}a^{1-(-3)}b^{-4-2}$

$= 7^2a^4b^{-6}$

$= \dfrac{49a^4}{b^6}$

59. $\left(\dfrac{a^{-5}b}{ab^3}\right)^{-4} = \dfrac{a^{-5(-4)}b^{-4}}{a^{-4}b^{3(-4)}}$

$= \dfrac{a^{20}b^{-4}}{a^{-4}b^{-12}}$

$= a^{20-(-4)}b^{-4-(-12)}$

$= a^{24}b^8$

61. $\dfrac{(xy^3)^5}{(xy)^{-4}} = \dfrac{x^5y^{3(5)}}{x^{-4}y^{-4}}$

$= \dfrac{x^5y^{15}}{x^{-4}y^{-4}}$

$= x^{5-(-4)}y^{15-(-4)}$

$= x^9y^{19}$

63. $\dfrac{(-2xy^{-3})^{-3}}{(xy^{-1})^{-1}} = \dfrac{(-2)^{-3}x^{-3}y^9}{x^{-1}y^1}$

$= (-2)^{-3}x^{-3-(-1)}y^{9-1}$

$= (-2)^{-3}x^{-2}y^8$

$= \dfrac{y^8}{(-2)^3x^2}$

$= -\dfrac{y^8}{8x^2}$

65. $\dfrac{6x^2y^3}{-7xy^5} = -\dfrac{6}{7}x^{2-1}y^{3-5} = -\dfrac{6}{7}x^1y^{-2} = -\dfrac{6x}{7y^2}$

67. $\dfrac{(a^4b^{-7})^{-5}}{(5a^2b^{-1})^{-2}} = \dfrac{(a^4)^{-5}(b^{-7})^{-5}}{5^{-2}(a^2)^{-2}(b^{-1})^{-2}}$

$= \dfrac{a^{-20}b^{35}}{5^{-2}a^{-4}b^2}$

$= 5^2a^{-20-(-4)}b^{35-2}$

$= 5^2a^{-16}b^{33}$

$= \dfrac{25b^{33}}{a^{16}}$

69. $78,000 = 7.8 \times 10^4$

71. $0.00000167 = 1.67 \times 10^{-6}$

73. $0.00635 = 6.35 \times 10^{-3}$

75. $1,160,000 = 1.16 \times 10^6$

77. $2,000,000,000 = 2 \times 10^9$

79. $1,212,000,000 = 1.212 \times 10^9$

81. $8.673 \times 10^{-10} = 0.0000000008673$

83. $3.3 \times 10^{-2} = 0.033$

85. $2.032 \times 10^4 = 20,320$

87. $7.0 \times 10^8 = 700,000,000$

89. $9.460 \times 10^{12} = 9,460,000,000,000$

91. The longest bar corresponds to Yahoo! sites, with approximately 130,000,000 or 1.3×10^8 visits.

93. $1,000,000,000 = 1 \times 10^9$

95. $5.7 \times 10^7 = 57,000,000$

97.
$$\begin{aligned}(1.2 \times 10^{-3})(3 \times 10^{-2}) &= (1.2 \cdot 3) \times (10^{-3} \cdot 10^{-2}) \\ &= 3.6 \times 10^{-5} \\ &= 0.000036\end{aligned}$$

99.
$$\begin{aligned}(4 \times 10^{-10})(7 \times 10^{-9}) &= (4 \cdot 7) \times (10^{-10} \cdot 10^{-9}) \\ &= 28 \times 10^{-19} \\ &= 2.8 \times 10^{-18} \\ &= 0.0000000000000000028\end{aligned}$$

101.
$$\begin{aligned}\frac{8 \times 10^{-1}}{16 \times 10^5} &= \frac{8}{16} \times 10^{-1-5} \\ &= 0.5 \times 10^{-6} \\ &= 5 \times 10^{-7} \\ &= 0.0000005\end{aligned}$$

103.
$$\begin{aligned}\frac{1.4 \times 10^{-2}}{7 \times 10^{-8}} &= \frac{1.4}{7} \times 10^{-2-(-8)} \\ &= 0.2 \times 10^6 \\ &= 2.0 \times 10^5 \\ &= 200,000\end{aligned}$$

105. $\frac{5x^7}{3x^4} = \frac{5}{3} \cdot x^{7-4} = \frac{5x^3}{3}$

107. $\frac{15z^4y^3}{21zy} = \frac{15}{21} z^{4-1} y^{3-1} = \frac{5z^3y^2}{7}$

109.
$$\begin{aligned}\frac{1}{y}(5y^2 - 6y + 5) &= \frac{1}{y}(5y^2) + \frac{1}{y}(-6y) + \frac{1}{y}(5) \\ &= 5y - 6 + \frac{5}{y}\end{aligned}$$

111. $\left(\frac{3x^{-2}}{z}\right)^3 = \frac{3^3 x^{-6}}{z^3} = \frac{27}{x^6 z^3}$

The volume is $\frac{27}{x^6 z^3}$ cubic inches.

113.
$$\begin{aligned}(2a^3)^3 a^4 + a^5 a^8 &= 2^3 (a^3)^3 a^4 + a^{5+8} \\ &= 8a^9 a^4 + a^{13} \\ &= 8a^{13} + a^{13} \\ &= 9a^{13}\end{aligned}$$

115. $x^{-5} = \frac{1}{x^5}$

117. Answers may vary

119. **a.** $9.7 \times 10^{-2} = 0.097$
$1.3 \times 10^1 = 130$
1.3×10^1 is larger.

b. $8.6 \times 10^5 = 860,000$
$4.4 \times 10^7 = 44,000,000$
4.4×10^7 is larger.

c. $6.1 \times 10^{-2} = 0.061$
$5.6 \times 10^{-4} = 0.00056$
6.1×10^{-2} is larger.

121. a. $5^{-1} = \frac{1}{5}$

$5^{-2} = \frac{1}{5^2} = \frac{1}{25}$

$\frac{1}{5} > \frac{1}{25}$

The statement is false.

b. $\left(\frac{1}{5}\right)^{-1} = \frac{1^{-1}}{5^{-1}} = \frac{5}{1} = 5$

$\left(\frac{1}{5}\right)^{-2} = \frac{1^{-2}}{5^{-2}} = \frac{5^2}{1^2} = 25$

$5 < 25$

The statement is true.

c. The statement is false, since the statement in part a is false.

123. $(x^{-3s})^3 = x^{-3s \cdot 3} = x^{-9s} = \frac{1}{x^{9s}}$

125. $a^{4m+1} \cdot a^4 = a^{4m+1+4} = a^{4m+5}$

127. $(6.785 \times 10^{-4})(4.68 \times 10^{10}) = 31,753,800$

129. $t = \frac{d}{r}$

$$t = \frac{93,000,000}{1.86 \times 10^5}$$

$$= \frac{9.3 \times 10^7}{1.86 \times 10^5}$$

$$= \frac{9.3}{1.86} \times \frac{10^7}{10^5}$$

$$= 5 \times 10^2$$

$$= 500$$

It takes the light of the sun 500 seconds to reach Earth.

Section 5.6

Practice Exercises

1. $\frac{8t^3 + 4t^2}{4t^2} = \frac{8t^3}{4t^2} + \frac{4t^2}{4t^2} = 2t + 1$

Check: $4t^2(2t+1) = 4t^2(2t) + 4t^2(1)$

$= 8t^3 + 4t^2$

2. $\frac{16x^6 + 20x^3 - 12x}{4x^2} = \frac{16x^6}{4x^2} + \frac{20x^3}{4x^2} - \frac{12x}{4x^2}$

$= 4x^4 + 5x - \frac{3}{x}$

Check: $4x^2\left(4x^4 + 5x - \frac{3}{x}\right)$

$= 4x^2(4x^4) + 4x^2(5x) - 4x^2\left(\frac{3}{x}\right)$

$= 16x^6 + 20x^3 - 12x$

3. $\frac{15x^4y^4 - 10xy + y}{5xy} = \frac{15x^4y^4}{5xy} - \frac{10xy}{5xy} + \frac{y}{5xy}$

$= 3x^3y^3 - 2 + \frac{1}{5x}$

Check: $5xy\left(3x^3y^3 - 2 + \frac{1}{5x}\right)$

$= 5xy(3x^3y^3) - 5xy(2) + 5xy\left(\frac{1}{5x}\right)$

$= 15x^4y^4 - 10xy + y$

4.

$$\begin{array}{r} x+3 \\ x+2 \overline{\smash{)} x^2+5x+6} \\ \underline{x^2+2x} \\ 3x+6 \\ \underline{3x+6} \\ 0 \end{array}$$

Check: $(x+2) \cdot (x+3) + 0 = x^2 + 5x + 6$

The quotient checks.

5.

$$\begin{array}{r} 2x+3 \\ 2x+1 \overline{\smash{)} 4x^2+8x-7} \\ \underline{4x^2+2x} \\ 6x-7 \\ \underline{6x+3} \\ -10 \end{array}$$

$\frac{4x^2 + 8x - 7}{2x+1} = 2x + 3 + \frac{-10}{2x+1}$

Check:

$(2x+1)(2x+3) + (-10) = (4x^2 + 8x + 3) - 10$

$= 4x^2 + 8x - 7$

The quotient checks.

6. Rewrite $11x-3+9x^3$ as $9x^3+0x^2+11x-3$.

$$
\begin{array}{r}
3x^2-2x+5 \\
3x+2\,\big)\,\overline{9x^3+0x^2+11x-3} \\
\underline{9x^3+6x^2} \\
-6x^2+11x \\
\underline{-6x^2-4x} \\
15x-3 \\
\underline{15x+10} \\
-13
\end{array}
$$

$$\frac{11x-3+9x^3}{3x+2}=3x^2-2x+5+\frac{-13}{3x+2}$$

7. Rewrite x^2+2 as x^2+0x+2.

$$
\begin{array}{r}
3x^2-2x-9 \\
x^2+0x+2\,\big)\,\overline{3x^4-2x^3-3x^2+x+4} \\
\underline{3x^4+0x^3+6x^2} \\
-2x^3-9x^2+x \\
\underline{-2x^3+0x^2-4x} \\
-9x^2+5x+4 \\
\underline{-9x^2+0x-18} \\
5x+22
\end{array}
$$

$$\frac{3x^4-2x^3-3x^2+x+4}{x^2+2}=3x^2-2x-9+\frac{5x+22}{x^2+2}$$

Vocabulary and Readiness Check

1. In $6\overline{)18}$ (quotient 3), the 18 is the <u>dividend</u>, the 3 is the <u>quotient</u> and the 6 is the <u>divisor</u>.

2. In $x+1\overline{)x^2+3x+2}$ (quotient $x+2$), the $x+1$ is the <u>divisor</u>, the x^2+3x+2 is the <u>dividend</u> and the $x+2$ is the <u>quotient</u>.

3. $\dfrac{a^6}{a^4}=a^{6-4}=2$

4. $\dfrac{p^8}{p^3}=p^{8-3}=p^5$

5. $\dfrac{y^2}{y}=\dfrac{y^2}{y^1}=y^{2-1}=y$

6. $\dfrac{a^3}{a}=\dfrac{a^3}{a^1}=a^{3-1}=a^2$

Exercise Set 5.6

1. $\dfrac{12x^4+3x^2}{x}=\dfrac{12x^4}{x}+\dfrac{3x^2}{x}=12x^3+3x$

3. $\dfrac{20x^3-30x^2+5x+5}{5}=\dfrac{20x^3}{5}-\dfrac{30x^2}{5}+\dfrac{5x}{5}+\dfrac{5}{5}$
$=4x^3-6x^2+x+1$

5. $\dfrac{15p^3+18p^2}{3p}=\dfrac{15p^3}{3p}+\dfrac{18p^2}{3p}=5p^2+6p$

7. $\dfrac{-9x^4+18x^5}{6x^5}=\dfrac{-9x^4}{6x^5}+\dfrac{18x^5}{6x^5}=-\dfrac{3}{2x}+3$

9. $\dfrac{-9x^5+3x^4-12}{3x^3}=\dfrac{-9x^5}{3x^3}+\dfrac{3x^4}{3x^3}-\dfrac{12}{3x^3}$
$=-3x^2+x-\dfrac{4}{x^3}$

11. $\dfrac{4x^4-6x^3+7}{-4x^4}=\dfrac{4x^4}{-4x^4}-\dfrac{6x^3}{-4x^4}+\dfrac{7}{-4x^4}$
$=-1+\dfrac{3}{2x}-\dfrac{7}{4x^4}$

13.

$$
\begin{array}{r}
x+1 \\
x+3\,\big)\,\overline{x^2+4x+3} \\
\underline{x^2+3x} \\
x+3 \\
\underline{x+3} \\
0
\end{array}
$$

$$\frac{x^2+4x+3}{x+3}=x+1$$

15.

$$
\begin{array}{r}
2x+3 \\
x+5\,\big)\,\overline{2x^2+13x+15} \\
\underline{2x^2+10x} \\
3x+15 \\
\underline{3x+15} \\
0
\end{array}
$$

$$\frac{2x^2+13x+15}{x+5}=2x+3$$

17.
$$\begin{array}{r|l} & 2x+1 \\ \hline x-4 & 2x^2-7x+3 \\ & \underline{2x^2-8x} \\ & \qquad x+3 \\ & \qquad \underline{x-4} \\ & \qquad\quad 7 \end{array}$$

$$\frac{2x^2-7x+3}{x-4}=2x+1+\frac{7}{x-4}$$

19.
$$\begin{array}{r|l} & 3a^2-3a+1 \\ \hline 3a+2 & 9a^3-3a^2-3a+4 \\ & \underline{9a^3+6a^2} \\ & \quad -9a^2-3a \\ & \quad \underline{-9a^2-6a} \\ & \qquad\quad 3a+4 \\ & \qquad\quad \underline{3a+2} \\ & \qquad\qquad 2 \end{array}$$

$$\frac{9a^3-3a^2-3a+4}{3a+2}=3a^2-3a+1+\frac{2}{3a+2}$$

21.
$$\begin{array}{r|l} & 4x+3 \\ \hline 2x+1 & 8x^2+10x+1 \\ & \underline{8x^2\ +4x} \\ & \qquad 6x+1 \\ & \qquad \underline{6x+3} \\ & \qquad\quad -2 \end{array}$$

$$\frac{8x^2+10x+1}{2x+1}=4x+3-\frac{2}{2x+1}$$

23.
$$\begin{array}{r|l} & 2x^2+6x-5 \\ \hline x-2 & 2x^3+2x^2\ -17x+8 \\ & \underline{2x^3-4x^2} \\ & \quad 6x^2-17x \\ & \quad \underline{6x^2-12x} \\ & \qquad\quad -5x\ +8 \\ & \qquad\quad \underline{-5x+10} \\ & \qquad\qquad -2 \end{array}$$

$$\frac{2x^3+2x^2-17x+8}{x-2}=2x^2+6x-5-\frac{2}{x-2}$$

25. Rewrite x^2-36 as $x^2+0x-36$.
$$\begin{array}{r|l} & x+6 \\ \hline x-6 & x^2+0x-36 \\ & \underline{x^2-6x} \\ & \qquad 6x-36 \\ & \qquad \underline{6x-36} \\ & \qquad\qquad 0 \end{array}$$

$$\frac{x^2-36}{x-6}=x+6$$

27. Rewrite x^3-27 as $x^3+0x^2+0x-27$.
$$\begin{array}{r|l} & x^2+3x+9 \\ \hline x-3 & x^3+0x^2+0x-27 \\ & \underline{x^3-3x^2} \\ & \qquad 3x^2+0x \\ & \qquad \underline{3x^2-9x} \\ & \qquad\qquad 9x-27 \\ & \qquad\qquad \underline{9x-27} \\ & \qquad\qquad\qquad 0 \end{array}$$

$$\frac{x^3-27}{x-3}=x^2+3x+9$$

29. Rewrite $1-3x^2$ as $-3x^2+0x+1$.
$$\begin{array}{r|l} & -3x+6 \\ \hline x+2 & -3x^2+0x\ \ +1 \\ & \underline{-3x^2-6x} \\ & \qquad 6x+\ 1 \\ & \qquad \underline{6x+12} \\ & \qquad\quad -11 \end{array}$$

$$\frac{1-3x^2}{x+2}=-3x+6-\frac{11}{x+2}$$

31. Rewrite $-4b+4b^2-5$ as $4b^2-4b-5$.
$$\begin{array}{r|l} & 2b-1 \\ \hline 2b-1 & 4b^2-4b-5 \\ & \underline{4b^2-2b} \\ & \qquad -2b-5 \\ & \qquad \underline{-2b+1} \\ & \qquad\quad -6 \end{array}$$

$$\frac{-4b+4b^2-5}{2b-1}=2b-1-\frac{6}{2b-1}$$

33. $$\frac{a^2b^2 - ab^3}{ab} = \frac{a^2b^2}{ab} - \frac{ab^3}{ab} = ab - b^2$$

35.
$$\begin{array}{r}
4x+9 \\
2x-3\overline{)8x^2+6x-27} \\
\underline{8x^2-12x} \\
18x-27 \\
\underline{18x-27} \\
0
\end{array}$$

$$\frac{8x^2+6x-27}{2x-3} = 4x+9$$

37. $$\frac{2x^2y+8x^2y^2-xy^2}{2xy} = \frac{2x^2y}{2xy} + \frac{8x^2y^2}{2xy} - \frac{xy^2}{2xy} = x + 4xy - \frac{y}{2}$$

39.
$$\begin{array}{r}
2b^2+b+2 \\
b+4\overline{)2b^3+9b^2+6b-4} \\
\underline{2b^3+8b^2} \\
b^2+6b \\
\underline{b^2+4b} \\
2b-4 \\
\underline{2b+8} \\
-12
\end{array}$$

$$\frac{2b^3+9b^2+6b-4}{b+4} = 2b^2+b+2-\frac{12}{b+4}$$

41.
$$\begin{array}{r}
5x-2 \\
x+6\overline{)5x^2+28x-10} \\
\underline{5x^2+30x} \\
-2x-10 \\
\underline{-2x-12} \\
2
\end{array}$$

$$\frac{5x^2+28x-10}{x+6} = 5x-2+\frac{2}{x+6}$$

43. $$\frac{10x^3-24x^2-10x}{10x} = \frac{10x^3}{10x} - \frac{24x^2}{10x} - \frac{10x}{10x} = x^2 - \frac{12x}{5} - 1$$

45.
$$\begin{array}{r}
6x-1 \\
x+3\overline{)6x^2+17x-4} \\
\underline{6x^2+18x} \\
-x-4 \\
\underline{-x-3} \\
-1
\end{array}$$

$$\frac{6x^2+17x-4}{x+3} = 6x-1-\frac{1}{x+3}$$

47.
$$\begin{array}{r}
6x-1 \\
5x-2\overline{)30x^2-17x+2} \\
\underline{30x^2-12x} \\
-5x+2 \\
\underline{-5x+2} \\
0
\end{array}$$

$$\frac{30x^2-17x+2}{5x-2} = 6x-1$$

49. $$\frac{3x^4-9x^3+12}{-3x} = \frac{3x^4}{-3x} - \frac{9x^3}{-3x} + \frac{12}{-3x} = -x^3+3x^2-\frac{4}{x}$$

51.
$$\begin{array}{r}
x^2+3x+9 \\
x+3\overline{)x^3+6x^2+18x+27} \\
\underline{x^3+3x^2} \\
3x^2+18x \\
\underline{3x^2+9x} \\
9x+27 \\
\underline{9x+27} \\
0
\end{array}$$

$$\frac{x^3+6x^2+18x+27}{x+3} = x^2+3x+9$$

53. Rewrite y^3+3y^2+4 as y^3+3y^2+0y+4.

$$\begin{array}{r} y^2+5y+10 \\ y-2\overline{)\,y^3+3y^2\quad+0y+4} \\ \underline{y^3-2y^2\qquad\qquad} \\ 5y^2\ +0y\qquad \\ \underline{5y^2-10y\qquad} \\ 10y\ +4 \\ \underline{10y-20} \\ 24 \end{array}$$

$$\frac{y^3+3y^2+4}{y-2}=y^2+5y+10+\frac{24}{y-2}$$

55. Rewrite $5-6x^2$ as $-6x^2+0x+5$.

$$\begin{array}{r} -6x-12 \\ x-2\overline{)-6x^2\ +0x\ +5} \\ \underline{-6x^2+12x\qquad} \\ -12x\ +5 \\ \underline{-12x+24} \\ -19 \end{array}$$

$$\frac{5-6x^2}{x-2}=-6x-12-\frac{19}{x-2}$$

57. Rewrite x^5+x^2 as $x^5+0x^4+0x^3+x^2$.

$$\begin{array}{r} x^3-x^2+x \\ x^2+x\overline{)\,x^5+0x^4+0x^3+x^2} \\ \underline{x^5\ +x^4\qquad\qquad\qquad} \\ -x^4+0x^3\qquad \\ \underline{-x^4\ -x^3\qquad} \\ x^3+x^2 \\ \underline{x^3+x^2} \\ 0 \end{array}$$

$$\frac{x^5+x^2}{x^2+x}=x^3-x^2+x$$

59. $2a(a^2+1)=2a(a^2)+2a(1)=2a^3+2a$

61. $2x(x^2+7x-5)=2x(x^2)+2x(7x)+2x(-5)$
$=2x^3+14x^2-10x$

63. $-3xy(xy^2+7x^2y+8)$
$=-3xy(xy^2)-3xy(7x^2y)-3xy(8)$
$=-3x^2y^3-21x^3y^2-24xy$

65. $9ab(ab^2c+4bc-8)$
$=9ab(ab^2c)+9ab(4bc)+9ab(-8)$
$=9a^2b^3c+36ab^2c-72ab$

67. The longest guitar corresponds to The Rolling Stones (2005).

69. The 2005 concert tour of U2 grossed approximately $139 million.

71. $P=4s,\ s=\frac{P}{4}$

$$\frac{12x^3+4x-16}{4}=\frac{12x^3}{4}+\frac{4x}{4}-\frac{16}{4}$$
$$=3x^3+x-4$$

Each side is $(3x^3+x-4)$ feet.

73. $\frac{a+7}{7}=\frac{a}{7}+\frac{7}{7}=\frac{a}{7}+1$;
choice c

75. Answers may vary

77. $A=l\cdot w,\ w=\frac{A}{l}$

$$w=\frac{49x^2+70x-200}{7x+20}$$

$$\begin{array}{r} 7x-10 \\ 7x+20\overline{)\,49x^2\ +70x-200} \\ \underline{49x^2+140x\qquad} \\ -70x-200 \\ \underline{-70x-200} \\ 0 \end{array}$$

The width is $(7x-10)$ inches.

79. $\frac{25y^{11b}+5y^{6b}-20y^{3b}+100y^b}{5y^b}$
$=\frac{25y^{11b}}{5y^b}+\frac{5y^{6b}}{5y^b}-\frac{20y^{3b}}{5y^b}+\frac{100y^b}{5y^b}$
$=5y^{10b}+y^{5b}-4y^{2b}+20$

The Bigger Picture

1. $-5.7+(-0.23)=-5.93$

2. $\frac{1}{2}-\frac{9}{10}=\frac{5}{10}-\frac{9}{10}=\frac{-4}{10}=-\frac{2}{5}$

3. $(-5x^2y^3)(-x^7y) = (-5\cdot -1)(x^2\cdot x^7)(y^3\cdot y)$
$= 5x^9y^4$

4. $2^{-3}a^{-7}a^3 = \frac{1}{2^3}\cdot\frac{1}{a^7}\cdot a^3 = \frac{1}{8}\cdot\frac{a^3}{a^7} = \frac{1}{8}\cdot a^{-4} = \frac{1}{8a^4}$

5. $(7y^3 - 6y + 2) - (y^3 + 2y^2 + 2)$
$= 7y^3 - 6y + 2 - y^3 - 2y^2 - 2$
$= 6y^3 - 2y^2 - 6y$

6. $(9y^2 - 3y) - (y^2 + 7) = 9y^2 - 3y - y^2 - 7$
$= 8y^2 - 3y - 7$

7. $(x-3)(4x^2 - x + 7)$
$= x(4x^2) + x(-x) + x(7) - 3(4x^2) - 3(-x) - 3(7)$
$= 4x^3 - x^2 + 7x - 12x^2 + 3x - 21$
$= 4x^3 - 13x^2 + 10x - 21$

8. $(6m-5)^2 = (6m)^2 - 2(6m)(5) + 5^2$
$= 36m^2 - 60m + 25$

9. $\frac{20n^2 - 5n + 10}{5n} = \frac{20n^2}{5n} - \frac{5n}{5n} + \frac{10}{5n} = 4n - 1 + \frac{2}{n}$

10.
$$\begin{array}{r} 2x-6 \\ 3x-1\overline{)6x^2-20x+20} \\ \underline{6x^2\ \ -2x} \\ -18x+20 \\ \underline{-18x\ \ +6} \\ 14 \end{array}$$

$\frac{6x^2 - 20x + 20}{3x-1} = 2x - 6 + \frac{14}{3x-1}$

11. $-6x = 3.6$
$x = \frac{3.6}{-6}$
$x = -0.6$

12. $-6x < 3.6$
$x > \frac{3.6}{-6}$
$x > -0.6$
$(-0.6, \infty)$

13. $6x + 6 \ge 8x + 2$
$6x \ge 8x - 4$
$-2x \ge -4$
$x \le \frac{-4}{-2}$
$x \le 2$
$(-\infty, 2]$

14. $7y + 3(y-1) = 4(y+1) - 3$
$7y + 3y - 3 = 4y + 4 - 3$
$10y - 3 = 4y + 1$
$6y - 3 = 1$
$6y = 4$
$\frac{6y}{6} = \frac{4}{6}$
$y = \frac{2}{3}$

Section 5.7

Practice Exercises

1. Since $x - c = x - 1$, c is 1.

$$\begin{array}{r|rrrr} 1 & 4 & -3 & 6 & 5 \\ & & 4 & 1 & 7 \\ \hline & 4 & 1 & 7 & 12 \end{array}$$

$4x^2 + x + 7 + \frac{12}{x-1}$

2. Since $x - c = x + 3 = x - (-3)$, c is -3.

$$\begin{array}{r|rrrrr} -3 & 1 & 3 & -5 & 6 & 12 \\ & & -3 & 0 & 15 & -63 \\ \hline & 1 & 0 & -5 & 21 & -51 \end{array}$$

$x^3 - 5x + 21 - \frac{51}{x+3}$

3. a. $P(x) = x^3 - 5x - 2$
$P(2) = 2^3 - 5(2) - 2$
$= 8 - 10 - 2$
$= -4$

b. Since $x - c = x - 2$, c is 2.

$$\begin{array}{r|rrrr} 2 & 1 & 0 & -5 & -2 \\ & & 2 & 4 & -2 \\ \hline & 1 & 2 & -1 & -4 \end{array}$$

The remainder is -4.

4. $$\begin{array}{r|rrrrrr} 3 & 2 & -18 & 0 & 90 & 59 & 0 \\ & & 6 & -36 & -108 & -54 & 15 \\ \hline & 2 & -12 & -36 & -18 & 5 & 15 \end{array}$$

$P(3) = 15$

Exercise Set 5.7

1. $x - 5 = x - c$ where $c = 5$.

$$\begin{array}{r|rrr} 5 & 1 & 3 & -40 \\ & & 5 & 40 \\ \hline & 1 & 8 & 0 \end{array}$$

$$\frac{x^2 + 3x - 40}{x - 5} = x + 8$$

3. $x + 6 = x - c$ where $c = -6$.

$$\begin{array}{r|rrr} -6 & 1 & 5 & -6 \\ & & -6 & 6 \\ \hline & 1 & -1 & 0 \end{array}$$

$$\frac{x^2 + 5x - 6}{x + 6} = x - 1$$

5. $x - 2 = x - c$ where $c = 2$.

$$\begin{array}{r|rrrr} 2 & 1 & -7 & -13 & 5 \\ & & 2 & -10 & -46 \\ \hline & 1 & -5 & -23 & -41 \end{array}$$

$$\frac{x^3 - 7x^2 - 13x + 5}{x - 2} = x^2 - 5x - 23 - \frac{41}{x - 2}$$

7. $x - 2 = x - c$ where $c = 2$.

$$\begin{array}{r|rrr} 2 & 4 & 0 & -9 \\ & & 8 & 16 \\ \hline & 4 & 8 & 7 \end{array}$$

$$\frac{4x^2 - 9}{x - 2} = 4x + 8 + \frac{7}{x - 2}$$

9. $P(x) = 3x^2 - 4x - 1$

a. $$\begin{aligned} P(2) &= 3(2)^2 - 4(2) - 1 \\ &= 3(4) - 8 - 1 \\ &= 12 - 8 - 1 \\ &= 3 \end{aligned}$$

b. $$\begin{array}{r|rrr} 2 & 3 & -4 & -1 \\ & & 6 & 4 \\ \hline & 3 & 2 & 3 \end{array}$$

Thus, $P(2) = 3$.

11. $$\begin{aligned} P(x) &= 4x^4 + 7x^2 + 9x - 1 \\ &= 4x^4 + 0x^3 + 7x^2 + 9x - 1 \end{aligned}$$

a. $$\begin{aligned} P(-2) &= 4(-2)^4 + 7(-2)^2 + 9(-2) - 1 \\ &= 4(16) + 7(4) - 18 - 1 \\ &= 64 + 28 - 18 - 1 \\ &= 73 \end{aligned}$$

b. $$\begin{array}{r|rrrrr} -2 & 4 & 0 & 7 & 9 & -1 \\ & & -8 & 16 & -46 & 74 \\ \hline & 4 & -8 & 23 & -37 & 73 \end{array}$$

Thus, $P(-2) = 73$.

13. $$\begin{aligned} P(x) &= x^5 + 3x^4 + 3x - 7 \\ &= x^5 + 3x^4 + 0x^3 + 0x^2 + 3x - 7 \end{aligned}$$

a. $$\begin{aligned} P(-1) &= (-1)^5 + 3(-1)^4 + 3(-1) - 7 \\ &= -1 + 3 - 3 - 7 \\ &= -8 \end{aligned}$$

b. $$\begin{array}{r|rrrrrr} -1 & 1 & 3 & 0 & 0 & 3 & -7 \\ & & -1 & -2 & 2 & -2 & -1 \\ \hline & 1 & 2 & -2 & 2 & 1 & -8 \end{array}$$

Thus, $P(-1) = -8$.

15. $x^3 - 3x^2 + 2 = x^3 - 3x^2 + 0x + 2$

$$\begin{array}{r|rrrr} 3 & 1 & -3 & 0 & 2 \\ & & 3 & 0 & 0 \\ \hline & 1 & 0 & 0 & 2 \end{array}$$

$$\frac{x^3 - 3x^2 + 2}{x - 3} = x^2 + \frac{2}{x - 3}$$

17. $x+1=x-(-1)$

$$\begin{array}{r|rrr} -1 & 6 & 13 & 8 \\ & & -6 & -7 \\ \hline & 6 & 7 & 1 \end{array}$$

$$\frac{6x^2+13x+8}{x+1}=6x+7+\frac{1}{x+1}$$

19.

$$\begin{array}{r|rrrrr} 5 & 2 & -13 & 16 & -9 & 20 \\ & & 10 & -15 & 5 & -20 \\ \hline & 2 & -3 & 1 & -4 & 0 \end{array}$$

$$\frac{2x^4-13x^3+16x^2-9x+20}{x-5}=2x^3-3x^2+x-4$$

21. $3x^2-15=3x^2+0x-15$; $x+3=x-(-3)$

$$\begin{array}{r|rrr} -3 & 3 & 0 & -15 \\ & & -9 & 27 \\ \hline & 3 & -9 & 12 \end{array}$$

$$\frac{3x^2-15}{x+3}=3x-9+\frac{12}{x+3}$$

23.

$$\begin{array}{r|rrrr} \frac{1}{2} & 3 & -6 & 4 & 5 \\ & & \frac{3}{2} & -\frac{9}{4} & \frac{7}{8} \\ \hline & 3 & -\frac{9}{2} & \frac{7}{4} & \frac{47}{8} \end{array}$$

$$\frac{3x^3-6x^2+4x+5}{x-\frac{1}{2}}=3x^2-\frac{9}{2}x+\frac{7}{4}+\frac{47}{8\left(x-\frac{1}{2}\right)}$$

25.

$$\begin{array}{r|rrrr} \frac{1}{3} & 3 & 2 & -4 & 1 \\ & & 1 & 1 & -1 \\ \hline & 3 & 3 & -3 & 0 \end{array}$$

$$\frac{3x^3+2x^2-4x+1}{x-\frac{1}{3}}=3x^2+3x-3$$

27. $7x^2-4x+12+3x^3=3x^3+7x^2-4x+12$

$x+1=x-(-1)$

$$\begin{array}{r|rrrr} -1 & 3 & 7 & -4 & 12 \\ & & -3 & -4 & 8 \\ \hline & 3 & 4 & -8 & 20 \end{array}$$

$$\frac{7x^2-4x+12+3x^3}{x+1}=3x^2+4x-8+\frac{20}{x+1}$$

29. $x^3-1=x^3+0x^2+0x-1$

$$\begin{array}{r|rrrr} 1 & 1 & 0 & 0 & -1 \\ & & 1 & 1 & 1 \\ \hline & 1 & 1 & 1 & 0 \end{array}$$

$$\frac{x^3-1}{x-1}=x^2+x+1$$

31. $x^2-36=x^2+0x-36$; $x+6=x-(-6)$

$$\begin{array}{r|rrr} -6 & 1 & 0 & -36 \\ & & -6 & 36 \\ \hline & 1 & -6 & 0 \end{array}$$

$$\frac{x^2-36}{x+6}=x-6$$

33.

$$\begin{array}{r|rrrr} 1 & 1 & 3 & -7 & 4 \\ & & 1 & 4 & -3 \\ \hline & 1 & 4 & -3 & 1 \end{array}$$

Thus, $P(1)=1$.

35.

$$\begin{array}{r|rrrr} -3 & 3 & -7 & -2 & 5 \\ & & -9 & 48 & -138 \\ \hline & 3 & -16 & 46 & -133 \end{array}$$

Thus, $P(-3)=-133$.

37.

$$\begin{array}{r|rrrrr} -1 & 4 & 0 & 1 & 0 & -2 \\ & & -4 & 4 & -5 & 5 \\ \hline & 4 & -4 & 5 & -5 & 3 \end{array}$$

Thus, $P(-1)=3$.

39.
$$\begin{array}{r|rrrrr} \frac{1}{3} & 2 & 0 & -3 & 0 & -2 \\ & & \frac{2}{3} & \frac{2}{9} & -\frac{25}{27} & -\frac{25}{81} \\ \hline & 2 & \frac{2}{3} & -\frac{25}{9} & -\frac{25}{27} & -\frac{187}{181} \end{array}$$

Thus, $P\left(\frac{1}{3}\right) = -\frac{187}{81}$.

41.
$$\begin{array}{r|rrrrrr} \frac{1}{2} & 1 & 1 & -1 & 0 & 0 & 3 \\ & & \frac{1}{2} & \frac{3}{4} & -\frac{1}{8} & -\frac{1}{16} & -\frac{1}{32} \\ \hline & 1 & \frac{3}{2} & -\frac{1}{4} & -\frac{1}{8} & -\frac{1}{16} & \frac{95}{32} \end{array}$$

Thus, $P\left(\frac{1}{2}\right) = \frac{95}{32}$.

43. Answers may vary

45.
$$\begin{aligned} 7x + 2 &= x - 3 \\ 7x - x &= -3 - 2 \\ 6x &= -5 \\ x &= -\frac{5}{6} \end{aligned}$$

The solution is $-\frac{5}{6}$.

47.
$$\begin{aligned} \frac{x}{3} - 5 &= 13 \\ 3\left(\frac{x}{3} - 5\right) &= (13) \cdot 3 \\ x - 15 &= 39 \\ x &= 54 \end{aligned}$$

The solution is 54.

49. $2^3 = 2 \cdot 2 \cdot 2 = 8$

51. $(-2)^5 = (-2)(-2)(-2)(-2)(-2) = -32$

53. $3 \cdot 4^2 = 3 \cdot 16 = 48$

55. Let $x = -5$.
$x^2 = (-5)^2 = (-5)(-5) = 25$

57. Let $x = -1$.
$2x^3 = 2(-1)^3 = 2(-1) = -2$

59. $(5x^2 - 3x + 2) \div (x + 2)$ is a candidate for synthetic division since $x + 2$ is in the form $x - c$, where $c = -2$.

61. $(x^7 - 2) \div (x^5 + 1)$ is not a candidate for synthetic division since $x^5 + 1$ does not have the form $x - c$.

63. $A = bh$ so $h = \frac{A}{b} = \frac{x^4 - 23x^2 + 9x - 5}{x + 5}$

$$\begin{array}{r|rrrrr} -5 & 1 & 0 & -23 & 9 & -5 \\ & & -5 & 25 & -10 & 5 \\ \hline & 1 & -5 & 2 & -1 & 0 \end{array}$$

The height is $(x^3 - 5x^2 + 2x - 1)$ cm.

65.
$$\begin{array}{r l} & x^3 + \frac{5}{3}x^2 + \frac{5}{3}x + \frac{8}{3} \\ x - 1 & \overline{)\, x^4 + \frac{2}{3}x^3 + 0x^2 + x + 0} \\ & \underline{x^4 - x^3} \\ & \quad \frac{5}{3}x^3 - 0x^2 \\ & \quad \underline{\frac{5}{3}x^3 - \frac{5}{3}x^2} \\ & \qquad \frac{5}{3}x^2 + x \\ & \qquad \underline{\frac{5}{3}x^2 - \frac{5}{3}x} \\ & \qquad\quad \frac{8}{3}x + 0 \\ & \qquad\quad \underline{\frac{8}{3}x - \frac{8}{3}} \\ & \qquad\qquad \frac{8}{3} \end{array}$$

Answer: $x^3 + \frac{5}{3}x^2 + \frac{5}{3}x + \frac{8}{3} + \frac{8}{3(x-1)}$

67.
$$\begin{array}{r|rrrr} -3 & 1 & 3 & 4 & 12 \\ & & -3 & 0 & -12 \\ \hline & 1 & 0 & 4 & 0 \end{array}$$

Remainder = 0 and
$(x + 3)(x^2 + 4) = x^3 + 3x^2 + 4x + 12$

69. $P(c)$ is equal to the remainder when $P(x)$ is divided by $x - c$. Therefore, $P(c) = 0$.

71. Multiply $(x^2 - x + 10)$ by $(x+3)$ and add the remainder, –2.
$(x^2 - x + 10)(x+3) - 2$
$= (x^3 + 3x^2 - x^2 - 3x + 10x + 30) - 2$
$= x^3 + 2x^2 + 7x + 28$

73. Answers may vary

Chapter 5 Vocabulary Check

1. A term is a number or the product of numbers and variables raised to powers.
2. The FOIL method may be used when multiplying two binomials.
3. A polynomial with exactly 3 terms is called a trinomial.
4. The degree of a polynomial is the greatest degree of any term of the polynomial.
5. A polynomial with exactly 2 terms is called a binomial.
6. The coefficient of a term is its numerical factor.
7. The degree of a term is the sum of the exponents on the variables in the term.
8. A polynomial with exactly 1 term is called a monomial.
9. Monomials, binomials, and trinomials are all examples of polynomials.

Chapter 5 Review

1. In 7^9, the base is 7 and the exponent is 9.
2. In $(-5)^4$, the base is –5 and the exponent is 4.
3. In -5^4, the base is 5 and the exponent is 4.
4. In x^6, the base is x and the exponent is 6.
5. $8^3 = 8 \cdot 8 \cdot 8 = 512$
6. $(-6)^2 = (-6)(-6) = 36$
7. $-6^2 = -6 \cdot 6 = -36$
8. $-4^3 - 4^0 = -64 - 1 = -65$
9. $(3b)^0 = 1$
10. $\frac{8b}{8b} = 1$
11. $y^2 \cdot y^7 = y^{2+7} = y^9$
12. $x^9 \cdot x^5 = x^{9+5} = x^{14}$
13. $(2x^5)(-3x^6) = (2 \cdot -3)(x^5 \cdot x^6) = -6x^{11}$
14. $(-5y^3)(4y^4) = (-5 \cdot 4)(y^3 \cdot y^4) = -20y^7$
15. $(x^4)^2 = x^{4 \cdot 2} = x^8$
16. $(y^3)^5 = y^{3 \cdot 5} = y^{15}$
17. $(3y^6)^4 = 3^4(y^6)^4 = 81y^{24}$
18. $2^3(x^3)^3 = 8x^9$
19. $\frac{x^9}{x^4} = x^{9-4} = x^5$
20. $\frac{z^{12}}{z^5} = z^{12-5} = z^7$
21. $\frac{a^5b^4}{ab} = a^{5-1}b^{4-1} = a^4b^3$
22. $\frac{x^4y^6}{xy} = x^{4-1}y^{6-1} = x^3y^5$
23. $\frac{12xy^6}{3x^4y^{10}} = \frac{12}{3}x^{1-3}y^{6-10} = 4x^{-2}y^{-4} = \frac{4}{x^2y^4}$
24. $\frac{2x^7y^8}{8xy^2} = \frac{2}{8}x^{7-1}y^{8-2} = \frac{x^6y^6}{4}$
25. $5a^7(2a^4)^3 = 5a^7(2^3)(a^4)^3$
$= (5 \cdot 8)(a^7 \cdot a^{12})$
$= 40a^{19}$

26. $(2x)^2(9x) = (2^2 \cdot x^2)(9x)$
$= (4 \cdot 9)(x^2 \cdot x)$
$= 36x^3$

27. $(-5a)^0 + 7^0 + 8^0 = 1+1+1 = 3$

28. $8x^0 + 9^0 = 8(1) + 1 = 9$

29. $\left(\frac{3x^4}{4y}\right)^3 = \frac{3^3 x^{4\cdot 3}}{4^3 y^3} = \frac{27x^{12}}{64y^3}$, choice b.

30. $\left(\frac{5a^6}{b^3}\right)^2 = \frac{5^2 a^{6\cdot 2}}{b^{3\cdot 2}} = \frac{25a^{12}}{b^6}$, choice c.

31. The degree of $-5x^4y^3$ is 4 + 3 = 7.

32. The degree of $10x^3y^2z$ is 3 + 2 + 1 = 6.

33. The degree of $35a^5bc^2$ is 5 + 1 + 2 = 8.

34. The degree of 95*xyz* is 1 + 1 + 1 = 3.

35. The degree is 5 because y^5 is the term with the highest degree.

36. The degree is 2 because $9y^2$ is the term with the highest degree.

37. The degree is 5 because $-28x^2y^3$ is the term with the highest degree.

38. The degree is 6 because $6x^2y^2z^2$ is the term with the highest degree.

39. a.

Term	Numerical Coefficient	Degree of Term
x^2y^2	1	4
$5x^2$	5	2
$-7y^2$	−7	2
$11xy$	11	2
−1	−1	0

b. The degree is 4.

40. $2x^2 + 20x$:
$x = 1$: $2(1)^2 + 20(1) = 22$
$x = 3$: $2(3)^2 + 20(3) = 78$
$x = 5.1$: $2(5.1)^2 + 20(5.1) = 154.02$
$x = 10$: $2(10)^2 + 20(10) = 400$

41. $6a^2 + 4a + 9a^2 = (6+9)a^2 + 4a$
$= 15a^2 + 4a$

42. $21x^2 + 3x + x^2 + 6 = (21+1)x^2 + 3x + 6$
$= 22x^2 + 3x + 6$

43. $4a^2b - 3b^2 - 8q^2 - 10a^2b + 7q^2$
$= (4a^2b - 10a^2b) - 3b^2 + (-8q^2 + 7q^2)$
$= -6a^2b - 3b^2 - q^2$

44. $2s^{14} + 3s^{13} + 12s^{12} - s^{10}$ cannot be combined.

45. $(3x^2 + 2x + 6) + (5x^2 + x)$
$= 3x^2 + 2x + 6 + 5x^2 + x$
$= 8x^2 + 3x + 6$

46. $(2x^5 + 3x^4 + 4x^3 + 5x^2) + (4x^2 + 7x + 6)$
$= 2x^5 + 3x^4 + 4x^3 + 5x^2 + 4x^2 + 7x + 6$
$= 2x^5 + 3x^4 + 4x^3 + 9x^2 + 7x + 6$

47. $(-5y^2 + 3) - (2y^2 + 4) = -5y^2 + 3 - 2y^2 - 4$
$= -7y^2 - 1$

48. $(3x^2 - 7xy + 7y^2) - (4x^2 - xy + 9y^2)$
$= 3x^2 - 7xy + 7y^2 - 4x^2 + xy - 9y^2$
$= -x^2 - 6xy - 2y^2$

49. $(7x - 14y) - (3x - y) = 7x - 14y - 3x + y$
$= 4x - 13y$

50. $[(x^2 + 7x + 9) + (x^2 + 4)] - (4x^2 + 8x - 7)$
$= x^2 + 7x + 9 + x^2 + 4 - 4x^2 - 8x + 7$
$= -2x^2 - x + 20$

51. $P(x) = 9x^2 - 7x + 8$

$$\begin{aligned} P(6) &= 9(6)^2 - 7(6) + 8 \\ &= 9(36) - 42 + 8 \\ &= 324 - 42 + 8 \\ &= 290 \end{aligned}$$

52. $P(x) = 9x^2 - 7x + 8$

$$\begin{aligned} P(-2) &= 9(-2)^2 - 7(-2) + 8 \\ &= 9(4) + 14 + 8 \\ &= 36 + 14 + 8 \\ &= 58 \end{aligned}$$

53. $(x^2y + 5) + (2x^2y - 6x + 1) + (x^2y + 5) + (2x^2y - 6x + 1)$

$$\begin{aligned} &= x^2y + 2x^2y + x^2y + 2x^2y - 6x - 6x + 5 + 1 + 5 + 1 \\ &= 6x^2y - 12x + 12 \end{aligned}$$

The perimeter is $(6x^2y - 12x + 12)$ cm.

54. Let $x = 8$.

$$\begin{aligned} f(8) &= 754(8)^2 - 228(8) + 80{,}134 \\ &= 126{,}566 \end{aligned}$$

Revenues from software sales in 2009 are predicted to be $126,566 million.

55. $4(2a + 7) = 4(2a) + 4(7) = 8a + 28$

56. $9(6a - 3) = 9(6a) - 9(3) = 54a - 27$

57. $-7x(x^2 + 5) = -7(x^2) - 7x(5) = -7x^3 - 35x$

58. $-8y(4y^2 - 6) = -8y(4y^2) - 8y(-6)$

$$= -32y^3 + 48y$$

59. $(3a^3 - 4a + 1)(-2a)$

$$\begin{aligned} &= 3a^3(-2a) - 4a(-2a) + 1(-2a) \\ &= -6a^4 + 8a^2 - 2a \end{aligned}$$

60. $(6b^3 - 4b + 2)(7b) = 6b^3(7b) - 4b(7b) + 2(7b)$

$$= 42b^4 - 28b^2 + 14b$$

61. $(2x + 2)(x - 7) = 2x^2 - 14x + 2x - 14$

$$= 2x^2 - 12x - 14$$

62. $(2x - 5)(3x + 2) = 6x^2 + 4x - 15x - 10$

$$= 6x^2 - 11x - 10$$

63. $(x - 9)^2 = (x - 9)(x - 9)$

$$\begin{aligned} &= x^2 - 9x - 9x + 81 \\ &= x^2 - 18x + 81 \end{aligned}$$

64. $(x - 12)^2 = (x - 12)(x - 12)$

$$\begin{aligned} &= x^2 - 12x - 12x + 144 \\ &= x^2 - 24x + 144 \end{aligned}$$

65. $(4a - 1)(a + 7) = 4a^2 + 28a - a - 7$

$$= 4a^2 + 27a - 7$$

66. $(6a - 1)(7a + 3) = 42a^2 + 18a - 7a - 3$

$$= 42a^2 + 11a - 3$$

67. $(5x + 2)^2 = (5x + 2)(5x + 2)$

$$\begin{aligned} &= 25x^2 + 10x + 10x + 4 \\ &= 25x^2 + 20x + 4 \end{aligned}$$

68. $(3x + 5)^2 = (3x + 5)(3x + 5)$

$$\begin{aligned} &= 9x^2 + 15x + 15x + 25 \\ &= 9x^2 + 30x + 25 \end{aligned}$$

69. $(x + 7)(x^3 + 4x - 5)$

$$\begin{aligned} &= x(x^3 + 4x - 5) + 7(x^3 + 4x - 5) \\ &= x^4 + 4x^2 - 5x + 7x^3 + 28x - 35 \\ &= x^4 + 7x^3 + 4x^2 + 23x - 35 \end{aligned}$$

70. $(x + 2)(x^5 + x + 1) = x(x^5 + x + 1) + 2(x^5 + x + 1)$

$$\begin{aligned} &= x^6 + x^2 + x + 2x^5 + 2x + 2 \\ &= x^6 + 2x^5 + x^2 + 3x + 2 \end{aligned}$$

71. $(x^2 + 2x + 4)(x^2 + 2x - 4)$

$$\begin{aligned} &= x^2(x^2 + 2x - 4) + 2x(x^2 + 2x - 4) + 4(x^2 + 2x - 4) \\ &= x^4 + 2x^3 - 4x^2 + 2x^3 + 4x^2 - 8x + 4x^2 + 8x - 16 \\ &= x^4 + 4x^3 + 4x^2 - 16 \end{aligned}$$

72. $(x^3+4x+4)(x^3+4x-4)$
$=x^3(x^3+4x-4)+4x(x^3+4x-4)$
$\quad +4(x^3+4x-4)$
$=x^6+4x^4-4x^3+4x^4+16x^2-16x+4x^3$
$\quad +16x-16$
$=x^6+8x^4+16x^2-16$

73. $(x+7)^3=(x+7)(x+7)(x+7)$
$=(x^2+7x+7x+49)(x+7)$
$=(x^2+14x+49)(x+7)$
$=(x^2+14x+49)x+(x^2+14x+49)7$
$=x^3+14x^2+49x+7x^2+98x+343$
$=x^3+21x^2+147x+343$

74. $(2x-5)^3$
$=(2x-5)(2x-5)(2x-5)$
$=(4x^2-10x-10x+25)(2x-5)$
$=(4x^2-20x+25)(2x-5)$
$=(4x^2-20x+25)(2x)+(4x^2-20x+25)(-5)$
$=8x^3-40x^2+50x-20x^2+100x-125$
$=8x^3-60x^2+150x-125$

75. $(x+7)^2=x^2+2(x)(7)+7^2=x^2+14x+49$

76. $(x-5)^2=x^2-2(x)(5)+5^2=x^2-10x+25$

77. $(3x-7)^2=(3x)^2-2(3x)(7)+7^2$
$=9x^2-42x+49$

78. $(4x+2)^2=(4x)^2+2(4x)(2)+2^2$
$=16x^2+16x+4$

79. $(5x-9)^2=(5x)^2-2(5x)(9)+9^2$
$=25x^2-90x+81$

80. $(5x+1)(5x-1)=(5x)^2-1^2=25x^2-1$

81. $(7x+4)(7x-4)=(7x)^2-4^2=49x^2-16$

82. $(a+2b)(a-2b)=a^2-(2b)^2=a^2-4b^2$

83. $(2x-6)(2x+6)=(2x)^2-6^2=4x^2-36$

84. $(4a^2-2b)(4a^2+2b)=(4a^2)^2-(2b)^2$
$=16a^4-4b^2$

85. $(3x-1)^2=(3x)^2-2(3x)(1)+1^2$
$=9x^2-6x+1$
The area is $(9x^2-6x+1)$ square meters.

86. $(5x+2)(x-1)=5x^2-5x+2x-2$
$=5x^2-3x-2$
The area is $(5x^2-3x-2)$ square miles.

87. $7^{-2}=\frac{1}{7^2}=\frac{1}{49}$

88. $-7^{-2}=-\frac{1}{7^2}=-\frac{1}{49}$

89. $2x^{-4}=\frac{2}{x^4}$

90. $(2x)^{-4}=\frac{1}{(2x)^4}=\frac{1}{16x^4}$

91. $\left(\frac{1}{5}\right)^{-3}=\frac{1^{-3}}{5^{-3}}=\frac{5^3}{1^3}=125$

92. $\left(\frac{-2}{3}\right)^{-2}=\frac{(-2)^{-2}}{3^{-2}}=\frac{3^2}{(-2)^2}=\frac{9}{4}$

93. $2^0+2^{-4}=1+\frac{1}{2^4}=\frac{16}{16}+\frac{1}{16}=\frac{17}{16}$

94. $6^{-1}-7^{-1}=\frac{1}{6}-\frac{1}{7}=\frac{7}{42}-\frac{6}{42}=\frac{1}{42}$

95. $\frac{x^5}{x^{-3}}=x^{5-(-3)}=x^8$

96. $\frac{z^4}{z^{-4}}=z^{4-(-4)}=z^8$

97. $\frac{r^{-3}}{r^{-4}}=r^{-3-(-4)}=r$

98. $\dfrac{y^{-2}}{y^{-5}} = y^{-2-(-5)} = y^3$

99. $\left(\dfrac{bc^{-2}}{bc^{-3}}\right)^4 = \dfrac{b^4c^{-8}}{b^4c^{-12}} = b^{4-4}c^{-8-(-12)} = c^4$

100. $\left(\dfrac{x^{-3}y^{-4}}{x^{-2}y^{-5}}\right)^{-3} = \dfrac{x^9y^{12}}{x^6y^{15}}$
$= x^{9-6}y^{12-15}$
$= x^3y^{-3}$
$= \dfrac{x^3}{y^3}$

101. $\dfrac{x^{-4}y^{-6}}{x^2y^7} = x^{-4-2}y^{-6-7}$
$= x^{-6}y^{-13}$
$= \dfrac{1}{x^6y^{13}}$

102. $\dfrac{a^5b^{-5}}{a^{-5}b^5} = a^{5-(-5)}b^{-5-5} = a^{10}b^{-10} = \dfrac{a^{10}}{b^{10}}$

103. $a^{6m}a^{5m} = a^{6m+5m} = a^{11m}$

104. $\dfrac{(x^{5+h})^3}{x^5} = \dfrac{x^{3(5+h)}}{x^5}$
$= \dfrac{x^{15+3h}}{x^5}$
$= x^{15+3h-5}$
$= x^{10+3h}$

105. $(3xy^{2z})^3 = 3^3x^3y^{2z(3)} = 27x^3y^{6z}$

106. $a^{m+2}a^{m+3} = a^{(m+2)+(m+3)} = a^{2m+5}$

107. $0.00027 = 2.7\times10^{-4}$

108. $0.8868 = 8.868\times10^{-1}$

109. $80{,}800{,}000 = 8.08\times10^7$

110. $868{,}000 = 8.68\times10^5$

111. $91{,}000{,}000 = 9.1\times10^7$

112. $150{,}000 = 1.5\times10^5$

113. $8.67\times10^5 = 867{,}000$

114. $3.86\times10^{-3} = 0.00386$

115. $8.6\times10^{-4} = 0.00086$

116. $8.936\times10^5 = 893{,}600$

117. $1.43128\times10^{15} = 1{,}431{,}280{,}000{,}000{,}000$

118. $1\times10^{-10} = 0.0000000001$

119. $(8\times10^4)(2\times10^{-7}) = (8\cdot2)\times(10^4\cdot10^{-7})$
$= 16\times10^{-3}$
$= 0.016$

120. $\dfrac{8\times10^4}{2\times10^{-7}} = \dfrac{8}{2}\times10^{4-(-7)}$
$= 4\times10^{11}$
$= 400{,}000{,}000{,}000$

121. $\dfrac{x^2+21x+49}{7x^2} = \dfrac{x^2}{7x^2}+\dfrac{21x}{7x^2}+\dfrac{49}{7x^2}$
$= \dfrac{1}{7}+\dfrac{3}{x}+\dfrac{7}{x^2}$

122. $\dfrac{5a^3b-15ab^2+20ab}{-5ab} = \dfrac{5a^3b}{-5ab}-\dfrac{15ab^2}{-5ab}+\dfrac{20ab}{-5ab}$
$= -a^2+3b-4$

123.

$$\begin{array}{r|l} & a+1 \\ \hline a-2 & a^2 - a + 4 \\ & \underline{a^2 - 2a} \\ & \quad a+4 \\ & \quad \underline{a-2} \\ & \quad\quad 6 \end{array}$$

$(a^2-a+4)\div(a-2) = a+1+\dfrac{6}{a-2}$

124.
$$\begin{array}{r}
4x \\
x+5 \overline{) 4x^2+20x+7} \\
\underline{4x^2+20x} \\
7
\end{array}$$

$$(4x^2+20x+7) \div (x+5) = 4x + \frac{7}{x+5}$$

125.
$$\begin{array}{r}
a^2+3a+8 \\
a-2 \overline{) a^3+a^2+2a+6} \\
\underline{a^3-2a^2} \\
3a^2+2a \\
\underline{3a^2-6a} \\
8a+6 \\
\underline{8a-16} \\
22
\end{array}$$

$$\frac{a^3+a^2+2a+6}{a-2} = a^2+3a+8+\frac{22}{a-2}$$

126.
$$\begin{array}{r}
3b^2-4b \\
3b-2 \overline{) 9b^3-18b^2+8b-1} \\
\underline{9b^3-6b^2} \\
-12b^2+8b \\
\underline{-12b^2+8b} \\
-1
\end{array}$$

$$\frac{9b^3-18b^2+8b-1}{3b-2} = 3b^2-4b-\frac{1}{3b-2}$$

127.
$$\begin{array}{r}
2x^3-x^2+2 \\
2x-1 \overline{) 4x^4-4x^3+x^2+4x-3} \\
\underline{4x^4-2x^3} \\
-2x^3+x^2 \\
\underline{-2x^3+x^2} \\
4x-3 \\
\underline{4x-2} \\
-1
\end{array}$$

$$\frac{4x^4-4x^3+x^2+4x-3}{2x-1} = 2x^3-x^2+2-\frac{1}{2x-1}$$

128. Rewrite $-10x^2-x^3-21x+18$ as $-x^3-10x^2-21x+18$.

$$\begin{array}{r}
-x^2-16x-117 \\
x-6 \overline{) -x^3-10x^2-21x+18} \\
\underline{-x^3+6x^2} \\
-16x^2-21x \\
\underline{-16x^2+96x} \\
-117x+18 \\
\underline{-117x+702} \\
-684
\end{array}$$

$$\frac{-10x^2-x^3-21x+18}{x-6} = -x^2-16x-117-\frac{684}{x-6}$$

129.
$$\frac{15x^3-3x^2+60}{3x^2} = \frac{15x^3}{3x^2}-\frac{3x^2}{3x^2}+\frac{60}{3x^2} = 5x-1+\frac{20}{x^2}$$

The width is $\left(5x-1+\frac{20}{x^2}\right)$ feet.

130.
$$\frac{21a^3b^6+3a-3}{3} = \frac{21a^3b^6}{3}+\frac{3a}{3}-\frac{3}{3} = 7a^3b^6+a-1$$

The length of a side is $(7a^3b^6+a-1)$ units.

131. $3x^3+12x-4 = 3x^3+0x^2+12x-4$

2	3	0	12	−4
		6	12	48
	3	6	24	44

$$\frac{3x^3+12x-4}{x-2} = 3x^2+6x+24+\frac{44}{x-2}$$

132. $x+\frac{3}{2} = x-\left(-\frac{3}{2}\right)$

$-\frac{3}{2}$	3	2	−4	−1
		$-\frac{9}{2}$	$\frac{15}{4}$	$\frac{3}{8}$
	3	$-\frac{5}{2}$	$-\frac{1}{4}$	$-\frac{5}{8}$

$$\frac{3x^3+2x^2-4x-1}{x+\frac{3}{2}} = 3x^2-\frac{5}{2}x-\frac{1}{4}-\frac{5}{8\left(x+\frac{3}{2}\right)}$$

133. $x^5 - 1 = x^5 + 0x^4 + 0x^3 + 0x^2 + 0x - 1;$
$x + 1 = x - (-1)$

$$\begin{array}{r|rrrrrr} -1 & 1 & 0 & 0 & 0 & 0 & -1 \\ & & -1 & 1 & -1 & 1 & -1 \\ \hline & 1 & -1 & 1 & -1 & 1 & -2 \end{array}$$

$$\frac{x^5 - 1}{x+1} = x^4 - x^3 + x^2 - x + 1 - \frac{2}{x+1}$$

134. $x^3 - 81 = x^3 + 0x^2 + 0x - 81$

$$\begin{array}{r|rrrr} 3 & 1 & 0 & 0 & -81 \\ & & 3 & 9 & 27 \\ \hline & 1 & 3 & 9 & -54 \end{array}$$

$$\frac{x^3 - 81}{x-3} = x^2 + 3x + 9 - \frac{54}{x-3}$$

135. $x^3 - x^2 + 3x^4 - 2 = 3x^4 + x^3 - x^2 + 0x - 2$

$$\begin{array}{r|rrrrr} 4 & 3 & 1 & -1 & 0 & -2 \\ & & 12 & 52 & 204 & 816 \\ \hline & 3 & 13 & 51 & 204 & 814 \end{array}$$

$$\frac{x^3 - x^2 + 3x^4 - 2}{x-4} = 3x^3 + 13x^2 + 51x + 204 + \frac{814}{x-4}$$

136. $3x^4 - 2x^2 + 10 = 3x^4 + 0x^3 - 2x^2 + 0x + 10$
$x + 2 = x - (-2)$

$$\begin{array}{r|rrrrr} -2 & 3 & 0 & -2 & 0 & 10 \\ & & -6 & 12 & -20 & 40 \\ \hline & 3 & -6 & 10 & -20 & 50 \end{array}$$

$$\frac{3x^4 - 2x^2 + 10}{x+2} = 3x^3 - 6x^2 + 10x - 20 + \frac{50}{x+2}$$

137. $P(x) = 3x^5 + 0x^4 + 0x^3 + 0x^2 - 9x + 7$

$$\begin{array}{r|rrrrrr} 4 & 3 & 0 & 0 & 0 & -9 & 7 \\ & & 12 & 48 & 192 & 768 & 3036 \\ \hline & 3 & 12 & 48 & 192 & 759 & 3043 \end{array}$$

Thus, $P(4) = 3043$.

138. $P(x) = 3x^5 + 0x^4 + 0x^3 + 0x^2 - 9x + 7$

$$\begin{array}{r|rrrrrr} -5 & 3 & 0 & 0 & 0 & -9 & 7 \\ & & -15 & 75 & -375 & 1875 & -9330 \\ \hline & 3 & -15 & 75 & -375 & 1866 & -9323 \end{array}$$

Thus, $P(-5) = -9323$.

139. $\left(-\frac{1}{2}\right)^3 = \left(-\frac{1}{2}\right)\left(-\frac{1}{2}\right)\left(-\frac{1}{2}\right) = -\frac{1}{8}$

140. $(4xy^2)(x^3y^5) = 4(x \cdot x^3)(y^2 \cdot y^5)$
$= 4x^{1+3}y^{2+5}$
$= 4x^4y^7$

141. $\frac{18x^9}{27x^3} = \frac{18}{27}x^{9-3} = \frac{2x^6}{3}$

142. $\left(\frac{3a^4}{b^2}\right)^3 = \frac{3^3(a^4)^3}{(b^2)^3} = \frac{27a^{12}}{b^6}$

143. $(2x^{-4}y^3)^{-4} = 2^{-4}(x^{-4})^{-4}(y^3)^{-4}$
$= \frac{1}{2^4}x^{16}y^{-12}$
$= \frac{x^{16}}{16y^{12}}$

144. $\frac{a^{-3}b^6}{9^{-1}a^{-5}b^{-2}} = 9a^{-3-(-5)}b^{6-(-2)} = 9a^2b^8$

145. $(6x + 2) + (5x - 7) = 6x + 2 + 5x - 7 = 11x - 5$

146. $(-y^2 - 4) + (3y^2 - 6) = -y^2 - 4 + 3y^2 - 6$
$= 2y^2 - 10$

147. $(8y^2 - 3y + 1) - (3y^2 + 2) = 8y^2 - 3y^2 - 3y + 1 - 2$
$= 5y^2 - 3y - 1$

148. $(5x^2 + 2x - 6) - (-x - 4) = 5x^2 + 2x - 6 + x + 4$
$= 5x^2 + 3x - 2$

149. $4x(7x^2 + 3) = 4x(7x^2) + 4x(3)$
$= 28x^3 + 12x$

150. $(2x+5)(3x-2) = 6x^2 - 4x + 15x - 10$
$= 6x^2 + 11x - 10$

151. $(x-3)(x^2+4x-6)$
$= x(x^2+4x-6) - 3(x^2+4x-6)$
$= x^3 + 4x^2 - 6x - 3x^2 - 12x + 18$
$= x^3 + x^2 - 18x + 18$

152. $(7x-2)(4x-9) = 28x^2 - 63x - 8x + 18$
$= 28x^2 - 71x + 18$

153. $(5x+4)^2 = (5x)^2 + 2(5x)(4) + 4^2$
$= 25x^2 + 40x + 16$

154. $(6x+3)(6x-3) = (6x)^2 - (3)^2 = 36x^2 - 9$

155. $\dfrac{8a^4 - 2a^3 + 4a - 5}{2a^3} = \dfrac{8a^4}{2a^3} - \dfrac{2a^3}{2a^3} + \dfrac{4a}{2a^3} - \dfrac{5}{2a^3}$
$= 4a - 1 + \dfrac{2}{a^2} - \dfrac{5}{2a^3}$

156.
$$\begin{array}{r|l} & x-3 \\ \hline x+5 & x^2+2x+10 \\ & \underline{x^2+5x} \\ & \quad -3x+10 \\ & \quad \underline{-3x-15} \\ & \qquad 25 \end{array}$$

$\dfrac{x^2+2x+10}{x+5} = x - 3 + \dfrac{25}{x+5}$

157.
$$\begin{array}{r|l} & 2x^2+7x+5 \\ \hline 2x-3 & 4x^3+8x^2-11x+4 \\ & \underline{4x^3-6x^2} \\ & \quad 14x^2-11x \\ & \quad \underline{14x^2-21x} \\ & \qquad 10x+4 \\ & \qquad \underline{10x-15} \\ & \qquad\quad 19 \end{array}$$

$\dfrac{4x^3+8x^2-11x+4}{2x-3} = 2x^2 + 7x + 5 + \dfrac{19}{2x-3}$

Chapter 5 Test

1. $2^5 = 2\cdot 2\cdot 2\cdot 2\cdot 2 = 32$

2. $(-3)^4 = (-3)(-3)(-3)(-3) = 81$

3. $-3^4 = -3\cdot 3\cdot 3\cdot 3 = -81$

4. $4^{-3} = \dfrac{1}{4^3} = \dfrac{1}{64}$

5. $(3x^2)(-5x^9) = (3)(-5)(x^2\cdot x^9) = -15x^{11}$

6. $\dfrac{y^7}{y^2} = y^{7-2} = y^5$

7. $\dfrac{r^{-8}}{r^{-3}} = r^{-8-(-3)} = r^{-5} = \dfrac{1}{r^5}$

8. $\left(\dfrac{x^2y^3}{x^3y^{-4}}\right)^2 = \dfrac{x^4y^6}{x^6y^{-8}}$
$= x^{4-6}y^{6-(-8)}$
$= x^{-2}y^{14}$
$= \dfrac{y^{14}}{x^2}$

9. $\left(\dfrac{6^2x^{-4}y^{-1}}{6^3x^{-3}y^7}\right) = 6^{2-3}x^{-4-(-3)}y^{-1-7}$
$= 6^{-1}x^{-1}y^{-8}$
$= \dfrac{1}{6xy^8}$

10. $563{,}000 = 5.63\times 10^5$

11. $0.0000863 = 8.63\times 10^{-5}$

12. $1.5\times 10^{-3} = 0.0015$

13. $6.23\times 10^4 = 62{,}300$

14. $(1.2\times 10^5)(3\times 10^{-7}) = (1.2)(3)\times 10^{5-7}$
$= 3.6\times 10^{-2}$
$= 0.036$

15. a.

Term	Numerical Coefficient	Degree of Term
$4xy^2$	4	3
$7xyz$	7	3
x^3y	1	4
-2	-2	0

b. The degree is 4.

16. $5x^2+4xy-7x^2+11+8xy$
$=(5x^2-7x^2)+(4xy+8xy)+11$
$=-2x^2+12xy+11$

17. $(8x^3+7x^2+4x-7)+(8x^3-7x-6)$
$=8x^3+7x^2+4x-7+8x^3-7x-6$
$=16x^3+7x^2-3x-13$

18.
$$\begin{array}{r} 5x^3+x^2+5x-2 \\ \underline{-(8x^3-4x^2+x-7)} \end{array}$$

$$\begin{array}{r} 5x^3+x^2+5x-2 \\ \underline{-8x^3+4x^2-x+7} \\ -3x^3+5x^2+4x+5 \end{array}$$

19. $[(8x^2+7x+5)+(x^3-8)]-(4x+2)$
$=8x^2+7x+5+x^3-8-4x-2$
$=x^3+8x^2+3x-5$

20. $(3x+7)(x^2+5x+2)$
$=3x(x^2+5x+2)+7(x^2+5x+2)$
$=3x^3+15x^2+6x+7x^2+35x+14$
$=3x^3+22x^2+41x+14$

21. $3x^2(2x^2-3x+7)$
$=3x^2(2x^2)+3x^2(-3x)+3x^2(7)$
$=6x^4-9x^3+21x^2$

22. $(x+7)(3x-5)=3x^2-5x+21x-35$
$=3x^2+16x-35$

23. $\left(3x-\frac{1}{5}\right)\left(3x+\frac{1}{5}\right)=(3x)^2-\left(\frac{1}{5}\right)^2=9x^2-\frac{1}{25}$

24. $(4x-2)^2=(4x)^2-2(4x)(2)+2^2$
$=16x^2-16x+4$

25. $(x^2-9b)(x^2+9b)=(x^2)^2-(9b)^2=x^4-81b^2$

26. $-16t^2+1001$
$t=0$: $-16(0)^2+1001=1001$ ft
$t=1$: $-16(1)^2+1001=985$ ft
$t=3$: $-16(3)^2+1001=857$ ft
$t=5$: $-16(5)^2+1001=601$ ft

27. $(2x+3)(2x-3)=(2x)^2-(3)^2=4x^2-9$
The area is $(4x^2-9)$ square inches.

28. $\frac{4x^2+24xy-7x}{8xy}=\frac{4x^2}{8xy}+\frac{24xy}{8xy}-\frac{7x}{8xy}$
$=\frac{x}{2y}+3-\frac{7}{8y}$

29.
$$\begin{array}{r} x+2 \\ x+5\overline{)x^2+7x+10} \\ \underline{x^2+5x} \\ 2x+10 \\ \underline{2x+10} \\ 0 \end{array}$$

$\frac{x^2+7x+10}{x+5}=x+2$

30. Rewrite $27x^3-8$ as $27x^3+0x^2+0x-8$.

$$\begin{array}{r} 9x^2-6x+4 \\ 3x+2\overline{)27x^3+0x^2+0x-8} \\ \underline{27x^3+18x^2} \\ -18x^2+0x \\ \underline{-18x^2-12x} \\ 12x-8 \\ \underline{12x+8} \\ -16 \end{array}$$

$\frac{27x^3-8}{3x+2}=9x^2-6x+4-\frac{16}{3x+2}$

31. $h(t) = -16t^2 + 96t + 880$

a. $h(1) = -16(1)^2 + 96(1) + 880$
$= -16 + 96 + 880$
$= 960$
The height of the pebble is 960 feet when $t = 1$.

b. $h(5.1) = -16(5.1)^2 + 96(5.1) + 880$
$= -16(26.01) + 489.6 + 880$
$= -416.16 + 489.6 + 880$
$= 953.44$
The height of the pebble is 953.44 feet when $t = 5.1$.

c. The pebble hits the ground when $h(t) = 0$.
$0 = -16t^2 + 96t + 880$
$0 = -16(t^2 - 6t - 55)$
$0 = -16(t + 5)(t - 11)$
$t + 5 = 0$ or $t - 11 = 0$
$t = -5$ $\quad t = 11$
Since the time cannot be negative, the pebble hits the ground when $t = 11$ seconds.

32. $4x^4 - 3x^3 - x - 1 = 4x^4 - 3x^3 + 0x^2 - x - 1$
$x + 3 = x - (-3)$

-3	4	-3	0	-1	-1
		-12	45	-135	408
	4	-15	45	-136	407

$\frac{4x^4 - 3x^3 - x - 1}{x + 3}$
$= 4x^3 - 15x^2 + 45x - 136 + \frac{407}{x + 3}$

33. $P(x) = 4x^4 + 0x^3 + 7x^2 - 2x - 5$

-2	4	0	7	-2	-5
		-8	16	-46	96
	4	-8	23	-48	91

Thus, $P(-2) = 91$.

Chapter 5 Cumulative Review

1. a. $8 \geq 8$ is true since $8 = 8$.

b. $8 \leq 8$ is true since $8 = 8$.

c. $23 \leq 0$ is false.

d. $23 \geq 0$ is true

2. a. $|-7.2| = 7.2$

b. $|0| = 0$

c. $\left|-\frac{1}{2}\right| = \frac{1}{2}$

3. a. $\frac{4}{5} \div \frac{5}{16} = \frac{4}{5} \cdot \frac{16}{5} = \frac{64}{25}$

b. $\frac{7}{10} \div 14 = \frac{7}{10} \div \frac{14}{1} = \frac{7}{10} \cdot \frac{1}{14} = \frac{7}{10 \cdot 7 \cdot 2} = \frac{1}{20}$

c. $\frac{3}{8} \div \frac{3}{10} = \frac{3}{8} \cdot \frac{10}{3} = \frac{3 \cdot 2 \cdot 5}{2 \cdot 4 \cdot 3} = \frac{5}{4}$

4. a. $\frac{3}{4} \cdot \frac{7}{21} = \frac{3 \cdot 7}{4 \cdot 3 \cdot 7} = \frac{1}{4}$

b. $\frac{1}{2} \cdot 4\frac{5}{6} = \frac{1}{2} \cdot \frac{29}{6} = \frac{29}{12} = 2\frac{5}{12}$

5. a. $3^2 = 3 \cdot 3 = 9$

b. $5^3 = 5 \cdot 5 \cdot 5 = 125$

c. $2^4 = 2 \cdot 2 \cdot 2 \cdot 2 = 16$

d. $7^1 = 7$

e. $\left(\frac{3}{7}\right)^2 = \left(\frac{3}{7}\right)\left(\frac{3}{7}\right) = \frac{9}{49}$

6. Let $x = 5$ and $y = 1$.
$\frac{2x - 7y}{x^2} = \frac{2(5) - 7(1)}{5^2}$
$= \frac{10 - 7}{25}$
$= \frac{3}{25}$

7. a. $-3+(-7)=-10$

b. $-1+(-20)=-21$

c. $-2+(-10)=-12$

8. $$\begin{aligned}8+3(2\cdot 6-1)&=8+3(12-1)\\&=8+3(11)\\&=8+33\\&=41\end{aligned}$$

9. $-4-8=-4+(-8)=-12$

10. $x=1$
$$\begin{aligned}5x^2+2&=x-8\\5(1)^2+2&\stackrel{?}{=}1-8\\5+2&\stackrel{?}{=}-7\\7&\stackrel{?}{=}-7 \quad \text{False}\end{aligned}$$
$x=1$ is not a solution.

11. a. The reciprocal of 22 is $\frac{1}{22}$.

b. The reciprocal of $\frac{3}{16}$ is $\frac{16}{3}$.

c. The reciprocal of -10 is $-\frac{1}{10}$.

d. The reciprocal of $-\frac{9}{13}$ is $-\frac{13}{9}$.

12. a. $7-40=7+(-40)=-33$

b. $-5-(-10)=-5+10=5$

13. a. $5+(4+6)=(5+4)+6$

b. $(-1\cdot 2)\cdot 5=-1\cdot(2\cdot 5)$

14. $\frac{4(-3)+(-8)}{5+(-5)}=\frac{-12+(-8)}{0}$ is undefined.

15. a. $$\begin{aligned}10+(x+12)&=10+(12+x)\\&=(10+12)+x\\&=22+x\end{aligned}$$

b. $-3(7x)=(-3\cdot 7)x=-21x$

16. $$\begin{aligned}-2(x+3y-z)&=-2(x)+(-2)(3y)-(-2)(z)\\&=-2x-6y+2z\end{aligned}$$

17. a. $5(x+2)=5x+5(2)=5x+10$

b. $$\begin{aligned}&-2(y+0.3z-1)\\&=-2(y)+(-2)(0.3z)-(-2)(1)\\&=-2y-0.6z+2y\end{aligned}$$

c. $$\begin{aligned}&-(x+y-2z+6)\\&=-1(x+y-2z+6)\\&=-1(x)+(-1)(y)-(-1)(2z)+(-1)(6)\\&=-x-y+2z-6\end{aligned}$$

18. $$\begin{aligned}2(6x-1)-(x-7)&=12x-2-x+7\\&=11x+5\end{aligned}$$

19. $$\begin{aligned}x-7&=10\\x-7+7&=10+7\\x&=17\end{aligned}$$

20. Let x = a number.
$(x+7)-2x$

21. $$\begin{aligned}\frac{5}{2}x&=15\\\frac{2}{5}\cdot\frac{5}{2}x&=\frac{2}{5}\cdot 15\\x&=6\end{aligned}$$

22. $$\begin{aligned}2x+\frac{1}{8}&=x-\frac{3}{8}\\x+\frac{1}{8}&=-\frac{3}{8}\\x&=-\frac{4}{8}\\x&=-\frac{1}{2}\end{aligned}$$

23. Let x = a number.
$$\begin{aligned}7+2x&=x-3\\7+x&=-3\\x&=-10\end{aligned}$$
The number is -10.

24. $$\begin{aligned}10&=5j-2\\12&=5j\\\frac{12}{5}&=j\end{aligned}$$

25. Let x = a number.

$$2(x+4) = 4x-12$$
$$2x+8 = 4x-12$$
$$-2x+8 = -12$$
$$-2x = -20$$
$$x = 10$$

The number is 10.

26.

$$\frac{7x+5}{3} = x+3$$
$$3\left(\frac{7x+5}{3}\right) = 3(x+3)$$
$$7x+5 = 3x+9$$
$$4x+5 = 9$$
$$4x = 4$$
$$x = 1$$

27. Let x = the width and $3x - 2$ = the length.

$$2L+2W = P$$
$$2(3x-2)+2x = 28$$
$$6x-4+2x = 28$$
$$8x-4 = 28$$
$$8x = 32$$
$$x = 4$$

$3x - 2 = 3(4) - 2 = 10$

The width is 4 feet and the length is 10 feet.

28. $x < 5$, $(-\infty, 5)$

5

29.

$$F = \frac{9}{5}C+32$$
$$F-32 = \frac{9}{5}C$$
$$\frac{5}{9}(F-32) = C$$
$$\frac{5F-160}{9} = C$$

30. a. $x = -1$ is a vertical line and the slope is undefined.

b. $y = 7$ is a horizontal line and the slope is zero.

31. $2 < x \le 4$

2 4

32. $m = \frac{y_2 - y_1}{x_2 - x_1} = \frac{2}{20} = \frac{1}{10} \cdot 100\% = 10\%$

33. $3x + y = 12$

a. $(0, \ )$: $3(0)+y = 12$

$y = 12$, $(0, 12)$

b. $(\ , 6)$: $3x+6 = 12$

$3x = 6$

$x = 2$, $(2, 6)$

c. $(-1, \ )$: $3(-1)+y = 12$

$-3+y = 12$

$y = 15$, $(-1, 15)$

34. $\begin{cases} 3x+2y = -8 \\ 2x-6y = -9 \end{cases}$

Multiply the first equation by 3 and add.

$$\begin{array}{rl} 9x+6y &= -24 \\ 2x-6y &= -9 \\ \hline 11x \quad &= -33 \\ x &= -3 \end{array}$$

Replace x with -3 in the first equation.

$$3(-3)+2y = -8$$
$$-9+2y = -8$$
$$2y = 1$$
$$y = \frac{1}{2}$$

The solution to the system is $\left(-3, \frac{1}{2}\right)$.

35. $2x + y = 5$

x	y
0	5
$\frac{5}{2}$	0

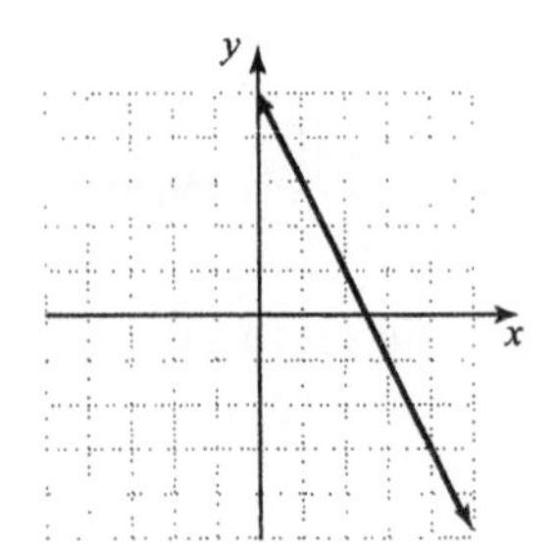

36. $\begin{cases} x = -3y + 3 \\ 2x + 9y = 5 \end{cases}$

Replace x with $-3y + 3$ in the second equation.

$$\begin{aligned} 2(-3y+3)+9y &= 5 \\ -6y+6+9y &= 5 \\ 3y+6 &= 5 \\ 3y &= -1 \\ y &= -\frac{1}{3} \end{aligned}$$

Replace y with $-\frac{1}{3}$ in the first equation.

$$x = -3\left(-\frac{1}{3}\right) + 3 = 1 + 3 = 4$$

The solution to the system is $\left(4, -\frac{1}{3}\right)$.

37.

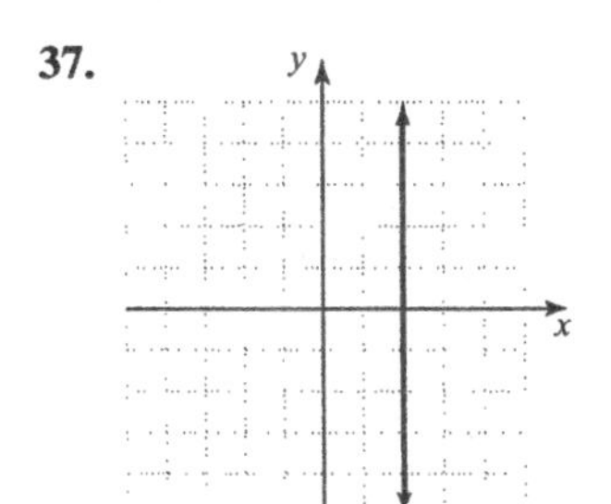

38. a. $(-5)^2 = (-5)(-5) = 25$

b. $-5^2 = -(5)(5) = -25$

c. $2 \cdot 5^2 = 2 \cdot 5 \cdot 5 = 50$

39. $x = 5$ is a vertical line and the slope is undefined.

40. $\dfrac{(z^2)^3 \cdot z^7}{z^9} = \dfrac{z^6 \cdot z^7}{z^9} = z^{6+7-9} = z^4$

41.
$$\begin{aligned} &(2x^3 + 8x^2 - 6x) - (2x^3 - x^2 + 1) \\ &= 2x^3 + 8x^2 - 6x - 2x^3 + x^2 - 1 \\ &= 2x^3 - 2x^3 + 8x^2 + x^2 - 6x - 1 \\ &= 9x^2 - 6x - 1 \end{aligned}$$

42. $(5y^2 - 6) - (y^2 + 2) = 5y^2 - 6 - y^2 - 2 = 4y^2 - 8$

43. $(2x^2)(-3x^5) = (2 \cdot -3)(x^2 \cdot x^5) = -6x^{2+5} = -6x^7$

44. $-x^2$

a. $-(2)^2 = -4$

b. $-(-2)^2 = -4$

45.
$$\begin{aligned} &(11x^3 - 12x^2 + x - 3) + (x^3 - 10x + 5) \\ &= 11x^3 - 12x^2 + x - 3 + x^3 - 10x + 5 \\ &= 11x^3 + x^3 - 12x^2 + x - 10x - 3 + 5 \\ &= 12x^3 - 12x^2 - 9x + 2 \end{aligned}$$

46.
$$\begin{aligned} (10x^2 - 3)(10x^2 + 3) &= (10x^2)^2 - 3^2 \\ &= 100x^4 - 9 \end{aligned}$$

47.
$$\begin{aligned} (2x - y)^2 &= (2x)^2 - 2(2x)(y) + (y)^2 \\ &= 4x^2 - 4xy + y^2 \end{aligned}$$

48.
$$\begin{aligned} (10x^2 + 3)^2 &= (10x^2)^2 + 2(10x^2)(3) + 3^2 \\ &= 100x^4 + 60x^2 + 9 \end{aligned}$$

49. $\dfrac{6m^2 + 2m}{2m} = \dfrac{6m^2}{2m} + \dfrac{2m}{2m} = 3m + 1$

50. a. $5^{-1} = \dfrac{1}{5}$

b. $7^{-2} = \dfrac{1}{7^2} = \dfrac{1}{49}$

Chapter 6

Section 6.1

Practice Exercises

1. a. $36 = 2 \cdot 2 \cdot 3 \cdot 3 = 2^2 \cdot 3^2$
$42 = 2 \cdot 3 \cdot 7$
$\text{GCF} = 2 \cdot 3 = 6$

b. $35 = 5 \cdot 7$
$44 = 2 \cdot 2 \cdot 11$
$\text{GCF} = 1$

c. $12 = 2 \cdot 2 \cdot 3 = 2^2 \cdot 3$
$16 = 2 \cdot 2 \cdot 2 \cdot 2 = 2^4$
$40 = 2 \cdot 2 \cdot 2 \cdot 5 = 2^3 \cdot 5$
$\text{GCF} = 2^2 = 4$

2. a. The GCF is y^4 since 4 is the smallest exponent to which y is raised.

b. The GCF is x^1 or x, since 1 is the smallest exponent on x.

3. a. $5y^4 = 5 \cdot y^4$
$15y^2 = 3 \cdot 5 \cdot y^2$
$-20y^3 = -1 \cdot 2 \cdot 2 \cdot 5 \cdot y^3$
$\text{GCF} = 5 \cdot y^2 = 5y^2$

b. $4x^2 = 2 \cdot 2 \cdot x^2$
$x^3 = x^3$
$3x^8 = 3 \cdot x^8$
$\text{GCF} = x^2$

c. The GCF of a^4, a^3, and a^2 is a^2.
The GCF of b^2, b^5, and b^3 is b^2.
Thus, the GCF of a^4b^2, a^3b^5, and a^2b^3 is a^2b^2.

4. a. $4t + 12$; GCF $= 4$
$4t + 12 = 4 \cdot t + 4 \cdot 3 = 4(t+3)$

b. $y^8 + y^4$; GCF $= y^4$
$y^8 + y^4 = y^4 \cdot y^4 + y^4 \cdot 1 = y^4(y^4 + 1)$

5. $-8b^6 + 16b^4 - 8b^2$
$= -8b^2(b^4) - 8b^2(-2b^2) - 8b^2(1)$
$= -8b^2(b^4 - 2b^2 + 1)$ or $8b^2(-b^4 + 2b^2 - 1)$

6. $5x^4 - 20x = 5x(x^3 - 4)$

7. $\frac{5}{9}z^5 + \frac{1}{9}z^4 - \frac{2}{9}z^3 = \frac{1}{9}z^3(5z^2 + z - 2)$

8. $8a^2b^4 - 20a^3b^3 + 12ab^3 = 4ab^3(2ab - 5a^2 + 3)$

9. $8(y-2) + x(y-2) = (y-2)(8+x)$

10. $7xy^3(p+q) - (p+q) = 7xy^3(p+q) - 1(p+q)$
$= (p+q)(7xy^3 - 1)$

11. $xy + 3y + 4x + 12 = (xy + 3y) + (4x + 12)$
$= y(x+3) + 4(x+3)$
$= (x+3)(y+4)$
Check: $(x+3)(y+4) = xy + 3y + 4x + 12$

12. $2xy + 3y^2 - 2x - 3y = (2xy + 3y^2) + (-2x - 3y)$
$= y(2x + 3y) - 1(2x + 3y)$
$= (2x + 3y)(y - 1)$

13. $7a^3 + 5a^2 + 7a + 5 = (7a^3 + 5a^2) + (7a + 5)$
$= a^2(7a + 5) + 1(7a + 5)$
$= (7a + 5)(a^2 + 1)$

14. $4xy + 15 - 12x - 5y = 4xy - 12x - 5y + 15$
$= (4xy - 12x) + (-5y + 15)$
$= 4x(y-3) - 5(y-3)$
$= (y-3)(4x-5)$

15. $9y - 18 + y^3 - 4y^2 = 9(y-2) + y^2(y-4)$
There is no common binomial factor, so it cannot be factored by grouping.

16. $3xy - 3ay - 6ax + 6a^2 = 3(xy - ay - 2ax + 2a^2)$
$= 3[y(x-a) - 2a(x-a)]$
$= 3(x-a)(y-2a)$

Vocabulary and Readiness Check

1. Since $5 \cdot 4 = 20$, the numbers 5 and 4 are called <u>factors</u> of 20.

2. The greatest common factor of a list of integers is the largest integer that is a factor of all the integers in the list.

3. The greatest common factor of a list of common variables raised to powers is the variable raised to the least exponent in the list.

4. The process of writing a polynomial as a product is called factoring.

5. $7(x + 3) + y(x + 3)$ is a sum, not a product. The statement is false.

6. $3x^3 + 6x + x^2 + 2 = 3x(x^2 + 2) + (x^2 + 2)$
$= (x^2 + 2)(3x + 1)$
The statement is false.

7. $14 = 2 \cdot 7$

8. $15 = 3 \cdot 5$

9. The GCF of 18 and 3 is 3.

10. The GCF of 7 and 35 is 7.

11. The GCF of 20 and 15 is 5.

12. The GCF of 6 and 15 is 3.

Exercise Set 6.1

1. $32 = 2 \cdot 2 \cdot 2 \cdot 2 \cdot 2 = 2^5$
$36 = 2 \cdot 2 \cdot 3 \cdot 3 = 2^2 \cdot 3^2$
$\text{GCF} = 2 \cdot 2 = 4$

3. $18 = 2 \cdot 3 \cdot 3 = 2 \cdot 3^2$
$42 = 2 \cdot 3 \cdot 7$
$84 = 2 \cdot 2 \cdot 3 \cdot 7 = 2^2 \cdot 3 \cdot 7$
$\text{GCF} = 2 \cdot 3 = 6$

5. $24 = 2 \cdot 2 \cdot 2 \cdot 3 = 2^3 \cdot 3$
$14 = 2 \cdot 7$
$21 = 3 \cdot 7$
$\text{GCF} = 1$

7. The GCF of y^2, y^4, and y^7 is y^2.

9. The GCF of z^7, z^9, and z^{11} is z^7.

11. The GCF of x^{10}, x, and x^3 is x.
The GCF of y^2, y^2, and y^3 is y^2.
Thus the GCF of $x^{10}y^2$, xy^2, and x^3y^3 is xy^2.

13. $14x = 2 \cdot 7 \cdot x$
$21 = 3 \cdot 7$
$\text{GCF} = 7$

15. $12y^4 = 2 \cdot 2 \cdot 3 \cdot y^4$
$20y^3 = 2 \cdot 2 \cdot 5 \cdot y^3$
$\text{GCF} = 2 \cdot 2 \cdot y^3 = 4y^3$

17. $-10x^2 = -1 \cdot 2 \cdot 5 \cdot x^2$
$15x^3 = 3 \cdot 5 \cdot x^3$
$\text{GCF} = 5 \cdot x^2 = 5x^2$

19. $12x^3 = 2 \cdot 2 \cdot 3 \cdot x^3$
$-6x^4 = -1 \cdot 2 \cdot 3 \cdot x^4$
$3x^5 = 3 \cdot x^5$
$\text{GCF} = 3 \cdot x^3 = 3x^3$

21. $-18x^2y = -1 \cdot 2 \cdot 3 \cdot 3 \cdot x^2 \cdot y$
$9x^3y^3 = 3 \cdot 3 \cdot x^3 \cdot y^3$
$36x^3y = 2 \cdot 2 \cdot 3 \cdot 3 \cdot x^3 \cdot y$
$\text{GCF} = 3 \cdot 3 \cdot x^2 \cdot y = 9x^2y$

23. $20a^6b^2c^8 = 2 \cdot 2 \cdot 5 \cdot a^6 \cdot b^2 \cdot c^8$
$50a^7b = 2 \cdot 5 \cdot 5 \cdot a^7 \cdot b$
$\text{GCF} = 2 \cdot 5 \cdot a^6 \cdot b = 10a^6b$

25. $3a + 6 = 3(a + 2)$

27. $30x - 15 = 15(2x - 1)$

29. $x^3 + 5x^2 = x^2(x + 5)$

31. $6y^4 + 2y^3 = 2y^3(3y + 1)$

33. $4x - 8y + 4 = 4(x - 2y + 1)$

35. $6x^3 - 9x^2 + 12x = 3x(2x^2 - 3x + 4)$

37. $a^7b^6 - a^3b^2 + a^2b^5 - a^2b^2$
$= a^2b^2(a^5b^4 - a + b^3 - 1)$

39. $8x^5+16x^4-20x^3+12=4(2x^5+4x^4-5x^3+3)$

41. $\frac{1}{3}x^4+\frac{2}{3}x^3-\frac{4}{3}x^5+\frac{1}{3}x$
$=\frac{1}{3}x(x^3+2x^2-4x^4+1)$

43. $y(x^2+2)+3(x^2+2)=(x^2+2)(y+3)$

45. $z(y+4)-3(y+4)=(y+4)(z-3)$

47. $r(z^2-6)+(z^2-6)=r(z^2-6)+1(z^2-6)$
$=(z^2-6)(r+1)$

49. $-2x-14=-2(x+7)$

51. $-2x^5+x^7=-x^5(2-x^2)$

53. $-6a^4+9a^3-3a^2=-3a^2(2a^2-3a+1)$

55. $x^3+2x^2+5x+10=x^2(x+2)+5(x+2)$
$=(x+2)(x^2+5)$

57. $5x+15+xy+3y=5(x+3)+y(x+3)$
$=(x+3)(5+y)$

59. $6x^3-4x^2+15x-10=2x^2(3x-2)+5(3x-2)$
$=(3x-2)(2x^2+5)$

61. $5m^3+6mn+5m^2+6n$
$=m(5m^2+6n)+1(5m^2+6n)$
$=(5m^2+6n)(m+1)$

63. $2y-8+xy-4x=2(y-4)+x(y-4)$
$=(y-4)(2+x)$

65. $2x^3-x^2+8x-4=x^2(2x-1)+4(2x-1)$
$=(2x-1)(x^2+4)$

67. $4x^2-8xy-3x+6y=4x(x-2y)-3(x-2y)$
$=(x-2y)(4x-3)$

69. $5q^2-4pq-5q+4p=q(5q-4p)-1(5q-4p)$
$=(5q-4p)(q-1)$

71. $2x^4+5x^3+2x^2+5x=x(2x^3+5x^2+2x+5)$
$=x[x^2(2x+5)+1(2x+5)]$
$=x(2x+5)(x^2+1)$

73. $12x^2y-42x^2-4y+14$
$=2(6x^2y-21x^2-2y+7)$
$=2[3x^2(2y-7)-1(2y-7)]$
$=2(2y-7)(3x^2-1)$

75. $32xy^2-18x^2=2x(16y-9x)$

77. $y(x+2)-3(x+2)=(x+2)(y-3)$

79. $14x^3y+7x^2y-7xy=7xy(2x^2+x-1)$

81. $28x^3-7x^2+12x-3=7x^2(4x-1)+3(4x-1)$
$=(4x-1)(7x^2+3)$

83. $-40x^8y^6-16x^9y^5=-8x^8y^5(5y+2x)$

85. $6a^2+9ab^2+6ab+9b^3$
$=3(2a^2+3ab^2+2ab+3b^3)$
$=3[a(2a+3b^2)+b(2a+3b^2)]$
$=3(2a+3b^2)(a+b)$

87. $(x+2)(x+5)=x^2+5x+2x+10=x^2+7x+10$

89. $(b+1)(b-4)=b^2-4b+b-4=b^2-3b-4$

	Two Numbers	Their Product	Their Sum
91.	2, 6	12	8
93.	−1, −8	8	−9
95.	−2, 5	−10	3

97. **a.** $8\cdot a-24=8a-24$

b. $8(a-3)=8a-24$

c. $4(2a-12)=8a-48$

d. $8\cdot a-2\cdot 12=8a-24$

The answer is b.

99. $(x+5)(x+y)$ is factored.

101. $3x(a + 2b) + 2(a + 2b)$ is not factored.

103. Answers may vary

105. Answers may vary

107. a. $-8x^2 + 50x + 3020 = -8(2)^2 + 50(2) + 3020$
$= -32 + 100 + 3020$
$= 3088$

In 2005, 3088 thousand, or 3,088,000, students graduated from U.S. high schools.

b. Let $x = 2007 - 2003 = 4$

$-8x^2 + 50x + 3020 = -8(4)^2 + 50(4) + 3020$
$= -128 + 200 + 3020$
$= 3092$

We predict 3092 thousand, or 3,092,000, students to graduate from U.S. high schools in 2007.

c. $-8x^2 + 50x + 3020 = -2(4x^2 - 25x - 1510)$

109. length of side of square $= 2x$
shaded area = square's area − circle's area
$= (2x)^2 - \pi x^2$
$= 4x^2 - \pi x^2$
$= x^2(4 - \pi)$

111. Area $= 5x^5 - 5x^2 = 5x^2(x^3 - 1)$

Since the width is $5x^2$ units, the length is $(x^3 - 1)$ units.

113. $x^{2n} + 6x^n + 10x^n + 60 = x^n(x^n + 6) + 10(x^n + 6)$
$= (x^n + 6)(x^n + 10)$

115. $12x^{2n} - 10x^n - 30x^n + 25$
$= 2x^n(6x^n - 5) - 5(6x^n - 5)$
$= (6x^n - 5)(2x^n - 5)$

Section 6.2

Practice Problems

1.

Positive Factors of 6	Sum of Factors
1, 6	7
2, 3	5

$x^2 + 5x + 6 = (x + 2)(x + 3)$

2.

Negative Factors of 70	Sum of Factors
−1, −70	−71
−2, −35	−37
−5, −14	−19
−7, −10	−17

$x^2 - 17x + 70 = (x - 7)(x - 10)$

3.

Factors of −14	Sum of Factors
−1, 14	13
1, −14	−13
−2, 7	5
2, −7	−5

$x^2 + 5x - 14 = (x - 2)(x + 7)$

4. The first term of each binomial is p. Then look for two numbers whose product is −63 and whose sum is −2.

$p^2 - 2p - 63 = (p - 9)(p + 7)$

5. The first term of each binomial is b. Then look for two numbers whose product is 1 and whose sum is 5. There are no such numbers.

$b^2 + 5b + 1$ is a prime polynomial.

6. The first term of each polynomial is x. Then look for two terms whose product is $12y^2$ and whose sum is $7y$.

$x^2 + 7xy + 12y^2 = (x + 3y)(x + 4y)$

7. The first term of each polynomial is x^2. Then look for two numbers whose product is 12 and whose sum is 13.

$x^4 + 13x^2 + 12 = (x^2 + 1)(x^2 + 12)$

8. $48 - 14x + x^2 = x^2 - 14x + 48$

The first term of each binomial is x. Then look for two factors whose product is 48 and whose sum is −14.

$x^2 - 14x + 48 = (x - 6)(x - 8)$

9. $4x^2 - 24x + 36 = 4(x^2 - 6x + 9)$
The first term of each binomial is x. Then look for two factors whose product is 9 and whose sum is −6.
$4(x^2 - 6x + 9) = 4(x-3)(x-3)$ or $4(x-3)^2$

10. $3y^4 - 18y^3 - 21y^2 = 3y^2(y^2 - 6y - 7)$
The first term of each binomial is y. Then look for two factors whose product is −7 and whose sum is −6.
$3y^2(y^2 - 6y - 7) = 3y^2(y-7)(y+1)$

Vocabulary and Readiness Check

1. The statement is true.

2. The statement is true.

3. Since $4x - 12 = 4(x - 3)$, the statement is false.

4. $(x+2y)^2 = (x+2y)(x+2y) \neq (x+2y)(x+y)$
The statement is false.

5. $x^2 + 9x + 20 = (x+4)(x\underline{+5})$

6. $x^2 + 12x + 35 = (x+5)(x\underline{+7})$

7. $x^2 - 7x + 12 = (x-4)(x\underline{-3})$

8. $x^2 - 13x + 22 = (x-2)(x\underline{-11})$

9. $x^2 + 4x + 4 = (x+2)(x\underline{+2})$

10. $x^2 + 10x + 24 = (x+6)(x\underline{+4})$

Exercise Set 6.2

1. $x^2 + 7x + 6 = (x+6)(x+1)$

3. $y^2 - 10y + 9 = (y-9)(y-1)$

5. $x^2 - 6x + 9 = (x-3)(x-3)$ or $(x-3)^2$

7. $x^2 - 3x - 18 = (x-6)(x+3)$

9. $x^2 + 3x - 70 = (x+10)(x-7)$

11. $x^2 + 5x + 2$ is a prime polynomial.

13. $x^2 + 8xy + 15y^2 = (x+5y)(x+3y)$

15. $a^4 - 2a^2 - 15 = (a^2 - 5)(a^2 + 3)$

17. $13 + 14m + m^2 = m^2 + 14m + 13 = (m+13)(m+1)$

19. $10t - 24 + t^2 = t^2 + 10t - 24 = (t-2)(t+12)$

21. $a^2 - 10ab + 16b^2 = (a-2b)(a-8b)$

23. $2z^2 + 20z + 32 = 2(z^2 + 10z + 16)$
$= 2(z+8)(z+2)$

25. $2x^3 - 18x^2 + 40x = 2x(x^2 - 9x + 20)$
$= 2x(x-5)(x-4)$

27. $x^2 - 3xy - 4y^2 = (x-4y)(x+y)$

29. $x^2 + 15x + 36 = (x+12)(x+3)$

31. $x^2 - x - 2 = (x-2)(x+1)$

33. $r^2 - 16r + 48 = (r-12)(r-4)$

35. $x^2 + xy - 2y^2 = (x+2y)(x-y)$

37. $3x^2 + 9x - 30 = 3(x^2 + 3x - 10) = 3(x+5)(x-2)$

39. $3x^2 - 60x + 108 = 3(x^2 - 20x + 36)$
$= 3(x-18)(x-2)$

41. $x^2 - 18x - 144 = (x-24)(x+6)$

43. $r^2 - 3r + 6$ is a prime polynomial.

45. $x^2 - 8x + 15 = (x-5)(x-3)$

47. $6x^3 + 54x^2 + 120x = 6x(x^2 + 9x + 20)$
$= 6x(x+4)(x+5)$

49. $4x^2y + 4xy - 12y = 4y(x^2 + x - 3)$

51. $x^2 - 4x - 21 = (x-7)(x+3)$

53. $x^2+7xy+10y^2=(x+5y)(x+2y)$

55. $64+24t+2t^2=2t^2+24t+64$
$=2(t^2+12t+32)$
$=2(t+8)(t+4)$

57. $x^3-2x^2-24x=x(x^2-2x-24)$
$=x(x-6)(x+4)$

59. $2t^5-14t^4+24t^3=2t^3(t^2-7t+12)$
$=2t^3(t-4)(t-3)$

61. $5x^3y-25x^2y^2-120xy^3=5xy(x^2-5xy-24y^2)$
$=5xy(x-8y)(x+3y)$

63. $162-45m+3m^2=3m^2-45m+162$
$=3(m^2-15m+54)$
$=3(m-9)(m-6)$

65. $-x^2+12x-11=-1(x^2-12x+11)$
$=-1(x-11)(x-1)$

67. $\frac{1}{2}y^2-\frac{9}{2}y-11=\frac{1}{2}(y^2-9y-22)$
$=\frac{1}{2}(y-11)(y+2)$

69. $x^3y^2+x^2y-20x=x(x^2y^2+xy-20)$
$=x(xy-4)(xy+5)$

71. $(2x+1)(x+5)=2x^2+10x+x+5$
$=2x^2+11x+5$

73. $(5y-4)(3y-1)=15y^2-5y-12y+4$
$=15y^2-17y+4$

75. $(a+3b)(9a-4b)=9a^2-4ab+27ab-12b^2$
$=9a^2+23ab-12b^2$

77. $(x-3)(x+8)=x^2+8x-3x-3(8)=x^2+5x-24$

79. Answers may vary

81. $P=2l+2w$
$l=x^2+10x$ and $w=4x+33$, so
$P=2(x^2+10x)+2(4x+33)$
$=2x^2+20x+8x+66$
$=2x^2+28x+66$
$=2(x^2+14x+33)$
$=2(x+11)(x+3)$
The perimeter of the rectangle is given by the polynomial $2x^2+28x+66$ which factors as $2(x+11)(x+3)$.

83. $-16t^2+64t+80=-16(t^2-4t-5)$
$=-16(t-5)(t+1)$

85. $x^2+\frac{1}{2}x+\frac{1}{16}=\left(x+\frac{1}{4}\right)\left(x+\frac{1}{4}\right)$ or $\left(x+\frac{1}{4}\right)^2$

87. $z^2(x+1)-3z(x+1)-70(x+1)$
$=(x+1)(z^2-3z-70)$
$=(x+1)(z-10)(z+7)$

89. $x^{2n}+8x^n-20=(x^n+10)(x^n-2)$

91. c must be the product of positive numbers that sum to 8.
$8=1+7;\ 1\cdot 7=7$
$8=2+6;\ 2\cdot 6=12$
$8=3+5;\ 3\cdot 5=15$
$8=4+4;\ 4\cdot 4=16$
t^2+8t+c is factorable when c is 7, 12, 15, or 16.

93. c must be the product of negative numbers that sum to -16.
$-16=-1+(-15);\ -1\cdot -15=15$
$-16=-2+(-14);\ -2\cdot -14=28$
$-16=-3+(-13);\ -3\cdot -13=39$
$-16=-4+(-12);\ -4\cdot -12=48$
$-16=-5+(-11);\ -5\cdot -11=55$
$-16=-6+(-10);\ -6\cdot -10=60$
$-16=-7+(-9);\ -7\cdot -9=63$
$-16=-8+(-8);\ -8\cdot -8=64$
$n^2-16n+c$ is factorable when c is 15, 28, 39, 48, 55, 60, 63, or 64.

95. b must be the sum of positive numbers which have a product of 20.
$20 = 1 \cdot 20;\ 1 + 20 = 21$
$20 = 2 \cdot 10;\ 2 + 10 = 12$
$20 = 4 \cdot 5;\ 4 + 5 = 9$
$y^2 + by + 20$ is factorable when b is 9, 12, or 21.

97. b must be the positive sum of a positive number and a negative number which have a product of -14.
$-14 = 14 \cdot -1;\ 14 + (-1) = 13$
$-14 = 7 \cdot -2;\ 7 + (-2) = 5$
$x^2 + bx - 14$ is factorable when b is 5 or 13.

Section 6.3

Practice Exercises

1. Factors of $2x^2$: $2x^2 = 2x \cdot x$
Factors of 15: $15 = 1 \cdot 15$, $15 = 3 \cdot 5$
Try possible combinations.
Factored form: $2x^2 + 11x + 15 = (2x+5)(x+3)$

2. Factors of $15x^2$: $15x^2 = 15x \cdot x$, $15x^2 = 5x \cdot 3x$
Factors of 8: $8 = -1 \cdot -8$, $8 = -2 \cdot -4$
Try possible combinations.
Factored form: $15x^2 - 22x + 8 = (5x-4)(3x-2)$

3. Factors of $4x^2$: $4x^2 = 4x \cdot x$, $4x^2 = 2x \cdot 2x$
Factors of -3: $-3 = -1 \cdot 3$, $-3 = 1 \cdot -3$
Try possible combinations.
Factored form: $4x^2 + 11x - 3 = (4x-1)(x+3)$

4. Factors of $21x^2$: $21x^2 = 21x \cdot x$, $21x^2 = 3x \cdot 7x$
Factors of
$-2y^2$: $-2y^2 = -2y \cdot y$, $-2y^2 = 2y \cdot -y$
Try possible combinations.
Factored form:
$21x^2 + 11xy - 2y^2 = (7x-y)(3x+2y)$

5. Factors of $2x^4$: $2x^4 = 2x^2 \cdot x^2$
Factors of -7: $-7 = -7 \cdot 1$, $-7 = 7 \cdot -1$
Try possible combinations.
$2x^4 - 5x^2 - 7 = (2x^2 - 7)(x^2 + 1)$

6. $3x^3 + 17x^2 + 10x = x(3x^2 + 17x + 10)$
Factors of $3x^2$: $3x^2 = 3x \cdot x$
Factors of 10: $10 = 1 \cdot 10$, $10 = 2 \cdot 5$
Try possible combinations:
$$3x^3 + 17x^2 + 10x = x(3x^2 + 17x + 10) = x(3x+2)(x+5)$$

7. $-8x^2 + 2x + 3 = -1(8x^2 - 2x - 3) = -1(4x-3)(2x+1)$

8. $x^2 = (x)^2$ and $49 = 7^2$
Is $2 \cdot x \cdot 7 = 14x$ the middle term? Yes.
$x^2 + 14x + 49 = (x+7)^2$

9. $4x^2 = (2x)^2$ and $9y^2 = (3y)^2$
Is $2 \cdot 2x \cdot 3y = 12xy$ the middle term? No.
Try other possibilities.
$4x^2 + 20xy + 9y^2 = (2x+9y)(2x+y)$

10. $36n^4 = (6n^2)^2$ and $1 = 1^2$
Is $2 \cdot 6n^2 \cdot 1 = 12n^2$ the middle term? Yes, the opposite of the middle term.
$36n^4 - 12n^2 + 1 = (6n^2 - 1)^2$

11.
$$12x^3 - 84x^2 + 147x = 3x(4x^2 - 28x + 49) = 3x[(2x)^2 - 2 \cdot 2x \cdot 7 + 7^2] = 3x(2x-7)^2$$

Vocabulary and Readiness Check

1. A perfect square trinomial is a trinomial that is the square of a binomial.

2. The term $25y^2$ written as a square is $(5y)^2$.

3. The expression $x^2 + 10xy + 25y^2$ is called a perfect square trinomial.

4. The factorization $(x + 5y)(x + 5y)$ may also be written as $(x+5y)^2$.

5. no

6. yes

7. $64 = 8^2$

8. $9 = 3^2$

9. $121a^2 = (11a)^2$

10. $81b^2 = (9b)^2$

11. $36p^4 = (6p^2)^2$

12. $4q^4 = (2q^2)^2$

Exercise Set 6.3

1. $5x^2 + 22x + 8 = (5x + 2)(x + 4)$

3. $50x^2 + 15x - 2 = (5x + 2)(10x - 1)$

5. $25x^2 - 20x + 4 = (5x - 2)(5x - 2)$

7. $2x^2 + 13x + 15 = (2x + 3)(x + 5)$

9. $8y^2 - 17y + 9 = (y - 1)(8y - 9)$

11. $2x^2 - 9x - 5 = (2x + 1)(x - 5)$

13. $20r^2 + 27r - 8 = (4r - 1)(5r + 8)$

15. $10x^2 + 31x + 3 = (10x + 1)(x + 3)$

17. $2m^2 + 17m + 10$ is prime.

19. $6x^2 - 13xy + 5y^2 = (3x - 5y)(2x - y)$

21. $15m^2 - 16m - 15 = (3m - 5)(5m + 3)$

23. $12x^3 + 11x^2 + 2x = x(12x^2 + 11x + 2)$
$= x(3x + 2)(4x + 1)$

25. $21b^2 - 48b - 45 = 3(7b^2 - 16b - 15)$
$= 3(7b + 5)(b - 3)$

27. $7z + 12z^2 - 12 = 12z^2 + 7z - 12 = (3z + 4)(4z - 3)$

29. $6x^2y^2 - 2xy^2 - 60y^2 = 2y^2(3x^2 - x - 30)$
$= 2y^2(3x - 10)(x + 3)$

31. $4x^2 - 8x - 21 = (2x - 7)(2x + 3)$

33. $-x^2 + 2x + 24 = -1(x^2 - 2x - 24)$
$= -1(x - 6)(x + 4)$

35. $4x^3 - 9x^2 - 9x = x(4x^2 - 9x - 9)$
$= x(4x + 3)(x - 3)$

37. $24x^2 - 58x + 9 = (4x - 9)(6x - 1)$

39. $x^2 + 22x + 121 = x^2 + 2 \cdot x \cdot 11 + 11^2 = (x + 11)^2$

41. $x^2 - 16x + 64 = x^2 - 2 \cdot x \cdot 8 + 8^2 = (x - 8)^2$

43. $16a^2 - 24a + 9 = (4a)^2 - 2 \cdot 4a \cdot 3 + 3^2 = (4a - 3)^2$

45. $x^4 + 4x^2 + 4 = (x^2)^2 + 2 \cdot x^2 \cdot 2 + 2^2 = (x^2 + 2)^2$

47. $2n^2 - 28n + 98 = 2(n^2 - 14n + 49)$
$= 2(n^2 - 2 \cdot n \cdot 7 + 7^2)$
$= 2(n - 7)^2$

49. $16y^2 + 40y + 25 = (4y)^2 + 2 \cdot 4y \cdot 5 + 5^2$
$= (4y + 5)^2$

51. $2x^2 - 7x - 99 = (2x + 11)(x - 9)$

53. $24x^2 + 41x + 12 = (8x + 3)(3x + 4)$

55. $3a^2 + 10ab + 3b^2 = (3a + b)(a + 3b)$

57. $-9x + 20 + x^2 = x^2 - 9x + 20 = (x - 4)(x - 5)$

59. $p^2 + 12pq + 36q^2 = p^2 + 2 \cdot p \cdot 6q + (6q)^2$
$= (p + 6q)^2$

61. $x^2y^2 - 10xy + 25 = (xy)^2 - 2 \cdot xy \cdot 5 + 5^2$
$= (xy - 5)^2$

63. $40a^2b + 9ab - 9b = b(40a^2 + 9a - 9)$
$= b(8a - 3)(5a + 3)$

65. $30x^3 + 38x^2 + 12x = 2x(15x^2 + 19x + 6)$
$= 2x(3x + 2)(5x + 3)$

67. $6y^3 - 8y^2 - 30y = 2y(3y^2 - 4y - 15)$
$= 2y(3y + 5)(y - 3)$

69. $10x^4 + 25x^3y - 15x^2y^2 = 5x^2(2x^2 + 5xy - 3y^2)$
$= 5x^2(2x - y)(x + 3y)$

71. $-14x^2 + 39x - 10 = -1(14x^2 - 39x + 10)$
$= -1(2x - 5)(7x - 2)$

73. $16p^4 - 40p^3 + 25p^2 = p^2(16p^2 - 40p + 25)$
$= p^2[(4p)^2 - 2 \cdot 4p \cdot 5 + 5^2]$
$= p^2(4p - 5)^2$

75. $x + 3x^2 - 2 = 3x^2 + x - 2 = (3x - 2)(x + 1)$

77. $8x^2 + 6xy - 27y^2 = (4x + 9y)(2x - 3y)$

79. $1 + 6x^2 + x^4 = x^4 + 6x^2 + 1$ is prime.

81. $9x^2 - 24xy + 16y^2 = (3x)^2 - 2 \cdot 3x \cdot 4y + (4y)^2$
$= (3x - 4y)^2$

83. $18x^2 - 9x - 14 = (6x - 7)(3x + 2)$

85. $-27t + 7t^2 - 4 = 7t^2 - 27t - 4 = (7t + 1)(t - 4)$

87. $49p^2 - 7p - 2 = (7p + 1)(7p - 2)$

89. $m^3 + 18m^2 + 81m = m(m^2 + 18m + 81)$
$= m(m^2 + 2 \cdot m \cdot 9 + 9^2)$
$= m(m + 9)^2$

91. $5x^2y^2 + 20xy + 1$ is prime.

93. $6a^5 + 37a^3b^2 + 6ab^4 = a(6a^4 + 37a^2b^2 + 6b^4)$
$= a(6a^2 + b^2)(a^2 + 6b^2)$

95. $(x - 2)(x + 2) = x^2 + 2x - 2x - 4 = x^2 - 4$

97. $(a + 3)(a^2 - 3a + 9)$
$= a^3 - 3a^2 + 9a + 3a^2 - 9a + 27$
$= a^3 + 27$

99. Look for the tallest graph. The income range is \$75,000 and above.

101. Answers may vary

103. no

105. Answers may vary

107. $P = (3x^2 + 1) + (6x + 4) + (x^2 + 15x)$
$= 3x^2 + 1 + 6x + 4 + x^2 + 15x$
$= 4x^2 + 21x + 5$
$= (4x + 1)(x + 5)$

109. $4x^2 + 2x + \frac{1}{4} = (2x)^2 + 2 \cdot 2x \cdot \frac{1}{2} + \left(\frac{1}{2}\right)^2$
$= \left(2x + \frac{1}{2}\right)^2$

111. $4x^2(y - 1)^2 + 10x(y - 1)^2 + 25(y - 1)^2$
$= (y - 1)^2(4x^2 + 10x + 25)$

113. $16 = 4^2$; $2 \cdot x \cdot 4 = 8x$; 8

115. $(a + b)^2 = a^2 + 2ab + b^2$

117. $b = 2$: $3x^2 + 2x - 5 = (3x + 5)(x - 1)$
$b = 14$: $3x^2 + 14x - 5 = (3x - 1)(x + 5)$

119. $c = 2$: $5x^2 + 7x + 2 = (5x + 2)(x + 1)$

121. $-12x^3y^2 + 3x^2y^2 + 15xy^2$
$= -3xy^2(4x^2 - x - 5)$
$= -3xy^2(4x - 5)(x + 1)$

123. $4x^2(y - 1)^2 + 20x(y - 1)^2 + 25(y - 1)^2$
$= (y - 1)^2(4x^2 + 20x + 25)$
$= (y - 1)^2[(2x)^2 + 2 \cdot 2x \cdot 5 + 5^2]$
$= (y - 1)^2(2x + 5)^2$

125. $3x^{2n} + 17x^n + 10 = (3x^n + 2)(x^n + 5)$

127. Answers may vary

Section 6.4

Practice Exercises

1.

Factors of $ac = 60$	Sum of Factors
1, 60	61 ← correct sum $b = 61$.
2, 30	32
3, 20	23
4, 15	19
5, 12	17
6, 10	16

$$\begin{aligned} 5x^2 + 61x + 12 &= 5x^2 + 1x + 60x + 12 \\ &= x(5x+1) + 12(5x+1) \\ &= (5x+1)(x+12) \end{aligned}$$

2.

Factors of $ac = 60$	Sum of Factors
−1, −60	−61
−2, −30	−32
−3, −20	−23
−4, −15	−19 ← Correct sum $b = -19$
−5, −12	−17
−6, −10	−60

$$\begin{aligned} 12x^2 - 19x + 5 &= 12x^2 - 15x - 4x + 5 \\ &= 3x(4x-5) - 1(4x-5) \\ &= (4x-5)(3x-1) \end{aligned}$$

3. $30x^2 - 14x - 4 = 2(15x^2 - 7x - 2)$

Find two numbers whose product is $ac = 15(-2) = -30$ and whose sum is b, −7. The numbers are −10 and 3.

$$\begin{aligned} 2(15x^2 - 7x - 2) &= 2(15x^2 - 10x + 3x - 2) \\ &= 2[5x(3x-2) + 1(3x-2)] \\ &= 2(3x-2)(5x+1) \end{aligned}$$

4. $40m^4 + 5m^3 - 35m^2 = 5m^2(8m^2 + m - 7)$

Find two numbers whose product is $ac = 8(-7) = -56$ and whose sum is b, 1. The numbers are 8 and −7.

$$\begin{aligned} 5m^2(8m^2 + m - 7) &= 5m^2(8m^2 + 8m - 7m - 7) \\ &= 5m^2[8m(m+1) - 7(m+1)] \\ &= 5m^2(m+1)(8m-7) \end{aligned}$$

5. Find two numbers whose product is $ac = 16 \cdot 9 = 144$ and whose sum is b, 24. The numbers are 12 and 12.

$$\begin{aligned} 16x^2 + 24x + 9 &= 16x^2 + 12x + 12x + 9 \\ &= 4x(4x+3) + 3(4x+3) \\ &= (4x+3)(4x+3) \\ &= (4x+3)^2 \end{aligned}$$

Exercise Set 6.4

1. $x^2 + 3x + 2x + 6 = x(x+3) + 2(x+3)$
$= (x+3)(x+2)$

3. $y^2 + 8y - 2y - 16 = y(y+8) - 2(y+8)$
$= (y+8)(y-2)$

5. $8x^2 - 5x - 24x + 15 = x(8x-5) - 3(8x-5)$
$= (8x-5)(x-3)$

7. $5x^4 - 3x^2 + 25x^2 - 15 = x^2(5x^2-3) + 5(5x^2-3)$
$= (5x^2-3)(x^2+5)$

9. a. $9 \cdot 2 = 18; 9 + 2 = 11; 9, 2$

b. $11x = 9x + 2x$

c.
$$\begin{aligned} 6x^2 + 11x + 3 &= 6x^2 + 9x + 2x + 3 \\ &= 3x(2x+3) + 1(2x+3) \\ &= (3x+1)(2x+3) \end{aligned}$$

11. a. $-20 \cdot (-3) = 60; -20 + (-3) = -23; -20, -3$

b. $-23x = -20x - 3x$

c.
$$\begin{aligned} 15x^2 - 23x + 4 &= 15x^2 - 20x - 3x + 4 \\ &= 5x(3x-4) - 1(3x-4) \\ &= (3x-4)(5x-1) \end{aligned}$$

13. $ac = 21 \cdot 2 = 42$; $b = 17$; two numbers: 14, 3

$$\begin{aligned}21y^2 + 17y + 2 &= 21y^2 + 14y + 3y + 2\\ &= 7y(3y+2) + 1(3y+2)\\ &= (3y+2)(7y+1)\end{aligned}$$

15. $ac = 7 \cdot (-11) = -77$; $b = -4$;
two numbers: -11, 7

$$\begin{aligned}7x^2 - 4x - 11 &= 7x^2 - 11x + 7x - 11\\ &= x(7x-11) + 1(7x-11)\\ &= (7x-11)(x+1)\end{aligned}$$

17. $ac = 10 \cdot 2 = 20$; $b = -9$; two numbers: -4, -5

$$\begin{aligned}10x^2 - 9x + 2 &= 10x^2 - 4x - 5x + 2\\ &= 2x(5x-2) - 1(5x-2)\\ &= (5x-2)(2x-1)\end{aligned}$$

19. $ac = 2 \cdot 5 = 10$; $b = -7$; two numbers: -5, -2

$$\begin{aligned}2x^2 - 7x + 5 &= 2x^2 - 5x - 2x + 5\\ &= x(2x-5) - 1(2x-5)\\ &= (2x-5)(x-1)\end{aligned}$$

21. $12x + 4x^2 + 9 = 4x^2 + 12x + 9$
$ac = 4 \cdot 9 = 36$; $b = 12$; two numbers: 6, 6

$$\begin{aligned}4x^2 + 12x + 9 &= 4x^2 + 6x + 6x + 9\\ &= 2x(2x+3) + 3(2x+3)\\ &= (2x+3)(2x+3)\\ &= (2x+3)^2\end{aligned}$$

23. $ac = 4 \cdot (-21) = -84$; $b = -8$;
two numbers: 6, -14

$$\begin{aligned}4x^2 - 8x - 21 &= 4x^2 + 6x - 14x - 21\\ &= 2x(2x+3) - 7(2x+3)\\ &= (2x+3)(2x-7)\end{aligned}$$

25. $ac = 10 \cdot 12 = 120$; $b = -23$;
two numbers: -8, -15

$$\begin{aligned}10x^2 - 23x + 12 &= 10x^2 - 8x - 15x + 12\\ &= 2x(5x-4) - 3(5x-4)\\ &= (5x-4)(2x-3)\end{aligned}$$

27. $2x^3 + 13x^2 + 15x = x(2x^2 + 13x + 15)$
$ac = 2 \cdot 15 = 30$; $b = 13$; two numbers: 3, 10

$$\begin{aligned}x(2x^2 + 13x + 15) &= x(2x^2 + 3x + 10x + 15)\\ &= x[x(2x+3) + 5(2x+3)]\\ &= x(2x+3)(x+5)\end{aligned}$$

29. $16y^2 - 34y + 18 = 2(8y^2 - 17y + 9)$
$ac = 8(9) = 72$; $b = -17$; two numbers: -9, -8

$$\begin{aligned}2(8y^2 - 17y + 9) &= 2(8y^2 - 9y - 8y + 9)\\ &= 2[y(8y-9) - 1(8y-9)]\\ &= 2(8y-9)(y-1)\end{aligned}$$

31. $-13x + 6 + 6x^2 = 6x^2 - 13x + 6$
$ac = 6 \cdot 6 = 36$; $b = -13$; two numbers: -9, -4

$$\begin{aligned}6x^2 - 13x + 6 &= 6x^2 - 9x - 4x + 6\\ &= 3x(2x-3) - 2(2x-3)\\ &= (2x-3)(3x-2)\end{aligned}$$

33. $54a^2 - 9a - 30 = 3(18a^2 - 3a - 10)$
$ac = 18(-10) = -180$; $b = -3$;
two numbers: 12, -15

$$\begin{aligned}3(18a^2 - 3a - 10) &= 3(18a^2 + 12a - 15a - 10)\\ &= 3[6a(3a+2) - 5(3a+2)]\\ &= 3(3a+2)(6a-5)\end{aligned}$$

35. $20a^3 + 37a^2 + 8a = a(20a^2 + 37a + 8)$
$ac = 20(8) = 160$; $b = 37$; two numbers: 5, 32

$$\begin{aligned}a(20a^2 + 37a + 8) &= a(20a^2 + 5a + 32a + 8)\\ &= a[5a(4a+1) + 8(4a+1)]\\ &= a(4a+1)(5a+8)\end{aligned}$$

37. $12x^3 - 27x^2 - 27x = 3x(4x^2 - 9x - 9)$
$ac = 4(-9) = -36$; $b = -9$; two numbers: 3, -12

$$\begin{aligned}3x(4x^2 - 9x - 9) &= 3x(4x^2 + 3x - 12x - 9)\\ &= 3x[x(4x+3) - 3(4x+3)]\\ &= 3x(4x+3)(x-3)\end{aligned}$$

39. $3x^2y + 4xy^2 + y^3 = y(3x^2 + 4xy + y^2)$
$ac = 3 \cdot 1 = 3$; $b = 4$; two numbers: 1, 3

$$\begin{aligned}y(3x^2 + 4xy + y^2) &= y(3x^2 + xy + 3xy + y^2)\\ &= y[x(3x+y) + y(3x+y)]\\ &= y(3x+y)(x+y)\end{aligned}$$

41. $ac = 20 \cdot 1 = 20$; $b = 7$; there are no two numbers.
$20z^2 + 7z + 1$ is prime.

43. $5x^2+50xy+125y^2=5(x^2+10xy+25y^2)$
$ac=1\cdot 25=25$; $b=10$; two numbers: 5, 5
$$5(x^2+10xy+25y^2)=5(x^2+5xy+5xy+25y^2)$$
$$=5[x(x+5y)+5y(x+5y)]$$
$$=5(x+5y)(x+5y)$$
$$=5(x+5y)^2$$

45. $24a^2-6ab-30b^2=6(4a^2-ab-5b^2)$
$ac=4\cdot(-5)=-20$; $b=-1$; two numbers: 4, −5
$$6(4a^2-ab-5b^2)=6(4a^2+4ab-5ab-5b^2)$$
$$=6[4a(a+b)-5b(a+b)]$$
$$=6(a+b)(4a-5b)$$

47. $15p^4+31p^3q+2p^2q^2=p^2(15p^2+31pq+2q^2)$
$ac=15(2)=30$; $b=31$; two numbers: 1, 30
$$p^2(15p^2+31pq+2q^2)$$
$$=p^2(15p^2+pq+30pq+2q^2)$$
$$=p^2[p(15p+q)+2q(15p+q)]$$
$$=p^2(15p+q)(p+2q)$$

49. $162a^4-72a^2+8=2(81a^4-36a^2+4)$
$ac=81\cdot 4=324$; $b=-36$;
two numbers: −18, −18
$$2(81a^4-36a^2+4)$$
$$=2(81a^4-18a^2-18a^2+4)$$
$$=2[9a^2(9a^2-2)-2(9a^2-2)]$$
$$=2(9a^2-2)(9a^2-2)$$
$$=2(9a^2-2)^2$$

51. $35+12x+x^2=x^2+12x+35$
$ac=1\cdot 35=35$; $b=12$; two numbers: 5, 7
$$x^2+12x+35=x^2+5x+7x+35$$
$$=x(x+5)+7(x+5)$$
$$=(x+5)(x+7)$$

53. $6-11x+5x^2=5x^2-11x+6$
$ac=5\cdot 6=30$; $b=-11$; two numbers: −6, −5
$$5x^2-11x+6=5x^2-6x-5x+6$$
$$=x(5x-6)-1(5x-6)$$
$$=(5x-6)(x-1)$$

55. $(x-2)(x+2)=x^2-2^2=x^2-4$

57. $(y+4)(y+4)=y^2+2\cdot y\cdot 4+4^2=y^2+8y+16$

59. $(9z+5)(9z-5)=(9z)^2-5^2=81z^2-25$

61. $(x-3)(x^2+3x+9)=x^3-3^3=x^3-27$

63. $5(2x^2+9x+9)=10x^2+45x+45$
$ac=2\cdot 9=18$; $b=9$; two numbers: 3, 6
$$5(2x^2+9x+9)=5(2x^2+3x+6x+9)$$
$$=5[x(2x+3)+3(2x+3)]$$
$$=15(2x+3)(x+3)$$

65. $x^{2n}+2x^n+3x^n+6=x^n(x^n+2)+3(x^n+2)$
$$=(x^n+2)(x^n+3)$$

67. $ac=3\cdot(-35)=-105$; $b=16$;
two numbers: −5, 21
$$3x^{2n}+16x^n-35=3x^{2n}-5x^n+21x^n-35$$
$$=x^n(3x^n-5)+7(3x^n-5)$$
$$=(3x^n-5)(x^n+7)$$

69. Answers may vary

Section 6.5

Practice Exercises

1. $x^2-81=x^2-9^2=(x+9)(x-9)$

2. a. $9x^2-1=(3x)^2-1^2=(3x+1)(3x-1)$

b. $36a^2-49b^2=(6a)^2-(7b)^2$
$$=(6a+7b)(6a-7b)$$

c. $p^2-\dfrac{25}{36}=p^2-\left(\dfrac{5}{6}\right)^2=\left(p+\dfrac{5}{6}\right)\left(p-\dfrac{5}{6}\right)$

3. $p^4-q^{10}=(p^2)^2-(q^5)^2=(p^2+q^5)(p^2-q^5)$

4. a. $z^4-81=(z^2)^2-9^2$
$$=(z^2+9)(z^2-9)$$
$$=(z^2+9)(z+3)(z-3)$$

b. m^2+49 is a prime polynomial.

5. $36y^3-25y=y(36y^2-25)$
$$=y[(6y)^2-5^2]$$
$$=y(6y+5)(6y-5)$$

6. $80y^4 - 5 = 5(16y^2 - 1)$
$= 5[(4y)^2 - 1^2]$
$= 5(4y+1)(4y-1)$

7. $-9x^2 + 100 = -1(9x^2 - 100)$
$= -1[(3x)^2 - 10^2]$
$= -1(3x+10)(3x-10)$

8. $x^3 + 64 = x^3 + 4^3$
$= (x+4)(x^2 - x \cdot 4 + 4^2)$
$= (x+4)(x^2 - 4x + 16)$

9. $x^3 - 125 = x^3 - 5^3$
$= (x-5)(x^2 + x \cdot 5 + 5^2)$
$= (x-5)(x^2 + 5x + 25)$

10. $27y^3 + 1 = (3y)^3 + 1^3$
$= (3y+1)[(3y)^2 - 3y \cdot 1 + 1^2]$
$= (3y+1)(9y^2 - 3y + 1)$

11. $32x^3 - 500y^3$
$= 4(8x^3 - 125y^3)$
$= 4[(2x)^3 - (5y)^3]$
$= 4(2x-5y)[(2x)^2 + 2x \cdot 5y + (5y)^2]$
$= 4(2x-5y)(4x^2 + 10xy + 25y^2)$

Calculator Explorations

x	$x^2 - 2x + 1$	$x^2 - 2x - 1$	$(x-1)^2$
5	16	14	16
−3	16	14	16
2.7	2.89	0.89	2.89
−12.1	171.61	169.61	171.61
0	1	−1	1

Vocabulary and Readiness Check

1. The expression $x^3 - 27$ is called a <u>difference of two cubes</u>.

2. The expression $x^2 - 49$ is called a <u>difference of two squares</u>.

3. The expression $z^3 + 1$ is called a <u>sum of two cubes</u>.

4. The binomial $y^2 + 9$ is prime. The statement is false.

5. $64 = 8^2$

6. $100 = 10^2$

7. $49x^2 = (7x)^2$

8. $25y^4 = (5y^2)^2$

9. $64 = 4^3$

10. $1 = 1^3$

11. $8y^3 = (2y)^3$

12. $x^6 = (x^2)^3$

Exercise Set 6.5

1. $x^2 - 4 = x^2 - 2^2 = (x+2)(x-2)$

3. $81p^2 - 1 = (9p)^2 - 1^2 = (9p+1)(9p-1)$

5. $25y^2 - 9 = (5y)^2 - 3^2 = (5y+3)(5y-3)$

7. $121m^2 - 100n^2 = (11m)^2 - (10n)^2$
$= (11m + 10n)(11m - 10n)$

9. $x^2y^2 - 1 = (xy)^2 - 1^2 = (xy+1)(xy-1)$

11. $x^2 - \frac{1}{4} = x^2 - \left(\frac{1}{2}\right)^2 = \left(x + \frac{1}{2}\right)\left(x - \frac{1}{2}\right)$

13. $-4r^2 + 1 = -1(4r^2 - 1)$
$= -1[(2r)^2 - 1^2]$
$= -1(2r+1)(2r-1)$

15. $16r^2 + 1$ is the sum of two squares, $(4r)^2 + 1^2$, not the difference of two squares. $16r^2 + 1$ is a prime polynomial.

17. $-36+x^2=-1(36-x^2)$
$=-1(6^2-x^2)$
$=-1(6+x)(6-x)$ or $(-6+x)(6+x)$

19. $m^4-1=(m^2)^2-1^2$
$=(m^2+1)(m^2-1)$
$=(m^2+1)(m+1)(m-1)$

21. $m^4-n^{18}=(m^2)^2-(n^9)^2$
$=(m^2+n^9)(m^2-n^9)$

23. $x^3+125=x^3+5^3$
$=(x+5)(x^2-x\cdot 5+5^2)$
$=(x+5)(x^2-5x+25)$

25. $8a^3-1=(2a)^3-1^3$
$=(2a-1)[(2a)^2+2a\cdot 1+1^2]$
$=(2a-1)(4a^2+2a+1)$

27. $m^3+27n^3=m^3+(3n)^3$
$=(m+3n)[m^2-m\cdot 3n+(3n)^2]$
$=(m+3n)(m^2-3mn+9n^2)$

29. $5k^3+40=5(k^3+8)$
$=5(k^3+2^3)$
$=5(k+2)[k^2-k\cdot 2+2^2]$
$=5(k+2)(k^2-2k+4)$

31. $x^3y^3-64=(xy)^3-4^3$
$=(xy-4)[(xy)^2+xy\cdot 4+4^2]$
$=(xy-4)(x^2y^2+4xy+16)$

33. $250r^3-128t^3=2(125r^3-64t^3)$
$=2[(5r)^3-(4t)^3]$
$=2(5r-4t)[(5r)^2+5r\cdot 4t+(4t)^2]$
$=2(5r-4t)(25r^2+20rt+16t^2)$

35. $r^2-64=r^2-8^2=(r+8)(r-8)$

37. $x^2-169y^2=x^2-(13y)^2=(x+13y)(x-13y)$

39. $27-t^3=3^3-t^3$
$=(3-t)(3^2+3\cdot t+t^2)$
$=(3-t)(9+3t+t^2)$

41. $18r^2-8=2(9r^2-4)$
$=2[(3r)^2-2^2]$
$=2(3r+2)(3r-2)$

43. $9xy^2-4x=x(9y^2-4)$
$=x[(3y)^2-2^2]$
$=x(3y+2)(3y-2)$

45. $8m^3+64=8(m^3+8)$
$=8(m^3+2^3)$
$=8(m+2)(m^2-m\cdot 2+2^2)$
$=8(m+2)(m^2-2m+4)$

47. $xy^3-9xyz^2=xy(y^2-9z^2)$
$=xy[y^2-(3z)^2]$
$=xy(y+3z)(y-3z)$

49. $36x^2-64y^2=4(9x^2-16y^2)$
$=4[(3x)^2-(4y)^2]$
$=4(3x+4y)(3x-4y)$

51. $144-81x^2=9(16-9x^2)$
$=9[4^2-(3x)^2]$
$=9(4+3x)(4-3x)$

53. $x^3y^3-z^6=(xy)^3-(z^2)^3$
$=(xy-z^2)[(xy)^2+xy\cdot z^2+(z^2)^2]$
$=(xy-z^2)(x^2y^2+xyz^2+z^4)$

55. $49-\frac{9}{25}m^2=7^2-\left(\frac{3}{5}m\right)^2=\left(7+\frac{3}{5}m\right)\left(7-\frac{3}{5}m\right)$

57. $t^3+343=t^3+7^3$
$=(t+7)(t^2-t\cdot 7+7^2)$
$=(t+7)(t^2-7t+49)$

59. $n^3-49n=n(n^2+49)$

61. $x^6 - 81x^2 = x^2(x^4 - 81)$
$= x^2[(x^2)^2 - 9^2]$
$= x^2(x^2 + 9)(x^2 - 9)$
$= x^2(x^2 + 9)(x + 3)(x - 3)$

63. $64p^3q - 81pq^3 = pq(64p^2 - 81q^2)$
$= pq[(8p)^2 - (9q)^2]$
$= pq(8p + 9q)(8p - 9q)$

65. $27x^2y^3 + xy^2 = xy^2(27xy + 1)$

67. $125a^4 - 64ab^3$
$= a(125a^3 - 64b^3)$
$= a[(5a)^3 - (4b)^3]$
$= a(5a - 4b)[(5a)^2 + 5a \cdot 4b + (4b)^2]$
$= a(5a - 4b)(25a^2 + 20ab + 16b^2)$

69. $16x^4 - 64x^2 = 16x^2(x^2 - 4)$
$= 16x^2(x^2 - 2^2)$
$= 16x^2(x + 2)(x - 2)$

71. $x - 6 = 0$
$x - 6 + 6 = 0 + 6$
$x = 6$

73. $2m + 4 = 0$
$2m + 4 - 4 = 0 - 4$
$2m = -4$
$\frac{2m}{2} = \frac{-4}{2}$
$m = -2$

75. $5z - 1 = 0$
$5z - 1 + 1 = 0 + 1$
$5z = 1$
$\frac{5z}{5} = \frac{1}{5}$
$z = \frac{1}{5}$

77. Let $x = 2003 - 2000 = 3$.
$-1.2x^2 + 4x + 80 = -1.2(3)^2 + 4(3) + 80$
$= -10.8 + 12 + 80$
$= 81.2$
81.2% of college students had credit cards in 2003.

79. $-1.2x^2 + 4x + 80 = -4(0.3x^2 - x - 20)$

81. $(x + 2)^2 - y^2 = (x + 2 + y)(x + 2 - y)$

83. $a^2(b - 4) - 16(b - 4) = (b - 4)(a^2 - 16)$
$= (b - 4)(a^2 - 4^2)$
$= (b - 4)(a + 4)(a - 4)$

85. $(x^2 + 6x + 9) - 4y^2 = (x + 3)^2 - 4y^2$
$= (x + 3)^2 - (2y)^2$
$= [(x + 3) + 2y][(x + 3) - 2y]$
$= (x + 3 + 2y)(x + 3 - 2y)$

87. $x^{2n} - 100 = (x^n)^2 - 10^2 = (x^n + 10)(x^n - 10)$

89. $x + 6$ since
$(x + 6)(x - 6) = x^2 - 6x + 6x - 36$
$= x^2 - 36$
$= x^2 - 6^2$

91. Answers may vary

93. **a.** Let $t = 2$.
$841 - 16t^2 = 841 - 16(2)^2$
$= 841 - 16(4)$
$= 841 - 64$
$= 777$
After 2 seconds, the height of the object is 777 feet.

b. Let $t = 5$.
$841 - 16t^2 = 841 - 16(5)^2$
$= 841 - 16(25)$
$= 841 - 400$
$= 441$
After 5 seconds the height of the object is 441 feet.

c. When the object hits the ground, its height is zero feet. Thus, to find the time, t, when the object's height is zero feet above the ground, we set the expression $841 - 16t^2$ equal to 0 and solve for t.

$$841-16t^2=0$$
$$841-16t^2+16t^2=0+16t^2$$
$$841=16t^2$$
$$\frac{841}{16}=\frac{16t^2}{16}$$
$$52.5625=t^2$$
$$\sqrt{52.5625}=\sqrt{t^2}$$
$$7.25=t$$

Thus, the object will hit the ground after approximately 7 seconds.

d. $841-16t^2=29^2-(4t)^2=(29+4t)(29-4t)$

95. a. Let $t=3$.
$1600-16t^2=1600-16(3)^2=1456$
After 3 seconds the height is 1456 feet.

b. Let $t=7$.
$1600-16t^2=1600-16(7)^2=816$
After 7 seconds the height is 816 feet.

c. When it hits the ground, the height is 0.
Let $0=1600-16t^2$.
$$16t^2=1600$$
$$t^2=100$$
$$t=\sqrt{100}$$
$$t=10$$
Thus, it will hit the ground after 10 seconds.

d.
$$1600-16t^2=16(100-t^2)$$
$$=16(10^2-t^2)$$
$$=16(10+t)(10-t)$$

Integrated Review

Practice Exercises

1. $6x^2-11x+3$
$ac=6\cdot3=18$; $b=-11$; two numbers: −2, −9
$$6x^2-11x+3=6x^2-2x-9x+3$$
$$=2x(3x-1)-3(3x-1)$$
$$=(3x-1)(2x-3)$$

2.
$$3x^3+x^2-12x-4=(3x^3+x^2)+(-12x-4)$$
$$=x^2(3x+1)-4(3x+1)$$
$$=(3x+1)(x^2-4)$$
$$=(3x+1)(x+2)(x-2)$$

3.
$$27x^2-3y^2=3(9x^2-y^2)$$
$$=3[(3x)^2-y^2]$$
$$=3(3x+y)(3x-y)$$

4.
$$8a^3+b^3=(2a)^3+b^3$$
$$=(2a+b)[(2a)^2-2a\cdot b+b^2]$$
$$=(2a+b)(4a^2-2ab+b^2)$$

5.
$$60x^3y^2-66x^2y^2-36xy^2$$
$$=6xy^2(10x^2-11x-6)$$
$$=6xy^2(5x+2)(2x-3)$$

Integrated Review

1. $x^2+2xy+y^2=(x+y)(x+y)=(x+y)^2$

2. $x^2-2xy+y^2=(x-y)(x-y)=(x-y)^2$

3. $a^2+11a-12=(a+12)(a-1)$

4. $a^2-11a+10=(a-10)(a-1)$

5. $a^2-a-6=(a-3)(a+2)$

6. $a^2-2a+1=(a-1)(a-1)=(a-1)^2$

7. $x^2+2x+1=(x+1)(x+1)=(x+1)^2$

8. $x^2+x-2=(x+2)(x-1)$

9. $x^2+4x+3=(x+3)(x+1)$

10. $x^2+x-6=(x+3)(x-2)$

11. $x^2+7x+12=(x+4)(x+3)$

12. $x^2+x-12=(x+4)(x-3)$

13. $x^2+3x-4=(x+4)(x-1)$

14. $x^2-7x+10=(x-5)(x-2)$

15. $x^2+2x-15=(x+5)(x-3)$

16. $x^2+11x+30=(x+6)(x+5)$

17. $x^2 - x - 30 = (x-6)(x+5)$

18. $x^2 + 11x + 24 = (x+8)(x+3)$

19. $2x^2 - 98 = 2(x^2 - 49)$
$= 2(x^2 - 7^2)$
$= 2(x+7)(x-7)$

20. $3x^2 - 75 = 3(x^2 - 25)$
$= 3(x^2 - 5^2)$
$= 3(x+5)(x-5)$

21. $x^2 + 3x + xy + 3y = x(x+3) + y(x+3)$
$= (x+3)(x+y)$

22. $3y - 21 + xy - 7x = 3(y-7) + x(y-7)$
$= (y-7)(3+x)$

23. $x^2 + 6x - 16 = (x+8)(x-2)$

24. $x^2 - 3x - 28 = (x-7)(x+4)$

25. $4x^3 + 20x^2 - 56x = 4x(x^2 + 5x - 14)$
$= 4x(x+7)(x-2)$

26. $6x^3 - 6x^2 - 120x = 6x(x^2 - x - 20)$
$= 6x(x-5)(x+4)$

27. $12x^2 + 34x + 24 = 2(6x^2 + 17x + 12)$
$= 2(6x^2 + 9x + 8x + 12)$
$= 2[3x(2x+3) + 4(2x+3)]$
$= 2(2x+3)(3x+4)$

28. $8a^2 + 6ab - 5b^2 = 8a^2 + 10ab - 4ab - 5b^2$
$= 2a(4a+5b) - b(4a+5b)$
$= (4a+5b)(2a-b)$

29. $4a^2 - b^2 = (2a)^2 - b^2 = (2a+b)(2a-b)$

30. $28 - 13x - 6x^2 = 28 - 21x + 8x - 6x^2$
$= 7(4-3x) + 2x(4-3x)$
$= (4-3x)(7+2x)$

31. $20 - 3x - 2x^2 = 20 - 8x + 5x - 2x^2$
$= 4(5-2x) + x(5-2x)$
$= (5-2x)(4+x)$

32. $x^2 - 2x + 4$ is a prime polynomial.

33. $a^2 + a - 3$ is a prime polynomial.

34. $6y^2 + y - 15 = 6y^2 + 10y - 9y - 15$
$= 2y(3y+5) - 3(3y+5)$
$= (3y+5)(2y-3)$

35. $4x^2 - x - 5 = 4x^2 - 5x + 4x - 5$
$= x(4x-5) + 1(4x-5)$
$= (4x-5)(x+1)$

36. $x^2y - y^3 = y(x^2 - y^2) = y(x-y)(x+y)$

37. $4t^2 + 36 = 4(t^2 + 9)$

38. $x^2 + x + xy + y = x(x+1) + y(x+1)$
$= (x+1)(x+y)$

39. $ax + 2x + a + 2 = x(a+2) + 1(a+2)$
$= (a+2)(x+1)$

40. $18x^3 - 63x^2 + 9x = 9x(2x^2 - 7x + 1)$

41. $12a^3 - 24a^2 + 4a = 4a(3a^2 - 6a + 1)$

42. $x^2 + 14x - 32 = (x+16)(x-2)$

43. $x^2 - 14x - 48$ is prime.

44. $16a^2 - 56ab + 49b^2 = (4a)^2 - 2(4a)(7b) + (7b)^2$
$= (4a - 7b)^2$

45. $25p^2 - 70pq + 49q^2 = (5p)^2 - 2(5p)(7q) + (7q)^2$
$= (5p - 7q)^2$

46. $7x^2 + 24xy + 9y^2 = 7x^2 + 3xy + 21xy + 9y^2$
$= x(7x+3y) + 3y(7x+3y)$
$= (7x+3y)(x+3y)$

47. $125 - 8y^3 = 5^3 - (2y)^3$
$= (5-2y)[5^2 + 5 \cdot 2y + (2y)^2]$
$= (5-2y)(25 + 10y + 4y^2)$

48. $64x^3+27=(4x)^3+3^3$
$=(4x+3)[(4x)^2-4x\cdot 3+3^2]$
$=(4x+3)(16x^2-12x+9)$

49. $-x^2-x+30=-1(x^2+x-30)=-(x+6)(x-5)$

50. $-x^2+6x-8=-1(x^2-6x+8)=-(x-2)(x-4)$

51. $14+5x-x^2=(7-x)(2+x)$

52. $3-2x-x^2=(3+x)(1-x)$

53. $3x^4y+6x^3y-72x^2y=3x^2y(x^2+2x-24)$
$=3x^2y(x+6)(x-4)$

54. $2x^3y+8x^2y^2-10xy^3=2xy(x^2+4xy-5y^2)$
$=2xy(x+5y)(x-y)$

55. $5x^3y^2-40x^2y^3+35xy^4=5xy^2-8xy+7y^2)$
$=5xy^2(x-7y)(x-y)$

56. $4x^4y-8x^3y-60x^2y=4x^2y(x^2-2x-15)$
$=4x^2y(x-5)(x+3)$

57. $12x^3y+243xy=3xy(4x^2+81)$

58. $6x^3y^2+8xy^2=2xy^2(3x^2+4)$

59. $4-x^2=2^2-x^2=(2+x)(2-x)$

60. $9-y^2=3^2-y^2=(3+y)(3-y)$

61. $3rs-s+12r-4=s(3r-1)+4(3r-1)$
$=(3r-1)(s+4)$

62. $x^3-2x^2+3x-6=x^2(x-2)+3(x-2)$
$=(x-2)(x^2+3)$

63. $4x^2-8xy-3x+6y=4x(x-2y)-3(x-2y)$
$=(x-2y)(4x-3)$

64. $4x^2-2xy-7yz+14xz$
$=2x(2x-y)+7z(-y+2x)$
$=(2x-y)(2x+7z)$

65. $6x^2+18xy+12y^2=6(x^2+3xy+2y^2)$
$=6(x+2)(x+y)$

66. $12x^2+46xy-8y^2=2(6x^2+23xy-4y^2)$
$=2(6x^2+24xy-xy-4y^2)$
$=2[6x(x+4y)-y(x+4y)]$
$=2(x+4y)(6x-y)$

67. $xy^2-4x+3y^2-12=x(y^2-4)+3(y^2-4)$
$=(y^2-4)(x+3)$
$=(y^2-2^2)(x+3)$
$=(y+2)(y-2)(x+3)$

68. $x^2y^2-9x^2+3y^2-27=x^2(y^2-9)+3(y^2-9)$
$=(y^2-9)(x^2+3)$
$=(y^2-3^2)(x^2+3)$
$=(y-3)(y+3)(x^2+3)$

69. $5(x+y)+x(x+y)=(x+y)(5+x)$

70. $7(x-y)+y(x-y)=(x-y)(7+y)$

71. $14t^2-9t+1=14t^2-7t-2t+1$
$=7t(2t-1)-1(2t-1)$
$=(2t-1)(7t-1)$

72. $3t^2-5t+1$ is a prime polynomial.

73. $3x^2+2x-5=3x^2+5x-3x-5$
$=x(3x+5)-1(3x+5)$
$=(3x+5)(x-1)$

74. $7x^2+19x-6=7x^2+21x-2x-6$
$=7x(x+3)-2(x+3)$
$=(x+3)(7x-2)$

75. $x^2+9xy-36y^2=(x+12y)(x-3y)$

76. $3x^2+10xy-8y^2=3x^2-2xy+12xy-8y^2$
$=x(3x-2y)+4y(3x-2y)$
$=(3x-2y)(x+4y)$

77. $1-8ab-20a^2b^2=1-10ab+2ab-20a^2b^2$
$=1(1-10ab)+2ab(1-10ab)$
$=(1-10ab)(1+2ab)$

78. $1-7ab-60a^2b^2 = 1-12ab+5ab-60a^2b^2$
$= 1(1-12ab)+5ab(1-12ab)$
$= (1-12ab)(1+5ab)$

79. $9-10x^2+x^4 = (9-x^2)(1-x^2)$
$= (3^2-x^2)(1^2-x^2)$
$= (3+x)(3-x)(1+x)(1-x)$

80. $36-13x^2+x^4 = (9-x^2)(4-x^2)$
$= (3^2-x^2)(2^2-x^2)$
$= (3+x)(3-x)(2+x)(2-x)$

81. $x^4-14x^2-32 = (x^2+2)(x^2-16)$
$= (x^2+2)(x^2-4^2)$
$= (x^2+2)(x+4)(x-4)$

82. $x^4-22x^2-75 = (x^2+3)(x^2-25)$
$= (x^2+3)(x^2-5^2)$
$= (x^2+3)(x+5)(x-5)$

83. $x^2-23x+120 = (x-15)(x-8)$

84. $y^2+22y+96 = (y+16)(y+6)$

85. $6x^3-28x^2+16x = 2x(3x^2-14x+8)$
$= 2x(3x-2)(x-4)$

86. $6y^3-8y^2-30y = 2y(3y^2-4y-15)$
$= 2y(3y+5)(y-3)$

87. $27x^3-125y^3 = (3x)^3-(5y)^3$
$= (3x-5y)[(3x)^2+3x\cdot 5y+(5y)^2]$
$= (3x-5y)(9x^2+15xy+25y^2)$

88. $216y^3-z^3 = (6y)^3-z^3$
$= (6y-z)[(6y)^2+6y\cdot z+z^2]$
$= (6y-z)(36y^2+6yz+z^2)$

89. $x^3y^3+8z^3 = (xy)^3+(2z)^3$
$= (xy+2z)[(xy)^2-xy\cdot 2z+(2z)^2]$
$= (xy+2z)(x^2y^2-2xyz+4z^2)$

90. $27a^3b^3+8 = (3ab)^3+2^3$
$= (3ab+2)[(3ab)^2-3ab\cdot 2+2^2]$
$= (3ab+2)(9a^2b^2-6ab+4)$

91. $2xy-72x^3y = 2xy(1-36x^2)$
$= 2xy[1^2-(6x)^2]$
$= 2xy(1+6x)(1-6x)$

92. $2x^3-18x = 2x(x^2-9)$
$= 2x(x^2-3^2)$
$= 2x(x+3)(x-3)$

93. $x^3+6x^2-4x-24 = x^2(x+6)-4(x+6)$
$= (x+6)(x^2-4)$
$= (x+6)(x^2-2^2)$
$= (x+6)(x+2)(x-2)$

94. $x^3-2x^2-36x+72 = x^2(x-2)-36(x-2)$
$= (x-2)(x^2-36)$
$= (x-2)(x^2-6^2)$
$= (x-2)(x+6)(x-6)$

95. $6a^3+10a^2 = 2a^2(3a+5)$

96. $4n^2-6n = 2n(2n-3)$

97. $a^2(a+2)+2(a+2) = (a+2)(a^2+2)$

98. $a-b+x(a-b) = (a-b)(1+x)$

99. $x^3-28+7x^2-4x = x^3+7x^2-28-4x$
$= x^2(x+7)-4(7+x)$
$= (x+7)(x^2-4)$
$= (x+7)(x^2-2^2)$
$= (x+7)(x+2)(x-2)$

100. $a^3-45-9a+5a^2 = a^3+5a^2-9a-45$
$= a^2(a+5)-9(a+5)$
$= (a+5)(a^2-9)$
$= (a+5)(a^2-3^2)$
$= (a+5)(a+3)(a-3)$

101. $(x-y)^2-z^2 = (x-y+z)(x-y-z)$

102. $(x+2y)^2-9=(x+2y)^2-3^2$
$=(x+2y+3)(x+2y-3)$

103. $81-(5x+1)^2=9^2-(5x+1)^2$
$=[9+(5x+1)][9-(5x+1)]$
$=(9+5x+1)(9-5x-1)$

104. $b^2-(4a+c)^2$
$=[b+(4a+c)][b-(4a+c)]$
$=(b+4a+c)(b-4a-c)$

105. Answers may vary

106. Yes; $9x^2+81y^2=9(x^2+9y^2)$

107. a, c

Section 6.6

Practice Exercises

1. $(x+4)(x-5)=0$
$x+4=0$ or $x-5=0$
$x=-4$ $\quad x=5$
Check:
Let $x=-4$.
$(x+4)(x-5)=0$
$(-4+4)(-4-5)\stackrel{?}{=}0$
$0(-9)=0$ True
Let $x=5$.
$(x+4)(x-5)=0$
$(5+4)(5-5)\stackrel{?}{=}0$
$9(0)=0$ True
The solutions are –4 and 5.

2. $x(7x-6)=0$
$x=0$ or $7x-6=0$
$7x=6$
$x=\frac{6}{7}$
Check:
Let $x=0$.
$x(7x-6)=0$
$0(7\cdot 0-6)\stackrel{?}{=}0$
$0(-6)=0$ True
Let $x=\frac{6}{7}$.
$x(7x-6)=0$
$\frac{6}{7}\left(7\cdot\frac{6}{7}-6\right)\stackrel{?}{=}0$
$\frac{6}{7}(6-6)\stackrel{?}{=}0$
$\frac{6}{7}(0)=0$ True
The solutions are 0 and $\frac{6}{7}$.

3. $x^2-8x-48=0$
$(x+4)(x-12)=0$
$x+4=0$ or $x-12=0$
$x=-4$ $\quad x=12$
Check:
Let $x=-4$.
$x^2-8x-48=0$
$(-4)^2-8(-4)-48\stackrel{?}{=}0$
$16+32-48\stackrel{?}{=}0$
$48-48\stackrel{?}{=}0$
$0=0$ True
Let $x=12$.
$x^2-8x-48=0$
$12^2-8\cdot 12-48\stackrel{?}{=}0$
$144-96-48\stackrel{?}{=}0$
$48-48\stackrel{?}{=}0$
$0=0$ True
The solutions are –4 and 12.

4. $9x^2-24x=-16$
$9x^2-24x+16=0$
$(3x-4)(3x-4)=0$
$3x-4=0$
$3x=4$
$x=\frac{4}{3}$
The solution is $\frac{4}{3}$.

5.
$$\begin{aligned} x(3x+7) &= 6 \\ 3x^2+7x &= 6 \\ 3x^2+7x-6 &= 0 \\ (3x-2)(x+3) &= 0 \end{aligned}$$
$3x-2=0$ or $x+3=0$
$3x=2$ $\quad x=-3$
$x=\frac{2}{3}$

The solutions are $\frac{2}{3}$ and -3.

6.
$$\begin{aligned} -3x^2-6x+72 &= 0 \\ -3(x^2+2x-24) &= 0 \\ -3(x+6)(x-4) &= 0 \end{aligned}$$
$x+6=0$ or $x-4=0$
$x=-6$ $\quad x=4$
The solutions are -6 and 4.

7.
$$\begin{aligned} 7x^3-63x &= 0 \\ 7x(x^2-9) &= 0 \\ 7x(x+3)(x-3) &= 0 \end{aligned}$$
$7x=0$ or $x+3=0$ or $x-3=0$
$x=0$ $\quad x=-3$ $\quad x=3$
The solutions are 0, -3, and 3.

8. $(3x-2)(2x^2-13x+15)=0$
$(3x-2)(2x-3)(x-5)=0$
$3x-2=0$ or $2x-3=0$ or $x-5=0$
$3x=2$ $\quad 2x=3$ $\quad x=5$
$x=\frac{2}{3}$ $\quad x=\frac{3}{2}$

The solutions are $\frac{2}{3}, \frac{3}{2}$, and 5.

9. $5x^3+5x^2-30x=0$
$5x(x^2+x-6)=0$
$5x(x+3)(x-2)=0$
$5x=0$ or $x+3=0$ or $x-2=0$
$x=0$ $\quad x=-3$ $\quad x=2$
The solutions are 0, -3, and 2.

10. $y=x^2-6x+8$
$0=x^2-6x+8$
$0=(x-4)(x-2)$
$x-4=0$ or $x-2=0$
$x=4$ $\quad x=2$
The x-intercepts of the graph of $y=x^2-6x+8$ are (2, 0) and (4, 0).

Calculator Explorations

1. -0.9, 2.2

2. -2.5, 3.5

3. no real solution

4. no real solution

5. -1.8, 2.8

6. -0.9, 0.3

Vocabulary and Readiness Check

1. An equation that can be written in the form $ax^2+bx+c=0$, (with $a \neq 0$), is called a quadratic equation.

2. If the product of two numbers is 0, then at least one of the numbers must be 0.

3. The solutions to $(x-3)(x+5)=0$ are 3, -5.

4. If $a \cdot b = 0$, then $a=0$ or $b=0$.

5. 3, 7

6. 5, 2

7. -8, -6

8. -2, -3

9. -1, 3

10. 1, -2

Exercise Set 6.6

1. $(x-2)(x+1)=0$
$x-2=0$ or $x+1=0$
$x=2$ $\quad x=-1$
The solutions are 2 and -1.

3. $(x+9)(x+17)=0$
$x+9=0$ or $x+17=0$
$x=-9$ $\quad x=-17$
The solutions are -9 and -17.

5. $x(x+6)=0$
$x=0$ or $x+6=0$
$x=-6$
The solutions are 0 and -6.

7. $3x(x-8)=0$
$3x=0$ or $x-8=0$
$x=0$ $\quad x=8$
The solutions are 0 and 8.

9. $(2x+3)(4x-5)=0$
$2x+3=0$ or $4x-5=0$
$2x=-3$ $\quad 4x=5$
$x=-\frac{3}{2}$ $\quad x=\frac{5}{4}$
The solutions are $-\frac{3}{2}$ and $\frac{5}{4}$.

11. $(2x-7)(7x+2)=0$
$2x-7=0$ or $7x+2=0$
$2x=7$ $\quad 7x=-2$
$x=\frac{7}{2}$ $\quad x=-\frac{2}{7}$
The solutions are $\frac{7}{2}$ and $-\frac{2}{7}$.

13. $\left(x-\frac{1}{2}\right)\left(x+\frac{1}{3}\right)=0$
$x-\frac{1}{2}=0$ or $x+\frac{1}{3}=0$
$x=\frac{1}{2}$ $\quad x=-\frac{1}{3}$
The solutions are $\frac{1}{2}$ and $-\frac{1}{3}$.

15. $(x+0.2)(x+1.5)=0$
$x+0.2=0$ or $x+1.5=0$
$x=-0.2$ $\quad x=-1.5$
The solutions are –0.2 and –1.5

17. Answers may vary. Possible answer:
If $x=6$ and $x=-1$ are the solutions, then
$x=6$ or $x=-1$
$x-6=0$ $\quad x+1=0$
$(x-6)(x+1)=0$

19. $x^2-13x+36=0$
$(x-9)(x-4)=0$
$x-9=0$ or $x-4=0$
$x=9$ $\quad x=4$
The solutions are 9 and 4.

21. $x^2+2x-8=0$
$(x+4)(x-2)=0$
$x+4=0$ or $x-2=0$
$x=-4$ $\quad x=2$
The solutions are –4 and 2.

23. $x^2-7x=0$
$x(x-7)=0$
$x=0$ or $x-7=0$
$x=7$
The solutions are 0 and 7.

25. $x^2-4x=32$
$x^2-4x-32=0$
$(x-8)(x+4)=0$
$x-8=0$ or $x+4=0$
$x=8$ $\quad x=-4$
The solutions are 8 and –4.

27. $x^2=16$
$x^2-16=0$
$(x+4)(x-4)=0$
$x+4=0$ or $x-4=0$
$x=-4$ $\quad x=4$
The solutions are –4 and 4.

29. $(x+4)(x-9)=4x$
$x^2-5x-36=4x$
$x^2-9x-36=0$
$(x-12)(x+3)=0$
$x-12=0$ or $x+3=0$
$x=12$ $\quad x=-3$
The solutions are 12 and –3.

31. $x(3x-1)=14$
$3x^2-x=14$
$3x^2-x-14=0$
$(3x-7)(x+2)=0$
$3x-7=0$ or $x+2=0$
$3x=7$ $\quad x=-2$
$x=\frac{7}{3}$
The solutions are $\frac{7}{3}$ and –2.

33.
$$-3x^2+75=0$$
$$-3(x^2-25)=0$$
$$-3(x+5)(x-5)=0$$
$$x+5=0 \quad \text{or} \quad x-5=0$$
$$x=-5 \qquad\qquad x=5$$
The solutions are –5 and 5.

35.
$$24x^2+44x=8$$
$$24x^2+44x-8=0$$
$$4(6x^2+11x-2)=0$$
$$4(6x-1)(x+2)=0$$
$$6x-1=0 \quad \text{or} \quad x+2=0$$
$$6x=1 \qquad\qquad x=-2$$
$$x=\frac{1}{6}$$
The solutions are $\frac{1}{6}$ and –2.

37.
$$x^3-12x^2+32x=0$$
$$x(x^2-12x+32)=0$$
$$x(x-8)(x-4)=0$$
$$x=0 \quad \text{or} \quad x-8=0 \quad \text{or} \quad x-4=0$$
$$x=8 \qquad\qquad x=4$$
The solutions are 0, 8, and 4.

39.
$$(4x-3)(16x^2-24x+9)=0$$
$$(4x-3)(4x-3)^2=0$$
$$(4x-3)^3=0$$
$$4x-3=0$$
$$4x=3$$
$$x=\frac{3}{4}$$
The solution is $\frac{3}{4}$.

41.
$$4x^3-x=0$$
$$x(4x^2-1)=0$$
$$x(2x+1)(2x-1)=0$$
$$x=0 \quad \text{or} \quad 2x+1=0 \quad \text{or} \quad 2x-1=0$$
$$2x=-1 \qquad\qquad 2x=1$$
$$x=-\frac{1}{2} \qquad\qquad x=\frac{1}{2}$$
The solutions are 0, $-\frac{1}{2}$, and $\frac{1}{2}$.

43.
$$32x^3-4x^2-6x=0$$
$$2x(16x^2-2x-3)=0$$
$$2x(2x-1)(8x+3)=0$$
$$2x=0 \quad \text{or} \quad 2x-1=0 \quad \text{or} \quad 8x+3=0$$
$$x=0 \qquad 2x=1 \qquad 8x=-3$$
$$x=\frac{1}{2} \qquad\qquad x=-\frac{3}{8}$$
The solutions are 0, $\frac{1}{2}$, and $-\frac{3}{8}$.

45.
$$(x+3)(x-2)=0$$
$$x+3=0 \quad \text{or} \quad x-2=0$$
$$x=-3 \qquad\qquad x=2$$
The solutions are –3 and 2.

47.
$$x^2+20x=0$$
$$x(x+20)=0$$
$$x=0 \quad \text{or} \quad x+20=0$$
$$x=-20$$
The solutions are 0 and –20.

49.
$$4(x-7)=6$$
$$4x-28=6$$
$$4x=34$$
$$x=\frac{34}{4}$$
$$x=\frac{17}{2}$$
The solution is $\frac{17}{2}$.

51.
$$4y^2-1=0$$
$$(2y+1)(2y-1)=0$$
$$2y+1=0 \quad \text{or} \quad 2y-1=0$$
$$2y=-1 \qquad\qquad 2y=1$$
$$y=-\frac{1}{2} \qquad\qquad y=\frac{1}{2}$$
The solutions are $-\frac{1}{2}$ and $\frac{1}{2}$.

53.
$$(2x+3)(2x^2-5x-3)=0$$
$$(2x+3)(2x+1)(x-3)=0$$
$$2x+3=0 \quad \text{or} \quad 2x+1=0 \quad \text{or} \quad x-3=0$$
$$2x=-3 \qquad 2x=-1 \qquad x=3$$
$$x=-\frac{3}{2} \qquad\qquad x=-\frac{1}{2}$$
The solutions are $-\frac{3}{2}$, $-\frac{1}{2}$, and 3.

55. $x^2 - 15 = -2x$
$x^2 + 2x - 15 = 0$
$(x+5)(x-3) = 0$
$x+5=0$ or $x-3=0$
$x=-5$ $\quad$ $x=3$
The solutions are −5 and 3.

57. $30x^2 - 11x - 30 = 0$
$(6x+5)(5x-6) = 0$
$6x+5=0$ or $5x-6=0$
$6x=-5$ $\quad$ $5x=6$
$x=-\frac{5}{6}$ $\quad$ $x=\frac{6}{5}$
The solutions are $-\frac{5}{6}$ and $\frac{6}{5}$.

59. $5x^2 - 6x - 8 = 0$
$(5x+4)(x-2) = 0$
$5x+4=0$ or $x-2=0$
$5x=-4$ $\quad$ $x=2$
$x=-\frac{4}{5}$
The solutions are $-\frac{4}{5}$ and 2.

61. $6y^2 - 22y - 40 = 0$
$2(3y^2 - 11y - 20) = 0$
$2(3y+4)(y-5) = 0$
$3y+4=0$ or $y-5=0$
$3y=-4$ $\quad$ $y=5$
$y=-\frac{4}{3}$
The solutions are $-\frac{4}{3}$ and 5.

63. $(y-2)(y+3) = 6$
$y^2 + y - 6 = 6$
$y^2 + y - 12 = 0$
$(y+4)(y-3) = 0$
$y+4=0$ or $y-3=0$
$y=-4$ $\quad$ $y=3$
The solutions are −4 and 3.

65. $3x^3 + 19x^2 - 72x = 0$
$x(3x^2 + 19x - 72) = 0$
$x(3x-8)(x+9) = 0$
$x=0$ or $3x-8=0$ or $x+9=0$
$3x=8$ $\quad$ $x=-9$
$x=\frac{8}{3}$
The solutions are 0, $\frac{8}{3}$, and −9.

67. $x^2 + 14x + 49 = 0$
$(x+7)^2 = 0$
$x+7=0$
$x=-7$
The solution is −7.

69. $12y = 8y^2$
$0 = 8y^2 - 12y$
$0 = 4y(2y-3)$
$4y=0$ or $2y-3=0$
$y=0$ $\quad$ $2y=3$
$y=\frac{3}{2}$
The solutions are 0 and $\frac{3}{2}$.

71. $7x^3 - 7x = 0$
$7x(x^2 - 1) = 0$
$7x(x+1)(x-1) = 0$
$7x=0$ or $x+1=0$ or $x-1=0$
$x=0$ $\quad$ $x=-1$ $\quad$ $x=1$
The solutions are 0, −1, and 1.

73. $3x^2 + 8x - 11 = 13 - 6x$
$3x^2 + 14x - 24 = 0$
$(3x-4)(x+6) = 0$
$3x-4=0$ or $x+6=0$
$3x=4$ $\quad$ $x=-6$
$x=\frac{4}{3}$
The solutions are $\frac{4}{3}$ and −6.

75. $3x^2 - 20x = -4x^2 - 7x - 6$

$$7x^2 - 13x + 6 = 0$$
$$(7x - 6)(x - 1) = 0$$
$$7x - 6 = 0 \quad \text{or} \quad x - 1 = 0$$
$$7x = 6 \qquad x = 1$$
$$x = \frac{6}{7}$$

The solutions are $\frac{6}{7}$ and 1.

77. Let $y = 0$ and solve for x.

$$y = (3x + 4)(x - 1)$$
$$0 = (3x + 4)(x - 1)$$
$$3x + 4 = 0 \quad \text{or} \quad x - 1 = 0$$
$$3x = -4 \qquad x = 1$$
$$x = -\frac{4}{3}$$

The intercepts are $\left(-\frac{4}{3}, 0\right)$ and (1, 0).

79. Let $y = 0$ and solve for x.

$$y = x^2 - 3x - 10$$
$$0 = x^2 - 3x - 10$$
$$0 = (x - 5)(x + 2)$$
$$x - 5 = 0 \quad \text{or} \quad x + 2 = 0$$
$$x = 5 \qquad x = -2$$

The x-intercepts are (5, 0) and (–2, 0).

81. Let $y = 0$ and solve for x.

$$y = 2x^2 + 11x - 6$$
$$0 = 2x^2 + 11x - 6$$
$$0 = (2x - 1)(x + 6)$$
$$2x - 1 = 0 \quad \text{or} \quad x + 6 = 0$$
$$2x = 1 \qquad x = -6$$
$$x = \frac{1}{2}$$

The x-intercepts are $\left(\frac{1}{2}, 0\right)$ and (–6, 0).

83. e; x-intercepts are (–2, 0), (1, 0)

85. b; x-intercepts are (0, 0), (–3, 0)

87. c; $y = 2x^2 - 8 = 2(x - 2)(x + 2)$

x-intercepts are (2, 0), (–2, 0).

89.
$$\frac{3}{5} + \frac{4}{9} = \frac{3 \cdot 9}{5 \cdot 9} + \frac{4 \cdot 5}{9 \cdot 5} = \frac{27}{45} + \frac{20}{45} = \frac{27 + 20}{45} = \frac{47}{45}$$

91.
$$\frac{7}{10} - \frac{5}{12} = \frac{7 \cdot 6}{10 \cdot 6} - \frac{5 \cdot 5}{12 \cdot 5} = \frac{42}{60} - \frac{25}{60} = \frac{42 - 25}{60} = \frac{17}{60}$$

93. $\frac{7}{8} \div \frac{7}{15} = \frac{7}{8} \cdot \frac{15}{7} = \frac{15}{8}$

95. $\frac{4}{5} \cdot \frac{7}{8} = \frac{4 \cdot 7}{5 \cdot 8} = \frac{4 \cdot 7}{5 \cdot 2 \cdot 4} = \frac{7}{10}$

97. Didn't write the equation in standard form; standard form should be:

$$x(x - 2) = 8$$
$$x^2 - 2x = 8$$
$$x^2 - 2x - 8 = 0$$
$$(x - 4)(x + 2) = 0$$
$$x - 4 = 0 \quad \text{or} \quad x + 2 = 0$$
$$x = 4 \qquad x = -2$$

99. Answers may vary. Possible answer: If the solutions are $x = 5$ and $x = 7$, then, by the zero factor property,

$$x = 5 \quad \text{or} \quad x = 7$$
$$x - 5 = 0 \qquad x - 7 = 0$$
$$(x - 5)(x - 7) = 0$$
$$x^2 - 7x - 5x + 35 = 0$$
$$x^2 - 12x + 35 = 0$$

101. $y = -16x^2 + 20x + 300$

a.

time x	0	1	2	3	4	5	6
height y	300	304	276	216	124	0	−156

b. The compass strikes the ground after 5 seconds, when the height, y, is zero feet.

c. The maximum height was approximately 304 feet.

d.

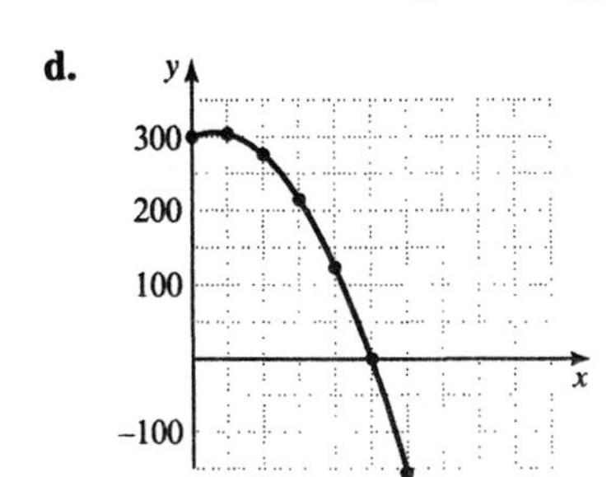

103. $(x-3)(3x+4) = (x+2)(x-6)$

$$3x^2 - 5x - 12 = x^2 - 4x - 12$$
$$2x^2 - x = 0$$
$$x(2x-1) = 0$$
$$2x - 1 = 0 \quad \text{or} \quad x = 0$$
$$x = \frac{1}{2}$$

The solutions are $\frac{1}{2}$ and 0.

105. $(2x-3)(x+8) = (x-6)(x+4)$

$$2x^2 + 13x - 24 = x^2 - 2x - 24$$
$$x^2 + 15x = 0$$
$$x(x+15) = 0$$
$$x + 15 = 0 \quad \text{or} \quad x = 0$$
$$x = -15$$

The solutions are −15 and 0.

The Bigger Picture

1. $-7 + (-27) = -34$

2. $\frac{(x^3)^4}{(x^{-2})^5} = \frac{x^{12}}{x^{-10}} = x^{12-(-10)} = x^{22}$

3. $(x^3 - 6x^2 + 2) - (5x^3 - 6)$
$= x^3 - 6x^2 + 2 - 5x^3 + 6$
$= x^3 - 5x^3 - 6x^2 + 2 + 6$
$= -4x^3 - 6x^2 + 8$

4. $\frac{3y^3 - 3y^2 + 9}{3y^2} = \frac{3y^3}{3y^2} - \frac{3y^2}{3y^2} + \frac{9}{3y^2} = y - 1 + \frac{3}{y^2}$

5. $10x^3 - 250x = 10x(x^2 - 25)$
$= 10x(x^2 - 5^2)$
$= 10x(x+5)(x-5)$

6. $x^2 - 36x + 35 = (x-1)(x-35)$

7. $6xy + 15x - 6y - 15 = 3(2xy + 5x - 2y - 5)$
$= 3[x(2y+5) - 1(2y+5)]$
$= 3(2y+5)(x-1)$

8. $5xy^2 - 2xy - 7x = x(5y^2 - 2y - 7)$
$= x(5y-7)(y+1)$

9. $(x-5)(2x+1) = 0$
$x - 5 = 0$ or $2x + 1 = 0$
$x = 5$ $\quad x = -\frac{1}{2}$
The solutions are 5 and $-\frac{1}{2}$.

10. $5x - 5 = 0$
$5x = 5$
$x = 1$
The solution is 1.

11. $x(x-12) = 28$
$x^2 - 12x = 28$
$x^2 - 12x - 28 = 0$
$(x+2)(x-14) = 0$
$x + 2 = 0$ or $x - 14 = 0$
$x = -2$ $\quad x = 14$
The solutions are −2 and 14.

12. $7(x-3) + 2(5x+1) = 14$
$7x - 21 + 10x + 2 = 14$
$17x - 19 = 14$
$17x = 33$
$x = \frac{33}{17}$
The solution is $\frac{33}{17}$.

Section 6.7

Practice Exercises

1. Find t when $h = 0$.
$h = -16t^2 + 64$
$0 = -16t^2 + 64$
$0 = -16(t^2 - 4)$
$0 = -16(t-2)(t+2)$
$t - 2 = 0$ or $t + 2 = 0$
$t = 2$ $\quad t = -2$
Since time cannot be negative, the diver will reach the pool in 2 seconds.

2. Let x = the number.
$x^2 - 8x = 48$
$x^2 - 8x - 48 = 0$
$(x-12)(x+4) = 0$
$x - 12 = 0$ or $x + 4 = 0$
$x = 12$ $\quad x = -4$
There are two numbers. They are −4 and 12.

3. Let x = height, then $3x - 1$ = base.
$A = \frac{1}{2}bh$
$210 = \frac{1}{2}(3x-1)(x)$
$420 = (3x-1)(x)$
$420 = 3x^2 - x$
$0 = 3x^2 - x - 420$
$0 = (3x+35)(x-12)$
$3x + 35 = 0$ or $x - 12 = 0$
$x = -\frac{35}{3}$ $\quad x = 12$
Since height cannot be negative, the height is 12 feet and the base is 3(12) − 1 = 35 feet.

4. Let x = first integer, then
$x + 1$ = next consecutive integer.

$$x(x+1) = x + (x+1) + 41$$
$$x^2 + x = 2x + 42$$
$$x^2 - x - 42 = 0$$
$$(x-7)(x+6) = 0$$
$$x - 7 = 0 \quad \text{or} \quad x + 6 = 0$$
$$x = 7 \qquad\qquad x = -6$$

The numbers are 7 and 8 or −6 and −5.

5. Let x = first leg, then $2x - 1$ = second leg, and $2x + 1$ = hypotenuse.

$$x^2 + (2x-1)^2 = (2x+1)^2$$
$$x^2 + 4x^2 - 4x + 1 = 4x^2 + 4x + 1$$
$$x^2 - 8x = 0$$
$$x(x-8) = 0$$
$$x = 0 \quad \text{or} \quad x - 8 = 0$$
$$x = 8$$

Since the length cannot be 0, the legs have lengths 8 units and $2(8) - 1 = 15$ units and the hypotenuse has length $2(8) + 1 = 17$ units.

Exercise Set 6.7

1. Let x = the width, then $x + 4$ = the length.

3. Let x = the first odd integer, then $x + 2$ = the next consecutive odd integer.

5. Let x = the base, then $4x + 1$ = the height.

7. Let x = the length of one side.

$$A = x^2$$
$$121 = x^2$$
$$0 = x^2 - 121$$
$$0 = x^2 - 11^2$$
$$0 = (x+11)(x-11)$$
$$x + 11 = 0 \quad \text{or} \quad x - 11 = 0$$
$$x = -11 \qquad\qquad x = 11$$

Since the length cannot be negative, the sides are 11 units long.

9. The perimeter is the sum of the lengths of the sides.

$$120 = (x+5) + (x^2 - 3x) + (3x - 8) + (x + 3)$$
$$120 = x + 5 + x^2 - 3x + 3x - 8 + x + 3$$
$$120 = x^2 + 2x$$
$$0 = x^2 + 2x - 120$$
$$x^2 + 2x - 120 = 0$$
$$(x+12)(x-10) = 0$$
$$x + 12 = 0 \quad \text{or} \quad x - 10 = 0$$
$$x = -12 \qquad\qquad x = 10$$

Since the dimensions cannot be negative, the lengths of the sides are:
$10 + 5 = 15$ cm, $10^2 - 3(10) = 70$ cm, $3(10) - 8 = 22$ cm, and $10 + 3 = 13$ cm.

11. $x + 5$ = the base and $x - 5$ = the height.

$$A = bh$$
$$96 = (x+5)(x-5)$$
$$96 = x^2 - 25$$
$$0 = x^2 - 121$$
$$x^2 - 121 = 0$$
$$(x+11)(x-11) = 0$$
$$x + 11 = 0 \quad \text{or} \quad x - 11 = 0$$
$$x = -11 \qquad\qquad x = 11$$

Since the dimensions cannot be negative, $x = 11$. The base is $11 + 5 = 16$ miles, and the height is $11 - 5 = 6$ miles.

13. Find t when $h = 0$.

$$h = -16t^2 + 64t + 80$$
$$0 = -16t^2 + 64t + 80$$
$$0 = -16(t^2 - 4t - 5)$$
$$0 = -16(t-5)(t+1)$$
$$t - 5 = 0 \quad \text{or} \quad t + 1 = 0$$
$$t = 5 \qquad\qquad t = -1$$

Since the time t cannot be negative, the object hits the ground after 5 seconds.

15. Let x = the width then $2x - 7$ = the length.

$$A = lw$$
$$30 = (2x-7)(x)$$
$$30 = 2x^2 - 7x$$
$$0 = 2x^2 - 7x - 30$$
$$0 = (2x+5)(x-6)$$

$2x+5=0$ or $x-6=0$

$x=-\frac{5}{2}$ $x=6$

Since the dimensions cannot be negative, the width is 6 cm and the length is 2(6) – 7 = 5 cm.

17. Let $n = 12$.

$D=\frac{1}{2}n(n-3)$

$D=\frac{1}{2}\cdot 12(12-3)=6(9)=54$

A polygon with 12 sides has 54 diagonals.

19. Let $D = 35$ and solve for n.

$D=\frac{1}{2}n(n-3)$

$35=\frac{1}{2}n(n-3)$

$70=n^2-3n$

$0=n^2-3n-70$

$0=(n-10)(n+7)$

$n-10=0$ or $n+7=0$

$n=10$ $n=-7$

The polygon has 10 sides.

21. Let x = the unknown number.

$x+x^2=132$

$x^2+x-132=0$

$(x+12)(x-11)=0$

$x+12=0$ or $x-11=0$

$x=-12$ $x=11$

The two numbers are –12 and 11.

23. Let x = the first room number, then $x + 1$ = next room number.

$x(x+1)=210$

$x^2+x=210$

$x^2+x-210=0$

$(x-14)(x+15)=0$

$x-14=0$ or $x+15=0$

$x=14$ $x=-15$

Since the room number is not negative, the room numbers are 14 and 15.

25. Let x = hypotenuse, then $x - 1$ = height.

$a^2+b^2=c^2$

$5^2+(x-1)^2=x^2$

$25+x^2-2x+1=x^2$

$26-2x=0$

$26=2x$

$13=x$

The length of the ladder is 13 feet.

27. Let x = the length of a side of the original square. Then $x + 3$ = the length of a side of the larger square.

$64=(x+3)^2$

$64=x^2+6x+9$

$0=x^2+6x-55$

$0=(x+11)(x-5)$

$x+11=0$ or $x-5=0$

$x=-11$ $x=5$

Since the length cannot be negative, the sides of the original square are 5 inches long.

29. Let x = the length of the shorter leg. Then $x + 4$ = the length of the longer leg and $x + 8$ = the length of the hypotenuse. By the Pythagorean theorem,

$x^2+(x+4)^2=(x+8)^2$

$x^2+x^2+8x+16=x^2+16x+64$

$x^2-8x-48=0$

$(x-12)(x+4)=0$

$x-12=0$ or $x+4=0$

$x=12$ $x=-4$

Since the length cannot be negative, the sides of the triangle are 12 mm, 12 + 4 = 16 mm, and 12 + 8 = 20 mm.

31. Let x = the height of the triangle, then $2x$ = the base.

$A=\frac{1}{2}bh$

$100=\frac{1}{2}(2x)(x)$

$100=x^2$

$0=x^2-100$

$0=(x+10)(x-10)$

$x+10=0$ or $x-10=0$

$x=-10$ $x=10$

Since the height cannot be negative, the height of the triangle is 10 km.

33. Let x = the length of the shorter leg, then $x + 12$ = the length of the longer leg and $2x - 12$ = the length of the hypotenuse.
By the Pythagorean theorem,

$$x^2 + (x+12)^2 = (2x-12)^2$$
$$x^2 + x^2 + 24x + 144 = 4x^2 - 48x + 144$$
$$0 = 2x^2 - 72x$$
$$0 = 2x(x-36)$$
$$2x = 0 \quad \text{or} \quad x - 36 = 0$$
$$x = 0 \qquad x = 36$$

Since the length cannot be zero feet, the shorter leg is 36 feet long.

35. Find t when $h = 0$.

$$h = -16t^2 + 1444$$
$$0 = -16t^2 + 1444$$
$$0 = -4(4t^2 - 361)$$
$$0 = -4(2t-19)(2t+19)$$
$$2t - 19 = 0 \quad \text{or} \quad 2t + 19 = 0$$
$$t = \frac{19}{2} \qquad t = -\frac{19}{2}$$

Since time cannot be negative, the object reaches the ground in $\frac{19}{2} = 9.5$ seconds.

37. Let $P = 100$ and $A = 144$.

$$A = P(1+r)^2$$
$$144 = 100(1+r)^2$$
$$144 = 100 + 200r + 100r^2$$
$$0 = 100r^2 + 200r - 44$$
$$0 = 4(25r^2 + 50r - 11)$$
$$0 = 4(5r-1)(5r+11)$$
$$5r - 1 = 0 \quad \text{or} \quad 5r + 11 = 0$$
$$5r = 1 \qquad 5r = -11$$
$$r = \frac{1}{5} \qquad r = -\frac{11}{5}$$
$$r = 0.2 \qquad r = -2.2$$

Since the interest rate cannot be negative $r = 0.2$ and the rate is 20%.

39. Let x = the length and $x - 7$ = the width.

$$A = lw$$
$$120 = (x-7)(x)$$
$$120 = x^2 - 7x$$
$$0 = x^2 - 7x - 120$$
$$0 = (x+8)(x-15)$$
$$x + 8 = 0 \quad \text{or} \quad x - 15 = 0$$
$$x = -8 \qquad x = 15$$

Since the length cannot be negative, the length is 15 miles. The width is $15 - 7 = 8$ miles.

41. Let $C = 9500$.

$$C = x^2 - 15x + 50$$
$$9500 = x^2 - 15x + 50$$
$$0 = x^2 - 15x - 9450$$
$$0 = (x+90)(x-105)$$
$$x + 90 = 0 \quad \text{or} \quad x - 105 = 0$$
$$x = -90 \qquad x = 105$$

Since the number of units cannot be negative the solution is 105 units.

43. In 1940, the size of the average farm was about 175 acres.

45. In 1940, there were approximately 6.25 million farms.

47. The lines appear to intersect in 1966.

49. Answers may vary

51. $\frac{24}{32} = \frac{2 \cdot 2 \cdot 2 \cdot 3}{2 \cdot 2 \cdot 2 \cdot 2 \cdot 2} = \frac{3}{4}$

53. $\frac{15}{27} = \frac{3 \cdot 5}{3 \cdot 3 \cdot 3} = \frac{5}{9}$

55. $\frac{45}{50} = \frac{3 \cdot 3 \cdot 5}{2 \cdot 5 \cdot 5} = \frac{9}{10}$

57. Let x = the length of a side of the square. Then x = the width of the rectangle and $x + 6$ = the length of the rectangle.
The area of the square is x^2. The area of the rectangle is $x(x+6) = x^2 + 6x$.

$$176 = x^2 + (x^2 + 6x)$$
$$176 = 2x^2 + 6x$$
$$0 = 2x^2 + 6x - 176$$
$$0 = 2(x^2 + 3x - 88)$$
$$0 = 2(x+11)(x-8)$$
$$x + 11 = 0 \quad \text{or} \quad x - 8 = 0$$
$$x = -11 \qquad x = 8$$

Since the length cannot be negative, the side of the square is 8 meters.

59. Let x = the first number, then $25 - x$ = the other number.

$$x^2 + (25-x)^2 = 325$$
$$x^2 + 625 - 50x + x^2 = 325$$
$$2x^2 - 50x + 625 = 325$$
$$2x^2 - 50x + 300 = 0$$
$$2(x^2 - 25x + 150) = 0$$
$$2(x-15)(x-10) = 0$$
$$x - 15 = 0 \quad \text{or} \quad x - 10 = 0$$
$$x = 15 \qquad\qquad x = 10$$

The numbers are 15 and 10.

61. Pool: width = x and length = $x + 6$
Total Area: width = $x + 8$ and length = $x + 14$

$$\text{Total area} = 576 + \text{Pool area}$$
$$(x+14)(x+8) = 576 + (x+6)(x)$$
$$x^2 + 22x + 112 = 576 + x^2 + 6x$$
$$16x + 112 = 576$$
$$16x = 464$$
$$x = 29$$
$$x + 6 = 29 + 6 = 35$$

The pool has length 35 meters and width 29 meters.

63. Answers may vary

Chapter 6 Vocabulary Check

1. An equation that can be written in the form $ax^2 + bx + c = 0$ (with a not 0) is called a quadratic equation.
2. Factoring is the process of writing an expression as a product.
3. The greatest common factor of a list of terms is the product of all common factors.
4. A trinomial that is the square of some binomial is called a perfect square trinomial.
5. The expression $a^2 - b^2$ is called a difference of two squares.
6. The expression $a^3 - b^3$ is called a difference of two cubes.
7. The expression $a^3 + b^3$ is called a sum of two cubes.
8. By the zero factor property, if the product of two numbers is 0, then at least one of the numbers must be 0.

Chapter 6 Review

1. $6x^2 - 15x = 3x(2x-5)$
2. $2x^3y - 6x^2y^2 - 8xy^3 = 2xy(x^2 - 3xy - 4y^2) = 2xy(x-4y)(x+y)$
3. $20x^2 + 12x = 4x(5x+3)$
4. $6x^2y^2 - 3xy^3 = 3xy^2(2x-y)$
5. $-8x^3y + 6x^2y^2 = -2x^2y(4x-3y)$
6. $3x(2x+3) - 5(2x+3) = (2x+3)(3x-5)$
7. $5x(x+1) - (x+1) = (x+1)(5x-1)$
8. $3x^2 - 3x + 2x - 2 = 3x(x-1) + 2(x-1) = (x-1)(3x+2)$
9. $6x^2 + 10x - 3x - 5 = 2x(3x+5) - 1(3x+5) = (3x+5)(2x-1)$
10. $3a^2 + 9ab + 3b^2 + ab = 3a(a+3b) + b(3b+a) = (a+3b)(3a+b)$
11. $x^2 + 6x + 8 = (x+4)(x+2)$
12. $x^2 - 11x + 24 = (x-8)(x-3)$
13. $x^2 + x + 2$ is a prime polynomial.
14. $x^2 - 5x - 6 = (x-6)(x+1)$
15. $x^2 + 2x - 8 = (x+4)(x-2)$
16. $x^2 + 4xy - 12y^2 = (x+6y)(x-2y)$
17. $x^2 + 8xy + 15y^2 = (x+5y)(x+3y)$
18. $3x^2y + 6xy^2 + 3y^3 = 3y(x^2 + 2xy + y^2) = 3y(x+y)(x+y) = 3y(x+y)^2$

19. $72-18x-2x^2=2(36-9x-x^2)$
$=2(3-x)(12+x)$

20. $32+12x-4x^2=4(8+3x-x^2)$

21. $2x^2+11x-6=(2x-1)(x+6)$

22. $4x^2-7x+4$ is a prime polynomial.

23. $4x^2+4x-3=4x^2+6x-2x-3$
$=2x(2x+3)-1(2x+3)$
$=(2x+3)(2x-1)$

24. $6x^2+5xy-4y^2=6x^2+8xy-3xy-4y^2$
$=2x(3x+4y)-y(3x+4y)$
$=(3x+4y)(2x-y)$

25. $6x^2-25xy+4y^2=(6x-y)(x-4y)$

26. $18x^2-60x+50=2(9x^2-30x+25)$
$=2[(3x)^2-2\cdot 3x\cdot 5+5^2]$
$=2(3x-5)^2$

27. $2x^2-23xy-39y^2=2x^2-26xy+3xy-39y^2$
$=2x(x-13y)+3y(x-13y)$
$=(x-13y)(2x+3y)$

28. $4x^2-28xy+49y^2=[(2x)^2-2\cdot 2x\cdot 7y+(7y)^2]$
$=(2x-7y)^2$

29. $18x^2-9xy-20y^2=18x^2-24xy+15xy-20y^2$
$=6x(3x-4y)+5y(3x-4y)$
$=(3x-4y)(6x+5y)$

30. $36x^3y+24x^2y^2-45xy^3$
$=3xy(12x^2+8xy-15y^2)$
$=3xy(12x^2+18xy-10xy-15y^2)$
$=3xy[6x(2x+3y)-5y(2x+3y)]$
$=3xy(2x+3y)(6x-5y)$

31. $4x^2-9=(2x)^2-3^2=(2x+3)(2x-3)$

32. $9t^2-25s^2=(3t)^2-(5s)^2=(3t+5s)(3t-5s)$

33. $16x^2+y^2$ is a prime polynomial.

34. $x^3-8y^3=x^3-(2y)^3$
$=(x-2y)[x^2+x\cdot 2y+(2y)^2]$
$=(x-2y)(x^2+2xy+4y^2)$

35. $8x^3+27=(2x)^3+3^3$
$=(2x+3)[(2x)^2-2x\cdot 3+3^2]$
$=(2x+3)(4x^2-6x+9)$

36. $2x^3+8x=2x(x^2+4)$

37. $54-2x^3y^3=2(27-x^3y^3)$
$=2[3^3-(xy)^3]$
$=2(3-xy)[3^2+3\cdot xy+(xy)^2]$
$=2(3-xy)(9+3xy+x^2y^2)$

38. $9x^2-4y^2=(3x)^2-(2y)^2=(3x-2y)(3x+2y)$

39. $16x^4-1=(4x^2)^2-1^2$
$=(4x^2+1)(4x^2-1)$
$=(4x^2+1)[(2x)^2-1^2]$
$=(4x^2+1)(2x+1)(2x-1)$

40. x^4+16 is a prime polynomial.

41. $(x+6)(x-2)=0$
$x+6=0$ or $x-2=0$
$x=-6$ $x=2$
The solutions are −6 and 2.

42. $3x(x+1)(7x-2)=0$
$3x=0$ or $x+1=0$ or $7x-2=0$
$x=0$ $x=-1$ $7x=2$
$x=\frac{2}{7}$

The solutions are 0, −1, and $\frac{2}{7}$.

43. $4(5x+1)(x+3)=0$
$5x+1=0$ or $x+3=0$
$5x=-1$ $x=-3$
$x=-\frac{1}{5}$

The solutions are $-\frac{1}{5}$ and −3.

44. $x^2+8x+7=0$
$(x+7)(x+1)=0$
$x+7=0$ or $x+1=0$
$x=-7$ $\quad$ $x=-1$
The solutions are −7 and −1.

45. $x^2-2x-24=0$
$(x-6)(x+4)=0$
$x-6=0$ or $x+4=0$
$x=6$ $\quad$ $x=-4$
The solutions are 6 and −4.

46. $x^2+10x=-25$
$x^2+10x+25=0$
$(x+5)(x+5)=0$
$x+5=0$ or $x+5=0$
$x=-5$ $\quad$ $x=-5$
The solution is −5.

47. $x(x-10)=-16$
$x^2-10x=-16$
$x^2-10x+16=0$
$(x-8)(x-2)=0$
$x-8=0$ or $x-2=0$
$x=8$ $\quad$ $x=2$
The solutions are 8 and 2.

48. $(3x-1)(9x^2+3x+1)=0$
$3x-1=0$ or $9x^2+3x+1=0$
$9x^2+3x+1$ is a prime polynomial.
$3x-1=0$
$3x=1$
$x=\frac{1}{3}$
The solution is $\frac{1}{3}$.

49. $56x^2-5x-6=0$
$56x^2+16x-21x-6=0$
$8x(7x+2)-3(7x+2)=0$
$(7x+2)(8x-3)=0$
$7x+2=0$ or $8x-3=0$
$7x=-2$ $\quad$ $8x=3$
$x=-\frac{2}{7}$ $\quad$ $x=\frac{3}{8}$
The solutions are $-\frac{2}{7}$ and $\frac{3}{8}$.

50. $20x^2-7x-6=0$
$(4x-3)(5x+2)=0$
$4x-3=0$ or $5x+2=0$
$4x=3$ $\quad$ $5x=-2$
$x=\frac{3}{4}$ $\quad$ $x=-\frac{2}{5}$
The solutions are $\frac{3}{4}$ and $-\frac{2}{5}$.

51. $5(3x+2)=4$
$15x+10=4$
$15x=-6$
$x=-\frac{6}{15}=-\frac{2}{5}$
The solution is $-\frac{2}{5}$.

52. $6x^2-3x+8=0$
The equation has no real solution.

53. $12-5t=-3$
$-5t=-15$
$t=3$
The solution is 3.

54. $5x^3+20x^2+20x=0$
$5x(x^2+4x+4)=0$
$5x(x+2)(x+2)=0$
$x+2=0$ or $5x=0$
$x=-2$ $\quad$ $x=0$
The solutions are −2 and 0.

55. $4t^3-5t^2-21t=0$
$t(4t^2-5t-21)=0$
$t(4t+7)(t-3)=0$
$t=0$ or $4t+7=0$ or $t-3=0$
$4t=-7$ $\quad$ $t=3$
$t=-\frac{7}{4}$
The solutions are 0, $-\frac{7}{4}$, and 3.

56. Answers may vary. Possible answer:
$(x-4)(x-5)=0$
$x^2-9x+20=0$

57. a. $7\neq 2\cdot 5$

b. $10 = 2 \cdot 5$

$$\begin{aligned} P &= 2l + 2w \\ &= 2(10) + 2(5) \\ &= 20 + 10 \\ &= 30 \neq 24 \end{aligned}$$

c. $8 = 2 \cdot 4$

$P = 2l + 2w = 2(8) + 2(4) = 16 + 8 = 24$

d. $10 \neq 2 \cdot 2$

Choice **c** gives the correct dimensions.

58. a. $3 \cdot 8 + 1 = 25 \neq 10$

b. $3 \cdot 4 + 1 = 13$

$A = lw = 13(4) = 52 \neq 80$

c. $3 \cdot 4 + 1 = 13 \neq 20$

d. $3 \cdot 5 + 1 = 16$

$A = lw = 5(16) = 80$

Choice **d** gives the correct dimensions.

59.

$$\begin{aligned} x^2 &= 81 \\ x^2 - 81 &= 0 \\ (x-9)(x+9) &= 0 \end{aligned}$$

$x - 9 = 0$ or $x + 9 = 0$

$x = 9$ $\quad x = -9$

Since length is not negative, the length of the side is 9 units.

60. $(2x+3)+(3x+1)+(x^2-3x)+(x+3) = 47$

$$\begin{aligned} x^2 + 3x + 7 &= 47 \\ x^2 + 3x - 40 &= 0 \\ (x-5)(x+8) &= 0 \end{aligned}$$

$x - 5 = 0$ or $x + 8 = 0$

$x = 5$ $\quad x = -8$

Length is not negative, so $x = 5$. The lengths are:

$x + 3 = 5 + 3 = 8$ units

$2x + 3 = 2(5) + 3 = 13$ units

$3x + 1 = 3(5) + 1 = 16$ units

$x^2 - 3x = 5^2 - 3(5) = 10$ units

61. Let x = the width of the flag. Then $2x - 15$ = the length of the flag.

$$\begin{aligned} A &= lw \\ 500 &= (2x-15)(x) \\ 500 &= 2x^2 - 15x \\ 0 &= 2x^2 - 15x - 500 \\ 0 &= (2x+25)(x-20) \end{aligned}$$

$2x + 25 = 0$

$2x = -25$ or $x - 20 = 0$

$x = -\frac{25}{2}$ $\quad x = 20$

Since the dimensions cannot be negative, the width is 20 inches and the length is $2(20) - 15 = 25$ inches.

62. Let x = the height of the sail, then $4x$ = the base of the sail.

$$\begin{aligned} A &= \frac{1}{2}bh \\ 162 &= \frac{1}{2}(4x)(x) \\ 162 &= 2x^2 \\ 0 &= 2x^2 - 162 \\ 0 &= 2(x^2 - 81) \\ 0 &= 2(x+9)(x-9) \end{aligned}$$

$x + 9 = 0$ or $x - 9 = 0$

$x = -9$ $\quad x = 9$

Since the dimensions cannot be negative, the height is 9 yards and the base is $4 \cdot 9 = 36$ yards.

63. Let x = the first integer. Then $x + 1$ = the next consecutive integer.

$$\begin{aligned} x(x+1) &= 380 \\ x^2 + x &= 380 \\ x^2 + x - 380 &= 0 \\ (x+20)(x-19) &= 0 \end{aligned}$$

$x + 20 = 0$ or $x - 19 = 0$

$x = -20$ $\quad x = 19$

The integers are 19 and 20.

64. a. Let $h = 2800$ and solve for t.

$$\begin{aligned} h &= -16t^2 + 440t \\ 2800 &= -16t^2 + 440t \\ 0 &= -16t^2 + 440t - 2800 \\ 0 &= -8(2t^2 - 55t + 350) \\ 0 &= -8(2t-35)(t-10) \end{aligned}$$

$$2t-35=0 \quad \text{or} \quad t-10=0$$
$$2t=35 \qquad t=10$$
$$t=\frac{35}{2}$$
$$t=17.5$$

The solutions are 17.5 sec and 10 sec. There are two answers because the rocket reaches a height of 2800 feet on its way up and on its way back down.

b. Let $h = 0$ and solve for t.

$$h=-16t^2+440t$$
$$0=-16t^2+440t$$
$$0=-8t(2t-55)$$
$$-8t=0 \quad \text{or} \quad 2t-55=0$$
$$t=0 \qquad 2t=55$$
$$t=\frac{55}{2}$$
$$t=27.5$$

$t = 0$ is when the rocket is launched, so it reaches the ground again after 27.5 seconds.

65. Find t when $h = 0$.

$$h=-16t^2+625$$
$$0=-16t^2+625$$
$$0=-1(16t^2-625)$$
$$0=-1[(4t)^2-(25)^2]$$
$$0=-1(4t+25)(4t-25)$$
$$4t+25=0 \quad \text{or} \quad 4t-25=0$$
$$t=-\frac{25}{4} \qquad t=\frac{25}{4}$$

Since time cannot be negative, the object reaches the ground after $\frac{25}{4} = 6.25$ seconds.

66. Let x = the length of the longer leg, then $x - 8$ = the length of the shorter leg and $x + 8$ = the length of the hypotenuse. By the Pythagorean theorem,

$$x^2+(x-8)^2=(x+8)^2$$
$$x^2+x^2-16x+64=x^2+16x+64$$
$$x^2-32x=0$$
$$x(x-32)=0$$
$$x=0 \quad \text{or} \quad x-32=0$$
$$x=32$$

Since the length cannot be 0 cm, the length of the longer leg is 32 cm.

67. $7x - 63 = 7(x - 9)$

68. $11x(4x - 3) - 6(4x - 3) = (4x - 3)(11x - 6)$

69. $m^2-\frac{4}{25}=m^2-\left(\frac{2}{5}\right)^2=\left(m+\frac{2}{5}\right)\left(m-\frac{2}{5}\right)$

70. $3x^3-4x^2+6x-8=x^2(3x-4)+2(3x-4)$
$=(3x-4)(x^2+2)$

71. $xy+2x-y-2=x(y+2)-1(y+2)$
$=(y+2)(x-1)$

72. $2x^2+2x-24=2(x^2+x-12)=2(x+4)(x-3)$

73. $3x^3-30x^2+27x=3x(x^2-10x+9)$
$=3x(x-9)(x-1)$

74. $4x^2-81=(2x)^2-9^2=(2x+9)(2x-9)$

75. $2x^2-18=2(x^2-9)$
$=2(x^2-3^2)$
$=2(x+3)(x-3)$

76. $16x^2-24x+9=(4x)^2-2\cdot 4x\cdot 3+3^2$
$=(4x-3)^2$

77. $5x^2+20x+20=5(x^2+4x+4)$
$=5(x^2+2\cdot x\cdot 2+2^2)$
$=5(x+2)^2$

78. $2x^2+5x-12=(2x-3)(x+4)$

79. $4x^2y-6xy^2=2xy(2x-3y)$

80. $8x^2-15x-x^3=-x(-8x+15+x^2)$
$=-x(x^2-8x+15)$
$=-x(x-5)(x-3)$

81. $125x^3+27=(5x)^3+3^3$
$=(5x+3)[(5x)^2-5x\cdot 3+3^2]$
$=(5x+3)(25x^2-15x+9)$

82. $24x^2-3x-18=3(8x^2-x-6)$

83. $(x+7)^2-y^2=[(x+7)+y][(x+7)-y]$
$=(x+7+y)(x+7-y)$

84. $x^2(x+3)-4(x+3)=(x+3)(x^2-4)$
$=(x+3)(x^2-2^2)$
$=(x+3)(x-2)(x+2)$

85. $54a^3b-2b=2b(27a^3-1)$
$=2b[(3a)^3-1^3]$
$=2b(3a-1)[(3a)^2+3a\cdot 1+1^2]$
$=2b(3a-1)(9a^2+3a+1)$

86. To factor $x^2+2x-48$, think of two numbers whose product is $\underline{-48}$ and whose sum if $\underline{2}$.

87. The first step is to factor out the GCF, 3.

88. $(x^2-2)+(x^2-4x)+(3x^2-5x)$
$=x^2+x^2+3x^2-4x-5x-2$
$=5x^2-9x-2$
$=(5x+1)(x-2)$

89. $2(2x^2+3)+2(6x^2-14x)$
$=4x^2+6+12x^2-28x$
$=16x^2-28x+6$
$=2(8x^2-14x+3)$
$=2(4x-1)(2x-3)$

90. $2x^2-x-28=0$
$(2x+7)(x-4)=0$
$2x+7=0$ or $x-4=0$
$x=-\frac{7}{2}$ $\quad x=4$
The solutions are $-\frac{7}{2}$ and 4.

91. $x^2-2x=15$
$x^2-2x-15=0$
$(x+3)(x-5)=0$
$x+3=0$ or $x-5=0$
$x=-3$ $\quad x=5$
The solutions are -3 and 5.

92. $2x(x+7)(x+4)=0$
$2x=0$ or $x+7=0$ or $x+4=0$
$x=0$ $\quad x=-7$ $\quad x=-4$
The solutions are 0, -7, and -4.

93. $x(x-5)=-6$
$x^2-5x=-6$
$x^2-5x+6=0$
$(x-3)(x-2)=0$
$x-3=0$ or $x-2=0$
$x=3$ $\quad x=2$
The solutions are 3 and 2.

94. $x^2=16x$
$x^2-16x=0$
$x(x-16)=0$
$x=0$ or $x-16=0$
$x=16$
The solutions are 0 and 16.

95. $(x^2+3)+(4x+5)+2x=48$
$x^2+6x+8=48$
$x^2+6x-40=0$
$(x-4)(x+10)=0$
$x-4=0$ or $x+10=0$
$x=4$ $\quad x=-10$
Since the length cannot be negative, $x=4$.
The lengths are:
$x^2+3=4^2+3=19$ inches
$4x+5=4\cdot 4+5=21$ inches
$2x=2\cdot 4=8$ inches

96. Let x = length, then $x-4$ = width.
$A=lw$
$12=x(x-4)$
$12=x^2-4x$
$0=x^2-4x-12$
$0=(x-6)(x+2)$
$x-6=0$ or $x+2=0$
$x=6$ $\quad x=-2$
Since length cannot be negative, the length is 6 inches and the width is $6-4=2$ inches.

97. Find t when $h=0$.
$h=-16t^2+729$
$0=-16t^2+729$
$0=-(16t^2-729)$
$0=-[(4t)^2-(27)^2]$
$0=-(4t+27)(4t-27)$

$4t+27=0$ or $4t-27=0$

$t=-\frac{27}{4}$ $\quad t=\frac{27}{4}$

Since time cannot be negative, the object reaches the ground in $\frac{27}{4}=6.75$ seconds.

98. Area of large figure – Area of circle

$= [(6x)(5x)-2x^2]-\pi x^2$
$= 30x^2-2x^2-\pi x^2$
$= 28x^2-\pi x^2$
$= x^2(28-\pi)$

Chapter 6 Test

1. $x^2+11x+28=(x+7)(x+4)$

2. $49-m^2=(7^2-m^2)=(7-m)(7+m)$

3. $y^2+22y+121=y^2+2\cdot y\cdot 11+11^2=(y+11)^2$

4. $4(a+3)-y(a+3)=(a+3)(4-y)$

5. x^2+4 is the sum of two perfect squares (not the difference). The polynomial is prime.

6. $y^2-8y-48=(y-12)(y+4)$

7. x^2+x-10 is a prime polynomial.

8. $9x^3+39x^2+12x=3x(3x^2+13x+4)$
$= 3x(3x+1)(x+4)$

9. $3a^2+3ab-7a-7b=3a(a+b)-7(a+b)$
$= (a+b)(3a-7)$

10. $3x^2-5x+2=(3x-2)(x-1)$

11. $x^2+14xy+24y^2=(x+12y)(x+2y)$

12. $180-5x^2=5(36-x^2)$
$= 5(6^2-x^2)$
$= 5(6+x)(6-x)$

13. $6t^2-t-5=(6t+5)(t-1)$

14. $xy^2-7y^2-4x+28=y^2(x-7)-4(x-7)$
$= (x-7)(y^2-4)$
$= (x-7)(y^2-2^2)$
$= (x-7)(y+2)(y-2)$

15. $x-x^5=x(1-x^4)$
$= x[1-(x^2)^2]$
$= x(1+x^2)(1-x^2)$
$= x(1+x^2)(1^2-x^2)$
$= x(1+x^2)(1+x)(1-x)$

16. $-xy^3-x^3y=xy(y^2+x^2)$

17. $64x^3-1=(4x)^3-1^3$
$= (4x-1)[(4x)^2+4x\cdot 1+1^2]$
$= (4x-1)(16x^2+4x+1)$

18. $8y^3-64=8(y^3-8)$
$= 8(y^3-2^3)$
$= 8(y-2)(y^2+y\cdot 2+2^2)$
$= 8(y-2)(y^2+2y+4)$

19. $(x-3)(x+9)=0$

$x-3=0$ or $x+9=0$

$x=3$ $\quad x=-9$

The solutions are 3 and –9.

20. $x^2+5x=14$

$x^2+5x-14=0$

$(x+7)(x-2)=0$

$x+7=0$ or $x-2=0$

$x=-7$ $\quad x=2$

The solutions are –7 and 2.

21. $x(x+6)=7$

$x^2+6x=7$

$x^2+6x-7=0$

$(x+7)(x-1)=0$

$x+7=0$ or $x-1=0$

$x=-7$ $\quad x=1$

The solutions are –7 and 1.

22. $3x(2x-3)(3x+4)=0$

$3x=0$ or $2x-3=0$ or $3x+4=0$

$x=0$ $\quad 2x=3$ $\quad 3x=-4$

$x=\frac{3}{2}$ $\quad x=-\frac{4}{3}$

The solutions are 0, $\frac{3}{2}$, and $-\frac{4}{3}$.

23. $5t^3-45t=0$

$5t(t^2-9)=0$

$5t(t+3)(t-3)=0$

$5t=0$ or $t+3=0$ or $t-3=0$

$t=0$ $\quad t=-3$ $\quad t=3$

The solutions are 0, –3, and 3.

24. $t^2-2t-15=0$

$(t-5)(t+3)=0$

$t-5=0$ or $t+3=0$

$t=5$ $\quad t=-3$

The solutions are 5 and –3.

25. $6x^2=15x$

$6x^2-15x=0$

$3x(2x-5)=0$

$3x=0$ or $2x-5=0$

$x=0$ $\quad 2x=5$

$x=\frac{5}{2}$

The solutions are 0 and $\frac{5}{2}$.

26. Let x = the altitude of the triangle, then $x+9$ = the base.

$A=\frac{1}{2}bh$

$68=\frac{1}{2}(x+9)(x)$

$136=x^2+9x$

$0=x^2+9x-136$

$0=(x+17)(x-8)$

$x+17=0$ or $x-8=0$

$x=-17$ $\quad x=8$

Since the length of the base cannot be negative, the base is 8 + 9 = 17 feet.

27. Let x = the first number, then $17-x$ = the other number.

$x^2+(17-x)^2=145$

$x^2+289-34x+x^2=145$

$2x^2-34x+144=0$

$2(x^2-17x+72)=0$

$2(x-9)(x-8)=0$

$x-9=0$ or $x-8=0$

$x=9$ $\quad x=8$

The numbers are 8 and 9.

28. Find t when $h=0$.

$h=-16t^2+784$

$0=-16t^2+784$

$0=-16(t^2-49)$

$0=-16(t+7)(t-7)$

$t+7=0$ or $t-7=0$

$t=-7$ $\quad t=7$

Since the time cannot be negative, the object reaches the ground after 7 seconds.

29. Let x = length of the shorter leg, then $x+10$ = length of hypotenuse, and $x+5$ = length of longer leg.

$x^2+(x+5)^2=(x+10)^2$

$x^2+x^2+10x+25=x^2+20x+100$

$x^2-10x-75=0$

$(x+5)(x-15)=0$

$x+5=0$ or $x-15=0$

$x=-5$ $\quad x=15$

Since length cannot be negative, the lengths of the triangle sides are:
shorter leg = 15 cm, longer leg = 20 cm, hypotenuse = 25 cm.

Chapter 6 Cumulative Review

1. a. $9\le 11$

b. $8>1$

c. $3\ne 4$

2. a. $|-5|>|-3|$

b. $|0|<|-2|$

3. a. $\frac{42}{49}=\frac{6\cdot 7}{7\cdot 7}=\frac{6}{7}$

b. $\frac{11}{27}=\frac{11}{3\cdot3\cdot3}=\frac{11}{27}$

c. $\frac{88}{20}=\frac{4\cdot22}{4\cdot5}=\frac{22}{5}$

4. Let $x = 20$ and $y = 10$.
$\frac{x}{y}+5x=\frac{20}{10}+5(20)=2+100=102$

5. $\frac{8+2\cdot3}{2^2-1}=\frac{8+6}{4-1}=\frac{14}{3}$

6. Let $x = -20$ and $y = 10$.
$\frac{x}{y}+5x=\frac{-20}{10}+5(-20)=-2-100=-102$

7. a. $3+(-7)+(-8)=3+(-15)=-12$

b. $$\begin{aligned}[7+(-10)]+[-2+|-4|]&=-3+(-2+4)\\&=-3+2\\&=-1\end{aligned}$$

8. Let $x = -20$ and $y = -10$.
$\frac{x}{y}+5x=\frac{-20}{-10}+5(-20)=2-100=-98$

9. a. $(-6)(4)=-24$

b. $2(-1)=-2$

c. $(-5)(-10)=50$

10. $5-2(3x-7)=5-6x+14=-6x+19$

11. a. $7x-3x=(7-3)x=4x$

b. $10y^2+y^2=(10+1)y^2=11y^2$

c. $8x^2+2x-3x=8x^2+(2-3)x=8x^2-x$

12. $$\begin{aligned}0.8y+0.2(y-1)&=1.8\\0.8y+0.2y-0.2&=1.8\\1.0y-0.2&=1.8\\y&=2.0\end{aligned}$$

13. $$\begin{aligned}\frac{y}{7}&=20\\7\left(\frac{y}{7}\right)&=7(20)\\y&=140\end{aligned}$$

14. $$\begin{aligned}\frac{x}{-7}&=-4\\-7\left(\frac{x}{-7}\right)&=-7(-4)\\x&=28\end{aligned}$$

15. $$\begin{aligned}-3x&=33\\\frac{-3x}{-3}&=\frac{33}{-3}\\x&=-11\end{aligned}$$

16. $$\begin{aligned}-\frac{2}{3}x&=-22\\\left(-\frac{3}{2}\right)\left(-\frac{2}{3}\right)x&=\left(-\frac{3}{2}\right)(-22)\\x&=33\end{aligned}$$

17. $$\begin{aligned}8(2-t)&=-5t\\16-8t&=-5t\\16-8t+5t&=-5t+5t\\16-3t&=0\\16-16-3t&=-16\\-3t&=-16\\\frac{-3t}{-3}&=\frac{-16}{-3}\\t&=\frac{16}{3}\end{aligned}$$

18. $$\begin{aligned}-z&=\frac{7z+3}{5}\\5(-z)&=5\left(\frac{7z+3}{5}\right)\\-5z&=7z+3\\-5z-7z&=7z-7z+3\\-12z&=3\\\frac{-12z}{-12}&=\frac{3}{-12}\\z&=-\frac{1}{4}\end{aligned}$$

19. Let x = the length of the shorter piece and $3x$ = the length of the longer piece.

$x+3x=48$

$4x=48$

$x=12$

$3x=3(12)=36$

The pieces are 12 inches and 36 inches in length.

20. $3x+9\le 5(x-1)$

$3x+9\le 5x-5$

$-2x+9\le -5$

$2x\le -14$

$\frac{-2x}{-2}\ge\frac{-14}{-2}$

$x\ge 7$, $[7, \infty)$

21. $y=-\frac{1}{3}x+2$

x	y
0	2
−3	3
3	1

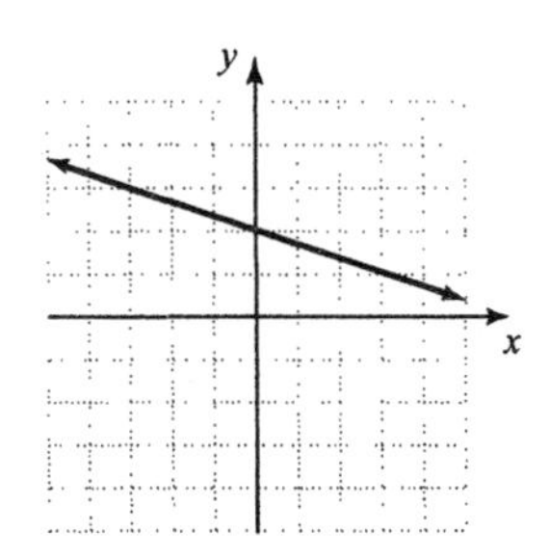

22. $-7x-8y=-9$

$(-1, 2)$: $-7(-1)-8(2)\stackrel{?}{=}-9$

$7-16\stackrel{?}{=}-9$

$-9=-9$ True

$(-1, 2)$ is a solution of the equation.

23. $3x-4y=4$

$-4y=-3x+4$

$y=\frac{3}{4}x-1$

$y=mx+b$

slope $=\frac{3}{4}$; y-intercept = $(0, -1)$

24. $(5, -6)$ and $(5, 2)$

$m=\frac{y_2-y_1}{x_2-x_1}=\frac{2-(-6)}{5-5}=\frac{8}{0}$

The slope is undefined.

25. a. If $x=5$, $2x^3=2(5)^3=2(125)=250$.

b. If $x=-3$, $\frac{9}{x^2}=\frac{9}{(-3)^2}=\frac{9}{9}=1$.

26. $7x-3y=2$

$-3y=-7x+2$

$y=\frac{-7x}{-3}+\frac{2}{-3}$

$y=\frac{7}{3}x-\frac{2}{3}$

$y=mx+b$

slope $=\frac{7}{3}$; y-intercept $=\left(0, -\frac{2}{3}\right)$

27. a. $3x^2$ has degree 2.

b. -2^3x^5 has degree 5.

c. y has degree 1.

d. $12x^2yz^3$ has degree 2 + 1 + 3 = 6.

e. 5 has degree 0.

28. Vertical line has equation $x=c$.

Point (0, 7)

$x=0$

29. $(2x^3+8x^2-6x)-(2x^3-x^2+1)$

$=2x^3+8x^2-6x-2x^3+x^2-1$

$=9x^2-6x-1$

30. $m=4$, $b=\frac{1}{2}$

$y=mx+b$

$y=4x+\frac{1}{2}$

$2y=8x+1$

$8x-2y=-1$

31. $(3x+2)(2x-5)$
$= 3x(2x)+3x(-5)+2(2x)+2(-5)$
$= 6x^2-15x+4x-10$
$= 6x^2-11x-10$

32. (–4, 0) and (6, –1)
$m = \frac{y_2-y_1}{x_2-x_1} = \frac{-1-0}{6-(-4)} = -\frac{1}{10}$
$m = -\frac{1}{10}$, point (–4, 0)
$y-y_1 = m(x-x_1)$
$y-0 = -\frac{1}{10}[x-(-4)]$
$y = -\frac{1}{10}x - \frac{4}{10}$
$10y = -x-4$
$x+10y = -4$

33. $(3y+1)^2 = (3y)^2+2(3y)(1)+1^2 = 9y^2+6y+1$

34. $\begin{cases} -x+3y=18 \\ -3x+2y=19 \end{cases}$
Multiply the first equation by –3.
$3x-9y=-54$
$-3x+2y=19$
$-7y=-35$
$y=5$
Substitute 5 for y in the first equation.
$-x+3(5)=18$
$-x+15=18$
$-x=3$
$x=-3$
The solution to the system is (–3, 5).

35. a. $3^{-2} = \frac{1}{3^2} = \frac{1}{9}$

b. $2x^{-3} = \frac{2}{x^3}$

c. $2^{-1}+4^{-1} = \frac{1}{2}+\frac{1}{4}$
$= \frac{1\cdot 2}{2\cdot 2}+\frac{1}{4}$
$= \frac{2}{4}+\frac{1}{4}$
$= \frac{2+1}{4}$
$= \frac{3}{4}$

d. $(-2)^{-4} = \frac{1}{(-2)^4} = \frac{1}{16}$

e. $\frac{1}{y^{-4}} = y^4$

f. $\frac{1}{7^{-2}} = 7^2 = 49$

36. $\frac{(5a^7)^2}{a^5} = \frac{5^2a^{14}}{a^5} = 25a^{14-5} = 25a^9$

37. a. $367,000,000 = 3.67\times 10^8$

b. $0.000003 = 3.0\times 10^{-6}$

c. $20,520,000,000 = 2.052\times 10^{10}$

d. $0.00085 = 8.5\times 10^{-4}$

38. $(3x-7y)^2 = (3x)^2-2(3x)(7y)+(7y)^2$
$= 9x^2-42xy+49y^2$

39.
$$\begin{array}{r} x+4 \\ x+3 \overline{)\, x^2+7x+12} \\ \underline{x^2+3x} \\ 4x+12 \\ \underline{4x+12} \\ 0 \end{array}$$

$\frac{x^2+7x+12}{x+3} = x+4$

40. $\dfrac{(xy)^{-3}}{(x^5y^6)^3} = \dfrac{x^{-3}y^{-3}}{x^{15}y^{18}}$
$= x^{-3-15}y^{-3-18}$
$= x^{-18}y^{-21}$
$= \dfrac{1}{x^{18}y^{21}}$

41. a. x^3, x^7, x^5: GCF $= x^3$

b. y, y^4, y^7: GCF $= y$

42. $z^3 + 7z + z^2 + 7 = z(z^2 + 7) + 1(z^2 + 7)$
$= (z^2 + 7)(z + 1)$

43. $x^2 + 7x + 12 = (x + 4)(x + 3)$

44. $2x^3 + 2x^2 - 84x = 2x(x^2 + x - 42)$
$= 2x(x + 7)(x - 6)$

45. $8x^2 - 22x + 5 = 8x^2 - 20x - 2x + 5$
$= 4x(2x - 5) - 1(2x - 5)$
$= (2x - 5)(4x - 1)$

46. $-4x^2 - 23x + 6 = -1(4x^2 + 23x - 6)$
$= -(4x^2 - x + 24x - 6)$
$= -[x(4x - 1) + 6(4x - 1)]$
$= -(4x - 1)(x + 6)$

47. $25a^2 - 9b^2 = (5a)^2 - (3b)^2 = (5a + 3b)(5a - 3b)$

48. $9xy^2 - 16x = x(9y^2 - 16)$
$= x[(3y)^2 - 4^2]$
$= x(3y + 4)(3y - 4)$

49. $(x - 3)(x + 1) = 0$
$x - 3 = 0$ or $x + 1 = 0$
$x = 3$ $\quad x = -1$
The solutions are 3 and -1.

50. $x^2 - 13x = -36$
$x^2 - 13x + 36 = 0$
$(x - 9)(x - 4) = 0$
$x - 9 = 0$ or $x - 4 = 0$
$x = 9$ $\quad x = 4$
The solutions are 9 and 4.

Chapter 7

Section 7.1

Practice Exercises

1. a. The denominator of $f(x)$ is never 0.
Domain: $\{x|x$ is a real number$\}$

b. Undefined values when
$x+3=0$, or $x=-3$
Domain: $\{x|x$ is a real number and $x \neq -3\}$

c. Undefined values when
$$x^2-5x+6=0$$
$$(x-3)(x-2)=0$$
$$x-3=0 \quad \text{or} \quad x-2=0$$
$$x=3 \quad \text{or} \quad x=2$$
Domain:
$\{x|x$ is a real number and $x \neq 2, x \neq 3\}$

2. a. $$\frac{5z^4}{10z^5-5z^4}=\frac{5z^4\cdot 1}{5z^4(2z-1)}=1\cdot\frac{1}{2z-1}=\frac{1}{2z-1}$$

b. $$\frac{5x^2+13x+6}{6x^2+7x-10}=\frac{(5x+3)(x+2)}{(6x-5)(x+2)}=\frac{5x+3}{6x-5}\cdot 1=\frac{5x+3}{6x-5}$$

3. a. $$\frac{x+3}{3+x}=\frac{x+3}{x+3}=1$$

b. $$\frac{3-x}{x-3}=\frac{-1(-3+x)}{x-3}=\frac{-1(x-3)}{x-3}=\frac{-1}{1}=-1$$

4. $$\frac{20-5x^2}{x^2+x-6}=\frac{5(4-x^2)}{(x+3)(x-2)}=\frac{5(2+x)(2-x)}{(x+3)(x-2)}=\frac{5(2+x)\cdot(-1)(x-2)}{(x+3)(x-2)}=-\frac{5(2+x)}{x+3}$$

5. a. $$\frac{x^3+64}{4+x}=\frac{(x+4)(x^2-4x+16)}{x+4}=x^2-4x+16$$

b. $$\frac{5z^2+10}{z^3-3z^2+2z-6}=\frac{5(z^2+2)}{(z^3-3z^2)+(2z-6)}=\frac{5(z^2+2)}{z^2(z-3)+2(z-3)}=\frac{5(z^2+2)}{(z-3)(z^2+2)}=\frac{5}{z-3}$$

6. $$-\frac{x+3}{6x-11}=\frac{-(x+3)}{6x-11}=\frac{-x-3}{6x-11}$$
Also,
$$-\frac{x+3}{6x-11}=\frac{x+3}{-(6x-11)}=\frac{x+3}{-6x+11} \text{ or } \frac{x+3}{11-6x}$$
Thus, some equivalent forms of $-\frac{x+3}{6x-11}$ are
$\frac{-(x+3)}{6x-11}, \frac{-x-3}{6x-11}, \frac{x+3}{-(6x-11)}, \frac{x+3}{-6x+11}$, and $\frac{x+3}{11-6x}$.

7. a. $$C(100)=\frac{3.2(100)+400}{100}=\frac{720}{100}=7.2$$
\$7.20 per tee shirt

b. $$C(1000)=\frac{3.2(1000)+400}{1000}=\frac{3600}{1000}=3.6$$
\$3.60 per tee shirt

Graphing Calculator Explorations

1. $$x^2-4=0$$
$$(x+2)(x-2)=0$$
$$x+2=0 \quad \text{or} \quad x-2=0$$
$$x=-2 \quad \text{or} \quad x=2$$
Domain: $\{x|x$ is a real number and $x \neq -2, x \neq 2\}$

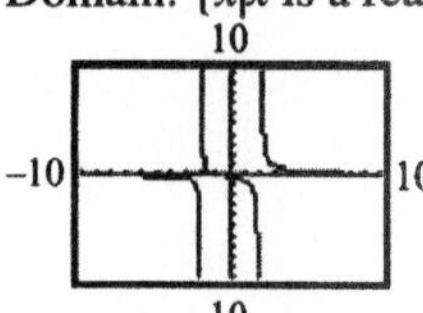

2. $x^2 - 9 = 0$

$(x+3)(x-3) = 0$

$x+3=0$ or $x-3=0$

$x=-3$ or $x=3$

Domain: $\{x|x \text{ is a real number and } x \neq -3, x \neq 3\}$

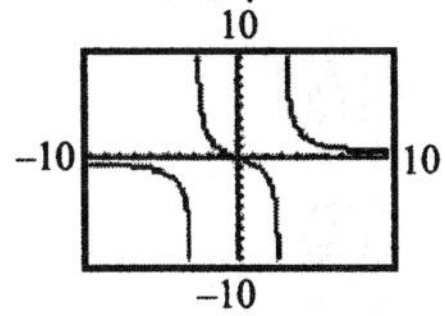

3. $2x^2+7x-4=0$

$(2x-1)(x+4)=0$

$2x-1=0$ or $x+4=0$

$2x=1$ or $x=-4$

$x=\frac{1}{2}$

Domain:

$\left\{x \middle| x \text{ is a real number and } x \neq -4, x \neq \frac{1}{2}\right\}$

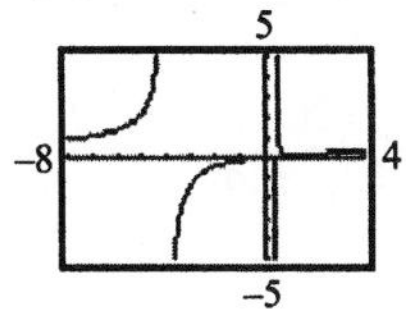

4. $4x^2-19x-5=0$

$(4x+1)(x-5)=0$

$4x+1=0$ or $x-5=0$

$4x=-1$ or $x=5$

$x=-\frac{1}{4}$

Domain:

$\left\{x \middle| x \text{ is a real number and } x \neq -\frac{1}{4}, x \neq 5\right\}$

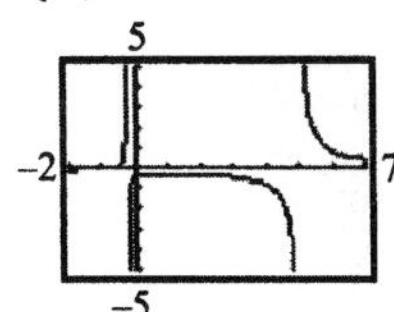

Vocabulary and Readiness Check

1. A <u>rational</u> expression is an expression that can be written as the quotient $\frac{P}{Q}$ of two polynomials P and Q as long as $Q \neq 0$.

2. A rational expression is undefined if the denominator is <u>0</u>.

3. The <u>domain</u> of the rational function $f(x)=\frac{2}{x}$ is $\{x|x \text{ is a real number and } x \neq 0\}$.

4. A rational expression is <u>simplified</u> if the numerator and denominator have no common factors other than 1 or −1.

5. The expression $\frac{x^2+2}{2+x^2}$ simplifies to <u>1</u>.

6. The expression $\frac{y-z}{z-y}$ simplifies to <u>−1</u>.

7. For a rational expression, $-\frac{a}{b}=\frac{-a}{\underline{b}}=\frac{a}{\underline{-b}}$.

8. The statement $\frac{a-6}{a+2}=\frac{-(a-6)}{-(a+2)}=\frac{-a+6}{-a-2}$ is true.

9. No, $\frac{x}{x+7}$ cannot be simplified.

10. Yes, $\frac{3+x}{x+3}$ can be simplified because $3+x=x+3$.

11. Yes, $\frac{5-x}{x-5}$ can be simplified because $5-x=-1(x-5)$.

12. No, $\frac{x+2}{x+8}$ cannot be simplified.

Exercise Set 7.1

1. 4 is never 0, so the domain of $f(x)=\frac{5x-7}{4}$ is $\{x|x \text{ is a real number}\}$.

3. $2t=0$

$t=0$

The domain of $s(t)=\frac{t^2+1}{2t}$ is $\{t|t \text{ is a real number and } t \neq 0\}$.

5. $7-x=0$
$7=x$

The domain of $f(x)=\frac{3x}{7-x}$ is $\{x|x \text{ is a real number and } x\neq 7\}$.

7. $3x-1=0$
$3x=1$
$x=\frac{1}{3}$

The domain of $f(x)=\frac{x}{3x-1}$ is $\left\{x\middle|x \text{ is a real number and } x\neq\frac{1}{3}\right\}$.

9. $x^3+x^2-2x=0$
$x(x^2+x-2)=0$
$x(x+2)(x-1)=0$
$x=0$ or $x+2=0$ or $x-1=0$
$x=0$ or $x=-2$ or $x=1$

The domain of $R(x)=\frac{3+2x}{x^3+x^2-2x}$ is $\{x|x \text{ is a real number and } x\neq -2, x\neq 0, x\neq 1\}$.

11. $x^2-4=0$
$(x+2)(x-2)=0$
$x+2=0$ or $x-2=0$
$x=-2$ or $x=2$

The domain of $C(x)=\frac{x+3}{x^2-4}$ is $\{x|x \text{ is a real number and } x\neq 2, x\neq -2\}$.

13. $-\frac{x-10}{x+8}=\frac{-(x-10)}{x+8}=\frac{-x+10}{x+8}$ or $\frac{10-x}{x+8}$
$-\frac{x-10}{x+8}=\frac{x-10}{-(x+8)}=\frac{x-10}{-x-8}$

15. $-\frac{5y-3}{y-12}=\frac{-(5y-3)}{y-12}=\frac{-5y+3}{y-12}$ or $\frac{3-5y}{y-12}$
$-\frac{5y-3}{y-12}=\frac{5y-3}{-(y-12)}=\frac{5y-3}{-y+12}$ or $\frac{5y-3}{12-y}$

17. $\frac{x+7}{7+x}=\frac{x+7}{x+7}=1$

19. $\frac{x-7}{7-x}=\frac{x-7}{-1(x-7)}=\frac{1}{-1}=-1$

21. $\frac{2}{8x+16}=\frac{2}{8(x+2)}=\frac{2(1)}{2(4)(x+2)}=\frac{1}{4(x+2)}$

23. $\frac{-5a-5b}{a+b}=\frac{-5(a+b)}{a+b}=-5$

25. $\frac{7x+35}{x^2+5x}=\frac{7(x+5)}{x(x+5)}=\frac{7}{x}$

27. $\frac{x+5}{x^2-4x-45}=\frac{x+5}{(x+5)(x-9)}=\frac{1}{x-9}$

29. $\frac{5x^2+11x+2}{x+2}=\frac{(5x+1)(x+2)}{x+2}=5x+1$

31. $\frac{x^3+7x^2}{x^2+5x-14}=\frac{x^2(x+7)}{(x+7)(x-2)}=\frac{x^2}{x-2}$

33. $\frac{2x^2-8}{4x-8}=\frac{2(x^2-4)}{4(x-2)}=\frac{2(x+2)(x-2)}{2\cdot 2(x-2)}=\frac{x+2}{2}$

35. $\frac{4-x^2}{x-2}=\frac{(-1)(x^2-4)}{x-2}$
$=-\frac{(x+2)(x-2)}{x-2}$
$=-(x+2)$ or $-x-2$

37. $\frac{11x^2-22x^3}{6x-12x^2}=\frac{11x^2(1-2x)}{6x(1-2x)}=\frac{11x}{6}$

39. $\frac{x^2+xy+2x+2y}{x+2}=\frac{x(x+y)+2(x+y)}{x+2}$
$=\frac{(x+y)(x+2)}{x+2}$
$=x+y$

41. $\frac{x^3+8}{x+2}=\frac{(x+2)(x^2-2x+4)}{x+2}=x^2-2x+4$

43. $\frac{x^3-1}{1-x}=\frac{(x-1)(x^2+x+1)}{-1(x-1)}$
$=-1(x^2+x+1)$
$=-x^2-x-1$

45. $\frac{2xy+5x-2y-5}{3xy+4x-3y-4}=\frac{x(2y+5)-1(2y+5)}{x(3y+4)-1(3y+4)}$
$=\frac{(2y+5)(x-1)}{(3y+4)(x-1)}$
$=\frac{2y+5}{3y+4}$

47. $\frac{3x^2-5x-2}{6x^3+2x^2+3x+1}=\frac{(3x+1)(x-2)}{2x^2(3x+1)+1(3x+1)}$
$=\frac{(3x+1)(x-2)}{(3x+1)(2x^2+1)}$
$=\frac{x-2}{2x^2+1}$

49. $\frac{9x^2-15x+25}{27x^3+125}=\frac{9x^2-15x+25}{(3x+5)(9x^2-15x+25)}$
$=\frac{1}{3x+5}$

51. $f(x)=\frac{x+8}{2x-1}$
$f(2)=\frac{2+8}{2(2)-1}=\frac{10}{4-1}=\frac{10}{3}$
$f(0)=\frac{0+8}{2(0)-1}=\frac{8}{0-1}=\frac{8}{-1}=-8$
$f(-1)=\frac{-1+8}{2(-1)-1}=\frac{7}{-2-1}=\frac{7}{-3}=-\frac{7}{3}$

53. $g(x)=\frac{x^2+8}{x^3-25x}$
$g(3)=\frac{3^2+8}{3^3-25(3)}=\frac{9+8}{27-75}=\frac{17}{-48}=-\frac{17}{48}$
$g(-2)=\frac{(-2)^2+8}{(-2)^3-25(-2)}=\frac{4+8}{-8+50}=\frac{12}{42}=\frac{2}{7}$
$g(1)=\frac{1^2+8}{1^3-25(1)}=\frac{1+8}{1-25}=\frac{9}{-24}=-\frac{3}{8}$

55. $R(x)=\frac{1000x^2}{x^2+4}$

a. $R(1)=\frac{1000\cdot 1^2}{1^2+4}=\frac{1000}{5}=200$
The revenue at the end of the first year is \$200 million.

b. $R(2)=\frac{1000\cdot 2^2}{2^2+4}$
$=\frac{1000\cdot 4}{4+4}$
$=\frac{4000}{8}$
$=500$
The revenue at the end of the second year is \$500 million.

c. The revenue during the second year is equal to the revenue at the end of the second year minus the revenue at the end of the first year.
$500-200=300$
The revenue during the second year is \$300 million.

d. $x^2+4=0$
$x^2=-4$
This equation has no solutions. Thus there are no values of x that would make the denominator equal to zero and the function undefined. The domain of $R(x)$ is $\{x|x \text{ is a real number}\}$.

57. Let $D = 1000$ and $A = 8$.
$C=\frac{DA}{A+12}=\frac{1000(8)}{8+12}=\frac{8000}{20}=400$
The child should receive 400 mg.

59. $C=\frac{100W}{L}$; $W=5, L=6.4$
$C=\frac{100(5)}{6.4}=\frac{500}{6.4}=78.125$
The skull is medium.

61. $C=\frac{250x+10,000}{x}$

a. $x=100$: $C=\frac{250(100)+10,000}{100}=\$350$

b. $x=1000$: $C=\frac{250(1000)+10,000}{1000}=\$260$

c. Decrease: Answers may vary.

63. $\frac{1}{3}\cdot\frac{9}{11}=\frac{1\cdot 9}{3\cdot 11}=\frac{3\cdot 3}{3\cdot 11}=\frac{3}{11}$

65. $\frac{1}{3} \div \frac{1}{4} = \frac{1}{3} \cdot \frac{4}{1} = \frac{4}{3}$

67. $\frac{13}{20} \div \frac{2}{9} = \frac{13}{20} \cdot \frac{9}{2} = \frac{13 \cdot 9}{20 \cdot 2} = \frac{117}{40}$

69. $\frac{5a-15}{5} = \frac{5(a-3)}{5} = a-3$
The statement is correct.

71. $\frac{1+2}{1+3} = \frac{3}{4}$
The statement is incorrect.

73. No; answers may vary

75. Answers may vary

77. a. $\frac{x+5}{5+x} = \frac{x+5}{x+5} = 1$

b. $\frac{x-5}{5-x} = \frac{x-5}{-(x-5)} = -1$

c. $\frac{x+5}{x-5}$ neither

d. $\frac{-x-5}{x+5} = \frac{-(x+5)}{x+5} = -1$

e. $\frac{x-5}{-x+5} = \frac{x-5}{-(x-5)} = -1$

f. $\frac{-5+x}{x-5} = \frac{x-5}{x-5} = 1$

79. $f(x) = \frac{20x}{100-x}$

x	0	10	30	50	70	90	95	99
y	0	$\frac{20}{9}$	$\frac{60}{7}$	20	$\frac{140}{3}$	180	380	1980

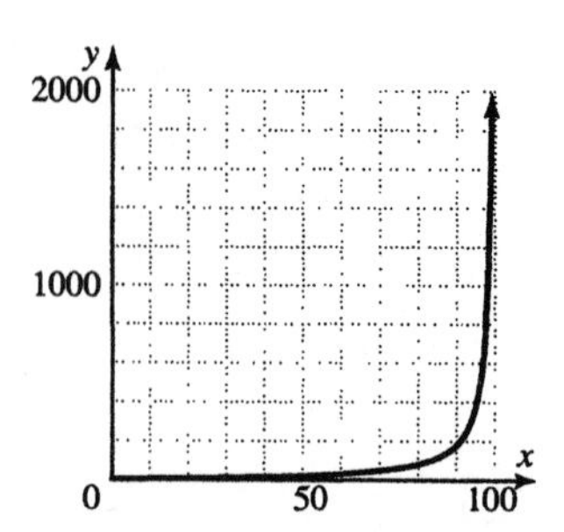

81. $y = \frac{x^2-16}{x-4} = \frac{(x+4)(x-4)}{x-4} = x+4,\ x \neq 4$

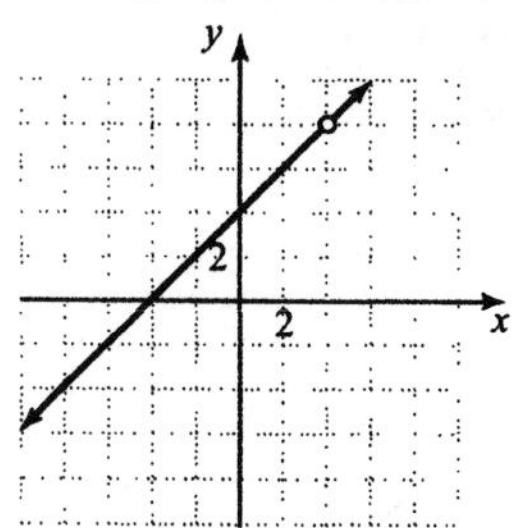

83. $y = \frac{x^2-6x+8}{x-2} = \frac{(x-2)(x-4)}{x-2} = x-4,\ x \neq 2$

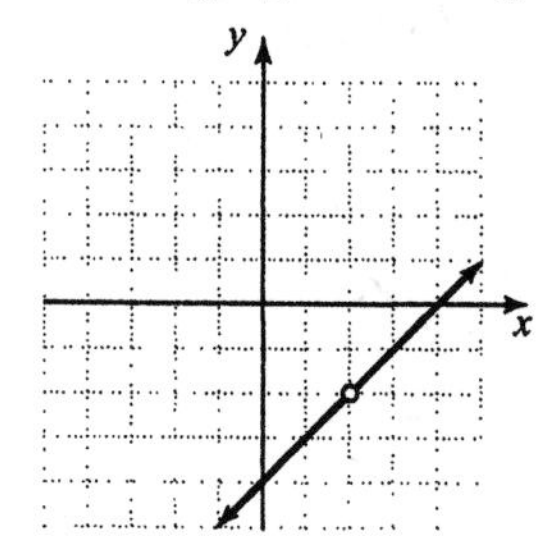

Section 7.2

Practice Exercises

1. a. $\frac{4a}{5} \cdot \frac{3}{b^2} = \frac{4a \cdot 3}{5 \cdot b^2} = \frac{12a}{5b^2}$

b. $\frac{-3p^4}{q^2} \cdot \frac{2q^3}{9p^4} = \frac{-3p^4 \cdot 2q^3}{q^2 \cdot 9p^4}$
$= \frac{-1 \cdot 3 \cdot p^4 \cdot 2 \cdot q \cdot q^2}{q^2 \cdot 3 \cdot 3 \cdot p^4}$
$= -\frac{2q}{3}$

2. $\frac{x^2-x}{5x}\cdot\frac{15}{x^2-1}=\frac{x(x-1)}{5x}\cdot\frac{3\cdot 5}{(x+1)(x-1)}$
$=\frac{x(x-1)\cdot 3\cdot 5}{5x\cdot(x+1)(x-1)}$
$=\frac{3}{x+1}$

3. $\frac{6-3x}{6x+6x^2}\cdot\frac{3x^2-2x-5}{x^2-4}$
$=\frac{3(2-x)}{2\cdot 3\cdot x(1+x)}\cdot\frac{(x+1)(3x-5)}{(x+2)(x-2)}$
$=\frac{3(2-x)(x+1)(3x-5)}{2\cdot 3x(1+x)(x+2)(x-2)}$
$=\frac{-1(x-2)(x+1)(3x-5)}{2x(x+1)(x+2)(x-2)}$
$=-\frac{3x-5}{2x(x+2)}$

4. $\frac{5a^3b^2}{24}\div\frac{10a^5}{6}=\frac{5a^3b^2}{24}\cdot\frac{6}{10a^5}$
$=\frac{5a^3b^2\cdot 6}{4\cdot 6\cdot 2\cdot 5\cdot a^2\cdot a^3}$
$=\frac{b^2}{8a^2}$

5. $\frac{(3x+1)(x-5)}{3}\div\frac{4x-20}{9}$
$=\frac{(3x+1)(x-5)}{3}\cdot\frac{9}{4x-20}$
$=\frac{(3x+1)(x-5)\cdot 3\cdot 3}{3\cdot 4(x-5)}$
$=\frac{3(3x+1)}{4}$

6. $\frac{10x-2}{x^2-9}\div\frac{5x^2-x}{x+3}=\frac{10x-2}{x^2-9}\cdot\frac{x+3}{5x^2-x}$
$=\frac{2(5x-1)(x+3)}{(x+3)(x-3)\cdot x(5x-1)}$
$=\frac{2}{x(x-3)}$

7. $\frac{3x^2-11x-4}{2x-8}\div\frac{9x+3}{6}=\frac{3x^2-11x-4}{2x-8}\cdot\frac{6}{9x+3}$
$=\frac{(3x+1)(x-4)\cdot 2\cdot 3}{2(x-4)\cdot 3(3x+1)}$
$=\frac{1}{1}$ or 1

8. a. $\frac{y+9}{8x}\cdot\frac{y+9}{2x}=\frac{(y+9)\cdot(y+9)}{8x\cdot 2x}=\frac{(y+9)^2}{16x^2}$

b. $\frac{y+9}{8x}\div\frac{y+9}{2}=\frac{y+9}{8x}\cdot\frac{2}{y+9}$
$=\frac{(y+9)\cdot 2}{2\cdot 4\cdot x\cdot(y+9)}$
$=\frac{1}{4x}$

c. $\frac{35x-7x^2}{x^2-25}\cdot\frac{x^2+3x-10}{x^2+4x}$
$=\frac{7x(5-x)}{(x+5)(x-5)}\cdot\frac{(x-2)(x+5)}{x(x+4)}$
$=\frac{7x\cdot(-1)(x-5)\cdot(x-2)(x+5)}{(x+5)(x-5)\cdot x(x+4)}$
$=-\frac{7(x-2)}{x+4}$

Vocabulary and Readiness Check

1. The expressions $\frac{x}{2y}$ and $\frac{2y}{x}$ are called reciprocals.

2. $\frac{a}{b}\cdot\frac{c}{d}=\frac{a\cdot c}{b\cdot d}$ or $\frac{ac}{bd}$

3. $\frac{a}{b}\div\frac{c}{d}=\frac{a\cdot d}{b\cdot c}$ or $\frac{ad}{bc}$

4. $\frac{x}{7}\cdot\frac{x}{6}=\frac{x^2}{42}$

5. $\frac{x}{7}\div\frac{x}{6}=\frac{6}{7}$

Exercise Set 7.2

1. $\frac{3x}{y^2}\cdot\frac{7y}{4x}=\frac{3\cdot x\cdot 7\cdot y}{y\cdot y\cdot 4\cdot x}=\frac{21}{4y}$

3. $\frac{8x}{2}\cdot\frac{x^5}{4x^2}=\frac{8x\cdot x^5}{2\cdot 4x^2}=\frac{2\cdot 4\cdot x\cdot x\cdot x^4}{2\cdot 4\cdot x\cdot x}=x^4$

5. $-\frac{5a^2b}{30a^2b^2}\cdot b^3=\frac{5a^2b\cdot b^3}{30a^2b^2}$
$=-\frac{5\cdot a^2\cdot b\cdot b\cdot b^2}{5\cdot 6\cdot a^2\cdot b^2}$
$=-\frac{b\cdot b}{6}$
$=-\frac{b^2}{6}$

7. $\frac{x}{2x-14}\cdot\frac{x^2-7x}{5}=\frac{x\cdot(x^2-7x)}{(2x-14)\cdot 5}$
$=\frac{x\cdot x(x-7)}{2(x-7)\cdot 5}$
$=\frac{x\cdot x}{2\cdot 5}$
$=\frac{x^2}{10}$

9. $\frac{6x+6}{5}\cdot\frac{10}{36x+36}=\frac{(6x+6)\cdot 10}{5\cdot(36x+36)}$
$=\frac{6(x+1)\cdot 2\cdot 5}{5\cdot 36(x+1)}$
$=\frac{6\cdot 5\cdot 2\cdot(x+1)}{6\cdot 5\cdot 2\cdot 3\cdot(x+1)}$
$=\frac{1}{3}$

11. $\frac{(m+n)^2}{m-n}\cdot\frac{m}{m^2+mn}=\frac{(m+n)(m+n)\cdot m}{(m-n)\cdot m(m+n)}=\frac{m+n}{m-n}$

13. $\frac{x^2-25}{x^2-3x-10}\cdot\frac{x+2}{x}=\frac{(x^2-25)\cdot(x+2)}{(x^2-3x-10)\cdot x}$
$=\frac{(x-5)(x+5)\cdot(x+2)}{(x-5)(x+2)\cdot x}$
$=\frac{x+5}{x}$

15. $\frac{x^2+6x+8}{x^2+x-20}\cdot\frac{x^2+2x-15}{x^2+8x+16}$
$=\frac{(x+2)(x+4)}{(x+5)(x-4)}\cdot\frac{(x+5)(x-3)}{(x+4)(x+4)}$
$=\frac{(x+2)(x+4)\cdot(x+5)(x-3)}{(x+5)(x-4)\cdot(x+4)(x+4)}$
$=\frac{(x+2)(x-3)}{(x-4)(x+4)}$

17. $\frac{5x^7}{2x^5}\div\frac{15x}{4x^3}=\frac{5x^7}{2x^5}\cdot\frac{4x^3}{15x}$
$=\frac{5\cdot x^2\cdot x^5\cdot 2\cdot 2\cdot x\cdot x^2}{2\cdot x^5\cdot 3\cdot 5\cdot x}$
$=\frac{2x^4}{3}$

19. $\frac{8x^2}{y^3}\div\frac{4x^2y^3}{6}=\frac{8x^2}{y^3}\cdot\frac{6}{4x^2y^3}=\frac{2\cdot 4\cdot x^2\cdot 6}{y^3\cdot 4x^2y^3}=\frac{12}{y^6}$

21. $\frac{(x-6)(x+4)}{4x}\div\frac{2x-12}{8x^2}$
$=\frac{(x-6)(x+4)}{4x}\cdot\frac{8x^2}{2x-12}$
$=\frac{(x-6)(x+4)\cdot 2\cdot 4\cdot x\cdot x}{4x\cdot 2(x-6)}$
$=x(x+4)$

23. $\frac{3x^2}{x^2-1}\div\frac{x^5}{(x+1)^2}=\frac{3x^2}{x^2-1}\cdot\frac{(x+1)^2}{x^5}$
$=\frac{3x^2\cdot(x+1)(x+1)}{(x-1)(x+1)\cdot x^2\cdot x^3}$
$=\frac{3(x+1)}{x^3(x-1)}$

25. $\frac{m^2-n^2}{m+n}\div\frac{m}{m^2+nm}=\frac{m^2-n^2}{m+n}\cdot\frac{m^2+nm}{m}$
$=\frac{(m-n)(m+n)\cdot m(m+n)}{(m+n)\cdot m}$
$=(m-n)(m+n)$
$=m^2-n^2$

27. $\frac{x+2}{7-x} \div \frac{x^2-5x+6}{x^2-9x+14} = \frac{x+2}{7-x} \cdot \frac{x^2-9x+14}{x^2-5x+6}$
$= \frac{(x+2)\cdot(x-7)(x-2)}{-1(x-7)\cdot(x-3)(x-2)}$
$= -\frac{x+2}{x-3}$

29. $\frac{x^2+7x+10}{x-1} \div \frac{x^2+2x-15}{x-1}$
$= \frac{x^2+7x+10}{x-1} \cdot \frac{x-1}{x^2+2x-15}$
$= \frac{(x+5)(x+2)\cdot(x-1)}{(x-1)\cdot(x+5)(x-3)}$
$= \frac{x+2}{x-3}$

31. $\frac{5x-10}{12} \div \frac{4x-8}{8} = \frac{5x-10}{12} \cdot \frac{8}{4x-8}$
$= \frac{5(x-2)\cdot 2\cdot 4}{6\cdot 2\cdot 4(x-2)}$
$= \frac{5}{6}$

33. $\frac{x^2+5x}{8} \cdot \frac{9}{3x+15} = \frac{x(x+5)\cdot 3\cdot 3}{8\cdot 3(x+5)} = \frac{3x}{8}$

35. $\frac{7}{6p^2+q} \div \frac{14}{18p^2+3q} = \frac{7}{6p^2+q} \cdot \frac{18p^2+3q}{14}$
$= \frac{7\cdot 3(6p^2+q)}{(6p^2+q)\cdot 7\cdot 2}$
$= \frac{3}{2}$

37. $\frac{3x+4y}{x^2+4xy+4y^2} \cdot \frac{x+2y}{2} = \frac{(3x+4y)\cdot(x+2y)}{(x+2y)(x+2y)\cdot 2}$
$= \frac{3x+4y}{2(x+2y)}$

39. $\frac{(x+2)^2}{x-2} \div \frac{x^2-4}{2x-4} = \frac{(x+2)^2}{x-2} \cdot \frac{2x-4}{x^2-4}$
$= \frac{(x+2)(x+2)\cdot 2(x-2)}{(x-2)\cdot(x+2)(x-2)}$
$= \frac{2(x+2)}{x-2}$

41. $\frac{x^2-4}{24x} \div \frac{2-x}{6xy} = \frac{x^2-4}{24x} \cdot \frac{6xy}{2-x}$
$= \frac{(x+2)(x-2)\cdot 6x\cdot y}{4\cdot 6x\cdot(-1)(x-2)}$
$= -\frac{y(x+2)}{4}$

43. $\frac{a^2+7a+12}{a^2+5a+6} \cdot \frac{a^2+8a+15}{a^2+5a+4}$
$= \frac{(a+3)(a+4)\cdot(a+5)(a+3)}{(a+3)(a+2)\cdot(a+4)(a+1)}$
$= \frac{(a+5)(a+3)}{(a+2)(a+1)}$

45. $\frac{5x-20}{3x^2+x} \cdot \frac{3x^2+13x+4}{x^2-16}$
$= \frac{5(x-4)}{x(3x+1)} \cdot \frac{(3x+1)(x+4)}{(x+4)(x-4)}$
$= \frac{5(x-4)\cdot(3x+1)(x+4)}{x(3x+1)\cdot(x+4)(x-4)}$
$= \frac{5}{x}$

47. $\frac{8n^2-18}{2n^2-5n+3} \div \frac{6n^2+7n-3}{n^2-9n+8}$
$= \frac{8n^2-18}{2n^2-5n+3} \cdot \frac{n^2-9n+8}{6n^2+7n-3}$
$= \frac{2(2n+3)(2n-3)\cdot(n-8)(n-1)}{(n-1)(2n-3)\cdot(2n+3)(3n-1)}$
$= \frac{2(n-8)}{3n-1}$

49. $\frac{x^2-9}{2x} \div \frac{x+3}{8x^4} = \frac{x^2-9}{2x} \cdot \frac{8x^4}{x+3}$
$= \frac{(x+3)(x-3)}{2x} \cdot \frac{8x^4}{x+3}$
$= \frac{2x\cdot 4x^3\cdot(x+3)(x-3)}{2x\cdot(x+3)}$
$= 4x^3(x-3)$

51. $\frac{a^2+ac+ba+bc}{a-b} \div \frac{a+c}{a+b}$
$= \frac{a(a+c)+b(a+c)}{a-b} \cdot \frac{a+b}{a+c}$
$= \frac{(a+c)(a+b)}{a-b} \cdot \frac{a+b}{a+c}$
$= \frac{(a+c)\cdot(a+b)\cdot(a+b)}{(a-b)\cdot(a+c)}$
$= \frac{(a+b)^2}{a-b}$

53. $\frac{3x^2+8x+5}{x^2+8x+7} \cdot \frac{x+7}{x^2+4} = \frac{(3x+5)(x+1)}{(x+7)(x+1)} \cdot \frac{x+7}{x^2+4}$
$= \frac{(3x+5)\cdot(x+1)\cdot(x+7)}{(x+7)\cdot(x+1)\cdot(x^2+4)}$
$= \frac{3x+5}{x^2+4}$

55. $\frac{x^3+8}{x^2-2x+4} \cdot \frac{4}{x^2-4}$
$= \frac{(x+2)(x^2-2x+4)}{x^2-2x+4} \cdot \frac{4}{(x+2)(x-2)}$
$= \frac{4\cdot(x+2)\cdot(x^2-2x+4)}{(x+2)(x-2)(x^2-2x+4)}$
$= \frac{4}{x-2}$

57. $\frac{a^2-ab}{6a^2+6ab} \div \frac{a^3-b^3}{a^2-b^2}$
$= \frac{a^2-ab}{6a^2+6ab} \cdot \frac{a^2-b^2}{a^3-b^3}$
$= \frac{a(a-b)}{6a(a+b)} \cdot \frac{(a-b)(a+b)}{(a-b)(a^2+ab+b^2)}$
$= \frac{a\cdot(a-b)\cdot(a-b)\cdot(a+b)}{6\cdot a\cdot(a+b)\cdot(a-b)\cdot(a^2+ab+b^2)}$
$= \frac{a-b}{6(a^2+ab+b^2)}$

59. $\frac{1}{5}+\frac{4}{5}=\frac{5}{5}=1$

61. $\frac{9}{9}-\frac{19}{9}=-\frac{10}{9}$

63. $\frac{6}{5}+\left(\frac{1}{5}-\frac{8}{5}\right)=\frac{6}{5}+\left(-\frac{7}{5}\right)=-\frac{1}{5}$

65. $x-2y=6$

x	y
0	−3
6	0

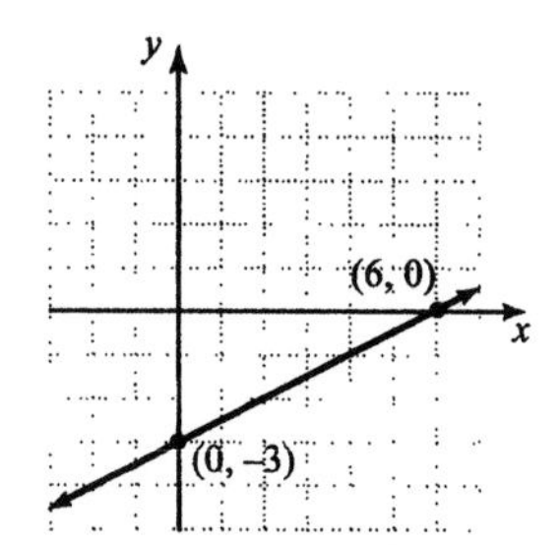

67. $\frac{4}{a}\cdot\frac{1}{b}=\frac{4\cdot 1}{a\cdot b}=\frac{4}{ab}$
The statement is true.

69. $\frac{x}{5}\cdot\frac{x+3}{4}=\frac{x\cdot(x+3)}{5\cdot 4}=\frac{x^2+3x}{20}$
The statement is false.

71. Area = length · width
$\frac{x+5}{9x}\cdot\frac{2x}{x^2-25}=\frac{(x+5)\cdot 2\cdot x}{9\cdot x\cdot(x+5)(x-5)}=\frac{2}{9(x-5)}$
The area of the rectangle is $\frac{2}{9(x-5)}$ square feet.

73. $\left(\frac{x^2-y^2}{x^2+y^2} \div \frac{x^2-y^2}{3x}\right)\cdot\frac{x^2+y^2}{6}$
$= \frac{x^2-y^2}{x^2+y^2}\cdot\frac{3x}{x^2-y^2}\cdot\frac{x^2+y^2}{6}$
$= \frac{(x^2-y^2)\cdot 3x\cdot(x^2+y^2)}{(x^2+y^2)\cdot(x^2-y^2)\cdot 2\cdot 3}$
$= \frac{x}{2}$

75. $\left(\frac{2a+b}{b^2}\cdot\frac{3a^2-2ab}{ab+2b^2}\right)\div\frac{a^2-3ab+2b^2}{5ab-10b^2}$

$$=\frac{2a+b}{b^2}\cdot\frac{3a^2-2ab}{ab+2b^2}\cdot\frac{5ab-10b^2}{a^2-3ab+2b^2}$$
$$=\frac{(2a+b)\cdot(3a^2-2ab)\cdot(5ab-10b^2)}{b^2\cdot(ab+2b^2)\cdot(a^2-3ab+2b^2)}$$
$$=\frac{(2a+b)\cdot a(3a-2b)\cdot 5b(a-2b)}{b^2\cdot b(a+2b)\cdot(a-2b)(a-b)}$$
$$=\frac{5a(2a+b)(3a-2b)}{b^2(a+2b)(a-b)}$$

77. Answers may vary

Section 7.3

Practice Exercises

1. $\frac{7a}{4b}+\frac{a}{4b}=\frac{7a+a}{4b}=\frac{8a}{4b}=\frac{2a}{b}$

2. $\frac{3x}{3x-2}-\frac{2}{3x-2}=\frac{3x-2}{3x-2}=\frac{1}{1}$ or 1

3. $\frac{4x^2+15x}{x+3}-\frac{8x+15}{x+3}=\frac{(4x^2+15x)-(8x+15)}{x+3}$

$$=\frac{4x^2+15x-8x-15}{x+3}$$
$$=\frac{4x^2+7x-15}{x+3}$$
$$=\frac{(x+3)(4x-5)}{x+3}$$
$$=4x-5$$

4. a. Find the prime factorization of each denominator.
$14=2\cdot 7$
$21=3\cdot 7$
The greatest number of times that the factor 2 appears is 1. The greatest number of times that the factor 3 appears is 1. The greatest number of times that the factor 7 appears is 1.
$\text{LCD}=2^1\cdot 3^1\cdot 7^1=42$

b. Factor each denominator.
$9y=3\cdot 3\cdot y=3^2\cdot y$
$15y^3=3\cdot 5\cdot y^3$
The greatest number of times that the factor 3 appears is 2. The greatest number of times that the factor 5 appears is 1. The greatest number of times that the factor y appears is 3.
$\text{LCD}=3^2\cdot 5^1\cdot y^3=9\cdot 5\cdot y^3=45y^3$

5. a. The denominators $y-5$ and $y-4$ are completely factored already. The factor $y-5$ appears once and the factor $y-4$ appears once.
$\text{LCD}=(y-5)(y-4)$

b. The denominators a and $a+2$ cannot be factored further. The factor a appears once and the factor $a+2$ appears once.
$\text{LCD}=a(a+2)$

6. Factor each denominator.
$(2x-1)^2=(2x-1)^2$
$6x-3=3(2x-1)$
The greatest number that the factor $2x-1$ appears in any one denominator is 2.
The greatest number of times that the factor 3 appears is 1.
$\text{LCD}=3(2x-1)^2$

7. Factor each denominator.
$x^2+5x+4=(x+1)(x+4)$
$x^2-16=(x-4)(x+4)$
$\text{LCD}=(x+1)(x+4)(x-4)$

8. The denominators $3-x$ and $x-3$ are opposites. That is, $3-x=-1(x-3)$. Use $x-3$ or $3-x$ as the LCD.
$\text{LCD}=x-3$ or $\text{LCD}=3-x$

9. a. Since $5y(7xy)=35xy^2$, multiply by 1 in the form of $\frac{7xy}{7xy}$.

$$\frac{3x}{5y}=\frac{3x}{5y}\cdot 1=\frac{3x}{5y}\cdot\frac{7xy}{7xy}=\frac{3x(7xy)}{5y(7xy)}=\frac{21x^2y}{35xy^2}$$

b. First, factor the denominator on the right.
$$\frac{9x}{4x+7}=\frac{}{2(4x+7)}$$
To obtain the denominator on the right from the denominator on the left, multiply by 1 in the form of $\frac{2}{2}$.

$$\frac{9x}{4x+7} = \frac{9x}{4x+7} \cdot \frac{2}{2}$$
$$= \frac{9x \cdot 2}{(4x+7) \cdot 2}$$
$$= \frac{18x}{2(4x+7)} \text{ or } \frac{18x}{8x+14}$$

10. First, factor the denominator $x^2 - 2x - 15$ as $(x + 3)(x - 5)$. If we multiply the original denominator $(x + 3)(x - 5)$ by $x - 2$, the result is the new denominator $(x - 2)(x + 3)(x - 5)$. Thus, we multiply by 1 in the form $\frac{x-2}{x-2}$.

$$\frac{3}{x^2-2x-15} = \frac{3}{(x+3)(x-5)}$$
$$= \frac{3}{(x+3)(x-5)} \cdot \frac{x-2}{x-2}$$
$$= \frac{3(x-2)}{(x+3)(x-5)(x-2)}$$
$$= \frac{3x-6}{(x-2)(x+30(x-5)}$$

Vocabulary and Readiness Check

1. $\frac{7}{11} + \frac{2}{11} = \underline{\frac{9}{11}}$

2. $\frac{7}{11} - \frac{2}{11} = \underline{\frac{5}{11}}$

3. $\frac{a}{b} + \frac{c}{b} = \underline{\frac{a+c}{b}}$

4. $\frac{a}{b} - \frac{c}{b} = \underline{\frac{a-c}{b}}$

5. $\frac{5}{x} - \frac{6+x}{x} = \underline{\frac{5-(6+x)}{x}}$

Exercise Set 7.3

1. $\frac{a+1}{13} + \frac{8}{13} = \frac{a+1+8}{13} = \frac{a+9}{13}$

3. $\frac{4m}{3n} + \frac{5m}{3n} = \frac{4m+5m}{3n} = \frac{9m}{3n} = \frac{3m}{n}$

5. $\frac{4m}{m-6} - \frac{24}{m-6} = \frac{4m-24}{m-6} = \frac{4(m-6)}{m-6} = 4$

7. $\frac{9}{3+y} + \frac{y+1}{3+y} = \frac{9+y+1}{3+y} = \frac{10+y}{3+y}$

9. $\frac{5x^2+4x}{x-1} - \frac{6x+3}{x-1} = \frac{5x^2+4x-(6x+3)}{x-1}$
$$= \frac{5x^2+4x-6x-3}{x-1}$$
$$= \frac{5x^2-2x-3}{x-1}$$
$$= \frac{(5x+3)(x-1)}{x-1}$$
$$= 5x+3$$

11. $\frac{4a}{a^2+2a-15} - \frac{12}{a^2+2a-15} = \frac{4a-12}{a^2+2a-15}$
$$= \frac{4(a-3)}{(a+5)(a-3)}$$
$$= \frac{4}{a+5}$$

13. $\frac{2x+3}{x^2-x-30} - \frac{x-2}{x^2-x-30} = \frac{2x+3-(x-2)}{x^2-x-30}$
$$= \frac{2x+3-x+2}{x^2-x-30}$$
$$= \frac{x+5}{x^2-x-30}$$
$$= \frac{x+5}{(x-6)(x+5)}$$
$$= \frac{1}{x-6}$$

15. $\frac{2x+1}{x-3} + \frac{3x+6}{x-3} = \frac{2x+1+3x+6}{x-3} = \frac{5x+7}{x-3}$

17. $\frac{2x^2}{x-5} - \frac{25+x^2}{x-5} = \frac{2x^2-(25+x^2)}{x-5}$
$$= \frac{2x^2-25-x^2}{x-5}$$
$$= \frac{x^2-25}{x-5}$$
$$= \frac{(x+5)(x-5)}{x-5}$$
$$= x+5$$

19. $\frac{5x+4}{x-1}-\frac{2x+7}{x-1}=\frac{5x+4-(2x+7)}{x-1}$
$=\frac{5x+4-2x-7}{x-1}$
$=\frac{3x-3}{x-1}$
$=\frac{3(x-1)}{x-1}$
$=3$

21. $2x=2\cdot x$
$4x^3=2^2\cdot x^3$
$\text{LCD}=2^2\cdot x^3=4x^3$

23. $8x=2^3\cdot x$
$2x+4=2(x+2)$
$\text{LCD}=2^3\cdot x\cdot(x+2)=8x(x+2)$

25. $x+3=x+3$
$x-2=x-2$
$\text{LCD}=(x+3)(x-2)$

27. $x+6=x+6$
$3x+18=3(x+6)$
$\text{LCD}=3(x+6)$

29. $(x-6)^2=(x-6)^2$
$5x-30=5(x-6)$
$\text{LCD}=5(x-6)^2$

31. $3x+3=3\cdot(x+1)$
$2x^2+4x+2=2(x^2+2x+1)=2\cdot(x+1)^2$
$\text{LCD}=2\cdot 3(x+1)^2=6(x+1)^2$

33. $x-8=x-8$
$8-x=-(x-8)$
$\text{LCD}=x-8 \text{ or } 8-x$

35. $x^2+3x-4=(x-1)(x+4)$
$x^2+2x-3=(x-1)(x+3)$
$\text{LCD}=(x-1)(x+4)(x+3)$

37. $3x^2+4x+1=(3x+1)(x+1)$
$2x^2-x-1=(x-1)(2x+1)$
$\text{LCD}=(3x+1)(x+1)(x-1)(2x+1)$

39. $x^2-16=(x+4)(x-4)$
$2x^3-8x^2=2x^2(x-4)$
$\text{LCD}=2x^2(x+4)(x-4)$

41. $\frac{3}{2x}=\frac{3(2x)}{2x(2x)}=\frac{6x}{4x^2}$

43. $\frac{6}{3a}=\frac{6(4b^2)}{3a(4b^2)}=\frac{24b^2}{12ab^2}$

45. $\frac{9}{2x+6}=\frac{9}{2(x+3)}=\frac{9(y)}{2(x+3)(y)}=\frac{9y}{2y(x+3)}$

47. $\frac{9a+2}{5a+10}=\frac{9a+2}{5(a+2)}=\frac{(9a+2)(b)}{5(a+2)(b)}=\frac{9ab+2b}{5b(a+2)}$

49. $\frac{x}{x^3+6x^2+8x}=\frac{x}{x(x+4)(x+2)}$
$=\frac{x(x+1)}{x(x+4)(x+2)(x+1)}$
$=\frac{x^2+x}{x(x+4)(x+2)(x+1)}$

51. $\frac{9y-1}{15x^2-30}=\frac{(9y-1)(2)}{(15x^2-30)2}=\frac{18y-2}{30x^2-60}$

53. $\frac{5x}{7}+\frac{9x}{7}=\frac{5x+9x}{7}=\frac{14x}{7}=\frac{2x}{1}=2x$

55. $\frac{x+3}{4}\div\frac{2x-1}{4}=\frac{x+3}{4}\cdot\frac{4}{2x-1}$
$=\frac{(x+3)\cdot 4}{4\cdot(2x-1)}$
$=\frac{x+3}{2x-1}$

57. $\frac{x^2}{x-6}-\frac{5x+6}{x-6}=\frac{x^2-(5x+6)}{x-6}$
$=\frac{x^2-5x-6}{x-6}$
$=\frac{(x+1)(x-6)}{x-6}$
$=x+1$

59. $\dfrac{-2x}{x^3-8x}+\dfrac{3x}{x^3-8x}=\dfrac{-2x+3x}{x^3-8x}$
$=\dfrac{x}{x(x^2-8)}$
$=\dfrac{1}{x^2-8}$

61. $\dfrac{12x-6}{x^2+3x}\cdot\dfrac{4x^2+13x+3}{4x^2-1}$
$=\dfrac{6(2x-1)\cdot(x+3)(4x+1)}{x(x+3)\cdot(2x+1)(2x-1)}$
$=\dfrac{6(4x+1)}{x(2x+1)}$

63. LCD = 21
$\dfrac{2}{3}+\dfrac{5}{7}=\dfrac{2(7)}{3(7)}+\dfrac{5(3)}{7(3)}=\dfrac{14}{21}+\dfrac{15}{21}=\dfrac{29}{21}$

65. $6=2\cdot 3$
$4=2^2$
$\text{LCD}=2^2\cdot 3=12$
$\dfrac{2}{6}-\dfrac{3}{4}=\dfrac{2(2)}{6(2)}-\dfrac{3(3)}{4(3)}=\dfrac{4}{12}-\dfrac{9}{12}=\dfrac{4-9}{12}=-\dfrac{5}{12}$

67. $12=2\cdot 2\cdot 3=2^2\cdot 3$
$20=2\cdot 2\cdot 5=2^2\cdot 5$
$\text{LCD}=2^2\cdot 3\cdot 5=60$
$\dfrac{1}{12}+\dfrac{3}{20}=\dfrac{1(5)}{12(5)}+\dfrac{3(3)}{20(3)}=\dfrac{5}{60}+\dfrac{9}{60}=\dfrac{14}{60}=\dfrac{7}{30}$

69. $4a-20=4(a-5)$
$(a-5)^2=(a-5)^2$
$\text{LCD}=4(a-5)^2$
The correct choice is d.

71. $\dfrac{3}{x}+\dfrac{y}{x}=\dfrac{3+y}{x}$
The correct choice is c.

73. $\dfrac{3}{x}\cdot\dfrac{y}{x}=\dfrac{3\cdot y}{x\cdot x}=\dfrac{3y}{x^2}$
The correct choice is b.

75. $\dfrac{5}{2-x}=\dfrac{5(-1)}{(2-x)(-1)}=-\dfrac{5}{x-2}$

77. $-\dfrac{7+x}{2-x}=\dfrac{7+x}{(-1)(2-x)}=\dfrac{7+x}{x-2}$

79. $P=\dfrac{5}{x-2}+\dfrac{5}{x-2}+\dfrac{5}{x-2}+\dfrac{5}{x-2}$
$=\dfrac{5+5+5+5}{x-2}$
$=\dfrac{20}{x-2}$
The perimeter is $\dfrac{20}{x-2}$ meters.

81. Answers may vary

83. $88=2^3\cdot 11$
$4332=2^3\cdot 3\cdot 19^2$
$\text{LCM}=2^3\cdot 3\cdot 11\cdot 19^2=95{,}304$
They will align again in 95,304 Earth days.

85. Answers may vary

87. Answers may vary

Section 7.4

Practice Exercises

1. a. Since $5=5$ and $15=3\cdot 5$, the $\text{LCD}=3\cdot 5=15$.
$\dfrac{2x}{5}-\dfrac{6x}{15}=\dfrac{2x(3)}{5(3)}-\dfrac{6x}{15}$
$=\dfrac{6x}{15}-\dfrac{6x}{15}$
$=\dfrac{6x-6x}{15}$
$=\dfrac{0}{15}$
$=0$

b. Since $8a=2^3\cdot a$ and $12a^2=2^2\cdot 3\cdot a^2$, the $\text{LCD}=2^3\cdot 3\cdot a^2=24a^2$.
$\dfrac{7}{8a}+\dfrac{5}{12a^2}=\dfrac{7(3a)}{8a(3a)}+\dfrac{5(2)}{12a^2(2)}$
$=\dfrac{21a}{24a^2}+\dfrac{10}{24a^2}$
$=\dfrac{21a+10}{24a^2}$

2. Since $x^2 - 25 = (x+5)(x-5)$, the LCD $= (x + 5)(x - 5)$.

$$\frac{12x}{x^2-25} - \frac{6}{x+5} = \frac{12x}{(x+5)(x-5)} - \frac{6(x-5)}{(x+5)(x-5)}$$
$$= \frac{12x-6(x-5)}{(x+5)(x-5)}$$
$$= \frac{12x-6x+30}{(x+5)(x-5)}$$
$$= \frac{6x+30}{(x+5)(x-5)}$$
$$= \frac{6(x+5)}{(x+5)(x-5)}$$
$$= \frac{6}{x-5}$$

3. The LCD is $5y(y + 1)$.

$$\frac{3}{5y} + \frac{2}{y+1} = \frac{3(y+1)}{5y(y+1)} + \frac{2(5y)}{(y+1)(5y)}$$
$$= \frac{3(y+1)+2(5y)}{5y(y+1)}$$
$$= \frac{3y+3+10y}{5y(y+1)}$$
$$= \frac{13y+3}{5y(y+1)}$$

4. $x - 5$ and $5 - x$ are opposites. Write the denominator $5 - x$ as $-(x - 5)$ and simplify.

$$\frac{6}{x-5} - \frac{7}{5-x} = \frac{6}{x-5} - \frac{7}{-(x-5)}$$
$$= \frac{6}{x-5} - \frac{-7}{x-5}$$
$$= \frac{6-(-7)}{x-5}$$
$$= \frac{13}{x-5}$$

5. Note that 2 is the same as $\frac{2}{1}$. The LCD of $\frac{2}{1}$ and $\frac{b}{b+3}$ is $b + 3$.

$$2 + \frac{b}{b+3} = \frac{2}{1} + \frac{b}{b+3}$$
$$= \frac{2(b+3)}{1(b+3)} + \frac{b}{b+3}$$
$$= \frac{2(b+3)+b}{b+3}$$
$$= \frac{2b+6+b}{b+3}$$
$$= \frac{3b+6}{b+3} \text{ or } \frac{3(b+2)}{b+3}$$

6. First, factor the denominators.

$$\frac{5}{2x^2+3x} - \frac{3x}{4x+6} = \frac{5}{x(2x+3)} - \frac{3x}{2(2x+3)}$$

The LCD is $2x(2x + 3)$.

$$\frac{5}{2x^2+3x} - \frac{3x}{4x+6} = \frac{5(2)}{x(2x+3)(2)} - \frac{3x(x)}{2(2x+3)(x)}$$
$$= \frac{10-3x^2}{2x(2x+3)}$$

7. First, factor the denominators.

$x^2 + 7x + 12 = (x+4)(x+3)$

$x^2 - 9 = (x+3)(x-3)$

LCD $= (x + 4)(x + 3)(x - 3)$

$$\frac{2x}{x^2+7x+12} + \frac{3x}{x^2-9}$$
$$= \frac{2x}{(x+4)(x+3)} + \frac{3x}{(x+3)(x-3)}$$
$$= \frac{2x(x-3)}{(x+4)(x+3)(x-3)} + \frac{3x(x+4)}{(x+3)(x-3)(x+4)}$$
$$= \frac{2x(x-3)+3x(x+4)}{(x+4)(x+3)(x-3)}$$
$$= \frac{2x^2-6x+3x^2+12x}{(x+4)(x+3)(x-3)}$$
$$= \frac{5x^2+6x}{(x+4)(x+3)(x-3)} \text{ or } \frac{x(5x+6)}{(x+4)(x+3)(x-3)}$$

Vocabulary and Readiness Check

1. The first step to perform on $\frac{3}{4} - \frac{y}{4}$ is to subtract the numerators and place the difference over the common denominator; choice d.

2. The first step to perform on $\frac{2}{a} \cdot \frac{3}{a+6}$ is to multiply the numerators and multiply the denominators; choice c.

3. The first step to perform on $\frac{x+1}{x} \div \frac{x-1}{x}$ is to multiply the first rational expression by the reciprocal of the second rational expression; choice a.

4. The first step to perform on $\frac{9}{x-2} - \frac{x}{x+2}$ is to find the LCD and write each expression as an equivalent expression with the LCD as denominator; choice b.

Exercise Set 7.4

1. $\text{LCD} = 2 \cdot 3 \cdot x = 6x$

$$\frac{4}{2x} + \frac{9}{3x} = \frac{4(3)}{2x(3)} + \frac{9(2)}{3x(2)} = \frac{12}{6x} + \frac{18}{6x} = \frac{30}{6x} = \frac{5(6)}{6x} = \frac{5}{x}$$

3. $\text{LCD} = 5b$

$$\frac{15a}{b} - \frac{6b}{5} = \frac{15a(5)}{b(5)} - \frac{6b(b)}{5(b)} = \frac{75a}{5b} - \frac{6b^2}{5b} = \frac{75a - 6b^2}{5b}$$

5. $\text{LCD} = 2x^2$

$$\frac{3}{x} + \frac{5}{2x^2} = \frac{3(2x)}{x(2x)} + \frac{5}{2x^2} = \frac{6x}{2x^2} + \frac{5}{2x^2} = \frac{6x+5}{2x^2}$$

7. $2x + 2 = 2(x + 1)$
$\text{LCD} = 2(x + 1)$

$$\frac{6}{x+1} + \frac{10}{2x+2} = \frac{6}{x+1} + \frac{10}{2(x+1)} = \frac{6(2)}{(x+1)2} + \frac{10}{2(x+1)} = \frac{12}{2(x+1)} + \frac{10}{2(x+1)} = \frac{12+10}{2(x+1)} = \frac{22}{2(x+1)} = \frac{2(11)}{2(x+1)} = \frac{11}{x+1}$$

9. $x^2 - 4 = (x+2)(x-2)$
$\text{LCD} = (x + 2)(x - 2)$

$$\frac{3}{x+2} - \frac{2x}{x^2-4} = \frac{3(x-2)}{(x+2)(x-2)} - \frac{2x}{(x+2)(x-2)} = \frac{3(x-2) - 2x}{(x+2)(x-2)} = \frac{3x - 6 - 2x}{(x+2)(x-2)} = \frac{x-6}{(x+2)(x-2)}$$

11. $\text{LCD} = 4x(x - 2)$

$$\frac{3}{4x} + \frac{8}{x-2} = \frac{3(x-2)}{4x(x-2)} + \frac{8(4x)}{(x-2)(4x)} = \frac{3x-6}{4x(x-2)} + \frac{32x}{4x(x-2)} = \frac{3x - 6 + 32x}{4x(x-2)} = \frac{35x-6}{4x(x-2)}$$

13. $3 - x = -(x - 3)$

$$\frac{6}{x-3} + \frac{8}{3-x} = \frac{6}{x-3} + \frac{8}{-(x-3)} = \frac{6}{x-3} + \frac{-8}{x-3} = \frac{6+(-8)}{x-3} = -\frac{2}{x-3}$$

15. $3-x=-(x-3)$

$$\frac{9}{x-3}+\frac{9}{3-x}=\frac{9}{x-3}+\frac{9}{-(x-3)}$$
$$=\frac{9}{x-3}+\frac{-9}{x-3}$$
$$=\frac{9+(-9)}{x-3}$$
$$=\frac{0}{x-3}$$
$$=0$$

17. $1-x^2=-(x^2-1)$

$$\frac{-8}{x^2-1}-\frac{7}{1-x^2}=\frac{8}{-(x^2-1)}-\frac{7}{1-x^2}$$
$$=\frac{8}{1-x^2}-\frac{7}{1-x^2}$$
$$=\frac{8-7}{1-x^2}$$
$$=\frac{1}{1-x^2} \text{ or } -\frac{1}{x^2-1}$$

19. LCD $= x$

$$\frac{5}{x}+2=\frac{5}{x}+\frac{2}{1}=\frac{5}{x}+\frac{2(x)}{1(x)}=\frac{5+2x}{x}$$

21. LCD $= x-2$

$$\frac{5}{x-2}+6=\frac{5}{x-2}+\frac{6}{1}$$
$$=\frac{5}{x-2}+\frac{6(x-2)}{1(x-2)}$$
$$=\frac{5}{x-2}+\frac{6x-12}{x-2}$$
$$=\frac{5+6x-12}{x-2}$$
$$=\frac{6x-7}{x-2}$$

23. LCD $= y+3$

$$\frac{y+2}{y+3}-2=\frac{y+2}{y+3}-\frac{2}{1}$$
$$=\frac{y+2}{y+3}-\frac{2(y+3)}{y+3}$$
$$=\frac{y+2}{y+3}-\frac{2y+6}{y+3}$$
$$=\frac{y+2-(2y+6)}{y+3}$$
$$=\frac{y+2-2y-6}{y+3}$$
$$=\frac{-y-4}{y+3}$$
$$=\frac{-(y+4)}{y+3}$$
$$=-\frac{y+4}{y+3}$$

25. LCD $= 4x$

$$\frac{-x+2}{x}-\frac{x-6}{4x}=\frac{(-x+2)(4)}{x(4)}-\frac{x-6}{4x}$$
$$=\frac{4(-x+2)-(x-6)}{4x}$$
$$=\frac{-4x+8-x+6}{4x}$$
$$=\frac{-5x+14}{4x} \text{ or } -\frac{5x-14}{4x}$$

27. $$\frac{5x}{x+2}-\frac{3x-4}{x+2}=\frac{5x-(3x-4)}{x+2}$$
$$=\frac{5x-3x+4}{x+2}$$
$$=\frac{2x+4}{x+2}$$
$$=\frac{2(x+2)}{x+2}$$
$$=2$$

29. LCD $= 21$

$$\frac{3x^4}{7}-\frac{4x^2}{21}=\frac{3x^4(3)}{7(3)}-\frac{4x^2}{21}$$
$$=\frac{3(3x^4)-4x^2}{21}$$
$$=\frac{9x^4-4x^2}{21}$$

31. LCD $= (x+3)^2$

$$\frac{1}{x+3}-\frac{1}{(x+3)^2}=\frac{1(x+3)}{(x+3)(x+3)}-\frac{1}{(x+3)^2}$$
$$=\frac{x+3}{(x+3)^2}-\frac{1}{(x+3)^2}$$
$$=\frac{x+3-1}{(x+3)^2}$$
$$=\frac{x+2}{(x+3)^2}$$

33. LCD $= 5b(b-1)$

$$\frac{4}{5b}+\frac{1}{b-1}=\frac{4(b-1)}{5b(b-1)}+\frac{1(5b)}{(b-1)(5b)}$$
$$=\frac{4b-4}{5b(b-1)}+\frac{5b}{5b(b-1)}$$
$$=\frac{4b-4+5b}{5b(b-1)}$$
$$=\frac{9b-4}{5b(b-1)}$$

35. LCD $= m$

$$\frac{2}{m}+1=\frac{2}{m}+\frac{1}{1}=\frac{2}{m}+\frac{1(m)}{1(m)}=\frac{2+m}{m}$$

37. LCD $= (x-7)(x-2)$

$$\frac{2x}{x-7}-\frac{x}{x-2}=\frac{2x(x-2)}{(x-7)(x-2)}-\frac{x(x-7)}{(x-2)(x-7)}$$
$$=\frac{2x(x-2)-x(x-7)}{(x-7)(x-2)}$$
$$=\frac{2x^2-4x-x^2+7x}{(x-7)(x-2)}$$
$$=\frac{x^2+3x}{(x-7)(x-2)} \text{ or } \frac{x(x+3)}{(x-7)(x-2)}$$

39. $2x-1=-(1-2x)$

$$\frac{6}{1-2x}-\frac{4}{2x-1}=\frac{6}{1-2x}-\frac{4}{-(1-2x)}$$
$$=\frac{6}{1-2x}-\frac{-4}{1-2x}$$
$$=\frac{6-(-4)}{1-2x}$$
$$=\frac{10}{1-2x}$$

41. LCD $= (x-1)(x+1)^2$

$$\frac{7}{(x+1)(x-1)}+\frac{8}{(x+1)^2}$$
$$=\frac{7(x+1)}{(x+1)(x-1)(x+1)}+\frac{8(x-1)}{(x+1)^2(x-1)}$$
$$=\frac{7x+7}{(x+1)^2(x-1)}+\frac{8x-8}{(x+1)^2(x-1)}$$
$$=\frac{7x+7+8x-8}{(x+1)^2(x-1)}$$
$$=\frac{15x-1}{(x+1)^2(x-1)}$$

43. $x^2-1=(x+1)(x-1)$

$x^2-2x+1=(x-1)^2$

LCD $= (x+1)(x-1)^2$

$$\frac{x}{x^2-1}-\frac{2}{x^2-2x+1}$$
$$=\frac{x(x-1)}{(x-1)(x+1)(x-1)}-\frac{2(x+1)}{(x-1)^2(x+1)}$$
$$=\frac{x^2-x}{(x-1)^2(x+1)}-\frac{2x+2}{(x-1)^2(x+1)}$$
$$=\frac{x^2-x-(2x+2)}{(x-1)^2(x+1)}$$
$$=\frac{x^2-x-2x-2}{(x-1)^2(x+1)}$$
$$=\frac{x^2-3x-2}{(x-1)^2(x+1)}$$

45. $2a+6=2(a+3)$

LCD $= 2(a+3)$

$$\frac{3a}{2a+6}-\frac{a-1}{a+3}=\frac{3a}{2(a+3)}-\frac{(a-1)(2)}{(a+3)(2)}$$
$$=\frac{3a}{2(a+3)}-\frac{2a-2}{2(a+3)}$$
$$=\frac{3a-(2a-2)}{2(a+3)}$$
$$=\frac{3a-2a+2}{2(a+3)}$$
$$=\frac{a+2}{2(a+3)}$$

47. LCD $= (2y+3)^2$

$$\frac{y-1}{2y+3}+\frac{3}{(2y+3)^2}=\frac{(y-1)(2y+3)}{(2y+3)(2y+3)}+\frac{3}{(2y+3)^2}$$
$$=\frac{(y-1)(2y+3)+3}{(2y+3)^2}$$
$$=\frac{2y^2+y-3+3}{(2y+3)^2}$$
$$=\frac{2y^2+y}{(2y+3)^2} \text{ or } \frac{y(2y+1)}{(2y+3)^2}$$

49. $2-x=-(x-2)$
$2x-4=2(x-2)$
LCD $= 2(x-2)$

$$\frac{5}{2-x}+\frac{x}{2x-4}=\frac{5}{-(x-2)}+\frac{x}{2(x-2)}$$
$$=\frac{-5}{x-2}+\frac{x}{2(x-2)}$$
$$=\frac{-5(2)}{(x-2)(2)}+\frac{x}{2(x-2)}$$
$$=\frac{-10}{2(x-2)}+\frac{x}{2(x-2)}$$
$$=\frac{x-10}{2(x-2)}$$

51. $x^2+6x+9=(x+3)^2$
LCD $= (x+3)^2$

$$\frac{15}{x^2+6x+9}+\frac{2}{x+3}=\frac{15}{(x+3)^2}+\frac{2(x+3)}{(x+3)(x+3)}$$
$$=\frac{15+2(x+3)}{(x+3)^2}$$
$$=\frac{15+2x+6}{(x+3)^2}$$
$$=\frac{2x+21}{(x+3)^2}$$

53. $x^2-5x-6=(x-3)(x-2)$
LCD $= (x-3)(x-2)$

$$\frac{13}{x^2-5x+6}-\frac{5}{x-3}$$
$$=\frac{13}{(x-3)(x-2)}-\frac{5(x-2)}{(x-3)(x-2)}$$
$$=\frac{13-(5x-10)}{(x-3)(x-2)}$$
$$=\frac{13-5x+10}{(x-3)(x-2)}$$
$$=\frac{-5x+23}{(x-3)(x-2)}$$

55. $m^2-100=(m+10)(m-10)$
LCD $= 2(m+10)(m-10)$

$$\frac{70}{m^2-100}+\frac{7}{2(m+10)}$$
$$=\frac{70(2)}{(m+10)(m-10)(2)}+\frac{7(m-10)}{2(m+10)(m-10)}$$
$$=\frac{70(2)+7(m-10)}{2(m+10)(m-10)}$$
$$=\frac{140+7m-70}{2(m+10)(m-10)}$$
$$=\frac{7m+70}{2(m+10)(m-10)}$$
$$=\frac{7(m+10)}{2(m+10)(m-10)}$$
$$=\frac{7}{2(m-10)}$$

57. $x^2-5x-6=(x-6)(x+1)$
$x^2-4x-5=(x-5)(x+1)$
LCD $= (x-6)(x+1)(x-5)$

$$\frac{x+8}{x^2-5x-6}+\frac{x+1}{x^2-4x-5}$$
$$=\frac{(x+8)(x-5)}{(x-6)(x+1)(x-5)}+\frac{(x+1)(x-6)}{(x-5)(x+1)(x-6)}$$
$$=\frac{x^2+3x-40+x^2-5x-6}{(x-6)(x+1)(x-5)}$$
$$=\frac{2x^2-2x-46}{(x-6)(x+1)(x-5)}$$
$$\text{or } \frac{2(x^2-x-23)}{(x-6)(x+1)(x-5)}$$

59. $4n^2-12n+8=4(n-1)(n-2)$

$3n^2-6n=3n(n-2)$

$\text{LCD}=4\cdot 3n(n-1)(n-2)=12n(n-1)(n-2)$

$$\frac{5}{4n^2-12n+8}-\frac{3}{3n^2-6n}$$
$$=\frac{5(3n)}{4(n-1)(n-2)(3n)}-\frac{3(4)(n-1)}{3n(n-2)(4)(n-1)}$$
$$=\frac{5(3n)-3(4)(n-1)}{12n(n-1)(n-2)}$$
$$=\frac{15n-12n+12}{12n(n-1)(n-2)}$$
$$=\frac{3n+12}{12n(n-1)(n-2)}$$
$$=\frac{3(n+4)}{12n(n-1)(n-2)}$$
$$=\frac{n+4}{4n(n-1)(n-2)}$$

61. $\frac{15x}{x+8}\cdot\frac{2x+16}{3x}=\frac{15x}{x+8}\cdot\frac{2(x+8)}{3x}$
$$=\frac{2\cdot 5\cdot 3x\cdot(x+8)}{3x\cdot(x+8)}$$
$$=10$$

63. $\frac{8x+7}{3x+5}-\frac{2x-3}{3x+5}=\frac{8x+7-(2x-3)}{3x+5}$
$$=\frac{8x+7-2x+3}{3x+5}$$
$$=\frac{6x+10}{3x+5}$$
$$=\frac{2(3x+5)}{3x+5}$$
$$=2$$

65. $\frac{5a+10}{18}\div\frac{a^2-4}{10a}=\frac{5a+10}{18}\cdot\frac{10a}{a^2-4}$
$$=\frac{5(a+2)\cdot 2\cdot 5a}{2\cdot 9\cdot(a-2)(a+2)}$$
$$=\frac{25a}{9(a-2)}$$

67. $x^2-3x+2=(x-2)(x-1)$

$\text{LCD}=(x-2)(x-1)$

$$\frac{5}{x^2-3x+2}+\frac{1}{x-2}$$
$$=\frac{5}{(x-2)(x-1)}+\frac{1}{(x-2)}\cdot\frac{(x-1)}{(x-1)}$$
$$=\frac{5}{(x-2)(x-1)}+\frac{x-1}{(x-2)(x-1)}$$
$$=\frac{5+x-1}{(x-2)(x-1)}$$
$$=\frac{x+4}{(x-2)(x-1)}$$

69.
$$3x+5=7$$
$$3x+5-5=7-5$$
$$3x=2$$
$$\frac{3x}{3}=\frac{2}{3}$$
$$x=\frac{2}{3}$$

71.
$$2x^2-x-1=0$$
$$(2x+1)(x-1)=0$$
$$2x+1=0 \quad\text{or}\quad x-1=0$$
$$2x=-1 \qquad x=1$$
$$x=-\frac{1}{2}$$

The solutions are $x=-\frac{1}{2}$ and $x=1$.

73.
$$4(x+6)+3=-3$$
$$4x+24+3=-3$$
$$4x+27=-3$$
$$4x=-30$$
$$x=\frac{-30}{4}=-\frac{15}{2}$$

75. $x^2-1=(x+1)(x-1)$

LCD = $x(x+1)(x-1)$

$$\frac{3}{x}-\frac{2x}{x^2-1}+\frac{5}{x+1}=\frac{3(x+1)(x-1)}{x(x+1)(x-1)}-\frac{2x(x)}{(x+1)(x-1)(x)}+\frac{5(x)(x-1)}{(x+1)(x)(x-1)}$$
$$=\frac{3(x+1)(x-1)-2x(x)+5x(x-1)}{x(x+1)(x-1)}$$
$$=\frac{3x^2-3-2x^2+5x^2-5x}{x(x+1)(x-1)}$$
$$=\frac{6x^2-5x-3}{x(x+1)(x-1)}$$

77. $x^2-4=(x+2)(x-2)$

$x^2-4x+4=(x-2)^2$

$x^2-x-6=(x-3)(x+2)$

LCD = $(x+2)(x-2)^2(x-3)$

$$\frac{5}{x^2-4}+\frac{2}{x^2-4x+4}-\frac{3}{x^2-x-6}=\frac{5(x-2)(x-3)}{(x-2)(x+2)(x-2)(x-3)}+\frac{2(x+2)(x-3)}{(x-2)^2(x+2)(x-3)}-\frac{3(x-2)^2}{(x-3)(x+2)(x-2)^2}$$
$$=\frac{5(x^2-5x+6)}{(x-2)^2(x+2)(x-3)}+\frac{2(x^2-x-6)}{(x-2)^2(x+2)(x-3)}-\frac{3(x^2-4x+4)}{(x-2)^2(x+2)(x-3)}$$
$$=\frac{5x^2-25x+30}{(x-2)^2(x+2)(x-3)}+\frac{2x^2-2x-12}{(x-2)^2(x+2)(x-3)}-\frac{3x^2-12x+12}{(x-2)^2(x+2)(x-3)}$$
$$=\frac{5x^2-25x+30+2x^2-2x-12-3x^2+12x-12}{(x-2)^2(x+2)(x-3)}$$
$$=\frac{4x^2-15x+6}{(x-2)^2(x+2)(x-3)}$$

79. $x^2+9x+14=(x+2)(x+7)$

$x^2+10x+21=(x+3)(x+7)$

$x^2+5x+6=(x+2)(x+3)$

LCD = $(x+2)(x+7)(x+3)$

$$\frac{9}{x^2+9x+14}-\frac{3x}{x^2+10x+21}+\frac{x+4}{x^2+5x+6}=\frac{9(x+3)}{(x+2)(x+7)(x+3)}-\frac{3x(x+2)}{(x+3)(x+7)(x+2)}+\frac{(x+4)(x+7)}{(x+2)(x+3)(x+7)}$$
$$=\frac{9(x+3)-3x(x+2)+(x+4)(x+7)}{(x+2)(x+7)(x+3)}$$
$$=\frac{9x+27-3x^2-6x+x^2+11x+28}{(x+2)(x+7)(x+3)}$$
$$=\frac{-2x^2+14x+55}{(x+2)(x+7)(x+3)}$$

81. The length of the other board is $\left(\frac{3}{x+4}-\frac{1}{x-4}\right)$ inches.

LCD = $(x+4)(x-4)$

$$\frac{3}{x+4}-\frac{1}{x-4}=\frac{3(x-4)}{(x+4)(x-4)}-\frac{1(x+4)}{(x-4)(x+4)}$$
$$=\frac{3(x-4)-(x+4)}{(x+4)(x-4)}$$
$$=\frac{3x-12-x-4}{(x+4)(x-4)}$$
$$=\frac{2x-16}{(x+4)(x-4)}$$

The length of the other board is $\frac{2x-16}{(x+4)(x-4)}$ inches.

83. $1-\frac{G}{P}=\frac{1}{1}-\frac{G}{P}=\frac{1(P)}{1(P)}-\frac{G}{P}=\frac{P-G}{P}$

85. Answers may vary

87. $90°-\left(\frac{40}{x}\right)°=\left(90-\frac{40}{x}\right)°$

LCD = x

$$\left(90\cdot\frac{x}{x}-\frac{40}{x}\right)°=\left(\frac{90x}{x}-\frac{40}{x}\right)°=\left(\frac{90x-40}{x}\right)°$$

89. Answers may vary

The Bigger Picture

1. $-8.6+(-9.1)=-17.7$

2. $(-8.6)(-9.1)=78.26$

3. $14-(-14)=14+14=28$

4. $3x^4-7+x^4-x^2-10=3x^4+x^4-x^2-7-10$
$=4x^4-x-17$

5. $\frac{5x^2-5}{25x+25}=\frac{5(x+1)(x-1)}{5\cdot5(x+1)}=\frac{x-1}{5}$

6.
$$\frac{7x}{x^2+4x+3}\div\frac{x}{2x+6}=\frac{7x}{x^2+4x+3}\cdot\frac{2x+6}{x}$$
$$=\frac{7\cdot x\cdot2\cdot(x+3)}{(x+3)(x+1)\cdot x}$$
$$=\frac{14}{x+1}$$

7. $9=3\cdot3=3^2$
$6=2\cdot3$
LCD $=2\cdot3^2$

$$\frac{2}{9}-\frac{5}{6}=\frac{2(2)}{9(2)}-\frac{5(3)}{6(3)}=\frac{4}{18}-\frac{15}{18}=-\frac{11}{18}$$

8. $9=3\cdot3=3^2$
LCD $=3^2\cdot5$

$$\frac{x}{9}-\frac{x+3}{5}=\frac{x(5)}{9(5)}-\frac{(x+3)(9)}{5(9)}$$
$$=\frac{5x-9(x+3)}{45}$$
$$=\frac{5x-9x-27}{45}$$
$$=\frac{-4x-27}{45}\text{ or }-\frac{4x+27}{45}$$

9. $9x^3-2x^2-11x=x(9x^2-2x-11)$
$=x(9x-11)(x+1)$

10. $12xy-21x+4y-7=3x(4y-7)+1(4y-7)$
$=(4y-7)(3x+1)$

11.
$7x-14=5x+10$
$2x-14=10$
$2x=24$
$x=12$

12.
$$\frac{-x+2}{5}<\frac{3}{10}$$
$$10\left(\frac{-x+2}{5}\right)<10\left(\frac{3}{10}\right)$$
$$2(-x+2)<3$$
$$-2x+4<3$$
$$-2x<-1$$
$$\frac{-2x}{-2}>\frac{-1}{-2}$$
$$x>\frac{1}{2}$$
$$\left(\frac{1}{2},\infty\right)$$

13.
$$\begin{aligned} 1+4(x+4) &= 3^2+x \\ 1+4x+16 &= 9+x \\ 4x+17 &= 9+x \\ 3x+17 &= 9 \\ 3x &= -8 \\ x &= -\frac{8}{3} \end{aligned}$$

14.
$$\begin{aligned} x(x-2) &= 24 \\ x^2-2x &= 24 \\ x^2-2x-24 &= 0 \\ (x+4)(x-6) &= 0 \end{aligned}$$
$x+4=0$ or $x-6=0$
$x=-4$ $\quad x=6$
The solutions are $x=-4, 6$.

Section 7.5

Practice Exercises

1. The LCD of 3, 5, and 15 is 15.
$$\begin{aligned} \frac{x}{3}+\frac{4}{5} &= \frac{12}{5} \\ 15\left(\frac{x}{3}+\frac{4}{5}\right) &= 15\left(\frac{2}{15}\right) \\ 15\left(\frac{x}{3}\right)+15\left(\frac{4}{5}\right) &= 15\left(\frac{2}{15}\right) \\ 5\cdot x+12 &= 2 \\ 5x &= -10 \\ x &= -2 \end{aligned}$$
Check:
$$\begin{aligned} \frac{x}{3}+\frac{4}{5} &= \frac{2}{15} \\ \frac{-2}{3}+\frac{4}{5} &\stackrel{?}{=} \frac{2}{15} \\ \frac{2}{15} &= \frac{2}{15} \quad \text{True} \end{aligned}$$
This number checks, so the solution is −2.

2. The LCD of 4, 3, and 12 is 12.
$$\begin{aligned} \frac{x+4}{4}-\frac{x-3}{3} &= \frac{11}{12} \\ 12\left(\frac{x+4}{4}-\frac{x-3}{3}\right) &= 12\left(\frac{11}{12}\right) \\ 12\left(\frac{x+4}{4}\right)-12\left(\frac{x-3}{3}\right) &= 12\left(\frac{11}{12}\right) \\ 3(x+4)-4(x-3) &= 11 \\ 3x+12-4x+12 &= 11 \\ -x+24 &= 11 \\ -x &= -13 \\ x &= 13 \end{aligned}$$
Check:
$$\begin{aligned} \frac{x+4}{4}-\frac{x-3}{3} &= \frac{11}{12} \\ \frac{13+4}{4}-\frac{13-3}{3} &\stackrel{?}{=} \frac{11}{12} \\ \frac{17}{4}-\frac{10}{3} &\stackrel{?}{=} \frac{11}{12} \\ \frac{11}{12} &= \frac{11}{12} \quad \text{True} \end{aligned}$$
The solution is 13.

3. In this equation, 0 cannot be a solution. The LCD is x.
$$\begin{aligned} 8+\frac{7}{x} &= x+2 \\ x\left(8+\frac{7}{x}\right) &= x(x+2) \\ x(8)+x\left(\frac{7}{x}\right) &= x\cdot x+x\cdot 2 \\ 8x+7 &= x^2+2x \\ 0 &= x^2-6x-7 \\ 0 &= (x+1)(x-7) \end{aligned}$$
$x+1=0$ or $x-7=0$
$x=-1$ $\quad x=7$
Neither −1 nor 7 makes the denominator in the original equation equal to 0.
Check:
$x=-1$
$$\begin{aligned} 8+\frac{7}{x} &= x+2 \\ 8+\frac{7}{-1} &\stackrel{?}{=} -1+2 \\ 8+(-7) &\stackrel{?}{=} 1 \\ 1 &= 1 \quad \text{True} \end{aligned}$$

$x = 7$

$8 + \frac{7}{7} \stackrel{?}{=} 7 + 2$

$8 + 1 \stackrel{?}{=} 9$

$9 = 9$ True

Both −1 and 7 are solutions.

4. $x^2 - 5x - 14 = (x+2)(x-7)$

The LCD is $(x + 2)(x - 7)$.

$$\frac{6x}{x^2 - 5x - 14} - \frac{3}{x+2} = \frac{1}{x-7}$$

$$(x+2)(x-7)\left(\frac{6x}{x^2 - 5x - 14} - \frac{3}{x+2}\right) = (x+2)(x-7)\left(\frac{1}{x-7}\right)$$

$$(x+2)(x-7)\cdot\frac{6x}{x^2 - 5x - 14} - (x+2)(x-7)\cdot\frac{3}{x+2} = (x+2)(x-7)\cdot\frac{1}{x-7}$$

$$6x - 3(x-7) = x + 2$$

$$6x - 3x + 21 = x + 2$$

$$3x + 21 = x + 2$$

$$2x = -19$$

$$x = -\frac{19}{2}$$

Check by replacing x with $-\frac{19}{2}$ in the original equation. The solution is $-\frac{19}{2}$.

5. The LCD is $x - 2$.

$$\frac{7}{x-2} = \frac{3}{x-2} + 4$$

$$(x-2)\left(\frac{7}{x-2}\right) = (x-2)\left(\frac{3}{x-2} + 4\right)$$

$$(x-2)\cdot\frac{7}{x-2} = (x-2)\cdot\frac{3}{x-2} + (x-2)\cdot 4$$

$$7 = 3 + 4x - 8$$

$$7 = 4x - 5$$

$$12 = 4x$$

$$3 = x$$

Check by replacing x with 3 in the original equation. The solution is 3.

6. From the denominators in the equation, 5 can't be a solution. The LCD is $x - 5$.

$$x + \frac{x}{x-5} = \frac{5}{x-5} - 7$$

$$(x-5)\left(x + \frac{x}{x-5}\right) = (x-5)\left(\frac{5}{x-5} - 7\right)$$

$$(x-5)(x) + (x-5)\left(\frac{x}{x-5}\right) = (x-5)\left(\frac{5}{x-5}\right) - (x-5)(7)$$

$$x^2 - 5x + x = 5 - 7x + 35$$

$$x^2 - 4x = 40 - 7x$$

$$x^2 + 3x - 40 = 0$$

$$(x+8)(x-5) = 0$$

$x+8=0 \quad \text{or} \quad x-5=0$
$x=-8 \qquad\qquad x=5$

Since 5 can't be a solution, check by replacing x with −8 in the original equation. The only solution is −8.

7. The LCD is abx.

$$\frac{1}{a}+\frac{1}{b}=\frac{1}{x}$$
$$abx\left(\frac{1}{a}+\frac{1}{b}\right)=abx\left(\frac{1}{x}\right)$$
$$abx\left(\frac{1}{a}\right)+abx\left(\frac{1}{b}\right)=abx\cdot\frac{1}{x}$$
$$bx+ax=ab$$
$$ax=ab-bx$$
$$ax=b(a-x)$$
$$\frac{ax}{a-x}=b$$

Calculator Explorations

1. $y_1=\frac{x-4}{2}-\frac{x-3}{9},\ y_2=\frac{5}{18}$

Use INTERSECT

10

−10 10

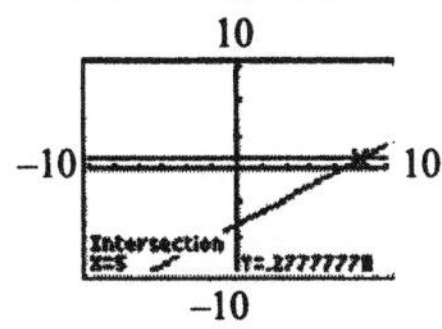

−10

The solution of the equation is 5.

2. $y_1=3-\frac{6}{x},\ y_2=x+8$

Use INTERSECT

10

−10 10

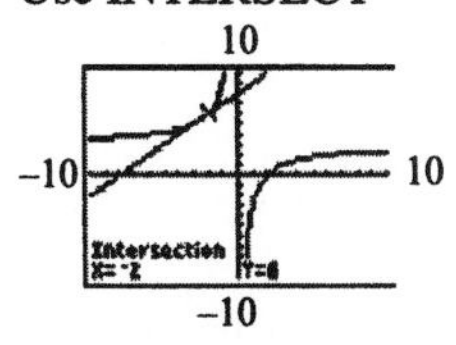

−10

One solution is −3.

10

−10 10

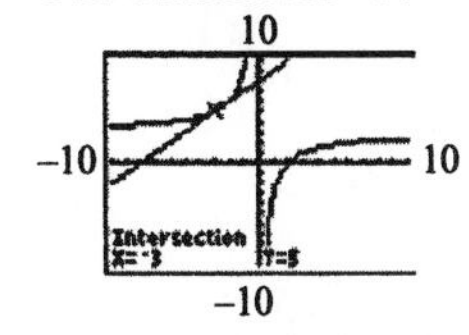

−10

The other solution is −2.

3. $y_1=\frac{2x}{x-4},\ y_2=\frac{8}{x-4}+1$

10

−10 10

−10

Using TRACE and ZOOM, it is clear that the curves never intersect. The equation has no solution.

4. $y_1=x+\frac{14}{x-2},\ y_2=\frac{7x}{x-2}+1$

Use INTERSECT

15

−5 15

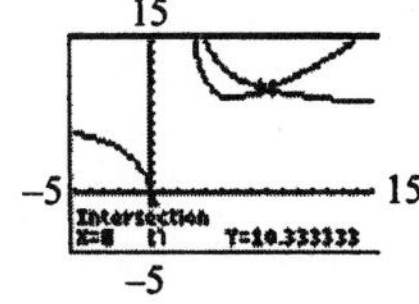

−5

The solution is 8.

Exercise Set 7.5

1. The LCD is 5.

$$\frac{x}{5}+3=9$$
$$5\left(\frac{x}{5}+3\right)=5(9)$$
$$5\left(\frac{x}{5}\right)+5(3)=5(9)$$
$$x+15=45$$
$$x=30$$

Check: $\frac{x}{5}+3=9$

$$\frac{30}{5}+3\stackrel{?}{=}9$$
$$6+3\stackrel{?}{=}9$$
$$9=9 \quad \text{True}$$

The solution is 30.

3. The LCD is 12.

$$\frac{x}{2}+\frac{5x}{4}=\frac{x}{12}$$
$$12\left(\frac{x}{2}+\frac{5x}{4}\right)=12\left(\frac{x}{12}\right)$$
$$12\left(\frac{x}{2}\right)+12\left(\frac{5x}{4}\right)=12\left(\frac{x}{12}\right)$$
$$6x+15x=x$$
$$21x=x$$
$$20x=0$$
$$x=0$$

Check: $\frac{x}{2}+\frac{5x}{4}=\frac{x}{12}$

$$\frac{0}{2}+\frac{5\cdot 0}{4}\stackrel{?}{=}\frac{0}{12}$$
$$0+\frac{0}{4}\stackrel{?}{=}0$$
$$0=0 \quad \text{True}$$

The solution is 0.

5. The LCD is x.

$$2-\frac{8}{x}=6$$
$$x\left(2-\frac{8}{x}\right)=x(6)$$
$$x\cdot 2-x\cdot\frac{8}{x}=x\cdot 6$$
$$2x-8=6x$$
$$-8=4x$$
$$-2=x$$

Check: $2-\frac{8}{x}=6$

$$2-\frac{8}{-2}\stackrel{?}{=}6$$
$$2-(-4)\stackrel{?}{=}6$$
$$6=6 \quad \text{True}$$

The solution is −2.

7. The LCD is x.

$$2+\frac{10}{x}=x+5$$
$$x\left(2+\frac{10}{x}\right)=x(x+5)$$
$$x(2)+x\left(\frac{10}{x}\right)=x(x+5)$$
$$2x+10=x^2+5x$$
$$0=x^2+3x-10$$
$$0=(x+5)(x-2)$$

$x+5=0$ or $x-2=0$

$x=-5$ $\quad$ $x=2$

Check:

$x=-5$: $\quad 2+\frac{10}{x}=x+5$

$$2+\frac{10}{-5}\stackrel{?}{=}-5+5$$
$$2+(-2)\stackrel{?}{=}-5+5$$
$$0=0 \quad \text{True}$$

$x=2$: $\quad 2+\frac{10}{x}=x+5$

$$2+\frac{10}{2}\stackrel{?}{=}2+5$$
$$2+5\stackrel{?}{=}2+5$$
$$7=7 \quad \text{True}$$

Both −5 and 2 are solutions.

9. The LCD is 10.

$$\frac{a}{5}=\frac{a-3}{2}$$
$$10\left(\frac{a}{5}\right)=10\left(\frac{a-3}{2}\right)$$
$$2a=5(a-3)$$
$$2a=5a-15$$
$$-3a=-15$$
$$a=5$$

Check: $\frac{a}{5}=\frac{a-3}{2}$

$$\frac{5}{5}\stackrel{?}{=}\frac{5-3}{2}$$
$$\frac{5}{5}\stackrel{?}{=}\frac{2}{2}$$
$$1=1 \quad \text{True}$$

The solution is 5.

11. The LCD is 10.

$$\frac{x-3}{5}+\frac{x-2}{2}=\frac{1}{2}$$
$$10\left(\frac{x-3}{5}+\frac{x-2}{2}\right)=10\left(\frac{1}{2}\right)$$
$$10\left(\frac{x-3}{5}\right)+10\left(\frac{x-2}{2}\right)=10\left(\frac{1}{2}\right)$$
$$2(x-3)+5(x-2)=5$$
$$2x-6+5x-10=5$$
$$7x-16=5$$
$$7x=21$$
$$x=3$$

Check: $\frac{x-3}{5}+\frac{x-2}{2}=\frac{1}{2}$

$$\frac{3-3}{5}+\frac{3-2}{2}\stackrel{?}{=}\frac{1}{2}$$
$$\frac{0}{5}+\frac{1}{2}\stackrel{?}{=}\frac{1}{2}$$
$$0+\frac{1}{2}\stackrel{?}{=}\frac{1}{2}$$
$$\frac{1}{2}=\frac{1}{2}\quad\text{True}$$

The solution is 3.

13. The LCD is $2a-5$.

$$\frac{3}{2a-5}=-1$$
$$(2a-5)\left(\frac{3}{2a-5}\right)=(2a-5)(-1)$$
$$3=-2a+5$$
$$-2=-2a$$
$$1=a$$

Check: $\frac{3}{2a-5}=-1$

$$\frac{3}{2(1)-5}\stackrel{?}{=}-1$$
$$\frac{3}{-3}\stackrel{?}{=}-1$$
$$-1=-1\quad\text{True}$$

The solution is 1.

15. The LCD is $y-4$.

$$\frac{4y}{y-4}+5=\frac{5y}{y-4}$$
$$(y-4)\left(\frac{4y}{y-4}+5\right)=(y-4)\left(\frac{5y}{y-4}\right)$$
$$(y-4)\left(\frac{4y}{y-4}\right)+(y-4)(5)=(y-4)\left(\frac{5y}{y-4}\right)$$
$$4y+5y-20=5y$$
$$9y-20=5y$$
$$4y-20=0$$
$$4y=20$$
$$y=5$$

Check: $\frac{4y}{y-4}+5=\frac{5y}{y-4}$

$$\frac{4(5)}{5-4}+5\stackrel{?}{=}\frac{5(5)}{5-4}$$
$$\frac{20}{1}+5\stackrel{?}{=}\frac{25}{1}$$
$$25=25\quad\text{True}$$

The solution is 5.

17. The LCD is $a - 3$.

$$2+\frac{3}{a-3}=\frac{a}{a-3}$$
$$(a-3)\left(2+\frac{3}{a-3}\right)=(a-3)\left(\frac{a}{a-3}\right)$$
$$(a-3)(2)+(a-3)\left(\frac{3}{a-3}\right)=a$$
$$2a-6+3=a$$
$$2a-3=a$$
$$-3=a-2a$$
$$-3=-a$$
$$\frac{-3}{-1}=a$$
$$3=a$$

When a is 3, a denominator equals zero. The equation has no solution.

19. $x^2-9=(x+3)(x-3)$

The LCD is $(x + 3)(x - 3)$.

$$\frac{1}{x+3}+\frac{6}{x^2-9}=1$$
$$(x+3)(x-3)\left(\frac{1}{x+3}+\frac{6}{(x+3)(x-3)}\right)=(x+3)(x-3)(1)$$
$$(x+3)(x-3)\cdot\frac{1}{x+3}+(x+3)(x-3)\cdot\frac{6}{(x+3)(x-3)}=(x+3)(x-3)\cdot 1$$
$$x-3+6=x^2-9$$
$$x+3=x^2-9$$
$$0=x^2-x-12$$
$$0=(x+3)(x-4)$$

$x+3=0$ or $x-4=0$

$x=-3$ $\qquad x=4$

When x is -3, a denominator equals zero. Check $x = 4$.

Check: $\frac{1}{x+3}+\frac{6}{x^2-9}=1$

$$\frac{1}{4+3}+\frac{6}{4^2-9}\stackrel{?}{=}1$$
$$\frac{1}{7}+\frac{6}{7}\stackrel{?}{=}1$$
$$1=1 \quad \text{True}$$

The solution is 4.

21. The LCD is $y + 4$.

$$\frac{2y}{y+4}+\frac{4}{y+4}=3$$
$$(y+4)\left(\frac{2y}{y+4}+\frac{4}{y+4}\right)=(y+4)(3)$$
$$(y+4)\cdot\frac{2y}{y+4}+(y+4)\cdot\frac{4}{y+4}=(y+4)\cdot 3$$
$$2y+4=3y+12$$
$$4=y+12$$
$$-8=y$$

Check: $\frac{2y}{y+4}+\frac{4}{y+4}=3$

$$\frac{2(-8)}{-8+4}+\frac{4}{-8+4}\stackrel{?}{=}3$$
$$\frac{-16}{-4}+\frac{4}{-4}\stackrel{?}{=}3$$
$$4-1\stackrel{?}{=}3$$
$$3=3 \quad \text{True}$$

The solution is -8.

23. The LCD is $(x + 2)(x - 2)$.

$$\frac{2x}{x+2}-2=\frac{x-8}{x-2}$$
$$(x+2)(x-2)\left(\frac{2x}{x+2}-2\right)=(x+2)(x-2)\left(\frac{x-8}{x-2}\right)$$
$$(x+2)(x-2)\cdot\frac{2x}{x+2}-(x+2)(x-2)(2)=(x+2)(x-2)\cdot\frac{x-8}{x-2}$$
$$2x(x-2)-2(x^2-4)=(x+2)(x-8)$$
$$2x^2-4x-2x^2+8=x^2-6x-16$$
$$-4x+8=x^2-6x-16$$
$$0=x^2-2x-24$$
$$0=(x+4)(x-6)$$

$x+4=0$ or $x-6=0$

$x=-4$ $\quad x=6$

Check $x=-4$: $\frac{2x}{x+2}-2=\frac{x-8}{x-2}$

$$\frac{2(-4)}{-4+2}-2\stackrel{?}{=}\frac{-4-8}{-4-2}$$
$$\frac{-8}{-2}-2\stackrel{?}{=}\frac{-12}{-6}$$
$$4-2\stackrel{?}{=}2$$
$$2=2 \quad \text{True}$$

Check $x = 6$: $\frac{2x}{x+2} - 2 = \frac{x-8}{x-2}$

$$\frac{2(6)}{6+2} - 2 \stackrel{?}{=} \frac{6-8}{6-2}$$
$$\frac{12}{8} - 2 \stackrel{?}{=} \frac{-2}{4}$$
$$\frac{3}{2} - 2 \stackrel{?}{=} -\frac{1}{2}$$
$$-\frac{1}{2} = -\frac{1}{2} \quad \text{True}$$

The solutions are –4 an 6.

25. The LCD is $2y$.

$$\frac{2}{y} + \frac{1}{2} = \frac{5}{2y}$$
$$2y\left(\frac{2}{y} + \frac{1}{2}\right) = 2y\left(\frac{5}{2y}\right)$$
$$2y\left(\frac{2}{y}\right) + 2y\left(\frac{1}{2}\right) = 2y\left(\frac{5}{2y}\right)$$
$$4 + y = 5$$
$$y = 1$$

The solution is 1.

27. The LCD is $(a - 6)(a - 1)$.

$$\frac{a}{a-6} = \frac{-2}{a-1}$$
$$(a-6)(a-1)\left(\frac{a}{a-6}\right) = (a-6)(a-1)\left(\frac{-2}{a-1}\right)$$
$$a(a-1) = -2(a-6)$$
$$a^2 - a = -2a + 12$$
$$a^2 + a - 12 = 0$$
$$(a+4)(a-3) = 0$$
$$a + 4 = 0 \quad \text{or} \quad a - 3 = 0$$
$$a = -4 \qquad\qquad a = 3$$

The solutions are –4 and 3.

29. The LCD is $6x$.

$$\frac{11}{2x} + \frac{2}{3} = \frac{7}{2x}$$
$$6x\left(\frac{11}{2x} + \frac{2}{3}\right) = 6x\left(\frac{7}{2x}\right)$$
$$6x \cdot \frac{11}{2x} + 6x \cdot \frac{2}{3} = 6x \cdot \frac{7}{2x}$$
$$33 + 4x = 21$$
$$4x = -12$$
$$x = -3$$

The solution is –3.

31. The LCD is $(x + 2)(x - 2)$.

$$\frac{2}{x-2}+1=\frac{x}{x+2}$$
$$(x+2)(x-2)\left(\frac{2}{x-2}+1\right)=(x+2)(x-2)\left(\frac{x}{x+2}\right)$$
$$(x+2)(x-2)\cdot\frac{2}{x-2}+(x+2)(x-2)\cdot 1=(x+2)(x-2)\cdot\frac{x}{x+2}$$
$$2(x+2)+(x+2)(x-2)=x(x-2)$$
$$2x+4+x^2-4=x^2-2x$$
$$x^2+2x=x^2-2x$$
$$2x=-2x$$
$$4x=0$$
$$x=0$$

The solution is 0.

33. The LCD is 6.

$$\frac{x+1}{3}-\frac{x-1}{6}=\frac{1}{6}$$
$$6\left(\frac{x+1}{3}-\frac{x-1}{6}\right)=6\left(\frac{1}{6}\right)$$
$$6\left(\frac{x+1}{3}\right)-6\left(\frac{x-1}{6}\right)=6\left(\frac{1}{6}\right)$$
$$2(x+1)-(x-1)=1$$
$$2x+2-x+1=1$$
$$x+3=1$$
$$x=-2$$

The solution is -2.

35. The LCD is $6(t - 4)$.

$$\frac{t}{t-4}=\frac{t+4}{6}$$
$$6(t-4)\left(\frac{t}{t-4}\right)=6(t-4)\left(\frac{t+4}{6}\right)$$
$$6t=(t-4)(t+4)$$
$$6t=t^2-16$$
$$0=t^2-6t-16$$
$$0=(t-8)(t+2)$$

$t+2=0$ or $t-8=0$

$t=-2$ $\quad$ $t=8$

The solutions are -2 and 8.

37. $2y + 2 = 2(y + 1)$

$4y + 4 = 2 \cdot 2(y + 1)$

The LCD is $4(y + 1)$.

$$\frac{y}{2y+2}+\frac{2y-16}{4y+4}=\frac{2y-3}{y+1}$$
$$4(y+1)\left(\frac{y}{2(y+1)}+\frac{2y-16}{4(y+1)}\right)=4(y+1)\left(\frac{2y-3}{y+1}\right)$$
$$4(y+1)\left(\frac{y}{2(y+1)}\right)+4(y+1)\left(\frac{2y-16}{4(y+1)}\right)=4(y+1)\left(\frac{2y-3}{y+1}\right)$$
$$2y+2y-16=4(2y-3)$$
$$4y-16=8y-12$$
$$-4y=4$$
$$y=-1$$

In the original equation, -1 makes a denominator 0. This equation has no solution.

39. $r^2+5r-14=(r+7)(r-2)$
The LCD is $(r+7)(r-2)$.

$$\frac{4r-4}{r^2+5r-14}+\frac{2}{r+7}=\frac{1}{r-2}$$
$$(r+7)(r-2)\left(\frac{4r-4}{(r+7)(r-2)}+\frac{2}{r+7}\right)=(r+7)(r-2)\left(\frac{1}{r-2}\right)$$
$$(r+7)(r-2)\left(\frac{4r-4}{(r+7)(r-2)}\right)+(r+7)(r-2)\left(\frac{2}{r+7}\right)=(r+7)(r-2)\left(\frac{1}{r-2}\right)$$
$$4r-4+2(r-2)=(r+7)(1)$$
$$4r-4+2r-4=r+7$$
$$6r-8=r+7$$
$$5r=15$$
$$r=3$$

The solution is 3.

41. $x^2+x-6=(x+3)(x-2)$
The LCD is $(x+3)(x-2)$.

$$\frac{x+1}{x+3}=\frac{x^2-11x}{x^2+x-6}-\frac{x-3}{x-2}$$
$$(x+3)(x-2)\left(\frac{x+1}{x+3}\right)=(x+3)(x-2)\left(\frac{x^2-11x}{(x+3)(x-2)}-\frac{x-3}{x-2}\right)$$
$$(x+3)(x-2)\cdot\frac{x+1}{x+3}=(x+3)(x-2)\cdot\frac{x^2-11x}{(x+3)(x-2)}-(x+3)(x-2)\cdot\frac{x-3}{x-2}$$
$$(x-2)(x+1)=x^2-11x-(x+3)(x-3)$$
$$x^2-x-2=x^2-11x-(x^2-9)$$
$$x^2-x-2=x^2-11x-x^2+9$$
$$x^2-x-2=-11x+9$$
$$x^2+10x-11=0$$
$$(x+11)(x-1)=0$$
$$x+11=0 \quad \text{or} \quad x-1=0$$
$$x=-11 \qquad\qquad x=1$$

The solutions are -11 and 1.

43.
$$R=\frac{E}{I}$$
$$I(R)=I\left(\frac{E}{I}\right)$$
$$IR=E$$
$$I=\frac{E}{R}$$

45.
$$T=\frac{2U}{B+E}$$
$$(B+E)(T)=(B+E)\left(\frac{2U}{B+E}\right)$$
$$BT+ET=2U$$
$$BT=2U-ET$$
$$B=\frac{2U-ET}{T}$$

47.
$$B=\frac{705w}{h^2}$$
$$h^2(B)=h^2\left(\frac{705w}{h^2}\right)$$
$$Bh^2=705w$$
$$\frac{Bh^2}{705}=w$$

49.
$$N=R+\frac{V}{G}$$
$$G(N)=G\left(R+\frac{V}{G}\right)$$
$$GN=GR+V$$
$$GN-GR=V$$
$$G(N-R)=V$$
$$G=\frac{V}{N-R}$$

51.
$$\frac{C}{\pi r}=2$$
$$\pi r\left(\frac{C}{\pi r}\right)=\pi r(2)$$
$$C=2\pi r$$
$$\frac{C}{2\pi}=\frac{2\pi r}{2\pi}$$
$$\frac{C}{2\pi}=r$$

53.
$$\frac{1}{y}+\frac{1}{3}=\frac{1}{x}$$
$$3xy\left(\frac{1}{y}+\frac{1}{3}\right)=3xy\left(\frac{1}{x}\right)$$
$$3xy\cdot\frac{1}{y}+3xy\cdot\frac{1}{3}=3xy\cdot\frac{1}{x}$$
$$3x+xy=3y$$
$$x(3+y)=3y$$
$$x=\frac{3y}{3+y}$$

55. The reciprocal of x is $\frac{1}{x}$.

57. The reciprocal of x, added to the reciprocal of 2 is $\frac{1}{x}+\frac{1}{2}$.

59. If a tank is filled in 3 hours, then $\frac{1}{3}$ of the tank is filled in one hour.

61. The graph crosses the x-axis at $x = 2$. It crosses the y-axis at $y = -2$. The x-intercept is (2, 0) and the y-intercept is (0, –2).

63. The graph crosses the x-axis at $x = -4$, $x = -2$ and $x = 3$. It crosses the y-axis at $y = 4$. The x-intercepts are (–4, 0), (–2, 0) and (3, 0), and the y-intercept is (0, 4).

65. Answers may vary

67. expression
$$\frac{1}{x}+\frac{5}{9}=\frac{1(9)}{x(9)}+\frac{5x}{9x}=\frac{5x+9}{9x}$$

69. equation
$$\frac{5}{x-1}-\frac{2}{x}=\frac{5}{x(x-1)}$$
$$x(x-1)\left(\frac{5}{x-1}\right)-x(x-1)\left(\frac{2}{x}\right)=x(x-1)\left(\frac{5}{x(x-1)}\right)$$
$$5x-2(x-1)=5$$
$$5x-2x+2=5$$
$$3x=3$$
$$x=1$$
1 makes a denominator zero. There is no solution.

71.
$$\frac{20x}{3}+\frac{32x}{6}=180$$
$$6\left(\frac{20x}{3}+\frac{32x}{6}\right)=6(180)$$
$$6\left(\frac{20x}{3}\right)+6\left(\frac{32x}{6}\right)=6(180)$$
$$40x+32x=1080$$
$$72x=1080$$
$$\frac{72x}{72}=\frac{1080}{72}$$
$$x=15$$
$$\frac{20x}{3}=\frac{20(15)}{3}=100$$
$$\frac{32x}{6}=\frac{32(15)}{6}=80$$
The angles are 100° and 80°.

73.
$$\frac{150}{x}+\frac{450}{x}=90$$
$$x\left(\frac{150}{x}+\frac{450}{x}\right)=x(90)$$
$$x\left(\frac{150}{x}\right)+x\left(\frac{450}{x}\right)=x(90)$$
$$150+450=90x$$
$$600=90x$$
$$\frac{600}{90}=\frac{90x}{90}$$
$$\frac{20}{3}=x$$
$$\frac{150}{x}=\frac{150}{\frac{20}{3}}=150\left(\frac{3}{20}\right)=\frac{45}{2}=22.5$$
$$\frac{450}{x}=\frac{450}{\frac{20}{3}}=450\left(\frac{3}{20}\right)=\frac{135}{2}=67.5$$
The angles are 22.5° and 67.5°.

75.

$$\frac{5}{a^2+4a+3}+\frac{2}{a^2+a-6}-\frac{3}{a^2-a-2}=0$$

$$\frac{5}{(a+3)(a+1)}+\frac{2}{(a+3)(a-2)}-\frac{3}{(a-2)(a+1)}=0$$

$$(a+3)(a+1)(a-2)\left(\frac{5}{(a+3)(a+1)}+\frac{2}{(a+3)(a-2)}-\frac{3}{(a-2)(a+1)}\right)=(a+3)(a+1)(a-2)(0)$$

$$(a+3)(a+1)(a-2)\left(\frac{5}{(a+3)(a+1)}\right)+(a+3)(a+1)(a-2)\left(\frac{2}{(a+3)(a-2)}\right)$$

$$-(a+3)(a+1)(a-2)\left(\frac{3}{(a-2)(a+1)}\right)=0$$

$$5(a-2)+2(a+1)-3(a+3)=0$$

$$5a-10+2a+2-3a-9=0$$

$$4a-17=0$$

$$4a=17$$

$$a=\frac{17}{4}$$

The solution is $\frac{17}{4}$.

Integrated Review

1. expression

$$\frac{1}{x}+\frac{2}{3}=\frac{1(3)}{x(3)}+\frac{2(x)}{3(x)}=\frac{3}{3x}+\frac{2x}{3x}=\frac{3+2x}{3x}$$

2. expression

$$\frac{3}{a}+\frac{5}{6}=\frac{3(6)}{a(6)}+\frac{5(a)}{6(a)}=\frac{18}{6a}+\frac{5a}{6a}=\frac{18+5a}{6a}$$

3. equation

$$\frac{1}{x}+\frac{2}{3}=\frac{3}{x}$$

$$3x\left(\frac{1}{x}+\frac{2}{3}\right)=3x\left(\frac{3}{x}\right)$$

$$3x\left(\frac{1}{x}\right)+3x\left(\frac{2}{3}\right)=3x\left(\frac{3}{x}\right)$$

$$3+2x=9$$

$$2x=6$$

$$x=3$$

The solution is 3.

4. equation

$$\frac{3}{a}+\frac{5}{6}=1$$
$$6a\left(\frac{3}{a}+\frac{5}{6}\right)=6a(1)$$
$$6a\left(\frac{3}{a}\right)+6a\left(\frac{5}{6}\right)=6a$$
$$18+5a=6a$$
$$18=a$$

The solution is 18.

5. expression

$$\frac{2}{x-1}-\frac{1}{x}=\frac{2(x)}{(x-1)(x)}-\frac{1(x-1)}{x(x-1)}$$
$$=\frac{2x-(x-1)}{x(x-1)}$$
$$=\frac{x+1}{x(x-1)}$$

6. expression

$$\frac{4}{x-3}-\frac{1}{x}=\frac{4(x)}{(x-3)(x)}-\frac{1(x-3)}{x(x-3)}$$
$$=\frac{4x-(x-3)}{x(x-3)}$$
$$=\frac{4x-x+3}{x(x-3)}$$
$$=\frac{3x+3}{x(x-3)}$$
$$=\frac{3(x+1)}{x(x-3)}$$

7. equation

$$\frac{2}{x+1}-\frac{1}{x}=1$$
$$x(x+1)\left(\frac{2}{x+1}-\frac{1}{x}\right)=x(x+1)(1)$$
$$x(x+1)\left(\frac{2}{x+1}\right)-x(x+1)\left(\frac{1}{x}\right)=x(x+1)$$
$$2x-(x+1)=x(x+1)$$
$$2x-x-1=x^2+x$$
$$x-1=x^2+x$$
$$-1=x^2$$

There is no real number solution.

8. equation

$$\frac{4}{x-3}-\frac{1}{x}=\frac{6}{x(x-3)}$$
$$x(x-3)\left(\frac{4}{x-3}-\frac{1}{x}\right)=x(x-3)\left(\frac{6}{x(x-3)}\right)$$
$$x(x-3)\left(\frac{4}{x-3}\right)-x(x-3)\left(\frac{1}{x}\right)=6$$
$$4x-(x-3)=6$$
$$4x-x+3=6$$
$$3x+3=6$$
$$3x=3$$
$$x=1$$

The solution is 1.

9. expression

$$\frac{15x}{x+8}\cdot\frac{2x+16}{3x}=\frac{15x\cdot(2x+16)}{(x+8)\cdot 3x}$$
$$=\frac{3\cdot 5\cdot x\cdot 2\cdot(x+8)}{(x+8)\cdot 3\cdot x}$$
$$=5\cdot 2$$
$$=10$$

10. expression

$$\frac{9z+5}{15}\cdot\frac{5z}{81z^2-25}=\frac{(9z+5)\cdot 5z}{15\cdot(81z^2-25)}$$
$$=\frac{(9z+5)\cdot 5\cdot z}{5\cdot 3\cdot(9z+5)(9z-5)}$$
$$=\frac{z}{3(9z-5)}$$

11. expression

$$\frac{2x+1}{x-3}+\frac{3x+6}{x-3}=\frac{2x+1+3x+6}{x-3}=\frac{5x+7}{x-3}$$

12. expression

$$\frac{4p-3}{2p+7}+\frac{3p+8}{2p+7}=\frac{4p-3+3p+8}{2p+7}=\frac{7p+5}{2p+7}$$

13. equation

$$\frac{x+5}{7}=\frac{8}{2}$$
$$14\left(\frac{x+5}{7}\right)=14\left(\frac{8}{2}\right)$$
$$2(x+5)=56$$
$$2x+10=56$$
$$2x=46$$
$$x=23$$

The solution is 23.

14. equation

$$\frac{1}{2}=\frac{x-1}{8}$$
$$8\left(\frac{1}{2}\right)=8\left(\frac{x-1}{8}\right)$$
$$4=x-1$$
$$5=x$$

The solution is 5.

15. expression

$$\frac{5a+10}{18}\div\frac{a^2-4}{10a}=\frac{5a+10}{18}\cdot\frac{10a}{a^2-4}$$
$$=\frac{5(a+2)\cdot 2\cdot 5\cdot a}{2\cdot 9(a+2)(a-2)}$$
$$=\frac{5\cdot 5\cdot a}{9(a-2)}$$
$$=\frac{25a}{9(a-2)}$$

16. expression

$$\frac{9}{x^2-1}+\frac{12}{3x+3}$$
$$=\frac{9(3)}{(x+1)(x-1)(3)}+\frac{12(x-1)}{3(x+1)(x-1)}$$
$$=\frac{27+12x-12}{3(x-1)(x+1)}$$
$$=\frac{15+12x}{3(x+1)(x-1)}$$
$$=\frac{3(5+4x)}{3(x+1)(x-1)}$$
$$=\frac{4x+5}{(x+1)(x-1)}$$

17. expression

$$\frac{x+2}{3x-1}+\frac{5}{(3x-1)^2}=\frac{(x+2)(3x-1)}{(3x-1)(3x-1)}+\frac{5}{(3x-1)^2}$$
$$=\frac{3x^2+5x-2+5}{(3x-1)^2}$$
$$=\frac{3x^2+5x+3}{(3x-1)^2}$$

18. expression

$$\frac{4}{(2x-5)^2}+\frac{x+1}{2x-5}=\frac{4}{(2x-5)^2}+\frac{(x+1)(2x-5)}{(2x-5)(2x-5)}$$
$$=\frac{4+2x^2-3x-5}{(2x-5)^2}$$
$$=\frac{2x^2-3x-1}{(2x-5)^2}$$

19. expression

$$\frac{x-7}{x}-\frac{x+2}{5x}=\frac{(x-7)(5)}{x(5)}-\frac{x+2}{5x}$$
$$=\frac{5x-35-x-2}{5x}$$
$$=\frac{4x-37}{5x}$$

20. equation

$$\frac{9}{x^2-4}+\frac{2}{x+2}=\frac{-1}{x-2}$$
$$(x^2-4)\left(\frac{9}{x^2-4}\right)+(x^2-4)\left(\frac{2}{x+2}\right)=(x^2-4)\left(\frac{-1}{x-2}\right)$$
$$9+(x-2)(2)=(x+2)(-1)$$
$$9+2x-4=-x-2$$
$$2x+5=-x-2$$
$$3x+5=-2$$
$$3x=-7$$
$$x=-\frac{7}{3}$$

The solution is $-\frac{7}{3}$.

21. equation

$$\frac{3}{x+3}=\frac{5}{x^2-9}-\frac{2}{x-3}$$
$$(x^2-9)\left(\frac{3}{x+3}\right)=(x^2-9)\left(\frac{5}{x^2-9}\right)-(x^2-9)\left(\frac{2}{x-3}\right)$$
$$(x-3)(3)=5-(x+3)(2)$$
$$3x-9=5-2x-6$$
$$3x-9=-2x-1$$
$$5x-9=-1$$
$$5x=8$$
$$x=\frac{8}{5}$$

The solution is $\frac{8}{5}$.

22. expression

$$\frac{10x-9}{x}-\frac{x-4}{3x}=\frac{(10x-9)(3)}{x(3)}-\frac{x-4}{3x}$$
$$=\frac{30x-27-x+4}{3x}$$
$$=\frac{29x-23}{3x}$$

Section 7.6

Practice Exercises

1. Solve the equation as a rational equation.

$$\frac{36}{x}=\frac{4}{11}$$
$$11x\cdot\frac{36}{x}=11x\cdot\frac{4}{11}$$
$$11\cdot 36=x\cdot 4$$
$$396=4x$$
$$\frac{396}{4}=\frac{4x}{4}$$
$$99=x$$

Solve the proportion using cross products.

$$\frac{36}{x}=\frac{4}{11}$$
$$36\cdot 11=x\cdot 4$$
$$396=4x$$
$$\frac{396}{4}=\frac{4x}{4}$$
$$99=x$$

Check: Both methods give a solution of 99. To check, substitute 99 for x in the original proportion. The solution is 99.

2.
$$\frac{3x+2}{9}=\frac{x-1}{2}$$
$$2(3x+2)=9(x-1)$$
$$6x+4=9x-9$$
$$6x=9x-13$$
$$-3x=-13$$
$$\frac{-3x}{-3}=\frac{-13}{-3}$$
$$x=\frac{13}{3}$$

Check: Verify that $\frac{13}{3}$ is the solution.

3. Let x = price of seven 2-liter bottles of Diet Pepsi.

$$\frac{4 \text{ bottles}}{7 \text{ bottles}} = \frac{\text{price of 4 bottles}}{\text{price of 7 bottles}}$$

$$\frac{4}{7} = \frac{5.16}{x}$$
$$4x = 7(5.16)$$
$$4x = 36.12$$
$$x = 9.03$$

Check: Verify that 4 bottles is to 7 bottles as \$5.16 is to \$9.03.
Seven 2-liter bottles of Diet Pepsi cost \$9.03.

4. Since the triangles are similar, their corresponding sides are in proportion.

$$\frac{20}{8} = \frac{15}{x}$$
$$20x = 8 \cdot 15$$
$$20x = 120$$
$$x = 6$$

Check: To check, replace x with 6 in the original proportion and see that a true statement results.
The missing length is 6 meters.

5. Let x = the unknown number.

In words	the quotient of x and 5	minus	$\frac{3}{2}$	is	the quotient of x and 10
	↓	↓	↓	↓	↓
Translate:	$\frac{x}{5}$	$-$	$\frac{3}{2}$	$=$	$\frac{x}{10}$

The LCD is 10.

$$10\left(\frac{x}{5} - \frac{3}{2}\right) = 10\left(\frac{x}{10}\right)$$
$$10\left(\frac{x}{5}\right) - 10\left(\frac{3}{2}\right) = 10\left(\frac{x}{10}\right)$$
$$2x - 15 = x$$
$$x - 15 = 0$$
$$x = 15$$

Check: To check, verify that "the quotient of 15 and 5 minus $\frac{3}{2}$ is the quotient of 15 and 10," or $\frac{15}{5} - \frac{3}{2} = \frac{15}{10}$.

6. Let x = the time in hours it takes Cindy and Mary to complete the job together. Then $\frac{1}{x}$ = the part of the job they complete in 1 hour.

	Hours to Complete Total Job	Part of Job Completed in 1 Hour
Cindy	3	$\frac{1}{3}$
Mary	4	$\frac{1}{4}$
Together	x	$\frac{1}{x}$

The part of the job Cindy completes in 1 hour, added to the part of the job Mary completes in 1 hour is equal to the part of the job they complete together in 1 hour.

$$\frac{1}{3}+\frac{1}{4}=\frac{1}{x}$$
$$12x\left(\frac{1}{3}\right)+12x\left(\frac{1}{4}\right)=12x\left(\frac{1}{x}\right)$$
$$4x+3x=12$$
$$7x=12$$
$$x=\frac{12}{7} \text{ or } 1\frac{5}{7}$$

Check: The proposed solution is reasonable since $1\frac{5}{7}$ hours is more than half of Cindy's time and less than half of Mary's time. Check $1\frac{5}{7}$ hours in the originally stated problem.

Cindy and Mary can complete the garden planting in $1\frac{5}{7}$ hours.

7. Let x = the speed of the bus. Then since the car's speed is 15 mph faster than that of the bus, the speed of the car is $x + 15$.
Since distance = rate · time, or $d = r \cdot t$, then $t=\frac{d}{r}$.
The bus travels 180 miles in the same time that the car travels 240 miles.

	Distance =	Rate	· Time
Bus	180	x	$\frac{180}{x}$
Car	240	$x+15$	$\frac{240}{x+15}$

Since the car and the bus traveled the same amount of time, $\frac{180}{x}=\frac{240}{x+15}$.

$$\frac{180}{x}=\frac{240}{x+15}$$
$$180(x+15)=240x$$
$$180x+2700=240x$$
$$2700=60x$$
$$45=x$$

The speed of the bus is 45 miles per hour. The speed of the car must then be $x + 15$ or 60 miles per hour.
Check: Find the time it takes the car to travel 240 miles and the time it takes the bus to travel 180 miles.

Car: $t=\frac{d}{r}=\frac{240}{60}=4$ hours

Bus: $t=\frac{d}{r}=\frac{180}{45}=4$ hours

Since the times are the same, the proposed solution is correct. The speed of the bus is 45 miles per hour and the speed of the car is 60 miles per hour.

Vocabulary and Readiness Check

1. If both people work together, they can complete the job in less time than either person working alone. That is, in less than 5 hours; choice c.

2. If both inlet pipes are on, they can fill the pond in less time than either pipe alone. That is, in less than 25 hours; choice a.

Exercise Set 7.6

1.
$$\frac{2}{3}=\frac{x}{6}$$
$$12=3x$$
$$4=x$$

3. $\frac{x}{10}=\frac{5}{9}$
$9x=50$
$x=\frac{50}{9}$

5. $\frac{x+1}{2x+3}=\frac{2}{3}$
$3(x+1)=2(2x+3)$
$3x+3=4x+6$
$3=x+6$
$-3=x$

7. $\frac{9}{5}=\frac{12}{3x+2}$
$9(3x+2)=5(12)$
$27x+18=60$
$27x=42$
$x=\frac{42}{27}=\frac{14}{9}$

9. Let x = the elephant's weight on Pluto.
$\frac{100}{3}=\frac{4100}{x}$
$100x=3(4100)$
$100x=12,300$
$x=123$
The elephant's weight is 123 pounds.

11. Let x = the number of calories in 43.2 grams.
$\frac{110}{28.8}=\frac{x}{43.2}$
$110(43.2)=28.8x$
$4752=28.8x$
$165=x$
There are 165 calories in 43.2 grams.

13. $\frac{16}{10}=\frac{34}{y}$
$16y=340$
$y=21.25$

15. $\frac{28}{20}=\frac{8}{y}$
$28y=160$
$y=\frac{160}{28}=\frac{40}{7}$
$y=5\frac{5}{7}$ feet

17. $3\cdot\frac{1}{x}=9\cdot\frac{1}{6}$
$\frac{3}{x}=\frac{9}{6}$
$6x\left(\frac{3}{x}\right)=6x\left(\frac{9}{6}\right)$
$18=9x$
$x=2$
The unknown number is 2.

19. $\frac{3+2x}{x+1}=\frac{3}{2}$
$2(x+1)\left(\frac{3+2x}{x+1}\right)=2(x+1)\left(\frac{3}{2}\right)$
$2(3+2x)=3(x+1)$
$6+4x=3x+3$
$x=-3$
The unknown number is -3.

21. Let x be the number of hours for the two surveyors to survey the roadbed together.

	Hours to Complete Total Job	Part of Job Completed in 1 Hour
Experienced	4	$\frac{1}{4}$
Apprentice	5	$\frac{1}{5}$
Together	x	$\frac{1}{x}$

$\frac{1}{4}+\frac{1}{5}=\frac{1}{x}$
$20x\left(\frac{1}{4}\right)+20x\left(\frac{1}{5}\right)=20x\left(\frac{1}{x}\right)$
$5x+4x=20$
$9x=20$
$x=\frac{20}{9}$ or $2\frac{2}{9}$

The experienced surveyor and apprentice surveyor, working together, can survey the road in $2\frac{2}{9}$ hours.

23. Let x be the number of minutes it takes the belts working together.

	Minutes to Complete Total Job	Part of Job Completed in 1 Minute
Larger belt	2	$\frac{1}{2}$
Smaller belt	6	$\frac{1}{6}$
Both belts	x	$\frac{1}{x}$

$$\frac{1}{2}+\frac{1}{6}=\frac{1}{x}$$
$$6x\left(\frac{1}{2}\right)+6x\left(\frac{1}{6}\right)=6x\left(\frac{1}{x}\right)$$
$$3x+x=6$$
$$4x=6$$
$$x=\frac{6}{4}=\frac{3}{2}=1\frac{1}{2}$$

Both belts together can move the cans to the storage area in $1\frac{1}{2}$ minutes.

25. Let r be the jogger's rate. Then, since distance = rate · time, or $d = r \cdot t$, then $t=\frac{d}{r}$.

	Distance =	Rate	· Time
Trip to Park	12	r	$\frac{12}{r}$
Return Trip	18	r	$\frac{18}{r}$

Since her time on the return trip is 1 hour longer than on the trip to the park, $\frac{18}{r}=\frac{12}{r}+1$.

$$r\left(\frac{18}{r}\right)=r\left(\frac{12}{r}\right)+r(1)$$
$$18=12+r$$
$$6=r$$

She jogs at 6 miles per hour.

27. Let r be his speed on the first portion. Then his speed on the cooldown portion is $r-2$.

	Distance =	Rate	· Time
1st portion	20	r	$\frac{20}{r}$
Cooldown portion	16	$r-2$	$\frac{16}{r-2}$

$$\frac{20}{r}=\frac{16}{r-2}$$
$$20(r-2)=16r$$
$$20r-40=16r$$
$$-40=-4r$$
$$r=10$$

and $r-2=10-2=8$

His speed was 10 miles per hour during the first portion and 8 miles per hour during the cooldown portion.

29. Let x = the minimum floor space needed by 40 students.

$$\frac{1}{9}=\frac{40}{x}$$
$$1x=9(40)$$
$$x=360$$

40 students need 360 square feet.

31.
$$\frac{1}{4}=\frac{x}{8}$$
$$8\left(\frac{1}{4}\right)=8\left(\frac{x}{8}\right)$$
$$2=x$$

The unknown number is 2.

33. Let x be the amount of time it takes Marcus and Tony working together.

	Hours to Complete Total Job	Part of Job Completed in 1 Hour
Marcus	6	$\frac{1}{6}$
Tony	4	$\frac{1}{4}$
Together	x	$\frac{1}{x}$

$$\frac{1}{6}+\frac{1}{4}=\frac{1}{x}$$
$$12x\left(\frac{1}{6}\right)+12x\left(\frac{1}{4}\right)=12x\left(\frac{1}{x}\right)$$
$$2x+3x=12$$
$$5x=12$$
$$x=\frac{12}{5}=2\frac{2}{5}$$
$$45\left(\frac{12}{5}\right)=108$$

Together Marcus and Tony work for $2\frac{2}{5}$ hours at \$45 per hour. The labor estimate should be \$108.

35. Let w be the speed of the wind.

	Distance =	Rate ·	Time
With wind	400	$230+w$	$\frac{400}{230+w}$
Against wind	336	$230-w$	$\frac{336}{230-w}$

Since the time with the wind is the same as the time against the wind, $\frac{336}{230-w}=\frac{400}{230+w}$.

$$\frac{336}{230-w}=\frac{400}{230+w}$$
$$336(230+w)=400(230-w)$$
$$77,280+336w=92,000-400w$$
$$736w=14,720$$
$$w=20$$

The speed of the wind is 20 miles per hour.

37. $\frac{y}{25}=\frac{3}{2}$

$$y\cdot 2=25\cdot 3$$
$$y\cdot 2=75$$
$$y=\frac{75}{2}$$
$$y=37\frac{1}{2}$$

The unknown length is $37\frac{1}{2}$ feet.

39. Let x = the number of rushing yards in one game.

$$\frac{x}{1}=\frac{4045}{12}$$
$$12x=1(4045)$$
$$12x=4045$$
$$x\approx 337$$

Ken averaged 337 yards per game.

41.
$$\frac{2}{x-3}-\frac{4}{x+3}=8\cdot\frac{1}{x^2-9}$$
$$(x-3)(x+3)\left(\frac{2}{x-3}-\frac{4}{x+3}\right)=(x-3)(x+3)\left(\frac{8}{x^2-9}\right)$$
$$(x-3)(x+3)\left(\frac{2}{x-3}\right)-(x-3)(x+3)\left(\frac{4}{x+3}\right)=8$$
$$2(x+3)-4(x-3)=8$$
$$2x+6-4x+12=8$$
$$-2x=-10$$
$$x=5$$

The unknown number is 5.

43. Let r be the rate of the plane in still air.

	Distance =	Rate	· Time
With wind	630	$r+35$	$\frac{630}{r+35}$
Against wind	455	$r-35$	$\frac{455}{r-35}$

$$\frac{630}{r+35}=\frac{455}{r-35}$$
$$630(r-35)=455(r+35)$$
$$630r-22,050=455r+15,925$$
$$175r=37,975$$
$$r=217$$

The speed in still air is 217 mph.

45. Let x = the number of gallons of water needed.
$$\frac{8}{2}=\frac{36}{x}$$
$$8x=2(36)$$
$$8x=72$$
$$x=9$$

Nine gallons of water are needed for the entire box.

47.

	r ×	t =	d
With wind	$16+x$	$\frac{48}{16+x}$	48
Into Wind	$16-x$	$\frac{16}{16-x}$	16

Since the times are the same, $\frac{48}{16+x}=\frac{16}{16-x}$.

$$\frac{48}{16+x} = \frac{16}{16-x}$$
$$48(16-x) = 16(16+x)$$
$$768-48x = 256+16x$$
$$512 = 64x$$
$$8 = x$$
The rate of the wind is 8 miles per hour.

49. Let x be the slower speed. Then $x + 40$ is the faster speed.

	r	× t	= d
Slower	x	$\frac{70}{x}$	70
Faster	$x+40$	$\frac{300}{x+40}$	300

Since the time spent at the faster speed was twice that spent at the slower speed, $\frac{300}{x+4} = 2\left(\frac{70}{x}\right)$.

$$\frac{300}{x+40} = \frac{140}{x}$$
$$300x = 140(x+40)$$
$$300x = 140x+5600$$
$$160x = 5600$$
$$x = 35$$
$$x + 40 = 35 + 40 = 75$$
The slower speed was 35 miles per hour and the faster speed was 75 miles per hour.

51. Let x be the amount of time it takes the second worker to do the job alone.

	Hours to Complete Total Job	Part of Job Completed in 1 Hour
Custodian	3	$\frac{1}{3}$
2nd Worker	x	$\frac{1}{x}$
Together	$1\frac{1}{2}$ or $\frac{3}{2}$	$\frac{2}{3}$

$$\frac{1}{3}+\frac{1}{x} = \frac{2}{3}$$
$$3x\left(\frac{1}{3}\right)+3x\left(\frac{1}{x}\right) = 3x\left(\frac{2}{3}\right)$$
$$x+3 = 2x$$
$$3 = x$$
It takes the second worker 3 hours to do the job alone.

53. Let x be the missing dimension.
$$\frac{x}{8} = \frac{20}{6}$$
$$6x = 8\cdot 20$$
$$x = \frac{160}{6}$$
$$x = \frac{80}{3} = 26\frac{2}{3}$$
The side is $26\frac{2}{3}$ feet long.

55. $\frac{3}{2} = \frac{324}{x}$
$$3\cdot x = 2\cdot 324$$
$$3x = 648$$
$$x = \frac{648}{3} = 216$$
There should be 216 other nuts in the can.

57. Let x be the speed of the plane in still air.

	r	× t	= d
With wind	$x+30$	$\frac{2160}{x+30}$	2160
Against Wind	$x-30$	$\frac{1920}{x-30}$	1920

Since the times are the same, $\frac{2160}{x+30} = \frac{1920}{x-30}$.

$$\frac{2160}{x+30} = \frac{1920}{x-30}$$
$$2160(x-30) = 1920(x+30)$$
$$2160x-64,800 = 1920x+57,600$$
$$240x = 122,400$$
$$x = 510$$
The speed of the plane in still air is 510 miles per hour.

59. Let x be the number of hours it would take the third pipe alone to fill the pool.

	Hours to Complete Total Job	Part of Job Completed in 1 Hour
1st Pipe	20	$\frac{1}{20}$
2nd Pipe	15	$\frac{1}{15}$
3rd Pipe	x	$\frac{1}{x}$
3 Pipes Together	6	$\frac{1}{6}$

$$\frac{1}{20}+\frac{1}{15}+\frac{1}{x}=\frac{1}{6}$$
$$60x\left(\frac{1}{20}\right)+60x\left(\frac{1}{15}\right)+60x\left(\frac{1}{x}\right)=60x\left(\frac{1}{6}\right)$$
$$3x+4x+60=10x$$
$$7x+60=10x$$
$$60=3x$$
$$20=x$$

It takes the third pipe 20 hours to fill the pool.

61. Let m be the speed of the motorcycle. Then the speed of the car is $m + 10$.

	r	$\times$ t	$=$ d
Motorcycle	m	$\frac{240}{m}$	240
Car	$m + 10$	$\frac{280}{m+10}$	280

Since the times are the same, $\frac{240}{m}=\frac{280}{m+10}$.

$$\frac{240}{m}=\frac{280}{m+10}$$
$$240(m+10)=280m$$
$$240m+2400=280m$$
$$2400=40m$$
$$60=m$$

$m + 10 = 60 + 10 = 70$

The speed of the motorcycle is 60 miles per hour and the speed of the car is 70 miles per hour.

63. Let x be the amount of time it takes the third cook alone to prepare the pies.

	Time	In one hour
First cook	6	$\frac{1}{6}$
Second cook	7	$\frac{1}{7}$
Third cook	x	$\frac{1}{x}$
Together	2	$\frac{1}{2}$

$$\frac{1}{6}+\frac{1}{7}+\frac{1}{x}=\frac{1}{2}$$
$$42x\left(\frac{1}{6}+\frac{1}{7}+\frac{1}{x}\right)=42x\left(\frac{1}{2}\right)$$
$$42x\left(\frac{1}{6}\right)+42x\left(\frac{1}{7}\right)+42x\left(\frac{1}{x}\right)=21x$$
$$7x+6x+42=21x$$
$$13x+42=21x$$
$$42=21x-13x$$
$$42=8x$$
$$\frac{42}{8}=x$$
$$\frac{21}{4}=x$$
$$5\frac{1}{4}=x$$

The third cook can prepare the pies in $5\frac{1}{4}$ hours.

65. Let x be the number of minutes it takes the second pump to fill the tank. Then it takes $3x$ minutes for the first pump to fill the tank.

	Minutes to Complete Total Job	Part of Job Completed in 1 Minute
1st Pump	$3x$	$\frac{1}{3x}$
2nd Pump	x	$\frac{1}{x}$
Together	21	$\frac{1}{21}$

$$\frac{1}{3x}+\frac{1}{x}=\frac{1}{21}$$
$$21x\left(\frac{1}{3x}\right)+21x\left(\frac{1}{x}\right)=21x\left(\frac{1}{21}\right)$$
$$7+21=x$$
$$28=x,\ 3x=3(28)=84$$

The 1st pump takes 28 minutes and the 2nd takes 84 minutes.

67. $\frac{9}{12}=\frac{3.75}{x}$
$9x=45$
$x=5$
The missing length is 5.

69. $\frac{16}{24}=\frac{9}{x}$
$16x=216$
$x=13.5$
The missing length is 13.5.

71. $(-2, 5), (4, -3)$
$m=\frac{-3-5}{4-(-2)}=\frac{-8}{6}=-\frac{4}{3}$
Since the slope is negative, the line moves downward.

73. $\frac{\frac{3}{4}+\frac{1}{4}}{\frac{3}{8}+\frac{13}{8}}=\frac{\frac{3+1}{4}}{\frac{3+13}{8}}=\frac{\frac{4}{4}}{\frac{16}{8}}=\frac{1}{2}$

75. $\frac{\frac{2}{5}+\frac{1}{5}}{\frac{7}{10}+\frac{7}{10}}=\frac{\frac{2+1}{5}}{\frac{7+7}{10}}$
$=\frac{\frac{3}{5}}{\frac{14}{10}}$
$=\frac{3}{5}\div\frac{14}{10}$
$=\frac{3}{5}\cdot\frac{10}{14}$
$=\frac{3\cdot 2\cdot 5}{5\cdot 2\cdot 7}$
$=\frac{3}{7}$

77. The capacity in 2001 was approximately 4400 megawatts. The capacity in 2003 was approximately 6400 megawatts. The increase in capacity was approximately $6400-4400=2000$ megawatts.

79. The capacity in 2007 was approximately 14,630 megawatts, or 14.63(1000 megawatts).
$14.63(560{,}000)\approx 8{,}190{,}000$
In 2007, the number of megawatts generate4d from wind would serve the electricity needs of 8,190,000 people.

81. Answers may vary

83. None; answers may vary

85. $\frac{1}{6}x+\frac{1}{12}x+\frac{1}{7}x+5+\frac{1}{2}x+4=x$
$\frac{1}{6}x+\frac{1}{12}x+\frac{1}{7}x+\frac{1}{2}x+9=x$
$84\left(\frac{1}{6}x+\frac{1}{12}x+\frac{1}{7}x+\frac{1}{2}x+9\right)=84x$
$14x+7x+12x+42x+756=84x$
$75x+756=84x$
$756=9x$
$\frac{756}{9}=\frac{9x}{9}$
$84=x$
He died when he was 84 years old.

87. $4+\frac{1}{2}x+\frac{1}{6}x+3+\frac{1}{10}x=x$
$30\left(7+\frac{1}{2}x+\frac{1}{6}x+\frac{1}{10}x\right)=(30)(x)$
$30\cdot 7+30\left(\frac{1}{2}x\right)+30\left(\frac{1}{6}x\right)+30\left(\frac{1}{10}x\right)=30x$
$210+15x+5x+3x=30x$
$210+23x=30x$
$210=30x-23x$
$210=7x$
$30=x$
You are 30 years old.

89. Let d be the distance that the giraffe runs before the hyena catches it. Then the hyena runs $d+0.5$ miles.

	Distance =	Rate ·	Time
Hyena	$d+0.5$	40	$\frac{d+0.5}{40}$
Giraffe	d	32	$\frac{d}{32}$

$$\frac{d+0.5}{40}=\frac{d}{32}$$
$$32(d+0.5)=40d$$
$$32d+16=40d$$
$$16=8d$$
$$2=d,\ \frac{d}{32}=\frac{2}{32}=\frac{1}{16}$$

It will take the hyena $\frac{1}{16}$ hour or 3.75 minutes to overtake the giraffe.

The Bigger Picture

1. $(3x-2)(4x^2-x-5)$
$=3x(4x^2-x-5)-2(4x^2-x-5)$
$=12x^3-3x^2-15x-8x^2+2x+10$
$=12x^3-11x^2-13x+10$

2. $(2x-y)^2=(2x)^2-2(2x)(y)+y^2$
$=4x^2-4xy+y^2$

3. $8y^3-20y^5=4y^3(2-5y^2)$

4. $9m^2-11mn+2n^2=9m^2-2mn-9mn+2n^2$
$=m(9m-2n)-n(9m-2n)$
$=(9m-2n)(m-n)$

5. $\frac{7}{x}=\frac{9}{x-10}$
$7(x-10)=9x$
$7x-70=9x$
$-70=2x$
$-35=x$

6. $\frac{7}{x}+\frac{9}{x-10}=\frac{7(x-10)}{x(x-10)}+\frac{9(x)}{(x-10)x}$
$=\frac{7(x-10)+9x}{x(x-10)}$
$=\frac{7x-70+9x}{x(x-10)}$
$=\frac{16x-70}{x(x-10)}$ or $\frac{2(8x-35)}{x(x-10)}$

7. $(-3x^5)\left(\frac{1}{2}x^7\right)(8x)=\left(-3\cdot\frac{1}{2}\cdot 8\right)(x^5\cdot x^7\cdot x^1)$
$=-12x^{5+7+1}$
$=-12x^{13}$

8. $5x-1=|-4|+|-5|$
$5x-1=4+5$
$5x-1=9$
$5x=10$
$x=2$

9. $\frac{8-12}{12\div 3\cdot 2}=\frac{-4}{4\cdot 2}=\frac{-4}{8}=-\frac{1}{2}$

10. $-2(3y-4)\le 5y-7-7y-1$
$-6y+8\le -2y-8$
$8\le 4y-8$
$16\le 4y$
$4\le y$
$[4,\infty)$

11. $\frac{7}{x}+\frac{5}{2x+3}=\frac{-2}{x}$
$x(2x+3)\left(\frac{7}{x}+\frac{5}{2x+3}\right)=x(2x+3)\left(\frac{-2}{x}\right)$
$x(2x+3)\cdot\frac{7}{x}+x(2x+3)\cdot\frac{5}{2x+3}=x(2x+3)\cdot\frac{-2}{x}$
$7(2x+3)+5x=-2(2x+3)$
$14x+21+5x=-4x-6$
$19x+21=-4x-6$
$23x+21=-6$
$23x=-27$
$x=-\frac{27}{23}$

12. $\frac{(a^{-3}b^2)^{-5}}{ab^4}=\frac{a^{(-3)(-5)}b^{2(-5)}}{ab^4}=\frac{a^{15}b^{-10}}{ab^4}=\frac{a^{14}}{b^{14}}$

Section 7.7

Practice Exercises

1. a. $\frac{\frac{5k}{36m}}{\frac{15k}{9}}=\frac{5k}{36m}\div\frac{15k}{9}$
$=\frac{5k}{36m}\cdot\frac{9}{15k}$
$=\frac{5k\cdot 9}{36m\cdot 15k}$
$=\frac{1}{12m}$

b. $\dfrac{\frac{8x}{x-4}}{\frac{3}{x+4}} = \dfrac{8x}{x-4} \div \dfrac{3}{x+4}$
$= \dfrac{8x}{x-4} \cdot \dfrac{x+4}{3}$
$= \dfrac{8x(x+4)}{3(x-4)}$

c. $\dfrac{\frac{5}{a}+\frac{b}{a^2}}{\frac{5a}{b^2}+\frac{1}{b}} = \dfrac{\frac{5\cdot a}{a\cdot a}+\frac{b}{a^2}}{\frac{5a}{b^2}+\frac{1\cdot b}{b\cdot b}}$
$= \dfrac{\frac{5a+b}{a^2}}{\frac{5a+b}{b^2}}$
$= \dfrac{5a+b}{a^2} \cdot \dfrac{b^2}{5a+b}$
$= \dfrac{b^2(5a+b)}{a^2(5a+b)}$
$= \dfrac{b^2}{a^2}$

2. a. The LCD is $(x-4)(x+4)$.
$\dfrac{\frac{8x}{x-4}}{\frac{3}{x+4}} = \dfrac{\left(\frac{8x}{x-4}\right)\cdot(x-4)(x+4)}{\left(\frac{3}{x+4}\right)\cdot(x-4)(x+4)}$
$= \dfrac{8x(x+4)}{3(x-4)}$

b. The LCD is a^2b^2.
$\dfrac{\frac{b}{a^2}+\frac{1}{a}}{\frac{a}{b^2}+\frac{1}{b}} = \dfrac{\left(\frac{b}{a^2}+\frac{1}{a}\right)\cdot a^2b^2}{\left(\frac{a}{b^2}+\frac{1}{b}\right)\cdot a^2b^2}$
$= \dfrac{\frac{b}{a^2}\cdot a^2b^2+\frac{1}{a}\cdot a^2b^2}{\frac{a}{b^2}\cdot a^2b^2+\frac{1}{b}\cdot a^2b^2}$
$= \dfrac{b^3+ab^2}{a^3+a^2b}$
$= \dfrac{b^2(b+a)}{a^2(a+b)}$
$= \dfrac{b^2}{a^2}$

3. $\dfrac{3x^{-1}+x^{-2}y^{-1}}{y^{-2}+xy^{-1}} = \dfrac{\frac{3}{x}+\frac{1}{x^2y}}{\frac{1}{y^2}+\frac{x}{y}}$

The LCD is x^2y^2.
$= \dfrac{\left(\frac{3}{x}+\frac{1}{x^2y}\right)\cdot x^2y^2}{\left(\frac{1}{y^2}+\frac{x}{y}\right)\cdot x^2y^2}$
$= \dfrac{\frac{3}{x}\cdot x^2y^2+\frac{1}{x^2y}\cdot x^2y^2}{\frac{1}{y^2}\cdot x^2y^2+\frac{x}{y}\cdot x^2y^2}$
$= \dfrac{3xy^2+y}{x^2+x^3y}$ or $\dfrac{y(3xy+1)}{x^2(1+xy)}$

4. $\dfrac{(3x)^{-1}-2}{5x^{-1}+2} = \dfrac{\frac{1}{3x}-2}{\frac{5}{x}+2}$
$= \dfrac{\left(\frac{1}{3x}-2\right)\cdot 3x}{\left(\frac{5}{x}+2\right)\cdot 3x}$
$= \dfrac{\frac{1}{3x}\cdot 3x-2\cdot 3x}{\frac{5}{x}\cdot 3x+2\cdot 3x}$
$= \dfrac{1-6x}{15+6x}$

Vocabulary and Readiness Check

1. $\dfrac{\frac{7}{x}}{\frac{1}{x}+\frac{z}{x}} = \dfrac{x\left(\frac{7}{x}\right)}{x\left(\frac{1}{x}\right)+x\left(\frac{z}{x}\right)} = \underline{\dfrac{7}{1+z}}$

2. $\dfrac{\frac{x}{4}}{\frac{x^2}{2}+\frac{1}{4}} = \dfrac{4\left(\frac{x}{4}\right)}{4\left(\frac{x^2}{2}\right)+4\left(\frac{1}{4}\right)} = \underline{\dfrac{x}{2x^2+1}}$

3. $x^{-2} = \underline{\dfrac{1}{x^2}}$

4. $y^{-3} = \underline{\dfrac{1}{y^3}}$

5. $2x^{-1} = \underline{\dfrac{2}{x}}$

6. $(2x)^{-1} = \underline{\dfrac{1}{2x}}$

7. $(9y)^{-1} = \dfrac{1}{9y}$

8. $9y^{-2} = \dfrac{9}{y^2}$

Exercise Set 7.7

1. $\dfrac{\frac{10}{3x}}{\frac{5}{6x}} = \dfrac{10}{3x} \cdot \dfrac{6x}{5} = \dfrac{60x}{15x} = 4$

3. $\dfrac{1+\frac{2}{5}}{2+\frac{3}{5}} = \dfrac{5\left(1+\frac{2}{5}\right)}{5\left(2+\frac{3}{5}\right)} = \dfrac{5+2}{10+3} = \dfrac{7}{13}$

5. $\dfrac{\frac{4}{x-1}}{\frac{x}{x-1}} = \dfrac{4}{x-1} \cdot \dfrac{x-1}{x} = \dfrac{4}{x}$

7. $\dfrac{1-\frac{2}{x}}{x+\frac{4}{9x}} = \dfrac{9x\left(1-\frac{2}{x}\right)}{9x\left(x+\frac{4}{9x}\right)} = \dfrac{9x-18}{9x^2+4} = \dfrac{9(x-2)}{9x^2+4}$

9.
$$\frac{\frac{4x^2-y^2}{xy}}{\frac{2}{y}-\frac{1}{x}} = \frac{\left(\frac{4x^2-y^2}{xy}\right)\cdot xy}{\left(\frac{2}{y}-\frac{1}{x}\right)\cdot xy}$$
$$= \frac{4x^2-y^2}{2x-y}$$
$$= \frac{(2x-y)(2x+y)}{2x-y}$$
$$= 2x+y$$

11. $\dfrac{\frac{x+1}{3}}{\frac{2x-1}{6}} = \dfrac{x+1}{3} \cdot \dfrac{6}{2x-1} = \dfrac{2(x+1)}{2x-1}$

13. $\dfrac{\frac{2}{x}+\frac{3}{x^2}}{\frac{4}{x^2}-\frac{9}{x}} = \dfrac{\left(\frac{2}{x}+\frac{3}{x^2}\right)x^2}{\left(\frac{4}{x^2}-\frac{9}{x}\right)x^2} = \dfrac{2x+3}{4-9x}$

15.
$$\frac{\frac{1}{x}+\frac{2}{x^2}}{x+\frac{8}{x^2}} = \frac{x^2\left(\frac{1}{x}+\frac{2}{x^2}\right)}{x^2\left(x+\frac{8}{x^2}\right)}$$
$$= \frac{x+2}{x^3+8}$$
$$= \frac{x+2}{(x+2)(x^2-2x+4)}$$
$$= \frac{1}{x^2-2x+4}$$

17.
$$\frac{\frac{4}{5-x}+\frac{5}{x-5}}{\frac{2}{x}+\frac{3}{x-5}} = \frac{-\frac{4}{x-5}+\frac{5}{x-5}}{\frac{2(x-5)+3x}{x(x-5)}}$$
$$= \frac{\frac{1}{x-5}}{\frac{2x-10+3x}{x(x-5)}}$$
$$= \frac{1}{x-5} \cdot \frac{x(x-5)}{5x-10}$$
$$= \frac{x}{5x-10} \text{ or } \frac{x}{5(x-2)}$$

19.
$$\frac{\frac{x+2}{x}-\frac{2}{x-1}}{\frac{x+1}{x}+\frac{x+1}{x-1}} = \frac{\frac{(x+2)(x-1)-2x}{x(x-1)}}{\frac{(x+1)(x-1)+(x+1)(x)}{x(x-1)}}$$
$$= \frac{\frac{x^2+x-2-2x}{x(x-1)}}{\frac{x^2-1+x^2+x}{x(x-1)}}$$
$$= \frac{x^2-x-2}{x(x-1)} \cdot \frac{x(x-1)}{2x^2+x-1}$$
$$= \frac{(x-2)(x+1)}{x(x-1)} \cdot \frac{x(x-1)}{(2x-1)(x+1)}$$
$$= \frac{x-2}{2x-1}$$

21.
$$\frac{\frac{2}{x}+3}{\frac{4}{x^2}-9} = \frac{\left(\frac{2}{x}+3\right)\cdot x^2}{\left(\frac{4}{x^2}-9\right)\cdot x^2}$$
$$= \frac{2x+3x^2}{4-9x^2}$$
$$= \frac{x(2+3x)}{(2+3x)(2-3x)}$$
$$= \frac{x}{2-3x}$$

23. $\dfrac{1-\frac{x}{y}}{\frac{x^2}{y^2}-1}=\dfrac{\left(1-\frac{x}{y}\right)\cdot y^2}{\left(\frac{x^2}{y^2}-1\right)\cdot y^2}$
$=\dfrac{y^2-xy}{x^2-y^2}$
$=\dfrac{y(y-x)}{(x+y)(x-y)}$
$=\dfrac{-y(x-y)}{(x+y)(x-y)}$
$=-\dfrac{y}{x+y}$

25. $\dfrac{\frac{-2x}{x-y}}{\frac{y}{x^2}}=\dfrac{-2x}{x-y}\cdot\dfrac{x^2}{y}=-\dfrac{2x^3}{y(x-y)}$

27. $\dfrac{\frac{2}{x}+\frac{1}{x^2}}{\frac{y}{x^2}}=\dfrac{\left(\frac{2}{x}+\frac{1}{x^2}\right)x^2}{\left(\frac{y}{x^2}\right)x^2}=\dfrac{2x+1}{y}$

29. $\dfrac{\frac{x}{9}-\frac{1}{x}}{1+\frac{3}{x}}=\dfrac{\left(\frac{x}{9}-\frac{1}{x}\right)\cdot 9x}{\left(1+\frac{3}{x}\right)\cdot 9x}$
$=\dfrac{x^2-9}{9x+27}$
$=\dfrac{(x+3)(x-3)}{9(x+3)}$
$=\dfrac{x-3}{9}$

31. $\dfrac{\frac{x-1}{x^2-4}}{1+\frac{1}{x-2}}=\dfrac{\frac{x-1}{x^2-4}}{\frac{x-2+1}{x-2}}=\dfrac{\frac{x-1}{x^2-4}}{\frac{x-1}{x-2}}$
$=\dfrac{x-1}{x^2-4}\cdot\dfrac{x-2}{x-1}$
$=\dfrac{x-1}{(x+2)(x-2)}\cdot\dfrac{x-2}{x-1}$
$=\dfrac{1}{x+2}$

33. $\dfrac{\frac{2}{x+5}+\frac{4}{x+3}}{\frac{3x+13}{x^2+8x+15}}=\dfrac{\frac{2}{x+5}+\frac{4}{x+3}}{\frac{3x+13}{(x+5)(x+3)}}$
$=\dfrac{\left(\frac{2}{x+5}+\frac{4}{x+3}\right)(x+5)(x+3)}{\frac{3x+13}{(x+5)(x+3)}(x+5)(x+3)}$
$=\dfrac{2(x+3)+4(x+5)}{3x+13}$
$=\dfrac{2x+6+4x+20}{3x+13}$
$=\dfrac{6x+26}{3x+13}$
$=\dfrac{2(3x+13)}{3x+13}$
$=2$

35. $\dfrac{x^{-1}}{x^{-2}+y^{-2}}=\dfrac{\frac{1}{x}}{\frac{1}{x^2}+\frac{1}{y^2}}$
$=\dfrac{x^2y^2\left(\frac{1}{x}\right)}{x^2y^2\left(\frac{1}{x^2}+\frac{1}{y^2}\right)}$
$=\dfrac{xy^2}{y^2+x^2}$
$=\dfrac{xy^2}{x^2+y^2}$

37. $\dfrac{2a^{-1}+3b^{-2}}{a^{-1}-b^{-1}}=\dfrac{\frac{2}{a}+\frac{3}{b^2}}{\frac{1}{a}-\frac{1}{b}}$
$=\dfrac{ab^2\left(\frac{2}{a}+\frac{3}{b^2}\right)}{ab^2\left(\frac{1}{a}-\frac{1}{b}\right)}$
$=\dfrac{2b^2+3a}{b^2-ab}$
$=\dfrac{2b^2+3a}{b(b-a)}$

39. $\dfrac{1}{x-x^{-1}}=\dfrac{1}{x-\frac{1}{x}}$
$=\dfrac{x(1)}{x\left(x-\frac{1}{x}\right)}$
$=\dfrac{x}{x^2-1}$
$=\dfrac{x}{(x+1)(x-1)}$

41. $\dfrac{a^{-1}+1}{a^{-1}-1}=\dfrac{\frac{1}{a}+1}{\frac{1}{a}-1}=\dfrac{a\left(\frac{1}{a}+1\right)}{a\left(\frac{1}{a}-1\right)}=\dfrac{1+a}{1-a}$

43. $\dfrac{3x^{-1}+(2y)^{-1}}{x^{-2}}=\dfrac{\frac{3}{x}+\frac{1}{2y}}{\frac{1}{x^2}}$
$=\dfrac{2x^2y\left(\frac{3}{x}+\frac{1}{2y}\right)}{2x^2y\left(\frac{1}{x^2}\right)}$
$=\dfrac{6xy+x^2}{2y}$
$=\dfrac{x(x+6y)}{2y}$

45. $\dfrac{2a^{-1}+(2a)^{-1}}{a^{-1}+2a^{-2}}=\dfrac{\frac{2}{a}+\frac{1}{2a}}{\frac{1}{a}+\frac{2}{a^2}}$
$=\dfrac{2a^2\left(\frac{2}{a}+\frac{1}{2a}\right)}{2a^2\left(\frac{1}{a}+\frac{2}{a^2}\right)}$
$=\dfrac{4a+a}{2a+4}$
$=\dfrac{5a}{2(a+2)}$

47. $\dfrac{5x^{-1}+2y^{-1}}{x^{-2}y^{-2}}=\dfrac{\frac{5}{x}+\frac{2}{y}}{\frac{1}{x^2y^2}}$
$=\dfrac{x^2y^2\left(\frac{5}{x}+\frac{2}{y}\right)}{x^2y^2\left(\frac{1}{x^2y^2}\right)}$
$=5xy^2+2x^2y$
$=xy(5y+2x)$

49. $\dfrac{5x^{-1}-2y^{-1}}{25x^{-2}-4y^{-2}}=\dfrac{\frac{5}{x}-\frac{2}{y}}{\frac{25}{x^2}-\frac{4}{y^2}}$
$=\dfrac{x^2y^2\left(\frac{5}{x}-\frac{2}{y}\right)}{x^2y^2\left(\frac{25}{x^2}-\frac{4}{y^2}\right)}$
$=\dfrac{5xy^2-2x^2y}{25y^2-4x^2}$
$=\dfrac{xy(5y-2x)}{(5y+2x)(5y-2x)}$
$=\dfrac{xy}{5y+2x}$ or $\dfrac{xy}{2x+5y}$

51. $\dfrac{3x^3y^2}{12x}=\dfrac{3x\cdot x^2y^2}{3x\cdot 4}=\dfrac{x^2y^2}{4}$

53. $\dfrac{144x^5y^5}{-16x^2y}=\dfrac{16x^2y\cdot 9x^3y^4}{16x^2y\cdot(-1)}=-9x^3y^4$

55. $P(x)=-x^2$
$P(-3)=-(-3)^2=-9$

57. $\dfrac{\frac{x+1}{9}}{\frac{y-2}{5}}=\dfrac{x+1}{9}\div\dfrac{y-2}{5}=\dfrac{x+1}{9}\cdot\dfrac{5}{y-2}$
Both a and c are equivalent to the original expression.

59. $\dfrac{a}{1-\frac{s}{770}}=\dfrac{770(a)}{770\left(1-\frac{s}{770}\right)}=\dfrac{770a}{770-s}$

61. $\dfrac{\frac{1}{x}}{\frac{3}{y}}=\dfrac{1}{x}\div\dfrac{3}{y}=\dfrac{1}{x}\cdot\dfrac{y}{3}$
Both a and b are equivalent to the original expression.

63. $\dfrac{1}{1+(1+x)^{-1}}=\dfrac{1}{1+\frac{1}{1+x}}$
$=\dfrac{(1+x)\cdot 1}{(1+x)\left(1+\frac{1}{1+x}\right)}$
$=\dfrac{1+x}{1+x+1}$
$=\dfrac{1+x}{2+x}$

65. $\dfrac{x}{1-\frac{1}{1+\frac{1}{x}}} = \dfrac{x}{1-\frac{1}{\frac{x+1}{x}}}$
$= \dfrac{x}{1-\frac{x}{x+1}}$
$= \dfrac{(x+1)(x)}{(x+1)\left(1-\frac{x}{x+1}\right)}$
$= \dfrac{x(x+1)}{x+1-x}$
$= \dfrac{x(x+1)}{1}$
$= x(x+1)$

67. $\dfrac{\frac{2}{y^2}-\frac{5}{xy}-\frac{3}{x^2}}{\frac{2}{y^2}+\frac{7}{xy}+\frac{3}{x^2}} = \dfrac{x^2y^2\left(\frac{2}{y^2}-\frac{5}{xy}-\frac{3}{x^2}\right)}{x^2y^2\left(\frac{2}{y^2}+\frac{7}{xy}+\frac{3}{x^2}\right)}$
$= \dfrac{2x^2-5xy-3y^2}{2x^2+7xy+3y^2}$
$= \dfrac{(2x+y)(x-3y)}{(2x+y)(x+3y)}$
$= \dfrac{x-3y}{x+3y}$

69. $\dfrac{3(a+1)^{-1}+4a^{-2}}{(a^3+a^2)^{-1}} = \dfrac{\frac{3}{a+1}+\frac{4}{a^2}}{\frac{1}{a^3+a^2}}$
$= \dfrac{\frac{3a^2+4(a+1)}{a^2(a+1)}}{\frac{1}{a^2(a+1)}}$
$= \dfrac{3a^2+4a+4}{a^2(a+1)} \cdot \dfrac{a^2(a+1)}{1}$
$= 3a^2+4a+4$

71. $f(x) = \dfrac{1}{x}$

a. $f(a+h) = \dfrac{1}{a+h}$

b. $f(a) = \dfrac{1}{a}$

c. $\dfrac{f(a+h)-f(a)}{h} = \dfrac{\frac{1}{a+h}-\frac{1}{a}}{h}$

d. $\dfrac{\frac{1}{a+h}-\frac{1}{a}}{h} = \dfrac{a(a+h)\left(\frac{1}{a+h}-\frac{1}{a}\right)}{a(a+h)\cdot h}$
$= \dfrac{a-(a+h)}{ah(a+h)}$
$= \dfrac{-h}{ah(a+h)}$
$= \dfrac{-1}{a(a+h)}$

73. $f(x) = \dfrac{3}{x+1}$

a. $f(a+h) = \dfrac{3}{a+h+1}$

b. $f(a) = \dfrac{3}{a+1}$

c. $\dfrac{f(a+h)-f(a)}{h} = \dfrac{\frac{3}{a+h+1}-\frac{3}{a+1}}{h}$

d. $\dfrac{\frac{3}{a+h+1}-\frac{3}{a+1}}{h}$
$= \dfrac{\left(\frac{3}{a+h+1}-\frac{3}{a+1}\right)\cdot(a+h+1)(a+1)}{h\cdot(a+h+1)(a+1)}$
$= \dfrac{3(a+1)-3(a+h+1)}{h(a+h+1)(a+1)}$
$= \dfrac{3a+3-3a-3h-3}{h(a+h+1)(a+1)}$
$= \dfrac{-3h}{h(a+h+1)(a+1)}$
$= \dfrac{-3}{(a+h+1)(a+1)}$

Chapter 7 Vocabulary Check

1. A ratio is the quotient of two numbers.

2. $\dfrac{x}{2} = \dfrac{7}{16}$ is an example of a proportion.

3. If $\dfrac{a}{b} = \dfrac{c}{d}$, then ad and bc are called cross products.

4. A rational expression is an expression that can be written in the form $\frac{P}{Q}$, where P and Q are polynomials and Q is not 0.

5. In a complex fraction, the numerator or denominator or both may contain fractions.

6. The domain of the rational function $f(x) = \frac{1}{x-3}$ is $\{x|x \text{ is a real number } x \neq 3\}$.

7. The reciprocal of $\frac{9}{7}$ is $\frac{7}{9}$.

Chapter 7 Review

1. 7 is never 0 so the domain of $f(x) = \frac{3-5x}{7}$ is $\{x|x \text{ is a real number}\}$.

2. 11 is never 0 so the domain of $g(x) = \frac{2x+4}{11}$ is $\{x|x \text{ is a real number}\}$.

3. $x-5=0$
   $x=5$
   The domain of $F(x) = \frac{-3x^2}{x-5}$ is $\{x|x \text{ is a real number and } x \neq 5\}$.

4. $3x-12=0$
   $3x=12$
   $x=4$
   The domain of $h(x) = \frac{4x}{3x-12}$ is $\{x|x \text{ is a real number and } x \neq 4\}$.

5. $x^2+8x=0$
   $x(x+8)=0$
   $x=0$ or $x+8=0$
   $x=0$ or $x=-8$
   The domain of $f(x) = \frac{x^3+2}{x^2+8x}$ is $\{x|x \text{ is a real number and } x \neq 0, x \neq -8\}$.

6. $3x^2-48=0$
   $3(x^2-16)=0$
   $3(x+4)(x-4)=0$
   $x+4=0$ or $x-4=0$
   $x=-4$ or $x=4$
   The domain of $G(x) = \frac{20}{3x^2-48}$ is $\{x|x \text{ is a real number and } x \neq -4, x \neq 4\}$.

7. $\frac{x-12}{12-x} = \frac{x-12}{-(x-12)} = -1$

8. $\frac{5x-15}{25x-75} = \frac{5(x-3)}{25(x-3)} = \frac{5}{25} = \frac{1}{5}$

9. $\frac{2x}{2x^2-2x} = \frac{2x}{2x(x-1)} = \frac{1}{x-1}$

10. $\frac{x+7}{x^2-49} = \frac{x+7}{(x-7)(x+7)} = \frac{1}{x-7}$

11. $\frac{2x^2+4x-30}{x^2+x-20} = \frac{2(x^2+2x-15)}{(x+5)(x-4)}$
    $= \frac{2(x+5)(x-3)}{(x+5)(x-4)}$
    $= \frac{2(x-3)}{x-4}$

12. $C(x) = \frac{35x+4200}{x}$

    a. $C(50) = \frac{35(50)+4200}{50}$
       $= \frac{1750+4200}{50}$
       $= \frac{5950}{50}$
       $=119$
       The average cost is \$119.

    b. $C(100) = \frac{35(100)+4200}{100}$
       $= \frac{3500+4200}{100}$
       $= \frac{7700}{100}$
       $=77$
       The average cost is \$77.

c. It will decrease.

13. $\frac{x^2+xa+xb+ab}{x^2-xc+bx-bc}=\frac{x(x+a)+b(x+a)}{x(x-c)+b(x-c)}$
$=\frac{(x+a)(x+b)}{(x-c)(x+b)}$
$=\frac{x+a}{x-c}$

14. $\frac{x^2+5x-2x-10}{x^2-3x-2x+6}=\frac{x(x+5)-2(x+5)}{x(x-3)-2(x-3)}$
$=\frac{(x+5)(x-2)}{(x-3)(x-2)}$
$=\frac{x+5}{x-3}$

15. $\frac{4-x}{x^3-64}=-\frac{x-4}{x^3-64}$
$=-\frac{x-4}{(x-4)(x^2+4x+16)}$
$=-\frac{1}{x^2+4x+16}$

16. $\frac{x^2-4}{x^3+8}=\frac{(x+2)(x-2)}{(x+2)(x^2-2x+4)}=\frac{x-2}{x^2-2x+4}$

17. $\frac{15x^3y^2}{z}\cdot\frac{z}{5xy^3}=\frac{15x^3y^2\cdot z}{z\cdot 5xy^3}$
$=\frac{3\cdot 5\cdot x^2\cdot x\cdot y^2\cdot z}{z\cdot 5\cdot x\cdot y^2\cdot y}$
$=\frac{3x^2}{y}$

18. $\frac{-y^3}{8}\cdot\frac{9x^2}{y^3}=-\frac{y^3\cdot 9x^2}{8\cdot y^3}=-\frac{9x^2}{8}$

19. $\frac{x^2-9}{x^2-4}\cdot\frac{x-2}{x+3}=\frac{(x^2-9)\cdot(x-2)}{(x^2-4)\cdot(x+3)}$
$=\frac{(x-3)(x+3)(x-2)}{(x+2)(x-2)(x+3)}$
$=\frac{x-3}{x+2}$

20. $\frac{2x+5}{x-6}\cdot\frac{2x}{-x+6}=\frac{2x+5}{x-6}\cdot\frac{2x}{-(x-6)}$
$=\frac{2x+5}{x-6}\cdot\frac{-2x}{x-6}$
$=\frac{(2x+5)\cdot(-2x)}{(x-6)\cdot(x-6)}$
$=\frac{-2x(2x+5)}{(x-6)^2}$

21. $\frac{x^2-5x-24}{x^2-x-12}\div\frac{x^2-10x+16}{x^2+x-6}$
$=\frac{x^2-5x-24}{x^2-x-12}\cdot\frac{x^2+x-6}{x^2-10x+16}$
$=\frac{(x-8)(x+3)\cdot(x+3)(x-2)}{(x-4)(x+3)\cdot(x-8)(x-2)}$
$=\frac{x+3}{x-4}$

22. $\frac{4x+4y}{xy^2}\div\frac{3x+3y}{x^2y}=\frac{4x+4y}{xy^2}\cdot\frac{x^2y}{3x+3y}$
$=\frac{4(x+y)\cdot x\cdot x\cdot y}{x\cdot y\cdot y\cdot 3(x+y)}$
$=\frac{4x}{3y}$

23. $\frac{x^2+x-42}{x-3}\cdot\frac{(x-3)^2}{x+7}$
$=\frac{(x+7)(x-6)\cdot(x-3)(x-3)}{(x-3)\cdot(x+7)}$
$=(x-6)(x-3)$

24. $\frac{2a+2b}{3}\cdot\frac{a-b}{a^2-b^2}=\frac{2(a+b)\cdot(a-b)}{3\cdot(a+b)(a-b)}=\frac{2}{3}$

25. $\frac{2x^2-9x+9}{8x-12}\div\frac{x^2-3x}{2x}=\frac{2x^2-9x+9}{8x-12}\cdot\frac{2x}{x^2-3x}$
$=\frac{(2x-3)(x-3)\cdot 2x}{4(2x-3)\cdot x(x-3)}$
$=\frac{2}{4}$
$=\frac{1}{2}$

26. $\frac{x^2-y^2}{x^2+xy}\div\frac{3x^2-2xy-y^2}{3x^2+6x}$
$=\frac{x^2-y^2}{x^2+xy}\cdot\frac{3x^2+6x}{3x^2-2xy-y^2}$
$=\frac{(x-y)(x+y)\cdot 3x(x+2)}{x(x+y)\cdot(3x+y)(x-y)}$
$=\frac{3(x+2)}{3x+y}$

27. $\frac{x-y}{4}\div\frac{y^2-2y-xy+2x}{16x+24}$
$=\frac{x-y}{4}\cdot\frac{16x+24}{y^2-2y-xy+2x}$
$=\frac{x-y}{4}\cdot\frac{8(2x+3)}{y(y-2)-x(y-2)}$
$=\frac{x-y}{4}\cdot\frac{8(2x+3)}{(y-2)(y-x)}$
$=-\frac{y-x}{4}\cdot\frac{8(2x+3)}{(y-2)(y-x)}$
$=-\frac{2\cdot 4(y-x)(2x+3)}{4(y-2)(y-x)}$
$=-\frac{2(2x+3)}{y-2}$

28. $\frac{5+x}{7}\div\frac{xy+5y-3x-15}{7y-35}$
$=\frac{5+x}{7}\cdot\frac{7y-35}{xy+5y-3x-15}$
$=\frac{(5+x)\cdot 7(y-5)}{7\cdot(x+5)(y-3)}$
$=\frac{y-5}{y-3}$

29. $\frac{x}{x^2+9x+14}+\frac{7}{x^2+9x+14}=\frac{x+7}{x^2+9x+14}$
$=\frac{x+7}{(x+7)(x+2)}$
$=\frac{1}{x+2}$

30. $\frac{x}{x^2+2x-15}+\frac{5}{x^2+2x-15}=\frac{x+5}{x^2+2x-15}$
$=\frac{x+5}{(x+5)(x-3)}$
$=\frac{1}{x-3}$

31. $\frac{4x-5}{3x^2}-\frac{2x+5}{3x^2}=\frac{4x-5-(2x+5)}{3x^2}$
$=\frac{4x-5-2x-5}{3x^2}$
$=\frac{2x-10}{3x^2}$

32. $\frac{9x+7}{6x^2}-\frac{3x+4}{6x^2}=\frac{9x+7-(3x+4)}{6x^2}$
$=\frac{9x+7-3x-4}{6x^2}$
$=\frac{6x+3}{6x^2}$
$=\frac{3(2x+1)}{3\cdot 2x^2}$
$=\frac{2x+1}{2x^2}$

33. $2x=2\cdot x$
$7x=7\cdot x$
$\text{LCD}=2\cdot 7\cdot x=14x$

34. $x^2-5x-24=(x-8)(x+3)$
$x^2+11x+24=(x+8)(x+3)$
$\text{LCD}=(x-8)(x+3)(x+8)$

35. $\frac{5}{7x}=\frac{5}{7x}\cdot\frac{2x^2y}{2x^2y}=\frac{5\cdot 2x^2y}{7x\cdot 2x^2y}=\frac{10x^2y}{14x^3y}$

36. $\frac{9}{4y}=\frac{9}{4y}\cdot\frac{4y^2x}{4y^2x}=\frac{9\cdot 4y^2x}{4y\cdot 4y^2x}=\frac{36y^2x}{16y^3x}$

37. $\frac{x+2}{x^2+11x+18}=\frac{x+2}{(x+9)(x+2)}$
$=\frac{(x+2)(x-5)}{(x+9)(x+2)(x-5)}$
$=\frac{x^2-3x-10}{(x+2)(x-5)(x+9)}$

38. $\frac{3x-5}{x^2+4x+4}=\frac{3x-5}{(x+2)^2}$
$=\frac{(3x-5)(x+3)}{(x+2)^2(x+3)}$
$=\frac{3x^2+4x-15}{(x+2)^2(x+3)}$

39. $\frac{4}{5x^2}-\frac{6}{y}=\frac{4(y)}{5x^2(y)}-\frac{6(5x^2)}{y(5x^2)}=\frac{4y-30x^2}{5x^2y}$

40.
$$\begin{aligned}\frac{2}{x-3}-\frac{4}{x-1}&=\frac{2(x-1)}{(x-3)(x-1)}-\frac{4(x-3)}{(x-1)(x-3)}\\&=\frac{2(x-1)-4(x-3)}{(x-3)(x-1)}\\&=\frac{2x-2-4x+12}{(x-3)(x-1)}\\&=\frac{-2x+10}{(x-3)(x-1)}\end{aligned}$$

41.
$$\begin{aligned}\frac{4}{x+3}-2&=\frac{4}{x+3}-\frac{2(x+3)}{x+3}\\&=\frac{4-2(x+3)}{x+3}\\&=\frac{4-2x-6}{x+3}\\&=\frac{-2x-2}{x+3}\end{aligned}$$

42.
$$\begin{aligned}&\frac{3}{x^2+2x-8}+\frac{2}{x^2-3x+2}\\&=\frac{3}{(x+4)(x-2)}+\frac{2}{(x-1)(x-2)}\\&=\frac{3(x-1)}{(x+4)(x-2)(x-1)}+\frac{2(x+4)}{(x-1)(x-2)(x+4)}\\&=\frac{3(x-1)+2(x+4)}{(x+4)(x-2)(x-1)}\\&=\frac{3x-3+2x+8}{(x+4)(x-2)(x-1)}\\&=\frac{5x+5}{(x+4)(x-2)(x-1)}\end{aligned}$$

43.
$$\begin{aligned}\frac{2x-5}{6x+9}-\frac{4}{2x^2+3x}&=\frac{2x-5}{3(2x+3)}-\frac{4}{x(2x+3)}\\&=\frac{(2x-5)(x)}{3(2x+3)(x)}-\frac{4(3)}{x(2x+3)(3)}\\&=\frac{2x^2-5x-12}{3x(2x+3)}\\&=\frac{(2x+3)(x-4)}{3x(2x+3)}\\&=\frac{x-4}{3x}\end{aligned}$$

44.
$$\begin{aligned}\frac{x-1}{x^2-2x+1}-\frac{x+1}{x-1}&=\frac{x-1}{(x-1)^2}-\frac{x+1}{x-1}\\&=\frac{1}{x-1}-\frac{x+1}{x-1}\\&=\frac{1-(x+1)}{x-1}\\&=\frac{1-x-1}{x-1}\\&=\frac{-x}{x-1}\\&=-\frac{x}{x-1}\end{aligned}$$

45. $P=2l+2w$

$$\begin{aligned}P&=2\left(\frac{x}{8}\right)+2\left(\frac{x+2}{4x}\right)\\&=\frac{x}{4}+\frac{2(x+2)}{4x}\\&=\frac{x\cdot x}{4\cdot x}+\frac{2x+4}{4x}\\&=\frac{x^2+2x+4}{4x}\end{aligned}$$

$A=l\cdot w$

$A=\frac{x}{8}\cdot\frac{x+2}{4x}=\frac{x\cdot(x+2)}{8\cdot 4x}=\frac{x+2}{32}$

The perimeter is $\frac{x^2+2x+4}{4x}$ units and the area is $\frac{x+2}{32}$ square units.

46.
$$\begin{aligned}P&=\frac{3x}{4x-4}+\frac{2x}{3x-3}+\frac{x}{x-1}\\&=\frac{3x}{4(x-1)}+\frac{2x}{3(x-1)}+\frac{x}{x-1}\\&=\frac{3x(3)}{4(x-1)(3)}+\frac{2x(4)}{3(x-1)(4)}+\frac{x(12)}{(x-1)(12)}\\&=\frac{9x+8x+12x}{12(x-1)}\\&=\frac{29x}{12(x-1)}\end{aligned}$$

$A=\frac{1}{2}\cdot b\cdot h$

$A=\frac{1}{2}\cdot\frac{x}{x-1}\cdot\frac{6y}{5}=\frac{1\cdot x\cdot 2\cdot 3y}{2\cdot(x-1)\cdot 5}=\frac{3xy}{5(x-1)}$

The perimeter is $\frac{29x}{12(x-1)}$ units and the area is $\frac{3xy}{5(x-1)}$ square units.

47.

$$\frac{n}{10} = 9 - \frac{n}{5}$$
$$10\left(\frac{n}{10}\right) = 10\left(9 - \frac{n}{5}\right)$$
$$10\left(\frac{n}{10}\right) = 10(9) - 10\left(\frac{n}{5}\right)$$
$$n = 90 - 2n$$
$$3n = 90$$
$$n = 30$$

48.

$$\frac{2}{x+1} - \frac{1}{x-2} = -\frac{1}{2}$$
$$2(x+1)(x-2)\left(\frac{2}{x+1} - \frac{1}{x-2}\right) = 2(x+1)(x-2)\left(-\frac{1}{2}\right)$$
$$2(x+1)(x-2)\left(\frac{2}{x+1}\right) - 2(x+1)(x-2)\left(\frac{1}{x-2}\right) = 2(x+1)(x-2)\left(-\frac{1}{2}\right)$$
$$4(x-2) - 2(x+1) = -(x+1)(x-2)$$
$$4x - 8 - 2x - 2 = -(x^2 - x - 2)$$
$$2x - 10 = -x^2 + x + 2$$
$$x^2 + x - 12 = 0$$
$$(x+4)(x-3) = 0$$

$x + 4 = 0$ or $x - 3 = 0$

$x = -4$ $\quad\quad x = 3$

49.

$$\frac{y}{2y+2} + \frac{2y-16}{4y+4} = \frac{y-3}{y+1}$$
$$\frac{y}{2(y+1)} + \frac{2y-16}{4(y+1)} = \frac{y-3}{y+1}$$
$$4(y+1)\left(\frac{y}{2(y+1)} + \frac{2y-16}{4(y+1)}\right) = 4(y+1)\left(\frac{y-3}{y+1}\right)$$
$$4(y+1)\left(\frac{y}{2(y+1)}\right) + 4(y+1)\left(\frac{2y-16}{4(y+1)}\right) = 4(y+1)\left(\frac{y-3}{y+1}\right)$$
$$2y + 2y - 16 = 4(y-3)$$
$$4y - 16 = 4y - 12$$
$$-16 = -12 \quad \text{False}$$

This equation has no solution.

50.
$$\frac{2}{x-3}-\frac{4}{x+3}=\frac{8}{x^2-9}$$
$$(x-3)(x+3)\left(\frac{2}{x-3}-\frac{4}{x+3}\right)=(x-3)(x+3)\left(\frac{8}{(x-3)(x+3)}\right)$$
$$(x-3)(x+3)\left(\frac{2}{x-3}\right)-(x-3)(x+3)\left(\frac{4}{x+3}\right)=8$$
$$2(x+3)-4(x-3)=8$$
$$2x+6-4x+12=8$$
$$-2x+18=8$$
$$-2x=-10$$
$$x=5$$

51.
$$\frac{x-3}{x+1}-\frac{x-6}{x+5}=0$$
$$(x+1)(x+5)\left(\frac{x-3}{x+1}-\frac{x-6}{x+5}\right)=(x+1)(x+5)(0)$$
$$(x+1)(x+5)\left(\frac{x-3}{x+1}\right)-(x+1)(x+5)\left(\frac{x-6}{x+5}\right)=0$$
$$(x+5)(x-3)-(x+1)(x-6)=0$$
$$x^2+2x-15-(x^2-5x-6)=0$$
$$x^2+2x-15-x^2+5x+6=0$$
$$7x-9=0$$
$$7x=9$$
$$x=\frac{9}{7}$$

52.
$$x+5=\frac{6}{x}$$
$$x(x+5)=x\left(\frac{6}{x}\right)$$
$$x^2+5x=6$$
$$x^2+5x-6=0$$
$$(x+6)(x-1)=0$$
$$x+6=0 \quad \text{or} \quad x-1=0$$
$$x=-6 \qquad x=1$$

53.
$$\frac{4A}{5b}=x^2$$
$$4A=5bx^2$$
$$\frac{4A}{5x^2}=\frac{5bx^2}{5x^2}$$
$$\frac{4A}{5x^2}=b$$

54. $\frac{x}{7}+\frac{y}{8}=10$

$56\left(\frac{x}{7}\right)+56\left(\frac{y}{8}\right)=56(10)$

$8x+7y=560$

$7y=560-8x$

$y=\frac{560-8x}{7}$

55. $\frac{x}{2}=\frac{12}{4}$

$4x=24$

$x=6$

56. $\frac{20}{1}=\frac{x}{25}$

$500=x$

57. $\frac{2}{x-1}=\frac{3}{x+3}$

$2(x+3)=3(x-1)$

$2x+6=3x-3$

$6=x-3$

$9=x$

58. $\frac{4}{y-3}=\frac{2}{y-3}$

$4(y-3)=2(y-3)$

$4y-12=2y-6$

$2y-12=-6$

$2y=6$

$y=3$

$y = 3$ doesn't check, so this equation has no solution.

59. Let x = the number of parts processed in 45 minutes.

$\frac{300}{20}=\frac{x}{45}$

$13,500=20x$

$675=x$

675 parts can be processed in 45 minutes.

60. Let x = the charge for 3 hours.

$\frac{90.00}{8}=\frac{x}{3}$

$270.00=8x$

$33.75=x$

He charges $33.75 for 3 hours.

61. $5\cdot\frac{1}{x}=\frac{3}{2}\cdot\frac{1}{x}+\frac{7}{6}$

$\frac{5}{x}=\frac{3}{2x}+\frac{7}{6}$

$6x\left(\frac{5}{x}\right)=6x\left(\frac{3}{2x}\right)+6x\left(\frac{7}{6}\right)$

$30=9+7x$

$21=7x$

$x=3$

The unknown number is 3.

62. $\frac{1}{x}=\frac{1}{4-x}$

$4-x=x$

$4=2x$

$2=x$

The unknown number is 2.

63. Let r be the rate of the faster car. Then the rate of the slower car is $r - 10$.

	Distance =	Rate ·	Time
Fast car	90	r	$\frac{90}{r}$
Slow car	60	$r-10$	$\frac{60}{r-10}$

$\frac{90}{r}=\frac{60}{r-10}$

$90(r-10)=60r$

$90r-900=60r$

$-900=-30r$

$30=r$

$r-10=30-10=20$

The rate of the fast car is 30 miles per hour and the rate of the slower car is 20 miles per hour.

64. Let r be the speed of the boat in still water.

	Distance =	Rate ·	Time
Upstream	48	$r-4$	$\frac{48}{r-4}$
Downstream	72	$r+4$	$\frac{72}{r+4}$

$$\begin{aligned}\frac{48}{r-4} &= \frac{72}{r+4}\\ 48(r+4) &= 72(r-4)\\ 48r+192 &= 72r-288\\ 480 &= 24r\\ r &= 20\end{aligned}$$

The speed of the boat in still water is 20 miles per hour.

65. Let x be the time it takes Maria working alone.

	Hours to Complete Total Job	Part of Job Completed in 1 Hour
Mark	7	$\frac{1}{7}$
Maria	x	$\frac{1}{x}$
Together	5	$\frac{1}{5}$

$$\begin{aligned}\frac{1}{7}+\frac{1}{x} &= \frac{1}{5}\\ 35x\left(\frac{1}{7}\right)+35x\left(\frac{1}{x}\right) &= 35x\left(\frac{1}{5}\right)\\ 5x+35 &= 7x\\ 35 &= 2x\\ x &= \frac{35}{2} \text{ or } 17\frac{1}{2}\end{aligned}$$

It takes Maria $17\frac{1}{2}$ hours to complete the job alone.

66. Let x be the number of days it takes the pipes to fill the pond together.

	Days to Complete Total Job	Part of Job Completed in 1 Day
Pipe A	20	$\frac{1}{20}$
Pipe B	15	$\frac{1}{15}$
Together	x	$\frac{1}{x}$

$$\begin{aligned}\frac{1}{20}+\frac{1}{25} &= \frac{1}{x}\\ 60x\left(\frac{1}{20}\right)+60x\left(\frac{1}{15}\right) &= 60x\left(\frac{1}{x}\right)\\ 3x+4x &= 60\\ 7x &= 60\\ x &= \frac{60}{7} = 8\frac{4}{7}\end{aligned}$$

Both pipes fill the pond in $8\frac{4}{7}$ days.

67. $$\begin{aligned}\frac{2}{3} &= \frac{10}{x}\\ 2x &= 30\\ x &= 15\end{aligned}$$

The missing length is 15.

68. $$\begin{aligned}\frac{12}{4} &= \frac{18}{x}\\ 12x &= 72\\ x &= 6\end{aligned}$$

The missing length is 6.

69. $$\frac{\frac{5x}{27}}{-\frac{10xy}{21}} = \frac{5x}{27}\cdot -\frac{21}{10xy} = -\frac{5x\cdot 3\cdot 7}{3\cdot 9\cdot 5\cdot 2\cdot x\cdot y} = -\frac{7}{18y}$$

70. $$\frac{\frac{3}{5}+\frac{2}{7}}{\frac{1}{5}+\frac{5}{6}} = \frac{\frac{21}{35}+\frac{10}{35}}{\frac{6}{30}+\frac{25}{30}} = \frac{\frac{31}{35}}{\frac{31}{30}} = \frac{31}{35}\cdot\frac{30}{31} = \frac{31\cdot 5\cdot 6}{5\cdot 7\cdot 31} = \frac{6}{7}$$

71. $$\frac{3-\frac{1}{y}}{2-\frac{1}{y}} = \frac{y\left(3-\frac{1}{y}\right)}{y\left(2-\frac{1}{y}\right)} = \frac{y(3)-y\left(\frac{1}{y}\right)}{y(2)-y\left(\frac{1}{y}\right)} = \frac{3y-1}{2y-1}$$

72. $$\begin{aligned}\frac{\frac{6}{x+2}+4}{\frac{8}{x+2}-4} &= \frac{(x+2)\left(\frac{6}{x+2}+4\right)}{(x+2)\left(\frac{8}{x+2}-4\right)}\\ &= \frac{(x+2)\left(\frac{6}{x+2}\right)+(x+2)(4)}{(x+2)\left(\frac{8}{x+2}\right)-(x+2)(4)}\\ &= \frac{6+4x+8}{8-4x-8}\\ &= \frac{4x+14}{-4x}\\ &= -\frac{2(2x+7)}{2\cdot 2x}\\ &= -\frac{2x+7}{2x}\end{aligned}$$

73. $\dfrac{\frac{x-3}{x+3}+\frac{x+3}{x-3}}{\frac{x-3}{x+3}-\frac{x+3}{x-3}}=\dfrac{(x+3)(x-3)\left(\frac{x-3}{x+3}+\frac{x+3}{x-3}\right)}{(x+3)(x-3)\left(\frac{x-3}{x+3}-\frac{x+3}{x-3}\right)}$

$=\dfrac{(x-3)^2+(x+3)^2}{(x-3)^2-(x+3)^2}$

$=\dfrac{x^2-6x+9+x^2+6x+9}{x^2-6x+9-(x^2+6x+9)}$

$=\dfrac{2x^2+18}{x^2-6x+9-x^2-6x-9}$

$=\dfrac{2(x^2+9)}{-12x}$

$=-\dfrac{x^2+9}{6x}$

74. $\dfrac{\frac{3}{x-1}-\frac{2}{1-x}}{\frac{2}{x-1}-\frac{2}{x}}=\dfrac{\frac{3}{x-1}+\frac{2}{x-1}}{\frac{2}{x-1}-\frac{2}{x}}$

$=\dfrac{\frac{5}{x-1}}{\frac{2}{x-1}-\frac{2}{x}}$

$=\dfrac{x(x-1)\frac{5}{x-1}}{x(x-1)\left(\frac{2}{x-1}-\frac{2}{x}\right)}$

$=\dfrac{5x}{2x-2(x-1)}$

$=\dfrac{5x}{2x-2x+2}$

$=\dfrac{5x}{2}$

75. $\dfrac{x+y^{-1}}{\frac{x}{y}}=\dfrac{x+\frac{1}{y}}{\frac{x}{y}}=\dfrac{y\left(x+\frac{1}{y}\right)}{x\left(\frac{x}{y}\right)}=\dfrac{xy+1}{x}$

76. $\dfrac{x-xy^{-1}}{\frac{1+x}{y}}=\dfrac{x-\frac{x}{y}}{\frac{1+x}{y}}=\dfrac{y\left(x-\frac{x}{y}\right)}{y\left(\frac{1+x}{y}\right)}=\dfrac{xy-x}{1+x}$

77. $\dfrac{4x+12}{8x^2+24x}=\dfrac{4(x+3)}{2\cdot4\cdot x(x+3)}=\dfrac{1}{2x}$

78. $\dfrac{x^3-6x^2+9x}{x^2+4x-21}=\dfrac{x(x-3)^2}{(x+7)(x-3)}=\dfrac{x(x-3)}{x+7}$

79. $\dfrac{x^2+9x+20}{x^2-25}\cdot\dfrac{x^2-9x+20}{x^2+8x+16}$

$=\dfrac{(x+4)(x+5)\cdot(x-4)(x-5)}{(x+5)(x-5)\cdot(x+4)(x+4)}$

$=\dfrac{x-4}{x+4}$

80. $\dfrac{x^2-x-72}{x^2-x-30}\div\dfrac{x^2+6x-27}{x^2-9x+18}$

$=\dfrac{x^2-x-72}{x^2-x-30}\cdot\dfrac{x^2-9x+18}{x^2+6x-27}$

$=\dfrac{(x-9)(x+8)\cdot(x-3)(x-6)}{(x+5)(x-6)\cdot(x+9)(x-3)}$

$=\dfrac{(x-9)(x+8)}{(x+5)(x+9)}$

81. $\dfrac{x}{x^2-36}+\dfrac{6}{x^2-36}=\dfrac{x+6}{x^2-36}$

$=\dfrac{x+6}{(x+6)(x-6)}$

$=\dfrac{1}{x-6}$

82. $\dfrac{5x-1}{4x}-\dfrac{3x-2}{4x}=\dfrac{5x-1-(3x-2)}{4x}$

$=\dfrac{5x-1-3x+2}{4x}$

$=\dfrac{2x+1}{4x}$

83. $\dfrac{4}{3x^2+8x-3}+\dfrac{2}{3x^2-7x+2}$

$=\dfrac{4}{(x+3)(3x-1)}+\dfrac{2}{(x-2)(3x-1)}$

$=\dfrac{4(x-2)}{(x+3)(3x-1)(x-2)}+\dfrac{2(x+3)}{(x-2)(3x-1)(x+3)}$

$=\dfrac{4(x-2)+2(x+3)}{(x+3)(3x-1)(x-2)}$

$=\dfrac{4x-8+2x+6}{(x+3)(3x-1)(x-2)}$

$=\dfrac{6x-2}{(x+3)(3x-1)(x-2)}$

$=\dfrac{2(3x-1)}{(x+3)(3x-1)(x-2)}$

$=\dfrac{2}{(x+3)(x-2)}$

84. $\frac{3x}{x^2+9x+14}-\frac{6x}{x^2+4x-21}$
$=\frac{3x}{(x+7)(x+2)}-\frac{6x}{(x+7)(x-3)}$
$=\frac{3x(x-3)}{(x+7)(x+2)(x-3)}-\frac{6x(x+2)}{(x+7)(x-3)(x+2)}$
$=\frac{3x(x-3)-6x(x+2)}{(x+7)(x+2)(x-3)}$
$=\frac{3x^2-9x-6x^2-12x}{(x+7)(x+2)(x-3)}$
$=\frac{-3x^2-21x}{(x+7)(x+2)(x-3)}$
$=\frac{-3x(x+7)}{(x+7)(x+2)(x-3)}$
$=-\frac{3x}{(x+2)(x-3)}$

85.
$$\frac{4}{a-1}+2=\frac{3}{a-1}$$
$$(a-1)\left(\frac{4}{a-1}\right)+(a-1)(2)=(a-1)\left(\frac{3}{a-1}\right)$$
$$4+2(a-1)=3$$
$$4+2a-2=3$$
$$2+2a=3$$
$$2a=1$$
$$a=\frac{1}{2}$$

86.
$$\frac{x}{x+3}+4=\frac{x}{x+3}$$
$$(x+3)\left(\frac{x}{x+3}\right)+(x+3)(4)=(x+3)\left(\frac{x}{x+3}\right)$$
$$x+4(x+3)=x$$
$$x+4x+12=x$$
$$5x+12=x$$
$$12=-4x$$
$$-3=x$$
Since $x=-3$ makes a denominator 0, the solution does not check. This equation has no solution.

87.
$$\frac{2x}{3}-\frac{1}{6}=\frac{x}{2}$$
$$6\left(\frac{2x}{3}\right)-6\left(\frac{1}{6}\right)=6\left(\frac{x}{2}\right)$$
$$4x-1=3x$$
$$-1=-x$$
$$1=x$$
The unknown number is 1.

88. Let x be the number of days it takes them to paint the house working together.

	Days to Complete Total Job	Part of Job Completed in 1 Day
Mr. Crocker	3	$\frac{1}{3}$
Son	4	$\frac{1}{4}$
Together	x	$\frac{1}{x}$

$$\frac{1}{3}+\frac{1}{4}=\frac{1}{x}$$
$$12x\left(\frac{1}{3}\right)+12x\left(\frac{1}{4}\right)=12x\left(\frac{1}{x}\right)$$
$$4x+3x=12$$
$$7x=12$$
$$x=\frac{12}{7}\text{ or }1\frac{5}{7}$$
Working together, Mr. Crocker and his son can paint the house in $1\frac{5}{7}$ days.

89. $\frac{5}{3}=\frac{10}{x}$
$5x=30$
$x=6$
The missing length is 6.

90. $\frac{6}{18}=\frac{4}{x}$
$6x=72$
$x=12$
The missing length is 12.

91. $\frac{\frac{1}{4}}{\frac{1}{3}+\frac{1}{2}}=\frac{12\left(\frac{1}{4}\right)}{12\left(\frac{1}{3}+\frac{1}{2}\right)}=\frac{12\left(\frac{1}{4}\right)}{12\left(\frac{1}{3}\right)+12\left(\frac{1}{2}\right)}=\frac{3}{4+6}=\frac{3}{10}$

92. $\frac{4+\frac{2}{x}}{6+\frac{3}{x}} = \frac{x\left(4+\frac{2}{x}\right)}{x\left(6+\frac{3}{x}\right)}$
$= \frac{x(4)+x\left(\frac{2}{x}\right)}{x(6)+x\left(\frac{3}{x}\right)}$
$= \frac{4x+2}{6x+3}$
$= \frac{2(2x+1)}{3(2x+1)}$
$= \frac{2}{3}$

93. $\frac{y^{-2}}{1-y^{-2}} = \frac{\frac{1}{y^2}}{1-\frac{1}{y^2}} = \frac{y^2\left(\frac{1}{y^2}\right)}{y^2\left(1-\frac{1}{y^2}\right)} = \frac{1}{y^2-1}$

94. $\frac{4+x^{-1}}{3+x^{-1}} = \frac{4+\frac{1}{x}}{3+\frac{1}{x}} = \frac{x\left(4+\frac{1}{x}\right)}{x\left(3+\frac{1}{x}\right)} = \frac{4x+1}{3x+1}$

Chapter 7 Test

1. The rational expression is undefined when
$x^2+4x+3=0$
$(x+3)(x+1)=0$
$x+3=0$ or $x+1=0$
$x=-3$ $x=-1$
The domain is
$\{x|x \text{ is a real number}, x=-1, x\neq-3\}$.

2. a. $C=\frac{100x+3000}{x}$
$=\frac{100(200)+3000}{200}$
$=\frac{20,000+3000}{200}$
$=\frac{23,000}{200}$
$=115$
The average cost per desk is \$115.

b. $C=\frac{100x+3000}{x}$
$=\frac{100(1000)+3000}{1000}$
$=\frac{100,000+3000}{1000}$
$=\frac{103,000}{1000}$
$=103$
The average cost per desk is \$103.

3. $\frac{3x-6}{5x-10}=\frac{3(x-2)}{5(x-2)}=\frac{3}{5}$

4. $\frac{x+6}{x^2+12x+36}=\frac{x+6}{(x+6)^2}=\frac{1}{x+6}$

5. $\frac{x+3}{x^3+27}=\frac{x+3}{(x+3)(x^2-3x+9)}=\frac{1}{x^2-3x+9}$

6. $\frac{2m^3-2m^2-12m}{m^2-5m+6}=\frac{2m(m^2-m-6)}{(m-3)(m-2)}$
$=\frac{2m(m-3)(m+2)}{(m-3)(m-2)}$
$=\frac{2m(m+2)}{m-2}$

7. $\frac{ay+3a+2y+6}{ay+3a+5y+15}=\frac{(y+3)(a+2)}{(y+3)(a+5)}=\frac{a+2}{a+5}$

8. $\frac{y-x}{x^2-y^2}=\frac{-(x-y)}{(x-y)(x+y)}=-\frac{1}{x+y}$

9. $\frac{3}{x-1}\cdot(5x-5)=\frac{3}{x-1}\cdot5(x-1)=\frac{3\cdot5(x-1)}{x-1}=15$

10. $\frac{y^2-5y+6}{2y+4}\cdot\frac{y+2}{2y-6}=\frac{(y-3)(y-2)\cdot(y+2)}{2(y+2)\cdot2(y-3)}$
$=\frac{y-2}{4}$

11. $\frac{15x}{2x+5}-\frac{6-4x}{2x+5}=\frac{15x-(6-4x)}{2x+5}$
$=\frac{15x-6+4x}{2x+5}$
$=\frac{19x-6}{2x+5}$

12. $\frac{5a}{a^2-a-6}-\frac{2}{a-3}$
$=\frac{5a}{(a-3)(a+2)}-\frac{2(a+2)}{(a-3)(a+2)}$
$=\frac{5a-2(a+2)}{(a-3)(a+2)}$
$=\frac{5a-2a-4}{(a-3)(a+2)}$
$=\frac{3a-4}{(a-3)(a+2)}$

13. $\frac{6}{x^2-1}+\frac{3}{x+1}=\frac{6}{(x+1)(x-1)}+\frac{3(x-1)}{(x+1)(x-1)}$
$=\frac{6+3x-3}{(x+1)(x-1)}$
$=\frac{3x+3}{(x+1)(x-1)}$
$=\frac{3(x+1)}{(x+1)(x-1)}$
$=\frac{3}{x-1}$

14. $\frac{x^2-9}{x^2-3x}\div\frac{xy+5x+3y+15}{2x+10}$
$=\frac{x^2-9}{x^2-3x}\cdot\frac{2x+10}{xy+5x+3y+15}$
$=\frac{(x-3)(x+3)\cdot 2(x+5)}{x(x-3)\cdot(x+3)(y+5)}$
$=\frac{2(x+5)}{x(y+5)}$

15. $\frac{x+2}{x^2+11x+18}+\frac{5}{x^2-3x-10}=\frac{x+2}{(x+9)(x+2)}+\frac{5}{(x-5)(x+2)}$
$=\frac{(x+2)(x-5)}{(x+9)(x+2)(x-5)}+\frac{5(x+9)}{(x-5)(x+2)(x+9)}$
$=\frac{(x+2)(x-5)+5(x+9)}{(x+9)(x+2)(x-5)}$
$=\frac{x^2-3x-10+5x+45}{(x+9)(x+2)(x-5)}$
$=\frac{x^2+2x+35}{(x+9)(x+2)(x-5)}$

16.
$$\frac{4}{y}-\frac{5}{3}=-\frac{1}{5}$$
$$15y\left(\frac{4}{y}-\frac{5}{3}\right)=15y\left(-\frac{1}{5}\right)$$
$$15y\left(\frac{4}{y}\right)-15y\left(\frac{5}{3}\right)=15y\left(-\frac{1}{5}\right)$$
$$60-25y=-3y$$
$$60=22y$$
$$\frac{60}{22}=y$$
$$y=\frac{30}{11}$$

17.
$$\frac{5}{y+1}=\frac{4}{y+2}$$
$$5(y+2)=4(y+1)$$
$$5y+10=4y+4$$
$$y=-6$$

18.
$$\frac{a}{a-3}=\frac{3}{a-3}-\frac{3}{2}$$
$$2(a-3)\left(\frac{a}{a-3}\right)=2(a-3)\left(\frac{3}{a-3}-\frac{3}{2}\right)$$
$$2a=2(a-3)\left(\frac{3}{a-3}\right)-2(a-3)\left(\frac{3}{2}\right)$$
$$2a=6-3(a-3)$$
$$2a=6-3a+9$$
$$2a=15-3a$$
$$5a=15$$
$$a=3$$

In the original equation, 3 makes a denominator 0. This equation has no solution.

19.
$$x-\frac{14}{x-1}=4-\frac{2x}{x-1}$$
$$(x-1)\left(x-\frac{14}{x-1}\right)=(x-1)\left(4-\frac{2x}{x-1}\right)$$
$$x(x-1)-14=4(x-1)-2x$$
$$x^2-x-14=4x-4-2x$$
$$x^2-x-14=2x-4$$
$$x^2-3x-10=0$$
$$(x-5)(x+2)=0$$
$$x-5=0 \quad \text{or} \quad x+2=0$$
$$x=5 \qquad\qquad x=-2$$

20.

$$\begin{aligned}
\frac{10}{x^2-25} &= \frac{3}{x+5}+\frac{1}{x-5} \\
\frac{10}{(x+5)(x-5)} &= \frac{3}{x+5}+\frac{1}{x-5} \\
(x+5)(x-5)\left(\frac{10}{(x+5)(x-5)}\right) &= (x+5)(x-5)\left(\frac{3}{x+5}\right)+(x+5)(x-5)\left(\frac{1}{x-5}\right) \\
10 &= 3(x-5)+1(x+5) \\
10 &= 3x-15+x+5 \\
10 &= 4x-10 \\
20 &= 4x \\
5 &= x
\end{aligned}$$

In the original equation 5 makes a denominator 0. This equation has no solution.

21. $$\frac{\frac{5x^2}{yz^2}}{\frac{10x}{z^3}} = \frac{5x^2}{yz^2}\cdot\frac{z^3}{10x} = -\frac{5\cdot x\cdot x\cdot z\cdot z^2}{y\cdot z^2\cdot 2\cdot 5\cdot x} = \frac{xz}{2y}$$

22.

$$\begin{aligned}
\frac{5-\frac{1}{y^2}}{\frac{1}{y}+\frac{2}{y^2}} &= \frac{y^2\left(5-\frac{1}{y^2}\right)}{y^2\left(\frac{1}{y}+\frac{2}{y^2}\right)} \\
&= \frac{y^2(5)-y^2\left(\frac{1}{y^2}\right)}{y^2\left(\frac{1}{y}\right)+y^2\left(\frac{2}{y^2}\right)} \\
&= \frac{5y^2-1}{y+2}
\end{aligned}$$

23. Let x = the number of defective bulbs.

$$\begin{aligned}
\frac{85}{3} &= \frac{510}{x} \\
85x &= 1530 \\
x &= 18
\end{aligned}$$

Expect to find 18 defective bulbs.

24. $x+5\cdot\frac{1}{x}=6$

$x+\frac{5}{x}=6$

$x\left(x+\frac{5}{x}\right)=x(6)$

$x(x)+x\left(\frac{5}{x}\right)=x(6)$

$x^2+5=6x$

$x^2-6x+5=0$

$(x-5)(x-1)=0$

$x-5=0$ or $x-1=0$

$x=5$ $\quad x=1$

The unknown number is 5 or 1.

25. Let r be the speed of the boat in still water.

	Distance	= Rate	· Time
Upstream	14	$r-2$	$\frac{14}{r-2}$
Downstream	16	$r+2$	$\frac{16}{r+2}$

$\frac{14}{r-2}=\frac{16}{r+2}$

$14(r+2)=16(r-2)$

$14r+28=16r-32$

$60=2r$

$r=30$

The speed of the boat in still water is 30 miles per hour.

26. Let x be the number of hours it takes to fill the tank using both pipes.

	Hours to Complete Total Job	Part of Job Completed in 1 Hour
1st Pipe	12	$\frac{1}{12}$
2nd Pipe	15	$\frac{1}{15}$
Together	x	$\frac{1}{x}$

$\frac{1}{12}+\frac{1}{15}=\frac{1}{x}$

$60x\left(\frac{1}{12}\right)+60x\left(\frac{1}{15}\right)=60x\left(\frac{1}{x}\right)$

$5x+4x=60$

$9x=60$

$x=\frac{60}{9}=\frac{20}{3}=6\frac{2}{3}$

Together, the pipes can fill the tank in $6\frac{2}{3}$ hours.

27. $\frac{8}{x}=\frac{10}{15}$

$8(15)=10x$

$120=10x$

$12=x$

The missing length is 12.

Chapter 7 Cumulative Review

1. a. $\frac{15}{x}=4$

b. $12-3=x$

c. $4x+17\neq 21$

d. $3x<48$

2. a. $12-x=-45$

b. $12x=-45$

c. $x-10=2x$

3. Let x = the amount invested at 9% for one year.

	Principal	· Rate =	Interest
9%	x	0.09	$0.09x$
7%	$20{,}000-x$	0.07	$0.07(20{,}000-x)$
Total	20,000		1550

$0.09x+0.07(20{,}000-x)=1550$

$0.09x+1400-0.07x=1550$

$0.02x+1400=1550$

$0.02x=150$

$x=7500$

$20{,}000-x=20{,}000-7500=12{,}500$

He invested \$7500 at 9% and \$12,500 at 7%.

4. Let x be the number of bankruptcies in 1994 then $2x - 80,000$ is the number in 2002.

$$x + 2x - 80,000 = 2,290,000$$
$$3x - 80,000 = 2,290,000$$
$$3x = 2,370,000$$
$$x = 790,000$$

$2x - 80,000 = 2(790,000) - 80,000 = 1,500,000$

There were 790,000 bankruptcies in 1994 and 1,500,000 in 2002.

5. $x - 3y = 6$

x	y
0	–2
6	0

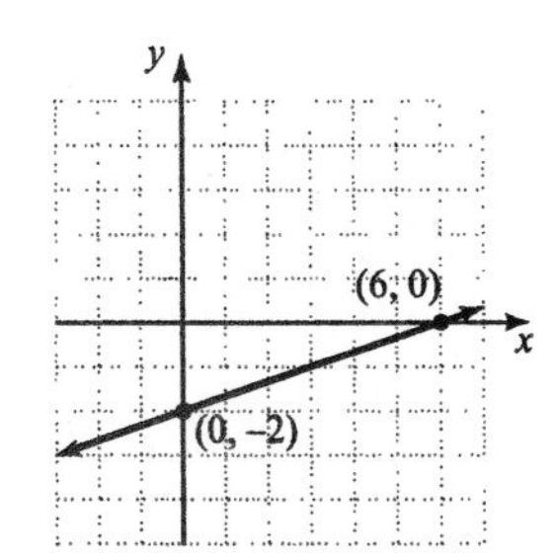

6. $7x + 2y = 9$

$$2y = -7x + 9$$
$$y = -\frac{7}{2}x + \frac{9}{2}$$
$$y = mx + b$$
$$m = -\frac{7}{2}$$

7. a. $4^2 \cdot 4^5 = 4^{2+5} = 4^7$

b. $x^4 \cdot x^6 = x^{4+6} = x^{10}$

c. $y^3 \cdot y = y^{3+1} = y^4$

d. $y^3 \cdot y^2 \cdot y^7 = y^{3+2+7} = y^{12}$

e. $(-5)^7 \cdot (-5)^8 = (-5)^{7+8} = (-5)^{15}$

f. $a^2 \cdot b^2 = a^2b^2$

8. a. $\frac{x^9}{x^7} = x^{9-7} = x^2$

b. $\frac{x^{19}y^5}{xy} = x^{19-1} \cdot y^{5-1} = x^{18}y^4$

c. $(x^5y^2)^3 = x^{5\cdot3}y^{2\cdot3} = x^{15}y^6$

d. $(-3a^2b)(5a^3b) = -15a^{2+3}b^{1+1} = -15a^5b^2$

9. $[(8z+11)+(9z-2)]-(5z-7)$

$$= 8z + 11 + 9z - 2 - 5z + 7$$
$$= 12z + 16$$

10. $(x+1)-(9x^2-6x+2) = x+1-9x^2+6x-2$

$$= -9x^2 + 7x - 1$$

11. $(3a+b)^3$

$$= (3a+b)(3a+b)^2$$
$$= (3a+b)[(3a)^2 + 2(3a)(b) + (b)^2]$$
$$= (3a+b)(9a^2 + 6ab + b^2)$$
$$= 27a^3 + 18a^2b + 3ab^2 + 9a^2b + 6ab^2 + b^3$$
$$= 27a^3 + 27a^2b + 9ab^2 + b^3$$

12. $(2x+1)(5x^2-x+2)$

$$= 2x(5x^2 - x + 2) + 1(5x^2 - x + 2)$$
$$= 10x^3 - 2x^2 + 4x + 5x^2 - x + 2$$
$$= 10x^3 + 3x^2 + 3x + 2$$

13. a. $(t+2)^2 = (t)^2 + 2(t)(2) + (2)^2 = t^2 + 4t + 4$

b. $(p-q)^2 = (p)^2 - 2(p)(q) + (q)^2$

$$= p^2 - 2pq + q^2$$

c. $(2x+5)^2 = (2x)^2 + 2(2x)(5) + (5)^2$

$$= 4x^2 + 20x + 25$$

d. $(x^2-7y)^2 = (x^2)^2 - 2(x^2)(7y) + (7y)^2$

$$= x^4 - 14x^2y + 49y^2$$

14. a. $(x+9)^2 = (x)^2 + 2(x)(9) + (9)^2$

$$= x^2 + 18x + 81$$

b. $(2x+1)(2x-1) = (2x)^2 - (1)^2 = 4x^2 - 1$

c. $8x(x^2+1)(x^2-1) = 8x[(x^2)^2 - (1)^2]$

$$= 8x[x^4 - 1]$$
$$= 8x^5 - 8x$$

15. a. $\frac{1}{x^{-3}} = x^3$

b. $\frac{1}{3^{-4}} = 3^4 = 81$

c. $\frac{p^{-4}}{q^{-9}} = \frac{q^9}{p^4}$

d. $\frac{5^{-3}}{2^{-5}} = \frac{2^5}{5^3} = \frac{32}{125}$

16. a. $5^{-3} = \frac{1}{5^3} = \frac{1}{125}$

b. $\frac{9}{x^{-7}} = 9x^7$

c. $\frac{11^{-1}}{7^{-2}} = \frac{7^2}{11^1} = \frac{49}{11}$

17.
$$\begin{array}{r|l} & 4x^2 - 4x + 6 \\ 2x+3 & 8x^3 + 4x^2 + 0x + 7 \\ & \underline{8x^3 + 12x^2} \\ & -8x^2 + 0x \\ & \underline{-8x^2 - 12x} \\ & 12x + 7 \\ & \underline{12x + 18} \\ & -11 \end{array}$$

$$\frac{4x^2 + 7 + 8x^3}{2x+3} = 4x^2 - 4x + 6 - \frac{11}{2x+3}$$

18.
$$\begin{array}{r|l} & 4x^2 + 16x + 55 \\ x-4 & 4x^3 + 0x^2 - 9x + 2 \\ & \underline{4x^3 - 16x^2} \\ & 16x^2 - 9x \\ & \underline{16x^2 - 64x} \\ & 55x + 2 \\ & \underline{55x - 220} \\ & 222 \end{array}$$

$$\frac{4x^3 - 9x + 2}{x-4} = 4x^2 + 16x + 55 + \frac{222}{x-4}$$

19. a. $28 = 2 \cdot 2 \cdot 7$
$40 = 2 \cdot 2 \cdot 2 \cdot 5$
$\text{GCF} = 2^2 = 4$

b. $55 = 5 \cdot 11$
$21 = 3 \cdot 7$
$\text{GCF} = 1$

c. $15 = 3 \cdot 5$
$18 = 2 \cdot 3 \cdot 3$
$66 = 2 \cdot 3 \cdot 11$
$\text{GCF} = 3$

20. $9x^2 = 3 \cdot 3 \cdot x^2$
$6x^3 = 2 \cdot 3 \cdot x^3$
$21x^5 = 3 \cdot 7 \cdot x^5$
$\text{GCF} = 3x^2$

21. $-9a^5 + 18a^2 - 3a = -3a(3a^4 - 6a + 1)$

22. $7x^6 - 7x^5 + 7x^4 = 7x^4(x^2 - x + 1)$

23. $3m^2 - 24m - 60 = 3(m^2 - 8m - 20)$
$= 3(m^2 - 10m + 2m - 20)$
$= 3[m(m-10) + 2(m-10)]$
$= 3(m-10)(m+2)$

24. $-2a^2 + 10a + 12 = -2(a^2 - 5a - 6)$
$= -2(a+1)(a-6)$

25. $3x^2 + 11x + 6 = 3x^2 + 2x + 9x + 6$
$= x(3x+2) + 3(3x+2)$
$= (3x+2)(x+3)$

26. $10m^2 - 7m + 1 = 10m^2 - 2m - 5m + 1$
$= 2m(5m-1) - 1(5m-1)$
$= (2m-1)(5m-1)$

27. $x^2 + 12x + 36 = x^2 + 2 \cdot x \cdot 6 + 6^2 = (x+6)^2$

28. $4x^2 + 12x + 9 = (2x)^2 + 2(2x)(3) + (3)^2$
$= (2x+3)^2$

29. $x^2 + 4$ is a prime polynomial.

30. $x^2 - 4 = (x)^2 - (2)^2 = (x+2)(x-2)$

31. $x^3+8=x^3+2^3$
$=(x+2)(x^2-x\cdot 2+2^2)$
$=(x+2)(x^2-2x+4)$

32. $27y^3-1=(3y)^3-(1)^3$
$=(3y-1)[(3y)^2+3y(1)+(1)^2]$
$=(3y-1)(9y^2+3y+1)$

33. $2x^3+3x^2-2x-3=x^2(2x+3)-1(2x+3)$
$=(2x+3)(x^2-1)$
$=(2x+3)(x^2-1^2)$
$=(2x+3)(x+1)(x-1)$

34. $3x^3+5x^2-12x-20=x^2(3x+5)-4(3x+5)$
$=(3x+5)(x^2-4)$
$=(3x+5)(x^2-2^2)$
$=(3x+5)(x+2)(x-2)$

35. $12m^2-3n^2=3(4m^2-n^2)$
$=3[(2m)^2-(n)^2]$
$=3(2m+n)(2m-n)$

36. $x^5-x=x(x^4-1)$
$=x[(x^2)^2-1^2]$
$=x(x^2+1)(x^2-1)$
$=x(x^2+1)(x+1)(x-1)$

37. $x(2x-7)=4$
$2x^2-7x=4$
$2x^2-7x-4=0$
$2x^2-8x+x-4=0$
$2x(x-4)+1(x-4)=0$
$(x-4)(2x+1)=0$
$2x+1=0$ or $x-4=0$
$2x=-1$ $\quad x=4$
$x=-\frac{1}{2}$

38. $3x^2+5x=2$
$3x^2+5x-2=0$
$3x^2+6x-x-2=0$
$3x(x+2)-1(x+2)=0$
$(x+2)(3x-1)=0$
$3x-1=0$ or $x+2=0$
$3x=1$ $\quad x=-2$
$x=\frac{1}{3}$

39. $y=x^2-5x+4$
$0=x^2-5x+4$
$0=(x-4)(x-1)$
$x-1=0$ or $x-4=0$
$x=1$ $\quad x=4$
The x-intercepts are (1, 0) and (4, 0).

40. $y=x^2-x-6$
$0=x^2-x-6$
$0=(x-3)(x+2)$
$x+2=0$ or $x-3=0$
$x=-2$ $\quad x=3$
The x-intercepts are (–2, 0) and (3, 0).

41. Let x = the base and $2x-2$ = the height.
$A=\frac{1}{2}bh$
$30=\frac{1}{2}x(2x-2)$
$30=\frac{1}{2}(2x)(x-1)$
$30=x(x-1)$
$30=x^2-x$
$0=x^2-x-30$
$0=(x+5)(x-6)$
$x-6=0$ or $x+5=0$
$x=6$ $\quad x=-5$
Length cannot be negative, so $x=6$.
$2x-2=2(6)-2=10$
The base is 6 meters and the height is 10 meters.

42. Let x = the base and $3x+5$ = the height.
$A=bh$
$182=x(3x+5)$
$182=3x^2+5x$
$0=3x^2+5x-182$
$0=3x^2+26x-21x-182$
$0=x(3x+26)-7(3x+26)$
$0=(x-7)(3x+26)$

$x-7=0$ or $3x+26=0$

$x=7$ $\quad x=-\frac{26}{3}$

Length cannot be negative so $x=7$.
$3x+5=3(7)+5=26$
The base is 7 ft and the height is 26 ft.

43. $\frac{18-2x^2}{x^2-2x-3}=\frac{2(9-x^2)}{(x+1)(x-3)}$
$=\frac{2(3+x)(3-x)}{(x+1)(x-3)}$
$=\frac{-2(3+x)(x-3)}{(x+1)(x-3)}$
$=-\frac{2(3+x)}{x+1}$

44. $\frac{2x^2-50}{4x^4-20x^3}=\frac{2(x^2-25)}{4x^3(x-5)}$
$=\frac{2(x+5)(x-5)}{4x^3(x-5)}$
$=\frac{x+5}{2x^3}$

45. $\frac{6x+2}{x^2-1}\div\frac{3x^2+x}{x-1}=\frac{6x+2}{x^2-1}\cdot\frac{x-1}{3x^2+x}$
$=\frac{2(3x+1)}{(x+1)(x-1)}\cdot\frac{x-1}{x(3x+1)}$
$=\frac{2}{x(x+1)}$

46. $\frac{6x^2-18x}{3x^2-2x}\cdot\frac{15x-10}{x^2-10}=\frac{6x(x-3)\cdot 5(3x-2)}{x(3x-2)\cdot(x+3)(x-3)}$
$=\frac{30}{x+3}$

47. $\frac{(2x)^{-1}+1}{2x^{-1}-1}=\frac{\frac{1}{2x}+1}{\frac{2}{x}-1}$
$=\frac{2x\left(\frac{1}{2x}+1\right)}{2x\left(\frac{2}{x}-1\right)}$
$=\frac{1+2x}{4-2x}$
$=\frac{1+2x}{2(2-x)}$

48. $\frac{\frac{m}{3}+\frac{n}{6}}{\frac{m+n}{12}}=\frac{12}{12}\cdot\frac{\frac{m}{3}+\frac{n}{6}}{\frac{m+n}{12}}$
$=\frac{12\left(\frac{m}{3}\right)+12\left(\frac{n}{6}\right)}{12\left(\frac{m+n}{12}\right)}$
$=\frac{4m+2n}{m+n}$ or $\frac{2(2m+n)}{m+n}$

Chapter 8

Section 8.1

Practice Exercises

1. $f(x) = 4x,\ g(x) = 4x - 3$

x	$f(x)$	$g(x)$
0	0	−3
−1	−4	−7
1	4	1

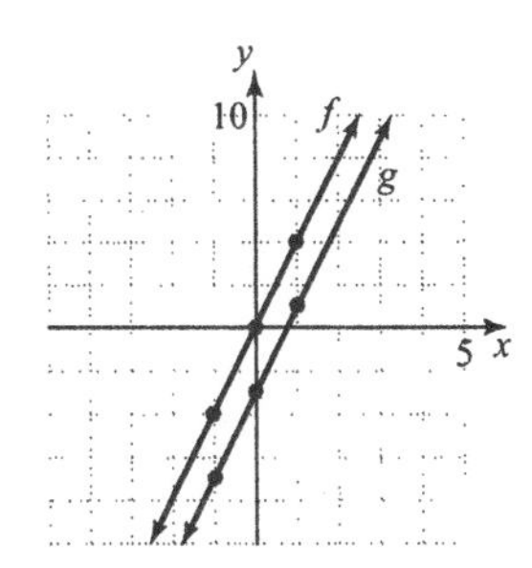

2. $f(x) = -2x,\ g(x) = -2x + 5$

x	$f(x)$	$g(x)$
0	0	5
−1	2	7
1	−2	3
2	−4	1

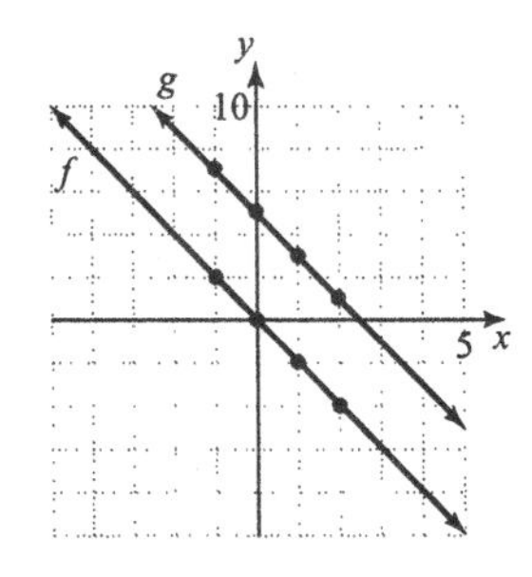

3. Use the slope-intercept form with $m = -4$ and $b = -3$.
$y = mx + b$
$y = -4x + (-3)$
$y = -4x - 3$
$f(x) = -4x - 3$

4. First find the slope.
$$m = \frac{0-2}{2-(-1)} = \frac{-2}{3} = -\frac{2}{3}$$
Use the slope and one of the points in the point-slope form. We use (2, 0).
$$y - y_1 = m(x - x_1)$$
$$y - 0 = -\frac{2}{3}(x-2)$$
$$y = -\frac{2}{3}x + \frac{4}{3}$$
$$f(x) = -\frac{2}{3}x + \frac{4}{3}$$

5. A horizontal line has an equation of the form $y = b$. Since the line contains the point (6, −2), the equation is $y = -2$ or $f(x) = -2$.

6. Solve the given equation for y.
$$3x + 4y = 1$$
$$4y = -3x + 1$$
$$y = -\frac{3}{4}x + \frac{1}{4}$$
The slope of this line is $-\frac{3}{4}$, so the slope of any line parallel to it is also $-\frac{3}{4}$. Use this slope and the point (8, −3) in the point-slope form.
$$y - y_1 = m(x - x_1)$$
$$y - (-3) = -\frac{3}{4}(x - 8)$$
$$4(y + 3) = -3(x - 8)$$
$$4y + 12 = -3x + 24$$
$$3x + 4y = 12$$

7. Solve the given equation for y.
$$3x + 4y = 1$$
$$4y = -3x + 1$$
$$y = -\frac{3}{4}x + \frac{1}{4}$$
The slope of this line is $-\frac{3}{4}$, so the slope of any line perpendicular to it is the negative reciprocal of $-\frac{3}{4}$, or $\frac{4}{3}$. Use this slope and the point

(8, −3) in the point-slope form.

$$y - y_1 = m(x - x_1)$$
$$y - (-3) = \frac{4}{3}(x - 8)$$
$$3(y + 3) = 4(x - 8)$$
$$3y + 9 = 4x - 32$$
$$3y = 4x - 41$$
$$y = \frac{4}{3}x - \frac{41}{3}$$
$$f(x) = \frac{4}{3}x - \frac{41}{3}$$

Graphing Calculator Explorations

1. $x = 3.5y$

$y = \frac{x}{3.5}$

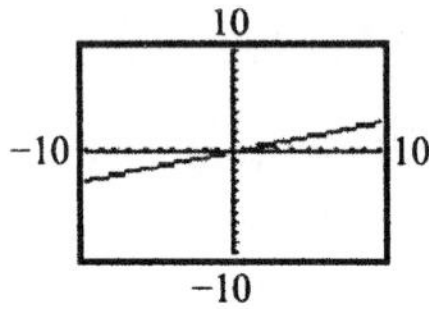

2. $-2.7y = x$

$y = \frac{x}{-2.7} = -\frac{x}{2.7}$

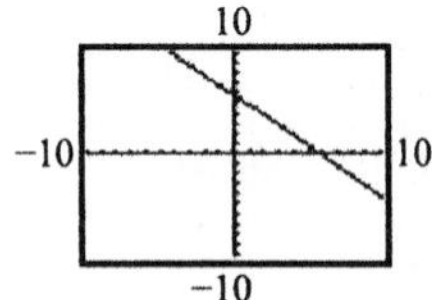

3. $5.78x + 2.31y = 10.98$

$2.31y = -5.78x + 10.98$

$y = -\frac{5.78}{2.31}x + \frac{10.98}{2.31}$

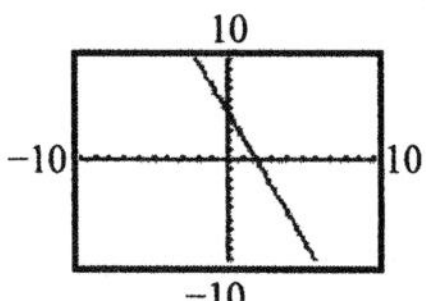

4. $-7.22x + 3.89y = 12.57$

$3.89y = 7.22x + 12.57$

$y = \frac{7.22}{3.89}x + \frac{12.57}{3.89}$

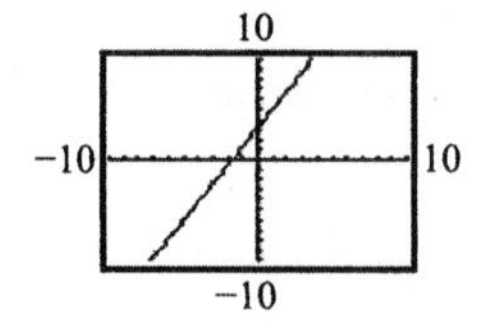

5. $y - |x| = 3.78$

$y = |x| + 3.78$

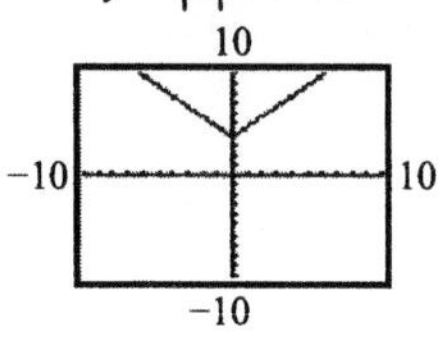

6. $3y - 5x^2 = 6x - 4$

$3y = 5x^2 + 6x - 4$

$y = \frac{5}{3}x^2 + 2x - \frac{4}{3}$

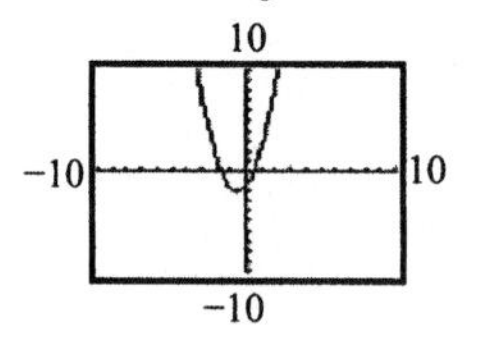

7. $y - 5.6x^2 = 7.7x + 1.5$

$y = 5.6x^2 + 7.7x + 1.5$

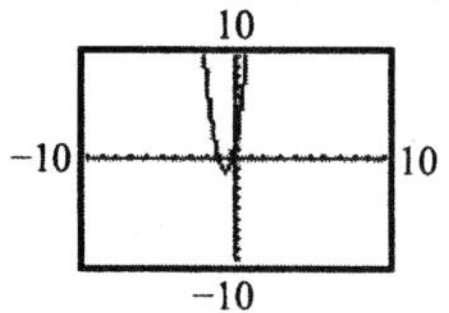

8. $y + 2.6|x| = -3.2$

$y = -2.6|x| - 3.2$

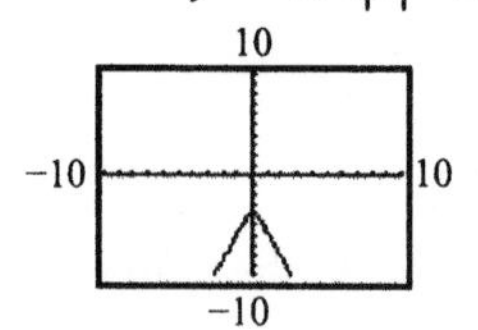

Vocabulary and Readiness Check

1. A linear function can be written in the form $f(x) = mx + b$.

2. In the form $f(x) = mx + b$, the y-intercept is <u>(0, b)</u> and the slope is <u>m</u>.

3. $m = -4$, $b = 12$ so y-intercept is (0, 12).

4. $m = \frac{2}{3}$, $b = -\frac{7}{2}$ so y-intercept is $\left(0, -\frac{7}{2}\right)$.

5. $m = 5$, $b = 0$ so y-intercept is (0, 0).

6. $m = -1$, $b = 0$ so y-intercept is (0, 0).

7. The lines both have slope 12 and they have different y-intercepts, (0, 6) and (0, −2), so they are parallel.

8. The lines both have slope −5 and they have different y-intercepts, (0, 8) and (0, −8), so they are parallel.

9. The line have slopes −9 and $\frac{3}{2}$. The slopes are not equal and their product is not −1, so the lines are neither parallel nor perpendicular.

10. The line have slopes 2 and $\frac{1}{2}$. The slopes are not equal and their product is not −1, so the lines are neither parallel nor perpendicular.

Exercise Set 8.1

1. $f(x) = -2x$

x	0	−1	1
y	0	2	−2

Plot the points to obtain the graph.

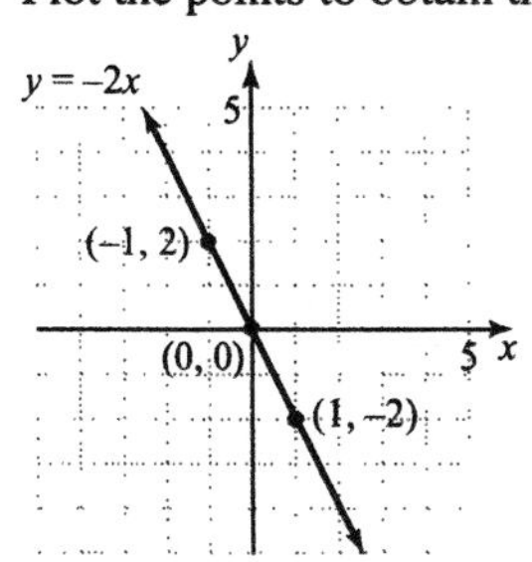

3. $f(x) = -2x + 3$

x	0	1	−1
y	3	1	5

Plot the points to obtain the graph.

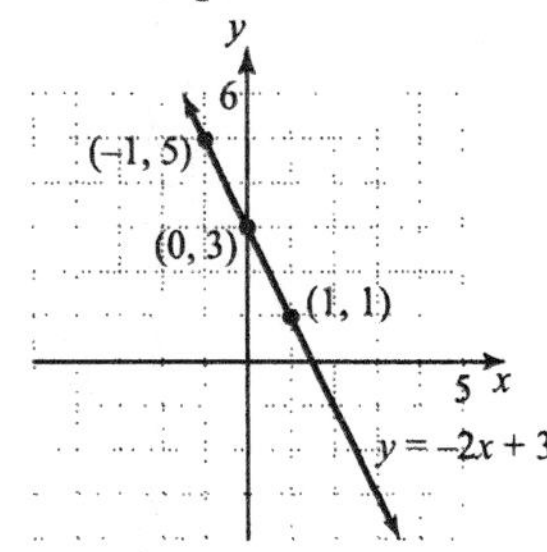

5. $f(x) = \frac{1}{2}x$

x	0	2	−2
y	0	1	−1

Plot the points to obtain the graph.

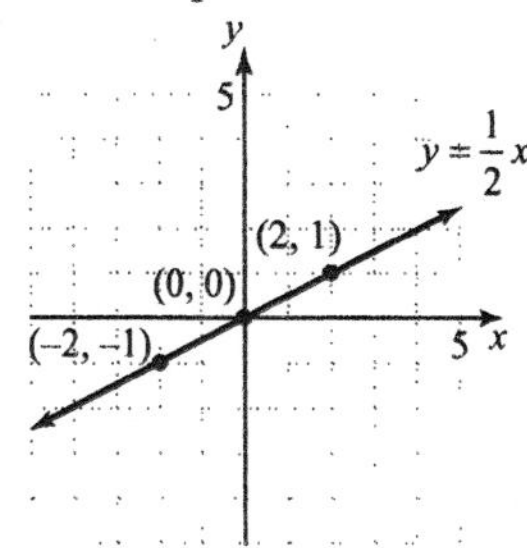

7. $f(x) = \frac{1}{2}x - 4$

x	0	2	4
y	−4	−3	−2

Plot the points to obtain the graph.

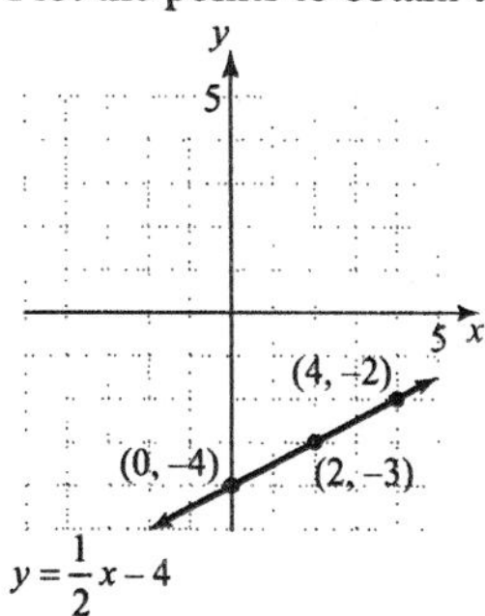

9. The graph of $f(x) = 5x - 3$ is the graph of $f(x) = 5x$ shifted down 3 units. The correct graph is C.

11. The graph of $f(x) = 5x + 1$ is the graph of $f(x) = 5x$ shifted up 1 unit. The correct graph is D.

13. $m = -1, b = 1$
$$y = mx + b$$
$$y = -1x + 1$$
$$y = -x + 1$$
$$f(x) = -x + 1$$

15. $m = 2,\ b = \frac{3}{4}$
$$y = mx + b$$
$$y = 2x + \frac{3}{4}$$
$$f(x) = 2x + \frac{3}{4}$$

17. $m = \frac{2}{7},\ b = 0$
$$y = mx + b$$
$$y = \frac{2}{7}x + 0$$
$$y = \frac{2}{7}x$$
$$f(x) = \frac{2}{7}x$$

19.
$$y - y_1 = m(x - x_1)$$
$$y - 2 = 3(x - 1)$$
$$y - 2 = 3x - 3$$
$$y = 3x - 1$$
$$f(x) = 3x - 1$$

21.
$$y - y_1 = m(x - x_1)$$
$$y - (-3) = -2(x - 1)$$
$$y + 3 = -2x + 2$$
$$y = -2x - 1$$
$$f(x) = -2x - 1$$

23.
$$y - y_1 = m(x - x_1)$$
$$y - 2 = \frac{1}{2}[x - (-6)]$$
$$y - 2 = \frac{1}{2}(x + 6)$$
$$y - 2 = \frac{1}{2}x + 3$$
$$y = \frac{1}{2}x + 5$$
$$f(x) = \frac{1}{2}x + 5$$

25.
$$y - y_1 = m(x - x_1)$$
$$y - 0 = -\frac{9}{10}[x - (-3)]$$
$$y = -\frac{9}{10}(x + 3)$$
$$y = -\frac{9}{10}x - \frac{27}{10}$$
$$f(x) = -\frac{9}{10}x - \frac{27}{10}$$

27. $m = \frac{6 - 0}{4 - 2} = \frac{6}{2} = 3$
$$y - 0 = 3(x - 2)$$
$$y = 3x - 6$$
$$f(x) = 3x - 6$$

29. $m = \frac{13 - 5}{-6 - (-2)} = \frac{8}{-4} = -2$
$$y - 5 = -2[x - (-2)]$$
$$y - 5 = -2(x + 2)$$
$$y - 5 = -2x - 4$$
$$y = -2x + 1$$
$$f(x) = -2x + 1$$

31. $m = \frac{-3 - (-4)}{-4 - (-2)} = \frac{1}{-2} = -\frac{1}{2}$
$$y - (-4) = -\frac{1}{2}[x - (-2)]$$
$$y + 4 = -\frac{1}{2}(x + 2)$$
$$2y + 8 = -(x + 2)$$
$$2y + 8 = -x - 2$$
$$2y = -x - 10$$
$$y = -\frac{1}{2}x - 5$$
$$f(x) = -\frac{1}{2}x - 5$$

33. $m = \frac{-9-(-8)}{-6-(-3)} = \frac{-1}{-3} = \frac{1}{3}$

$$y-(-8) = \frac{1}{3}[x-(-3)]$$
$$y+8 = \frac{1}{3}(x+3)$$
$$3y+24 = x+3$$
$$3y = x-21$$
$$y = \frac{1}{3}x-7$$
$$f(x) = \frac{1}{3}x-7$$

35. $m = \frac{\frac{7}{10}-\frac{4}{10}}{-\frac{1}{5}-\frac{3}{5}} = \frac{\frac{3}{10}}{-\frac{4}{5}} = \frac{3}{10}\left(-\frac{5}{4}\right) = -\frac{3}{8}$

$$y-\frac{4}{10} = -\frac{3}{8}\left(x-\frac{3}{5}\right)$$
$$y-\frac{4}{10} = -\frac{3}{8}x+\frac{9}{40}$$
$$y = -\frac{3}{8}x+\frac{5}{8}$$
$$f(x) = -\frac{3}{8}x+\frac{5}{8}$$

37.
$$y = mx+b$$
$$-4 = 0(-2)+b$$
$$-4 = b$$
$$y = -4$$
$$f(x) = -4$$

39. Every horizontal line is in the form $y = c$. Since the line passes through the point (0, 5), its equation is $y = 5$ or $f(x) = 5$.

41. $y = 4x-2$ so $m = 4$

$$y-8 = 4(x-3)$$
$$y-8 = 4x-12$$
$$y = 4x-4$$
$$f(x) = 4x-4$$

43. $3y = x-6$ or $y = \frac{1}{3}x-2$ so

$m = \frac{1}{3}$ and $m_{\perp} = -3$

$$y-(-5) = -3(x-2)$$
$$y+5 = -3x+6$$
$$y = -3x+1$$
$$f(x) = -3x+1$$

45. $3x+2y = 5$

$$2y = -3x+5$$
$$y = -\frac{3}{2}x+\frac{5}{2} \text{ so } m = -\frac{3}{2}$$
$$y-(-3) = -\frac{3}{2}[x-(-2)]$$
$$2(y+3) = -3(x+2)$$
$$2y+6 = -3(x+2)$$
$$2y+6 = -3x-6$$
$$y = -\frac{3}{2}x-6$$
$$f(x) = -\frac{3}{2}x-6$$

47.
$$y-3 = 2[x-(-2)]$$
$$y-3 = 2(x+2)$$
$$y-3 = 2x+4$$
$$2x-y = -7$$

49. $m = \frac{2-6}{5-1} = \frac{-4}{4} = -1$

$$y-6 = -1(x-1)$$
$$y-6 = -x+1$$
$$y = -x+7$$
$$f(x) = -x+7$$

51.
$$y = -\frac{1}{2}x+11$$
$$f(x) = -\frac{1}{2}x+11$$

53. $m = \frac{-6-(-4)}{0-(-7)} = \frac{-2}{7} = -\frac{2}{7}$

$$y = -\frac{2}{7}x-6$$
$$7y = -2x-42$$
$$2x+7y = -42$$

55.
$$y-0 = -\frac{4}{3}[x-(-5)]$$
$$3y = -4(x+5)$$
$$3y = -4x-20$$
$$4x+3y = -20$$

57. Every horizontal line is in the form $y = c$. Since the line passes through the point (–2, –10), its equation is $y = -10$ or $f(x) = -10$.

59. $2x+4y=8$
$$4y=-2x+8$$
$$y=-\frac{1}{2}x+2 \text{ so } m=-\frac{1}{2}$$
$$y-(-2)=-\frac{1}{2}(x-6)$$
$$2(y+2)=-(x-6)$$
$$2y+4=-x+6$$
$$x+2y=2$$

61. Lines with slopes of 0 are horizontal. Every horizontal line is in the form $y=c$. Since the line passes through (–9, 12), its equation is $y = 12$ or $f(x) = 12$.

63. $8x-y=9$
$$y=8x-9 \text{ so } m=8$$
$$y-1=8(x-6)$$
$$y-1=8x-48$$
$$8x-y=47$$

65. A line perpendicular to $y = 9$ will have the form $x=c$. Since the line passes through the point (5, –6), its equation is $x = 5$.

67. $m=\frac{-5-(-8)}{-6-2}=\frac{3}{-8}=-\frac{3}{8}$
$$y-(-8)=-\frac{3}{8}(x-2)$$
$$8(y+8)=-3(x-2)$$
$$8y+64=-3x+6$$
$$y=-\frac{3}{8}x-\frac{29}{4}$$
$$f(x)=-\frac{3}{8}x-\frac{29}{4}$$

69. $2x-7\le 21$
$$2x\le 28$$
$$x\le 14$$
$$(-\infty, 14]$$

71. $5(x-2)\ge 3(x-1)$
$$5x-10\ge 3x-3$$
$$2x\ge 7$$
$$x\ge\frac{7}{2}$$
$$\left[\frac{7}{2}, \infty\right)$$

73. $\frac{x}{2}+\frac{1}{4}<\frac{1}{8}$
$$8\left(\frac{x}{2}+\frac{1}{4}\right)<8\left(\frac{1}{8}\right)$$
$$4x+2<1$$
$$4x<-1$$
$$x<-\frac{1}{4}$$
$$\left(-\infty, -\frac{1}{4}\right)$$

75. (0, 3), (1, 1)
$$m=\frac{1-3}{1-0}=\frac{-2}{1}=-2$$
$$b=3$$
$$y=-2x+3$$
$$f(x)=-2x+3$$

77. (–2, 1), (4, 5)
$$m=\frac{5-1}{4-(-2)}=\frac{4}{6}=\frac{2}{3}$$
$$y-1=\frac{2}{3}(x+2)$$
$$y-1=\frac{2}{3}x+\frac{4}{3}$$
$$y=\frac{2}{3}x+\frac{7}{3}$$
$$f(x)=\frac{2}{3}x+\frac{7}{3}$$

79. a. (1, 30,000), (4, 66,000)
$$m=\frac{66,000-30,000}{4-1}=12,000$$
$$y-30,000=12,000(x-1)$$
$$y=12,000x+18,000$$
$$P(x)=12,000x+18,000$$

b. $P(7)=12,000(7)+18,000$
$$=\$102,000$$

c. $126,000=12,000x+18,000$
$$x=\frac{126,000-18,000}{12,000}$$
$$x=9 \text{ years}$$

81. a. (3, 10,000), (5, 8000)

$$m = \frac{8000 - 10,000}{5 - 3} = -1000$$

$$y - 10,000 = -1000(x - 3)$$
$$y - 10,000 = -1000x + 3,000$$
$$y = -1000x + 13,000$$

b. $y = -1000(3.5) + 13,000$

$y = 9500$

9500 Fun Noodles

83. a. We have two ordered pairs, (0, 150,900) and (5, 222,000). Find the slope.

$$m = \frac{222,000 - 150,900}{5 - 0} = \frac{71,100}{5} = 14,220$$

Use the slope and the y-intercept, (0, 150,900) to write the equation.

$y = mx + b$

$y = 14,220x + 150,900$

b. The year 2010 corresponds to $x = 9$.

$$y = 14,220(9) + 150,900 = 127,980 + 150,900 = 278,880$$

We predict that the median existing home price will be $278,880 in 2010.

c. The slope is 14,220. Every year the median price of a home increases by $14,220.

85. a. (0, 387), (10, 589)

$$m = \frac{589 - 387}{10 - 0} = \frac{202}{10} = 20.2$$

$$y - 387 = 20.2(x - 0)$$
$$y = 20.2x + 387$$

b. $x = 2013 - 2004 = 9$

$y = 20.2(9) + 387$

$= 568.8$ thousand people

87. Since any vertical line intersects any horizontal line in a right angle, the statement is true.

89. $m = \frac{1-(-1)}{-5-3} = \frac{2}{-8} = -\frac{1}{4}$ so $m_{\perp} = 4$

$$M((3, -1), (5, 1)) = \left(\frac{3-5}{2}, \frac{-1+1}{2}\right) = (1, 0)$$

$$y - 0 = 4[x - (-1)]$$
$$y = 4(x + 1)$$
$$y = 4x + 4$$
$$-4x + y = 4$$

91. $m = \frac{-4-6}{-22-(-2)} = \frac{-10}{-20} = \frac{1}{2}$ so $m_{\perp} = -2$

$$M((-2, 6), (-22, -4)) = \left(\frac{-2-22}{2}, \frac{6-4}{2}\right) = (-12, 1)$$

$$y - 1 = -2[x - (-12)]$$
$$y - 1 = -2(x + 12)$$
$$y - 1 = -2x - 24$$
$$2x + y = -23$$

93. $m = \frac{7-3}{-4-2} = \frac{4}{-6} = -\frac{2}{3}$ so $m_{\perp} = \frac{3}{2}$

$$M((2, 3), (-4, 7)) = \left(\frac{2-4}{2}, \frac{3+7}{2}\right) = (-1, 5)$$

$$y - 5 = \frac{3}{2}[x - (-1)]$$
$$2(y - 5) = 3(x + 1)$$
$$2y - 10 = 3x + 3$$
$$3x - 2y = -13$$

95. Answers may vary

Section 8.2

Practice Exercises

1. a. To find $f(1)$, find the y-value when $x = 1$. We see from the graph that when $x = 1$, y or $f(x) = -3$. Thus, $f(1) = -3$.

b. $f(0) = -2$ from the ordered pair $(0, -2)$.

c. $g(-2) = 3$ from the ordered pair $(-2, 3)$.

d. $g(0) = 1$ from the ordered pair $(0, 1)$.

e. To find x-values such that $f(x) = 1$, we are looking for any ordered pairs on the graph of f whose $f(x)$ or y-value is 1. They are $(-1, 1)$ and $(3, 1)$. Thus, $f(-1) = 1$ and $f(3) = 1$. The x-values are -1 and 3.

f. Find ordered pairs on the graph of g whose $g(x)$ or y-value is -2. There is one such ordered pair, $(-3, -2)$. Thus, $g(-3) = -2$. The only x-value is -3.

2. Find the year 2003 and move upward until you reach the graph. From the point on the graph, move horizontally to the left until the other axis is reached. In 2003, approximately $35 billion was spent.

3. Find $f(2012)$.
$$f(x) = 2.602x - 5178$$
$$f(2012) = 2.602(2012) - 5178$$
$$= 57.224$$
We predict that $57.224 billion will be spent in 2012.

4. a. $\sqrt{121} = 11$ since 11 is positive and $11^2 = 121$.

b. $\sqrt{\frac{1}{16}} = \frac{1}{4}$ since $\left(\frac{1}{4}\right)^2 = \frac{1}{16}$.

c. $-\sqrt{64} = -8$

d. $\sqrt{-64}$ is not a real number.

e. $\sqrt{100} = 10$ since $10^2 = 100$.

5. $f(x) = 2x^2$
This equation is not linear because of the x^2 term. Its graph is not a line.
If $x = -3$, then $f(-3) = 2(-3)^2$, or 18.
If $x = -2$, then $f(-2) = 2(-2)^2$, or 8.
If $x = -1$, then $f(-1) = 2(-1)^2$, or 2.
If $x = 0$, then $f(0) = 2(0)^2$, or 0.
If $x = 1$, then $f(1) = 2(1)^2$, or 2.
If $x = 2$, then $f(2) = 2(2)^2$, or 8.
If $x = 3$, then $f(3) = 2(3)^2$, or 18.

x	y or $f(x)$
−3	18
−2	8
−1	2
0	0
1	2
2	8
3	18

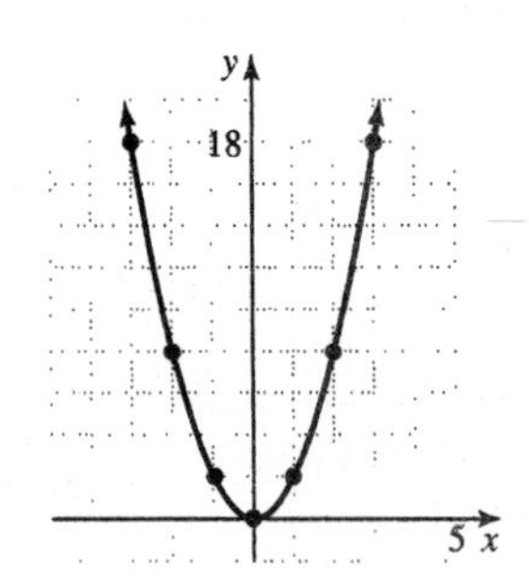

6. $f(x) = -|x|$
This equation is not linear because it cannot be written in the form $Ax + By = C$. Its graph is not a line.
If $x = -3$, then $f(-3) = -|-3|$, or -3.
If $x = -2$, then $f(-2) = -|-2|$, or -2.
If $x = -1$, then $f(-1) = -|-1|$, or -1.
If $x = 0$, then $f(0) = -|0|$, or 0.
If $x = 1$, then $f(1) = -|1|$, or -1.
If $x = 2$, then $f(2) = -|2|$, or -2.
If $x = 3$, then $f(3) = -|3|$, or -3.

x	y or $f(x)$
−3	−3
−2	−2
−1	−1
0	0
1	−1
2	−2
3	−3

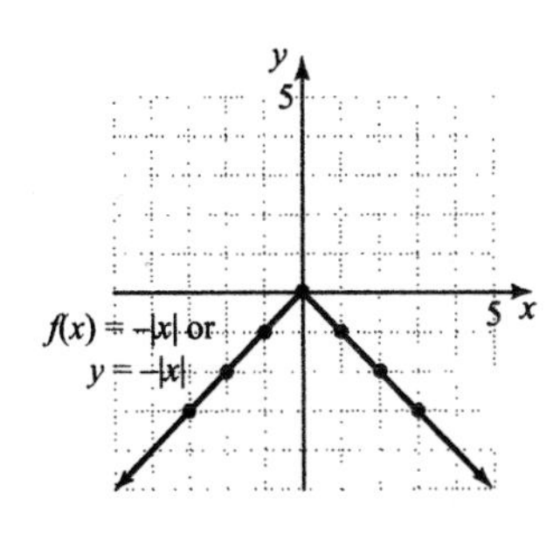

7. $f(x) = \sqrt{x} + 1$

   This equation is not linear because it cannot be written in the form $Ax + By = C$. Its graph is not a line.

   If $x = 0$, then $f(0) = \sqrt{0} + 1$, or 1.

   If $x = 1$, then $f(1) = \sqrt{1} + 1$, or 2.

   If $x = 4$, then $f(4) = \sqrt{4} + 1$, or 3.

   If $x = 9$, then $f(9) = \sqrt{9} + 1$, or 4.

x	y or $f(x)$
0	1
1	2
4	3
9	4

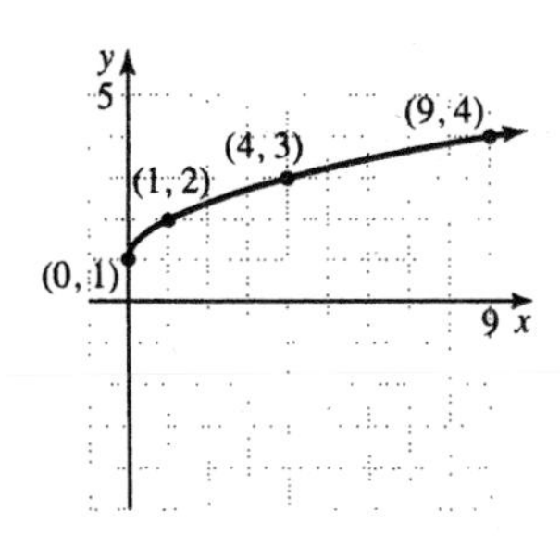

Graphing Calculator Explorations

1.

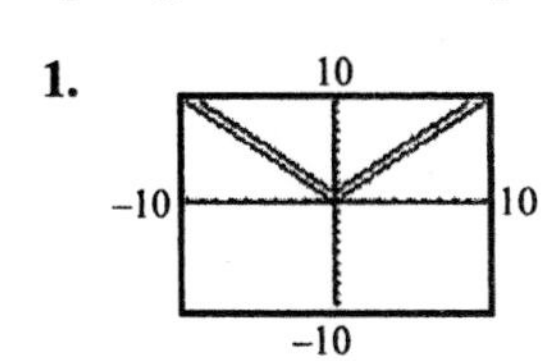

2.

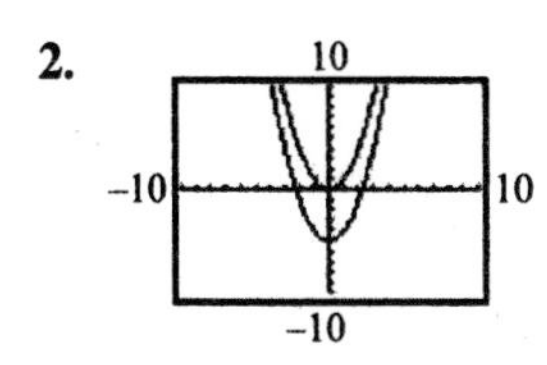

3.

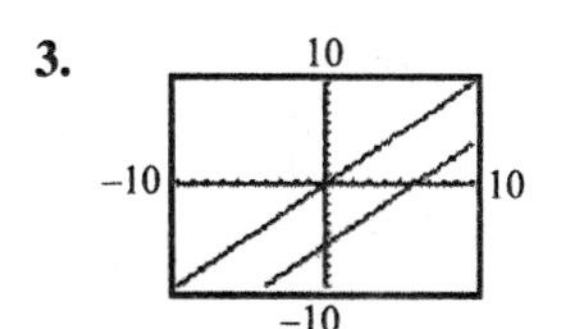

4.

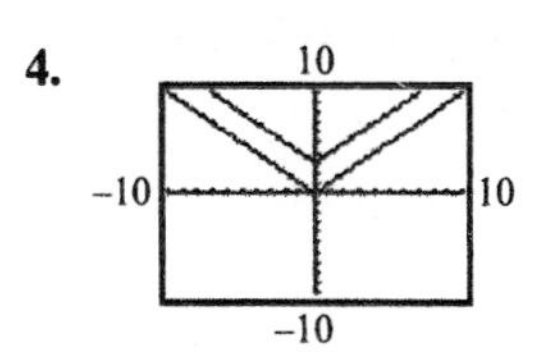

5.

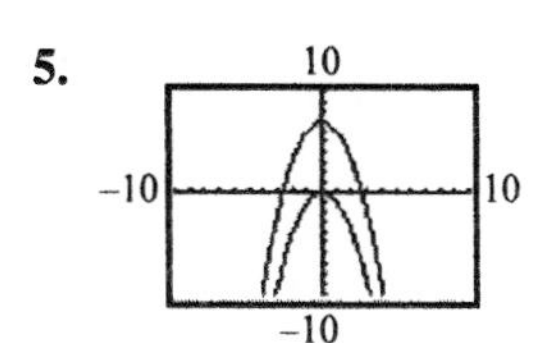

6.

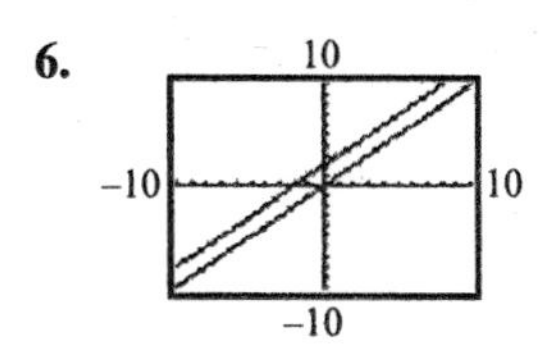

Vocabulary and Readiness Check

1. The graph of $y = |x|$ looks V-shaped.

2. The graph of $y = x^2$ is a parabola.

3. If $f(-2) = 1.7$, the corresponding ordered pair is $(-2, 1.7)$.

4. If $f(x) = x^2$, then $f(-3) = 9$.

Exercise Set 8.2

1. To find $f(1)$, find the y-value when $x = 1$. We see from the graph that when $x = 1$, y or $f(x)$ is 0. Thus, $f(1) = 0$.

3. To find $f(-1)$, find the y-value when $x = -1$. We see from the graph that when $x = -1$, y or $f(x)$ is -4. Thus, $f(-1) = -4$.

5. To find x-values such that $f(x) = 4$, find any ordered pairs on the graph with $f(x)$- or y-value of 4. The only such point is (3, 4). Thus, $f(3) = 4$ and the x-value is 3.

7. If $f(1) = -10$, then $y = -10$ when $x = 1$. The ordered pair is $(1, -10)$.

9. If $g(4) = 56$, then $y = 56$ when $x = 4$. The ordered pair is (4, 56).

11. The ordered pair (−1, −2) is on the graph of f. Thus, $f(-1) = -2$.

13. The ordered pair (2, 0) is on the graph of g. Thus, $g(2) = 0$.

15. There are two ordered pairs on the graph of f with a y-value of −5, (−4, −5) and (0, −5). The x-values are −4 and 0.

17. To the right of the y-axis, there is one ordered pair on the graph of g with a y-value of 4, (3, 4). The x-value is 3.

19. $\sqrt{49} = 7,$ since $7^2 = 49.$

21. $-\sqrt{\frac{4}{9}} = -\frac{2}{3},$ since $\left(\frac{2}{3}\right)^2 = \frac{4}{9}.$

23. $\sqrt{64} = 8,$ since $8^2 = 64.$

25. $\sqrt{81} = 9,$ since $9^2 = 81.$

27. $\sqrt{-100}$ is not a real number.

29. $f(x) = x^2 + 3$

x	y or $f(x)$
−2	7
−1	4
0	3
1	4
2	7

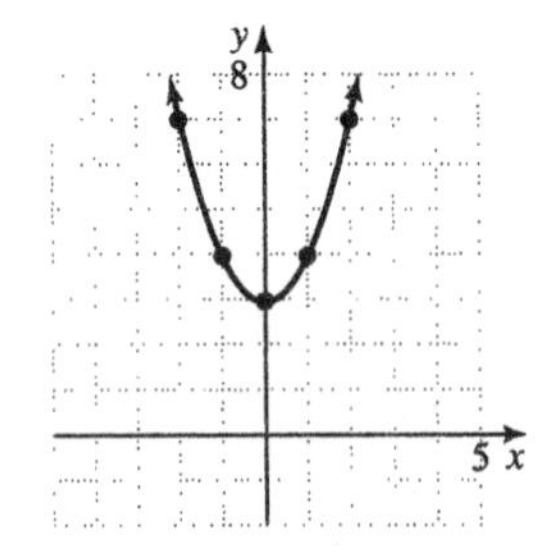

31. $h(x) = |x| - 2$

x	y or $h(x)$
−2	0
−1	−1
0	−2
1	−1
2	0

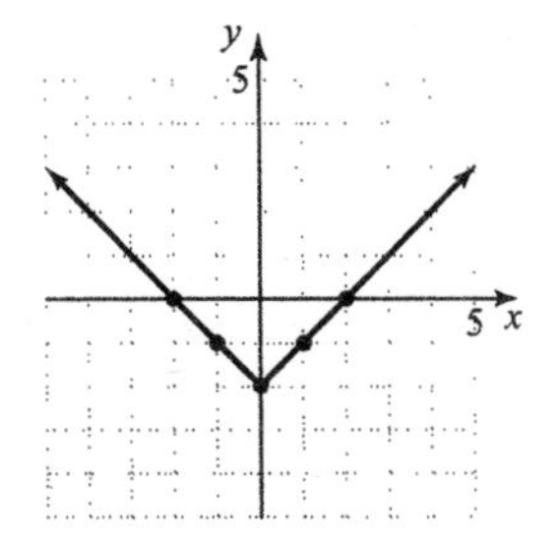

33. $g(x) = 2x^2$

x	y or $g(x)$
−2	8
−1	2
0	0
1	2
2	8

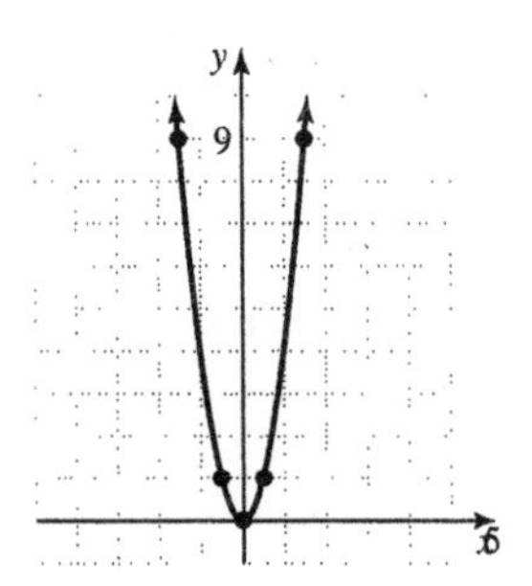

35. $f(x) = 5x - 1$

x	y or $f(x)$
−1	−6
0	−1
1	4

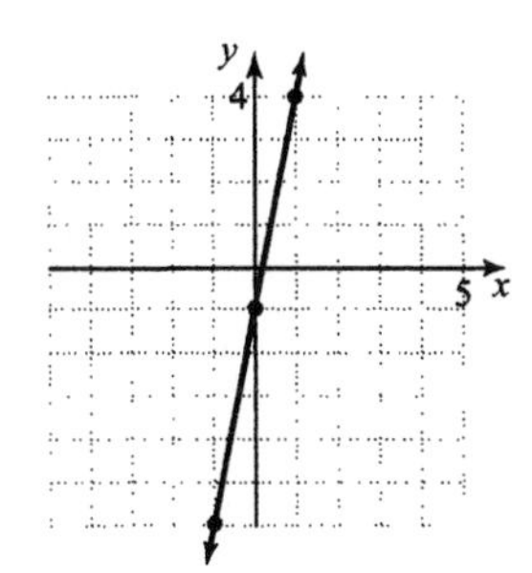

37. $f(x) = \sqrt{x+1}$

x	y or $f(x)$
−1	0
0	1
3	2
8	3

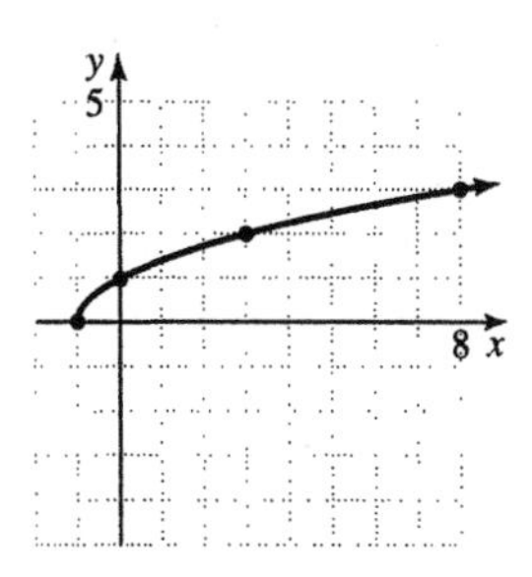

39. $g(x) = -2|x|$

x	y or $g(x)$
−2	−4
−1	−2
0	0
1	−2
2	−4

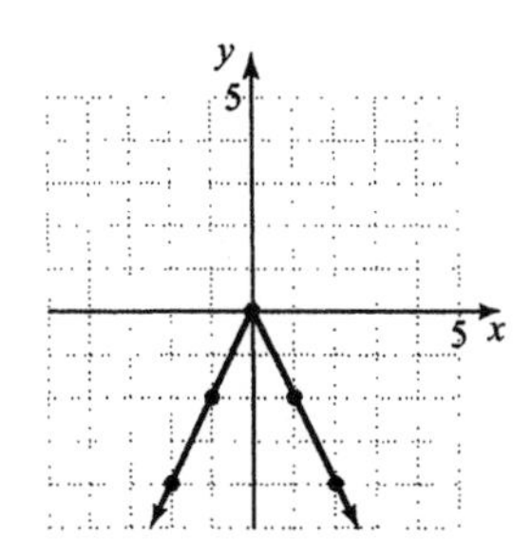

41. $h(x) = \sqrt{x} + 2$

x	y or $h(x)$
0	2
1	3
4	4
9	5

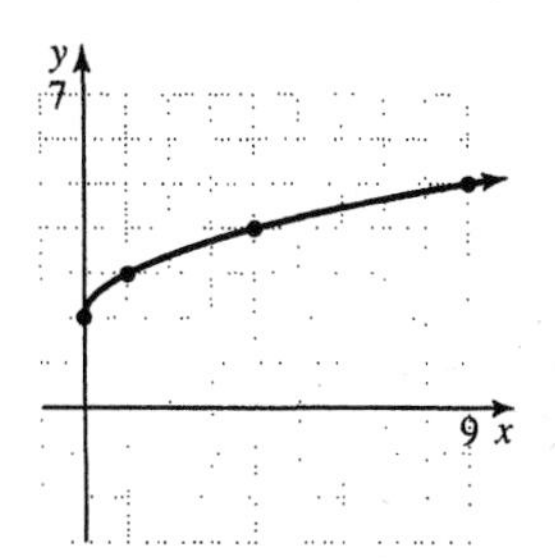

43. **a.** Find the year 1996 and move upward until you reach the graph. From the point on the graph, move horizontally to the left until the other axis is reached. In 1996, approximately $17 billion was spent.

b. Find $f(1996)$.
$f(x) = 2.602x - 5178$
$f(1996) = 2.602(1996) - 5178$
$= 15.592$
Approximately $15.592 billion was spent in 1996.

45. Since 2012 is 12 years after 2000, find $f(12)$.
$f(x) = 0.42x + 10.5$
$f(12) = 0.42(12) + 10.5$
$= 15.54$
We predict that diamond production will be $15.54 billion in 2012.

47. $A(r) = \pi r^2$
$A(5) = \pi(5)^2 = 25\pi$ square centimeters

49. $V(x) = x^3$
$V(14) = (14)^3 = 2744$ cubic inches

51. $H(f) = 2.59f + 47.24$
$H(46) = 2.59(46) + 47.24$
$= 166.38$ centimeters

53. $D(x) = \frac{136}{25}x$

$D(30) = \frac{136}{25}(30) = 163.2$ milligrams

55. Infinite number
The reason is that a graph is a function as long as it passes the vertical line test. So it does not matter if the equation of the graph takes on the value 0 many times.

57. $3(x-2)+5x = 6x-16$

$3x-6+5x = 6x-16$

$8x-6 = 6x-16$

$2x = -10$

$x = -5$

The solution is -5.

59. $3x+\frac{2}{5} = \frac{1}{10}$

$30x+4 = 1$

$30x = -3$

$x = -\frac{1}{10}$

The solution is $-\frac{1}{10}$.

61. Look for the graph where the only nonzero y-values are 40 and 60. The answer is b.

63. Look for the graph where all the y-values are between 10 and 30. The answer is c.

65. The first segment in the graph with y-coordinate greater than 0.25 begins in February 1991. Thus, 1991 is the first year that the price of a first-class stamp rose above \$0.25.

67. Answers may vary

69. $y = x^2 - 4x + 7$

x	y
0	7
1	4
2	3
3	4
4	7

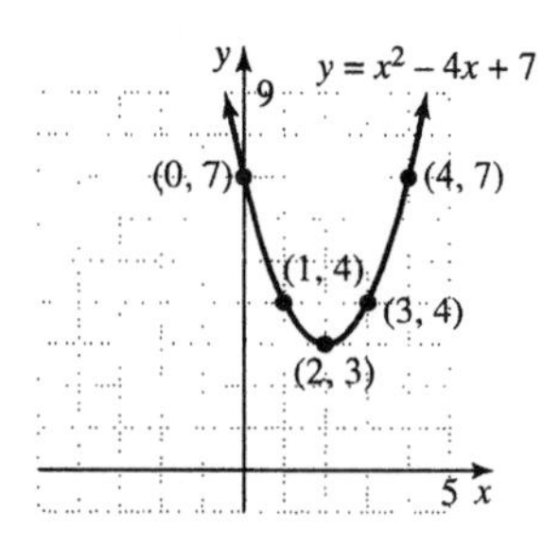

71. $f(x) = [x]$

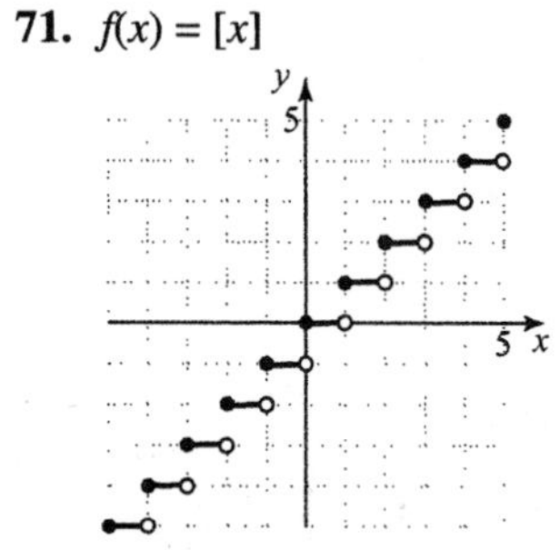

Integrated Review

1. $f(x) = 3x-5$

$y = 3x-5$

$y = mx+b$

The slope is $m = 3$; the y-intercept is $(0, b) = (0, -5)$.

2. $f(x) = \frac{5}{2}x - \frac{7}{2}$

$y = \frac{5}{2}x - \frac{7}{2}$

$y = mx+b$

The slope is $m = \frac{5}{2}$; the y-intercept is $(0, b) = \left(0, -\frac{7}{2}\right)$.

3. $f(x) = 8x - 6$: slope 8; y-intercept $(0, -6)$
$g(x) = 8x + 6$: slope 8; y-intercept $(0, 6)$
The lines have the same slope and different y-intercepts, so they are parallel.

4. $f(x) = \frac{2}{3}x+1$: slope $\frac{2}{3}$; y-intercept $(0, 1)$

$2y+3x = 1$

$2y = -3x+1$

$y = -\frac{3}{2}x+\frac{1}{2}$

slope $-\frac{3}{2}$; y-intercept $\left(0, \frac{1}{2}\right)$

The slopes of the lines are not equal and their product is -1, so the lines are perpendicular.

5. (1, 6), (5, 2)

$m = \frac{2-6}{5-1} = \frac{-4}{4} = -1$

$y - y_1 = m(x - x_1)$
$y - 6 = -1(x - 1)$
$y - 6 = -x + 1$
$y = -x + 7$
$f(x) = -x + 7$

6. (2, −8), (−6, −5)

$m = \frac{-5-(-8)}{-6-2} = \frac{3}{-8} = -\frac{3}{8}$

$y - y_1 = m(x - x_1)$
$y - (-8) = -\frac{3}{8}(x - 2)$
$y + 8 = -\frac{3}{8}x + \frac{3}{4}$
$y = -\frac{3}{8}x - \frac{29}{4}$
$f(x) = -\frac{3}{8}x - \frac{29}{4}$

7. $3x - y = 5$
$3x - 5 = y$

The line $3x - y = 5$ has slope 3. A parallel line will also have slope 3.

$y - y_1 = m(x - x_1)$; (−1, −5)
$y - (-5) = 3[x - (-1)]$
$y + 5 = 3(x + 1)$
$y + 5 = 3x + 3$
$y = 3x - 2$
$f(x) = 3x - 2$

8. $4x - 5y = 10$
$-5y = -4x + 10$
$y = \frac{4}{5}x - 2$

The line $4x - 5y = 10$ has slope $\frac{4}{5}$.

A perpendicular line has slope $-\frac{5}{4}$.

$y = mx + b$; (0, 4)
$y = -\frac{5}{4}x + 4$
$f(x) = -\frac{5}{4}x + 4$

9. $4x + y = \frac{2}{3}$
$y = -4x + \frac{2}{3}$

The line $4x + y = \frac{2}{3}$ has slope −4. A perpendicular line has slope $\frac{1}{4}$.

$y - y_1 = m(x - x_1)$; (2, −3)
$y - (-3) = \frac{1}{4}(x - 2)$
$y + 3 = \frac{1}{4}x - \frac{1}{2}$
$y = \frac{1}{4}x - \frac{7}{2}$
$f(x) = \frac{1}{4}x - \frac{7}{2}$

10. $5x + 2y = 2$
$2y = -5x + 2$
$y = -\frac{5}{2}x + 1$

The line $5x + 2y = 2$ has slope $-\frac{5}{2}$. A parallel line will also have slope $-\frac{5}{2}$.

$y - y_1 = m(x - x_1)$; (−1, 0)
$y - 0 = -\frac{5}{2}[x - (-1)]$
$y = -\frac{5}{2}(x + 1)$
$y = -\frac{5}{2}x - \frac{5}{2}$
$f(x) = -\frac{5}{2}x - \frac{5}{2}$

11. $f(x) = 4x - 2$
Linear

x	0	$\frac{1}{2}$	1
y or $f(x)$	−2	0	2

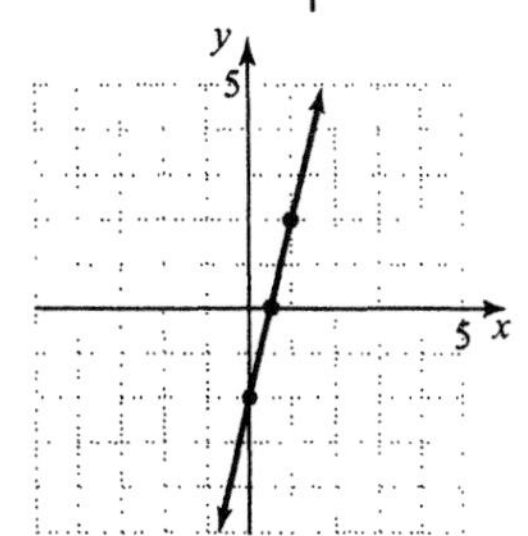

12. $f(x) = 6x - 5$
Linear

x	0	$\frac{1}{2}$	1
y or $f(x)$	−5	−2	1

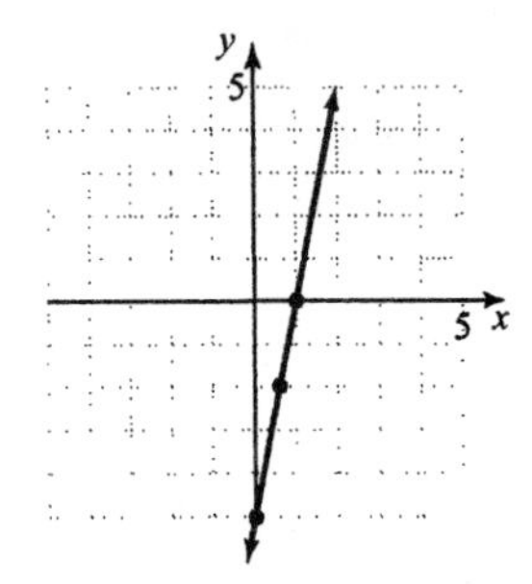

13. $g(x) = |x| + 3$
Not linear

x	−2	−1	0	1	2
y or $g(x)$	5	4	3	4	5

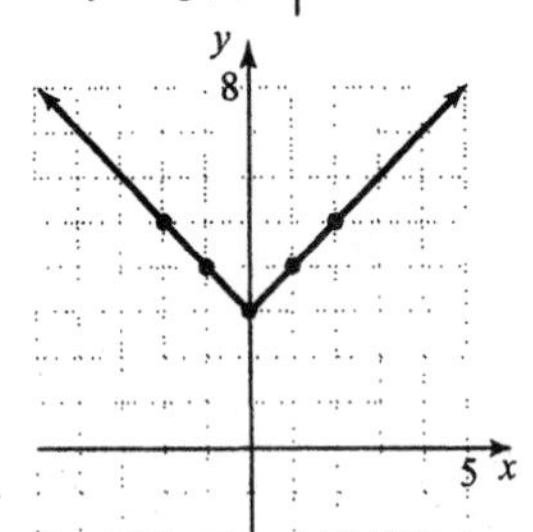

14. $h(x) = |x| + 2$
Not linear

x	−2	−1	0	1	2
y or $h(x)$	4	3	2	3	4

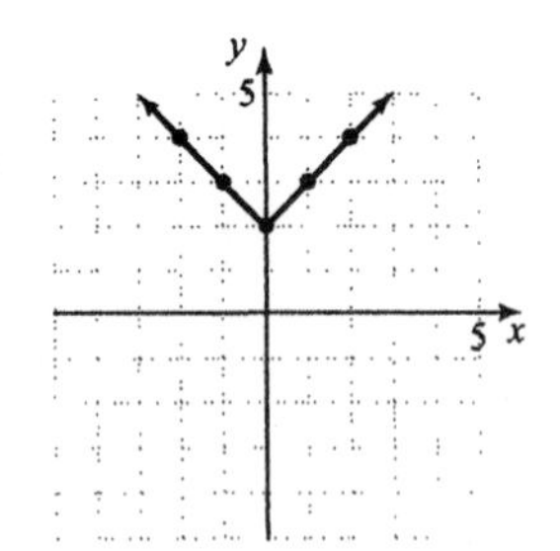

15. $f(x) = 2x^2$
Not linear

x	−2	−1	0	1	2
y or $f(x)$	8	2	0	2	8

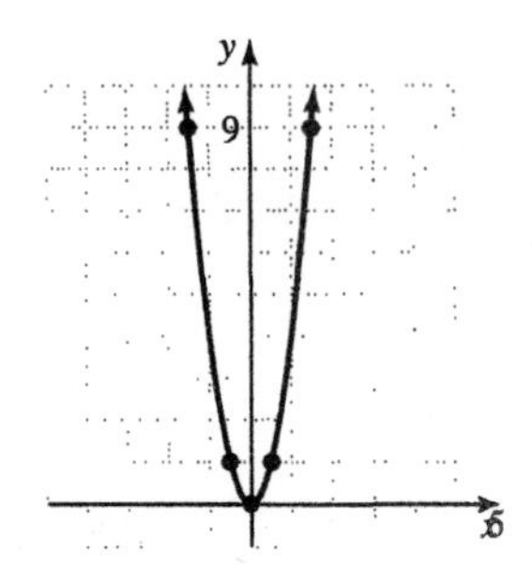

16. $F(x) = 3x^2$
Not linear

x	−2	−1	0	1	2
y or $F(x)$	12	3	0	3	12

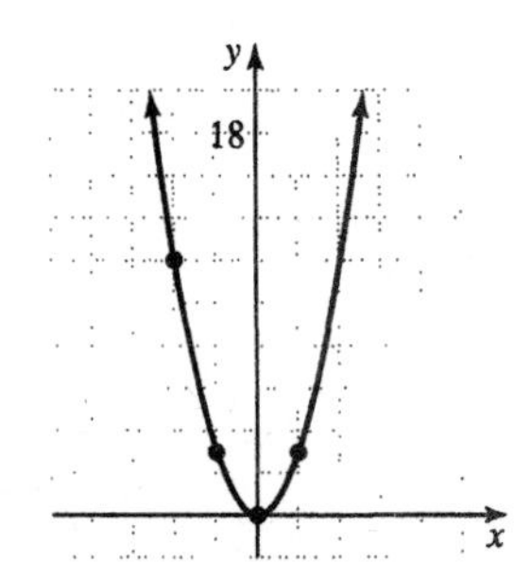

17. $h(x) = x^2 - 3$
Not linear

x	−2	−1	0	1	2
y or $h(x)$	1	−2	−3	−2	1

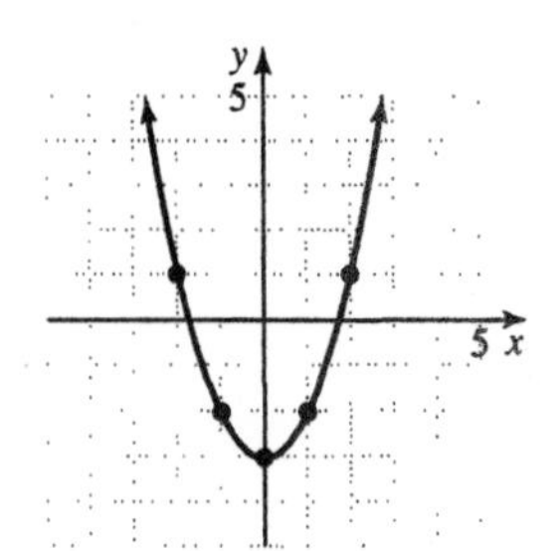

18. $G(x) = x^2 + 3$
Not linear

x	-2	-1	0	1	2
y or $G(x)$	7	4	3	4	7

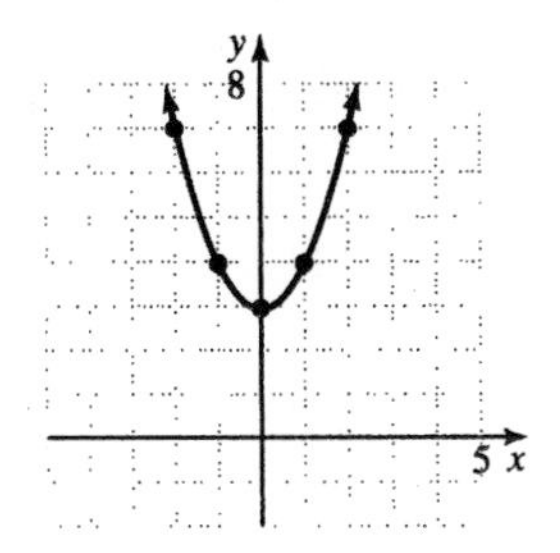

19. $F(x) = -2x$
Linear

x	-1	0	1
y or $F(x)$	2	0	-2

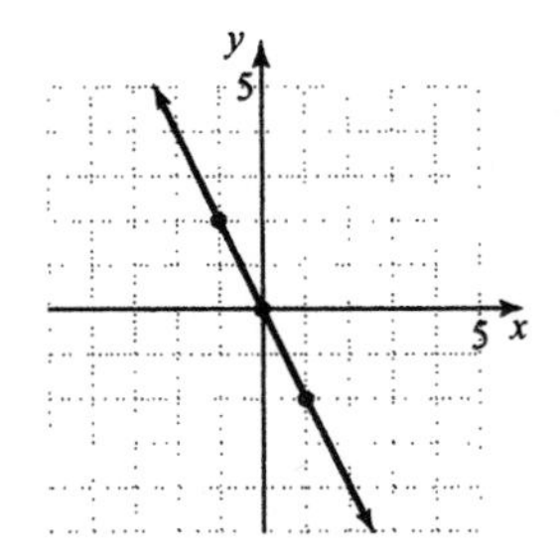

20. $H(x) = -3x$
Linear

x	-1	0	1
y or $H(x)$	3	0	-3

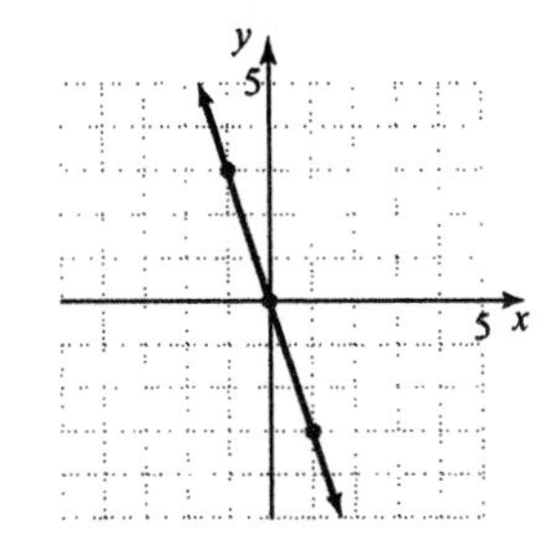

21. $G(x) = |x + 2|$
Not linear

x	-4	-3	-2	-1	0
y or $G(x)$	2	1	0	1	2

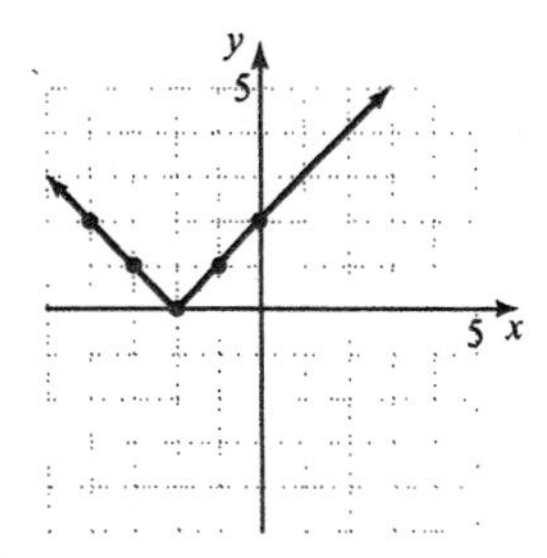

22. $g(x) = |x - 1|$
Not linear

x	-1	0	1	2	3
y or $g(x)$	2	1	0	1	2

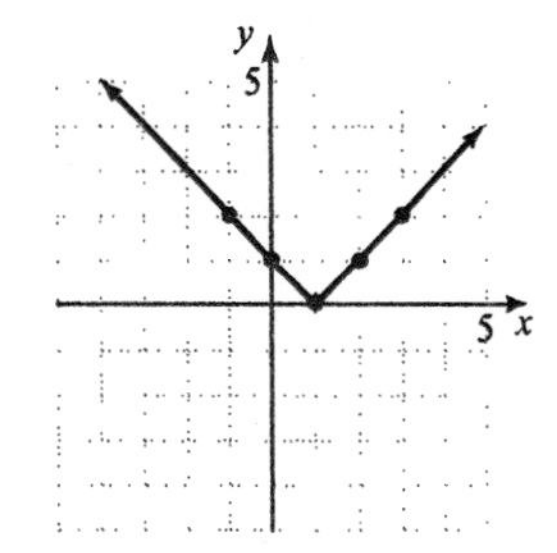

23. $f(x) = \frac{1}{3}x - 1$
Linear

x	-3	0	3
y or $f(x)$	-2	-1	0

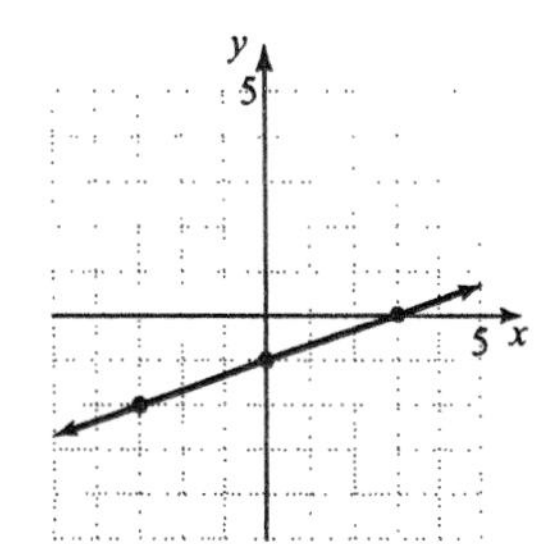

24. $f(x) = \frac{1}{2}x - 3$

Linear

x	−2	0	2
y or $f(x)$	−4	−3	−2

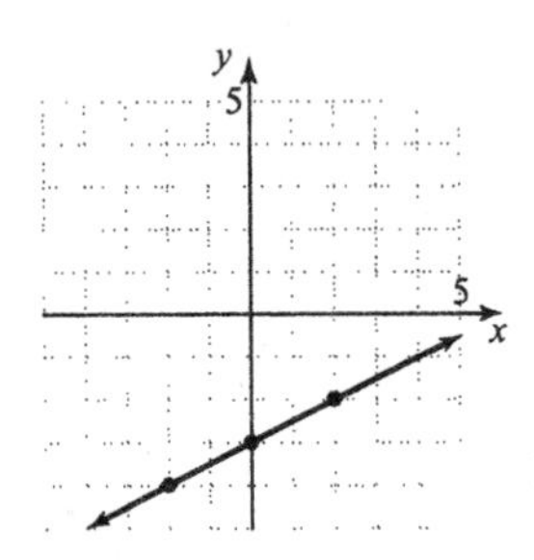

25. $g(x) = -\frac{3}{2}x + 1$

Linear

x	−2	0	2
y or $g(x)$	4	1	−2

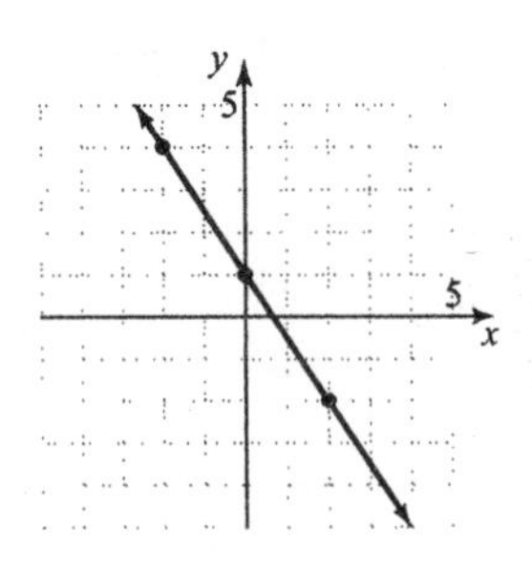

26. $G(x) = -\frac{2}{3}x + 1$

Linear

x	−3	0	3
y or $G(x)$	3	1	−1

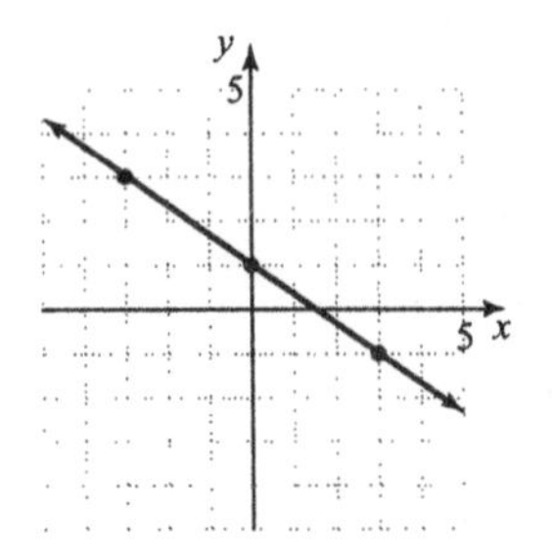

Section 8.3

Practice Exercises

1. $f(x) = \begin{cases} -4x - 2 & \text{if } x \le 0 \\ x + 1 & \text{if } x > 0 \end{cases}$

Since $4 > 0$, $f(4) = 4 + 1 = 5$.

Since $-2 \le 0$, $f(-2) = -4(-2) - 2 = 8 - 2 = 6$.

Since $0 \le 0$, $f(0) = -4(0) - 2 = 0 - 2 = -2$.

2. $f(x) = \begin{cases} -4x - 2 & \text{if } x \le 0 \\ x + 1 & \text{if } x > 0 \end{cases}$

For $x \le 0$:

x	$f(x)$
−2	6
−1	2
0	−2

For $x > 0$:

x	$f(x)$
1	2
2	3
3	4

Graph a closed circle at (0, −2). Graph an open circle at (0, 1), which is found by substituting 0 for x in $f(x) = x + 1$.

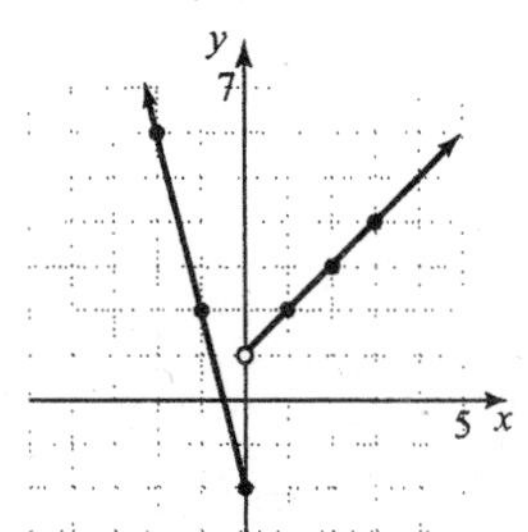

3. $f(x) = x^2$ and $g(x) = x^2 - 3$

The graph of $g(x) = x^2 - 3$ is the graph of $f(x) = x^2$ moved downward 3 units.

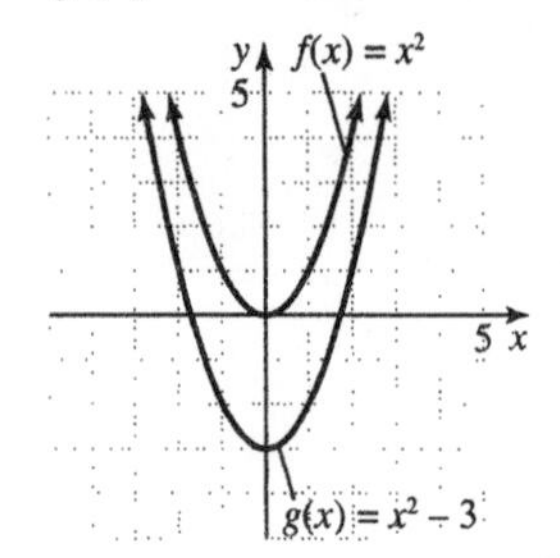

4. $f(x)=\sqrt{x}$ and $g(x)=\sqrt{x}+1$

The graph of $g(x)=\sqrt{x}+1$ is the graph of $f(x)=\sqrt{x}$ moved upward 1 unit.

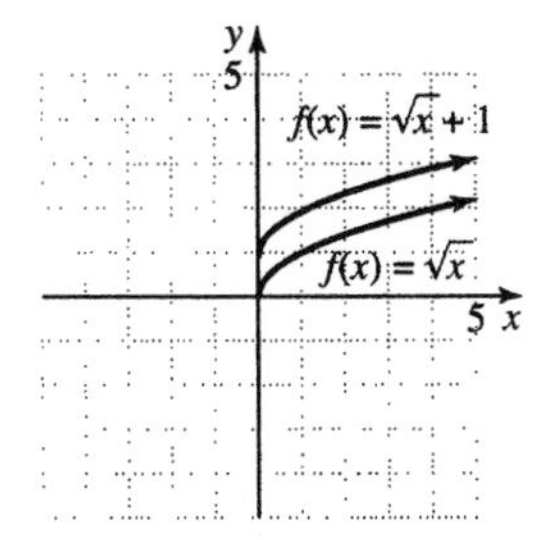

5. $f(x)=|x|$ and $g(x)=|x-3|$

x	$f(x)$	$g(x)$
−2	2	5
−1	1	4
0	0	3
1	1	2
2	2	1
3	3	0
4	4	1
5	5	2

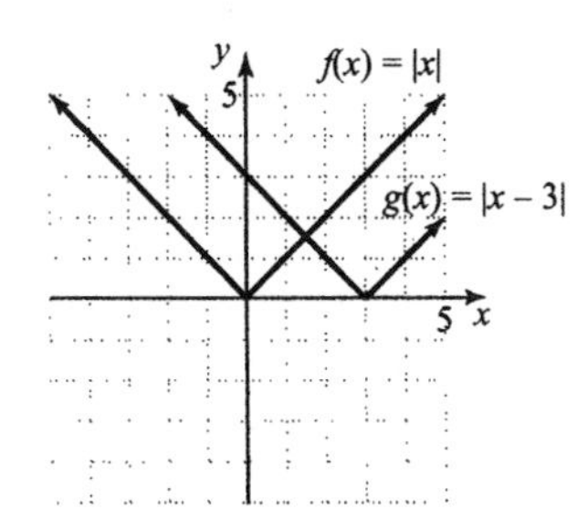

6. $f(x)=|x|$ and $g(x)=|x-2|+3$

The graph of $g(x)$ is the same as the graph of $f(x)$ shifted 2 units to the right and 3 units up.

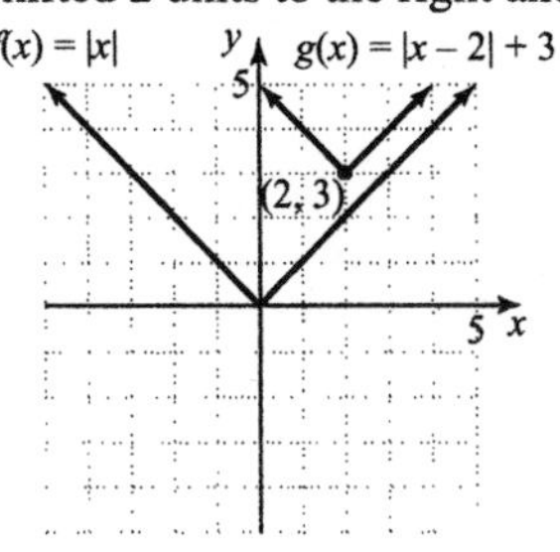

7. $h(x)=-(x+2)^2-1$

The graph of $h(x)=-(x+2)^2-1$ is the same as the graph of $f(x)=x^2$ reflected about the x-axis, then moved 2 units to the left and 1 unit downward.

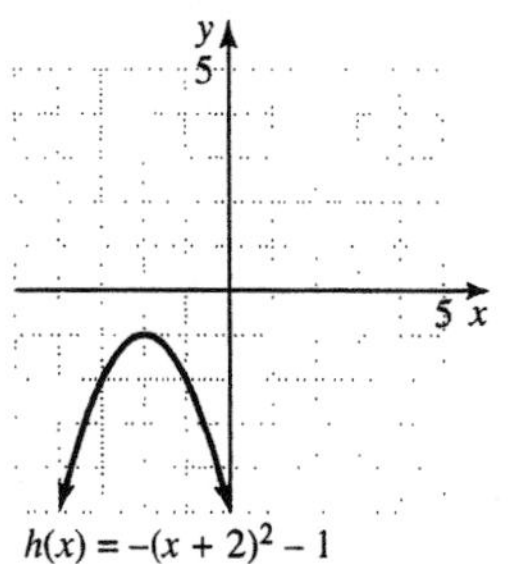

Vocabulary and Readiness Check

1. The graph that corresponds to $y=\sqrt{x}$ is C.
2. The graph that corresponds to $y=x^2$ is B.
3. The graph that corresponds to $y=x$ is D.
4. The graph that corresponds to $y=|x|$ is A.

Exercise Set 8.3

1. $f(x)=\begin{cases} 2x & \text{if } x<0 \\ x+1 & \text{if } x\geq 0 \end{cases}$

For $x<0$:

x	$f(x)$
−3	−6
−2	−4
−1	−2

For $x\geq 0$:

x	$f(x)$
0	1
1	2
2	3

Graph a closed circle at (0, 1). Graph an open circle at (0, 0), which is found by substituting 0 for x in $f(x)=2x$.

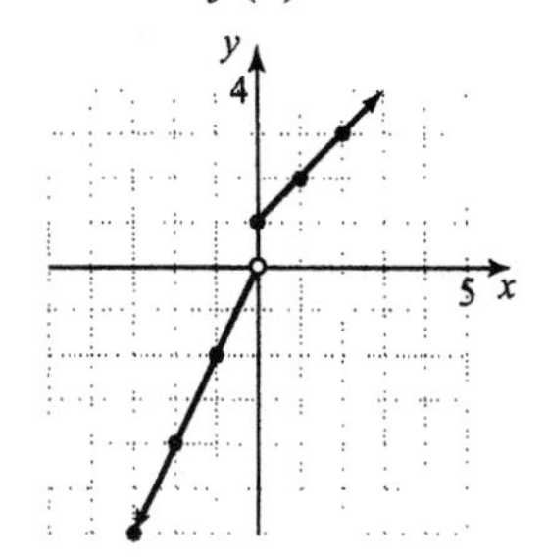

3. $f(x)=\begin{cases}4x+5 & \text{if } x\le 0\\ \frac{1}{4}x+2 & \text{if } x>0\end{cases}$

For $x \le 0$:

x	$f(x)$
−2	−3
−1	1
0	5

For $x > 0$:

x	$f(x)$
1	$2\frac{1}{4}$
2	$2\frac{1}{2}$
4	3

Graph a closed circle at (0, 5). Graph an open circle at (0, 2), which is found by substituting 0 for x in $f(x)=\frac{1}{4}x+2$.

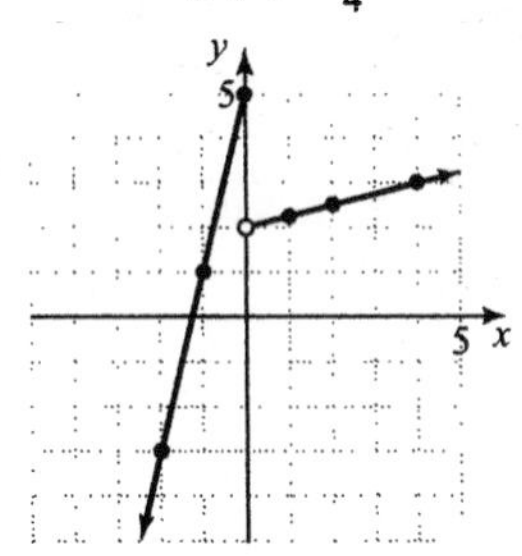

5. $g(x)=\begin{cases}-x & \text{if } x\le 1\\ 2x+1 & \text{if } x>1\end{cases}$

For $x \le 1$:

x	$g(x)$
−1	1
0	0
1	−1

For $x > 1$:

x	$g(x)$
2	5
3	7
4	9

Graph a closed circle at $(1, -1)$. Graph an open circle at $(1, 3)$, which is found by substituting 1 for x in $g(x)=2x+1$.

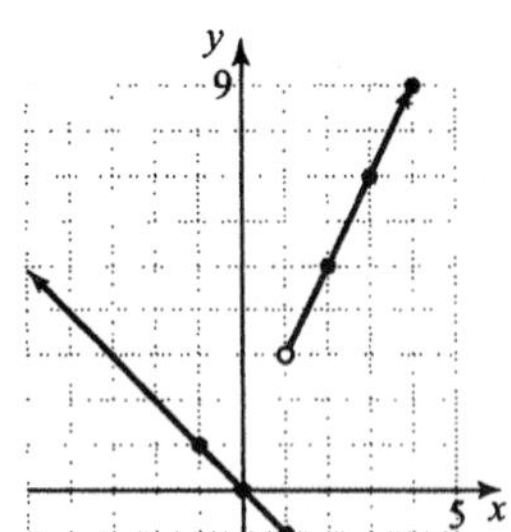

7. $f(x)=\begin{cases}5 & \text{if } x<-2\\ 3 & \text{if } x\ge -2\end{cases}$

For $x < -2$:

x	$f(x)$
−5	5
−4	5
−3	5

For $x \ge -2$:

x	$f(x)$
−2	3
−1	3
0	3

Graph a closed circle at (−2, 3). Graph an open circle at (−2, 5), which is found by substituting −2 for x in $f(x)=5$.

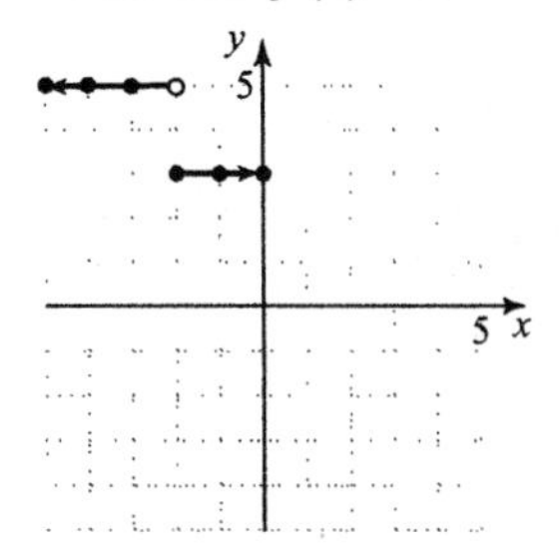

9. $f(x)=\begin{cases}-2x & \text{if } x\le 0\\ 2x+1 & \text{if } x>0\end{cases}$

For $x \le 0$:

x	$f(x)$
−1	2
0	0

For $x > 0$:

x	$f(x)$
1	3
2	5

Graph a closed circle at (0, 0). Graph an open circle at (0, 1), which is found by substituting 0 for x in $f(x)=2x+1$.

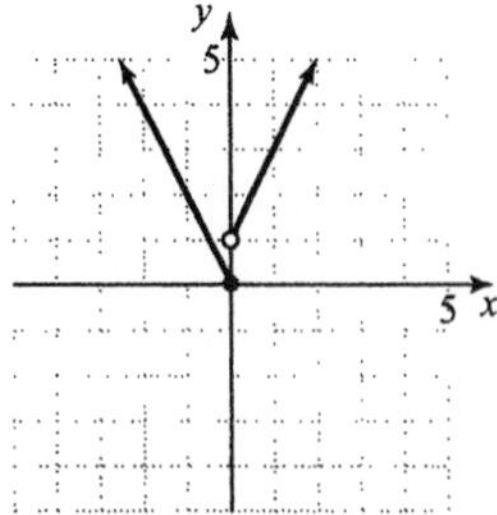

The function is defined for all real numbers, so the domain is $(-\infty, \infty)$. The function takes on all y-values greater than or equal to 0, so the range is $[0, \infty)$.

11. $h(x)=\begin{cases}5x-5 & \text{if } x<2\\ -x+3 & \text{if } x\geq 2\end{cases}$

For $x<2$:

x	$h(x)$
0	–5
1	0

For $x\geq 2$:

x	$h(x)$
2	1
3	0

Graph a closed circle at (2, 1). Graph an open circle at (2, 5), which is found by substituting 2 for x in $h(x)=5x-5$.

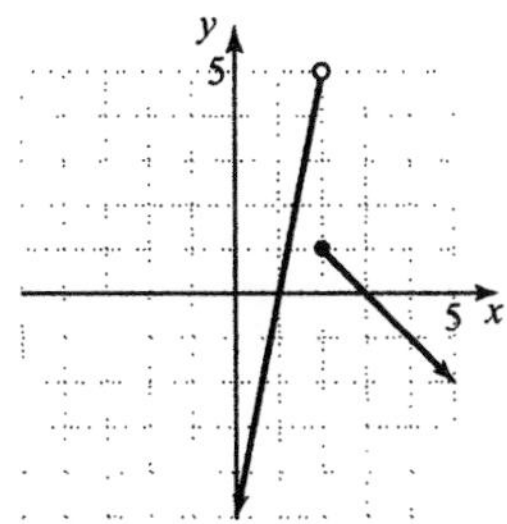

The function is defined for all real numbers, so the domain is $(-\infty, \infty)$. The function takes on all y-values less than 5, so the range is $(-\infty, 5)$.

13. $f(x)=\begin{cases}x+3 & \text{if } x<-1\\ -2x+4 & \text{if } x\geq -1\end{cases}$

For $x<-1$:

x	$f(x)$
–4	–1
–3	0
–2	1

For $x\geq -1$:

x	$f(x)$
–1	6
0	4
1	2

Graph a closed circle at (–1, 6). Graph an open circle at (–1, 2), which is found by substituting –1 for x in $f(x)=x+3$.

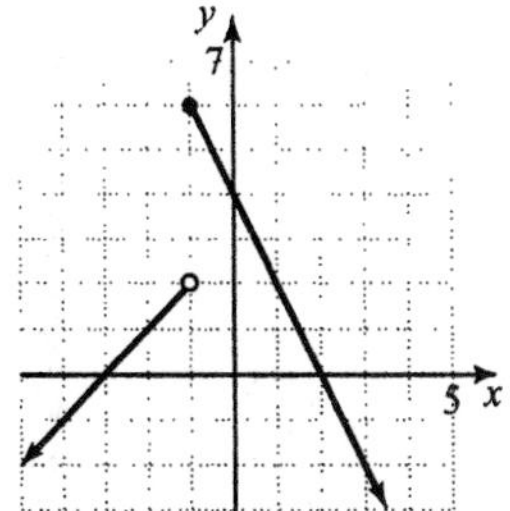

The function is defined for all real numbers, so the domain is $(-\infty, \infty)$. The function takes on all y-values less than or equal to 6, so the range is $(-\infty, 6]$.

15. $g(x)=\begin{cases}-2 & \text{if } x\leq 0\\ -4 & \text{if } x\geq 1\end{cases}$

For $x\leq 0$:

x	$g(x)$
–2	–2
–1	–2
0	–2

For $x\geq 1$:

x	$g(x)$
1	–4
2	–4
3	–4

Graph closed circles at (0, –2) and (1, –4).

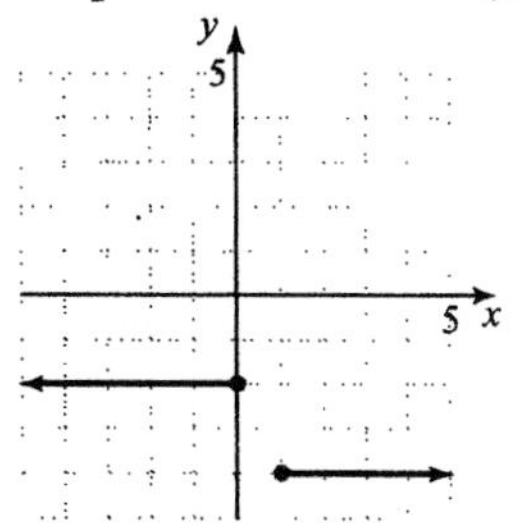

The function is defined for $x\leq 0$ or $x\geq 1$, so the domain is $(-\infty, 0]\cup[1, \infty)$. The function takes on two y-values, –2 and –4, so the range is $\{-2, -4\}$.

17. $f(x)=|x|+3$

The graph of $f(x)=|x|+3$ is the same as the graph of $y=|x|$ shifted up 3 units.

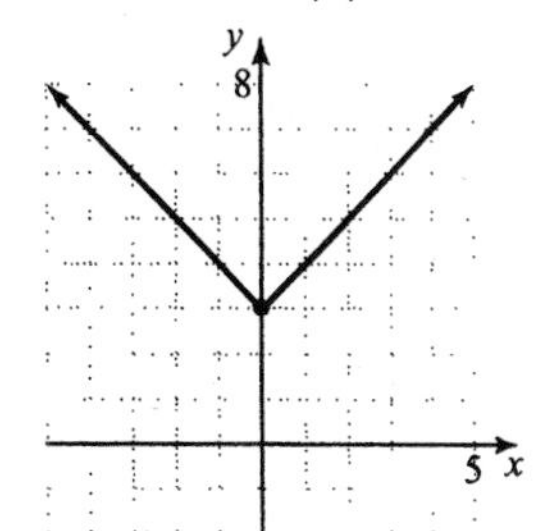

19. $f(x)=\sqrt{x}-2$

The graph of $f(x)=\sqrt{x}-2$ is the same as the graph of $y=\sqrt{x}$ shifted down 2 units.

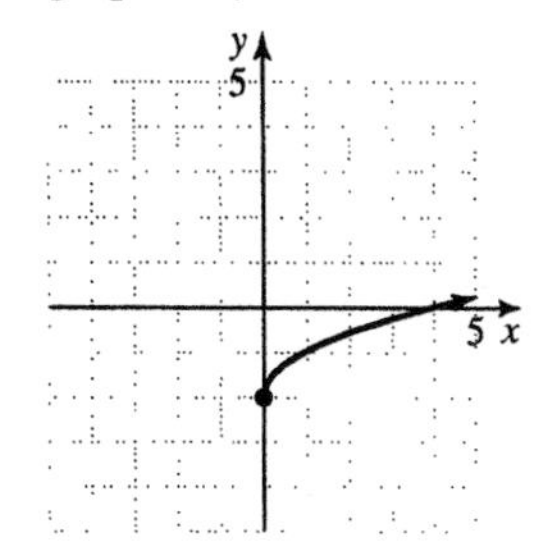

21. $f(x)=|x-4|$

The graph of $f(x)=|x-4|$ is the same as the graph of $y=|x|$ shifted right 4 units.

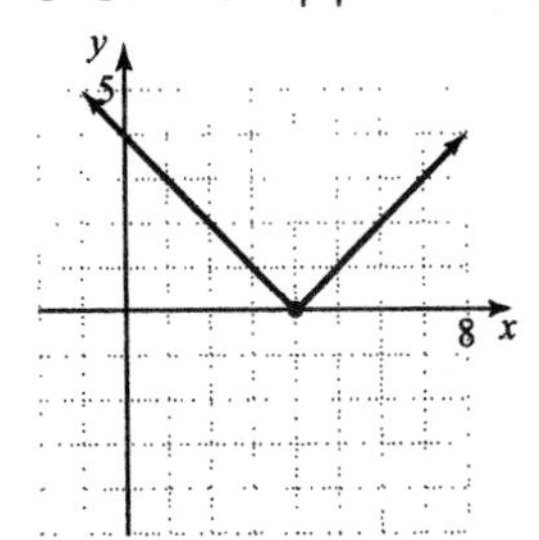

23. $f(x)=\sqrt{x+2}$

The graph of $f(x)=\sqrt{x+2}$ is the same as the graph of $y=\sqrt{x}$ shifted left 2 units.

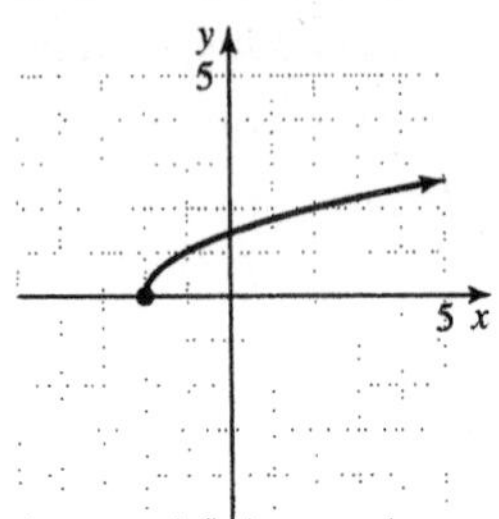

25. $y=(x-4)^2$

The graph of $y=(x-4)^2$ is the same as the graph of $y=x^2$ shifted right 4 units.

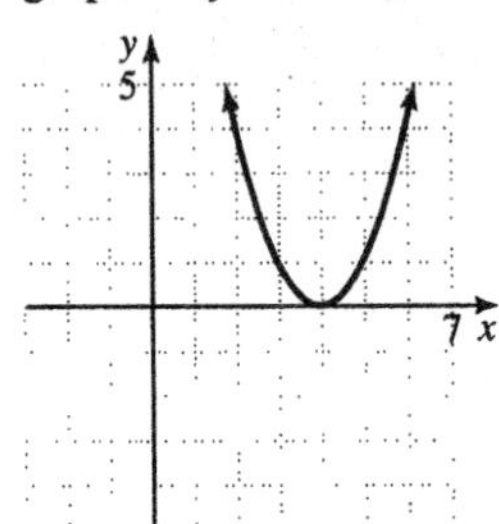

27. $f(x)=x^2+4$

The graph of $f(x)=x^2+4$ is the same as the graph of $y=x^2$ shifted up 4 units.

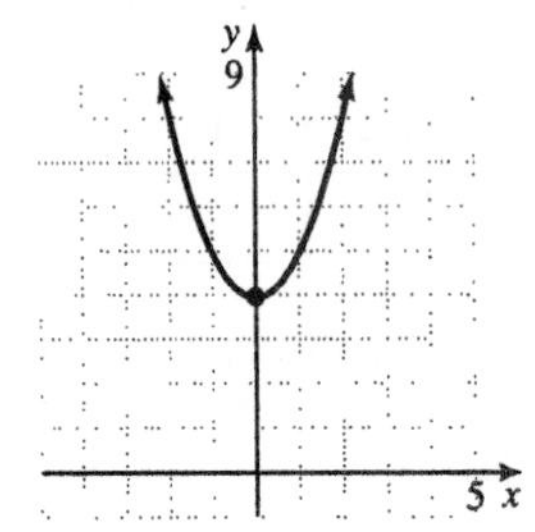

29. $f(x)=\sqrt{x-2}+3$

The graph of $f(x)=\sqrt{x-2}+3$ is the same as the graph of $y=\sqrt{x}$ shifted right 2 units and up 3 units.

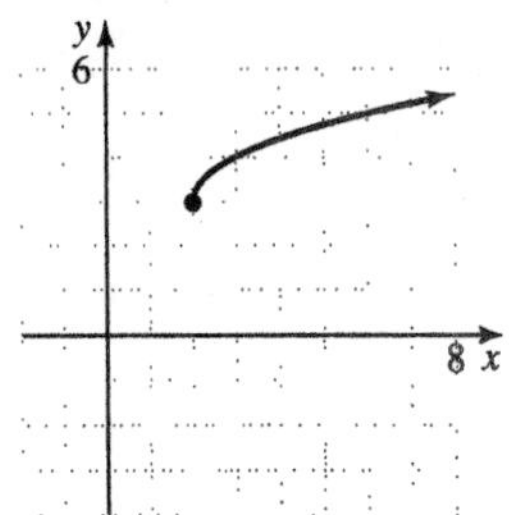

31. $f(x)=|x-1|+5$

The graph of $f(x)=|x-1|+5$ is the same as the graph of $y=|x|$ shifted right 1 unit and up 5 units.

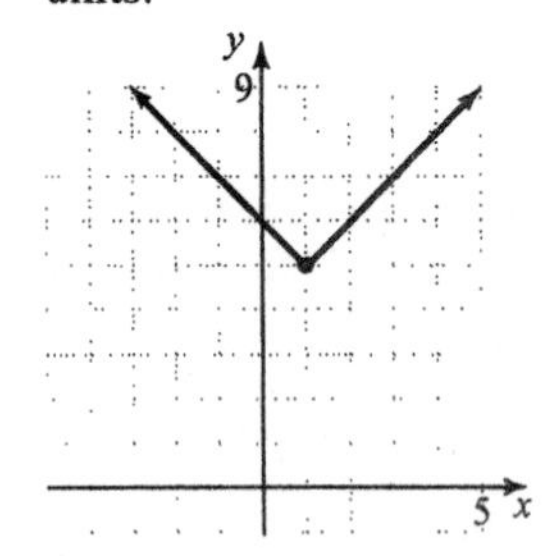

33. $f(x) = \sqrt{x+1} + 1$

The graph of $f(x) = \sqrt{x+1} + 1$ is the same as the graph of $y = \sqrt{x}$ shifted left 1 unit and up 1 unit.

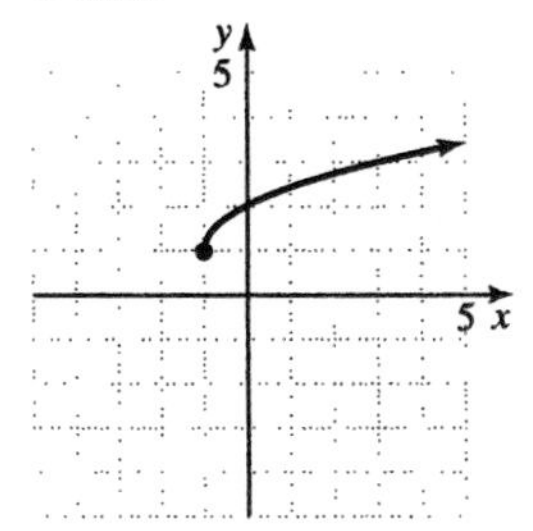

35. $f(x) = |x+3| - 1$

The graph of $f(x) = |x+3| - 1$ is the same as the graph of $y = |x|$ shifted left 3 units and down 1 unit.

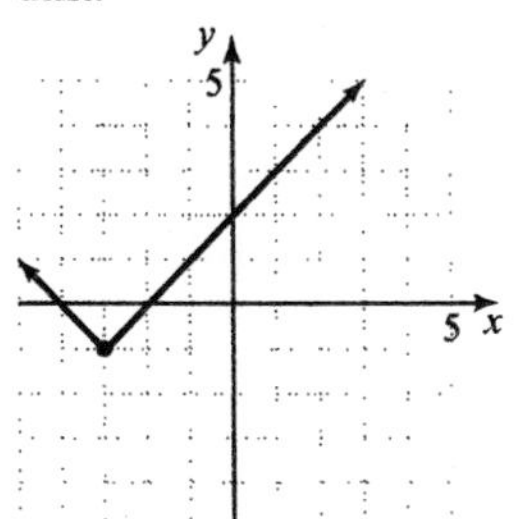

37. $g(x) = (x-1)^2 - 1$

The graph of $g(x) = (x-1)^2 - 1$ is the same as the graph of $y = x^2$ shifted right 1 unit and down 1 unit.

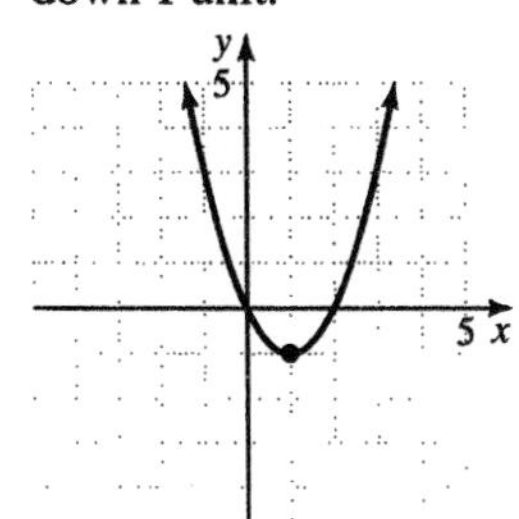

39. $f(x) = (x+3)^2 - 2$

The graph of $f(x) = (x+3)^2 - 2$ is the same as the graph of $y = x^2$ shifted left 3 units and down 2 units.

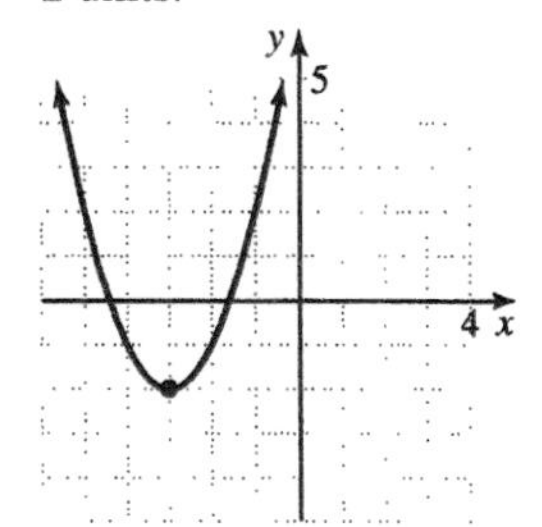

41. $f(x) = -(x-1)^2$

The graph of $f(x) = -(x-1)^2$ is the same as the graph of $y = x^2$ reflected about the x-axis and then shifted right 1 unit.

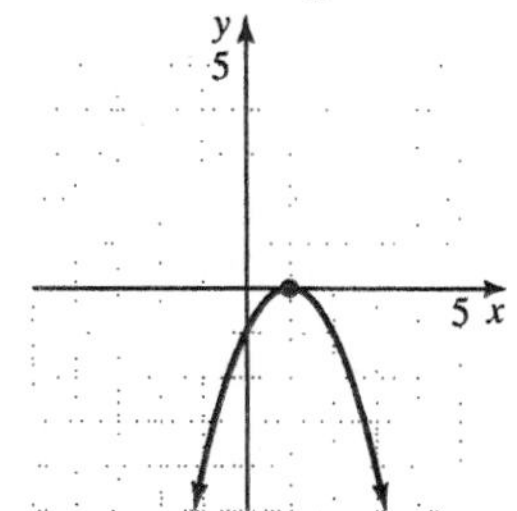

43. $h(x) = -\sqrt{x} + 3$

The graph of $h(x) = -\sqrt{x} + 3$ is the same as the graph of $y = \sqrt{x}$ reflected about the x-axis and then shifted up 3 units.

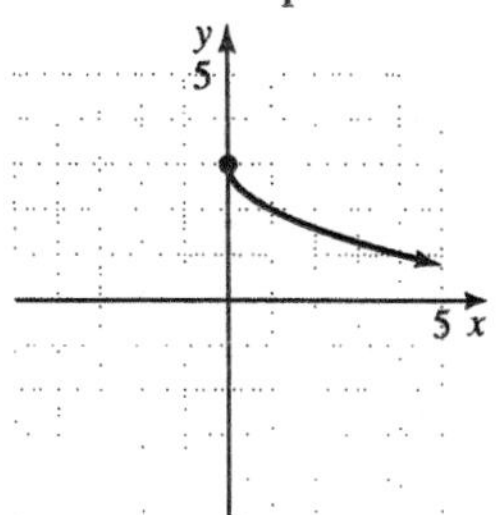

45. $h(x) = -|x+2| + 3$

The graph of $h(x) = -|x+2| + 3$ is the same as the graph of $y = |x|$ reflected about the x-axis and then shifted left 2 units and up 3 units.

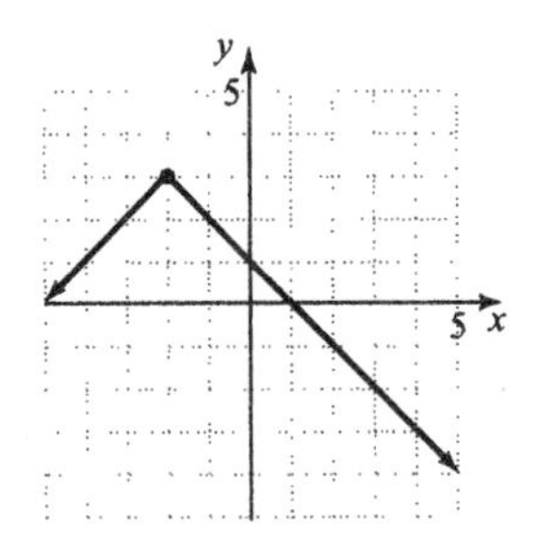

47. $f(x) = (x-3)+2$
Since the function can be simplified to $f(x) = x-1$, we see that its graph is a line with slope $m = 1$ and y-intercept $(0, -1)$.

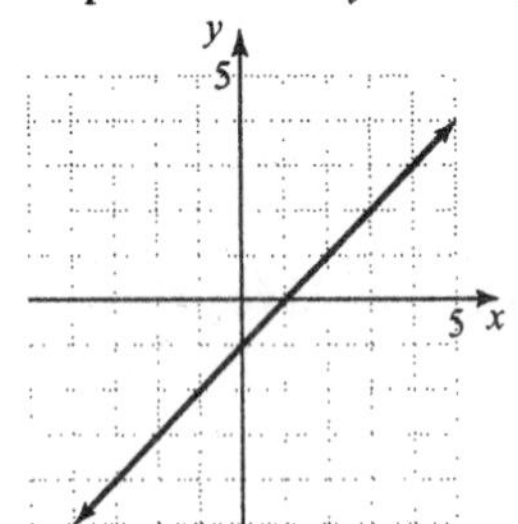

49. The graph of $y = -1$ is a horizontal line with y-intercept $(0, -1)$. The correct graph is A.

51. The graph of $x = 3$ is a vertical line with x-intercept $(3, 0)$. The correct graph is D.

53. Answers may vary

55. $f(x) = \begin{cases} -\frac{1}{2}x & \text{if } x \le 0 \\ x+1 & \text{if } 0 < x \le 2 \\ 2x-1 & \text{if } x > 2 \end{cases}$

Some points for $x \le 0$: $(-4, 2)$, $(-2, 1)$, $(0, 0)$
Closed dot at $(0, 0)$
Some points for $0 < x \le 2$: $(1, 2)$, $(2, 3)$
Open dot at $(0, 1)$, closed dot at $(2, 3)$
Some points for $x > 2$: $(3, 5)$, $(4, 7)$
There would be an open dot at $(2, 3)$ except that it gets filled by the middle piece of the graph.

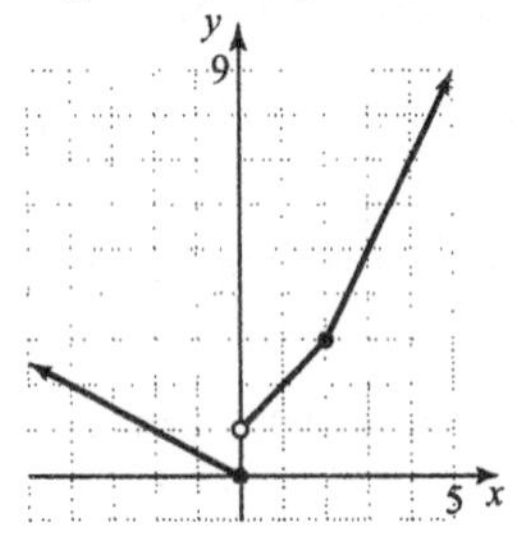

57. $f(x) = \sqrt{x-2}+3$
The function is defined when $x-2 \ge 0$, or $x \ge 2$, so the domain is $[2, \infty)$. The function takes on all y-values greater than or equal to 3, so the range is $[3, \infty)$.

59. $h(x) = -|x+2|+3$
The function is defined for all real numbers, so the domain is $(-\infty, \infty)$. The function takes on all y-values less than or equal to 3, so the range is $(-\infty, 3]$.

61. $f(x) = 5\sqrt{x-20}+1$
The function is defined when $x-20 \ge 0$, or $x \ge 20$, so the domain is $[20, \infty)$.

63. $h(x) = 5|x-20|+1$
The function is defined for all real numbers, so the domain is $(-\infty, \infty)$.

65. $g(x) = 9-\sqrt{x+103}$
The function is defined when $x+103 \ge 0$, or $x \ge -103$, so the domain is $[-103, \infty)$.

67. $f(x) = \begin{cases} |x| & \text{if } x \le 0 \\ x^2 & \text{if } x > 0 \end{cases}$

For $x \le 0$:

x	$f(x)$
−2	2
−1	1
0	0

For $x > 0$:

x	$f(x)$
1	1
2	4
3	9

Graph a closed circle at $(0, 0)$. The graph of $f(x) = x^2$ for $x > 0$ also approaches the point $(0, 0)$.

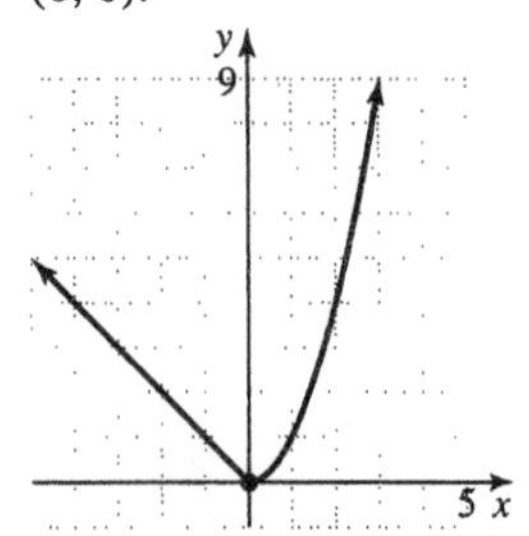

The function is defined for all real numbers, so the domain is $(-\infty, \infty)$. The function takes on all y-values greater than or equal to 0, so the range is $[0, \infty)$.

69. $g(x) = \begin{cases} |x-2| & \text{if } x < 0 \\ -x^2 & \text{if } x \geq 0 \end{cases}$

For $x < 0$:

x	$g(x)$
−3	5
−2	4
−1	3

For $x \geq 0$:

x	$g(x)$
0	0
1	−1
2	−4

Graph an open circle at (0, 2). Graph a closed circle at (0, 0).

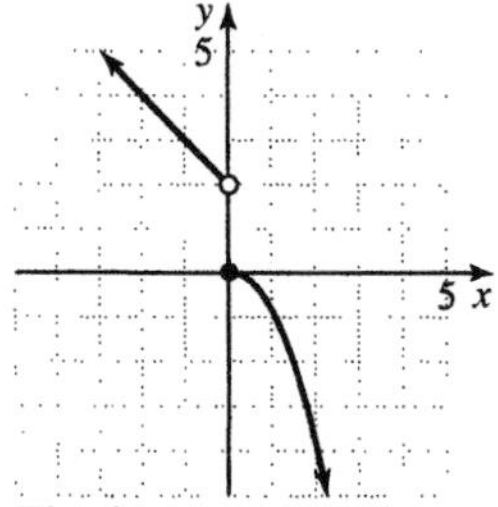

The function is defined for all real numbers, so the domain is $(-\infty, \infty)$. The function takes on all y-values such $y > 2$ or $y \leq 0$, so the range is $(-\infty, 0] \cup (2, \infty)$.

Section 8.4

Practice Exercises

1. $y = kx$
$20 = k(15)$
$\frac{4}{3} = k$
$k = \frac{4}{3};\ y = \frac{4}{3}x$

2. $d = kw$
$9 = k(36)$
$\frac{1}{4} = k$
$d = \frac{1}{4}w$
$d = \frac{1}{4}(75)$
$d = \frac{75}{4}$ inches or $18\frac{3}{4}$ inches

3. $b = \frac{k}{a}$
$5 = \frac{k}{9}$
$k = 45;\ b = \frac{45}{a}$

4. $P = \frac{k}{V}$
$350 = \frac{k}{2.8}$
$980 = k$
$P = \frac{980}{V}$
$P = \frac{980}{1.5}$
$P = 653\frac{1}{3}$ kilopascals

5. $A = kap$

6. $y = \frac{k}{x^3}$
$\frac{1}{2} = \frac{k}{2^3}$
$\frac{1}{2} = \frac{k}{8}$
$4 = k$
$k = 4;\ y = \frac{4}{x^3}$

7. $y = \frac{kz}{x^3}$
$15 = \frac{k \cdot 5}{3^3}$
$81 = k$
$k = 81;\ y = \frac{81z}{x^3}$

Vocabulary and Readiness Check

1. $y = 5x$ represents direct variation.

2. $y = \frac{700}{x}$ represents inverse variation.

3. $y = 5xz$ represents joint variation.

4. $y = \frac{1}{2}abc$ represents joint variation.

5. $y = \frac{9.1}{x}$ represents inverse variation.

6. $y = 2.3x$ represents direct variation.

7. $y = \frac{2}{3}x$ represents direct variation.

8. $y = 3.1st$ represents joint variation.

Exercise Set 8.4

1. $y = kx$
$4 = k(20)$
$k = \frac{1}{5}$
$y = \frac{1}{5}x$

3. $y = kx$
$6 = k(4)$
$k = \frac{3}{2}$
$y = \frac{3}{2}x$

5. $y = kx$
$7 = k\left(\frac{1}{2}\right)$
$k = 14$
$y = 14x$

7. $y = kx$
$0.2 = k(0.8)$
$k = 0.25$
$y = 0.25x$

9. $W = kr^3$
$1.2 = k \cdot 2^3$
$k = \frac{1.2}{8} = 0.15$
$W = 0.15r^3$
$= 0.15(3)^3$
$= 0.15(27)$
$= 4.05$
The ball weighs 4.05 pounds.

11. $P = kN$
$260,000 = k(442,000)$
$k = \frac{10}{17}$
$P = \frac{10}{17}N$
$P = \frac{10}{17}(348,000) \approx 204,706$
St. Louis produces 204,706 tons.

13. $y = \frac{k}{x}$
$6 = \frac{k}{5}$
$k = 30$
$y = \frac{30}{x}$

15. $y = \frac{k}{x}$
$100 = \frac{k}{7}$
$k = 700$
$y = \frac{700}{x}$

17. $y = \frac{k}{x}$
$\frac{1}{8} = \frac{k}{16}$
$k = 2$
$y = \frac{2}{x}$

19. $y = \frac{k}{x}$
$0.2 = \frac{k}{0.7}$
$k = 0.14$
$y = \frac{0.14}{x}$

21. $R = \frac{k}{T}$
$45 = \frac{k}{6}$
$k = 270$
$R = \frac{270}{5} = 54$
The car's speed is 54 mph.

23. $I = \frac{k}{R}$

$40 = \frac{k}{270}$

$k = 10{,}800$

$I = \frac{10{,}800}{R} = \frac{10{,}800}{150} = 72$

The current is 72 amps.

25. $I_1 = \frac{k}{d^2}$

Replace d by $2d$.

$I_2 = \frac{k}{(2d)^2} = \frac{k}{4d^2} = \frac{1}{4}I_1$

Thus, the intensity is divided by 4.

27. $x = kyz$

29. $r = kst^3$

31. $y = kx^3$

$9 = k(3)^3$

$9 = 27k$

$k = \frac{1}{3}$

$y = \frac{1}{3}x^3$

33. $y = k\sqrt{x}$

$0.4 = k\sqrt{4}$

$0.4 = 2k$

$\frac{0.4}{2} = k$

$0.2 = k$

$y = 0.2\sqrt{x}$

35. $y = \frac{k}{x^2}$

$0.052 = \frac{k}{5^2}$

$k = 1.3$

$y = \frac{1.3}{x^2}$

37. $y = kxz^3$

$120 = k(5)(2^3)$

$120 = k(5)(8)$

$120 = 40k$

$3 = k$

$y = 3xz^3$

39. $\text{Weight} = \frac{kwh^2}{l}$

$12 = \frac{k\left(\frac{1}{2}\right)\left(\frac{1}{3}\right)^2}{10}$

$120 = k \cdot \frac{1}{2} \cdot \frac{1}{9}$

$k = 2160$

$\text{Weight} = \frac{2160wh^2}{l} = \frac{2160\left(\frac{2}{3}\right)\left(\frac{1}{2}\right)^2}{16} = 22.5$

The beam can support 22.5 tons.

41. $V = kr^2h$

$32\pi = k(4)^2(6)$

$32\pi = k(16)(6)$

$32\pi = 96k$

$\frac{32\pi}{96} = k$

$\frac{\pi}{3} = k$

$V = \frac{\pi}{3}r^2h$

$V = \frac{\pi}{3}(3)^2(5)$

$V = 15\pi$

The volume is 15π cubic inches.

43. $I = \frac{k}{x^2}$

$80 = \frac{k}{2^2}$

$k = 320$

$I = \frac{320}{x^2}$

$5 = \frac{320}{x^2}$

$5x^2 = 320$

$x^2 = 64$

$x = 8$

The source is 8 feet from the light source.

45. y varies directly as x is written as $y = kx$.

47. a varies inversely as b is written as $a = \frac{k}{b}$.

49. y varies jointly as x and z is written as $y = kxz$.

51. y varies inversely as x^3 is written as $y = \frac{k}{x^3}$.

53. y varies directly as x and inversely as p^2 is written as $y = \frac{kx}{p^2}$.

55. $r = 4$ in.
$C = 2\pi r = 2\pi(4) = 8\pi$ in.
$A = \pi r^2 = \pi(4)^2 = 16\pi$ sq in.

57. $r = 9$ cm
$C = 2\pi r = 2\pi(9) = 18\pi$ cm
$A = \pi r^2 = \pi(9)^2 = 81\pi$ sq cm

59. $|-1.2| = 1.2$

61. $-|7| = -7$

63. $-\left|-\frac{1}{2}\right| = -\frac{1}{2}$

65. $\left(\frac{2}{3}\right)^3 = \left(\frac{2}{3}\right)\left(\frac{2}{3}\right)\left(\frac{2}{3}\right) = \frac{8}{27}$

67. $y = \frac{2}{3}x$ is an example of direct variation; a.

69. $y = 9ab$ is an example of joint variation; c.

71. $H_1 = ks^3$
$H_2 = k(2s)^3 = 8(ks^3) = 8H_1$
It is multiplied by 8.

73. $y_1 = kx$
$y_2 = k(2x) = 2(kx) = 2y_1$
It is multiplied by 2.

75.

x	$\frac{1}{4}$	$\frac{1}{2}$	1	2	4
$y = \frac{3}{x}$	12	6	3	$\frac{3}{2}$	$\frac{3}{4}$

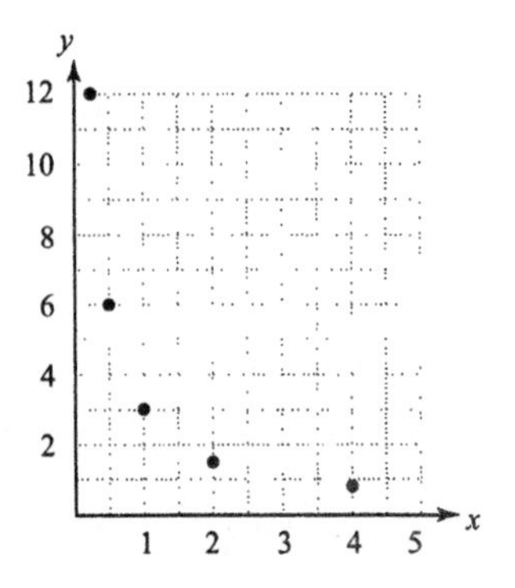

77.

x	$\frac{1}{4}$	$\frac{1}{2}$	1	2	4
$y = \frac{1}{2x}$	2	1	$\frac{1}{2}$	$\frac{1}{4}$	$\frac{1}{8}$

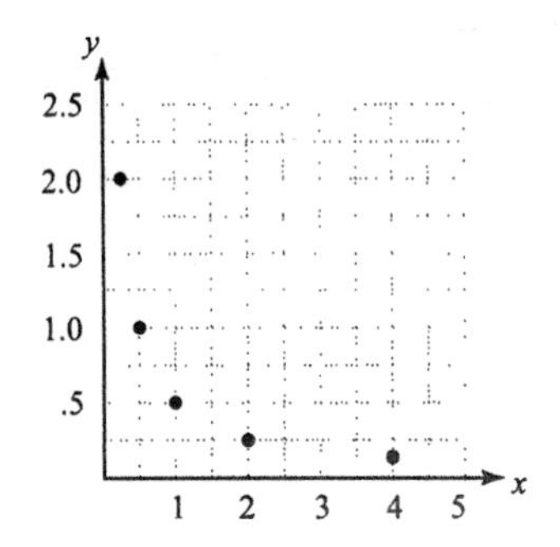

Chapter 8 Vocabulary Check

1. Parallel lines have the same slope and different y-intercepts.
2. Slope-intercept form of a linear equation in two variables is $y = mx + b$.
3. A function is a relation in which each first component in the ordered pairs corresponds to exactly one second component.
4. In the equation $y = 4x - 2$, the coefficient of x is the slope of its corresponding graph.
5. Two lines are perpendicular if the product of their slopes is -1.
6. A linear function is a function that can be written in the form $f(x) = mx + b$.
7. In the equation $y = kx$, y varies directly as x.
8. In the equation $y = \frac{k}{x}$, y varies inversely as x.

9. In the equation $y = kxz$, y varies jointly as x and z.

Chapter 8 Review

1. $f(x) = x$ or $y = x$
$m = 1,\ b = 0$

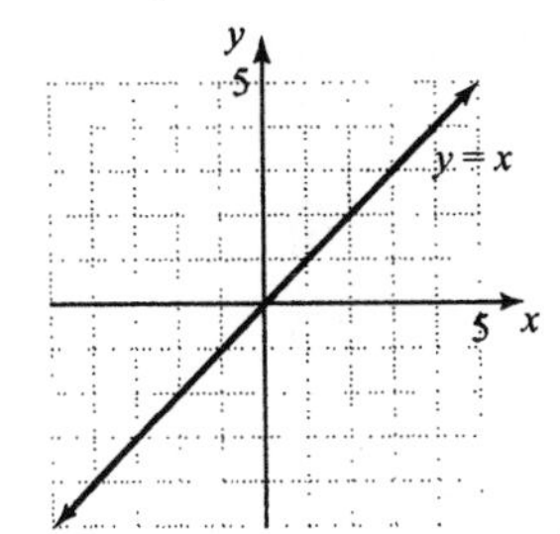

2. $f(x) = -\frac{1}{3}x$ or $y = -\frac{1}{3}x$
$m = -\frac{1}{3},\ b = 0$

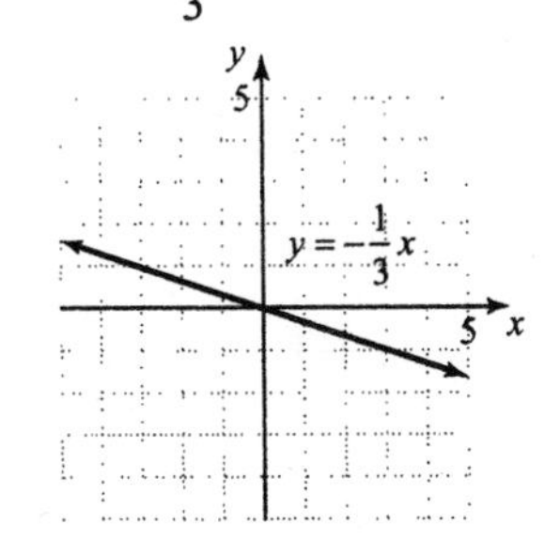

3. $g(x) = 4x - 1$ or $y = 4x - 1$
$m = 4,\ b = -1$

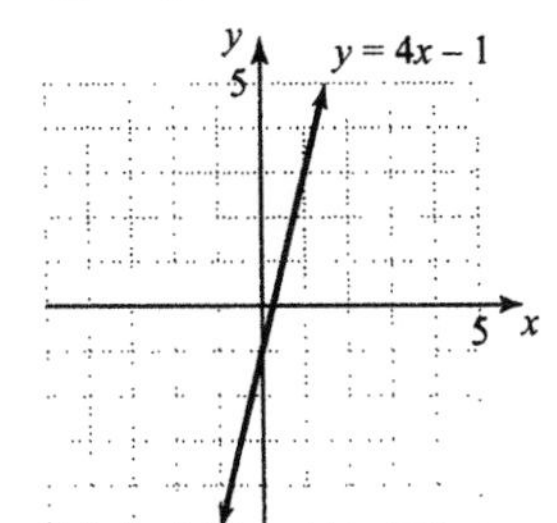

4. $F(x) = -\frac{2}{3}x + 2$ or $y = -\frac{2}{3}x + 2$
$m = -\frac{2}{3},\ b = 2$

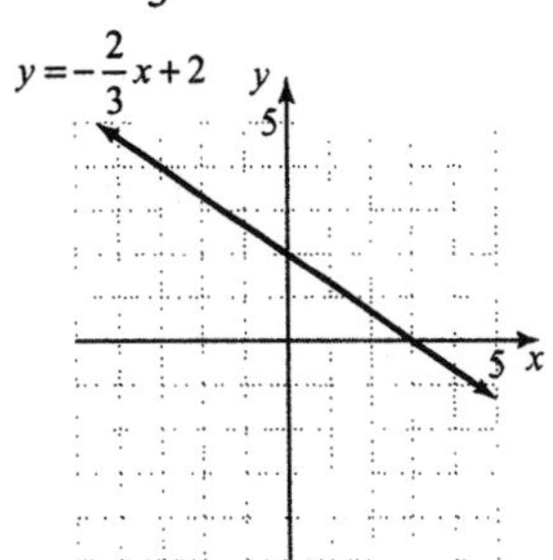

5. $f(x) = 3x + 1$
The y-intercept should be (0, 1). The correct graph is C.

6. $f(x) = 3x - 2$
The y-intercept should be (0, −2). The correct graph is A.

7. $f(x) = 3x + 2$
The y-intercept should be (0, 2). The correct graph is B.

8. $f(x) = 3x - 5$
The y-intercept should be (0, −5). The correct graph is D.

9. $f(x) = \frac{2}{5}x - \frac{4}{3}$
slope $m = \frac{2}{5}$; y-intercept $(0, b) = \left(0, -\frac{4}{3}\right)$

10. $f(x) = -\frac{2}{7}x + \frac{3}{2}$
slope $m = -\frac{2}{7}$; y-intercept $(0, b) = \left(0, \frac{3}{2}\right)$

11.
$$\begin{aligned} y - y_1 &= m(x - x_1) \\ y - (-2) &= 2(x - 5) \\ y + 2 &= 2x - 10 \\ 2x - y &= 12 \end{aligned}$$

12.
$$\begin{aligned} y - y_1 &= m(x - x_1) \\ y - 5 &= 3[x - (-3)] \\ y - 5 &= 3(x + 3) \\ y - 5 &= 3x + 9 \\ 3x - y &= -14 \end{aligned}$$

13. $m = \dfrac{-8-3}{-4-(-5)} = \dfrac{-11}{1} = -11$

$$\begin{aligned} y - y_1 &= m(x - x_1) \\ y - 3 &= -11[x - (-5)] \\ y - 3 &= -11(x+5) \\ y - 3 &= -11x - 55 \\ 11x + y &= -52 \end{aligned}$$

14. $m = \dfrac{-2-(-1)}{-4-(-6)} = \dfrac{-1}{2} = -\dfrac{1}{2}$

$$\begin{aligned} y - y_1 &= m(x - x_1) \\ y - (-1) &= -\frac{1}{2}[x - (-6)] \\ 2(y+1) &= -(x+6) \\ 2y + 2 &= -x - 6 \\ x + 2y &= -8 \end{aligned}$$

15. $y = 8$ has slope $= 0$
A line parallel to $y = 8$ has slope $= 0$.
$y = -5$

16. $x = 4$ has undefined slope.
A line perpendicular to $x = 4$ has slope $= 0$ and is therefore horizontal.
$y = 3$

17. Every horizontal line is in the form $y = c$. Since the line passes through the point $(3, -1)$, its equation is $y = -1$ or $f(x) = -1$.

18.

$$\begin{aligned} y &= mx + b \\ y &= -\frac{2}{3}x + 4 \\ f(x) &= -\frac{2}{3}x + 4 \end{aligned}$$

19.

$$\begin{aligned} y &= mx + b \\ y &= -x - 2 \\ f(x) &= -x - 2 \end{aligned}$$

20. $6x + 3y = 5$

$$\begin{aligned} 3y &= -6x + 5 \\ y &= -2x + \frac{5}{3} \text{ so } m = -2 \\ y - y_1 &= m(x - x_1) \\ y - (-6) &= -2(x-2) \\ y + 6 &= -2x + 4 \\ y &= -2x - 2 \\ f(x) &= -2x - 2 \end{aligned}$$

21. $3x + 2y = 8$

$$\begin{aligned} 2y &= -3x + 8 \\ y &= -\frac{3}{2}x + 4 \text{ so } m = -\frac{3}{2} \\ y - y_1 &= m(x - x_1) \\ y - (-2) &= -\frac{3}{2}[x - (-4)] \\ 2(y+2) &= -3(x+4) \\ 2y + 4 &= -3x - 12 \\ 2y &= -3x - 16 \\ y &= -\frac{3}{2}x - 8 \\ f(x) &= -\frac{3}{2}x - 8 \end{aligned}$$

22. $4x + 3y = 5$

$$\begin{aligned} 3y &= -4x + 5 \\ y &= -\frac{4}{3}x + \frac{5}{3} \end{aligned}$$

so $m = -\dfrac{4}{3}$ and $m_\perp = \dfrac{3}{4}$

$$\begin{aligned} y - y_1 &= m(x - x_1) \\ y - (-1) &= \frac{3}{4}[x - (-6)] \\ 4(y+1) &= 3(x+6) \\ 4y + 4 &= 3x + 18 \\ 4y &= 3x + 14 \\ y &= \frac{3}{4}x + \frac{7}{2} \\ f(x) &= \frac{3}{4}x + \frac{7}{2} \end{aligned}$$

23. $2x - 3y = 6$

$$\begin{aligned} -3y &= -2x + 6 \\ y &= \frac{2}{3}x - 2 \end{aligned}$$

so $m = \dfrac{2}{3}$ and $m_\perp = -\dfrac{3}{2}$

$$\begin{aligned} y - y_1 &= m(x - x_1) \\ y - 5 &= -\frac{3}{2}[x - (-4)] \\ 2(y-5) &= -3(x+4) \\ 2y - 10 &= -3x - 12 \\ 2y &= -3x - 2 \\ y &= -\frac{3}{2}x - 1 \\ f(x) &= -\frac{3}{2}x - 1 \end{aligned}$$

24. a. Use ordered pairs (0, 71) and (5, 82)

$$m = \frac{82-71}{5-0} = \frac{11}{5} = 2.2 \text{ and } b = 71$$

$y = 2.2x + 71$

b. $x = 2009 - 2000 = 9$

$y = 2.2(9) + 71 = 90.8$

About 91% of US drivers will be wearing seat belts.

25. a. Use ordered pairs (0, 43) and (22, 60)

$$m = \frac{60-43}{22-0} = \frac{17}{22} \text{ and } b = 43$$

$$y = \frac{17}{22}x + 43$$

b. $x = 2010 - 1998 = 12$

$$y = \frac{17}{22}(12) + 43 \approx 52.3$$

There will be about 52 million people reporting arthritis.

26. $-x + 3y = 2$

$3y = x + 2$

$$y = \frac{1}{3}x + \frac{2}{3}$$

$$m = \frac{1}{3}, b = \frac{2}{3}$$

$6x - 18y = 3$

$-18y = -6x + 3$

$$y = \frac{1}{3}x - \frac{1}{6}$$

$$m = \frac{1}{3}, b = -\frac{1}{6}$$

The slopes are the same and the y-intercepts are different, so the lines are parallel.

27. When $x = -1$, y or $f(x)$ is 0, so $f(-1) = 0$.

28. When $x = 1$, y or $f(x)$ is -2, so $f(1) = -2$.

29. The x-values that correspond to a y-value of 1 are -2 and 4, so $f(x) = 1$ for $x = -2$ and $x = 4$.

30. The x-values that correspond to a y-value of -1 are 0 and 2, so $f(x) = -1$ for $x = 0$ and $x = 2$.

31. $f(x) = 3x$; Linear

x	-1	0	1
y	-3	0	3

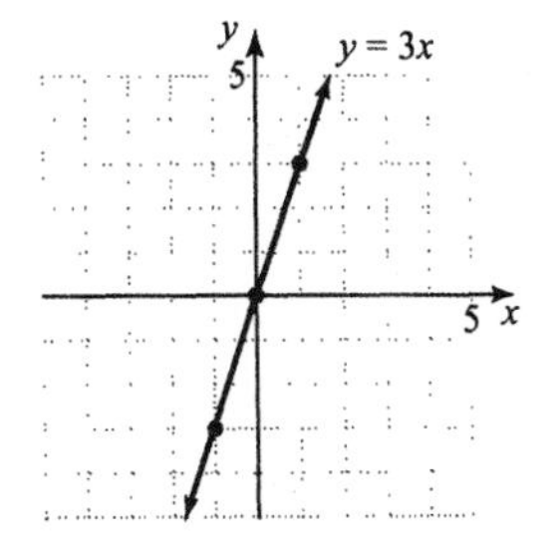

32. $f(x) = 5x$; Linear

x	-1	0	1
y	-5	0	5

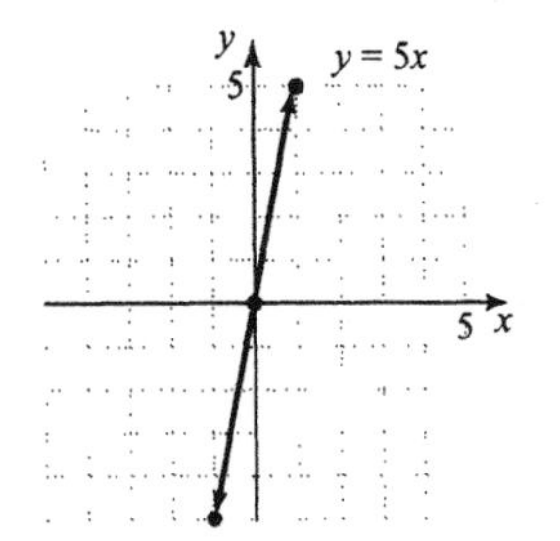

33. $g(x) = |x| + 4$; Nonlinear

x	-3	-2	-1	0	1	2	3
y	7	6	5	4	5	6	7

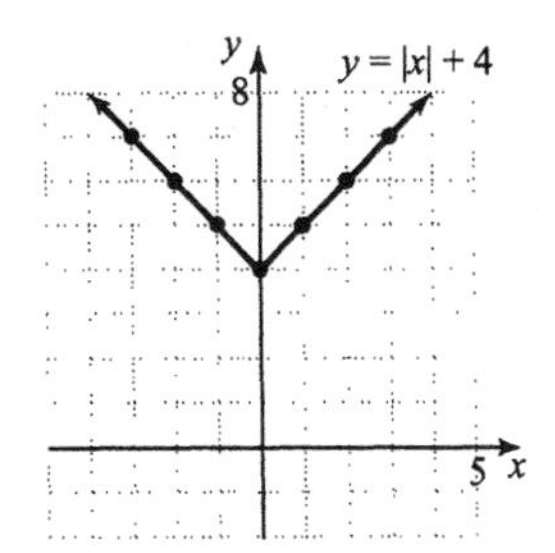

34. $h(x) = x^2 + 4$; Nonlinear

x	-3	-2	-1	0	1	2	3
y	13	8	5	4	5	8	13

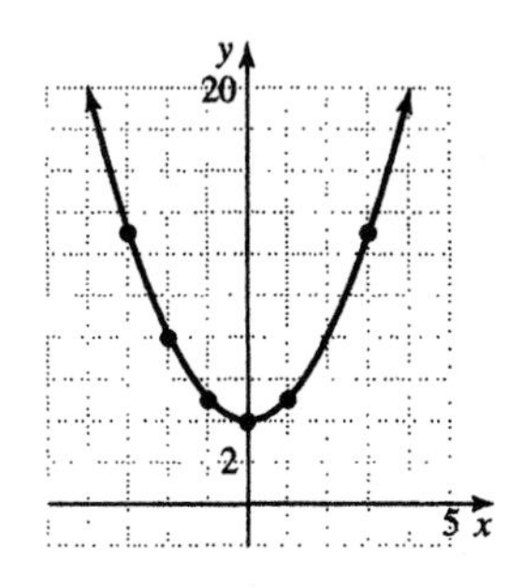

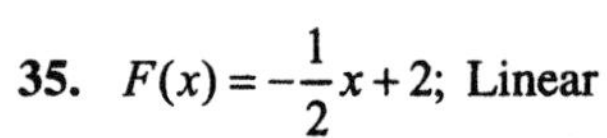

35. $F(x) = -\frac{1}{2}x + 2$; Linear

Find three ordered pair solutions, or find *x*- and *y*-intercepts, or find *m* and *b*.

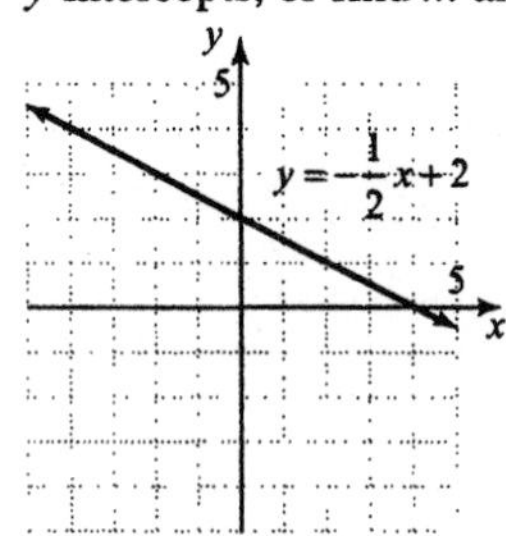

36. $G(x) = -x + 5$; Linear

Find three ordered pair solutions, or find *x*- and *y*-intercepts, or find *m* and *b*.

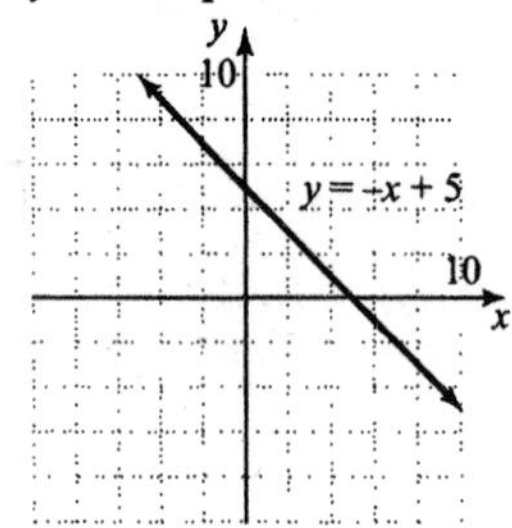

37. $y = -1.36x$; Linear

Find three ordered pair solutions, or find *x*- and *y*-intercepts, or find *m* and *b*.

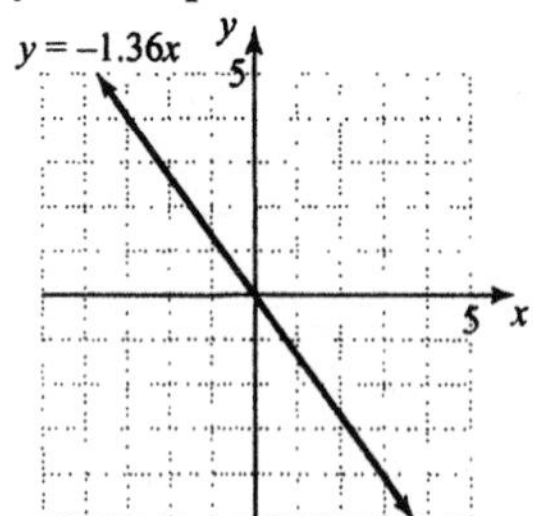

38. $y = 2.1x + 5.9$; Linear

Find three ordered pair solutions, or find *x*- and *y*-intercepts, or find *m* and *b*.

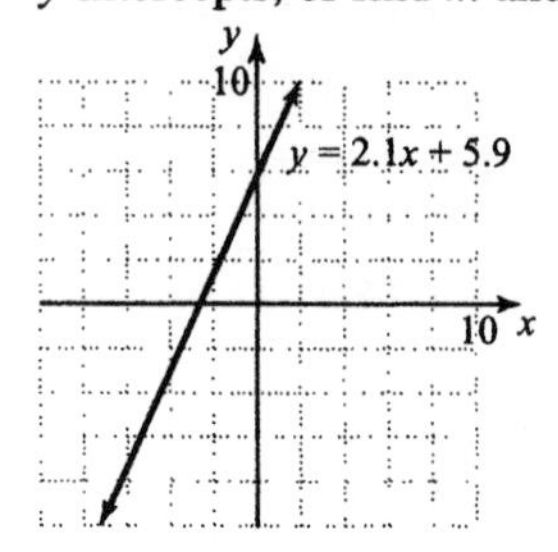

39. $H(x) = (x-2)^2$; Nonlinear

x	0	1	2	3	4
y	4	1	0	1	4

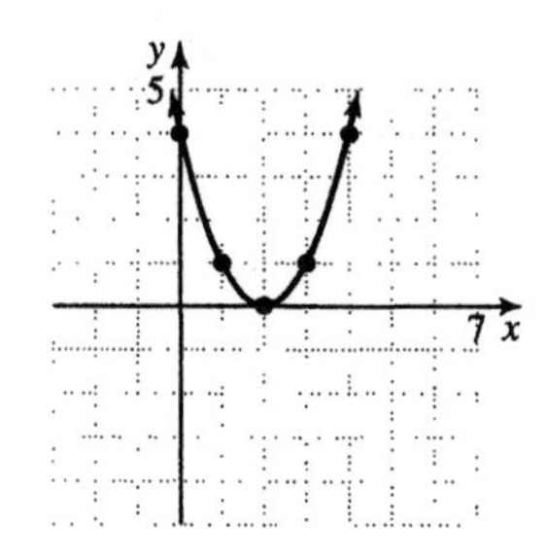

40. $f(x) = -|x - 3|$; Nonlinear

x	1	2	3	4	5
y	−2	−1	0	−1	−2

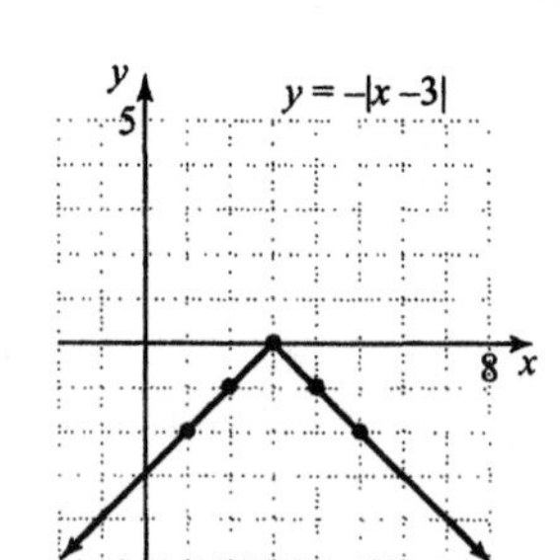

41. $g(x)=\begin{cases} -\frac{1}{5}x & \text{if } x \le -1 \\ -4x+2 & \text{if } x > -1 \end{cases}$

For $x \le -1$:

x	$g(x)$
–5	1
–3	$\frac{3}{5}$
–1	$\frac{1}{5}$

For $x > -1$:

x	$g(x)$
0	2
1	–2
2	–6

Graph a closed circle at $\left(-1, \frac{1}{5}\right)$. Graph an open circle at (–1, 6), which is found by substituting –1 for x in $g(x) = -4x + 2$.

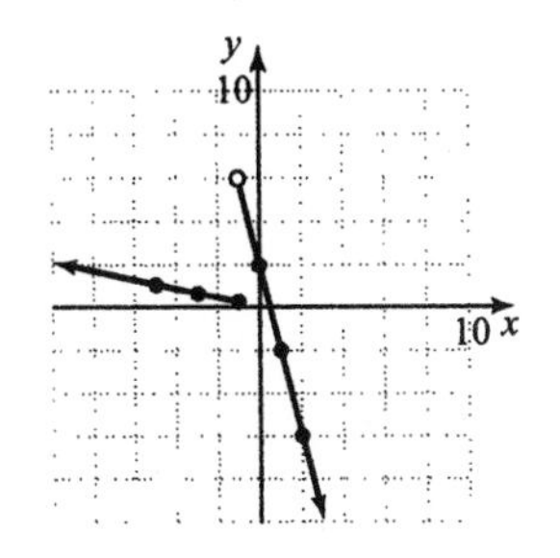

42. $f(x)=\begin{cases} -3x & \text{if } x < 0 \\ x-3 & \text{if } x \ge 0 \end{cases}$

For $x < 0$:

x	$f(x)$
–3	9
–2	6
–1	3

For $x \ge 0$:

x	$f(x)$
0	–3
1	–2
2	–1

Graph a closed circle at (0, –3). Graph an open circle at (0, 0), which is found by substituting 0 for x in $f(x) = -3x$.

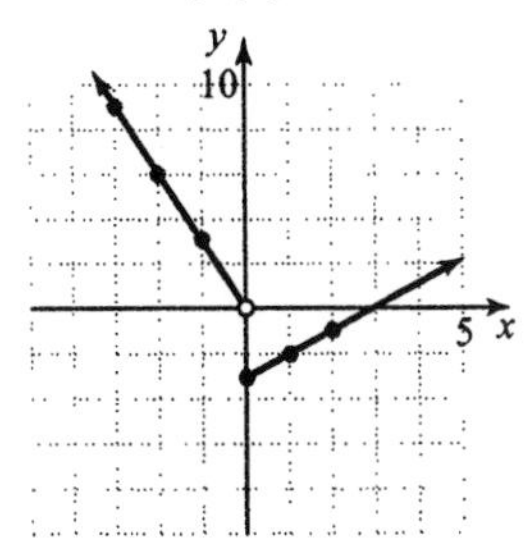

43. $f(x)=\sqrt{x-4}$

The graph of $f(x)=\sqrt{x-4}$ is the same as the graph of $y=\sqrt{x}$ shifted right 4 units.

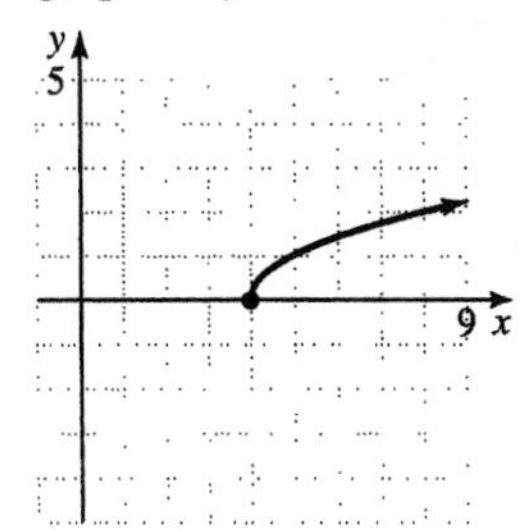

44. $y=\sqrt{x}-4$

The graph of $f(x)=\sqrt{x}-4$ is the same as the graph of $y=\sqrt{x}$ shifted down 4 units.

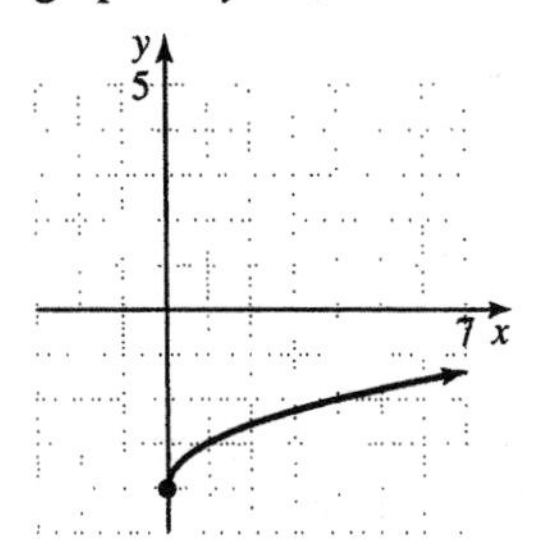

45. $h(x)=-(x+3)^2-1$

The graph of $h(x)=-(x+3)^2-1$ is the same as the graph of $y=x^2$ reflected about the x-axis and then shifted left 3 units and down 1 unit.

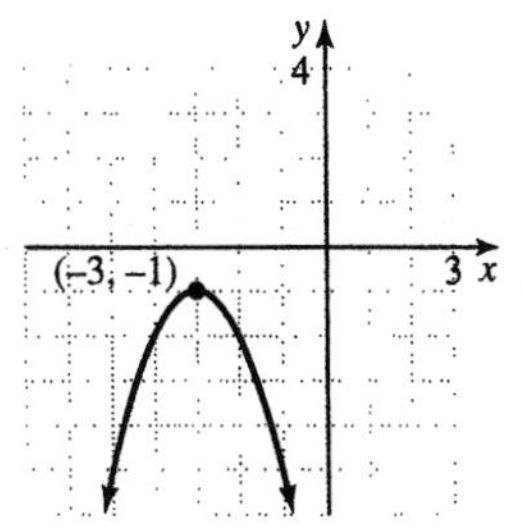

46. $g(x)=|x-2|-2$

The graph of $g(x)=|x-2|-2$ is the same as the graph of $y=|x|$ shifted right 2 units and down 2 units.

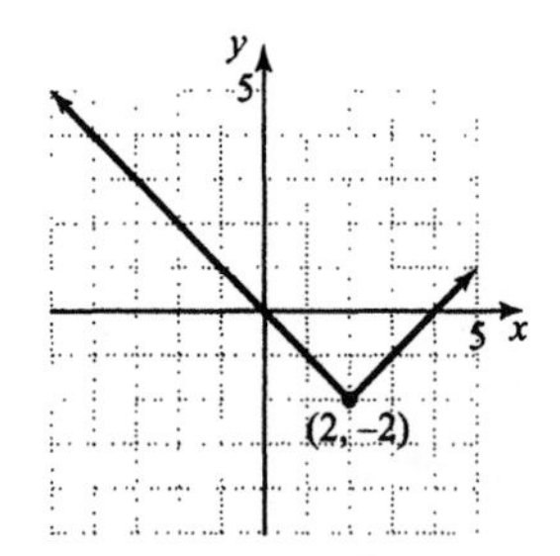

47. $A = kB$

$6 = k(14)$

$k = \frac{6}{14} = \frac{3}{7}$

$A = \frac{3}{7}B$

$A = \frac{3}{7}(21) = 9$

48. $C = \frac{k}{D}$

$12 = \frac{k}{8}$

$96 = k$

$C = \frac{96}{D}$

$C = \frac{96}{24} = 4$

49. $P = \frac{k}{V}$

$1250 = \frac{k}{2}$

$k = 2500$

$P = \frac{2500}{V}$

$800 = \frac{2500}{V}$

$800V = 2500$

$V = 3.125$

When the pressure is 800 kilopascals, the volume is 3.125 cubic meters.

50. $A = kr^2$

$36\pi = k(3)^2$

$36\pi = 9k$

$4\pi = k$

$A = 4\pi r^2$

$A = 4\pi(4)^2 = 64\pi$

When the radius is 4 inches, the surface area is 64π square inches.

51. Slope 0; through $\left(-4, \frac{9}{2}\right)$

A line with slope 0 is horizontal, and a horizontal line has an equation of the form $y = b$, where b is the y-coordinate of any point on the line. The equation is $y = \frac{9}{2}$ or $f(x) = \frac{9}{2}$.

52. Slope $\frac{3}{4}$; through (−8, −4)

$y - y_1 = m(x - x_1)$

$y - (-4) = \frac{3}{4}(x - (-8))$

$y + 4 = \frac{3}{4}(x + 8)$

$4(y + 4) = 3(x + 8)$

$4y + 16 = 3x + 24$

$4y = 3x + 8$

$y = \frac{3}{4}x + 2$

$f(x) = \frac{3}{4}x + 2$

53. Through (−3, 8) and (−2, 3)

Find the slope.

$m = \frac{3-8}{-2-(-3)} = \frac{-5}{1} = -5$

Use the slope and one of the points in the point-slope form. We use (−2, 3).

$y - y_1 = m(x - x_1)$

$y - 3 = -5(x - (-2))$

$y - 3 = -5(x + 2)$

$y - 3 = -5x - 10$

$y = -5x - 7$

$f(x) = -5x - 7$

54. Through (−6, 1); parallel to $y = -\frac{3}{2}x + 11$

The slope of a line parallel to $y = -\frac{3}{2}x + 11$ will have the same slope, $-\frac{3}{2}$.

$$y - y_1 = m(x - x_1)$$
$$y - 1 = -\frac{3}{2}(x - (-6))$$
$$y - 1 = -\frac{3}{2}(x + 6)$$
$$2(y - 1) = -3(x + 6)$$
$$2y - 2 = -3x - 18$$
$$2y = -3x - 16$$
$$y = -\frac{3}{2}x - 8$$
$$f(x) = -\frac{3}{2}x - 8$$

55. Through (−5, 7); perpendicular to $5x - 4y = 10$
Find the slope of $5x - 4y = 10$.

$$5x - 4y = 10$$
$$-4y = -5x + 10$$
$$y = \frac{5}{4}x - \frac{5}{2}$$

The slope is $\frac{5}{4}$. The slope of any line perpendicular to this line is the negative reciprocal of $\frac{5}{4}$, or $-\frac{4}{5}$.

$$y - y_1 = m(x - x_1)$$
$$y - 7 = -\frac{4}{5}(x - (-5))$$
$$y - 7 = -\frac{4}{5}(x + 5)$$
$$5(y - 7) = -4(x + 5)$$
$$5y - 35 = -4x - 20$$
$$5y = -4x + 15$$
$$y = -\frac{4}{5}x + 3$$
$$f(x) = -\frac{4}{5}x + 3$$

56. $g(x) = \begin{cases} 4x - 3 & \text{if } x \le 1 \\ 2x & \text{if } x > 1 \end{cases}$

For $x \le 1$:

x	$g(x)$
−1	−7
0	−3
1	1

For $x > 1$:

x	$g(x)$
2	4
3	6
4	8

Graph a closed circle at (1, 1). Graph an open circle at (1, 2), which is found by substituting 1 for x in $g(x) = 2x$.

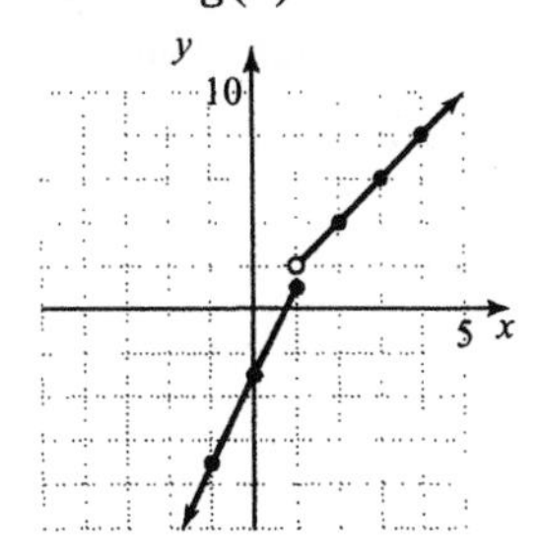

57. $f(x) = \begin{cases} x - 2 & \text{if } x \le 0 \\ -\frac{x}{3} & \text{if } x \ge 3 \end{cases}$

For $x \le 0$:

x	$f(x)$
−2	−4
−1	−3
0	−2

For $x \ge 3$:

x	$f(x)$
3	−1
4	$-\frac{4}{3}$
6	−2

Graph closed circles at (0, −2) and (3, −1).

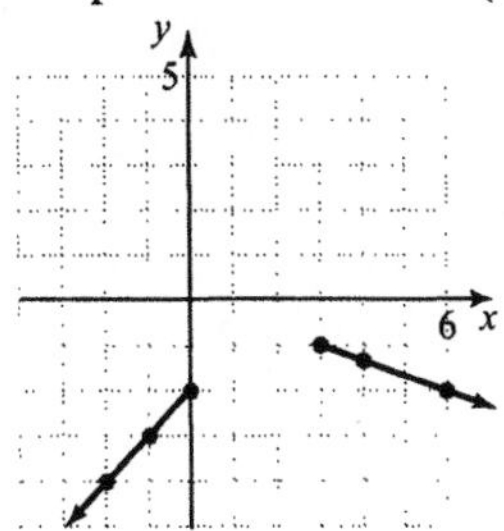

58. $f(x) = |x + 1| - 3$

The graph of $f(x) = |x + 1| - 3$ is the same as the graph of $y = |x|$ shifted left 1 unit and down 3 units.

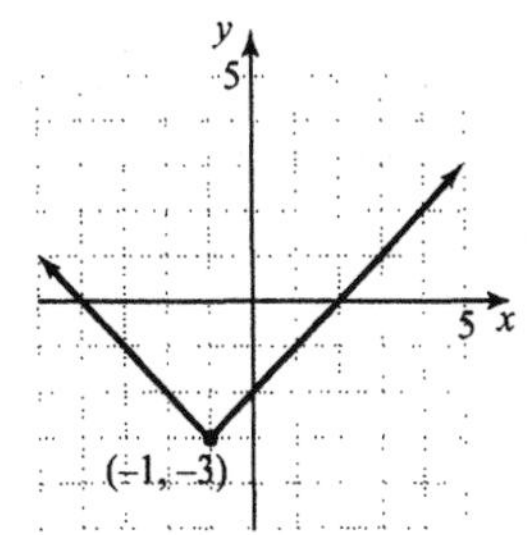

59. $f(x) = \sqrt{x-2}$

The graph of $f(x) = \sqrt{x-2}$ is the same as the graph of $y = \sqrt{x}$ shifted right 2 units.

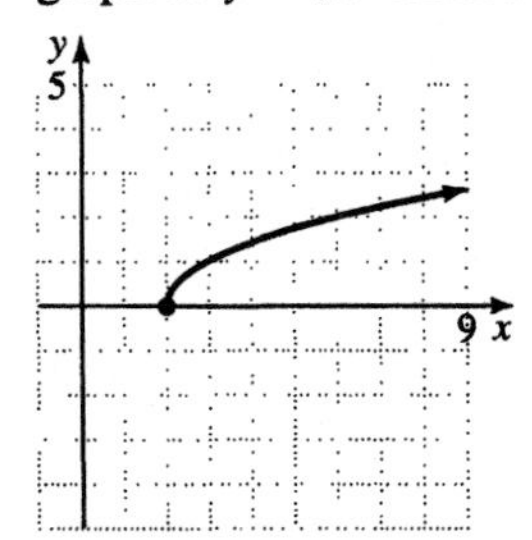

60.
$$y = \frac{k}{x}$$
$$14 = \frac{k}{6}$$
$$84 = k$$
$$y = \frac{84}{x}$$
$$y = \frac{84}{21} = 4$$

Chapter 8 Test

1. When $x = 1$, y or $f(x)$ is 3, so $f(1) = 3$.

2. When $x = -3$, y or $f(x)$ is -5, so $f(-3) = -5$.

3. The x-values that correspond to a y-value of 0 are -2 and 2, so $f(x) = 0$ for $x = -2$ and $x = 2$.

4. The x-value that corresponds to a y-value of 4 is 0, so $f(x) = 4$ for $x = 0$.

5.
$$2x - 3y = -6$$
$$-3y = -2x - 6$$
$$y = \frac{2}{3}x + 2$$
$$m = \frac{2}{3},\ b = 2$$

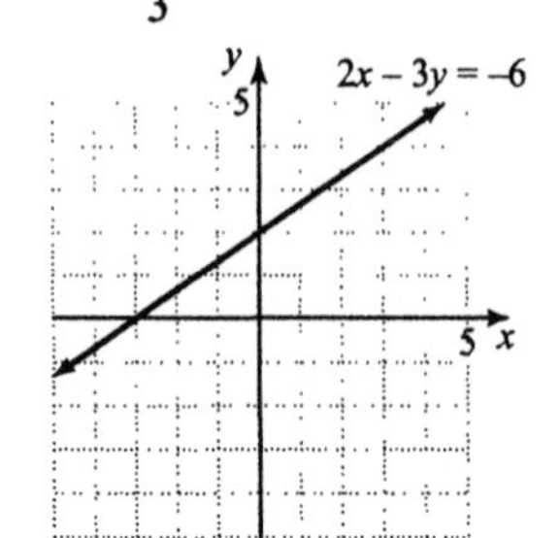

6. $f(x) = \frac{2}{3}x$ or $y = \frac{2}{3}x$

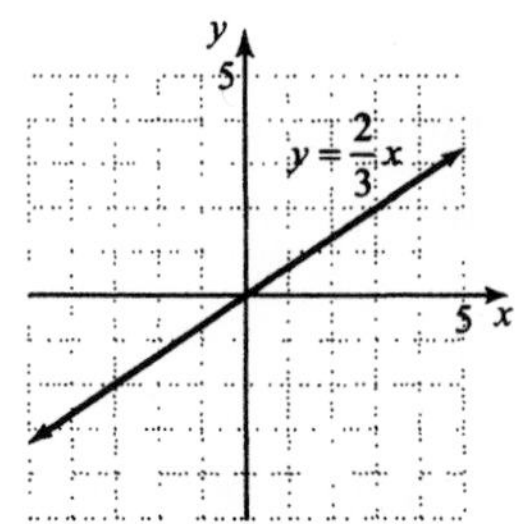

7. Horizontal; through $(2, -8)$
A horizontal line has an equation of the form $y = b$, where b is the y-coordinate of any point on the line. The equation is $y = -8$.

8.
$$y - y_1 = m(x - x_1)$$
$$y - (-1) = -3(x - 4)$$
$$y + 1 = -3x + 12$$
$$3x + y = 11$$

9.
$$y - y_1 = m(x - x_1)$$
$$y - (-2) = 5(x - 0)$$
$$y + 2 = 5x$$
$$5x - y = 2$$

10.
$$m = \frac{-3-(-2)}{6-4} = \frac{-1}{2} = -\frac{1}{2}$$
$$y - y_1 = m(x - x_1)$$
$$y - (-2) = -\frac{1}{2}(x - 4)$$
$$2(y + 2) = -(x - 4)$$
$$2y + 4 = -x + 4$$
$$2y = -x$$
$$y = -\frac{1}{2}x$$
$$f(x) = -\frac{1}{2}x$$

11.
$$3x - y = 4$$
$$y = 3x - 4$$
$$m = 3 \text{ so } m_{\perp} = -\frac{1}{3}$$
$$y - y_1 = m(x - x_1)$$
$$y - 2 = -\frac{1}{3}[x - (-1)]$$
$$3(y - 2) = -(x + 1)$$
$$3y - 6 = -x - 1$$
$$3y = -x + 5$$
$$y = -\frac{1}{3}x + \frac{5}{3}$$
$$f(x) = -\frac{1}{3}x + \frac{5}{3}$$

12. $2y + x = 3$

$$2y = -x + 3$$

$$y = -\frac{1}{2}x + 3 \text{ so } m = -\frac{1}{2}$$

$$y - y_1 = m(x - x_1)$$

$$y - (-2) = -\frac{1}{2}(x - 3)$$

$$2(y + 2) = -(x - 3)$$

$$2y + 4 = -x + 3$$

$$2y = -x - 1$$

$$y = -\frac{1}{2}x - \frac{1}{2}$$

$$f(x) = -\frac{1}{2}x - \frac{1}{2}$$

13. $2x - 5y = 8$

$$-5y = -2x + 8$$

$$y = \frac{2}{5}x - \frac{8}{5} \text{ so } m_1 = \frac{2}{5}$$

$$m_2 = \frac{-1-4}{-1-1} = \frac{-5}{-2} = \frac{5}{2}$$

Therefore, lines L_1 and L_2 are neither parallel nor perpendicular since their slopes are not equal and the product of their slopes is not -1.

14. Domain: $(-\infty, \infty)$
Range: $\{5\}$
Function since it passes the vertical line test.

15. Domain: $\{-2\}$
Range: $(-\infty, \infty)$
Not a function since it fails the vertical line test.

16. Domain: $(-\infty, \infty)$
Range: $[0, \infty)$
Function since it passes the vertical line test.

17. Domain: $(-\infty, \infty)$
Range: $(-\infty, \infty)$
Function since it passes the vertical line test.

18. $f(x) = 1031x + 25{,}193$

a. $x = 0$
$f(0) = 1031(0) + 25{,}193 = 25{,}193$
The average earnings in 2000 were $25,193.

b. $x = 2007 - 2000 = 7$
$f(7) = 1031(7) + 25{,}193 = 32{,}410$
The average earnings in 2007 were $32,410.

c. $40{,}000 \le 1031x + 25{,}193$

$$14{,}807 \le 1031x$$

$$14.4 \le x$$

$2000 + 15 = 2015$
The average earnings will be greater than $40,000 in 2015.

d. slope = 1031; the yearly earnings for high school graduates increases $1031 per year.

e. (0, 25,193); the yearly earnings for a high school graduate in 2000 were $25,193.

19. $f(x) = \begin{cases} -\frac{1}{2}x & \text{if } x \le 0 \\ 2x - 3 & \text{if } x > 0 \end{cases}$

For $x \le 0$:

x	$f(x)$
−4	2
−2	1
0	0

For $x > 0$:

x	$f(x)$
1	−1
2	1
3	3

Graph a closed circle at (0, 0). Graph an open circle at (0, −3), which is found by substituting 0 for x in $f(x) = 2x - 3$.

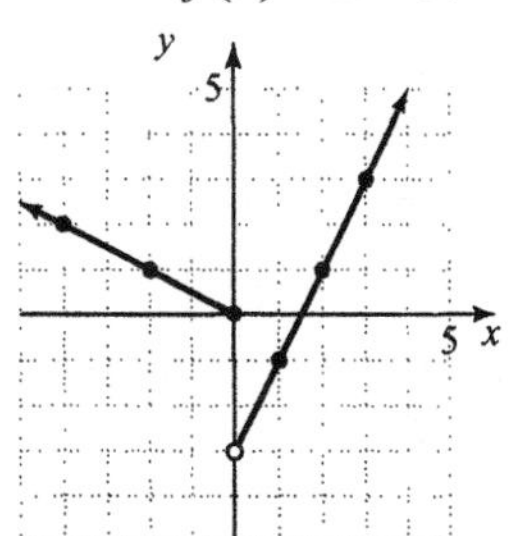

Domain: $(-\infty, \infty)$;
range: $(-3, \infty)$

20. $f(x) = (x - 4)^2$

The graph of $f(x) = (x - 4)^2$ is the same as the graph of $y = x^2$ shifted right 4 units.

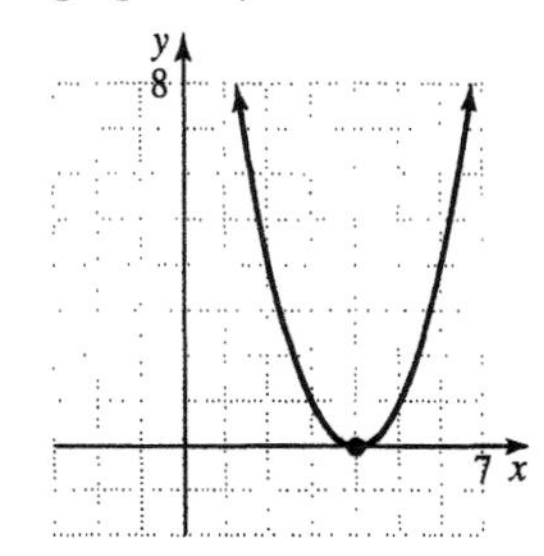

21. $g(x) = -|x+2|-1$

The graph of $g(x) = -|x+2|-1$ is the same as the graph of $y = |x|$ reflected about the x-axis and then shifted left 2 units and down 1 unit.

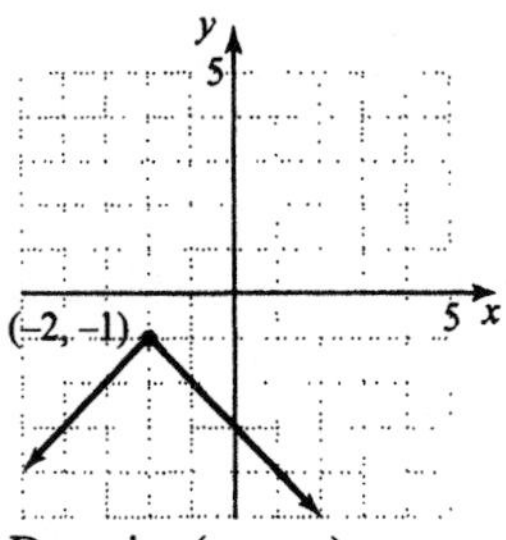

Domain: $(-\infty, \infty)$; range: $(-\infty, -1]$

22. $h(x) = \sqrt{x} - 1$

The graph of $h(x) = \sqrt{x} - 1$ is the same as the graph of $y = \sqrt{x}$ shifted down 1 unit.

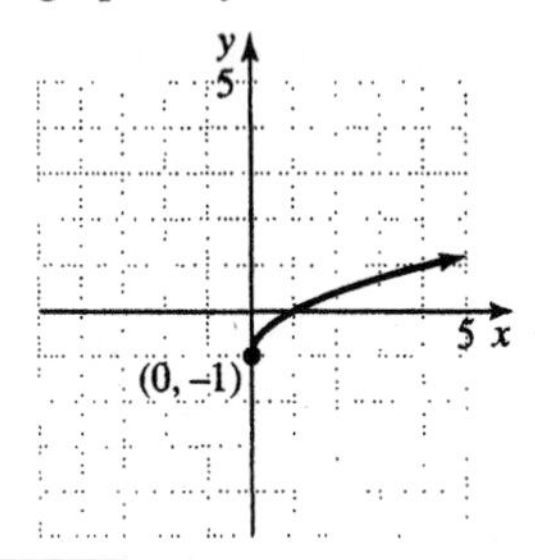

23. $W = \frac{k}{V}$

$20 = \frac{k}{12}$

$240 = k$

$W = \frac{240}{V}$

$W = \frac{240}{15} = 16$

24. $Q = kRS^2$

$24 = k(3)(4)^2$

$24 = 48k$

$\frac{1}{2} = k$

$Q = \frac{1}{2}RS^2$

$Q = \frac{1}{2}(2)(3)^2 = \frac{1}{2}(2)(9) = 9$

25. $s = k\sqrt{d}$

$160 = k\sqrt{400}$

$160 = 20k$

$8 = k$

$s = 8\sqrt{d}$

$128 = 8\sqrt{d}$

$16 = \sqrt{d}$

$16^2 = \left(\sqrt{d}\right)^2$

$256 = d$

The cliff is 256 feet tall.

Chapter 8 Cumulative Review

1. $3[4+2(10-1)] = 3[4+2(9)]$
$= 3[4+18]$
$= 3[22]$
$= 66$

2. $5[3+6(8-5)] = 5[3+6(3)]$
$= 5[3+18]$
$= 5[21]$
$= 105$

3. Let $x = 2$ and $y = -5$.

a. $\frac{x-y}{12+x} = \frac{2-(-5)}{12+2} = \frac{2+5}{14} = \frac{7}{14} = \frac{1}{2}$

b. $x^2 - 3y = 2^2 - 3(-5) = 4 + 15 = 19$

4. Let $x = 2$ and $y = -5$.

a. $\frac{x+y}{3y} = \frac{2+(-5)}{3(-5)} = \frac{-3}{-15} = \frac{1}{5}$

b. $y^2 - x = (-5)^2 - 2 = 25 - 2 = 23$

5. $-3x = 33$

$\frac{-3x}{-3} = \frac{33}{-3}$

$x = -11$

6. $\frac{2}{3}y = 7$

$\frac{3}{2}\left(\frac{2}{3}y\right) = \frac{3}{2}(7)$

$y = \frac{21}{2}$

7.
$$\begin{aligned}8(2-t)&=-5t\\16-8t&=-5t\\16-8t+8t&=-5t+8t\\16&=3t\\\frac{16}{3}&=\frac{3t}{3}\\\frac{16}{3}&=t\end{aligned}$$

8.
$$\begin{aligned}5x-9&=5x-29\\5x-5x-9&=5x-5x-29\\-9&=-29\end{aligned}$$
This is a false statement, so the equation has no solution.

9.
$$\begin{aligned}y&=mx+b\\y-b&=mx\end{aligned}$$
$\frac{y-b}{m}=x$ or $x=\frac{y-b}{m}$

10.
$$\begin{aligned}y&=7x-2\\y+2&=7x\end{aligned}$$
$\frac{y+2}{7}=x$ or $x=\frac{y+2}{7}$

11.
$$\begin{aligned}-4x+7&\ge-9\\-4x+7-7&\ge-9-7\\-4x&\ge-16\\\frac{-4x}{-4}&\le\frac{-16}{-4}\\x&\le4\end{aligned}$$
$(-\infty, 4]$

12.
$$\begin{aligned}-5x-6&<3x+1\\-5x-3x-6&<3x-3x+1\\-8x-6&<1\\-8x-6+6&<1+6\\-8x&<7\\\frac{-8x}{-8}&>\frac{7}{-8}\\x&>-\frac{7}{8}\end{aligned}$$
$\left(-\frac{7}{8}, \infty\right)$

13. $(-1, 7), (2, 2)$
$$m=\frac{2-7}{2-(-1)}=\frac{-5}{3}=-\frac{5}{3}$$
The slope of a perpendicular line is $\frac{3}{5}$.

14. $(0, 7), (-1, 0)$
$$m=\frac{0-7}{-1-0}=\frac{-7}{-1}=7$$
The slope of a parallel line is 7.

15. $g(x)=x^2-3$

a. $g(2)=2^2-3=4-3=1$

b. $g(-2)=(-2)^2-3=4-3=1$

c. $g(0)=0^2-3=0-3=-3$

16. $f(x)=3-x^2$

a. $f(2)=3-2^2=3-4=-1$

b. $f(-2)=3-(-2)^2=3-4=-1$

c. $f(0)=3-0^2=3-0=3$

17. $\begin{cases}2x+y=10\\x=y+2\end{cases}$

Substitute $y + 2$ for x in the first equation.
$$\begin{aligned}2(y+2)+y&=10\\2y+4+y&=10\\4+3y&=10\\3y&=6\\y&=2\end{aligned}$$
Let $y = 2$ in the second equation.
$x = 2 + 2 = 4$
The solution is (4, 2).

18. $\begin{cases}3y=x+10\\2x+5y=24\end{cases}$

Solve the first equation for x.
$x = 3y - 10$
Substitute $3y - 10$ for x in the second equation.
$$\begin{aligned}2(3y-10)+5y&=24\\6y-20+5y&=24\\11y-20&=24\\11y&=44\\y&=4\end{aligned}$$
Let $y = 4$ in $x = 3y - 10$.
$x = 3(4) - 10 = 12 - 10 = 2$
The solution is (2, 4).

19. $\begin{cases} -x - \frac{y}{2} = \frac{5}{2} \\ -\frac{x}{2} + \frac{y}{4} = 0 \end{cases}$

Multiply the first equation by 2 and the second equation by 4, then add.

$$\begin{aligned} -2x - y &= 5 \\ -2x + y &= 0 \\ \hline -4x \quad &= 5 \\ x &= -\frac{5}{4} \end{aligned}$$

Let $x = -\frac{5}{4}$ in the first equation.

$$\begin{aligned} -\left(-\frac{5}{4}\right) - \frac{y}{2} &= \frac{5}{2} \\ \frac{5}{4} - \frac{y}{2} &= \frac{5}{2} \\ -\frac{y}{2} &= \frac{10}{4} - \frac{5}{4} \\ -\frac{y}{2} &= \frac{5}{4} \\ -2\left(-\frac{y}{2}\right) &= -2\left(\frac{5}{4}\right) \\ y &= -\frac{5}{2} \end{aligned}$$

The solution is $\left(-\frac{5}{4}, -\frac{5}{2}\right)$.

20. $\begin{cases} \frac{x}{2} + y = \frac{5}{6} \\ 2x - y = \frac{5}{6} \end{cases}$

Multiply both equations by 6, then add.

$$\begin{aligned} 3x + 6y &= 5 \\ 12x - 6y &= 5 \\ \hline 15x \quad &= 10 \\ x &= \frac{2}{3} \end{aligned}$$

Let $x = \frac{2}{3}$ in the first equation.

$$\begin{aligned} \frac{\frac{2}{3}}{2} + y &= \frac{5}{6} \\ \frac{1}{3} + y &= \frac{5}{6} \\ y &= \frac{5}{6} - \frac{2}{6} \\ y &= \frac{3}{6} = \frac{1}{2} \end{aligned}$$

The solution is $\left(\frac{2}{3}, \frac{1}{2}\right)$.

21.
$$\begin{array}{r} x+4 \\ x+3\overline{\smash{)}x^2+7x+12} \\ \underline{x^2+3x} \\ 4x+12 \\ \underline{4x+12} \\ 0 \end{array}$$

$$\frac{x^2+7x+12}{x+3} = x+4$$

22. $\frac{5x^2y - 6xy + 2}{6xy} = \frac{5x^2y}{6xy} - \frac{6xy}{6xy} + \frac{2}{6xy}$
$= \frac{5x}{6} - 1 + \frac{1}{3xy}$

23. a. $6t + 18 = 6 \cdot t + 6 \cdot 3 = 6(t + 3)$

b. $y^5 - y^7 = y^5 \cdot 1 - y^5 \cdot y^2 = y^5(1 - y^2)$

24. a. $5y - 20 = 5 \cdot y - 5 \cdot 4 = 5(y - 4)$

b. $z^{10} - z^3 = z^3 \cdot z^7 - z^3 \cdot 1 = z^3(z^7 - 1)$

25. $x^2 + 4x - 12 = (x - 2)(x + 6)$

26. $x^2 - 10x + 21 = (x - 7)(x - 3)$

27. $10x^2 - 13xy - 3y^2 = 10x^2 - 15xy + 2xy - 3y^2$
$= 5x(2x - 3y) + y(2x - 3y)$
$= (2x - 3y)(5x + y)$

28. $12a^2 + 5ab - 2b^2 = 12a^2 - 3ab + 8ab - 2b^2$
$= 3a(4a - b) + 2b(4a - b)$
$= (4a - b)(3a + 2b)$

29. $x^3+8=x^3+2^3$
$=(x+2)(x^2-x\cdot 2+2^2)$
$=(x+2)(x^2-2x+4)$

30. $y^3-27=y^3-3^3$
$=(y-3)(y^2+y\cdot 3+3^2)$
$=(y-3)(y^2+3y+9)$

31. $x^2-9x-22=0$
$(x+2)(x-11)=0$
$x+2=0$ or $x-11=0$
$x=-2$ $\quad x=11$
The solutions are –2 and 11.

32. $y^2-5y=-6$
$y^2-5y+6=0$
$(y-2)(y-3)=0$
$y-2=0$ or $y-3=0$
$y=2$ $\quad y=3$
The solutions are 2 and 3.

33. a. $\dfrac{2x^2}{10x^3-2x^2}=\dfrac{2x^2}{2x^2(5x-1)}=\dfrac{1}{5x-1}$

b. $\dfrac{9x^2+13x+4}{8x^2+x-7}=\dfrac{(9x+4)(x+1)}{(8x-7)(x+1)}=\dfrac{9x+4}{8x-7}$

34. a. $\dfrac{33x^4y^2}{3xy}=\dfrac{33}{3}x^{4-1}y^{2-1}=11x^3y^1=11x^3y$

b. $\dfrac{9y}{90y^2+9y}=\dfrac{9y}{9y(10y+1)}=\dfrac{1}{10y+1}$

35. $\dfrac{3x+3}{5x-5x^2}\cdot\dfrac{2x^2+x-3}{4x^2-9}$
$=\dfrac{3(x+1)\cdot(2x+3)(x-1)}{-5x(x-1)\cdot(2x+3)(2x-3)}$
$=-\dfrac{3(x+1)}{5x(2x-3)}$

36. $\dfrac{2x}{x-6}-\dfrac{x+6}{x-6}=\dfrac{2x-(x+6)}{x-6}$
$=\dfrac{2x-x-6}{x-6}$
$=\dfrac{x-6}{x-6}$
$=1$

37. $\dfrac{3x^2+2x}{x-1}-\dfrac{10x-5}{x-1}=\dfrac{3x^2+2x-(10x-5)}{x-1}$
$=\dfrac{3x^2+2x-10x+5}{x-1}$
$=\dfrac{3x^2-8x+5}{x-1}$
$=\dfrac{(3x-5)(x-1)}{x-1}$
$=3x-5$

38. $\dfrac{9}{y^2}-4y=\dfrac{9}{y^2}-\dfrac{4y}{1}\cdot\dfrac{y^2}{y^2}=\dfrac{9}{y^2}-\dfrac{4y^3}{y^2}=\dfrac{9-4y^3}{y^2}$

39. $3-\dfrac{6}{x}=x+8$
$x\left(3-\dfrac{6}{x}\right)=x(x+8)$
$3x-6=x^2+8x$
$0=x^2+5x+6$
$0=(x+3)(x+2)$
$x+3=0$ or $x+2=0$
$x=-3$ $\quad x=-2$
The solutions are –3 and –2.

40. $\dfrac{x}{2}+\dfrac{x}{5}=\dfrac{x-7}{20}$
$20\left(\dfrac{x}{2}+\dfrac{x}{5}\right)=20\left(\dfrac{x-7}{20}\right)$
$10x+4x=x-7$
$14x=x-7$
$13x=-7$
$x=-\dfrac{7}{13}$
The solution is $-\dfrac{7}{13}$.

41. (4, 0), (–4, –5)

$$m = \frac{-5-0}{-4-4} = \frac{-5}{-8} = \frac{5}{8}$$

$$y - y_1 = m(x - x_1)$$

$$y - 0 = \frac{5}{8}(x-4)$$

$$y = \frac{5}{8}x - \frac{5}{2}$$

$$f(x) = \frac{5}{8}x - \frac{5}{2}$$

42. (–1, 3), (–2, 7)

$$m = \frac{7-3}{-2-(-1)} = \frac{4}{-2+1} = \frac{4}{-1} = -4$$

$$y - y_1 = m(x - x_1)$$

$$y - 3 = -4[x-(-1)]$$

$$y - 3 = -4(x+1)$$

$$y - 3 = -4x - 4$$

$$y = -4x - 1$$

$$f(x) = -4x - 1$$

Chapter 9

Section 9.1

Practice Exercises

1. $A = \{1, 3, 5, 7, 9\}$ and $B = \{1, 2, 3, 4\}$
The numbers 1 and 3 are in sets A and B.
The intersection is $\{1, 3\}$. $A \cap B = \{1, 3\}$.

2.
$$\begin{aligned} x+3<8 \quad &\text{and} \quad 2x-1<3 \\ x<5 \quad &\text{and} \quad 2x<4 \\ x<5 \quad &\text{and} \quad x<2 \end{aligned}$$

$\{x|x<5\}$, $(-\infty, 5)$

5

$\{x|x<2\}$, $(-\infty, 2)$

2

$\{x|x<5 \text{ and } x<2\} = \{x|x<2\}$

2

The solution set is $(-\infty, 2)$.

3.
$$\begin{aligned} 4x \le 0 \quad &\text{and} \quad 3x+2>8 \\ x \le 0 \quad &\text{and} \quad 3x>6 \\ x \le 0 \quad &\text{and} \quad x>2 \end{aligned}$$

$\{x|x \le 0\}$, $(-\infty, 0]$

0

$\{x|x>2\}$, $(2, \infty)$

2

$\{x|4x \le 0 \text{ and } 3x+2>8\} = \{\ \}$ or $\varnothing$

4.
$$\begin{aligned} 3 &< 5-x < 9 \\ 3-5 &< 5-x-5 < 9-5 \\ -2 &< -x < 4 \\ \frac{-2}{-1} &> \frac{-x}{-1} > \frac{4}{-1} \\ 2 &> x > -4 \end{aligned}$$
or $-4 < x < 2$
The solution set is $(-4, 2)$.

5.
$$\begin{aligned} -4 &\le \frac{x}{2}-1 \le 3 \\ 2(-4) &\le 2\left(\frac{x}{2}-1\right) \le 2(3) \\ -8 &\le x-2 \le 6 \\ -8+2 &\le x-2+2 \le 6+2 \\ -6 &\le x \le 8 \end{aligned}$$
The solution set is $[-6, 8]$.

6. $A = \{1, 3, 5, 7, 9\}$ and $B = \{2, 3, 4, 5, 6\}$.
The numbers that are in either set or both sets are $\{1, 2, 3, 4, 5, 6, 7, 9\}$. This set is the union, $A \cup B$.

7.
$$\begin{aligned} 8x+5 \le 8 \quad &\text{or} \quad x-1 \ge 2 \\ 8x \le 3 \quad &\text{or} \quad x \ge 3 \\ x \le \frac{3}{8} \quad &\text{or} \quad x \ge 3 \end{aligned}$$

$\left\{x \middle| x \le \frac{3}{8}\right\}, \left(-\infty, \frac{3}{8}\right]$

$\frac{3}{8}$

$\{x|x \ge 3\}$, $[3, \infty)$

3

$\left\{x \middle| x \le \frac{3}{8} \text{ or } x \ge 3\right\} = \left(-\infty, \frac{3}{8}\right] \cup [3, \infty)$

$\frac{3}{8}$ 3

The solution set is $\left(-\infty, \frac{3}{8}\right] \cup [3, \infty)$.

8.
$$\begin{aligned} -3x-2>-8 \quad &\text{or} \quad 5x>0 \\ -3x>-6 \quad &\text{or} \quad x>0 \\ x<2 \quad &\text{or} \quad x>0 \end{aligned}$$

$\{x|x<2\}$, $(-\infty, 2)$

2

$\{x|x>0\}$, $(0, \infty)$

0

$\{x|x<2 \text{ or } x>0\}$, $(-\infty, \infty)$

The solution set is $(-\infty, \infty)$.

Vocabulary and Readiness Check

1. Two inequalities joined by the words "and" or "or" are called compound inequalities.
2. The word and means intersection.
3. The word or means union.
4. The symbol $\cap$ means intersection.
5. The symbol $\cup$ represents union.

6. The symbol $\varnothing$ is the empty set.

7. The inequality $-2 \le x < 1$ means $-2 \le x$ and $x < 1$.

8. $\{x|x < 0 \text{ and } x > 0\} = \varnothing$.

Exercise Set 9.1

1. $C \cup D = \{2, 3, 4, 5, 6, 7\}$

3. $A \cap D = \{4, 6\}$

5. $A \cup B = \{..., -2, -1, 0, 1, ...\}$

7. $B \cap D = \{5, 7\}$

9. $B \cup C = \{x|x \text{ is an odd integer or } x = 2 \text{ or } x = 4\}$

11. $A \cap C = \{2, 4\}$

13. $x < 1$ and $x > -3$
$-3 < x < 1$
$(-3, 1)$

15. $x \le -3$ and $x \ge -2$
$\varnothing$

17. $x < -1$ and $x < 1$
$x < -1$
$(-\infty, -1)$

19. $x + 1 \ge 7$ and $3x - 1 \ge 5$
$x \ge 6$ and $3x \ge 6$
$x \ge 2$

$x \ge 6$
$[6, \infty)$

21. $4x + 2 \le -10$ and $2x \le 0$
$4x \le -12$ and $x \le 0$
$x \le -3$

$x \le -3$
$(-\infty, -3]$

23. $-2x < -8$ and $x - 5 < 5$
$x > 4$ and $x < 10$
$(4, 10)$

25. $5 < x - 6 < 11$
$11 < x < 17$
$(11, 17)$

27. $-2 \le 3x - 5 \le 7$
$3 \le 3x \le 12$
$1 \le x \le 4$
$[1, 4]$

29. $1 \le \frac{2}{3}x + 3 \le 4$
$-2 \le \frac{2}{3}x \le 1$
$-3 \le x \le \frac{3}{2}$
$\left[-3, \frac{3}{2}\right]$

31. $-5 \le \frac{-3x+1}{4} \le 2$
$4(-5) \le 4\left(\frac{-3x+1}{4}\right) \le 4(2)$
$-20 \le -3x + 1 \le 8$
$-21 \le -3x \le 7$
$7 \ge x \ge -\frac{7}{3}$
$-\frac{7}{3} \le x \le 7$
$\left[-\frac{7}{3}, 7\right]$

33. $x < 4$ or $x < 5$
$(-\infty, 5)$

35. $x \le -4$ or $x \ge 1$
$(-\infty, -4] \cup [1, \infty)$

37. $x > 0$ or $x < 3$
$(-\infty, \infty)$

39. $-2x \le -4$ or $5x - 20 \ge 5$
$x \ge 2$ or $5x \ge 25$
$x \ge 5$

$x \ge 2$
$[2, \infty)$

41. $x+4<0$ or $6x>-12$
$x<-4$ or $x>-2$
$(-\infty, -4) \cup (-2, \infty)$

43. $3(x-1)<12$ or $x+7>10$
$x-1<4$ or $x>3$
$x<5$
$(-\infty, \infty)$

45. $x<\frac{2}{3}$ and $x>-\frac{1}{2}$
$-\frac{1}{2}<x<\frac{2}{3}$
$\left(-\frac{1}{2}, \frac{2}{3}\right)$

47. $x<\frac{2}{3}$ or $x>-\frac{1}{2}$
$(-\infty, \infty)$

49. $0 \le 2x-3 \le 9$
$3 \le 2x \le 12$
$\frac{3}{2} \le x \le 6$
$\left[\frac{3}{2}, 6\right]$

51. $\frac{1}{2}<x-\frac{3}{4}<2$
$4\left(\frac{1}{2}\right)<4\left(x-\frac{3}{4}\right)<4(2)$
$2<4x-3<8$
$5<4x<11$
$\frac{5}{4}<x<\frac{11}{4}$
$\left(\frac{5}{4}, \frac{11}{4}\right)$

53. $x+3 \ge 3$ and $x+3 \le 2$
$x \ge 0$ and $x \le -1$
No solution exists.
$\varnothing$

55. $3x \ge 5$ or $-\frac{5}{8}x-6>1$
$x \ge \frac{5}{3}$ or $-\frac{5}{8}x>7$
$x<-\frac{56}{5}$
$\left(-\infty, -\frac{56}{5}\right) \cup \left[\frac{5}{3}, \infty\right)$

57. $0<\frac{5-2x}{3}<5$
$0<5-2x<15$
$\frac{-5}{-2}>\frac{-2x}{-2}>\frac{10}{-2}$
$\frac{5}{2}>x>-5$
$-5<x<\frac{5}{2}$
$\left(-5, \frac{5}{2}\right)$

59. $-6<3(x-2) \le 8$
$-6<3x-6 \le 8$
$0<3x \le 14$
$0<x<\frac{14}{3}$
$\left(0, \frac{14}{3}\right]$

61. $-x+5>6$ and $1+2x \le -5$
$-x>1$ and $2x \le -6$
$x<-1$ and $x \le -3$
$x \le -3$
$(-\infty, -3]$

63. $3x+2 \le 5$ or $7x>29$
$3x \le 3$ or $x>\frac{29}{7}$
$x \le 1$ or $x>\frac{29}{7}$
$(-\infty, 1] \cup \left(\frac{29}{7}, \infty\right)$

65. $5-x>7$ and $2x+3 \ge 13$
$-x>2$ and $2x \ge 10$
$x<-2$ and $x \ge 5$
No solution exists.
$\varnothing$

67. $-\frac{1}{2} \le \frac{4x-1}{6} < \frac{5}{6}$

$6\left(-\frac{1}{2}\right) \le 6\left(\frac{4x-1}{6}\right) < 6\left(\frac{5}{6}\right)$

$-3 \le 4x-1 < 5$

$-2 \le 4x < 6$

$-\frac{1}{2} \le x < \frac{3}{2}$

$\left[-\frac{1}{2}, \frac{3}{2}\right)$

69. $\frac{1}{15} < \frac{8-3x}{15} < \frac{4}{5}$

$15\left(\frac{1}{15}\right) < 15\left(\frac{8-3x}{15}\right) < 15\left(\frac{4}{5}\right)$

$1 < 8-3x < 12$

$-7 < -3x < 4$

$-\frac{4}{3} < x < \frac{7}{3}$

$\left(-\frac{4}{3}, \frac{7}{3}\right)$

71. $0.3 < 0.2x - 0.9 < 1.5$

$1.2 < 0.2x < 2.4$

$6 < x < 12$

$(6, 12)$

73. $|-7| - |19| = 7 - 19 = -12$

75. $-(-6) - |-10| = 6 - 10 = -4$

77. $|x| = 7$

$x = -7, 7$

79. $|x| = 0$

$x = 0$

81. The years that the consumption of bottled water was greater than 20 gallons per person were 2003, 2004, and 2005. The years that consumption of diet soda was greater than 14 gallons per person were 2003, 2004, and 2005. The years in common are 2003, 2004, and 2005.

83. The number of jobs in 2000 was greater than 1800 thousand for registered nurses. The predicted number of jobs in 2012 is greater than 1800 thousand for post secondary teachers and registered nurses. The intersection of these is registered nurses.

85. $-29 \le C \le 35$

$-29 \le \frac{5}{9}(F-32) \le 35$

$-52.5 \le F - 32 \le 63$

$-20.2 \le F \le 95$

$-20.2° \le F \le 95°$

87. $70 \le \frac{68+65+75+78+2x}{6} \le 79$

$420 \le 286 + 2x \le 474$

$134 \le 2x \le 188$

$67 \le x \le 94$

If Christian scores between 67 and 94 inclusive on his final exam, he will receive a C in the course.

89. $2x - 3 < 3x + 1 < 4x - 5$

$2x-3 < 3x+1$ and $3x+1 < 4x-5$

$-x < 4$ and $-x < -6$

$x > -4$ and $x > 6$

$x > 6$

$(6, \infty)$

91. $-3(x-2) \le 3 - 2x \le 10 - 3x$

$-3x+6 \le 3-2x$ and $3-2x \le 10-3x$

$-x \le -3$ and $x \le 7$

$x \ge 3$

$3 \le x \le 7$

$[3, 7]$

93. $5x - 8 < 2(2 + x) < -2(1 + 2x)$

$5x-8 < 4+2x$ and $4+2x < -2-4x$

$3x < 12$ and $6x < -6$

$x < 4$ and $x < -1$

$x < -1$

$(-\infty, -1)$

The Bigger Picture

1. $-\frac{1}{2} - \left(-\frac{3}{8}\right) = -\frac{1}{2} + \frac{3}{8} = -\frac{4}{8} + \frac{3}{8} = -\frac{1}{8}$

2. $(8xy - 7y^2) - (4xy - y^2) = 8xy - 7y^2 - 4xy + y^2$

$= 4xy - 6y^2$

3. $\dfrac{x+2}{xy-z^2}-\dfrac{x+1}{xy-z^2}=\dfrac{x+2-(x+1)}{xy-z^2}$
$=\dfrac{x+2-x-1}{xy-z^2}$
$=\dfrac{1}{xy-z^2}$

4. $\dfrac{x^3-8}{x-2}\cdot\dfrac{x^2-4}{x-2}$
$=\dfrac{(x-2)(x^2+2x+4)\cdot(x+2)(x-2)}{(x-2)\cdot(x-2)}$
$=(x+2)(x^2+2x+4)$

5. $x-2\le 1$ and $3x-1\ge -4$
$x\le 3$ and $3x\ge -3$
$x\ge -1$

$-1\le x\le 3$
$[-1, 3]$

6. $-2<x-1<5$
$-2+1<x-1+1<5+1$
$-1<x<6$
$(-1, 6)$

7. $-2x+2.5=-7.7$
$-2x=-10.2$
$x=5.1$

8. $-5x>20$
$\dfrac{-5x}{-5}<\dfrac{20}{-5}$
$x<-4$
$(-\infty, -4)$

9. $x\le -3$ or $x\le -5$
$x\le -3$
$(-\infty, -3]$

10. $5x<-10$ or $3x-4>2$
$x<-2$ or $3x>6$
$x>2$

$(-\infty, -2)\cup(2, \infty)$

11. $\dfrac{5t}{2}-\dfrac{3t}{4}=7$
$4\left(\dfrac{5t}{2}-\dfrac{3t}{4}\right)=4(7)$
$2(5t)-3t=28$
$10t-3t=28$
$7t=28$
$t=4$

12. $5(x-3)+x+2\ge 3(x+2)+2x$
$5x-15+x+2\ge 3x+6+2x$
$6x-13\ge 5x+6$
$6x-5x\ge 13+6$
$x\ge 19$

$[19, \infty)$

Section 9.2

Practice Exercises

1. $|q|=7$
$q=7$ or $q=-7$
The solution set is $\{-7, 7\}$.

2. $|2x-3|=5$
$2x-3=5$ or $2x-3=-5$
$2x=8$ or $2x=-2$
$x=4$ or $x=-1$
The solution set is $\{-1, 4\}$.

3. $\left|\dfrac{x}{5}+1\right|=15$

$\dfrac{x}{5}+1=15$ or $\dfrac{x}{5}+1=-15$
$\dfrac{x}{5}=14$ or $\dfrac{x}{5}=-16$
$x=70$ or $x=-80$
The solutions are -80 and 70.

4. $|3x|+8=14$
$|3x|=6$
$3x=6$ or $3x=-6$
$x=2$ or $x=-2$
The solutions are -2 and 2.

5. $|z|=0$
The solution is 0.

6. $3|z|+9=7$

$3|z|=-2$

$|z|=-\frac{2}{3}$

The absolute value of a number is never negative, so there is no solution. The solution set is { } or ∅.

7. $\left|\frac{5x+3}{4}\right|=-8$

The absolute value of a number is never negative, so there is no solution. The solution set is { } or ∅.

8. $|2x+4|=|3x-1|$

$2x+4=3x-1$ or $2x+4=-(3x-1)$

$-x+4=-1$ $\quad$ $2x+4=-3x+1$

$-x=-5$ $\quad$ $5x+4=1$

$x=5$ $\quad$ $5x=-3$

$x=-\frac{3}{5}$

The solutions are $-\frac{3}{5}$ and 5.

9. $|x-2|=|8-x|$

$x-2=8-x$ or $x-2=-(8-x)$

$2x-2=8$ $\quad$ $x-2=-8+x$

$2x=10$ $\quad$ $-2=-8$ False

$x=5$

The solution is 5.

Vocabulary and Readiness Check

1. $|x-2|=5$

C. $x-2=5$ or $x-2=-5$

2. $|x-2|=0$

A. $x-2=0$

3. $|x-2|=|x+3|$

B. $x-2=x+3$ or $x-2=-(x+3)$

4. $|x+3|=5$

E. $x+3=5$ or $x+3=-5$

5. $|x+3|=-5$

D. ∅

Exercise Set 9.2

1. $|x|=7$

$x=7$ or $x=-7$

3. $|3x|=12.6$

$3x=12.6$ or $3x=-12.6$

$x=4.2$ or $x=-4.2$

5. $|2x-5|=9$

$2x-5=9$ or $2x-5=-9$

$2x=14$ or $2x=-4$

$x=7$ or $x=-2$

7. $\left|\frac{x}{2}-3\right|=1$

$\frac{x}{2}-3=1$ or $\frac{x}{2}-3=-1$

$2\left(\frac{x}{2}-3\right)=2(1)$ or $2\left(\frac{x}{2}-3\right)=2(-1)$

$x-6=2$ or $x-6=-2$

$x=8$ or $x=4$

9. $|z|+4=9$

$|z|=5$

$z=-5$ or $z=-5$

11. $|3x|+5=14$

$|3x|=9$

$3x=9$ or $3x=-9$

$x=3$ or $x=-3$

13. $|2x|=0$

$2x=0$

$x=0$

15. $|4n+1|+10=4$

$|4n+1|=-6$ which is impossible.

The solution set is ∅.

17. $|5x-1|=0$

$5x-1=0$

$5x=1$

$x=\frac{1}{5}$

19. $|x|=5$

21. $|5x-7|=|3x+11|$
$5x-7=3x+11$ or $5x-7=-(3x+11)$
$2x=18$ or $5x-7=-3x-11$
$x=9$ or $8x=-4$
$x=-\frac{1}{2}$

23. $|z+8|=|z-3|$
$z+8=z-3$ or $z+8=-(z-3)$
$8=-3$ or $z+8=-z+3$
$2z=-5$
$z=-\frac{5}{2}$
The only solution is $-\frac{5}{2}$.

25. Answers may vary

27. $|x|=4$
$x=4$ or $x=-4$

29. $|y|=0$; $y=0$

31. $|z|=-2$ is impossible. The solution set is $\varnothing$.

33. $|7-3x|=7$
$7-3x=7$ or $7-3x=-7$
$-3x=0$ or $-3x=-14$
$x=0$ or $x=\frac{14}{3}$

35. $|6x|-1=11$
$|6x|=12$
$6x=12$ or $6x=-12$
$x=2$ or $x=-2$

37. $|4p|=-8$ is impossible. The solution set is $\varnothing$.

39. $|x-3|+3=7$
$|x-3|=4$
$x-3=4$ or $x-3=-4$
$x=7$ or $x=-1$

41. $\left|\frac{z}{4}+5\right|=-7$ is impossible. The solution set is $\varnothing$.

43. $|9v-3|=-8$ is impossible. The solution set is $\varnothing$.

45. $|8n+1|=0$
$8n+1=0$
$8n=-1$
$n=-\frac{1}{8}$

47. $|1-6c|-7=-3$
$|1-6c|=4$
$1-6c=4$ or $1-6c=-4$
$6c=3$ or $6c=-5$
$c=\frac{1}{2}$ or $c=-\frac{5}{6}$

49. $|5x+1|=11$
$5x+1=11$ or $5x+1=-11$
$5x=10$ or $5x=-12$
$x=2$ or $x=-\frac{12}{5}$

51. $|4x-2|=|-10|$
$|4x-2|=10$
$4x-2=10$ or $4x-2=-10$
$4x=12$ or $4x=-8$
$x=3$ or $x=-2$

53. $|5x+1|=|4x-7|$
$5x+1=4x-7$ or $5x+1=-(4x-7)$
$x=-8$ or $5x+1=-4x+7$
$9x=6$
$x=\frac{2}{3}$

55. $|6+2x|=-|-7|$
$|6+2x|=-7$ which is impossible. The solution set is $\varnothing$.

57. $|2x-6|=|10-2x|$
$2x-6=10-2x$ or $2x-6=-(10-2x)$
$4x=16$ or $2x-6=-10+2x$
$x=4$ or $-6=-10$
$-6=-10$ is impossible. The only solution is 4.

59. $\left|\frac{2x-5}{3}\right|=7$
$\frac{2x-5}{3}=7$ or $\frac{2x-5}{3}=-7$
$2x-5=21$ or $2x-5=-21$
$2x=26$ or $2x=-16$
$x=13$ or $x=-8$

61. $2+|5n|=17$
$|5n|=15$
$5n=15$ or $5n=-15$
$n=3$ or $n=-3$

63. $\left|\frac{2x-1}{3}\right|=|-5|$
$\left|\frac{2x-1}{3}\right|=5$
$\frac{2x-1}{3}=5$ or $\frac{2x-1}{3}=-5$
$2x-1=15$ or $2x-1=-15$
$2x=16$ or $2x=-14$
$x=8$ or $x=-7$

65. $|2y-3|=|9-4y|$
$2y-3=9-4y$ or $2y-3=-(9-4y)$
$6y=12$ or $2y-3=-9+4y$
$y=2$ or $-2y=-6$
$y=3$

67. $\left|\frac{3n+2}{8}\right|=|-1|$
$\left|\frac{3n+2}{8}\right|=1$
$\frac{3n+2}{8}=1$ or $\frac{3n+2}{8}=-1$
$3n+2=8$ or $3n+2=-8$
$3n=6$ or $3n=-10$
$n=2$ or $n=-\frac{10}{3}$

69. $|x+4|=|7-x|$
$x+4=7-x$ or $x+4=-(7-x)$
$2x=3$ or $x+4=-7+x$
$x=\frac{3}{2}$ or $4=-7$
$4=-7$ is impossible. The only solution is $\frac{3}{2}$.

71. $\left|\frac{8c-7}{3}\right|=-|-5|$
$\left|\frac{8c-7}{3}\right|=-5$ which is impossible.
The solution set is $\varnothing$.

73. Answers may vary

75. 34% of cheese consumption came from cheddar cheese.

77. $33\%\cdot(120 \text{ pounds}) = 0.33(120 \text{ pounds})$
$= 39.6 \text{ pounds}$
We might expect they consumed 39.6 pounds.

79. $|x|\ge -2$
Answers may vary; 0, 1, 2, 3, 4, for example

81. $|y|<0$
No solution

83. $|x-7|=2$

85. $|2x-1|=4$

87. $|ax+b|=c$

a. one solution if $c=0$

b. no solutions if c is a negative number

c. two solutions if c is a positive number

Section 9.3

Practice Exercises

1. $|x|<2$
The solution set of this inequality contains all numbers whose distance from 0 is less than 2. The solution set is $(-2, 2)$.

2. $|b+1|<3$
$-3<b+1<3$
$-3-1<b+1-1<3-1$
$-4<b<2$
$(-4, 2)$

3. $|3x-2|+5\le 9$
$|3x-2|\le 9-5$
$|3x-2|\le 4$
$-4\le 3x-2\le 4$
$-4+2\le 3x-2+2\le 4+2$
$-2\le 3x\le 6$
$-\frac{2}{3}\le x\le 2$
$\left[-\frac{2}{3}, 2\right]$

4. $\left|3x+\frac{5}{8}\right|<-4$

The absolute value of a number is always nonnegative and can never be less than −4. The solution set is { } or ∅.

5. $|y+4|\ge 6$

$$\begin{aligned} y+4 &\le -6 \quad \text{or} \quad y+4\ge 6 \\ y+4-4 &\le -6-4 \quad \text{or} \quad y+4-4\ge 6-4 \\ y &\le -10 \quad \text{or} \quad y\ge 2 \end{aligned}$$

$(-\infty, -10] \cup [2, \infty)$

6.

$$\begin{aligned} |4x+3|+5 &> 3 \\ |4x+3|+5-5 &> 3-5 \\ |4x+3| &> -2 \end{aligned}$$

The absolute value of any number is always nonnegative and thus is always greater than −2.

$(-\infty, \infty)$

7.

$$\begin{aligned} \left|\frac{x}{2}-3\right|-5 &> -2 \\ \left|\frac{x}{2}-3\right|-5+5 &> -2+5 \\ \left|\frac{x}{2}-3\right| &> 3 \end{aligned}$$

$$\begin{aligned} \frac{x}{2}-3 &< -3 \quad \text{or} \quad \frac{x}{2}-3>3 \\ 2\left(\frac{x}{2}-3\right) &< 2(-3) \quad \text{or} \quad 2\left(\frac{x}{2}-3\right)>2(3) \\ x-6 &< -6 \quad \text{or} \quad x-6>6 \\ x &< 0 \quad \text{or} \quad x>12 \end{aligned}$$

$(-\infty, 0) \cup (12, \infty)$

8.

$$\begin{aligned} \left|\frac{3(x-2)}{5}\right| &\le 0 \\ \frac{3(x-2)}{5} &= 0 \\ 5\left[\frac{3(x-2)}{5}\right] &= 5(0) \\ 3(x-2) &= 0 \\ 3x-6 &= 0 \\ 3x &= 6 \\ x &= 2 \end{aligned}$$

The solution set is {2}.

Vocabulary and Readiness Check

1. D
2. E
3. C
4. B
5. A

Exercise Set 9.3

1. $|x|\le 4$

$-4\le x\le 4$

$[-4, 4]$

3. $|x-3|<2$

$-2<x-3<2$

$1<x<5$

$(1, 5)$

5. $|x+3|<2$

$-2<x+3<2$

$-5<x<-1$

$(-5, -1)$

7. $|2x+7|\le 3$

$-13\le 2x+7\le 13$

$-20\le 2x\le 6$

$-10\le x\le 3$

$[-10, 3]$

9. $|x|+7\le 12$

$|x|\le 5$

$-5\le x\le 5$

$[-5, 5]$

11. $|3x-1|<-5$

No real solutions; ∅

13. $|x-6|-7 \le -1$
$|x-6| \le 6$
$-6 \le x-6 \le 6$
$0 \le x \le 12$
$[0, 12]$

15. $|x| > 3$
$x < -3$ or $x > 3$
$(-\infty, -3) \cup (3, \infty)$

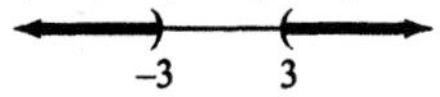

17. $|x+10| \ge 14$
$x+10 \le -14$ or $x+10 \ge 14$
$x \le -24$ or $x \ge 4$
$(-\infty, -24] \cup [4, \infty)$

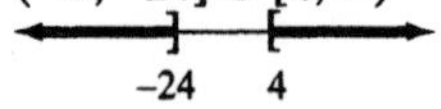

19. $|x|+2 > 6$
$|x| > 4$
$x < -4$ or $x > 4$
$(-\infty, -4) \cup (4, \infty)$

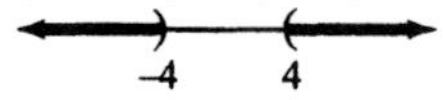

21. $|5x| > -4$
All real numbers
$(-\infty, \infty)$

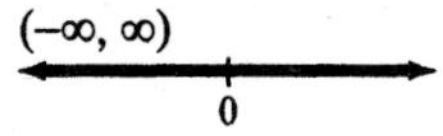

23. $|6x-8|+3 > 7$
$|6x-8| > 4$
$6x-8 < -4$ or $6x-8 > 4$
$6x < 4$ or $6x > 12$
$x < \frac{2}{3}$ or $x > 2$
$\left(-\infty, \frac{2}{3}\right) \cup (2, \infty)$

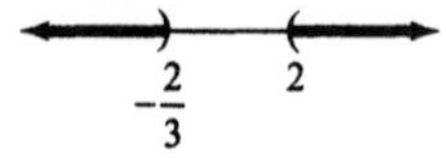

25. $|x| \le 0$
$|x| = 0$
$x = 0$

27. $|8x+3| > 0$ only excludes $|8x+3| = 0$
$8x+3 = 0$
$8x = -3$
$x = -\frac{3}{8}$
All real numbers except $-\frac{3}{8}$.
$\left(-\infty, -\frac{3}{8}\right) \cup \left(-\frac{3}{8}, \infty\right)$

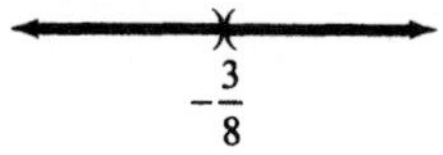

29. $|x| \le 2$
$-2 \le x \le 2$
$[-2, 2]$

31. $|y| > 1$
$y < -1$ or $y > 1$
$(-\infty, -1) \cup (1, \infty)$

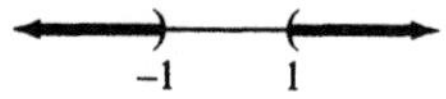

33. $|x-3| < 8$
$-8 < x-3 < 8$
$-5 < x < 11$
$(-5, 11)$

35. $|0.6x-3| > 0.6$
$0.6x-3 < -0.6$ or $0.6x-3 > 0.6$
$0.6x < 2.4$ or $0.6x > 3.6$
$x < 4$ or $x > 6$
$(-\infty, 4) \cup (6, \infty)$

37. $5+|x| \le 2$
$|x| \le -3$
No real solution
$\varnothing$

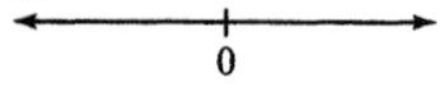

39. $|x| > -4$
All real numbers
$(-\infty, \infty)$

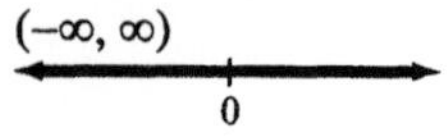

41. $|2x-7| \le 11$
$-11 \le 2x-7 \le 11$
$-4 \le 2x \le 18$
$-2 \le x \le 9$
$[-2, 9]$

43. $|x+5|+2 \ge 8$
$|x+5| \ge 6$
$x+5 \le -6$ or $x+5 \ge 6$
$x \le -11$ or $x \ge 1$
$(-\infty, -11] \cup [1, \infty)$

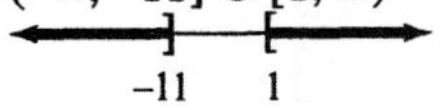

45. $|x| > 0$ only excludes $|x| = 0$, or $x = 0$.
All real numbers except $x = 0$
$(-\infty, 0) \cup (0, \infty)$

47. $9+|x| > 7$
$|x| > -2$
All real numbers
$(-\infty, \infty)$

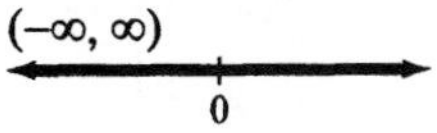

49. $6+|4x-1| \le 9$
$|4x-1| \le 3$
$-3 \le 4x-1 \le 3$
$-2 \le 4x \le 4$
$-\frac{1}{2} \le x \le 1$
$\left[-\frac{1}{2}, 1\right]$

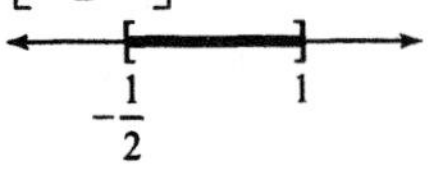

51. $\left|\frac{2}{3}x+1\right| > 1$
$\frac{2}{3}x+1 < -1$ or $\frac{2}{3}x+1 > 1$
$\frac{2}{3}x < -2$ or $\frac{2}{3}x > 0$
$x < -3$ or $x > 0$
$(-\infty, -3) \cup (0, \infty)$

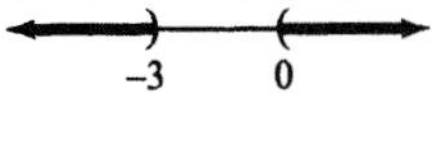

53. $|5x+3| < -6$
No real solution
$\varnothing$

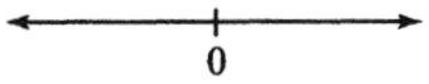

55. $\left|\frac{8x-3}{4}\right| \le 0$
$\frac{8x-3}{4} = 0$
$8x-3 = 0$
$8x = 3$
$x = \frac{3}{8}$
$\left\{\frac{3}{8}\right\}$

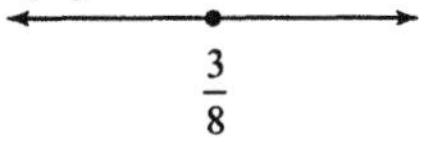

57. $|1+3x|+4 < 5$
$|1+3x| < 1$
$-1 < 1+3x < 1$
$-2 < 3x < 0$
$-\frac{2}{3} < x < 0$
$\left(-\frac{2}{3}, 0\right)$

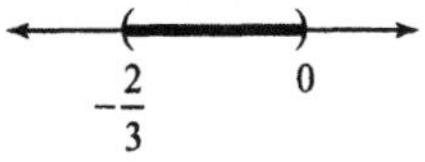

59. $\left|\frac{x+6}{3}\right| > 2$
$\frac{x+6}{3} < -2$ or $\frac{x+6}{3} > 2$
$x+6 < -6$ or $x+6 > 6$
$x < -12$ or $x > 0$
$(-\infty, -12) \cup (0, \infty)$

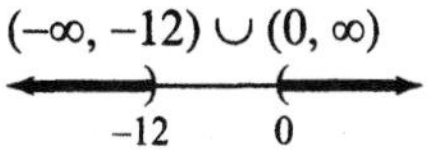

61. $-15+|2x-7| \le -6$
$|2x-7| \le 9$
$-9 \le 2x-7 \le 9$
$-2 \le 2x \le 16$
$-1 \le x \le 8$
$[-1, 8]$

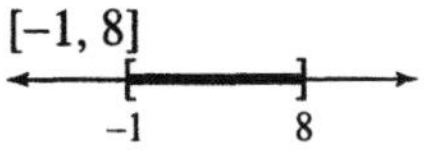

63. $\left|2x+\frac{3}{4}\right|-7\le -2$
$\left|2x+\frac{3}{4}\right|\le 5$
$-5\le 2x+\frac{3}{4}\le 5$
$-20\le 8x+3\le 20$
$-23\le 8x\le 17$
$-\frac{23}{8}\le x\le \frac{17}{8}$
$\left[-\frac{23}{8}, \frac{17}{8}\right]$

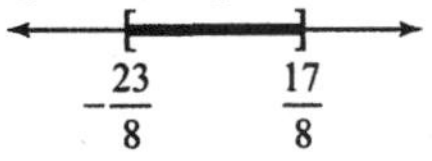

65. $|2x-3|<7$
$-7<2x-3<7$
$-4<2x<10$
$-2<x<5$
$(-2, 5)$

67. $|2x-3|=7$
$2x-3=7$ or $2x-3=-7$
$2x=10$ or $2x=-4$
$x=5$ or $x=-2$

69. $|x-5|\ge 12$
$x-5\le -12$ or $x-5\ge 12$
$x\le -7$ or $x\ge 17$
$(-\infty, -7]\cup[17, \infty)$

71. $|9+4x|=0$
$9+4x=0$
$4x=-9$
$x=-\frac{9}{4}$

73. $|2x+1|+4<7$
$|2x+1|<3$
$-3<2x+1<3$
$-4<2x<2$
$-2<x<1$
$(-2, 1)$

75. $|3x-5|+4=5$
$|3x-5|=1$
$3x-5=1$ or $3x-5=-1$
$3x=6$ or $3x=4$
$x=2$ or $x=\frac{4}{3}$

77. $|x+11|=-1$ is impossible. The solution set is $\varnothing$.

79. $\left|\frac{2x-1}{3}\right|=6$
$\frac{2x-1}{3}=6$ or $\frac{2x-1}{3}=-6$
$2x-1=18$ or $2x-1=-18$
$2x=19$ or $2x=-17$
$x=\frac{19}{2}$ or $x=-\frac{17}{2}$

81. $\left|\frac{3x-5}{6}\right|>5$
$\frac{3x-5}{6}<-5$ or $\frac{3x-5}{6}>5$
$3x-5<-30$ or $3x-5>30$
$3x<-25$ or $3x>35$
$x<-\frac{25}{3}$ or $x>\frac{35}{3}$
$\left(-\infty, -\frac{25}{3}\right)\cup\left(\frac{35}{3}, \infty\right)$

83. $3x-4y=12$
$3(2)-4y=12$
$6-4y=12$
$-4y=6$
$y=-\frac{3}{2}=-1.5$

85. $3x-4y=12$
$3x-4(-3)=12$
$3x+12=12$
$3x=0$
$x=0$

87. $|x|<7$

89. $|x|\le 5$

91. Answers may vary

93. $|3.5 - x| < 0.05$

$-0.05 < 3.5 - x < 0.05$

$-3.55 < -x < -3.45$

$3.55 > x > 3.45$

$3.45 < x < 3.55$

The Bigger Picture

1. $9x - 14 = 11x + 2$

$9x - 11x = 14 + 2$

$-2x = 16$

$x = -8$

2. $|x - 4| = 17$

$x - 4 = -17$ or $x - 4 = 17$

$x = -13$ or $x = 21$

3. $x - 1 \le 5$ or $3x - 2 \le 10$

$x \le 6$ or $3x \le 12$

$x \le 6$ or $x \le 4$

$(-\infty, 6]$

4. $-x < 7$ and $4x \le 20$

$x > -7$ and $x \le 5$

$(-7, 5]$

5. $|x - 2| = |x + 15|$

$x - 2 = x + 15$ or $x - 2 = -(x + 15)$

$-2 = 15$ False

$x - 2 = -x - 15$

$2x - 2 = -15$

$2x = -13$

$x = -\frac{13}{2}$

The only solution is $-\frac{13}{2}$.

6. $9y - 6y + 1 = 4y + 10 - y + 3$

$3y + 1 = 3y + 13$

$1 = 13$

$\varnothing$

7. $1.5x - 3 = 1.2x - 18$

$1.5x - 1.2x = 3 - 18$

$0.3x = -15$

$x = -50$

8. $\frac{7x+1}{8} - 3 = x + \frac{2x+1}{4}$

$8\left(\frac{7x+1}{8} - 3\right) = 8\left(x + \frac{2x+1}{4}\right)$

$7x + 1 - 8 \cdot 3 = 8x + 2(2x + 1)$

$7x + 1 - 24 = 8x + 4x + 2$

$7x - 23 = 12x + 2$

$7x - 12x = 2 + 23$

$-5x = 25$

$x = -5$

9. $|5x + 2| - 10 \le -3$

$|5x + 2| \le 7$

$-7 \le 5x + 2 \le 7$

$-9 \le 5x \le 5$

$-\frac{9}{5} \le x \le 1$

$\left[-\frac{9}{5}, 1\right]$

10. $|x + 11| > 2$

$x + 11 > 2$ or $x + 11 < -2$

$x > -9$ or $x < -13$

$(-\infty, -13) \cup (-9, \infty)$

11. $|9x + 2| - 1 = 24$

$|9x + 2| = 25$

$9x + 2 = -25$ or $9x + 2 = 25$

$9x = -27$ or $9x = 23$

$x = -3$ or $x = \frac{23}{9}$

12. $\left|\frac{3x-1}{2}\right| = |2x+5|$

$$\frac{3x-1}{2} = -(2x+5) \quad \text{or} \quad \frac{3x-1}{2} = 2x+5$$
$$2\left(\frac{3x-1}{2}\right) = 2[-(2x+5)] \quad \text{or} \quad 2\left[\frac{3x-1}{2}\right] = 2(2x+5)$$
$$3x-1 = -4x-10 \quad \text{or} \quad 3x-1 = 4x+10$$
$$3x+4x = -10+1 \quad \text{or} \quad 3x-4x = 10+1$$
$$7x = -9 \quad \text{or} \quad -x = 11$$
$$x = -\frac{9}{7} \quad \text{or} \quad x = -11$$

Integrated Review

1. $x < 7$ and $x > -5$ is $-5 < x < 7$. The solution set is $(-5, 7)$.

-5 7

2. $x < 7$ or $x > -5$
The solution set is $(-\infty, \infty)$.

3. $|4x-3| = 1$

$$4x-3 = 1 \quad \text{or} \quad 4x-3 = -1$$
$$4x = 4 \quad \text{or} \quad 4x = 2$$
$$x = 1 \quad \text{or} \quad x = \frac{1}{2}$$

The solutions are 1 and $\frac{1}{2}$.

4. $|2x+1| < 5$
$-5 < 2x+1 < 5$
$-6 < 2x < 4$
$-3 < x < 2$
The solution set is $(-3, 2)$.

5. $|6x| - 9 \geq -3$
$|6x| \geq 6$
$6x \leq -6$ or $6x \geq 6$
$x \leq -1$ or $x \geq 1$
The solution set is $(-\infty, -1] \cup [1, \infty)$.

6. $|x-7| = |2x+11|$

$x-7 = 2x+11$ or $x-7 = -(2x+11)$

$-7 = x+11$ or $x-7 = -2x-11$

$-18 = x$ or $3x-7 = -11$

$3x = -4$

$-18 = x$ or $x = -\frac{4}{3}$

The solutions are -18 and $-\frac{4}{3}$.

7. $-5 \le \frac{3x-8}{2} \le 2$

$-10 \le 3x-8 \le 4$

$-2 \le 3x \le 12$

$-\frac{2}{3} \le x \le 4$

The solution set is $\left[-\frac{2}{3}, 4\right]$.

8. $|9x-1| = -3$

The absolute value of a number cannot be negative. There is no solution, or $\varnothing$.

9. $3x+2 \le 5$ or $-3x \ge 0$

$3x \le 3$ or $\frac{-3x}{-3} \le \frac{0}{-3}$

$x \le 1$ or $x \le 0$

The solution set is $(-\infty, 1]$.

10. $3x+2 \le 5$ and $-3x \ge 0$

$3x \le 3$ and $\frac{-3x}{-3} \le \frac{0}{-3}$

$x \le 1$ and $x \le 0$

The solution set is $(-\infty, 0]$.

11. $|3-x|-5 \le -2$

$|3-x| \le 3$

$-3 \le 3-x \le 3$

$-6 \le -x \le 0$

$\frac{-6}{-1} \ge \frac{-x}{-1} \ge \frac{0}{-1}$

$6 \ge x \ge 0$

$0 \le x \le 6$

The solution set is $[0, 6]$.

12. $\left|\frac{4x+1}{5}\right| = |-1|$

$\frac{4x+1}{5} = 1$ or $\frac{4x+1}{5} = -1$

$4x+1 = 5$ or $4x+1 = -5$

$4x = 4$ or $4x = -6$

$x = 1$ or $x = \frac{-6}{4} = -\frac{3}{2}$

The solutions are 1 and $-\frac{3}{2}$.

13. $|2x+1| = 5$

$2x+1 = 5$ or $2x+1 = -5$

This is statement B.

14. $|2x+1| < 5$

$-5 < 2x+1 < 5$

This is statement E.

15. $|2x+1| > 5$

$2x+1 < -5$ or $2x+1 > 5$

This is statement A.

16. $x < 3$ or $x < 5$ is $x < 5$. This is statement C.

17. $x < 3$ and $x < 5$ is $x < 3$. This is statement D.

Section 9.4

Practice Exercises

1.

2.

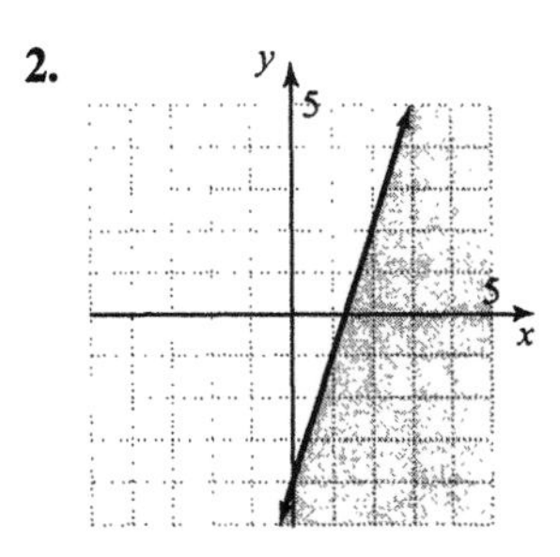

3.

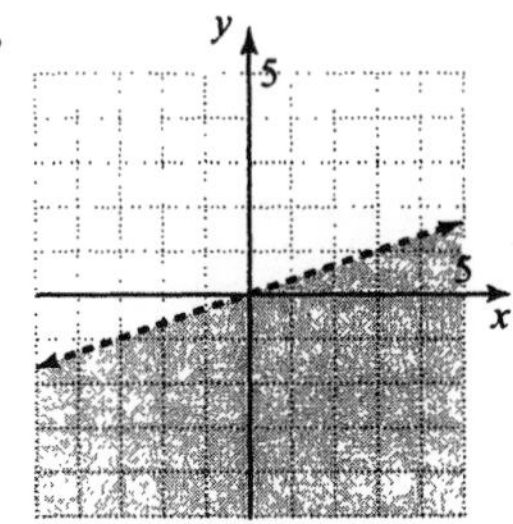

4.

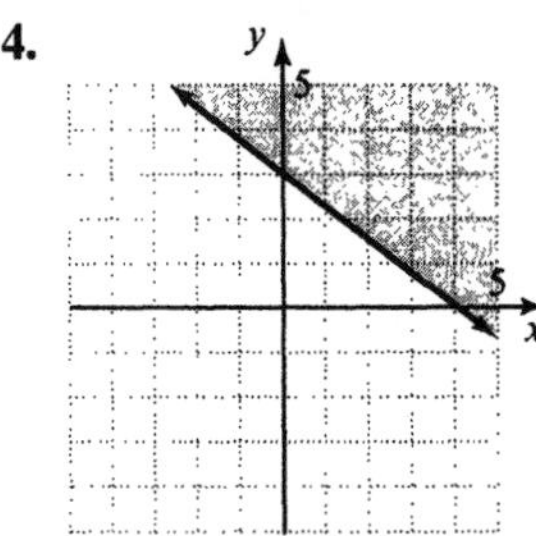

5.

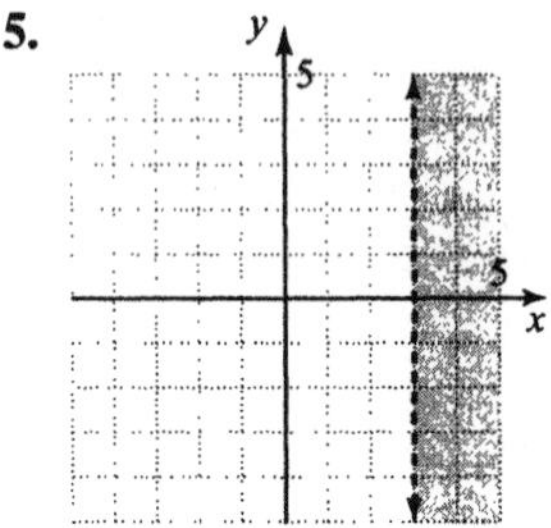

6. $\begin{cases} 4x \le y \\ x + 3y \ge 9 \end{cases}$

Graph $4x \le y$ with a solid line.
Test (1, 0)

$$4(1) \overset{?}{\le} 0$$

False
Shade above.
Graph $x + 3y \ge 9$ with a solid line.
Test (0, 0)

$$0 + 3(0) \overset{?}{\ge} 9$$

False
Shade above.
The solution of the system is the darker shaded region and includes parts of both boundary lines.

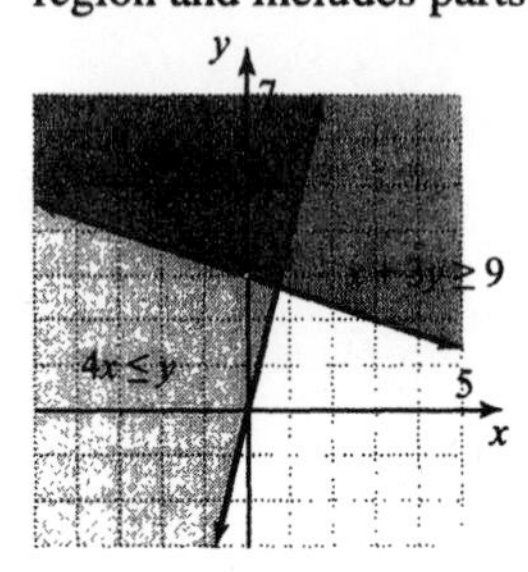

7. $\begin{cases} x - y > 4 \\ x + 3y < -4 \end{cases}$

Graph both inequalities using dashed lines. The solution of the system is the darker shaded region which does not include any of the boundary lines.

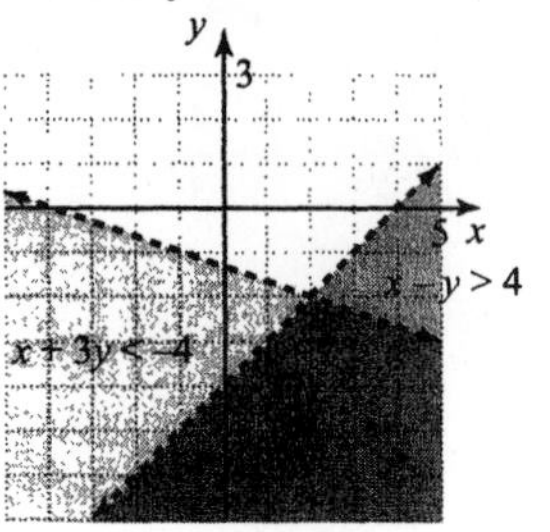

8. $\begin{cases} y \le 6 \\ -2x + 5y > 10 \end{cases}$

Graph both inequalities. The solution of the system is the darker shaded region.

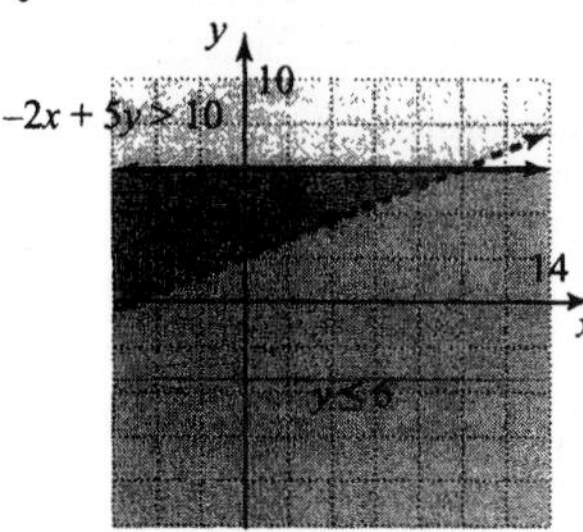

Vocabulary and Readiness Check

1. The statement $5x - 6y < 7$ is an example of a linear inequality in two variables.

2. A boundary line divides a plane into two regions called half-planes.

3. The graph of $5x - 6y < 7$ does not include its corresponding boundary line. The statement is false.

4. When graphing a linear inequality to determine which side of the boundary line to shade, choose a point *not* on the boundary line. The statement is true.

5. The boundary line for the inequality $5x - 6y < 7$ is the graph of $5x - 6y = 7$. The statement is true.

6. The graph of $y < 3$ is

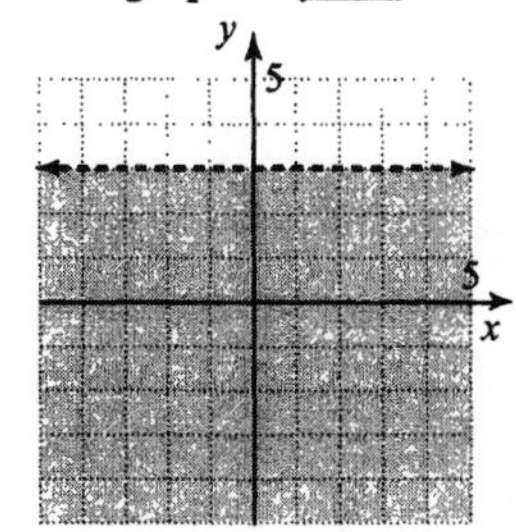

7. Yes, since the inequality is ≥, the graph includes the boundary line.

8. No, since the inequality is >, the graph does not include the boundary line.

9. Yes, since the inequality is ≥, the graph includes the boundary line.

10. No, since the inequality is >, the graph does not include the boundary line.

11. $x + y > -5$, (0, 0)

$$0 + 0 \overset{?}{>} -5$$
$$0 \overset{?}{>} -5$$

Yes, (0, 0) is a solution.

12. $2x + 3y < 10$, (0, 0)

$$2(0) + 3(0) \overset{?}{<} 10$$
$$0 \overset{?}{<} 10$$

Yes, (0, 0) is a solution.

13. $x - y \le -1$, (0,)

$$0 - 0 \overset{?}{\le} -1$$
$$0 \overset{?}{\le} -1$$

No, (0, 0) is not a solution.

14. $\frac{2}{3}x + \frac{5}{6}y > 4$, (0, 0)

$$\frac{2}{3}(0) + \frac{5}{6}(0) \overset{?}{>} 4$$
$$0 \overset{?}{>} 4$$

No, (0, 0) is not a solution.

Exercise Set 9.4

1. $x - y > 3$

$$(2, -1),\ 2 - (-1) \overset{?}{>} 3$$
$$2 + 1 \overset{?}{>} 3$$
$$3 \overset{?}{>} 3, \text{ False}$$

(2, −1) is not a solution.

$$(5, 1),\ 5 - 1 \overset{?}{>} 3$$
$$4 \overset{?}{>} 3, \text{ True}$$

(5, 1) is a solution.

3. $3x - 5y \le -4$

$$(-1, -1),\ 3(-1) - 5(-1) \overset{?}{\le} -4$$
$$-3 + 5 \overset{?}{\le} -4$$
$$2 \overset{?}{\le} -4, \text{ False}$$

(−1, −1) is not a solution.

$$(4, 0),\ 3(4) - 5(0) \overset{?}{\le} -4$$
$$12 - 0 \overset{?}{\le} -4$$
$$12 \overset{?}{\le} -4, \text{ False}$$

(4, 0) is not a solution.

5. $x < -y$

$$(0, 2),\ 0 \overset{?}{<} -2, \text{ False}$$

(0, 2) is not a solution.

$$(-5, 1),\ -5 \overset{?}{<} -1, \text{ True}$$

(−5, 1) is a solution.

7. $x + y \le 1$

Test (0, 0)

$$0 + 0 \overset{?}{\le} 1, \text{ True}$$

Shade below.

9. $2x + y > -4$
Test (0, 0)
$2(0) + 0 \overset{?}{>} -4$
True
Shade above.

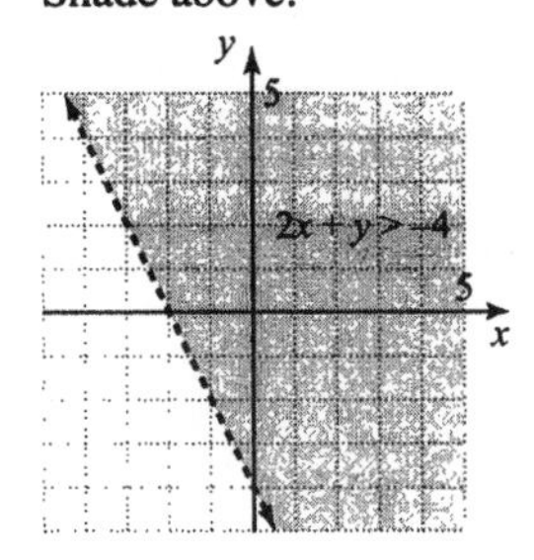

11. $x + 6y \le -6$
Test (0, 0)
$0 + 6(0) \overset{?}{\le} -6$
False
Shade below.

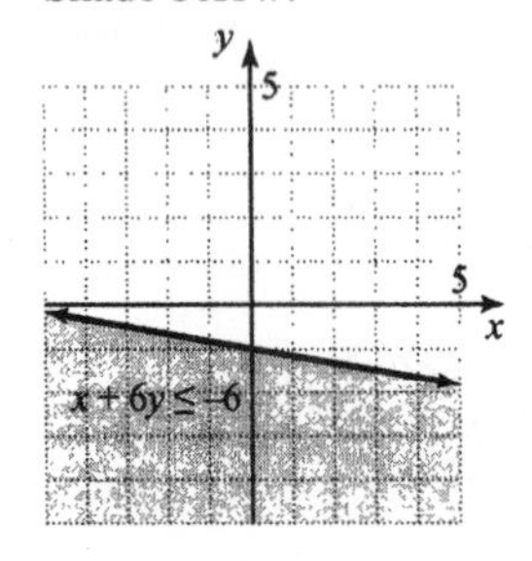

13. $2x + 5y > -10$
Test (0, 0)
$2(0) + 5(0) \overset{?}{>} -10$
True
Shade above.

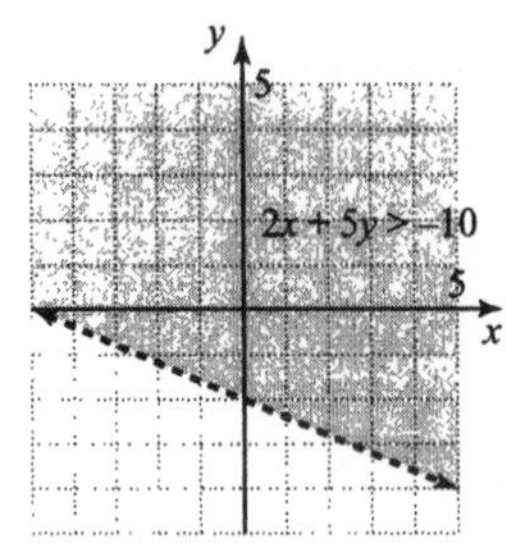

15. $x + 2y \le 3$
Test (0, 0)
$0 + 2(0) \overset{?}{\le} 3$
True
Shade below.

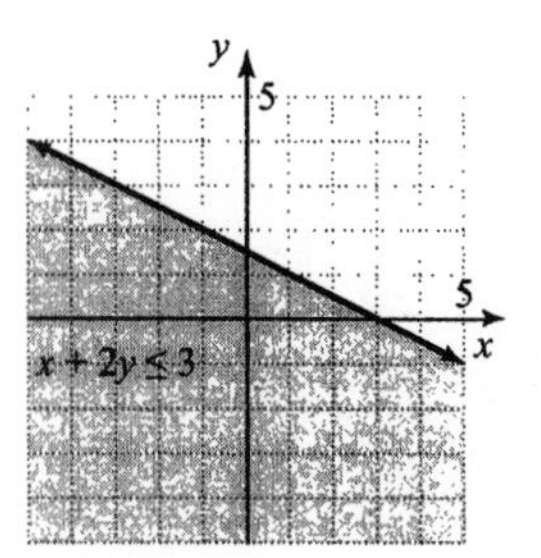

17. $2x + 7y > 5$
Test (0, 0)
$2(0) + 7(0) \overset{?}{>} 5$
False
Shade above.

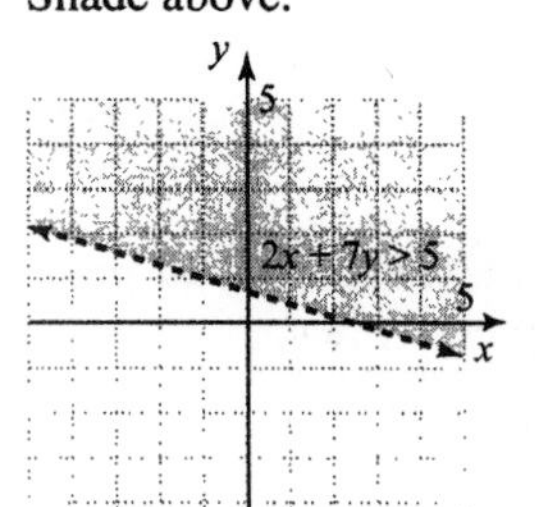

19. $x - 2y \ge 3$
Test (0, 0)
$(0) - 2(0) \overset{?}{\ge} 3$
False
Shade below.

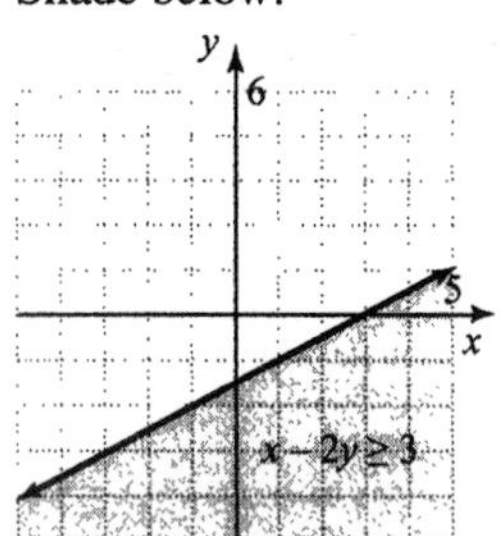

21. $5x + y < 3$
Test (0, 0)
$5(0) + 0 \overset{?}{<} 3$
True
Shade below.

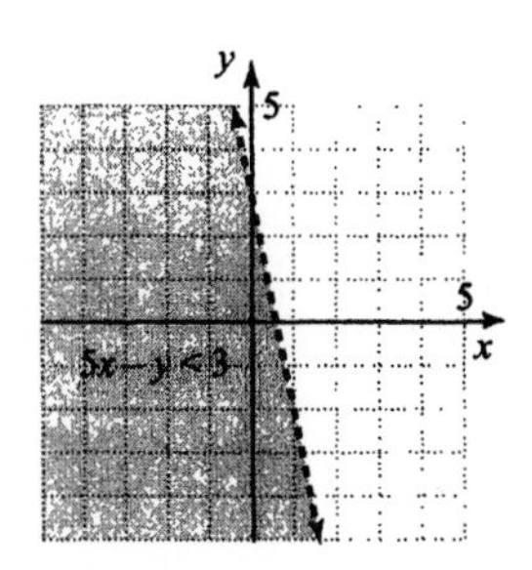

23. $4x + y < 8$
Test (0, 0)

$4(09) + 0 \overset{?}{<} 8$

True
Shade below.

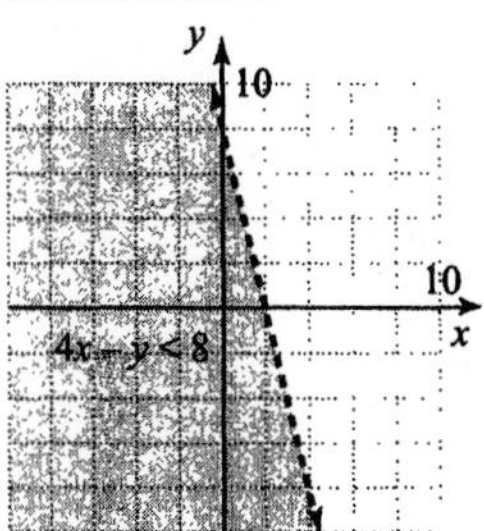

25. $y \geq 2x$
Test (1, 0)

$0 \overset{?}{\geq} 2(1)$

False
Shade above.

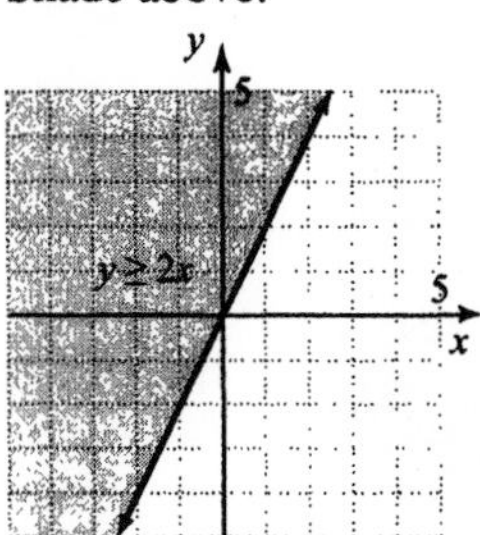

27. $x \geq 0$
Shade right.

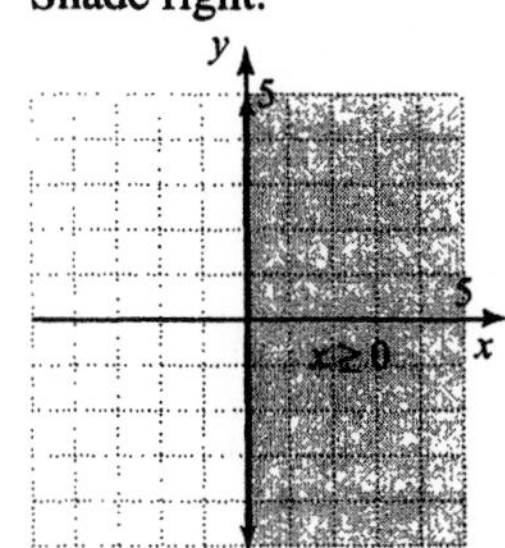

29. $y \leq -3$
Shade below.

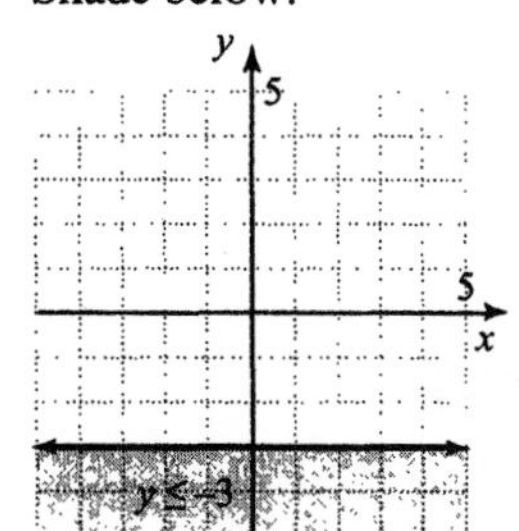

31. $2x - 7y > 0$
Test (1, 0)

$2(1) - 7(0) \overset{?}{>} 0$

True
Shade below.

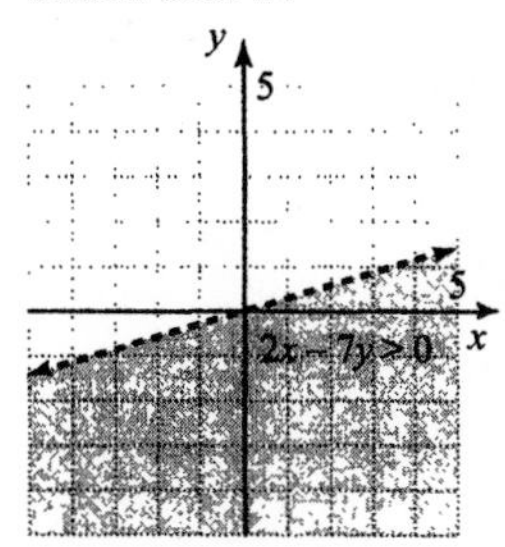

33. $3x - 7y \geq 0$
Test (1, 0)

$3(1) - 7(0) \overset{?}{\geq} 0$

True
Shade below.

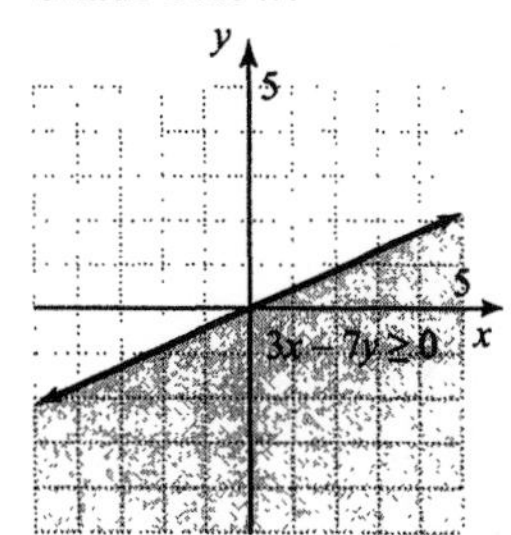

35. $x > y$
Test (0, 1)

$0 \overset{?}{>} 1$

False
Shade below.

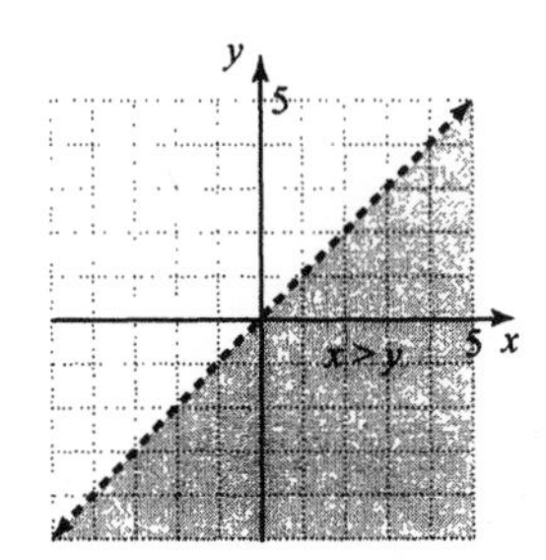

37. $x - y \le 6$
Test (0, 0)
$0 - 0 \overset{?}{\le} 6$
True
Shade above.

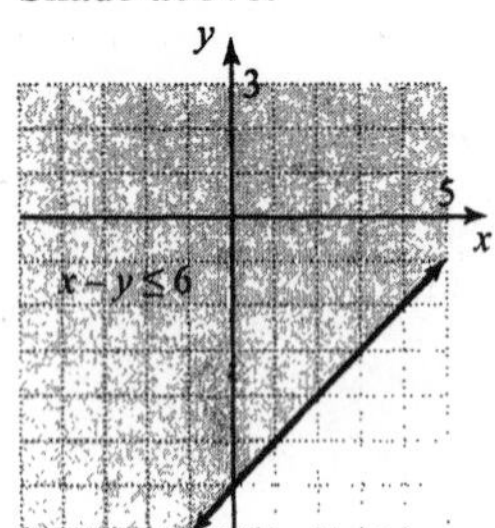

39. $-\frac{1}{4}y + \frac{1}{3}x > 1$
Test (0, 0)
$-\frac{1}{4}(0) + \frac{1}{3}(0) \overset{?}{>} 1$
False
Shade below.

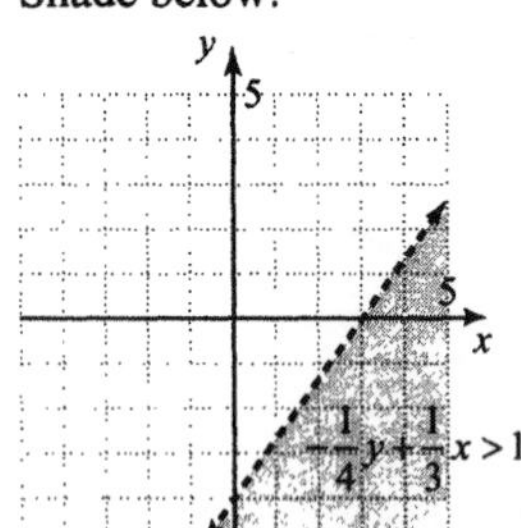

41. $-x < 0.4y$
Test (1, 0)
$-(1) \overset{?}{<} 0$
True
Shade above.

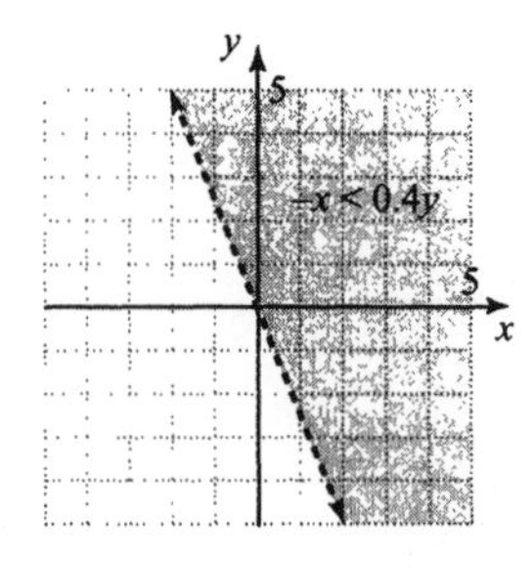

43. e

45. c

47. f

49. $\begin{cases} y \ge x+1 \\ y \ge 3-x \end{cases}$

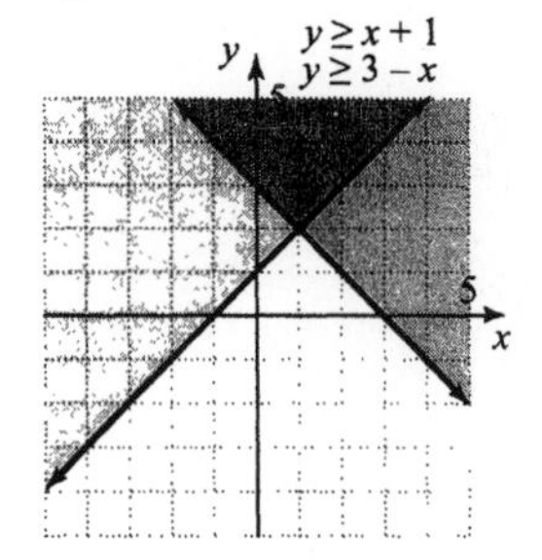

51. $\begin{cases} y < 3x-4 \\ y \le x+2 \end{cases}$

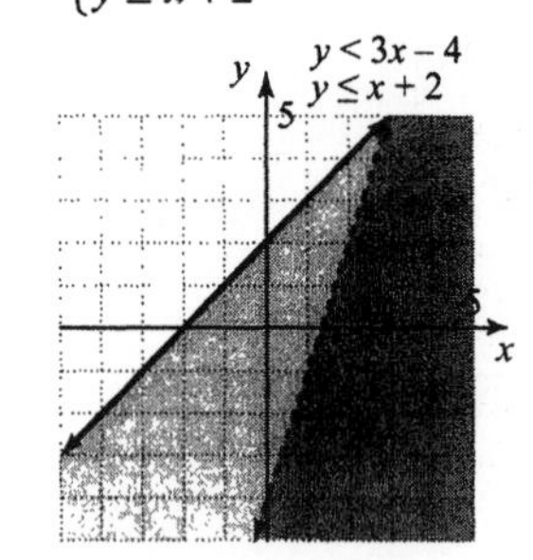

53. $\begin{cases} y \le -2x-2 \\ y \ge x+4 \end{cases}$

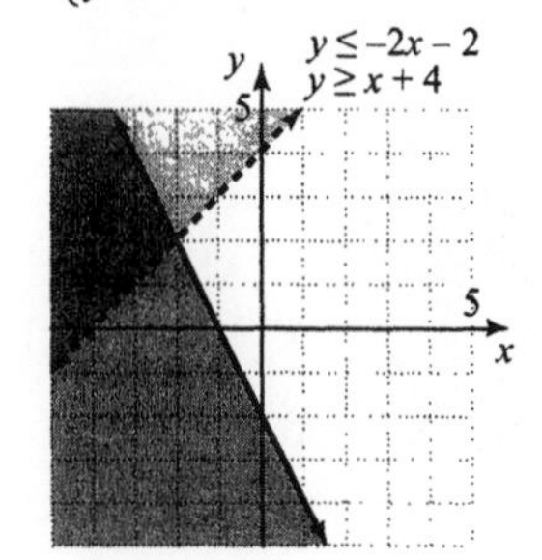

55. $\begin{cases} y \geq -x+2 \\ y \leq 2x+5 \end{cases}$

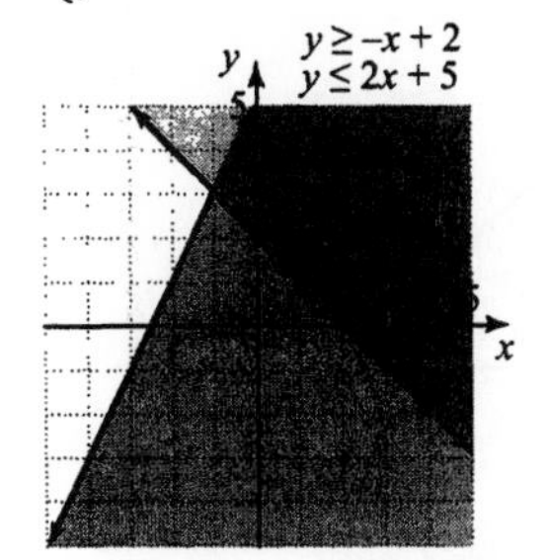

57. $\begin{cases} x \geq 3y \\ x+3y \leq 6 \end{cases}$

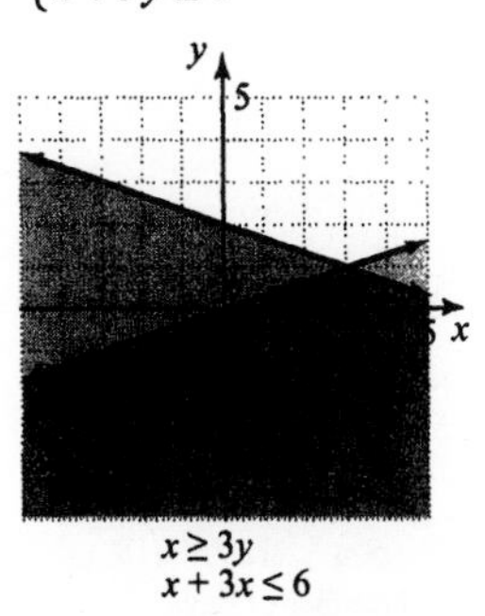

59. $\begin{cases} y+2x \geq 0 \\ 5x-3y \leq 12 \end{cases}$

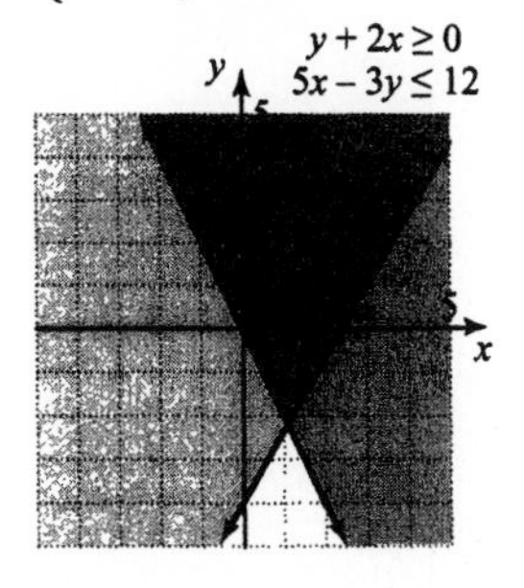

61. $\begin{cases} 3x-4y \geq -6 \\ 2x+y \leq 7 \end{cases}$

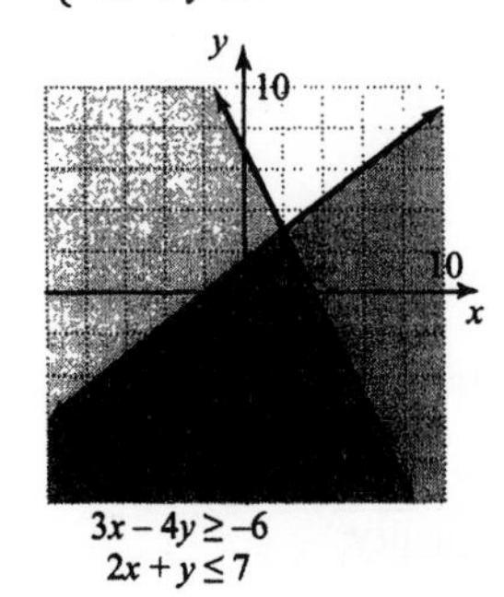

63. $\begin{cases} x \leq 2 \\ y \geq -3 \end{cases}$

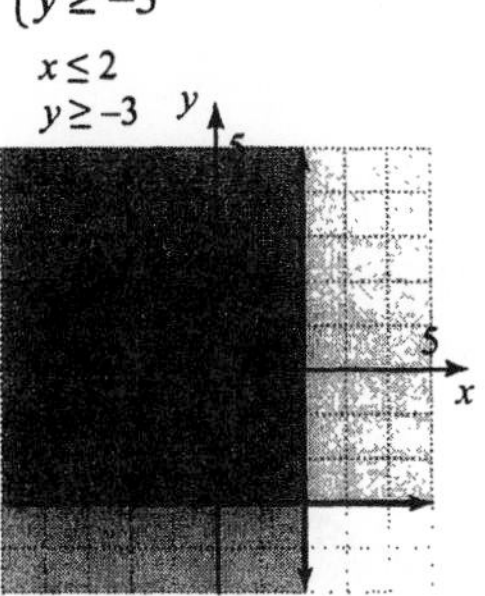

65. $\begin{cases} y \geq 1 \\ x < -3 \end{cases}$

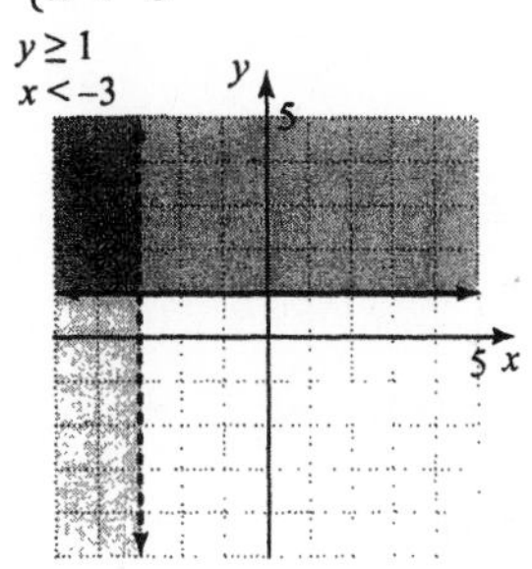

67. $\begin{cases} 2x+3y < -8 \\ x \geq -4 \end{cases}$

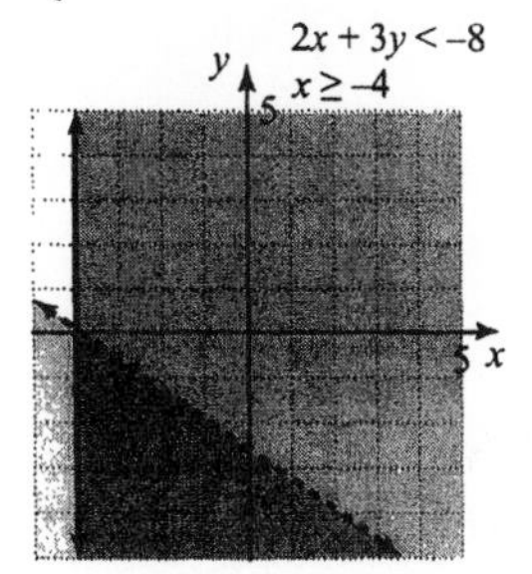

69. $\begin{cases} 2x-5y \leq 9 \\ y \leq -3 \end{cases}$

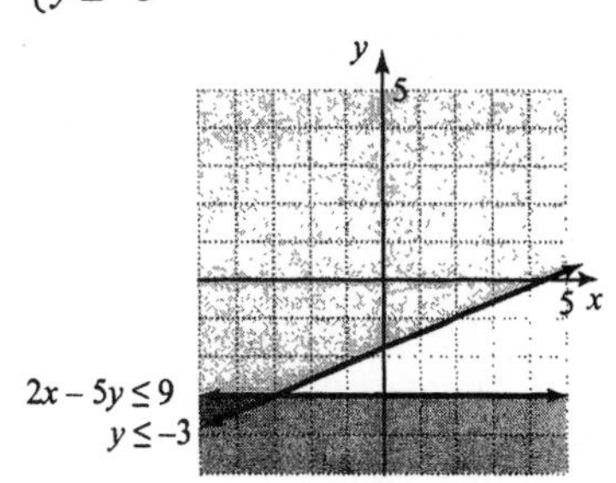

71. $\begin{cases} y \geq \frac{1}{2}x + 2 \\ y \leq \frac{1}{2}x - 3 \end{cases}$

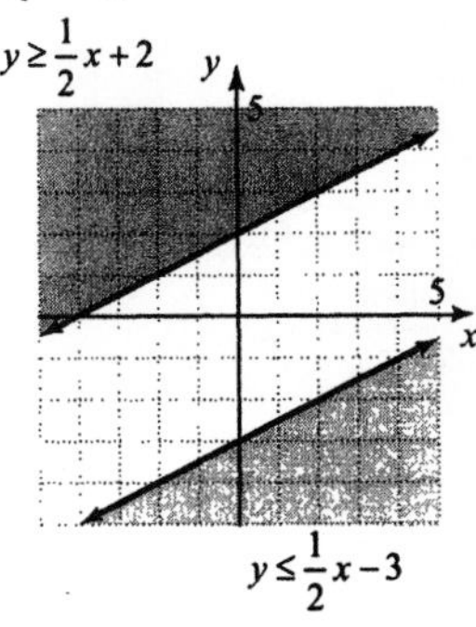

73. Let $x = -5$.
$x^2 = (-5)(-5) = 25$

75. Let $x = -1$.
$2x^3 = 2(-1)(-1)(-1) = -2$

77. $3x + 4y < 8$; (1,1)
$3(1) + 4(1) < 8$
$3 + 4 < 8$
$7 < 8$ True
(1, 1) is included in the graph.

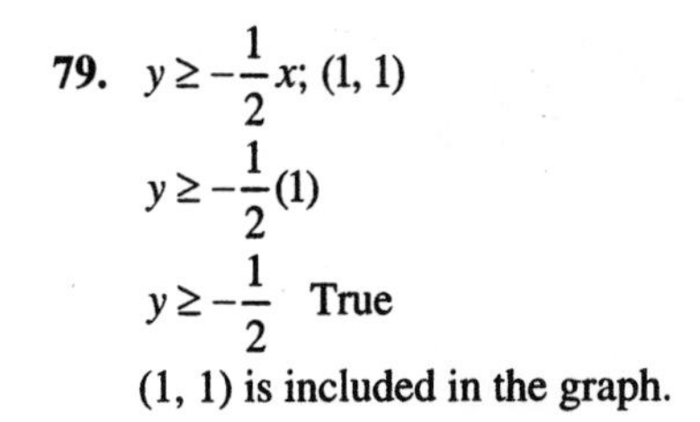

79. $y \geq -\frac{1}{2}x$; (1, 1)
$y \geq -\frac{1}{2}(1)$
$y \geq -\frac{1}{2}$ True
(1, 1) is included in the graph.

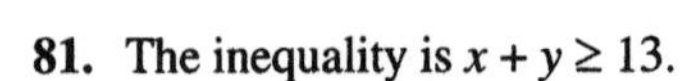

81. The inequality is $x + y \geq 13$.

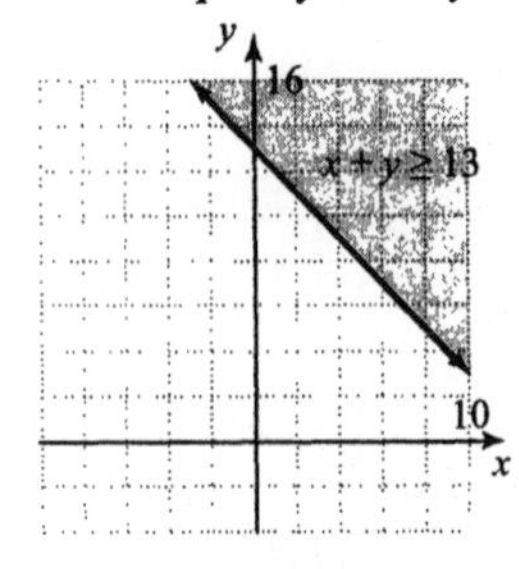

83. Answers may vary

85.

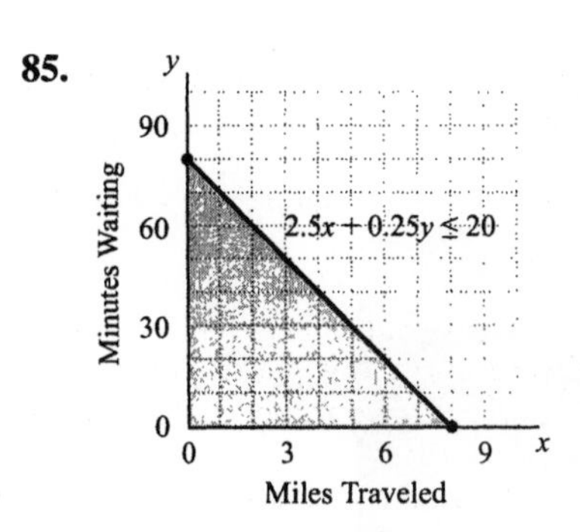

87. Answers may vary

89. **a.** $30x + 0.15y \leq 500$

b.

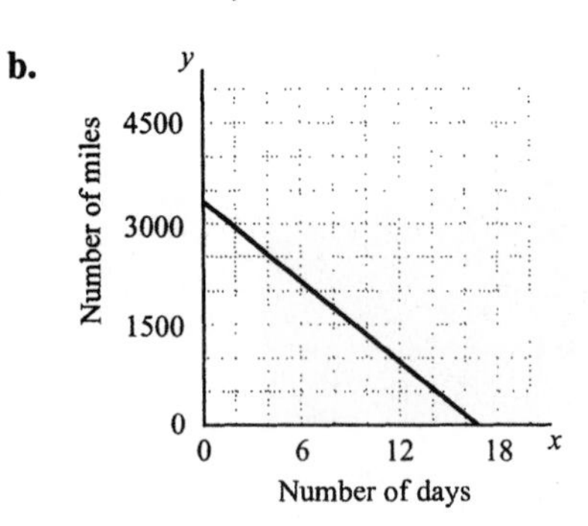

c. Answers may vary

91. C

93. D

95. $\begin{cases} 2x - y \leq 6 \\ x \geq 3 \\ y > 2 \end{cases}$

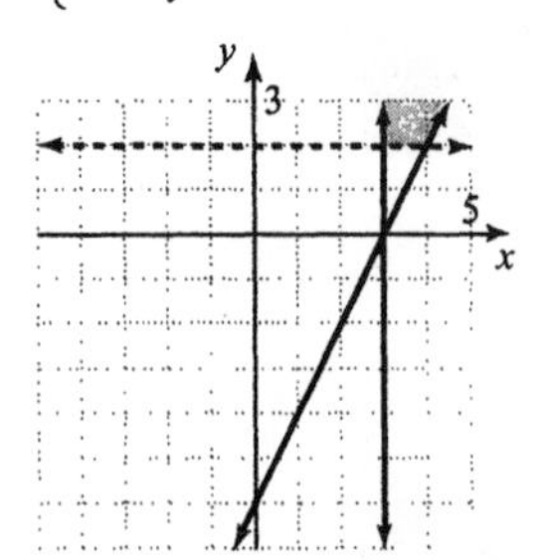

97. Answers may vary

Chapter 9 Vocabulary Check

1. The statement "$x < 5$ or $x > 7$" is called a compound inequality.

2. The intersection of two sets is the set of all elements common to both sets.

3. The union of two sets is the set of all elements that belong to either of the sets.

4. A number's distance from 0 is called its absolute value.

5. When a variable in an equation is replaced by a number and the resulting equation is true, then that number is called a solution of the equation.

6. Two or more linear inequalities are called a system of linear inequalities.

Chapter 9 Review

1. $-3 < 4(2x-1) < 12$
$-3 < 8x-4 < 12$
$1 < 8x < 16$
$\frac{1}{8} < x < 2$
$\left(\frac{1}{8}, 2\right)$

2. $-2 \le 8+5x < -1$
$-10 \le 5x \le -9$
$-2 \le x \le -\frac{9}{5}$
$\left[-2, \frac{9}{5}\right)$

3. $\frac{1}{6} < \frac{4x-3}{3} \le \frac{4}{5}$
$30\left(\frac{1}{6}\right) < 30\left(\frac{4x-3}{3}\right) \le 30\left(\frac{4}{5}\right)$
$5 < 10(4x-3) \le 24$
$5 < 40x-30 \le 24$
$35 < 40x < 54$
$\frac{7}{8} < x \le \frac{27}{20}$
$\left(\frac{7}{8}, \frac{27}{20}\right]$

4. $-6 < x-(3-4x) < -3$
$-6 < x-3+4x < -3$
$-6 < 5x-3 < -3$
$-3 < 5x < 0$
$-\frac{3}{5} < x < 0$
$\left(-\frac{3}{5}, 0\right)$

5. $3x-5 > 6$ or $-x < -5$
$3x > 11$ or $x > 5$
$x > \frac{11}{3}$ or $x > 5$
$x > \frac{11}{3}$
$\left(\frac{11}{3}, \infty\right)$

6. $x \le 2$ and $x > -5$
$-5 < x \le 2$
$(-5, 2]$

7. $|8-x| = 3$
$8-x = 3$ or $8-x = -3$
$-x = -5$ or $-x = -11$
$x = 5$ or $x = 11$

8. $|x-7| = 9$
$x-7 = 9$ or $x-7 = -9$
$x = 16$ or $x = -2$

9. $|-3x+4| = 7$
$-3x+4 = 7$ or $-3x+4 = -7$
$-3x = 3$ or $-3x = -11$
$x = -1$ or $x = \frac{11}{3}$

10. $|2x+9| = 9$
$2x+9 = 9$ or $2x+9 = -9$
$2x = 0$ or $2x = -18$
$x = 0$ or $x = -9$

11. $5+|6x+1| = 5$
$|6x+1| = 0$
$6x+1 = 0$
$6x = -1$
$x = -\frac{1}{6}$

12. $|3x-2|+6 = 10$
$|3x-2| = 4$
$3x-2 = 4$ or $3x-2 = -4$
$3x = 6$ or $3x = -2$
$x = 2$ or $x = -\frac{2}{3}$

13. $|5-6x|+8=3$
$|5-6x|=-5$
The solution set is $\varnothing$.

14. $-5=|4x-3|$
The solution set is $\varnothing$.

15. $\left|\dfrac{3x-7}{4}\right|=2$
$\dfrac{3x-7}{4}=2$ or $\dfrac{3x-7}{4}=-2$
$3x-7=8$ or $3x-7=-8$
$3x=15$ or $3x=-1$
$x=5$ or $x=-\dfrac{1}{3}$

16. $-8=|x-3|-10$
$2=|x-3|$
$x-3=2$ or $x-3=-2$
$x=5$ or $x=1$

17. $|6x+1|=|15+4x|$
$6x+1=15+4x$ or $6x+1=-(15+4x)$
$2x=14$ or $6x+1=-15-4x$
$x=7$ or $10x=-16$
$x=-\dfrac{8}{5}$

18. $|x-3|=|x+5|$
$x-3=x+5$ or $x-3=-(x+5)$
$-3=5$ False or $x-3=-x-5$
$2x-3=-5$
$2x=-2$
$x=-1$

19. $|5x-1|<9$
$-9<5x-1<9$
$-8<5x<10$
$-\dfrac{8}{5}<x<2$
$\left(-\dfrac{8}{5}, 2\right)$
[number line: open interval from $-\frac{8}{5}$ to 2]

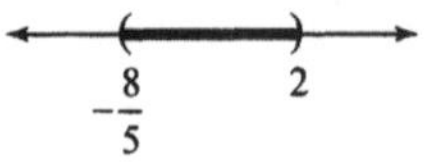

20. $|6+4x|\geq 10$
$6+4x\leq -10$ or $6+4x\geq 10$
$4x\leq -16$ or $4x\geq 4$
$x\leq -4$ or $x\geq 1$
$(-\infty, -4]\cup[1, \infty)$
[number line: shaded to the left of −4 (closed) and right of 1 (closed)]

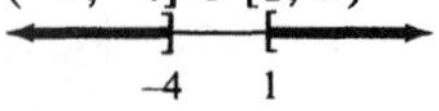

21. $|3x|-8>1$
$|3x|>9$
$3x<-9$ or $3x>9$
$x<-3$ or $x>3$
$(-\infty, -3)\cup(3, \infty)$
[number line: shaded to the left of −3 (open) and right of 3 (open)]

22. $9+|5x|<24$
$|5x|<15$
$-15<5x<15$
$-3<x<3$
$(-3, 3)$
[number line: open interval from −3 to 3]

23. $|6x-5|\leq -1$
The solution set is $\varnothing$.
[number line: empty, 0 marked]

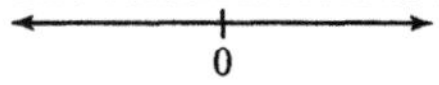

24. $|6x-5|\leq 5$
$-5\leq 6x-5\leq 5$
$0\leq 6x\leq 10$
$\dfrac{0}{6}\leq x\leq \dfrac{10}{6}$
$0\leq x\leq \dfrac{5}{3}$
$\left[0, \dfrac{5}{3}\right]$
[number line: closed interval from 0 to $\frac{5}{3}$]

25. $\left|3x+\frac{2}{5}\right| \ge 4$

$$3x+\frac{2}{5} \le -4 \quad \text{or} \quad 3x+\frac{2}{5} \ge 4$$
$$5\left(3x+\frac{2}{5}\right) \le 5(-4) \quad \text{or} \quad 5\left(3x+\frac{2}{5}\right) \ge 5(4)$$
$$15x+2 \le -20 \quad \text{or} \quad 15x+2 \ge 20$$
$$15x \le -22 \quad \text{or} \quad 15x \ge 18$$
$$x \le -\frac{22}{15} \quad \text{or} \quad x \ge \frac{6}{5}$$
$$\left(-\infty, -\frac{22}{15}\right] \cup \left[\frac{6}{5}, \infty\right)$$

$-\frac{22}{15}$ $\frac{6}{5}$

26. $|5x-3| > 2$

$$5x-3 < -2 \quad \text{or} \quad 5x-3 > 2$$
$$5x < 1 \quad \text{or} \quad 5x > 5$$
$$x < \frac{1}{5} \quad \text{or} \quad x > 1$$
$$\left(-\infty, \frac{1}{5}\right) \cup (1, \infty)$$

$\frac{1}{5}$ 1

27. $\left|\frac{x}{3}+6\right| - 8 > -5$

$$\left|\frac{x}{3}+6\right| > 3$$
$$\frac{x}{3}+6 < -3 \quad \text{or} \quad \frac{x}{3}+6 > 3$$
$$\frac{x}{3} < -9 \quad \text{or} \quad \frac{x}{3} > -3$$
$$x < -27 \quad \text{or} \quad x > -9$$
$$(-\infty, -27) \cup (-9, \infty)$$

−27 −9

28. $\left|\frac{4(x-1)}{7}\right| + 10 < 2$

$$\left|\frac{4(x-1)}{7}\right| < -8$$

The solution set is ∅.

0

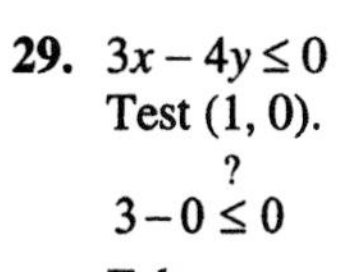

29. $3x-4y \le 0$
Test (1, 0).

$$3-0 \overset{?}{\le} 0$$

False
Shade above.

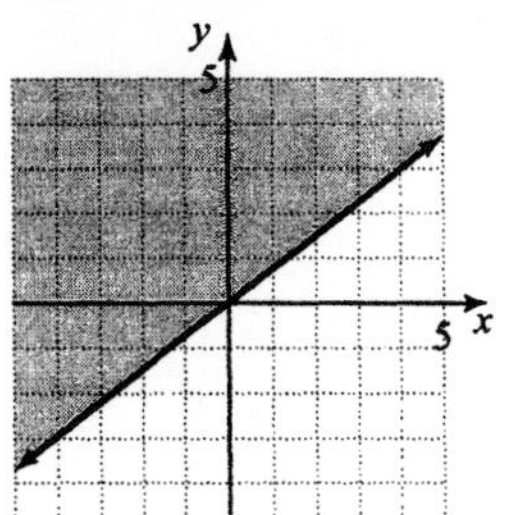

30. $3x-4y \ge 0$
Test (1, 0).

$$3-0 \overset{?}{\ge} 0$$

True
Shade below.

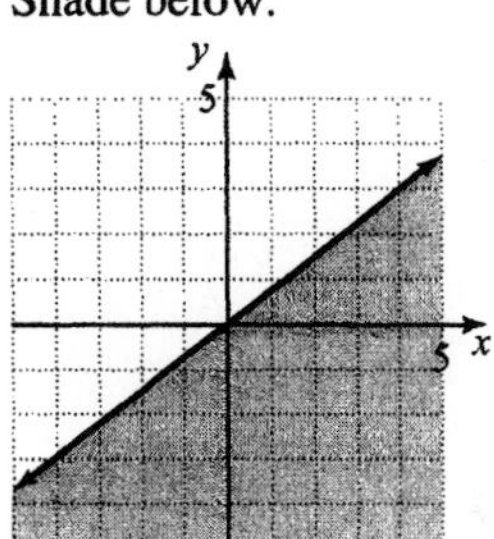

31. $x+6y < 6$
Test (0, 0)

$$0+6(0) \overset{?}{<} 6$$

True
Shade below.

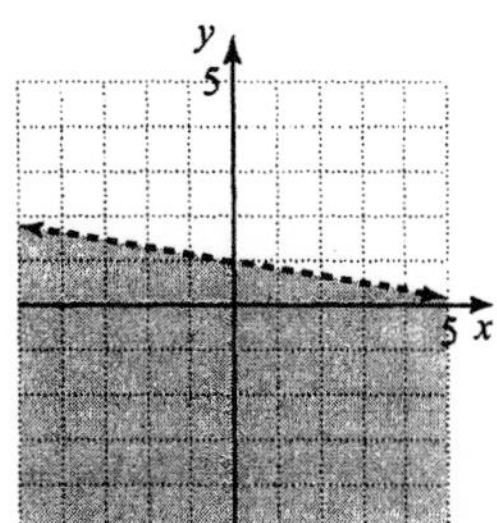

32. $y \le -4$
Shade below.

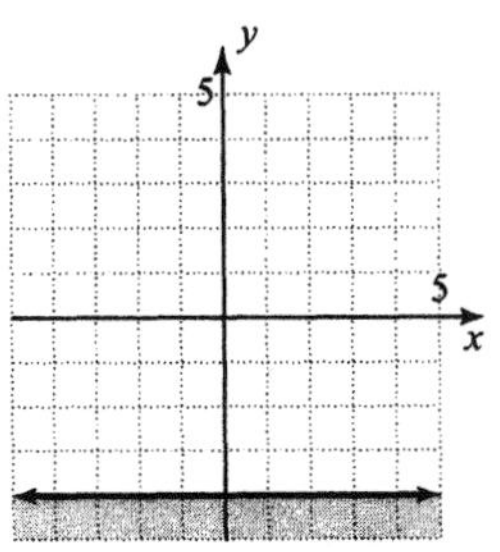

33. $y \ge -7$
Shade above.

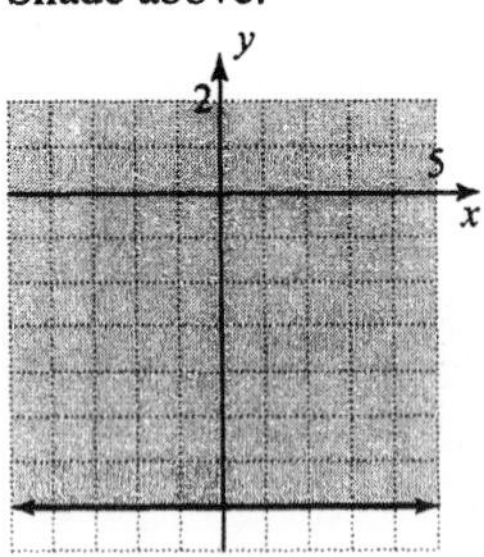

34. $x \ge -y$
Test (1, 0)
$1 \overset{?}{\ge} 0$
True
Shade above.

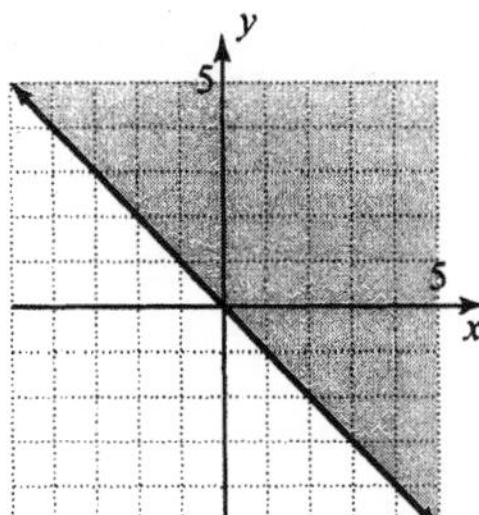

35. $\begin{cases} y \ge 2x - 3 \\ y \le -2x + 1 \end{cases}$

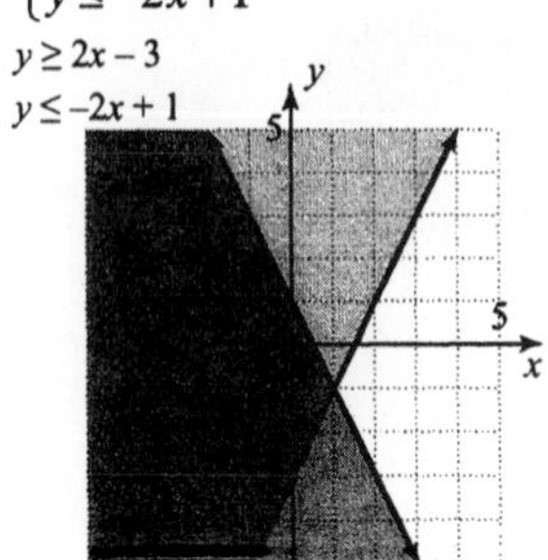

36. $\begin{cases} y \le -3x - 3 \\ y \le 2x + 7 \end{cases}$

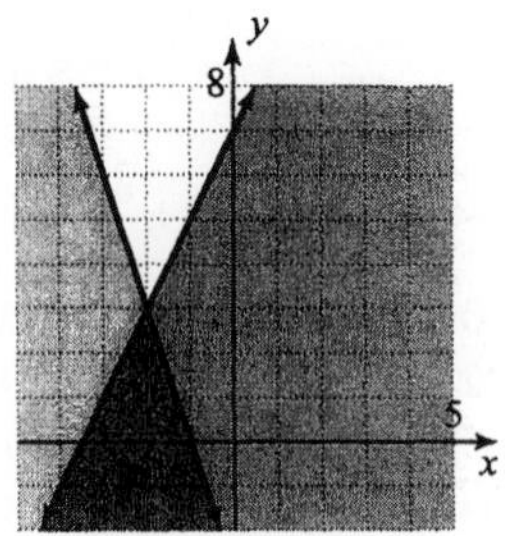

37. $\begin{cases} x + 2y > 0 \\ x - y \le 6 \end{cases}$

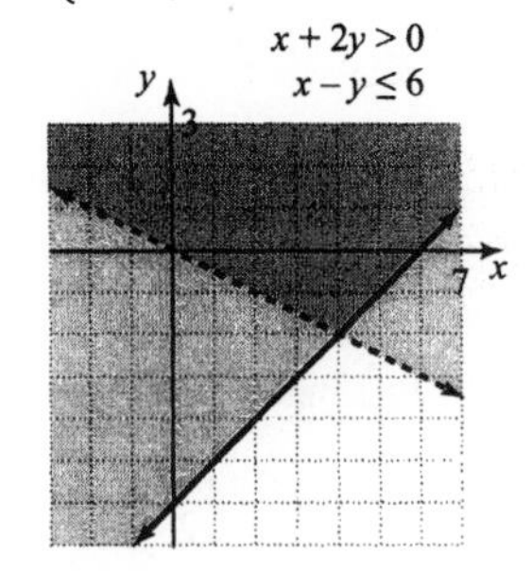

38. $\begin{cases} 4x - y \le 0 \\ 3x - 2y \ge -5 \end{cases}$

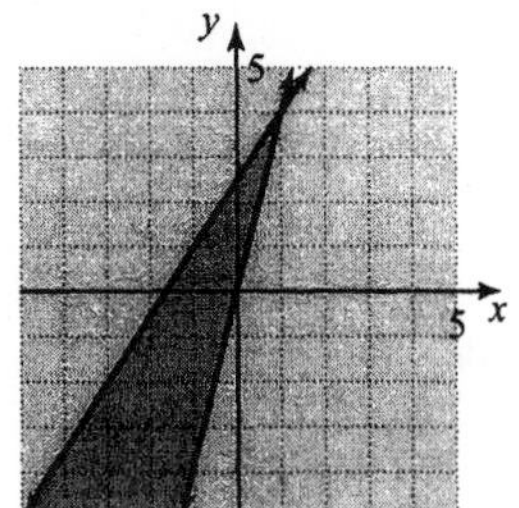

39. $\begin{cases} 3x - 2y \le 4 \\ 2x + y \ge 5 \end{cases}$

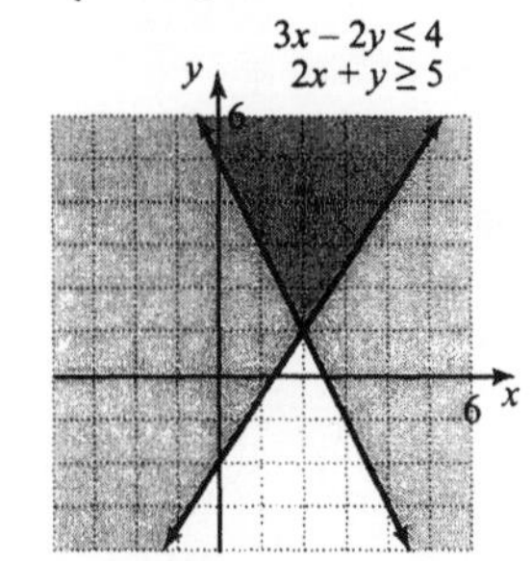

40. $\begin{cases} -2x+3y>-7 \\ x \geq -2 \end{cases}$

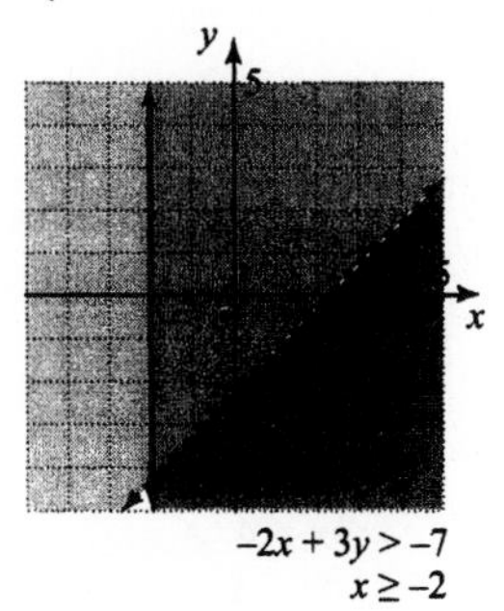

41. $0 \leq \frac{2(3x+4)}{5} \leq 3$

$5(0) \leq 5\left[\frac{2(3x+4)}{5}\right] \leq 5(3)$

$0 \leq 2(3x+4) \leq 15$

$0 \leq 6x+8 \leq 15$

$-8 \leq 6x \leq 7$

$-\frac{4}{3} \leq x \leq \frac{7}{6}$

$\left[-\frac{4}{3}, \frac{7}{6}\right]$

42. $x \leq 2$ or $x > -5$

$(-\infty, \infty)$

43. $-2x \leq 6$ and $-2x+3 < -7$

$x \geq -3$ and $-2x < -10$

$x \geq -3$ and $x > 5$

$x > 5$

$(5, \infty)$

44. $|7x| - 26 = -5$

$|7x| = 21$

$7x = 21$ or $7x = -21$

$x = 3$ or $x = -3$

45. $\left|\frac{9-2x}{5}\right| = -3$

The solution set is $\varnothing$.

46. $|x-3| = |7+2x|$

$x-3 = 7+2x$ or $x-3 = -(7+2x)$

$-10 = x$ or $x-3 = -7-2x$

$3x = -4$

$x = -\frac{4}{3}$

47. $|6x-5| \geq -1$

Since $|6x-5|$ is nonnegative for all numbers x, the solution set is $(-\infty, \infty)$.

48. $\left|\frac{4x-3}{5}\right| < 1$

$-1 < \frac{4x-3}{5} < 1$

$-5 < 4x-3 < 5$

$-2 < 4x < 8$

$-\frac{1}{2} < x < 2$

$\left(-\frac{1}{2}, 2\right)$

49. $-x \leq y$

Test (1, 0)

$-1 \overset{?}{\leq} 0$

True

Shade above.

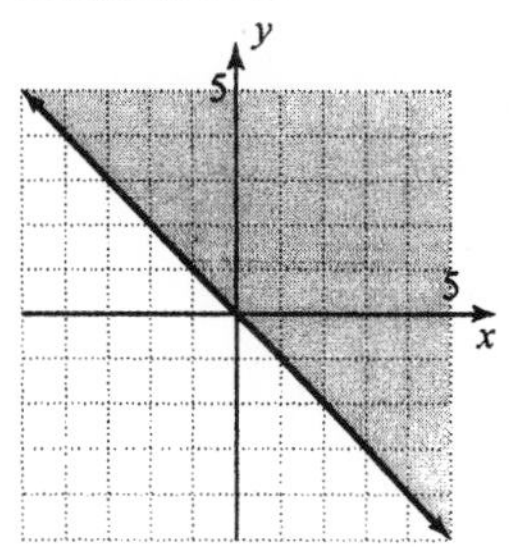

50. $x+y > -2$

Test (0, 0)

$0+0 \overset{?}{>} -2$

True

Shade above.

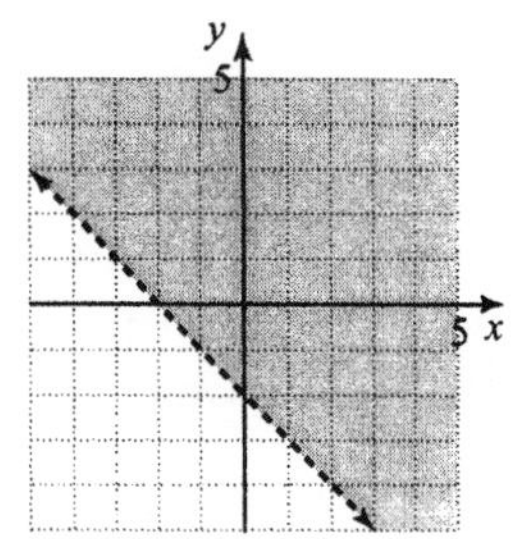

51. 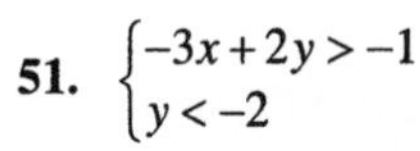

$$\begin{cases} -3x+2y>-1 \\ y<-2 \end{cases}$$

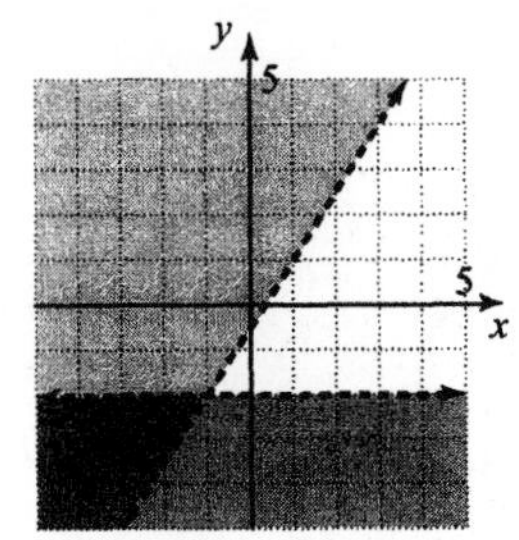

$-3x+2y>-1$
$y<-2$

52. $\begin{cases} x-2y\ge 7 \\ x+y\le -5 \end{cases}$

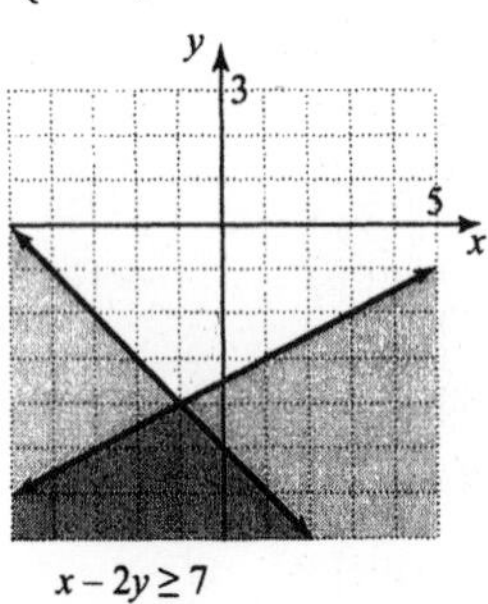

$x-2y\ge 7$
$x+y\le -5$

Chapter 9 Test

1. $|6x-5|-3=-2$
$|6x-5|=1$
$6x-5=1$ or $6x-5=-1$
$6x=6$ or $6x=4$
$x=1$ or $x=\frac{2}{3}$

2. $|8-2t|=-6$
No solution, $\varnothing$

3. $|x-5|=|x+2|$
$x-5=x+2$ or $x-5=-(x+2)$
$-5=2$ False or $x-5=-x-2$
$2x=3$
$x=\frac{3}{2}$

Since $-5=2$ is not possible, the only solution is $\frac{3}{2}$.

4. $-3<2(x-3)\le 4$
$-3<2x-6\le 4$
$3<2x\le 10$
$\frac{3}{2}<x\le 5$
$\left(\frac{3}{2}, 5\right]$

5. $|3x+1|>5$
$3x+1<-5$ or $3x+1>5$
$3x<-6$ or $3x>4$
$x<-2$ or $x>\frac{4}{3}$
$(-\infty, -2)\cup\left(\frac{4}{3}, \infty\right)$

6. $|x-5|-4<-2$
$|x-5|<2$
$-2<x-5<2$
$3<x<7$
$(3, 7)$

7. $x\le -2$ and $x\le -5$
$(-\infty, -5]$

8. $x\le -2$ or $x\le -5$
$(-\infty, -2]$

9. $-x>1$ and $3x+3\ge x-3$
$\frac{-x}{-1}<\frac{1}{-1}$ and $2x\ge -6$
$x<-1$ and $x\ge -3$
$-3\le x<-1$
$[-3, -1)$

10. $6x+1>5x+4$ or $1-x>-4$
$x>3$ or $5>x$
$(-\infty, \infty)$

11. $\left|\frac{5x-7}{2}\right|=4$
$\frac{5x-7}{2}=4$ or $\frac{5x-7}{2}=-4$
$5x-7=8$ or $5x-7=-8$
$5x=15$ or $5x=-1$
$x=3$ or $x=-\frac{1}{5}$

12. $\left|17x-\frac{1}{5}\right|>-2$

The solution set is $(-\infty, \infty)$ since an absolute value is never negative.

13.

$$-1 \le \frac{2x-5}{3} < 2$$
$$3(-1) \le 3\left(\frac{2x-5}{3}\right) < 3(2)$$
$$-3 \le 2x-5 < 6$$
$$-3+5 \le 2x-5+5 < 6+5$$
$$2 \le 2x < 11$$
$$\frac{2}{2} \le \frac{2x}{2} < \frac{11}{2}$$
$$1 \le x < \frac{11}{2}$$

$\left[1, \frac{11}{2}\right)$

14. $y > -4x$

Test (1, 0)

$0 \overset{?}{>} -4(1)$

True

Shade above.

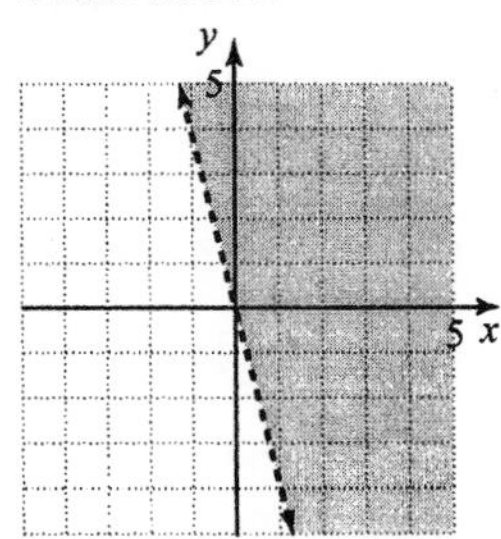

15. $2x - 3y > -6$

Test (0, 0)

$2(0) - 3(0) \overset{?}{>} -6$

True

Shade below.

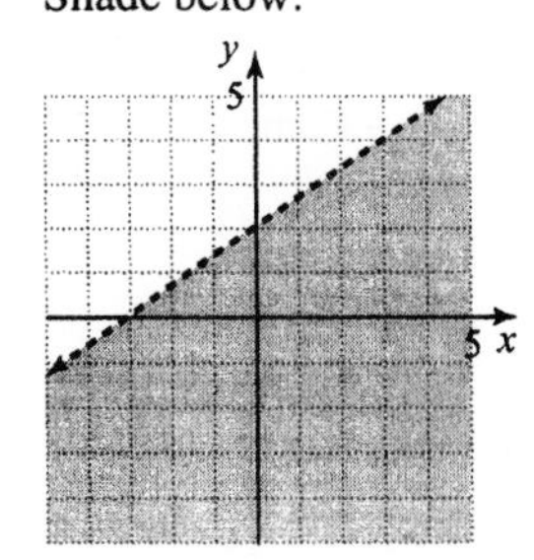

16. $\begin{cases} y + 2x \le 4 \\ y \ge 2 \end{cases}$

$y + 2x \le 4$
$y \ge 2$

17. $\begin{cases} 2y - x \ge 1 \\ x + y \ge -4 \end{cases}$

$2y - x \ge 1$
$x + y \ge -4$

Chapter 9 Cumulative Review

1. Let $x = 2$ and $y = -5$.

a. $\frac{x-y}{12+x} = \frac{2-(-5)}{12+2} = \frac{2+5}{14} = \frac{7}{14} = \frac{1}{2}$

b. $x^2 - 3y = 2^2 - 3(-5) = 4 + 15 = 19$

2. Let $x = -4$ and $y = 7$.

a. $\frac{x-y}{7-x} = \frac{-4-7}{7-(-4)} = \frac{-11}{7+4} = \frac{-11}{11} = -1$

b. $x^2 + 2y = (-4)^2 + 2 \cdot 7 = 16 + 14 = 30$

3. a. $\frac{(-12)(-3)+3}{-7-(-2)} = \frac{36+3}{-7+2} = \frac{39}{-5} = -\frac{39}{5}$

b. $\frac{2(-3)^2 - 20}{-5+4} = \frac{2 \cdot 9 - 20}{-1} = \frac{18-20}{-1} = \frac{-2}{-1} = 2$

4. a. $\frac{4(-3)-(-6)}{-8+4} = \frac{-12+6}{-8+4} = \frac{-6}{-4} = \frac{3}{2}$

b. $\frac{3+(-3)(-2)^3}{-1-(-4)} = \frac{3+(-3)(-8)}{-1+4}$
$= \frac{3+24}{3}$
$= \frac{27}{3}$
$= 9$

5. a. $2x+3x+5+2 = (2+3)x+(5+2) = 5x+7$

b. $-5a-3+a+2 = -5a+a-3+2 = -4a-1$

c. $4y-3y^2$ cannot be simplified.

d. $2.3x + 5x - 6 = 7.3x - 6$

e. $-\frac{1}{2}b+b = \left(-\frac{1}{2}+1\right)b = \frac{1}{2}b$

6. a. $4x - 3 + 7 - 5x = 4x - 5x - 3 + 7 = -x + 4$

b. $-6y+3y-8+8y = -6y+3y+8y-8$
$= 5y-8$

c. $2+8.1a+a-6 = 8.1a+a+2-6 = 9.1a-4$

d. $2x^2-2x$ cannot be simplified.

7. $2x+3x-5+7 = 10x+3-6x-4$
$5x+2 = 4x-1$
$x+2 = -1$
$x = -3$

8. $6y-11+4+2y = 8+15y-8y$
$8y-7 = 7y+8$
$y-7 = 8$
$y = 15$

9. $y = 3x$
$x = -1$: $y = 3(-1) = -3$
$y = 0$: $0 = 3x$
$\frac{0}{3} = \frac{3x}{3}$
$0 = x$
$y = -9$: $-9 = 3x$
$\frac{-9}{3} = \frac{3x}{3}$
$-3 = x$

x	y
–1	–3
0	0
–3	–9

10. $2x + y = 6$
$x = 0$: $2(0) + y = 6$
$y = 6$
$y = -2$: $2x + (-2) = 6$
$2x - 2 = 6$
$2x = 8$
$x = 4$
$x = 3$: $2(3) + y = 6$
$6 + y = 6$
$y = 0$

x	y
0	6
4	–2
3	0

11. a. x-intercept: (–3, 0); y-intercept: (0, 2)

b. x-intercepts: (–4, 0) and (–1, 0);
y-intercept: (0, 1)

c. x-intercept: (0, 0); y-intercept: (0, 0)

d. x-intercept: (2, 0); no y-intercept

e. x-intercepts: (–1, 0) and (3, 0);
y-intercepts: (0, –1) and (0, 2)

12. a. x-intercept: (4, 0); y-intercept: (0, 1)

b. x-intercepts: (–2, 0), (0, 0), and (3, 0);
y-intercept: (0, 0)

c. no x-intercept; y-intercept: (0, –3)

d. x-intercepts: (–3, 0) and (3, 0);
y-intercepts: (0, –3) and (0, 3)

13. $y = -\frac{1}{5}x+1$: $m = -\frac{1}{5}$, $b = 1$
$2x+10y = 30$
$10y = -2x+30$
$y = -\frac{1}{5}x+3$: $m = -\frac{1}{5}$, $b = 3$
The slopes are the same, but the y-intercepts are different, so the lines are parallel.

14. $y = 3x + 7$: $m = 3$, $b = 7$

$$x + 3y = -15$$
$$3y = -x - 15$$
$$y = -\frac{1}{3}x - 5\text{: } m = -\frac{1}{3},\ b = -5$$

The product of the slopes is -1, so the lines are perpendicular.

15. y-intercept $(0, -3)$: $b = -3$; $m = \frac{1}{4}$

$$y = mx + b$$
$$y = \frac{1}{4}x + (-3) \text{ or } y = \frac{1}{4}x - 3$$

16. y-intercept $(0, 4)$: $b = 4$; $m = -2$

$$y = mx + b$$
$$y = -2x + 4$$

17. The line $y = 5$ is vertical. A parallel line will also be vertical. The vertical line passing through $(-2, -3)$ has equation $y = -3$.

18. $y = 2x + 4$: $m = 2$

A perpendicular line has slope $m = -\frac{1}{2}$.

$$(x_1, y_1) = (1, 5)$$
$$y - y_1 = m(x - x_1)$$
$$y - 5 = -\frac{1}{2}(x - 1)$$
$$y - 5 = -\frac{1}{2}x + \frac{1}{2}$$
$$y = -\frac{1}{2}x + \frac{11}{2}$$

19. a, b, and c are functions since they represent non-vertical lines.

20. a, c, and d are functions since they represent non-vertical lines.

21. $\begin{cases} 2x - 3y = 6 \\ x = 2y \end{cases}$

a. Let $x = 12$ and $y = 6$.

$$2x - 3y = 6 \qquad\qquad x = 2y$$
$$2(12) - 3(6) \stackrel{?}{=} 6 \qquad\qquad 12 \stackrel{?}{=} 2(6)$$
$$24 - 18 \stackrel{?}{=} 6 \qquad\qquad 12 = 12 \quad \text{True}$$
$$6 = 6 \quad \text{True}$$

$(12, 6)$ is a solution.

b. Let $x = 0$ and $y = -2$.

$$2x - 3y = 6 \qquad\qquad x = 2y$$
$$2(0) - 3(-2) \stackrel{?}{=} 6 \qquad\qquad 0 \stackrel{?}{=} 2(-2)$$
$$0 + 6 \stackrel{?}{=} 6 \qquad\qquad 0 = -4 \quad \text{False}$$
$$6 = 6 \quad \text{True}$$

$(0, -2)$ is not a solution.

22. $\begin{cases} 2x + y = 4 \\ x + y = 2 \end{cases}$

a. Let $x = 1$ and $y = 1$.

$$2x + y = 4$$
$$2(1) + 1 \stackrel{?}{=} 4$$
$$2 + 1 \stackrel{?}{=} 4$$
$$3 = 4 \quad \text{False}$$

$(1, 1)$ is not a solution.

b. Let $x = 2$ and $y = 0$.

$$2x + y = 4 \qquad\qquad x + y = 2$$
$$2(2) + 0 \stackrel{?}{=} 4 \qquad\qquad 2 + 0 \stackrel{?}{=} 2$$
$$4 + 0 \stackrel{?}{=} 4 \qquad\qquad 2 = 2 \quad \text{True}$$
$$4 = 4 \quad \text{True}$$

$(2, 0)$ is a solution.

23. $(11x^3 - 12x^2 + x - 3) + (x^3 - 10x + 5)$

$$= 11x^3 + x^3 - 12x^2 + x - 10x - 3 + 5$$
$$= 12x^3 - 12x^2 - 9x + 2$$

24. $4a^2 + 3a - 2a^2 + 7a - 5$

$$= 4a^2 - 2a^2 + 3a + 7a - 5$$
$$= 2a^2 + 10a - 5$$

25. $x^2 + 7yx + 6y^2 = (x + 6y)(x + y)$

26. $3x^2 + 15x + 18 = 3(x^2 + 5x + 6) = 3(x + 2)(x + 3)$

27. $\frac{3x^3y^7}{40} \div \frac{4x^3}{y^2} = \frac{3x^3y^7}{40} \cdot \frac{y^2}{4x^3} = \frac{3y^9}{160}$

28. $\frac{12x^2y^3}{5} \div \frac{3y^2}{x} = \frac{12x^2y^3}{5} \cdot \frac{x}{3y^2} = \frac{4x^3y}{5}$

29. $\frac{2y}{2y - 7} - \frac{7}{2y - 7} = \frac{2y - 7}{2y - 7} = 1$

30. $\frac{-4x^2}{x + 1} - \frac{4x}{x + 1} = \frac{-4x^2 - 4x}{x + 1} = \frac{-4x(x + 1)}{x + 1} = -4x$

31. $\frac{2x}{x^2+2x+1}+\frac{x}{x^2-1}=\frac{2x}{(x+1)^2}+\frac{x}{(x+1)(x-1)}$
$=\frac{2x(x-1)+x(x+1)}{(x+1)^2(x-1)}$
$=\frac{2x^2-2x+x^2+x}{(x+1)^2(x-1)}$
$=\frac{3x^2-x}{(x+1)^2(x-1)}$
$=\frac{x(3x-1)}{(x+1)^2(x-1)}$

32. $\frac{3x}{x^2+5x+6}+\frac{1}{x^2+2x-3}$
$=\frac{3x}{(x+2)(x+3)}+\frac{1}{(x+3)(x-1)}$
$=\frac{3x(x-1)+1(x+2)}{(x+2)(x+3)(x-1)}$
$=\frac{3x^2-3x+x+2}{(x+2)(x+3)(x-1)}$
$=\frac{3x^2-2x+2}{(x+2)(x+3)(x-1)}$

33. $\frac{x}{2}+\frac{8}{3}=\frac{1}{6}$
$6\left(\frac{x}{2}+\frac{8}{3}\right)=6\left(\frac{1}{6}\right)$
$3x+16=1$
$3x=-15$
$x=-5$

34. $\frac{1}{21}+\frac{x}{7}=\frac{5}{3}$
$21\left(\frac{1}{21}+\frac{x}{7}\right)=21\left(\frac{5}{3}\right)$
$1+3x=35$
$3x=34$
$x=\frac{34}{3}$

35. $\begin{cases}2x+y=7\\2y=-4x\end{cases}$

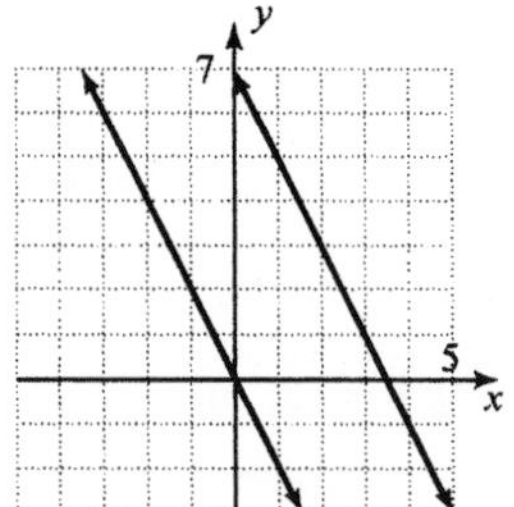

The system has no solution.

36. $\begin{cases}y=x+2\\2x+y=5\end{cases}$

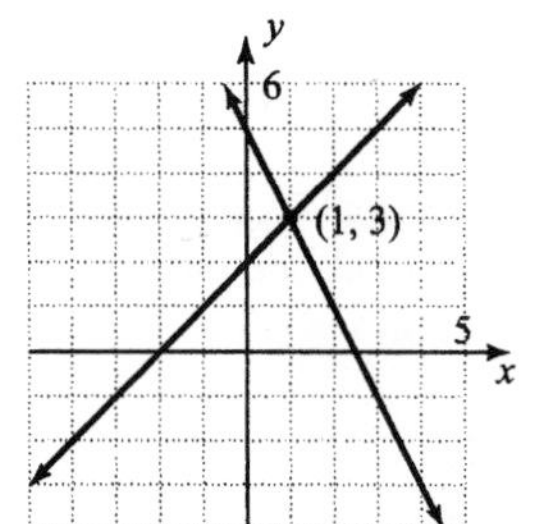

The solution is (1, 3).

37. $\begin{cases}7x-3y=-14\\-3x+y=6\end{cases}$

Solve the second equation for y.
$y = 3x + 6$
Substitute $3x + 6$ for y in the first equation.
$7x-3(3x+6)=-14$
$7x-9x-18=-14$
$-2x-18=-14$
$-2x=4$
$x=-2$
Let $x = -2$ in $y = 3x + 6$.
$y = 3(-2) + 6 = -6 + 6 = 0$
The solution is (–2, 0).

38. $\begin{cases}5x+y=3\\y=-5x\end{cases}$

Substitute $-5x$ for y in the first equation.
$5x+(-5x)=3$
$0=3$
This is a false statement, so the system has no solution.

39. $\begin{cases} 3x-2y=2 \\ -9x+6y=-6 \end{cases}$

Multiply the first equation by 3, then add.

$$\begin{array}{r} 9x-6y=6 \\ -9x+6y=-6 \\ \hline 0=0 \end{array}$$

This is a true statement, so the system has an infinite number of solutions.

40. $\begin{cases} -2x+y=7 \\ 6x-3y=-21 \end{cases}$

Multiply the first equation by 3, then add.

$$\begin{array}{r} -6x+3y=21 \\ 6x-3y=-21 \\ \hline 0=0 \end{array}$$

This is a true statement, so the system has an infinite number of solutions.

41. $\begin{cases} -3x+4y<12 \\ x\ge 2 \end{cases}$

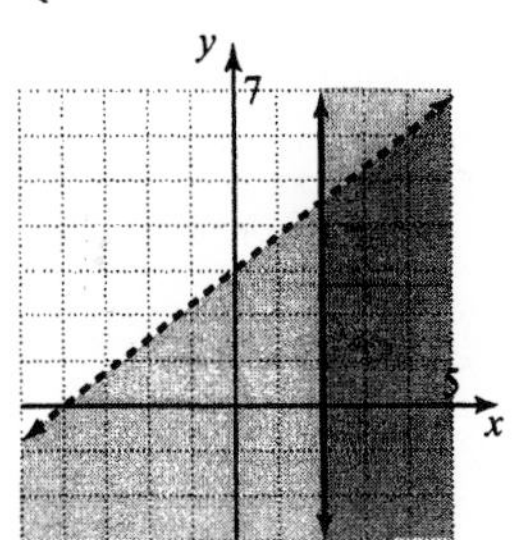

$-3x+4y<12$
$x\ge 2$

42. $\begin{cases} 2x-y\le 6 \\ y\ge 2 \end{cases}$

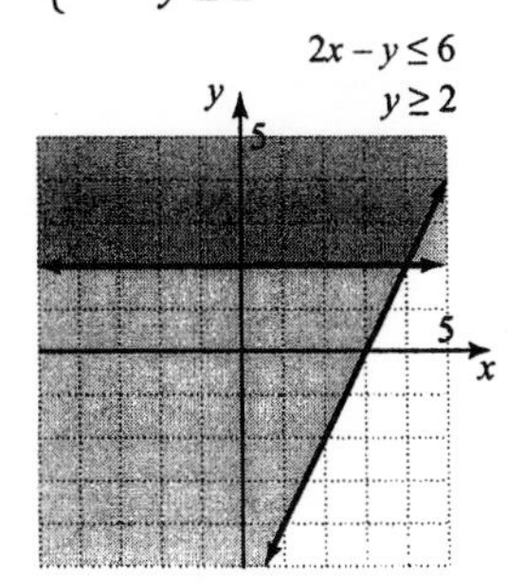

43. a. $\left(\frac{st}{2}\right)^4=\frac{s^4t^4}{2^4}=\frac{s^4t^4}{16}$

b. $(9y^5z^7)^2=9^2(y^5)^2(z^7)^2=81y^{10}z^{14}$

c. $\left(\frac{-5x^2}{y^3}\right)^2=\frac{(-5)^2(x^2)^2}{(y^3)^2}=\frac{25x^4}{y^6}$

44. a. $\left(\frac{-6x}{y^3}\right)^3=\frac{(-6)^3x^3}{(y^3)^3}=\frac{-216x^3}{y^9}=-\frac{216x^3}{y^9}$

b.
$$\begin{aligned}\frac{a^2b^7}{(2b^2)^5}&=\frac{a^2b^7}{2^5(b^2)^5}\\&=\frac{a^2b^7}{32b^{10}}\\&=\frac{a^2}{32}b^{7-10}\\&=\frac{a^2}{32}b^{-3}\\&=\frac{a^2}{32b^3}\end{aligned}$$

c. $\frac{(3y)^2}{y^2}=\frac{3^2y^2}{y^2}=9y^{2-2}=9y^0=9$

d.
$$\begin{aligned}\frac{(x^2y^4)^2}{xy^3}&=\frac{(x^2)^2(y^4)^2}{xy^3}\\&=\frac{x^4y^8}{xy^3}\\&=x^{4-1}y^{8-3}\\&=x^3y^5\end{aligned}$$

45.
$$\begin{aligned}(5x-1)(2x^2+15x+18)&=0\\(5x-1)(2x+3)(x+6)&=0\end{aligned}$$

$5x-1=0$ or $2x+3=0$ or $x+6=0$

$5x=1 \qquad 2x=-3 \qquad x=-6$

$x=\frac{1}{5} \qquad x=-\frac{3}{2}$

The solutions are -6, $-\frac{3}{2}$, and $\frac{1}{5}$.

46. $(x+1)(2x^2-3x-5)=0$
$(x+1)(2x-5)(x+1)=0$
$x+1=0$ or $2x-5=0$
$x=-1$ $\quad 2x=5$
$x=\frac{5}{2}$

The solutions are -1 and $\frac{5}{2}$.

47. $\frac{45}{x}=\frac{5}{7}$
$45\cdot 7=5x$
$315=5x$
$\frac{315}{5}=\frac{5x}{5}$
$63=x$

48. $\frac{2x+7}{3}=\frac{x-6}{2}$
$2(2x+7)=3(x-6)$
$4x+14=3x-18$
$x+14=-18$
$x=-32$

Chapter 10

Section 10.1

Practice Exercises

1. a. $\sqrt{49} = 7$ because $7^2 = 49$ and 7 is not negative.

   b. $\sqrt{\frac{0}{1}} = \sqrt{0} = 0$ because $0^2 = 0$ and 0 is not negative.

   c. $\sqrt{\frac{16}{81}} = \frac{4}{9}$ because $\left(\frac{4}{9}\right)^2 = \frac{16}{81}$ and $\frac{4}{9}$ is not negative.

   d. $\sqrt{0.64} = 0.8$ because $(0.8)^2 = 0.64$.

   e. $\sqrt{z^8} = z^4$ because $(z^4)^2 = z^8$.

   f. $\sqrt{16b^4} = 4b^2$ because $(4b^2)^2 = 16b^4$.

   g. $-\sqrt{36} = -6$. The negative in front of the radical indicates the negative square root of 36.

   h. $\sqrt{-36}$ is not a real number.

2. $\sqrt{45} \approx 6.708$
   Since $36 < 45 < 49$, then $\sqrt{36} < \sqrt{45} < \sqrt{49}$, or $6 < \sqrt{45} < 7$. The approximation is between 6 and 7 and thus is reasonable.

3. a. $\sqrt[3]{-1} = -1$ because $(-1)^3 = -1$.

   b. $\sqrt[3]{27} = 3$ because $3^3 = 27$.

   c. $\sqrt[3]{\frac{27}{64}} = \frac{3}{4}$ because $\left(\frac{3}{4}\right)^3 = \frac{27}{64}$.

   d. $\sqrt[3]{x^{12}} = x^4$ because $(x^4)^3 = x^{12}$.

   e. $\sqrt[3]{-8x^3} = -2x$ because $(-2x)^3 = -8x^3$.

4. a. $\sqrt[4]{10{,}000} = 10$ because $10^4 = 10{,}000$ and 10 is positive.

   b. $\sqrt[5]{-1} = -1$ because $(-1)^5 = -1$.

   c. $-\sqrt{81} = -9$ because -9 is the opposite of $\sqrt{81}$.

   d. $\sqrt[4]{-625}$ is not a real number. There is no real number that, when raised to the fourth power, is -625.

   e. $\sqrt[3]{27x^9} = 3x^3$ because $(3x^3)^3 = 27x^9$.

5. a. $\sqrt{(-4)^2} = |-4| = 4$

   b. $\sqrt{x^{14}} = |x^7|$

   c. $\sqrt[4]{(x+7)^4} = |x+7|$

   d. $\sqrt[3]{(-7)^3} = -7$

   e. $\sqrt[5]{(3x-5)^5} = 3x-5$

   f. $\sqrt{49x^2} = 7|x|$

   g. $\sqrt{x^2+4x+4} = \sqrt{(x+2)^2} = |x+2|$

6. $f(x) = \sqrt{x+5},\ g(x) = \sqrt[3]{x-3}$

   a. $f(11) = \sqrt{11+5} = \sqrt{16} = 4$

   b. $f(-1) = \sqrt{-1+5} = \sqrt{4} = 2$

   c. $g(11) = \sqrt[3]{11-3} = \sqrt[3]{8} = 2$

   d. $g(-5) = \sqrt[3]{-5-3} = \sqrt[3]{-8} = -2$

7. $h(x) = \sqrt{x+2}$
   Find the domain.
   $x+2 \geq 0$
   $x \geq -2$
   The domain of $h(x)$ is $\{x \mid x \geq -2\}$.

x	$h(x)=\sqrt{x+2}$
−2	0
−1	1
1	$\sqrt{1+2}=\sqrt{3}\approx 1.7$
2	2
7	3

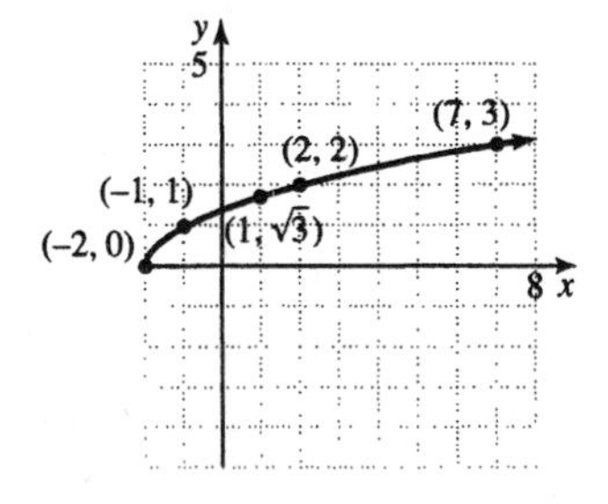

8. $f(x)=\sqrt[3]{x}-4$
The domain is the set of all real numbers.

x	$f(x)=\sqrt[3]{x}-4$
0	−4
1	−3
−1	−5
6	$\sqrt[3]{6}-4\approx 1.8-4=-2.2$
−6	$\sqrt[3]{-6}-4\approx -1.8-4=-5.8$
8	−2
−8	−6

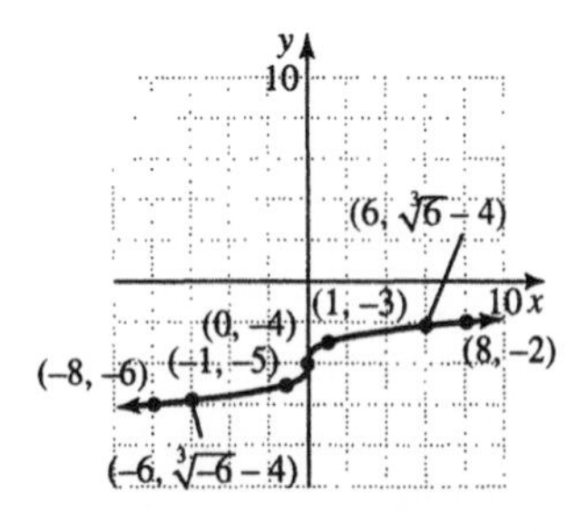

Vocabulary and Readiness Check

1. In the expression $\sqrt[n]{a}$, the n is called the index, the $\sqrt{\ }$ is called the radical sign, and a is called the radicand.

2. If $\sqrt{a}$ is the positive square root of a, $a \neq 0$, then $-\sqrt{a}$ is the negative square root of a.

3. The square root of a negative number is not a real number.

4. Numbers such as 1, 4, 9, and 25 are called perfect squares where numbers such as 1, 8, 27, and 125 are called perfect cubes.

5. The domain of the function $f(x)=\sqrt{x}$ is $[0, \infty)$.

6. The domain of the function $f(x)=\sqrt[3]{x}$ is $(-\infty, \infty)$.

7. If $f(16) = 4$, the corresponding ordered pair is $(16, 4)$.

8. If $g(-8) = -2$, the corresponding ordered pair is $(-8, -2)$.

9. The radical that is not a real number is $\sqrt{-10}$, choice d.

10. The radicals that simplify to 3 are $\sqrt{9}$ and $\sqrt[3]{27}$, choices a and c.

11. The radical that simplifies to −3 is $\sqrt[3]{-27}$, choice d.

12. The radical that does not simplify to a whole number is $\sqrt{8}$, choice c.

Exercise Set 10.1

1. $\sqrt{100}=10$ because $10^2=100$.

3. $\sqrt{\frac{1}{4}}=\frac{1}{2}$ because $\left(\frac{1}{2}\right)^2=\frac{1}{4}$.

5. $\sqrt{0.0001}=0.01$ because $(0.01)^2=0.0001$.

7. $-\sqrt{36}=-1\cdot\sqrt{36}=-1\cdot 6=-6$

9. $\sqrt{x^{10}}=x^5$ because $(x^5)^2=x^{10}$.

11. $\sqrt{16y^6}=4y^3$ because $(4y^3)^2=16y^6$.

13. $\sqrt{7} \approx 2.646$
Since $4 < 7 < 9$, then $\sqrt{4} < \sqrt{7} < \sqrt{9}$, or $2 < \sqrt{7} < 3$. The approximation is between 2 and 3 and thus is reasonable.

15. $\sqrt{38} \approx 6.164$
Since $36 < 38 < 49$, then $\sqrt{36} < \sqrt{38} < \sqrt{49}$, or $6 < \sqrt{38} < 7$. The approximation is between 6 and 7 and thus is reasonable.

17. $\sqrt{200} \approx 14.142$
Since $196 < 200 < 225$, then $\sqrt{196} < \sqrt{200} < \sqrt{225}$, or $14 < \sqrt{200} < 15$. The approximation is between 14 and 15 and thus is reasonable.

19. $\sqrt[3]{64} = 4$ because $4^3 = 64$.

21. $\sqrt[3]{\frac{1}{8}} = \frac{1}{2}$ because $\left(\frac{1}{2}\right)^3 = \frac{1}{8}$.

23. $\sqrt[3]{-1} = -1$ because $(-1)^3 = -1$.

25. $\sqrt[3]{x^{12}} = x^4$ because $(x^4)^3 = x^{12}$.

27. $\sqrt[3]{-27x^9} = -3x^3$ because $(-3x^3)^3 = -27x^9$.

29. $-\sqrt[4]{16} = -2$ because $2^4 = 16$.

31. $\sqrt[4]{-16}$ is not a real number. There is no real number that, when raised to the fourth power, is -16.

33. $\sqrt[5]{-32} = -2$ because $(-2)^5 = -32$.

35. $\sqrt[5]{x^{20}} = x^4$ because $(x^4)^5 = x^{20}$.

37. $\sqrt[6]{64x^{12}} = 2x^2$ because $(2x^2)^6 = 64x^{12}$.

39. $\sqrt{81x^4} = 9x^2$ because $(9x^2)^2 = 81x^4$.

41. $\sqrt[4]{256x^8} = 4x^2$ because $(4x^2)^4 = 256x^8$.

43. $\sqrt{(-8)^2} = |-8| = 8$

45. $\sqrt[3]{(-8)^3} = -8$

47. $\sqrt{4x^2} = 2|x|$

49. $\sqrt[3]{x^3} = x$

51. $\sqrt{(x-5)^2} = |x-5|$

53. $\sqrt{x^2+4x+4} = \sqrt{(x+2)^2} = |x+2|$

55. $-\sqrt{121} = -11$

57. $\sqrt[3]{8x^3} = 2x$

59. $\sqrt{y^{12}} = y^6$

61. $\sqrt{25a^2b^{20}} = 5ab^{10}$

63. $\sqrt[3]{-27x^{12}y^9} = -3x^4y^3$

65. $\sqrt[4]{a^{16}b^4} = a^4b$

67. $\sqrt[5]{-32x^{10}y^5} = -2x^2y$

69. $\sqrt{\frac{25}{49}} = \frac{5}{7}$

71. $\sqrt{\frac{x^2}{4y^2}} = \frac{x}{2y}$

73. $-\sqrt[3]{\frac{z^{21}}{27x^3}} = -\frac{z^7}{3x}$

75. $\sqrt[4]{\frac{x^4}{16}} = \frac{x}{2}$

77. $f(x) = \sqrt{2x+3}$
$f(0) = \sqrt{2(0)+3} = \sqrt{3}$

79. $g(x) = \sqrt[3]{x-8}$
$g(7) = \sqrt[3]{7-8} = \sqrt[3]{-1} = -1$

81. $g(x) = \sqrt[3]{x-8}$
$g(-19) = \sqrt[3]{-19-8} = \sqrt[3]{-27} = -3$

83. $f(x) = \sqrt{2x+3}$
$f(2) = \sqrt{2(2)+3} = \sqrt{7}$

85. $f(x) = \sqrt{x} + 2$
$x \geq 0$
Domain: $[0, \infty)$

x	$f(x) = \sqrt{x} + 2$
0	$\sqrt{0} + 2 = 2$
1	$\sqrt{1} + 2 = 3$
3	$\sqrt{3} + 2 \approx 3.7$
4	$\sqrt{4} + 2 = 4$

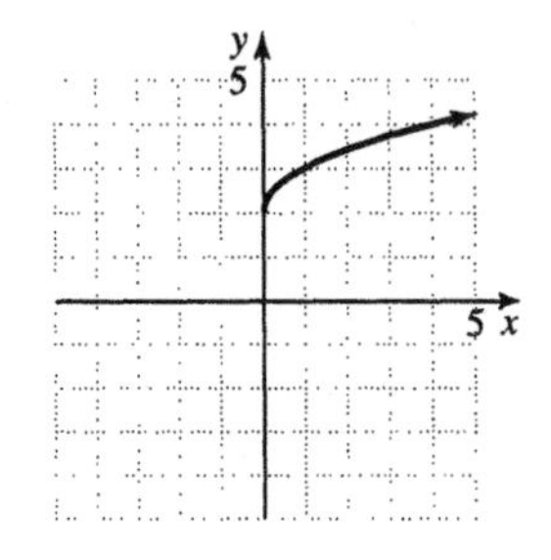

87. $f(x) = \sqrt{x-3}$
$x - 3 \geq 0$
$x \geq 3$
Domain: $[3, \infty)$

x	$f(x) = \sqrt{x-3}$
3	$\sqrt{3-3} = \sqrt{0} = 0$
4	$\sqrt{4-3} = \sqrt{1} = 1$
7	$\sqrt{7-3} = \sqrt{4} = 2$
12	$\sqrt{12-3} = \sqrt{9} = 3$

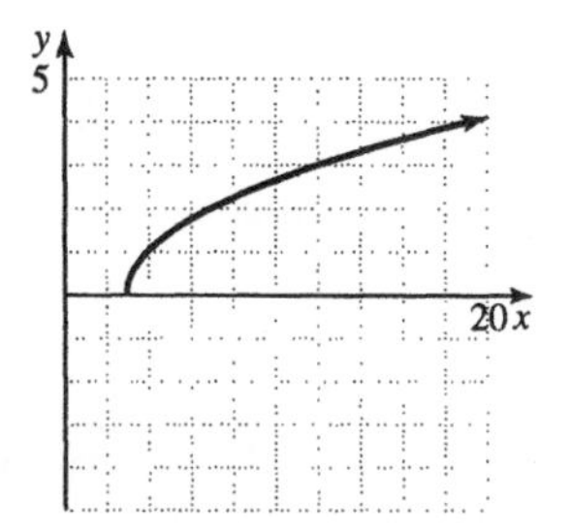

89. $f(x) = \sqrt[3]{x} + 1$
Domain: $(-\infty, \infty)$

x	$f(x) = \sqrt[3]{x} + 1$
−4	$\sqrt[3]{-4} + 1 \approx -0.6$
−1	$\sqrt[3]{-1} + 1 = 0$
0	$\sqrt[3]{0} + 1 = 1$
1	$\sqrt[3]{1} + 1 = 2$
4	$\sqrt[3]{4} + 1 \approx 2.6$

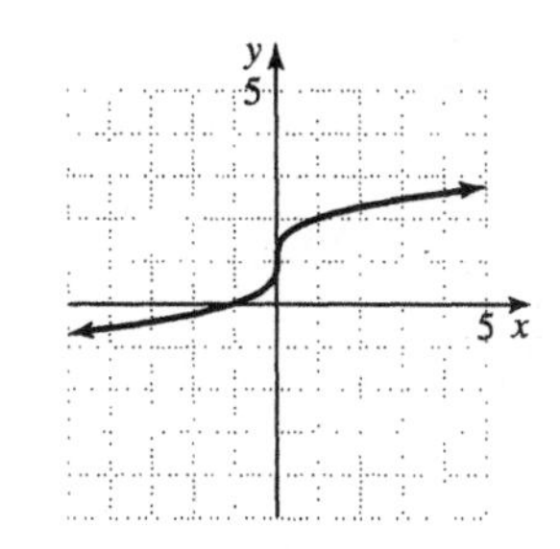

91. $g(x) = \sqrt[3]{x-1}$
Domain: $(-\infty, \infty)$

x	$g(x) = \sqrt[3]{x-1}$
1	$\sqrt[3]{1-1} = \sqrt[3]{0} = 0$
2	$\sqrt[3]{2-1} = \sqrt[3]{1} = 1$
0	$\sqrt[3]{0-1} = \sqrt[3]{-1} = -1$
9	$\sqrt[3]{9-1} = \sqrt[3]{8} = 2$
−7	$\sqrt[3]{-7-1} = \sqrt[3]{-8} = -2$

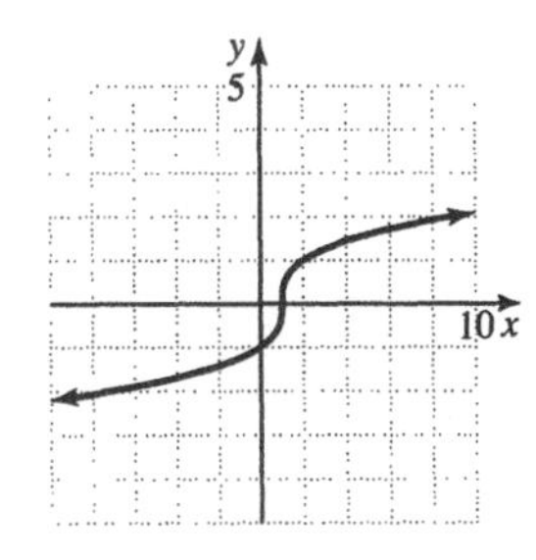

93. $(-2x^3y^2)^5 = (-2)^5 x^{3\cdot5} y^{2\cdot5} = -32x^{15}y^{10}$

95. $(-3x^2y^3z^5)(20x^5y^7) = -3(20)x^{2+5}y^{3+7}z^5$
$= -60x^7y^{10}z^5$

97. $\dfrac{7x^{-1}y}{14(x^5y^2)^{-2}} = \dfrac{7x^{-1}y}{14x^{-10}y^{-4}} = \dfrac{x^9y^5}{2}$

99. $\sqrt{-17}$ is not a real number.

101. $\sqrt[10]{-17}$ is not a real number.

103. Answers may vary

105. $144 < 160 < 169$ so $\sqrt{144} < \sqrt{160} < \sqrt{169}$, or $12 < \sqrt{160} < 13$. Thus $\sqrt{160}$ is between 12 and 13. Therefore, the answer is **b.**

107. $\sqrt{30} \approx 5$, $\sqrt{10} \approx 3$, and $\sqrt{90} \approx 10$ so $P = \sqrt{30} + \sqrt{10} + \sqrt{90} \approx 5 + 3 + 10 = 18$.
Therefore, the answer is **b.**

109. $B = \sqrt{\dfrac{hw}{3131}} = \sqrt{\dfrac{66\cdot135}{3131}}$
$= \sqrt{\dfrac{8910}{3131}}$
≈ 1.69 sq meters

111. Answers may vary

113. $f(x) = \sqrt{x} + 2$

10
−10 10
−10

Domain: $[0, \infty)$

115. $f(x) = \sqrt[3]{x} + 1$

10
−10 10
−10

Domain: $(-\infty, \infty)$

Section 10.2

Practice Exercises

1. a. $36^{1/2} = \sqrt{36} = 6$

b. $1000^{1/3} = \sqrt[3]{1000} = 10$

c. $x^{1/5} = \sqrt[5]{x}$

d. $1^{1/4} = \sqrt[4]{1} = 1$

e. $-64^{1/2} = -\sqrt{64} = -8$

f. $(125x^9)^{1/3} = \sqrt[3]{125x^9} = 5x^3$

g. $(3x)^{1/4} = \sqrt[4]{3x}$

2. a. $16^{3/2} = \left(\sqrt{16}\right)^3 = 4^3 = 64$

b. $-1^{3/5} = -\left(\sqrt[5]{1}\right)^3 = -(1)^3 = -1$

c. $-(81)^{3/4} = -\left(\sqrt[4]{81}\right)^3 = -(3)^3 = -27$

d. $\left(\dfrac{1}{25}\right)^{3/2} = \left(\sqrt{\dfrac{1}{25}}\right)^3 = \left(\dfrac{1}{5}\right)^3 = \dfrac{1}{125}$

e. $(3x+2)^{5/9} = \sqrt[9]{(3x+2)^5}$

3. a. $9^{-3/2} = \dfrac{1}{9^{3/2}} = \dfrac{1}{\left(\sqrt{9}\right)^3} = \dfrac{1}{3^3} = \dfrac{1}{27}$

b. $(-64)^{-2/3} = \dfrac{1}{(-64)^{2/3}} = \dfrac{1}{\left(\sqrt[3]{-64}\right)^2} = \dfrac{1}{(-4)^2} = \dfrac{1}{16}$

4. a. $y^{2/3} \cdot y^{8/3} = y^{(2/3+8/3)} = y^{10/3}$

b. $x^{3/5} \cdot x^{1/4} = x^{3/5+1/4} = x^{12/20+5/20} = x^{17/20}$

c. $\dfrac{9^{2/7}}{9^{9/7}} = 9^{2/7-9/7} = 9^{-7/7} = 9^{-1} = \dfrac{1}{9}$

d. $b^{4/9} \cdot b^{-2/9} = b^{4/9+(-2/9)} = b^{2/9}$

e. $\dfrac{\left(3x^{1/4}y^{-2/3}\right)^4}{x^4y} = \dfrac{3^4(x^{1/4})^4(y^{-2/3})^4}{x^4y}$
$= \dfrac{81xy^{-8/3}}{x^4y}$
$= 81x^{1-4}y^{-8/3-3/3}$
$= 81x^{-3}y^{-11/3}$
$= \dfrac{81}{x^3y^{11/3}}$

5. a. $x^{3/5}(x^{1/3} - x^2) = x^{3/5}x^{1/3} - x^{3/5}x^2$
$= x^{(3/5+1/3)} - x^{(3/5+2)}$
$= x^{(9/15+5/15)} - x^{(3/5+10/5)}$
$= x^{14/15} - x^{13/5}$

b. $(x^{1/2} + 6)(x^{1/2} - 2)$
$= x^{2/2} - 2x^{1/2} + 6x^{1/2} - 12$
$= x + 4x^{1/2} - 12$

6. $2x^{-1/5} - 7x^{4/5} = (x^{-1/5})(2) - (x^{-1/5})(7x^{5/5})$
$= x^{-1/5}(2 - 7x)$

7. a. $\sqrt[9]{x^3} = x^{3/9} = x^{1/3} = \sqrt[3]{x}$

b. $\sqrt[4]{36} = 36^{1/4} = (6^2)^{1/4} = 6^{2/4} = 6^{1/2} = \sqrt{6}$

c. $\sqrt[8]{a^4b^2} = (a^4b^2)^{1/8}$
$= a^{4/8}b^{2/8}$
$= a^{2/4}b^{1/4}$
$= (a^2b)^{1/4}$
$= \sqrt[4]{a^2b}$

8. a. $\sqrt[3]{x} \cdot \sqrt[4]{x} = x^{1/3} \cdot x^{1/4}$
$= x^{1/3+1/4}$
$= x^{4/12+3/12}$
$= x^{7/12}$
$= \sqrt[12]{x^7}$

b. $\dfrac{\sqrt[3]{y}}{\sqrt[5]{y}} = \dfrac{y^{1/3}}{y^{1/5}}$
$= y^{1/3-1/5}$
$= y^{5/15-3/15}$
$= y^{2/15}$
$= \sqrt[15]{y^2}$

c. $\sqrt[3]{5} \cdot \sqrt{3} = 5^{1/3} \cdot 3^{1/2}$
$= 5^{2/6} \cdot 3^{3/6}$
$= (5^2 \cdot 3^3)^{1/6}$
$= \sqrt[6]{5^2 \cdot 3^3}$
$= \sqrt[6]{675}$

Vocabulary and Readiness Check

1. It is true that $9^{-1/2}$ is a positive number.

2. It is false that $9^{-1/2}$ is a whole number.

3. It is true that $\dfrac{1}{a^{-m/n}} = a^{m/n}$ (where $a^{m/n}$ is a nonzero real number).

4. To simplify $x^{2/3} \cdot x^{1/5}$, <u>add</u> the exponents.

5. To simplify $(x^{2/3})^{1/5}$, <u>multiply</u> the exponents.

6. To simplify $\dfrac{x^{2/3}}{x^{1/5}}$, <u>subtract</u> the exponents.

7. $4^{1/2} = 2$, A

8. $-4^{1/2} = -2$, B

9. $(-4)^{1/2}$ is not a real number, C

10. $8^{1/3} = 2$, A

11. $-8^{1/3} = -2$, B

12. $(-8)^{1/3} = -2$, B

Exercise Set 10.2

1. $49^{1/2} = \sqrt{49} = 7$

3. $27^{1/3} = \sqrt[3]{27} = 3$

5. $\left(\frac{1}{16}\right)^{1/4} = \sqrt[4]{\frac{1}{16}} = \frac{1}{2}$

7. $169^{1/2} = \sqrt{169} = 13$

9. $2m^{1/3} = 2\sqrt[3]{m}$

11. $(9x^4)^{1/2} = \sqrt{9x^4} = 3x^2$

13. $(-27)^{1/3} = \sqrt[3]{-27} = -3$

15. $-16^{1/4} = -\sqrt[4]{16} = -2$

17. $16^{3/4} = \left(\sqrt[4]{16}\right)^3 = 2^3 = 8$

19. $(-64)^{2/3} = \left(\sqrt[3]{-64}\right)^2 = (-4)^2 = 16$

21. $(-16)^{3/4} = \left(\sqrt[4]{-16}\right)^3$ is not a real number.

23. $(2x)^{3/5} = \sqrt[5]{(2x)^3}$ or $\left(\sqrt[5]{2x}\right)^3$

25. $(7x+2)^{2/3} = \sqrt[3]{(7x+2)^2}$ or $\left(\sqrt[3]{7x+2}\right)^2$

27. $\left(\frac{16}{9}\right)^{3/2} = \left(\sqrt{\frac{16}{9}}\right)^3 = \left(\frac{4}{3}\right)^3 = \frac{64}{27}$

29. $8^{-4/3} = \frac{1}{8^{4/3}} = \frac{1}{\left(\sqrt[3]{8}\right)^4} = \frac{1}{2^4} = \frac{1}{16}$

31. $(-64)^{-2/3} = \frac{1}{(-64)^{2/3}} = \frac{1}{\left(\sqrt[3]{-64}\right)^2} = \frac{1}{(-4)^2} = \frac{1}{16}$

33. $(-4)^{-3/2} = \frac{1}{(-4)^{3/2}} = \frac{1}{\left(\sqrt{-4}\right)^3}$ is not a real number.

35. $x^{-1/4} = \frac{1}{x^{1/4}}$

37. $\frac{1}{a^{-2/3}} = a^{2/3}$

39. $\frac{5}{7x^{-3/4}} = \frac{5x^{3/4}}{7}$

41. $a^{2/3}a^{5/3} = a^{2/3+5/3} = a^{7/3}$

43. $x^{-2/5} \cdot x^{7/5} = x^{-\frac{2}{5}+\frac{7}{5}} = x^{5/5} = x$

45. $3^{1/4} \cdot 3^{3/8} = 3^{\frac{1}{4}+\frac{3}{8}} = 3^{\frac{2}{8}+\frac{3}{8}} = 3^{5/8}$

47. $\frac{y^{1/3}}{y^{1/6}} = y^{\frac{1}{3}-\frac{1}{6}} = y^{\frac{2}{6}-\frac{1}{6}} = y^{1/6}$

49.
$$\begin{aligned}(4u^2)^{3/2} &= 4^{3/2}u^{2(3/2)} \\ &= \left(\sqrt{4}\right)^3 u^3 \\ &= 2^3 u^3 \\ &= 8u^3\end{aligned}$$

51. $\frac{b^{1/2}b^{3/4}}{-b^{1/4}} = -b^{\frac{1}{2}+\frac{3}{4}-\frac{1}{4}} = -b^{\frac{2}{4}+\frac{3}{4}-\frac{1}{4}} = -b^1 = -b$

53.
$$\begin{aligned}\frac{(x^3)^{1/2}}{x^{7/2}} &= \frac{x^{3/2}}{x^{7/2}} \\ &= x^{3/2-7/2} \\ &= x^{-2} \\ &= \frac{1}{x^2}\end{aligned}$$

55. $\frac{(3x^{1/4})^3}{x^{1/12}} = \frac{3^3 x^{3/4}}{x^{1/12}}$
$= 27x^{\frac{3}{4}-\frac{1}{12}}$
$= 27x^{\frac{9}{12}-\frac{1}{12}}$
$= 27x^{8/12}$
$= 27x^{2/3}$

57. $\frac{(y^3 z)^{1/6}}{y^{-1/2} z^{1/3}} = \frac{y^{3/6} z^{1/6}}{y^{-1/2} z^{1/3}}$
$= y^{3/6-(-1/2)} z^{1/6-1/3}$
$= y^{1/2+1/2} z^{1/6-2/6}$
$= y^1 z^{-1/6}$
$= \frac{y}{z^{1/6}}$

59. $\frac{(x^3 y^2)^{1/4}}{(x^{-5} y^{-1})^{-1/2}} = \frac{x^{3/4} y^{2/4}}{x^{5/2} y^{1/2}}$
$= x^{\frac{3}{4}-\frac{5}{2}} y^{\frac{2}{4}-\frac{1}{2}}$
$= x^{\frac{3}{4}-\frac{10}{4}} y^{\frac{1}{2}-\frac{1}{2}}$
$= x^{-7/4}$
$= \frac{1}{x^{7/4}}$

61. $y^{1/2}(y^{1/2} - y^{2/3}) = y^{1/2} y^{1/2} - y^{1/2} y^{2/3}$
$= y^{1/2+1/2} - y^{1/2+2/3}$
$= y^1 - y^{7/6}$
$= y - y^{7/6}$

63. $x^{2/3}(x-2) = x \cdot x^{2/3} - 2x^{2/3}$
$= x^{1+2/3} - 2x^{2/3}$
$= x^{5/3} - 2x^{2/3}$

65. $(2x^{1/3}+3)(2x^{1/3}-3) = (2x^{1/3})^2 - 3^2$
$= 2^2(x^{1/3})^2 - 9$
$= 4x^{2/3} - 9$

67. $x^{8/3} + x^{10/3} = x^{8/3}(1) + x^{8/3}(x^{2/3})$
$= x^{8/3}(1 + x^{2/3})$

69. $x^{2/5} - 3x^{1/5} = x^{1/5}(x^{1/5}) - x^{1/5}(3)$
$= x^{1/5}(x^{1/5} - 3)$

71. $5x^{-1/3} + x^{2/3} = x^{-1/3}(5) + x^{-1/3}(x^{3/3})$
$= x^{-1/3}(5 + x)$

73. $\sqrt[6]{x^3} = x^{3/6} = x^{1/2} = \sqrt{x}$

75. $\sqrt[6]{4} = 4^{1/6} = (2^2)^{1/6} = 2^{1/3} = \sqrt[3]{2}$

77. $\sqrt[4]{16x^2} = (16x^2)^{1/4}$
$= 16^{1/4} x^{2/4} = 2x^{1/2} = 2\sqrt{x}$

79. $\sqrt[8]{x^4 y^4} = (x^4 y^4)^{1/8}$
$= x^{4/8} y^{4/8}$
$= x^{1/2} y^{1/2}$
$= (xy)^{1/2}$
$= \sqrt{xy}$

81. $\sqrt[12]{a^8 b^4} = a^{8/12} b^{4/12}$
$= a^{2/3} b^{1/3}$
$= (a^2 b)^{1/3}$
$= \sqrt[3]{a^2 b}$

83. $\sqrt[4]{(x+3)^2} = (x+3)^{2/4} = (x+3)^{1/2} = \sqrt{x+3}$

85. $\sqrt[3]{y} \cdot \sqrt[5]{y^2} = y^{1/3} \cdot y^{2/5}$
$= y^{\frac{1}{3}+\frac{2}{5}}$
$= y^{\frac{5}{15}+\frac{6}{15}}$
$= y^{11/15}$
$= \sqrt[15]{y^{11}}$

87. $\frac{\sqrt[3]{b^2}}{\sqrt[4]{b}} = \frac{b^{2/3}}{b^{1/4}} = b^{\frac{2}{3}-\frac{1}{4}} = b^{\frac{8}{12}-\frac{3}{12}} = b^{5/12} = \sqrt[12]{b^5}$

89. $\sqrt[3]{x} \cdot \sqrt[4]{x} \cdot \sqrt[8]{x^3} = x^{1/3} \cdot x^{1/4} \cdot x^{3/8}$
$= x^{8/24} \cdot x^{6/24} \cdot x^{9/24}$
$= x^{23/24}$
$= \sqrt[24]{x^{23}}$

91. $\frac{\sqrt[3]{a^2}}{\sqrt[6]{a}} = \frac{a^{2/3}}{a^{1/6}}$
$= a^{\frac{2}{3}-\frac{1}{6}} = a^{\frac{4}{6}-\frac{1}{6}} = a^{3/6} = a^{1/2} = \sqrt{a}$

93. $\sqrt{3}\cdot\sqrt[3]{4}=3^{1/2}\cdot 4^{1/3}$
$=3^{3/6}\cdot 4^{2/6}$
$=(3^3\cdot 4^2)^{1/6}$
$=(27\cdot 16)^{1/6}$
$=(432)^{1/6}$
$=\sqrt[6]{432}$

95. $\sqrt[5]{7}\cdot\sqrt[3]{y}=7^{1/5}\cdot y^{1/3}$
$=7^{3/15}\cdot y^{5/15}$
$=(7^3\cdot y^5)^{1/15}$
$=(343y^5)^{1/15}$
$=\sqrt[15]{343y^5}$

97. $\sqrt{5r}\cdot\sqrt[3]{s}=(5r)^{1/2}\cdot s^{1/3}$
$=(5r)^{3/6}\cdot s^{2/6}$
$=[(5r)^3\cdot s^2]^{1/6}$
$=(125r^3s^2)^{1/6}$
$=\sqrt[6]{125r^3s^2}$

99. $75=25\cdot 3$ where 25 is a perfect square.

101. $48=4\cdot 12$ or $16\cdot 3$ where both 4 and 16 are perfect squares.

103. $16=8\cdot 2$ where 8 is a perfect cube.

105. $54=27\cdot 2$ where 27 is a perfect cube.

107. $B(w)=70w^{3/4}$
$B(60)=70(60)^{3/4}$
≈ 1509 calories

109. $f(x)=33.3x^{4/5}$
$f(10)=33.3(10)^{4/5}$
≈ 210.1 million subscriptions

111. $\square\cdot a^{2/3}=a^{3/3}$
$\square=\dfrac{a^{3/3}}{a^{2/3}}$
$\square=a^{3/3-2/3}$
$\square=a^{1/3}$

113. $\dfrac{\square}{x^{-2/5}}=x^{3/5}$
$x^{-2/5}\left(\dfrac{\square}{x^{-2/5}}\right)=x^{3/5}\cdot x^{-2/5}$
$\square=x^{3/5-2/5}$
$\square=x^{1/5}$

115. $8^{1/4}\approx 1.6818$

117. $18^{3/5}\approx 5.6645$

119. $\dfrac{\sqrt{t}}{\sqrt{u}}=\dfrac{t^{1/2}}{u^{1/2}}$

Section 10.3

Practice Exercises

1. a. $\sqrt{5}\cdot\sqrt{7}=\sqrt{5\cdot 7}=\sqrt{35}$

b. $\sqrt{13}\cdot\sqrt{z}=\sqrt{13z}$

c. $\sqrt[4]{125}\cdot\sqrt[4]{5}=\sqrt[4]{125\cdot 5}=\sqrt[4]{625}=5$

d. $\sqrt[3]{5y}\cdot\sqrt[3]{3x^2}=\sqrt[3]{5y\cdot 3x^2}=\sqrt[3]{15x^2y}$

e. $\sqrt{\dfrac{5}{m}}\cdot\sqrt{\dfrac{t}{2}}=\sqrt{\dfrac{5}{m}\cdot\dfrac{t}{2}}=\sqrt{\dfrac{5t}{2m}}$

2. a. $\sqrt{\dfrac{36}{49}}=\dfrac{\sqrt{36}}{\sqrt{49}}=\dfrac{6}{7}$

b. $\sqrt{\dfrac{z}{16}}=\dfrac{\sqrt{z}}{\sqrt{16}}=\dfrac{\sqrt{z}}{4}$

c. $\sqrt[3]{\dfrac{125}{8}}=\dfrac{\sqrt[3]{125}}{\sqrt[3]{8}}=\dfrac{5}{2}$

d. $\sqrt[4]{\dfrac{5}{81x^8}}=\dfrac{\sqrt[4]{5}}{\sqrt[4]{81x^8}}=\dfrac{\sqrt[4]{5}}{3x^2}$

3. a. $\sqrt{98}=\sqrt{49\cdot 2}=\sqrt{49}\cdot\sqrt{2}=7\sqrt{2}$

b. $\sqrt[3]{54}=\sqrt[3]{27\cdot 2}=\sqrt[3]{27}\cdot\sqrt[3]{2}=3\sqrt[3]{2}$

c. The largest perfect square factor of 35 is 1, so $\sqrt{35}$ cannot be simplified further.

d. $\sqrt[4]{243} = \sqrt[4]{81 \cdot 3} = \sqrt[4]{81} \cdot \sqrt[4]{3} = 3\sqrt[4]{3}$

4. a. $\sqrt{36z^7} = \sqrt{36z^6 \cdot z} = \sqrt{36z^6} \cdot \sqrt{z} = 6z^3\sqrt{z}$

b. $\sqrt[3]{32p^4q^7} = \sqrt[3]{8 \cdot 4 \cdot p^3 \cdot p \cdot q^6 \cdot q}$
$= \sqrt[3]{8p^3q^6 \cdot 4pq}$
$= \sqrt[3]{8p^3q^6} \cdot \sqrt[3]{4pq}$
$= 2pq^2\sqrt[3]{4pq}$

c. $\sqrt[4]{16x^{15}} = \sqrt[4]{16 \cdot x^{12} \cdot x^3}$
$= \sqrt[4]{16x^{12}} \cdot \sqrt[4]{x^3}$
$= 2x^3\sqrt[4]{x^3}$

5. a. $\dfrac{\sqrt{80}}{\sqrt{5}} = \sqrt{\dfrac{80}{5}} = \sqrt{16} = 4$

b. $\dfrac{\sqrt{98z}}{3\sqrt{2}} = \dfrac{1}{3} \cdot \sqrt{\dfrac{98z}{2}}$
$= \dfrac{1}{3} \cdot \sqrt{49z}$
$= \dfrac{1}{3} \cdot \sqrt{49} \cdot \sqrt{z}$
$= \dfrac{1}{3} \cdot 7 \cdot \sqrt{z}$
$= \dfrac{7}{3}\sqrt{z}$

c. $\dfrac{5\sqrt[3]{40x^5y^7}}{\sqrt[3]{5y}} = 5 \cdot \sqrt[3]{\dfrac{40x^5y^7}{5y}}$
$= 5 \cdot \sqrt[3]{8x^5y^6}$
$= 5 \cdot \sqrt[3]{8x^3y^6 \cdot x^2}$
$= 5 \cdot \sqrt[3]{8x^3y^6} \cdot \sqrt[3]{x^2}$
$= 5 \cdot 2xy^2 \cdot \sqrt[3]{x^2}$
$= 10xy^2\sqrt[3]{x^2}$

d. $\dfrac{3\sqrt[5]{64x^9y^8}}{\sqrt[5]{x^{-1}y^2}} = 3 \cdot \sqrt[5]{\dfrac{64x^9y^8}{x^{-1}y^2}}$
$= 3 \cdot \sqrt[5]{64x^{10}y^6}$
$= 3 \cdot \sqrt[5]{32 \cdot x^{10} \cdot y^5 \cdot 2 \cdot y}$
$= 3 \cdot \sqrt[5]{32x^{10}y^5} \cdot \sqrt[5]{2y}$
$= 3 \cdot 2x^2y \cdot \sqrt[5]{2y}$
$= 6x^2y\sqrt[5]{2y}$

6. Let $(x_1, y_1) = (-3, 7)$ and $(x_2, y_2) = (-2, 3)$.
$d = \sqrt{(x_2 - x_1)^2 + (y_2 - y_1)^2}$
$= \sqrt{[-2-(-3)]^2 + (3-7)^2}$
$= \sqrt{(1)^2 + (-4)^2}$
$= \sqrt{1+16}$
$= \sqrt{17} \approx 4.123$
The distance between the two points is exactly $\sqrt{17}$ units, or approximately 4.123 units.

7. Let $(x_1, y_1) = (5, -2)$ and $(x_2, y_2) = (8, -6)$.
$\text{midpoint} = \left(\dfrac{x_1 + x_2}{2}, \dfrac{y_1 + y_2}{2}\right)$
$= \left(\dfrac{5+8}{2}, \dfrac{-2+(-6)}{2}\right)$
$= \left(\dfrac{13}{2}, \dfrac{-8}{2}\right)$
$= \left(\dfrac{13}{2}, -4\right)$
The midpoint of the segment is $\left(\dfrac{13}{2}, -4\right)$.

Vocabulary and Readiness Check

1. The midpoint of a line segment is a point exactly halfway between the two endpoints of the line segment.
2. The distance formula is $d = \sqrt{(x_2 - x_1)^2 + (y_2 - y_1)^2}$.
3. The midpoint formula is $\left(\dfrac{x_1 + x_2}{2}, \dfrac{y_1 + y_2}{2}\right)$.
4. The statement $\sqrt[n]{a} \cdot \sqrt[n]{b} = \sqrt[n]{ab}$ is true.
5. The statement $\sqrt[3]{7} \cdot \sqrt[3]{11} = \sqrt[3]{18}$ is false.

6. The statement $\sqrt[3]{7}\cdot\sqrt{11}=\sqrt{77}$ is <u>false</u>.

7. The statement $\sqrt{x^7y^8}=\sqrt{x^7}\cdot\sqrt{y^8}$ is <u>true</u>.

8. The statement $\frac{\sqrt[n]{a}}{\sqrt[n]{b}}=\sqrt[n]{\frac{a}{b}}$ is <u>true</u>.

9. The statement $\frac{\sqrt[3]{12}}{\sqrt[3]{4}}=\sqrt[3]{8}$ is <u>false</u>.

10. The statement $\frac{\sqrt[n]{x^7}}{\sqrt[n]{x}}=\sqrt[n]{x^6}$ is <u>true</u>.

Exercise Set 10.3

1. $\sqrt{7}\cdot\sqrt{2}=\sqrt{7\cdot 2}=\sqrt{14}$

3. $\sqrt[4]{8}\cdot\sqrt[4]{2}=\sqrt[4]{8\cdot 2}=\sqrt[4]{16}=2$

5. $\sqrt[3]{4}\cdot\sqrt[3]{9}=\sqrt[3]{4\cdot 9}=\sqrt[3]{36}$

7. $\sqrt{2}\cdot\sqrt{3x}=\sqrt{2\cdot 3x}=\sqrt{6x}$

9. $\sqrt{\frac{7}{x}}\cdot\sqrt{\frac{2}{y}}=\sqrt{\frac{7}{x}\cdot\frac{2}{y}}=\sqrt{\frac{14}{xy}}$

11. $\sqrt[4]{4x^3}\cdot\sqrt[4]{5}=\sqrt[4]{4x^3\cdot 5}=\sqrt[4]{20x^3}$

13. $\sqrt{\frac{6}{49}}=\frac{\sqrt{6}}{\sqrt{49}}=\frac{\sqrt{6}}{7}$

15. $\sqrt{\frac{2}{49}}=\frac{\sqrt{2}}{\sqrt{49}}=\frac{\sqrt{2}}{7}$

17. $\sqrt[4]{\frac{x^3}{16}}=\frac{\sqrt[4]{x^3}}{\sqrt[4]{16}}=\frac{\sqrt[4]{x^3}}{2}$

19. $\sqrt[3]{\frac{4}{27}}=\frac{\sqrt[3]{4}}{\sqrt[3]{27}}=\frac{\sqrt[3]{4}}{3}$

21. $\sqrt[4]{\frac{8}{x^8}}=\frac{\sqrt[4]{8}}{\sqrt[4]{x^8}}=\frac{\sqrt[4]{8}}{x^2}$

23.
$$\sqrt[3]{\frac{2x}{81y^{12}}}=\frac{\sqrt[3]{2x}}{\sqrt[3]{81y^{12}}}=\frac{\sqrt[3]{2x}}{\sqrt[3]{27y^{12}}\cdot\sqrt[3]{3}}=\frac{\sqrt[3]{2x}}{3y^4\sqrt[3]{3}}$$

25. $\sqrt{\frac{x^2y}{100}}=\frac{\sqrt{x^2y}}{\sqrt{100}}=\frac{\sqrt{x^2}\sqrt{y}}{10}=\frac{x\sqrt{y}}{10}$

27. $\sqrt{\frac{5x^2}{4y^2}}=\frac{\sqrt{5x^2}}{\sqrt{4y^2}}=\frac{\sqrt{5}\sqrt{x^2}}{2y}=\frac{x\sqrt{5}}{2y}$

29. $-\sqrt[3]{\frac{z^7}{27x^3}}=-\frac{\sqrt[3]{z^7}}{\sqrt[3]{27x^3}}=-\frac{\sqrt[3]{z^6\cdot z}}{3x}=-\frac{z^2\sqrt[3]{z}}{3x}$

31. $\sqrt{32}=\sqrt{16\cdot 2}=\sqrt{16}\cdot\sqrt{2}=4\sqrt{2}$

33. $\sqrt[3]{192}=\sqrt[3]{64\cdot 3}=\sqrt[3]{64}\cdot\sqrt[3]{3}=4\sqrt[3]{3}$

35. $5\sqrt{75}=5\sqrt{25\cdot 3}=5\sqrt{25}\cdot\sqrt{3}=5(5)\sqrt{3}=25\sqrt{3}$

37. $\sqrt{24}=\sqrt{4\cdot 6}=\sqrt{4}\cdot\sqrt{6}=2\sqrt{6}$

39. $\sqrt{100x^5}=\sqrt{100x^4\cdot x}=\sqrt{100x^4}\cdot\sqrt{x}=10x^2\sqrt{x}$

41. $\sqrt[3]{16y^7}=\sqrt[3]{8y^6\cdot 2y}=\sqrt[3]{8y^6}\cdot\sqrt[3]{2y}=2y^2\sqrt[3]{2y}$

43. $\sqrt[4]{a^8b^7}=\sqrt[4]{a^8b^4\cdot b^3}=\sqrt[4]{a^8b^4}\cdot\sqrt[4]{b^3}=a^2b\sqrt[4]{b^3}$

45. $\sqrt{y^5}=\sqrt{y^4\cdot y}=\sqrt{y^4}\cdot\sqrt{y}=y^2\sqrt{y}$

47.
$$\sqrt{25a^2b^3}=\sqrt{25a^2b^2\cdot b}=\sqrt{25a^2b^2}\cdot\sqrt{b}=5ab\sqrt{b}$$

49.
$$\sqrt[5]{-32x^{10}y}=\sqrt[5]{-32x^{10}\cdot y}=\sqrt[5]{-32x^{10}}\cdot\sqrt[5]{y}=-2x^2\sqrt[5]{y}$$

51. $\sqrt[3]{50x^{14}} = \sqrt[3]{x^{12} \cdot 50x^2}$
$= \sqrt[3]{x^{12}} \cdot \sqrt[3]{50x^2}$
$= x^4 \sqrt[3]{50x^2}$

53. $-\sqrt{32a^8b^7} = -\sqrt{16a^8b^6 \cdot 2b}$
$= -\sqrt{16a^8b^6} \cdot \sqrt{2b}$
$= -4a^4b^3\sqrt{2b}$

55. $\sqrt{9x^7y^9} = \sqrt{9x^6y^8 \cdot xy}$
$= \sqrt{9x^6y^8} \cdot \sqrt{xy}$
$= 3x^3y^4\sqrt{xy}$

57. $\sqrt[3]{125r^9s^{12}} = 5r^3s^4$

59. $\dfrac{\sqrt{14}}{\sqrt{7}} = \sqrt{\dfrac{14}{7}} = \sqrt{2}$

61. $\dfrac{\sqrt[3]{24}}{\sqrt[3]{3}} = \sqrt[3]{\dfrac{24}{3}} = \sqrt[3]{8} = 2$

63. $\dfrac{5\sqrt[4]{48}}{\sqrt[4]{3}} = 5\sqrt[4]{\dfrac{48}{3}} = 5\sqrt[4]{16} = 5(2) = 10$

65. $\dfrac{\sqrt{x^5y^3}}{\sqrt{xy}} = \sqrt{\dfrac{x^5y^3}{xy}} = \sqrt{x^4y^2} = x^2y$

67. $\dfrac{8\sqrt[3]{54m^7}}{\sqrt[3]{2m}} = 8\sqrt[3]{\dfrac{54m^7}{2m}}$
$= 8\sqrt[3]{27m^6}$
$= 8(3m^2)$
$= 24m^2$

69. $\dfrac{3\sqrt{100x^2}}{2\sqrt{2x^{-1}}} = \dfrac{3}{2}\sqrt{\dfrac{100x^2}{2x^{-1}}}$
$= \dfrac{3}{2}\sqrt{50x^3}$
$= \dfrac{3}{2}\sqrt{25x^2 \cdot 2x}$
$= \dfrac{3}{2}(5x)\sqrt{2x}$
$= \dfrac{15x}{2}\sqrt{2x}$

71. $\dfrac{\sqrt[4]{96a^{10}b^3}}{\sqrt[4]{3a^2b^3}} = \sqrt[4]{\dfrac{96a^{10}b^3}{3a^2b^3}}$
$= \sqrt[4]{32a^8}$
$= \sqrt[4]{16a^8 \cdot 2}$
$= 2a^2\sqrt[4]{2}$

73. (5, 1), (8, 5)
$d = \sqrt{(8-5)^2 + (5-1)^2}$
$= \sqrt{3^2 + 4^2}$
$= \sqrt{9+16}$
$= \sqrt{25}$
$= 5$ units

75. (–3, 2), (1, –3)
$d = \sqrt{[1-(-3)]^2 + (-3-2)^2}$
$= \sqrt{4^2 + (-5)^2}$
$= \sqrt{16+25}$
$= \sqrt{41} \approx 6.403$ units

77. (–9, 4), (–8, 1)
$d = \sqrt{[-8-(-9)]^2 + (1-4)^2}$
$= \sqrt{1^2 + (-3)^2}$
$= \sqrt{1+9}$
$= \sqrt{10} \approx 3.162$ units

79. $\left(0, -\sqrt{2}\right), \left(\sqrt{3}, 0\right)$
$d = \sqrt{\left(\sqrt{3}-0\right)^2 + \left[0-\left(-\sqrt{2}\right)\right]^2}$
$= \sqrt{\left(\sqrt{3}\right)^2 + \left(\sqrt{2}\right)^2}$
$= \sqrt{3+2}$
$= \sqrt{5} \approx 2.236$ units

81. (1.7, –3.6), (–8.6, 5.7)
$d = \sqrt{(-8.6-1.7)^2 + [5.7-(-3.6)]^2}$
$= \sqrt{(-10.3)^2 + (9.3)^2}$
$= \sqrt{192.58} \approx 13.877$ units

83. (6, –8), (2, 4)
$\left(\dfrac{6+2}{2}, \dfrac{-8+4}{2}\right) = \left(\dfrac{8}{2}, \dfrac{-4}{2}\right) = (4, -2)$
The midpoint of the segment is (4, –2).

85. $(-2, -1), (-8, 6)$

$$\left(\frac{-2+(-8)}{2}, \frac{-1+6}{2}\right) = \left(\frac{-10}{2}, \frac{5}{2}\right) = \left(-5, \frac{5}{2}\right)$$

The midpoint of the segment is $\left(-5, \frac{5}{2}\right)$.

87. $(7, 3), (-1, -3)$

$$\left(\frac{7+(-1)}{2}, \frac{3+(-3)}{2}\right) = \left(\frac{6}{2}, \frac{0}{2}\right) = (3, 0)$$

The midpoint of the segment is (3, 0).

89. $\left(\frac{1}{2}, \frac{3}{8}\right), \left(-\frac{3}{2}, \frac{5}{8}\right)$

$$\left(\frac{\frac{1}{2}+\left(-\frac{3}{2}\right)}{2}, \frac{\frac{3}{8}+\frac{5}{8}}{2}\right) = \left(\frac{-1}{2}, \frac{1}{2}\right)$$

The midpoint of the segment is $\left(-\frac{1}{2}, \frac{1}{2}\right)$.

91. $\left(\sqrt{2}, 3\sqrt{5}\right), \left(\sqrt{2}, -2\sqrt{5}\right)$

$$\left(\frac{\sqrt{2}+\sqrt{2}}{2}, \frac{3\sqrt{5}+\left(-2\sqrt{5}\right)}{2}\right) = \left(\frac{2\sqrt{2}}{2}, \frac{\sqrt{5}}{2}\right) = \left(\sqrt{2}, \frac{\sqrt{5}}{2}\right)$$

The midpoint of the segment is $\left(\sqrt{2}, \frac{\sqrt{5}}{2}\right)$.

93. $(4.6, -3.5), (7.8, -9.8)$

$$\left(\frac{4.6+7.8}{2}, \frac{-3.5+(-9.8)}{2}\right) = \left(\frac{12.4}{2}, \frac{-13.3}{2}\right) = (6.2, -6.65)$$

The midpoint of the segment is (6.2, –6.65).

95. $6x + 8x = (6 + 8)x = 14x$

97. $(2x+3)(x-5) = 2x^2 - 10x + 3x - 15$
$= 2x^2 - 7x - 15$

99. $9y^2 - 8y^2 = (9-8)y^2 = 1y^2 = y^2$

101. $-3(x+5) = -3x - 3(5) = -3x - 15$

103. $(x-4)^2 = x^2 - 2(x)(4) + 4^2$
$= x^2 - 8x + 16$

105. $\frac{\sqrt[3]{64}}{\sqrt{64}} = \frac{4}{8} = \frac{1}{2}$

107. $\sqrt[5]{x^{35}} = x^7$

109. $\sqrt[4]{a^{12}b^4c^{20}} = a^3bc^5$

111. $\sqrt[3]{z^{32}} = \sqrt[3]{z^{30} \cdot z^2} = \sqrt[3]{z^{30}} \cdot \sqrt[3]{z^2} = z^{10}\sqrt[3]{z^2}$

113. $\sqrt[7]{q^{17}r^{40}s^7} = \sqrt[7]{q^{14} \cdot q^3 \cdot r^{35} \cdot r^5 \cdot s^7}$
$= \sqrt[7]{q^{14}r^{35}s^7 \cdot q^3r^5}$
$= q^2r^5s\sqrt[7]{q^3r^5}$

115. $r = \sqrt{\frac{A}{4\pi}} = \sqrt{\frac{32.17}{4\pi}} \approx \sqrt{2.56} = 1.6$

The radius of a standard zorb is 1.6 meters.

117. $F(x) = 0.6\sqrt{49 - x^2}$

a. $F(3) = 0.6\sqrt{49 - 3^2}$
$= 0.6\sqrt{49 - 9}$
$= 0.6\sqrt{40} \approx 3.8$ times

b. $F(5) = 0.6\sqrt{49 - 5^2}$
$= 0.6\sqrt{49 - 25}$
$= 0.6\sqrt{24} \approx 2.9$ times

c. Answers may vary

Section 10.4

Practice Exercises

1. a. $3\sqrt{17} + 5\sqrt{17} = (3+5)\sqrt{17} = 8\sqrt{17}$

b. $7\sqrt[3]{5z} - 12\sqrt[3]{5z} = (7-12)\sqrt[3]{5z} = -5\sqrt[3]{5z}$

c. $3\sqrt{2} + 5\sqrt[3]{2}$

This expression cannot be simplified since $3\sqrt{2}$ and $5\sqrt[3]{2}$ do not contain like radicals.

2. a. $\sqrt{24}+3\sqrt{54}=\sqrt{4\cdot 6}+3\sqrt{9\cdot 6}$
$=\sqrt{4}\cdot\sqrt{6}+3\cdot\sqrt{9}\cdot\sqrt{6}$
$=2\cdot\sqrt{6}+3\cdot 3\cdot\sqrt{6}$
$=2\sqrt{6}+9\sqrt{6}$
$=11\sqrt{6}$

b. $\sqrt[3]{24}-4\sqrt[3]{81}+\sqrt[3]{3}$
$=\sqrt[3]{8}\cdot\sqrt[3]{3}-4\cdot\sqrt[3]{27}\cdot\sqrt[3]{3}+\sqrt[3]{3}$
$=2\cdot\sqrt[3]{3}-4\cdot 3\cdot\sqrt[3]{3}+\sqrt[3]{3}$
$=2\sqrt[3]{3}-12\sqrt[3]{3}+\sqrt[3]{3}$
$=-9\sqrt[3]{3}$

c. $\sqrt{75x}-3\sqrt{27x}+\sqrt{12x}$
$=\sqrt{25}\cdot\sqrt{3x}-3\cdot\sqrt{9}\cdot\sqrt{3x}+\sqrt{4}\cdot\sqrt{3x}$
$=5\cdot\sqrt{3x}-3\cdot 3\cdot\sqrt{3x}+2\cdot\sqrt{3x}$
$=5\sqrt{3x}-9\sqrt{3x}+2\sqrt{3x}$
$=-2\sqrt{3x}$

d. $\sqrt{40}+\sqrt[3]{40}=\sqrt{4}\cdot\sqrt{10}+\sqrt[3]{8}\cdot\sqrt[3]{5}$
$=2\sqrt{10}+2\sqrt[3]{5}$

e. $\sqrt[3]{81x^4}+\sqrt[3]{3x^4}=\sqrt[3]{27x^3}\cdot\sqrt[3]{3x}+\sqrt[3]{x^3}\cdot\sqrt[3]{3x}$
$=3x\sqrt[3]{3x}+x\sqrt[3]{3x}$
$=4x\sqrt[3]{3x}$

3. a. $\frac{\sqrt{28}}{3}-\frac{\sqrt{7}}{4}=\frac{2\sqrt{7}}{3}-\frac{\sqrt{7}}{4}$
$=\frac{2\sqrt{7}\cdot 4}{3\cdot 4}-\frac{\sqrt{7}\cdot 3}{4\cdot 3}$
$=\frac{8\sqrt{7}}{12}-\frac{3\sqrt{7}}{12}$
$=\frac{5\sqrt{7}}{12}$

b. $\sqrt[3]{\frac{6y}{64}}+3\sqrt[3]{6y}=\frac{\sqrt[3]{6y}}{\sqrt[3]{64}}+3\sqrt[3]{6y}$
$=\frac{\sqrt[3]{6y}}{4}+3\sqrt[3]{6y}$
$=\frac{\sqrt[3]{6y}}{4}+\frac{3\sqrt[3]{6y}\cdot 4}{4}$
$=\frac{\sqrt[3]{6y}}{4}+\frac{12\sqrt[3]{6y}}{4}$
$=\frac{13\sqrt[3]{6y}}{4}$

4. a. $\sqrt{5}(2+\sqrt{15})=\sqrt{5}(2)+\sqrt{5}(\sqrt{15})$
$=2\sqrt{5}+\sqrt{5\cdot 15}$
$=2\sqrt{5}+\sqrt{5\cdot 5\cdot 3}$
$=2\sqrt{5}+5\sqrt{3}$

b. $(\sqrt{2}-\sqrt{5})(\sqrt{6}+2)$
$=\sqrt{2}\cdot\sqrt{6}+\sqrt{2}\cdot 2-\sqrt{5}\cdot\sqrt{6}-\sqrt{5}\cdot 2$
$=\sqrt{2\cdot 2\cdot 3}+2\sqrt{2}-\sqrt{30}-2\sqrt{5}$
$=2\sqrt{3}+2\sqrt{2}-\sqrt{30}-2\sqrt{5}$

c. $(3\sqrt{z}-4)(2\sqrt{z}+3)$
$=3\sqrt{z}(2\sqrt{z})+3\sqrt{z}(3)-4(2\sqrt{z})-4(3)$
$=6\cdot z+9\sqrt{z}-8\sqrt{z}-12$
$=6z+\sqrt{z}-12$

d. $(\sqrt{6}-3)^2=(\sqrt{6}-3)(\sqrt{6}-3)$
$=\sqrt{6}(\sqrt{6})-\sqrt{6}(3)-3(\sqrt{6})-3(-3)$
$=6-3\sqrt{6}-3\sqrt{6}+9$
$=6-6\sqrt{6}+9$
$=15-6\sqrt{6}$

e. $(\sqrt{5x}+3)(\sqrt{5x}-3)$
$=\sqrt{5x}\cdot\sqrt{5x}-3\sqrt{5x}+3\sqrt{5x}-3\cdot 3$
$=5x-9$

f. $(\sqrt{x+2}+3)^2=(\sqrt{x+2})^2+2\cdot\sqrt{x+2}\cdot 3+3^2$
$=x+2+6\sqrt{x+2}+9$
$=x+11+6\sqrt{x+2}$

Vocabulary and Readiness Check

1. The terms $\sqrt{7}$ and $\sqrt[3]{7}$ are unlike terms.
2. The terms $\sqrt[3]{x^2y}$ and $\sqrt[3]{yx^2}$ are like terms.
3. The terms $\sqrt[3]{abc}$ and $\sqrt[3]{cba}$ are like terms.
4. The terms $2x\sqrt{5}$ and $2x\sqrt{10}$ are unlike terms.
5. $2\sqrt{3}+4\sqrt{3}=6\sqrt{3}$
6. $5\sqrt{7}+3\sqrt{7}=8\sqrt{7}$
7. $8\sqrt{x}-\sqrt{x}=7\sqrt{x}$
8. $3\sqrt{y}-\sqrt{y}=2\sqrt{y}$

9. $7\sqrt[3]{x}+\sqrt[3]{x}=\underline{8\sqrt[3]{x}}$

10. $8\sqrt[3]{z}+\sqrt[3]{z}=\underline{9\sqrt[3]{z}}$

11. $\sqrt{11}+\sqrt[3]{11}=\underline{\sqrt{11}+\sqrt[3]{11}}$

12. $9\sqrt{13}-\sqrt[4]{13}=\underline{9\sqrt{13}-\sqrt[4]{13}}$

13. $8\sqrt[3]{2x}+3\sqrt[3]{2x}-\sqrt[3]{2x}=\underline{10\sqrt[3]{2x}}$

14. $8\sqrt[3]{2x}+3\sqrt[3]{2x^2}-\sqrt[3]{2x}=\underline{7\sqrt[3]{2x}+3\sqrt[3]{2x^2}}$

Exercise Set 10.4

1.
$$\begin{aligned}\sqrt{8}-\sqrt{32}&=\sqrt{4\cdot 2}-\sqrt{16\cdot 2}\\&=\sqrt{4}\cdot\sqrt{2}-\sqrt{16}\cdot\sqrt{2}\\&=2\sqrt{2}-4\sqrt{2}\\&=-2\sqrt{2}\end{aligned}$$

3.
$$\begin{aligned}2\sqrt{2x^3}+4x\sqrt{8x}&=2\sqrt{x^2\cdot 2x}+4x\sqrt{4\cdot 2x}\\&=2\sqrt{x^2}\cdot\sqrt{2x}+4x\sqrt{4}\cdot\sqrt{2x}\\&=2x\sqrt{2x}+4x(2)\sqrt{2x}\\&=2x\sqrt{2x}+8x\sqrt{2x}\\&=10x\sqrt{2x}\end{aligned}$$

5.
$$\begin{aligned}&2\sqrt{50}-3\sqrt{125}+\sqrt{98}\\&=2\sqrt{25\cdot 2}-3\sqrt{25\cdot 5}+\sqrt{49\cdot 2}\\&=2\sqrt{25}\cdot\sqrt{2}-3\sqrt{25}\cdot\sqrt{5}+\sqrt{49}\cdot\sqrt{2}\\&=2(5)\sqrt{2}-3(5)\sqrt{5}+7\sqrt{2}\\&=10\sqrt{2}-15\sqrt{5}+7\sqrt{2}\\&=17\sqrt{2}-15\sqrt{5}\end{aligned}$$

7.
$$\begin{aligned}\sqrt[3]{16x}-\sqrt[3]{54x}&=\sqrt[3]{8\cdot 2x}-\sqrt[3]{27\cdot 2x}\\&=\sqrt[3]{8}\cdot\sqrt[3]{2x}-\sqrt[3]{27}\cdot\sqrt[3]{2x}\\&=2\sqrt[3]{2x}-3\sqrt[3]{2x}\\&=-\sqrt[3]{2x}\end{aligned}$$

9.
$$\begin{aligned}&\sqrt{9b^3}-\sqrt{25b^3}+\sqrt{49b^3}\\&=\sqrt{9b^2\cdot b}-\sqrt{25b^2\cdot b}+\sqrt{49b^2\cdot b}\\&=\sqrt{9b^2}\cdot\sqrt{b}-\sqrt{25b^2}\cdot\sqrt{b}+\sqrt{49b^2}\cdot\sqrt{b}\\&=3b\sqrt{b}-5b\sqrt{b}+7b\sqrt{b}\\&=5b\sqrt{b}\end{aligned}$$

11.
$$\begin{aligned}\frac{5\sqrt{2}}{3}+\frac{2\sqrt{2}}{5}&=\frac{5(5\sqrt{2})+3(2\sqrt{2})}{3(5)}\\&=\frac{25\sqrt{2}+6\sqrt{2}}{15}\\&=\frac{31\sqrt{2}}{15}\end{aligned}$$

13.
$$\begin{aligned}\sqrt[3]{\frac{11}{8}}-\frac{\sqrt[3]{11}}{6}&=\frac{\sqrt[3]{11}}{\sqrt[3]{8}}-\frac{\sqrt[3]{11}}{6}\\&=\frac{\sqrt[3]{11}}{2}-\frac{\sqrt[3]{11}}{6}\\&=\frac{3\sqrt[3]{11}-\sqrt[3]{11}}{6}\\&=\frac{2\sqrt[3]{11}}{6}\\&=\frac{\sqrt[3]{11}}{3}\end{aligned}$$

15.
$$\begin{aligned}\frac{\sqrt{20x}}{9}+\sqrt{\frac{5x}{9}}&=\frac{\sqrt{4\cdot 5x}}{9}+\frac{\sqrt{5x}}{\sqrt{9}}\\&=\frac{2\sqrt{5x}}{9}+\frac{\sqrt{5x}}{3}\\&=\frac{2\sqrt{5x}+3\sqrt{5x}}{9}\\&=\frac{5\sqrt{5x}}{9}\end{aligned}$$

17.
$$\begin{aligned}7\sqrt{9}-7+\sqrt{3}&=7(3)-7+\sqrt{3}\\&=21-7+\sqrt{3}\\&=14+\sqrt{3}\end{aligned}$$

19.
$$\begin{aligned}2+3\sqrt{y^2}-6\sqrt{y^2}+5&=2+3y-6y+5\\&=7-3y\end{aligned}$$

21.
$$\begin{aligned}&3\sqrt{108}-2\sqrt{18}-3\sqrt{48}\\&=3\sqrt{36\cdot 3}-2\sqrt{9\cdot 2}-3\sqrt{16\cdot 3}\\&=3\sqrt{36}\cdot\sqrt{3}-2\sqrt{9}\cdot\sqrt{2}-3\sqrt{16}\cdot\sqrt{3}\\&=3(6)\sqrt{3}-2(3)\sqrt{2}-3(4)\sqrt{3}\\&=18\sqrt{3}-6\sqrt{2}-12\sqrt{3}\\&=6\sqrt{3}-6\sqrt{2}\end{aligned}$$

23. $-5\sqrt[3]{625}+\sqrt[3]{40} = -5\sqrt[3]{125\cdot 5}+\sqrt[3]{8\cdot 5}$
$= -5(5)\sqrt[3]{5}+2\sqrt[3]{5}$
$= -25\sqrt[3]{5}+2\sqrt[3]{5}$
$= -23\sqrt[3]{5}$

25. $\sqrt{9b^3}-\sqrt{25b^3}+\sqrt{16b^3}$
$= \sqrt{9b^2\cdot b}-\sqrt{25b^2\cdot b}+\sqrt{16b^2\cdot b}$
$= 3b\sqrt{b}-5b\sqrt{b}+4b\sqrt{b}$
$= 2b\sqrt{b}$

27. $5y\sqrt{8y}+2\sqrt{50y^3} = 5y\sqrt{4\cdot 2y}+2\sqrt{25y^2\cdot 2y}$
$= 5y(2)\sqrt{2y}+2(5y)\sqrt{2y}$
$= 10y\sqrt{2y}+10y\sqrt{2y}$
$= 20y\sqrt{2y}$

29. $\sqrt[3]{54xy^3}-5\sqrt[3]{2xy^3}+y\sqrt[3]{128x}$
$= \sqrt[3]{27y^3\cdot 2x}-5\sqrt[3]{y^3\cdot 2x}+y\sqrt[3]{64\cdot 2x}$
$= 3y\sqrt[3]{2x}-5y\sqrt[3]{2x}+4y\sqrt[3]{2x}$
$= 2y\sqrt[3]{2x}$

31. $6\sqrt[3]{11}+8\sqrt{11}-12\sqrt{11} = 6\sqrt[3]{11}-4\sqrt{11}$

33. $-2\sqrt[4]{x^7}+3\sqrt[4]{16x^7} = -2\sqrt[4]{x^4\cdot x^3}+3\sqrt[4]{16x^4\cdot x^3}$
$= -2x\sqrt[4]{x^3}+3(2x)\sqrt[4]{x^3}$
$= -2x\sqrt[4]{x^3}+6x\sqrt[4]{x^3}$
$= 4x\sqrt[4]{x^3}$

35. $\frac{4\sqrt{3}}{3}-\frac{\sqrt{12}}{3} = \frac{4\sqrt{3}}{3}-\frac{\sqrt{4\cdot 3}}{3}$
$= \frac{4\sqrt{3}-2\sqrt{3}}{3}$
$= \frac{2\sqrt{3}}{3}$

37. $\frac{\sqrt[3]{8x^4}}{7}+\frac{3x\sqrt[3]{x}}{7} = \frac{\sqrt[3]{8x^3\cdot x}}{7}+\frac{3x\sqrt[3]{x}}{7}$
$= \frac{2x\sqrt[3]{x}+3x\sqrt[3]{x}}{7}$
$= \frac{5x\sqrt[3]{x}}{7}$

39. $\sqrt{\frac{28}{x^2}}+\sqrt{\frac{7}{4x^2}} = \frac{\sqrt{28}}{\sqrt{x^2}}+\frac{\sqrt{7}}{\sqrt{4x^2}}$
$= \frac{\sqrt{4\cdot 7}}{x}+\frac{\sqrt{7}}{2x}$
$= \frac{2\sqrt{7}}{x}+\frac{\sqrt{7}}{2x}$
$= \frac{2(2\sqrt{7})+\sqrt{7}}{2x}$
$= \frac{4\sqrt{7}+\sqrt{7}}{2x}$
$= \frac{5\sqrt{7}}{2x}$

41. $\sqrt[3]{\frac{16}{27}}-\frac{\sqrt[3]{54}}{6} = \frac{\sqrt[3]{8\cdot 2}}{\sqrt[3]{27}}-\frac{\sqrt[3]{27\cdot 2}}{6}$
$= \frac{2\sqrt[3]{2}}{3}-\frac{3\sqrt[3]{2}}{6}$
$= \frac{2(2\sqrt[3]{2})-3\sqrt[3]{2}}{6}$
$= \frac{4\sqrt[3]{2}-3\sqrt[3]{2}}{6}$
$= \frac{\sqrt[3]{2}}{6}$

43. $-\frac{\sqrt[3]{2x^4}}{9}+\sqrt[3]{\frac{250x^4}{27}} = -\frac{\sqrt[3]{x^3\cdot 2x}}{9}+\frac{\sqrt[3]{125x^3\cdot 2x}}{\sqrt[3]{27}}$
$= \frac{-x\sqrt[3]{2x}}{9}+\frac{5x\sqrt[3]{2x}}{3}$
$= \frac{-x\sqrt[3]{2x}+3(5x\sqrt[3]{2x})}{9}$
$= \frac{-x\sqrt[3]{2x}+15x\sqrt[3]{2x}}{9}$
$= \frac{14x\sqrt[3]{2x}}{9}$

45. $P = 2\sqrt{12}+\sqrt{12}+2\sqrt{27}+3\sqrt{3}$
$= 2\sqrt{4\cdot 3}+\sqrt{4\cdot 3}+2\sqrt{9\cdot 3}+3\sqrt{3}$
$= 2(2)\sqrt{3}+2\sqrt{3}+2(3)\sqrt{3}+3\sqrt{3}$
$= 4\sqrt{3}+2\sqrt{3}+6\sqrt{3}+3\sqrt{3}$
$= 15\sqrt{3}$ inches

47. $\sqrt{7}\left(\sqrt{5}+\sqrt{3}\right) = \sqrt{7}\sqrt{5}+\sqrt{7}\sqrt{3}$
$= \sqrt{35}+\sqrt{21}$

49. $\left(\sqrt{5}-\sqrt{2}\right)^2=\left(\sqrt{5}\right)^2-2\sqrt{5}\sqrt{2}+\left(\sqrt{2}\right)^2$
$=5-2\sqrt{10}+2$
$=7-2\sqrt{10}$

51. $\sqrt{3x}\left(\sqrt{3}-\sqrt{x}\right)=\sqrt{3x}\sqrt{3}-\sqrt{3x}\sqrt{x}$
$=\sqrt{9x}-\sqrt{3x^2}$
$=3\sqrt{x}-x\sqrt{3}$

53. $\left(2\sqrt{x}-5\right)\left(3\sqrt{x}+1\right)$
$=2\sqrt{x}\left(3\sqrt{x}\right)+2\sqrt{x}\cdot1-5\left(3\sqrt{x}\right)-5(1)$
$=6x+2\sqrt{x}-15\sqrt{x}-5$
$=6x-13\sqrt{x}-5$

55. $\left(\sqrt[3]{a}-4\right)\left(\sqrt[3]{a}+5\right)$
$=\sqrt[3]{a}\left(\sqrt[3]{a}\right)+\sqrt[3]{a}\cdot5-4\sqrt[3]{a}-4(5)$
$=\sqrt[3]{a^2}+5\sqrt[3]{a}-4\sqrt[3]{a}-20$
$=\sqrt[3]{a^2}+\sqrt[3]{a}-20$

57. $6\left(\sqrt{2}-2\right)=6\sqrt{2}-6(2)=6\sqrt{2}-12$

59. $\sqrt{2}\left(\sqrt{2}+x\sqrt{6}\right)=\sqrt{2}\sqrt{2}+\sqrt{2}\left(x\sqrt{6}\right)$
$=2+x\sqrt{12}$
$=2+x\sqrt{4\cdot3}$
$=2+2x\sqrt{3}$

61. $\left(2\sqrt{7}+3\sqrt{5}\right)\left(\sqrt{7}-2\sqrt{5}\right)$
$=2\sqrt{7}\sqrt{7}+2\sqrt{7}\left(-2\sqrt{5}\right)+3\sqrt{5}\sqrt{7}+3\sqrt{5}\left(-2\sqrt{5}\right)$
$=2(7)-4\sqrt{35}+3\sqrt{35}-6(5)$
$=14-\sqrt{35}-30$
$=-16-\sqrt{35}$

63. $\left(\sqrt{x}-y\right)\left(\sqrt{x}+y\right)=\left(\sqrt{x}\right)^2-y^2=x-y^2$

65. $\left(\sqrt{3}+x\right)^2=\left(\sqrt{3}\right)^2+2\sqrt{3}\cdot x+x^2$
$=3+2x\sqrt{3}+x^2$

67. $\left(\sqrt{5x}-2\sqrt{3x}\right)\left(\sqrt{5x}-3\sqrt{3x}\right)$
$=\left(\sqrt{5x}\right)^2-\sqrt{5x}\left(3\sqrt{3x}\right)-2\sqrt{3x}\left(\sqrt{5x}\right)$
$\quad-2\sqrt{3x}\left(-3\sqrt{3x}\right)$
$=5x-3x\sqrt{15}-2x\sqrt{15}+6\cdot3x$
$=23x-5x\sqrt{15}$

69. $\left(\sqrt[3]{4}+2\right)\left(\sqrt[3]{2}-1\right)$
$=\sqrt[3]{4}\left(\sqrt[3]{2}\right)+\sqrt[3]{4}\cdot(-1)+2\sqrt[3]{2}+2(-1)$
$=\sqrt[3]{8}-\sqrt[3]{4}+2\sqrt[3]{2}-2$
$=2-\sqrt[3]{4}+2\sqrt[3]{2}-2$
$=2\sqrt[3]{2}-\sqrt[3]{4}$

71. $\left(\sqrt[3]{x}+1\right)\left(\sqrt[3]{x^2}-\sqrt[3]{x}+1\right)$
$=\sqrt[3]{x}\left(\sqrt[3]{x^2}\right)-\sqrt[3]{x}\left(\sqrt[3]{x}\right)+\sqrt[3]{x}(1)$
$\quad+1\left(\sqrt[3]{x^2}\right)-1\left(\sqrt[3]{x}\right)+1(1)$
$=\sqrt[3]{x^3}-\sqrt[3]{x^2}+\sqrt[3]{x}+\sqrt[3]{x^2}-\sqrt[3]{x}+1$
$=x+1$

73. $\left(\sqrt{x-1}+5\right)^2=\left(\sqrt{x-1}\right)^2+2\sqrt{x-1}\cdot5+5^2$
$=(x-1)+10\sqrt{x-1}+25$
$=x+10\sqrt{x-1}+24$

75. $\left(\sqrt{2x+5}-1\right)^2=\left(\sqrt{2x+5}\right)^2-2\sqrt{2x+5}\cdot1+1^2$
$=(2x+5)-2\sqrt{2x+5}+1$
$=2x-2\sqrt{2x+5}+6$

77. $\dfrac{2x-14}{2}=\dfrac{2(x-7)}{2}=x-7$

79. $\dfrac{7x-7y}{x^2-y^2}=\dfrac{7(x-y)}{(x+y)(x-y)}=\dfrac{7}{x+y}$

81. $\dfrac{6a^2b-9ab}{3ab}=\dfrac{3ab(2a-3)}{3ab}=2a-3$

83. $\dfrac{-4+2\sqrt{3}}{6}=\dfrac{2\left(-2+\sqrt{3}\right)}{6}=\dfrac{-2+\sqrt{3}}{3}$

85. $P = 2l + 2w$
$= 2(3\sqrt{20}) + 2(\sqrt{125})$
$= 6\sqrt{4\cdot 5} + 2\sqrt{25\cdot 5}$
$= 6(2)\sqrt{5} + 2(5)\sqrt{5}$
$= 12\sqrt{5} + 10\sqrt{5}$
$= 22\sqrt{5}$ feet
$A = lw$
$= (3\sqrt{20})(\sqrt{125})$
$= 3\sqrt{4\cdot 5}\sqrt{25\cdot 5}$
$= 3(2)\sqrt{5}\cdot 5\sqrt{5}$
$= 30\cdot 5$
$= 150$ square feet

87. a. $\sqrt{3} + \sqrt{3} = 2\sqrt{3}$

b. $\sqrt{3}\cdot\sqrt{3} = \sqrt{9} = 3$

c. Answers may vary

89. Answer may vary

Section 10.5

Practice Exercises

1. a. $\frac{5}{\sqrt{3}} = \frac{5\cdot\sqrt{3}}{\sqrt{3}\cdot\sqrt{3}} = \frac{5\sqrt{3}}{3}$

b. $\frac{3\sqrt{25}}{\sqrt{4x}} = \frac{3(5)}{2\sqrt{x}} = \frac{15}{2\sqrt{x}} = \frac{15\cdot\sqrt{x}}{2\sqrt{x}\cdot\sqrt{x}} = \frac{15\sqrt{x}}{2x}$

c. $\sqrt[3]{\frac{2}{9}} = \frac{\sqrt[3]{2}}{\sqrt[3]{9}} = \frac{\sqrt[3]{2}\cdot\sqrt[3]{3}}{\sqrt[3]{3^2}\cdot\sqrt[3]{3}} = \frac{\sqrt[3]{6}}{3}$

2. $\sqrt{\frac{3z}{5y}} = \frac{\sqrt{3z}}{\sqrt{5y}} = \frac{\sqrt{3z}\cdot\sqrt{5y}}{\sqrt{5y}\cdot\sqrt{5y}} = \frac{\sqrt{15yz}}{5y}$

3. $\frac{\sqrt[3]{z^2}}{\sqrt[3]{27x^4}} = \frac{\sqrt[3]{z^2}}{\sqrt[3]{27x^3}\cdot\sqrt[3]{x}}$
$= \frac{\sqrt[3]{z^2}}{3x\sqrt[3]{x}}$
$= \frac{\sqrt[3]{z^2}\cdot\sqrt[3]{x^2}}{3x\sqrt[3]{x}\cdot\sqrt[3]{x^2}}$
$= \frac{\sqrt[3]{z^2x^2}}{3x\sqrt[3]{x^3}}$
$= \frac{\sqrt[3]{x^2z^2}}{3x^2}$

4. a. $\frac{5}{3\sqrt{5}+2} = \frac{5(3\sqrt{5}-2)}{(3\sqrt{5}+2)(3\sqrt{5}-2)}$
$= \frac{5(3\sqrt{5}-2)}{(3\sqrt{5})^2 - 2^2}$
$= \frac{5(3\sqrt{5}-2)}{45-4}$
$= \frac{5(3\sqrt{5}-2)}{41}$

b. $\frac{\sqrt{2}+5}{\sqrt{3}-\sqrt{5}} = \frac{(\sqrt{2}+5)(\sqrt{3}+\sqrt{5})}{(\sqrt{3}-\sqrt{5})(\sqrt{3}+\sqrt{5})}$
$= \frac{\sqrt{2}\sqrt{3}+\sqrt{2}\sqrt{5}+5\sqrt{3}+5\sqrt{5}}{(\sqrt{3})^2-(\sqrt{5})^2}$
$= \frac{\sqrt{6}+\sqrt{10}+5\sqrt{3}+5\sqrt{5}}{3-5}$
$= \frac{\sqrt{6}+\sqrt{10}+5\sqrt{3}+5\sqrt{5}}{-2}$

c. $\frac{3\sqrt{x}}{2\sqrt{x}+\sqrt{y}} = \frac{3\sqrt{x}(2\sqrt{x}-\sqrt{y})}{(2\sqrt{x}+\sqrt{y})(2\sqrt{x}-\sqrt{y})}$
$= \frac{6\sqrt{x^2}-3\sqrt{xy}}{(2\sqrt{x})^2-(\sqrt{y})^2}$
$= \frac{6x-3\sqrt{xy}}{4x-y}$

5. $\frac{\sqrt{32}}{\sqrt{80}} = \frac{\sqrt{16\cdot 2}}{\sqrt{16\cdot 5}} = \frac{4\sqrt{2}}{4\sqrt{5}} = \frac{\sqrt{2}}{\sqrt{5}} = \frac{\sqrt{2}\cdot\sqrt{2}}{\sqrt{5}\cdot\sqrt{2}} = \frac{2}{\sqrt{10}}$

6. $\frac{\sqrt[3]{5b}}{\sqrt[3]{2a}} = \frac{\sqrt[3]{5b}\cdot\sqrt[3]{25b^2}}{\sqrt[3]{2a}\cdot\sqrt[3]{25b^2}} = \frac{\sqrt[3]{125b^3}}{\sqrt[3]{50ab^2}} = \frac{5b}{\sqrt[3]{50ab^2}}$

7. $\frac{\sqrt{x}-3}{4} = \frac{(\sqrt{x}-3)(\sqrt{x}+3)}{4(\sqrt{x}+3)}$
$= \frac{(\sqrt{x})^2-(3)^2}{4(\sqrt{x}+3)}$
$= \frac{x-9}{4(\sqrt{x}+3)}$

Vocabulary and Readiness Check

1. The conjugate of $a+b$ is $a-b$.

2. The process of writing an equivalent expression, but without a radical in the denominator is called rationalizing the denominator.

3. The process of writing an equivalent expression, but without a radical in the numerator is called rationalizing the numerator.

4. To rationalize the denominator of $\frac{5}{\sqrt{3}}$, we multiply by $\frac{\sqrt{3}}{\sqrt{3}}$.

5. The conjugate of $\sqrt{2}+x$ is $\sqrt{2}-x$.

6. The conjugate of $\sqrt{3}+y$ is $\sqrt{3}-y$.

7. The conjugate of $5-\sqrt{a}$ is $5+\sqrt{a}$.

8. The conjugate of $6-\sqrt{b}$ is $6+\sqrt{b}$.

9. The conjugate of $-7\sqrt{5}+8\sqrt{x}$ is $-7\sqrt{5}-8\sqrt{x}$.

10. The conjugate of $-9\sqrt{2}-6\sqrt{y}$ is $-9\sqrt{2}+6\sqrt{y}$.

Exercise Set 10.5

1. $\frac{\sqrt{2}}{\sqrt{7}} = \frac{\sqrt{2}\cdot\sqrt{7}}{\sqrt{7}\cdot\sqrt{7}} = \frac{\sqrt{14}}{\sqrt{49}} = \frac{\sqrt{14}}{7}$

3. $\sqrt{\frac{1}{5}} = \frac{\sqrt{1}}{\sqrt{5}} = \frac{1\cdot\sqrt{5}}{\sqrt{5}\cdot\sqrt{5}} = \frac{\sqrt{5}}{5}$

5. $\sqrt{\frac{4}{x}} = \frac{\sqrt{4}}{\sqrt{x}} = \frac{2\cdot\sqrt{x}}{\sqrt{x}\cdot\sqrt{x}} = \frac{2\sqrt{x}}{\sqrt{x^2}} = \frac{2\sqrt{x}}{x}$

7. $\frac{4}{\sqrt[3]{3}} = \frac{4\cdot\sqrt[3]{9}}{\sqrt[3]{3}\cdot\sqrt[3]{9}} = \frac{4\sqrt[3]{9}}{\sqrt[3]{27}} = \frac{4\sqrt[3]{9}}{3}$

9. $\frac{3}{\sqrt{8x}} = \frac{3\cdot\sqrt{2x}}{\sqrt{8x}\cdot\sqrt{2x}} = \frac{3\sqrt{2x}}{\sqrt{16x^2}} = \frac{3\sqrt{2x}}{4x}$

11. $\frac{3}{\sqrt[3]{4x^2}} = \frac{3\cdot\sqrt[3]{2x}}{\sqrt[3]{4x^2}\cdot\sqrt[3]{2x}} = \frac{3\sqrt[3]{2x}}{\sqrt[3]{8x^3}} = \frac{3\sqrt[3]{2x}}{2x}$

13. $\frac{9}{\sqrt{3a}} = \frac{9\cdot\sqrt{3a}}{\sqrt{3a}\cdot\sqrt{3a}} = \frac{9\sqrt{3a}}{3a} = \frac{3\sqrt{3a}}{a}$

15. $\frac{3}{\sqrt[3]{2}} = \frac{3\cdot\sqrt[3]{4}}{\sqrt[3]{2}\cdot\sqrt[3]{4}} = \frac{3\sqrt[3]{4}}{\sqrt[3]{8}} = \frac{3\sqrt[3]{4}}{2}$

17. $\frac{2\sqrt{3}}{\sqrt{7}} = \frac{2\sqrt{3}\cdot\sqrt{7}}{\sqrt{7}\cdot\sqrt{7}} = \frac{2\sqrt{21}}{\sqrt{49}} = \frac{2\sqrt{21}}{7}$

19. $\sqrt{\frac{2x}{5y}} = \frac{\sqrt{2x}}{\sqrt{5y}} = \frac{\sqrt{2x}\cdot\sqrt{5y}}{\sqrt{5y}\cdot\sqrt{5y}} = \frac{\sqrt{10xy}}{5y}$

21. $\sqrt[3]{\frac{3}{5}} = \frac{\sqrt[3]{3}}{\sqrt[3]{5}}\cdot\frac{\sqrt[3]{25}}{\sqrt[3]{25}} = \frac{\sqrt[3]{75}}{5}$

23. $\sqrt{\frac{3x}{50}} = \frac{\sqrt{3x}}{\sqrt{50}}$
$= \frac{\sqrt{3x}}{5\sqrt{2}}$
$= \frac{\sqrt{3x}\cdot\sqrt{2}}{5\sqrt{2}\cdot\sqrt{2}}$
$= \frac{\sqrt{6x}}{5\cdot 2}$
$= \frac{\sqrt{6x}}{10}$

25. $\frac{1}{\sqrt{12z}} = \frac{1}{\sqrt{4\cdot 3z}} = \frac{1}{2\sqrt{3z}}\cdot\frac{\sqrt{3z}}{\sqrt{3z}} = \frac{\sqrt{3z}}{6z}$

27. $\frac{\sqrt[3]{2y^2}}{\sqrt[3]{9x^2}} = \frac{\sqrt[3]{2y^2}\cdot\sqrt[3]{3x}}{\sqrt[3]{9x^2}\cdot\sqrt[3]{3x}} = \frac{\sqrt[3]{6xy^2}}{3x}$

29. $\sqrt[4]{\frac{81}{8}} = \frac{\sqrt[4]{81}}{\sqrt[4]{8}} = \frac{3 \cdot \sqrt[4]{2}}{\sqrt[4]{8} \cdot \sqrt[4]{2}} = \frac{3\sqrt[4]{2}}{\sqrt[4]{16}} = \frac{3\sqrt[4]{2}}{2}$

31. $\sqrt[4]{\frac{16}{9x^7}} = \frac{\sqrt[4]{16}}{\sqrt[4]{9x^7}} = \frac{2 \cdot \sqrt[4]{9x}}{\sqrt[4]{9x^7} \cdot \sqrt[4]{9x}} = \frac{2\sqrt[4]{9x}}{\sqrt[4]{81x^8}} = \frac{2\sqrt[4]{9x}}{3x^2}$

33.
$$\frac{5a}{\sqrt[5]{8a^9b^{11}}} = \frac{5a \cdot \sqrt[5]{4ab^4}}{\sqrt[5]{8a^9b^{11}} \cdot \sqrt[5]{4ab^4}}$$
$$= \frac{5a\sqrt[5]{4ab^4}}{\sqrt[5]{32a^{10}b^{15}}}$$
$$= \frac{5a\sqrt[5]{4ab^4}}{2a^2b^3}$$

35.
$$\frac{6}{2-\sqrt{7}} = \frac{6(2+\sqrt{7})}{(2-\sqrt{7})(2+\sqrt{7})}$$
$$= \frac{6(2+\sqrt{7})}{2^2-(\sqrt{7})^2}$$
$$= \frac{6(2+\sqrt{7})}{4-7}$$
$$= \frac{6(2+\sqrt{7})}{-3}$$
$$= -2(2+\sqrt{7})$$

37.
$$\frac{-7}{\sqrt{x}-3} = \frac{-7(\sqrt{x}+3)}{(\sqrt{x}-3)(\sqrt{x}+3)}$$
$$= \frac{-7(\sqrt{x}+3)}{(\sqrt{x})^2-(3)^2}$$
$$= \frac{-7(\sqrt{x}+3)}{x-9} \text{ or } \frac{7(\sqrt{x}+3)}{9-x}$$

39.
$$\frac{\sqrt{2}-\sqrt{3}}{\sqrt{2}+\sqrt{3}} = \frac{(\sqrt{2}-\sqrt{3})(\sqrt{2}-\sqrt{3})}{(\sqrt{2}+\sqrt{3})(\sqrt{2}-\sqrt{3})}$$
$$= \frac{(\sqrt{2})^2 - 2\sqrt{2}\sqrt{3} + (\sqrt{3})^2}{(\sqrt{2})^2-(\sqrt{3})^2}$$
$$= \frac{2-2\sqrt{6}+3}{2-3}$$
$$= \frac{5-2\sqrt{6}}{-1}$$
$$= -5+2\sqrt{6}$$

41.
$$\frac{\sqrt{a}+1}{2\sqrt{a}-\sqrt{b}}$$
$$= \frac{(\sqrt{a}+1)(2\sqrt{a}+\sqrt{b})}{(2\sqrt{a}-\sqrt{b})(2\sqrt{a}+\sqrt{b})}$$
$$= \frac{\sqrt{a} \cdot 2\sqrt{a} + \sqrt{a}\sqrt{b} + 1 \cdot 2\sqrt{a} + 1 \cdot \sqrt{b}}{(2\sqrt{a})^2 - (\sqrt{b})^2}$$
$$= \frac{2a+\sqrt{ab}+2\sqrt{a}+\sqrt{b}}{4a-b}$$

43.
$$\frac{8}{1+\sqrt{10}} = \frac{8(1-\sqrt{10})}{(1+\sqrt{10})(1-\sqrt{10})}$$
$$= \frac{8(1-\sqrt{10})}{1^2-(\sqrt{10})^2}$$
$$= \frac{8(1-\sqrt{10})}{1-10}$$
$$= -\frac{8(1-\sqrt{10})}{9}$$

45. $\frac{\sqrt{x}}{\sqrt{x}+\sqrt{y}} = \frac{\sqrt{x}(\sqrt{x}-\sqrt{y})}{(\sqrt{x}+\sqrt{y})(\sqrt{x}-\sqrt{y})}$

$= \frac{\sqrt{x}(\sqrt{x}-\sqrt{y})}{(\sqrt{x})^2-(\sqrt{y})^2}$

$= \frac{\sqrt{x}(\sqrt{x}-\sqrt{y})}{x-y}$

$= \frac{\sqrt{x}\sqrt{x}-\sqrt{x}\sqrt{y}}{x-y}$

$= \frac{x-\sqrt{xy}}{x-y}$

47. $\frac{2\sqrt{3}+\sqrt{6}}{4\sqrt{3}-\sqrt{6}} = \frac{(2\sqrt{3}+\sqrt{6})(4\sqrt{3}+\sqrt{6})}{(4\sqrt{3}-\sqrt{6})(4\sqrt{3}+\sqrt{6})}$

$= \frac{8\cdot 3+2\sqrt{18}+4\sqrt{18}+6}{(4\sqrt{3})^2-(\sqrt{6})^2}$

$= \frac{30+6\sqrt{18}}{16\cdot 3-6}$

$= \frac{30+6(3)\sqrt{2}}{42}$

$= \frac{30+18\sqrt{2}}{42}$

$= \frac{6(5+3\sqrt{2})}{42}$

$= \frac{5+3\sqrt{2}}{7}$

49. $\sqrt{\frac{5}{3}} = \frac{\sqrt{5}}{\sqrt{3}} = \frac{\sqrt{5}\cdot\sqrt{5}}{\sqrt{3}\cdot\sqrt{5}} = \frac{\sqrt{25}}{\sqrt{15}} = \frac{5}{\sqrt{15}}$

51. $\sqrt{\frac{18}{5}} = \frac{\sqrt{18}}{\sqrt{5}}$

$= \frac{\sqrt{9}\cdot\sqrt{2}}{\sqrt{5}}$

$= \frac{3\sqrt{2}}{\sqrt{5}}$

$= \frac{3\sqrt{2}\cdot\sqrt{2}}{\sqrt{5}\cdot\sqrt{2}}$

$= \frac{3\cdot 2}{\sqrt{10}}$

$= \frac{6}{\sqrt{10}}$

53. $\frac{\sqrt{4x}}{7} = \frac{2\sqrt{x}}{7} = \frac{2\sqrt{x}\cdot\sqrt{x}}{7\cdot\sqrt{x}} = \frac{2\sqrt{x^2}}{7\sqrt{x}} = \frac{2x}{7\sqrt{x}}$

55. $\frac{\sqrt[3]{5y^2}}{\sqrt[3]{4x}} = \frac{\sqrt[3]{5y^2}\cdot\sqrt[3]{5^2y}}{\sqrt[3]{4x}\cdot\sqrt[3]{5^2y}} = \frac{\sqrt[3]{5^3y^3}}{\sqrt[3]{100xy}} = \frac{5y}{\sqrt[3]{100xy}}$

57. $\sqrt{\frac{2}{5}} = \frac{\sqrt{2}}{\sqrt{5}} = \frac{\sqrt{2}\cdot\sqrt{2}}{\sqrt{5}\cdot\sqrt{2}} = \frac{\sqrt{4}}{\sqrt{10}} = \frac{2}{\sqrt{10}}$

59. $\frac{\sqrt{2x}}{11} = \frac{\sqrt{2x}\cdot\sqrt{2x}}{11\cdot\sqrt{2x}} = \frac{\sqrt{4x^2}}{11\sqrt{2x}} = \frac{2x}{11\sqrt{2x}}$

61. $\sqrt[3]{\frac{7}{8}} = \frac{\sqrt[3]{7}}{\sqrt[3]{8}}$

$= \frac{\sqrt[3]{7}}{2}$

$= \frac{\sqrt[3]{7}\cdot\sqrt[3]{7^2}}{2\cdot\sqrt[3]{7^2}}$

$= \frac{\sqrt[3]{7^3}}{2\sqrt[3]{49}}$

$= \frac{7}{2\sqrt[3]{49}}$

63. $\frac{\sqrt[3]{3x^5}}{10} = \frac{\sqrt[3]{x^3\cdot 3x^2}}{10}$

$= \frac{x\sqrt[3]{3x^2}}{10}$

$= \frac{x\sqrt[3]{3x^2}\cdot\sqrt[3]{3^2x}}{10\cdot\sqrt[3]{3^2x}}$

$= \frac{x\sqrt[3]{3^3x^3}}{10\sqrt[3]{9x}}$

$= \frac{x\cdot 3x}{10\sqrt[3]{9x}}$

$= \frac{3x^2}{10\sqrt[3]{9x}}$

65. $\sqrt{\dfrac{18x^4y^6}{3z}} = \dfrac{\sqrt{18x^4y^6}}{\sqrt{3z}}$
$= \dfrac{\sqrt{9x^4y^6 \cdot 2}}{\sqrt{3z}}$
$= \dfrac{3x^2y^3\sqrt{2}}{\sqrt{3z}}$
$= \dfrac{3x^2y^3\sqrt{2} \cdot \sqrt{2}}{\sqrt{3z} \cdot \sqrt{2}}$
$= \dfrac{3x^2y^3 \cdot 2}{\sqrt{6z}}$
$= \dfrac{6x^2y^3}{\sqrt{6z}}$

67. Answers may vary

69. $\dfrac{2-\sqrt{11}}{6} = \dfrac{(2-\sqrt{11})(2+\sqrt{11})}{6(2+\sqrt{11})}$
$= \dfrac{4-11}{12+6\sqrt{11}}$
$= \dfrac{-7}{12+6\sqrt{11}}$

71. $\dfrac{2-\sqrt{7}}{-5} = \dfrac{(2-\sqrt{7})(2+\sqrt{7})}{-5(2+\sqrt{7})}$
$= \dfrac{4-7}{-5(2+\sqrt{7})}$
$= \dfrac{-3}{-5(2+\sqrt{7})}$
$= \dfrac{3}{5(2+\sqrt{7})}$
$= \dfrac{3}{10+5\sqrt{7}}$

73. $\dfrac{\sqrt{x}+3}{\sqrt{x}} = \dfrac{(\sqrt{x}+3)(\sqrt{x}-3)}{\sqrt{x}(\sqrt{x}-3)}$
$= \dfrac{\sqrt{x^2}-9}{\sqrt{x^2}-3\sqrt{x}}$
$= \dfrac{x-9}{x-3\sqrt{x}}$

75. $\dfrac{\sqrt{2}-1}{\sqrt{2}+1} = \dfrac{(\sqrt{2}-1)(\sqrt{2}+1)}{(\sqrt{2}+1)(\sqrt{2}+1)}$
$= \dfrac{\sqrt{4}-1}{\sqrt{4}+2\sqrt{2}+1}$
$= \dfrac{2-1}{2+2\sqrt{2}+1}$
$= \dfrac{1}{3+2\sqrt{2}}$

77. $\dfrac{\sqrt{x}+1}{\sqrt{x}-1} = \dfrac{(\sqrt{x}+1)(\sqrt{x}-1)}{(\sqrt{x}-1)(\sqrt{x}-1)}$
$= \dfrac{\sqrt{x^2}-1}{\sqrt{x^2}-2\sqrt{x}+1}$
$= \dfrac{x-1}{x-2\sqrt{x}+1}$

79. $2x-7 = 3(x-4)$
$2x-7 = 3x-12$
$-x-7 = -12$
$-x = -5$
$x = 5$
The solution is 5.

81. $(x-6)(2x+1) = 0$
$x-6 = 0$ or $2x+1 = 0$
$x = 6$ or $2x = -1$
$x = -\dfrac{1}{2}$
The solutions are $-\dfrac{1}{2}$, 6.

83. $x^2 - 8x = -12$
$x^2 - 8x + 12 = 0$
$(x-6)(x-2) = 0$
$x-6 = 0$ or $x-2 = 0$
$x = 6$ or $x = 2$
The solutions are 2, 6.

85. $\dfrac{9}{\sqrt[3]{5}} = \dfrac{9}{\sqrt[3]{5}} \cdot \dfrac{\sqrt[3]{25}}{\sqrt[3]{25}} = \dfrac{9\sqrt[3]{25}}{\sqrt[3]{125}} = \dfrac{9\sqrt[3]{25}}{5}$
The smallest number is $\sqrt[3]{25}$.

87. $r = \sqrt{\dfrac{A}{4\pi}}$
$= \dfrac{\sqrt{A}}{\sqrt{4\pi}}$
$= \dfrac{\sqrt{A}}{2\sqrt{\pi}}$
$= \dfrac{\sqrt{A} \cdot \sqrt{\pi}}{2\sqrt{\pi} \cdot \sqrt{\pi}}$
$= \dfrac{\sqrt{A\pi}}{2\pi}$

89. Answers may vary

The Bigger Picture

1. $\sqrt{56} = \sqrt{4 \cdot 14} = \sqrt{4} \cdot \sqrt{14} = 2\sqrt{14}$

2. $\sqrt{\dfrac{20x^5}{49}} = \dfrac{\sqrt{20x^5}}{\sqrt{49}}$
$= \dfrac{\sqrt{4x^4 \cdot 5x}}{7}$
$= \dfrac{\sqrt{4x^4} \cdot \sqrt{5x}}{7}$
$= \dfrac{2x^2\sqrt{5x}}{7}$

3. $(-5x^{12}y^{-3})(3x^{-7}y^{14}) = -5 \cdot 3x^{12-7}y^{-3+14}$
$= -15x^5y^{11}$

4. $\sqrt{\dfrac{10}{11}} = \dfrac{\sqrt{10}}{\sqrt{11}} = \dfrac{\sqrt{10} \cdot \sqrt{11}}{\sqrt{11} \cdot \sqrt{11}} = \dfrac{\sqrt{110}}{11}$

5. $\dfrac{8}{\sqrt{5}-1} = \dfrac{8\left(\sqrt{5}+1\right)}{\left(\sqrt{5}-1\right)\left(\sqrt{5}+1\right)}$
$= \dfrac{8\left(\sqrt{5}+1\right)}{5-1}$
$= \dfrac{8\left(\sqrt{5}+1\right)}{4}$
$= 2\left(\sqrt{5}+1\right)$ or $2\sqrt{5}+2$

6. $\frac{1}{2}(6x^2-4) + \frac{1}{3}(6x^2-9) - 14$
$= 3x^2 - 2 + 2x^2 - 3 - 14$
$= 5x^2 - 19$

7. $\dfrac{\sqrt{13}}{\sqrt{2x^5}} = \dfrac{\sqrt{13}}{\sqrt{x^4 \cdot 2x}}$
$= \dfrac{\sqrt{13}}{x^2\sqrt{2x}}$
$= \dfrac{\sqrt{13}}{x^2\sqrt{2x}} \cdot \dfrac{\sqrt{2x}}{\sqrt{2x}}$
$= \dfrac{\sqrt{26x}}{2x^3}$

8. $\dfrac{y}{y^2+1} - \dfrac{2y-6}{y^2+1} = \dfrac{y-(2y-6)}{y^2+1}$
$= \dfrac{y-2y+6}{y^2+1}$
$= \dfrac{-y+6}{y^2+1}$

9. $\dfrac{5x^3+20x}{10x-10y} \div \dfrac{2x^2+8}{x^2-y^2} = \dfrac{5x^3+20x}{10x-10y} \cdot \dfrac{x^2-y^2}{2x^2+8}$
$= \dfrac{5x(x^2+4) \cdot (x+y)(x-y)}{2 \cdot 5(x-y) \cdot 2(x^2+4)}$
$= \dfrac{x(x+y)}{4}$

10. $\sqrt[3]{16y^{20}} = \sqrt[3]{8y^{18} \cdot 2y^2}$
$= \sqrt[3]{2^3y^{18}}\sqrt[3]{2y^2}$
$= 2y^6\sqrt[3]{2y^2}$

Integrated Review

1. $\sqrt{81} = 9$ because $9^2 = 81$.

2. $\sqrt[3]{-8} = -2$ because $(-2)^3 = -8$.

3. $\sqrt[4]{\dfrac{1}{16}} = \dfrac{1}{2}$ because $\left(\dfrac{1}{2}\right)^4 = \dfrac{1}{16}$.

4. $\sqrt{x^6} = x^3$ because $(x^3)^2 = x^6$.

5. $\sqrt[3]{y^9} = y^3$ because $(y^3)^3 = y^9$.

6. $\sqrt{4y^{10}} = 2y^5$ because $(2y^5)^2 = 4y^{10}$.

7. $\sqrt[5]{-32y^5} = -2y$ because $(-2y)^5 = -32y^5$.

8. $\sqrt[4]{81b^{12}} = 3b^3$ because $(3b^3)^4 = 81b^{12}$.

9. $36^{1/2} = \sqrt{36} = 6$

10. $(3y)^{1/4} = \sqrt[4]{3y}$

11. $64^{-2/3} = \dfrac{1}{\left(\sqrt[3]{64}\right)^2} = \dfrac{1}{4^2} = \dfrac{1}{16}$

12. $(x+1)^{3/5} = \sqrt[5]{(x+1)^3}$

13. $y^{-1/6} \cdot y^{7/6} = y^{-\frac{1}{6}+\frac{7}{6}} = y^{6/6} = y$

14.
$$\begin{aligned}\frac{(2x^{1/3})^4}{x^{5/6}} &= 16x^{4/3}x^{-5/6}\\ &= 16x^{\frac{8}{6}-\frac{5}{6}}\\ &= 16x^{3/6}\\ &= 16x^{1/2}\end{aligned}$$

15. $\dfrac{x^{1/4}x^{3/4}}{x^{-1/4}} = x^{\frac{1}{4}+\frac{3}{4}+\frac{1}{4}} = x^{5/4}$

16. $4^{1/3} \cdot 4^{2/5} = 4^{\frac{1}{3}+\frac{2}{5}} = 4^{\frac{5}{15}+\frac{6}{15}} = 4^{11/15}$

17. $\sqrt[3]{8x^6} = (8x^6)^{1/3} = (2^3x^6)^{1/3} = 2^{3/3}x^{6/3} = 2x^2$

18.
$$\begin{aligned}\sqrt[12]{a^9b^6} &= (a^9b^6)^{1/12}\\ &= a^{9/12}b^{6/12}\\ &= a^{3/4}b^{1/2}\\ &= a^{3/4}b^{2/4}\\ &= (a^3b^2)^{1/4}\\ &= \sqrt[4]{a^3b^2}\end{aligned}$$

19. $\sqrt[4]{x} \cdot \sqrt{x} = x^{1/4} \cdot x^{1/2} = x^{\frac{1}{4}+\frac{2}{4}} = x^{3/4} = \sqrt[4]{x^3}$

20.
$$\begin{aligned}\sqrt{5} \cdot \sqrt[3]{2} &= 5^{1/2} \cdot 2^{1/3}\\ &= 5^{3/6} \cdot 2^{2/6}\\ &= (5^3 \cdot 2^2)^{1/6}\\ &= \sqrt[6]{5^3 \cdot 2^2}\\ &= \sqrt[6]{500}\end{aligned}$$

21. $\sqrt{40} = \sqrt{4}\sqrt{10} = 2\sqrt{10}$

22. $\sqrt[4]{16x^7y^{10}} = \sqrt[4]{16x^4y^8}\sqrt[4]{x^3y^2} = 2xy^2\sqrt[4]{x^3y^2}$

23. $\sqrt[3]{54x^4} = \sqrt[3]{27x^3}\sqrt[3]{2x} = 3x\sqrt[3]{2x}$

24. $\sqrt[5]{-64b^{10}} = \sqrt[5]{-32b^{10}}\sqrt[5]{2} = -2b^2\sqrt[5]{2}$

25. $\sqrt{5} \cdot \sqrt{x} = \sqrt{5x}$

26. $\sqrt[3]{8x} \cdot \sqrt[3]{8x^2} = \sqrt[3]{64x^3} = 4x$

27.
$$\begin{aligned}\frac{\sqrt{98y^6}}{\sqrt{2y}} &= \sqrt{\frac{98y^6}{2y}}\\ &= \sqrt{49y^5}\\ &= \sqrt{49y^4} \cdot \sqrt{y}\\ &= 7y^2\sqrt{y}\end{aligned}$$

28.
$$\begin{aligned}\frac{\sqrt[4]{48a^9b^3}}{\sqrt[4]{ab^3}} &= \sqrt[4]{\frac{48a^9b^3}{ab^3}}\\ &= \sqrt[4]{48a^8}\\ &= \sqrt[4]{16a^8} \cdot \sqrt[4]{3}\\ &= 2a^2\sqrt[4]{3}\end{aligned}$$

29.
$$\begin{aligned}\sqrt{20} - \sqrt{75} + 5\sqrt{7} &= \sqrt{4}\sqrt{5} - \sqrt{25}\sqrt{3} + 5\sqrt{7}\\ &= 2\sqrt{5} - 5\sqrt{3} + 5\sqrt{7}\end{aligned}$$

30.
$$\begin{aligned}\sqrt[3]{54y^4} - y\sqrt[3]{16y} &= \sqrt[3]{27y^3}\sqrt[3]{2y} - y\sqrt[3]{8}\sqrt[3]{2y}\\ &= 3y\sqrt[3]{2y} - 2y\sqrt[3]{2y}\\ &= y\sqrt[3]{2y}\end{aligned}$$

31. $\sqrt{3}\left(\sqrt{5} - \sqrt{2}\right) = \sqrt{3}\sqrt{5} - \sqrt{3}\sqrt{2} = \sqrt{15} - \sqrt{6}$

32.
$$\begin{aligned}\left(\sqrt{7} + \sqrt{3}\right)^2 &= \left(\sqrt{7}\right)^2 + 2\sqrt{7}\sqrt{3} + \left(\sqrt{3}\right)^2\\ &= 7 + 2\sqrt{21} + 3\\ &= 10 + 2\sqrt{21}\end{aligned}$$

33.
$$\begin{aligned}\left(2x - \sqrt{5}\right)\left(2x + \sqrt{5}\right) &= (2x)^2 - \left(\sqrt{5}\right)^2\\ &= 4x^2 - 5\end{aligned}$$

34.
$$\begin{aligned}\left(\sqrt{x+1} - 1\right)^2 &= \left(\sqrt{x+1}\right)^2 - 2\left(\sqrt{x+1}\right) + 1^2\\ &= x + 1 - 2\sqrt{x+1} + 1\\ &= x + 2 - 2\sqrt{x+1}\end{aligned}$$

35. $\sqrt{\frac{7}{3}}=\frac{\sqrt{7}}{\sqrt{3}}=\frac{\sqrt{7}}{\sqrt{3}}\cdot\frac{\sqrt{3}}{\sqrt{3}}=\frac{\sqrt{21}}{3}$

36. $\frac{5}{\sqrt[3]{2x^2}}=\frac{5}{\sqrt[3]{2x^2}}\cdot\frac{\sqrt[3]{4x}}{\sqrt[3]{4x}}=\frac{5\sqrt[3]{4x}}{\sqrt[3]{8x^3}}=\frac{5\sqrt[3]{4x}}{2x}$

37. $\frac{\sqrt{3}-\sqrt{7}}{2\sqrt{3}+\sqrt{7}}$

$=\frac{\sqrt{3}-\sqrt{7}}{2\sqrt{3}+\sqrt{7}}\cdot\frac{\left(2\sqrt{3}-\sqrt{7}\right)}{\left(2\sqrt{3}-\sqrt{7}\right)}$

$=\frac{\sqrt{3}\left(2\sqrt{3}\right)-\sqrt{3}\sqrt{7}-\sqrt{7}\left(2\sqrt{3}\right)+\sqrt{7}\sqrt{7}}{\left(2\sqrt{3}\right)^2-\left(\sqrt{7}\right)^2}$

$=\frac{6-\sqrt{21}-2\sqrt{21}+7}{12-7}$

$=\frac{13-3\sqrt{21}}{5}$

38. $\sqrt{\frac{7}{3}}=\frac{\sqrt{7}}{\sqrt{3}}=\frac{\sqrt{7}}{\sqrt{3}}\cdot\frac{\sqrt{7}}{\sqrt{7}}=\frac{7}{\sqrt{21}}$

39. $\sqrt[3]{\frac{9y}{11}}=\frac{\sqrt[3]{9y}}{\sqrt[3]{11}}=\frac{\sqrt[3]{9y}}{\sqrt[3]{11}}\cdot\frac{\sqrt[3]{3y^2}}{\sqrt[3]{3y^2}}=\frac{\sqrt[3]{27y^3}}{\sqrt[3]{31y^2}}=\frac{3y}{\sqrt[3]{33y^2}}$

40. $\frac{\sqrt{x}-2}{\sqrt{x}}=\frac{\sqrt{x}-2}{\sqrt{x}}\cdot\frac{\sqrt{x}+2}{\sqrt{x}+2}$

$=\frac{\left(\sqrt{x}\right)^2-2^2}{\sqrt{x}\sqrt{x}+2\sqrt{x}}$

$=\frac{x-4}{x+2\sqrt{x}}$

Section 10.6

Practice Exercises

1. $\sqrt{3x-5}=7$

$\left(\sqrt{3x-5}\right)^2=7^2$

$3x-5=49$

$3x=54$

$x=18$

Check:

$\sqrt{3x-5}=7$

$\sqrt{3(18)-5}\stackrel{?}{=}7$

$\sqrt{54-5}\stackrel{?}{=}7$

$\sqrt{49}\stackrel{?}{=}7$

$7=7$

The solution is 18.

2. $\sqrt{3-2x}-4x=0$

$\sqrt{3-2x}-4x+4x=0+4x$

$\sqrt{3-2x}=4x$

$\left(\sqrt{3-2x}\right)^2=(4x)^2$

$3-2x=16x^2$

$16x^2+2x-3=0$

$(8x-3)(2x+1)=0$

$8x-3=0$ or $2x+1=0$

$x=\frac{3}{8}$ or $x=-\frac{1}{2}$

Check $\frac{3}{8}$:

$\sqrt{3-2x}-4x=0$

$\sqrt{3-2\left(\frac{3}{8}\right)}-4\left(\frac{3}{8}\right)\stackrel{?}{=}0$

$\sqrt{\frac{24}{8}-\frac{6}{8}}-\frac{3}{2}\stackrel{?}{=}0$

$\sqrt{\frac{9}{4}}-\frac{3}{2}\stackrel{?}{=}0$

$\frac{3}{2}-\frac{3}{2}=0$

Check $-\frac{1}{2}$:

$\sqrt{3-2x}-4x=0$

$\sqrt{3-2\left(-\frac{1}{2}\right)}-4\left(-\frac{1}{2}\right)\stackrel{?}{=}0$

$\sqrt{3+1}+2\stackrel{?}{=}0$

$2+2\stackrel{?}{=}0$

$4\neq 0$

$-\frac{1}{2}$ does not check, so the only solution is $\frac{3}{8}$.

3. $\sqrt[3]{x-2}+1=3$
$\sqrt[3]{x-2}=2$
$\left(\sqrt[3]{x-2}\right)^3=2^3$
$x-2=8$
$x=10$
Check:
$\sqrt[3]{x-2}+1=3$
$\sqrt[3]{10-2}+1\stackrel{?}{=}3$
$\sqrt[3]{8}+1\stackrel{?}{=}3$
$2+1=3$
The solution is 10.

4. $\sqrt{16+x}=x-4$
$\left(\sqrt{16+x}\right)^2=(x-4)^2$
$16+x=x^2-8x+16$
$x^2-9x=0$
$x(x-9)=0$
$x=0$ or $x-9=0$
$x=9$
Check 0:
$\sqrt{16+x}=x-4$
$\sqrt{16+0}\stackrel{?}{=}0-4$
$\sqrt{16}\stackrel{?}{=}-4$
$4\neq-4$
Check 9:
$\sqrt{16+x}=x-4$
$\sqrt{16+9}\stackrel{?}{=}9-4$
$\sqrt{25}\stackrel{?}{=}5$
$5=5$
0 does not check, so the only solution is 9.

5. $\sqrt{8x+1}+\sqrt{3x}=2$
$\sqrt{8x+1}=2-\sqrt{3x}$
$\left(\sqrt{8x+1}\right)^2=\left(2-\sqrt{3x}\right)^2$
$8x+1=4-4\sqrt{3x}+3x$
$4\sqrt{3x}=3-5x$
$\left(4\sqrt{3x}\right)^2=(3-5x)^2$
$16(3x)=9-30x+25x^2$
$25x^2-78x+9=0$
$(25x-3)(x-3)=0$
$25x-3=0$ or $x-3=0$
$x=\frac{3}{25}$ or $x=3$

Check $\frac{3}{25}$:
$\sqrt{8x+1}+\sqrt{3x}=2$
$\sqrt{8\left(\frac{3}{25}\right)+1}+\sqrt{3\left(\frac{3}{25}\right)}\stackrel{?}{=}2$
$\sqrt{\frac{24}{25}+\frac{25}{25}}+\sqrt{\frac{9}{25}}\stackrel{?}{=}2$
$\sqrt{\frac{49}{25}}+\sqrt{\frac{9}{25}}\stackrel{?}{=}2$
$\frac{7}{5}+\frac{3}{5}\stackrel{?}{=}2$
$\frac{10}{5}=2$
Check 3:
$\sqrt{8x+1}+\sqrt{3x}=2$
$\sqrt{8(3)+1}+\sqrt{3(3)}\stackrel{?}{=}2$
$\sqrt{25}+\sqrt{9}\stackrel{?}{=}2$
$5+3\neq2$
3 does not check, so the only solution is $\frac{3}{25}$.

6. $a^2+b^2=c^2$
$a^2+6^2=12^2$
$a^2+36=144$
$a^2=108$
$a=\pm\sqrt{108}=\pm\sqrt{36\cdot3}=\pm6\sqrt{3}$
Since a is a length, we will use the positive value only. The unknown leg is $6\sqrt{3}$ meters long.

7. Consider the base of the tank, and the plastic divider in the diagonal. Use the Pythagorean theorem to find l.

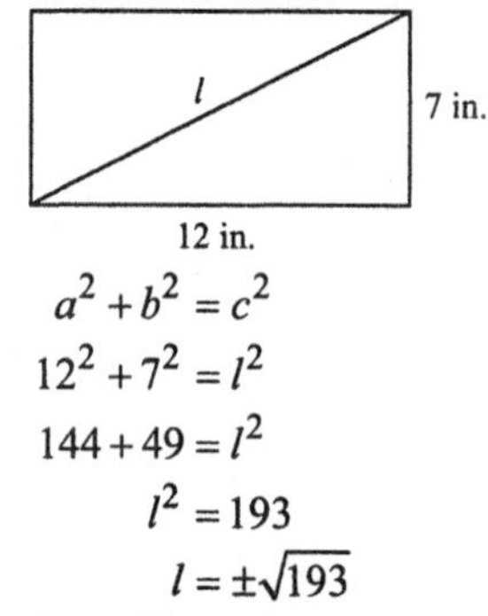

$a^2+b^2=c^2$
$12^2+7^2=l^2$
$144+49=l^2$
$l^2=193$
$l=\pm\sqrt{193}$
We will use the positive value because l represents length. The divider must be $\sqrt{193}\approx13.89$ inches long.

Graphing Calculator Explorations

1.

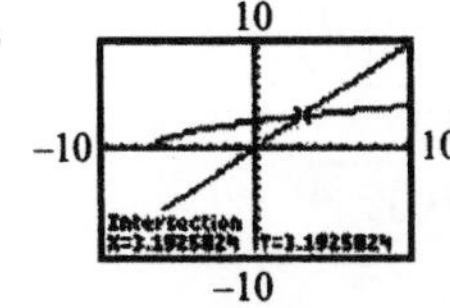

The solution is 3.19.

2.

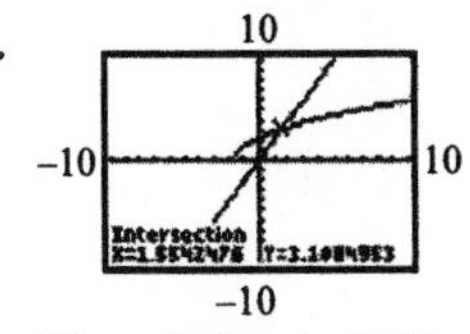

The solution is 1.55.

3.

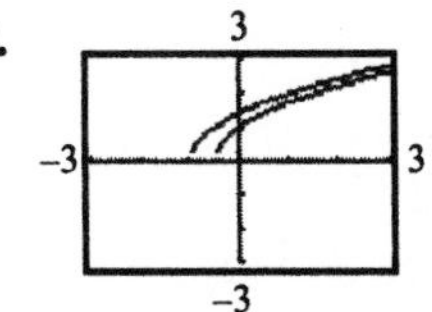

There is no solution. The solution set is ∅.

4.

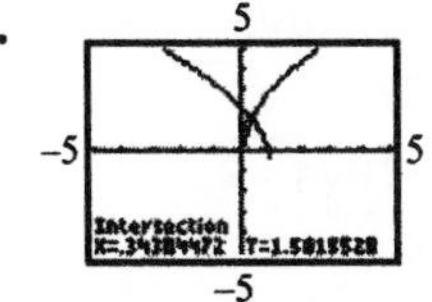

The solution is 0.34.

5.

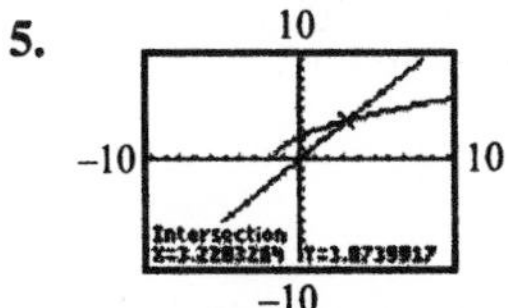

The solution is 3.23.

6.

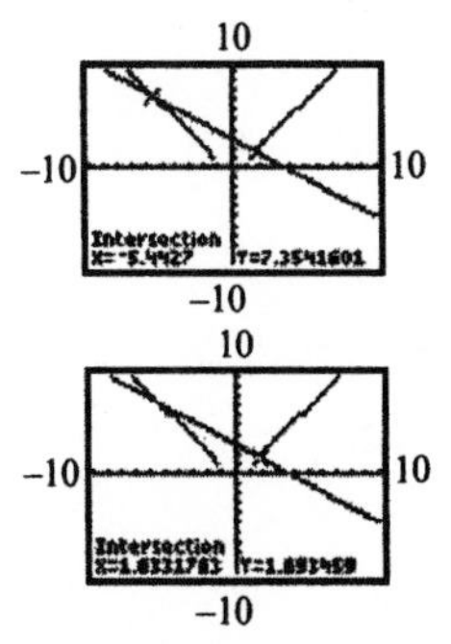

The solutions are −5.44 and 7.35.

Vocabulary and Readiness Check

1. A proposed solution that is not a solution of the original equation is called an <u>extraneous solution</u>.

2. The Pythagorean Theorem states that $a^2 + b^2 = c^2$ where a and b are the lengths of the <u>legs</u> of a <u>right</u> triangle and c is the length of the <u>hypotenuse</u>.

3. The square of $x - 5$, or $(x-5)^2 = \underline{x^2 - 10x + 25}$.

4. The square of $4-\sqrt{7x}$, or

$$\left(4-\sqrt{7x}\right)^2 = \underline{16 - 8\sqrt{7x} + 7x}.$$

Exercise Set 10.6

1.
$$\sqrt{2x} = 4$$
$$\left(\sqrt{2x}\right)^2 = 4^2$$
$$2x = 16$$
$$x = 8$$
The solution is 8.

3.
$$\sqrt{x-3} = 2$$
$$\left(\sqrt{x-3}\right)^2 = 2^2$$
$$x - 3 = 4$$
$$x = 7$$
The solution is 7.

5. $\sqrt{2x} = -4$

No solution since a principle square root does not yield a negative number.

7.
$$\sqrt{4x-3} - 5 = 0$$
$$\sqrt{4x-3} = 5$$
$$\left(\sqrt{4x-3}\right)^2 = 5^2$$
$$4x - 3 = 25$$
$$4x = 28$$
$$x = 7$$
The solution is 7.

9. $\sqrt{2x-3}-2=1$
$$\sqrt{2x-3}=3$$
$$\left(\sqrt{2x-3}\right)^2=3^2$$
$$2x-3=9$$
$$2x=12$$
$$x=6$$
The solution is 6.

11. $\sqrt[3]{6x}=-3$
$$\left(\sqrt[3]{6x}\right)^3=(-3)^3$$
$$6x=-27$$
$$x=-\frac{27}{6}=-\frac{9}{2}$$
The solution is $-\frac{9}{2}$.

13. $\sqrt[3]{x-2}-3=0$
$$\sqrt[3]{x-2}=3$$
$$\left(\sqrt[3]{x-2}\right)^3=3^3$$
$$x-2=27$$
$$x=29$$
The solution is 29.

15. $\sqrt{13-x}=x-1$
$$\left(\sqrt{13-x}\right)^2=(x-1)^2$$
$$13-x=x^2-2x+1$$
$$0=x^2-x-12$$
$$0=(x-4)(x+3)$$
$$x-4=0 \text{ or } x+3=0$$
$$x=4 \text{ or } x=-3$$
We discard –3 as extraneous. The solution is 4.

17. $x-\sqrt{4-3x}=-8$
$$x+8=\sqrt{4-3x}$$
$$(x+8)^2=\left(\sqrt{4-3x}\right)^2$$
$$x^2+16x+64=4-3x$$
$$x^2+19x+60=0$$
$$(x+4)(x+15)=0$$
$$x+4=0 \text{ or } x+15=0$$
$$x=-4 \text{ or } x=-15$$
We discard –15 as extraneous. The solution is –4.

19. $\sqrt{y+5}=2-\sqrt{y-4}$
$$\left(\sqrt{y+5}\right)^2=\left(2-\sqrt{y-4}\right)^2$$
$$y+5=4-4\sqrt{y-4}+(y-4)$$
$$y+5=y-4\sqrt{y-4}$$
$$5=-4\sqrt{y-4}$$
$$5^2=\left(-4\sqrt{y-4}\right)^2$$
$$25=16(y-4)$$
$$25=16y-64$$
$$89=16y$$
$$\frac{89}{16}=y$$
We discard $\frac{89}{16}$ as extraneous. There is no solution.

21. $\sqrt{x-3}+\sqrt{x+2}=5$
$$\sqrt{x-3}=5-\sqrt{x+2}$$
$$\left(\sqrt{x-3}\right)^2=\left(5-\sqrt{x+2}\right)^2$$
$$x-3=25-10\sqrt{x+2}+(x+2)$$
$$x-3=27-10\sqrt{x+2}+x$$
$$-30=-10\sqrt{x+2}$$
$$3=\sqrt{x+2}$$
$$3^2=\left(\sqrt{x+2}\right)^2$$
$$9=x+2$$
$$7=x$$
The solution is 7.

23. $\sqrt{3x-2}=5$
$$\left(\sqrt{3x-2}\right)^2=5^2$$
$$3x-2=25$$
$$3x=27$$
$$x=9$$
The solution is 9.

25. $-\sqrt{2x}+4=-6$
$$10=\sqrt{2x}$$
$$10^2=\left(\sqrt{2x}\right)^2$$
$$100=2x$$
$$50=x$$
The solution is 50.

27. $\sqrt{3x+1}+2=0$

$\sqrt{3x+1}=-2$

No solution since a principle square root does not yield a negative number.

29. $\sqrt[4]{4x+1}-2=0$

$\sqrt[4]{4x+1}=2$

$\left(\sqrt[4]{4x+1}\right)^4=2^4$

$4x+1=16$

$4x=15$

$x=\frac{15}{4}$

The solution is $\frac{15}{4}$.

31. $\sqrt{4x-3}=7$

$\left(\sqrt{4x-3}\right)^2=7^2$

$4x-3=49$

$4x=52$

$x=13$

The solution is 13.

33. $\sqrt[3]{6x-3}-3=0$

$\sqrt[3]{6x-3}=3$

$\left(\sqrt[3]{6x-3}\right)^3=3^3$

$6x-3=27$

$6x=30$

$x=5$

The solution is 5.

35. $\sqrt[3]{2x-3}-2=-5$

$\sqrt[3]{2x-3}=-3$

$\left(\sqrt[3]{2x-3}\right)^3=(-3)^3$

$2x-3=-27$

$2x=-24$

$x=-12$

The solution is –12.

37. $\sqrt{x+4}=\sqrt{2x-5}$

$\left(\sqrt{x+4}\right)^2=\left(\sqrt{2x-5}\right)^2$

$x+4=2x-5$

$-x=-9$

$x=9$

The solution is 9.

39. $x-\sqrt{1-x}=-5$

$x+5=\sqrt{1-x}$

$(x+5)^2=\left(\sqrt{1-x}\right)^2$

$x^2+10x+25=1-x$

$x^2+11x+24=0$

$(x+8)(x+3)=0$

$x+8=0$ or $x+3=0$

$x=-8$ or $x=-3$

We discard –8 as extraneous. The solution is –3.

41. $\sqrt[3]{-6x-1}=\sqrt[3]{-2x-5}$

$\left(\sqrt[3]{-6x-1}\right)^3=\left(\sqrt[3]{-2x-5}\right)^3$

$-6x-1=-2x-5$

$-4x=-4$

$x=1$

The solution is 1.

43. $\sqrt{5x-1}-\sqrt{x}+2=3$

$\sqrt{5x-1}=\sqrt{x}+1$

$\left(\sqrt{5x-1}\right)^2=\left(\sqrt{x}+1\right)^2$

$5x-1=x+2\sqrt{x}+1$

$4x-2=2\sqrt{x}$

$2x-1=\sqrt{x}$

$(2x-1)^2=\left(\sqrt{x}\right)^2$

$4x^2-4x+1=x$

$4x^2-5x+1=0$

$(4x-1)(x-1)=0$

$4x-1=0$ or $x-1=0$

$4x=1$ or $x=1$

$x=\frac{1}{4}$

We discard $\frac{1}{4}$ as extraneous. The solution is 1.

45. $\sqrt{2x-1}=\sqrt{1-2x}$

$\left(\sqrt{2x-1}\right)^2=\left(\sqrt{1-2x}\right)^2$

$2x-1=1-2x$

$4x=2$

$x=\frac{2}{4}=\frac{1}{2}$

The solution is $\frac{1}{2}$.

47. $\sqrt{3x+4}-1=\sqrt{2x+1}$
$\sqrt{3x+4}=\sqrt{2x+1}+1$
$\left(\sqrt{3x+4}\right)^2=\left(\sqrt{2x+1}+1\right)^2$
$3x+4=(2x+1)+2\sqrt{2x+1}+1$
$3x+4=2x+2+2\sqrt{2x+1}$
$x+2=2\sqrt{2x+1}$
$(x+2)^2=\left(2\sqrt{2x+1}\right)^2$
$x^2+4x+4=4(2x+1)$
$x^2+4x+4=8x+4$
$x^2-4x=0$
$x(x-4)=0$
$x=0$ or $x-4=0$
$x=4$
The solutions are 0 and 4.

49. $\sqrt{y+3}-\sqrt{y-3}=1$
$\sqrt{y+3}=1+\sqrt{y-3}$
$\left(\sqrt{y+3}\right)^2=\left(1+\sqrt{y-3}\right)^2$
$y+3=1+2\sqrt{y-3}+(y-3)$
$y+3=-2+2\sqrt{y-3}+y$
$5=2\sqrt{y-3}$
$(5)^2=\left(2\sqrt{y-3}\right)^2$
$25=4(y-3)$
$25=4y-12$
$37=4y$
$\frac{37}{4}=y$
The solution is $\frac{37}{4}$.

51. Let c = length of the hypotenuse.
$6^2+3^2=c^2$
$36+9=c^2$
$45=c^2$
$\sqrt{45}=\sqrt{c^2}$
$\sqrt{9\cdot 5}=c$
$3\sqrt{5}=c$ so $c=3\sqrt{5}$ feet

53. Let b = length of the unknown leg.
$3^2+b^2=7^2$
$9+b^2=49$
$b^2=40$
$\sqrt{b^2}=\sqrt{40}$
$b=\sqrt{4\cdot 10}$
$b=2\sqrt{10}$ meters

55. Let b = length of the unknown leg.
$9^2+b^2=\left(11\sqrt{5}\right)^2$
$81+b^2=121\cdot 5$
$81+b^2=605$
$b^2=524$
$\sqrt{b^2}=\sqrt{524}$
$b=\sqrt{4\cdot 131}$
$b=2\sqrt{131}\approx 22.9$ meters

57. Let c = length of the hypotenuse.
$7^2+(7.2)^2=c^2$
$49+51.84=c^2$
$100.84=c^2$
$100.84=\sqrt{c^2}$
$c=\sqrt{100.84}\approx 10.0$ mm

59. Let c = amount of cable needed.
$15^2+8^2=c^2$
$225+64=c^2$
$289=c^2$
$\sqrt{289}=\sqrt{c^2}$
$17=c$
Thus, 17 feet of cable is needed.

61. Let c = length of the ladder.
$12^2+5^2=c^2$
$144+25=c^2$
$169=c^2$
$\sqrt{169}=\sqrt{c^2}$
$13=c$
A 13-foot ladder is needed.

63.
$$r = \sqrt{\frac{A}{4\pi}}$$
$$1080 = \sqrt{\frac{A}{4\pi}}$$
$$(1080)^2 = \left(\sqrt{\frac{A}{4\pi}}\right)^2$$
$$1{,}166{,}400 = \frac{A}{4\pi}$$
$$14{,}657{,}415 \approx A$$
The surface area is approximately 14,657,415 square miles.

65.
$$v = \sqrt{2gh}$$
$$80 = \sqrt{2(32)h}$$
$$(80)^2 = \left(\sqrt{64h}\right)^2$$
$$6400 = 64h$$
$$100 = h$$
The object fell 100 feet.

67.
$$S = 2\sqrt{I} - 9$$
$$11 = 2\sqrt{I} - 9$$
$$20 = 2\sqrt{I}$$
$$10 = \sqrt{I}$$
$$10^2 = \left(\sqrt{I}\right)^2$$
$$100 = I$$
The estimated IQ is 100.

69.
$$P = 2\pi\sqrt{\frac{l}{32}}$$
$$= 2\pi\sqrt{\frac{2}{32}}$$
$$= 2\pi\sqrt{\frac{1}{16}}$$
$$= 2\pi\left(\frac{1}{4}\right)$$
$$= \frac{\pi}{2} \text{ sec} \approx 1.57 \text{ sec}$$

71.
$$P = 2\pi\sqrt{\frac{l}{32}}$$
$$4 = 2\pi\sqrt{\frac{l}{32}}$$
$$\frac{4}{2\pi} = \sqrt{\frac{l}{32}}$$
$$\left(\frac{2}{\pi}\right)^2 = \left(\sqrt{\frac{l}{32}}\right)^2$$
$$\frac{4}{\pi^2} = \frac{l}{32}$$
$$l = 32\left(\frac{4}{\pi^2}\right) \approx 12.97 \text{ feet}$$

73. Answers may vary

75.
$$s = \frac{1}{2}(6+10+14) = \frac{1}{2}(30) = 15$$
$$A = \sqrt{s(s-a)(s-b)(s-c)}$$
$$= \sqrt{15(15-6)(15-10)(15-14)}$$
$$= \sqrt{15(9)(5)(1)}$$
$$= \sqrt{675}$$
$$= \sqrt{225 \cdot 3}$$
$$= 15\sqrt{3} \text{ sq mi} \approx 25.98 \text{ sq mi.}$$

77. Answers may vary

79.
$$D(h) = 111.7\sqrt{h}$$
$$80 = 111.7\sqrt{h}$$
$$\frac{80}{111.7} = \sqrt{h}$$
$$\left(\frac{80}{111.7}\right)^2 = \left(\sqrt{h}\right)^2$$
$$0.5129483389 \approx h$$
$$h \approx 0.51 \text{ km}$$

81. Function; no vertical line intersects the graph more than one time.

83. Function; no vertical line intersects the graph more than one time.

85. Not a function; the y-axis is an example of a vertical line that intersects the graph more than one time.

87.
$$\frac{\frac{x}{6}}{\frac{2x}{3} + \frac{1}{2}} = \frac{\left(\frac{x}{6}\right)6}{\left(\frac{2x}{3} + \frac{1}{2}\right)6} = \frac{x}{4x+3}$$

89. $\dfrac{\frac{z}{5}+\frac{1}{10}}{\frac{z}{20}-\frac{z}{5}}=\dfrac{\left(\frac{z}{5}+\frac{1}{10}\right)20}{\left(\frac{z}{20}-\frac{z}{5}\right)20}$

$=\dfrac{4z+2}{z-4z}$

$=\dfrac{4z+2}{-3z}$

$=-\dfrac{4z+2}{3z}$

91. $\sqrt{5x-1}+4=7$

$\sqrt{5x-1}=3$

$\left(\sqrt{5x-1}\right)^2=3^2$

$5x-1=9$

$5x=10$

$x=2$

93. $\sqrt{\sqrt{x+3}+\sqrt{x}}=\sqrt{3}$

$\left(\sqrt{\sqrt{x+3}+\sqrt{x}}\right)^2=\left(\sqrt{3}\right)^2$

$\sqrt{x+3}+\sqrt{x}=3$

$\sqrt{x+3}=3-\sqrt{x}$

$\left(\sqrt{x+3}\right)^2=\left(3-\sqrt{x}\right)^2$

$x+3=9-6\sqrt{x}+x$

$-6=-6\sqrt{x}$

$(-6)^2=\left(-6\sqrt{x}\right)^2$

$36=36x$

$1=x$

95. a. Answers may vary

b. Answers may vary

97. $\sqrt{(x^2-x)+7}=2(x^2-x)-1$

Let $t=x^2-x$. Then

$\sqrt{t+7}=2t-1$

$\left(\sqrt{t+7}\right)^2=(2t-1)^2$

$t+7=4t^2-4t+1$

$0=4t^2-5t-6$

$0=(4t+3)(t-2)$

$t=-\dfrac{3}{4}$ or $t=2$

Replace t with x^2-x.

$x^2-x=-\dfrac{3}{4}$ or $x^2-x=2$

$4x^2-4x+3=0$ which has no real solutions

$x^2-x-2=0$

$(x-2)(x+1)=0$

$x=2$ or $x=-1$

The solutions are -1 and 2.

99. $x^2+6x=4\sqrt{x^2+6x}$

Let $t=x^2+6x$. Then

$t=4\sqrt{t}$

$t^2=\left(4\sqrt{t}\right)^2$

$t^2=16t$

$t^2-16t=0$

$t(t-16)=0$

$t=0$ or $t=16$

Replace t with x^2+6x.

$x^2+6x=0$

$x(x+6)=0$

$x=0$ or $x=-6$

or

$x^2+6x=16$

$x^2+6x-16=0$

$(x+8)(x-2)=0$

$x=-8$ or $x=2$

The solutions are -8, -6, 0, and 2.

The Bigger Picture

1. $\dfrac{x}{4}+\dfrac{x+18}{20}=\dfrac{x-5}{5}$

$20\left(\dfrac{x}{4}\right)+20\left(\dfrac{x+18}{20}\right)=20\left(\dfrac{x-5}{5}\right)$

$5x+(x+18)=4(x-5)$

$6x+18=4x-20$

$2x=-38$

$x=-19$

The solution set is $\{-19\}$.

2. $|3x-5|=10$

$3x-5=-10$ or $3x-5=10$

$3x=-5$ $\quad$ $3x=15$

$x=-\dfrac{5}{3}$ $\quad$ $x=5$

The solution set is $\left\{-\dfrac{5}{3}, 5\right\}$.

3. $2x^2 - x = 45$
$2x^2 - x - 45 = 0$
$(2x+9)(x-5) = 0$
$2x+9 = 0$ or $x-5 = 0$
$x = -\frac{9}{2}$ $\quad x = 5$

The solution set is $\left\{-\frac{9}{2}, 5\right\}$.

4. $-6 \le -5x - 1 \le 10$
$-5 \le -5x \le 11$
$1 \ge x \ge -\frac{11}{5}$
$-\frac{11}{5} \le x \le 1$

The solution is $\left[-\frac{11}{5}, 1\right]$.

5. $4(x-1)+3x > 1+2(x-6)$
$4x-4+3x > 1+2x-12$
$7x-4 > 2x-11$
$5x > -7$
$x > -\frac{7}{5}$

The solution is $\left(-\frac{7}{5}, \infty\right)$.

6. $\sqrt{x}+14 = x-6$
$\sqrt{x} = x-20$
$\left(\sqrt{x}\right)^2 = (x-20)^2$
$x = x^2 - 40x + 400$
$0 = x^2 - 41x + 400$
$0 = (x-25)(x-16)$
$x-25 = 0$ or $x-16 = 0$
$x = 25$ $\quad x = 16$

Discard 16 as an extraneous solution. The solution set is {25}.

7. $x \ge 10$ or $-x < 5$
$x \ge 10$ or $x > -5$

The solution is $(-5, \infty)$.

8. $\sqrt{3x-1}+4 = 1$
$\sqrt{3x-1} = -3$

There is no real number whose square root is negative. The solution set is ∅.

9. $|x-2| > 15$
$x-2 < -15$ or $x-2 > 15$
$x < -13$ or $x > 17$

The solution is $(-\infty, -13) \cup (17, \infty)$.

10. $5x - 4[x - 2(3x+1)] = 25$
$5x - 4(x - 6x - 2) = 25$
$5x - 4(-5x - 2) = 25$
$5x + 20x + 8 = 25$
$25x = 17$
$x = \frac{17}{25}$

The solution set is $\left\{\frac{17}{25}\right\}$.

Section 10.7

Practice Exercises

1. a. $\sqrt{-4} = \sqrt{-1 \cdot 4} = \sqrt{-1} \cdot \sqrt{4} = i \cdot 2$, or $2i$

b. $\sqrt{-7} = \sqrt{-1(7)} = \sqrt{-1} \cdot \sqrt{7} = i\sqrt{7}$

c. $-\sqrt{-18} = -\sqrt{-1 \cdot 18}$
$= -\sqrt{-1} \cdot \sqrt{9 \cdot 2}$
$= -i \cdot 3\sqrt{2}$
$= -3i\sqrt{2}$

2. a. $\sqrt{-5} \cdot \sqrt{-6} = i\sqrt{5}\left(i\sqrt{6}\right)$
$= i^2\sqrt{30}$
$= -1\sqrt{30}$
$= -\sqrt{30}$

b. $\sqrt{-9} \cdot \sqrt{-1} = 3i \cdot i = 3i^2 = 3(-1) = -3$

c. $\sqrt{125} \cdot \sqrt{-5} = 5\sqrt{5}\left(i\sqrt{5}\right)$
$= 5i\left(\sqrt{5}\sqrt{5}\right)$
$= 5i(5)$
$= 25i$

d. $\frac{\sqrt{-27}}{\sqrt{3}} = \frac{i\sqrt{27}}{\sqrt{3}} = i\sqrt{9} = 3i$

3. a. $(3-5i)+(-4+i) = (3-4)+(-5+1)i$
$= -1 - 4i$

b. $4i-(3-i)=4i-3+i$
$=-3+(4+1)i$
$=-3+5i$

c. $(-5-2i)-(-8)=-5-2i+8$
$=(-5+8)-2i$
$=3-2i$

4. a. $-4i\cdot 5i=-20i^2=-20(-1)=20$

b. $5i(2+i)=5i\cdot 2+5i\cdot i$
$=10i+5i^2$
$=10i+5(-1)$
$=10i-5$
$=-5+10i$

c. $(2+3i)(6-i)=2(6)-2(i)+3i(6)-3i(i)$
$=12-2i+18i-3i^2$
$=12+16i-3(-1)$
$=12+16i+3$
$=15+16i$

d. $(3-i)^2=(3-i)(3-i)$
$=3(3)-3(i)-3(i)+i^2$
$=9-6i+(-1)$
$=8-6i$

e. $(9+2i)(9-2i)=9(9)-9(2i)+2i(9)-2i(2i)$
$=81-18i+18i-4i^2$
$=81-4(-1)$
$=81+4$
$=85$

5. a. $\frac{4-i}{3+i}=\frac{(4-i)(3-i)}{(3+i)(3-i)}$
$=\frac{4(3)-4(i)-3(i)+i^2}{3^2-i^2}$
$=\frac{12-7i-1}{9+1}$
$=\frac{11-7i}{10}$
$=\frac{11}{10}-\frac{7i}{10}$ or $\frac{11}{10}-\frac{7}{10}i$

b. $\frac{5}{2i}=\frac{5(-2i)}{2i(-2i)}$
$=\frac{-10i}{-4i^2}$
$=\frac{-10i}{-4(-1)}$
$=\frac{-10i}{4}$
$=\frac{-5i}{2}$
$=0-\frac{5i}{2}$ or $0-\frac{5}{2}i$

6. a. $i^9=i^4\cdot i^4\cdot i=1\cdot 1\cdot i=i$

b. $i^{16}=(i^4)^4=1^4=1$

c. $i^{34}=i^{32}\cdot i^2=(i^4)^8\cdot i^2=1^8(-1)=-1$

d. $i^{-24}=\frac{1}{i^{24}}=\frac{1}{(i^4)^6}=\frac{1}{(1)^6}=\frac{1}{1}=1$

Vocabulary and Readiness Check

1. A complex number is one that can be written in the form $a+bi$ where a and b are real numbers.

2. In the complex number system, i denotes the imaginary unit.

3. $i^2=\underline{-1}$

4. $i=\underline{\sqrt{-1}}$

5. A complex number, $a+bi$, is a real number if $b=0$.

6. A complex number, $a+bi$, is a pure imaginary number if $a=0$ and $b\neq 0$.

7. $\sqrt{-81}=9i$

8. $\sqrt{-49}=7i$

9. $\sqrt{-7}=i\sqrt{7}$

10. $\sqrt{-3}=i\sqrt{3}$

11. $-\sqrt{16}=-4$

12. $-\sqrt{4} = -2$

13. $\sqrt{-64} = 8i$

14. $\sqrt{-100} = 10i$

Exercise Set 10.7

1. $\sqrt{-24} = \sqrt{-1 \cdot 24} = \sqrt{-1}\sqrt{4 \cdot 6} = i \cdot 2\sqrt{6} = 2i\sqrt{6}$

3. $-\sqrt{-36} = -\sqrt{-1 \cdot 36} = -\sqrt{-1}\sqrt{36} = -i \cdot 6 = -6i$

5.
$$\begin{aligned} 8\sqrt{-63} &= 8\sqrt{-1 \cdot 63} \\ &= 8\sqrt{-1}\sqrt{9 \cdot 7} \\ &= 8i \cdot 3\sqrt{7} \\ &= 24i\sqrt{7} \end{aligned}$$

7. $-\sqrt{54} = -\sqrt{9 \cdot 6} = -3\sqrt{6}$

9.
$$\begin{aligned} \sqrt{-2} \cdot \sqrt{-7} &= i\sqrt{2} \cdot i\sqrt{7} \\ &= i^2\sqrt{14} \\ &= (-1)\sqrt{14} \\ &= -\sqrt{14} \end{aligned}$$

11.
$$\begin{aligned} \sqrt{-5} \cdot \sqrt{-10} &= i\sqrt{5} \cdot i\sqrt{10} \\ &= i^2\sqrt{50} \\ &= (-1)\sqrt{25 \cdot 2} \\ &= -5\sqrt{2} \end{aligned}$$

13. $\sqrt{16} \cdot \sqrt{-1} = 4i$

15. $\dfrac{\sqrt{-9}}{\sqrt{3}} = \dfrac{i\sqrt{9}}{\sqrt{3}} = i\sqrt{\dfrac{9}{3}} = i\sqrt{3}$

17. $\dfrac{\sqrt{-80}}{\sqrt{-10}} = \dfrac{i\sqrt{80}}{i\sqrt{10}} = \sqrt{\dfrac{80}{10}} = \sqrt{8} = \sqrt{4 \cdot 2} = 2\sqrt{2}$

19.
$$\begin{aligned} (4-7i)+(2+3i) &= (4+2)+(-7+3)i \\ &= 6+(-4)i \\ &= 6-4i \end{aligned}$$

21.
$$\begin{aligned} (6+5i)-(8-i) &= 6+5i-8+i \\ &= (6-8)+(5+1)i \\ &= -2+6i \end{aligned}$$

23.
$$\begin{aligned} 6-(8+4i) &= 6-8-4i \\ &= (6-8)-4i \\ &= -2-4i \end{aligned}$$

25. $-10i \cdot -4i = 40i^2 = 40(-1) = -40$

27.
$$\begin{aligned} 6i(2-3i) &= 12i-18i^2 \\ &= 12i-18(-1) \\ &= 18+12i \end{aligned}$$

29.
$$\begin{aligned} &\left(\sqrt{3}+2i\right)\left(\sqrt{3}-2i\right) \\ &= \sqrt{3} \cdot \sqrt{3} - \sqrt{3} \cdot 2i + \sqrt{3} \cdot 2i - 4i^2 \\ &= 3-4(-1) \\ &= 3+4 \\ &= 7 \end{aligned}$$

31.
$$\begin{aligned} (4-2i)^2 &= (4-2i)(4-2i) \\ &= 16-4 \cdot 2i - 4 \cdot 2i + 4i^2 \\ &= 16-8i-8i+4(-1) \\ &= 16-16i-4 \\ &= 12-16i \end{aligned}$$

33. $\dfrac{4}{i} = \dfrac{4(-i)}{i(-i)} = \dfrac{-4i}{-i^2} = \dfrac{-4i}{-(-1)} = -4i$

35.
$$\begin{aligned} \frac{7}{4+3i} &= \frac{7(4-3i)}{(4+3i)(4-3i)} \\ &= \frac{28-21i}{4^2-9i^2} \\ &= \frac{28-21i}{16+9} \\ &= \frac{28-21i}{25} \\ &= \frac{28}{25}-\frac{21}{25}i \end{aligned}$$

37.
$$\begin{aligned} \frac{3+5i}{1+i} &= \frac{(3+5i)(1-i)}{(1+i)(1-i)} \\ &= \frac{3-3i+5i-5i^2}{1^2-i^2} \\ &= \frac{3+2i+5}{1+1} \\ &= \frac{8+2i}{2} \\ &= \frac{8}{2}+\frac{2}{2}i \\ &= 4+i \end{aligned}$$

39. $\frac{5-i}{3-2i}=\frac{(5-i)(3+2i)}{(3-2i)(3+2i)}$
$=\frac{15+10i-3i-2i^2}{3^2-4i^2}$
$=\frac{15+7i+2}{9+4}$
$=\frac{17+7i}{13}$
$=\frac{17}{13}+\frac{7}{13}i$

41. $(7i)(-9i)=-63i^2=-63(-1)=63$

43. $(6-3i)-(4-2i)=6-3i-4+2i=2-i$

45. $-3i(-1+9i)=3i-27i^2$
$=3i-27(-1)$
$=27+3i$

47. $\frac{4-5i}{2i}=\frac{4-5i}{2i}\cdot\frac{-2i}{-2i}$
$=\frac{-8i+10i^2}{-4i^2}$
$=\frac{-10-8i}{4}$
$=\frac{-10}{4}-\frac{8}{4}i$
$=-\frac{5}{2}-2i$

49. $(4+i)(5+2i)=20+8i+5i+2i^2$
$=20+13i+2(-1)$
$=20+13i-2$
$=18+13i$

51. $(6-2i)(3+i)=18+6i-6i-2i^2$
$=18+2$
$=20$

53. $(8-3i)+(2+3i)=8-3i+2+3i=10$

55. $(1-i)(1+i)=1+i-i-i^2=1+1=2$

57. $\frac{16+15i}{-3i}=\frac{(16+15i)(3i)}{-3i(3i)}$
$=\frac{48i+45i^2}{-9i^2}$
$=\frac{-45+48i}{9}$
$=\frac{-45}{9}+\frac{48}{9}i$
$=-5+\frac{16}{3}i$

59. $(9+8i)^2=9^2+2(9)(8i)+(8i)^2$
$=81+144i+64i^2$
$=81+144i-64$
$=17+144i$

61. $\frac{2}{3+i}=\frac{2(3-i)}{(3+i)(3-i)}$
$=\frac{6-2i}{3^2-i^2}$
$=\frac{6-2i}{9+1}$
$=\frac{6-2i}{10}$
$=\frac{6}{10}-\frac{2}{10}i$
$=\frac{3}{5}-\frac{1}{5}i$

63. $(5-6i)-4i=5-6i-4i=5-10i$

65. $\frac{2-3i}{2+i}=\frac{(2-3i)(2-i)}{(2+i)(2-i)}$
$=\frac{4-2i-6i+3i^2}{2^2-i^2}$
$=\frac{4-8i-3}{4+1}$
$=\frac{1-8i}{5}$
$=\frac{1}{5}-\frac{8}{5}i$

67. $(2+4i)+(6-5i)=2+4i+6-5i=8-i$

69. $\left(\sqrt{3}+2i\right)\left(\sqrt{3}-2i\right)=\left(\sqrt{3}\right)^2-(2i)^2$
$=3-4i^2$
$=3-4(-1)$
$=7$

71. $(4-2i)^2 = 16-2\cdot4\cdot2i+4i^2$
$= 16-16i+4(-1)$
$= 16-4-16i$
$= 12-16i$

73. $i^8 = (i^4)^2 = 1^2 = 1$

75. $i^{21} = i^{20}\cdot i = (i^4)^5\cdot i = 1^5\cdot i = i$

77. $i^{11} = i^8\cdot i^3 = (i^4)^2\cdot i^3 = 1^2\cdot(-i) = -i$

79. $i^{-6} = \frac{1}{i^6} = \frac{1}{i^4\cdot i^2} = \frac{1}{1\cdot(-1)} = -1$

81. $(2i)^6 = 2^6 i^6 = 64i^4\cdot i^2 = 64(1)(-1) = -64$

83. $(-3i)^5 = (-3)^5 i^5 = -243i^4\cdot i = -243(1)i = -243i$

85. $x+50°+90° = 180°$
$x+140° = 180°$
$x = 40°$

87.

1	1	−6	3	−4
		1	−5	−2
	1	−5	−2	−6

Answer: $x^2-5x-2-\frac{6}{x-1}$

89. 5 people

91. $5+9=14$ people

93. $\frac{5 \text{ people}}{30 \text{ people}} = \frac{1}{6} \approx 0.1666$
About 16.7% of the people reported an average checking balance of \$201 to \$300.

95. $i^3 - i^4 = -i-1 = -1-i$

97. $i^6+i^8 = i^4\cdot i^2+(i^4)^2 = 1(-1)+1^2 = -1+1 = 0$

99. $2+\sqrt{-9} = 2+i\sqrt{9} = 2+3i$

101. $\frac{6+\sqrt{-18}}{3} = \frac{6+i\sqrt{9\cdot2}}{3}$
$= \frac{6+3i\sqrt{2}}{3}$
$= \frac{6}{3}+\frac{3\sqrt{2}}{3}i$
$= 2+i\sqrt{2}$

103. $\frac{5-\sqrt{-75}}{10} = \frac{5-i\sqrt{25\cdot3}}{10}$
$= \frac{5-5i\sqrt{3}}{10}$
$= \frac{5}{10}-\frac{5\sqrt{3}}{10}i$
$= \frac{1}{2}-\frac{\sqrt{3}}{2}i$

105. Answers may vary

107. $\left(8-\sqrt{-4}\right)-\left(2+\sqrt{-16}\right) = (8-2i)-(2+4i)$
$= 8-2i-2-4i$
$= 6-6i$

109. $x^2+2x = -2$
$(-1+i)^2+2(-1+i) = -2$
$(1-2i+i^2)-2+2i = -2$
$1-1-2 = -2$
$-2 = -2$, which is true.
Yes, $-1+i$ is a solution.

Chapter 10 Vocabulary Check

1. The conjugate of $\sqrt{3}+2$ is $\sqrt{3}-2$.
2. The principal square root of a nonnegative number a is written as $\sqrt{a}$.
3. The process of writing a radical expression as an equivalent expression but without a radical in the denominator is called rationalizing the denominator.
4. The imaginary unit written i, is the number whose square is -1.
5. The cube root of a number is written as $\sqrt[3]{a}$.
6. In the notation $\sqrt[n]{a}$, n is called the index and a is called the radicand.

7. Radicals with the same index and the same radicand are called like radicals.

8. A complex number is a number that can be written in the form $a + bi$, where a and b are real numbers.

9. The distance formula is $d = \sqrt{(x_2 - x_1)^2 + (y_2 - y_1)^2}$.

10. The midpoint formula is $\left(\frac{x_1 + x_2}{2}, \frac{y_1 + y_2}{2}\right)$.

Chapter 10 Review

1. $\sqrt{81} = 9$ because $9^2 = 81$.

2. $\sqrt[4]{81} = 3$ because $3^4 = 81$.

3. $\sqrt[3]{-8} = -2$ because $(-2)^3 = -8$.

4. $\sqrt[4]{-16}$ is not a real number.

5. $-\sqrt{\frac{1}{49}} = -\frac{1}{7}$ because $\left(\frac{1}{7}\right)^2 = \frac{1}{49}$.

6. $\sqrt{x^{64}} = x^{32}$ because $(x^{32})^2 = x^{32\cdot 2} = x^{64}$.

7. $-\sqrt{36} = -6$ because $6^2 = 36$.

8. $\sqrt[3]{64} = 4$ because $4^3 = 64$.

9. $\sqrt[3]{-a^6b^9} = \sqrt[3]{-1}\sqrt[3]{a^6}\sqrt[3]{b^9}$
$= -1a^2b^3$
$= -a^2b^3$

10. $\sqrt{16a^4b^{12}} = \sqrt{16}\sqrt{a^4}\sqrt{b^{12}} = 4a^2b^6$

11. $\sqrt[5]{32a^5b^{10}} = \sqrt[5]{32}\sqrt[5]{a^5}\sqrt[5]{b^{10}} = 2ab^2$

12. $\sqrt[5]{-32x^{15}y^{20}} = \sqrt[5]{-32}\sqrt[5]{x^{15}}\sqrt[5]{y^{20}} = -2x^3y^4$

13. $\sqrt{\frac{x^{12}}{36y^2}} = \frac{\sqrt{x^{12}}}{\sqrt{36y^2}} = \frac{x^6}{6y}$

14. $\sqrt[3]{\frac{27y^3}{z^{12}}} = \frac{\sqrt[3]{27y^3}}{\sqrt[3]{z^{12}}} = \frac{3y}{z^4}$

15. $\sqrt{(-x)^2} = |-x|$

16. $\sqrt[4]{(x^2-4)^4} = |x^2 - 4|$

17. $\sqrt[3]{(-27)^3} = -27$

18. $\sqrt[5]{(-5)^5} = -5$

19. $-\sqrt[5]{x^5} = -x$

20. $\sqrt[4]{16(2y+z)^{12}} = \sqrt[4]{16}\sqrt[4]{(2y+z)^{12}} = 2\left|(2y+z)^3\right|$

21. $\sqrt{25(x-y)^{10}} = \sqrt{25}\sqrt{(x-y)^{10}}$
$= 5\left|(x-y)^5\right|$

22. $\sqrt[5]{-y^5} = \sqrt[5]{-1}\sqrt[5]{y^5} = -1y = -y$

23. $\sqrt[9]{-x^9} = \sqrt[9]{-1}\sqrt[9]{x^9} = -1x = -x$

24. $f(x) = \sqrt{x} + 3$
$x \geq 0$
Domain: $[0, \infty)$

x	0	1	4	9
$f(x)$	3	4	5	6

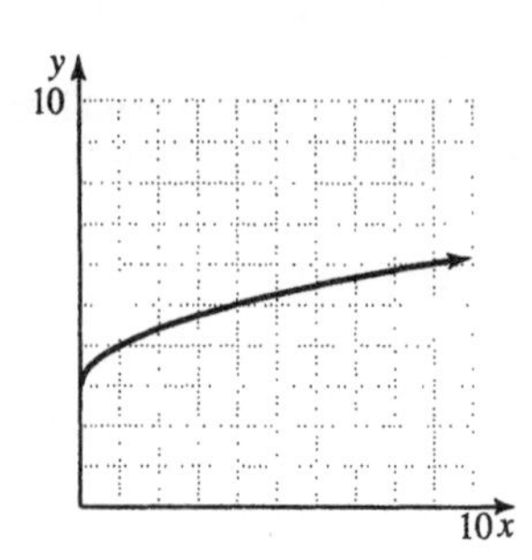

25. $g(x) = \sqrt[3]{x-3}$
Domain: $(-\infty, \infty)$

x	-5	2	3	4	11
$g(x)$	-2	-1	0	1	2

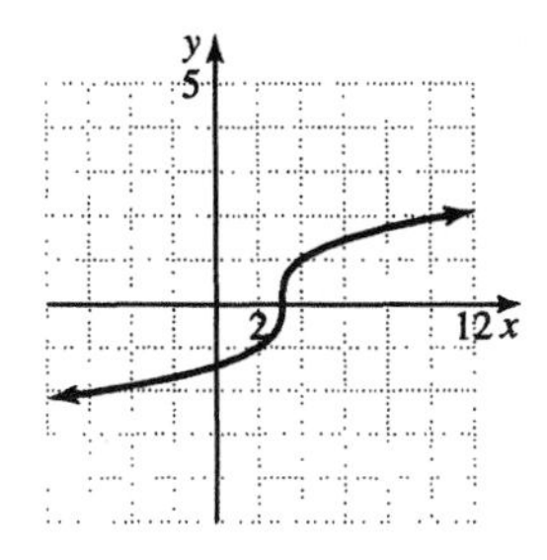

26. $\left(\frac{1}{81}\right)^{1/4} = \frac{1}{81^{1/4}} = \frac{1}{\sqrt[4]{81}} = \frac{1}{3}$

27. $\left(-\frac{1}{27}\right)^{1/3} = -\frac{1}{27^{1/3}} = -\frac{1}{\sqrt[3]{27}} = -\frac{1}{3}$

28. $(-27)^{-1/3} = \frac{1}{(-27)^{1/3}} = \frac{1}{\sqrt[3]{-27}} = \frac{1}{-3} = -\frac{1}{3}$

29. $(-64)^{-1/3} = \frac{1}{(-64)^{1/3}} = \frac{1}{\sqrt[3]{-64}} = \frac{1}{-4} = -\frac{1}{4}$

30. $-9^{3/2} = -\left(\sqrt{9}\right)^3 = -3^3 = -27$

31. $64^{-1/3} = \frac{1}{64^{1/3}} = \frac{1}{\sqrt[3]{64}} = \frac{1}{4}$

32. $(-25)^{5/2} = \left(\sqrt{-25}\right)^5$ is not a real number, since there is no real number whose square is -25.

33.
$$\left(\frac{25}{49}\right)^{-3/2} = \frac{1}{\left(\frac{25}{49}\right)^{3/2}} = \frac{1}{\left(\sqrt{\frac{25}{49}}\right)^3} = \frac{1}{\left(\frac{5}{7}\right)^3} = \frac{1}{\frac{125}{343}} = \frac{343}{125}$$

34. $\left(\frac{8}{27}\right)^{-2/3} = \frac{1}{\left(\frac{8}{27}\right)^{2/3}} = \frac{1}{\left(\sqrt[3]{\frac{8}{27}}\right)^2} = \frac{1}{\left(\frac{2}{3}\right)^2} = \frac{1}{\frac{4}{9}} = \frac{9}{4}$

35. $\left(-\frac{1}{36}\right)^{-1/4} = \frac{1}{\left(-\frac{1}{36}\right)^{1/4}} = \frac{1}{\sqrt[4]{-\frac{1}{36}}}$ is not a real number, since there is no real number whose 4th power is $-\frac{1}{36}$.

36. $\sqrt[3]{x^2} = (x^2)^{1/3} = x^{2/3}$

37.
$$\sqrt[5]{5x^2y^3} = (5x^2y^3)^{1/5} = 5^{1/5}(x^2)^{1/5}(y^3)^{1/5} = 5^{1/5}x^{2/5}y^{3/5}$$

38. $y^{4/5} = (y^4)^{1/5} = \sqrt[5]{y^4}$

39. $5(xy^2z^5)^{1/3} = 5\sqrt[3]{xy^2z^5}$

40. $(x+2y)^{-1/2} = \frac{1}{(x+2y)^{1/2}} = \frac{1}{\sqrt{x+2y}}$

41. $a^{1/3}a^{4/3}a^{1/2} = a^{\frac{1}{3}+\frac{4}{3}+\frac{1}{2}} = a^{\frac{2}{6}+\frac{8}{6}+\frac{3}{6}} = a^{13/6}$

42. $\frac{b^{1/3}}{b^{4/3}} = b^{1/3-4/3} = b^{-3/3} = b^{-1} = \frac{1}{b}$

43. $(a^{1/2}a^{-2})^3 = (a^{1/2-2})^3$
$= (a^{1/2-4/2})^3$
$= (a^{-3/2})^3$
$= a^{-9/2}$
$= \dfrac{1}{a^{9/2}}$

44. $(x^{-3}y^6)^{1/3} = (x^{-3})^{1/3}(y^6)^{1/3} = x^{-1}y^2 = \dfrac{y^2}{x}$

45. $\left(\dfrac{b^{3/4}}{a^{-1/2}}\right)^8 = (a^{1/2}b^{3/4})^8$
$= (a^{1/2})^8(b^{3/4})^8$
$= a^4b^6$

46. $\dfrac{x^{1/4}x^{-1/2}}{x^{2/3}} = x^{1/4+(-1/2)-2/3}$
$= x^{\frac{3}{12}-\frac{6}{12}-\frac{8}{12}}$
$= x^{-11/12}$
$= \dfrac{1}{x^{11/12}}$

47. $\left(\dfrac{49c^{5/3}}{a^{-1/4}b^{5/6}}\right)^{-1} = \dfrac{49^{-1}c^{-5/3}}{a^{1/4}b^{-5/6}} = \dfrac{b^{5/6}}{49a^{1/4}c^{5/3}}$

48. $a^{-1/4}(a^{5/4} - a^{9/4}) = a^{-1/4}(a^{5/4}) - a^{-1/4}(a^{9/4})$
$= a^{-1/4+5/4} - a^{-1/4+9/4}$
$= a^{4/4} - a^{8/4}$
$= a - a^2$

49. $\sqrt{20} \approx 4.472$

50. $\sqrt[3]{-39} \approx -3.391$

51. $\sqrt[4]{726} \approx 5.191$

52. $56^{1/3} \approx 3.826$

53. $-78^{3/4} \approx -26.246$

54. $105^{-2/3} \approx 0.045$

55. $\sqrt[3]{2} \cdot \sqrt{7} = 2^{1/3} \cdot 7^{1/2}$
$= 2^{2/6} \cdot 7^{3/6}$
$= (2^2 \cdot 7^3)^{1/6}$
$= \sqrt[6]{4 \cdot 343}$
$= \sqrt[6]{1372}$

56. $\sqrt[3]{3} \cdot \sqrt[4]{x} = 3^{1/3} \cdot x^{1/4}$
$= 3^{4/12} \cdot x^{3/12}$
$= (3^4 \cdot x^3)^{1/12}$
$= \sqrt[12]{81x^3}$

57. $\sqrt{3} \cdot \sqrt{8} = \sqrt{24} = \sqrt{4 \cdot 6} = 2\sqrt{6}$

58. $\sqrt[3]{7y} \cdot \sqrt[3]{x^2z} = \sqrt[3]{7y \cdot x^2z} = \sqrt[3]{7x^2yz}$

59. $\dfrac{\sqrt{44x^3}}{\sqrt{11x}} = \sqrt{\dfrac{44x^3}{11x}} = \sqrt{4x^2} = 2x$

60. $\dfrac{\sqrt[4]{a^6b^{13}}}{\sqrt[4]{a^2b}} = \sqrt[4]{\dfrac{a^6b^{13}}{a^2b}} = \sqrt[4]{a^4b^{12}} = ab^3$

61. $\sqrt{60} = \sqrt{4 \cdot 15} = 2\sqrt{15}$

62. $-\sqrt{75} = -\sqrt{25 \cdot 3} = -5\sqrt{3}$

63. $\sqrt[3]{162} = \sqrt[3]{27 \cdot 6} = 3\sqrt[3]{6}$

64. $\sqrt[3]{-32} = \sqrt[3]{-8 \cdot 4} = -2\sqrt[3]{4}$

65. $\sqrt{36x^7} = \sqrt{36x^6 \cdot x} = 6x^3\sqrt{x}$

66. $\sqrt[3]{24a^5b^7} = \sqrt[3]{8a^3b^6 \cdot 3a^2b} = 2ab^2\sqrt[3]{3a^2b}$

67. $\sqrt{\dfrac{p^{17}}{121}} = \dfrac{\sqrt{p^{17}}}{\sqrt{121}} = \dfrac{\sqrt{p^{16} \cdot p}}{11} = \dfrac{p^8\sqrt{p}}{11}$

68. $\sqrt[3]{\dfrac{y^5}{27x^6}} = \dfrac{\sqrt[3]{y^5}}{\sqrt[3]{27x^6}} = \dfrac{\sqrt[3]{y^3y^2}}{\sqrt[3]{27x^6}} = \dfrac{y\sqrt[3]{y^2}}{3x^2}$

69. $\sqrt[4]{\dfrac{xy^6}{81}} = \dfrac{\sqrt[4]{xy^6}}{\sqrt[4]{81}} = \dfrac{\sqrt[4]{y^4 \cdot xy^2}}{3} = \dfrac{y\sqrt[4]{xy^2}}{3}$

70. $\sqrt{\frac{2x^3}{49y^4}} = \frac{\sqrt{2x^3}}{\sqrt{49y^4}} = \frac{\sqrt{x^2 \cdot 2x}}{7y^2} = \frac{x\sqrt{2x}}{7y^2}$

71. $r = \sqrt{\frac{A}{\pi}}$

a. $r = \sqrt{\frac{25}{\pi}} = \frac{\sqrt{25}}{\sqrt{\pi}} = \frac{5}{\sqrt{\pi}}$ meters, or

$r = \frac{5}{\sqrt{\pi}} = \frac{5\sqrt{\pi}}{\sqrt{\pi}\sqrt{\pi}} = \frac{5\sqrt{\pi}}{\pi}$ meters

b. $r = \sqrt{\frac{104}{\pi}} \approx 5.75$ inches

72. $(x_1, y_1) = (-6, 3), (x_2, y_2) = (8, 4)$

$$\begin{aligned} d &= \sqrt{(x_2 - x_1)^2 + (y_2 - y_1)^2} \\ &= \sqrt{(8+6)^2 + (4-3)^2} \\ &= \sqrt{196+1} \\ &= \sqrt{197} \approx 14.036 \text{ units} \end{aligned}$$

73. $(x_1, y_1) = (-4, -6), (x_2, y_2) = (-1, 5)$

$$\begin{aligned} d &= \sqrt{(x_2 - x_1)^2 + (y_2 - y_1)^2} \\ &= \sqrt{(-1+4)^2 + (5+6)^2} \\ &= \sqrt{9+121} \\ &= \sqrt{130} \approx 11.402 \text{ units} \end{aligned}$$

74. $(x_1, y_1) = (-1, 5), (x_2, y_2) = (2, -3)$

$$\begin{aligned} d &= \sqrt{(x_2 - x_1)^2 + (y_2 - y_1)^2} \\ &= \sqrt{(2+1)^2 + (-3-5)^2} \\ &= \sqrt{9+64} \\ &= \sqrt{73} \approx 8.544 \text{ units} \end{aligned}$$

75. $(x_1, y_1) = \left(-\sqrt{2}, 0\right), (x_2, y_2) = \left(0, -4\sqrt{6}\right)$

$$\begin{aligned} d &= \sqrt{(x_2 - x_1)^2 + (y_2 - y_1)^2} \\ &= \sqrt{\left(0+\sqrt{2}\right)^2 + \left(-4\sqrt{6}-0\right)^2} \\ &= \sqrt{2+96} \\ &= \sqrt{98} \\ &= 7\sqrt{2} \approx 9.899 \text{ units} \end{aligned}$$

76. $(x_1, y_1) = \left(-\sqrt{5}, -\sqrt{11}\right)$,

$(x_2, y_2) = \left(-\sqrt{5}, -3\sqrt{11}\right)$

$$\begin{aligned} d &= \sqrt{(x_2 - x_1)^2 + (y_2 - y_1)^2} \\ &= \sqrt{\left(-\sqrt{5}+\sqrt{5}\right)^2 + \left(-3\sqrt{11}+\sqrt{11}\right)^2} \\ &= \sqrt{0+44} \\ &= \sqrt{44} \\ &= 2\sqrt{11} \approx 6.633 \text{ units} \end{aligned}$$

77. $(x_1, y_1) = (7.4, -8.6), (x_2, y_2) = (-1.2, 5.6)$

$$\begin{aligned} d &= \sqrt{(-1.2-7.4)^2 + (5.6+8.6)^2} \\ &= \sqrt{(-8.6)^2 + (14.2)^2} \\ &= \sqrt{73.96+201.64} \\ &= \sqrt{275.6} \approx 16.601 \text{ units} \end{aligned}$$

78. $(x_1, y_1) = (2, 6), (x_2, y_2) = (-12, 4)$

$$\begin{aligned} \text{midpoint} &= \left(\frac{x_1 + x_2}{2}, \frac{y_1 + y_2}{2}\right) \\ &= \left(\frac{2-12}{2}, \frac{6+4}{2}\right) \\ &= \left(\frac{-10}{2}, \frac{10}{2}\right) \\ &= (-5, 5) \end{aligned}$$

79. $(x_1, y_1) = (-6, -5), (x_2, y_2) = (-9, 7)$

$$\begin{aligned} \text{midpoint} &= \left(\frac{x_1 + x_2}{2}, \frac{y_1 + y_2}{2}\right) \\ &= \left(\frac{-6-9}{2}, \frac{-5+7}{2}\right) \\ &= \left(\frac{-15}{2}, \frac{2}{2}\right) \\ &= \left(-\frac{15}{2}, 1\right) \end{aligned}$$

80. $(x_1, y_1) = (4, -6), (x_2, y_2) = (-15, 2)$

$$\begin{aligned} \text{midpoint} &= \left(\frac{x_1 + x_2}{2}, \frac{y_1 + y_2}{2}\right) \\ &= \left(\frac{4-15}{2}, \frac{-6+2}{2}\right) \\ &= \left(\frac{-11}{2}, \frac{-4}{2}\right) \\ &= \left(-\frac{11}{2}, -2\right) \end{aligned}$$

81. $(x_1, y_1) = \left(0, -\frac{3}{8}\right), (x_2, y_2) = \left(\frac{1}{10}, 0\right)$

$$\begin{aligned}\text{midpoint} &= \left(\frac{x_1+x_2}{2}, \frac{y_1+y_2}{2}\right)\\ &= \left(\frac{0+\frac{1}{10}}{2}, \frac{-\frac{3}{8}+0}{2}\right)\\ &= \left(\frac{1}{20}, -\frac{3}{16}\right)\end{aligned}$$

82. $(x_1, y_1) = \left(\frac{3}{4}, -\frac{1}{7}\right), (x_2, y_2) = \left(-\frac{1}{4}, -\frac{3}{7}\right)$

$$\begin{aligned}\text{midpoint} &= \left(\frac{x_1+x_2}{2}, \frac{y_1+y_2}{2}\right)\\ &= \left(\frac{\frac{3}{4}-\frac{1}{4}}{2}, \frac{-\frac{1}{7}-\frac{3}{7}}{2}\right)\\ &= \left(\frac{\frac{1}{2}}{2}, \frac{-\frac{4}{7}}{2}\right)\\ &= \left(\frac{1}{4}, -\frac{2}{7}\right)\end{aligned}$$

83. $(x_1, y_1) = \left(\sqrt{3}, -2\sqrt{6}\right), (x_2, y_2) = \left(\sqrt{3}, -4\sqrt{6}\right)$

$$\begin{aligned}\text{midpoint} &= \left(\frac{x_1+x_2}{2}, \frac{y_1+y_2}{2}\right)\\ &= \left(\frac{\sqrt{3}+\sqrt{3}}{2}, \frac{-2\sqrt{6}-4\sqrt{6}}{2}\right)\\ &= \left(\frac{2\sqrt{3}}{2}, \frac{-6\sqrt{6}}{2}\right)\\ &= \left(\sqrt{3}, -3\sqrt{6}\right)\end{aligned}$$

84.
$$\begin{aligned}&2\sqrt{50}-3\sqrt{125}+\sqrt{98}\\ &= 2\sqrt{25\cdot 2}-3\sqrt{25\cdot 5}+\sqrt{49\cdot 2}\\ &= 2\cdot 5\sqrt{2}-3\cdot 5\sqrt{5}+7\sqrt{2}\\ &= 10\sqrt{2}-15\sqrt{5}+7\sqrt{2}\\ &= 17\sqrt{2}-15\sqrt{5}\end{aligned}$$

85.
$$\begin{aligned}x\sqrt{75xy}-\sqrt{27x^3y} &= x\sqrt{25\cdot 3xy}-\sqrt{9x^2\cdot 3xy}\\ &= x\cdot 5\sqrt{3xy}-3x\sqrt{3xy}\\ &= 2x\sqrt{3xy}\end{aligned}$$

86.
$$\begin{aligned}\sqrt[3]{128}+\sqrt[3]{250} &= \sqrt[3]{64\cdot 2}+\sqrt[3]{125\cdot 2}\\ &= 4\sqrt[3]{2}+5\sqrt[3]{2}\\ &= 9\sqrt[3]{2}\end{aligned}$$

87.
$$\begin{aligned}3\sqrt[4]{32a^5}-a\sqrt[4]{162a} &= 3\sqrt[4]{16a^4\cdot 2a}-a\sqrt[4]{81\cdot 2a}\\ &= 3\cdot 2a\sqrt[4]{2a}-3a\sqrt[4]{2a}\\ &= 6a\sqrt[4]{2a}-3a\sqrt[4]{2a}\\ &= 3a\sqrt[4]{2a}\end{aligned}$$

88. $\frac{5}{\sqrt{4}}+\frac{\sqrt{3}}{3} = \frac{5}{2}+\frac{\sqrt{3}}{3} = \frac{5\cdot 3+2\sqrt{3}}{6} = \frac{15+2\sqrt{3}}{6}$

89.
$$\begin{aligned}\sqrt{\frac{8}{x^2}}-\sqrt{\frac{50}{16x^2}} &= \frac{\sqrt{8}}{\sqrt{x^2}}-\frac{\sqrt{50}}{\sqrt{16x^2}}\\ &= \frac{\sqrt{4\cdot 2}}{x}-\frac{\sqrt{25\cdot 2}}{4x}\\ &= \frac{2\sqrt{2}\cdot 4}{x\cdot 4}-\frac{5\sqrt{2}}{4x}\\ &= \frac{8\sqrt{2}-5\sqrt{2}}{4x}\\ &= \frac{3\sqrt{2}}{4x}\end{aligned}$$

90.
$$\begin{aligned}&2\sqrt{32x^2y^3}-xy\sqrt{98y}\\ &= 2\sqrt{16x^2y^2\cdot 2y}-xy\sqrt{49\cdot 2y}\\ &= 2\cdot 4xy\sqrt{2y}-xy\cdot 7\sqrt{2y}\\ &= 8xy\sqrt{2y}-7xy\sqrt{2y}\\ &= xy\sqrt{2y}\end{aligned}$$

91.
$$\begin{aligned}&2a\sqrt[4]{32b^5}-3b\sqrt[4]{162a^4b}+\sqrt[4]{2a^4b^5}\\ &= 2a\sqrt[4]{16b^4\cdot 2b}-3b\sqrt[4]{81a^4\cdot 2b}+\sqrt[4]{a^4b^4\cdot 2b}\\ &= 2a\cdot 2b\sqrt[4]{2b}-3b\cdot 3a\sqrt[4]{2b}+ab\sqrt[4]{2b}\\ &= 4ab\sqrt[4]{2b}-9ab\sqrt[4]{2b}+ab\sqrt[4]{2b}\\ &= -4ab\sqrt[4]{2b}\end{aligned}$$

92.
$$\begin{aligned}\sqrt{3}\left(\sqrt{27}-\sqrt{3}\right) &= \sqrt{3}\left(\sqrt{9\cdot 3}-\sqrt{3}\right)\\ &= \sqrt{3}\left(3\sqrt{3}-\sqrt{3}\right)\\ &= \sqrt{3}\left(2\sqrt{3}\right)\\ &= 2\sqrt{9}\\ &= 2(3)\\ &= 6\end{aligned}$$

93. $\left(\sqrt{x}-3\right)^2=\left(\sqrt{x}\right)^2-2\cdot\sqrt{x}\cdot 3+3^2=x-6\sqrt{x}+9$

94. $\left(\sqrt{5}-5\right)\left(2\sqrt{5}+2\right)=2\sqrt{25}+2\sqrt{5}-10\sqrt{5}-10$
$=2(5)-8\sqrt{5}-10$
$=10-8\sqrt{5}-10$
$=-8\sqrt{5}$

95. $\left(2\sqrt{x}-3\sqrt{y}\right)\left(2\sqrt{x}+3\sqrt{y}\right)$
$=\left(2\sqrt{x}\right)^2-\left(3\sqrt{y}\right)^2$
$=2^2\left(\sqrt{x}\right)^2-3^2\left(\sqrt{y}\right)^2$
$=4x-9y$

96. $\left(\sqrt{a}+3\right)\left(\sqrt{a}-3\right)=\left(\sqrt{a}\right)^2-(3)^2=a-9$

97. $\left(\sqrt[3]{a}+2\right)^2=\left(\sqrt[3]{a}\right)^2+2\cdot\sqrt[3]{a}\cdot 2+2^2$
$=\sqrt[3]{a^2}+4\sqrt[3]{a}+4$

98. $\left(\sqrt[3]{5x}+9\right)\left(\sqrt[3]{5x}-9\right)=\left(\sqrt[3]{5x}\right)^2-9^2$
$=\sqrt[3]{(5x)^2}-81$
$=\sqrt[3]{25x^2}-81$

99. $\left(\sqrt[3]{a}+4\right)\left(\sqrt[3]{a^2}-4\sqrt[3]{a}+16\right)$
$=\left(\sqrt[3]{a}\right)\left(\sqrt[3]{a^2}\right)-4\cdot\left(\sqrt[3]{a}\right)^2+16\sqrt[3]{a}+4\sqrt[3]{a^2}$
$-16\sqrt[3]{a}+64$
$=\sqrt[3]{a^3}-4\sqrt[3]{a^2}+4\sqrt[3]{a^2}+64$
$=a+64$

100. $\dfrac{3}{\sqrt{7}}=\dfrac{3\cdot\sqrt{7}}{\sqrt{7}\cdot\sqrt{7}}=\dfrac{3\sqrt{7}}{7}$

101. $\sqrt{\dfrac{x}{12}}=\dfrac{\sqrt{x}}{\sqrt{12}}$
$=\dfrac{\sqrt{x}}{\sqrt{4\cdot 3}}$
$=\dfrac{\sqrt{x}}{2\sqrt{3}}$
$=\dfrac{\sqrt{x}\cdot\sqrt{3}}{2\sqrt{3}\cdot\sqrt{3}}$
$=\dfrac{\sqrt{3x}}{2\cdot 3}$
$=\dfrac{\sqrt{3x}}{6}$

102. $\dfrac{5}{\sqrt[3]{4}}=\dfrac{5\cdot\sqrt[3]{2}}{\sqrt[3]{4}\cdot\sqrt[3]{2}}=\dfrac{5\sqrt[3]{2}}{\sqrt[3]{8}}=\dfrac{5\sqrt[3]{2}}{2}$

103. $\sqrt{\dfrac{24x^5}{3y^2}}=\sqrt{\dfrac{8x^5}{y^2}}$
$=\dfrac{\sqrt{8x^5}}{\sqrt{y^2}}$
$=\dfrac{\sqrt{4x^4\cdot 2x}}{y}$
$=\dfrac{2x^2\sqrt{2x}}{y}$

104. $\sqrt[3]{\dfrac{15x^6y^7}{z^2}}=\dfrac{\sqrt[3]{15x^6y^7}}{\sqrt[3]{z^2}}$
$=\dfrac{\sqrt[3]{15x^6y^7}\cdot\sqrt[3]{z}}{\sqrt[3]{z^2}\cdot\sqrt[3]{z}}$
$=\dfrac{\sqrt[3]{15x^6y^7z}}{\sqrt[3]{z^3}}$
$=\dfrac{\sqrt[3]{15x^6y^6\cdot yz}}{z}$
$=\dfrac{x^2y^2\sqrt[3]{15yz}}{z}$

105. $\frac{5}{2-\sqrt{7}}=\frac{5(2+\sqrt{7})}{(2-\sqrt{7})(2+\sqrt{7})}$
$=\frac{5(2+\sqrt{7})}{2^2-(\sqrt{7})^2}$
$=\frac{10+5\sqrt{7}}{4-7}$
$=\frac{10+5\sqrt{7}}{-3}$
$=-\frac{10+5\sqrt{7}}{3}$

106. $\frac{3}{\sqrt{y}-2}=\frac{3(\sqrt{y}+2)}{(\sqrt{y}-2)(\sqrt{y}+2)}$
$=\frac{3(\sqrt{y}+2)}{(\sqrt{y})^2-2^2}$
$=\frac{3\sqrt{y}+6}{y-4}$

107. $\frac{\sqrt{2}-\sqrt{3}}{\sqrt{2}+\sqrt{3}}=\frac{(\sqrt{2}-\sqrt{3})(\sqrt{2}-\sqrt{3})}{(\sqrt{2}+\sqrt{3})(\sqrt{2}-\sqrt{3})}$
$=\frac{2-\sqrt{2}\sqrt{3}-\sqrt{3}\sqrt{2}+3}{(\sqrt{2})^2-(\sqrt{3})^2}$
$=\frac{5-\sqrt{6}-\sqrt{6}}{2-3}$
$=\frac{5-2\sqrt{6}}{-1}$
$=-5+2\sqrt{6}$

108. $\frac{\sqrt{11}}{3}=\frac{\sqrt{11}\cdot\sqrt{11}}{3\cdot\sqrt{11}}=\frac{11}{3\sqrt{11}}$

109. $\sqrt{\frac{18}{y}}=\frac{\sqrt{18}}{\sqrt{y}}=\frac{3\sqrt{2}}{\sqrt{y}}=\frac{3\sqrt{2}\cdot\sqrt{2}}{\sqrt{y}\cdot\sqrt{2}}=\frac{3\cdot 2}{\sqrt{2y}}=\frac{6}{\sqrt{2y}}$

110. $\frac{\sqrt[3]{9}}{7}=\frac{\sqrt[3]{9}\cdot\sqrt[3]{3}}{7\cdot\sqrt[3]{3}}=\frac{\sqrt[3]{27}}{7\sqrt[3]{3}}=\frac{3}{7\sqrt[3]{3}}$

111. $\sqrt{\frac{24x^5}{3y^2}}=\sqrt{\frac{8x^5}{y^2}}$
$=\frac{\sqrt{4x^4\cdot 2x}}{\sqrt{y^2}}$
$=\frac{2x^2\sqrt{2x}}{y}$
$=\frac{2x^2\sqrt{2x}\cdot\sqrt{2x}}{y\cdot\sqrt{2x}}$
$=\frac{2x^2\cdot 2x}{y\sqrt{2x}}=\frac{4x^3}{y\sqrt{2x}}$

112. $\sqrt[3]{\frac{xy^2}{10z}}=\frac{\sqrt[3]{xy^2}}{\sqrt[3]{10z}}$
$=\frac{\sqrt[3]{xy^2}\cdot\sqrt[3]{x^2y}}{\sqrt[3]{10z}\cdot\sqrt[3]{x^2y}}$
$=\frac{\sqrt[3]{x^3y^3}}{\sqrt[3]{10x^2yz}}$
$=\frac{xy}{\sqrt[3]{10x^2yz}}$

113. $\frac{\sqrt{x}+5}{-3}=\frac{(\sqrt{x}+5)(\sqrt{x}-5)}{-3(\sqrt{x}-5)}$
$=\frac{(\sqrt{x})^2-5^2}{-3\sqrt{x}+15}$
$=\frac{x-25}{-3\sqrt{x}+15}$

114. $\sqrt{y-7}=5$
$(\sqrt{y-7})^2=5^2$
$y-7=25$
$y=32$
The solution is 32.

115. $\sqrt{2x}+10=4$
$\sqrt{2x}=-6$
No solution exists since the principle square root of a number is not negative.

116. $\sqrt[3]{2x-6}=4$

$\left(\sqrt[3]{2x-6}\right)^3=4^3$

$2x-6=64$

$2x=70$

$x=35$

The solution is 35.

117. $\sqrt{x+6}=\sqrt{x+2}$

$\left(\sqrt{x+6}\right)^2=\left(\sqrt{x+2}\right)^2$

$x+6=x+2$

$6=2$, which is false.

There is no solution.

118. $2x-5\sqrt{x}=3$

$2x-3=5\sqrt{x}$

$(2x-3)^2=\left(5\sqrt{x}\right)^2$

$4x^2-12x+9=25x$

$4x^2-37x+9=0$

$(4x-1)(x-9)=0$

$4x-1=0$ or $x-9=0$

$4x=1$ or $x=9$

$x=\frac{1}{4}$

Discard the solution $\frac{1}{4}$ as extraneous. The solution is 9.

119. $\sqrt{x+9}=2+\sqrt{x-7}$

$\left(\sqrt{x+9}\right)^2=\left(2+\sqrt{x-7}\right)^2$

$x+9=4+4\sqrt{x-7}+(x-7)$

$x+9=x-3+4\sqrt{x-7}$

$12=4\sqrt{x-7}$

$3=\sqrt{x-7}$

$3^2=\left(\sqrt{x-7}\right)^2$

$9=x-7$

$16=x$

The solution is 16.

120. Let c = length of the hypotenuse.

$3^2+3^2=c^2$

$18=c^2$

$\sqrt{18}=\sqrt{c^2}$

$3\sqrt{2}=c$

The length is $3\sqrt{2}$ centimeters.

121. Let c = length of the hypotenuse.

$7^2+\left(8\sqrt{3}\right)^2=c^2$

$49+64\cdot 3=c^2$

$241=c^2$

$\sqrt{241}=\sqrt{c^2}$

$\sqrt{241}=c$

The length is $\sqrt{241}$ feet.

122. Let b = width of the lake.

$a^2+b^2=c^2$

$40^2+b^2=65^2$

$1600+b^2=4225$

$b^2=2625$

$\sqrt{b^2}=\sqrt{2625}$

$b=51.23475$

The width is about 51.2 feet.

123. Let c = length of the shortest pipe.

$a^2+b^2=c^2$

$3^2+3^2=c^2$

$18=c^2$

$\sqrt{18}=\sqrt{c^2}$

$4.24264=c$

The shortest possible pipe is 4.24 feet.

124. $\sqrt{-8}=i\sqrt{4\cdot 2}=2i\sqrt{2}$

125. $-\sqrt{-6}=-i\sqrt{6}$

126. $\sqrt{-4}+\sqrt{-16}=2i+4i=6i$

127. $\sqrt{-2}\cdot\sqrt{-5}=i\sqrt{2}\cdot i\sqrt{5}$

$=i^2\sqrt{10}$

$=-1\cdot\sqrt{10}$

$=-\sqrt{10}$

128. $(12-6i)+(3+2i)=(12+3)+(-6+2)i$
$=15+(-4)i$
$=15-4i$

129. $(-8-7i)-(5-4i)=-8-7i-5+4i$
$=-13-3i$

130. $(2i)^6=2^6i^6=64i^4\cdot i^2=64(1)(-1)=-64$

131. $-3i(6-4i)=-18i+12i^2$
$=-18i+12(-1)$
$=-12-18i$

132. $(3+2i)(1+i)=3+3i+2i+2i^2$
$=3+5i+2(-1)$
$=1+5i$

133. $(2-3i)^2=2^2+2\cdot 2\cdot(-3i)+(3i)^2$
$=4-12i+9i^2$
$=4-12i+9(-1)$
$=-5-12i$

134. $\left(\sqrt{6}-9i\right)\left(\sqrt{6}+9i\right)=\left(\sqrt{6}\right)^2-(9i)^2$
$=6-81i^2$
$=6+81$
$=87$

135. $\frac{2+3i}{2i}=\frac{(2+3i)\cdot(-2i)}{2i\cdot(-2i)}$
$=\frac{-4i-6i^2}{-4i^2}$
$=\frac{-4i+6}{4}$
$=\frac{6}{4}-\frac{4}{4}i$
$=\frac{3}{2}-i$

136. $\frac{1+i}{-3i}=\frac{(1+i)\cdot(3i)}{-3i\cdot(3i)}$
$=\frac{3i+3i^2}{-9i^2}$
$=\frac{3i-3}{9}$
$=\frac{-3}{9}-\frac{3}{9}i$
$=-\frac{1}{3}+\frac{1}{3}i$

137. $\sqrt[3]{x^3}=x$

138. $\sqrt{(x+2)^2}=|x+2|$

139. $-\sqrt{100}=-10$

140. $\sqrt[3]{-x^{12}y^3}=-x^4y$

141. $\sqrt[4]{\frac{y^{20}}{16x^{12}}}=\frac{\sqrt[4]{y^{20}}}{\sqrt[4]{16x^{12}}}=\frac{y^5}{2x^3}$

142. $9^{1/2}=\sqrt{9}=3$

143. $64^{-1/2}=\frac{1}{64^{1/2}}=\frac{1}{\sqrt{64}}=\frac{1}{8}$

144. $\left(\frac{27}{64}\right)^{-2/3}=\left(\frac{64}{27}\right)^{2/3}$
$=\left(\sqrt[3]{\frac{64}{27}}\right)^2$
$=\left(\frac{4}{3}\right)^2$
$=\frac{16}{9}$

145. $\frac{(x^{2/3}x^{-3})^3}{x^{-1/2}}=\frac{x^{6/3}x^{-9}}{x^{-1/2}}$
$=x^{2-9+\frac{1}{2}}$
$=x^{-13/2}$
$=\frac{1}{x^{13/2}}$

146. $\sqrt{200x^9}=\sqrt{100x^8\cdot 2x}=10x^4\sqrt{2x}$

147. $\sqrt{\frac{3n^3}{121m^{10}}}=\frac{\sqrt{3n^3}}{\sqrt{121m^{10}}}=\frac{\sqrt{n^2\cdot 3n}}{\sqrt{121m^{10}}}=\frac{n\sqrt{3n}}{11m^5}$

148. $3\sqrt{20}-7x\sqrt[3]{40}+3\sqrt[3]{5x^3}$
$=3\sqrt{4}\sqrt{5}-7x\sqrt[3]{8}\sqrt[3]{5}+3\sqrt[3]{x^3}\sqrt[3]{5}$
$=6\sqrt{5}-14x\sqrt[3]{5}+3x\sqrt[3]{5}$
$=6\sqrt{5}-11x\sqrt[3]{5}$

149. $\left(2\sqrt{x}-5\right)^2 = \left(2\sqrt{x}\right)^2 - 2(5)\left(2\sqrt{x}\right) + 5^2$
$= 4x - 20\sqrt{x} + 25$

150. $(x_1, y_1) = (-3, 5), (x_2, y_2) = (-8, 9)$

$$\begin{aligned} d &= \sqrt{(x_2-x_1)^2 + (y_2-y_1)^2} \\ &= \sqrt{(-8+3)^2 + (9-5)^2} \\ &= \sqrt{(-5)^2 + (4)^2} \\ &= \sqrt{25+16} \\ &= \sqrt{41} \end{aligned}$$

The distance is $\sqrt{41}$ units.

151. $(x_1, y_1) = (-3, 8), (x_2, y_2) = (11, 24)$

$$\begin{aligned} \text{midpoint} &= \left(\frac{x_1+x_2}{2}, \frac{y_1+y_2}{2}\right) \\ &= \left(\frac{-3+11}{2}, \frac{8+24}{2}\right) \\ &= \left(\frac{8}{2}, \frac{32}{2}\right) \\ &= (4, 16) \end{aligned}$$

152. $\frac{7}{\sqrt{13}} = \frac{7}{\sqrt{13}} \cdot \frac{\sqrt{13}}{\sqrt{13}} = \frac{7\sqrt{13}}{13}$

153. $\frac{2}{\sqrt{x}+3} = \frac{2}{\sqrt{x}+3} \cdot \frac{\sqrt{x}-3}{\sqrt{x}-3} = \frac{2\sqrt{x}-6}{x-9}$

154. $\sqrt{x}+2 = x$

$$\begin{aligned} \sqrt{x} &= x-2 \\ \left(\sqrt{x}\right)^2 &= (x-2)^2 \\ x &= x^2 - 4x + 4 \\ 0 &= x^2 - 5x + 4 \\ 0 &= (x-4)(x-1) \end{aligned}$$

$x-4=0$ or $x-1=0$
$x=4$ $\quad$ $x=1$

Discard the extraneous solution 1. The solution set is $\{4\}$.

Chapter 10 Test

1. $\sqrt{216} = \sqrt{36 \cdot 6} = 6\sqrt{6}$

2. $-\sqrt[4]{x^{64}} = -x^{16}$

3. $\left(\frac{1}{125}\right)^{1/3} = \frac{1}{125^{1/3}} = \frac{1}{\sqrt[3]{125}} = \frac{1}{5}$

4. $\left(\frac{1}{125}\right)^{-1/3} = \frac{1}{\left(\frac{1}{125}\right)^{1/3}} = \frac{1}{\frac{1}{5}} = 5$

5.
$$\begin{aligned} \left(\frac{8x^3}{27}\right)^{2/3} &= \frac{(8x^3)^{2/3}}{27^{2/3}} \\ &= \frac{\left(\sqrt[3]{8x^3}\right)^2}{\left(\sqrt[3]{27}\right)^2} \\ &= \frac{(2x)^2}{3^2} \\ &= \frac{4x^2}{9} \end{aligned}$$

6. $\sqrt[3]{-a^{18}b^9} = \sqrt[3]{-1a^{18}b^9} = (-1)a^6b^3 = -a^6b^3$

7.
$$\begin{aligned} \left(\frac{64c^{4/3}}{a^{-2/3}b^{5/6}}\right)^{1/2} &= \left(\frac{64a^{2/3}c^{4/3}}{b^{5/6}}\right)^{1/2} \\ &= \frac{64^{1/2}(a^{2/3})^{1/2}(c^{4/3})^{1/2}}{(b^{5/6})^{1/2}} \\ &= \frac{\sqrt{64}a^{1/3}c^{2/3}}{b^{5/12}} \\ &= \frac{8a^{1/3}c^{2/3}}{b^{5/12}} \end{aligned}$$

8.
$$\begin{aligned} a^{-2/3}(a^{5/4} - a^3) &= a^{-2/3}a^{5/4} - a^{-2/3}a^3 \\ &= a^{-\frac{2}{3}+\frac{5}{4}} - a^{-\frac{2}{3}+3} \\ &= a^{-\frac{8}{12}+\frac{15}{12}} - a^{-\frac{2}{3}+\frac{9}{3}} \\ &= a^{7/12} - a^{7/3} \end{aligned}$$

9. $\sqrt[4]{(4xy)^4} = |4xy| = 4|xy|$

10. $\sqrt[3]{(-27)^3} = -27$

11. $\sqrt{\frac{9}{y}} = \frac{\sqrt{9}}{\sqrt{y}} = \frac{3}{\sqrt{y}} = \frac{3 \cdot \sqrt{y}}{\sqrt{y} \cdot \sqrt{y}} = \frac{3\sqrt{y}}{y}$

12. $\frac{4-\sqrt{x}}{4+2\sqrt{x}}=\frac{4-\sqrt{x}}{2(2+\sqrt{x})}$
$=\frac{(4-\sqrt{x})(2-\sqrt{x})}{2(2+\sqrt{x})(2-\sqrt{x})}$
$=\frac{8-4\sqrt{x}-2\sqrt{x}+x}{2\left[2^2-(\sqrt{x})^2\right]}$
$=\frac{8-6\sqrt{x}+x}{2(4-x)}$ or $\frac{8-6\sqrt{x}+x}{8-2x}$

13. $\frac{\sqrt[3]{ab}}{\sqrt[3]{ab^2}}=\sqrt[3]{\frac{ab}{ab^2}}$
$=\sqrt[3]{\frac{1}{b}}$
$=\frac{1}{\sqrt[3]{b}}$
$=\frac{1\cdot\sqrt[3]{b^2}}{\sqrt[3]{b}\cdot\sqrt[3]{b^2}}$
$=\frac{\sqrt[3]{b^2}}{b}$

14. $\frac{\sqrt{6}+x}{8}=\frac{(\sqrt{6}+x)(\sqrt{6}-x)}{8(\sqrt{6}-x)}$
$=\frac{(\sqrt{6})^2-x^2}{8(\sqrt{6}-x)}$
$=\frac{6-x^2}{8(\sqrt{6}-x)}$

15. $\sqrt{125x^3}-3\sqrt{20x^3}=\sqrt{25x^2\cdot 5x}-3\sqrt{4x^2\cdot 5x}$
$=5x\sqrt{5x}-3\cdot 2x\sqrt{5x}$
$=5x\sqrt{5x}-6x\sqrt{5x}$
$=-x\sqrt{5x}$

16. $\sqrt{3}(\sqrt{16}-\sqrt{2})=\sqrt{3}(4-\sqrt{2})$
$=4\sqrt{3}-\sqrt{3}\sqrt{2}$
$=4\sqrt{3}-\sqrt{6}$

17. $(\sqrt{x}+1)^2=(\sqrt{x})^2+2\sqrt{x}+1^2$
$=x+2\sqrt{x}+1$

18. $(\sqrt{2}-4)(\sqrt{3}+1)=\sqrt{2}\sqrt{3}+1\cdot\sqrt{2}-4\sqrt{3}-4$
$=\sqrt{6}+\sqrt{2}-4\sqrt{3}-4$

19. $(\sqrt{5}+5)(\sqrt{5}-5)=(\sqrt{5})^2-5^2$
$=5-25$
$=-20$

20. $\sqrt{561}\approx 23.685$

21. $386^{-2/3}\approx 0.019$

22. $x=\sqrt{x-2}+2$
$x-2=\sqrt{x-2}$
$(x-2)^2=(\sqrt{x-2})^2$
$x^2-4x+4=x-2$
$x^2-5x+6=0$
$(x-2)(x-3)=0$
$x=2$ or $x=3$
The solutions are 2 and 3.

23. $\sqrt{x^2-7}+3=0$
$\sqrt{x^2-7}=-3$
No solution exists since the principle square root of a number is not negative.

24. $\sqrt[3]{x+5}=\sqrt[3]{2x-1}$
$(\sqrt[3]{x+5})^3=(\sqrt[3]{2x-1})^3$
$x+5=2x-1$
$-x=-6$
$x=6$
The solution is 6.

25. $\sqrt{-2}=i\sqrt{2}$

26. $-\sqrt{-8}=-i\sqrt{4\cdot 2}=-2i\sqrt{2}$

27. $(12-6i)-(12-3i)=12-6i-12+3i=-3i$

28. $(6-2i)(6+2i)=6^2-(2i)^2$
$=36-4i^2$
$=36+4$
$=40$

29. $(4+3i)^2 = 4^2 + 2\cdot 4\cdot 3i + (3i)^2$
$= 16 + 24i + 9i^2$
$= 16 + 24i - 9$
$= 7 + 24i$

30. $\frac{1+4i}{1-i} = \frac{(1+4i)(1+i)}{(1-i)(1+i)}$
$= \frac{1+i+4i+4i^2}{1^2 - i^2}$
$= \frac{1+5i-4}{1-(-1)}$
$= \frac{-3+5i}{2}$
$= -\frac{3}{2} + \frac{5}{2}i$

31. $x^2 + x^2 = 5^2$
$2x^2 = 25$
$x^2 = \frac{25}{2}$
$\sqrt{x^2} = \sqrt{\frac{25}{2}}$
$x = \frac{5}{\sqrt{2}} = \frac{5\cdot\sqrt{2}}{\sqrt{2}\cdot\sqrt{2}} = \frac{5\sqrt{2}}{2}$

32. $g(x) = \sqrt{x+2}$
$x + 2 \geq 0$
$x \geq -2$
Domain: $[-2, \infty)$

x	−2	−1	2	7
$g(x)$	0	1	2	3

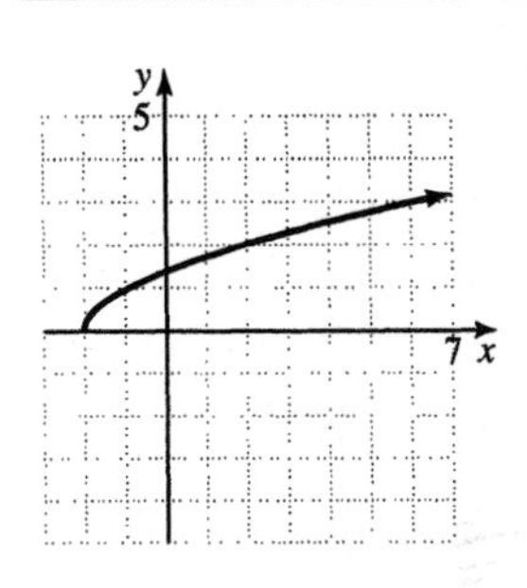

33. $(x_1, y_1) = (-6, 3), (x_2, y_2) = (-8, -7)$
$d = \sqrt{(-8-(-6))^2 + (-7-3)^2}$
$= \sqrt{(-2)^2 + (-10)^2}$
$= \sqrt{4+100}$
$= \sqrt{104}$
$= \sqrt{4\cdot 26}$
$= 2\sqrt{26}$
The distance is $2\sqrt{26}$ units.

34. $(x_1, y_1) = \left(-2\sqrt{5}, \sqrt{10}\right)$,
$(x_2, y_2) = \left(-\sqrt{5}, 4\sqrt{10}\right)$
$d = \sqrt{(x_2 - x_1)^2 + (y_2 - y_1)^2}$
$= \sqrt{\left(-\sqrt{5} + 2\sqrt{5}\right)^2 + \left(4\sqrt{10} - \sqrt{10}\right)^2}$
$= \sqrt{\left(\sqrt{5}\right)^2 + \left(3\sqrt{10}\right)^2}$
$= \sqrt{5+90}$
$= \sqrt{95}$
The distance is $\sqrt{95}$ units.

35. $(x_1, y_1) = (-2, -5), (x_2, y_2) = (-6, 12)$
$\text{midpoint} = \left(\frac{x_1+x_2}{2}, \frac{y_1+y_2}{2}\right)$
$= \left(\frac{-2-6}{2}, \frac{-5+12}{2}\right)$
$= \left(-\frac{8}{2}, \frac{7}{2}\right)$
$= \left(-4, \frac{7}{2}\right)$

36. $(x_1, y_1) = \left(-\frac{2}{3}, -\frac{1}{5}\right), (x_2, y_2) = \left(-\frac{1}{3}, \frac{4}{5}\right)$
$\text{midpoint} = \left(\frac{x_1+x_2}{2}, \frac{y_1+y_2}{2}\right)$
$= \left(\frac{-\frac{2}{3}-\frac{1}{3}}{2}, \frac{-\frac{1}{5}+\frac{4}{5}}{2}\right)$
$= \left(\frac{-\frac{3}{3}}{2}, \frac{\frac{3}{5}}{2}\right)$
$= \left(-\frac{1}{2}, \frac{3}{10}\right)$

37. $V(r) = \sqrt{2.5r}$
$V(300) = \sqrt{2.5(300)} = \sqrt{750} \approx 27$ mph

38. $V(r) = \sqrt{2.5r}$
$$30 = \sqrt{2.5r}$$
$$30^2 = \left(\sqrt{2.5r}\right)^2$$
$$900 = 2.5r$$
$$r = \frac{900}{2.5} = 360 \text{ feet}$$

Chapter 10 Cumulative Review

1. a. $\frac{(-12)(-3)+3}{-7-(-2)} = \frac{36+3}{-7+2} = \frac{39}{-5} = -\frac{39}{5}$

b. $\frac{2(-3)^2-20}{-5+4} = \frac{2\cdot 9-20}{-1} = \frac{18-20}{-1} = \frac{-2}{-1} = 2$

2. a. $2(x-3)+(5x+3) = 2x-6+5x+3$
$= 7x-3$

b. $4(3x+2)-3(5x-1) = 12x+8-15x+3$
$= -3x+11$

c. $7x+2(x-7)-3x = 7x+2x-14-3x$
$= 6x-14$

3. $\frac{x}{2}-1 = \frac{2}{3}x-3$
$$6\left(\frac{x}{2}-1\right) = 6\left(\frac{2}{3}x-3\right)$$
$$3x-6 = 4x-18$$
$$-6 = x-18$$
$$12 = x$$

4. $\frac{a-1}{2}+a = 2-\frac{2a+7}{8}$
$$8\left(\frac{a-1}{2}+a\right) = 8\left(2-\frac{2a+7}{8}\right)$$
$$4(a-1)+8a = 16-(2a+7)$$
$$4a-4+8a = 16-2a-7$$
$$12a-4 = 9-2a$$
$$14a = 13$$
$$a = \frac{13}{14}$$

5. Let x = the length of the shorter board. Then the longer board has length $3x$.
$$x+3x = 48$$
$$4x = 48$$
$$\frac{4x}{4} = \frac{48}{4}$$
$$x = 12$$
$$3x = 3(12) = 36$$
The pieces are 12 inches and 36 inches long.

6. Let r = their average speed.
$$t_{\text{going}} + t_{\text{returning}} = 4.5 \text{ hr}$$
$$\frac{121.5}{r} + \frac{121.5}{r} = 4.5$$
$$\frac{243}{r} = 4.5$$
$$243 = 4.5r$$
$$r = \frac{243}{4.5} = 54$$
Their average speed was 54 mph.

7. $\begin{cases} 3x-y=4 \\ x+2y=8 \end{cases}$
$$3x-y = 4$$
$$-y = -3x+4$$
$$y = 3x-4;\ m = 3$$
$$x+2y = 8$$
$$2y = -x+8$$
$$y = -\frac{1}{2}x+4;\ m = -\frac{1}{2}$$
Since the slopes are different, the lines intersect in one point. The system has one solution.

8. $|3x-2|+5 = 5$
$$|3x-2| = 0$$
$$3x-2 = 0$$
$$3x = 2$$
$$x = \frac{2}{3}$$

9. $\begin{cases} x+2y=7 \\ 2x+2y=13 \end{cases}$
Solve the first equation for x.
$x = -2y+7$
Substitute $-2y+7$ for x in the second equation.
$$2(-2y+7)+2y = 13$$
$$-4y+14+2y = 13$$
$$-2y+14 = 13$$
$$-2y = -1$$
$$y = \frac{1}{2}$$

Let $y = \frac{1}{2}$ in the equation $x = -2y + 7$.

$x = -2y + 7 = -2\left(\frac{1}{2}\right) + 7 = -1 + 7 = 6$

The solution is $\left(6, \frac{1}{2}\right)$.

10. $\left|\frac{x}{2} - 1\right| \le 0$

$\frac{x}{2} - 1 = 0$

$\frac{x}{2} = 1$

$x = 2$

11. $\begin{cases} 2x - y = 7 \\ 8x - 4y = 1 \end{cases}$

Multiply the first equation by −4, then add.

$-8x + 4y = -28$

$8x - 4y = 1$

$0 = -27$

This is a false statement, so the system has no solution.

12. $y = |x - 2|$

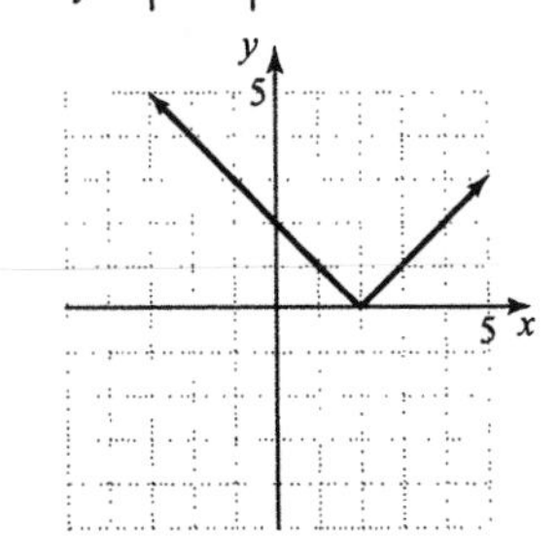

13. Let x be the amount of 30% alcohol solution and y the amount of 80% solution.

$\begin{cases} x + y = 70 \\ 0.30x + 0.80y = 0.50(70) \end{cases}$

$\begin{cases} x + y = 70 \\ 3x + 8y = 350 \end{cases}$

Multiply the first equation by −3, then add.

$-3x - 3y = -210$

$3x + 8y = 350$

$5y = 140$

$y = 28$

Let $y = 28$ in the first equation.

$x + y = 70$

$x + 28 = 70$

$x = 42$

She should mix 42 liters of 30% solution with 28 liters of 80% solution.

14. a. Domain: $(-\infty, 0]$, Range: $(-\infty, \infty)$

not a function

b. Domain: $(-\infty, \infty)$, Range: $(-\infty, \infty)$

function

c. Domain: $(-\infty, -2] \cup [2, \infty)$

Range: $(-\infty, \infty)$

not a function

15. $P(x) = 3x^2 - 2x - 5$

a. $P(1) = 3(1)^2 - 2(1) - 5$
$= 3(1) - 2(1) - 5$
$= 3 - 2 - 5$
$= -4$

b. $P(-2) = 3(-2)^2 - 2(-2) - 5$
$= 3(4) - (-4) - 5$
$= 12 + 4 - 5$
$= 11$

16. $f(x) = -2$

This is a horizontal line passing through (0, −2).

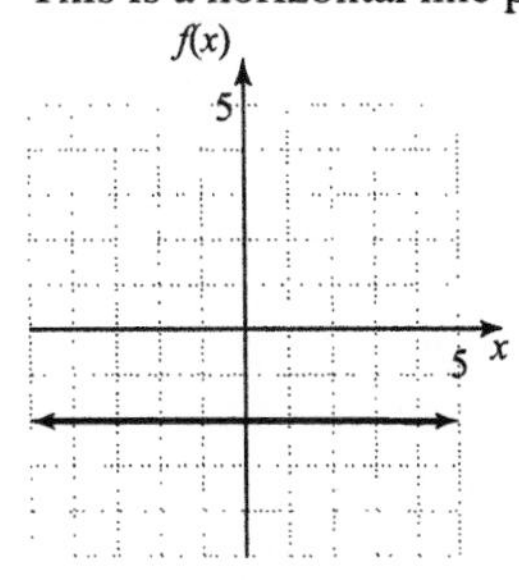

17. $\frac{6m^2 + 2m}{2m} = \frac{2m(3m + 1)}{2m} = 3m + 1$

18. $y = -3$ is a horizontal line. The slope is 0.

19. $\underline{3|}\,2 \quad -1 \quad -13 \quad 1$

$\quad\quad 6 \quad 15 \quad 6$

$2 \quad 5 \quad 2 \quad 7$

Answer: $2x^2+5x+2+\dfrac{7}{x-3}$

20. $\begin{cases} \dfrac{x}{6}-\dfrac{y}{2}=1 \\ \dfrac{x}{3}-\dfrac{y}{4}=2 \end{cases}$ or $\begin{cases} x-3y=6 & (1) \\ 4x-3y=24 & (2) \end{cases}$

Solve equation (1) for x.

$x-3y=6$

$x=3y+6$

Replace x with $3y+6$ in equation (2).

$4(3y+6)-3y=24$

$12y+24-3y=24$

$9y+24=24$

$9y=0$

$y=0$

Substitute 0 for y in $x=3y+6$.

$x=3(0)+6=0+6=6$

The solution is (6, 0).

21. $x^2+7yx+6y^2=(x+6y)(x+y)$

22. Let x = number of tee-shirts and y = number of shorts.

$\begin{cases} x+y=9 & (1) \\ 3.50x+4.25y=33.75 & (2) \end{cases}$

Solve equation (1) for y.

$x+y=9$

$y=9-x$

Substitute $9-x$ for y in equation (2).

$3.50x+4.25(9-x)=33.75$

$3.50x+38.25-4.25x=33.75$

$-0.75x+38.25=33.75$

$-0.75x=-4.5$

$x=\dfrac{-4.5}{-0.75}=6$

Replace x with 6 in $y=9-x$.

$y=9-6=3$

Nana bought 6 shirts and 3 shorts.

23. a. $\dfrac{x^3+8}{2+x}=\dfrac{x^3+2^3}{x+2}$

$=\dfrac{(x+2)(x^2-2x+4)}{x+2}$

$=x^2-2x+4$

b. $\dfrac{2y^2+2}{y^3-5y^2+y-5}=\dfrac{2(y^2+1)}{y^2(y-5)+1(y-5)}$

$=\dfrac{2(y^2+1)}{(y-5)(y^2+1)}$

$=\dfrac{2}{y-5}$

24. $\dfrac{0.0000035\times 4000}{0.28}=\dfrac{(3.5\times 10^{-6})\times(4\times 10^3)}{2.8\times 10^{-1}}$

$=\dfrac{3.5\times 4}{2.8}\times 10^{-6+3-(-1)}$

$=5\times 10^{-2}$

25. $\dfrac{3x^3y^7}{40}\div\dfrac{4x^3}{y^2}=\dfrac{3x^3y^7}{40}\cdot\dfrac{y^2}{4x^3}=\dfrac{3y^9}{160}$

26. $[(5x^2-3x+6)+(4x^2+5x-3)]-(2x-5)$

$=5x^2-3x+6+4x^2+5x-3-2x+5$

$=9x^2+8$

27. $\dfrac{2y}{2y-7}-\dfrac{7}{2y-7}=\dfrac{2y-7}{2y-7}=1$

28. a. $(y-2)(3y+4)=3y^2+4y-6y-8$

$=3y^2-2y-8$

b. $(3y-1)(2y^2+3y-1)$

$=6y^3+9y^2-3y-2y^2-3y+1$

$=6y^3+7y^2-6y+1$

29. $\dfrac{2x}{x^2+2x+1}+\dfrac{x}{x^2-1}=\dfrac{2x}{(x+1)^2}+\dfrac{x}{(x+1)(x-1)}$

$=\dfrac{2x(x-1)+x(x+1)}{(x+1)^2(x-1)}$

$=\dfrac{2x^2-2x+x^2+x}{(x+1)^2(x-1)}$

$=\dfrac{3x^2-x}{(x+1)^2(x-1)}$

$=\dfrac{x(3x-1)}{(x+1)^2(x-1)}$

30. $x^3 - x^2 + 4x - 4 = (x^3 - x^2) + (4x - 4)$
$= x^2(x-1) + 4(x-1)$
$= (x-1)(x^2+4)$

31. a. $\dfrac{\frac{5x}{x+2}}{\frac{10}{x-2}} = \dfrac{5x}{x+2} \cdot \dfrac{x-2}{10} = \dfrac{x(x-2)}{2(x+2)}$

b. $\dfrac{\frac{x}{y^2}+\frac{1}{y}}{\frac{y}{x^2}+\frac{1}{x}} = \dfrac{\left(\frac{x}{y^2}+\frac{1}{y}\right)x^2y^2}{\left(\frac{y}{x^2}+\frac{1}{x}\right)x^2y^2}$
$= \dfrac{x^3 + x^2y}{y^3 + xy^2}$
$= \dfrac{x^2(x+y)}{y^2(y+x)}$
$= \dfrac{x^2}{y^2}$

32. a. $\dfrac{a^3-8}{2-a} = \dfrac{a^3-2^3}{2-a}$
$= \dfrac{(a-2)(a^2+2a+4)}{-1(a-2)}$
$= -1(a^2+2a+4)$
$= -a^2 - 2a - 4$

b. $\dfrac{3a^2-3}{a^3+5a^2-a-5} = \dfrac{3(a^2-1)}{a^2(a+5)-1(a+5)}$
$= \dfrac{3(a^2-1)}{(a+5)(a^2-1)}$
$= \dfrac{3}{a+5}$

33. $\dfrac{x}{2} + \dfrac{8}{3} = \dfrac{1}{6}$
$6\left(\dfrac{x}{2}+\dfrac{8}{3}\right) = 6\left(\dfrac{1}{6}\right)$
$3x + 16 = 1$
$3x = -15$
$x = -5$

34. a. $\dfrac{3}{xy^2} - \dfrac{2}{3x^2y} = \dfrac{3\cdot 3x}{xy^2\cdot 3x} - \dfrac{2\cdot y}{3x^2y\cdot y}$
$= \dfrac{9x-2y}{3x^2y^2}$

b. $\dfrac{5x}{x+3} - \dfrac{2x}{x-3} = \dfrac{5x(x-3)-2x(x+3)}{(x+3)(x-3)}$
$= \dfrac{5x^2-15x-2x^2-6x}{(x+3)(x-3)}$
$= \dfrac{3x^2-21x}{(x+3)(x-3)}$
or $\dfrac{3x(x-7)}{(x+3)(x-3)}$

c. $\dfrac{x}{x-2} - \dfrac{5}{2-x} = \dfrac{x}{x-2} + \dfrac{5}{x-2} = \dfrac{x+5}{x-2}$

35. $\dfrac{x}{10} = \dfrac{3}{2}$
$2x = 10 \cdot 3$
$2x = 30$
$x = 15$
The missing length is 15 yards.

36. a. $\dfrac{\frac{y-2}{16}}{\frac{2y+3}{12}} = \dfrac{y-2}{16} \cdot \dfrac{12}{2y+3} = \dfrac{3(y-2)}{4(2y+3)}$

b. $\dfrac{\frac{x}{16}-\frac{1}{x}}{1-\frac{4}{x}} = \dfrac{\left(\frac{x}{16}-\frac{1}{x}\right)16x}{\left(1-\frac{4}{x}\right)16x}$
$= \dfrac{x^2-16}{16x-64}$
$= \dfrac{(x+4)(x-4)}{16(x-4)}$
$= \dfrac{x+4}{16}$

37. a. $\sqrt[3]{1} = 1$, since $1^3 = 1$.

b. $\sqrt[3]{-64} = \sqrt[3]{(-4)^3} = -4$

c. $\sqrt[3]{\dfrac{8}{125}} = \sqrt[3]{\left(\dfrac{2}{5}\right)^3} = \dfrac{2}{5}$

d. $\sqrt[3]{x^6} = \sqrt[3]{(x^2)^3} = x^2$

e. $\sqrt[3]{-27x^9} = \sqrt[3]{(-3x^3)^3} = -3x^3$

38.
$$\begin{array}{r|l} & x^2 \qquad\quad +3 \\ \hline x-2 & x^3-2x^2+3x-6 \\ & \underline{x^3-2x^2} \\ & \qquad\qquad 3x-6 \\ & \qquad\qquad \underline{3x-6} \\ & \qquad\qquad\quad 0 \end{array}$$

Answer: x^2+3

39. a. $16^{-3/4} = \frac{1}{16^{3/4}} = \frac{1}{\left(\sqrt[4]{16}\right)^3} = \frac{1}{(2)^3} = \frac{1}{8}$

b. $$\begin{aligned}(-27)^{-2/3} &= \frac{1}{(-27)^{2/3}} \\ &= \frac{1}{\left(\sqrt[3]{-27}\right)^2} \\ &= \frac{1}{(-3)^2} \\ &= \frac{1}{9}\end{aligned}$$

40.
$$\begin{array}{r|rrrr} 3 & 4 & -12 & -1 & 12 \\ & & 12 & 0 & -3 \\ \hline & 4 & 0 & -1 & 9 \end{array}$$

Answer: $4y^2-1+\frac{9}{y-3}$

41. $$\begin{aligned}\frac{\sqrt{x}+2}{5} &= \frac{\left(\sqrt{x}+2\right)\left(\sqrt{x}-2\right)}{5\left(\sqrt{x}-2\right)} \\ &= \frac{\left(\sqrt{x}\right)^2-2^2}{5\left(\sqrt{x}-2\right)} \\ &= \frac{x-4}{5\left(\sqrt{x}-2\right)}\end{aligned}$$

42.

$$\frac{28}{9-a^2} = \frac{2a}{a-3} + \frac{6}{a+3}$$
$$\frac{28}{-(a^2-9)} = \frac{2a}{a-3} + \frac{6}{a+3}$$
$$\frac{-28}{(a+3)(a-3)} = \frac{2a}{a-3} + \frac{6}{a+3}$$
$$(a+3)(a-3) \cdot \frac{-28}{(a+3)(a-3)} = (a+3)(a-3) \cdot \left(\frac{2a}{a-3} + \frac{6}{a+3}\right)$$
$$-28 = 2a(a+3) + 6(a-3)$$
$$-28 = 2a^2 + 6a + 6a - 18$$
$$0 = 2a^2 + 12a + 10$$
$$0 = 2(a^2 + 6a + 5)$$
$$0 = 2(a+5)(a+1)$$

$a = -5$ or $a = -1$

The solutions are –5 and –1.

43.

$$u = \frac{k}{w}$$
$$3 = \frac{k}{5}$$
$$k = 3(5) = 15$$
$$u = \frac{15}{w}$$

44.

$$y = kx$$
$$0.51 = k(3)$$
$$k = \frac{0.51}{3} = 0.17$$
$$y = 0.17x$$

Chapter 11

Section 11.1

Practice Exercises

1. $x^2 = 18$
   $x = \pm\sqrt{18}$
   $x = \pm 3\sqrt{2}$
   Check:
   Let $x = 3\sqrt{2}$.
   $x^2 = 18$
   $(3\sqrt{2})^2 \stackrel{?}{=} 18$
   $9 \cdot 2 \stackrel{?}{=} 18$
   $18 = 18$ True
   Let $x = -3\sqrt{2}$.
   $x^2 = 18$
   $(-3\sqrt{2})^2 \stackrel{?}{=} 18$
   $9 \cdot 2 \stackrel{?}{=} 18$
   $18 = 18$ True
   The solutions are $3\sqrt{2}$ and $-3\sqrt{2}$, or the solution set is $\{-3\sqrt{2}, 3\sqrt{2}\}$.

2. First we get the squared variable alone on one side of the equation.
   $3x^2 - 30 = 0$
   $3x^2 = 30$
   $x^2 = 10$
   $x = \pm\sqrt{10}$
   The solutions are $\sqrt{10}$ and $-\sqrt{10}$, or the solution set is $\{-\sqrt{10}, \sqrt{10}\}$.

3. $(x+3)^2 = 20$
   $x + 3 = \pm\sqrt{20}$
   $x + 3 = \pm 2\sqrt{5}$
   $x = -3 \pm 2\sqrt{5}$
   Check:
   $(x+3)^2 = 20$
   $(-3 + 2\sqrt{5} + 3)^2 \stackrel{?}{=} 20$
   $(2\sqrt{5})^2 \stackrel{?}{=} 20$
   $4 \cdot 5 \stackrel{?}{=} 20$
   $20 = 20$ True
   $(x+3)^2 = 20$
   $(-3 - 2\sqrt{5} + 3)^2 \stackrel{?}{=} 20$
   $(-2\sqrt{5})^2 \stackrel{?}{=} 20$
   $4 \cdot 5 \stackrel{?}{=} 20$
   $20 = 20$ True
   The solutions are $-3 + 2\sqrt{5}$ and $-3 - 2\sqrt{5}$.

4. $(5x-2)^2 = -9$
   $5x - 2 = \pm\sqrt{-9}$
   $5x - 2 = \pm 3i$
   $5x = 2 \pm 3i$
   $x = \dfrac{2 \pm 3i}{5}$
   The solutions are $\dfrac{2+3i}{5}$ and $\dfrac{2-3i}{5}$.

5. $b^2 + 4b = 3$
   Add the square of half the coefficient of b to both sides.
   $b^2 + 4b + \left(\dfrac{4}{2}\right)^2 = 3 + \left(\dfrac{4}{2}\right)^2$
   $b^2 + 4b + 4 = 7$
   $(b+2)^2 = 7$
   $b + 2 = \pm\sqrt{7}$
   $b = -2 \pm \sqrt{7}$
   The solutions are $-2 + \sqrt{7}$ and $-2 - \sqrt{7}$.

6. $p^2 - 3p + 1 = 0$
   Subtract 1 from both sides.
   $p^2 - 3p = -1$
   Add the square of half the coefficient of p to both sides.

$$p^2 - 3p + \left(\frac{-3}{2}\right)^2 = -1 + \left(\frac{-3}{2}\right)^2$$
$$p^2 - 3p + \frac{9}{4} = -1 + \frac{9}{4} = \frac{5}{4}$$
$$\left(p - \frac{3}{2}\right)^2 = \frac{5}{4}$$
$$p - \frac{3}{2} = \pm\frac{\sqrt{5}}{2}$$
$$p = \frac{3 \pm \sqrt{5}}{2}$$

The solutions are $\frac{3+\sqrt{5}}{2}$ and $\frac{3-\sqrt{5}}{2}$.

7. $3x^2 - 12x + 1 = 0$

Divide both sides by 3.

$$3x^2 - 12x + 1 = 0$$
$$x^2 - 4x + \frac{1}{3} = 0$$
$$x^2 - 4x = -\frac{1}{3}$$

Find the square of half of –4.

$$\left(\frac{-4}{2}\right)^2 = (-2)^2 = 4$$

Add 4 to both sides of the equation.

$$x^2 - 4x + 4 = -\frac{1}{3} + 4$$
$$(x-2)^2 = -\frac{1}{3} + \frac{12}{3} = \frac{11}{3}$$
$$x - 2 = \pm\sqrt{\frac{11}{3}} = \pm\frac{\sqrt{33}}{3}$$
$$x = \frac{6}{3} \pm \frac{\sqrt{33}}{3} = \frac{6 \pm \sqrt{33}}{3}$$

The solutions are $\frac{6+\sqrt{33}}{3}$ and $\frac{6-\sqrt{33}}{3}$.

8. $2x^2 - 5x + 7 = 0$

$$2x^2 - 5x = -7$$
$$x^2 - \frac{5}{2}x = -\frac{7}{2}$$

Since $\frac{1}{2}\left(-\frac{5}{2}\right) = -\frac{5}{4}$ and $\left(-\frac{5}{4}\right)^2 = \frac{25}{16}$, we add $\frac{25}{16}$ to both sides of the equation.

$$x^2 - \frac{5}{2}x + \frac{25}{16} = -\frac{7}{2} + \frac{25}{16}$$
$$\left(x - \frac{5}{4}\right)^2 = -\frac{56}{16} + \frac{25}{16} = -\frac{31}{16}$$
$$x - \frac{5}{4} = \pm\sqrt{-\frac{31}{16}}$$
$$x = \frac{5}{4} \pm \frac{i\sqrt{31}}{4} = \frac{5 \pm i\sqrt{31}}{4}$$

The solutions are $\frac{5+i\sqrt{31}}{4}$ and $\frac{5-i\sqrt{31}}{4}$.

9. $A = P(1+r)^t$; $A = 5618$, $P = 5000$, $t = 2$

$$A = P(1+r)^t$$
$$5618 = 5000(1+r)^2$$
$$1.1236 = (1+r)^2$$
$$\pm\sqrt{1.1236} = 1 + r$$
$$-1 \pm 1.06 = r$$
$$0.06 = r \text{ or } -2.06 = r$$

The rate cannot be negative, so we reject –2.06.

Check: $A = 5000(1+0.06)^2$
$$= 5000(1.06)^2$$
$$= 5000 \cdot 1.1236$$
$$= 5618$$

The interest rate is 6% compounded annually.

Graphing Calculator Explorations

1. –1.27, 6.27

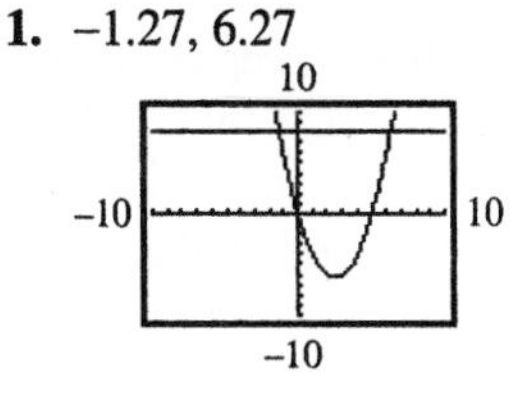

2. –3.45, 1.45

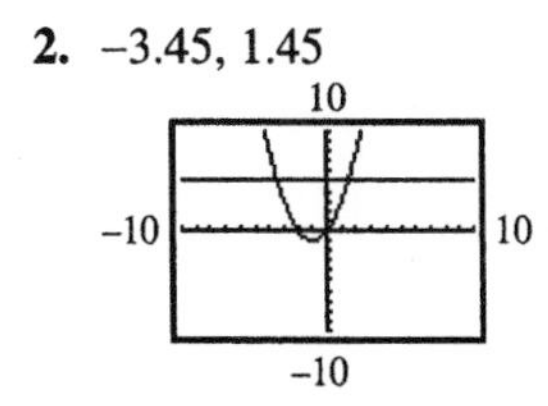

3. –1.10, 0.90

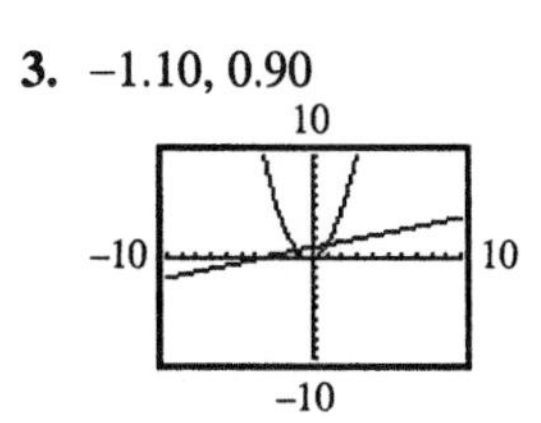

4. −1.54, 1.94

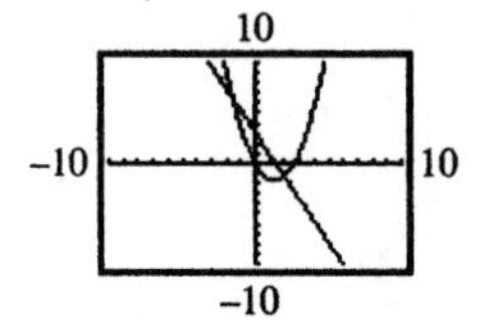

5. No real solutions, or ∅

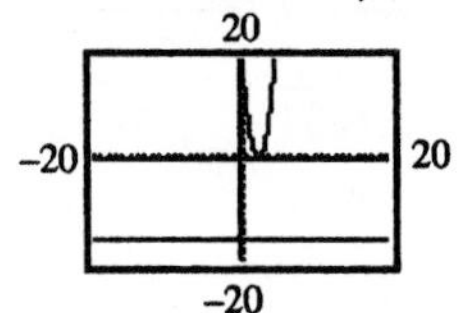

6. Answers may vary

Vocabulary and Readiness Check

1. By the square root property, if b is a real number, and $a^2 = b$, then $a = \pm\sqrt{b}$.

2. A quadratic equation can be written in the form $ax^2 + bx + c = 0$, $a \neq 0$.

3. The process of writing a quadratic equation so that one side is a perfect square trinomial is called completing the square.

4. A perfect square trinomial is one that can be factored as a binomial squared.

5. To solve $x^2 + 6x = 10$ by completing the square, add 9 to both sides.

6. To solve $x^2 + bx = c$ by completing the square, add $\left(\frac{b}{2}\right)^2$ to both sides.

7. $m^2 + 2m + 1$

8. $m^2 - 2m + 1$

9. $y^2 - 14y + 49$

10. $z^2 + z + \frac{1}{4}$

Exercise Set 11.1

1. $x^2 = 16$
$x = \pm\sqrt{16}$
$x = \pm 4$

3. $x^2 - 7 = 0$
$x^2 = 7$
$x = \pm\sqrt{7}$

5. $x^2 = 18$
$x = \pm\sqrt{18}$
$x = \pm\sqrt{9 \cdot 2}$
$x = \pm 3\sqrt{2}$

7. $3z^2 - 30 = 0$
$3z^2 = 30$
$z^2 = 10$
$z = \pm\sqrt{10}$

9. $(x+5)^2 = 9$
$x + 5 = \pm\sqrt{9}$
$x + 5 = \pm 3$
$x = -5 \pm 3$
$x = -8$ or $x = -2$

11. $(z-6)^2 = 18$
$z - 6 = \pm\sqrt{18}$
$z - 6 = \pm 3\sqrt{2}$
$z = 6 \pm 3\sqrt{2}$

13. $(2x-3)^2 = 8$
$2x - 3 = \pm\sqrt{8}$
$2x - 3 = \pm 2\sqrt{2}$
$2x = 3 \pm 2\sqrt{2}$
$x = \frac{3 \pm 2\sqrt{2}}{2}$

15. $x^2 + 9 = 0$
$x^2 = -9$
$x = \pm\sqrt{-9}$
$x = \pm 3i$

17. $x^2 - 6 = 0$
$$x^2 = 6$$
$$x = \pm\sqrt{6}$$

19. $2z^2 + 16 = 0$
$$2z^2 = -16$$
$$z^2 = -8$$
$$z = \pm\sqrt{-8}$$
$$z = \pm i\sqrt{8}$$
$$z = \pm 2i\sqrt{2}$$

21. $(x-1)^2 = -16$
$$x - 1 = \pm\sqrt{-16}$$
$$x - 1 = \pm 4i$$
$$x = 1 \pm 4i$$

23. $(z+7)^2 = 5$
$$z + 7 = \pm\sqrt{5}$$
$$z = -7 \pm \sqrt{5}$$

25. $(x+3)^2 = -8$
$$x + 3 = \pm\sqrt{-8}$$
$$x + 3 = \pm i\sqrt{8}$$
$$x + 3 = \pm 2i\sqrt{2}$$
$$x = -3 \pm 2i\sqrt{2}$$

27. $x^2 + 16x + \left(\frac{16}{2}\right)^2 = x^2 + 16x + 64$
$$= (x+8)^2$$

29. $z^2 - 12z + \left(\frac{-12}{2}\right)^2 = z^2 - 12z + 36$
$$= (z-6)^2$$

31. $p^2 + 9p + \left(\frac{9}{2}\right)^2 = p^2 + 9p + \frac{81}{4}$
$$= \left(p + \frac{9}{2}\right)^2$$

33. $x^2 + x + \left(\frac{1}{2}\right)^2 = x^2 + 16x + \frac{1}{4}$
$$= \left(x + \frac{1}{2}\right)^2$$

35. $x^2 + 8x = -15$
$$x^2 + 8x + \left(\frac{8}{2}\right)^2 = -15 + 16$$
$$x^2 + 8x + 16 = 1$$
$$(x+4)^2 = 1$$
$$x + 4 = \pm\sqrt{1}$$
$$x = -4 \pm 1$$
$$x = -5 \text{ or } x = -3$$

37. $x^2 + 6x + 2 = 0$
$$x^2 + 6x = -2$$
$$x^2 + 6x + \left(\frac{6}{2}\right)^2 = -2 + 9$$
$$x^2 + 6x + 9 = 7$$
$$(x+3)^2 = 7$$
$$x + 3 = \pm\sqrt{7}$$
$$x = -3 \pm \sqrt{7}$$

39. $x^2 + x - 1 = 0$
$$x^2 + x = 1$$
$$x^2 + x + \left(\frac{1}{2}\right)^2 = 1 + \frac{1}{4}$$
$$x^2 + x + \frac{1}{4} = \frac{5}{4}$$
$$\left(x + \frac{1}{2}\right)^2 = \frac{5}{4}$$
$$x + \frac{1}{2} = \pm\sqrt{\frac{5}{4}}$$
$$x = -\frac{1}{2} \pm \frac{\sqrt{5}}{2} = \frac{-1 \pm \sqrt{5}}{2}$$

41. $x^2 + 2x - 5 = 0$
$$x^2 + 2x = 5$$
$$x^2 + 2x + \left(\frac{2}{2}\right)^2 = 5 + 1$$
$$x^2 + 2x + 1 = 6$$
$$(x+1)^2 = 6$$
$$x + 1 = \pm\sqrt{6}$$
$$x = -1 \pm \sqrt{6}$$

43.
$$3p^2-12p+2=0$$
$$3p^2-12p=-2$$
$$p^2-4p=-\frac{2}{3}$$
$$p^2-4p+\left(\frac{-4}{2}\right)^2=-\frac{2}{3}+4$$
$$(p-2)^2=\frac{10}{3}$$
$$p-2=\pm\sqrt{\frac{10}{3}}$$
$$p-2=\pm\frac{\sqrt{10}\cdot\sqrt{3}}{\sqrt{3}\cdot\sqrt{3}}$$
$$p-2=\pm\frac{\sqrt{30}}{3}$$
$$p=2\pm\frac{\sqrt{30}}{3}=\frac{6\pm\sqrt{30}}{3}$$

45.
$$4y^2-12y-2=0$$
$$4y^2-12y=2$$
$$y^2-3y=\frac{1}{2}$$
$$y^2-3y+\left(\frac{-3}{2}\right)^2=\frac{1}{2}+\frac{9}{4}$$
$$y^2-3y+\frac{9}{4}=\frac{11}{4}$$
$$\left(y-\frac{3}{2}\right)^2=\frac{11}{4}$$
$$y-\frac{3}{2}=\pm\sqrt{\frac{11}{4}}$$
$$y=\frac{3}{2}\pm\frac{\sqrt{11}}{2}=\frac{3\pm\sqrt{11}}{2}$$

47.
$$2x^2+7x=4$$
$$x^2+\frac{7}{2}x=2$$
$$x^2+\frac{7}{2}x+\left(\frac{\frac{7}{2}}{2}\right)^2=2+\frac{49}{16}$$
$$x^2+\frac{7}{2}x+\frac{49}{16}=\frac{81}{16}$$
$$\left(x+\frac{7}{4}\right)^2=\frac{81}{16}$$
$$x+\frac{7}{4}=\pm\sqrt{\frac{81}{16}}$$
$$x=-\frac{7}{4}\pm\frac{9}{4}=\frac{-7\pm 9}{4}$$
$$x=-4,\frac{1}{2}$$

49.
$$x^2-4x-5=0$$
$$x^2-4x=5$$
$$x^2-4x+\left(\frac{-4}{2}\right)^2=5+4$$
$$x^2-4x+4=9$$
$$(x-2)^2=9$$
$$x-2=\pm\sqrt{9}$$
$$x=2\pm 3$$
$$x=-1, 5$$

51.
$$x^2+8x+1=0$$
$$x^2+8x=-1$$
$$x^2+8x+\left(\frac{8}{2}\right)^2=-1+16$$
$$x^2+8x+16=15$$
$$(x+4)^2=15$$
$$x+4=\pm\sqrt{15}$$
$$x=-4\pm\sqrt{15}$$

53.
$$3y^2+6y-4=0$$
$$3y^2+6y=4$$
$$y^2+2y=\frac{4}{3}$$
$$y^2+2y+\left(\frac{2}{2}\right)^2=\frac{4}{3}+1$$
$$y^2+2y+1=\frac{7}{3}$$
$$(y+1)^2=\frac{7}{3}$$
$$y+1=\pm\sqrt{\frac{7}{3}}$$
$$y+1=\pm\frac{\sqrt{7}\cdot\sqrt{3}}{\sqrt{3}\cdot\sqrt{3}}$$
$$y+1=\pm\frac{\sqrt{21}}{3}$$
$$y=-1\pm\frac{\sqrt{21}}{3}=\frac{-3\pm\sqrt{21}}{3}$$

55.
$$2x^2-3x-5=0$$
$$2x^2-3x=5$$
$$x^2-\frac{3}{2}x=\frac{5}{2}$$
$$x^2-\frac{3}{2}x+\left(\frac{\frac{3}{2}}{2}\right)^2=\frac{5}{2}+\frac{9}{16}$$
$$x^2-\frac{3}{2}x+\frac{9}{16}=\frac{49}{16}$$
$$\left(x-\frac{3}{4}\right)^2=\frac{49}{16}$$
$$x-\frac{3}{4}=\pm\sqrt{\frac{49}{16}}$$
$$x=\frac{3}{4}\pm\frac{7}{4}=\frac{3\pm 7}{4}$$
$$x=-1,\frac{5}{2}$$

57.
$$y^2+2y+2=0$$
$$y^2+2y=-2$$
$$y^2+2y+\left(\frac{2}{2}\right)^2=-2+1$$
$$y^2+2y+1=-1$$
$$(y+1)^2=-1$$
$$y+1=\pm\sqrt{-1}$$
$$y=-1\pm i$$

59.
$$x^2-6x+3=0$$
$$x^2-6x=-3$$
$$x^2-6x+\left(\frac{-6}{2}\right)^2=-3+9$$
$$x^2-6x+9=6$$
$$(x-3)^2=6$$
$$x-3=\pm\sqrt{6}$$
$$x=3\pm\sqrt{6}$$

61.
$$2a^2+8a=-12$$
$$a^2+4a=-6$$
$$a^2+4a+\left(\frac{4}{2}\right)^2=-6+4$$
$$a^2+4a+4=-2$$
$$(a+2)^2=-2$$
$$a+2=\pm\sqrt{-2}$$
$$a+2=\pm i\sqrt{2}$$
$$a=-2\pm i\sqrt{2}$$

63.
$$5x^2+15x-1=0$$
$$5x^2+15x=1$$
$$x^2+3x=\frac{1}{5}$$
$$x^2+3x+\left(\frac{3}{2}\right)^2=\frac{1}{5}+\frac{9}{4}$$
$$x^2+3x+\frac{9}{4}=\frac{49}{20}$$
$$\left(x+\frac{3}{2}\right)^2=\frac{49}{20}$$
$$x+\frac{3}{2}=\pm\sqrt{\frac{49}{20}}$$
$$x+\frac{3}{2}=\pm\frac{7}{\sqrt{20}}$$
$$x+\frac{3}{2}=\pm\frac{7}{2\sqrt{5}}$$
$$x+\frac{3}{2}=\pm\frac{7\cdot\sqrt{5}}{2\sqrt{5}\cdot\sqrt{5}}$$
$$x+\frac{3}{2}=\pm\frac{7\sqrt{5}}{10}$$
$$x=-\frac{3}{2}\pm\frac{7\sqrt{5}}{10}=\frac{-15\pm7\sqrt{5}}{10}$$

65.
$$2x^2-x+6=0$$
$$2x^2-x=-6$$
$$x^2-\frac{1}{2}x=-3$$
$$x^2-\frac{1}{2}x+\left(\frac{-\frac{1}{2}}{2}\right)^2=-3+\frac{1}{16}$$
$$x^2-\frac{1}{2}x+\frac{1}{16}=-\frac{47}{16}$$
$$\left(x-\frac{1}{4}\right)^2=-\frac{47}{16}$$
$$x-\frac{1}{4}=\pm\sqrt{-\frac{47}{16}}$$
$$x-\frac{1}{4}=\pm i\frac{\sqrt{47}}{4}$$
$$x=\frac{1}{4}\pm i\frac{\sqrt{47}}{4}=\frac{1\pm i\sqrt{47}}{4}$$

67.
$$x^2+10x+28=0$$
$$x^2+10x=-28$$
$$x^2+10x+\left(\frac{10}{2}\right)^2=-28+25$$
$$(x+5)^2=-3$$
$$x+5=\pm\sqrt{-3}$$
$$x=-5\pm i\sqrt{3}$$

69.
$$z^2+3z-4=0$$
$$z^2+3z=4$$
$$z^2+3z+\left(\frac{3}{2}\right)^2=4+\frac{9}{4}$$
$$z^2+3z+\frac{9}{4}=\frac{25}{4}$$
$$\left(z+\frac{3}{2}\right)^2=\frac{25}{4}$$
$$z+\frac{3}{2}=\pm\sqrt{\frac{25}{4}}$$
$$z=-\frac{3}{2}\pm\frac{5}{2}=\frac{-3\pm5}{2}$$
$$z=-4, 1$$

71.
$$2x^2-4x=-3$$
$$x^2-2x=-\frac{3}{2}$$
$$x^2-2x+\left(\frac{-2}{2}\right)^2=-\frac{3}{2}+1$$
$$x^2-2x+1=-\frac{1}{2}$$
$$(x-1)^2=-\frac{1}{2}$$
$$x-1=\pm\sqrt{-\frac{1}{2}}$$
$$x-1=\pm i\frac{1}{\sqrt{2}}$$
$$x-1=\pm i\frac{1\cdot\sqrt{2}}{\sqrt{2}\cdot\sqrt{2}}$$
$$x-1=\pm i\frac{\sqrt{2}}{2}$$
$$x=1\pm i\frac{\sqrt{2}}{2}=\frac{2\pm i\sqrt{2}}{2}$$

73.
$$3x^2+3x=5$$
$$x^2+x=\frac{5}{3}$$
$$x^2+x+\left(\frac{1}{2}\right)^2=\frac{5}{3}+\frac{1}{4}$$
$$x^2+x+\frac{1}{4}=\frac{23}{12}$$
$$\left(x+\frac{1}{2}\right)^2=\frac{23}{12}$$
$$x+\frac{1}{2}=\pm\sqrt{\frac{23}{12}}$$
$$x+\frac{1}{2}=\pm\frac{\sqrt{23}}{2\sqrt{3}}$$
$$x+\frac{1}{2}=\pm\frac{\sqrt{23}\cdot\sqrt{3}}{2\sqrt{3}\cdot\sqrt{3}}$$
$$x+\frac{1}{2}=\pm\frac{\sqrt{69}}{6}$$
$$x=-\frac{1}{2}\pm\frac{\sqrt{69}}{6}=\frac{-3\pm\sqrt{69}}{6}$$

75.
$$A=P(1+r)^t$$
$$4320=3000(1+r)^2$$
$$\frac{4320}{3000}=(1+r)^2$$
$$1.44=(1+r)^2$$
$$\pm\sqrt{1.44}=1+r$$
$$\pm1.2=1+r$$
$$-1\pm1.2=r$$
$$-2.2=r \text{ or } 0.2=r$$
Rate cannot be negative, so the rate is $r=0.2=20\%$.

77.
$$A=P(1+r)^t$$
$$16,224=15,000(1+r)^2$$
$$\frac{16,224}{15,000}=(1+r)^2$$
$$\pm\sqrt{1.0816}=1+r$$
$$\pm1.04=1+r$$
$$-1\pm1.04=r$$
$$0.04=r \text{ or } -2.04=r$$
Rate cannot be negative, so the rate is $r=0.04$, or 4%.

79. Answers may vary

81.
$$s(t)=16t^2$$
$$1053=16t^2$$
$$t^2=\frac{1053}{16}$$
$$t=\pm\sqrt{\frac{1053}{16}}$$
$$t\approx 8.11 \text{ or } -8.11 \text{ (disregard)}$$
It would take 8.11 seconds.

83.
$$s(t)=16t^2$$
$$725=16t^2$$
$$t^2=\frac{725}{16}$$
$$t=\pm\sqrt{\frac{725}{16}}$$
$$t\approx 6.73 \text{ or } -6.73 \text{ (disregard)}$$
It would take 6.73 seconds.

85. Simple; answers may vary

87. $\frac{3}{5}+\sqrt{\frac{16}{25}}=\frac{3}{5}+\frac{4}{5}=\frac{7}{5}$

89. $\frac{9}{10}-\sqrt{\frac{49}{100}}=\frac{9}{10}-\frac{7}{10}=\frac{2}{10}=\frac{1}{5}$

91. $\frac{10-20\sqrt{3}}{2}=\frac{10}{2}-\frac{20\sqrt{3}}{2}=5-10\sqrt{3}$

93.
$$\frac{12-8\sqrt{7}}{16}=\frac{12}{16}-\frac{8\sqrt{7}}{16}$$
$$=\frac{3}{4}-\frac{\sqrt{7}}{2}$$
$$=\frac{3}{4}-\frac{2\sqrt{7}}{4}$$
$$=\frac{3-2\sqrt{7}}{4}$$

95.
$$\sqrt{b^2-4ac}=\sqrt{(6)^2-4(1)(2)}$$
$$=\sqrt{36-8}$$
$$=\sqrt{28}$$
$$=\sqrt{4\cdot7}$$
$$=2\sqrt{7}$$

97. $\sqrt{b^2-4ac} = \sqrt{(-3)^2-4(1)(-1)}$
$= \sqrt{9+4}$
$= \sqrt{13}$

99. The solutions of $(y-5)^2 = -9$ are complex, but not real numbers; answers may vary

101. The solutions of $4x^2 = 17$ are real; answers may vary

103. The solutions of $(3m+2)^2 + 4 = 1$ are complex, but not real numbers; answers may vary

105. $y^2 + \_\_ + 9$

$$\left(\frac{b}{2}\right)^2 = 9$$
$$\frac{b}{2} = \pm\sqrt{9}$$
$$\frac{b}{2} = \pm 3$$
$$b = \pm 6$$

Answer: $\pm 6y$

107. $x^2 + \_\_ + \frac{1}{4}$

$$\left(\frac{b}{2}\right)^2 = \frac{1}{4}$$
$$\frac{b}{2} = \pm\sqrt{\frac{1}{4}}$$
$$\frac{b}{2} = \pm\frac{1}{2}$$
$$b = \pm 1$$

Answer: $\pm x$

109.
$$A = \pi r^2$$
$$36\pi = \pi r^2$$
$$r^2 = \frac{36\pi}{\pi}$$
$$r^2 = 36$$
$$r = \pm\sqrt{36}$$
$$r = 6 \text{ or } -6 \text{ (disregard)}$$

The radius is 6 inches.

111.
$$a^2 + b^2 = c^2$$
$$(4x)^2 + (3x)^2 = 27^2$$
$$16x^2 + 9x^2 = 729$$
$$25x^2 = 729$$
$$x^2 = \frac{729}{25}$$
$$x = \pm\sqrt{\frac{729}{25}} = \pm\frac{27}{5}$$
$$x = 5.4 \text{ or } -5.4 \text{ (disregard)}$$
$$3x = 3(5.4) = 16.2$$
$$4x = 4(5.4) = 21.6$$

The sides are 16.2 in. and 21.6 in.

113.
$$p = -x^2 + 15$$
$$7 = -x^2 + 15$$
$$x^2 = 8$$
$$x = \pm\sqrt{8}$$
$$x \approx \pm 2.828$$

Demand cannot be negative. Therefore, the demand is approximately 2.828 thousand (or 2828) units.

Section 11.2

Practice Exercises

1. $3x^2 - 5x - 2 = 0$

$a = 3, b = -5, c = -2$

$$x = \frac{-b \pm \sqrt{b^2-4ac}}{2a}$$
$$= \frac{-(-5) \pm \sqrt{(-5)^2 - 4(3)(-2)}}{2(3)}$$
$$= \frac{5 \pm \sqrt{25+24}}{6}$$
$$= \frac{5 \pm \sqrt{49}}{6}$$
$$= \frac{5 \pm 7}{6}$$
$$x = \frac{5+7}{6} = \frac{12}{6} = 2 \text{ or } x = \frac{5-7}{6} = \frac{-2}{6} = -\frac{1}{3}$$

The solutions are $-\frac{1}{3}$ and 2, or the solution set is $\left\{-\frac{1}{3}, 2\right\}$.

2. $3x^2 - 8x = 2$
Write in standard form.
$3x^2 - 8x - 2 = 0$
$a = 3, b = -8, c = -2$

$$x = \frac{-b \pm \sqrt{b^2 - 4ac}}{2a}$$
$$= \frac{-(-8) \pm \sqrt{(-8)^2 - 4(3)(-2)}}{2(3)}$$
$$= \frac{8 \pm \sqrt{64 + 24}}{6}$$
$$= \frac{8 \pm \sqrt{88}}{6}$$
$$= \frac{8 \pm 2\sqrt{22}}{6}$$
$$= \frac{4 \pm \sqrt{22}}{3}$$

The solutions are $\frac{4+\sqrt{22}}{3}$ and $\frac{4-\sqrt{22}}{3}$, or the solution set is $\left\{\frac{4+\sqrt{22}}{3}, \frac{4-\sqrt{22}}{3}\right\}$.

3. $\frac{1}{8}x^2 - \frac{1}{4}x - 2 = 0$
Multiply both sides of the equation by 8.
$$8\left(\frac{1}{8}x^2 - \frac{1}{4}x - 2\right) = 8 \cdot 0$$
$$x^2 - 2x - 16 = 0$$
Substitute $a = 1$, $b = -2$, and $c = -16$ into the quadratic formula and simplify.
$$x = \frac{-(-2) \pm \sqrt{(-2)^2 - 4(1)(-16)}}{2(1)}$$
$$= \frac{2 \pm \sqrt{4 + 64}}{2}$$
$$= \frac{2 \pm \sqrt{68}}{2}$$
$$= \frac{2 \pm 2\sqrt{17}}{2}$$
$$= 1 \pm \sqrt{17}$$
The solutions are $1 + \sqrt{17}$ or $1 - \sqrt{17}$.

4. $x = -2x^2 - 2$
The equation in standard form is $2x^2 + x + 2 = 0$. Thus, let $a = 2$, $b = 1$, and $c = 2$ in the quadratic formula.
$$x = \frac{-1 \pm \sqrt{1^2 - 4(2)(2)}}{2(2)}$$
$$= \frac{-1 \pm \sqrt{1 - 16}}{4}$$
$$= \frac{-1 \pm \sqrt{-15}}{4}$$
$$= \frac{-1 \pm i\sqrt{15}}{4}$$
The solutions are $\frac{-1+i\sqrt{15}}{4}$ and $\frac{-1-i\sqrt{15}}{4}$.

5. a. $x^2 - 6x + 9 = 0$
In $x^2 - 6x + 9$, $a = 1$, $b = -6$, and $c = 9$. Thus,
$b^2 - 4ac = (-6)^2 - 4(1)(9) = 36 - 36 = 0$
Since $b^2 - 4ac = 0$, this equation has one real solution.

b. $x^2 - 3x - 1 = 0$
In this equation, $a = 1$, $b = -3$, and $c = -1$.
$b^2 - 4ac = (-3)^2 - 4(1)(-1) = 9 + 4 = 13 > 0$
Since $b^2 - 4ac$ is positive, this equation has two real solutions.

c. $7x^2 + 11 = 0$
In this equation, $a = 7$, $b = 0$, and $c = 11$.
$b^2 - 4ac = 0^2 - 4(7)(11) = -308 < 0$
Since $b^2 - 4ac$ is negative, this equation has two complex but not real solutions.

6. By the Pythagorean theorem, we have
$$x^2 + (x+3)^2 = 15^2$$
$$x^2 + x^2 + 6x + 9 = 225$$
$$2x^2 + 6x - 216 = 0$$
$$x^2 + 3x - 108 = 0$$
Here, $a = 1$, $b = 3$, and $c = -108$. By the quadratic formula,

$$x = \frac{-3 \pm \sqrt{3^2 - 4(1)(-108)}}{2(1)}$$
$$= \frac{-3 \pm \sqrt{9+432}}{2}$$
$$= \frac{-3 \pm \sqrt{441}}{2}$$
$$= \frac{-3 \pm 21}{2}$$
$$x = \frac{-3+21}{2} = \frac{18}{2} = 9 \text{ or}$$
$$x = \frac{-3-21}{2} = \frac{-24}{2} = -12$$

The length can't be negative, so reject −12. The distance along the sidewalk is
$x + (x + 3) = 2x + 3 = 2(9) + 3 = 18 + 3 = 21$ feet
A person can save $21 - 15 = 6$ feet by cutting across the lawn.

7. $h = -16t^2 + 20t + 45$
At the ground, $h = 0$.
$0 = -16t^2 + 20t + 45$
Here, $a = -16$, $b = 20$, and $c = 45$. By the quadratic formula,

$$t = \frac{-20 \pm \sqrt{20^2 - 4(-16)(45)}}{2(-16)}$$
$$= \frac{-20 \pm \sqrt{400+2880}}{-32}$$
$$= \frac{-20 \pm \sqrt{3280}}{-32}$$
$$= \frac{20 \pm \sqrt{16 \cdot 205}}{32}$$
$$= \frac{20 \pm 4\sqrt{205}}{32}$$
$$= \frac{5 \pm \sqrt{205}}{8}$$
$$t = \frac{5+\sqrt{205}}{8} \approx 2.4 \text{ or } t = \frac{5-\sqrt{205}}{8} \approx -1.2$$

Since the time won't be negative, we reject −1.2. The rocket will strike the ground 2.4 seconds after launch.

Vocabulary and Readiness Check

1. The quadratic formula is $\underline{x = \frac{-b \pm \sqrt{b^2 - 4ac}}{2a}}$.

2. For $2x^2 + x + 1 = 0$, if $a = 2$, then $b = \underline{1}$ and $c = \underline{1}$.

3. For $5x^2 - 5x - 7 = 0$, if $a = 5$, then $b = \underline{-5}$ and $c = \underline{-7}$.

4. For $7x^2 - 4 = 0$, if $a = 7$, then $b = \underline{0}$ and $c = \underline{-4}$.

5. For $x^2 + 9 = 0$, if $c = 9$, then $a = \underline{1}$ and $b = \underline{0}$.

6. The correct simplified form of $\frac{5 \pm 10\sqrt{2}}{5}$ is $\underline{1 \pm 2\sqrt{2}}$. The answer is **c**.

Exercise Set 11.2

1. $m^2 + 5m - 6 = 0$
$a = 1, b = 5, c = -6$

$$m = \frac{-5 \pm \sqrt{(5)^2 - 4(1)(-6)}}{2(1)}$$
$$= \frac{-5 \pm \sqrt{25+24}}{2}$$
$$= \frac{-5 \pm \sqrt{49}}{2}$$
$$= \frac{-5 \pm 7}{2}$$
$$= -6 \text{ or } 1$$

The solutions are −6 and 1.

3. $2y = 5y^2 - 3$
$5y^2 - 2y - 3 = 0$
$a = 5, b = -2, c = -3$

$$y = \frac{2 \pm \sqrt{(-2)^2 - 4(5)(-3)}}{2(5)}$$
$$= \frac{2 \pm \sqrt{4+60}}{10}$$
$$= \frac{2 \pm \sqrt{64}}{10}$$
$$= \frac{2 \pm 8}{10}$$
$$= -\frac{3}{5} \text{ or } 1$$

The solutions are $-\frac{3}{5}$ and 1.

5. $x^2-6x+9=0$
$a=1, b=-6, c=9$

$$x=\frac{6\pm\sqrt{(-6)^2-4(1)(9)}}{2(1)}=\frac{6\pm\sqrt{36-36}}{2}=\frac{6\pm\sqrt{0}}{2}=\frac{6}{2}=3$$

The solution is 3.

7. $x^2+7x+4=0$
$a=1, b=7, c=4$

$$x=\frac{-7\pm\sqrt{(7)^2-4(1)(4)}}{2(1)}=\frac{-7\pm\sqrt{49-16}}{2}=\frac{-7\pm\sqrt{33}}{2}$$

The solutions are $\frac{-7+\sqrt{33}}{2}$ and $\frac{-7-\sqrt{33}}{2}$.

9. $8m^2-2m=7$
$8m^2-2m-7=0$
$a=8, b=-2, c=-7$

$$m=\frac{2\pm\sqrt{(-2)^2-4(8)(-7)}}{2(8)}=\frac{2\pm\sqrt{4+224}}{16}=\frac{2\pm\sqrt{228}}{16}=\frac{2\pm\sqrt{4\cdot 57}}{16}=\frac{2\pm 2\sqrt{57}}{16}=\frac{1\pm\sqrt{57}}{8}$$

The solutions are $\frac{1+\sqrt{57}}{8}$ and $\frac{1-\sqrt{57}}{8}$.

11. $3m^2-7m=3$
$3m^2-7m-3=0$
$a=3, b=-7, c=-3$

$$m=\frac{7\pm\sqrt{(-7)^2-4(3)(-3)}}{2(3)}=\frac{7\pm\sqrt{49+36}}{6}=\frac{7\pm\sqrt{85}}{6}$$

The solutions are $\frac{7+\sqrt{85}}{6}$ and $\frac{7-\sqrt{85}}{6}$.

13. $\frac{1}{2}x^2-x-1=0$
$x^2-2x-2=0$
$a=1, b=-2, c=-2$

$$x=\frac{2\pm\sqrt{(-2)^2-4(1)(-2)}}{2(1)}=\frac{2\pm\sqrt{4+8}}{2}=\frac{2\pm\sqrt{12}}{2}=\frac{2\pm 2\sqrt{3}}{2}=1\pm\sqrt{3}$$

The solutions are $1+\sqrt{3}$ and $1-\sqrt{3}$.

15. $\frac{2}{5}y^2+\frac{1}{5}y=\frac{3}{5}$
$2y^2+y-3=0$
$a=2, b=1, c=-3$

$$y=\frac{-1\pm\sqrt{(1)^2-4(2)(-3)}}{2(2)}=\frac{-1\pm\sqrt{1+24}}{4}=\frac{-1\pm\sqrt{25}}{4}=\frac{-1\pm 5}{4}=-\frac{3}{2}\text{ or }1$$

The solutions are $-\frac{3}{2}$ and 1.

17.
$$\frac{1}{3}y^2 = y + \frac{1}{6}$$
$$\frac{1}{3}y^2 - y - \frac{1}{6} = 0$$
$$2y^2 - 6y - 1 = 0$$
$$a = 2, b = -6, c = -1$$
$$y = \frac{6 \pm \sqrt{(-6)^2 - 4(2)(-1)}}{2(2)} = \frac{6 \pm \sqrt{36+8}}{4} = \frac{6 \pm \sqrt{44}}{4} = \frac{6 \pm 2\sqrt{11}}{4} = \frac{3 \pm \sqrt{11}}{2}$$

The solutions are $\frac{3+\sqrt{11}}{2}$ and $\frac{3-\sqrt{11}}{2}$.

19.
$$x^2 + 5x = -2$$
$$x^2 + 5x + 2 = 0$$
$$a = 1, b = 5, c = 2$$
$$x = \frac{-5 \pm \sqrt{(5)^2 - 4(1)(2)}}{2(1)} = \frac{-5 \pm \sqrt{25-8}}{2} = \frac{-5 \pm \sqrt{17}}{2}$$

The solutions are $\frac{-5+\sqrt{17}}{2}$ and $\frac{-5-\sqrt{17}}{2}$.

21.
$$(m+2)(2m-6) = 5(m-1) - 12$$
$$2m^2 - 6m + 4m - 12 = 5m - 5 - 12$$
$$2m^2 - 7m + 5 = 0$$
$$a = 2, b = -7, c = 5$$
$$m = \frac{7 \pm \sqrt{(-7)^2 - 4(2)(5)}}{2(2)} = \frac{7 \pm \sqrt{49-40}}{4} = \frac{7 \pm \sqrt{9}}{4} = \frac{7 \pm 3}{4} = 1 \text{ or } \frac{5}{2}$$

The solutions are 1 and $\frac{5}{2}$.

23. $x^2 + 6x + 13 = 0$
$$a = 1, b = 6, c = 13$$
$$x = \frac{-6 \pm \sqrt{(6)^2 - 4(1)(13)}}{2(1)} = \frac{-6 \pm \sqrt{36-52}}{2} = \frac{-6 \pm \sqrt{-16}}{2} = \frac{-6 \pm 4i}{2} = -3 \pm 2i$$

The solutions are $-3 + 2i$ and $-3 - 2i$.

25. $(x+5)(x-1) = 2$
$$x^2 + 4x - 5 = 2$$
$$x^2 + 4x - 7 = 0$$
$$a = 1, b = 4, c = -7$$
$$x = \frac{-4 \pm \sqrt{(4)^2 - 4(1)(-7)}}{2(1)} = \frac{-4 \pm \sqrt{16+28}}{2} = \frac{-4 \pm \sqrt{44}}{2} = \frac{-4 \pm 2\sqrt{11}}{2} = -2 \pm \sqrt{11}$$

The solutions are $-2 + \sqrt{11}$ and $-2 - \sqrt{11}$.

27.
$$6 = -4x^2 + 3x$$
$$4x^2 - 3x + 6 = 0$$
$$a = 4, b = -3, c = 6$$
$$x = \frac{3 \pm \sqrt{(-3)^2 - 4(4)(6)}}{2(4)} = \frac{3 \pm \sqrt{9 - 96}}{8} = \frac{3 \pm \sqrt{-87}}{8} = \frac{3 \pm i\sqrt{87}}{8}$$

The solutions are $\frac{3 + i\sqrt{87}}{8}$ and $\frac{3 - i\sqrt{87}}{8}$.

29.
$$\frac{x^2}{3} - x = \frac{5}{3}$$
$$x^2 - 3x = 5$$
$$x^2 - 3x - 5 = 0$$
$$a = 1, b = -3, c = -5$$
$$x = \frac{3 \pm \sqrt{(-3)^2 - 4(1)(-5)}}{2(1)} = \frac{3 \pm \sqrt{9 + 20}}{2} = \frac{3 \pm \sqrt{29}}{2}$$

The solutions are $\frac{3 + \sqrt{29}}{2}$ and $\frac{3 - \sqrt{29}}{2}$.

31. $10y^2 + 10y + 3 = 0$
$$a = 10, b = 10, c = 3$$
$$y = \frac{-10 \pm \sqrt{(10)^2 - 4(10)(3)}}{2(10)} = \frac{-10 \pm \sqrt{100 - 120}}{20} = \frac{-10 \pm \sqrt{-20}}{20} = \frac{-10 \pm i\sqrt{4 \cdot 5}}{20} = \frac{-10 \pm 2i\sqrt{5}}{20} = \frac{-5 \pm i\sqrt{5}}{10}$$

The solutions are $\frac{-5 + i\sqrt{5}}{10}$ and $\frac{-5 - i\sqrt{5}}{10}$.

33.
$$x(6x + 2) = 3$$
$$x(6x + 2) - 3 = 0$$
$$6x^2 + 2x - 3 = 0$$
$$a = 6, b = 2, c = -3$$
$$x = \frac{-2 \pm \sqrt{(2)^2 - 4(6)(-3)}}{2(6)} = \frac{-2 \pm \sqrt{4 + 72}}{12} = \frac{-2 \pm \sqrt{76}}{12} = \frac{-2 \pm \sqrt{4 \cdot 19}}{12} = \frac{-2 \pm 2\sqrt{19}}{12} = \frac{-1 \pm \sqrt{19}}{6}$$

The solutions are $\frac{-1 + \sqrt{19}}{6}$ and $\frac{-1 - \sqrt{19}}{6}$.

35. $\frac{2}{5}y^2 + \frac{1}{5}y + \frac{3}{5} = 0$
$$2y^2 + y + 3 = 0$$
$$a = 2, b = 1, c = 3$$
$$y = \frac{-1 \pm \sqrt{(1)^2 - 4(2)(3)}}{2(2)} = \frac{-1 \pm \sqrt{1 - 24}}{4} = \frac{-1 \pm \sqrt{-23}}{4} = \frac{-1 \pm i\sqrt{23}}{4}$$

The solutions are $\frac{-1 + i\sqrt{23}}{4}$ and $\frac{-1 - i\sqrt{23}}{4}$.

37. $\frac{1}{2}y^2 = y - \frac{1}{2}$

$y^2 = 2y - 1$

$y^2 - 2y + 1 = 0$

$a = 1, b = -2, c = 1$

$$y = \frac{2 \pm \sqrt{(-2)^2 - 4(1)(1)}}{2(1)} = \frac{2 \pm \sqrt{4-4}}{2} = \frac{2 \pm \sqrt{0}}{2} = \frac{2}{2} = 1$$

The solution is 1.

39. $(n-2)^2 = 2n$

$n^2 - 4n + 4 = 2n$

$n^2 - 6n + 4 = 0$

$a = 1, b = -6, c = 4$

$$n = \frac{6 \pm \sqrt{(-6)^2 - 4(1)(4)}}{2(1)} = \frac{6 \pm \sqrt{36-16}}{2} = \frac{6 \pm \sqrt{20}}{2} = \frac{6 \pm 2\sqrt{5}}{2} = 3 \pm \sqrt{5}$$

The solutions are $3+\sqrt{5}$ and $3-\sqrt{5}$.

41. $x^2 - 5 = 0$

$a = 1, b = 0, c = -5$

$b^2 - 4ac = 0^2 - 4(1)(-5) = 20 > 0$

Therefore, there are two real solutions.

43. $4x^2 + 12x = -9$

$4x^2 - 12x + 9 = 0$

$a = 4, b = -12, c = 9$

$b^2 - 4ac = (-12)^2 - 4(4)(9)$
$= 144 - 144$
$= 0$

Therefore, there is one real solution.

45. $3x = -2x^2 + 7$

$2x^2 + 3x - 7 = 0$

$a = 2, b = 3, c = -7$

$b^2 - 4ac = 3^2 - 4(2)(-7)$
$= 9 + 56$
$= 65 > 0$

Therefore, there are two real solutions.

47. $6 = 4x - 5x^2$

$5x^2 - 4x + 6 = 0$

$a = 5, b = -4, c = 6$

$b^2 - 4ac = (-4)^2 - 4(5)(6)$
$= 16 - 120$
$= -104 < 0$

Therefore, there are two complex but not real solutions.

49. $9x - 2x^2 + 5 = 0$

$-2x^2 + 9x + 5 = 0$

$a = -2, b = 9, c = 5$

$b^2 - 4ac = 9^2 - 4(-2)(5)$
$= 81 + 40$
$= 121 > 0$

Therefore, there are two real solutions.

51. $(x+8)^2 + x^2 = 36^2$

$(x^2 + 16x + 64) + x^2 = 1296$

$2x^2 + 16x - 1232 = 0$

$a = 2, b = 16, c = -1232$

$$x = \frac{-16 \pm \sqrt{(16)^2 - 4(2)(-1232)}}{2(2)} = \frac{-16 \pm \sqrt{10,112}}{4}$$

$x \approx 21$ or $x \approx -29$ (disregard)

$x + (x+8) = 21 + 21 + 8 = 50$

$50 - 36 = 14$

They save about 14 feet of walking distance.

53. Let x = length of leg. Then
$x + 2$ = length of hypotenuse.

$x^2 + x^2 = (x+2)^2$

$2x^2 = x^2 + 4x + 4$

$x^2 - 4x - 4 = 0$

$a = 1, b = -4, c = -4$

$$x = \frac{4 \pm \sqrt{(-4)^2 - 4(1)(-4)}}{2(1)}$$
$$= \frac{4 \pm \sqrt{32}}{2}$$
$$= \frac{4 \pm 4\sqrt{2}}{2}$$
$$= 2 \pm 2\sqrt{2} \text{ (disregard the negative)}$$
$$= 2 + 2\sqrt{2}$$

The sides measure $2+2\sqrt{2}$ cm, $2+2\sqrt{2}$ cm, and $4+2\sqrt{2}$ cm.

55. Let x = width; then $x + 10$ = length.
Area = length · width
$$400 = (x+10)x$$
$$0 = x^2 + 10x - 400$$
$$a = 1, b = 10, c = -400$$
$$x = \frac{-10 \pm \sqrt{(10)^2 - 4(1)(-400)}}{2(1)}$$
$$= \frac{-10 \pm \sqrt{1700}}{2}$$
$$= \frac{-10 \pm 10\sqrt{17}}{2}$$
$$= -5 \pm 5\sqrt{17}$$
Disregard the negative length. The width is $-5+5\sqrt{17}$ ft and the length is $5+5\sqrt{17}$ ft.

57. a. Let x = length.
$$x^2 + x^2 = 100^2$$
$$2x^2 - 10{,}000 = 0$$
$$a = 2, b = 0, c = -10{,}000$$
$$x = \frac{0 \pm \sqrt{(0)^2 - 4(2)(-10{,}000)}}{2(2)}$$
$$= \frac{\pm\sqrt{80{,}000}}{4}$$
$$= \frac{\pm 200\sqrt{2}}{4}$$
$$= \pm 50\sqrt{2}$$
Disregard the negative length. The side measures $50\sqrt{2}$ meters.

b. Area $= s^2$
$$= \left(50\sqrt{2}\right)^2$$
$$= 2500(2)$$
$$= 5000$$
The area is 5000 square meters.

59. Let w = width; then $w + 1.1$ = height.
Area = length · width
$$1439.9 = (w+1.1)w$$
$$0 = w^2 + 1.1w - 1439.9$$
$$a = 1, b = 1.1, c = -1439.9$$
$$w = \frac{-1.1 \pm \sqrt{(1.1)^2 - 4(1)(-1439.9)}}{2(1)}$$
$$= \frac{-1.1 \pm \sqrt{5760.81}}{2}$$
$$= 37.4 \text{ or } -38.5 \text{ (disregard)}$$
Its width is 37.4 ft and its height is 38.5 ft.

61. Let h = height. Then $2h + 4$ = base.
$$\text{Area} = \frac{1}{2}\text{base} \cdot \text{height}$$
$$42 = \frac{1}{2}(2h+4)h$$
$$42 = h^2 + 2h$$
$$0 = h^2 + 2h - 42$$
$$a = 1, b = 2, c = -42$$
$$h = \frac{-2 \pm \sqrt{(2)^2 - 4(1)(-42)}}{2(1)}$$
$$= \frac{-2 \pm \sqrt{172}}{2}$$
$$= \frac{-2 \pm 2\sqrt{43}}{2}$$
$$= -1 \pm \sqrt{43} \text{ (disregard the negative)}$$
$$\text{base} = 2\left(-1+\sqrt{43}\right) + 4 = 2 + 2\sqrt{43}$$
Height: $-1+\sqrt{43}$ cm
Base: $2+2\sqrt{43}$ cm

63. $h = -16t^2 + 20t + 1100$
$$0 = -16t^2 + 20t + 1100$$
$$a = -16, b = 20, c = 1100$$
$$t = \frac{-20 \pm \sqrt{(20)^2 - 4(-16)(1100)}}{2(-16)}$$
$$= \frac{-20 \pm \sqrt{70{,}800}}{-32}$$
$$\approx 8.9 \text{ or } -7.7 \text{ (disregard)}$$
It will take about 8.9 seconds.

65. $h = -16t^2 - 20t + 180$
$0 = -16t^2 - 20t + 180$
$a = -16, b = -20, c = 180$
$$t = \frac{20 \pm \sqrt{(-20)^2 - 4(-16)(180)}}{2(-16)} = \frac{20 \pm \sqrt{11,920}}{-32}$$
≈ 2.8 or -4.0 (disregard)
It will take about 2.8 seconds.

67. $\sqrt{5x-2} = 3$
$\left(\sqrt{5x-2}\right)^2 = 3^2$
$5x - 2 = 9$
$5x = 11$
$x = \frac{11}{5}$

69. $\frac{1}{x} + \frac{2}{5} = \frac{7}{x}$
$5x\left(\frac{1}{x} + \frac{2}{5}\right) = 5x\left(\frac{7}{x}\right)$
$5 + 2x = 35$
$2x = 30$
$x = 15$

71. $x^4 + x^2 - 20 = (x^2 + 5)(x^2 - 4)$
$= (x^2 + 5)(x + 2)(x - 2)$

73. $z^4 - 13z^2 + 36 = (z^2 - 9)(z^2 - 4)$
$= (z + 3)(z - 3)(z + 2)(z - 2)$

75. $x^2 = -10$
$x^2 + 10 = 0$
$a = 1, b = 0, c = 10$
The correct substitution is **b**.

77. $m^2 + 5m - 6 = 0$
$(m + 6)(m - 1) = 0$
$m + 6 = 0$ or $m - 1 = 0$
$m = -6$ or $m = 1$
The results are the same. Answers may vary.

79. $2x^2 - 6x + 3 = 0$
$a = 2, b = -6, c = 3$
$$x = \frac{6 \pm \sqrt{(-6)^2 - 4(2)(3)}}{2(2)} = \frac{6 \pm \sqrt{12}}{4}$$
≈ 0.6 or 2.4

81. From Sunday to Monday

83. Wednesday

85. $f(x) = 3x^2 - 18x + 56$
$f(4) = 3(4)^2 - 18(4) + 56 = 32$
This answers appears to agree with the graph.

87. $f(x) = 115x^2 + 711x + 3946$

a. $x = 2004 - 2000 = 4$
$f(4) = 115(4)^2 + 711(4) + 3946 = 8630$
There were 8630 stores.

b. Let $f(x) = 25,000$.
$25,000 = 115x^2 + 711x + 3946$
$0 = 115x^2 + 711x - 21,054$
$a = 115, b = 711, c = -21,054$
$$x = \frac{-711 \pm \sqrt{711^2 - 4(115)(-21,054)}}{2(115)} = \frac{-711 \pm \sqrt{10,190,361}}{230}$$
$\approx -3.0913 \pm 13.8793$
$x \approx 10.788 \approx 11$ or $x \approx -16.9706 \approx -17$
We are not concerned with the past, so we reject -17. There will be 25,000 Starbucks in 2011.

89. $$\frac{-b + \sqrt{b^2 - 4ac}}{2a} + \frac{-b - \sqrt{b^2 - 4ac}}{2a} = \frac{-b + \sqrt{b^2 - 4ac} - b - \sqrt{b^2 - 4ac}}{2a} = \frac{-2b}{2a} = -\frac{b}{a}$$

91. $3x^2 - \sqrt{12}x + 1 = 0$

$a = 3, b = -\sqrt{12}, c = 1$

$$x = \frac{\sqrt{12} \pm \sqrt{\left(-\sqrt{12}\right)^2 - 4(3)(1)}}{2(3)}$$
$$= \frac{\sqrt{12} \pm \sqrt{12 - 12}}{6}$$
$$= \frac{\sqrt{4 \cdot 3} \pm \sqrt{0}}{6}$$
$$= \frac{2\sqrt{3}}{6}$$
$$= \frac{\sqrt{3}}{3}$$

The solution is $\frac{\sqrt{3}}{3}$.

93. $x^2 + \sqrt{2}x + 1 = 0$

$a = 1, b = \sqrt{2}, c = 1$

$$x = \frac{-\sqrt{2} \pm \sqrt{\left(\sqrt{2}\right)^2 - 4(1)(1)}}{2(1)}$$
$$= \frac{-\sqrt{2} \pm \sqrt{2 - 4}}{2}$$
$$= \frac{-\sqrt{2} \pm \sqrt{-2}}{2}$$
$$= \frac{-\sqrt{2} \pm i\sqrt{2}}{2}$$

The solutions are $\frac{-\sqrt{2} + i\sqrt{2}}{2}$ and $\frac{-\sqrt{2} - i\sqrt{2}}{2}$.

95. $2x^2 - \sqrt{3}x - 1 = 0$

$a = 2, b = -\sqrt{3}, c = -1$

$$x = \frac{\sqrt{3} \pm \sqrt{\left(-\sqrt{3}\right)^2 - 4(2)(-1)}}{2(2)}$$
$$= \frac{\sqrt{3} \pm \sqrt{3 + 8}}{4}$$
$$= \frac{\sqrt{3} \pm \sqrt{11}}{4}$$

The solutions are $\frac{\sqrt{3} + \sqrt{11}}{4}$ and $\frac{\sqrt{3} - \sqrt{11}}{4}$.

97. Exercise 63:

1200

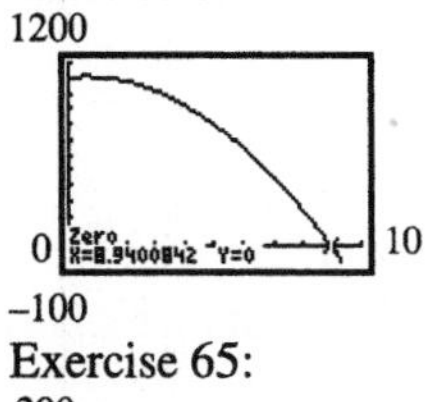

0 10

−100

Exercise 65:

200

0 4

−10

99. $y = 9x - 2x^2 + 5$

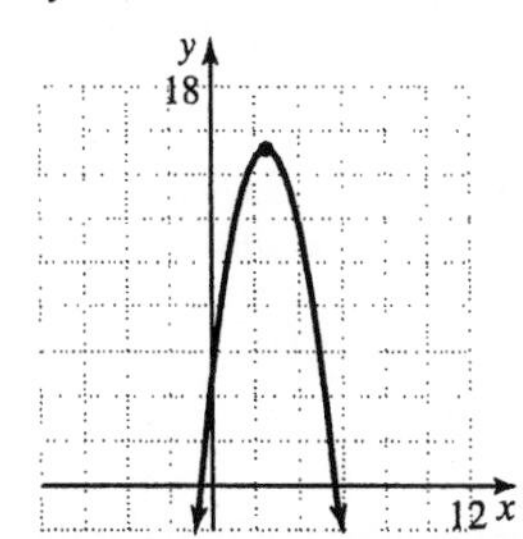

There are two x-intercepts. There are two real solutions.

Section 11.3

Practice Exercises

1. $x - \sqrt{x+1} - 5 = 0$

Get the radical alone on one side of the equation. Then square both sides.

$$x - \sqrt{x+1} - 5 = 0$$
$$x - 5 = \sqrt{x+1}$$
$$(x-5)^2 = x + 1$$
$$x^2 - 10x + 25 = x + 1$$
$$x^2 - 11x + 24 = 0$$
$$(x-8)(x-3) = 0$$
$$x - 8 = 0 \quad \text{or} \quad x - 3 = 0$$
$$x = 8 \quad \text{or} \quad x = 3$$

Check:

Let $x = 3$.

$$x - \sqrt{x+1} - 5 = 0$$
$$3 - \sqrt{3+1} - 5 \stackrel{?}{=} 0$$
$$-2 - \sqrt{4} \stackrel{?}{=} 0$$
$$-2 - 2 \stackrel{?}{=} 0$$
$$-4 = 0 \quad \text{False}$$

Let $x = 8$.

$$x-\sqrt{x+1}-5=0$$
$$8-\sqrt{8+1}-5 \stackrel{?}{=} 0$$
$$3-\sqrt{9} \stackrel{?}{=} 0$$
$$3-3 \stackrel{?}{=} 0$$
$$0=0 \quad \text{True}$$

The solution is 8 or the solution set is $\{8\}$.

2. $\dfrac{5x}{x+1}-\dfrac{x+4}{x}=\dfrac{3}{x(x+1)}$

x cannot be either -1 or 0, because these values cause denominators to equal zero. Multiply both sides of the equation by $x(x + 1)$.

$$x(x+1)\left(\frac{5x}{x+1}\right)-x(x+1)\left(\frac{x+4}{x}\right)=x(x+1)\left[\frac{3}{x(x+1)}\right]$$
$$5x^2-(x+1)(x+4)=3$$
$$5x^2-x^2-5x-4=3$$
$$4x^2-5x-7=0$$

Use the quadratic formula with $a = 4$, $b = -5$, and $c = -7$.

$$x=\frac{-(-5)\pm\sqrt{(-5)^2-4(4)(-7)}}{2(4)}=\frac{5\pm\sqrt{25+112}}{8}=\frac{5\pm\sqrt{137}}{8}$$

Neither proposed solution will make denominators 0. The solutions are $\dfrac{5+\sqrt{137}}{8}$ and $\dfrac{5-\sqrt{137}}{8}$ or the solution set is $\left\{\dfrac{5+\sqrt{137}}{8}, \dfrac{5-\sqrt{137}}{8}\right\}$.

3.
$$p^4-7p^2-144=0$$
$$(p^2+9)(p^2-16)=0$$
$$(p^2+9)(p+4)(p-4)=0$$
$$p^2+9=0 \quad \text{or} \quad p+4=0 \quad \text{or} \quad p-4=0$$
$$p^2=-9 \qquad p=-4 \qquad p=4$$
$$p=\pm\sqrt{-9}$$
$$p=\pm 3i$$

The solutions are 4, -4, $3i$, and $-3i$.

4. $(x+2)^2-2(x+2)-3=0$

Let $y = x + 2$.

$$y^2-2y-3=0$$
$$(y-3)(y+1)=0$$
$$y-3=0 \quad \text{or} \quad y+1=0$$
$$y=3 \qquad y=-1$$

Substitute $x + 2$ for y.

$$x+2=3 \quad \text{or} \quad x+2=-1$$
$$x=1 \qquad x=-3$$

Both 1 and -3 check. The solutions are 1 and -3.

5. $x^{2/3} - 5x^{1/3} + 4 = 0$

Let $m = x^{1/3}$.

$$m^2 - 5m + 4 = 0$$
$$(m-4)(m-1) = 0$$
$$m - 4 = 0 \quad \text{or} \quad m - 1 = 0$$
$$m = 4 \qquad\qquad m = 1$$

Since $m = x^{1/3}$, we have

$$x^{1/3} = 4 \quad \text{or} \quad x^{1/3} = 1$$
$$x = 4^3 = 64 \qquad x = 1^3 = 1$$

Both 64 and 1 check. The solutions are 64 and 1.

6. Let x = the time in hours it takes Steve to groom all the dogs. Then, $x - 1$ = the time it takes Katy to groom all the dogs.

The part of the job completed in one hour by Steve is $\frac{1}{x}$, and the part completed by Katy in one hour is $\frac{1}{x-1}$.

In one hour, $\frac{1}{4}$ of the job is completed. We have,

$$\frac{1}{x} + \frac{1}{x-1} = \frac{1}{4}$$
$$4x(x-1)\left(\frac{1}{x}\right) + 4x(x-1)\left(\frac{1}{x-1}\right) = 4x(x-1)\left(\frac{1}{4}\right)$$
$$4(x-1) + 4x = x(x-1)$$
$$4x - 4 + 4x = x^2 - x$$
$$0 = x^2 - 9x + 4$$

Use the quadratic formula with $a = 1$, $b = -9$, and $c = 4$.

$$x = \frac{-(-9) \pm \sqrt{(-9)^2 - 4(1)(4)}}{2(1)}$$
$$x = \frac{9 \pm \sqrt{81-16}}{2} = \frac{9 \pm \sqrt{65}}{2}$$
$$x \approx 8.53 \quad \text{or} \quad x \approx 0.47$$

Since $x - 1 = 0.47 - 1 = -0.53 < 0$, representing negative time worked, we reject 0.47. It takes Steve $\frac{9+\sqrt{65}}{2} \approx 8.5$ hours and Katy $\frac{9+\sqrt{65}}{2} - 1 = \frac{7+\sqrt{65}}{2} \approx 7.5$ hours to groom all the dogs when working alone.

7. Let x = the speed driven to Shanghai. Then $x + 50$ = the speed driven to Ningbo.

	distance =	*rate* ·	*time*
To Shanghai	36	x	$\frac{36}{x}$
To Ningbo	36	$x + 50$	$\frac{36}{x+50}$

The total travel time was 1.3 hours, so

$$\frac{36}{x}+\frac{36}{x+50}=1.3$$

$$x(x+50)\left(\frac{36}{x}\right)+x(x+50)\left(\frac{36}{x+50}\right)=1.3x(x+50)$$

$$36(x+50)+36x=1.3x^2+65x$$

$$36x+1800+36x=1.3x^2+65x$$

$$0=1.3x^2-7x-1800$$

Use the quadratic formula with $a = 1.3$, $b = -7$, and $c = -1800$.

$$x=\frac{-(-7)\pm\sqrt{(-7)^2-4(1.3)(-1800)}}{2(1.3)}=\frac{7\pm\sqrt{9409}}{2.6}$$

$$x=\frac{7+\sqrt{9409}}{2.6}=40 \quad \text{or} \quad x=\frac{7-\sqrt{9409}}{2.6}\approx -34.6$$

The speed is not negative, so reject –34.6. The speed to Shanghai was 40 km/hr and to Ningbo it was 40 + 50 = 90 km/hr.

Exercise Set 11.3

1.

$$2x=\sqrt{10+3x}$$

$$4x^2=10+3x$$

$$4x^2-3x-10=0$$

$$(4x+5)(x-2)=0$$

$$4x+5=0 \quad \text{or} \quad x-2=0$$

$$x=-\frac{5}{4} \quad \text{or} \quad x=2$$

Discard $-\frac{5}{4}$. The solution is 2.

3.

$$x-2\sqrt{x}=8$$

$$x-8=2\sqrt{x}$$

$$(x-8)^2=\left(2\sqrt{x}\right)^2$$

$$x^2-16x+64=4x$$

$$x^2-20x+64=0$$

$$(x-16)(x-4)=0$$

$$x-16=0 \quad \text{or} \quad x-4=0$$

$$x=16 \quad \text{or} \quad x=4 \text{ (discard)}$$

The solution is 16.

5. $\sqrt{9x} = x+2$

$$\left(\sqrt{9x}\right)^2 = (x+2)^2$$
$$9x = x^2 + 4x + 4$$
$$0 = x^2 - 5x + 4$$
$$0 = (x-4)(x-1)$$
$$x-4=0 \text{ or } x-1=0$$
$$x=4 \text{ or } x=1$$

The solutions are 1 and 4.

7. $\frac{2}{x} + \frac{3}{x-1} = 1$

Multiply each term by $x(x-1)$.

$$2(x-1)+3x = x(x-1)$$
$$2x-2+3x = x^2 - x$$
$$0 = x^2 - 6x + 2$$
$$x = \frac{6 \pm \sqrt{(-6)^2 - 4(1)(2)}}{2(1)}$$
$$= \frac{6 \pm \sqrt{28}}{2}$$
$$= \frac{6 \pm 2\sqrt{7}}{2} = 3 \pm \sqrt{7}$$

The solutions are $3+\sqrt{7}$ and $3-\sqrt{7}$.

9. $\frac{3}{x} + \frac{4}{x+2} = 2$

Multiply each term by $x(x+2)$.

$$3(x+2)+4x = 2x(x+2)$$
$$3x+6+4x = 2x^2 + 4x$$
$$0 = 2x^2 - 3x - 6$$
$$x = \frac{3 \pm \sqrt{(-3)^2 - 4(2)(-6)}}{2(2)}$$
$$= \frac{3 \pm \sqrt{57}}{4}$$

The solutions are $\frac{3+\sqrt{57}}{4}$ and $\frac{3-\sqrt{57}}{4}$.

11. $\frac{7}{x^2-5x+6} = \frac{2x}{x-3} - \frac{x}{x-2}$

$$\frac{7}{(x-3)(x-2)} = \frac{2x}{x-3} - \frac{x}{x-2}$$

Multiply each term by $(x-3)(x-2)$.

$$7 = 2x(x-2) - x(x-3)$$
$$7 = 2x^2 - 4x - x^2 + 3x$$
$$0 = x^2 - x - 7$$
$$x = \frac{1 \pm \sqrt{(-1)^2 - 4(1)(-7)}}{2(1)}$$
$$= \frac{1 \pm \sqrt{29}}{2}$$

The solutions are $\frac{1+\sqrt{29}}{2}$ and $\frac{2-\sqrt{29}}{2}$.

13. $p^4 - 16 = 0$

$$(p^2-4)(p^2+4) = 0$$
$$(p+2)(p-2)(p^2+4) = 0$$
$$p+2=0 \text{ or } p-2=0 \text{ or } p^2+4=0$$
$$p=-2 \text{ or } p=2 \text{ or } p^2=-4$$
$$p = \pm\sqrt{-4}$$
$$p = \pm 2i$$

The solutions are –2, 2, $-2i$, and $2i$.

15. $4x^4 + 11x^2 = 3$

$$4x^4 + 11x^2 - 3 = 0$$
$$(4x^2-1)(x^2+3) = 0$$
$$(2x+1)(2x-1)(x^2+3) = 0$$
$$2x+1=0 \text{ or } 2x-1=0 \text{ or } x^2+3=0$$
$$x = -\frac{1}{2} \text{ or } x = \frac{1}{2} \text{ or } x^2 = -3$$
$$x = \pm\sqrt{-3}$$
$$x = \pm i\sqrt{3}$$

The solutions are $-\frac{1}{2}, \frac{1}{2}, -i\sqrt{3}$, and $i\sqrt{3}$.

17. $z^4 - 13z^2 + 36 = 0$

$$(z^2-9)(z^2-4) = 0$$
$$(z+3)(z-3)(z+2)(z-2) = 0$$
$$z=-3,\ z=3,\ z=-2,\ z=2$$

The solutions are –3, 3, –2, and 2.

19. $x^{2/3} - 3x^{1/3} - 10 = 0$

Let $y = x^{1/3}$. Then $y^2 = x^{2/3}$ and

$$y^2 - 3y - 10 = 0$$
$$(y-5)(y+2) = 0$$
$$y-5=0 \text{ or } y+2=0$$
$$y=5 \text{ or } y=-2$$
$$x^{1/3}=5 \text{ or } x^{1/3}=-2$$
$$x=125 \text{ or } x=-8$$

The solutions are –8 and 125.

21. $(5n+1)^2+2(5n+1)-3=0$

Let $y = 5n + 1$. Then $y^2=(5n+1)^2$ and

$$y^2+2y-3=0$$
$$(y+3)(y-1)=0$$
$$y+3=0 \text{ or } y-1=0$$
$$y=-3 \text{ or } y=1$$
$$5n+1=-3 \text{ or } 5n+1=1$$
$$5n=-4 \text{ or } 5n=0$$
$$n=-\frac{4}{5} \text{ or } n=0$$

The solutions are $-\frac{4}{5}$ and 0.

23. $2x^{2/3}-5x^{1/3}=3$

Let $y=x^{1/3}$. Then $y^2=x^{2/3}$ and

$$2y^2-5y=3$$
$$2y^2-5y-3=0$$
$$(2y+1)(y-3)=0$$
$$2y+1=0 \text{ or } y-3=0$$
$$y=-\frac{1}{2} \text{ or } y=3$$
$$x^{1/3}=-\frac{1}{2} \text{ or } x^{1/3}=3$$
$$x=-\frac{1}{8} \text{ or } x=27$$

The solutions are $-\frac{1}{8}$ and 27.

25.
$$1+\frac{2}{3t-2}=\frac{8}{(3t-2)^2}$$
$$(3t-2)^2+2(3t-2)=8$$
$$(3t-2)^2+2(3t-2)-8=0$$

Let $y=3t-2$. Then $y^2=(3t-2)^2$ and

$$y^2+2y-8=0$$
$$(y+4)(y-2)=0$$
$$y+4=0 \text{ or } y-2=0$$
$$y=-4 \text{ or } y=2$$
$$3t-2=-4 \text{ or } 3t-2=2$$
$$3t=-2 \text{ or } 3t=4$$
$$t=-\frac{2}{3} \text{ or } t=\frac{4}{3}$$

The solutions are $-\frac{2}{3}$ and $\frac{4}{3}$.

27. $20x^{2/3}-6x^{1/3}-2=0$

Let $y=x^{1/3}$. Then $y^2=x^{2/3}$ and

$$20y^2-6y-2=0$$
$$2(10y^2-3y-1)=0$$
$$2(5y+1)(2y-1)=0$$
$$5y+1=0 \text{ or } 2y-1=0$$
$$y=-\frac{1}{5} \text{ or } y=\frac{1}{2}$$
$$x^{1/3}=-\frac{1}{5} \text{ or } x^{1/3}=\frac{1}{2}$$
$$x=-\frac{1}{125} \text{ or } x=\frac{1}{8}$$

The solutions are $\frac{1}{8}$ and $-\frac{1}{125}$.

29.
$$a^4-5a^2+6=0$$
$$(a^2-3)(a^2-2)=0$$
$$a^2-3=0 \text{ or } a^2-2=0$$
$$a^2=3 \text{ or } a^2=2$$
$$a=\pm\sqrt{3} \text{ or } a=\pm\sqrt{2}$$

The solutions are $-\sqrt{3}, \sqrt{3}, -\sqrt{2}$, and $\sqrt{2}$.

31. $\frac{2x}{x-2}+\frac{x}{x+3}=-\frac{5}{x+3}$

Multiply each term by $(x + 3)(x - 2)$.

$$2x(x+3)+x(x-2)=-5(x-2)$$
$$2x^2+6x+x^2-2x=-5x+10$$
$$3x^2+9x-10=0$$
$$x=\frac{-9\pm\sqrt{(9)^2-4(3)(-10)}}{2(3)}$$
$$=\frac{-9\pm\sqrt{201}}{6}$$

The solutions are $\frac{-9+\sqrt{201}}{6}$ and $\frac{-9-\sqrt{201}}{6}$.

33.
$$(p+2)^2=9(p+2)-20$$
$$(p+2)^2-9(p+2)+20=0$$

Let $x=p+2$. Then $x^2=(p+2)^2$ and

$$x^2-9x+20=0$$
$$(x-5)(x-4)=0$$
$$x=5 \text{ or } x=4$$
$$p+2=5 \text{ or } p+2=4$$
$$p=3 \text{ or } p=2$$

The solutions are 2 and 3.

35.
$$2x = \sqrt{11x+3}$$
$$(2x)^2 = \left(\sqrt{11x+3}\right)^2$$
$$4x^2 = 11x+3$$
$$4x^2 - 11x - 3 = 0$$
$$(4x+1)(x-3) = 0$$
$$x = -\frac{1}{4} \text{ (discard) or } x = 3$$
The solution is 3.

37. $x^{2/3} - 8x^{1/3} + 15 = 0$

Let $y = x^{1/3}$. Then $y^2 = x^{2/3}$ and
$$y^2 - 8y + 15 = 0$$
$$(y-5)(y-3) = 0$$
$$y = 5 \quad \text{or} \quad y = 3$$
$$x^{1/3} = 5 \quad \text{or} \quad x^{1/3} = 3$$
$$x = 125 \quad \text{or} \quad x = 27$$
The solutions are 27 and 125.

39.
$$y^3 + 9y - y^2 - 9 = 0$$
$$y(y^2+9) - 1(y^2+9) = 0$$
$$(y^2+9)(y-1) = 0$$
$$y^2 + 9 = 0 \quad \text{or} \quad y - 1 = 0$$
$$y^2 = -9 \quad \text{or} \quad y = 1$$
$$y = \pm\sqrt{-9}$$
$$y = \pm 3i$$
The solutions are 1, $-3i$, and $3i$.

41. $2x^{2/3} + 3x^{1/3} - 2 = 0$

Let $y = x^{1/3}$. Then $y^2 = x^{2/3}$ and
$$2y^2 + 3y - 2 = 0$$
$$(2y-1)(y+2) = 0$$
$$y = \frac{1}{2} \quad \text{or} \quad y = -2$$
$$x^{1/3} = \frac{1}{2} \quad \text{or} \quad x^{1/3} = -2$$
$$x = \frac{1}{8} \quad \text{or} \quad x = -8$$
The solutions are -8 and $\frac{1}{8}$.

43. $x^{-2} - x^{-1} - 6 = 0$

Let $y = x^{-1}$. Then $y^2 = x^{-2}$ and
$$y^2 - y - 6 = 0$$
$$(y-3)(y+2) = 0$$
$$y = 3 \quad \text{or} \quad y = -2$$
$$x^{-1} = 3 \quad \text{or} \quad x^{-1} = -2$$
$$\frac{1}{x} = 3 \quad \text{or} \quad \frac{1}{x} = -2$$
$$x = \frac{1}{3} \quad \text{or} \quad x = -\frac{1}{2}$$
The solutions are $-\frac{1}{2}$ and $\frac{1}{3}$.

45.
$$x - \sqrt{x} = 2$$
$$x - 2 = \sqrt{x}$$
$$(x-2)^2 = x$$
$$x^2 - 4x + 4 = x$$
$$x^2 - 5x + 4 = 0$$
$$(x-4)(x-1) = 0$$
$x = 4$ or $x = 1$ (discard)

The solution is 4.

47.
$$\frac{x}{x-1} + \frac{1}{x+1} = \frac{2}{x^2-1}$$
$$\frac{x}{x-1} + \frac{1}{x+1} = \frac{2}{(x+1)(x-1)}$$
$$x(x+1) + (x-1) = 2$$
$$x^2 + x + x - 1 = 2$$
$$x^2 + 2x - 3 = 0$$
$$(x+3)(x-1) = 0$$
$x = -3$ or $x = 1$ (discard)

The solution is -3.

49.
$$p^4 - p^2 - 20 = 0$$
$$(p^2-5)(p^2+4) = 0$$
$$p^2 - 5 = 0 \quad \text{or} \quad p^2 + 4 = 0$$
$$p^2 = 5 \quad \text{or} \quad p^2 = -4$$
$$p = \pm\sqrt{5} \quad \text{or} \quad p = \pm 2i$$
The solutions are $-\sqrt{5}, \sqrt{5}, -2i$, and $2i$.

51. $(x+3)(x^2-3x+9)=0$

$x+3=0$ or $x^2-3x+9=0$

$x=-3$ or

$$x=\frac{3\pm\sqrt{(-3)^2-4(1)(9)}}{2(1)}=\frac{3\pm\sqrt{-27}}{2}=\frac{3\pm 3i\sqrt{3}}{2}$$

The solutions are -3, $\frac{3+3i\sqrt{3}}{2}$, and $\frac{3-3i\sqrt{3}}{2}$.

53. $1=\frac{4}{x-7}+\frac{5}{(x-7)^2}$

$(x-7)^2-4(x-7)-5=0$

Let $y=x-7$. Then $y^2=(x-7)^2$ and

$y^2-4y-5=0$

$(y-5)(y+1)=0$

$y=5$ or $y=-1$

$x-7=5$ or $x-7=-1$

$x=12$ or $x=6$

The solutions are 6 and 12.

55. $27y^4+15y^2=2$

$27y^4+15y^2-2=0$

$(9y^2-1)(3y^2+2)=0$

$(3y+1)(3y-1)(3y^2+2)=0$

$y=-\frac{1}{3}$ or $y=\frac{1}{3}$ or $y^2=-\frac{2}{3}$

$y=\pm\sqrt{-\frac{2}{3}}$

$y=\pm\frac{i\sqrt{6}}{3}$

The solutions are $-\frac{1}{3}$, $\frac{1}{3}$, $-\frac{i\sqrt{6}}{3}$, and $\frac{i\sqrt{6}}{3}$.

57. Let x = speed on the first part. Then $x-1$ = speed on the second part.

$d=rt \Rightarrow t=\frac{d}{r}$

$t_{\text{on first part}}+t_{\text{on second part}}=1\frac{3}{5}$

$\frac{3}{x}+\frac{4}{x-1}=\frac{8}{5}$

$3\cdot 5(x-1)+4\cdot 5x=8x(x-1)$

$15x-15+20x=8x^2-8x$

$0=8x^2-43x+15$

$0=(8x-3)(x-5)$

$8x-3=0$ or $x-5=0$

$x=\frac{3}{8}$ or $x=5$

$x-1=4$

Discard $\frac{3}{8}$. Her speeds were 5 mph and 4 mph.

59. Let x = time for hose alone. Then $x-1$ = time for the inlet pipe alone.

$\frac{1}{x}+\frac{1}{x-1}=\frac{1}{8}$

$8(x-1)+8x=x(x-1)$

$8x-8+8x=x^2-x$

$0=x^2-17x+8$

$$x=\frac{17\pm\sqrt{(-17)^2-4(1)(8)}}{2(1)}=\frac{17\pm\sqrt{257}}{2}$$

$x\approx 0.5$ (discard) or $x\approx 16.5$

$x-1\approx 15.5$

Hose: 16.5 hrs; Inlet pipe: 15.5 hrs

61. Let x = original speed. Then $x+11$ = return speed.

$d=rt \Rightarrow t=\frac{d}{r}$

$t_{\text{return}}=t_{\text{original}}-1$

$\frac{330}{x+11}=\frac{330}{x}-1$

$330x=330(x+11)-x(x+11)$

$330x=330x+3630-x^2-11x$

$x^2+11x-3630=0$

$$x = \frac{-11 \pm \sqrt{(11)^2 - 4(1)(-3630)}}{2(1)}$$
$$= \frac{-11 \pm \sqrt{14,641}}{2}$$
$$= \frac{-11 \pm 121}{2} = 55 \text{ or } -66 \text{ (disregard)}$$
$x + 11 = 55 + 11 = 66$
Original speed: 55 mph
Return speed: 66 mph

63. Let x = time for son alone. Then $x - 1$ = time for dad alone.
$$\frac{1}{x} + \frac{1}{x-1} = \frac{1}{4}$$
$$4(x-1) + 4x = x(x-1)$$
$$4x - 4 + 4x = x^2 - x$$
$$0 = x^2 - 9x + 4$$
$$x = \frac{9 \pm \sqrt{(-9)^2 - 4(1)(4)}}{2(1)}$$
$$= \frac{9 \pm \sqrt{65}}{2}$$
≈ 0.5 (discard) or 8.5
It takes his son about 8.5 hours.

65. Let x = the number.
$$x(x-4) = 96$$
$$x^2 - 4x - 96 = 0$$
$$(x-12)(x+8) = 0$$
$x = 12$ or $x = -8$
The number is 12 or –8.

67. a. length $= x - 3 - 3 = x - 6 = (x - 6)$ cm

b. $V = lwh$
$300 = (x-6)(x-6) \cdot 3$

c. $300 = 3(x-6)^2$
$100 = x^2 - 12x + 36$
$0 = x^2 - 12x - 64$
$0 = (x-16)(x+4)$
$x = 16$ or $x = -4$ (discard)
The sheet is 16 cm by 16 cm.
Check: $V = 3(x-6)(x-6)$
$= 3(16-6)(16-6)$
$= 3(10)(10)$
$= 300$ cubic cm

69. Let x = length of the side of the square.
Area $= x^2$
$920 = x^2$
$\sqrt{920} = x$
Adding another radial line to a different corner would yield a right triangle with legs r and hypotenuse x.
$r^2 + r^2 = x^2$
$2r^2 = \left(\sqrt{920}\right)^2$
$2r^2 = 920$
$r^2 = 460$
$r = \pm\sqrt{460} = \pm 21.4476$
Disregard the negative. The smallest radius would be 22 feet.

71. $\frac{5x}{3} + 2 \le 7$
$\frac{5x}{3} \le 5$
$5x \le 15$
$x \le 3$
$(-\infty, 3]$

73. $\frac{y-1}{15} > -\frac{2}{5}$
$15\left(\frac{y-1}{15}\right) > 15\left(-\frac{2}{5}\right)$
$y - 1 > -6$
$y > -5$
$(-5, \infty)$

75. Domain: $\{x \mid x \text{ is a real number}\}$ or $(-\infty, \infty)$
Range: $\{y \mid y \text{ is a real number}\}$ or $(-\infty, \infty)$
It is a function.

77. Domain: $\{x \mid x \text{ is a real number}\}$ or $(-\infty, \infty)$
Range: $\{y \mid y \ge -1\}$ or $[-1, \infty)$
It is a function.

79.
$$y^3 + 9y - y^2 - 9 = 0$$
$$y(y^2 + 9) - (y^2 + 9) = 0$$
$$(y - 1)(y^2 + 9) = 0$$
$$y - 1 = 0 \quad \text{or} \quad y^2 + 9 = 0$$
$$y = 1 \qquad\qquad y^2 = -9$$
$$y = \pm\sqrt{-9} = \pm 3i$$

The solutions are 1, $3i$, and $-3i$.

81.
$$x^{-2} - x^{-1} - 6 = 0$$
$$1 - x - 6x^2 = 0$$
$$(1 - 3x)(1 + 2x) = 0$$
$$1 - 3x = 0 \quad \text{or} \quad 1 + 2x = 0$$
$$1 = 3x \qquad\qquad 2x = -1$$
$$\frac{1}{3} = x \qquad\qquad x = -\frac{1}{2}$$

The solutions are $\frac{1}{3}$ and $-\frac{1}{2}$.

83.
$$2x^3 = -54$$
$$x^3 = -27$$
$$x^3 + 27 = 0$$
$$(x + 3)(x^2 - 3x + 9) = 0$$
$$x + 3 = 0 \quad \text{or} \quad x^2 - 3x + 9 = 0$$
$$x = -3$$
$$x = \frac{-(-3) \pm \sqrt{(-3)^2 - 4(1)(9)}}{2(1)}$$
$$x = \frac{3 \pm \sqrt{9 - 36}}{2} = \frac{3 \pm \sqrt{-27}}{2}$$
$$x = \frac{3 \pm 3i\sqrt{3}}{2}$$

The solutions are -3, $\frac{3 + 3i\sqrt{3}}{2}$, and $\frac{3 - 3i\sqrt{3}}{2}$.

85. Answers may vary

87. a. Let x = Bourdais's fastest lap speed and $x + 0.55$ = Pagenaud's fastest lap speed.

Using $t = \frac{d}{r}$, we have

$$t_{\text{Bourdais}} = t_{\text{Pagenaud}} + 0.25$$
$$\frac{10,391}{x} = \frac{10,391}{x + 0.55} + 0.25$$
$$10,391(x + 0.55) = 10,391x + 0.25x(x + 0.55)$$
$$10,391x + 5715.05 = 10,391x + 0.25x^2 + 0.1375x$$
$$0 = 0.25x^2 + 0.1375x - 5715.05$$

$$x = \frac{-0.1375 \pm \sqrt{(0.1375)^2 - 4(0.25)(-5715.05)}}{2(0.25)}$$

Using the positive square root, $x \approx 150.92$ feet per second.

b. $x + 0.55 = 150.92 + 0.55$
$= 151.47$ feet per second

c. 5280 ft = 1 mile, and 3600 sec = 1 hr.

Bourdais: $\frac{150.92 \text{ ft}}{\text{sec}} \cdot \frac{3600 \text{ sec}}{\text{hr}} \cdot \frac{1 \text{ mile}}{5280 \text{ ft}} \approx 102.9$ mph

Fernandez: $\frac{151.47 \text{ ft}}{\text{sec}} \cdot \frac{3600 \text{ sec}}{\text{hr}} \cdot \frac{1 \text{ mile}}{5280 \text{ ft}} \approx 103.3$ mph

Integrated Review

1. $x^2 - 10 = 0$
$x^2 = 10$
$x = \pm\sqrt{10}$

2. $x^2 - 14 = 0$
$x^2 = 14$
$x = \pm\sqrt{14}$

3. $(x-1)^2 = 8$
$x - 1 = \pm\sqrt{8}$
$x - 1 = \pm 2\sqrt{2}$
$x = 1 \pm 2\sqrt{2}$

4. $(x+5)^2 = 12$
$x + 5 = \pm\sqrt{12}$
$x + 5 = \pm 2\sqrt{3}$
$x = -5 \pm 2\sqrt{3}$

5. $x^2 + 2x - 12 = 0$
$x^2 + 2x + \left(\frac{2}{2}\right)^2 = 12 + 1$
$x^2 + 2x + 1 = 13$
$(x+1)^2 = 13$
$x + 1 = \pm\sqrt{13}$
$x = -1 \pm \sqrt{13}$

6.
$$x^2-12x+11=0$$
$$x^2-12x+\left(\frac{-12}{2}\right)^2=-11+36$$
$$x^2-12x+36=25$$
$$(x-6)^2=\pm\sqrt{25}$$
$$x-6=\pm 5$$
$$x=6\pm 5$$
$$x=1 \text{ or } x=11$$

7.
$$3x^2+3x=5$$
$$x^2+x=\frac{5}{3}$$
$$x^2+x+\left(\frac{1}{2}\right)^2=\frac{5}{3}+\frac{1}{4}$$
$$x^2+x+\frac{1}{4}=\frac{23}{12}$$
$$\left(x+\frac{1}{2}\right)^2=\frac{23}{12}$$
$$x+\frac{1}{2}=\pm\sqrt{\frac{23}{12}}$$
$$x+\frac{1}{2}=\pm\frac{\sqrt{23}}{2\sqrt{3}}$$
$$x+\frac{1}{2}=\pm\frac{\sqrt{23}\cdot\sqrt{3}}{2\sqrt{3}\cdot\sqrt{3}}$$
$$x+\frac{1}{2}=\pm\frac{\sqrt{69}}{6}$$
$$x=-\frac{1}{2}\pm\frac{\sqrt{69}}{6}=\frac{-3\pm\sqrt{69}}{6}$$

8.
$$16y^2+16y=1$$
$$y^2+y=\frac{1}{16}$$
$$y^2+y+\left(\frac{1}{2}\right)^2=\frac{1}{16}+\frac{1}{4}$$
$$y^2+y+\frac{1}{4}=\frac{5}{16}$$
$$\left(y+\frac{1}{2}\right)^2=\frac{5}{16}$$
$$y+\frac{1}{2}=\pm\sqrt{\frac{5}{16}}$$
$$y+\frac{1}{2}=\pm\frac{\sqrt{5}}{4}$$
$$y=-\frac{1}{2}\pm\frac{\sqrt{5}}{4}=\frac{-2\pm\sqrt{5}}{4}$$

9.
$$2x^2-4x+1=0$$
$$a=2,\ b=-4,\ c=1$$
$$x=\frac{4\pm\sqrt{(-4)^2-4(2)(1)}}{2(2)}$$
$$=\frac{4\pm\sqrt{8}}{4}$$
$$=\frac{4\pm 2\sqrt{2}}{4}=\frac{2\pm\sqrt{2}}{2}$$

10.
$$\frac{1}{2}x^2+3x+2=0$$
$$x^2+6x+4=0$$
$$a=1,\ b=6,\ c=4$$
$$x=\frac{-6\pm\sqrt{(6)^2-4(1)(4)}}{2(1)}$$
$$=\frac{-6\pm\sqrt{20}}{2}$$
$$=\frac{-6\pm 2\sqrt{5}}{2}=-3\pm\sqrt{5}$$

11.
$$x^2+4x=-7$$
$$x^2+4x+7=0$$
$$a=1,\ b=4,\ c=7$$
$$x=\frac{-4\pm\sqrt{(4)^2-4(1)(7)}}{2(1)}$$
$$=\frac{-4\pm\sqrt{-12}}{2}$$
$$=\frac{-4\pm i\sqrt{4\cdot 3}}{2}$$
$$=\frac{-4\pm 2i\sqrt{3}}{2}=-2\pm i\sqrt{3}$$

12.
$$x^2+x=-3$$
$$x^2+x+3=0$$
$$a=1,\ b=1,\ c=3$$
$$x=\frac{-1\pm\sqrt{(1)^2-4(1)(3)}}{2(1)}$$
$$=\frac{-1\pm\sqrt{-11}}{2}$$
$$=\frac{-1\pm i\sqrt{11}}{2}$$

13. $x^2+3x+6=0$

$a=1, b=3, c=6$

$$x=\frac{-3\pm\sqrt{(3)^2-4(1)(6)}}{2(1)}=\frac{-3\pm\sqrt{-15}}{2}=\frac{-3\pm i\sqrt{15}}{2}$$

14. $2x^2+18=0$

$$2x^2=-18$$
$$x^2=-9$$
$$x=\pm\sqrt{-9}$$
$$x=\pm 3i$$

15. $x^2+17x=0$

$x(x+17)=0$

$x=0$ or $x+17=0$

$x=-17$

$x=0, -17$

16. $4x^2-2x-3=0$

$a=4, b=-2, c=-3$

$$x=\frac{2\pm\sqrt{(-2)^2-4(4)(-3)}}{2(4)}=\frac{2\pm\sqrt{52}}{8}=\frac{2\pm 2\sqrt{13}}{8}=\frac{1\pm\sqrt{13}}{4}$$

17. $(x-2)^2=27$

$$x-2=\pm\sqrt{27}$$
$$x-2=\pm 3\sqrt{3}$$
$$x=2\pm 3\sqrt{3}$$

18. $\frac{1}{2}x^2-2x+\frac{1}{2}=0$

$$x^2-4x+1=0$$
$$x^2-4x+\left(\frac{-4}{2}\right)^2=-1+4$$
$$x^2-4x+4=3$$
$$(x-2)^2=3$$
$$x-2=\pm\sqrt{3}$$
$$x=2\pm\sqrt{3}$$

19. $3x^2+2x=8$

$3x^2+2x-8=0$

$(3x-4)(x+2)=0$

$3x-4=0$ or $x+2=0$

$x=\frac{4}{3}$ or $x=-2$

20. $2x^2=-5x-1$

$2x^2+5x+1=0$

$a=2, b=5, c=1$

$$x=\frac{-5\pm\sqrt{(5)^2-4(2)(1)}}{2(2)}=\frac{-5\pm\sqrt{17}}{4}$$

21. $x(x-2)=5$

$$x^2-2x=5$$
$$x^2-2x+\left(\frac{-2}{2}\right)^2=5+1$$
$$x^2-2x+1=6$$
$$(x-1)^2=6$$
$$x-1=\pm\sqrt{6}$$
$$x=1\pm\sqrt{6}$$

22. $x^2-31=0$

$$x^2=31$$
$$x=\pm\sqrt{31}$$

23. $5x^2-55=0$

$$5x^2=55$$
$$x^2=11$$
$$x=\pm\sqrt{11}$$

24. $5x^2+55=0$
$5x^2=-55$
$x^2=-11$
$x=\pm\sqrt{-11}$
$x=\pm i\sqrt{11}$

25. $x(x+5)=66$
$x^2+5x=66$
$x^2+5x-66=0$
$(x+11)(x-6)=0$
$x+11=0$ or $x-6=0$
$x=-11$ or $x=6$

26. $5x^2+6x-2=0$
$a=5, b=6, c=-2$

$$x=\frac{-6\pm\sqrt{(6)^2-4(5)(-2)}}{2(5)}=\frac{-6\pm\sqrt{76}}{10}=\frac{-6\pm\sqrt{4\cdot19}}{10}=\frac{-6\pm2\sqrt{19}}{10}=\frac{-3\pm\sqrt{19}}{5}$$

27. $2x^2+3x=1$
$2x^2+3x-1=0$
$a=2, b=3, c=-1$

$$x=\frac{-3\pm\sqrt{(3)^2-4(2)(-1)}}{2(2)}=\frac{-3\pm\sqrt{17}}{4}$$

28. $a^2+b^2=c^2$
$x^2+x^2=20^2$
$2x^2=400$
$x^2=200$
$x=\pm\sqrt{200}$
$=\pm10\sqrt{2}\approx14.1421$
Disregard the negative. A side of the room is $10\sqrt{2}$ feet $\approx$ 14.1 feet.

29. Let x = time for Jack alone. Then $x-2$ = time for Lucy alone.

$$\frac{1}{x}+\frac{1}{x-2}=\frac{1}{4}$$
$4(x-2)+4x=x(x-2)$
$4x-8+4x=x^2-2x$
$0=x^2-10x+8$

$$x=\frac{10\pm\sqrt{(-10)^2-4(1)(8)}}{2(1)}=\frac{10\pm\sqrt{68}}{2}$$
≈ 9.1 or 0.9 (disregard)
$x-2=9.1-2=7.1$
It would take Jack 9.1 hours and Lucy 7.1 hours.

30. Let x = initial speed on treadmill. Then $x+1$ = speed increased.

$$t_{\text{initial}}+t_{\text{increased}}=\frac{4}{3}$$
$$\frac{5}{x}+\frac{2}{x+1}=\frac{4}{3}$$
$5\cdot3(x+1)+2\cdot3x=4x(x+1)$
$15x+15+6x=4x^2+4x$
$0=4x^2-17x-15$
$0=(4x+3)(x-5)$
$x=-\frac{4}{3}$ (disregard) or $x=5$
$x+1=5+1=6$
Initial speed: 5 mph
Increased speed: 6 mph

Section 11.4

Practice Exercises

1. $(x-4)(x+3)>0$
Solve the related equation, $(x-4)(x+3)=0$.
$(x-4)(x+3)=0$
$x-4=0$ or $x+3=0$
$x=4$ $x=-3$
Test points in the three regions separated by $x=4$ and $x=-3$.

Region	Test Point	$(x-4)(x+3)>0$ Result
A: $(-\infty, -3)$	-4	$(-8)(-1)>0$ True
B: $(-3, 4)$	0	$(-4)(3)>0$ False
C: $(4, \infty)$	5	$(1)(8)>0$ True

The points in regions A and C satisfy the inequality. The numbers 4 and -3 are not included in the solution since the inequality symbol is >. The solution set is $(-\infty, -3) \cup (4, \infty)$.

2. $x^2 - 8x \le 0$

Solve the related equation, $x^2 - 8x = 0$.

$x^2 - 8x = 0$
$x(x-8) = 0$
$x = 0$ or $x - 8 = 0$
$x = 8$

The numbers 0 and 8 separate the number line into three regions, A, B, and C. Test a point in each region.

Region	Test Point	$x^2 - 8x \le 0$ Result
A: $(-\infty, 0]$	-1	$1 + 8 \le 0$ False
B: $[0, 8]$	1	$1 - 8 \le 0$ True
C: $[8, \infty)$	9	$81 - 72 \le 0$ False

Values in region B satisfy the inequality. The numbers 0 and 8 are included in the solution since the inequality symbol is $\le$. The solution set is $[0, 8]$.

3. $(x+3)(x-2)(x+1) \le 0$
Solve $(x+3)(x-2)(x+1) = 0$ by inspection.
$x = -3$ or $x = 2$ or $x = -1$
These separate the number line into four regions. Test points in each region.

Region	Test Point	$(x+3)(x-2)(x+1) \le 0$ Result
A: $(-\infty, -3]$	-4	$(-1)(-6)(-3) \le 0$ True
B: $[-3, -1]$	-2	$(1)(-4)(-1) \le 0$ False
C: $[-1, 2]$	0	$(3)(-2)(1) \le 0$ True
D: $[2, \infty)$	3	$(6)(1)(4) \le 0$ False

The solution set is $(-\infty, -3] \cup [-1, 2]$. We include the numbers -3, -1, and 2 because the inequality symbol is $\le$.

4. $\frac{x-5}{x+4} \le 0$

$x + 4 = 0$
$x = -4$

$x = -4$ makes the denominator zero. Solve the related equation $\frac{x-5}{x+4} = 0$.

$\frac{x-5}{x+4} = 0$
$x - 5 = 0$
$x = 5$

Test points in the three regions separated by $x = -4$ and $x = 5$.

Region	Test Point	$\frac{x-5}{x+4} \le 0$ Result
A: $(-\infty, -4)$	-5	$\frac{-10}{-1} \le 0$ False
B: $(-4, 5]$	0	$\frac{-5}{4} \le 0$ True
C: $[5, \infty)$	6	$\frac{1}{10} \le 0$ False

The solution set is $(-4, 5]$. The interval includes 5 because 5 satisfies the original inequality. This interval does not include -4, because -4 would make the denominator zero.

5. $\frac{7}{x+3} < 5$

$x+3=0$

$x=-3$

$x=-3$ makes the denominator zero.

Solve $\frac{7}{x+3}=5.$

$$(x+3)\left(\frac{7}{x+3}\right)=5(x+3)$$
$$7=5x+15$$
$$-8=5x$$
$$-\frac{8}{5}=x$$

We use these two solutions to divide the number line into three regions and choose test points.

Region	Test Point	$\frac{7}{x+3}<5$ Result
$A: (-\infty, -3)$	-4	$\frac{7}{-1}<5$ True
$B: \left(-3, -\frac{8}{5}\right)$	-2	$\frac{7}{1}<5$ False
$C: \left(-\frac{8}{5}, \infty\right)$	0	$\frac{7}{3}<5$ True

The solution set is $(-\infty, -3)\cup\left(-\frac{8}{5}, \infty\right)$.

Vocabulary and Readiness Check

1. $[-7, 3)$

2. $(-1, 5]$

3. $(-\infty, 0]$

4. $(-\infty, -8]$

5. $(-\infty, -12)\cup[-10, \infty)$

6. $(-\infty, -3]\cup(4, \infty)$

Exercise Set 11.4

1. $(x+1)(x+5)>0$

$x+1=0$ or $x+5=0$

$x=-1$ or $x=-5$

Region	Test Point	$(x+1)(x+5)>0$ Result
$A: (-\infty, -5)$	-6	$(-5)(-1)>0$ True
$B: (-5, -1)$	-2	$(-1)(3)>0$ False
$C: (-1, \infty)$	0	$(1)(5)>0$ True

Solution: $(-\infty, -5)\cup(-1, \infty)$

3. $(x-3)(x+4)\le 0$

$x-3=0$ or $x+4=0$

$x=3$ or $x=-4$

Region	Test Point	$(x-3)(x+4)\le 0$ Result
$A: (-\infty, -4]$	-5	$(-8)(-1)\le 0$ False
$B: [-4, 3]$	0	$(-3)(4)\le 0$ True
$C: [3, \infty)$	4	$(1)(8)\le 0$ False

Solution: $[-4, 3]$

5. $x^2-7x+10\le 0$

$(x-5)(x-2)\le 0$

$x-5=0$ or $x-2=0$

$x=5$ or $x=2$

Region	Test Point	$(x-5)(x-2)\le 0$ Result
$A: (-\infty, 2]$	0	$(-5)(-2)\le 0$ False
$B: [2, 5]$	3	$(-2)(1)\le 0$ True
$C: [5, \infty)$	6	$(1)(4)\le 0$ False

Solution: $[2, 5]$

7. $3x^2 + 16 < -5$

$3x^2 + 16x + 5 < 0$

$(3x+1)(x+5) < 0$

$3x + 1 = 0$ or $x + 5 = 0$

$x = -\frac{1}{3}$ or $x = -5$

Region	Test Point	$(3x + 1)(x + 5) < 0$ Result
A: $(-\infty, -5)$	-6	$(-17)(-1) < 0$ False
B: $\left(-5, -\frac{1}{3}\right)$	-1	$(-2)(4) < 0$ True
C: $\left(-\frac{1}{3}, \infty\right)$	0	$(1)(5) < 0$ False

Solution: $\left(-5, -\frac{1}{3}\right)$

9. $(x - 6)(x - 4)(x - 2) > 0$

$x - 6 = 0$ or $x - 4 = 0$ or $x - 2 = 0$

$x = 6$ or $x = 4$ or $x = 2$

Region	Test Point	$(x - 6)(x - 4)(x - 2) > 0$ Result
A: $(-\infty, 2)$	0	$(-6)(-4)(-2) > 0$ False
B: $(2, 4)$	3	$(-3)(-1)(1) > 0$ True
C: $(4, 6)$	5	$(-1)(1)(3) > 0$ False
D: $(6, \infty)$	7	$(1)(3)(5) > 0$ True

Solution: $(2, 4) \cup (6, \infty)$

11. $x(x-1)(x+4) \le 0$

$x = 0$ or $x - 1 = 0$ or $x + 4 = 0$

$x = 1$ or $x = -4$

Region	Test Point	$x(x-1)(x+4) \le 0$ Result
A: $(-\infty, -4]$	-5	$-5(-6)(-1) \le 0$ True
B: $[-4, 0]$	-1	$-1(-2)(3) \le 0$ False
C: $[0, 1]$	$\frac{1}{2}$	$\frac{1}{2}\left(-\frac{1}{2}\right)\left(\frac{9}{2}\right) \le 0$ True
D: $[1, \infty)$	2	$2(1)(6) \le 0$ False

Solution: $(-\infty, -4] \cup [0, 1]$

13. $(x^2 - 9)(x^2 - 4) > 0$

$(x+3)(x-3)(x+2)(x-2) > 0$

$x + 3 = 0$ or $x - 3 = 0$ or $x + 2 = 0$ or $x - 2 = 0$

$x = -3$ or $x = 3$ or $x = -2$ or $x = 2$

Region	Test Point	$(x+3)(x-3)(x+2)(x-2) > 0$ Result
A: $(-\infty, -3)$	-4	$(-1)(-7)(-2)(-6) > 0$ True
B: $(-3, -2)$	$-\frac{5}{2}$	$\left(\frac{1}{2}\right)\left(-\frac{11}{2}\right)\left(-\frac{1}{2}\right)\left(-\frac{9}{2}\right) > 0$ False
C: $(-2, 2)$	0	$(3)(-3)(2)(-2) > 0$ True
D: $(2, 3)$	$\frac{5}{2}$	$\left(\frac{11}{2}\right)\left(-\frac{1}{2}\right)\left(\frac{9}{2}\right)\left(\frac{1}{2}\right) > 0$ False
E: $(3, \infty)$	4	$(7)(1)(6)(2) > 0$ True

Solution: $(-\infty, -3) \cup (-2, 2) \cup (3, \infty)$

15. $\frac{x+7}{x-2}<0$

$x+7=0$ or $x-2=0$

$x=-7$ or $x=2$

Region	Test Point	$\frac{x+7}{x-2}<0$ False
$A: (-\infty, -7)$	-8	$\frac{-1}{-10}<0$ False
$B: (-7, 2)$	0	$\frac{7}{-2}<0$ True
$C: (2, \infty)$	3	$\frac{10}{1}<0$ False

Solution: $(-7, 2)$

17. $\frac{5}{x+1}>0$

$x+1=0$

$x=-1$

Region	Test Point	$\frac{5}{x+1}>0$ Result
$A: (-\infty, -1)$	-2	$\frac{5}{-1}>0$ False
$B: (-1, \infty)$	0	$\frac{5}{1}>0$ True

Solution: $(-1, \infty)$

19. $\frac{x+1}{x-4}\geq 0$

$x+1=0$ or $x-4=0$

$x=-1$ or $x=4$

Region	Test Point	$\frac{x+1}{x-4}\geq 0$ Result
$A: (-\infty, -1]$	-2	$\frac{-1}{-6}\geq 0$ True
$B: [-1, 4)$	0	$\frac{1}{-4}\geq 0$ False
$C: (4, \infty)$	5	$\frac{6}{1}\geq 0$ True

Solution: $(-\infty, -1] \cup (4, \infty)$

21. $\frac{3}{x-2}<4$

The denominator is equal to 0 when $x-2=0$, or $x=2$.

$\frac{3}{x-2}=4$

$3=4x-8$

$11=4x$

$\frac{11}{4}=x$

Region	Test Point	$\frac{3}{x-2}<4$ Result
$A: (-\infty, 2)$	0	$\frac{3}{-2}<4$ True
$B: \left(2, \frac{11}{4}\right)$	$\frac{5}{2}$	$\frac{3}{\frac{1}{2}}=6<4$ False
$C: \left(\frac{11}{4}, \infty\right)$	4	$\frac{3}{2}<4$ True

Solution: $(-\infty, 2)\cup\left(\frac{11}{4}, \infty\right)$

23. $\frac{x^2+6}{5x} \geq 1$

The denominator is equal to 0 when $5x = 0$, or $x = 0$.

$$\frac{x^2+6}{5x} = 1$$
$$x^2+6 = 5x$$
$$x^2-5x+6 = 0$$
$$(x-2)(x-3) = 0$$
$$x-2=0 \quad \text{or} \quad x-3=0$$
$$x=2 \quad \text{or} \quad x=3$$

Region	Test Point	$\frac{x^2+6}{5x} \geq 1$ Result
A: $(-\infty, 0)$	-1	$\frac{7}{-5} \geq 1$ False
B: $(0, 2]$	1	$\frac{7}{5} \geq 1$ True
C: $[2, 3]$	$\frac{5}{2}$	$\frac{\frac{49}{4}}{\frac{25}{2}} = \frac{49}{50} \geq 1$ False
D: $[3, \infty)$	4	$\frac{22}{20} \geq 1$ True

Solution: $(0, 2] \cup [3, \infty)$

25. $(x-8)(x+7) > 0$

$$x-8=0 \quad \text{or} \quad x+7=0$$
$$x=8 \quad \text{or} \quad x=-7$$

Region	Test Point	$(x-8)(x+7) > 0$ Result
A: $(-\infty, -7)$	-8	$(-16)(-1) > 0$ True
B: $(-7, 8)$	0	$(-8)(7) > 0$ False
C: $(8, \infty)$	9	$(1)(16) > 0$ True

Solution: $(-\infty, -7) \cup (8, \infty)$

27. $(2x-3)(4x+5) \leq 0$

$$2x-3=0 \quad \text{or} \quad 4x+5=0$$
$$x=\frac{3}{2} \quad \text{or} \quad x=-\frac{5}{4}$$

Region	Test Point	$(2x-3)(4x+5) \leq 0$ Result
A: $\left(-\infty, -\frac{5}{4}\right]$	-2	$(-7)(-3) \leq 0$ False
B: $\left[-\frac{5}{4}, \frac{3}{2}\right]$	0	$(-3)(5) \leq 0$ True
C: $\left[\frac{3}{2}, \infty\right)$	2	$(1)(13) \leq 0$ False

Solution: $\left[-\frac{5}{4}, \frac{3}{2}\right]$

29. $x^2 > x$

$$x^2 - x > 0$$
$$x(x-1) > 0$$
$$x=0 \quad \text{or} \quad x-1=0$$
$$x=1$$

Region	Test Point	$x(x-1) > 0$ Result
A: $(-\infty, 0)$	-1	$-1(-2) > 0$ True
B: $(0, 1)$	$\frac{1}{2}$	$\frac{1}{2}\left(-\frac{1}{2}\right) > 0$ False
C: $(1, \infty)$	2	$2(1) > 0$ True

Solution: $(-\infty, 0) \cup (1, \infty)$

31. $(2x-8)(x+4)(x-6) \le 0$

$2x-8=0$ or $x+4=0$ or $x-6=0$

$x=4$ or $x=-4$ or $x=6$

Region	Test Point	$(2x-8)(x+4)(x-6) \le 0$ Result
A: $(-\infty, -4]$	-5	$(-18)(-1)(-11) \le 0$ True
B: $[-4, 4]$	0	$(-8)(4)(-6) \le 0$ False
C: $[4, 6]$	5	$(2)(9)(-1) \le 0$ True
D: $[6, \infty)$	7	$(6)(11)(1) \le 0$ False

Solution: $(-\infty, -4] \cup [4, 6]$

33.

$$6x^2-5x \ge 6$$
$$6x^2-5x-6 \ge 0$$
$$(3x+2)(2x-3) \ge 0$$

$3x+2=0$ or $2x-3=0$

$x=-\frac{2}{3}$ or $x=\frac{3}{2}$

Region	Test Point	$(3x+2)(2x-3) \ge 0$ Result
A: $\left(-\infty, -\frac{2}{3}\right]$	-1	$(-1)(-5) \ge 0$ True
B: $\left[-\frac{2}{3}, \frac{3}{2}\right]$	0	$(2)(-3) \ge 0$ False
C: $\left[\frac{3}{2}, \infty\right)$	2	$(8)(1) \ge 0$ True

Solution: $\left(-\infty, -\frac{2}{3}\right] \cup \left[\frac{3}{2}, \infty\right)$

35.

$$4x^3+16x^2-9x-36>0$$
$$4x^2(x+4)-9(x+4)>0$$
$$(x+4)(4x^2-9)>0$$
$$(x+4)(2x+3)(2x-3)>0$$

$x+4=0$ or $2x+3=0$ or $2x-3=0$

$x=-4$ or $x=-\frac{3}{2}$ or $x=\frac{3}{2}$

Region	Test Point	$(x+4)(2x+3)(2x-3)>0$
$A: (-\infty, -4)$	-5	$(-1)(-7)(-13)>0$ False
$B: \left(-4, -\frac{3}{2}\right)$	-3	$(1)(-3)(-9)>0$ True
$C: \left(-\frac{3}{2}, \frac{3}{2}\right)$	0	$(4)(3)(-3)>0$ False
$D: \left(\frac{3}{2}, \infty\right)$	4	$(8)(11)(5)>0$ True

Solution: $\left(-4, -\frac{3}{2}\right)\cup\left(\frac{3}{2}, \infty\right)$

37.

$$x^4-26x^2+25\ge 0$$
$$(x^2-25)(x^2-1)\ge 0$$
$$(x+5)(x-5)(x+1)(x-1)\ge 0$$
$$x=-5 \quad\text{or}\quad x=5 \quad\text{or}\quad x=-1 \quad\text{or}\quad x=1$$

Region	Test Point	$(x+5)(x-5)(x+1)(x-1)\ge 0$ Result
$A: (-\infty, -5]$	-6	$(-1)(-11)(-5)(-7)\ge 0$ True
$B: [-5, -1]$	-2	$(3)(-7)(-1)(-3)\ge 0$ False
$C: [-1, 1]$	0	$(5)(-5)(1)(-1)\ge 0$ True
$D: [1, 5]$	2	$(7)(-3)(3)(1)\ge 0$ False
$E: [5, \infty)$	6	$(11)(1)(7)(5)\ge 0$ True

Solution: $(-\infty, -5]\cup[-1, 1]\cup[5, \infty)$

39. $(2x-7)(3x+5) > 0$

$2x-7=0$ or $3x+5=0$

$x=\frac{7}{2}$ or $x=-\frac{5}{3}$

Region	Test Point	$(2x-7)(3x+5) > 0$
$A: \left(-\infty, -\frac{5}{3}\right)$	−2	$(-11)(-1) > 0$ True
$B: \left(-\frac{5}{3}, \frac{7}{2}\right)$	0	$(-7)(5) > 0$ False
$C: \left(\frac{7}{2}, \infty\right)$	4	$(1)(17) > 0$ True

Solution: $\left(-\infty, -\frac{5}{3}\right) \cup \left(\frac{7}{2}, \infty\right)$

41. $\frac{x}{x-10} < 0$

$x=0$ or $x-10=0$

$x=10$

Region	Test Point	$\frac{x}{x-10} < 0$ Result
$A: (-\infty, 0)$	−1	$\frac{-1}{-11} < 0$ False
$B: (0, 10)$	5	$\frac{5}{-5} < 0$ True
$C: (10, \infty)$	11	$\frac{11}{1} < 0$ False

Solution: (0, 10)

43. $\frac{x-5}{x+4} \geq 0$

$x-5=0$ or $x+4=0$

$x=5$ or $x=-4$

Region	Test Point	$\frac{x-5}{x+4} \geq 0$ Result
A: $(-\infty, -4)$	-5	$\frac{-10}{-1} \geq 0$ True
B: $(-4, 5]$	0	$\frac{-5}{4} \geq 0$ False
C: $[5, \infty)$	6	$\frac{1}{10} \geq 0$ True

Solution: $(-\infty, -4) \cup [5, \infty)$

45. $\frac{x(x+6)}{(x-7)(x+1)} \geq 0$

$x = 0$ or $x+6=0$ or $x-7=0$ or $x+1=0$

$x=-6$ or $x=7$ or $x=-1$

Region	Test Point	$\frac{x(x+6)}{(x-7)(x+1)} \geq 0$ Result
A: $(-\infty, -6]$	-7	$\frac{-7(-1)}{(-14)(-6)} \geq 0$ True
B: $[-6, -1)$	-3	$\frac{-3(3)}{(-10)(-2)} \geq 0$ False
C: $(-1, 0]$	$-\frac{1}{2}$	$\frac{-\frac{1}{2}\left(\frac{11}{2}\right)}{\left(-\frac{15}{2}\right)\left(\frac{1}{2}\right)} \geq 0$ True
D: $[0, 7)$	2	$\frac{2(8)}{(-5)(3)} \geq 0$ False
E: $(7, \infty)$	8	$\frac{8(14)}{(1)(9)} \geq 0$ True

Solution: $(-\infty, -6] \cup (-1, 0] \cup (7, \infty)$

47. $\frac{-1}{x-1} > -1$

The denominator is equal to 0 when $x - 1 = 0$, or $x = 1$.

$$\frac{-1}{x-1} = -1$$
$$-1 = -1(x-1)$$
$$-1 = -x+1$$
$$x = 2$$

Region	Test Point	$\frac{-1}{x-1} > -1$ Result
A: $(-\infty, 1)$	0	$\frac{-1}{-1} > -1$ True
B: $(1, 2)$	$\frac{3}{2}$	$\frac{-1}{\frac{1}{2}} = -2 > -1$ False
C: $(2, \infty)$	3	$\frac{-1}{2} > -1$ True

Solution: $(-\infty, 1) \cup (2, \infty)$

49. $\frac{x}{x+4} \le 2$

The denominator is equal to 0 when $x + 4 = 0$, or $x = -4$.

$$\frac{x}{x+4} = 2$$
$$x = 2x+8$$
$$-x = 8$$
$$x = -8$$

Region	Test Point	$\frac{x}{x+4} \le 2$ Result
A: $(-\infty, -8]$	−9	$\frac{-9}{-5} \le 2$ True
B: $[-8, -4)$	−6	$\frac{-6}{-2} \le 2$ False
C: $(-4, \infty)$	0	$\frac{0}{4} \le 2$ True

Solution: $(-\infty, -8] \cup (-4, \infty)$

51. $\dfrac{z}{z-5} \geq 2z$

The denominator is equal to 0 when $z-5=0$, or $z=5$.

$$\frac{z}{z-5} = 2z$$
$$z = 2z(z-5)$$
$$z = 2z^2 - 10z$$
$$0 = 2z^2 - 11z$$
$$0 = z(2z-11)$$
$$z = 0 \quad \text{or} \quad 2z - 11 = 0$$
$$z = \frac{11}{2}$$

Region	Test Point	$\dfrac{z}{z-5} \geq 2z$ Result
A: $(-\infty, 0]$	-1	$\dfrac{-1}{-6} \geq -2$ True
B: $[0, 5)$	1	$\dfrac{1}{-4} \geq 2$ False
C: $\left(5, \dfrac{11}{2}\right]$	$\dfrac{21}{4}$	$\dfrac{\left(\frac{21}{4}\right)}{\left(\frac{1}{4}\right)} \geq \dfrac{21}{2}$ $21 \geq \dfrac{21}{2}$ True
D: $\left[\dfrac{11}{2}, \infty\right)$	6	$\dfrac{6}{1} \geq 12$ False

Solution: $(-\infty, 0] \cup \left(5, \dfrac{11}{2}\right]$

53. $\dfrac{(x+1)^2}{5x} > 0$

The denominator is equal to 0 when $5x = 0$, or $x = 0$.

$$\frac{(x+1)^2}{5x} = 0$$
$$(x+1)^2 = 0$$
$$x + 1 = 0$$
$$x = -1$$

Region	Test Point	$\dfrac{(x+1)^2}{5x} > 0$ Result
A: $(-\infty, -1)$	-2	$\dfrac{1}{-10} > 0$ False
B: $(-1, 0)$	$-\dfrac{1}{2}$	$\dfrac{\left(\frac{1}{4}\right)}{\left(-\frac{5}{2}\right)} > 0$ False
C: $(0, \infty)$	1	$\dfrac{4}{5} > 0$ True

Solution: $(0, \infty)$

55. $g(x) = |x| + 2$

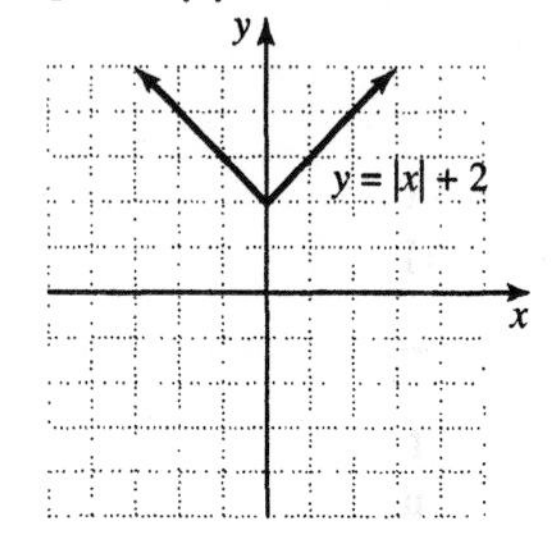

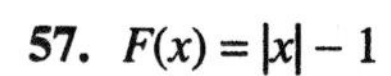

57. $F(x) = |x| - 1$

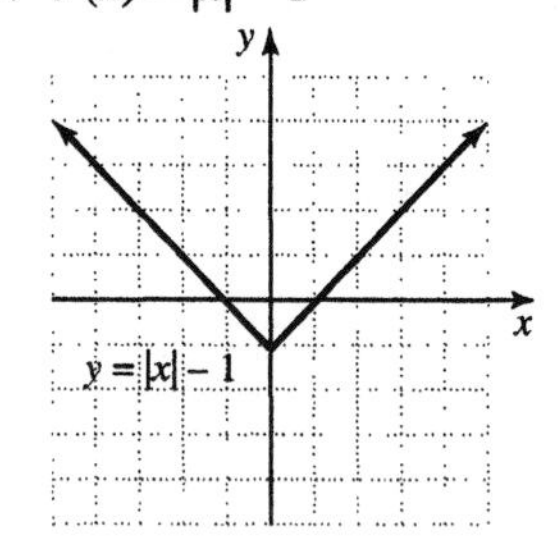

59. $F(x) = x^2 - 3$

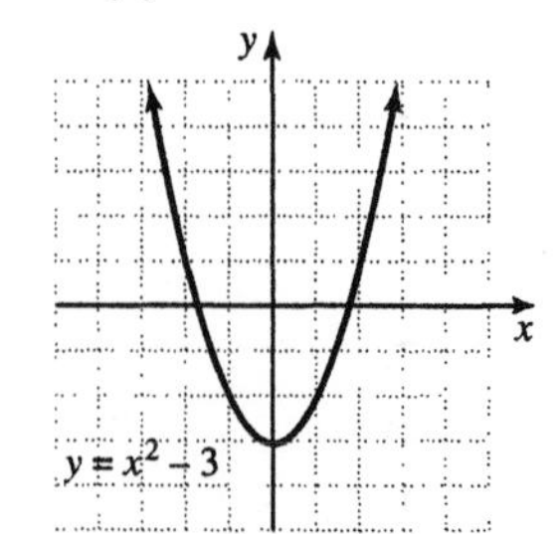

61. $H(x) = x^2 + 1$

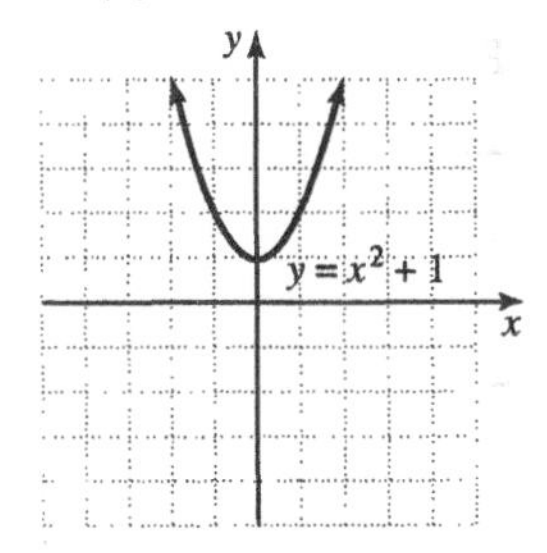

63. Answers may vary

65. Let x = the number. Then

$\frac{1}{x}$ = the reciprocal of the number.

$$x - \frac{1}{x} < 0$$
$$\frac{x^2 - 1}{x} < 0$$
$$\frac{(x+1)(x-1)}{x} < 0$$
$$x + 1 = 0 \quad \text{or} \quad x - 1 = 0 \quad \text{or} \quad x = 0$$
$$x = -1 \quad \text{or} \quad x = 1$$

Region	Test Point	$\frac{(x+1)(x-1)}{x} < 0$ Result
A: $(-\infty, -1)$	-2	$\frac{(-1)(-3)}{-2} < 0$ True
B: $(-1, 0)$	$-\frac{1}{2}$	$\frac{\left(\frac{1}{2}\right)\left(-\frac{3}{2}\right)}{\left(-\frac{1}{2}\right)} < 0$ False
C: $(0, 1)$	$\frac{1}{2}$	$\frac{\left(\frac{3}{2}\right)\left(-\frac{1}{2}\right)}{\left(\frac{1}{2}\right)} < 0$ True
D: $(1, \infty)$	2	$\frac{(3)(1)}{2} < 0$ False

Any number less than -1 or between 0 and 1 and its reciprocal satisfy the conditions.

67. $P(x) = -2x^2 + 26x - 44$

$$-2x^2 + 26x - 44 > 0$$
$$-2(x^2 + 13x - 22) > 0$$
$$-2(x - 11)(x - 2) > 0$$
$$x - 11 = 0 \quad \text{or} \quad x - 2 = 0$$
$$x = 11 \quad \text{or} \quad x = 2$$

Region	Test Point	$-2(x - 11)(x - 2) > 0$ Result
A: $(0, 2)$	1	$-2(-10)(-3) > 0$ False
B: $(2, 11)$	3	$-2(-8)(1) > 0$ True
C: $(11, \infty)$	12	$-2(1)(10) > 0$ False

The company makes a profit when x is between 2 and 11.

69.

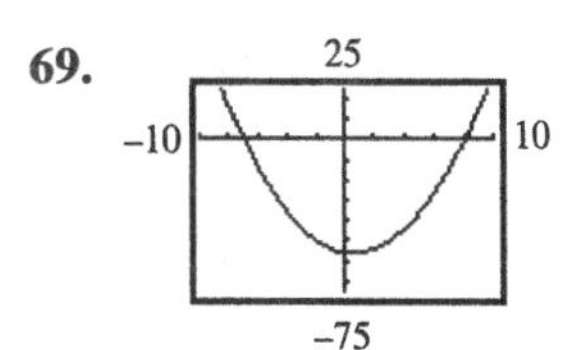

71.

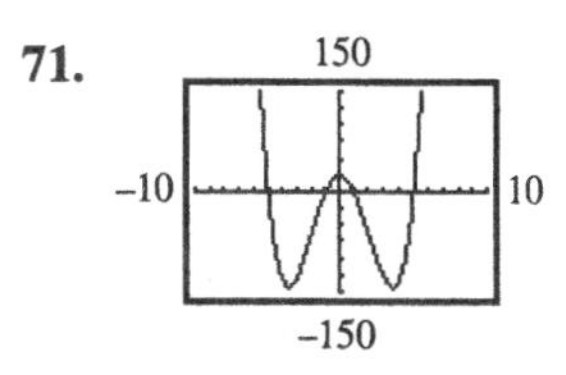

The Bigger Picture

1. $|x - 8| = |2x + 1|$

$$x - 8 = 2x + 1 \quad \text{or} \quad x - 8 = -(2x + 1)$$
$$-9 = x \qquad x - 8 = -2x - 1$$
$$3x = 7$$
$$x = \frac{7}{3}$$

2. $0 < -x + 7 < 3$

$$-7 < -x < -4$$
$$7 > x > 4$$
$$4 < x < 7$$

Solution: $(4, 7)$

3. $\sqrt{3x-11}+3=x$

$\sqrt{3x-11}=x-3$

$3x-11=(x-3)^2$

$3x-11=x^2-6x+9$

$0=x^2-9x+20$

$0=(x-4)(x-5)$

$x-4=0$ or $x-5=0$

$x=4$ $\quad$ $x=5$

The solutions are 4 and 5.

4. $x(3x+1)=1$

$3x^2+x-1=0$

$a=3, b=1, c=-1$

$$x=\frac{-1\pm\sqrt{1^2-4(3)(-1)}}{2(3)}=\frac{-1\pm\sqrt{1+12}}{6}=\frac{-1\pm\sqrt{13}}{6}$$

The solutions are $\frac{-1+\sqrt{13}}{6}$ and $\frac{-1-\sqrt{13}}{6}$.

5. $\frac{x+2}{x-7}\le 0$

$x-7=0$, so $x=7$ makes the denominator 0.

$\frac{x+2}{x-7}=0$

$x+2=0$

$x=-2$

Region	Test Point	$\frac{x+2}{x-7}\le 0$ Result
A: $(-\infty, -2]$	-3	$\frac{-1}{-10}\le 0$ False
B: $[-2, 7)$	0	$\frac{2}{-7}\le 0$
C: $(7, \infty)$	8	$\frac{10}{1}\le 0$ False

Solution: $[-2, 7)$

6. $x(x-6)+4=x^2-2(3-x)$

$x^2-6x+4=x^2-6+2x$

$-6x+4=-6+2x$

$10=8x$

$\frac{5}{4}=x$

The solution is $\frac{5}{4}$.

7. $x(5x-36)=-7$

$5x^2-36x+7=0$

$a=5, b=-36, c=7$

$$x=\frac{-(-36)\pm\sqrt{(-36)^2-4(5)(7)}}{2(5)}=\frac{36\pm\sqrt{1156}}{10}=\frac{36\pm 34}{10}$$

$x=\frac{36+34}{10}=\frac{70}{10}=7$ or $x=\frac{36-34}{10}=\frac{2}{10}=\frac{1}{5}$

The solutions are 7 and $\frac{1}{5}$.

8. $2x^2-4\ge 7x$

Solve $2x^2-4=7x$.

$2x^2-7x-4=0$

$(2x+1)(x-4)=0$

$2x+1=0$ or $x-4=0$

$2x=-1$ $\quad$ $x=4$

$x=-\frac{1}{2}$

Region	Test Point	$2x^2-4\ge 7x$ Result
A: $\left(-\infty, -\frac{1}{2}\right]$	-1	$2-4\ge -7$ True
B: $\left[-\frac{1}{2}, 4\right]$	0	$-4\ge 0$ False
C: $[4, \infty)$	5	$50-4\ge 35$ True

Solution: $\left(-\infty, -\frac{1}{2}\right]\cup[4, \infty)$

9. $\left|\frac{x-7}{3}\right| > 5$

$\frac{x-7}{3} > 5$ or $\frac{x-7}{3} < -5$

$x-7 > 15$ or $x-7 < -15$

$x > 22$ or $x < -8$

Solution: $(-\infty, -8) \cup (22, \infty)$

10. $2(x-5)+4 < 1+7(x-5)-x$

$2x-10+4 < 1+7x-35-x$

$2x-6 < 6x-34$

$28 < 4x$

$7 < x$

Solution: $(7, \infty)$

Section 11.5

Practice Exercises

1. $f(x) = x^2$ and $g(x) = x^2 - 4$

Construct a table of values for $f(x)$ and $g(x)$.

x	$f(x) = x^2$	$g(x) = x^2 - 4$
−2	4	0
−1	1	−3
0	0	−4
1	1	−3
2	4	0

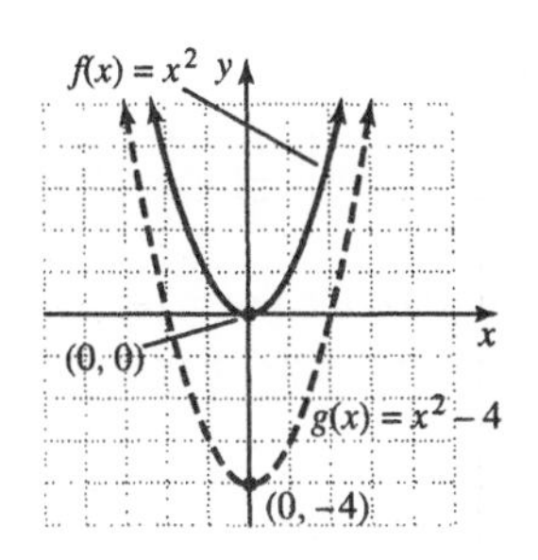

2. a. $f(x) = x^2 - 5$

The graph of $f(x)$ is obtained by shifting the graph of $y = x^2$ downward 5 units.

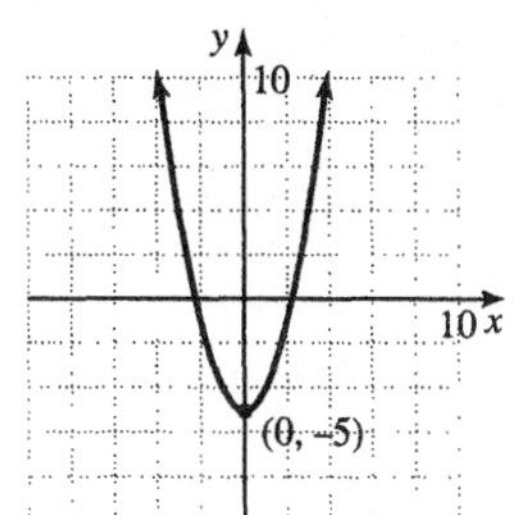

b. $g(x) = x^2 + 3$

The graph of $g(x)$ is obtained by shifting the graph of $y = x^2$ upward 3 units.

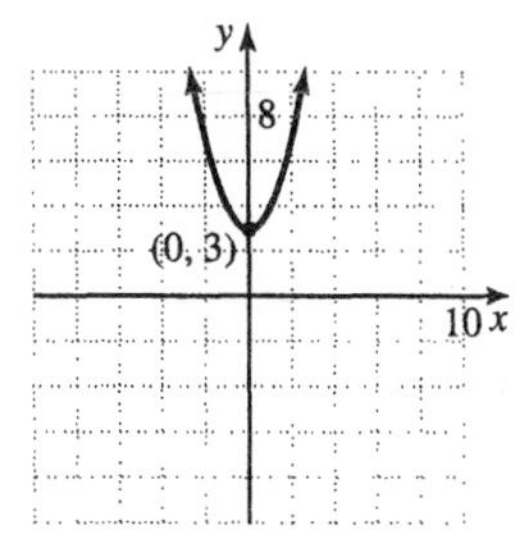

3. $f(x) = x^2$ and $g(x) = (x+6)^2$

Plot points. Notice that the graph of $g(x)$ is the graph of $f(x)$ shifted 6 units to the left.

x	$f(x) = x^2$	x	$g(x) = (x+6)^2$
−2	4	−8	4
−1	1	−7	1
0	0	−6	0
1	1	−5	1
2	4	−4	4

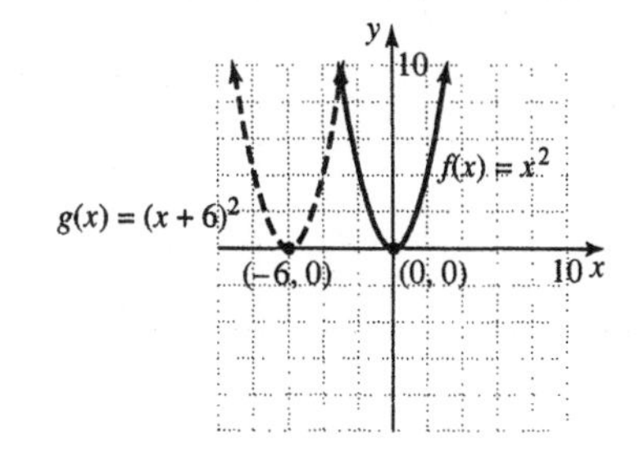

4. a. $G(x) = (x+4)^2$

The graph of $G(x)$ is obtained by shifting the graph of $y = x^2$ to the left 4 units.

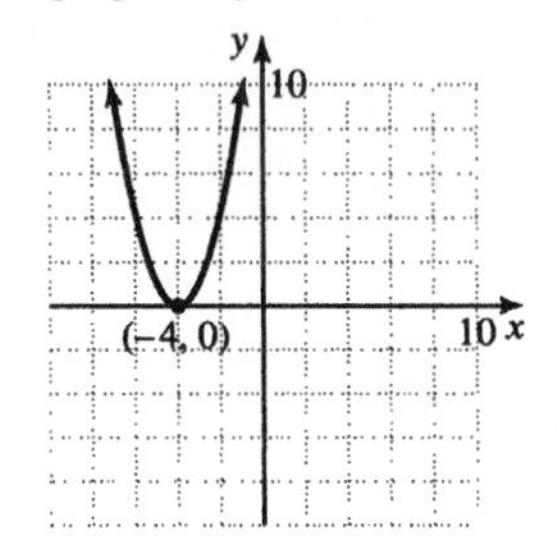

b. $H(x) = (x-7)^2$

The graph of $H(x)$ is obtained by shifting the graph of $y = x^2$ to the right 7 units.

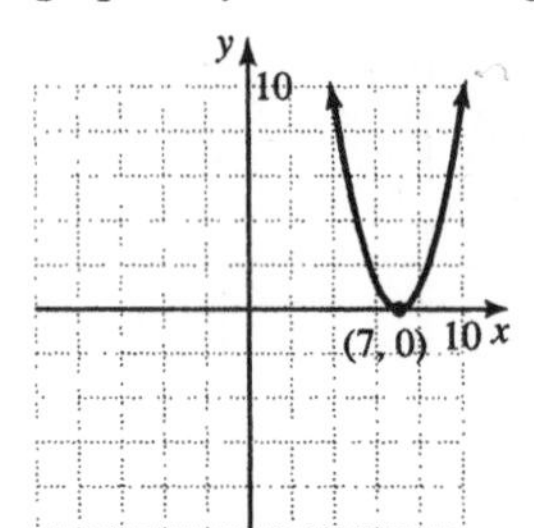

5. $f(x) = (x+2)^2 + 2$

The graph of $f(x)$ is the graph of $y = x^2$ shifted 2 units to the left and 2 units upward. The vertex is then (−2, 2), and the axis of symmetry is $x = -2$.

x	$f(x) = (x+2)^2 + 2$
−4	6
−3	3
−1	3
0	6

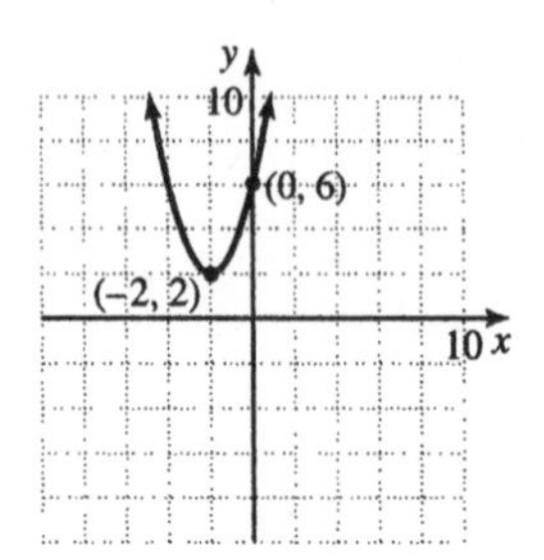

6. $f(x) = x^2$, $g(x) = 4x^2$, and $h(x) = \frac{1}{4}x^2$

Comparing tables of values, we see that for each x-value, the corresponding value of $g(x)$ is four times that of $f(x)$. Similarly, the value of $h(x)$ is one quarter the value of $f(x)$.

x	$f(x) = x^2$	$g(x) = 4x^2$	$h(x) = \frac{1}{4}x^2$
−2	4	16	1
−1	1	4	$\frac{1}{4}$
0	0	0	0
1	1	4	$\frac{1}{4}$
2	4	16	1

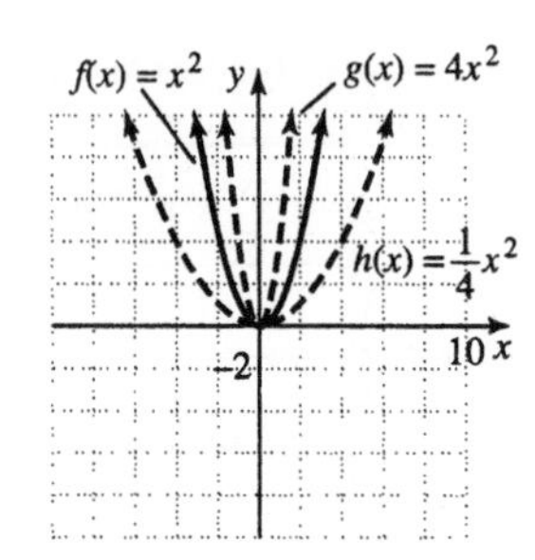

7. $f(x) = -\frac{1}{2}x^2$

Because $a = -\frac{1}{2}$, a negative value, this parabola opens downward. Since $\left|-\frac{1}{2}\right| = \frac{1}{2} < 1$, the parabola is wider than the graph of $y = x^2$. The vertex is (0, 0), and the axis of symmetry is the y-axis.

x	$f(x) = -\frac{1}{2}x^2$
−2	−2
−1	$-\frac{1}{2}$
0	0
1	$-\frac{1}{2}$
2	−2

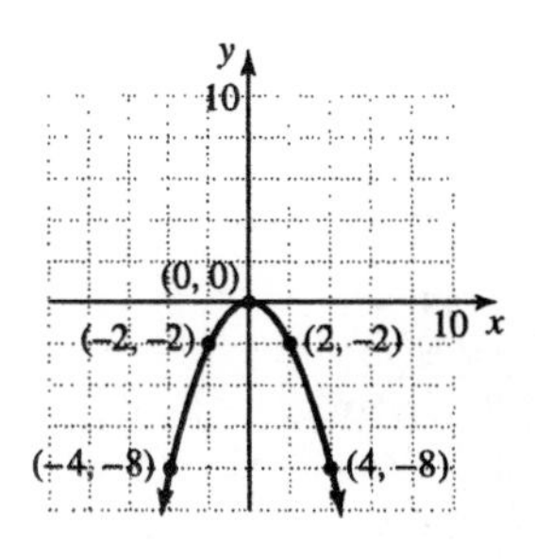

8. $h(x) = \frac{1}{3}(x-4)^2 - 3$

This graph is the same as $y = x^2$ shifted 4 units to the right and 3 units downward, and it is wider because a is $\frac{1}{3}$. The vertex is (4, −3), and the axis of symmetry is $x = 4$.

x	$h(x) = \frac{1}{3}(x-4)^2 - 3$
2	$-\frac{5}{3}$
3	$-\frac{8}{3}$
4	−3
5	$-\frac{8}{3}$
6	$-\frac{5}{3}$

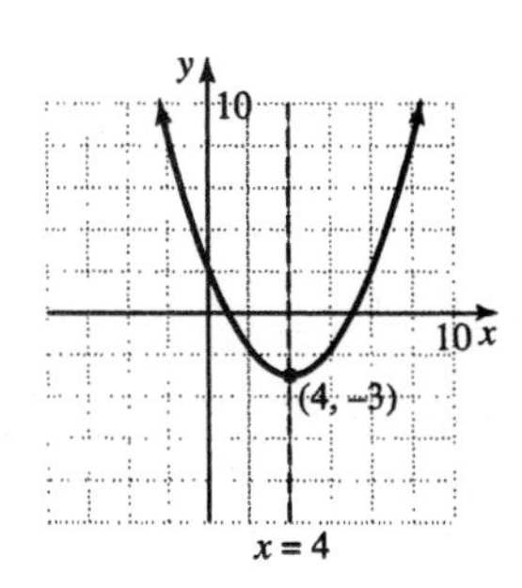

Graphing Calculator Explorations

1.

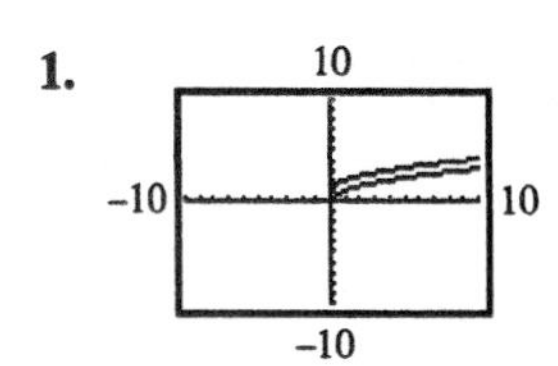

2.

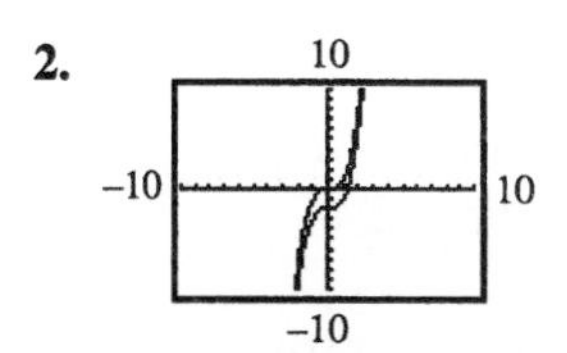

3.

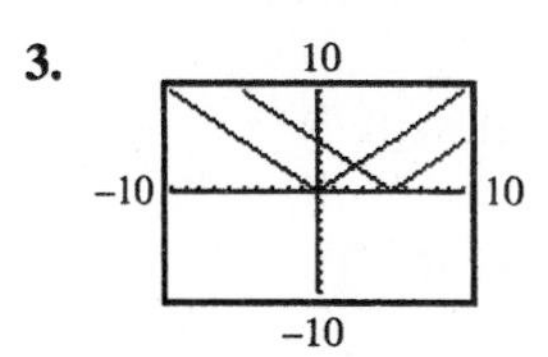

4.

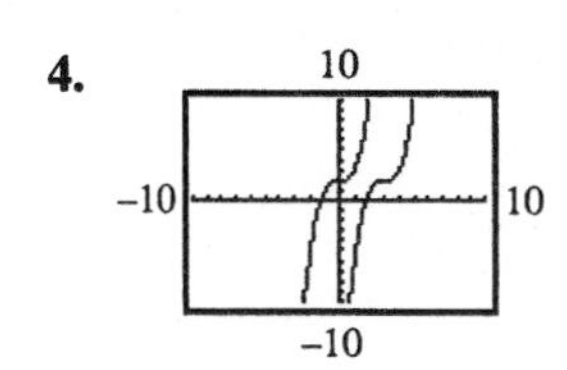

5.

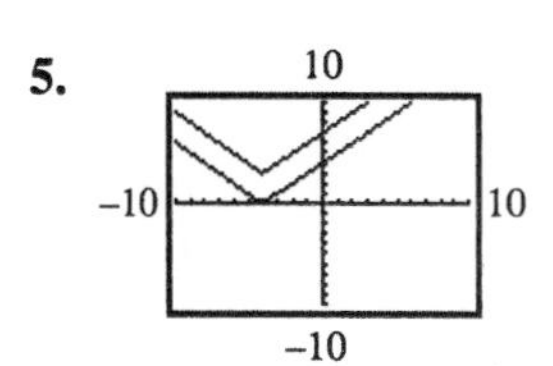

6.

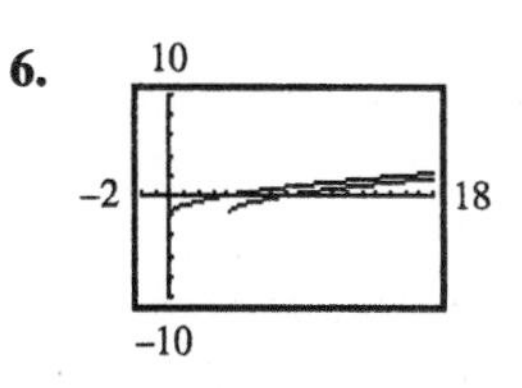

Vocabulary and Readiness Check

1. A quadratic function is one that can be written in the form $f(x) = ax^2 + bx + c,\ a \neq 0$.

2. The graph of a quadratic function is a parabola opening upward or downward.

3. If $a > 0$, the graph of the quadratic function opens upward.

4. If $a < 0$, the graph of the quadratic function opens downward.

5. The vertex of a parabola is the lowest point if $a > 0$.

6. The vertex of a parabola is the highest point if $a < 0$.

7. $f(x) = x^2$; vertex: (0, 0)

8. $f(x) = -5x^2$; vertex: (0, 0)

9. $g(x) = (x-2)^2$; vertex: (2, 0)

10. $g(x) = (x+5)^2$; vertex: (−5, 0)

11. $f(x) = 2x^2 + 3$; vertex: (0, 3)

12. $h(x) = x^2 - 1$; vertex: (0, −1)

13. $g(x) = (x+1)^2 + 5$; vertex: (−1, 5)

14. $h(x) = (x-10)^2 - 7$; vertex: (10, −7)

Exercise Set 11.5

1. $f(x) = x^2 - 1$

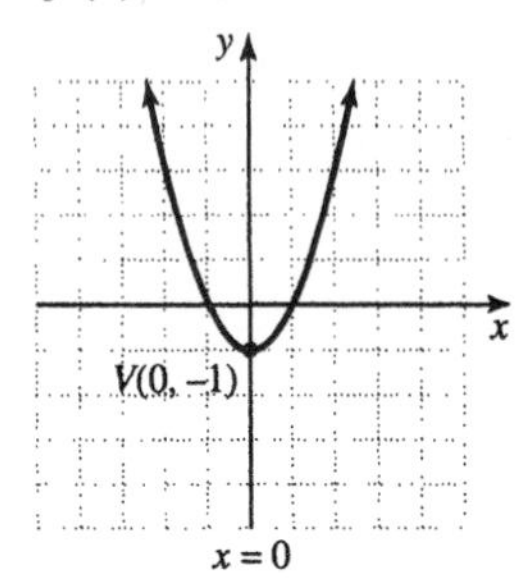

3. $h(x) = x^2 + 5$

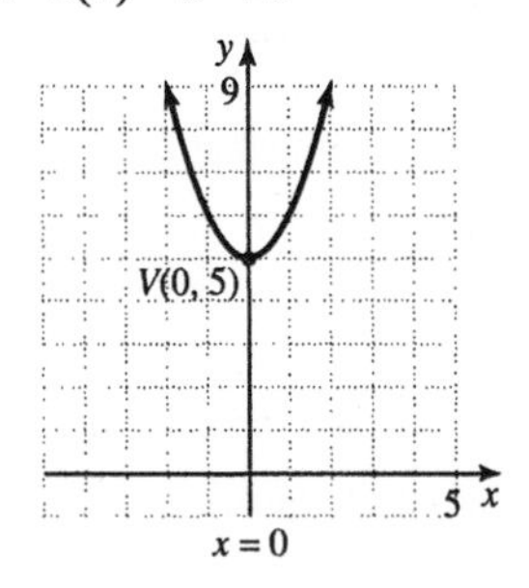

5. $g(x) = x^2 + 7$

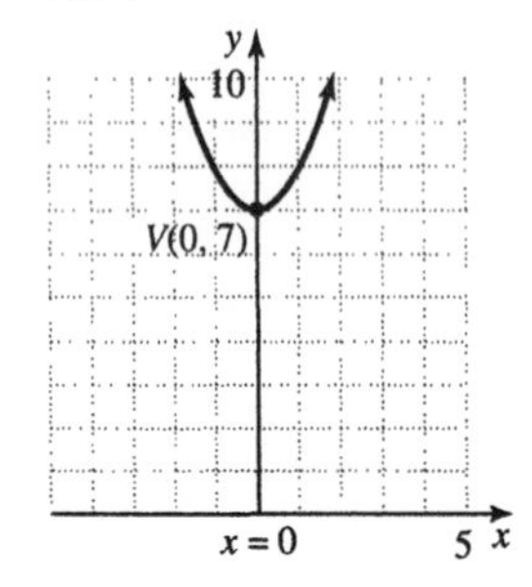

7. $f(x) = (x-5)^2$

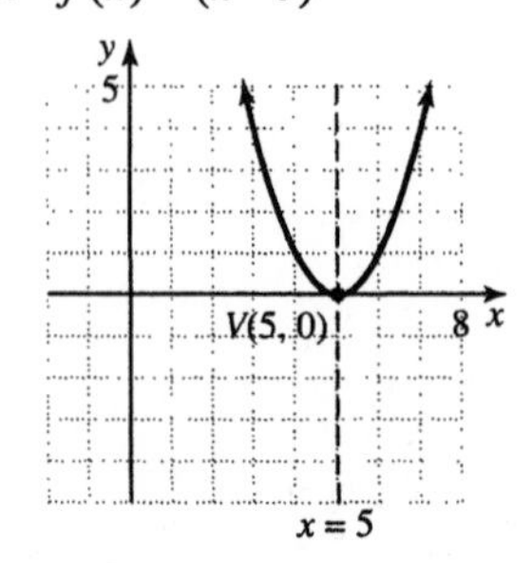

9. $h(x) = (x+2)^2$

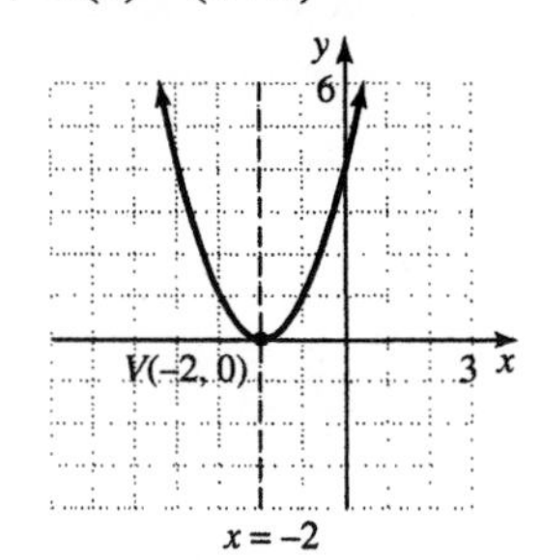

11. $G(x) = (x+3)^2$

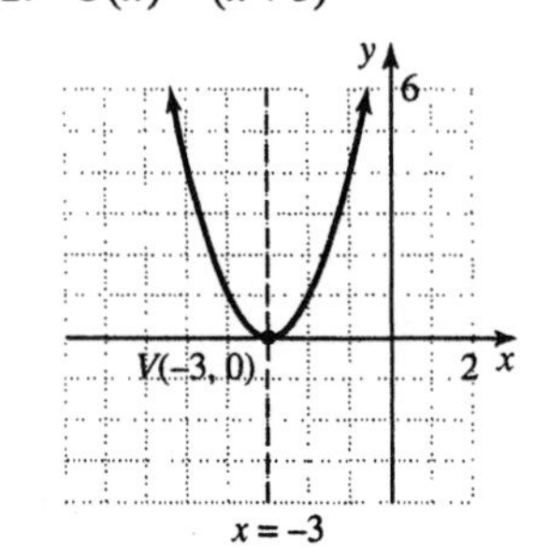

13. $f(x) = (x-2)^2 + 5$

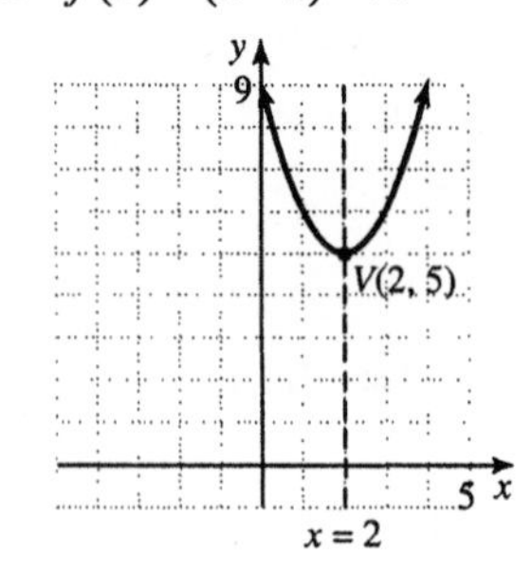

15. $h(x) = (x+1)^2 + 4$

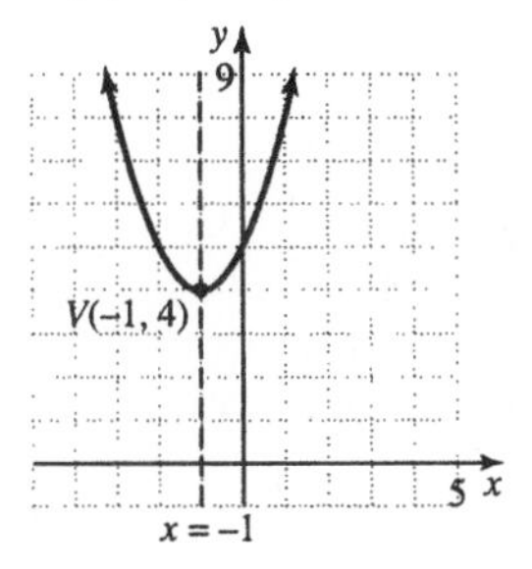

17. $g(x) = (x+2)^2 - 5$

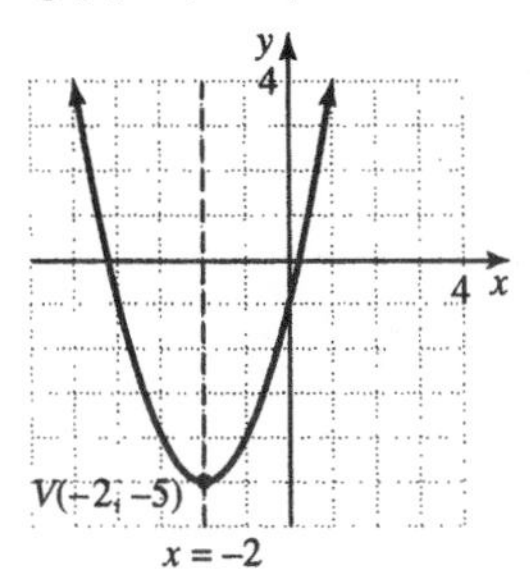

19. $g(x) = -x^2$

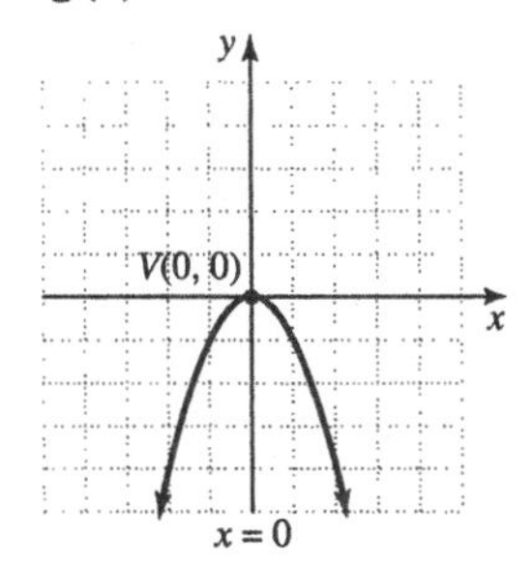

21. $h(x) = \frac{1}{3}x^2$

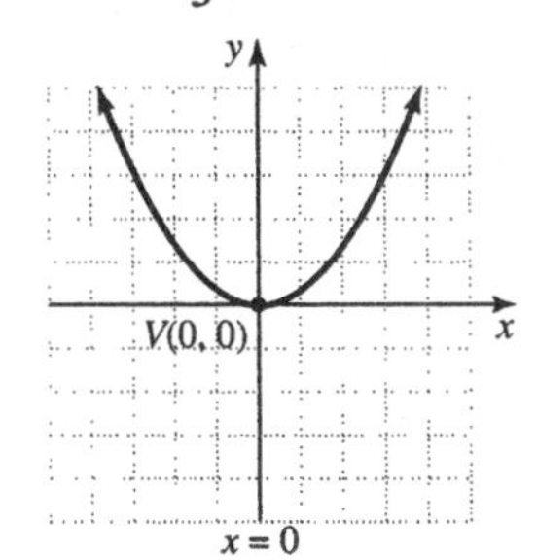

23. $H(x) = 2x^2$

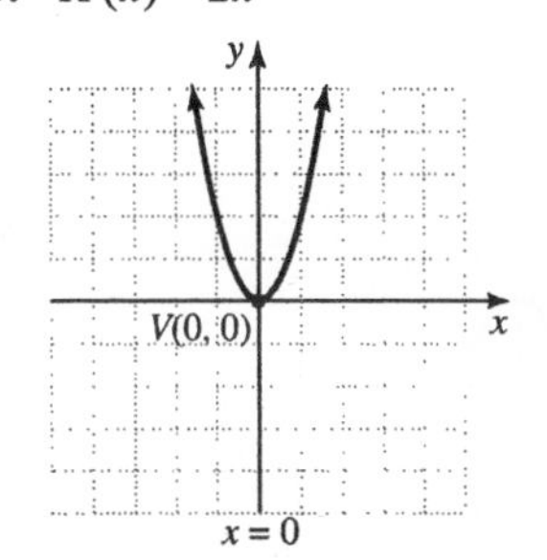

25. $f(x) = 2(x-1)^2 + 3$

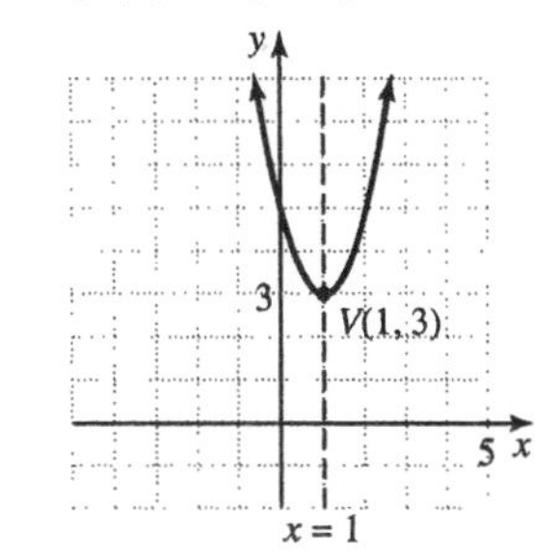

27. $h(x) = -3(x+3)^2 + 1$

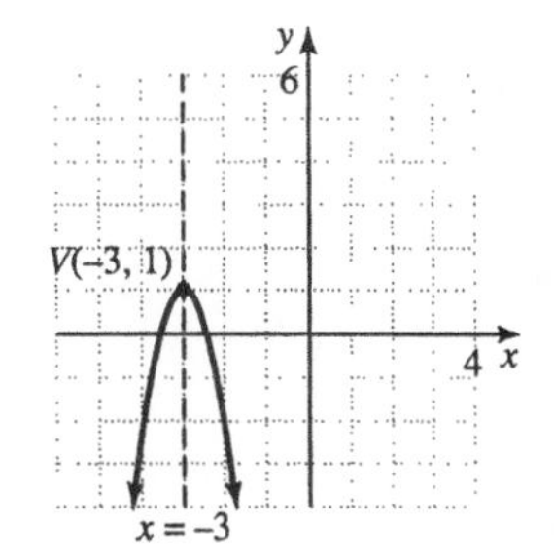

29. $H(x) = \frac{1}{2}(x-6)^2 - 3$

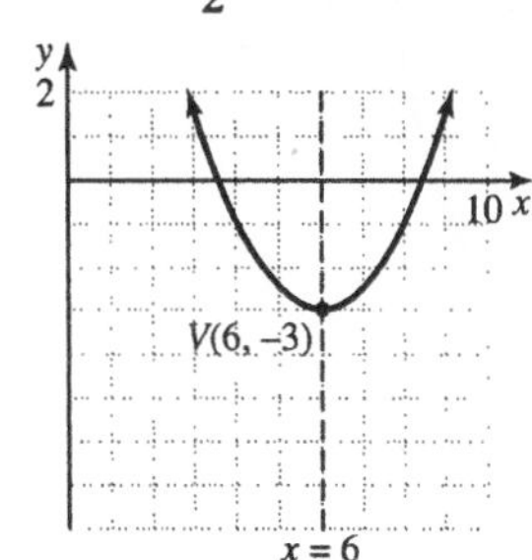

31. $f(x) = -(x-2)^2$

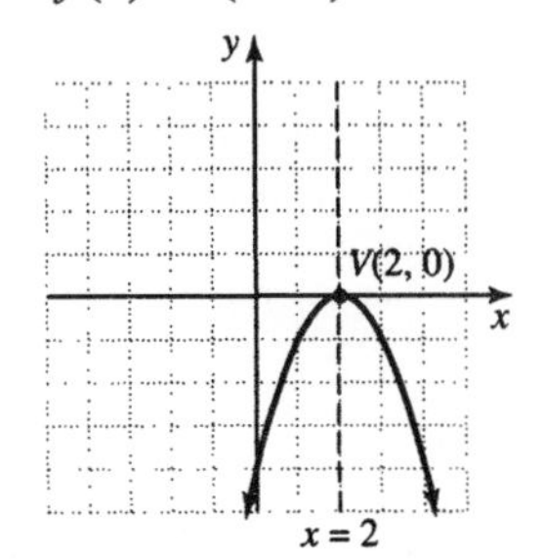

33. $F(x) = -x^2 + 4$

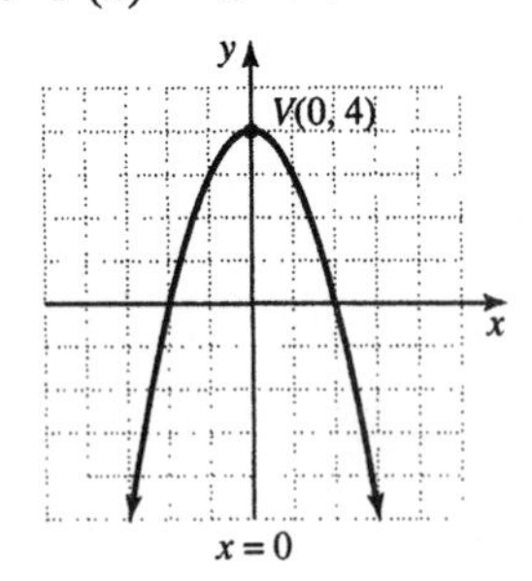

35. $F(x) = 2x^2 - 5$

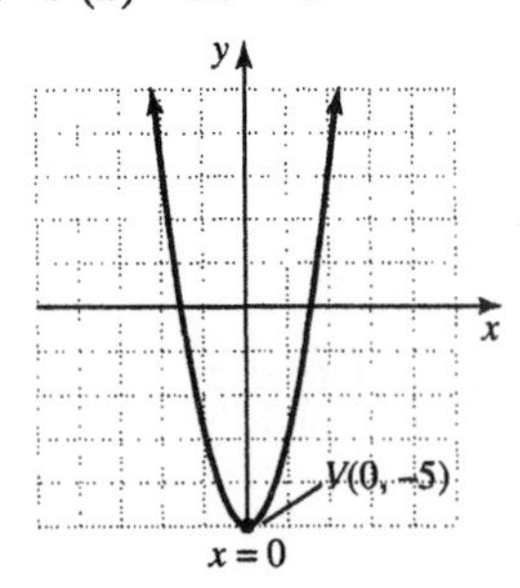

37. $h(x) = (x-6)^2 + 4$

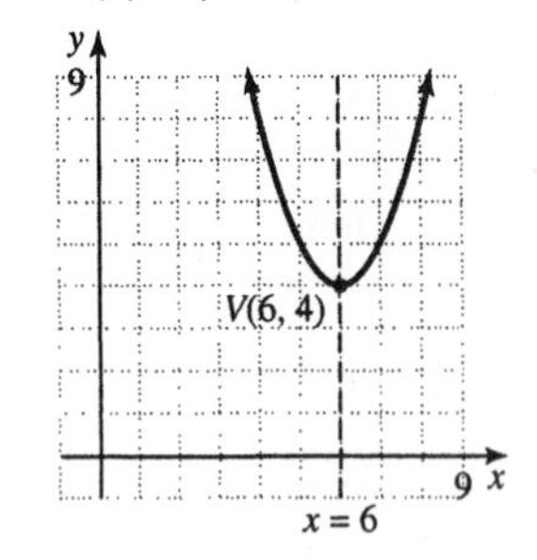

39. $F(x) = \left(x + \frac{1}{2}\right)^2 - 2$

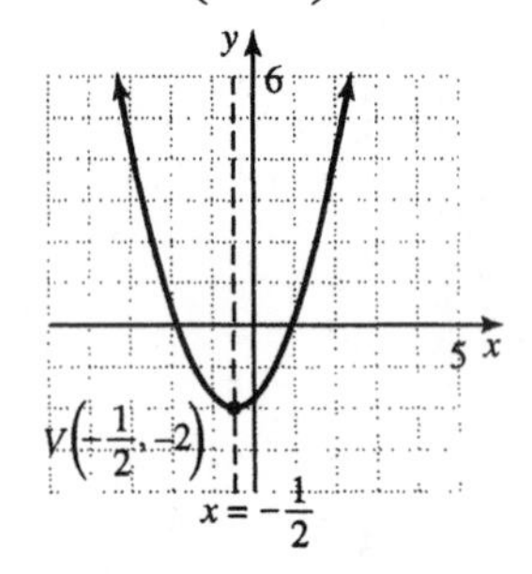

41. $F(x) = \frac{3}{2}(x+7)^2 + 1$

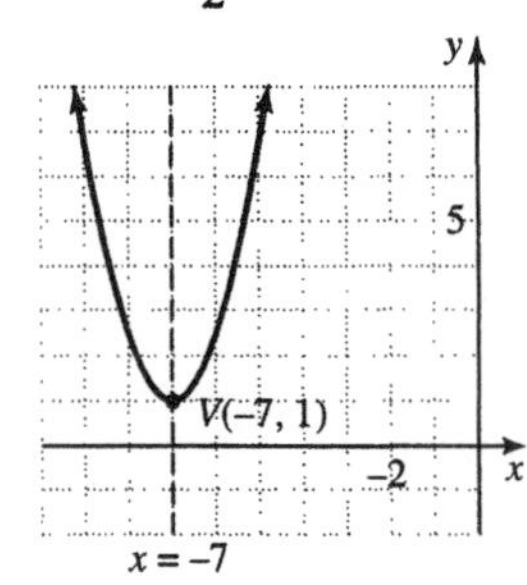

43. $f(x) = \frac{1}{4}x^2 - 9$

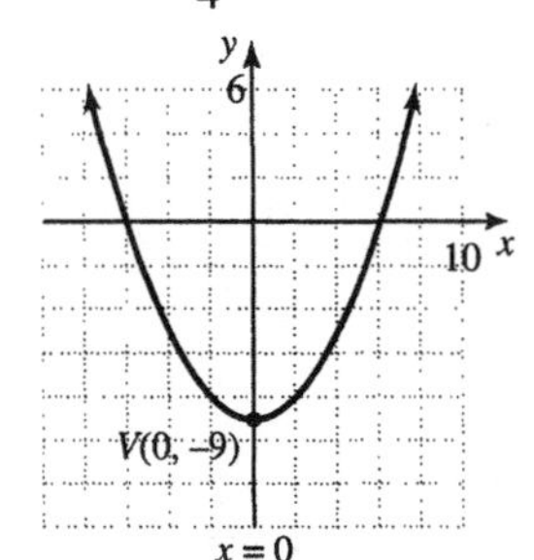

45. $G(x) = 5\left(x + \frac{1}{2}\right)^2$

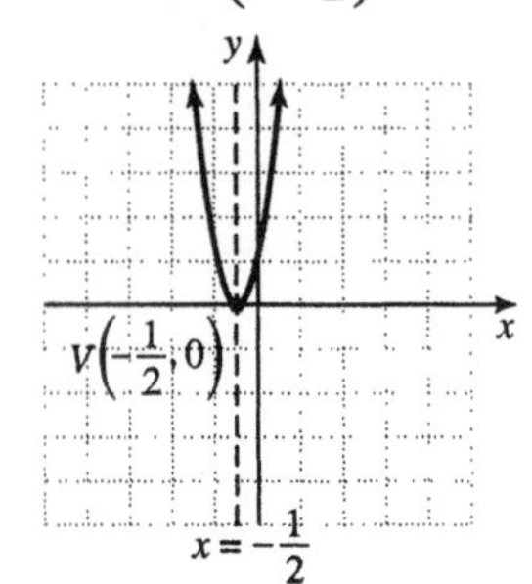

47. $h(x) = -(x-1)^2 - 1$

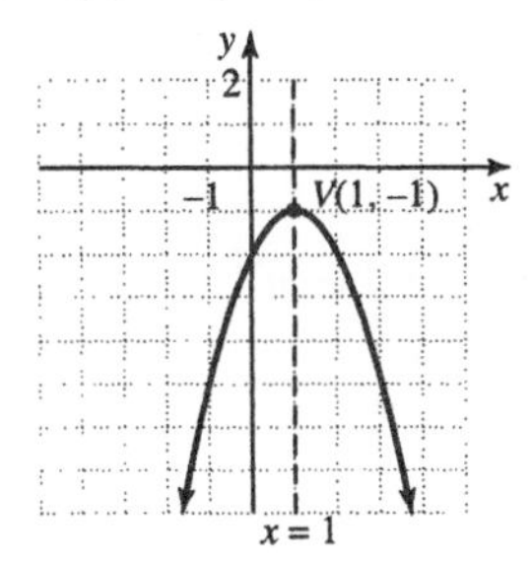

49. $g(x) = \sqrt{3}(x+5)^2 + \frac{3}{4}$

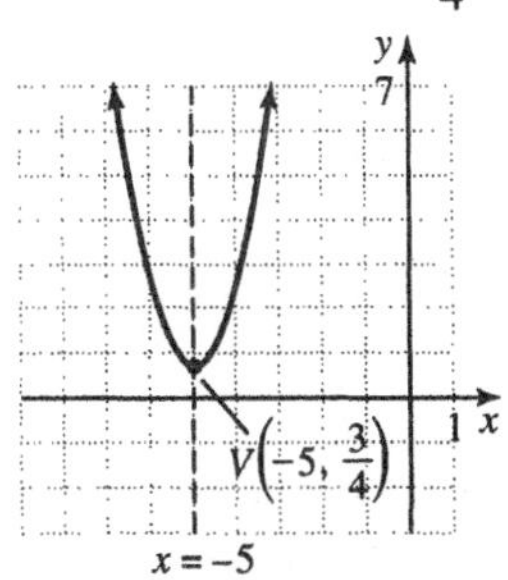

51. $h(x) = 10(x+4)^2 - 6$

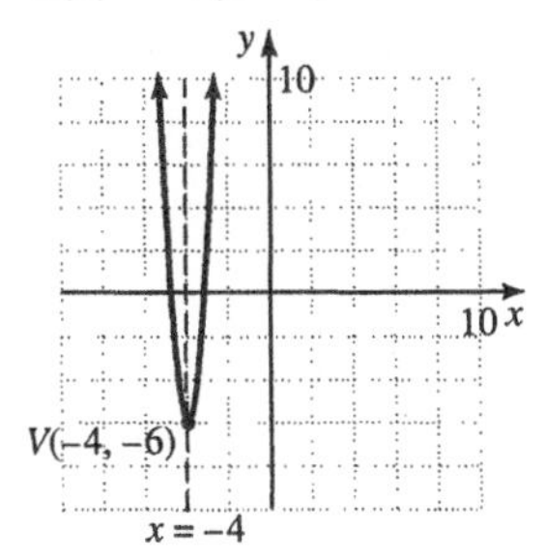

53. $f(x) = -2(x-4)^2 + 5$

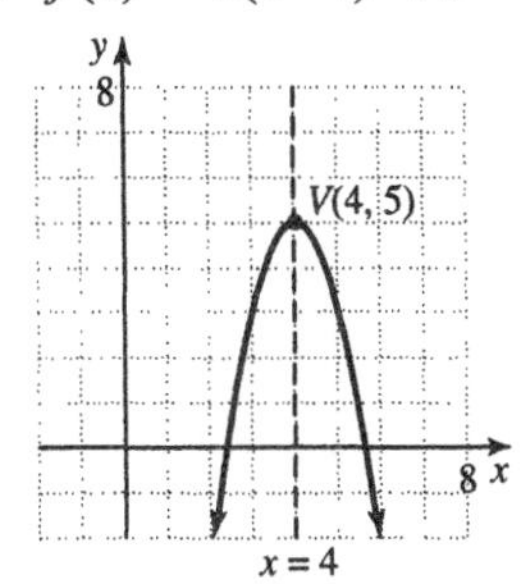

55. $x^2 + 8x$

$\left[\frac{1}{2}(8)\right]^2 = (4)^2 = 16$

$x^2 + 8x + 16$

57. $z^2 - 16z$

$\left[\frac{1}{2}(-16)\right]^2 = (-8)^2 = 64$

$z^2 - 16z + 64$

59. $y^2 + y$

$\left[\frac{1}{2}(1)\right]^2 = \left(\frac{1}{2}\right)^2 = \frac{1}{4}$

$y^2 + y + \frac{1}{4}$

61.
$$\begin{aligned} x^2 + 4x &= 12 \\ x^2 + 4x + \left(\frac{4}{2}\right)^2 &= 12 + 4 \\ x^2 + 4x + 4 &= 16 \\ (x+2)^2 &= 16 \\ x + 2 &= \pm\sqrt{16} \\ x + 2 &= \pm 4 \\ x &= -2 \pm 4 \\ x &= -6 \text{ or } 2 \end{aligned}$$

63.
$$\begin{aligned} z^2 + 10z - 1 &= 0 \\ z^2 + 10z &= 1 \\ z^2 + 10z + \left(\frac{10}{2}\right)^2 &= 1 + 25 \\ z^2 + 10z + 25 &= 26 \\ (z+5)^2 &= 26 \\ z + 5 &= \pm\sqrt{26} \\ z &= -5 \pm \sqrt{26} \end{aligned}$$

65.
$$\begin{aligned} z^2 - 8z &= 2 \\ z^2 - 8z + \left(\frac{-8}{2}\right)^2 &= 2 + 16 \\ z^2 - 8z + 16 &= 18 \\ (z-4)^2 &= 18 \\ z - 4 &= \pm\sqrt{18} \\ z - 4 &= \pm 3\sqrt{2} \\ z &= 4 \pm 3\sqrt{2} \end{aligned}$$

67. $f(x) = -213(x - 0.1)^2 + 3.6$

$a = -213 < 0$, so $f(x)$ opens downward.
The vertex is (0.1, 3.6). The correct answer is **c**.

69. $f(x) = 5(x-2)^2 + 3$

71. $f(x) = 5[x-(-3)]^2 + 6$
$= 5(x+3)^2 + 6$

73. $y = f(x)+1$

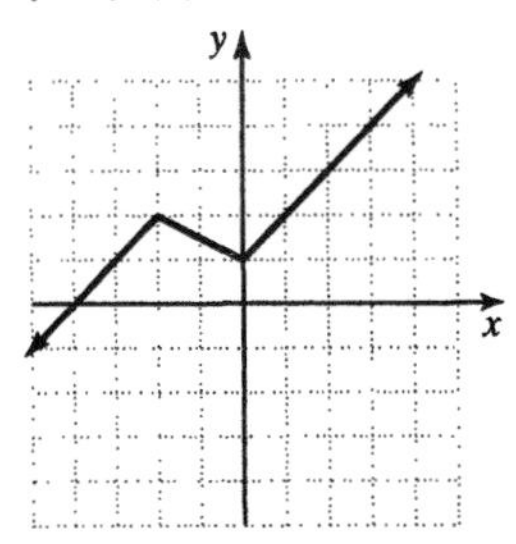

75. $y = f(x-3)$

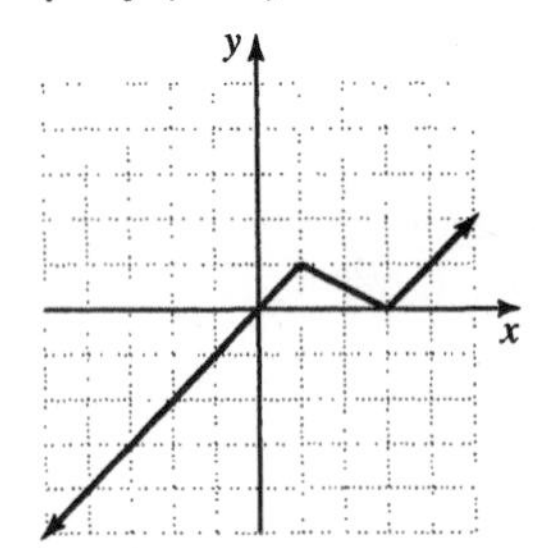

77. $y = f(x+2)+2$

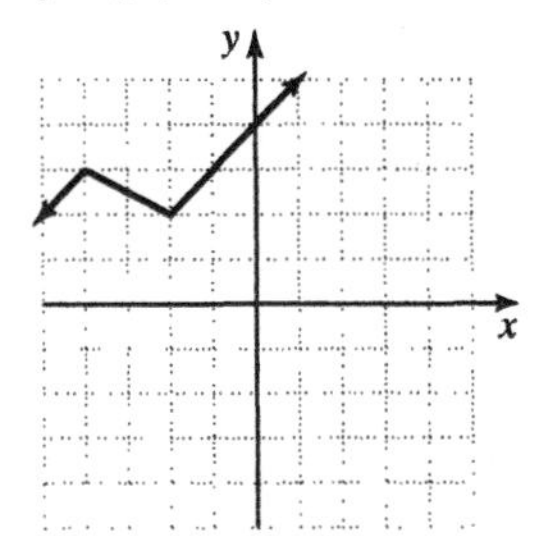

79. $f(x) = 668.7x^2 - 2990.7x + 938$

a. $x = 2004 - 1985 = 19$
$f(19) = 668.7(19)^2 - 2990.7(19) + 938$
$= 185{,}515.4$
There were approximately 185,515 thousand subscribers.

b. $x = 2007 - 1985 = 22$
$f(22) = 668.7(22)^2 - 2990.7(22) + 938$
$= 258{,}793.4$
There were approximately 258,793 thousand subscribers.

Section 11.6

Practice Exercises

1. $g(x) = x^2 - 2x - 3$

Write in the form $y = (x-h)^2 + k$ by completing the square.

$$y = x^2 - 2x - 3$$
$$y + 3 = x^2 - 2x$$
$$y + 3 + \left(\frac{-2}{2}\right)^2 = x^2 - 2x + \left(\frac{-2}{2}\right)^2$$
$$y + 4 = x^2 - 2x + 1$$
$$y = (x-1)^2 - 4$$

The vertex is at (1, –4).
Let $g(x) = 0$.
$0 = x^2 - 2x - 3$
$0 = (x-3)(x+1)$
$x - 3 = 0$ or $x + 1 = 0$
$x = 3$ $x = -1$
The x-intercepts are (3, 0) and (–1, 0).
Let $x = 0$.
$g(0) = 0^2 - 2(0) - 3 = -3$
The y-intercept is (0, –3).

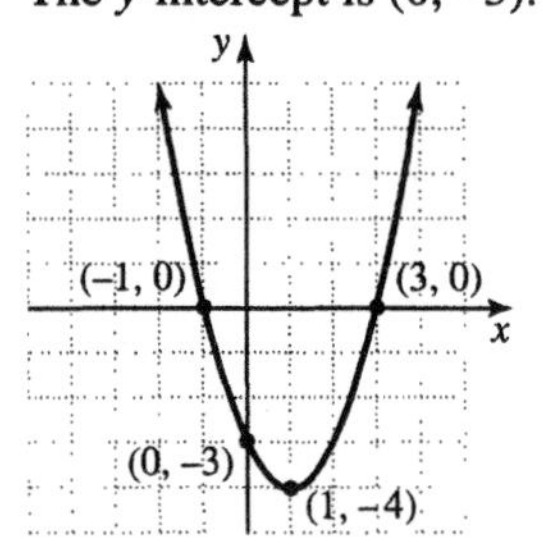

2. $g(x) = 4x^2 + 4x + 3$
Replace $g(x)$ with y and complete the square to write the equation in the form $y = a(x-h)^2 + k$.

$$y = 4x^2 + 4x + 3$$
$$y - 3 = 4x^2 + 4x = 4(x^2 + x)$$
$$y - 3 + 4\left(\frac{1}{2}\right)^2 = 4\left[x^2 + x + \left(\frac{1}{2}\right)^2\right]$$
$$y - 3 + 1 = 4\left(x^2 + x + \frac{1}{4}\right)$$
$$y = 4\left(x + \frac{1}{2}\right)^2 + 2$$

$a = 4$, $h = -\frac{1}{2}$, and $k = 2$.

The parabola opens upward with vertex $\left(-\frac{1}{2}, 2\right)$, and has an axis of symmetry $x = -\frac{1}{2}$.

Let $x = 0$.

$g(0) = 4(0)^2 + 4(0) + 3 = 3$

The y-intercept is (0, 3). There are no x-intercepts.

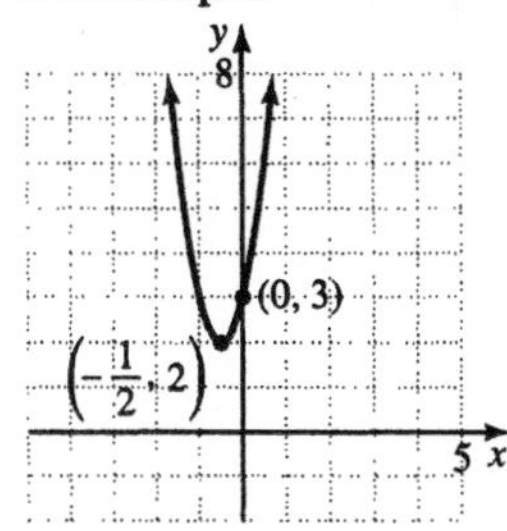

3. $g(x) = -x^2 + 5x + 6$

Write $g(x)$ in the form $a(x-h)^2 + k$ by completing the square. Replace $g(x)$ with y.

$$y = -x^2 + 5x + 6$$
$$y - 6 = -x^2 + 5x$$
$$y - 6 = -1(x^2 - 5x)$$
$$y - 6 - \left(\frac{-5}{2}\right)^2 = -1\left[x^2 - 5x + \left(\frac{-5}{2}\right)^2\right]$$
$$y - 6 - \frac{25}{4} = -1\left(x^2 - 5x + \frac{25}{4}\right)$$
$$y - \frac{49}{4} = -\left(x - \frac{5}{2}\right)^2$$
$$y = -\left(x - \frac{5}{2}\right)^2 + \frac{49}{4}$$

Since $a = -1$, the parabola opens downward with vertex $\left(\frac{5}{2}, \frac{49}{4}\right)$ and axis of symmetry $x = \frac{5}{2}$.

Let $x = 0$.

$y = -0^2 + 5(0) + 6 = 6$

The y-intercept is (0, 6). Let $y = 0$.

$0 = -x^2 + 5x + 6$

$0 = x^2 - 5x - 6$

$0 = (x-6)(x+1)$

$x - 6 = 0$ or $x + 1 = 0$

$x = 6$ $\quad$ $x = -1$

The x-intercepts are (6, 0) and (−1, 0).

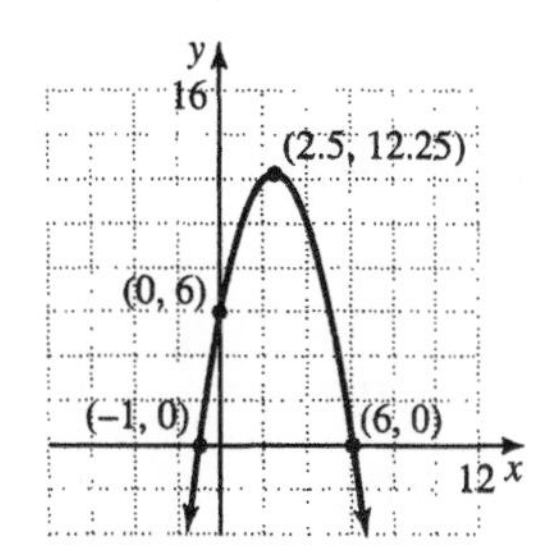

4. $g(x) = x^2 - 2x - 3$

$a = 1$, $b = -2$, and $c = -3$

$\frac{-b}{2a} = \frac{-(-2)}{2(1)} = \frac{2}{2} = 1$

The x-value of the vertex is 1.

$g(1) = 1^2 - 2(1) - 3 = 1 - 2 - 3 = -4$

The vertex is (1, −4).

5. $h(t) = -16t^2 + 24t$

Find the vertex of $h(t)$ to find its maximum value.

$a = -16$, $b = 24$, and $c = 0$

$\frac{-b}{2a} = \frac{-24}{2(-16)} = \frac{3}{4}$

The t-value of the vertex is $\frac{3}{4}$.

$$h\left(\frac{3}{4}\right) = -16\left(\frac{3}{4}\right)^2 + 24\left(\frac{3}{4}\right)$$
$$= -16\left(\frac{9}{16}\right) + 18$$
$$= -9 + 18$$
$$= 9$$

The vertex is $\left(\frac{3}{4}, 9\right)$. Thus, the ball reaches its maximum height of 9 feet in $\frac{3}{4}$ second.

Vocabulary and Readiness Check

1. If a quadratic function is in the form $f(x) = a(x-h)^2 + k$, the vertex of its graph is $\underline{(h, k)}$.

2. The graph of $f(x) = ax^2 + bx + c$, $a \neq 0$ is a parabola whose vertex has x-value of $\underline{\frac{-b}{2a}}$.

	Parabola Opens	*Vertex Location*	*Number of x-intercept(s)*	*Number of y-intercept(s)*
3.	up	Q I	0	1
4.	up	Q III	2	1
5.	down	Q II	2	1
6.	down	Q IV	0	1
7.	up	*x*-axis	1	1
8.	down	*x*-axis	1	1
9.	down	Q III	0	
10.	down	Q I	2	
11.	up	Q IV	2	
12.	up	Q II	0	

Exercise Set 11.6

1. $f(x) = x^2 + 8x + 7$

$-\frac{b}{2a} = \frac{-8}{2(1)} = -4$ and

$$\begin{aligned} f(-4) &= (-4)^2 + 8(-4) + 7 \\ &= 16 - 32 + 7 \\ &= -9 \end{aligned}$$

Thus, the vertex is (–4, –9).

3. $f(x) = -x^2 + 10x + 5$

$-\frac{b}{2a} = \frac{-10}{2(-1)} = 5$ and

$$\begin{aligned} f(5) &= -(5)^2 + 10(5) + 5 \\ &= -25 + 50 + 5 \\ &= 30 \end{aligned}$$

Thus, the vertex is (5, 30).

5. $f(x) = 5x^2 - 10x + 3$

$-\frac{b}{2a} = \frac{-(-10)}{2(5)} = 1$ and

$$\begin{aligned} f(1) &= 5(1)^2 - 10(1) + 3 \\ &= 5 - 10 + 3 \\ &= -2 \end{aligned}$$

Thus, the vertex is (1, –2).

7. $f(x) = -x^2 + x + 1$

$-\frac{b}{2a} = \frac{-1}{2(-1)} = \frac{1}{2}$ and

$$f\left(\frac{1}{2}\right) = -\left(\frac{1}{2}\right)^2 + \left(\frac{1}{2}\right) + 1$$
$$= -\frac{1}{4} + \frac{1}{2} + 1$$
$$= \frac{5}{4}$$

Thus, the vertex is $\left(\frac{1}{2}, \frac{5}{4}\right)$.

9. $f(x) = x^2 - 4x + 3$

$-\frac{b}{2a} = \frac{-(-4)}{2(1)} = 2$ and

$f(2) = (2)^2 - 4(2) + 3 = -1$

The vertex is (2, –1), so the graph is D.

11. $f(x) = x^2 - 2x - 3$

$-\frac{b}{2a} = \frac{-(-2)}{2(1)} = 1$ and

$f(1) = (1)^2 - 2(1) - 3 = -4$

The vertex is (1, –4), so the graph is B.

13. $f(x) = x^2 + 4x - 5$

$-\frac{b}{2a} = \frac{-4}{2(1)} = -2$ and

$f(-2) = (-2)^2 + 4(-2) - 5 = -9$

Thus, the vertex is (–2, –9).

The graph opens upward ($a = 1 > 0$).

$$x^2 + 4x - 5 = 0$$
$$(x+5)(x-1) = 0$$
$$x+5 = 0 \quad \text{or} \quad x-1 = 0$$
$$x = -5 \quad \text{or} \quad x = 1$$

x-intercepts: (–5, 0) and (1, 0).

$f(0) = -5$, so the y-intercept is (0, –5).

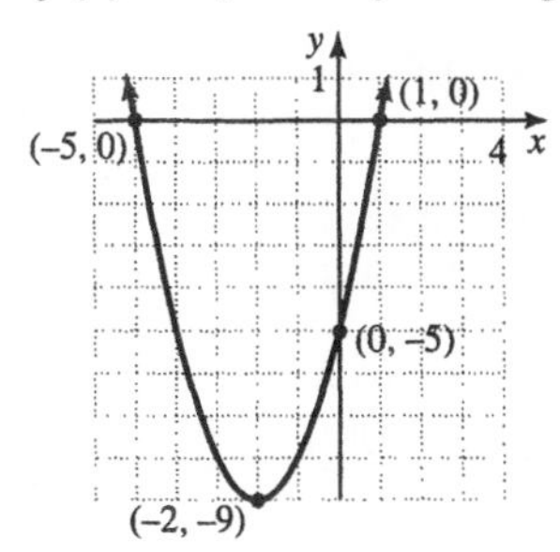

15. $f(x) = -x^2 + 2x - 1$

$-\frac{b}{2a} = \frac{-2}{2(-1)} = 1$ and

$f(1) = -(1)^2 + 2(1) - 1 = 0$

Thus, the vertex is (1, 0).

The graph opens downward ($a = -1 < 0$).

$$-x^2 + 2x - 1 = 0$$
$$x^2 - 2x + 1 = 0$$
$$(x-1)^2 = 0$$
$$x - 1 = 0$$
$$x = 1$$

x-intercept: (1, 0).

$f(0) = -1$, so the y-intercept is (0, –1).

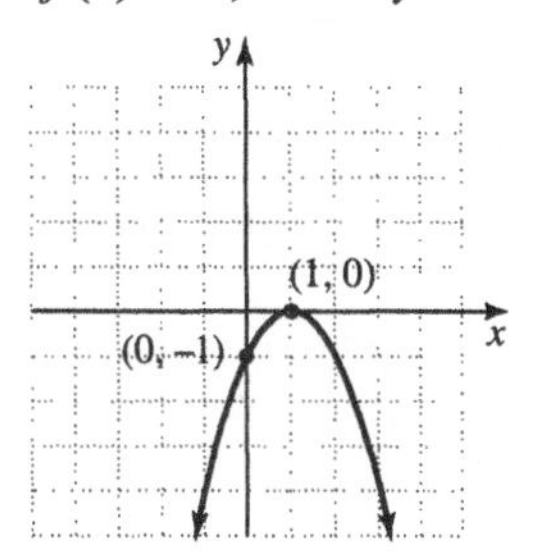

17. $f(x) = x^2 - 4$

$-\frac{b}{2a} = \frac{-0}{2(1)} = 0$ and

$f(0) = (0)^2 - 4 = -4$

Thus, the vertex is (0, –4).

The graph opens upward ($a = 1 > 0$).

$$x^2 - 4 = 0$$
$$(x+2)(x-2) = 0$$
$$x+2 = 0 \quad \text{or} \quad x-2 = 0$$
$$x = -2 \quad \text{or} \quad x = 2$$

x-intercepts: (–2, 0) and (2, 0).

$f(0) = -4$, so the y-intercept is (0, –4).

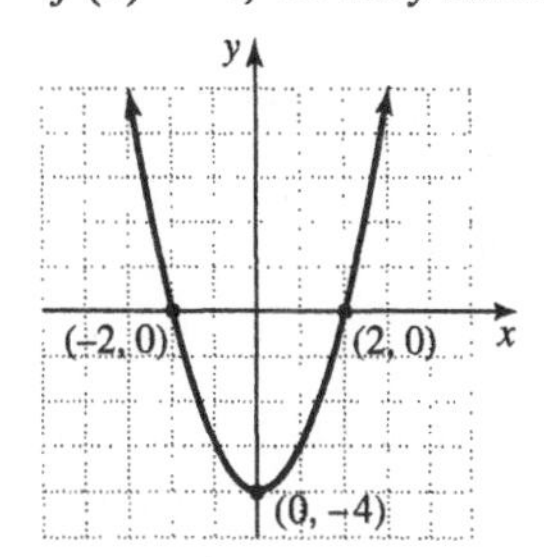

19. $f(x) = 4x^2 + 4x - 3$

$-\frac{b}{2a} = \frac{-4}{2(4)} = -\frac{1}{2}$ and

$f\left(-\frac{1}{2}\right) = 4\left(-\frac{1}{2}\right)^2 + 4\left(-\frac{1}{2}\right) - 3 = -4$

Thus, the vertex is $\left(-\frac{1}{2}, -4\right)$.

The graph opens upward ($a = 4 > 0$).

$4x^2 + 4x - 3 = 0$

$(2x+3)(2x-1) = 0$

$2x + 3 = 0$ or $2x - 1 = 0$

$x = -\frac{3}{2}$ or $x = \frac{1}{2}$

x-intercepts: $\left(-\frac{3}{2}, 0\right)$ and $\left(\frac{1}{2}, 0\right)$.

$f(0) = -3$, so the y-intercept is $(0, -3)$.

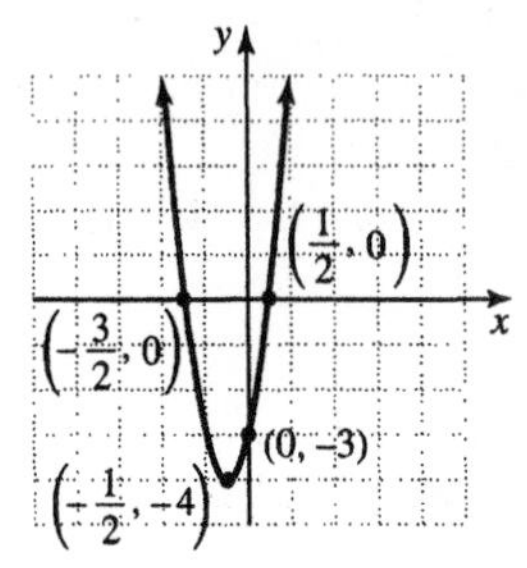

21. $f(x) = x^2 + 8x + 15$

$y = x^2 + 8x + 15$

$y - 15 = x^2 + 8x$

$y - 15 + 16 = x^2 + 8x + 16$

$y + 1 = (x+4)^2$

$y = (x+4)^2 - 1$

$f(x) = (x+4)^2 - 1$

Thus, the vertex is $(-4, -1)$.

The graph opens upward ($a = 1 > 0$).

$x^2 + 8x + 15 = 0$

$(x+5)(x+3) = 0$

$x + 5 = 0$ or $x + 3 = 0$

$x = -5$ or $x = -3$

x-intercepts: $(-5, 0)$ and $(-3, 0)$.

$f(0) = 15$, so the y-intercept is $(0, 15)$.

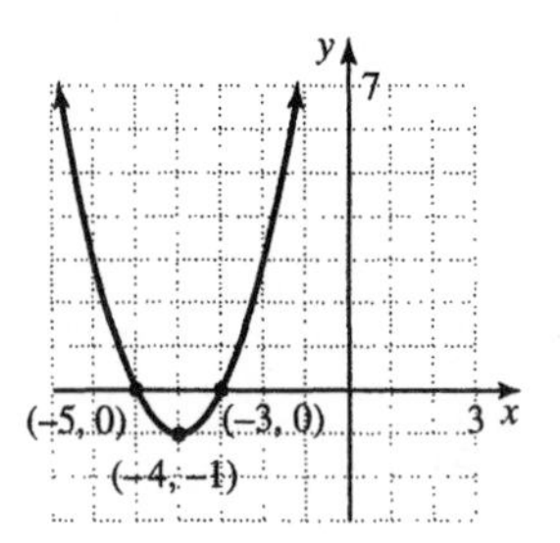

23. $f(x) = x^2 - 6x + 5$

$y = x^2 - 6x + 5$

$y - 5 = x^2 - 6x$

$y - 5 + 9 = x^2 - 6x + 9$

$y + 4 = (x-3)^2$

$y = (x-3)^2 - 4$

$f(x) = (x-3)^2 - 4$

Thus, the vertex is $(3, -4)$.

The graph opens upward ($a = 1 > 0$).

$x^2 - 6x + 5 = 0$

$(x-5)(x-1) = 0$

$x = 5$ or $x = 1$

x-intercepts: $(5, 0)$ and $(1, 0)$.

$f(0) = 5$, so the y-intercept is $(0, 5)$.

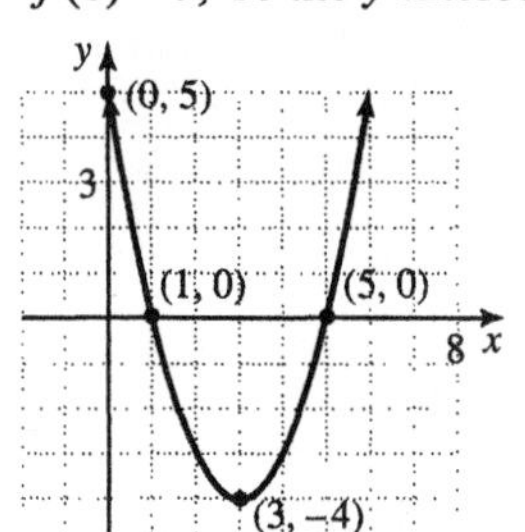

25. $f(x) = x^2 - 4x + 5$

$y = x^2 - 4x + 5$

$y - 5 = x^2 - 4x$

$y - 5 + 4 = x^2 - 4x + 4$

$y - 1 = (x-2)^2$

$y = (x-2)^2 + 1$

$f(x) = (x-2)^2 + 1$

Thus, the vertex is $(2, 1)$.

The graph opens upward ($a = 1 > 0$).

$x^2 - 4x + 5 = 0$

$$x = \frac{4 \pm \sqrt{(-4)^2 - 4(1)(5)}}{2(1)} = \frac{4 \pm \sqrt{-4}}{2}$$

which give non-real solutions.

Hence, there are no x-intercepts.

$f(0) = 5,$ so the y-intercept is (0, 5).

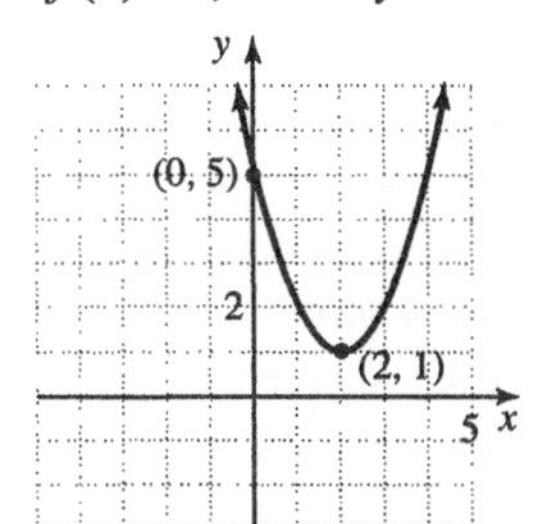

27.
$$f(x) = 2x^2 + 4x + 5$$
$$y = 2x^2 + 4x + 5$$
$$y - 5 = 2(x^2 + 2x)$$
$$y - 5 + 2(1) = 2(x^2 + 2x + 1)$$
$$y - 3 = 2(x+1)^2$$
$$y = 2(x+1)^2 + 3$$
$$f(x) = 2(x+1)^2 + 3$$
Thus, the vertex is (–1, 3).
The graph opens upward ($a = 2 > 0$).
$$2x^2 + 4x + 5 = 0$$
$$x = \frac{-4 \pm \sqrt{(4)^2 - 4(2)(5)}}{2(2)} = \frac{-4 \pm \sqrt{-24}}{4}$$
which give non-real solutions.
Hence, there are no x-intercepts.
$f(0) = 5,$ so the y-intercept is (0, 5).

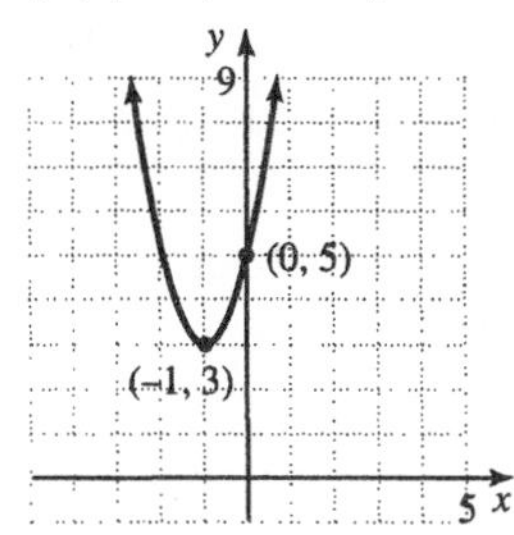

29.
$$f(x) = -2x^2 + 12x$$
$$y = -2(x^2 - 6x)$$
$$y + [-2(9)] = -2(x^2 - 6x + 9)$$
$$y - 18 = -2(x-3)^2$$
$$y = -2(x-3)^2 + 18$$
$$f(x) = -2(x-3)^2 + 18$$
Thus, the vertex is (3, 18).
The graph opens downward ($a = -2 < 0$).

$$-2x^2 + 12x = 0$$
$$-2x(x-6) = 0$$
$$x = 0 \text{ or } x - 6 = 0$$
$$x = 6$$
x-intercepts: (0, 0) and (6, 0)
$f(0) = 0,$ so the y-intercept is (0, 0).

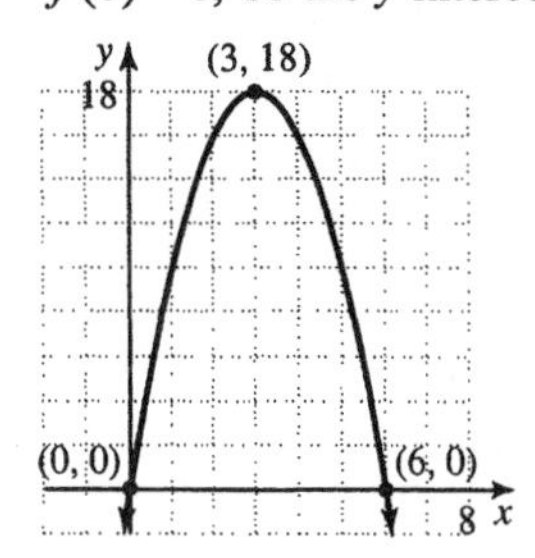

31. $f(x) = x^2 + 1$
$$x = -\frac{b}{2a} = -\frac{0}{2(1)} = 0$$
$$f(0) = (0)^2 + 1 = 1$$
Thus, the vertex is (0, 1).
The graph opens upward ($a = 1 > 0$).
$$x^2 + 1 = 0$$
$$x^2 = -1$$
which give non-real solutions.
Hence, there are no x-intercepts.
$f(0) = 1,$ so the y-intercept is (0, 1).

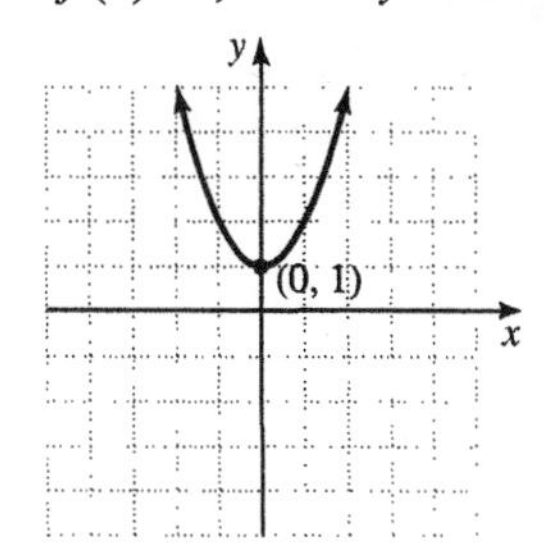

33.
$$f(x) = x^2 - 2x - 15$$
$$y = x^2 - 2x - 15$$
$$y + 15 = x^2 - 2x$$
$$y + 15 + 1 = x^2 - 2x + 1$$
$$y + 16 = (x-1)^2$$
$$y = (x-1)^2 - 16$$
$$f(x) = (x-1)^2 - 16$$
Thus, the vertex is (1, –16).
The graph opens upward ($a = 1 > 0$).

$x^2 - 2x - 15 = 0$
$(x-5)(x+3) = 0$
$x = 5$ or $x = -3$
x-intercepts: (–3, 0) and (5, 0).
$f(0) = -15$ so the y-intercept is (0, –15).

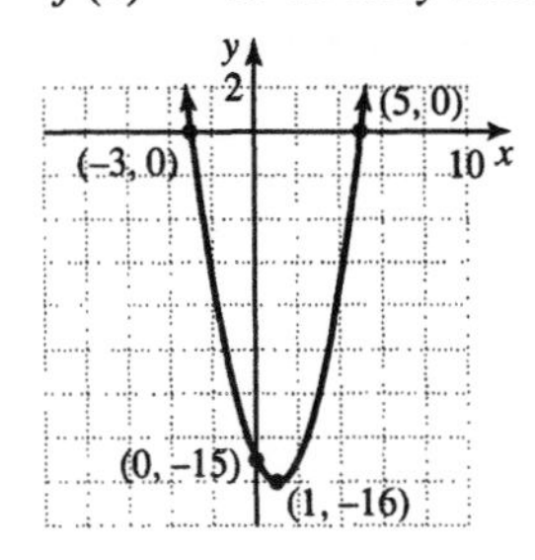

35. $f(x) = -5x^2 + 5x$

$x = -\frac{b}{2a} = \frac{-5}{2(-5)} = \frac{1}{2}$ and

$f\left(\frac{1}{2}\right) = -5\left(\frac{1}{2}\right)^2 + 5\left(\frac{1}{2}\right) = -\frac{5}{4} + \frac{5}{2} = \frac{5}{4}$

Thus, the vertex is $\left(\frac{1}{2}, \frac{5}{4}\right)$.

The graph opens downward ($a = -5 < 0$).

$-5x^2 + 5x = 0$
$-5x(x-1) = 0$
$x = 0$ or $x - 1 = 0$
$x = 1$

x-intercepts: (0, 0) and (1, 0)
$f(0) = 0$, so the y-intercept is (0, 0).

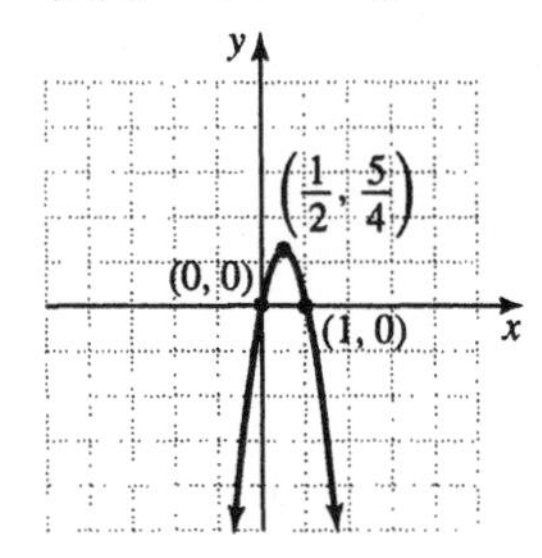

37. $f(x) = -x^2 + 2x - 12$

$x = -\frac{b}{2a} = \frac{-2}{2(-1)} = 1$ and

$f(1) = -(1)^2 + 2(1) - 12 = -11$

Thus, the vertex is (1, –11).
The graph opens downward ($a = -1 < 0$).

$-x^2 + 2x - 12 = 0$
$x^2 - 2x + 12 = 0$

$x = \frac{2 \pm \sqrt{(-2)^2 - 4(1)(12)}}{2(1)} = \frac{2 \pm \sqrt{-44}}{2}$

which yields non-real solutions.
Hence, there are no x-intercepts.
$f(0) = -12$ so the y-intercept is (0, –12).

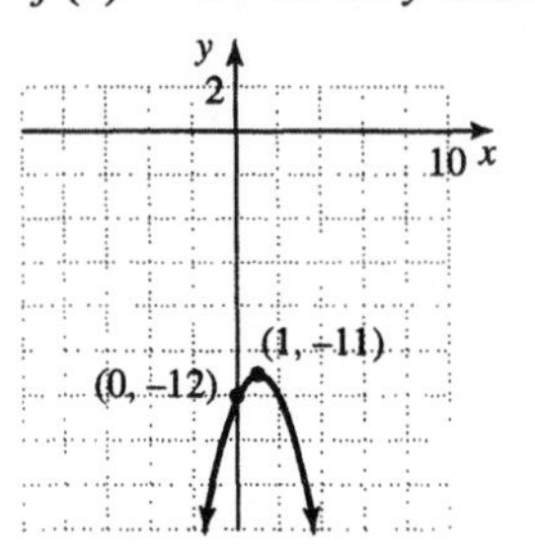

39. $f(x) = 3x^2 - 12x + 15$

$x = -\frac{b}{2a} = \frac{-(-12)}{2(3)} = \frac{12}{6} = 2$ and

$f(2) = 3(2)^2 - 12(2) + 15$
$= 12 - 24 + 15$
$= 3$

Thus, the vertex is (2, 3).
The graph opens upward ($a = 3 > 0$).

$3x^2 - 12x + 15 = 0$
$x^2 - 4x + 5 = 0$

$x = \frac{4 \pm \sqrt{(-4)^2 - 4(1)(5)}}{2(1)} = \frac{4 \pm \sqrt{-4}}{2}$

which yields non-real solutions.
Hence, there are no x-intercepts.
$f(0) = 15$, so the y-intercept is (0, 15).

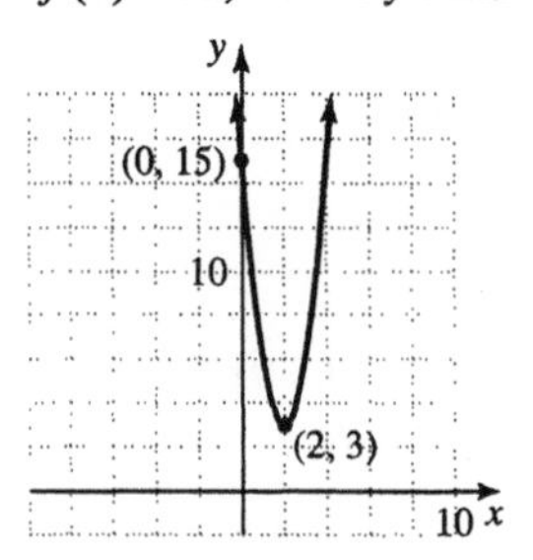

41. $f(x) = x^2 + x - 6$

$x = -\frac{b}{2a} = \frac{-1}{2(1)} = -\frac{1}{2}$ and

$$f\left(-\frac{1}{2}\right) = \left(-\frac{1}{2}\right)^2 + \left(-\frac{1}{2}\right) - 6 = \frac{1}{4} - \frac{1}{2} - 6 = -\frac{25}{4}$$

Thus, the vertex is $\left(-\frac{1}{2}, -\frac{25}{4}\right)$.

The graph opens upward ($a = 1 > 0$).

$x^2 + x - 6 = 0$
$(x+3)(x-2) = 0$
$x = -3$ or $x = 2$

x-intercepts: (–3, 0) and (2, 0).

$f(0) = -6$ so the y-intercept is (0, –6).

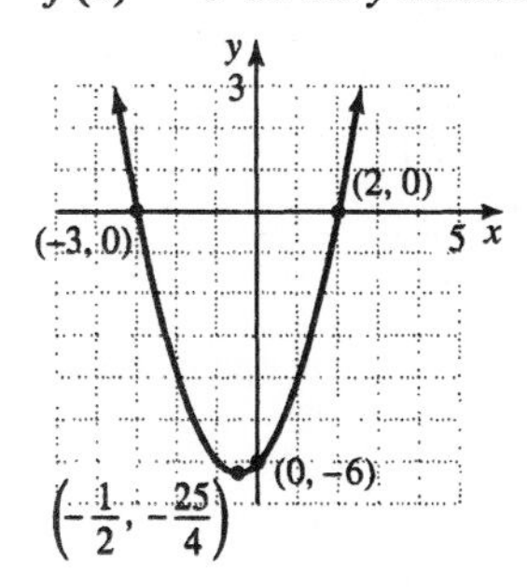

43. $f(x) = -2x^2 - 3x + 35$

$x = -\frac{b}{2a} = \frac{-(-3)}{2(-2)} = -\frac{3}{4}$ and

$$f\left(-\frac{3}{4}\right) = -2\left(-\frac{3}{4}\right)^2 - 3\left(-\frac{3}{4}\right) + 35 = -\frac{9}{8} + \frac{9}{4} + 35 = \frac{289}{8}$$

Thus, the vertex is $\left(-\frac{3}{4}, \frac{289}{8}\right)$.

The graph opens downward ($a = -2 < 0$).

$-2x^2 - 3x + 35 = 0$
$2x^2 + 3x - 35 = 0$
$(2x-7)(x+5) = 0$
$2x - 7 = 0$ or $x + 5 = 0$
$x = \frac{7}{2}$ or $x = -5$

x-intercepts: (–5, 0) and $\left(\frac{7}{2}, 0\right)$.

$f(0) = 35$ so the y-intercept is (0, 35).

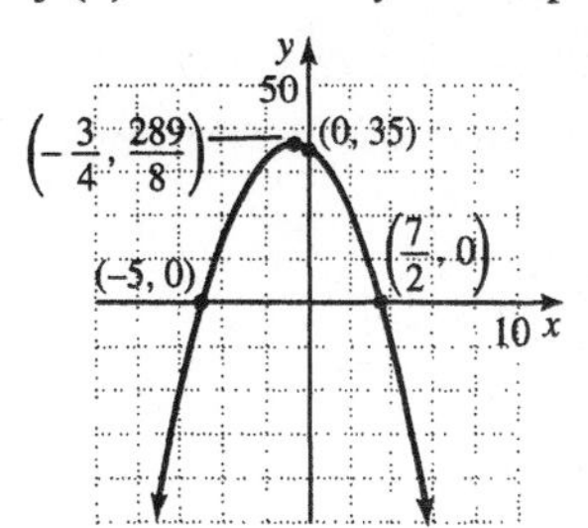

45. $h(t) = -16t^2 + 96t$

$t = -\frac{b}{2a} = \frac{-96}{2(-16)} = \frac{96}{32} = 3$ and

$$h(3) = -16(3)^2 + 96(3) = -144 + 288 = 144$$

The maximum height is 144 feet.

47. $C(x) = 2x^2 - 800x + 92{,}000$

a. $x = -\frac{b}{2a} = \frac{-(-800)}{2(2)} = 200$

200 bicycles are needed to minimize the cost.

b. $C(200) = 2(200)^2 - 800(200) + 92{,}000 = 12{,}000$

The minimum cost is \$12,000.

49. Let x = one number. Then $60 - x$ = the other number.

$$f(x) = x(60 - x) = 60x - x^2 = -x^2 + 60x$$

The maximum will occur at the vertex.

$x = -\frac{b}{2a} = \frac{-60}{2(-1)} = 30$

$60 - x = 60 - 30 = 30$

The numbers are 30 and 30.

51. Let x = one number. Then $10 + x$ = the other number.

$$f(x) = x(10 + x) = 10x + x^2 = x^2 + 10x$$

The minimum will occur at the vertex.

$x = -\frac{b}{2a} = \frac{-10}{2(1)} = -5$

$10 + x = 10 + (-5) = 5$

The numbers are –5 and 5.

53. Let x = width. Then $40 - x$ = the length.

Area = length · width

$A(x) = (40 - x)x$
$= 40x - x^2$
$= -x^2 + 40x$

The maximum will occur at the vertex.

$x = -\frac{b}{2a} = \frac{-40}{2(-1)} = 20$

$40 - x = 40 - 20 = 20$

The maximum area will occur when the length and width are 20 units each.

55. $f(x) = x^2 + 2$

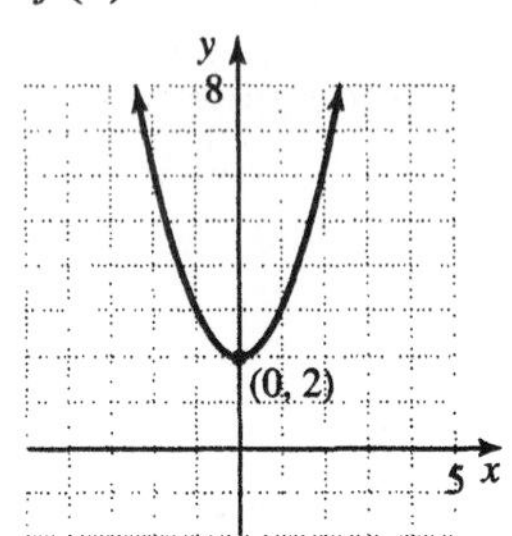

57. $g(x) = x + 2$

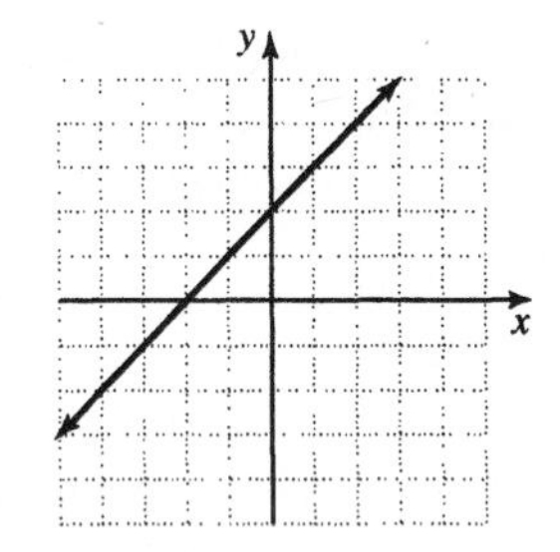

59. $f(x) = (x + 5)^2 + 2$

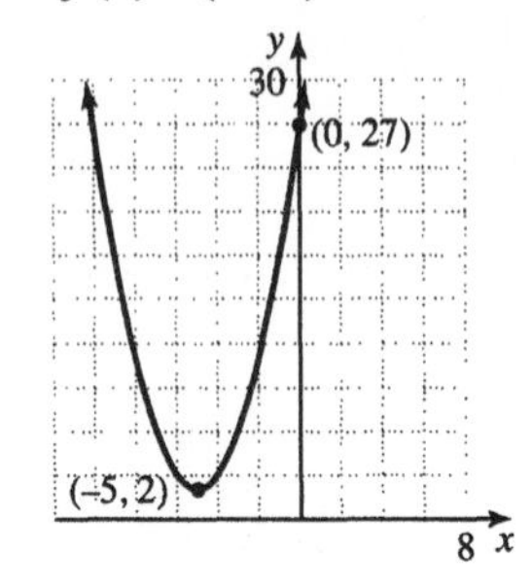

61. $f(x) = 3(x - 4)^2 + 1$

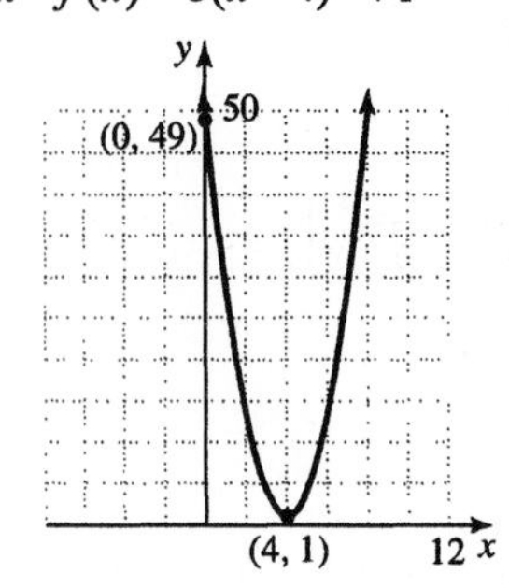

63. $f(x) = -(x - 4)^2 + \frac{3}{2}$

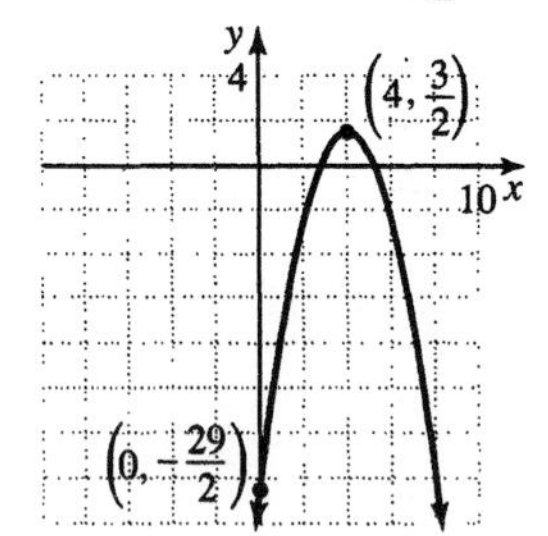

65. $f(x) = 2x^2 - 5$

Since $a = 2 > 0$, the graph opens upward; thus, $f(x)$ has a minimum value.

67. $F(x) = 3 - \frac{1}{2}x^2$

Since $a = -\frac{1}{2} < 0$, the graph opens downward; thus, $F(x)$ has a maximum value.

69. $f(x) = x^2 + 10x + 15$

$x = -\frac{b}{2a} = \frac{-10}{2(1)} = -5$ and

$f(-5) = (-5)^2 + 10(-5) + 15 = -10$

Thus, the vertex is (–5, –10).

The graph opens upward ($a = 1 > 0$).

$f(0) = 15$ so the y-intercept is (0, 15).

$x^2 + 10x + 15 = 0$

$$x = \frac{-10 \pm \sqrt{(10)^2 - 4(1)(15)}}{2(1)} = \frac{-10 \pm \sqrt{40}}{2} \approx -8.2 \text{ or } -1.8$$

The x-intercepts are approximately (–8.2, 0) and (–1.8, 0).

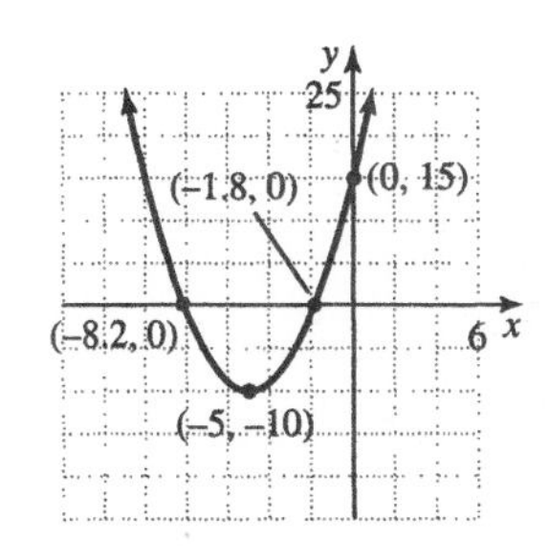

71. $f(x) = 3x^2 - 6x + 7$

$x = -\frac{b}{2a} = \frac{-(-6)}{2(3)} = 1$ and

$f(1) = 3(1)^2 - 6(1) + 7 = 4$

Thus, the vertex is (1, 4).

The graph opens upward ($a = 3 > 0$).

$f(0) = 7$ so the y-intercept is (0, 7).

$3x^2 - 6x + 7 = 0$

$x = \frac{6 \pm \sqrt{(-6)^2 - 4(3)(7)}}{2(3)} = \frac{6 \pm \sqrt{-48}}{6}$

which yields non-real solutions.

Hence, there are no x-intercepts.

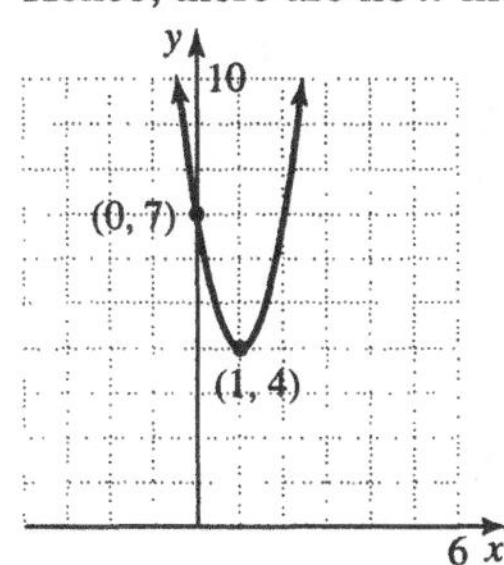

73. $f(x) = 2.3x^2 - 6.1x + 3.2$

$x = \frac{-(-6.1)}{2(2.3)} \approx 1.33$

$f(1.33) \approx -0.84$

minimum ≈ -0.84

Alternative solution:

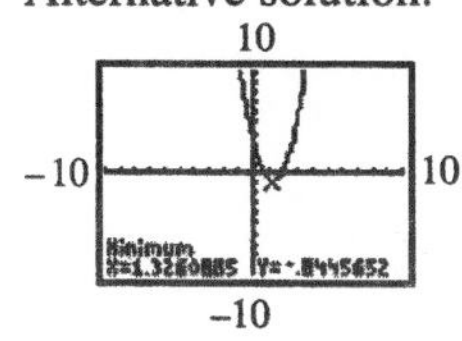

75. $f(x) = -1.9x^2 + 5.6x - 2.7$

$x = \frac{-5.6}{2(-1.9)} \approx 1.47$

$f(1.47) \approx 1.43$

maximum ≈ 1.43

Alternate solution:

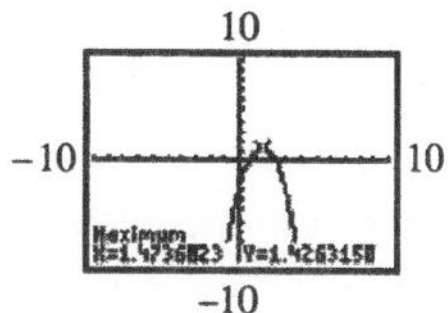

77. $f(x) = -96x^2 + 1018x + 28{,}824$

a. It will have a maximum; answer may vary (e.g., since $a = -96 < 0$).

b. $x = -\frac{b}{2a} = \frac{-1018}{2(-96)} \approx 5.3$

2000 + 5 = 2005

In the year 2005.

c. $f(5.3) = -96(5.3)^2 + 1018(5.3) + 28{,}824$
$= 31{,}522.76$

The maximum number of McDonald's is predicted to be about 31,523.

79.

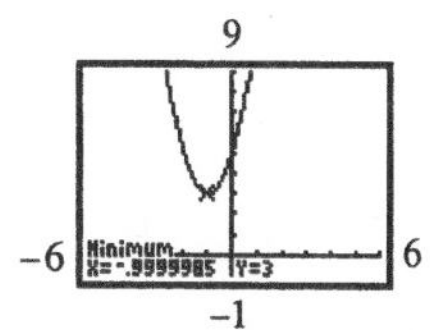

81.

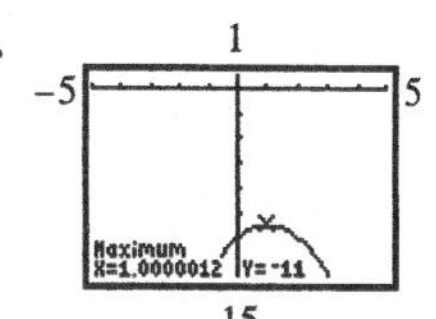

Chapter 11 Vocabulary Check

1. The discriminant helps us find the number and type of solutions of a quadratic equation.

2. If $a^2 = b$, then $a = \pm\sqrt{b}$.

3. The graph of $f(x) = ax^2 + bx + c$ where a is not 0 is a parabola whose vertex has x-value of $\frac{-b}{2a}$.

4. A quadratic inequality is an inequality that can be written so that one side is a quadratic expression and the other side is 0.

5. The process of writing a quadratic equation so that one side is a perfect square trinomial is called completing the square.

6. The graph of $f(x) = x^2 + k$ has vertex (0, k).

7. The graph of $f(x) = (x-h)^2$ has vertex (h, 0).

8. The graph of $f(x) = (x-h)^2 + k$ has vertex (h, k).

9. The formula $x = \frac{-b \pm \sqrt{b^2 - 4ac}}{2a}$ is called the quadratic formula.

10. A quadratic equation is one that can be written in the form $ax^2 + bx + c = 0$ where *a*, *b*, and *c* are real numbers and *a* is not 0.

Chapter 11 Review

1.
$$x^2 - 15x + 14 = 0$$
$$(x-14)(x-1) = 0$$
$$x - 14 = 0 \text{ or } x - 1 = 0$$
$$x = 14 \text{ or } x = 1$$
The solutions are 1 and 14.

2.
$$7a^2 = 29a + 30$$
$$7a^2 - 29a - 30 = 0$$
$$(7a+6)(a-5) = 0$$
$$7a + 6 = 0 \text{ or } a - 5 = 0$$
$$7a = -6 \text{ or } a = 5$$
$$a = -\frac{6}{7}$$
The solutions are $-\frac{6}{7}$ and 5.

3.
$$4m^2 = 196$$
$$m^2 = 49$$
$$m = \pm\sqrt{49}$$
$$m = \pm 7$$
The solutions are –7 and 7.

4.
$$(5x-2)^2 = 2$$
$$5x - 2 = \pm\sqrt{2}$$
$$5x = 2 \pm \sqrt{2}$$
$$x = \frac{2 \pm \sqrt{2}}{5}$$
The solutions are $\frac{2+\sqrt{2}}{5}$ and $\frac{2-\sqrt{2}}{5}$.

5.
$$z^2 + 3z + 1 = 0$$
$$z^2 + 3z = -1$$
$$z^2 + 3z + \left(\frac{3}{2}\right)^2 = -1 + \frac{9}{4}$$
$$\left(z + \frac{3}{2}\right)^2 = \frac{5}{4}$$
$$z + \frac{3}{2} = \pm\sqrt{\frac{5}{4}}$$
$$z + \frac{3}{2} = \pm\frac{\sqrt{5}}{2}$$
$$z = -\frac{3}{2} \pm \frac{\sqrt{5}}{2} = \frac{-3 \pm \sqrt{5}}{2}$$
The solutions are $\frac{-3+\sqrt{5}}{2}$ and $\frac{-3-\sqrt{5}}{2}$.

6.
$$(2x+1)^2 = x$$
$$4x^2 + 4x + 1 = x$$
$$4x^2 + 3x = -1$$
$$x^2 + \frac{3}{4}x = -\frac{1}{4}$$
$$x^2 + \frac{3}{4}x + \left(\frac{\frac{3}{4}}{2}\right)^2 = -\frac{1}{4} + \frac{9}{64}$$
$$\left(x + \frac{3}{8}\right)^2 = -\frac{7}{64}$$
$$x + \frac{3}{8} = \pm\sqrt{-\frac{7}{64}}$$
$$x + \frac{3}{8} = \pm\frac{i\sqrt{7}}{8}$$
$$x = -\frac{3}{8} \pm \frac{i\sqrt{7}}{8} = \frac{-3 \pm i\sqrt{7}}{8}$$
The solutions are $\frac{-3+i\sqrt{7}}{8}$ and $\frac{-3-i\sqrt{7}}{8}$.

7.
$$A = P(1+r)^2$$
$$2717 = 2500(1+r)^2$$
$$\frac{2717}{2500} = (1+r)^2$$
$$(1+r)^2 = 1.0868$$
$$1+r = \pm\sqrt{1.0868}$$
$$1+r = \pm 1.0425$$
$$r = -1 \pm 1.0425$$
$$= 0.0425 \text{ or } -2.0425 \text{ (disregard)}$$
The interest rate is 4.25%.

8. Let x = distance traveled.
$$a^2 + b^2 = c^2$$
$$x^2 + x^2 = (150)^2$$
$$2x^2 = 22{,}500$$
$$x^2 = 11{,}250$$
$$x = \pm 75\sqrt{2} \approx \pm 106.1$$
Disregard the negative. The ships each traveled $75\sqrt{2} \approx 106.1$ miles.

9. Two complex but not real solutions exist.

10. Two real solutions exist.

11. Two real solutions exist.

12. One real solution exists.

13. $x^2 - 16x + 64 = 0$
$a = 1, b = -16, c = 64$
$$x = \frac{16 \pm \sqrt{(-16)^2 - 4(1)(64)}}{2(1)}$$
$$= \frac{16 \pm \sqrt{256 - 256}}{2}$$
$$= \frac{16 \pm \sqrt{0}}{2}$$
$$= 8$$
The solution is 8.

14. $x^2 + 5x = 0$
$a = 1, b = 5, c = 0$
$$x = \frac{-5 \pm \sqrt{(5)^2 - 4(1)(0)}}{2(1)}$$
$$= \frac{-5 \pm \sqrt{25}}{2}$$
$$= \frac{-5 \pm 5}{2}$$
$$= 0 \text{ or } -5$$
The solutions are −5 and 0.

15.
$$2x^2 + 3x = 5$$
$$2x^2 + 3x - 5 = 0$$
$a = 2, b = 3, c = -5$
$$x = \frac{-3 \pm \sqrt{(3)^2 - 4(2)(-5)}}{2(2)}$$
$$= \frac{-3 \pm \sqrt{49}}{4}$$
$$= \frac{-3 \pm 7}{4}$$
$$= 1 \text{ or } -\frac{5}{2}$$
The solutions are $-\frac{5}{2}$ and 1.

16.
$$9a^2 + 4 = 2a$$
$$9a^2 - 2a + 4 = 0$$
$$a = \frac{2 \pm \sqrt{(-2)^2 - 4(9)(4)}}{2(9)}$$
$$= \frac{2 \pm \sqrt{-140}}{18}$$
$$= \frac{2 \pm i\sqrt{4 \cdot 35}}{18}$$
$$= \frac{2 \pm 2i\sqrt{35}}{18}$$
$$= \frac{1 \pm i\sqrt{35}}{9}$$
The solutions are $\frac{1+i\sqrt{35}}{9}$ and $\frac{1-i\sqrt{35}}{9}$.

17.
$$6x^2+7=5x$$
$$6x^2-5x+7=0$$
$$a=6, b=-5, c=7$$
$$x=\frac{5\pm\sqrt{(-5)^2-4(6)(7)}}{2(6)}$$
$$=\frac{5\pm\sqrt{25-168}}{12}$$
$$=\frac{5\pm\sqrt{-143}}{12}$$
$$=\frac{5\pm i\sqrt{143}}{12}$$
The solutions are $\frac{5+i\sqrt{143}}{12}$ and $\frac{5-i\sqrt{143}}{12}$.

18.
$$(2x-3)^2=x$$
$$4x^2-12x+9-x=0$$
$$4x^2-13x+9=0$$
$$a=4, b=-13, c=9$$
$$x=\frac{13\pm\sqrt{(-13)^2-4(4)(9)}}{2(4)}$$
$$=\frac{13\pm\sqrt{169-144}}{8}$$
$$=\frac{13\pm\sqrt{25}}{8}$$
$$=\frac{13\pm5}{8}$$
$$=\frac{9}{4} \text{ or } 1$$
The solutions are 1 and $\frac{9}{4}$.

19. $d(t)=-16t^2+30t+6$

a.
$$d(1)=-16(1)^2+30(1)+6$$
$$=-16+30+6$$
$$=20 \text{ feet}$$

b.
$$-16t^2+30t+6=0$$
$$8t^2-15t-3=0$$
$$a=8, b=-15, c=-3$$
$$t=\frac{15\pm\sqrt{(-15)^2-4(8)(-3)}}{2(8)}$$
$$=\frac{15\pm\sqrt{225+96}}{16}$$
$$=\frac{15\pm\sqrt{321}}{16}$$
Disregarding the negative, we have
$$t=\frac{15+\sqrt{321}}{16} \text{ seconds}$$
$$\approx 2.1 \text{ seconds.}$$

20. Let x = length of the legs. Then $x+6$ = length of the hypotenuse.
$$x^2+x^2=(x+6)^2$$
$$2x^2=x^2+12x+36$$
$$x^2-12x-36=0$$
$$a=1, b=-12, c=-36$$
$$x=\frac{12\pm\sqrt{(-12)^2-4(1)(-36)}}{2(1)}$$
$$=\frac{12\pm\sqrt{144+144}}{2}$$
$$=\frac{12\pm\sqrt{144\cdot2}}{2}$$
$$=\frac{12\pm12\sqrt{2}}{2}$$
$$=6\pm6\sqrt{2}$$
Disregard the negative. The length of each leg is $\left(6+6\sqrt{2}\right)$ cm.

21.
$$x^3 = 27$$
$$x^3 - 27 = 0$$
$$(x-3)(x^2+3x+9) = 0$$
$$x-3=0 \text{ or } x^2+3x+9=0$$
$$x = 3 \qquad a=1, b=3, c=9$$
$$x = \frac{-3 \pm \sqrt{(3)^2 - 4(1)(9)}}{2(1)} = \frac{-3 \pm \sqrt{9-36}}{2} = \frac{-3 \pm \sqrt{-27}}{2} = \frac{-3 \pm 3i\sqrt{3}}{2}$$

The solutions are 3, $\frac{-3+3i\sqrt{3}}{2}$, and $\frac{-3-3i\sqrt{3}}{2}$.

22.
$$y^3 = -64$$
$$y^3 + 64 = 0$$
$$(y+4)(y^2-4y+16) = 0$$
$$y+4=0 \text{ or } y^2-4y+16=0$$
$$y=-4 \qquad a=1, b=-4, c=16$$
$$y = \frac{4 \pm \sqrt{(-4)^2 - 4(1)(16)}}{2(1)} = \frac{4 \pm \sqrt{16-64}}{2} = \frac{4 \pm \sqrt{-48}}{2} = \frac{4 \pm 4i\sqrt{3}}{2} = 2 \pm 2i\sqrt{3}$$

The solutions are −4, $2+2i\sqrt{3}$, and $2-2i\sqrt{3}$.

23.
$$\frac{5}{x} + \frac{6}{x-2} = 3$$
$$x(x-2)\left(\frac{5}{x} + \frac{6}{x-2}\right) = 3x(x-2)$$
$$5(x-2)+6x = 3x^2 - 6x$$
$$5x - 10 + 6x = 3x^2 - 6x$$
$$0 = 3x^2 - 17x + 10$$
$$0 = (3x-2)(x-5)$$
$$3x-2=0 \text{ or } x-5=0$$
$$x = \frac{2}{3} \text{ or } x = 5$$

The solutions are $\frac{2}{3}$ and 5.

24.
$$x^4 - 21x^2 - 100 = 0$$
$$(x^2-25)(x^2+4) = 0$$
$$(x+5)(x-5)(x^2+4) = 0$$
$$x+5=0 \text{ or } x-5=0 \text{ or } x^2+4=0$$
$$x=-5 \text{ or } x=5 \text{ or } x^2=-4$$
$$x = \pm 2i$$

The solutions are −5, 5 −2*i*, and 2*i*.

25. $x^{2/3} - 6x^{1/3} + 5 = 0$

Let $y = x^{1/3}$. Then $y^2 = x^{2/3}$ and
$$y^2 - 6y + 5 = 0$$
$$(y-5)(y-1) = 0$$
$$y-5=0 \text{ or } y-1=0$$
$$y=5 \text{ or } y=1$$
$$x^{1/3}=5 \text{ or } x^{1/3}=1$$
$$x=125 \text{ or } x=1$$

The solutions are 1 and 125.

26. $5(x+3)^2-19(x+3)=4$

$5(x+3)^2-19(x+3)-4=0$

Let $y=x+3$. Then $y^2=(x+3)^2$ and

$5y^2-19y-4=0$

$(5y+1)(y-4)=0$

$5y+1=0$ or $y-4=0$

$y=-\frac{1}{5}$ or $y=4$

$x+3=-\frac{1}{5}$ or $x+3=4$

$x=-\frac{16}{5}$ or $x=1$

The solutions are $-\frac{16}{5}$ and 1.

27. $a^6-a^2=a^4-1$

$a^6-a^4-a^2+1=0$

$a^4(a^2-1)-1(a^2-1)=0$

$(a^2-1)(a^4-1)=0$

$(a+1)(a-1)(a^2+1)(a^2-1)=0$

$(a+1)(a-1)(a^2+1)(a+1)(a-1)=0$

$(a+1)^2(a-1)^2(a^2+1)=0$

$(a+1)^2=0$ or $(a-1)^2=0$ or $a^2+1=0$

$a+1=0$ or $a-1=0$ or $a^2=-1$

$a=-1$ or $a=1$ or $a=\pm i$

The solutions are –1, 1, –*i*, and *i*.

28. $y^{-2}+y^{-1}=20$

$\frac{1}{y^2}+\frac{1}{y}=20$

$1+y=20y^2$

$0=20y^2-y-1$

$0=(5y+1)(4y-1)$

$5y+1=0$ or $4y-1=0$

$y=-\frac{1}{5}$ or $y=\frac{1}{4}$

The solutions are $-\frac{1}{5}$ and $\frac{1}{4}$.

29. Let x = time for Jerome alone. Then $x-1$ = time for Tim alone.

$\frac{1}{x}+\frac{1}{x-1}=\frac{1}{5}$

$5(x-1)+5x=x(x-1)$

$5x-5+5x=x^2-x$

$0=x^2-11x+5$

$a=1, b=-11, c=5$

$x=\frac{11\pm\sqrt{(-11)^2-4(1)(5)}}{2(1)}$

$=\frac{11\pm\sqrt{101}}{2}$

≈ 0.475 (disregard) or 10.525

Jerome: 10.5 hours

Tim: 9.5 hours

30. Let x = the number; then $\frac{1}{x}$ = the reciprocal of the number.

$x-\frac{1}{x}=-\frac{24}{5}$

$5x\left(x-\frac{1}{x}\right)=5x\left(-\frac{24}{5}\right)$

$5x^2-5=-24x$

$5x^2+24x-5=0$

$(5x-1)(x+5)=0$

$5x-1=0$ or $x+5=0$

$x=\frac{1}{5}$ or $x=-5$

Disregard the positive value as extraneous. The number is –5.

31. $2x^2-50\le 0$

$2(x^2-25)\le 0$

$2(x+5)(x-5)\le 0$

$x+5=0$ or $x-5=0$

$x=-5$ or $x=5$

Region	Test Point	$2(x+5)(x-5)\le 0$ Result
A: $(-\infty, -5]$	–6	$2(-1)(-11)\le 0$ False
B: $[-5, 5]$	0	$2(5)(-5)\le 0$ True
C: $[5, \infty)$	6	$2(11)(1)\le 0$ False

Solution: $[-5, 5]$

32.

$$\frac{1}{4}x^2 < \frac{1}{16}$$
$$x^2 < \frac{1}{4}$$
$$x^2 - \frac{1}{4} < 0$$
$$\left(x+\frac{1}{2}\right)\left(x-\frac{1}{2}\right) < 0$$
$$x+\frac{1}{2}=0 \quad \text{or} \quad x-\frac{1}{2}=0$$
$$x=-\frac{1}{2} \quad \text{or} \quad x=\frac{1}{2}$$

Region	Test Point	$\left(x+\frac{1}{2}\right)\left(x-\frac{1}{2}\right) < 0$ Result
$A: \left(-\infty, -\frac{1}{2}\right)$	-1	$\left(-\frac{1}{2}\right)\left(-\frac{3}{2}\right) < 0$ False
$B: \left(-\frac{1}{2}, \frac{1}{2}\right)$	0	$\left(\frac{1}{2}\right)\left(-\frac{1}{2}\right) < 0$ True
$C: \left(\frac{1}{2}, \infty\right)$	1	$\left(\frac{3}{2}\right)\left(\frac{1}{2}\right) < 0$ False

Solution: $\left(-\frac{1}{2}, \frac{1}{2}\right)$

33. $\frac{x-5}{x-6}<0$

$x-5=0$ or $x-6=0$

$x=5$ or $x=6$

Region	Test Point	$\frac{x-5}{x-6}<0$ Result
A: $(-\infty, 5)$	0	$\frac{-5}{-6}<0$ False
B: $(5, 6)$	$\frac{11}{2}$	$\frac{\frac{1}{2}}{-\frac{1}{2}}<0$ True
C: $(6, \infty)$	7	$\frac{2}{1}<0$ False

Solution: $(5, 6)$

34. $(x^2-16)(x^2-1)>0$

$(x+4)(x-4)(x+1)(x-1)>0$

$x+4=0$ or $x-4=0$ or $x+1=0$ or $x-1=0$

$x=-4$ or $x=4$ or $x=-1$ or $x=1$

Region	Test Point	$(x+4)(x-4)(x+1)(x-1)>0$ Result
A: $(-\infty, -4)$	−5	$(-1)(-9)(-4)(-6)>0$ True
B: $(-4, -1)$	−2	$(2)(-6)(-1)(-3)>0$ False
C: $(-1, 1)$	0	$(4)(-4)(1)(-1)>0$ True
D: $(1, 4)$	2	$(6)(-2)(3)(1)>0$ False
E: $(4, \infty)$	5	$(9)(1)(6)(4)>0$ True

Solution: $(-\infty, -4)\cup(-1, 1)\cup(4, \infty)$

35. $\dfrac{(4x+3)(x-5)}{x(x+6)} > 0$

$4x+3=0,\ x-5=0,\ x=0,\ \text{or}\ x+6=0$

$x=-\dfrac{3}{4},\ x=5,\ x=0,\ \text{or}\ x=-6$

Region	Test Point	$\dfrac{(4x+3)(x-5)}{x(x+6)} > 0$ Result
A: $(-\infty, -6)$	-7	$\dfrac{(-25)(-12)}{-7(-1)} > 0$ True
B: $\left(-6, -\dfrac{3}{4}\right)$	-3	$\dfrac{(-9)(-8)}{-3(3)} > 0$ False
C: $\left(-\dfrac{3}{4}, 0\right)$	$-\dfrac{1}{2}$	$\dfrac{(1)\left(-\frac{11}{2}\right)}{-\frac{1}{2}\left(\frac{11}{2}\right)} > 0$ True
D: $(0, 5)$	1	$\dfrac{(7)(-4)}{1(7)} > 0$ False
E: $(5, \infty)$	6	$\dfrac{(27)(1)}{6(12)} > 0$ True

Solution: $(-\infty, -6) \cup \left(-\dfrac{3}{4}, 0\right) \cup (5, \infty)$

36. $(x+5)(x-6)(x+2) \le 0$

$x+5=0\ \text{or}\ x-6=0\ \text{or}\ x+2=0$

$x=-5\ \text{or}\ x=6\ \text{or}\ x=-2$

Region	Test Point	$(x+5)(x-6)(x+2) \le 0$ Result
A: $(-\infty, -5]$	-6	$(-1)(-12)(-4) \le 0$ True
B: $[-5, -2]$	-3	$(2)(-9)(-1) \le 0$ False
C: $[-2, 6]$	0	$(5)(-6)(2) \le 0$ True
D: $[6, \infty)$	7	$(12)(1)(9) \le 0$ False

Solution: $(-\infty, -5] \cup [-2, 6]$

37. $x^3+3x^2-25x-75>0$

$x^2(x+3)-25(x+3)>0$

$(x+3)(x^2-25)>0$

$(x+3)(x+5)(x-5)>0$

$x+3=0\ \text{or}\ x+5=0\ \text{or}\ x-5=0$

$x=-3\ \text{or}\ x=-5\ \text{or}\ x=5$

Region	Test Point	$(x+3)(x+5)(x-5) > 0$ Result
A: $(-\infty, -5)$	-6	$(-3)(-1)(-11) > 0$ False
B: $(-5, -3)$	-4	$(-1)(1)(-9) > 0$ True
C: $(-3, 5)$	0	$(3)(5)(-5) > 0$ False
D: $(5, \infty)$	6	$(9)(11)(1) > 0$ True

Solution: $(-5, -3) \cup (5, \infty)$

38. $\dfrac{x^2+4}{3x} \le 1$

The denominator equals 0 when $3x = 0$, or $x = 0$.

$$\frac{x^2+4}{3x}=1$$

$$x^2+4=3x$$

$$x^2-3x+4=0$$

$$x=\frac{3\pm\sqrt{(-3)^2-4(1)(4)}}{2(1)}=\frac{3\pm\sqrt{-7}}{2}$$

which yields non-real solutions.

Region	Test Point	$\frac{x^2+4}{3x} \le 1$ Result
$A: (-\infty, 0)$	-1	$\frac{5}{-3} \le 1$ True
$B: (0, \infty)$	1	$\frac{5}{3} \le 1$ False

Solution: $(\infty, 0)$

39. $\frac{(5x+6)(x-3)}{x(6x-5)} < 0$

$x = -\frac{6}{5}$ or $x = 3$ or $x = 0$ or $x = \frac{5}{6}$

Region	Test Point	$\frac{(5x+6)(x-3)}{x(6x-5)} < 0$ Result
$A: \left(-\infty, -\frac{6}{5}\right)$	-2	$\frac{(-4)(-5)}{-2(-17)} < 0$ False
$B: \left(-\frac{6}{5}, 0\right)$	-1	$\frac{(1)(-4)}{-1(-11)} < 0$ True
$C: \left(0, \frac{5}{6}\right)$	$\frac{1}{2}$	$\frac{\left(\frac{17}{2}\right)\left(-\frac{5}{2}\right)}{\frac{1}{2}(-2)} < 0$ False
$D: \left(\frac{5}{6}, 3\right)$	2	$\frac{(16)(-1)}{2(7)} < 0$ True
$E: (3, \infty)$	4	$\frac{(26)(1)}{4(19)} < 0$ False

Solution: $\left(-\frac{6}{5}, 0\right) \cup \left(\frac{5}{6}, 3\right)$

40. $\frac{3}{x-2} > 2$

The denominator is equal to 0 when $x - 2 = 0$, or $x = 2$.

$$\frac{3}{x-2} = 2$$
$$3 = 2(x-2)$$
$$3 = 2x - 4$$
$$7 = 2x$$
$$\frac{7}{2} = x$$

Region	Test Point	$\frac{3}{x-2} > 2$ Result
$A: (-\infty, 2)$	0	$\frac{3}{-2} > 2$ False
$B: \left(2, \frac{7}{2}\right)$	3	$\frac{3}{1} > 2$ True
$C: \left(\frac{7}{2}, \infty\right)$	5	$\frac{3}{3} > 2$ False

Solution: $\left(2, \frac{7}{2}\right)$

41. $f(x) = x^2 - 4$

Vertex: $(0, -4)$

Axis of symmetry: $x = 0$

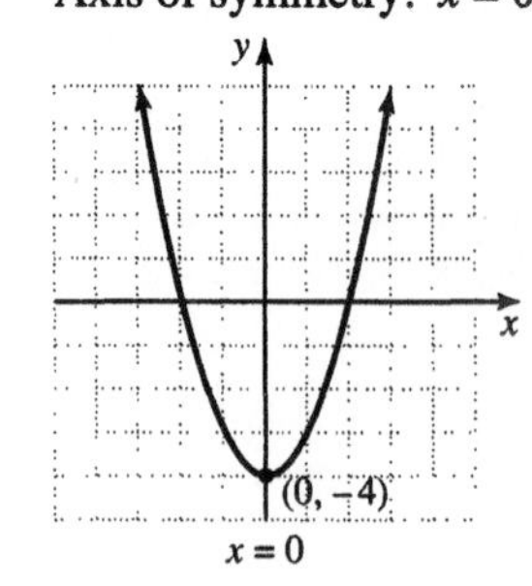

42. $g(x) = x^2 + 7$
Vertex: (0, 7)
Axis of symmetry: $x = 0$

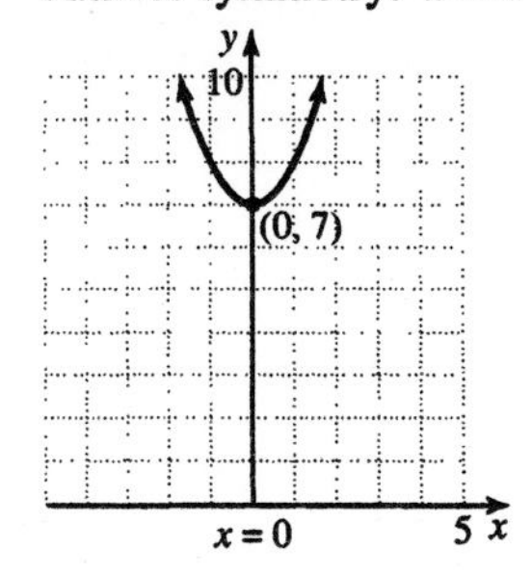

43. $H(x) = 2x^2$
Vertex: (0, 0)
Axis of symmetry: $x = 0$

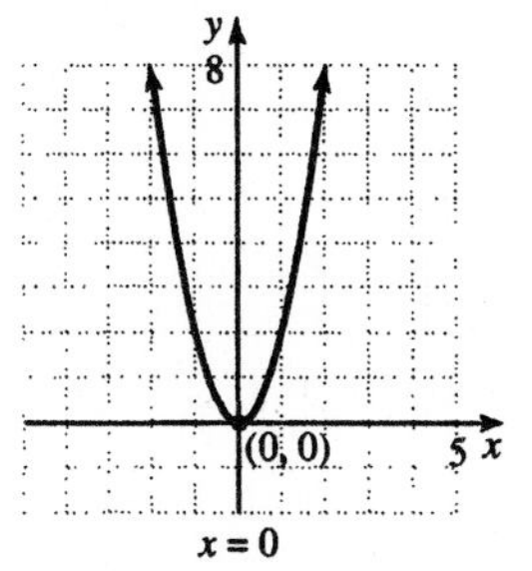

44. $h(x) = -\frac{1}{3}x^2$
Vertex: (0, 0)
Axis of symmetry: $x = 0$

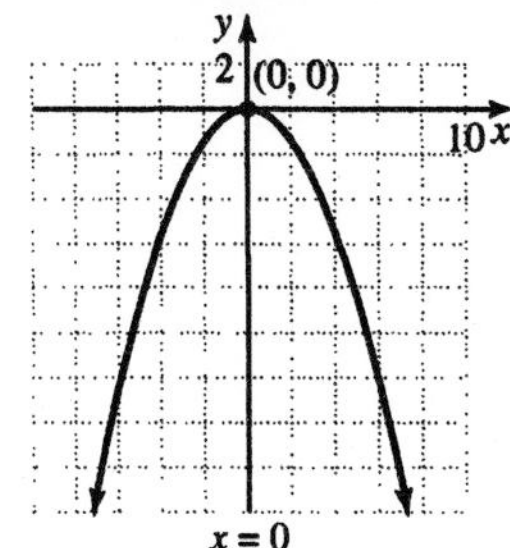

45. $F(x) = (x-1)^2$
Vertex: (1, 0)
Axis of symmetry: $x = 1$

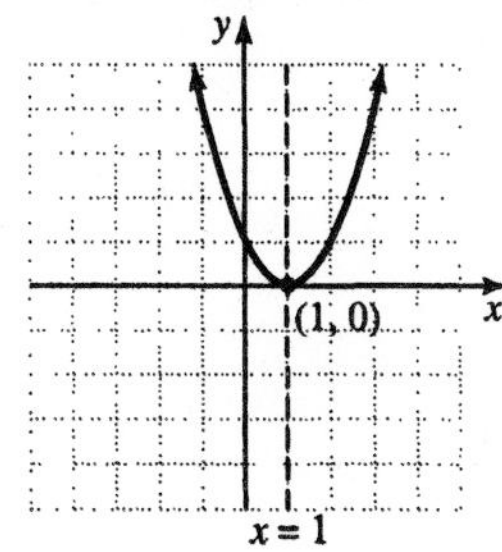

46. $G(x) = (x+5)^2$
Vertex: (–5, 0)
Axis of symmetry: $x = -5$

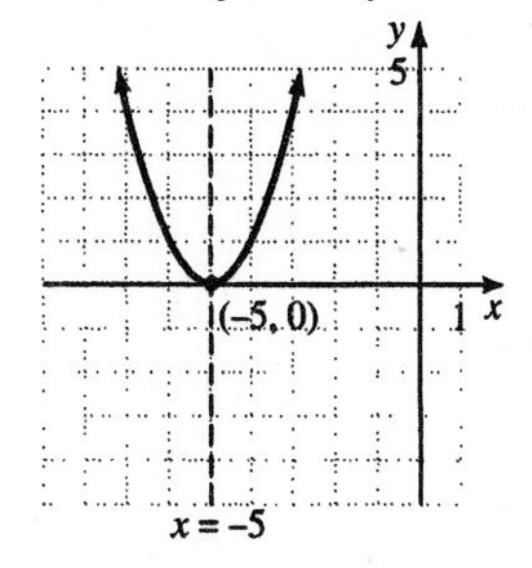

47. $f(x) = (x-4)^2 - 2$
Vertex: (4, –2)
Axis of symmetry: $x = 4$

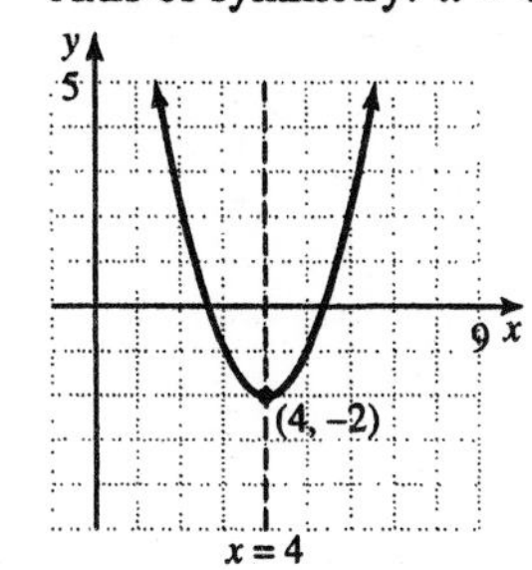

48. $f(x) = -3(x-1)^2 + 1$
Vertex: (1, 1)
Axis of symmetry: $x = 1$

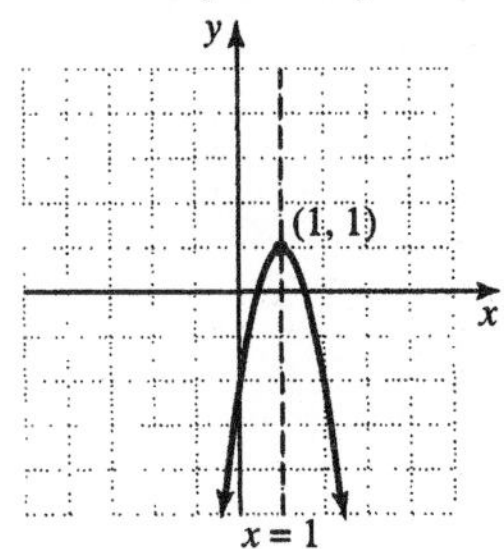

49. $f(x) = x^2 + 10x + 25$

$x = -\frac{b}{2a} = \frac{-10}{2(1)} = -5$

$f(-5) = (-5)^2 + 10(-5) + 25 = 0$

Vertex: (–5, 0)

$x^2 + 10x + 25 = 0$

$(x+5)^2 = 0$

$x + 5 = 0$

$x = -5$

x-intercept: (–5, 0)

$f(0) = 25$ so the y-intercept is (0, 25).

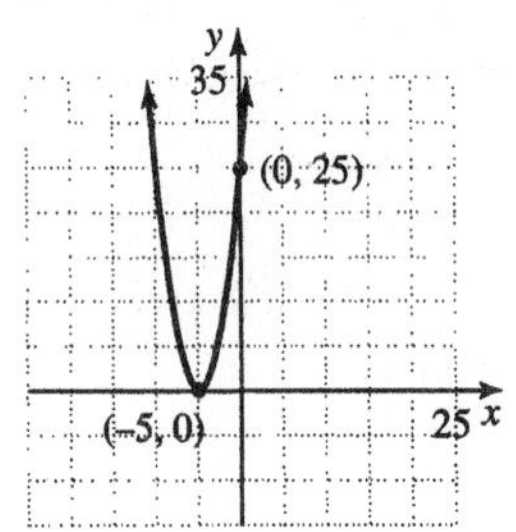

50. $f(x) = -x^2 + 6x - 9$

$x = -\frac{b}{2a} = \frac{-6}{2(-1)} = 3$

$f(3) = -(3)^2 + 6(3) - 9 = 0$

Vertex: (3, 0)
x-intercept: (3, 0)
$f(0) = -9$
y-intercept: (0, –9)

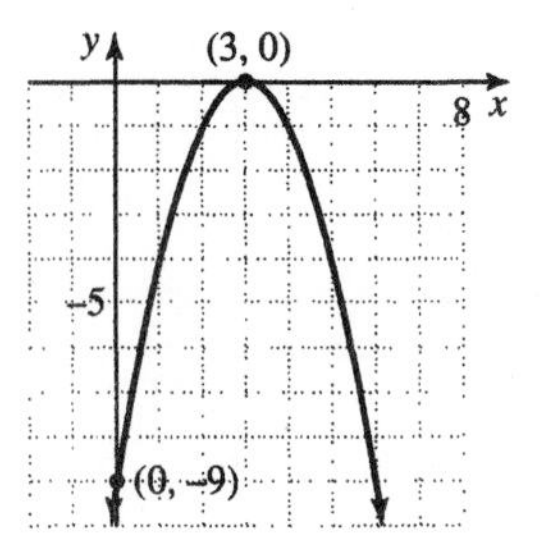

51. $f(x) = 4x^2 - 1$

$x = -\frac{b}{2a} = \frac{-0}{2(4)} = 0$

$f(0) = 4(0)^2 - 1 = -1$

Vertex: (0, –1)

$4x^2 - 1 = 0$

$(2x+1)(2x-1) = 0$

$x = -\frac{1}{2}$ or $x = \frac{1}{2}$

x-intercepts: $\left(-\frac{1}{2}, 0\right), \left(\frac{1}{2}, 0\right)$

$f(0) = -1$

y-intercept: (0, –1)

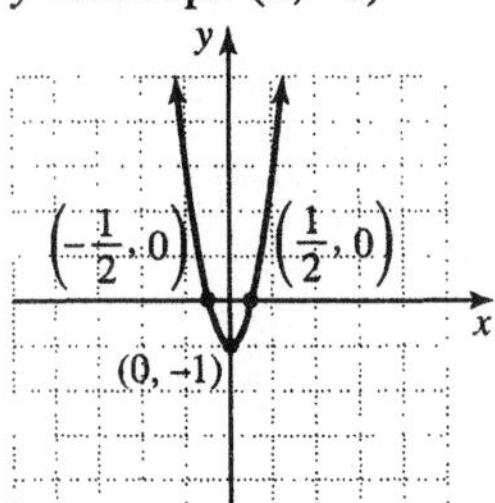

52. $f(x) = -5x^2 + 5$

$x = -\frac{b}{2a} = \frac{-0}{2(-5)} = 0$

$f(0) = -5(0)^2 + 5 = 5$

Vertex: (0, 5)

$-5x^2 + 5 = 0$

$-5x^2 = -5$

$x^2 = 1$

$x = \pm 1$

x-intercepts: (–1, 0), (1, 0)

$f(0) = 5$

y-intercept: (0, 5)

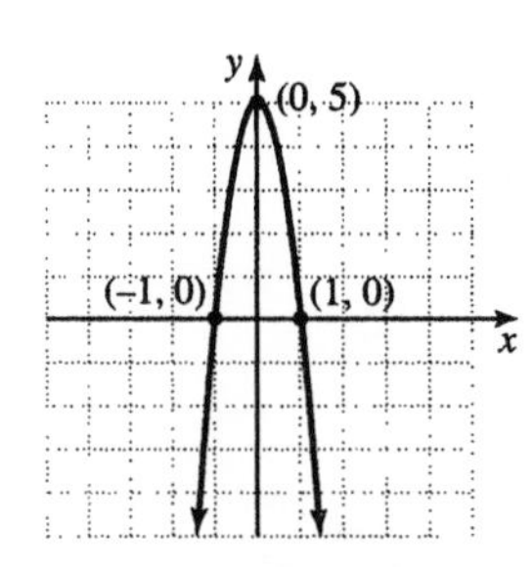

53. $f(x) = -3x^2 - 5x + 4$

$x = -\frac{b}{2a} = \frac{-(-5)}{2(-3)} = -\frac{5}{6}$

$f\left(-\frac{5}{6}\right) = -3\left(-\frac{5}{6}\right)^2 - 5\left(-\frac{5}{6}\right) + 4 = \frac{73}{12}$

Vertex: $\left(-\frac{5}{6}, \frac{73}{12}\right)$

The graph opens downward ($a = -3 < 0$).

$f(0) = 4 \Rightarrow$ y-intercept: (0, 4)

$-3x^2 - 5x + 4 = 0$

$x = \frac{5 \pm \sqrt{(-5)^2 - 4(-3)(4)}}{2(-3)}$

$= \frac{5 \pm \sqrt{73}}{-6}$

≈ -2.2573 or 0.5907

x-intercepts: (−2.3, 0), (0.6, 0)

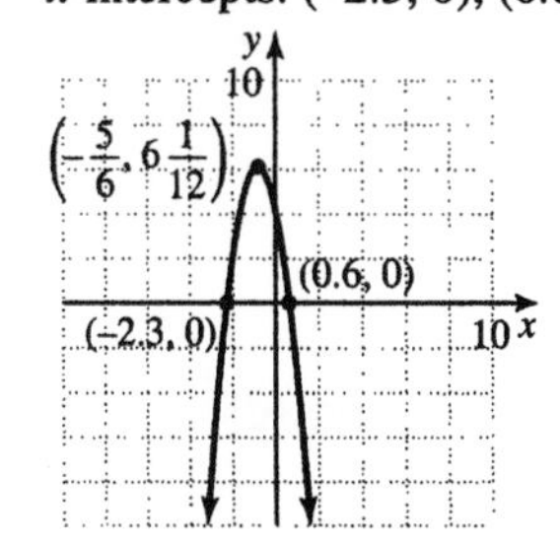

54. $h(t) = -16t^2 + 120t + 300$

a. $350 = -16t^2 + 120t + 300$

$16t^2 - 120t + 50 = 0$

$8t^2 - 60t + 25 = 0$

$a = 8, b = -60, c = 25$

$t = \frac{60 \pm \sqrt{(-60)^2 - 4(8)(25)}}{2(8)}$

$= \frac{60 \pm \sqrt{2800}}{16}$

≈ 0.4 second and 7.1 seconds

b. The object will be at 350 feet on the way up and on the way down.

55. Let x = one number; then

$420 - x$ = the other number.

Let $f(x)$ represent their product.

$f(x) = x(420 - x)$

$= 420x - x^2$

$= -x^2 + 420x$

$x = -\frac{b}{2a} = \frac{-420}{2(-1)} = 210;$

$420 - x = 420 - 210 = 210$

Therefore, the numbers are both 210.

56. $y = a(x - h)^2 + k$

vertex (−3, 7) gives $y = a(x + 3)^2 + 7$.

Passing through the origin gives

$0 = a(0 + 3)^2 + 7$

$-7 = 9a$

$-\frac{7}{9} = a$

Thus, $y = -\frac{7}{9}(x + 3)^2 + 7$.

57. $x^2 - x - 30 = 0$

$(x + 5)(x - 6) = 0$

$x + 5 = 0$ or $x - 6 = 0$

$x = -5$ or $x = 6$

The solutions are −5 and 6.

58.
$$10x^2 = 3x + 4$$
$$10x^2 - 3x - 4 = 0$$
$$(5x-4)(2x+1) = 0$$
$$5x - 4 = 0 \text{ or } 2x + 1 = 0$$
$$5x = 4 \text{ or } 2x = -1$$
$$x = \frac{4}{5} \text{ or } x = -\frac{1}{2}$$
The solutions are $-\frac{1}{2}$ and $\frac{4}{5}$.

59. $9y^2 = 36$
$$y^2 = 4$$
$$y = \pm\sqrt{4}$$
$$y = \pm 2$$
The solutions are –2 and 2.

60. $(9n+1)^2 = 9$
$$9n + 1 = \pm\sqrt{9}$$
$$9n + 1 = \pm 3$$
$$9n = -1 \pm 3$$
$$n = \frac{-1 \pm 3}{9} = \frac{2}{9}, -\frac{4}{9}$$
The solutions are $-\frac{4}{9}$ and $\frac{2}{9}$.

61.
$$x^2 + x + 7 = 0$$
$$x^2 + x = -7$$
$$x^2 + x + \left(\frac{1}{2}\right)^2 = -7 + \frac{1}{4}$$
$$\left(x + \frac{1}{2}\right)^2 = -\frac{27}{4}$$
$$x + \frac{1}{2} = \pm\sqrt{-\frac{27}{4}}$$
$$x + \frac{1}{2} = \pm\frac{i\sqrt{9 \cdot 3}}{2}$$
$$x + \frac{1}{2} = \pm\frac{3i\sqrt{3}}{2}$$
$$x = -\frac{1}{2} \pm \frac{3i\sqrt{3}}{2} = \frac{-1 \pm 3i\sqrt{3}}{2}$$
The solutions are $\frac{-1+3i\sqrt{3}}{2}$ and $\frac{-1-3i\sqrt{3}}{2}$.

62.
$$(3x-4)^2 = 10x$$
$$9x^2 - 24x + 16 = 10x$$
$$9x^2 - 34x = -16$$
$$x^2 - \frac{34}{9}x = -\frac{16}{9}$$
$$x^2 - \frac{34}{9}x + \left(\frac{-\frac{34}{9}}{2}\right)^2 = -\frac{16}{9} + \frac{289}{81}$$
$$\left(x - \frac{17}{9}\right)^2 = \frac{145}{81}$$
$$x - \frac{17}{9} = \pm\sqrt{\frac{145}{81}}$$
$$x - \frac{17}{9} = \pm\frac{\sqrt{145}}{9}$$
$$x = \frac{17 \pm \sqrt{145}}{9}$$
The solutions are $\frac{17+\sqrt{145}}{9}$ and $\frac{17-\sqrt{145}}{9}$.

63. $x^2 + 11 = 0$
$$a = 1, b = 0, c = 11$$
$$x = \frac{0 \pm \sqrt{(0)^2 - 4(1)(11)}}{2(1)}$$
$$= \frac{\pm\sqrt{-44}}{2}$$
$$= \frac{\pm 2i\sqrt{11}}{2}$$
$$= \pm i\sqrt{11}$$
The solutions are $-i\sqrt{11}$ and $i\sqrt{11}$.

64.
$$(5a-2)^2 - a = 0$$
$$25a^2 - 20a + 4 - a = 0$$
$$25a^2 - 21a + 4 = 0$$
$$a = \frac{21 \pm \sqrt{(-21)^2 - 4(25)(4)}}{2(25)}$$
$$= \frac{21 \pm \sqrt{441 - 400}}{50}$$
$$= \frac{21 \pm \sqrt{41}}{50}$$
The solutions are $\frac{21+\sqrt{41}}{50}$ and $\frac{21-\sqrt{41}}{50}$.

65. $\frac{7}{8}=\frac{8}{x^2}$

$7x^2=64$

$x^2=\frac{64}{7}$

$x=\pm\sqrt{\frac{64}{7}}$

$x=\pm\frac{8}{\sqrt{7}}=\pm\frac{8\cdot\sqrt{7}}{\sqrt{7}\cdot\sqrt{7}}=\pm\frac{8\sqrt{7}}{7}$

The solutions are $-\frac{8\sqrt{7}}{7}$ and $\frac{8\sqrt{7}}{7}$.

66. $x^{2/3}-6x^{1/3}=-8$

$x^{2/3}-6x^{1/3}+8=0$

Let $y=x^{1/3}$. Then $y^2=x^{2/3}$ and

$y^2-6y+8=0$

$(y-4)(y-2)=0$

$y-4=0$ or $y-2=0$

$y=4$ or $y=2$

$x^{1/3}=4$ or $x^{1/3}=2$

$x=64$ or $x=8$

The solutions are 8 and 64.

67. $(2x-3)(4x+5)\ge 0$

$2x-3=0$ or $4x+5=0$

$x=\frac{3}{2}$ or $x=-\frac{5}{4}$

Region	Test Point	$(2x-3)(4x+5)\ge 0$ Result
$A: \left(-\infty, -\frac{5}{4}\right]$	-2	$(-7)(-3)\ge 0$ True
$B: \left[-\frac{5}{4}, \frac{3}{2}\right]$	0	$(-3)(5)\ge 0$ False
$C: \left[\frac{3}{2}, \infty\right)$	3	$(3)(17)\ge 0$ True

Solution: $\left(-\infty, -\frac{5}{4}\right]\cup\left[\frac{3}{2}, \infty\right)$

68. $\frac{x(x+5)}{4x-3}\ge 0$

$x=0$ or $x+5=0$ or $4x-3=0$

$x=-5$ or $x=\frac{3}{4}$

Region	Test Point	$\frac{x(x+5)}{4x-3}\ge 0$ Result
$A: (-\infty, -5]$	-6	$\frac{-6(-1)}{-27}\ge 0$ False
$B: [-5, 0]$	-1	$\frac{-1(4)}{-7}\ge 0$ True
$C: \left[0, \frac{3}{4}\right)$	$\frac{1}{2}$	$\frac{\frac{1}{2}\left(\frac{11}{2}\right)}{-1}\ge 0$ False
$D: \left(\frac{3}{4}, \infty\right)$	1	$\frac{1(6)}{1}\ge 0$ True

Solution: $[-5, 0]\cup\left(\frac{3}{4}, \infty\right)$

69. $\frac{3}{x-2}>2$

The denominator is equal to 0 when $x-2=0$, or $x=2$.

$\frac{3}{x-2}=2$

$3=2(x-2)$

$3=2x-4$

$7=2x$

$\frac{7}{2}=x$

Region	Test Point	$\frac{3}{x-2} > 2$ Result
$A: (-\infty, 2)$	0	$\frac{3}{-2} > 2$ False
$B: \left(2, \frac{7}{2}\right)$	3	$\frac{3}{1} > 2$ True
$C: \left(\frac{7}{2}, \infty\right)$	5	$\frac{3}{3} > 2$ False

Solution: $\left(2, \frac{7}{2}\right)$

70. $y = 6.46x^2 + 1236.5x + 7289$

a. $x = 2000 - 1980 = 20$

$$y = 6.46(20)^2 + 1236.5(20) + 7289 = 34,603 \text{ thousand}$$

The passenger traffic was approximately 34,603,000.

b. Let $y = 60,000$.

$$60,000 = 6.46x^2 + 1236.5x + 7289$$
$$0 = 6.46x^2 + 1236.5x - 52,711$$
$$x = \frac{-1236.5 \pm \sqrt{1236.5^2 - 4(6.46)(-52,711)}}{2(6.46)}$$

Choosing the positive root, $x \approx 36$, we see that there will be 60,000,000 passengers in $1980 + 36 = 2016$.

Chapter 11 Test

1.
$$5x^2 - 2x = 7$$
$$5x^2 - 2x - 7 = 0$$
$$(5x-7)(x+1) = 0$$
$$5x - 7 = 0 \quad \text{or} \quad x + 1 = 0$$
$$x = \frac{7}{5} \quad \text{or} \quad x = -1$$

The solutions are -1 and $\frac{7}{5}$.

2. $(x+1)^2 = 10$
$$x + 1 = \pm\sqrt{10}$$
$$x = -1 \pm \sqrt{10}$$

The solutions are $-1+\sqrt{10}$ and $-1-\sqrt{10}$.

3. $m^2 - m + 8 = 0$

$a = 1, b = -1, c = 8$

$$m = \frac{1 \pm \sqrt{(-1)^2 - 4(1)(8)}}{2(1)} = \frac{1 \pm \sqrt{1-32}}{2} = \frac{1 \pm \sqrt{-31}}{2} = \frac{1 \pm i\sqrt{31}}{2}$$

The solutions are $\frac{1+i\sqrt{31}}{2}$ and $\frac{1-i\sqrt{31}}{2}$.

4. $u^2 - 6u + 2 = 0$

$a = 1, b = -6, c = 2$

$$u = \frac{-(-6) \pm \sqrt{(-6)^2 - 4(1)(2)}}{2(1)} = \frac{6 \pm \sqrt{36-8}}{2} = \frac{6 \pm \sqrt{28}}{2} = \frac{6 \pm 2\sqrt{7}}{2} = 3 \pm \sqrt{7}$$

The solutions are $3+\sqrt{7}$ and $3-\sqrt{7}$.

5.
$$7x^2 + 8x + 1 = 0$$
$$(7x+1)(x+1) = 0$$
$$7x + 1 = 0 \quad \text{or} \quad x + 1 = 0$$
$$7x = -1 \qquad x = -1$$
$$x = -\frac{1}{7}$$

The solutions are $-\frac{1}{7}$ and -1.

6. $y^2 - 3y = 5$

$y^2 - 3y - 5 = 0$

$a = 1, b = -3, c = -5$

$$y = \frac{3 \pm \sqrt{(-3)^2 - 4(1)(-5)}}{2(1)} = \frac{3 \pm \sqrt{9+20}}{2} = \frac{3 \pm \sqrt{29}}{2}$$

The solutions are $\frac{3+\sqrt{29}}{2}$ and $\frac{3-\sqrt{29}}{2}$.

7. $\frac{4}{x+2} + \frac{2x}{x-2} = \frac{6}{x^2-4}$

$\frac{4}{x+2} + \frac{2x}{x-2} = \frac{6}{(x+2)(x-2)}$

$4(x-2) + 2x(x+2) = 6$

$4x - 8 + 2x^2 + 4x = 6$

$2x^2 + 8x - 14 = 0$

$x^2 + 4x - 7 = 0$

$a = 1, b = 4, c = -7$

$$x = \frac{-4 \pm \sqrt{(4)^2 - 4(1)(-7)}}{2(1)} = \frac{-4 \pm \sqrt{16+28}}{2} = \frac{-4 \pm \sqrt{44}}{2} = \frac{-4 \pm 2\sqrt{11}}{2} = -2 \pm \sqrt{11}$$

The solutions are $-2+\sqrt{11}$ and $-2-\sqrt{11}$.

8. $x^5 + 3x^4 = x + 3$

$x^5 + 3x^4 - x - 3 = 0$

$x^4(x+3) - 1(x+3) = 0$

$(x+3)(x^4 - 1) = 0$

$(x+3)(x^2+1)(x^2-1) = 0$

$x+3 = 0$ or $x^2+1 = 0$ or $x^2 - 1 = 0$

$x = -3$ or $x^2 = -1$ or $x^2 = 1$

$x = \pm i$ or $x = \pm 1$

The solutions are –3, –1, 1, $-i$, and i.

9. $x^6 + 1 = x^4 + x^2$

$x^6 - x^4 - x^2 + 1 = 0$

$x^4(x^2-1) - (x^2-1) = 0$

$(x^4-1)(x^2-1) = 0$

$(x^2+1)(x^2-1)(x+1)(x-1) = 0$

$(x^2+1)(x+1)^2(x-1)^2 = 0$

$x^2 + 1 = 0$ or $x + 1 = 0$ or $x - 1 = 0$

$x^2 = -1$ $\quad x = -1$ $\quad x = 1$

$x = \pm i$

The solutions are $-i$, i, –1, and 1.

10. $(x+1)^2 - 15(x+1) + 56 = 0$

Let $y = x + 1$. Then $y^2 = (x+1)^2$ and

$y^2 - 15y + 56 = 0$

$(y-8)(y-7) = 0$

$y = 8$ or $y = 7$

$x + 1 = 8$ or $x + 1 = 7$

$x = 7$ or $x = 6$

The solutions are 6 and 7.

11. $x^2 - 6x = -2$

$x^2 - 6x + \left(\frac{-6}{2}\right)^2 = -2 + 9$

$x^2 - 6x + 9 = 7$

$(x-3)^2 = 7$

$x - 3 = \pm\sqrt{7}$

$x = 3 \pm \sqrt{7}$

The solutions are $3+\sqrt{7}$ and $3-\sqrt{7}$.

12.

$$2a^2+5=4a$$
$$2a^2-4a=-5$$
$$a^2-2a=-\frac{5}{2}$$
$$a^2-2a+\left(\frac{-2}{2}\right)^2=-\frac{5}{2}+1$$
$$a^2-2a+1=-\frac{3}{2}$$
$$(a-1)^2=-\frac{3}{2}$$
$$a-1=\pm\sqrt{-\frac{3}{2}}=\pm\frac{i\sqrt{3}}{\sqrt{2}}$$
$$a-1=\pm\frac{i\sqrt{6}}{2}$$
$$a=1\pm\frac{i\sqrt{6}}{2} \text{ or } \frac{2\pm i\sqrt{6}}{2}$$

The solutions are $\frac{2+i\sqrt{6}}{2}$ and $\frac{2-i\sqrt{6}}{2}$.

13.

$$2x^2-7x>15$$
$$2x^2-7x-15>0$$
$$(2x+3)(x-5)>0$$
$$2x+3=0 \quad \text{or } x-5=0$$
$$x=-\frac{3}{2} \quad \text{or} \quad x=5$$

Region	Test Point	$(2x+3)(x-5)>0$ Result
A: $\left(-\infty, -\frac{3}{2}\right)$	-2	$(-1)(-7)>0$ True
B: $\left(-\frac{3}{2}, 5\right)$	0	$(3)(-5)>0$ False
C: $(5, \infty)$	6	$(15)(1)>0$ True

Solution: $\left(-\infty, -\frac{3}{2}\right)\cup(5, \infty)$

14. $(x^2-16)(x^2-25) \geq 0$

$(x+4)(x-4)(x+5)(x-5) \geq 0$

$x+4=0$ or $x-4=0$ or $x+5=0$ or $x-5=0$

$x=-4$ or $x=4$ or $x=-5$ or $x=5$

Region	Test Point	$(x+4)(x-4)(x+5)(x-5) \geq 0$ Result
A: $(-\infty, -5]$	-6	$(-2)(-10)(-1)(-11) \geq 0$ True
B: $[-5, -4]$	$-\frac{9}{2}$	$\left(-\frac{1}{2}\right)\left(-\frac{17}{2}\right)\left(\frac{1}{2}\right)\left(-\frac{19}{2}\right) \geq 0$ False
C: $[-4, 4]$	0	$(4)(-4)(5)(-5) \geq 0$ True
D: $[4, 5]$	$\frac{9}{2}$	$\left(\frac{17}{2}\right)\left(\frac{1}{2}\right)\left(\frac{19}{2}\right)\left(-\frac{1}{2}\right) \geq 0$ False
E: $[5, \infty)$	6	$(10)(2)(11)(1) \geq 0$ True

Solution: $(-\infty, -5] \cup [-4, 4] \cup [5, \infty)$

15. $\frac{5}{x+3} < 1$

The denominator is equal to 0 when $x + 3 = 0$, or $x = -3$.

$\frac{5}{x+3} = 1$

$5 = x+3$ so $x = 2$

Region	Test Point	$\frac{5}{x+3} < 1$ Result
A: $(-\infty, -3)$	-4	$\frac{5}{-1} < 1$ True
B: $(-3, 2)$	0	$\frac{5}{3} < 1$ False
C: $(2, \infty)$	3	$\frac{5}{6} < 1$ True

Solution: $(-\infty, -3) \cup (2, \infty)$

16. $\dfrac{7x-14}{x^2-9} \le 0$

$\dfrac{7(x-2)}{(x+3)(x-3)} \le 0$

$x-2=0$ or $x+3=0$ or $x-3=0$

$x=2$ or $x=-3$ or $x=3$

Region	Test Point	$\dfrac{7(x-2)}{(x+3)(x-3)} \le 0$ Result
A: $(-\infty, -3)$	-4	$\dfrac{7(-6)}{(-1)(-7)} \le 0$ True
B: $(-3, 2]$	0	$\dfrac{7(-2)}{(3)(-3)} \le 0$ False
C: $[2, 3)$	$\frac{5}{2}$	$\dfrac{7\left(\frac{1}{2}\right)}{\left(\frac{11}{2}\right)\left(-\frac{1}{2}\right)} \le 0$ True
D: $(3, \infty)$	4	$\dfrac{7(2)}{(7)(1)} \le 0$ False

Solution: $(-\infty, -3) \cup [2, 3)$

17. $f(x) = 3x^2$

Vertex: (0, 0)

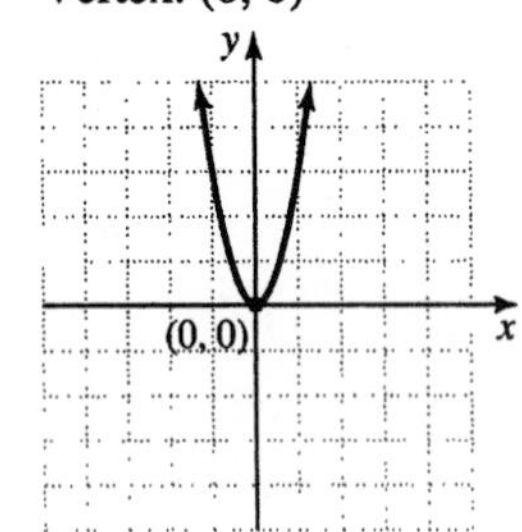

18. $G(x) = -2(x-1)^2 + 5$

Vertex: (1, 5)

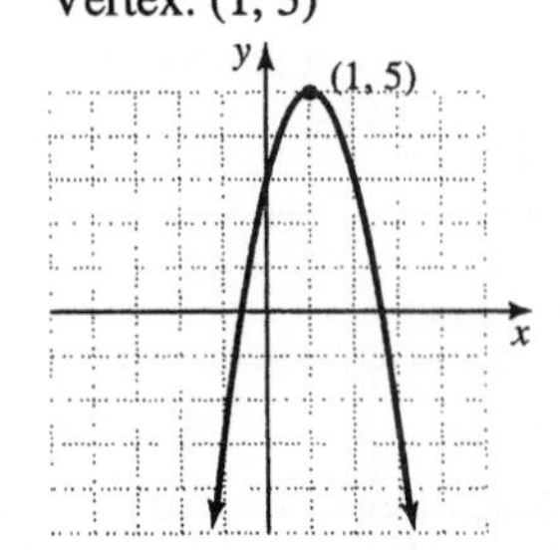

19. $h(x) = x^2 - 4x + 4$

$x = -\dfrac{b}{2a} = \dfrac{-(-4)}{2(1)} = 2$

$h(2) = (2)^2 - 4(2) + 4 = 0$

Vertex: (2, 0)

$h(0) = 4 \Rightarrow$ y-intercept: (0, 4)

x-intercept: (2, 0)

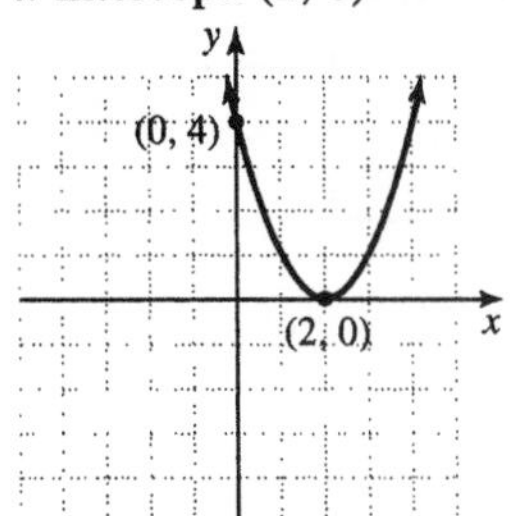

20. $F(x) = 2x^2 - 8x + 9$

$x = -\dfrac{b}{2a} = \dfrac{-(-8)}{2(2)} = 2$

$F(2) = 2(2)^2 - 8(2) + 9 = 1$

Vertex: (2, 1)

$F(0) = 9 \Rightarrow$ y-intercept: (0, 9)

$2x^2 - 8x + 9 = 0$

$a = 2, b = -8, c = 9$

$x = \dfrac{8 \pm \sqrt{(-8)^2 - 4(2)(9)}}{2(2)}$

$= \dfrac{8 \pm \sqrt{-8}}{4}$

which yields non-real solutons.
Therefore, there are no x-intercepts.

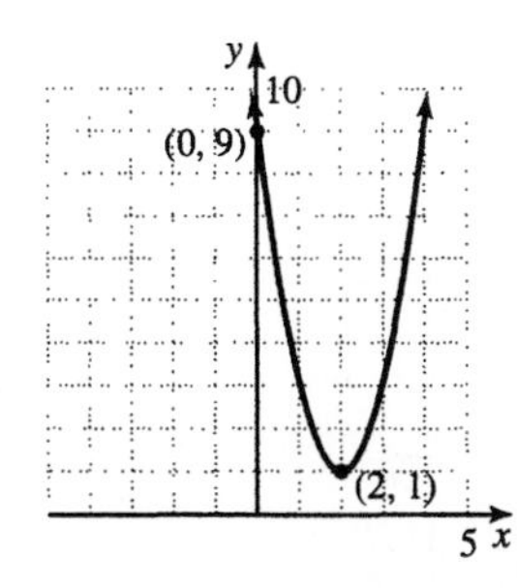

21. Let t = time for Sandy alone. Then $t-2$ = time for Dave alone.

$$\frac{1}{t}+\frac{1}{t-2}=\frac{1}{4}$$
$$4(t-2)+4t=t(t-2)$$
$$4t-8+4t=t^2-2t$$
$$0=t^2-10t+8$$
$$a=1, b=-10, c=8$$
$$t=\frac{10\pm\sqrt{(-10)^2-4(1)(8)}}{2(1)}$$
$$=\frac{10\pm\sqrt{68}}{2}$$
$$=\frac{10\pm 2\sqrt{17}}{2}$$
$$=5\pm\sqrt{17}$$
$$\approx 9.12 \text{ or } 0.88 \text{ (discard)}$$

It takes her about 9.12 hours.

22. $s(t)=-16t^2+32t+256$

a. $t=-\frac{b}{2a}=\frac{-32}{2(-16)}=1$

$s(1)=-16(1)^2+32(1)+256=272$

Vertex: (1, 272)

The maximum height is 272 feet.

b. $-16t^2+32t+256=0$

$$t^2-2t-16=0$$
$$a=1, b=-2, c=-16$$
$$t=\frac{2\pm\sqrt{(-2)^2-4(1)(-16)}}{2(1)}$$
$$=\frac{2\pm\sqrt{68}}{2}$$
$$=\frac{2\pm 2\sqrt{17}}{2}$$
$$=1\pm\sqrt{17}$$
$$\approx -3.12 \text{ and } 5.12$$

Disregard the negative. The stone will hit the water in about 5.12 seconds.

23.
$$a^2+b^2=c^2$$
$$x^2+(x+8)^2=(20)^2$$
$$x^2+(x^2+16x+64)=400$$
$$2x^2+16x-336=0$$
$$x^2+8x-168=0$$
$$a=1, b=8, c=-168$$
$$x=\frac{-8\pm\sqrt{(8)^2-4(1)(-168)}}{2(1)}$$
$$=\frac{-8\pm\sqrt{736}}{2}$$
$$\approx -17.565 \text{ or } 9.565$$

Disregard the negative.

$x\approx 9.6$

$x+8\approx 9.6+8=17.6$

$17.6+9.6=27.2$

$27.2-20=7.2$

They would save about 7 feet.

Chapter 11 Cumulative Review

1. Let $x = 2$ and $y = -5$.

a. $\frac{x-y}{12+x}=\frac{2-(-5)}{12+2}=\frac{2+5}{14}=\frac{7}{14}=\frac{1}{2}$

b. $x^2-3y=2^2-3(-5)=4+15=19$

2. $|3x-2|=-5$ which is impossible. Thus, there is no solution, or $\varnothing$.

3. a. $2x+3x+5+2=(2+3)x+(5+2)=5x+7$

b. $-5a-3+a+2=-5a+a-3+2=-4a-1$

c. $4y - 3y^2$ cannot be simplified.

d. $2.3x + 5x - 6 = 7.3x - 6$

e. $-\frac{1}{2}b + b = \left(-\frac{1}{2} + 1\right)b = \frac{1}{2}b$

4. $\begin{cases} -6x + y = 5 & (1) \\ 4x - 2y = 6 & (2) \end{cases}$

Multiply E1 by 2 and add to E2.

$$\begin{aligned} -12x + 2y &= 10 \\ 4x - 2y &= 6 \\ \hline -8x \quad &= 16 \\ x &= -2 \end{aligned}$$

Replace x with -2 in E1.

$$\begin{aligned} -6(-2) + y &= 5 \\ 12 + y &= 5 \\ y &= -7 \end{aligned}$$

The solution is $(-2, -7)$.

5. $\begin{cases} 2x + y = 7 \\ 2y = -4x \end{cases}$

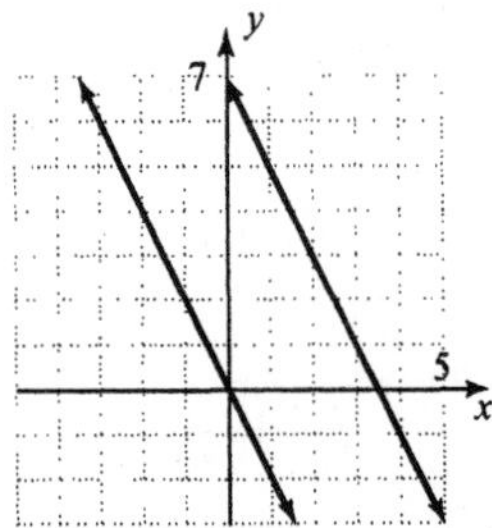

The system has no solution.

6. a. $(a^{-2}bc^3)^{-3} = (a^{-2})^{-3}b^{-3}(c^3)^{-3}$
$= a^6b^{-3}c^{-9}$
$= \dfrac{a^6}{b^3c^9}$

b. $\left(\dfrac{a^{-4}b^2}{c^3}\right)^{-2} = \dfrac{(a^{-4})^{-2}(b^2)^{-2}}{(c^3)^{-2}}$
$= \dfrac{a^8b^{-4}}{c^{-6}}$
$= \dfrac{a^8c^6}{b^4}$

c. $\left(\dfrac{3a^8b^2}{12a^5b^5}\right)^{-2} = \left(\dfrac{a^3}{4b^3}\right)^{-2}$
$= \dfrac{(a^3)^{-2}}{4^{-2}(b^3)^{-2}}$
$= \dfrac{4^2a^{-6}}{b^{-6}}$
$= \dfrac{16b^6}{a^6}$

7. $\begin{cases} 7x - 3y = -14 \\ -3x + y = 6 \end{cases}$

Solve the second equation for y.
$y = 3x + 6$
Substitute $3x + 6$ for y in the first equation.

$$\begin{aligned} 7x - 3(3x + 6) &= -14 \\ 7x - 9x - 18 &= -14 \\ -2x - 18 &= -14 \\ -2x &= 4 \\ x &= -2 \end{aligned}$$

Let $x = -2$ in $y = 3x + 6$.
$y = 3x + 6 = 3(-2) + 6 = -6 + 6 = 0$
The solution is $(-2, 0)$.

8. a. $(4a - 3)(7a - 2) = 28a^2 - 8a - 21a + 6$
$= 28a^2 - 29a + 6$

b. $(2a + b)(3a - 5b)$
$= 6a^2 - 10ab + 3ab - 5b^2$
$= 6a^2 - 7ab - 5b^2$

9. a. $\dfrac{x^5}{x^2} = x^{5-2} = x^3$

b. $\dfrac{4^7}{4^3} = 4^{7-3} = 4^4 = 256$

c. $\dfrac{(-3)^5}{(-3)^2} = (-3)^{5-2} = (-3)^3 = -27$

d. $\dfrac{s^2}{t^3}$ cannot be simplified.

e. $\dfrac{2x^5y^2}{xy} = 2x^{5-1}y^{2-1} = 2x^4y^1 = 2x^4y$

10. a. $9x^3+27x^2-15x=3x(3x^2+9x-5)$

b. $2x(3y-2)-5(3y-2)$
$=(3y-2)(2x-5)$

c. $2xy+6x-y-3=2x(y+3)-1(y+3)$
$=(y+3)(2x-1)$

11. $P(x)=2x^3-4x^2+5$

a. $P(2)=2(2)^3-4(2)^2+5$
$=2(8)-4(4)+5$
$=16-16+5$
$=5$

b.

2	2	−4	0	5
		4	0	0
	2	0	0	5

Thus, $P(2)=5$.

12. $x^2-2x-48=(x+6)(x-8)$

13. $(5x-1)(2x^2+15x+18)=0$
$(5x-1)(2x+3)(x+6)=0$
$5x-1=0$ or $2x+3=0$ or $x+6=0$
$5x=1$ $\quad 2x=-3$ $\quad x=-6$
$x=\frac{1}{5}$ $\quad x=-\frac{3}{2}$

The solutions are -6, $-\frac{3}{2}$, and $\frac{1}{5}$.

14. $2ax^2-12axy+18ay^2=2a(x^2-6xy+9y^2)$
$=2a(x-3y)(x-3y)$
$=2a(x-3y)^2$

15. $\frac{2x^2}{10x^3-2x^2}=\frac{2x^2}{2x^2(5x-1)}=\frac{1}{5x-1}$

16. $2(a^2+2)-8=-2a(a-2)-5$
$2a^2+4-8=-2a^2+4a-5$
$4a^2-4a+1=0$
$(2a-1)^2=0$
$2a-1=0$
$2a=1$
$a=\frac{1}{2}$

The solution is $\frac{1}{2}$.

17. $\frac{x^{-1}+2xy^{-1}}{x^{-2}-x^{-2}y^{-1}}=\frac{\frac{1}{x}+\frac{2x}{y}}{\frac{1}{x^2}-\frac{1}{x^2y}}$

$=\frac{\left(\frac{1}{x}+\frac{2x}{y}\right)x^2y}{\left(\frac{1}{x^2}-\frac{1}{x^2y}\right)x^2y}$

$=\frac{xy+2x^3}{y-1}$

18. $f(x)=x^2+x-12$
$x=-\frac{b}{2a}=-\frac{1}{2(1)}=-\frac{1}{2}$
$f\left(-\frac{1}{2}\right)=\left(-\frac{1}{2}\right)^2+\left(-\frac{1}{2}\right)-12$
$=\frac{1}{4}-\frac{1}{2}-12$
$=-\frac{49}{4}$

Vertex: $\left(-\frac{1}{2},-\frac{49}{4}\right)$

$x^2+x-12=0$
$(x+4)(x-3)=0$
$x+4=0$ or $x-3=0$
$x=-4$ $\quad x=3$

x-intercepts: $(-4, 0)$, $(3, 0)$

$f(0)=0^2+0-12=-12$

y-intercept: $(0, -12)$

19. $4m^2-4m+1=(2m)^2-2\cdot 2m\cdot 1+1^2=(2m-1)^2$

20. $\frac{x^2-4x+4}{2-x}=\frac{(x-2)^2}{-(x-2)}=\frac{x-2}{-1}=2-x$

21. Let x = the number.

$$x^2+3x=70$$
$$x^2+3x-70=0$$
$$(x+10)(x-7)=0$$
$$x+10=0 \quad \text{or} \quad x-7=0$$
$$x=-10 \qquad x=7$$

The number is −10 or 7.

22. $\dfrac{a+1}{a^2-6a+8}-\dfrac{3}{16-a^2}$

$$=\frac{a+1}{(a-4)(a-2)}-\frac{3}{(4+a)(4-a)}$$
$$=\frac{a+1}{(a-4)(a-2)}+\frac{3}{(4+a)(a-4)}$$
$$=\frac{(a+1)(a+4)+3(a-2)}{(a-4)(a-2)(a+4)}$$
$$=\frac{(a^2+4a+a+4)+3a-6}{(a-4)(a-2)(a+4)}$$
$$=\frac{a^2+8a-2}{(a-4)(a-2)(a+4)}$$

23. a. $\sqrt{25x^3}=\sqrt{25x^2\cdot x}=5x\sqrt{x}$

b. $\sqrt[3]{54x^6y^8}=\sqrt[3]{27x^6y^6\cdot 2y^2}$
$=3x^2y^2\sqrt[3]{2y^2}$

c. $\sqrt[4]{81z^{11}}=\sqrt[4]{81z^8\cdot z^3}=3z^2\sqrt[4]{z^3}$

24. $\dfrac{(2a)^{-1}+b^{-1}}{a^{-1}+(2b)^{-1}}=\dfrac{\frac{1}{2a}+\frac{1}{b}}{\frac{1}{a}+\frac{1}{2b}}$

$$=\frac{\left(\frac{1}{2a}+\frac{1}{b}\right)2ab}{\left(\frac{1}{a}+\frac{1}{2b}\right)2ab}$$
$$=\frac{b+2a}{2b+a}$$
$$=\frac{2a+b}{a+2b}$$

25. a. $\dfrac{2}{\sqrt{5}}=\dfrac{2\cdot\sqrt{5}}{\sqrt{5}\cdot\sqrt{5}}=\dfrac{2\sqrt{5}}{5}$

b. $\dfrac{2\sqrt{16}}{\sqrt{9x}}=\dfrac{2\cdot 4}{3\sqrt{x}}=\dfrac{8\cdot\sqrt{x}}{3\sqrt{x}\cdot\sqrt{x}}=\dfrac{8\sqrt{x}}{3x}$

c. $\sqrt[3]{\dfrac{1}{2}}=\dfrac{\sqrt[3]{1}}{\sqrt[3]{2}}=\dfrac{1}{\sqrt[3]{2}}=\dfrac{1\cdot\sqrt[3]{2^2}}{\sqrt[3]{2}\cdot\sqrt[3]{2^2}}=\dfrac{\sqrt[3]{4}}{2}$

26.

$$\begin{array}{r|l} & x^2-6x+8 \\ x+3 & x^3-3x^2-10x+24 \\ & \underline{x^3+3x^2} \\ & -6x^2-10x \\ & \underline{-6x^2-18x} \\ & 8x+24 \\ & \underline{8x+24} \\ & 0 \end{array}$$

Answer: x^2-6x+8

27. $\sqrt{2x+5}+\sqrt{2x}=3$

$$\sqrt{2x+5}=3-\sqrt{2x}$$
$$\left(\sqrt{2x+5}\right)^2=\left(3-\sqrt{2x}\right)^2$$
$$2x+5=9-6\sqrt{2x}+2x$$
$$-4=-6\sqrt{2x}$$
$$(-4)^2=\left(-6\sqrt{2x}\right)^2$$
$$16=36(2x)$$
$$16=72x$$
$$x=\frac{16}{72}=\frac{2}{9}$$

The solution is $\dfrac{2}{9}$.

28. $P(x)=4x^3-2x^2+3$

a. $P(-2)=4(-2)^3-2(-2)^2+3$
$=4(-8)-2(4)+3$
$=-32-8+3$
$=-37$

b.

$$\begin{array}{r|rrrr} -2 & 4 & -2 & 0 & 3 \\ & & -8 & 20 & -40 \\ \hline & 4 & -10 & 20 & -37 \end{array}$$

Thus, $P(-2)=-37$.

29. $\frac{x}{2}+\frac{8}{3}=\frac{1}{6}$

$6\left(\frac{x}{2}+\frac{8}{3}\right)=6\left(\frac{1}{6}\right)$

$3x+16=1$

$3x=-15$

$x=-5$

30. $\frac{x+3}{x^2+5x+6}=\frac{3}{2x+4}-\frac{1}{x+3}$

$\frac{x+3}{(x+3)(x+2)}=\frac{3}{2(x+2)}-\frac{1}{x+3}$

$2(x+3)=3(x+3)-2(x+2)$

$2x+6=3x+9-2x-4$

$2x+6=x+5$

$x=-1$

31. Let x = the number.

$\frac{x}{6}-\frac{5}{3}=\frac{x}{2}$

$6\left(\frac{x}{6}-\frac{5}{3}\right)=6\left(\frac{x}{2}\right)$

$x-10=3x$

$-10=2x$

$-5=x$

The number is -5.

32. Let t = time to roof the house together.

$\frac{1}{24}+\frac{1}{40}=\frac{1}{t}$

$120t\left(\frac{1}{24}+\frac{1}{40}\right)=120t\left(\frac{1}{t}\right)$

$5t+3t=120$

$8t=120$

$t=\frac{120}{8}=15$

It would take them 15 hours to roof the house working together.

33. $y=kx$

$5=k(30)$

$k=\frac{5}{30}=\frac{1}{6}$ and $y=\frac{1}{6}x$

34. $y=\frac{k}{x}$

$8=\frac{k}{14}$

$k=8(14)=112$ and $y=\frac{112}{x}$

35. a. $\sqrt{(-3)^2}=|-3|=3$

b. $\sqrt{x^2}=|x|$

c. $\sqrt[4]{(x-2)^4}=|x-2|$

d. $\sqrt[3]{(-5)^3}=-5$

e. $\sqrt[5]{(2x-7)^5}=2x-7$

f. $\sqrt{25x^2}=\sqrt{25}\cdot\sqrt{x^2}=5|x|$

g. $\sqrt{x^2+2x+1}=\sqrt{(x+1)^2}=|x+1|$

36. a. $\sqrt{(-2)^2}=|-2|=2$

b. $\sqrt{y^2}=|y|$

c. $\sqrt[4]{(a-3)^4}=|a-3|$

d. $\sqrt[3]{(-6)^3}=-6$

e. $\sqrt[5]{(3x-1)^5}=3x-1$

37. a. $\sqrt[8]{x^4}=x^{4/8}=x^{1/2}=\sqrt{x}$

b. $\sqrt[6]{25}=(25)^{1/6}$

$=(5^2)^{1/6}=5^{2/6}=5^{1/3}=\sqrt[3]{5}$

c. $\sqrt[4]{r^2s^6}=(r^2s^6)^{1/4}$

$=r^{2/4}s^{6/4}$

$=r^{1/2}s^{3/2}$

$=(rs^3)^{1/2}=\sqrt{rs^3}$

38. a. $\sqrt[4]{5^2}=5^{2/4}=5^{1/2}=\sqrt{5}$

b. $\sqrt[12]{x^3}=x^{3/12}=x^{1/4}=\sqrt[4]{x}$

c. $\sqrt[6]{x^2y^4} = (x^2y^4)^{1/6}$
$= x^{2/6}y^{4/6}$
$= x^{1/3}y^{2/3}$
$= (xy^2)^{1/3} = \sqrt[3]{xy^2}$

39. a. $\frac{2+i}{1-i} = \frac{(2+i)\cdot(1+i)}{(1-i)\cdot(1+i)}$
$= \frac{2+2i+1i+i^2}{1^2-i^2}$
$= \frac{2+3i-1}{1+1}$
$= \frac{1+3i}{2}$ or $\frac{1}{2}+\frac{3}{2}i$

b. $\frac{7}{3i} = \frac{7\cdot(-3i)}{3i\cdot(-3i)} = \frac{-21i}{-9i^2} = \frac{-21i}{9} = -\frac{7}{3}i$

40. a. $3i(5-2i) = 15i - 6i^2$
$= 15i + 6$
$= 6 + 15i$

b. $(6-5i)^2 = 6^2 - 2(6)(5i) + (5i)^2$
$= 36 - 60i + 25i^2$
$= 36 - 60i - 25$
$= 11 - 60i$

c. $\left(\sqrt{3}+2i\right)\left(\sqrt{3}-2i\right) = \left(\sqrt{3}\right)^2 - (2i)^2$
$= 3 - 4i^2$
$= 3 + 4$
$= 7$

41. $(x+1)^2 = 12$
$x+1 = \pm\sqrt{12}$
$x+1 = \pm 2\sqrt{3}$
$x = -1 \pm 2\sqrt{3}$
The solutions are $-1+2\sqrt{3}$ and $-1-2\sqrt{3}$.

42. $(y-1)^2 = 24$
$y-1 = \pm\sqrt{24}$
$y-1 = \pm 2\sqrt{6}$
$y = 1 \pm 2\sqrt{6}$
The solutions are $1+2\sqrt{6}$ and $1-2\sqrt{6}$.

43. $x - \sqrt{x} - 6 = 0$
Let $y = \sqrt{x}$. Then $y^2 = x$ and
$y^2 - y - 6 = 0$
$(y-3)(y+2) = 0$
$y-3 = 0$ or $y+2 = 0$
$y = 3$ or $y = -2$
$\sqrt{x} = 3$ or $\sqrt{x} = -2$ (can't happen)
$x = 9$
The solution is 9.

44. $m^2 = 4m + 8$
$m^2 - 4m - 8 = 0$
$a = 1, b = -4, c = -8$
$x = \frac{4 \pm \sqrt{(-4)^2 - 4(1)(-8)}}{2(1)}$
$= \frac{4 \pm \sqrt{16+32}}{2}$
$= \frac{4 \pm \sqrt{48}}{2}$
$= \frac{4 \pm 4\sqrt{3}}{2}$
$= 2 \pm 2\sqrt{3}$
The solutions are $2+2\sqrt{3}$ and $2-2\sqrt{3}$.

Chapter 12

Section 12.1

Practice Exercises

1. $f(x) = x + 2$; $g(x) = 3x + 5$

   a. $$\begin{aligned}(f+g)(x) &= f(x)+g(x)\\ &= (x+2)+(3x+5)\\ &= 4x+7\end{aligned}$$

   b. $$\begin{aligned}(f-g)(x) &= f(x)-g(x)\\ &= (x+2)-(3x+5)\\ &= x+2-3x-5\\ &= -2x-3\end{aligned}$$

   c. $$\begin{aligned}(f\cdot g)(x) &= f(x)\cdot g(x)\\ &= (x+2)(3x+5)\\ &= 3x^2+6x+5x+10\\ &= 3x^2+11x+10\end{aligned}$$

   d. $\left(\frac{f}{g}\right)(x) = \frac{f(x)}{g(x)} = \frac{x+2}{3x+5}$, where $x \neq -\frac{5}{3}$.

2. $f(x) = x^2 + 1$; $g(x) = 3x - 5$

   a. $(f\circ g)(4) = f(g(4)) = f(7) = 50$
   $(g\circ f)(4) = g(f(4)) = g(17) = 46$

   b. $$\begin{aligned}(f\circ g)(x) &= f(g(x))\\ &= f(3x-5)\\ &= (3x-5)^2+1\\ &= 9x^2-30x+26\end{aligned}$$
   $$\begin{aligned}(g\circ f)(x) &= g(f(x))\\ &= g(x^2+1)\\ &= 3(x^2+1)-5\\ &= 3x^2-2\end{aligned}$$

3. $f(x) = x^2 + 5$; $g(x) = x + 3$

   a. $$\begin{aligned}(f\circ g)(x) &= f(g(x))\\ &= f(x+3)\\ &= (x+3)^2+5\\ &= x^2+6x+14\end{aligned}$$

   b. $$\begin{aligned}(g\circ f)(x) &= g(f(x))\\ &= g(x^2+5)\\ &= (x^2+5)+3\\ &= x^2+8\end{aligned}$$

4. $f(x) = 3x$; $g(x) = x - 4$; $h(x) = |x|$

   a. $F(x) = |x - 4|$
   $$\begin{aligned}F(x) &= (h\circ g)(x)\\ &= h(g(x))\\ &= h(x-4)\\ &= |x-4|\end{aligned}$$

   b. $G(x) = 3x - 4$
   $$\begin{aligned}G(x) &= (g\circ f)(x)\\ &= g(f(x))\\ &= g(3x)\\ &= 3x-4\end{aligned}$$

Vocabulary and Readiness Check

1. C
2. E
3. F
4. A
5. D
6. B

Exercise Set 12.1

1. a. $(f+g)(x) = (x-7)+(2x+1) = 3x-6$

   b. $$\begin{aligned}(f-g)(x) &= (x-7)-(2x+1)\\ &= x-7-2x-1\\ &= -x-8\end{aligned}$$

   c. $(f\cdot g)(x) = (x-7)(2x+1) = 2x^2-13x-7$

   d. $\left(\frac{f}{g}\right)(x) = \frac{x-7}{2x+1}$, where $x \neq -\frac{1}{2}$.

3. a. $(f+g)(x) = (x^2+1)+5x = x^2+5x+1$

   b. $(f-g)(x) = (x^2+1)-5x = x^2-5x+1$

c. $(f \cdot g)(x) = (x^2+1)(5x) = 5x^3 + 5x$

d. $\left(\frac{f}{g}\right)(x) = \frac{x^2+1}{5x}$, where $x \neq 0$

5. a. $(f+g)(x) = \sqrt{x} + x + 5$

b. $(f-g)(x) = \sqrt{x} - (x+5) = \sqrt{x} - x - 5$

c. $(f \cdot g)(x) = \sqrt{x}(x+5)$
$= x\sqrt{x} + 5\sqrt{x}$

d. $\left(\frac{f}{g}\right)(x) = \frac{\sqrt{x}}{x+5}$; where $x \neq -5$.

7. a. $(f+g)(x) = -3x + 5x^2$ or $5x^2 - 3x$

b. $(f-g)(x) = -3x - 5x^2$ or $-5x^2 - 3x$

c. $(f \cdot g)(x) = (-3x)(5x^2) = -15x^3$

d. $\left(\frac{f}{g}\right)(x) = \frac{-3x}{5x^2}$
$= -\frac{3}{5x}$, where $x \neq 0$.

9. $(f \circ g)(2) = f(g(2))$
$= f(-4)$
$= (-4)^2 - 6(-4) + 2$
$= 16 + 24 + 2$
$= 42$

11. $(g \circ f)(-1) = g(f(-1))$
$= g(9)$
$= -2(9)$
$= -18$

13. $(g \circ h)(0) = g(h(0))$
$= g(0)$
$= -2(0)$
$= 0$

15. $(f \circ g)(x) = f(g(x))$
$= f(5x)$
$= (5x)^2 + 1$
$= 25x^2 + 1$

$(g \circ f)(x) = g(f(x))$
$= g(x^2+1)$
$= 5(x^2+1)$
$= 5x^2 + 5$

17. $(f \circ g)(x) = f(g(x))$
$= f(x+7)$
$= 2(x+7) - 3$
$= 2x + 14 - 3$
$= 2x + 11$
$(g \circ f)(x) = g(f(x))$
$= g(2x-3)$
$= (2x-3) + 7$
$= 2x + 4$

19. $(f \circ g)(x) = f(g(x))$
$= f(-2x)$
$= (-2x)^3 + (-2x) - 2$
$= -8x^3 - 2x - 2$
$(g \circ f)(x) = g(f(x))$
$= g(x^3 + x - 2)$
$= -2(x^3 + x - 2)$
$= -2x^3 - 2x + 4$

21. $(f \circ g)(x) = f(g(x))$
$= f(10x-3)$
$= |10x-3|$

$(g \circ f)(x) = g(f(x)) = g(|x|) = 10|x| - 3$

23. $(f \circ g)(x) = f(g(x)) = f(-5x+2) = \sqrt{-5x+2}$
$(g \circ f)(x) = g(f(x)) = g(\sqrt{x}) = -5\sqrt{x} + 2$

25. $H(x) = (g \circ h)(x)$
$= g(h(x))$
$= g(x^2+2)$
$= \sqrt{x^2+2}$

27. $F(x) = (h \circ f)(x)$
$= h(f(x))$
$= h(3x)$
$= (3x)^2 + 2$
$= 9x^2 + 2$

29. $G(x) = (f \circ g)(x)$
$= f(g(x))$
$= f\left(\sqrt{x}\right)$
$= 3\sqrt{x}$

31. Answers may vary. For example, $g(x) = x+2$ and $f(x) = x^2$.

33. Answers may vary. For example, $g(x) = x+5$ and $f(x) = \sqrt{x}+2$.

35. Answers may vary. For example, $g(x) = 2x-3$ and $f(x) = \frac{1}{x}$.

37. $x = y + 2$
$y = x - 2$

39. $x = 3y$
$y = \frac{x}{3}$

41. $x = -2y - 7$
$2y = -x - 7$
$y = -\frac{x+7}{2}$

43. $(f + g)(2) = f(2) + g(2) = 7 + (-1) = 6$

45. $(f \circ g)(2) = f(g(2)) = f(-1) = 4$

47. $(f \cdot g)(7) = f(7) \cdot g(7) = 1 \cdot 4 = 4$

49. $\left(\frac{f}{g}\right)(-1) = \frac{f(-1)}{g(-1)} = \frac{4}{-4} = -1$

51. Answers may vary.

53. Profit is equal to the revenue minus the cost; $P(x) = R(x) - C(x)$

Section 12.2

Practice Exercises

1. a. $f = \{(4, -3), (3, -4), (2, 7), (5, 0)\}$
f is one-to-one since each y-value corresponds to only one x-value.

b. $g = \{(8, 4), (-2, 0), (6, 4), (2, 6)\}$
g is not one-to-one because the y-value 4 in (8, 4) and (6, 4) corresponds to two different x-values.

c. $h = \{(2, 4), (1, 3), (4, 6), (-2, 4)\}$
h is not one-to-one because the y-value 4 in (2, 4) and (−2, 4) corresponds to two different x-values.

d.

Year	1950	1963	1968	1975	1997	2002
Federal Minimum Wage	\$0.75	\$1.25	\$1.60	\$2.10	\$5.15	\$5.15

This function is not one-to-one because the wage \$5.15 corresponds to two different years.

e. The function represented by the graph is not one-to-one because the y-value 2 in (2, 2) and (3, 2) corresponds to two different x-values.

f. The function represented by the diagram is not one-to-one because the score 509 corresponds to two different states.

2. Graphs **a**, **b**, and **c** all pass the vertical line test, so only these graphs are functions. But, of these, only **b** and **c** pass the horizontal line test, so only **b** and **c** are graphs of one-to-one functions.

3. $f(x) = \{(3, 4), (-2, 0), (2, 8), (6, 6)\}$
Switching the coordinates of each ordered pair gives $f^{-1}(x) = \{(4, 3), (0, -2), (8, 2), (6, 6)\}$

4. $f(x) = 6 - x$
Replace $f(x)$ with y.
$y = 6 - x$
Interchange x and y.
$x = 6 - y$
Solve for y.
$x = 6 - y$
$y = 6 - x$
Replace y with $f^{-1}(x)$.
$f^{-1}(x) = 6 - x$

5. $f(x) = 5x + 2$
Replace $f(x)$ with y.
$y = 5x + 2$
Interchange x and y.
$x = 5y + 2$
Solve for y.
$x = 5y + 2$
$x - 2 = 5y$
$\frac{x-2}{5} = y$

Replace y with $f^{-1}(x)$.

$f^{-1}(x) = \frac{x-2}{5}$

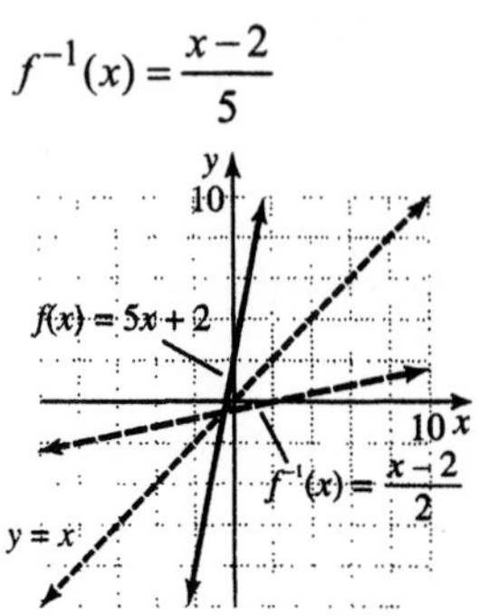

6. a. $f(x) = 2x - 3$

$y = 2x - 3$

$$x = 2y - 3$$
$$x + 3 = 2y$$
$$\frac{x+3}{2} = y$$
$$f^{-1}(x) = \frac{x+3}{2}$$

y 10 $f^{-1}(x) = \frac{x+3}{2}$ 10 x $f(x) = 2x - 3$ y = x

b. $f(x) = x^3$

$y = x^3$

$$x = y^3$$
$$\sqrt[3]{x} = y$$
$$f^{-1}(x) = \sqrt[3]{x}$$

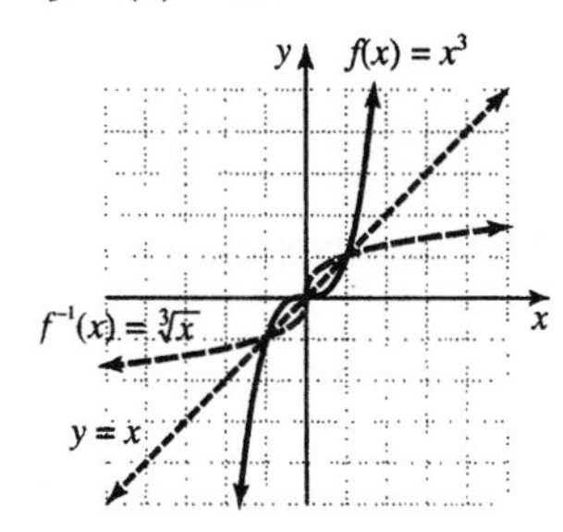

7. $f(x) = 4x - 1$; $f^{-1}(x) = \frac{x+1}{4}$

$$\begin{aligned}(f \circ f^{-1})(x) &= f(f^{-1}(x)) \\ &= f\left(\frac{x+1}{4}\right) \\ &= 4\left(\frac{x+1}{4}\right) - 1 \\ &= x + 1 - 1 \\ &= x\end{aligned}$$

$$\begin{aligned}(f^{-1} \circ f)(x) &= f^{-1}(f(x)) \\ &= f^{-1}(4x - 1) \\ &= \frac{(4x-1)+1}{4} \\ &= \frac{4x-1+1}{4} \\ &= \frac{4x}{4} \\ &= x\end{aligned}$$

Since $f \circ f^{-1} = x$ and $f^{-1} \circ f = x$, if $f(x) = 4x - 1$, $f^{-1}(x) = \frac{x+1}{4}$.

Vocabulary and Readiness Check

1. If $f(2) = 11$, the corresponding ordered pair is (2, 11).
2. The symbol f^{-1} means the inverse of f.
3. If (7, 3) is an ordered pair solution of $f(x)$, and $f(x)$ has an inverse, then an ordered pair solution of $f^{-1}(x)$ is (3, 7).
4. To tell whether a graph is the graph of a function, use the vertical line test.
5. To tell whether the graph of a function is also a one-to-one function, use the horizontal line test.
6. The graphs of f and f^{-1} are symmetric about the $y = x$ line.
7. Two functions are inverse of each other if $(f \circ f^{-1})(x) = x$ and $(f^{-1} \circ f)(x) = x$.

Exercise Set 12.2

1. $f = \{(-1,-1),(1,1),(0,2),(2,0)\}$ is a one-to-one function.

$f^{-1} = \{(-1,-1),(1,1),(2,0),(0,2)\}$

3. $h = \{(10,10)\}$ is a one-to-one function.

$h^{-1} = \{(10,10)\}$

5. $f = \{(11,12),(4,3),(3,4),(6,6)\}$
is a one-to-one function.
$f^{-1} = \{(12,11),(3,4),(4,3),(6,6)\}$

7. This function is not one-to-one because there are two pairs of two months with the same output: (January, 4.6) and (February, 4.6); (March, 4.4) and (April, 4.4).

9. This function is one-to-one.

Rank in Population (input)	1	19	35	4	48
State (output)	California	Maryland	Nevada	Florida	North Dakota

11. $f(x) = x^3 + 2$

a. $f(1) = 1^3 + 2 = 3$

b. $f^{-1}(3) = 1$

13. $f(x) = x^3 + 2$

a. $f(-1) = (-1)^3 + 2 = 1$

b. $f^{-1}(1) = -1$

15. The graph represents a one-to-one function because it passes the horizontal line test.

17. The graph does not represent a one-to-one function because it does not pass the horizontal line test.

19. The graph represents a one-to-one function because it passes the horizontal line test.

21. The graph does not represent a one-to-one function because it does not pass the horizontal line test.

23.
$$f(x) = x + 4$$
$$y = x + 4$$
$$x = y + 4$$
$$y = x - 4$$
$$f^{-1}(x) = x - 4$$

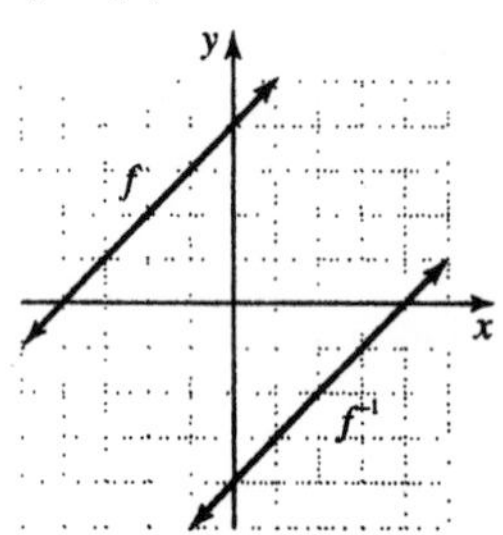

25.
$$\begin{aligned} f(x) &= 2x-3 \\ y &= 2x-3 \\ x &= 2y-3 \\ 2y &= x+3 \\ y &= \frac{x+3}{2} \\ f^{-1}(x) &= \frac{x+3}{2} \end{aligned}$$

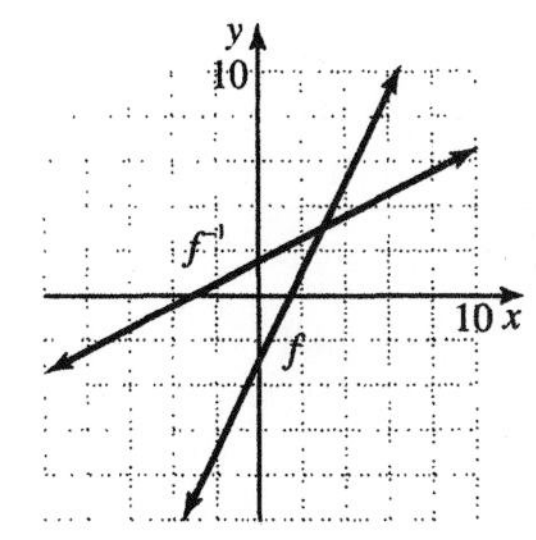

27.
$$\begin{aligned} f(x) &= \frac{1}{2}x-1 \\ y &= \frac{1}{2}x-1 \\ x &= \frac{1}{2}y-1 \\ \frac{1}{2}y &= x+1 \\ y &= 2x+2 \\ f^{-1}(x) &= 2x+2 \end{aligned}$$

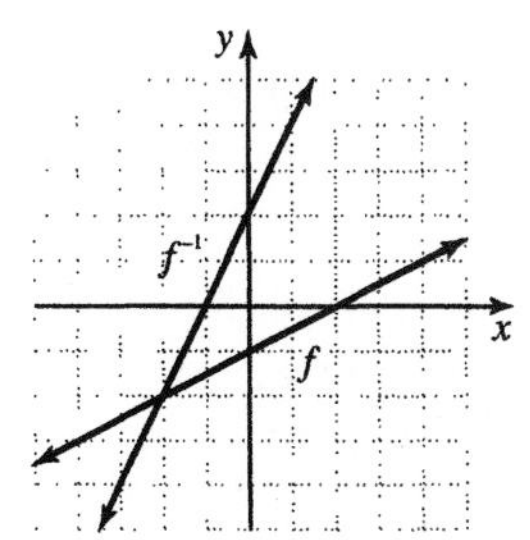

29.
$$\begin{aligned} f(x) &= x^3 \\ y &= x^3 \\ x &= y^3 \\ y &= \sqrt[3]{x} \\ f^{-1}(x) &= \sqrt[3]{x} \end{aligned}$$

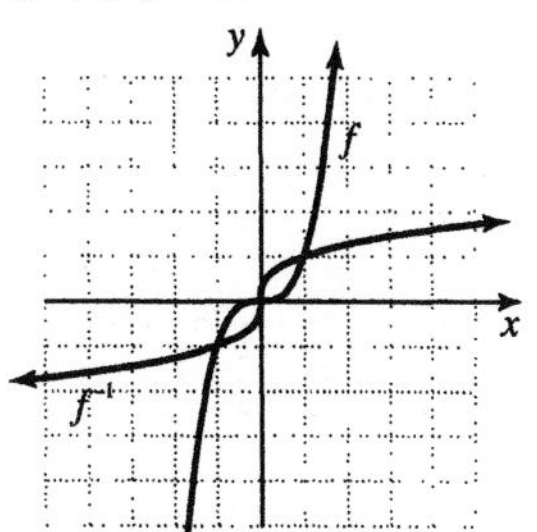

31.
$$\begin{aligned} f(x) &= 5x+2 \\ y &= 5x+2 \\ x &= 5y+2 \\ 5y &= x-2 \\ y &= \frac{x-2}{5} \\ f^{-1}(x) &= \frac{x-2}{5} \end{aligned}$$

33.
$$\begin{aligned} f(x) &= \frac{x-2}{5} \\ y &= \frac{x-2}{5} \\ x &= \frac{y-2}{5} \\ 5x &= y-2 \\ y &= 5x+2 \\ f^{-1}(x) &= 5x+2 \end{aligned}$$

35.
$$\begin{aligned} f(x) &= \sqrt[3]{x} \\ y &= \sqrt[3]{x} \\ x &= \sqrt[3]{y} \\ x^3 &= y \\ f^{-1}(x) &= x^3 \end{aligned}$$

37.
$$
\begin{aligned}
f(x) &= \frac{5}{3x+1}\\
y &= \frac{5}{3x+1}\\
x &= \frac{5}{3y+1}\\
3y+1 &= \frac{5}{x}\\
3y &= \frac{5}{x}-1\\
3y &= \frac{5-x}{x}\\
y &= \frac{5-x}{3x}\\
f^{-1}(x) &= \frac{5-x}{3x}
\end{aligned}
$$

39.
$$
\begin{aligned}
f(x) &= (x+2)^3\\
y &= (x+2)^3\\
x &= (y+2)^3\\
\sqrt[3]{x} &= y+2\\
\sqrt[3]{x}-2 &= y\\
f^{-1}(x) &= \sqrt[3]{x}-2
\end{aligned}
$$

41.

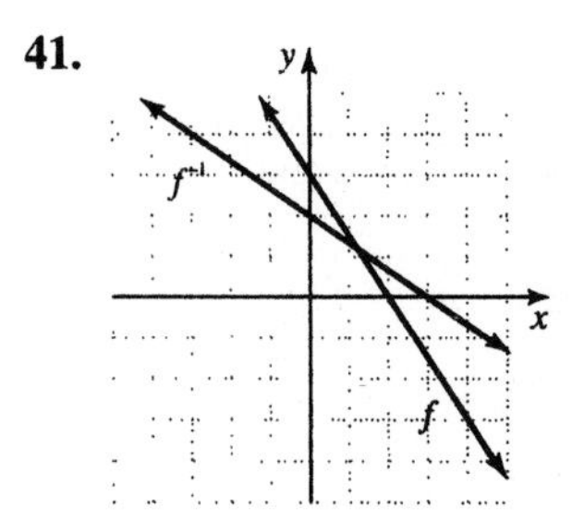

43.

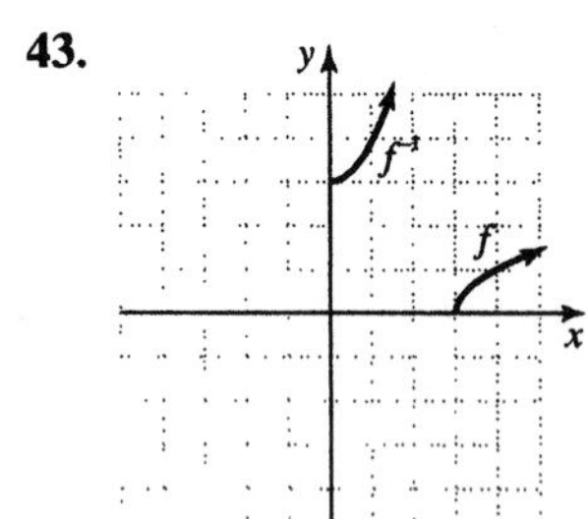

45.

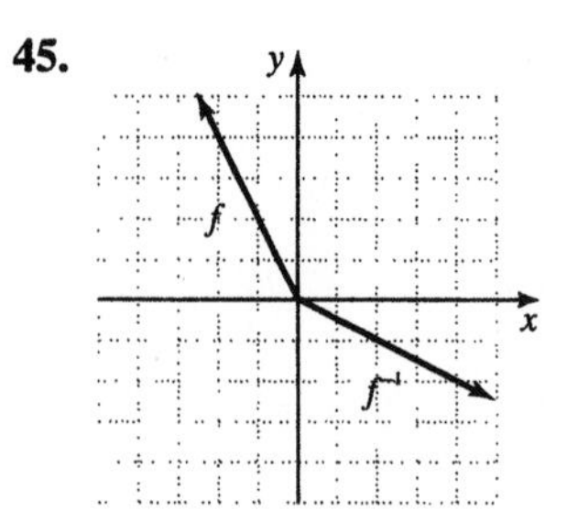

47.

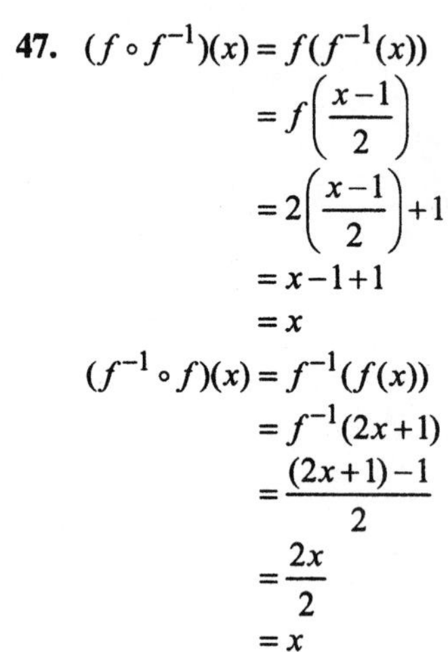

$$
\begin{aligned}
(f\circ f^{-1})(x) &= f(f^{-1}(x))\\
&= f\left(\frac{x-1}{2}\right)\\
&= 2\left(\frac{x-1}{2}\right)+1\\
&= x-1+1\\
&= x
\end{aligned}
$$

$$
\begin{aligned}
(f^{-1}\circ f)(x) &= f^{-1}(f(x))\\
&= f^{-1}(2x+1)\\
&= \frac{(2x+1)-1}{2}\\
&= \frac{2x}{2}\\
&= x
\end{aligned}
$$

49.
$$
\begin{aligned}
(f\circ f^{-1})(x) &= f(f^{-1}(x))\\
&= f\left(\sqrt[3]{x-6}\right)\\
&= \left(\sqrt[3]{x-6}\right)^3+6\\
&= x-6+6\\
&= x
\end{aligned}
$$

$$
\begin{aligned}
(f^{-1}\circ f)(x) &= f^{-1}(f(x))\\
&= f^{-1}(x^3+6)\\
&= \sqrt[3]{(x^3+6)-6}\\
&= \sqrt[3]{x^3}\\
&= x
\end{aligned}
$$

51. $25^{1/2} = \sqrt{25} = 5$

53. $16^{3/4} = \left(\sqrt[4]{16}\right)^3 = 2^3 = 8$

55. $9^{-3/2} = \frac{1}{9^{3/2}} = \frac{1}{\left(\sqrt{9}\right)^3} = \frac{1}{3^3} = \frac{1}{27}$

57. $f(x) = 3^x$
$f(2) = 3^2 = 9$

59. $f(x) = 3^x$

$f\left(\frac{1}{2}\right) = 3^{1/2} \approx 1.73$

61. $f(2) = 9$

a. $(2, 9)$

b. $(9, 2)$

63. a. $\left(-2, \frac{1}{4}\right), \left(-1, \frac{1}{2}\right), (0,1), (1,2), (2,5)$

b. $\left(\frac{1}{4}, -2\right), \left(\frac{1}{2}, -1\right), (1,0), (2,1), (5,2)$

c, d.

65. Answers may vary.

67.
$$f(x) = 3x + 1$$
$$y = 3x + 1$$
$$x = 3y + 1$$
$$x - 1 = 3y$$
$$y = \frac{x-1}{3}$$
$$f^{-1}(x) = \frac{x-1}{3}$$

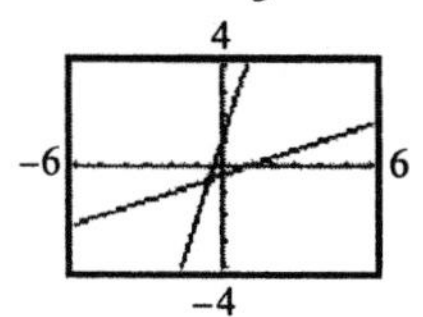

69. $f(x) = \sqrt[3]{x+1}$

$$y = \sqrt[3]{x+1}$$
$$x = \sqrt[3]{y+1}$$
$$x^3 = y+1$$
$$y = x^3 - 1$$
$$f^{-1}(x) = x^3 - 1$$

4
−6 6
−4

Section 12.3

Practice Exercises

1.

$f(x) = 2^x$	x	0	1	2	3	−1	−2
	$f(x)$	1	2	4	8	$\frac{1}{2}$	$\frac{1}{4}$

$g(x) = 7^x$	x	0	1	2	3	−1	−2
	$g(x)$	1	7	49	343	$\frac{1}{7}$	$\frac{1}{49}$

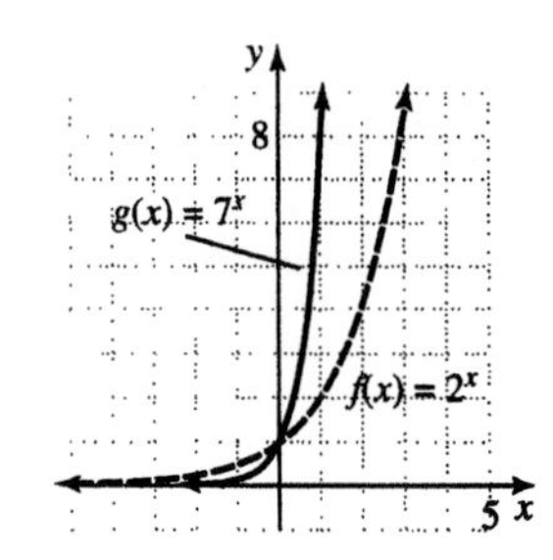

2.

$f(x) = \left(\frac{1}{3}\right)^x$	x	0	1	2	3	−1	−2
	$f(x)$	1	$\frac{1}{3}$	$\frac{1}{9}$	$\frac{1}{27}$	3	9

$g(x) = \left(\frac{1}{5}\right)^x$	x	0	1	2	3	−1	−2
	$g(x)$	1	$\frac{1}{5}$	$\frac{1}{25}$	$\frac{1}{125}$	5	25

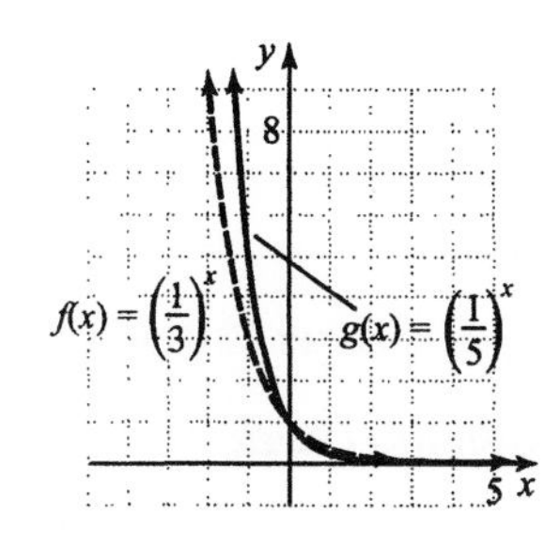

3. $f(x) = 2^{x+3}$

$y = 2^{x+3}$	x	0	−1	−2	−3	−4	−5
	y	8	4	2	1	$\frac{1}{2}$	$\frac{1}{4}$

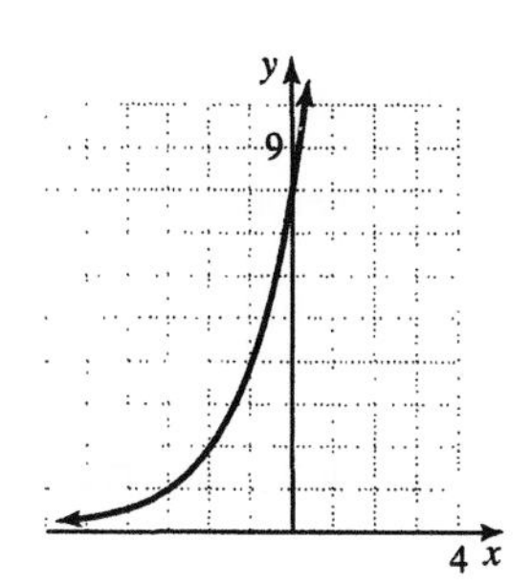

4. a. $3^x = 9$

Write 9 as a power of 3, $9 = 3^2$.

$3^x = 3^2$, thus, $x = 2$.

b. $8^x = 16$

Write 8 and 16 as powers of 2.

$8 = 2^3$ and $16 = 2^4$.

$$8^x = 16$$
$$(2^3)^x = 2^4$$
$$2^{3x} = 2^4$$
$$3x = 4$$
$$x = \frac{4}{3}$$

c. $125^x = 25^{x-2}$
Write 125 and 25 as powers of 5.
$125 = 5^3$ and $25 = 5^2$.
$$125^x = 25^{x-2}$$
$$(5^3)^x = (5^2)^{x-2}$$
$$5^{3x} = 5^{2x-4}$$
$$3x = 2x-4$$
$$x = -4$$

5. $P = \$3000$, $r = 7\% = 0.07$, $n = 2$, and $t = 4$.
$$A = P\left(1+\frac{r}{n}\right)^{nt}$$
$$A = 3000\left(1+\frac{0.07}{2}\right)^{2(4)}$$
$$= 3000(1.035)^8$$
$$\approx 3950.43$$
Thus, the amount A owed is approximately \$3950.43.

6. $p(n) = 100(2.7)^{-0.05n}$, $n = 10$ sheets of glass.
$$p(10) = 100(2.7)^{-0.05(10)} = 100(2.7)^{-0.5} \approx 60.86$$
Thus, approximately 60.86% of the light passes through.

Graphing Calculator Explorations

1.

90
1 3
80

The expected percent after 2 days is 81.98%.

2.

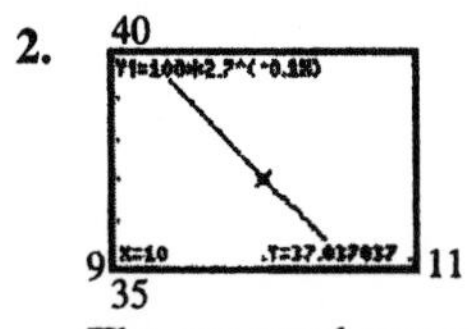

The expected percent after 10 days is 37.04%.

3.

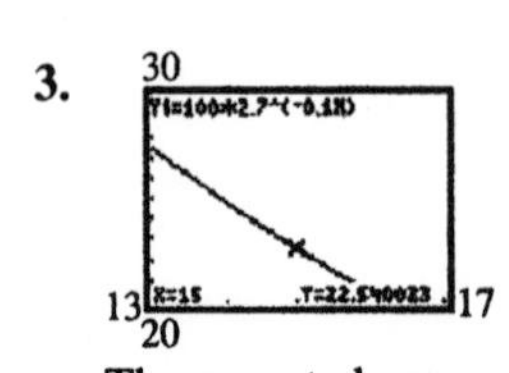

The expected percent after 15 days is 22.54%.

4.

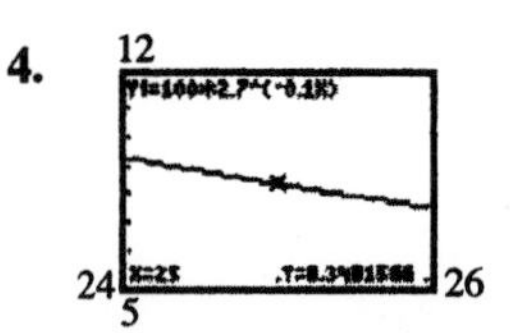

The expected percent after 25 days is 8.35%.

Vocabulary and Readiness Check

1. A function such as $f(x) = 2^x$ is an exponential function; **C**.

2. If $7^x = 7^y$, then $x = y$; **B**.

3. Yes, the function passes both the vertical- and horizontal-line tests.

4. The function has no x-intercept.

5. The function has a y-intercept of (0, 1).

6. The domain of this function, in interval notation, is $(-\infty, \infty)$.

7. The range of this function, in interval notation, is $(0, \infty)$.

Exercise Set 12.3

1. $y = 4^x$

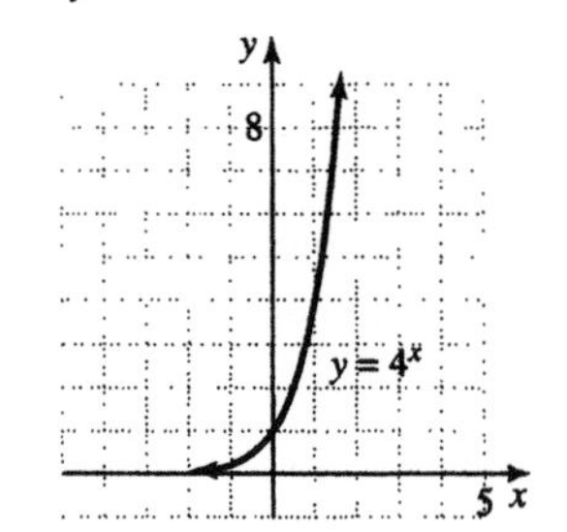

3. $y = 2^x + 1$

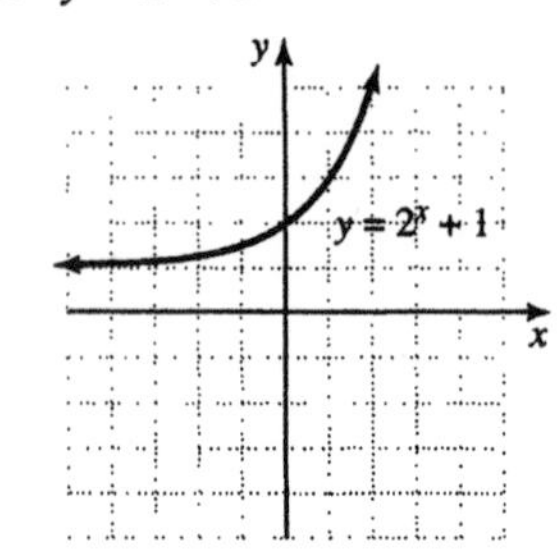

5. $y = \left(\frac{1}{4}\right)^x$

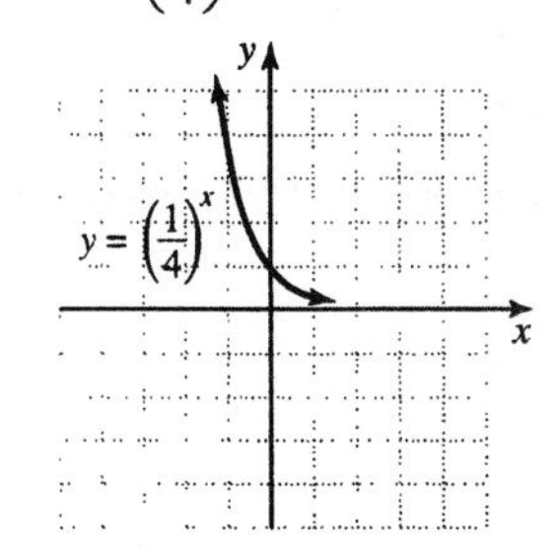

7. $y = \left(\frac{1}{2}\right)^x - 2$

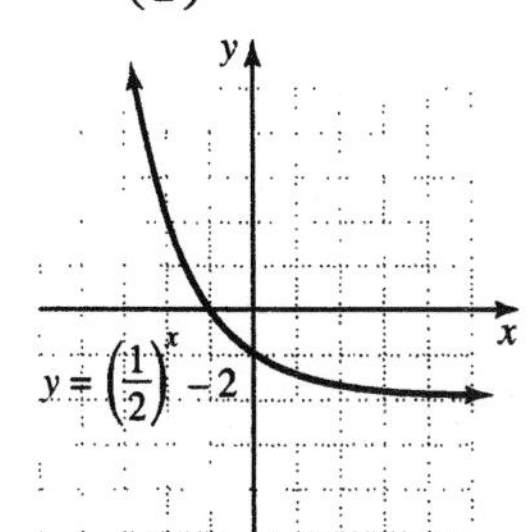

9. $y = -2^x$

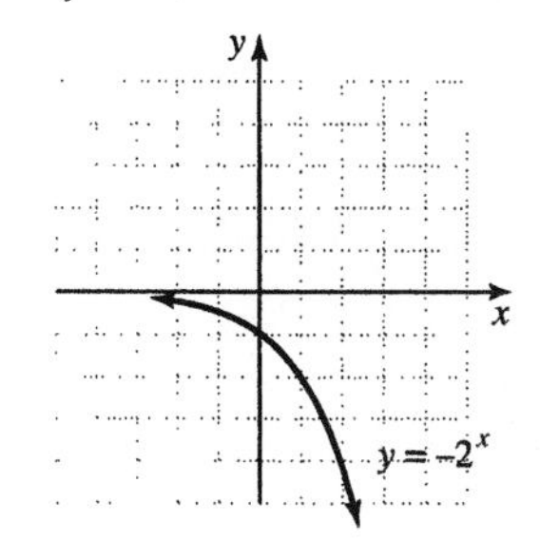

11. $y = -\left(\frac{1}{4}\right)^x$

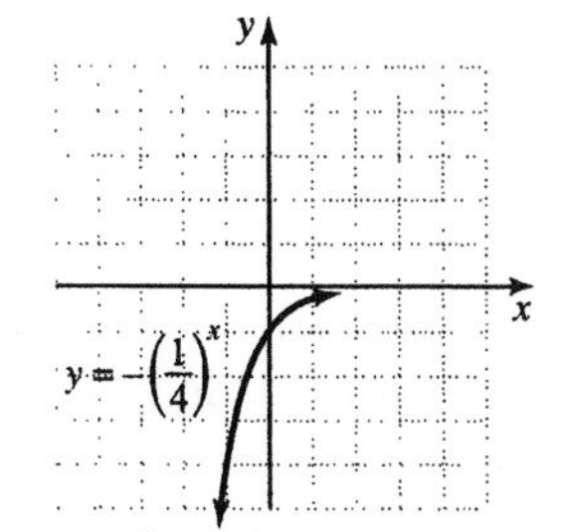

13. $f(x) = 2^{x+1}$

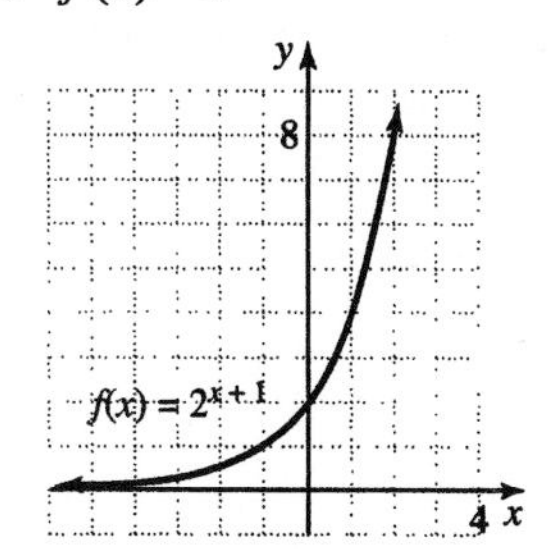

15. $f(x) = 4^{x-2}$

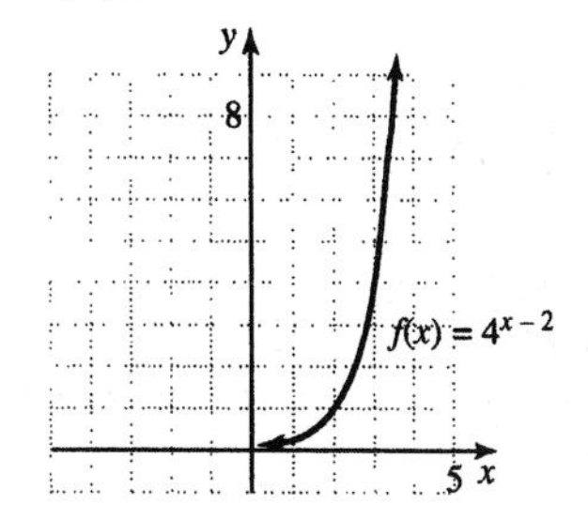

17. C

19. B

21. $3^x = 27$

$3^x = 3^3$

$x = 3$

The solution is 3.

23. $16^x = 8$

$(2^4)^x = 2^3$

$2^{4x} = 2^3$

$4x = 3$

$x = \frac{3}{4}$

The solution is $\frac{3}{4}$.

25. $32^{2x-3} = 2$

$(2^5)^{2x-3} = 2^1$

$2^{10x-15} = 2^1$

$10x - 15 = 1$

$10x = 16$

$x = \frac{8}{5}$

The solution is $\frac{8}{5}$.

27. $\frac{1}{4} = 2^{3x}$
$2^{-2} = 2^{3x}$
$3x = -2$
$x = -\frac{2}{3}$
The solution is $-\frac{2}{3}$.

29. $5^x = 625$
$5^x = 5^4$
$x = 4$
The solution is 4.

31. $4^x = 8$
$(2^2)^x = 2^3$
$2^{2x} = 2^3$
$2x = 3$
$x = \frac{3}{2}$
The solution is $\frac{3}{2}$.

33. $27^{x+1} = 9$
$(3^3)^{x+1} = 3^2$
$3^{3x+3} = 3^2$
$3x + 3 = 2$
$3x = -1$
$x = -\frac{1}{3}$
The solution is $-\frac{1}{3}$.

35. $81^{x-1} = 27^{2x}$
$(3^4)^{x-1} = (3^3)^{2x}$
$3^{4x-4} = 3^{6x}$
$4x - 4 = 6x$
$-4 = 2x$
$x = -2$
The solution is -2.

37. $y = 30(2.7)^{-0.004t}$, $t = 50$
$y = 30(2.7)^{-(0.004)(50)}$
$= 30(2.7)^{-0.2}$
≈ 24.6
Approximately 24.6 pounds of uranium will remain after 50 days.

39. $y = 260(2.7)^{0.025t}$, $t = 10$
$y = 260(2.7)^{0.025(10)}$
$= 260(2.7)^{0.25}$
≈ 333
There should be about 333 bison in the park in 10 years.

41. $y = 5(2.7)^{-0.15t}$, $t = 10$
$y = 5(2.7)^{-0.15(10)}$
$= 5(2.7)^{-1.5}$
≈ 1.1
After 10 seconds there will be about 1.1 grams.

43. $p(h) = 760(2.7)^{-0.145h}$

a. $p(1) = 760(2.7)^{-0.145(1)} \approx 658.1$
The pressure is 658.1 Pascals at a height of 1 kilometer.

b. $p(10) = 760(2.7)^{-0.145(10)}$
$= 760(2.7)^{-1.45}$
≈ 180.0
The pressure is 180.0 Pascals at a height of 10 kilometers.

45. $y = 84,949(1.096)^x$

a. $x = 2000 - 1995 = 5$
$y = 84,949(1.096)^5 \approx 134,342$
134,342 American students studied abroad in 2000.

b. $x = 2020 - 1995 = 25$
$y = 84,949(1.096)^{25} \approx 840,276$
840,276 American students would be studying abroad in 2020.

47. $A = P\left(1+\frac{r}{n}\right)^{nt}$

$t = 3$, $P = 6000$, $r = 0.08$, and $n = 12$

$$A = 6000\left(1+\frac{0.08}{12}\right)^{12(3)}$$
$$= 6000\left(1+\frac{0.08}{12}\right)^{36}$$
$$\approx 7621.42$$

Erica would owe \$7621.42 after 3 years.

49. $A = P\left(1+\frac{r}{n}\right)^{nt}$

$P = 2000$

$r = 0.06$, $n = 2$, and $t = 12$

$$A = 2000\left(1+\frac{0.06}{2}\right)^{2(12)}$$
$$= 2000(1.03)^{24}$$
$$\approx 4065.59$$

Janina has approximately \$4065.59 in her savings account.

51. $y = 18(1.24)^x$

$x = 2010 - 1994 = 16$

$y = 18(1.24)^{16} \approx 562$

There will be approximately 562 million cellular phone users in 2010.

53. $5x - 2 = 18$

$5x = 20$

$x = 4$

The solution is 4.

55. $3x - 4 = 3(x+1)$

$3x - 4 = 3x + 3$

$-4 = 3$

This is a false statement. The solution set is $\varnothing$.

57. $x^2 + 6 = 5x$

$x^2 - 5x + 6 = 0$

$(x-2)(x-3) = 0$

$x = 2$ or $x = 3$

The solutions are 2 and 3.

59. $2^x = 8$

$2^3 = 8$

$x = 3$

61. $5^x = \frac{1}{5}$

$5^{-1} = \frac{1}{5}$

$x = -1$

63. Answers may vary

65. $y = |3^x|$

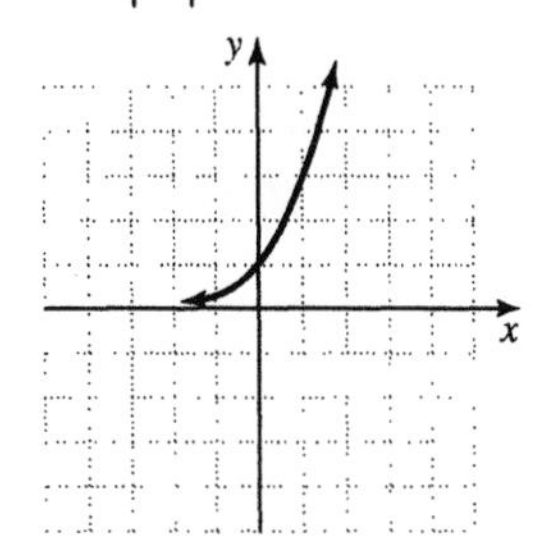

67. $y = 3^{|x|}$

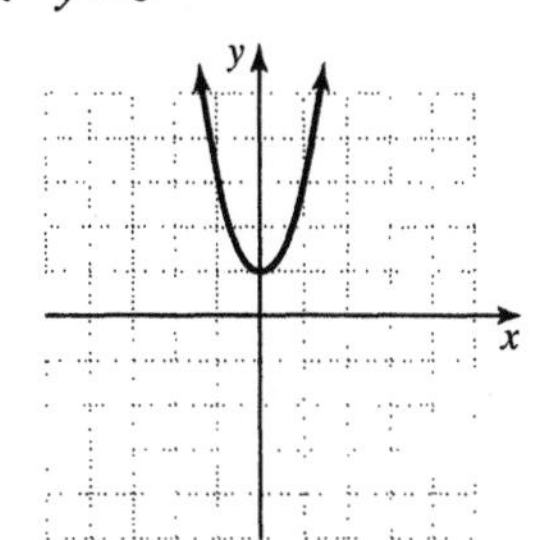

69.

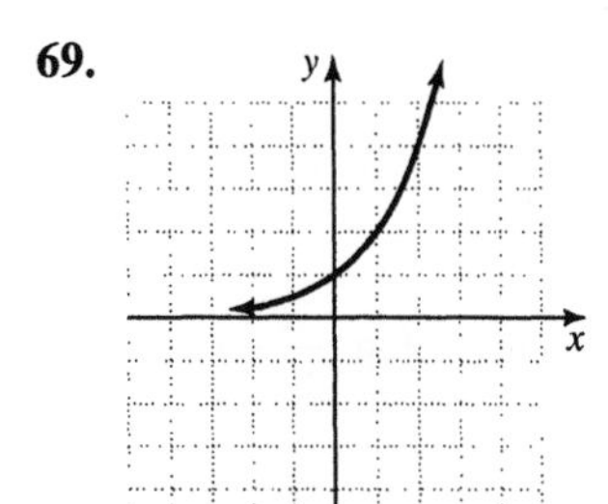

The graphs are the same, since $\left(\frac{1}{2}\right)^{-x} = 2^x$.

71.

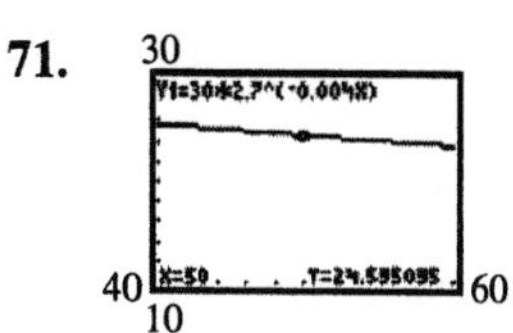

24.60 pounds remain after 50 days.

73.

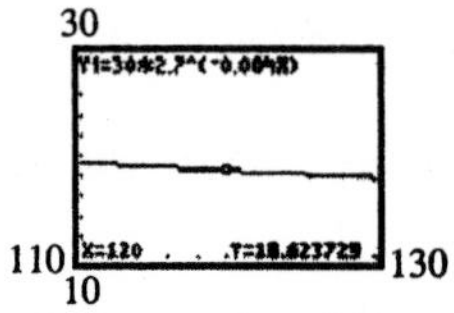

18.62 pounds will be available after 120 days.

75.

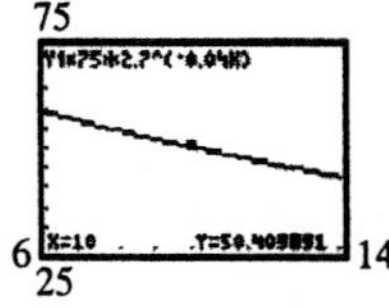

50.41 grams remain after 10 days.

Section 12.4

Practice Exercises

1. a. $\log_3 81 = 4$ means $3^4 = 81$.

b. $\log_5 \frac{1}{5} = -1$ means $5^{-1} = \frac{1}{5}$.

c. $\log_7 \sqrt{7} = \frac{1}{2}$ means $7^{1/2} = \sqrt{7}$.

d. $\log_{13} y = 4$ means $13^4 = y$.

2. a. $4^3 = 64$ means $\log_4 64 = 3$.

b. $6^{1/3} = \sqrt[3]{6}$ means $\log_6 \sqrt[3]{6} = \frac{1}{3}$.

c. $5^{-3} = \frac{1}{125}$ means $\log_5 \frac{1}{125} = -3$.

d. $\pi^7 = z$ means $\log_\pi z = 7$.

3. a. $\log_3 9 = 2$ because $3^2 = 9$.

b. $\log_2 \frac{1}{8} = -3$ because $2^{-3} = \frac{1}{8}$.

c. $\log_{49} 7 = \frac{1}{2}$ because $49^{1/2} = 7$.

4. a. $\log_5 \frac{1}{25} = x$

$\log_5 \frac{1}{25} = x$ means $5^x = \frac{1}{25}$. Solve

$5^x = \frac{1}{25}$.

$5^x = \frac{1}{25}$

$5^x = 5^{-2}$

Since the bases are the same, by the uniqueness of b^x, we have that $x = -2$. The solution is -2 or the solution set is $\{-2\}$.

b. $\log_x 8 = 3$

$x^3 = 8$

$x^3 = 2^3$

$x = 2$

c. $\log_6 x = 2$

$6^2 = x$

$36 = x$

d. $\log_{13} 1 = x$

$13^x = 1$

$13^x = 13^0$

$x = 0$

e. $\log_h 1 = x$

$h^x = 1$

$h^x = h^0$

$x = 0$

5. a. From Property 2, $\log_5 5^4 = 4$.

b. From Property 2, $\log_9 9^{-2} = -2$.

c. From Property 3, $6^{\log_6 5} = 5$.

d. From Property 3, $7^{\log_7 4} = 4$.

6. $y = \log_7 x$ means that $7^y = x$. Find some ordered pair solutions that satisfy $7^y = x$.

$x = 7^y$	y
1	0
7	1
$\frac{1}{7}$	−1
$\frac{1}{49}$	−2

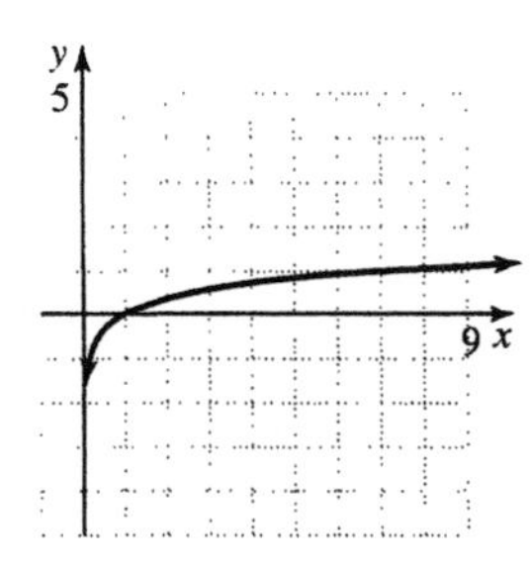

7. $y = \log_{1/4} x$ means that $\left(\frac{1}{4}\right)^y = x$. Find some ordered-pair solutions that satisfy $\left(\frac{1}{4}\right)^y = x$.

$x = \left(\frac{1}{4}\right)^y$	y
1	0
$\frac{1}{4}$	1
4	−1
16	−2

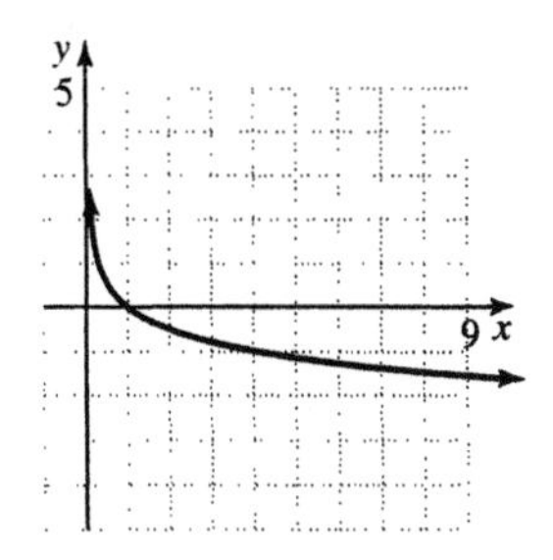

Vocabulary and Readiness Check

1. A function, such as $y = \log_2 x$ is a <u>logarithmic</u> function; **B**.

2. If $y = \log_2 x$, then $\underline{2^y = x}$; **C**.

3. Yes, the function passes both the horizontal- and vertical-line tests.

4. The function has an x-intercept of $\underline{(1, 0)}$.

5. The function has no y-intercept.

6. The domain of this function, in interval notation, is $\underline{(0, \infty)}$.

7. The range of this function, in interval notation, is $\underline{(-\infty, \infty)}$.

Exercise Set 12.4

1. $\log_6 36 = 2$
$6^2 = 36$

3. $\log_3 \frac{1}{27} = -3$
$3^{-3} = \frac{1}{27}$

5. $\log_{10} 1000 = 3$
$10^3 = 1000$

7. $\log_9 x = 4$
$9^4 = x$

9. $\log_\pi \frac{1}{\pi^2} = -2$
$\pi^{-2} = \frac{1}{\pi^2}$

11. $\log_7 \sqrt{7} = \frac{1}{2}$
$7^{1/2} = \sqrt{7}$

13. $\log_{0.7} 0.343 = 3$
$0.7^3 = 0.343$

15. $\log_3 \frac{1}{81} = -4$
$3^{-4} = \frac{1}{81}$

17. $2^4 = 16$
$\log_2 16 = 4$

19. $10^2 = 100$
$\log_{10} 100 = 2$

21. $\pi^3 = x$
$\log_\pi x = 3$

23. $10^{-1} = \frac{1}{10}$
$\log_{10} \frac{1}{10} = -1$

25. $4^{-2} = \frac{1}{16}$
$\log_4 \frac{1}{16} = -2$

27. $5^{1/2} = \sqrt{5}$
$\log_5 \sqrt{5} = \frac{1}{2}$

29. $\log_2 8 = 3$ since $2^3 = 8$.

31. $\log_3 \frac{1}{9} = -2$ since $3^{-2} = \frac{1}{9}$.

33. $\log_{25} 5 = \frac{1}{2}$ since $25^{1/2} = 5$.

35. $\log_{1/2} 2 = -1$ since $\left(\frac{1}{2}\right)^{-1} = 2$.

37. $\log_6 1 = 0$ since $6^0 = 1$.

39. $\log_{10} 100 = \log_{10} 10^2 = 2$

41. $\log_3 81 = \log_3 3^4 = 4$

43. $\log_4 \frac{1}{64} = \log_4 4^{-3} = -3$

45. $\log_3 9 = x$
$3^x = 9$
$3^x = 3^2$
$x = 2$

47. $\log_3 x = 4$
$x = 3^4 = 81$

49. $\log_x 49 = 2$
$x^2 = 49$
$x = \pm 7$
We discard the negative base.
$x = 7$

51. $\log_2 \frac{1}{8} = x$
$2^x = \frac{1}{8}$
$2^x = 2^{-3}$
$x = -3$

53. $\log_3 \frac{1}{27} = x$
$\frac{1}{27} = 3^x$
$3^{-3} = 3^x$
$-3 = x$

55. $\log_8 x = \frac{1}{3}$
$x = 8^{1/3} = 2$

57. $\log_4 16 = x$
$4^x = 16$
$4^x = 4^2$
$x = 2$

59. $\log_{3/4} x = 3$
$\left(\frac{3}{4}\right)^3 = x$
$\frac{27}{64} = x$

61. $\log_x 100 = 2$
$x^2 = 100$
$x = \pm 10$
We discard the negative base.
$x = 10$

63. $\log_2 2^4 = x$
$$2^x = 2^4$$
$$x = 4$$

65. $3^{\log_3 5} = x$
$$5 = x$$

67. $\log_x \frac{1}{7} = \frac{1}{2}$
$$x^{1/2} = \frac{1}{7}$$
$$x = \frac{1}{49}$$

69. $\log_5 5^3 = 3$

71. $2^{\log_2 3} = 3$

73. $\log_9 9 = 1$

75. $y = \log_3 x$
$y = 0$:
$\log_3 x = 0$
$$x = 3^0 = 1$$
$(1, 0)$ is the only x-intercept. No y-intercept exists.

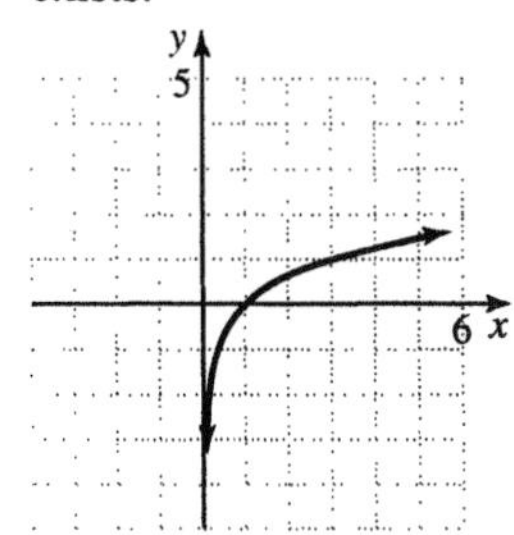

77. $f(x) = \log_{1/4} x$
$y = 0$: $0 = \log_{1/4} x$
$$x = \left(\frac{1}{4}\right)^0 = 1$$
$(1, 0)$ is the x-intercept. No y-intercept exists.

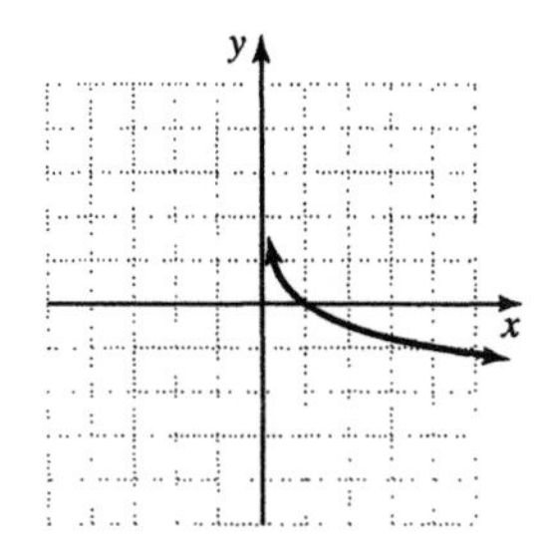

79. $f(x) = \log_5 x$
$y = 0$: $0 = \log_5 x$
$$x = 5^0 = 1$$
$(1, 0)$ is the x-intercept. No y-intercept exists.

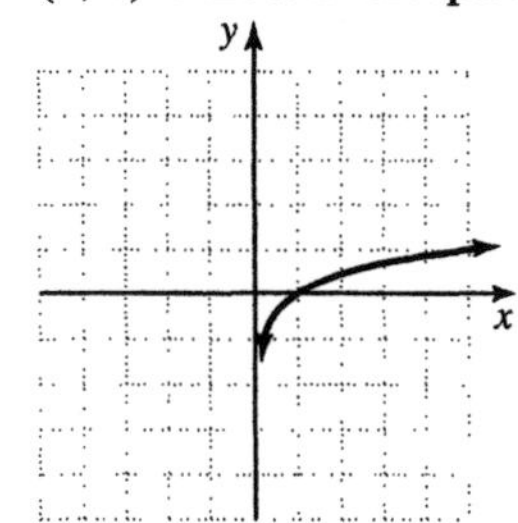

81. $f(x) = \log_{1/16} x$
$y = 0$:
$0 = \log_{1/6} x$
$$x = \left(\frac{1}{6}\right)^0 = 1$$
$(1, 0)$ is the x-intercept. No y-intercept exists.

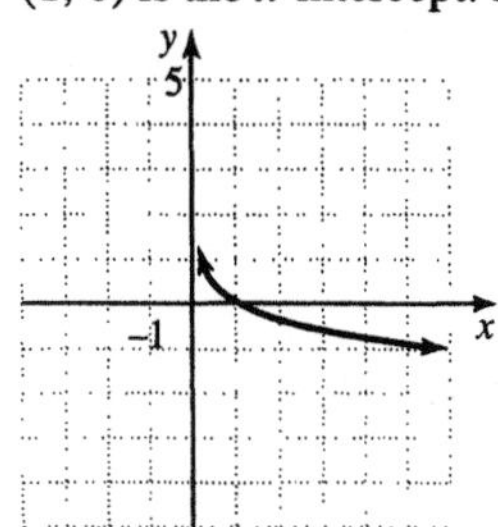

83. $\frac{x+3}{3+x} = \frac{x+3}{x+3} = 1$

85. $\frac{x^2 - 8x + 16}{2x - 8} = \frac{(x-4)^2}{2(x-4)} = \frac{x-4}{2}$

87. $\frac{2}{x} + \frac{3}{x^2} = \frac{2x}{x^2} + \frac{3}{x^2} = \frac{2x+3}{x^2}$

89. $\frac{m^2}{m+1} - \frac{1}{m+1} = \frac{m^2-1}{m+1}$
$= \frac{(m+1)(m-1)}{m+1}$
$= m-1$

91. $f(x) = \log_5 x$; $f^{-1}(x) = g(x) = 5^x$

a. (2, 25) implies $g(2) = 25$.

b. Since $f^{-1}(x) = g(x)$, (25, 2) is a solution of $f(x)$.

c. (25, 2) implies $f(25) = 2$.

93. Answers may vary

95. $\log_7(5x-2) = 1$
$5x - 2 = 7^1$
$5x = 9$
$x = \frac{9}{5}$

97. $\log_3(\log_5 125) = \log_3(3) = 1$

99. $y = 4^x$; $y = \log_4 x$

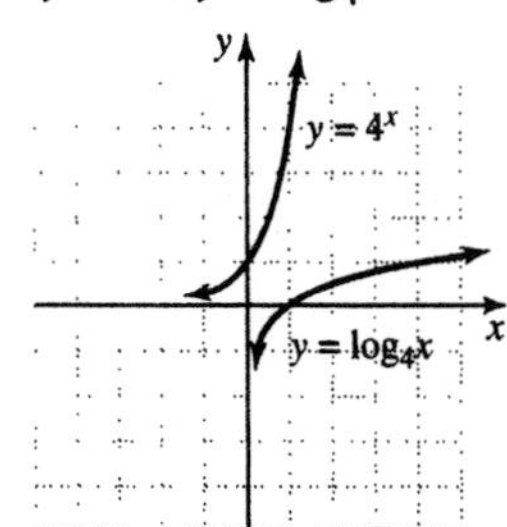

101. $y = \left(\frac{1}{3}\right)^x$; $y = \log_{1/3} x$

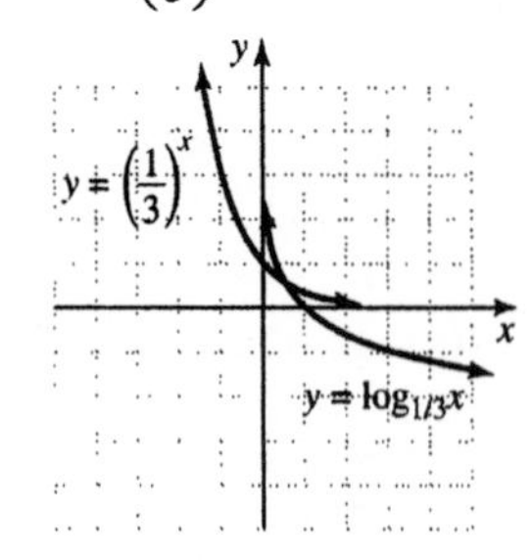

103. Answers may vary

105. $\log_{10}(1-k) = \frac{-0.3}{H}$, $H = 8$
$\log_{10}(1-k) = \frac{-0.3}{8} = -0.0375$
$1 - k = 10^{-0.0375}$
$1 - 10^{-0.0375} = k$
$k \approx 0.0827$
The rate of decay is 0.0827.

Section 12.5

Practice Exercises

1. a. $\log_8 5 + \log_8 3 = \log_8(5 \cdot 3) = \log_8 15$

b. $\log_2 \frac{1}{3} + \log_2 18 = \log_2\left(\frac{1}{3} \cdot 18\right) = \log_2 6$

c. $\log_5(x-1) + \log_5(x+1) = \log_5[(x+1)(x+1)]$
$= \log_5(x^2 - 1)$

2. a. $\log_5 18 - \log_5 6 = \log_5 \frac{18}{6} = \log_5 3$

b. $\log_6 x - \log_6 3 = \log_6 \frac{x}{3}$

c. $\log_4(x^2+1) - \log_4(x^2+3) = \log_4 \frac{x^2+1}{x^2+3}$

3. a. $\log_7 x^8 = 8\log_7 x$

b. $\log_5 \sqrt[4]{7} = \log_5 7^{1/4} = \frac{1}{4}\log_5 7$

4. a. $2\log_5 4 + 5\log_5 2 = \log_5 4^2 + \log_5 2^5$
$= \log_5 16 + \log_5 32$
$= \log_5(16 \cdot 32)$
$= \log_5 512$

b. $2\log_8 x - \log_8(x+3) = \log_8 x^2 - \log_8(x+3)$
$= \log_8 \frac{x^2}{x+3}$

c. $\log_7 12 + \log_7 5 - \log_7 4$
$= \log_7(12\cdot 5) - \log_7 4$
$= \log_7 60 - \log_7 4$
$= \log_7 \frac{60}{4}$
$= \log_7 15$

5. a. $\log_5 \frac{4\cdot 3}{7} = \log_5(4\cdot 3) - \log_5 7$
$= \log_5 4 + \log_5 3 - \log_5 7$

b. $\log_4 \frac{a^2}{b^5} = \log_4 a^2 - \log_4 b^5$
$= 2\log_4 a - 5\log_4 b$

6. $\log_b 5 = 0.83$ and $\log_b 3 = 0.56$

a. $\log_b 15 = \log_b(3\cdot 5)$
$= \log_b 3 + \log_b 5$
$= 0.56 + 0.83$
$= 1.39$

b. $\log_b 25 = \log_b 5^2 = 2\log_b 5 = 2(0.83) = 1.66$

c. $\log_b \sqrt{3} = \log_b 3^{1/2}$
$= \frac{1}{2}\log_b 3$
$= \frac{1}{2}(0.56)$
$= 0.28$

Vocabulary and Readiness Check

1. $\log_b 12 + \log_b 3 = \log_b(12\cdot 3) = \log_b \underline{36}$; **a.**

2. $\log_b 12 - \log_b 3 = \log_b \frac{12}{3} = \log_b \underline{4}$; **c.**

3. $7\log_b 2 = \underline{\log_b 2^7}$; **b.**

4. $\log_b 1 = \underline{0}$; **c.**

5. $b^{\log_b x} = \underline{x}$; **a.**

6. $\log_5 5^2 = \underline{2}$; **b.**

Exercise Set 12.5

1. $\log_5 2 + \log_5 7 = \log_5(2\cdot 7) = \log_5 14$

3. $\log_4 9 + \log_4 x = \log_4 9x$

5. $\log_6 x + \log_6(x+1) = \log_6[x(x+1)]$
$= \log_6(x^2 + x)$

7. $\log_{10} 5 + \log_{10} 2 + \log_{10}(x^2+2)$
$= \log_{10}\left[5\cdot 2\left(x^2+2\right)\right]$
$= \log_{10}\left(10x^2 + 20\right)$

9. $\log_5 12 - \log_5 4 = \log_5 \frac{12}{4} = \log_5 3$

11. $\log_3 8 - \log_3 2 = \log_3 \frac{8}{2} = \log_3 4$

13. $\log_2 x - \log_2 y = \log_2 \frac{x}{y}$

15. $\log_2(x^2+6) - \log(x^2+1) = \log_2 \frac{x^2+6}{x^2+1}$

17. $\log_3 x^2 = 2\log_3 x$

19. $\log_4 5^{-1} = -\log_4 5$

21. $\log_5 \sqrt{y} = \log_5 y^{1/2} = \frac{1}{2}\log_5 y$

23. $\log_2 5 + \log_2 x^3 = \log_2 5x^3$

25. $3\log_4 2 + \log_4 6 = \log_4 2^3 + \log_4 6$
$= \log_4 8 + \log_4 6$
$= \log_4(8\cdot 6)$
$= \log_4 48$

27. $3\log_5 x + 6\log_5 z = \log_5 x^3 + \log_5 z^6$
$= \log_5 x^3 z^6$

29. $\log_4 2 + \log_4 10 - \log_4 5 = \log_4(2\cdot 10) - \log_4 5$
$= \log_4 \frac{20}{5}$
$= \log_4 4$
$= 1$

31. $\log_7 6+\log_7 3-\log_7 4=\log_7(6\cdot 3)-\log_7 4$
$=\log_7 \frac{18}{4}$
$=\log_7 \frac{9}{2}$

33. $\log_{10} x-\log_{10}(x+1)+\log_{10}(x^2-2)$
$=\log_{10}\frac{x}{x+1}+\log_{10}(x^2-2)$
$=\log_{10}\frac{x(x^2-2)}{x+1}$
$=\log_{10}\frac{x^3-2x}{x+1}$

35. $3\log_2 x+\frac{1}{2}\log_2 x-2\log_2(x+1)$
$=\log_2 x^3+\log_2 x^{1/2}-\log_2(x+1)^2$
$=\log_2(x^3\cdot x^{1/2})-\log_2(x+1)^2$
$=\log_2 x^{7/2}-\log_2(x+1)^2$
$=\log_2\frac{x^{7/2}}{(x+1)^2}$

37. $2\log_8 x-\frac{2}{3}\log_8 x+4\log_8 x=\left(2-\frac{2}{3}+4\right)\log_8 x$
$=\frac{16}{3}\log_8 x$
$=\log_8 x^{16/3}$

39. $\log_3\frac{4y}{5}=\log_3 4y-\log_3 5$
$=\log_3 4+\log_3 y-\log_3 5$

41. $\log_4\frac{2}{9z}=\log_4 2-\log_4 9z$
$=\log_4 2-(\log_4 9+\log_4 z)$
$=\log_4 2-\log_4 9-\log_4 z$

43. $\log_2\frac{x^3}{y}=\log_2 x^3-\log_2 y$
$=3\log_2 x-\log_2 y$

45. $\log_b\sqrt{7x}=\log_b(7x)^{1/2}$
$=\frac{1}{2}\log_b(7x)$
$=\frac{1}{2}[\log_b 7+\log_b x]$
$=\frac{1}{2}\log_b 7+\frac{1}{2}\log_b x$

47. $\log_6 x^4y^5=\log_6 x^4+\log_6 y^5$
$=4\log_6 x+5\log_6 y$

49. $\log_5 x^3(x+1)=\log_5 x^3+\log_5(x+1)$
$=3\log_5 x+\log_5(x+1)$

51. $\log_6\frac{x^2}{x+3}=\log_6 x^2-\log_6(x+3)$
$=2\log_6 x-\log_6(x+3)$

53. $\log_b 15=\log_b(5\cdot 3)$
$=\log_b 5+\log_b 3$
$=0.7+0.5$
$=1.2$

55. $\log_b\frac{5}{3}=\log_b 5-\log_b 3=0.7-0.5=0.2$

57. $\log_b\sqrt{5}=\log_b 5^{1/2}=\frac{1}{2}\log_b 5=\frac{1}{2}(0.7)=0.35$

59. $\log_b 8=\log_b 2^3=3\log_b 2=3(0.43)=1.29$

61. $\log_b\frac{3}{9}=\log_b 3-\log_b 9$
$=\log_b 3-\log_b 3^2$
$=\log_b 3-2\log_b 3$
$=-\log_b 3$
$=-0.68$

63. $\log_b\sqrt{\frac{2}{3}}=\log_b\left(\frac{2}{3}\right)^{1/2}$
$=\frac{1}{2}\log_b\frac{2}{3}$
$=\frac{1}{2}(\log_b 2-\log_b 3)$
$=\frac{1}{2}(0.43-0.68)$
$=\frac{1}{2}(-0.25)$
$=-0.125$

65. $y = 10^x$ and $y = \log_{10} x$

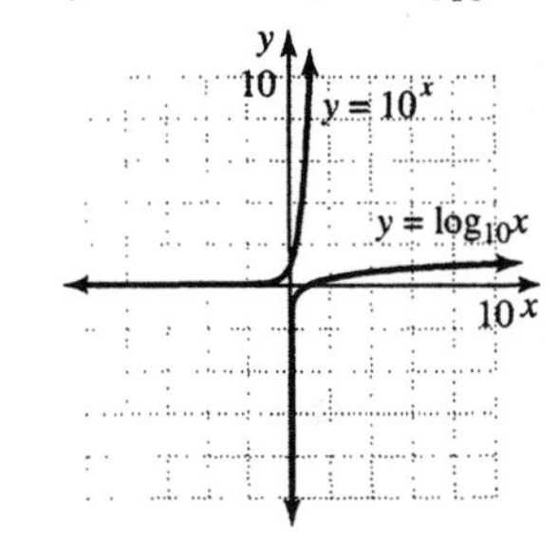

67.
$$\log_{10} \frac{1}{10} = x$$
$$\log_{10} 10^{-1} = x$$
$$-1 = x$$

69.
$$\log_7 \sqrt{7} = x$$
$$\log_7 7^{1/2} = x$$
$$\frac{1}{2} = x$$

71. $\log_9 \frac{21}{3} = \log_9 7 = \log_9 21 - \log_9 3$

The correct answers are **a** and **d**.

73. $\log_3(x+y) = \log_3 x + \log_3 y$ is false.

75. $\log_7 \frac{14}{8} = \log_7 14 - \log_7 8$ is true.

77. $(\log_3 6) \cdot (\log_3 4) = \log_3 24$ is false.

Integrated Review

1. $(f+g)(x) = x - 6 + x^2 + 1 = x^2 + x - 5$

2. $(f-g)(x) = x - 6 - (x^2 + 1) = -x^2 + x - 7$

3. $(f \cdot g)(x) = (x-6)(x^2+1) = x^3 - 6x^2 + x - 6$

4. $\left(\frac{f}{g}\right)(x) = \frac{x-6}{x^2+1}$

5. $(f \circ g)(x) = f(g(x)) = f(3x-1) = \sqrt{3x-1}$

6. $(g \circ f)(x) = g(f(x)) = g\left(\sqrt{x}\right) = 3\sqrt{x} - 1$

7. one-to-one; inverse:
$\{(6,-2),(8,4),(-6,2),(3,3)\}$

8. not one-to-one

9. not one-to-one

10. one-to-one

11. not one-to-one

12.
$$f(x) = 3x$$
$$y = 3x$$
$$x = 3y$$
$$y = \frac{x}{3}$$
$$f^{-1}(x) = \frac{x}{3}$$

13.
$$f(x) = x + 4$$
$$y = x + 4$$
$$x = y + 4$$
$$y = x - 4$$
$$f^{-1}(x) = x - 4$$

14.
$$f(x) = 5x - 1$$
$$y = 5x - 1$$
$$x = 5y - 1$$
$$5y = x + 1$$
$$y = \frac{x+1}{5}$$
$$f^{-1}(x) = \frac{x+1}{5}$$

15.
$$f(x) = 3x + 2$$
$$y = 3x + 2$$
$$x = 3y + 2$$
$$3y = x - 2$$
$$y = \frac{x-2}{3}$$
$$f^{-1}(x) = \frac{x-2}{3}$$

16. 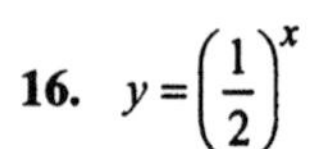

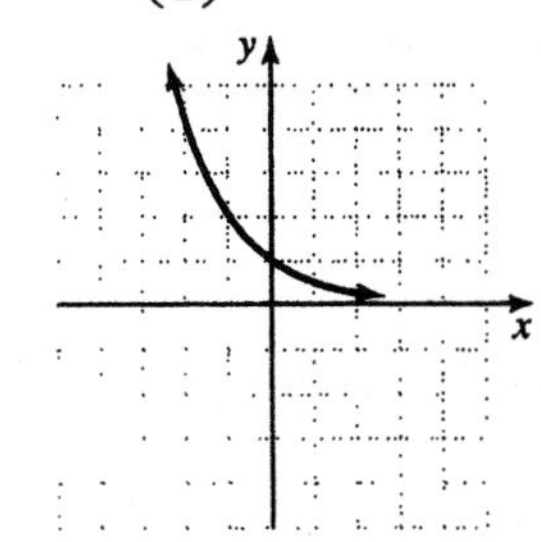

17. $y = 2^x + 1$

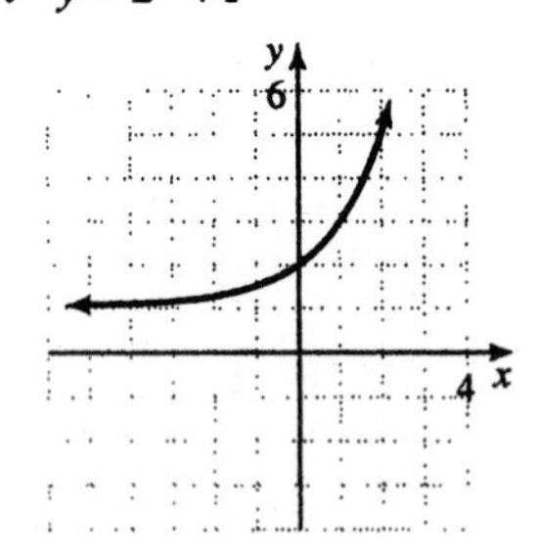

18. $y = \log_3 x$

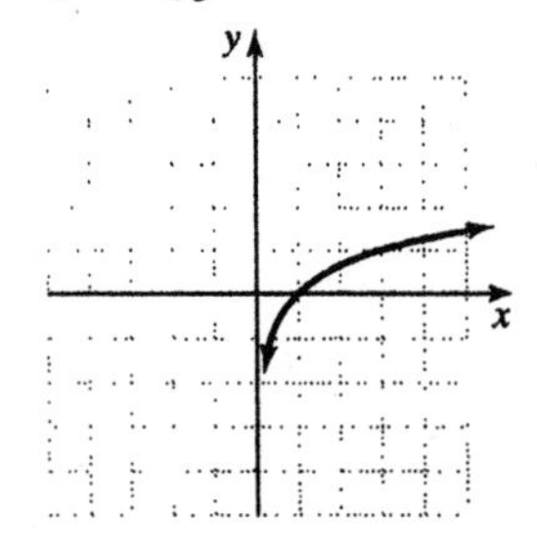

19. $y = \log_{1/3} x$

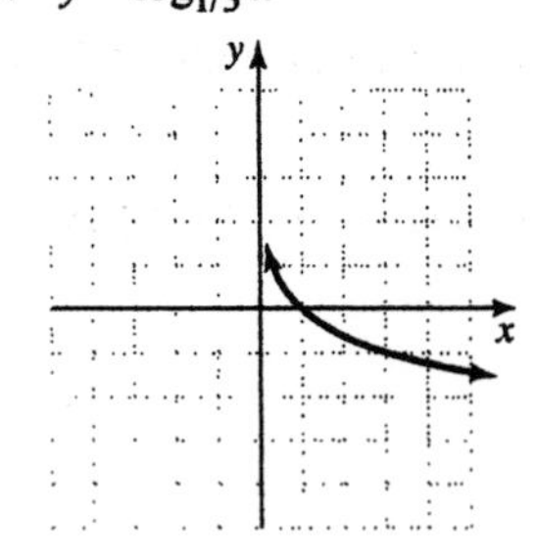

20. $2^x = 8$
$2^x = 2^3$
$x = 3$
The solution is 3.

21. $9 = 3^{x-5}$
$3^2 = 3^{x-5}$
$2 = x - 5$
$7 = x$
The solution is 7.

22. $4^{x-1} = 8^{x+2}$
$(2^2)^{x-1} = (2^3)^{x+2}$
$2^{2x-2} = 2^{3x+6}$
$2x - 2 = 3x + 6$
$-8 = x$
The solution is –8.

23. $25^x = 125^{x-1}$
$(5^2)^x = (5^3)^{x-1}$
$5^{2x} = 5^{3x-3}$
$2x = 3x - 3$
$3 = x$
The solution is 3.

24. $\log_4 16 = x$
$4^x = 16$
$4^x = 4^2$
$x = 2$
The solution is 2.

25. $\log_{49} 7 = x$
$49^x = 7$
$(7^2)^x = 7$
$7^{2x} = 7$
$2x = 1$
$x = \frac{1}{2}$
The solution is $\frac{1}{2}$.

26. $\log_2 x = 5$
$2^5 = x$
$32 = x$
The solution is 32.

27. $\log_x 64 = 3$
$x^3 = 64$
$x^3 = 4^3$
$x = 4$
The solution is 4.

28. $\log_x \frac{1}{125} = -3$

$x^{-3} = \frac{1}{125}$

$x^{-3} = 5^{-3}$

$x = 5$

The solution is 5.

29. $\log_3 x = -2$

$3^{-2} = x$

$x = \frac{1}{3^2} = \frac{1}{9}$

The solution is $\frac{1}{9}$.

30. $5\log_2 x = \log_2 x^5$

31. $x\log_2 5 = \log_2 5^x$

32. $3\log_5 x - 5\log_5 y = \log_5 x^3 - \log_5 y^5 = \log_5 \frac{x^3}{y^5}$

33. $9\log_5 x + 3\log_5 y = \log_5 x^9 + \log_5 y^3$

$= \log_5 x^9 y^3$

34. $\log_2 x + \log_2(x-3) - \log_2(x^2+4)$

$= \log_2[x(x-3)] - \log_2(x^2+4)$

$= \log_2(x^2-3x) - \log_2(x^2+4)$

$= \log_2 \frac{x^2-3x}{x^2+4}$

35. $\log_3 y - \log_3(y+2) + \log_3(y^3+11)$

$= \log_3 \frac{y}{y+2} + \log_3(y^3+11)$

$= \log_3 \frac{y(y^3+11)}{y+2}$

$= \log_3 \frac{y^4+11y}{y+2}$

36. $\log_7 \frac{9x^2}{y} = \log_7 9x^2 - \log_7 y$

$= \log_7 9 + \log_7 x^2 - \log_7 y$

$= \log_7 9 + 2\log_7 x - \log_7 y$

37. $\log_6 \frac{5y}{z^2} = \log_6 5y - \log_6 z^2$

$= \log_6 5 + \log_6 y - 2\log_6 z$

Section 12.6

Practice Exercises

1. To four decimal places, $\log 15 \approx 1.1761$.

2. a. $\log \frac{1}{100} = \log 10^{-2} = 2$

b. $\log 100{,}000 = \log 10^5 = 5$

c. $\log \sqrt[5]{10} = \log 10^{1/5} = \frac{1}{5}$

d. $\log 0.001 = \log 10^{-3} = -3$

3. $\log x = 3.4$

$x = 10^{3.4}$

$x \approx 2511.8864$

4. $a = 450$ micrometers

$T = 4.2$ seconds

$B = 3.6$

$R = \log\left(\frac{a}{T}\right) + B$

$= \log\left(\frac{450}{4.2}\right) + 3.6$

$\approx 2.0 + 3.6$

$= 5.6$

The earthquake had a magnitude of 5.6 on the Richter scale.

5. To four decimal places, $\ln 13 \approx 2.5649$.

6. a. $\ln e^4 = 4$

b. $\ln \sqrt[3]{e} = \ln e^{1/3} = \frac{1}{3}$

7. $\ln 5x = 8$

$e^8 = 5x$

$\frac{e^8}{5} = x$

$x = \frac{1}{5}e^8 \approx 596.1916$

8. $P = \$2400$
$r = 6\% = 0.06$
$t = 4$ years
$A = Pe^{rt} = 2400e^{0.06(4)} = 2400e^{0.24} \approx 3051.00$
The total amount of money owed is $3051.00.

9. $\log_8 5 = \frac{\log 5}{\log 8} \approx \frac{0.6989700043}{0.903089987} \approx 0.773976$

To four decimal places, $\log_8 5 \approx 0.7740$.

Vocabulary and Readiness Check

1. The base of log 7 is <u>10</u>; **c.**

2. The base of ln 7 is <u>*e*</u>; **a.**

3. $\log_{10} 10^7 = \underline{7}$; **b.**

4. $\log_7 1 = \underline{0}$; **d.**

5. $\log_e e^5 = \underline{5}$; **b.**

6. $\ln e^5 = \underline{5}$; **b.**

7. $\log_2 7 = \underline{\frac{\log 7}{\log 2} = \frac{\ln 7}{\ln 2}}$; **a** and **b.**

Exercise Set 12.6

1. $\log 8 \approx 0.9031$

3. $\log 2.31 \approx 0.3636$

5. $\ln 2 \approx 0.6931$

7. $\ln 0.0716 \approx -2.6367$

9. $\log 12.6 \approx 1.1004$

11. $\ln 5 \approx 1.6094$

13. $\log 41.5 \approx 1.6180$

15. Answers may vary

17. $\log 100 = \log 10^2 = 2$

19. $\log\left(\frac{1}{1000}\right) = \log 10^{-3} = -3$

21. $\ln e^2 = 2$

23. $\ln \sqrt[4]{e} = \ln e^{1/4} = \frac{1}{4}$

25. $\log 10^3 = 3$

27. $\ln e^{-7} = -7$

29. $\log 0.0001 = \log 10^{-4} = -4$

31. $\ln \sqrt{e} = \ln e^{1/2} = \frac{1}{2}$

33. $\ln 2x = 7$
$2x = e^7$
$x = \frac{1}{2}e^7 \approx 548.3166$

35. $\log x = 1.3$
$x = 10^{1.3} \approx 19.9526$

37. $\log 2x = 1.1$
$2x = 10^{1.1}$
$x = \frac{10^{1.1}}{2} \approx 6.2946$

39. $\ln x = 1.4$
$x = e^{1.4} \approx 4.0552$

41. $\ln(3x-4) = 2.3$
$3x - 4 = e^{2.3}$
$3x = 4 + e^{2.3}$
$x = \frac{4 + e^{2.3}}{3} \approx 4.6581$

43. $\log x = 2.3$
$x = 10^{2.3} \approx 199.5262$

45. $\ln x = -2.3$
$x = e^{-2.3} \approx 0.1003$

47. $\log(2x+1) = -0.5$
$2x + 1 = 10^{-0.5}$
$2x = 10^{-0.5} - 1$
$x = \frac{10^{-0.5} - 1}{2} \approx -0.3419$

49. $\ln 4x = 0.18$
$4x = e^{0.18}$
$x = \frac{e^{0.18}}{4} \approx 0.2993$

51. $\log_2 3 = \frac{\log 3}{\log 2} \approx 1.5850$

53. $\log_{1/2} 5 = \frac{\ln 5}{\ln\left(\frac{1}{2}\right)} \approx -2.3219$

55. $\log_4 9 = \frac{\ln 9}{\ln 4} \approx 1.5850$

57. $\log_3\left(\frac{1}{6}\right) = \frac{\log\left(\frac{1}{6}\right)}{\log 3} \approx -1.6309$

59. $\log_8 6 = \frac{\log 6}{\log 8} \approx 0.8617$

61. $R = \log\left(\frac{a}{T}\right) + B,\ a = 200,\ T = 1.6$
$B = 2.1$
$R = \log\left(\frac{200}{1.6}\right) + 2.1 \approx 4.2$
The earthquake measures 4.2 on the Richter scale.

63. $R = \log\left(\frac{a}{T}\right) + B,\ a = 400,\ T = 2.6$
$B = 3.1$
$R = \log\left(\frac{400}{2.6}\right) + 3.1 \approx 5.3$
The earthquake measures 5.3 on the Richter scale.

65. $A = Pe^{rt},\ t = 12,\ P = 1400,\ r = 0.08$
$A = 1400e^{(0.08)12} = 1400e^{0.96} \approx 3656.38$
Dana has \$3656.38 after 12 years.

67. $A = Pe^{rt},\ t = 4,\ P = 2000,\ r = 0.06$
$A = 2000e^{(0.06)4} = 2000e^{0.24} \approx 2542.50$
Barbara owes \$2542.50 at the end of 4 years.

69. $6x - 3(2 - 5x) = 6$
$6x - 6 + 15x = 6$
$21x = 12$
$x = \frac{12}{21} = \frac{4}{7}$
The solution is $\frac{4}{7}$.

71. $2x + 3y = 6x$
$3y = 4x$
$x = \frac{3y}{4}$

73. $x^2 + 7x = -6$
$x^2 + 7x + 6 = 0$
$(x + 6)(x + 1) = 0$
$x + 6 = 0$ or $x + 1 = 0$
$x = -6$ or $x = -1$
The solutions are −6 and −1.

75. $\begin{cases} x + 2y = -4 \\ 3x - y = 9 \end{cases}$
Multiply the second equation by 2, then add.
$x + 2y = -4$
$6x - 2y = 18$
$7x = 14$
$x = 2$
Replace x with 2 in the first equation.
$x + 2y = -4$
$2 + 2y = -4$
$2y = -6$
$y = -3$
The solution is (2, −3).

77. ln 50 is larger. Answers may vary

79. $f(x) = e^x$

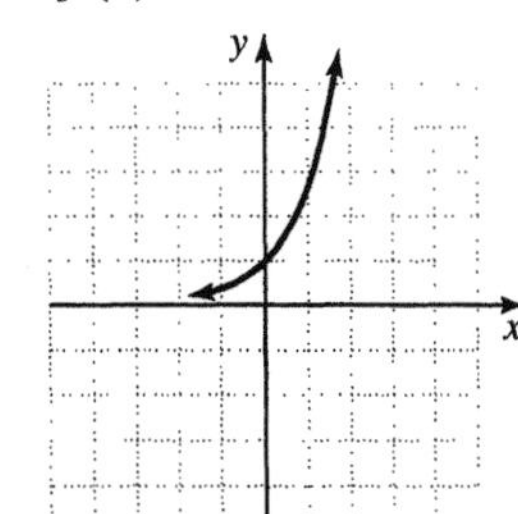

81. $f(x) = e^{-3x}$

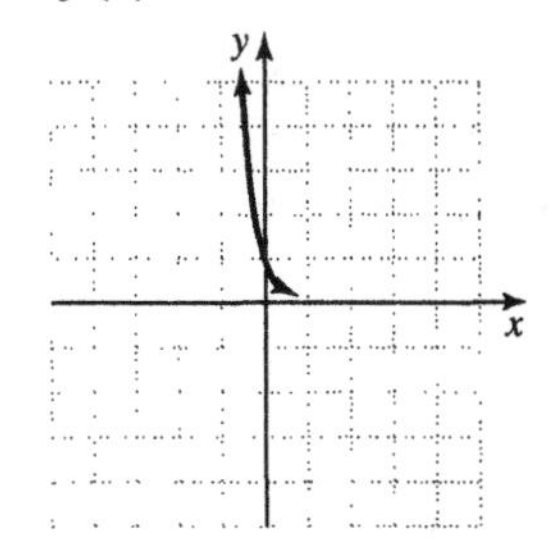

83. $f(x) = e^x + 2$

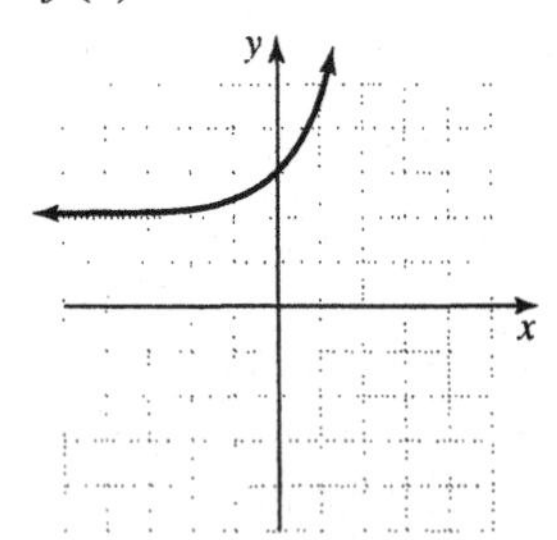

85. $f(x) = e^{x-1}$

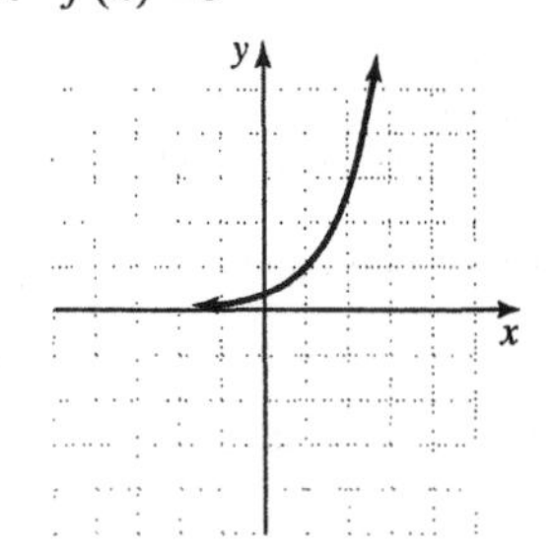

87. $f(x) = 3e^x$

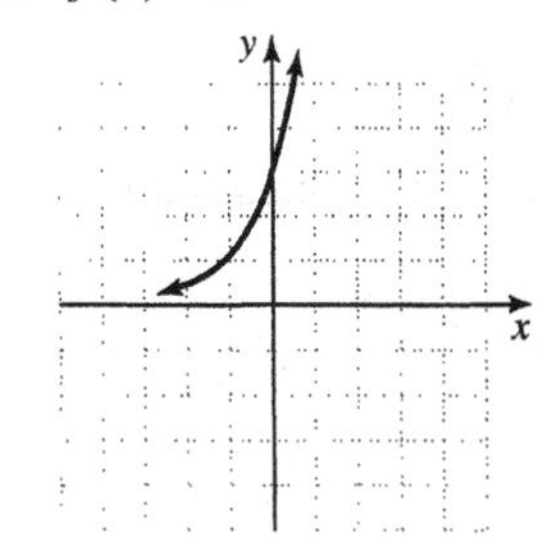

89. $f(x) = \ln x$

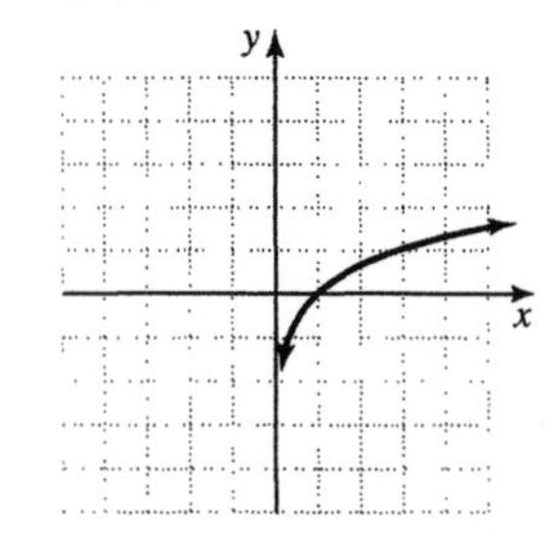

91. $f(x) = -2\log x$

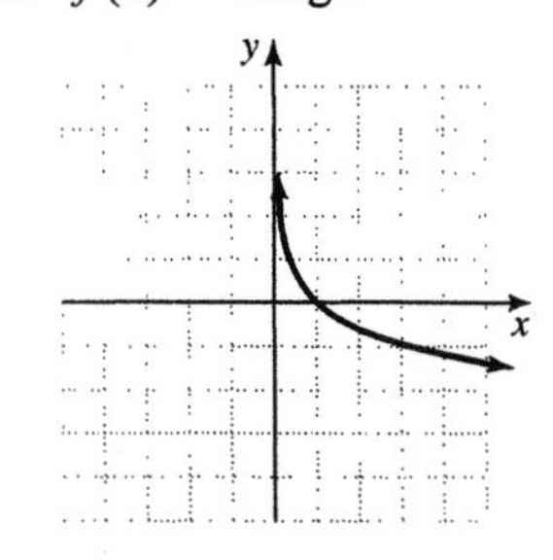

93. $f(x) = \log(x+2)$

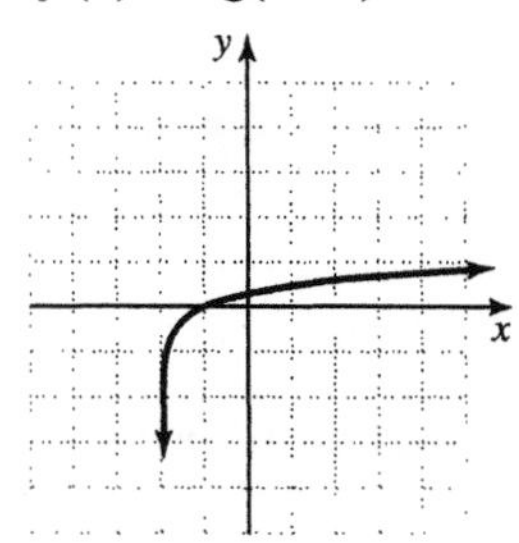

95. $f(x) = \ln x - 3$

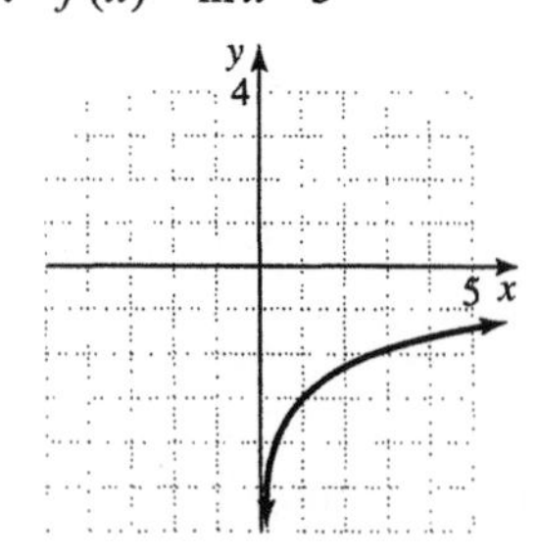

97. $f(x) = e^x$

$f(x) = e^x + 2$

$f(x) = e^x - 3$

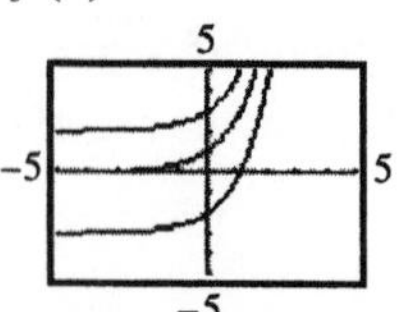

Answers may vary

Section 12.7

Practice Exercises

1.
$$5^x = 9$$
$$\log 5^x = \log 9$$
$$x \log 5 = \log 9$$
$$x = \frac{\log 9}{\log 5} \approx 1.3652$$
The solution is $\frac{\log 9}{\log 5}$, or approximately 1.3652.

2. $\log_2(x-1) = 5$
$$2^5 = x-1$$
$$32 = x-1$$
$$33 = x$$
Check: $\log_2(x-1) = 5$
$$\log_2(33-1) \stackrel{?}{=} 5$$
$$\log_2 32 \stackrel{?}{=} 5$$
$$2^5 = 32 \quad \text{True}$$
The solution is 33.

3. $\log_5 x + \log_5(x+4) = 1$
$$\log_5 x(x+4) = 1$$
$$\log_5(x^2+4x) = 1$$
$$5^1 = x^2 + 4x$$
$$0 = x^2 + 4x - 5$$
$$0 = (x+5)(x-1)$$
$$x+5 = 0 \quad \text{or} \quad x-1 = 0$$
$$x = -5 \qquad\qquad x = 1$$
Since $\log_5(-5)$ is undefined, -5 is rejected. The solution is 1.

4. $\log(x+3) - \log x = 1$
$$\log\frac{x+3}{x} = 1$$
$$10^1 = \frac{x+3}{x}$$
$$10x = x+3$$
$$9x = 3$$
$$x = \frac{1}{3}$$
The solution is $\frac{1}{3}$.

5. $y_0 = 60$; $t = 3$
$$y = y_0 e^{0.916t}$$
$$y = 60e^{0.916(3)} = 60e^{2.748} \approx 937$$
The population will be approximately 937 rabbits.

6. $P = \$3000$; $r = 7\% = 0.07$; $n = 12$;
$A = 2P = \$6000$
$$A = P\left(1+\frac{r}{n}\right)^{nt}$$
$$6000 = 3000\left(1+\frac{0.07}{12}\right)^{12t}$$
$$2 = \left(1+\frac{0.07}{12}\right)^{12t}$$
$$\log 2 = \log\left(1+\frac{0.07}{12}\right)^{12t}$$
$$\log 2 = 12t \log\left(1+\frac{0.07}{12}\right)$$
$$\frac{\log 2}{12\log\left(1+\frac{0.07}{12}\right)} = t$$
$$9.9 \approx t$$
It takes nearly 10 years to double.

Graphing Calculator Explorations

1. $Y_1 = 5000\left(1+\frac{0.05}{4}\right)^{4x}$, $Y_2 = 6000$

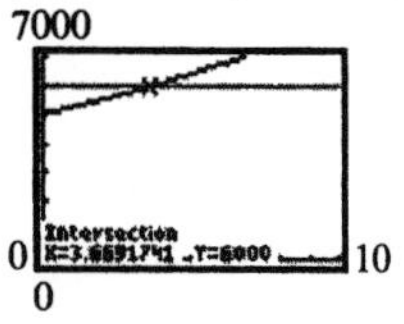

7000

0 10

0

It takes 3.67 years, or 3 years and 8 months.

2. $Y_1 = 1000\left(1+\frac{0.045}{365}\right)^{365x}$, $Y_2 = 2000$

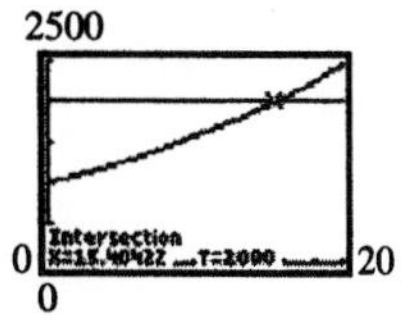

2500

0 20

0

It takes 15.40 years or 15 years and 5 months.

3. $Y_1 = 10,000\left(1+\frac{0.06}{12}\right)^{12x}$, $Y_2 = 40,000$

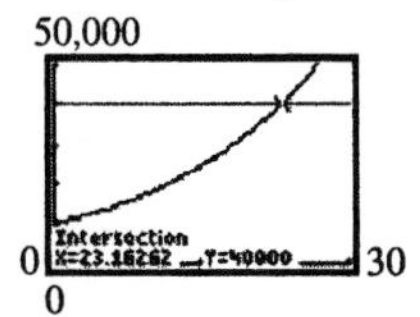

It takes 23.16 years or 23 years and 2 months.

4. $Y_1 = 500\left(1+\frac{0.04}{2}\right)^{2x}$, $Y_2 = 800$

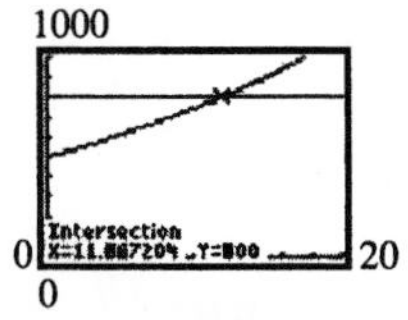

It takes 11.87 years or 11 years and 10 months.

Exercise Set 12.7

1.
$$\begin{aligned} 3^x &= 6 \\ \log 3^x &= \log 6 \\ x\log 3 &= \log 6 \\ x &= \frac{\log 6}{\log 3} \approx 1.6309 \end{aligned}$$

3.
$$\begin{aligned} 3^{2x} &= 3.8 \\ \log 3^{2x} &= \log 3.8 \\ 2x\log 3 &= \log 3.8 \\ x &= \frac{\log 3.8}{2\log 3} \approx 0.6076 \end{aligned}$$

5.
$$\begin{aligned} 2^{x-3} &= 5 \\ \log 2^{x-3} &= \log 5 \\ (x-3)\log 2 &= \log 5 \\ x-3 &= \frac{\log 5}{\log 2} \\ x &= 3+\frac{\log 5}{\log 2} \approx 5.3219 \end{aligned}$$

7.
$$\begin{aligned} 9^x &= 5 \\ \log 9^x &= \log 5 \\ x\log 9 &= \log 5 \\ x &= \frac{\log 5}{\log 9} \approx 0.7325 \end{aligned}$$

9.
$$\begin{aligned} 4^{x+7} &= 3 \\ \log 4^{x+7} &= \log 3 \\ (x+7)\log 4 &= \log 3 \\ x+7 &= \frac{\log 3}{\log 4} \\ x &= -7+\frac{\log 3}{\log 4} \approx -6.2075 \end{aligned}$$

11.
$$\begin{aligned} 7^{3x-4} &= 11 \\ \log 7^{3x-4} &= \log 11 \\ (3x-4)\log 7 &= \log 11 \\ 3x-4 &= \frac{\log 11}{\log 7} \\ 3x &= 4+\frac{\log 11}{\log 7} \\ x &= \frac{1}{3}\left(4+\frac{\log 11}{\log 7}\right) \approx 1.7441 \end{aligned}$$

13.
$$\begin{aligned} e^{6x} &= 5 \\ \ln e^{6x} &= \ln 5 \\ 6x &= \ln 5 \\ x &= \frac{\ln 5}{6} \approx 0.2682 \end{aligned}$$

15.
$$\begin{aligned} \log_2(x+5) &= 4 \\ x+5 &= 2^4 \\ x+5 &= 16 \\ x &= 11 \end{aligned}$$

17.
$$\begin{aligned} \log_3 x^2 &= 4 \\ x^2 &= 3^4 \\ x^2 &= 81 \\ x &= \pm 9 \end{aligned}$$

19.
$$\begin{aligned} \log_4 2+\log_4 x &= 0 \\ \log_4(2x) &= 0 \\ 2x &= 4^0 \\ 2x &= 1 \\ x &= \frac{1}{2} \end{aligned}$$

21. $\log_2 6 - \log_2 x = 3$

$$\log_2\left(\frac{6}{x}\right) = 3$$
$$\frac{6}{x} = 2^3$$
$$\frac{6}{x} = 8$$
$$8x = 6$$
$$x = \frac{3}{4}$$

23. $\log_4 x + \log_4(x+6) = 2$

$$\log_4 x(x+6) = 2$$
$$x(x+6) = 4^2$$
$$x^2 + 6x = 16$$
$$x^2 + 6x - 16 = 0$$
$$(x+8)(x-2) = 0$$

$x = -8$ or $x = 2$

We discard −8 as extraneous, the solution is 2.

25. $\log_5(x+3) - \log_5 x = 2$

$$\log_5\left(\frac{x+3}{x}\right) = 2$$
$$\frac{x+3}{x} = 5^2$$
$$\frac{x+3}{x} = 25$$
$$x + 3 = 25x$$
$$3 = 24x$$
$$x = \frac{1}{8}$$

27. $\log_3(x-2) = 2$

$$x - 2 = 3^2$$
$$x - 2 = 9$$
$$x = 11$$

29. $\log_4(x^2 - 3x) = 1$

$$x^2 - 3x = 4$$
$$x^2 - 3x - 4 = 0$$
$$(x-4)(x+1) = 0$$

$x = 4$ or $x = -1$

31. $\ln 5 + \ln x = 0$

$$\ln(5x) = 0$$
$$e^0 = 5x$$
$$1 = 5x$$
$$\frac{1}{5} = x$$

33. $3\log x - \log x^2 = 2$

$$3\log x - 2\log x = 2$$
$$\log x = 2$$
$$x = 10^2$$
$$x = 100$$

35. $\log_2 x + \log_2(x+5) = 1$

$$\log_2 x(x+5) = 1$$
$$x(x+5) = 2$$
$$x^2 + 5x - 2 = 0$$

$a = 1, b = 5, c = -2$

$$x = \frac{-5 \pm \sqrt{5^2 - 4(1)(-2)}}{2(1)}$$
$$x = \frac{-5 \pm \sqrt{33}}{2}$$

Discard $\frac{-5-\sqrt{33}}{2}$, the solution is $\frac{-5+\sqrt{33}}{2}$.

37. $\log_4 x - \log_4(2x-3) = 3$

$$\log_4\left(\frac{x}{2x-3}\right) = 3$$
$$\frac{x}{2x-3} = 4^3$$
$$x = 64(2x-3)$$
$$x = 128x - 192$$
$$192 = 127x$$
$$x = \frac{192}{127}$$

39. $\log_2 x + \log_2(3x+1) = 1$

$$\log_2 x(3x+1) = 1$$
$$x(3x+1) = 2$$
$$3x^2 + x - 2 = 0$$
$$(3x-2)(x+1) = 0$$

$3x - 2 = 0$ or $x + 1 = 0$

$x = \frac{2}{3}$ or $x = -1$

We discard −1 as extraneous, the solution is $\frac{2}{3}$.

41. $y = y_0 e^{0.043t}$, $y_0 = 83$, $t = 5$

$y = 83e^{0.043(5)} = 83e^{0.215} \approx 103$

There should be 103 wolves in 5 years.

43. $y = y_0 e^{0.023t}$, $y_0 = 294,380$, $t = 8$

$y = 294,380e^{0.023(8)} \approx 354,000$

There will be approximately 354,000 inhabitants in 2015.

45. $y = y_0 e^{-0.00033t}$

$y_0 = 82,400$, $y = 82,000$

$82,000 = 82,400e^{-0.00033t}$

$\frac{82,000}{82,400} = e^{-0.005t}$

$t = \frac{\ln\left(\frac{82,000}{82,400}\right)}{-0.00033} \approx 14.7$

It will take approximately 14.7 years to reach 82,000.

47. $A = P\left(1+\frac{r}{n}\right)^{nt}$, $P = 600$,

$A = 2(600) = 1200$, $r = 0.07$, $n = 12$

$1200 = 600\left(1+\frac{0.07}{12}\right)^{12t}$

$2 = \left(1+\frac{0.07}{12}\right)^{12t}$

$\log 2 = \log\left(1+\frac{0.07}{12}\right)^{12t}$

$\log 2 = 12t\log\left(1+\frac{0.07}{12}\right)$

$\frac{\log 2}{12\log\left(1+\frac{0.07}{12}\right)} = t$

$9.9 \approx t$

It takes approximately 9.9 years for the $600 to double.

49. $A = P\left(1+\frac{r}{n}\right)^{nt}$, $P = 1200$,

$A = P + I = 1200 + 200 = 1400$

$r = 0.09$, $n = 4$

$1400 = 1200\left(1+\frac{0.09}{4}\right)^{4t}$

$\frac{7}{6} = (1.0225)^{4t}$

$\log\frac{7}{6} = \log 1.0225^{4t}$

$\log\frac{7}{6} = 4t\log 1.0225$

$t = \frac{\log\frac{7}{6}}{4\log 1.0225}$

$t \approx 1.7$

It would take the investment approximately 1.7 years to earn $200.

51. $A = P\left(1+\frac{r}{n}\right)^{nt}$, $P = 1000$

$A = 2(1000) = 2000$, $r = 0.08$, $n = 2$

$2000 = 1000\left(1+\frac{0.08}{2}\right)^{2t}$

$2 = (1.04)^{2t}$

$\log 2 = \log 1.04^{2t}$

$\log 2 = 2t\log 1.04$

$t = \frac{\log 2}{2\log 1.04}$

$t \approx 8.8$

It takes 8.8 years to double.

53. $w = 0.00185h^{2.67}$, and $h = 35$

$w = 0.00185(35)^{2.67} \approx 24.5$

The expected weight of a boy 35 inches tall is 24.5 pounds.

55. $w = 0.00185h^{2.67}$, and $w = 85$

$85 = 0.00185h^{2.67}$

$\frac{85}{0.00185} = h^{2.67}$

$h = \left(\frac{85}{0.00185}\right)^{1/2.67} \approx 55.7$

The expected height of the boy is 55.7 inches.

57. $P = 14.7e^{-0.21x}$, $x = 1$

$P = 14.7e^{-0.21(1)}$

$= 14.7e^{-0.21}$

≈ 11.9

The average atmospheric pressure in Denver is approximately 11.9 pounds per square inch.

59. $P = 14.7e^{-0.21x}$, $P = 7.5$

$$7.5 = 14.7e^{-0.21x}$$
$$\frac{7.5}{14.7} = e^{-0.21x}$$
$$-0.21x = \ln\left(\frac{7.5}{14.7}\right)$$
$$x = -\frac{1}{0.21}\ln\left(\frac{7.5}{14.7}\right) \approx 3.2$$

The elevation of the jet is approximately 3.2 miles.

61. $t = \frac{1}{c}\ln\left(\frac{A}{A-N}\right)$

$$t = \frac{1}{0.09}\ln\left(\frac{75}{75-50}\right)$$
$$t = \frac{1}{0.09}\ln(3)$$
$$t \approx 12.21$$

It will take 12 weeks.

63. $t = \frac{1}{c}\ln\left(\frac{A}{A-N}\right)$

$$t = \frac{1}{0.07}\ln\left(\frac{210}{210-150}\right)$$
$$t = \frac{1}{0.07}\ln(3.5)$$
$$t \approx 17.9$$

It will take 18 weeks.

65. $\frac{x^2 - y + 2z}{3x} = \frac{(-2)^2 - 0 + 2(3)}{3(-2)}$

$$= \frac{4+6}{-6}$$
$$= \frac{10}{-6}$$
$$= -\frac{5}{3}$$

67. $\frac{3z - 4x + y}{x + 2z} = \frac{3(3) - 4(-2) + 0}{-2 + 2(3)} = \frac{9+8}{-2+6} = \frac{17}{4}$

69. $f(x) = 5x + 2$

$$y = 5x + 2$$
$$x = 5y + 2$$
$$\frac{x-2}{5} = y$$
$$f^{-1}(x) = \frac{x-2}{5}$$

71. $y = 6{,}123{,}106$; $y_0 = 5{,}130{,}632$; $t = 6$

$$y = y_0 e^{kt}$$
$$6{,}123{,}106 = 5{,}130{,}632e^{k(6)}$$
$$\frac{6{,}123{,}106}{5{,}130{,}632} = e^{6k}$$
$$\ln\frac{6{,}123{,}106}{5{,}130{,}632} = 6k$$
$$k = \frac{1}{6}\ln\frac{6{,}123{,}106}{5{,}130{,}632} \approx 0.029$$

The annual rate of population growth was approximately 2.9%.

73. Answers may vary

75. $Y_1 = e^{0.3x}$, $Y_2 = 8$

10

−10   10

Intersection X=6.9314718 Y=8

−10

$x \approx 6.93$

77. $Y_1 = 2\log(-5.6x + 1.3) + x + 1$, $Y_2 = 0$

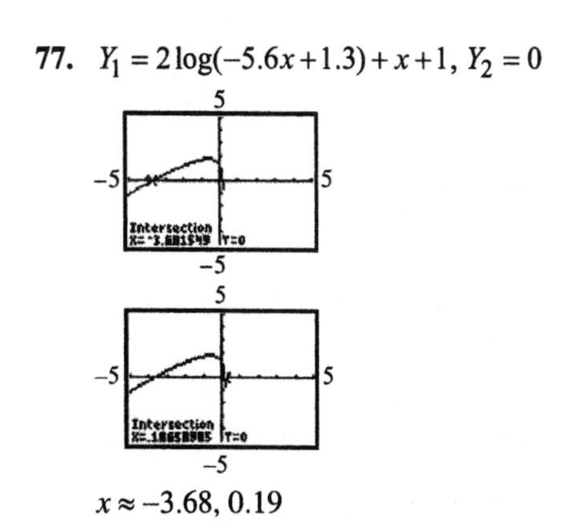

$x \approx -3.68, 0.19$

79. $Y_1 = 7^{3x-4} - 11, Y_2 = 0$

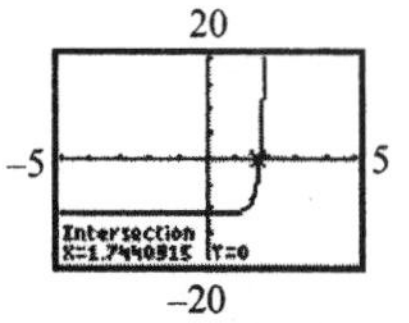

$x \approx 1.74$

81. $Y_1 = \ln 5 + \ln x, Y_2 = 0$

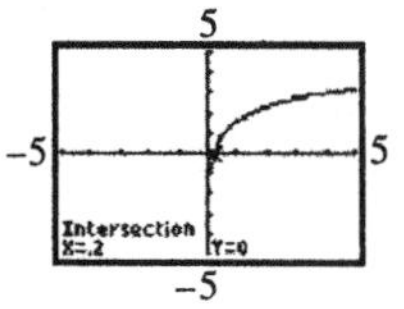

$x = 0.2 = \frac{1}{5}$

The Bigger Picture

1. $8^x = 2^{x-3}$
$2^{3x} = 2^{x-3}$
$3x = x - 3$
$2x = -3$
$x = -\frac{3}{2}$

2. $11^x = 5$
$\log 11^x = \log 5$
$x \log 11 = \log 5$
$x = \frac{\log 5}{\log 11} \approx 0.6712$

3. $-7x + 3 \le -5x + 13$
$-10 \le 2x$
$-5 \le x$
The solution is $[-5, \infty)$.

4. $-7 \le 3x + 6 \le 0$
$-13 \le 3x \le -6$
$-\frac{13}{3} \le x \le -2$
The solution is $\left[-\frac{13}{3}, -2\right]$.

5. $|5y + 3| < 3$
$-3 < 5y + 3 < 3$
$-6 < 5y < 0$
$-\frac{6}{5} < y < 0$
The solution is $\left(-\frac{6}{5}, 0\right)$.

6. $(x - 6)(5x + 1) = 0$
$x - 6 = 0$ or $5x + 1 = 0$
$x = 6$ $\quad$ $5x = -1$
$x = -\frac{1}{5}$
The solutions are 6 and $-\frac{1}{5}$.

7. $\log_{13} 8 + \log_{13}(x - 1) = 1$
$\log_{13}(8x - 8) = 1$
$13^1 = 8x - 8$
$21 = 8x$
$\frac{21}{8} = x$

8. $\left|\frac{3x-1}{4}\right| = 2$
$\frac{3x-1}{4} = 2$ or $\frac{3x-1}{4} = -2$
$3x - 1 = 8$ $\quad$ $3x - 1 = -8$
$3x = 9$ $\quad$ $3x = -7$
$x = 3$ $\quad$ $x = -\frac{7}{3}$
The solutions are 3 and $-\frac{7}{3}$.

9. $|7x + 1| > -2$ is a true statement for all x, so the solution is $(-\infty, \infty)$.

10. $x^2 = 4$
$x = \pm\sqrt{4} = \pm 2$
The solutions are 2 and -2.

11. $(x + 5)^2 = 3$
$x + 5 = \pm\sqrt{3}$
$x = -5 \pm \sqrt{3}$
The solutions are $-5 + \sqrt{3}$ and $-5 - \sqrt{3}$.

12. $\log_7(4x^2 - 27x) = 1$

$$7^1 = 4x^2 - 27x$$
$$0 = 4x^2 - 27x - 7$$
$$0 = (4x+1)(x-7)$$

$4x + 1 = 0$ or $x - 7 = 0$

$4x = -1$ $\quad x = 7$

$x = -\frac{1}{4}$

The solutions are $-\frac{1}{4}$ and 7.

Chapter 12 Vocabulary Check

1. For each one-to-one function, we can find its inverse function by switching the coordinates of the ordered pairs of the function.
2. The composition of functions f and g is $(f \circ g)(x) = f(g(x))$.
3. A function of the form $f(x) = b^x$ is called an exponential function if $b > 0$, b is not 1, and x is a real number.
4. The graphs of f and f^{-1} are symmetric about the line $y = x$.
5. Natural logarithms are logarithms to base e.
6. Common logarithms are logarithms to base 10.
7. To see whether a graph is the graph of a one-to-one function, apply the vertical line test to see if it is a function, and then apply the horizontal line test to see if it is a one-to-one function.
8. A logarithmic function is a function that can be defined by $f(x) = \log_b x$ where x is a positive real number, b is a constant positive real number, and b is not 1.

Chapter 12 Review

1. $(f+g)(x) = f(x) + g(x)$
$$= (x-5) + (2x+1)$$
$$= x - 5 + 2x + 1$$
$$= 3x - 4$$

2. $(f-g)(x) = f(x) - g(x)$
$$= (x-5) - (2x+1)$$
$$= x - 5 - 2x - 1$$
$$= -x - 6$$

3. $(f \cdot g)(x) = f(x) \cdot g(x)$
$$= (x-5)(2x+1)$$
$$= 2x^2 + x - 10x - 5$$
$$= 2x^2 - 9x - 5$$

4. $\left(\frac{g}{f}\right)(x) = \frac{g(x)}{f(x)} = \frac{2x+1}{x-5}, x \neq 5$

5. $(f \circ g)(x) = f(g(x))$
$$= f(x+1)$$
$$= (x+1)^2 - 2$$
$$= x^2 + 2x - 1$$

6. $(g \circ f)(x) = g(f(x))$
$$= g(x^2 - 2)$$
$$= x^2 - 2 + 1$$
$$= x^2 - 1$$

7. $(h \circ g)(2) = h(g(2)) = h(3) = 3^3 - 3^2 = 18$

8. $(f \circ f)(x) = f(f(x))$
$$= f(x^2 - 2)$$
$$= (x^2 - 2)^2 - 2$$
$$= x^4 - 4x^2 + 4 - 2$$
$$= x^4 - 4x^2 + 2$$

9. $(f \circ g)(-1) = f(g(-1)) = f(0) = 0^2 - 2 = -2$

10. $(h \circ h)(2) = h(h(2)) = h(4) = 4^3 - 4^2 = 48$

11. The function is one-to-one.
$h^{-1} = \{(14, -9), (8, 6), (12, -11), (15, 15)\}$

12. The function is not one-to-one.

13. The function is one-to-one.

Rank in Auto Thefts (Input)	2	4	1	3
U.S. Region (Output)	West	Midwest	South	Northeast

14. The function is not one-to-one.

15. $f(x)=\sqrt{x+2}$

a. $f(7)=\sqrt{7+2}=\sqrt{9}=3$

b. $f^{-1}(3)=7$

16. $f(x)=\sqrt{x+2}$

a. $f(-1)=\sqrt{-1+2}=\sqrt{1}=1$

b. $f^{-1}(1)=-1$

17. The graph does not represent a one-to-one function.

18. The graph does not represent a one-to-one function.

19. The graph does not represent a one-to-one function.

20. The graph represents a one-to-one function.

21.
$$\begin{aligned} f(x)&=x-9\\ y&=x-9\\ x&=y-9\\ y&=x+9\\ f^{-1}(x)&=x+9 \end{aligned}$$

22.
$$\begin{aligned} f(x)&=x+8\\ y&=x+8\\ x&=y+8\\ y&=x-8\\ f^{-1}(x)&=x-8 \end{aligned}$$

23.
$$\begin{aligned} f(x)&=6x+11\\ y&=6x+11\\ x&=6y+11\\ 6y&=x-11\\ y&=\frac{x-11}{6}\\ f^{-1}(x)&=\frac{x-11}{6} \end{aligned}$$

24.
$$\begin{aligned} f(x)&=12x\\ y&=12x\\ x&=12y\\ y&=\frac{x}{12}\\ f^{-1}(x)&=\frac{x}{12} \end{aligned}$$

25.
$$\begin{aligned} f(x)&=x^3-5\\ y&=x^3-5\\ x&=y^3-5\\ y^3&=x+5\\ y&=\sqrt[3]{x+5}\\ f^{-1}(x)&=\sqrt[3]{x+5} \end{aligned}$$

26.
$$\begin{aligned} f(x)&=\sqrt[3]{x+2}\\ y&=\sqrt[3]{x+2}\\ x&=\sqrt[3]{y+2}\\ x^3&=y+2\\ y&=x^3-2\\ f^{-1}(x)&=x^3-2 \end{aligned}$$

27.
$$\begin{aligned} g(x)&=\frac{12x-7}{6}\\ y&=\frac{12x-7}{6}\\ x&=\frac{12y-7}{6}\\ 6x&=12y-7\\ 12y&=6x+7\\ y&=\frac{6x+7}{12}\\ g^{-1}(x)&=\frac{6x+7}{12} \end{aligned}$$

28.
$$\begin{aligned} r(x)&=\frac{13}{2}x-4\\ y&=\frac{13}{2}x-4\\ x&=\frac{13}{2}y-4\\ x+4&=\frac{13}{2}y\\ y&=\frac{2(x+4)}{13}\\ r^{-1}(x)&=\frac{2(x+4)}{13} \end{aligned}$$

29. $y = g(x) = \sqrt{x}$
$x = \sqrt{y}$
$x^2 = y = g^{-1}(x),\ x \ge 0$

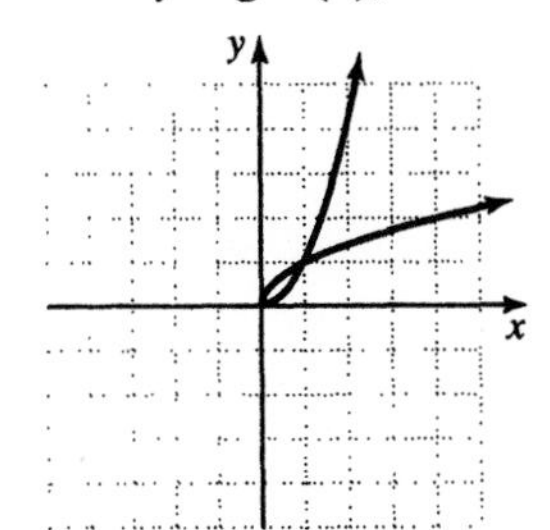

30. $h(x) = 5x - 5$
$y = 5x - 5$
$x = 5y - 5$
$5y = x + 5$
$y = \frac{x+5}{5}$
$h^{-1}(x) = \frac{x+5}{5}$

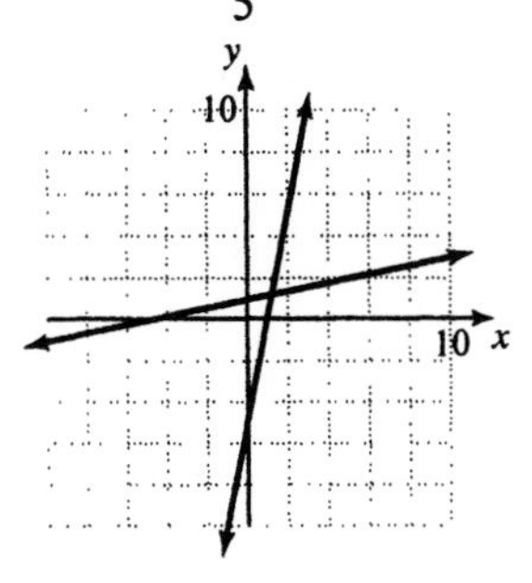

31. $f(x) = 2x - 3$
$y = 2x - 3$
$x = 2y - 3$
$y = \frac{x+3}{2}$
$f^{-1}(x) = \frac{x+3}{2}$

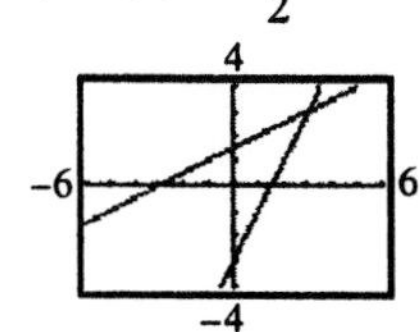

32. $4^x = 64$
$4^x = 4^3$
$x = 3$

33. $3^x = \frac{1}{9}$
$3^x = 3^{-2}$
$x = -2$

34. $2^{3x} = \frac{1}{16}$
$2^{3x} = 2^{-4}$
$3x = -4$
$x = -\frac{4}{3}$

35. $5^{2x} = 125$
$5^{2x} = 5^3$
$2x = 3$
$x = \frac{3}{2}$

36. $9^{x+1} = 243$
$(3^2)^{x+1} = 3^5$
$3^{2x+2} = 3^5$
$2x + 2 = 5$
$2x = 3$
$x = \frac{3}{2}$

37. $8^{3x-2} = 4$
$(2^3)^{3x-2} = 2^2$
$2^{9x-6} = 2^2$
$9x - 6 = 2$
$9x = 8$
$x = \frac{8}{9}$

38. $y = 3^x$

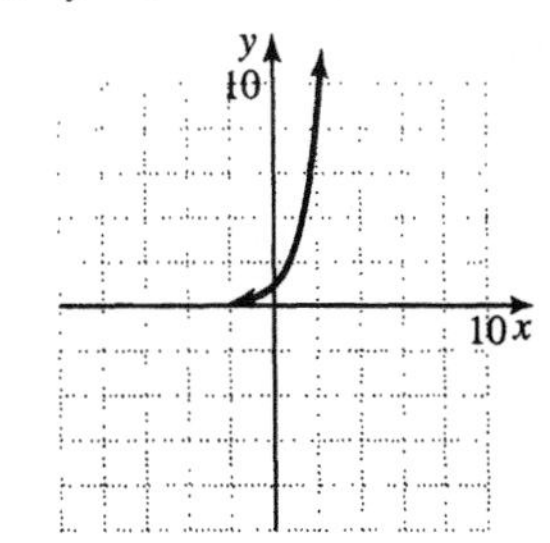

39. 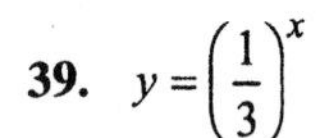 $y = \left(\frac{1}{3}\right)^x$

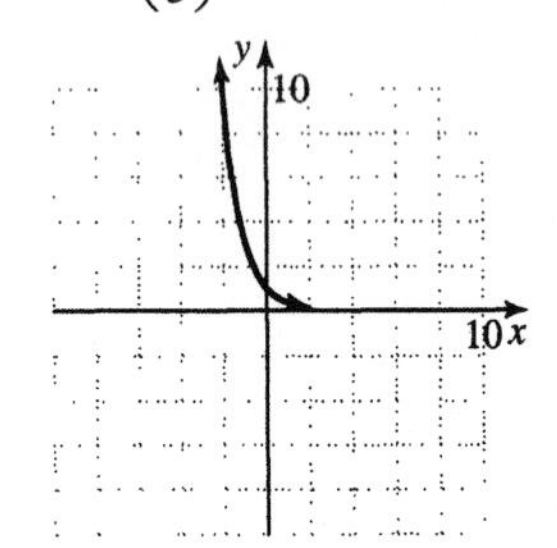

40. $y = 4 \cdot 2^x$

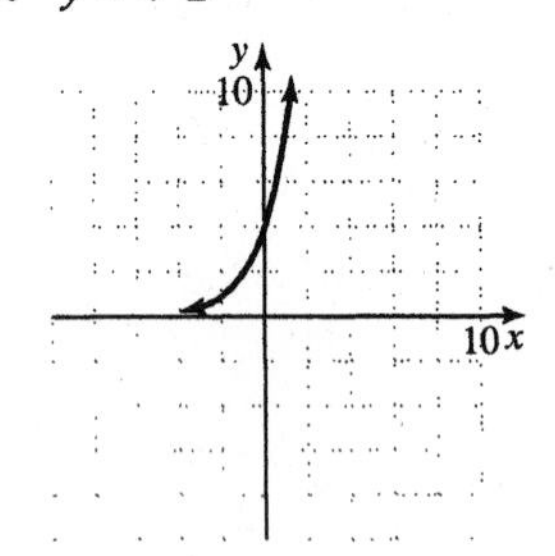

41. $y = 2^x + 4$

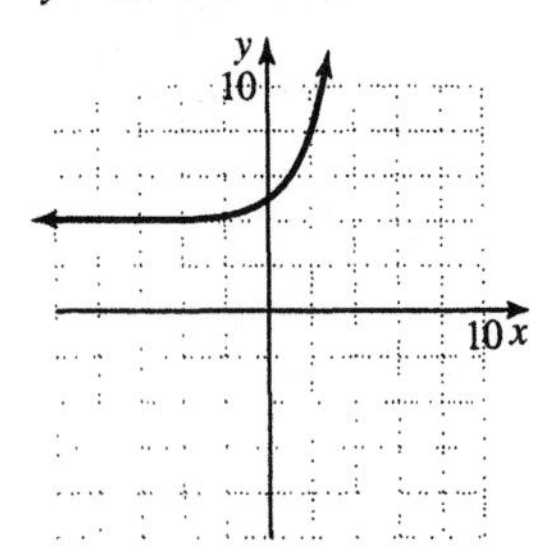

42. $A = P\left(1 + \frac{r}{n}\right)^{nt}$

$A = 1600\left(1 + \frac{0.09}{2}\right)^{(2)(7)}$

$A \approx 2963.11$

The amount accrued is \$2963.11.

43. $A = P\left(1 + \frac{r}{n}\right)^{nt}$

$A = 800\left(1 + \frac{0.07}{4}\right)^{(4)(5)}$

$A \approx 1131.82$

The certificate is worth \$1131.82 at the end of 5 years.

44. $y = 4 \cdot 2^x$

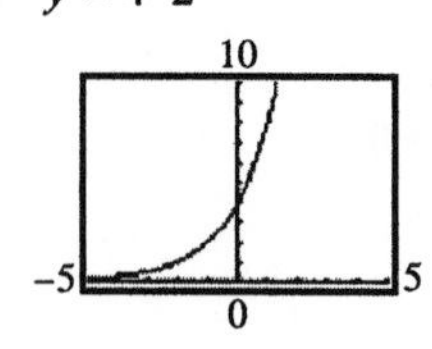

45. $49 = 7^2$

$\log_7 49 = 2$

46. $2^{-4} = \frac{1}{16}$

$\log_2 \frac{1}{16} = -4$

47. $\log_{1/2} 16 = -4$

$\left(\frac{1}{2}\right)^{-4} = 16$

48. $\log_{0.4} 0.064 = 3$

$0.4^3 = 0.064$

49. $\log_4 x = -3$

$x = 4^{-3} = \frac{1}{64}$

50. $\log_3 x = 2$

$x = 3^2 = 9$

51. $\log_3 1 = x$

$3^x = 1$

$3^x = 3^0$

$x = 0$

52. $\log_4 64 = x$

$4^x = 64$

$4^x = 4^3$

$x = 3$

53. $\log_x 64 = 2$

$x^2 = 64$

$x = \pm\sqrt{64} = \pm 8$

$x = 8$ since the base must be positive

54. $\log_x 81 = 4$

$x^4 = 81$

$x = \pm 3$

$x = 3$ since the base must be positive

55. $\log_4 4^5 = x$
$x = 5$

56. $\log_7 7^{-2} = x$
$x = -2$

57. $5^{\log_5 4} = x$
$x = 4$

58. $2^{\log_2 9} = x$
$9 = x$

59. $\log_2(3x-1) = 4$
$$\begin{aligned} 3x-1 &= 2^4 \\ 3x-1 &= 16 \\ 3x &= 17 \\ x &= \frac{17}{3} \end{aligned}$$

60. $\log_3(2x+5) = 2$
$$\begin{aligned} 2x+5 &= 3^2 \\ 2x+5 &= 9 \\ 2x &= 4 \\ x &= 2 \end{aligned}$$

61. $\log_4(x^2-3x) = 1$
$$\begin{aligned} x^2-3x &= 4 \\ x^2-3x-4 &= 0 \\ (x+1)(x-4) &= 0 \end{aligned}$$
$x = -1$ or $x = 4$

62. $\log_8(x^2+7x) = 1$
$$\begin{aligned} x^2+7x &= 8 \\ x^2+7x-8 &= 0 \\ (x+8)(x-1) &= 0 \end{aligned}$$
$x = -8$ or $x = 1$

63. $y = 2^x$ and $y = \log_2 x$

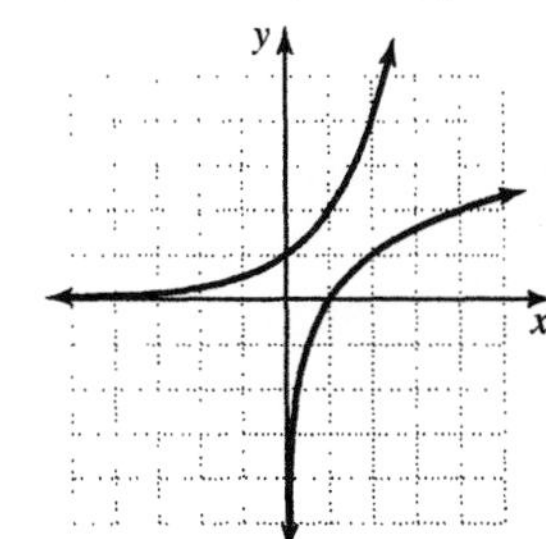

64. $y = \left(\frac{1}{2}\right)^x$ and $y = \log_{1/2} x$

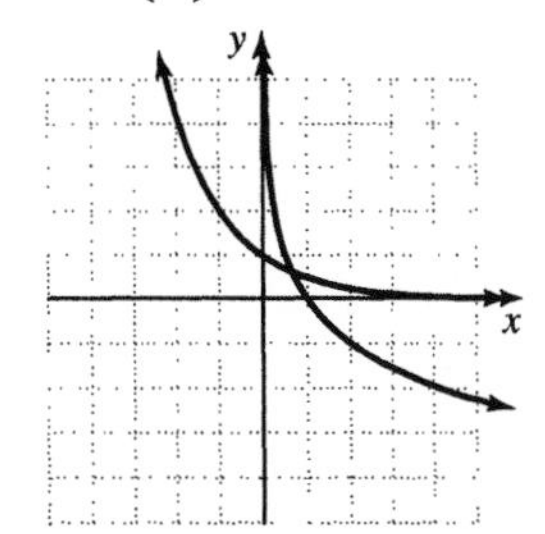

65. $\log_3 8 + \log_3 4 = \log_3(8\cdot 4) = \log_3 32$

66. $\log_2 6 + \log_2 3 = \log_2(6\cdot 3) = \log_2 18$

67. $\log_7 15 - \log_7 20 = \log_7 \frac{15}{20} = \log_7 \frac{3}{4}$

68. $\log 18 - \log 12 = \log \frac{18}{12} = \log \frac{3}{2}$

69. $\log_{11} 8 + \log_{11} 3 - \log_{11} 6 = \log_{11} \frac{(8)(3)}{6}$
$= \log_{11} 4$

70. $\log_5 14 + \log_5 3 - \log_5 21$
$$\begin{aligned} &= \log_5(14\cdot 3) - \log_5 21 \\ &= \log_5 \frac{42}{21} \\ &= \log_5 2 \end{aligned}$$

71. $2\log_5 x - 2\log_5(x+1) + \log_5 x$
$$\begin{aligned} &= \log_5 x^2 - \log_5(x+1)^2 + \log_5 x \\ &= \log_5 \frac{(x^2)(x)}{(x+1)^2} \\ &= \log_5 \frac{x^3}{(x+1)^2} \end{aligned}$$

72. $4\log_3 x - \log_3 x + \log_3(x+2)$
$$\begin{aligned} &= 3\log_3 x + \log_3(x+2) \\ &= \log_3 x^3 + \log_3(x+2) \\ &= \log_3\left[x^3(x+2)\right] \\ &= \log_3(x^4+2x^3) \end{aligned}$$

73. $\log_3 \frac{x^3}{x+2} = \log_3 x^3 - \log_3(x+2)$
$= 3\log_3 x - \log_3(x+2)$

74. $\log_4 \frac{x+5}{x^2} = \log_4(x+5) - \log_4 x^2$
$= \log_4(x+5) - 2\log_4 x$

75. $\log_2 \frac{3x^2 y}{z} = \log_2(3x^2 y) - \log_2 z$
$= \log_2 3 + \log_2 x^2 + \log_2 y - \log_2 z$
$= \log_2 3 + 2\log_2 x + \log_2 y - \log_2 z$

76. $\log_7 \frac{yz^3}{x} = \log_7(yz^3) - \log_7 x$
$= \log_7 y + \log_7 z^3 - \log_7 x$
$= \log_7 y + 3\log_7 z - \log_7 x$

77. $\log_b 50 = \log_b(5)(5)(2)$
$= \log_b(5) + \log_b(5) + \log_b(2)$
$= 0.83 + 0.83 + 0.36$
$= 2.02$

78. $\log_b \frac{4}{5} = \log_b 4 - \log_b 5$
$= \log_b 2^2 - \log_b 5$
$= 2\log_b 2 - \log_b 5$
$= 2(0.36) - 0.83$
$= 0.72 - 0.83$
$= -0.11$

79. $\log 3.6 \approx 0.5563$

80. $\log 0.15 \approx -0.8239$

81. $\ln 1.25 \approx 0.2231$

82. $\ln 4.63 \approx 1.5326$

83. $\log 1000 = \log 10^3 = 3$

84. $\log \frac{1}{10} = \log 10^{-1} = -1$

85. $\ln \frac{1}{e} = \ln e^{-1} = -1$

86. $\ln e^4 = 4$

87. $\ln(2x) = 2$
$2x = e^2$
$x = \frac{e^2}{2}$

88. $\ln(3x) = 1.6$
$3x = e^{1.6}$
$x = \frac{e^{1.6}}{3}$

89. $\ln(2x-3) = -1$
$2x - 3 = e^{-1}$
$x = \frac{e^{-1}+3}{2}$

90. $\ln(3x+1) = 2$
$3x + 1 = e^2$
$3x = e^2 - 1$
$x = \frac{e^2 - 1}{3}$

91. $\ln \frac{I}{I_0} = -kx$
$\ln \frac{0.03 I_0}{I_0} = -2.1x$
$\ln 0.03 = -2.1x$
$\frac{\ln 0.03}{-2.1} = x$
$x \approx 1.67$
The depth is 1.67 millimeters.

92. $\ln \frac{I}{I_0} = -kx$
$\ln \frac{0.02 I_0}{I_0} = -3.2x$
$\ln 0.02 = -3.2x$
$\frac{\ln 0.02}{-3.2} = x$
$x \approx 1.22$
2% of the original radioactivity will penetrate at a depth of approximately 1.22 millimeters.

93. $\log_5 1.6 = \frac{\log 1.6}{\log 5} \approx 0.2920$

94. $\log_3 4 = \frac{\log 4}{\log 3} \approx 1.2619$

95. $A = Pe^{rt}$

$A = 1450e^{(0.06)(5)}$

$A \approx 1957.30$

The accrued amount is \$1957.30.

96. $A = Pe^{rt}$

$A = 940e^{0.11(3)} = 940e^{0.33} \approx 1307.51$

The investment grows to \$1307.51.

97.

$$\begin{aligned} 3^{2x} &= 7 \\ \log 3^{2x} &= \log 7 \\ 2x \log 3 &= \log 7 \\ x &= \frac{\log 7}{2\log 3} \approx 0.8856 \end{aligned}$$

98.

$$\begin{aligned} 6^{3x} &= 5 \\ \log 6^{3x} &= \log 5 \\ 3x \log 6 &= \log 5 \\ x &= \frac{\log 5}{3\log 6} \approx 0.2994 \end{aligned}$$

99.

$$\begin{aligned} 3^{2x+1} &= 6 \\ \log 3^{2x+1} &= \log 6 \\ (2x+1)\log 3 &= \log 6 \\ 2x &= \frac{\log 6}{\log 3} - 1 \\ x &= \frac{1}{2}\left(\frac{\log 6}{\log 3} - 1\right) \approx 0.3155 \end{aligned}$$

100.

$$\begin{aligned} 4^{3x+2} &= 9 \\ \log 4^{3x+2} &= \log 9 \\ (3x+2)\log 4 &= \log 9 \\ 3x &= \frac{\log 9}{\log 4} - 2 \\ x &= \frac{1}{3}\left(\frac{\log 9}{\log 4} - 2\right) \approx -0.1383 \end{aligned}$$

101.

$$\begin{aligned} 5^{3x-5} &= 4 \\ \log 5^{3x-5} &= \log 4 \\ (3x-5)\log 5 &= \log 4 \\ 3x &= \frac{\log 4}{\log 5} + 5 \\ x &= \frac{1}{3}\left(\frac{\log 4}{\log 5} + 5\right) \approx 1.9538 \end{aligned}$$

102.

$$\begin{aligned} 8^{4x-2} &= 3 \\ \log 8^{4x-2} &= \log 3 \\ (4x-2)\log 8 &= \log 3 \\ 4x &= \frac{\log 3}{\log 8} + 2 \\ x &= \frac{1}{4}\left(\frac{\log 3}{\log 8} + 2\right) \approx 0.6321 \end{aligned}$$

103.

$$\begin{aligned} 2 \cdot 5^{x-1} &= 1 \\ \log(2 \cdot 5^{x-1}) &= \log 1 \\ \log 2 + (x-1)\log 5 &= 0 \\ (x-1)\log 5 &= -\log 2 \\ x &= -\frac{\log 2}{\log 5} + 1 \approx 0.5693 \end{aligned}$$

104.

$$\begin{aligned} 3 \cdot 4^{x+5} &= 2 \\ 4^{x+5} &= \frac{2}{3} \\ \log 4^{x+5} &= \log\frac{2}{3} \\ (x+5)\log 4 &= \log\frac{2}{3} \\ x &= \frac{\log\left(\frac{2}{3}\right)}{\log 4} - 5 \approx -5.2925 \end{aligned}$$

105.

$$\begin{aligned} \log_5 2 + \log_5 x &= 2 \\ \log_5 2x &= 2 \\ 2x &= 5^2 \\ 2x &= 25 \\ x &= \frac{25}{2} \end{aligned}$$

106.

$$\begin{aligned} \log_3 x + \log_3 10 &= 2 \\ \log_3(10x) &= 2 \\ 10x &= 3^2 \\ 10x &= 9 \\ x &= \frac{9}{10} \end{aligned}$$

107. $\log(5x) - \log(x+1) = 4$

$$\log\frac{5x}{x+1} = 4$$
$$\frac{5x}{x+1} = 10^4$$
$$\frac{5x}{x+1} = 10,000$$
$$5x = 10,000x + 10,000$$
$$x = -1.0005$$

no solution, or $\varnothing$

108. $\ln(3x) - \ln(x-3) = 2$

$$\ln\left(\frac{3x}{x-3}\right) = 2$$
$$\frac{3x}{x-3} = e^2$$
$$3x = e^2x - 3e^2$$
$$3x - e^2x = -3e^2$$
$$(3-e^2)x = -3e^2$$
$$x = \frac{3e^2}{e^2-3}$$

109. $\log_2 x + \log_2 2x - 3 = 1$

$$\log_2(x \cdot 2x) = 4$$
$$2x^2 = 2^4$$
$$2x^2 = 16$$
$$x^2 = 8$$
$$x = \pm 2\sqrt{2}$$

$-2\sqrt{2}$ is rejected since $\log_2\left(-2\sqrt{2}\right)$ is undefined. The solution is $2\sqrt{2}$.

110. $-\log_6(4x+7) + \log_6 x = 1$

$$\log_6\frac{x}{4x+7} = 1$$
$$\frac{x}{4x+7} = 6$$
$$x = 6(4x+7)$$
$$x = 24x + 42$$
$$x = -\frac{42}{23}$$

$-\frac{42}{23}$ is rejected since $\log_6\left(-\frac{42}{23}\right)$ is undefined.

There is no solution, or $\varnothing$.

111. $y = y_0e^{kt}$

$$y = 155,000e^{0.06(4)}$$
$$\approx 197,044$$

There will be 197,044 ducks after 4 weeks.

112. $y = y_0e^{kt}$

$$y = 2,971,650e^{-0.00129(8)}$$
$$= 2,971,650e^{-0.01032}$$
$$\approx 2,941,140$$

The population of Armenia in the year 2015 will be approximately 2,941,140.

113. $y = y_0e^{kt}$

$$1,500,000,000 = 1,321,851,888e^{0.00606t}$$
$$\frac{1,500,000,000}{1,321,851,888} = e^{0.00606t}$$
$$\ln\frac{1,500,000,000}{1,321,851,888} = \ln e^{0.00606t}$$
$$\ln\frac{1,500,000,000}{1,321,851,888} = 0.00606t$$
$$t = \frac{1}{0.00606}\ln\frac{1,500,000,000}{1,321,851,888}$$
$$t \approx 20.9$$

It will take approximately 20.9 years.

114. $y = y_0e^{kt}$

$$2(33,390,141) = 33,390,141e^{0.009t}$$
$$2 = e^{0.009t}$$
$$\ln 2 = \ln e^{0.009t}$$
$$\ln 2 = 0.009t$$
$$t = \frac{\ln 2}{0.009}$$
$$t \approx 77.0$$

It will take approximately 77.0 years.

115. $y = y_0e^{kt}$

$$2(24,821,286) = 24,821,286e^{0.018t}$$
$$2 = e^{0.018t}$$
$$\ln 2 = \ln e^{0.018t}$$
$$\ln 2 = 0.018t$$
$$t = \frac{\ln 2}{0.018}$$
$$t \approx 38.5$$

It will take approximately 38.5 years.

116.
$$A = P\left(1+\frac{r}{n}\right)^{nt}$$
$$10{,}000 = 5000\left(1+\frac{0.08}{4}\right)^{4t}$$
$$2 = (1.02)^{4t}$$
$$\log 2 = \log 1.02^{4t}$$
$$\log 2 = 4t \log 1.02$$
$$t = \frac{\log 2}{4 \log 1.02} \approx 8.8$$
It will take 8.8 years.

117.
$$A = P\left(1+\frac{r}{n}\right)^{nt}$$
$$10{,}000 = 6000\left(1+\frac{0.06}{12}\right)^{12t}$$
$$\frac{5}{3} = (1.005)^{12t}$$
$$\log\frac{5}{3} = \log 1.005^{12t}$$
$$\log\frac{5}{3} = 12t \log 1.005$$
$$t = \frac{1}{12}\left(\frac{\log\left(\frac{5}{3}\right)}{\log(1.005)}\right) \approx 8.5$$
It was invested for approximately 8.5 years.

118. $Y_1 = e^x, Y_2 = 2$

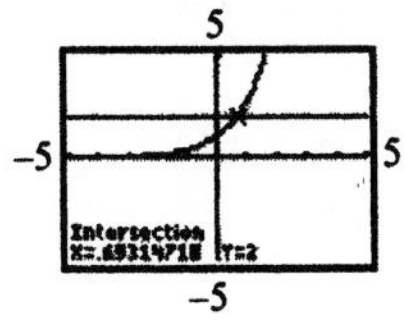

$x \approx 0.69$

119. $Y_1 = 10^{0.3x}, Y_2 = 7$

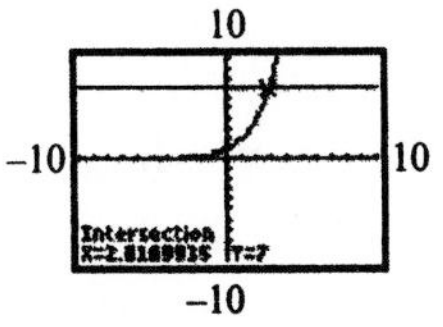

$x \approx 2.82$

120.
$$3^x = \frac{1}{81}$$
$$3^x = 3^{-4}$$
$$x = -4$$

121.
$$7^{4x} = 49$$
$$7^{4x} = 7^2$$
$$4x = 2$$
$$x = \frac{1}{2}$$

122.
$$8^{3x-2} = 32$$
$$(2^3)^{(3x-2)} = 2^5$$
$$2^{9x-6} = 2^5$$
$$9x - 6 = 5$$
$$9x = 11$$
$$x = \frac{11}{9}$$

123.
$$\log_4 4 = x$$
$$4^x = 4^1$$
$$x = 1$$

124.
$$\log_3 x = 4$$
$$3^4 = x$$
$$81 = x$$

125.
$$\log_5(x^2 - 4x) = 1$$
$$5^1 = x^2 - 4x$$
$$0 = x^2 - 4x - 5$$
$$0 = (x-5)(x+1)$$
$x - 5 = 0$ or $x + 1 = 0$

$x = 5$ $\quad$ $x = -1$

Both check, so the solutions are 5 and -1.

126.
$$\log_4(3x-1) = 2$$
$$4^2 = 3x - 1$$
$$16 + 1 = 3x$$
$$\frac{17}{3} = x$$

127.
$$\ln x = -3.2$$
$$e^{\ln x} = e^{-3.2}$$
$$x = e^{-3.2}$$

128.
$$\log_5 x + \log_5 10 = 2$$
$$\log_5(10x) = 2$$
$$5^2 = 10x$$
$$\frac{25}{10} = x$$
$$\frac{5}{2} = x$$

129. $\ln x - \ln 2 = 1$

$$\ln\frac{x}{2} = 1$$
$$e^{\ln\frac{x}{2}} = e^1$$
$$\frac{x}{2} = e$$
$$x = 2e$$

130. $\log_6 x - \log_6(4x+7) = 1$

$$\log_6\frac{x}{4x+7} = 1$$
$$6^1 = \frac{x}{4x+7}$$
$$24x + 42 = x$$
$$23x = -42$$
$$x = -\frac{42}{23}$$

$-\frac{42}{23}$ is rejected since $\log_6\left(-\frac{42}{23}\right)$ is undefined.

There is no solution, or ∅.

Chapter 12 Test

1. $f(x) = x$ and $g(x) = 2x - 3$
$(f \cdot g)(x) = f(x) \cdot g(x) = x(2x-3) = 2x^2 - 3x$

2. $f(x) = x$ and $g(x) = 2x - 3$

$$\begin{aligned}(f-g)(x) &= f(x) - g(x)\\ &= x - (2x-3)\\ &= -x+3\\ &= 3-x\end{aligned}$$

3. $(f \circ h)(0) = f(h(0)) = f(5) = 5$

4. $(g \circ f)(x) = g(f(x)) = g(x) = x - 7$

5.
$$\begin{aligned}(g \circ h)(x) &= g(h(x))\\ &= g(x^2 - 6x + 5)\\ &= x^2 - 6x + 5 - 7\\ &= x^2 - 6x - 2\end{aligned}$$

6. $f(x) = 7x - 14$, $f^{-1}(x) = \frac{x+14}{7}$

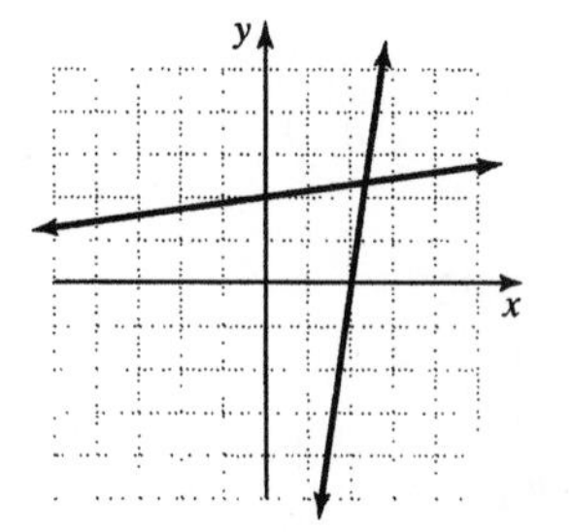

7. The graph represents a one-to-one function.

8. The graph does not represent a one-to-one function.

9. $y = 6 - 2x$ is one-to-one.

$$x = 6 - 2y$$
$$2y = -x + 6$$
$$y = \frac{-x+6}{2}$$
$$f^{-1}(x) = \frac{-x+6}{2}$$

10. $f = \{(0,0),(2,3),(-1,5)\}$ is one-to-one.
$f^{-1} = \{(0,0),(3,2),(5,-1)\}$

11. The function is not one-to-one.

12. $\log_3 6 + \log_3 4 = \log_3(6 \cdot 4) = \log_3 24$

13.
$$\begin{aligned}&\log_5 x + 3\log_5 x - \log_5(x+1)\\ &= 4\log_5 x - \log_5(x+1)\\ &= \log_5 x^4 - \log_5(x+1)\\ &= \log_5\frac{x^4}{x+1}\end{aligned}$$

14.
$$\begin{aligned}\log_6\frac{2x}{y^3} &= \log_6 2x - \log_6 y^3\\ &= \log_6 2 + \log_6 x - 3\log_6 y\end{aligned}$$

15.
$$\begin{aligned}\log_b\left(\frac{3}{25}\right) &= \log_b 3 - \log_b 25\\ &= \log_b 3 - \log_b 5^2\\ &= \log_b 3 - 2\log_b 5\\ &= 0.79 - 2(1.16)\\ &= -1.53\end{aligned}$$

16. $\log_7 8 = \frac{\ln 8}{\ln 7} \approx 1.0686$

17. $8^{x-1} = \frac{1}{64}$
$8^{x-1} = 8^{-2}$
$x - 1 = -2$
$x = -1$

18. $3^{2x+5} = 4$
$\log 3^{2x+5} = \log 4$
$(2x+5)\log 3 = \log 4$
$2x = \frac{\log 4}{\log 3} - 5$
$x = \frac{1}{2}\left(\frac{\log 4}{\log 3} - 5\right)$
$x \approx -1.8691$

19. $\log_3 x = -2$
$x = 3^{-2}$
$x = \frac{1}{9}$

20. $\ln\sqrt{e} = x$
$\ln e^{1/2} = x$
$\frac{1}{2} = x$

21. $\log_8(3x-2) = 2$
$3x - 2 = 8^2$
$3x - 2 = 64$
$3x = 66$
$x = \frac{66}{3} = 22$

22. $\log_5 x + \log_5 3 = 2$
$\log_5(3x) = 2$
$3x = 5^2$
$3x = 25$
$x = \frac{25}{3}$

23. $\log_4(x+1) - \log_4(x-2) = 3$
$\log_4 \frac{x+1}{x-2} = 3$
$\frac{x+1}{x-2} = 4^3$
$\frac{x+1}{x-2} = 64$
$x + 1 = 64x - 128$
$129 = 63x$
$\frac{129}{63} = x$
$\frac{43}{21} = x$

24. $\ln(3x+7) = 1.31$
$3x + 7 = e^{1.31}$
$3x = e^{1.31} - 7$
$x = \frac{e^{1.31} - 7}{3} \approx -1.0979$

25. $y = \left(\frac{1}{2}\right)^x + 1$

26. $y = 3^x$ and $y = \log_3 x$

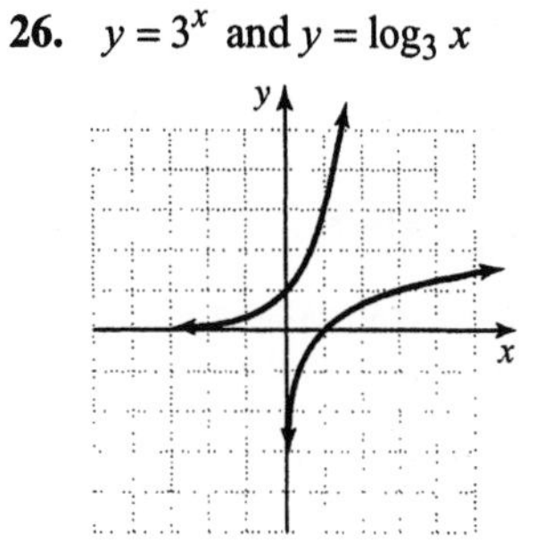

27. $A = \left(1+\frac{r}{n}\right)^{nt}, P = 4000, t = 3, r = 0.09,$ and $n = 12$

$$A = 4000\left(1+\frac{0.09}{12}\right)^{12(3)}$$
$$= 4000(1.0075)^{36}$$
$$\approx 5234.58$$

\$5234.58 will be in the account.

28. $A = \left(1+\frac{r}{n}\right)^{nt}$, $P = 2000$, $A = 3000$

$r = 0.07$, $n = 2$

$$3000 = 2000\left(1+\frac{0.07}{2}\right)^{2t}$$
$$1.5 = (1.035)^{2t}$$
$$\log 1.5 = \log 1.035^{2t}$$
$$\log 1.5 = 2t \log 1.035$$
$$t = \frac{\log 1.5}{2\log 1.035} \approx 5.9$$

It would take 6 years.

29. $y = y_0 e^{kt}$

$$y = 57{,}000e^{0.026(5)}$$
$$= 57{,}000e^{0.13}$$
$$\approx 64{,}913$$

There will be approximately 64,913 prairie dogs 5 years from now.

30.
$$y = y_0 e^{kt}$$
$$1000 = 400e^{0.062(t)}$$
$$2.5 = e^{0.062t}$$
$$\ln 2.5 = \ln e^{0.062t}$$
$$0.062t = \ln 2.5$$
$$t = \frac{\ln 2.5}{0.062} \approx 14.8$$

It will take the naturalists approximately 15 years to reach their goal.

31. $\log(1+k) = \frac{0.3}{D}, D = 56$

$$\log(1+k) = \frac{0.3}{56}$$
$$1+k = 10^{0.3/56}$$
$$k = -1+10^{0.3/56}$$
$$k \approx 0.012$$

The rate of population increase is approximately 1.2%.

Chapter 12 Cumulative Review

1. a. $\frac{4}{5} \div \frac{5}{16} = \frac{4}{5} \cdot \frac{16}{5} = \frac{4\cdot 16}{5\cdot 5} = \frac{64}{25}$

b. $\frac{7}{10} \div 14 = \frac{7}{10} \div \frac{14}{1} = \frac{7}{10} \cdot \frac{1}{14} = \frac{7\cdot 1}{10\cdot 2\cdot 7} = \frac{1}{20}$

c. $\frac{3}{8} \div \frac{3}{10} = \frac{3}{8} \cdot \frac{10}{3} = \frac{3\cdot 2\cdot 5}{2\cdot 4\cdot 3} = \frac{5}{4}$

2.
$$\frac{1}{3}(x-2) = \frac{1}{4}(x+1)$$
$$4(x-2) = 3(x+1)$$
$$4x-8 = 3x+3$$
$$x = 11$$

3. $f(x) = x^2$

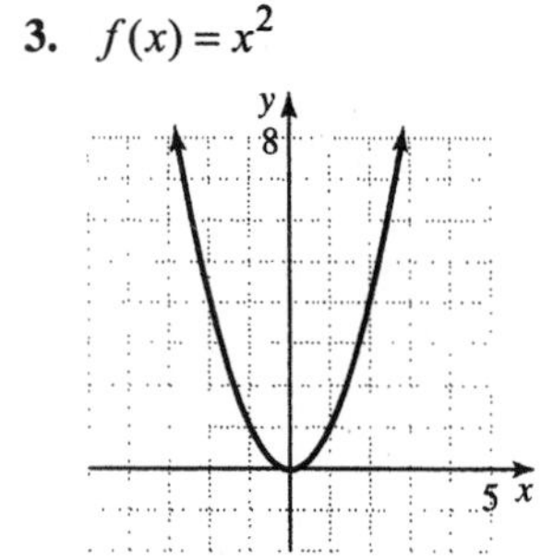

4. $y = f(x) = -3x + 4$, $m = -3$

Perpendicular line: $m = \frac{1}{3}$, through $(-2, 6)$

$$y - y_1 = m(x - x_1)$$
$$y - 6 = \frac{1}{3}[x - (-2)]$$
$$y - 6 = \frac{1}{3}x + \frac{2}{3}$$
$$y = \frac{1}{3}x + \frac{20}{3}$$
$$f(x) = \frac{1}{3}x + \frac{20}{3}$$

5. Equation 2 is twice the opposite of equation 1 and equation 3 is one-half of equation 1. Therefore, the system is dependent. The solution is $\{(x, y, z) | x - 5y - 2z = 6\}$.

6. The angles labeled $y°$ and $(x - 40)°$ are alternate interior angles, so $y = x - 40$. The angles labeled $x°$ and $y°$ are supplementary, so $x + y = 180$.

$$\begin{cases} y = x - 40 \\ x + y = 180 \end{cases}$$

Replace y with $x - 40$ in the second equation.

$$x+(x-40)=180$$
$$2x=220$$
$$x=110$$
$$y=x-40=110-40=70$$

7. a. $-3+[(-2-5)-2]=-3+[-7-2]$
$=-3+[-9]$
$=-12$

b. $2^3-|10|+[-6-(-5)]=2^3-|10|+[-6+5]$
$=2^3-|10|+(-1)$
$=8-10-1$
$=-2-1$
$=-3$

8. a. $(4a^3)^2=4^2(a^3)^2=16a^6$

b. $\left(-\frac{2}{3}\right)^3=\frac{(-2)^3}{3^3}=\frac{-8}{27}=-\frac{8}{27}$

c. $\left(\frac{4a^5}{b^3}\right)^3=\frac{4^3(a^5)^3}{(b^3)^3}=\frac{64a^{15}}{b^9}$

d. $\left(\frac{3^{-2}}{x}\right)^{-3}=\frac{(3^{-2})^{-3}}{x^{-3}}=\frac{3^6}{x^{-3}}=729x^3$

e. $(a^{-2}b^3c^{-4})^{-2}=(a^{-2})^{-2}(b^3)^{-2}(c^{-4})^{-2}$
$=a^4b^{-6}c^8$
$=\frac{a^4c^8}{b^6}$

9. a. $C(100)=\frac{2.6(100)+10,000}{100}$
$=102.60$

The cost is \$102.60 per disc for 100 discs.

b. $C(1000)=\frac{2.6(1000)+10,000}{1000}$
$=12.60$

The cost is \$12.60 per disc for 1000 discs.

10. a. $(3x-1)^2=(3x)^2-2(3x)(1)+1^2$
$=9x^2-6x+1$

b. $\left(\frac{1}{2}x+3\right)\left(\frac{1}{2}x-3\right)=\left(\frac{1}{2}x\right)^2-3^2$
$=\frac{1}{4}x^2-9$

c. $(2x-5)(6x+7)=12x^2+14x-30x-35$
$=12x^2-16x-35$

11. $12a-8a=10+2a-13-7$
$4a=2a-10$
$2a=-10$
$a=-5$

12. $\frac{5}{x-2}+\frac{3}{x^2+4x+4}-\frac{6}{x+2}$
$=\frac{5}{x-2}+\frac{3}{(x+2)^2}-\frac{6}{x+2}$
$=\frac{5(x+2)^2+3(x-2)-6(x-2)(x+2)}{(x-2)(x+2)(x+2)}$
$=\frac{-x^2+23x+38}{(x-2)(x+2)^2}$

13. $\frac{8x^2y^2-16xy+2x}{4xy}=\frac{8x^2y^2}{4xy}-\frac{16xy}{4xy}+\frac{2x}{4xy}$
$=2xy-4+\frac{1}{2y}$

14. a. $\frac{\frac{a}{5}}{\frac{a-1}{10}}=\frac{a}{5}\cdot\frac{10}{a-1}=\frac{2a}{a-1}$

b. $\frac{\frac{3}{2+a}+\frac{6}{2-a}}{\frac{5}{a+2}-\frac{1}{a-2}}=\frac{\frac{3}{a+2}-\frac{6}{a-2}}{\frac{5}{a+2}-\frac{1}{a-2}}$

Multiply the numerator and the denominator by $(a+2)(a-2)$.

$\frac{3(a-2)-6(a+2)}{5(a-2)-1(a+2)}=\frac{3a-6-6a-12}{5a-10-a-2}$
$=\frac{-3a-18}{4a-12}$

c. $\frac{x^{-1}+y^{-1}}{xy}=\frac{\frac{1}{x}+\frac{1}{y}}{xy}=\frac{\left(\frac{1}{x}+\frac{1}{y}\right)xy}{(xy)(xy)}=\frac{y+x}{x^2y^2}$

15. $3m^2-24m-60=3(m^2-8m-20)$
$=3(m+2)(m-10)$

16. $5x^2-85x+350=5(x^2-17x+70)$
$=5(x-10)(x-7)$

17. $\frac{3x^2+2x}{x-1}-\frac{10x-5}{x-1}=\frac{3x^2+2x-(10x-5)}{x-1}$
$=\frac{3x^2+2x-10x+5}{x-1}$
$=\frac{3x^2-8x+5}{x-1}$
$=\frac{(3x-5)(x-1)}{x-1}$
$=3x-5$

18.

2	8	−12	−7
		16	8
	8	4	1

Solution: $8x+4+\frac{1}{x-2}$

19. a. $\sqrt[4]{81}=\sqrt[4]{3^4}=3$

b. $\sqrt[5]{-243}=\sqrt[5]{(-3)^5}=-3$

c. $-\sqrt{25}=-\sqrt{5^2}=-5$

d. $\sqrt[4]{-81}$ is not a real number.

e. $\sqrt[3]{64x^3}=\sqrt[3]{4^3x^3}=4x$

20. $\frac{1}{a+5}=\frac{1}{3a+6}-\frac{a+2}{a^2+7a+10}$
$\frac{1}{a+5}=\frac{1}{3(a+2)}-\frac{a+2}{(a+2)(a+5)}$
$3(a+2)=a+5-3(a+2)$
$3a+6=a+5-3a-6$
$5a=-7$
$a=-\frac{7}{5}$

21. a. $\sqrt{x}\cdot\sqrt[4]{x}=x^{1/2}\cdot x^{1/4}=x^{3/4}=\sqrt[4]{x^3}$

b. $\frac{\sqrt{x}}{\sqrt[3]{x}}=\frac{x^{1/2}}{x^{1/3}}=x^{\frac{1}{2}-\frac{1}{3}}=x^{1/6}=\sqrt[6]{x}$

c. $\sqrt[3]{3}\cdot\sqrt{2}=3^{1/3}\cdot 2^{1/2}$
$=3^{2/6}\cdot 2^{3/6}$
$=9^{1/6}\cdot 8^{1/6}$
$=72^{1/6}$
$=\sqrt[6]{72}$

22. $y=kx$
$\frac{1}{2}=12k$
$k=\frac{1}{24},\ y=\frac{1}{24}x$

23. a. $\sqrt{3}\left(5+\sqrt{30}\right)=5\sqrt{3}+\sqrt{90}=5\sqrt{3}+3\sqrt{10}$

b. $\left(\sqrt{5}-\sqrt{6}\right)\left(\sqrt{7}+1\right)=\sqrt{35}+\sqrt{5}-\sqrt{42}-\sqrt{6}$

c. $\left(7\sqrt{x}+5\right)\left(3\sqrt{x}-\sqrt{5}\right)$
$=21x-7\sqrt{5x}+15\sqrt{x}-5\sqrt{5}$

d. $\left(4\sqrt{3}-1\right)^2$
$=\left(4\sqrt{3}\right)^2-2\left(4\sqrt{3}\right)(1)+1^2$
$=16\cdot 3-8\sqrt{3}+1$
$=49-8\sqrt{3}$

e. $\left(\sqrt{2x}-5\right)\left(\sqrt{2x}+5\right)=\left(\sqrt{2x}\right)^2-5^2$
$=2x-25$

f. $\left(\sqrt{x-3}+5\right)^2=\left(\sqrt{x-3}\right)^2+2\sqrt{x-3}(5)+5^2$
$=x-3+10\sqrt{x-3}+25$
$=x+22+10\sqrt{x-3}$

24. a. $\sqrt[4]{81}=\sqrt[4]{3^4}=3$

b. $\sqrt[3]{-27}=\sqrt[3]{(-3)^3}=-3$

c. $\sqrt{\frac{9}{64}}=\sqrt{\left(\frac{3}{8}\right)^2}=\frac{3}{8}$

d. $\sqrt[4]{x^{12}}=x^3$

e. $\sqrt[3]{-125y^6}=-5y^2$

25. $\frac{\sqrt[4]{x}}{\sqrt[4]{81y^5}} = \frac{\sqrt[4]{x}}{\sqrt[4]{81y^5}} \cdot \frac{\sqrt[4]{y^3}}{\sqrt[4]{y^3}} = \frac{\sqrt[4]{xy^3}}{3y^2}$

26. a. $a^{1/4}(a^{3/4} - a^8) = a^{4/4} - a^{33/4} = a - a^{39/4}$

b. $(x^{1/2} - 3)(x^{1/2} + 5)$
$= x^{2/2} + 5x^{1/2} - 3x^{1/2} - 15$
$= x + 2x^{1/2} - 15$

27.
$$\sqrt{4-x} = x-2$$
$$\left(\sqrt{4-x}\right)^2 = (x-2)^2$$
$$4 - x = x^2 - 4x + 4$$
$$0 = x^2 - 3x$$
$$0 = x(x-3)$$
$x = 0$ or $x - 3 = 0$
$x = 3$
$x = 0$ does not check, so the only solution is $x = 3$.

28. a. $\frac{\sqrt{54}}{\sqrt{6}} = \sqrt{\frac{54}{6}} = \sqrt{9} = 3$

b. $\frac{\sqrt{108a^2}}{3\sqrt{3}} = \frac{1}{3}\sqrt{\frac{108a^2}{3}}$
$= \frac{1}{3}\sqrt{36a^2}$
$= \frac{1}{3}(6a)$
$= 2a$

c. $\frac{3\sqrt[3]{81a^5b^{10}}}{\sqrt[3]{3b^4}} = 3\sqrt[3]{\frac{81a^5b^{10}}{3b^4}}$
$= 3\sqrt[3]{27a^5b^6}$
$= 9ab^2\sqrt[3]{a^2}$

29.
$$3x^2 - 9x + 8 = 0$$
$$x^2 - 3x + \frac{8}{3} = 0$$
$$x^2 - 3x = -\frac{8}{3}$$
$$x^2 - 3x + \left(\frac{-3}{2}\right)^2 = -\frac{8}{3} + \left(\frac{-3}{2}\right)^2$$
$$x^2 - 3x + \frac{9}{4} = -\frac{8}{3} + \frac{9}{4}$$
$$\left(x - \frac{3}{2}\right)^2 = -\frac{5}{12}$$
$$x - \frac{3}{2} = \pm\sqrt{-\frac{5}{12}}$$
$$x - \frac{3}{2} = \pm\frac{i\sqrt{5}}{2\sqrt{3}}$$
$$x - \frac{3}{2} = \pm\frac{i\sqrt{15}}{6}$$
$$x = \frac{3}{2} \pm \frac{i\sqrt{15}}{6}$$
$$= \frac{9}{6} \pm \frac{i\sqrt{15}}{6}$$
$$= \frac{9 \pm i\sqrt{15}}{6}$$
The solutions are $\frac{9 + i\sqrt{15}}{6}$ and $\frac{9 - i\sqrt{15}}{6}$.

30. a. $\frac{\sqrt{20}}{3} + \frac{\sqrt{5}}{4} = \frac{2\sqrt{5}}{3} + \frac{\sqrt{5}}{4}$
$= \frac{8\sqrt{5} + 3\sqrt{5}}{12}$
$= \frac{11\sqrt{5}}{12}$

b. $\sqrt[3]{\frac{24x}{27}} - \frac{\sqrt[3]{3x}}{2} = \frac{2\sqrt[3]{3x}}{3} - \frac{\sqrt[3]{3x}}{2}$
$= \frac{4\sqrt[3]{3x} - 3\sqrt[3]{3x}}{6}$
$= \frac{\sqrt[3]{3x}}{6}$

31. $\frac{3x}{x-2} - \frac{x+1}{x} = \frac{6}{x(x-2)}$

$3x(x) - (x+1)(x-2) = 6$

$3x^2 - x^2 + x + 2 = 6$

$2x^2 + x - 4 = 0$

$a = 2, b = 1, c = -4$

$x = \frac{-1 \pm \sqrt{1^2 - 4(2)(-4)}}{2(2)} = \frac{-1 \pm \sqrt{33}}{4}$

32. $\sqrt[3]{\frac{27}{m^4n^8}} = \frac{\sqrt[3]{27}}{\sqrt[3]{m^4n^8}}$

$= \frac{3}{mn^2\sqrt[3]{mn^2}}$

$= \frac{3 \cdot \sqrt[3]{m^2n}}{mn^2\sqrt[3]{mn^2} \cdot \sqrt[3]{m^2n}}$

$= \frac{3\sqrt[3]{m^2n}}{m^2n^3}$

33. $x^2 - 4x \le 0$

$x(x-4) = 0$

$x = 0, x = 4$

Region	Test Point	$x(x-4) \le 0$	Result
$x < 0$	$x = -1$	$(-1)(-5) \le 0$	False
$0 < x < 4$	$x = 2$	$2(-2) \le 0$	True
$x > 4$	$x = 5$	$5(1) \le 0$	False

Solution: [0, 4]

34. $c^2 = a^2 + b^2$

$8^2 = 4^2 + b^2$

$64 = 16 + b^2$

$48 = b^2$

$\pm 4\sqrt{3} = b$

$b > 0$ so the length is $4\sqrt{3}$ inches.

35. $F(x) = (x-3)^2 + 1$

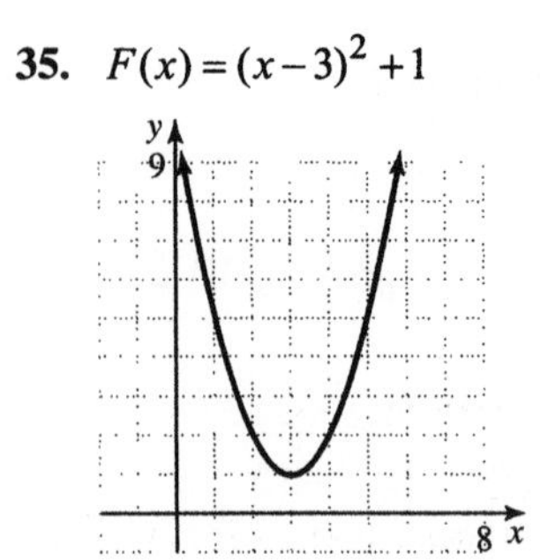

36. a. $i^8 = (i^2)^4 = (-1)^4 = 1$

b. $i^{21} = i(i^{20}) = i$

c. $i^{42} = i^2(i^{40}) = i^2 = -1$

d. $i^{-13} = \frac{1}{i^{13}} = \frac{1}{i(i^{12})} = \frac{1}{i} = \frac{i}{i^2} = -i$

37. $\frac{45}{x} = \frac{5}{7}$

$45 \cdot 7 = 5x$

$315 = 5x$

$63 = x$

38. $4x^2 + 8x - 1 = 0$

$x^2 + 2x - \frac{1}{4} = 0$

$x^2 + 2x = \frac{1}{4}$

$x^2 + 2x + \left(\frac{2}{2}\right)^2 = \frac{1}{4} + \left(\frac{2}{2}\right)^2$

$x^2 + 2x + 1 = \frac{1}{4} + 1$

$(x+1)^2 = \frac{5}{4}$

$x + 1 = \pm\sqrt{\frac{5}{4}}$

$x + 1 = \pm\frac{\sqrt{5}}{2}$

$x = -1 \pm \frac{\sqrt{5}}{2}$

$= \frac{-2 \pm \sqrt{5}}{2}$

The solutions are $\frac{-2+\sqrt{5}}{2}$ and $\frac{-2-\sqrt{5}}{2}$.

39. $f(x) = x + 3$

$y = x + 3$

$x = y + 3$

$y = x - 3$

$f^{-1}(x) = x - 3$

40. $\left(x - \frac{1}{2}\right)^2 = \frac{x}{2}$

$x^2 - x + \frac{1}{4} = \frac{1}{2}x$

$x^2 - \frac{3}{2}x + \frac{1}{4} = 0$

$4x^2 - 6x + 1 = 0$

$a = 4, b = -6, c = 1$

$x = \frac{-(-6) \pm \sqrt{(-6)^2 - 4(4)(1)}}{2(4)}$

$= \frac{6 \pm \sqrt{20}}{8}$

$= \frac{6 \pm 2\sqrt{5}}{8}$

$= \frac{3 \pm \sqrt{5}}{4}$

The solutions are $\frac{3+\sqrt{5}}{4}$ and $\frac{3-\sqrt{5}}{4}$.

41. a. $\log_4 16 = \log_4 4^2 = 2$

b. $\log_{10} \frac{1}{10} = \log_{10} 10^{-1} = -1$

c. $\log_9 3 = \log_9 9^{1/2} = \frac{1}{2}$

42. $f(x) = -(x+1)^2 + 1$

Vertex: (−1, 1)

Axis of symmetry: $x = -1$

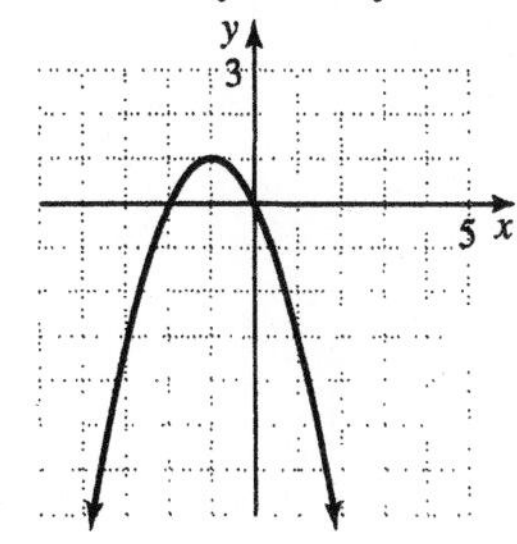

Chapter 13

Section 13.1

Practice Exercises

1. $x = \frac{1}{2}y^2$; $a = \frac{1}{2}$, $h = 0, k = 0$; vertex: (0, 0)

x	y
2	−2
$\frac{1}{2}$	−1
0	0
$\frac{1}{2}$	1
2	2

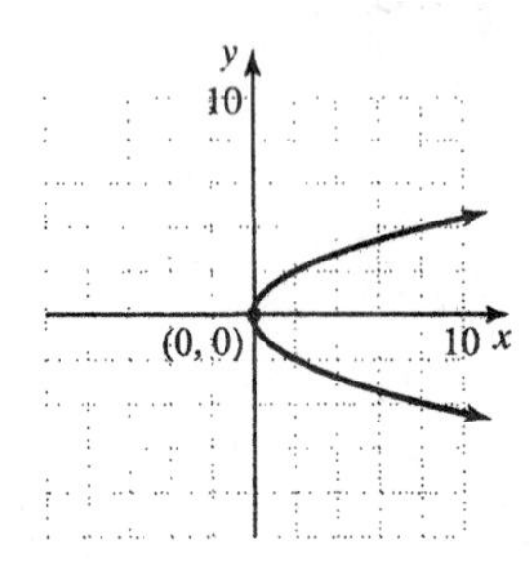

2. $x = -2(y+4)^2 - 1$; $a = -2, h = -1, k = -4$;
vertex: (−1, −4)

x	y
−9	−6
3	−5
−1	−4
−3	−3
−9	−2

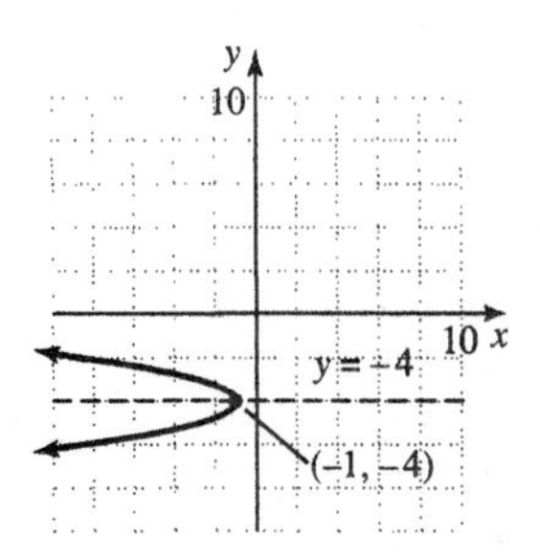

3.

$$\begin{aligned} y &= -x^2 + 4x + 6 \\ y - 6 &= -x^2 + 4x \\ y - 6 &= -(x^2 - 4x) \\ y - 6 - (+4) &= -(x^2 - 4x + 4) \\ y - 10 &= -(x-2)^2 \\ y &= -(x-2)^2 + 10 \end{aligned}$$

$a = -1, h = 2, k = 10$
vertex: (2, 10)

x	y
−1	1
0	6
1	9
2	10
3	9
4	6
5	1

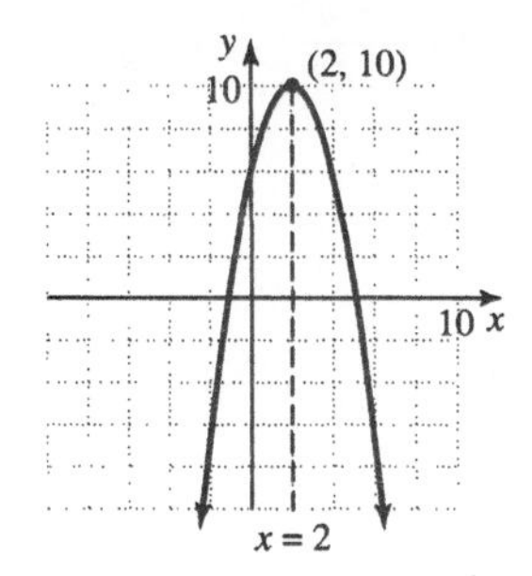

4. $x = 3y^2 + 6y + 4$
Find the vertex.

$$y = \frac{-b}{2a} = \frac{-6}{2(3)} = -1$$

$$x = 3(-1)^2 + 6(-1) + 4 = 3 - 6 + 4 = 1$$

vertex: (1, −1)
The axis of symmetry is the line $y = -1$.
Since $a > 0$, the parabola opens to the right.

$$x = 3(0)^2 + 6(0) + 4 = 4$$

The x-intercept is (4, 0).

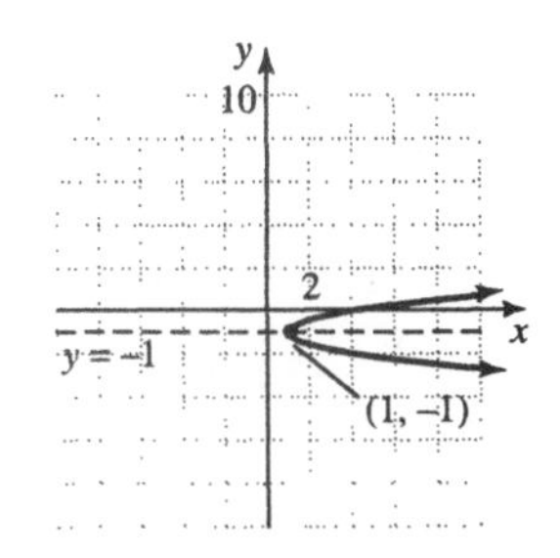

5. $x^2 + y^2 = 25$

$(x-0)^2 + (y-0)^2 = 5^2$

center: (0, 0); radius = 5

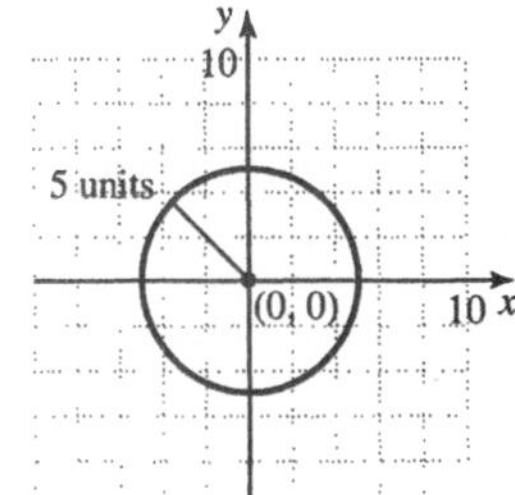

6. $(x-3)^2 + (y+2)^2 = 4$

$h = 3, k = -2, \ r = \sqrt{4} = 2$

center: (3, −2)

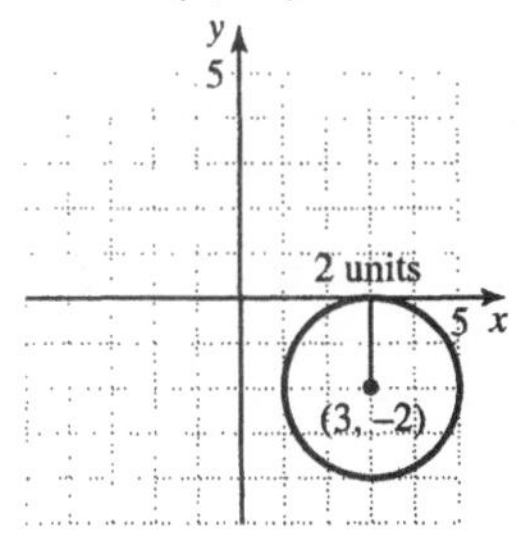

7. Center: (−2, −5); radius = 9

$(x-h)^2 + (y-k)^2 = r^2$

$h = -2$, $k = -5$, and $r = 9$.

The equation is $(x+2)^2 + (y+5)^2 = 81$.

8.

$$x^2 + y^2 + 6x - 2y = 6$$
$$(x^2 + 6x) + (y^2 - 2y) = 6$$
$$(x^2 + 6x + 9) + (y^2 - 2y + 1) = 6 + 9 + 1$$
$$(x+3)^2 + (y-1)^2 = 16$$

Center: (−3, 1); radius = $\sqrt{16} = 4$

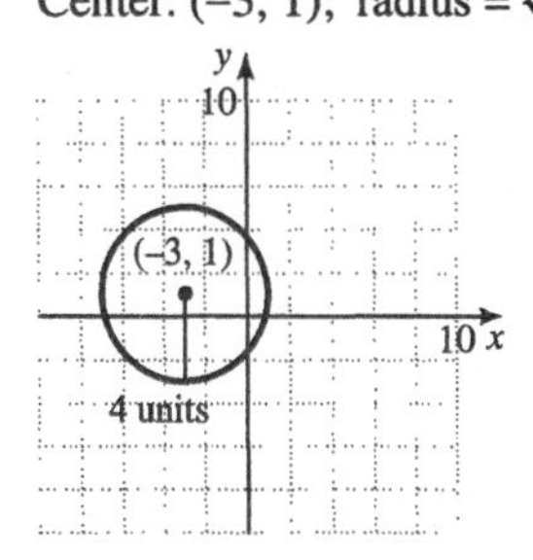

Graphing Calculator Explorations

1. $x^2 + y^2 = 55$

$y^2 = 55 - x^2$

$y = \pm\sqrt{55 - x^2}$

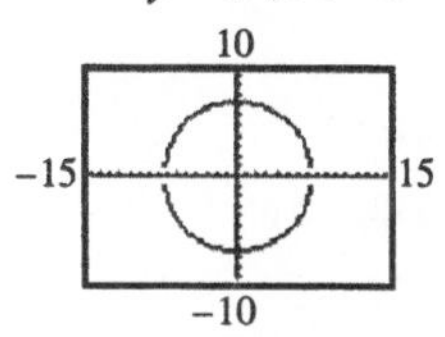

2. $x^2 + y^2 = 20$

$y^2 = 20 - x^2$

$y = \pm\sqrt{20 - x^2}$

10

−15 15

−10

3. $5x^2 + 5y^2 = 50$

$5y^2 = 50 - 5x^2$

$y^2 = 10 - x^2$

$y = \pm\sqrt{10 - x^2}$

10

−15 15

−10

4. $6x^2+6y^2=105$

$$6y^2=105-6x^2$$
$$y^2=17.5-x^2$$
$$y=\pm\sqrt{17.5-x^2}$$

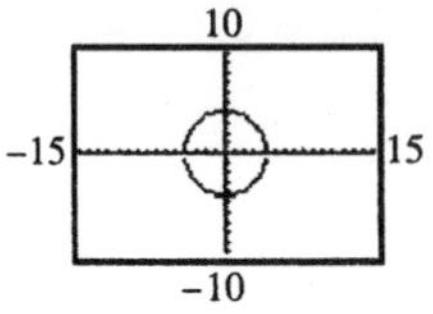

10, −15, 15, −10

5. $2x^2+2y^2-34=0$

$$2y^2=34-2x^2$$
$$y^2=17-x^2$$
$$y=\pm\sqrt{17-x^2}$$

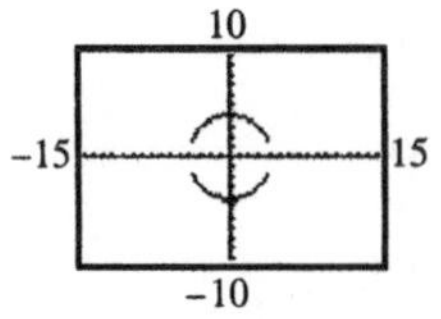

10, −15, 15, −10

6. $4x^2+4y^2-48=0$

$$4y^2=48-4x^2$$
$$y^2=12-x^2$$
$$y=\pm\sqrt{12-x^2}$$

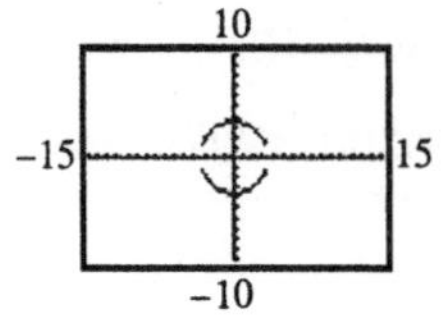

10, −15, 15, −10

7. $7x^2+7y^2-89=0$

$$7y^2=89-7x^2$$
$$y^2=\frac{89-7x^2}{7}$$
$$y=\pm\sqrt{\frac{89-7x^2}{7}}$$

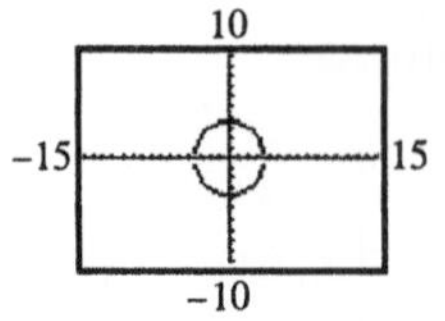

10, −15, 15, −10

8. $3x^2+3y^2-35=0$

$$3y^2=35-3x^2$$
$$y^2=\frac{35-3x^2}{3}$$
$$y=\pm\sqrt{\frac{35-3x^2}{3}}$$

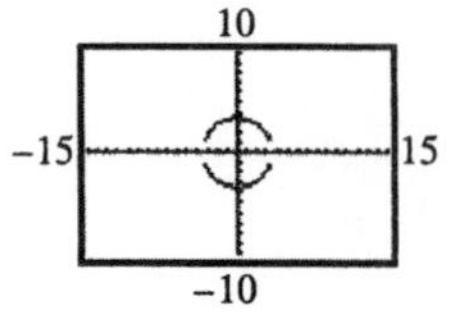

10, −15, 15, −10

Vocabulary and Readiness Check

1. The circle, parabola, ellipse, and hyperbola are called the conic sections.
2. For a parabola that opens upward the lowest point is the vertex.
3. A circle is the set of all points in a plane that are the same distance from a fixed point. The fixed point is called the center.
4. The midpoint of a diameter of a circle is the center.
5. The distance from the center of a circle to any point of the circle is called the radius.
6. Twice a circle's radius is its diameter.
7. $y=x^2-7x+5$; $a=1$, upward
8. $y=-x^2+16$; $a=-1$, downward
9. $x=-y^2-y+2$; $a=-1$, to the left
10. $x=3y^2+2y-5$; $a=3$, to the right
11. $y=-x^2+2x+1$; $a=-1$, downward
12. $x=-y^2+2y-6$; $a=-1$, to the left

Exercise Set 13.1

1. $x = 3y^2$

$x = 3(y-0)^2 + 0$

Vertex: (0, 0)

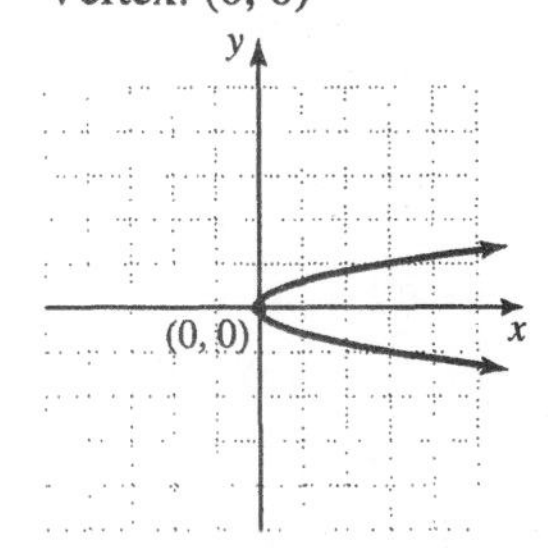

3. $x = (y-2)^2 + 3$

Vertex: (3, 2)

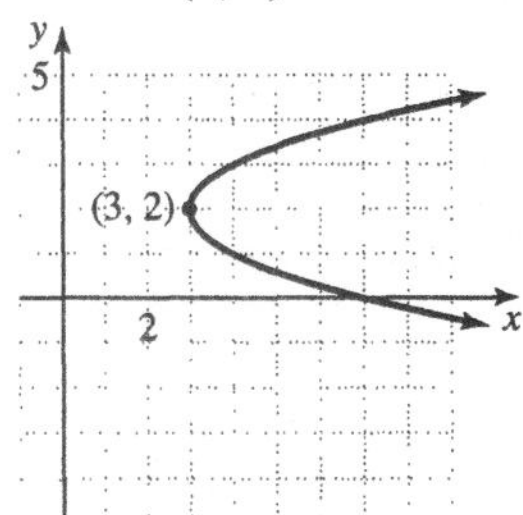

5. $y = 3(x-1)^2 + 5$

Vertex: (1, 5)

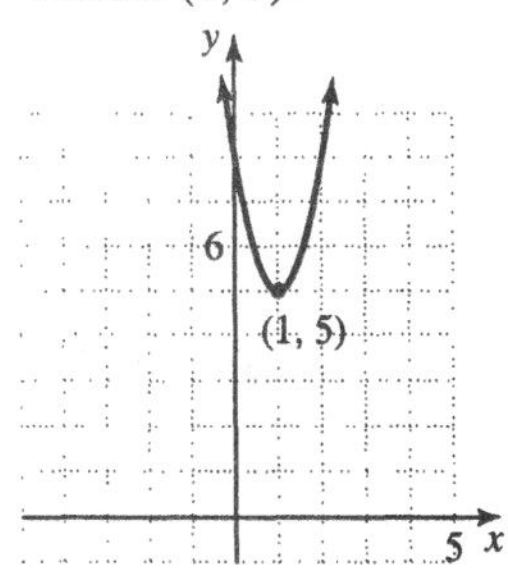

7.

$$x = y^2 + 6y + 8$$
$$x - 8 = y^2 + 6y$$
$$x - 8 + 9 = y^2 + 6y + 9$$
$$x + 1 = (y+3)^2$$
$$x = (y+3)^2 - 1$$

Vertex: (–1, –3)

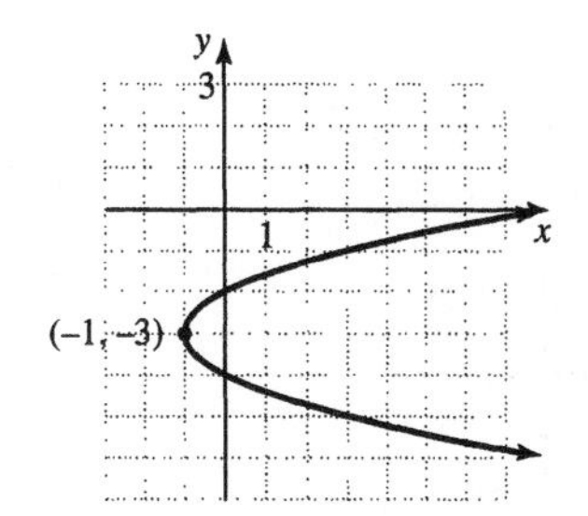

9.

$$y = x^2 + 10x + 20$$
$$y - 20 = x^2 + 10x$$
$$y - 20 + 25 = x^2 + 10x + 25$$
$$y + 5 = (x+5)^2$$
$$y = (x+5)^2 - 5$$

Vertex: (–5, –5)

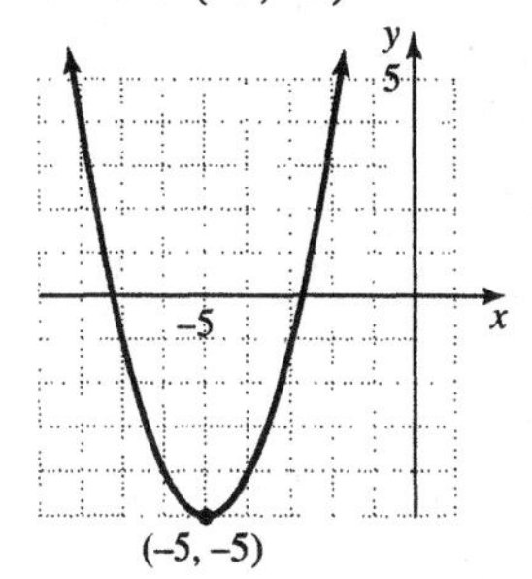

11.

$$x = -2y^2 + 4y + 6$$
$$x - 6 = -2(y^2 - 2y)$$
$$x - 6 + [-2(1)] = -2(y^2 - 2y + 1)$$
$$x - 8 = -2(y-1)^2$$
$$x = -2(y-1)^2 + 8$$

Vertex: (8, 1)

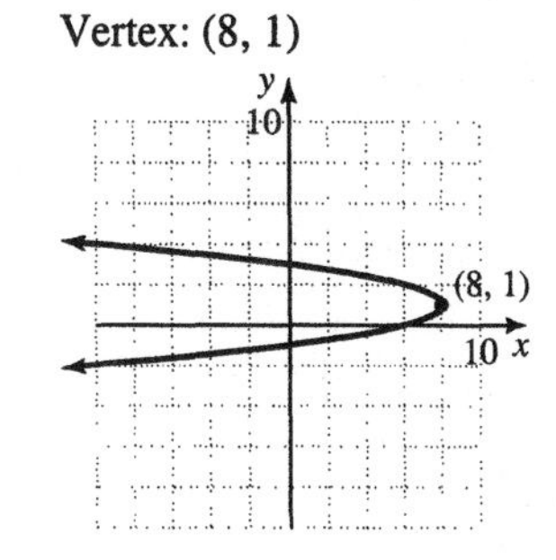

13. $x^2+y^2=9$

$(x-0)^2+(y-0)^2=3^2$

Center: (0, 0), radius $r=3$.

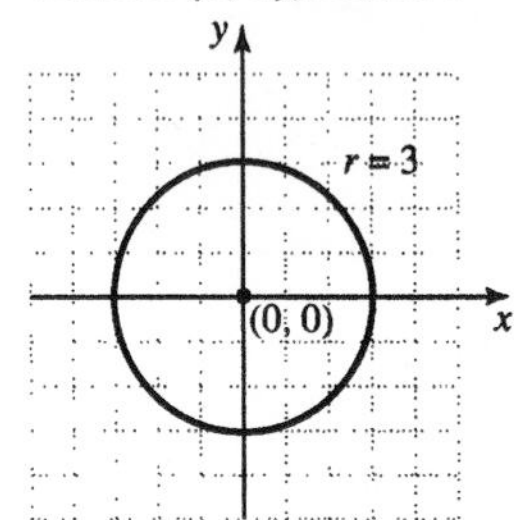

15. $x^2+(y-2)^2=1$

$(x-0)^2+(y-2)^2=1^2$

Center: (0, 2), radius $r=1$.

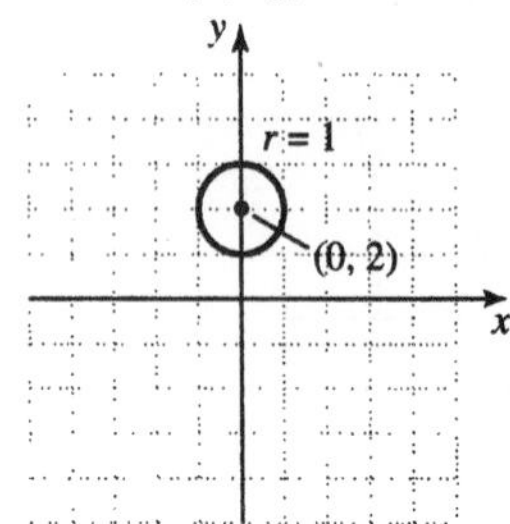

17. $(x-5)^2+(y+2)^2=1$

$(x-5)^2+(y+2)^2=1^2$

Center: (5, –2), radius $r=1$.

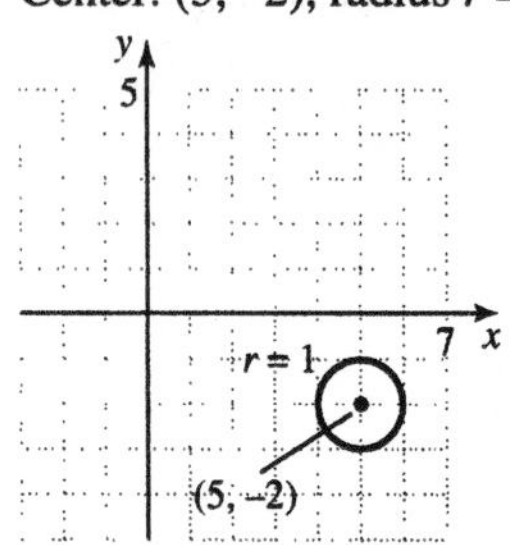

19.

$$x^2+y^2+6y=0$$
$$x^2+(y^2+6y)=0$$
$$x^2+(y^2+6y+9)=9$$
$$(x-0)^2+(y+3)^2=3^2$$

Center: (0, –3), radius $r=3$.

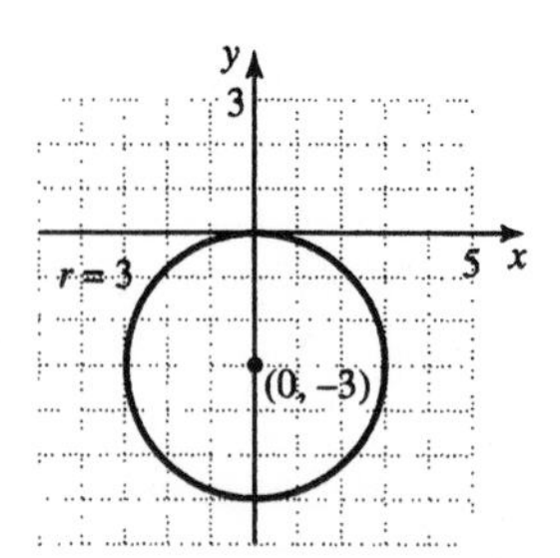

21.

$$x^2+y^2+2x-4y=4$$
$$(x^2+2x)+(y^2-4y)=4$$
$$(x^2+2x+1)+(y^2-4y+4)=4+1+4$$
$$(x+1)^2+(y-2)^2=9$$

Center: (–1, 2), radius $r=\sqrt{9}=3$.

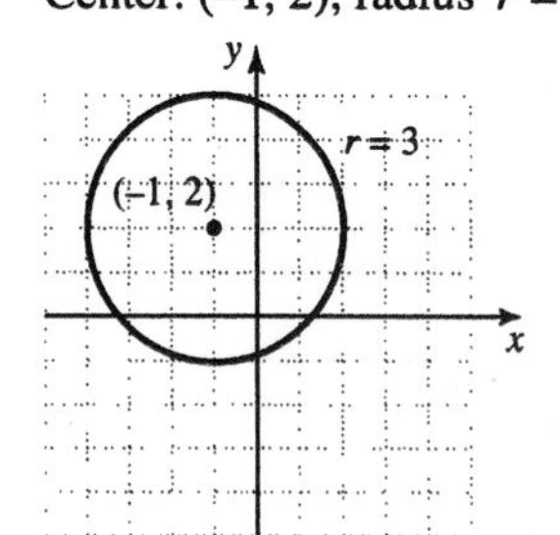

23.

$$x^2+y^2-4x-8y-2=0$$
$$(x^2-4x)+(y^2-8y)=2$$
$$(x^2-4x+4)+(y^2-8y+16)=2+4+16$$
$$(x-2)^2+(y-4)^2=22$$

Center: (2, 4), radius $r=\sqrt{22}$.

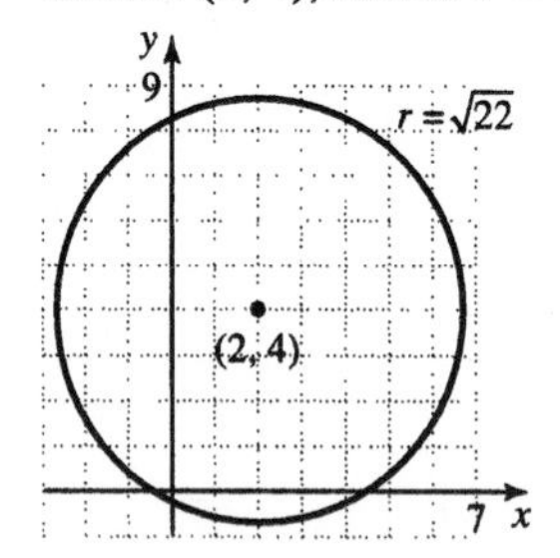

25. Center $(h, k)=(2, 3)$ and radius $r=6$.

$(x-h)^2+(y-k)^2=r^2$

$(x-2)^2+(y-3)^2=6^2$

$(x-2)^2+(y-3)^2=36$

27. Center $(h, k) = (0, 0)$ and radius $r = \sqrt{3}$.

$$(x-h)^2 + (y-k)^2 = r^2$$
$$(x-0)^2 + (y-0)^2 = \left(\sqrt{3}\right)^2$$
$$x^2 + y^2 = 3$$

29. Center $(h, k) = (-5, 4)$ and radius $r = 3\sqrt{5}$.

$$(x-h)^2 + (y-k)^2 = r^2$$
$$[x-(-5)]^2 + (y-4)^2 = \left(3\sqrt{5}\right)^2$$
$$(x+5)^2 + (y-4)^2 = 45$$

31. The radius is $\sqrt{10}$.

33. $x = y^2 - 3$

$x = (y-0)^2 - 3$

Vertex: $(-3, 0)$

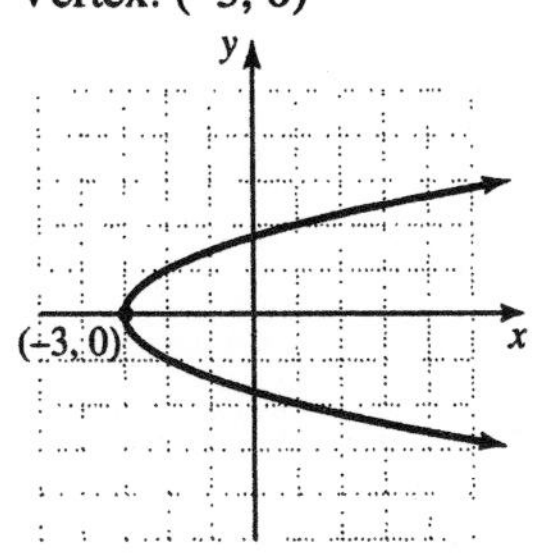

35. $y = (x-2)^2 - 2$

Vertex: $(2, -2)$

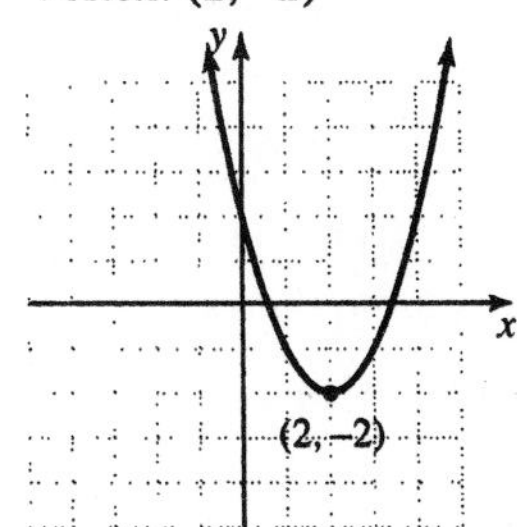

37. $x^2 + y^2 = 1$

Center: $(0, 0)$, radius $r = \sqrt{1} = 1$

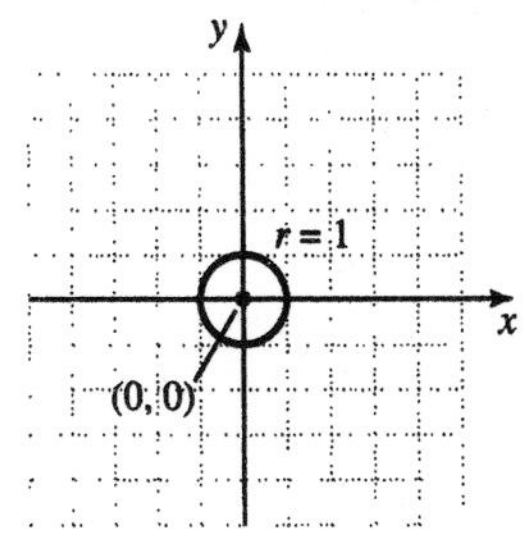

39. $x = (y+3)^2 - 1$

Vertex: $(-1, -3)$

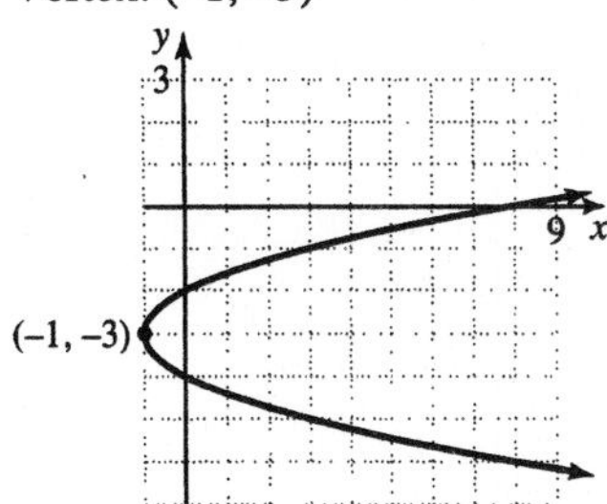

41. $(x-2)^2 + (y-2)^2 = 16$

Center: $(2, 2)$, radius $r = \sqrt{16} = 4$

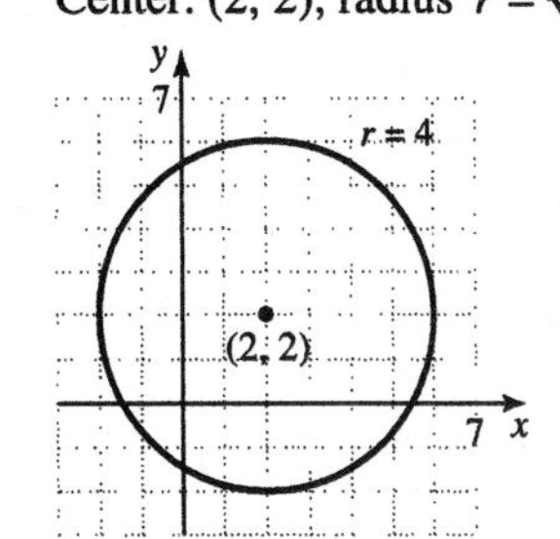

43. $x = -(y-1)^2$

Vertex: $(0, 1)$

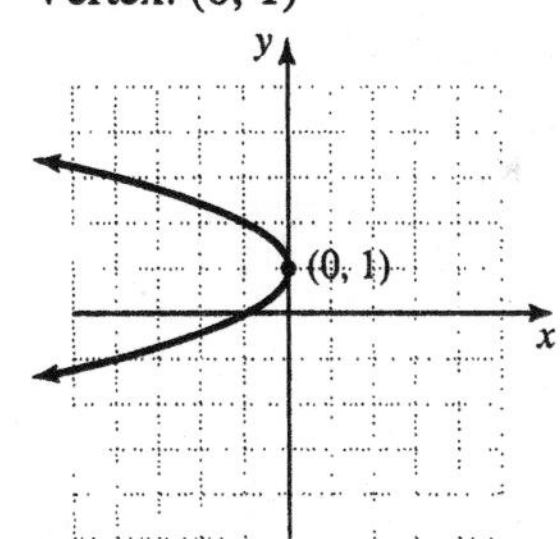

45. $(x-4)^2 + y^2 = 7$

Center: $(4, 0)$, radius $r = \sqrt{7}$

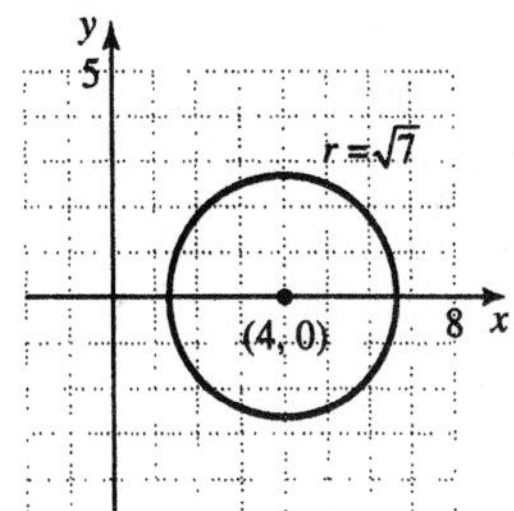

47. $y = 5(x+5)^2 + 3$

Vertex: (–5, 3)

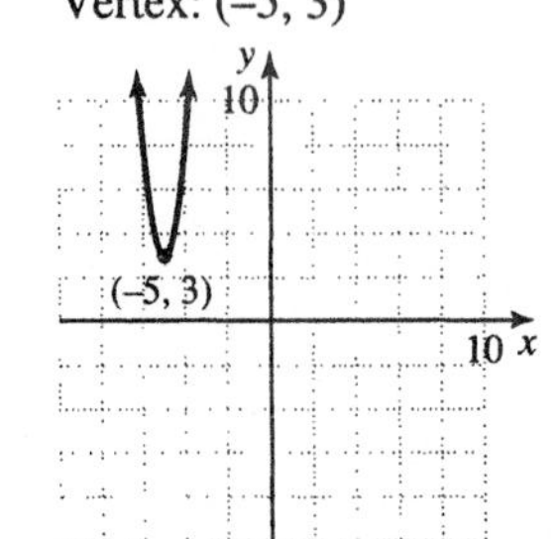

49.
$$\frac{x^2}{8} + \frac{y^2}{8} = 2$$
$$8\left(\frac{x^2}{8} + \frac{y^2}{8}\right) = 8(2)$$
$$x^2 + y^2 = 16$$

Center: (0, 0), radius $r = \sqrt{16} = 4$

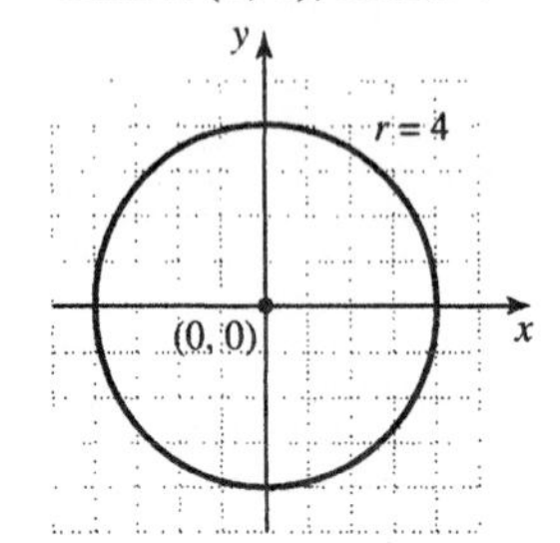

51.
$$y = x^2 + 7x + 6$$
$$y - 6 = x^2 + 7x$$
$$y - 6 + \frac{49}{4} = x^2 + 7x + \frac{49}{4}$$
$$y + \frac{25}{4} = \left(x + \frac{7}{2}\right)^2$$
$$y = \left(x + \frac{7}{2}\right)^2 - \frac{25}{4}$$

Vertex: $\left(-\frac{7}{2}, -\frac{25}{4}\right)$

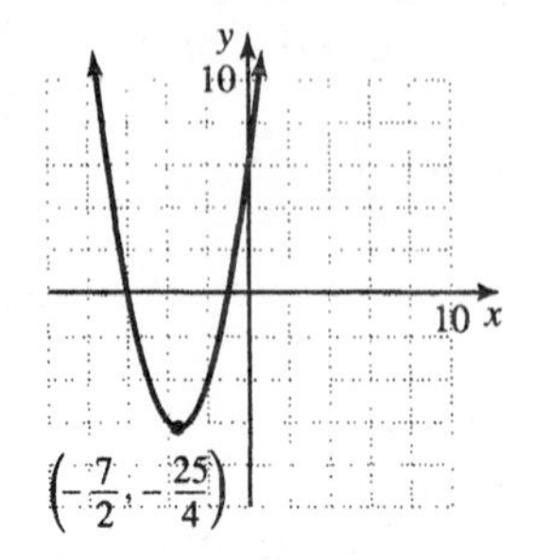

53.
$$x^2 + y^2 + 2x + 12y - 12 = 0$$
$$(x^2 + 2x) + (y^2 + 12y) = 12$$
$$(x^2 + 2x + 1) + (y^2 + 12y + 36) = 12 + 1 + 36$$
$$(x+1)^2 + (y+6)^2 = 49$$

Center: (–1, –6), radius $r = \sqrt{49} = 7$

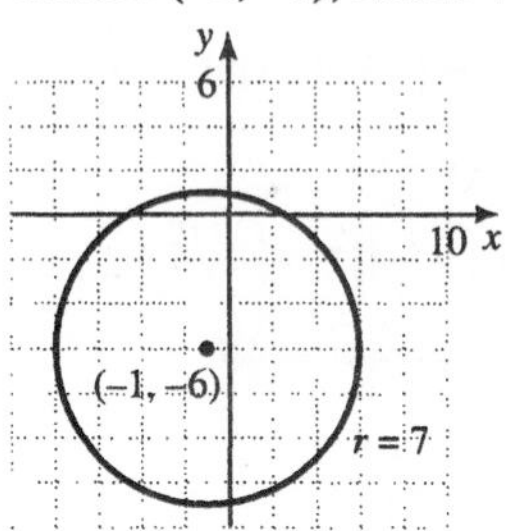

55.
$$x = y^2 + 8y - 4$$
$$x + 4 = y^2 + 8y$$
$$x + 4 + 16 = y^2 + 8y + 16$$
$$x + 20 = (y+4)^2$$
$$x = (y+4)^2 - 20$$

Vertex: (–20, –4)

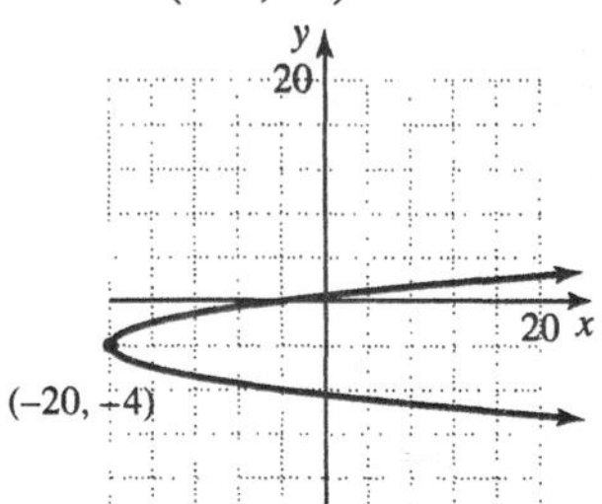

57.
$$x^2 - 10y + y^2 + 4 = 0$$
$$x^2 + (y^2 - 10y) = -4$$
$$x^2 + (y^2 - 10y + 25) = -4 + 25$$
$$x^2 + (y-5)^2 = 21$$

Center: (0, 5), radius $r = \sqrt{21}$

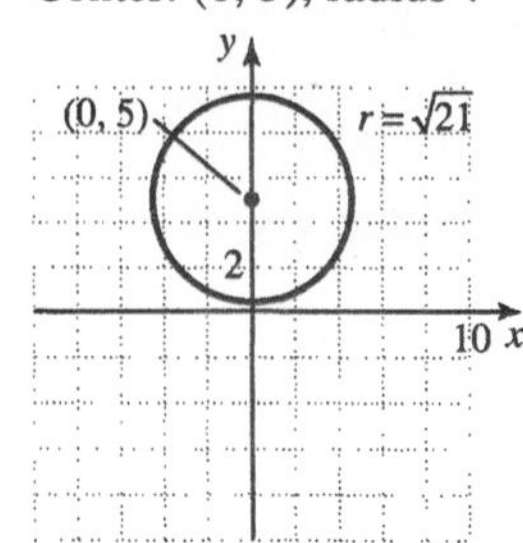

59.
$$x = -3y^2 + 30y$$
$$x = -3(y^2 - 10y)$$
$$x + [-3(25)] = -3(y^2 - 10y + 25)$$
$$x - 75 = -3(y - 5)^2$$
$$x = -3(y - 5)^2 + 75$$

Vertex: (75, 5)

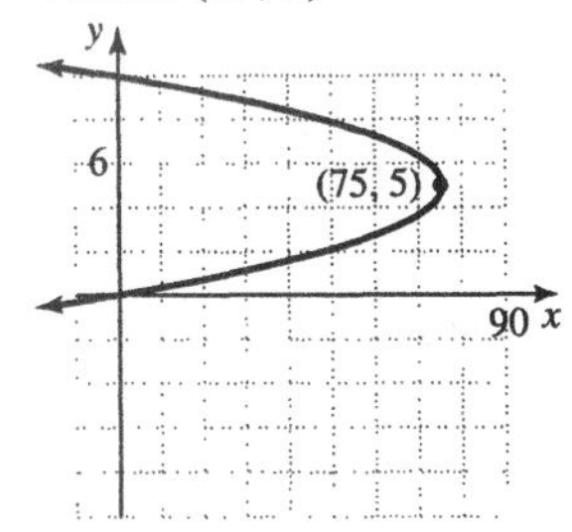

61. $5x^2 + 5y^2 = 25$
$$x^2 + y^2 = 5$$

Center: (0, 0), radius $r = \sqrt{5}$

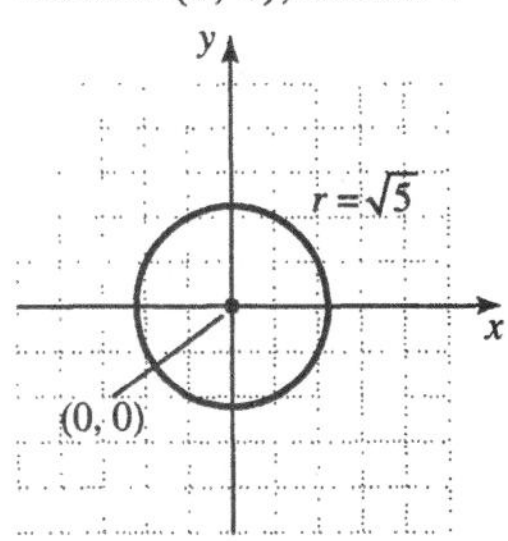

63.
$$y = 5x^2 - 20x + 16$$
$$y - 16 = 5(x^2 - 4x)$$
$$y - 16 + 5(4) = 5(x^2 - 4x + 4)$$
$$y + 4 = (x - 2)^2$$
$$y = (x - 2)^2 - 4$$

Vertex: (2, –4)

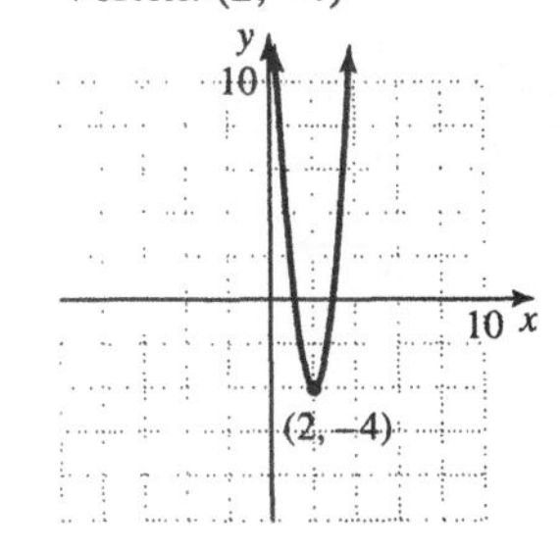

65. $y = -3x + 3$

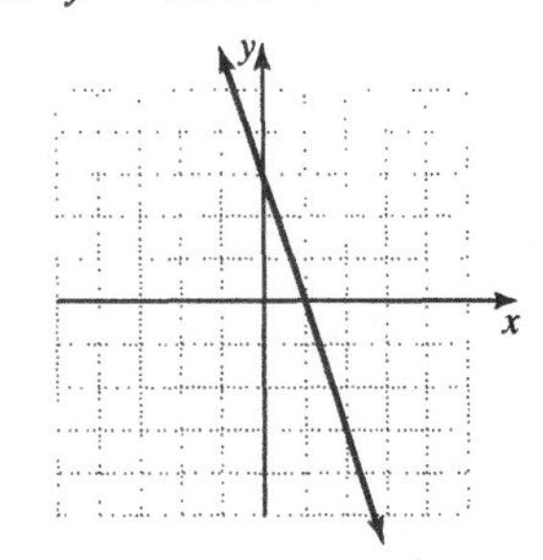

67. $x = -2$

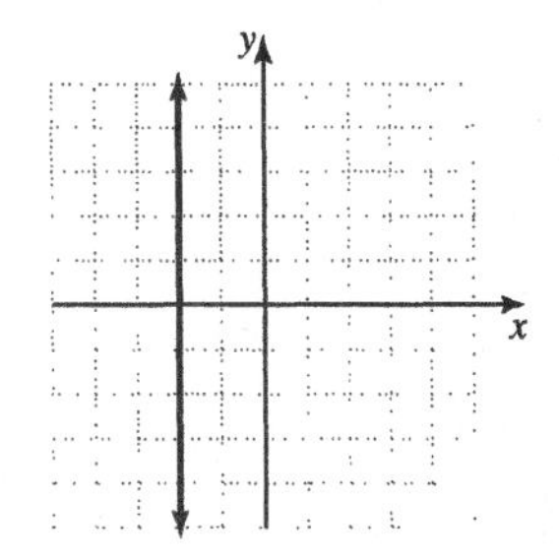

69. $\dfrac{\sqrt{5}}{\sqrt{8}} = \dfrac{\sqrt{5}}{2\sqrt{2}} = \dfrac{\sqrt{5} \cdot \sqrt{2}}{2\sqrt{2} \cdot \sqrt{2}} = \dfrac{\sqrt{10}}{2 \cdot 2} = \dfrac{\sqrt{10}}{4}$

71. $\dfrac{10}{\sqrt{5}} = \dfrac{10 \cdot \sqrt{5}}{\sqrt{5} \cdot \sqrt{5}} = \dfrac{10\sqrt{5}}{5} = 2\sqrt{5}$

73. **a.** radius $= \dfrac{1}{2}(\text{diameter}) = \dfrac{1}{2}(135 \text{ meters})$
$= 67.5$ meters

b. The wheel is at ground level or 0 meters.

c. The height of the center is equal to the radius or 67.5 meters.

d. The coordinates of the center are (0, 67.5).

e. $h = 0$, $k = 67.5$, and $r = 67.5$.
$$(x - h)^2 + (y - k)^2 = r^2$$
$$x^2 + (y - 67.5)^2 = 67.5^2$$

75. **a.** radius $= \dfrac{1}{2}(\text{diameter}) = \dfrac{1}{2}(153 \text{ meters})$
$= 76.5$ meters

b. height – diameter
$= 160 - 153$
$= 7$ meters from the ground

c. height from ground + radius
$= 7 + 76.5$
$= 83.5$ meters from the ground

d. Center: (0, 83.5)

e. $h = 0$, $k = 83.5$, and $r = 76.5$.

$$(x-h)^2 + (y-k)^2 = r^2$$
$$x^2 + (y-83.5)^2 = 76.5^2$$

77. diameter = 20 cm; center: (0, 0)

a.

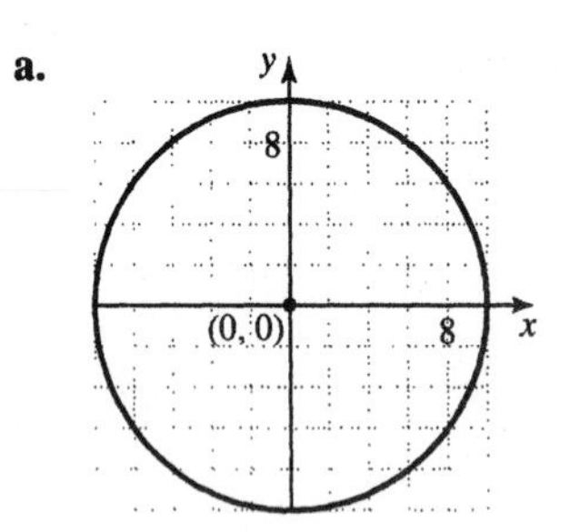

b. $h = 0$, $k = 0$, and $r = \frac{1}{2}(20) = 10$

$$(x-h)^2 + (y-k)^2 = r^2$$
$$x^2 + y^2 = 100$$

c. $h = 0$, $k = 0$, and the radius is the distance from the fountain or 5 feet.

$$x^2 + y^2 = 25$$

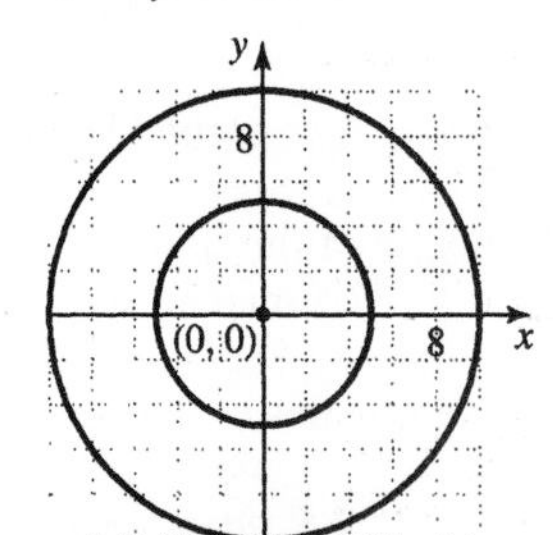

79. $5x^2 + 5y^2 = 25$

$$5y^2 = 25 - 5x^2$$
$$y^2 = 5 - x^2$$
$$y = \pm\sqrt{5 - x^2}$$

81. $y = 5x^2 - 20x + 16$

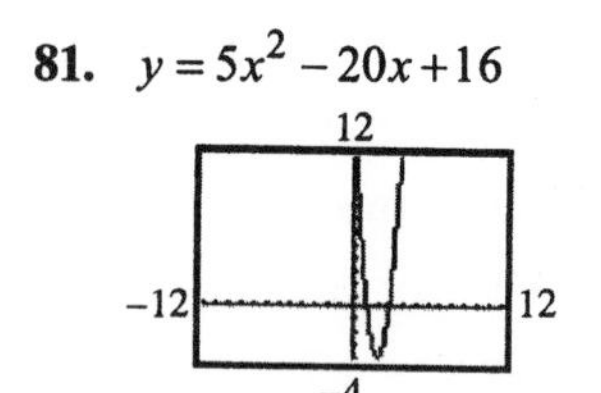

Section 13.2

Practice Exercises

1. $\frac{x^2}{25} + \frac{y^2}{4} = 1$

The equation is an ellipse with $a = 5$ and $b = 2$. The center is (0, 0). The x-intercepts are (5, 0) and (−5, 0). The y-intercepts are (2, 0) and (−2, 0).

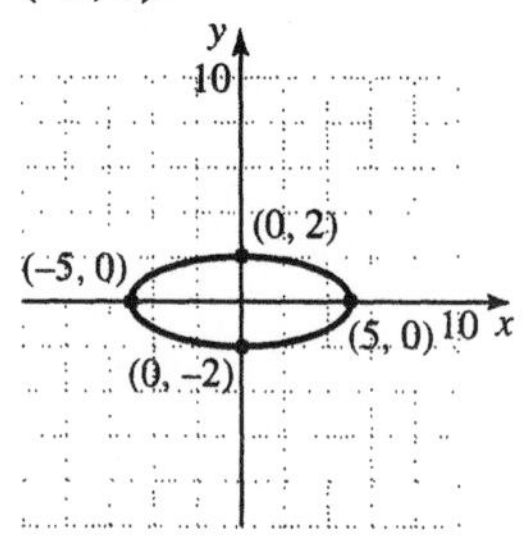

2. $9x^2 + 4y^2 = 36$

$$\frac{9x^2}{36} + \frac{4y^2}{36} = \frac{36}{36}$$
$$\frac{x^2}{4} + \frac{y^2}{9} = 1$$

This is an equation of an ellipse with $a = 2$ and $b = 3$. The ellipse has center (0, 0), x-intercepts (2, 0) and (−2, 0), and y-intercepts (3, 0) and (−3, 0).

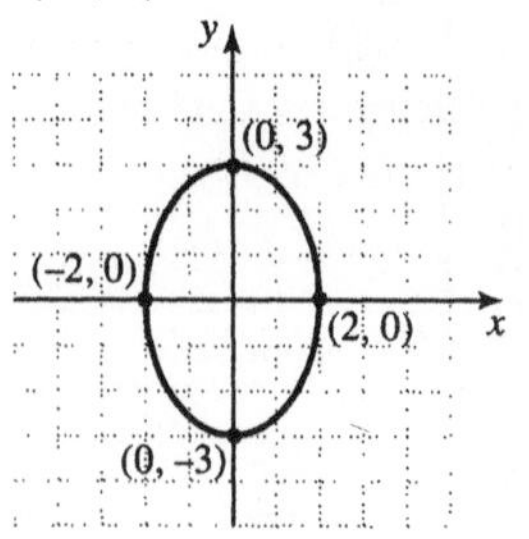

3. $\frac{(x-4)^2}{49} + \frac{(y+1)^2}{81} = 1$

This ellipse has center (4, −1).
$a = 7$ and $b = 9$.
Find four points on the ellipse.

$(4+7, -1) = (11, -1)$
$(4-7, -1) = (-3, -1)$
$(4, -1+9) = (4, 8)$
$(4, -1-9) = (4, -10)$

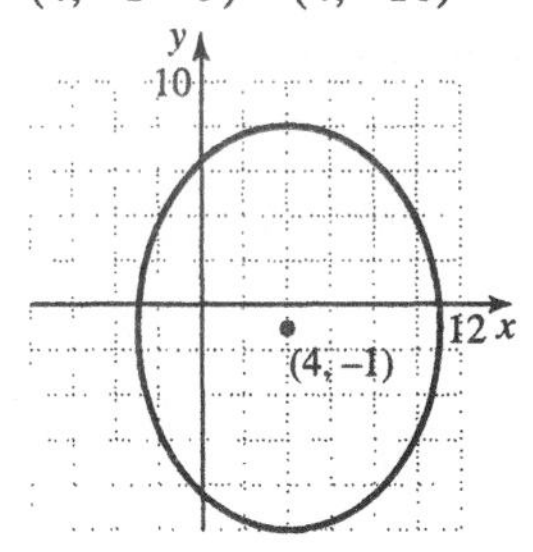

4. $\frac{x^2}{9} - \frac{y^2}{16} = 1$

This is a hyperbola with $a = 3$ and $b = 4$. It has center (0, 0) and x-intercepts (3, 0) and (−3, 0). The asymptotes pass through (3, 4), (3, −4), (−3, 4), and (−3, −4).

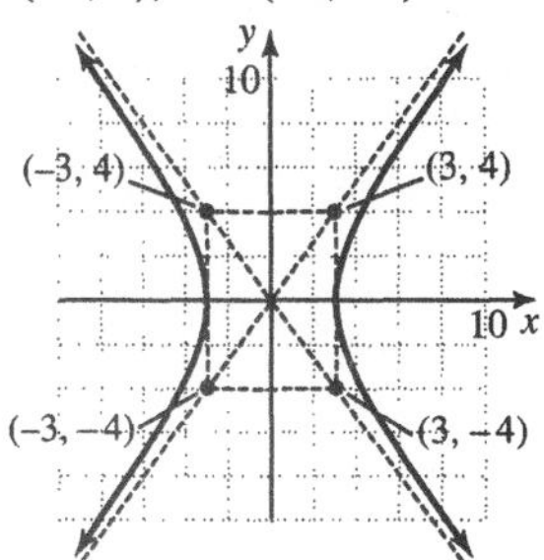

5. $9y^2 - 25x^2 = 225$

$$\frac{9y^2}{225} - \frac{25x^2}{225} = \frac{225}{225}$$

$$\frac{y^2}{25} - \frac{x^2}{9} = 1$$

This is a hyperbola with $a = 3$ and $b = 5$. The center is at (0, 0) with y-intercepts (0, 5) and (0, −5). The asymptotes pass through (3, 5), (3, −5), (−3, 5), and (−3, −5).

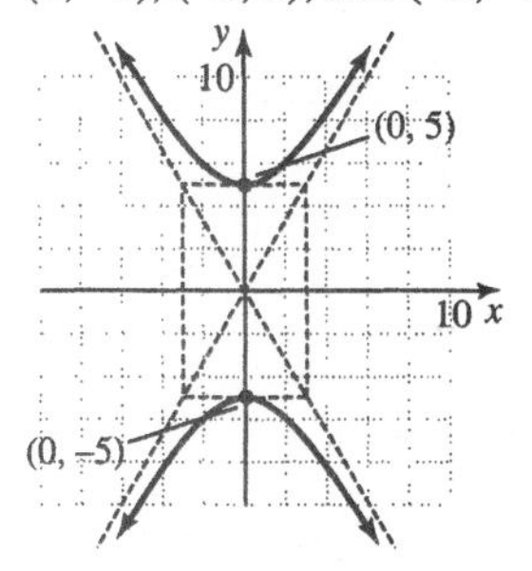

Graphing Calculator Explorations

1. $10x^2 + y^2 = 32$

$y^2 = 32 - 10x^2$

$y = \pm\sqrt{32 - 10x^2}$

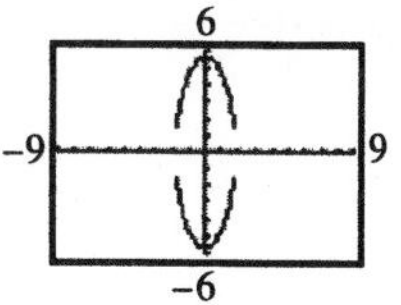

2. $x^2 + 6y^2 = 35$

$6y^2 = 35 - x^2$

$y^2 = \frac{35 - x^2}{6}$

$y = \pm\sqrt{\frac{35 - x^2}{6}}$

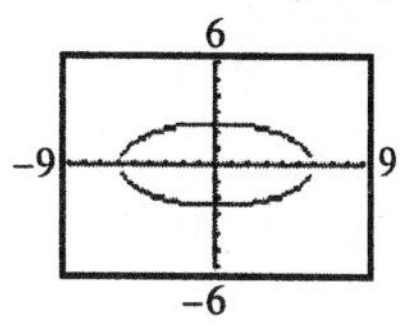

3. $20x^2 + 5y^2 = 100$

$5y^2 = 100 - 20x^2$

$y^2 = 20 - 4x^2$

$y = \pm\sqrt{20 - 4x^2}$

4. $4y^2 + 12x^2 = 48$

$4y^2 = 48 - 12x^2$

$y^2 = 12 - 3x^2$

$y = \pm\sqrt{12 - 3x^2}$

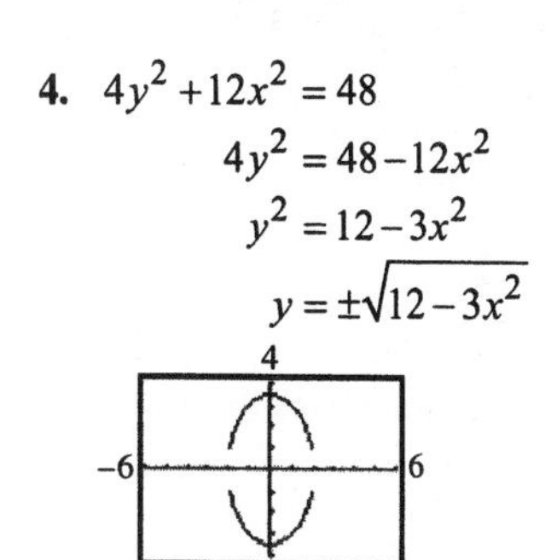

5. $7.3x^2 + 15.5y^2 = 95.2$
$$15.5y^2 = 95.2 - 7.3x^2$$
$$y^2 = \frac{95.2 - 7.3x^2}{15.5}$$
$$y = \pm\sqrt{\frac{95.2 - 7.3x^2}{15.5}}$$

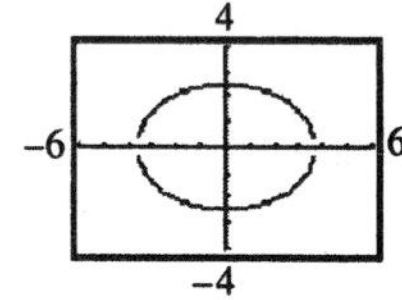

6. $18.8x^2 + 36.1y^2 = 205.8$
$$36.1y^2 = 205.8 - 18.8x^2$$
$$y^2 = \frac{205.8 - 18.8x^2}{36.1}$$
$$y = \pm\sqrt{\frac{205.8 - 18.8x^2}{36.1}}$$

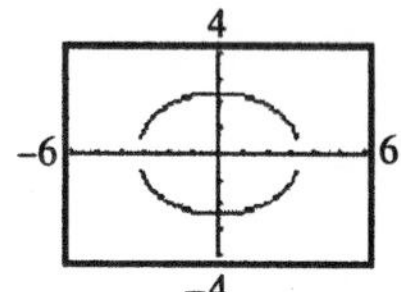

Vocabulary and Readiness Check

1. A hyperbola is the set of points in a plane such that the absolute value of the differences of their distances from two fixed points is constant.

2. An ellipse is the set of points in a plane such that the sum of their distances from two fixed points is constant.

3. The two fixed points are each called a focus.

4. The point midway between the foci is called the center.

5. The graph of $\frac{x^2}{a^2} - \frac{y^2}{b^2} = 1$ is a hyperbola with center (0, 0) and x-intercepts of $(a, 0)$ and $(-a, 0)$.

6. The graph of $\frac{x^2}{b^2} + \frac{y^2}{a^2} = 1$ is an ellipse with center (0, 0) and x-intercepts of $(b, 0)$ and $(-b, 0)$.

7. $\frac{x^2}{16} + \frac{y^2}{4} = 1$ is an ellipse.

8. $\frac{x^2}{16} - \frac{y^2}{4} = 1$ is a hyperbola.

9. $x^2 - 5y^2 = 3$ is a hyperbola.

10. $-x^2 + 5y^2 = 3$ or
$5y^2 - x^2 = 3$ is a hyperbola.

11. $-\frac{y^2}{25} + \frac{x^2}{36} = 1$ or
$\frac{x^2}{36} - \frac{y^2}{25} = 1$ is a hyperbola.

12. $\frac{y^2}{25} + \frac{x^2}{36} = 1$ is an ellipse.

Exercise Set 13.2

1. $\frac{x^2}{4} + \frac{y^2}{25} = 1$
$$\frac{x^2}{2^2} + \frac{y^2}{5^2} = 1$$
Center: (0, 0)
x-intercepts: (−2, 0), (2, 0)
y-intercepts: (0, −5), (0, 5)

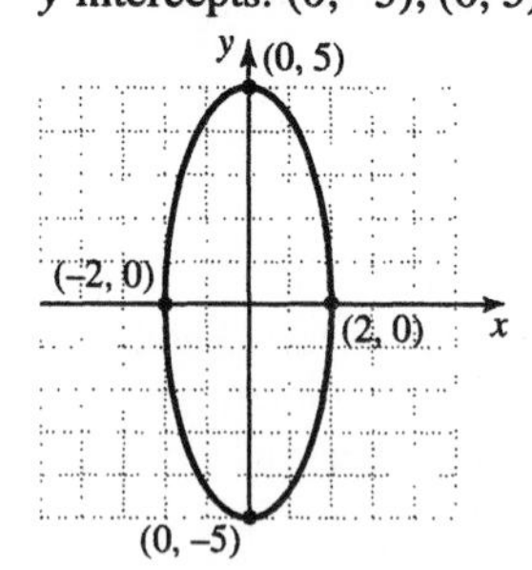

3. $\frac{x^2}{9} + y^2 = 1$
$$\frac{x^2}{3^2} + \frac{y^2}{1^2} = 1$$
Center: (0, 0)
x-intercepts: (−3, 0), (3, 0)
y-intercepts: (0, −1), (0, 1)

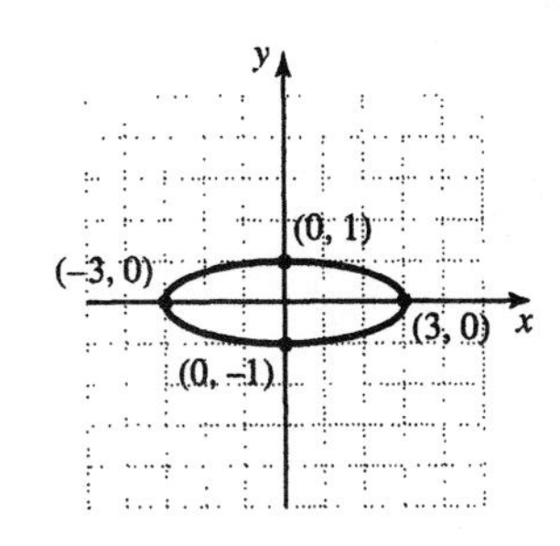

5. $9x^2 + y^2 = 36$

$$\frac{x^2}{4} + \frac{y^2}{36} = 1$$

$$\frac{x^2}{2^2} + \frac{y^2}{6^2} = 1$$

Center: (0, 0)
x-intercepts: (–2, 0), (2, 0)
y-intercepts: (0, –6), (0, 6)

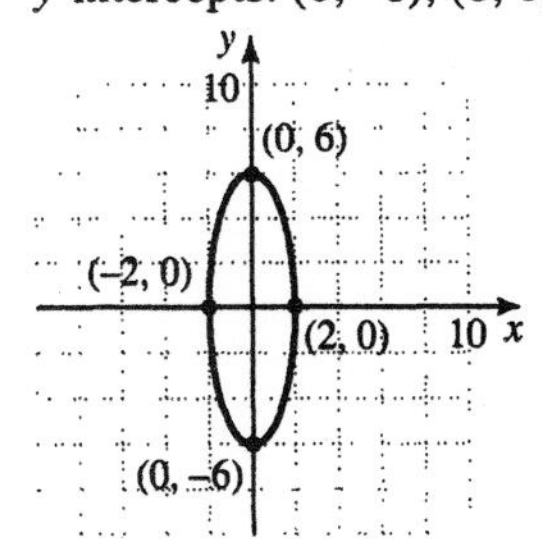

7. $4x^2 + 25y^2 = 100$

$$\frac{x^2}{25} + \frac{y^2}{4} = 1$$

$$\frac{x^2}{5^2} + \frac{y^2}{2^2} = 1$$

Center: (0, 0)
x-intercepts: (–5, 0), (5, 0)
y-intercepts: (0, –2), (0, 2)

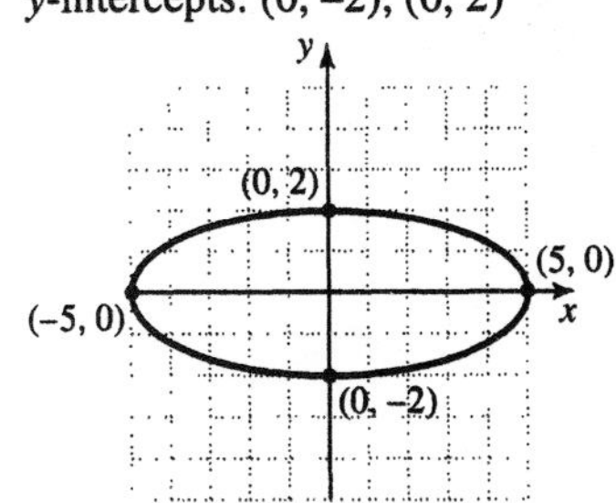

9. $\frac{(x+1)^2}{36} + \frac{(y-2)^2}{49} = 1$

$$\frac{(x+1)^2}{6^2} + \frac{(y-2)^2}{7^2} = 1$$

Center: (–1, 2)

Other points:
$(-1-6, 2) = (-7, 2)$
$(-1+6, 2) = (5, 2)$
$(-1, 2-7) = (-1, -5)$
$(-1, 2+7) = (-1, 9)$

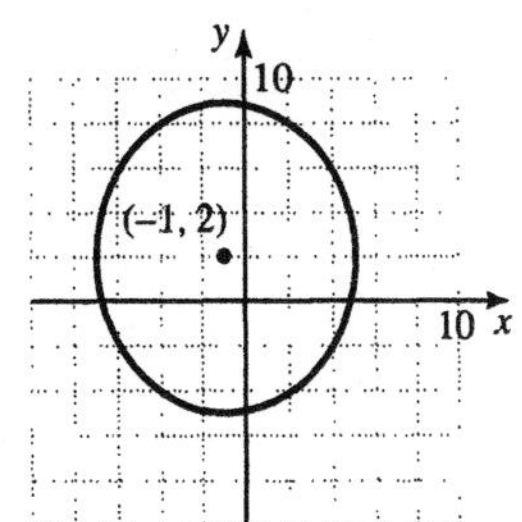

11. $\frac{(x-1)^2}{4} + \frac{(y-1)^2}{25} = 1$

$$\frac{(x-1)^2}{2^2} + \frac{(y-1)^2}{5^2} = 1$$

Center: (1, 1)
Other points:
$(1-2, 1) = (-1, 1)$
$(1+2, 1) = (3, 1)$
$(1, 1-5) = (1, -4)$
$(1, 1+5) = (1, 6)$

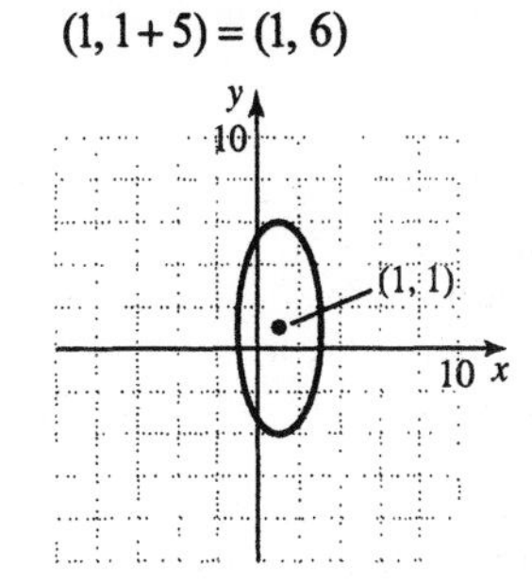

13. $\frac{x^2}{4} - \frac{y^2}{9} = 1$

$$\frac{x^2}{2^2} - \frac{y^2}{3^2} = 1$$

$a = 2, b = 3$

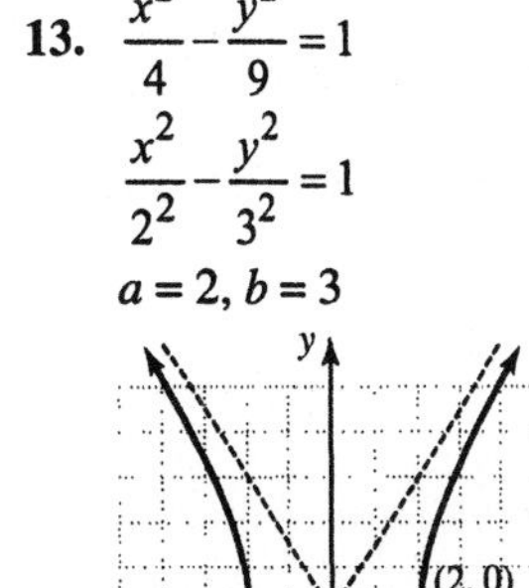

15. $\dfrac{y^2}{25}-\dfrac{x^2}{16}=1$

$\dfrac{y^2}{5^2}-\dfrac{x^2}{4^2}=1$

$a=4, b=5$

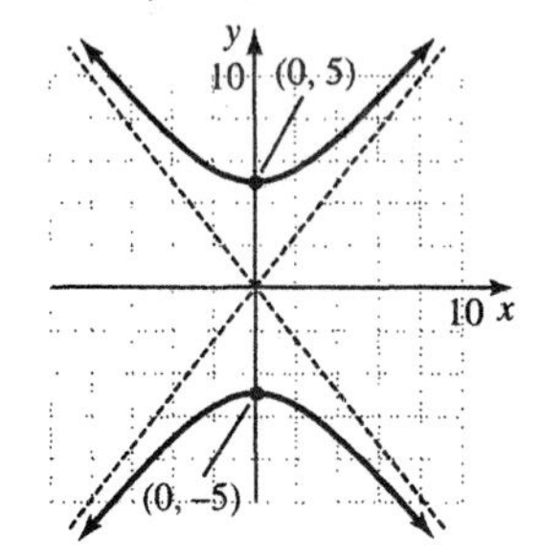

17. $x^2-4y^2=16$

$\dfrac{x^2}{16}-\dfrac{y^2}{4}=1$

$\dfrac{x^2}{4^2}-\dfrac{y^2}{2^2}=1$

$a=4, b=2$

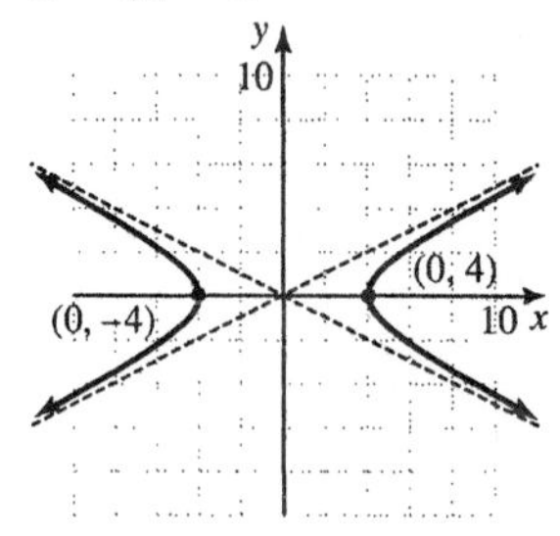

19. $16y^2-x^2=16$

$\dfrac{y^2}{1}-\dfrac{x^2}{16}=1$

$\dfrac{y^2}{1^2}-\dfrac{x^2}{4^2}=1$

$a=4, b=1$

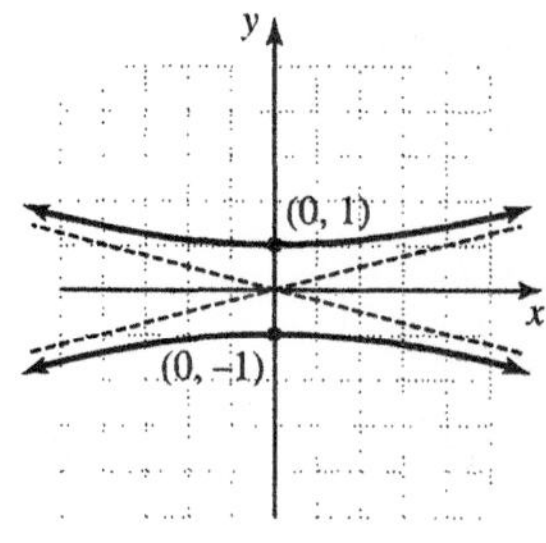

21. Answers may vary

23. $y=x^2+4$

Parabola; vertex (0, 4), opens upward

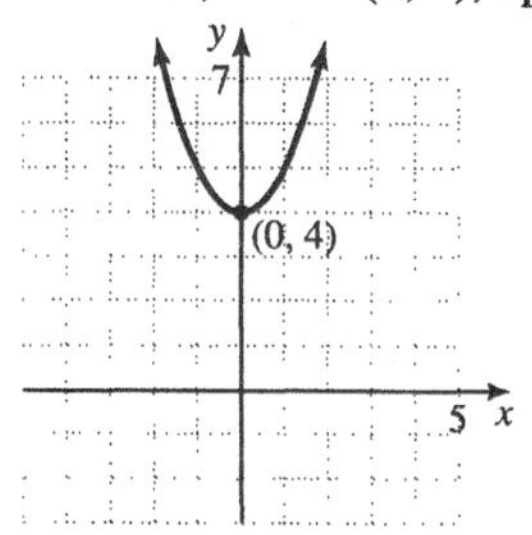

25. $\dfrac{x^2}{4}+\dfrac{y^2}{9}=1$

$\dfrac{x^2}{2^2}+\dfrac{y^2}{3^2}=1$

Ellipse; center: (0, 0)

x-intercepts: (−2, 0), (2, 0)

y-intercepts: (0, −3), (0, 3)

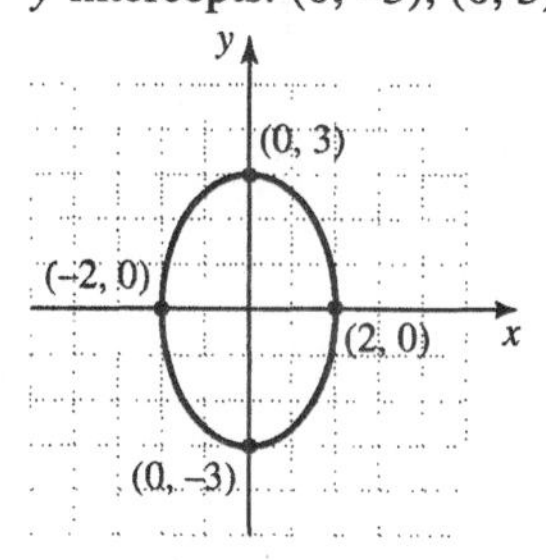

27. $\dfrac{x^2}{16}-\dfrac{y^2}{4}=1$

$\dfrac{x^2}{4^2}-\dfrac{y^2}{2^2}=1$

Hyperbola; center: (0, 0)

$a=4, b=2$

x-intercepts (−4, 0), (4, 0)

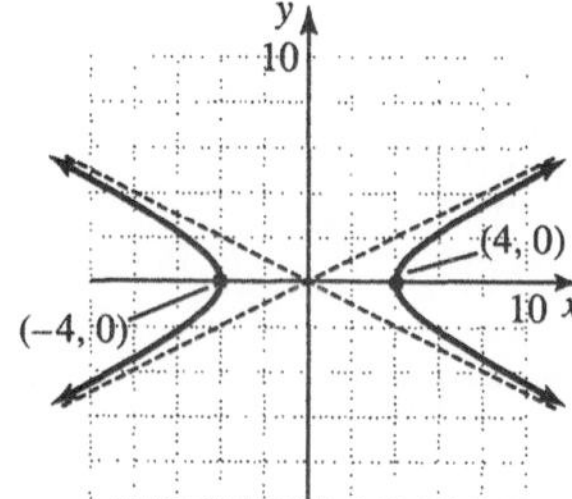

29. $x^2 + y^2 = 16$

Circle; center: (0, 0), radius: $r = \sqrt{16} = 4$

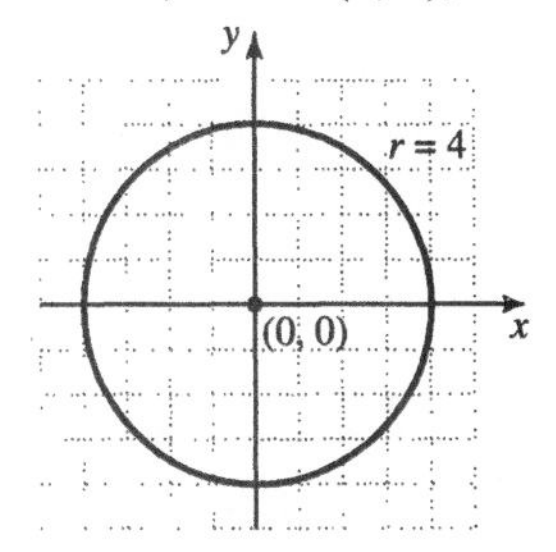

31. $x = -y^2 + 6y$

Parabola: $y = \dfrac{-b}{2a} = \dfrac{-6}{2(-1)} = 3$

$x = -(3)^2 + 6(3) = -9 + 18 = 9$

Vertex: (9, 3), opens to the left

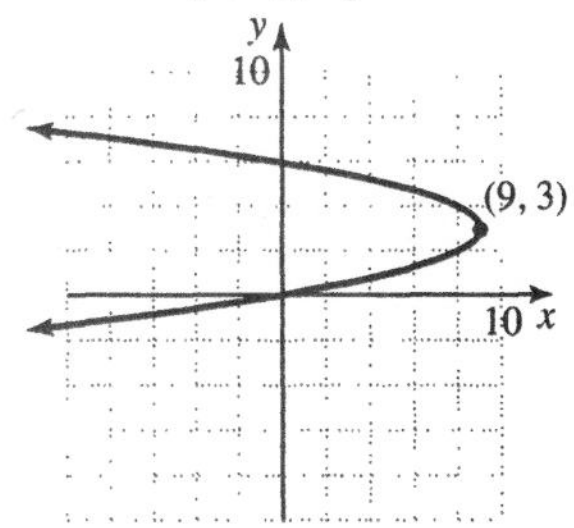

33. $9x^2 + 4y^2 = 36$

$$\frac{x^2}{4} + \frac{y^2}{9} = 1$$

$$\frac{x^2}{2^2} - \frac{y^2}{3^2} = 1$$

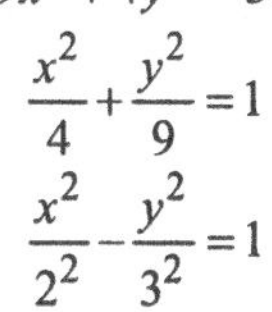

Ellipse; center: (0, 0)
$a = 2, b = 3$

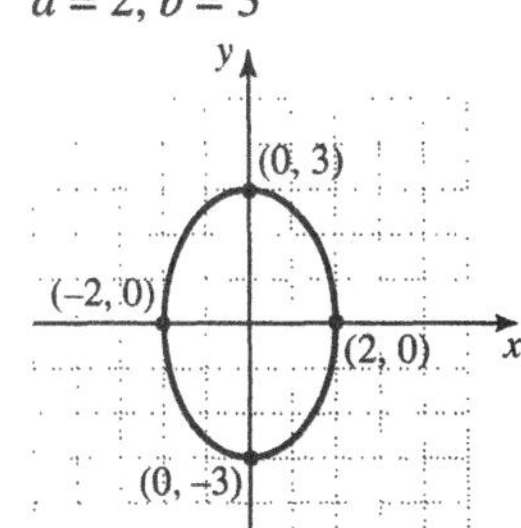

35.

$$y^2 = x^2 + 16$$

$$y^2 - x^2 = 16$$

$$\frac{y^2}{16} - \frac{x^2}{16} = 1$$

$$\frac{y^2}{4^2} - \frac{x^2}{4^2} = 1$$

Hyperbola; center: (0, 0)
$a = 4, b = 4$

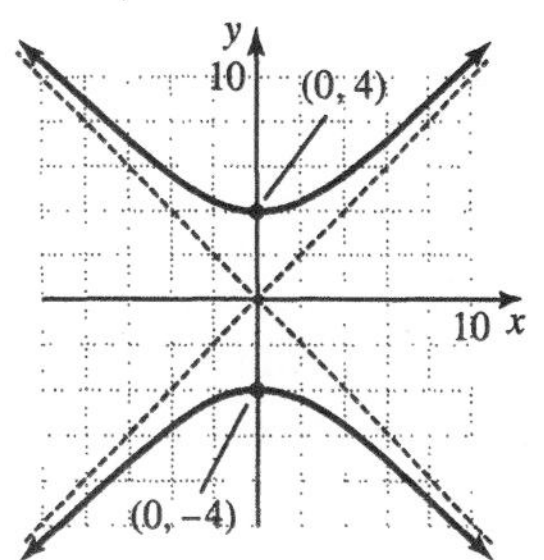

37. $y = -2x^2 + 4x - 3$

Parabola; $x = \dfrac{-b}{2a} = \dfrac{-4}{2(-2)} = 1$

$y = -2(1)^2 + 4(1) - 3 = -1$

Vertex: (1, −1)

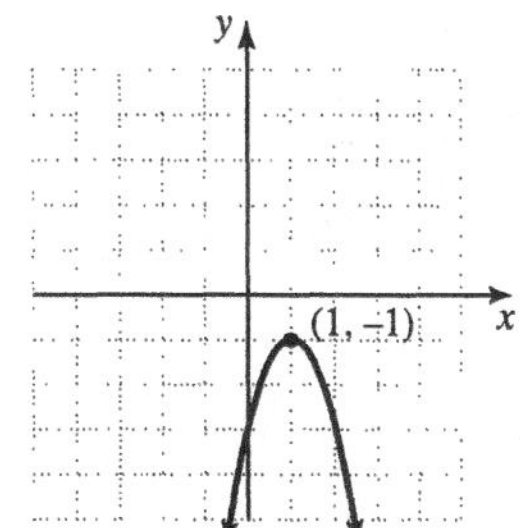

39. $x < 5$ or $x < 1$

$x < 5$

$(-\infty, 5)$

41. $2x - 1 \geq 7$ and $-3x \leq -6$

$2x \geq 8$ and $x \geq 2$

$x \geq 4$

$x \geq 4$

$[4, \infty)$

43. $2x^3 - 4x^3 = -2x^3$

45. $(-5x^2)(x^2) = -5x^{2+2} = -5x^4$

47. $\frac{x^2}{100}+\frac{y^2}{49}=1$

$\sqrt{100}=10,$ so the distance between the x-intercepts is 10 + 10 = 20 units.

$\sqrt{49}=7,$ so the distance between the y-intercepts is 7 + 7 = 14 units.

The distance between the x-intercepts is longer by 20 − 14 = 6 units.

49. $x^2+4y^2=36$

$\frac{x^2}{36}+\frac{y^2}{9}=1$

$\sqrt{36}=6,$ so the distance between the x-intercepts is 6 + 6 = 12 units.

$\sqrt{9}=3,$ so the distance between the y-intercepts is 3 + 3 = 6 units.

The distance between the x-intercepts is longer by 12 − 6 = 6 units.

51. Circles: B, F
Ellipses: C, E, H
Hyperbolas: A, D, G

53. A: $c^2=36+13=49;\ c=\sqrt{49}=7$

B: $c^2=4-4=0;\ c=\sqrt{0}=0$

C: $c^2=|25-16|=9;\ c=\sqrt{9}=3$

D: $c^2=39+25=64;\ c=\sqrt{64}=8$

E: $c^2=|81-17|=64;\ c=\sqrt{64}=8$

F: $c^2=|36-36|=0;\ c=\sqrt{0}=0$

G: $c^2=65+16=81;\ c=\sqrt{81}=9$

H: $c^2=|144-140|=4;\ c=\sqrt{4}=2$

55. A: $e=\frac{7}{6}$

B: $e=\frac{0}{2}=0$

C: $e=\frac{3}{5}$

D: $e=\frac{8}{5}$

E: $e=\frac{8}{9}$

F: $e=\frac{0}{6}=0$

G: $e=\frac{9}{4}$

H: $e=\frac{2}{12}=\frac{1}{6}$

57. They are equal to 0.

59. Answers may vary

61. $a=130{,}000{,}000 \Rightarrow a^2=(130{,}000{,}000)^2=1.69\times10^{16}$

$b=125{,}000{,}000 \Rightarrow b^2=(125{,}000{,}000)^2=1.5625\times10^{16}$

Thus, the equation is

$\frac{x^2}{1.69\times10^{16}}+\frac{y^2}{1.5625\times10^{16}}=1.$

63. $9x^2+4y^2=36$

$4y^2=36-9x^2$

$y^2=\frac{36-9x^2}{4}$

$y=\pm\sqrt{\frac{36-9x^2}{4}}=\pm\frac{\sqrt{36-9x^2}}{2}$

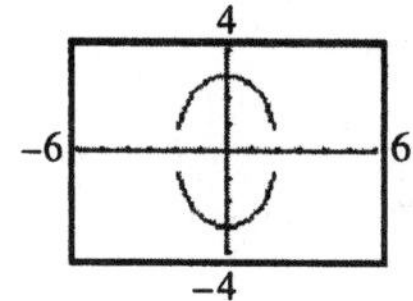

65. $\frac{(x-1)^2}{4}-\frac{(y+1)^2}{25}=1$

Center: (1, −1)
$a=2,\ b=5$

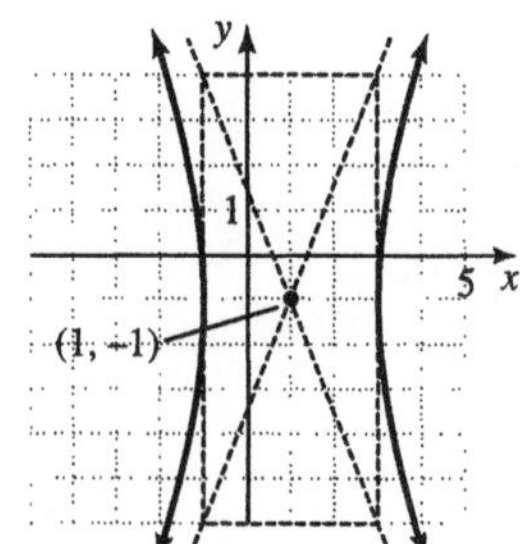

67. $\frac{y^2}{16}-\frac{(x+3)^2}{9}=1$

Center: (−3, 0)
$a=3,\ b=4$

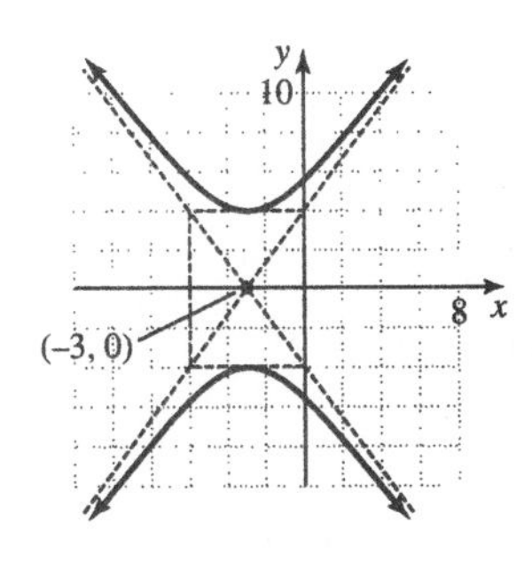

69. $\frac{(x+5)^2}{16} - \frac{(y+2)^2}{25} = 1$

Center: (−5, −2)

$a = 4, b = 5$

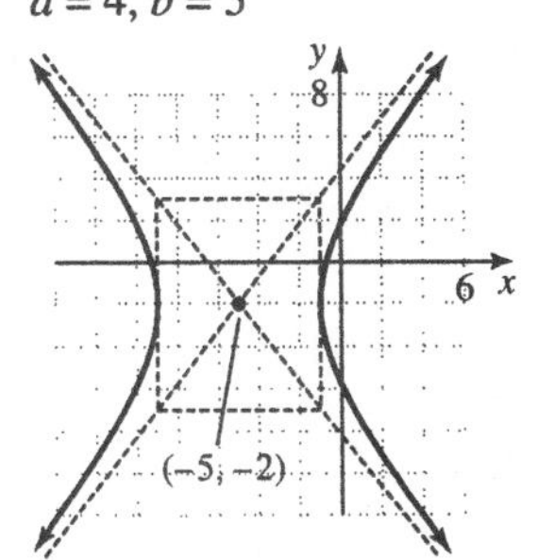

Integrated Review

1. $(x-7)^2 + (y-2)^2 = 4$

Circle; center: (7, 2),

radius: $r = \sqrt{4} = 2$

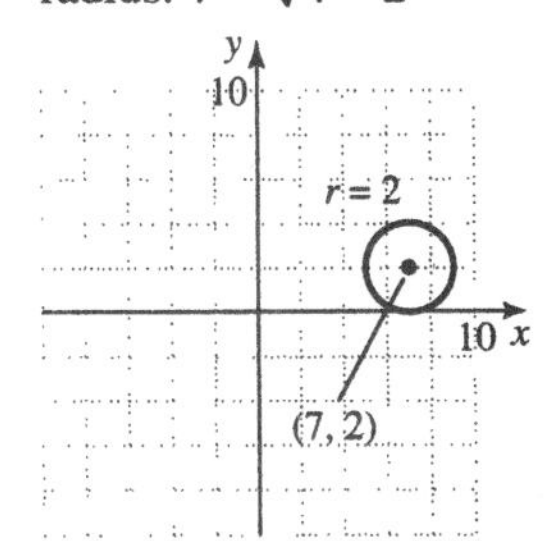

2. $y = x^2 + 4$

Parabola; vertex: (0, 4)

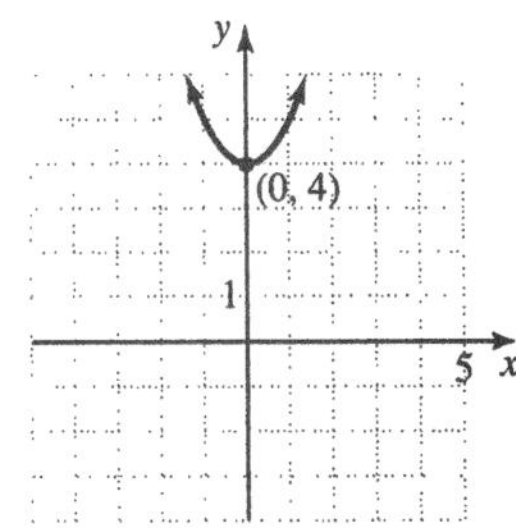

3. $y = x^2 + 12x + 36$

Parabola; $x = \frac{-b}{2a} = \frac{-12}{2(1)} = -6$

$y = (-6)^2 + 12(-6) + 36 = 0$

Vertex: (−6, 0)

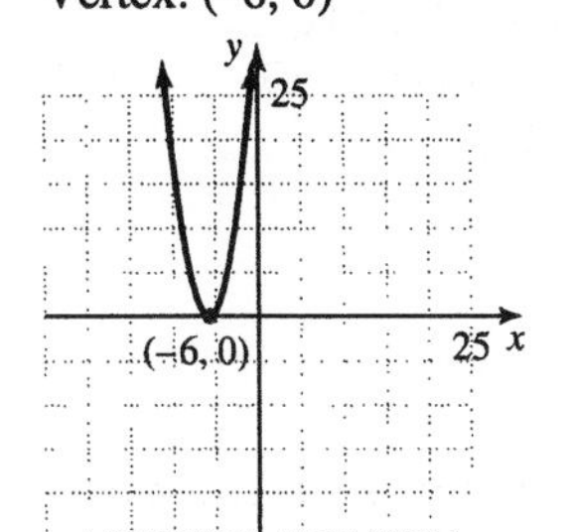

4. $\frac{x^2}{4} + \frac{y^2}{9} = 1$

Ellipse; center: (0, 0)

$a = 2, b = 3$

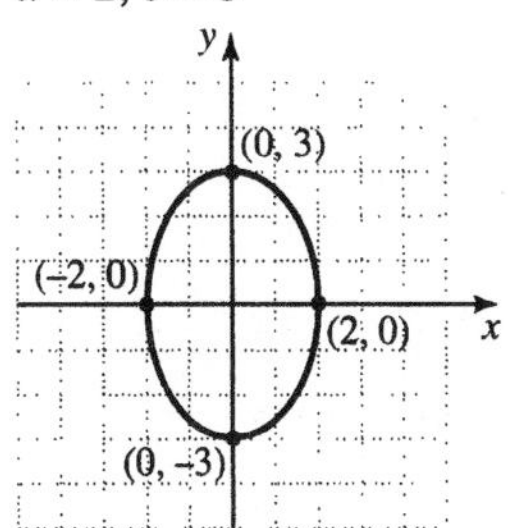

5. $\frac{y^2}{9} - \frac{x^2}{9} = 1$

Hyperbola; center: (0, 0)

$a = 3, b = 3$

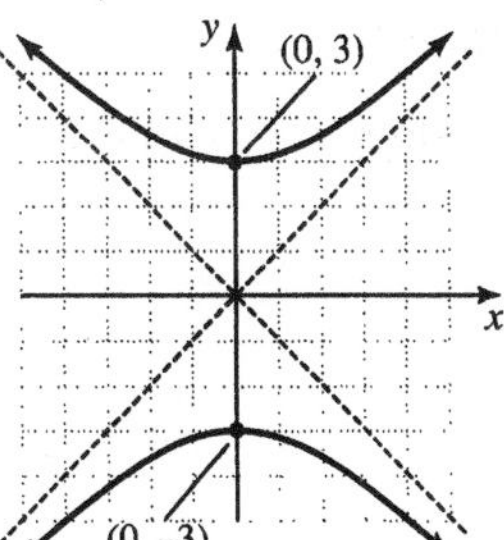

6. $\frac{x^2}{16} - \frac{y^2}{4} = 1$

Hyperbola; center: (0, 0)

$a = 4, b = 2$

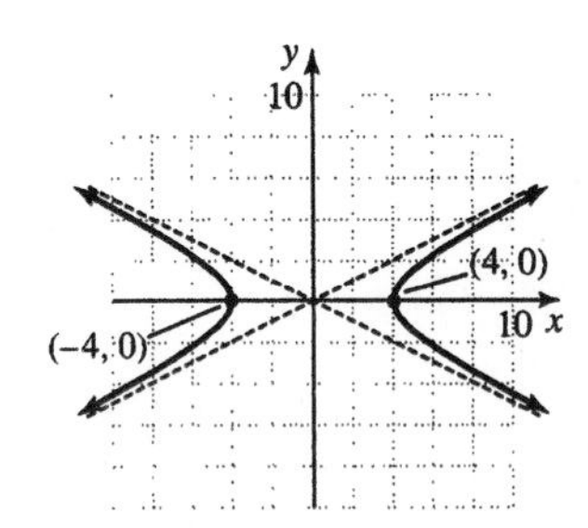

7. $\frac{x^2}{16}+\frac{y^2}{4}=1$

Ellipse; center: (0, 0)
$a=4, b=2$

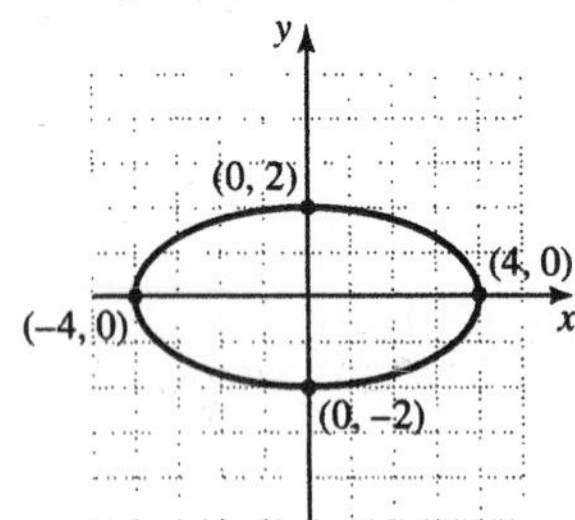

8. $x^2+y^2=16$

Circle; center: (0, 0)
radius: $r=\sqrt{16}=4$

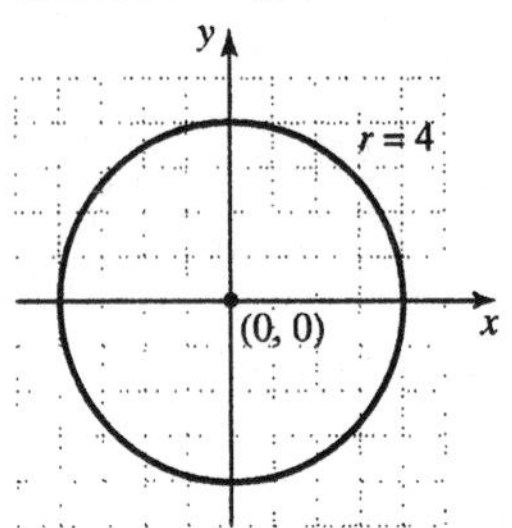

9. $x=y^2+4y-1$

Parabola; $y=\frac{-b}{2a}=\frac{-4}{2(1)}=-2$

$x=(-2)^2+4(-2)-1=-5$
Vertex: (−5, −2)

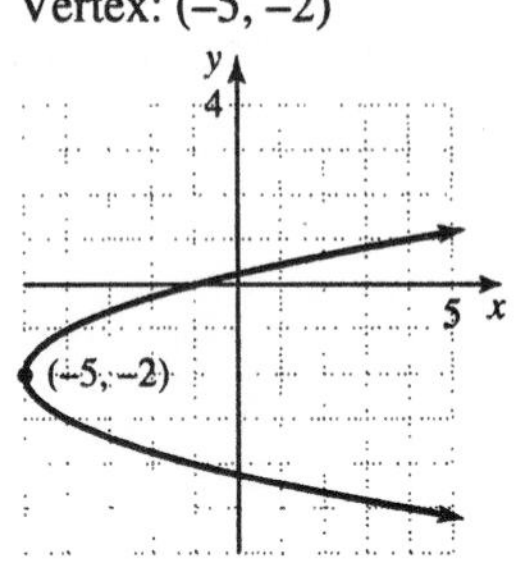

10. $x=-y^2+6y$

Parabola; $y=\frac{-b}{2a}=\frac{-6}{2(-1)}=3$

$x=-(3)^2+6(3)=9$
Vertex: (9, 3)

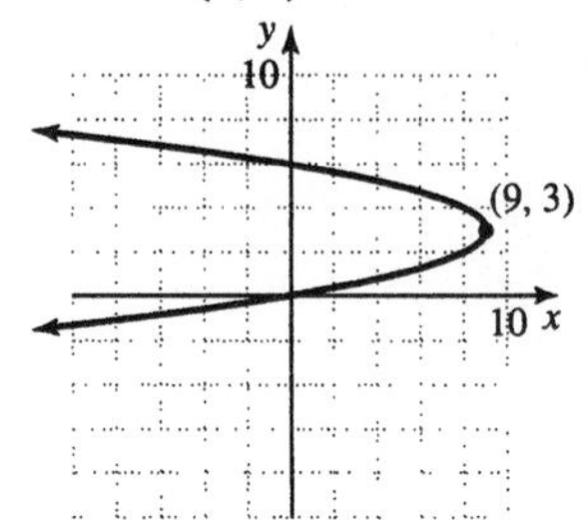

11. $9x^2-4y^2=36$

$\frac{x^2}{4}-\frac{y^2}{9}=1$

Hyperbola; center: (0, 0)
$a=2, b=3$

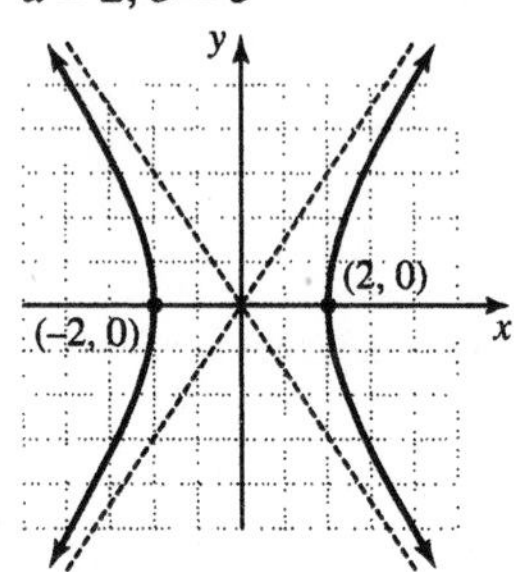

12. $9x^2+4y^2=36$

$\frac{x^2}{4}+\frac{y^2}{9}=1$

Ellipse; center: (0, 0)
$a=2, b=3$

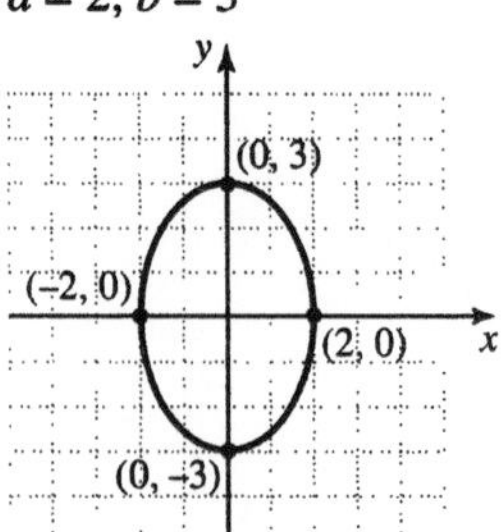

13. $\frac{(x-1)^2}{49}+\frac{(y+2)^2}{25}=1$

Ellipse; center: (1, −2),
$a=7, b=5$

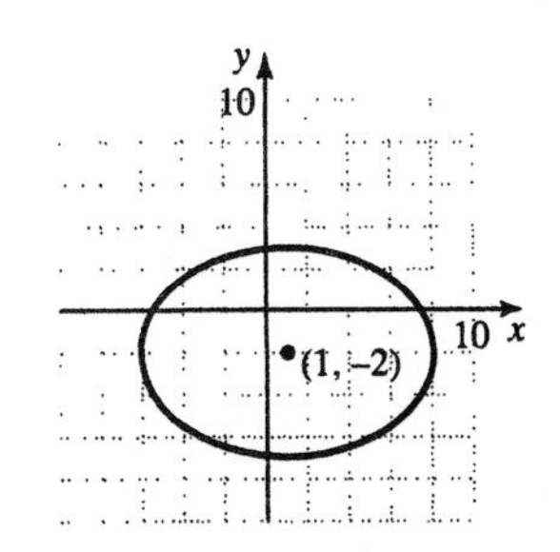

14. $y^2 = x^2 + 16$

$y^2 - x^2 = 16$

$\frac{y^2}{16} - \frac{x^2}{16} = 1$

Hyperbola; center: (0, 0)

$a = 4, b = 4$

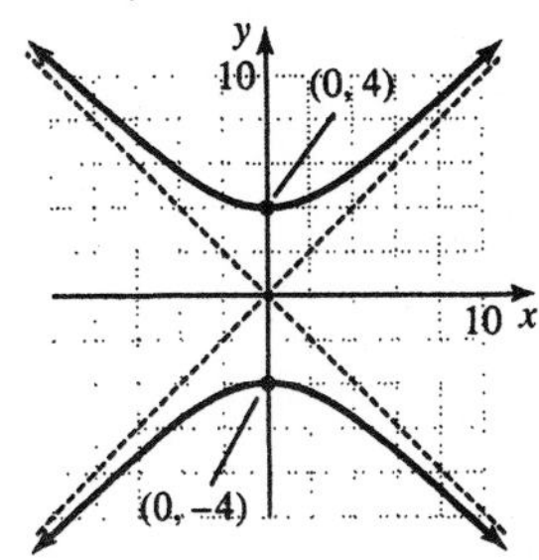

15. $\left(x+\frac{1}{2}\right)^2 + \left(y-\frac{1}{2}\right)^2 = 1$

Circle; center: $\left(-\frac{1}{2}, \frac{1}{2}\right)$, radius: $r = \sqrt{1} = 1$

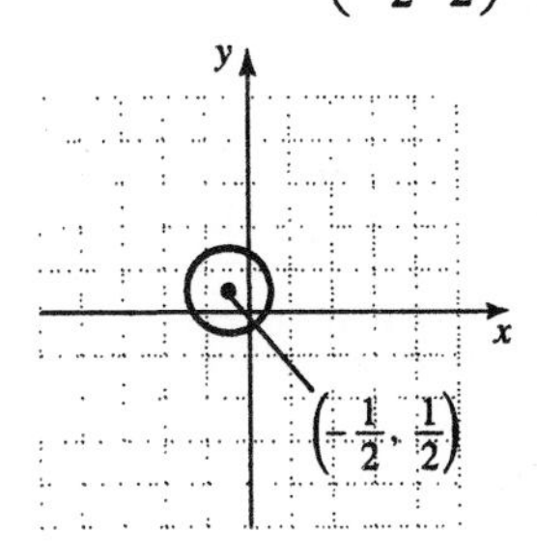

Section 13.3

Practice Exercises

1. $\begin{cases} x^2 - 4y = 4 \\ x + y = -1 \end{cases}$

Solve $x + y = -1$ for y.

$y = -x - 1$

Replace y with $-x - 1$ in the first equation and solve for x.

$x^2 - 4(-x-1) = 4$

$x^2 + 4x + 4 = 4$

$x^2 + 4x = 0$

$x(x+4) = 0$

$x = 0$ or $x = -4$

Let $x = 0$, $y = -0 - 1 = -1$

Let $x = -4$, $y = -(-4) - 1 = 3$

The solutions are (0, −1) and (−4, 3).

2. $\begin{cases} y = -\sqrt{x} \\ x^2 + y^2 = 20 \end{cases}$

Substitute $-\sqrt{x}$ for y in the second equation.

$x^2 + \left(-\sqrt{x}\right)^2 = 20$

$x^2 + x = 20$

$x^2 + x - 20 = 0$

$(x+5)(x-4) = 0$

$x = -5$ or $x = 4$

Let $x = -5$.

$y = -\sqrt{-5}$ Not a real number

Let $x = 4$.

$y = -\sqrt{4} = -2$

The solution is (4, −2).

3. $\begin{cases} x^2 + y^2 = 9 \\ x - y = 5 \end{cases}$

Solve the second equation for x.

$x = y + 5$

Let $x = y + 5$ in the first equation.

$(y+5)^2 + y^2 = 9$

$y^2 + 10y + 25 + y^2 = 9$

$2y^2 + 10y + 16 = 0$

$y^2 + 5y + 8 = 0$

By the quadratic formula,

$y = \frac{-5 \pm \sqrt{5^2 - 4(1)(8)}}{2(1)} = \frac{-5 \pm \sqrt{-7}}{2}$

$\sqrt{-7}$ is not a real number. There is no real solution, or ∅.

4. $\begin{cases} x^2+4y^2=16 \\ x^2-y^2=1 \end{cases}$

Add the opposite of the second equation to the first.

$$\begin{array}{r} x^2+4y^2=16 \\ -x^2+y^2=-1 \\ \hline 0+5y^2=15 \end{array}$$

$$y^2=3$$

$$y=\pm\sqrt{3}$$

Let $y=\sqrt{3}$.

$$x^2-\left(\sqrt{3}\right)^2=1$$
$$x^2-3=1$$
$$x^2=4$$
$$x=\pm 2$$

Let $y=-\sqrt{3}$.

$$x^2-\left(-\sqrt{3}\right)^2=1$$
$$x^2-3=1$$
$$x^2=4$$
$$x=\pm 2$$

The solutions are $\left(2, \sqrt{3}\right)$, $\left(2, -\sqrt{3}\right)$, $\left(-2, \sqrt{3}\right)$, and $\left(-2, -\sqrt{3}\right)$.

Exercise Set 13.3

1. $\begin{cases} x^2+y^2=25 & (1) \\ 4x+3y=0 & (2) \end{cases}$

Solve E2 for y.

$$3y=-4x$$
$$y=-\frac{4x}{3}$$

Substitute into E1.

$$x^2+\left(-\frac{4x}{3}\right)^2=25$$
$$x^2+\frac{16x^2}{9}=25$$
$$9\left(x^2+\frac{16x^2}{9}\right)=9(25)$$
$$9x^2+16x^2=225$$
$$25x^2=225$$
$$x^2=9$$
$$x=\pm\sqrt{9}=\pm 3$$

$$x=3: y=-\frac{4(3)}{3}=-4$$
$$x=-3: y=-\frac{4(-3)}{3}=4$$

The solutions are (3, –4) and (–3, 4).

3. $\begin{cases} x^2+4y^2=10 & (1) \\ y=x & (2) \end{cases}$

Substitute x for y in E1.

$$x^2+4x^2=10$$
$$5x^2=10$$
$$x^2=2$$
$$x=\pm\sqrt{2}$$

Substitute these values into E2.

$$x=\sqrt{2}: y=x=\sqrt{2}$$
$$x=-\sqrt{2}: y=x=-\sqrt{2}$$

The solutions are $\left(\sqrt{2}, \sqrt{2}\right)$ and $\left(-\sqrt{2}, -\sqrt{2}\right)$.

5. $\begin{cases} y^2=4-x & (1) \\ x-2y=4 & (2) \end{cases}$

Solve E2 for x.

$$x=2y+4$$

Substitute into E1.

$$y^2=4-(2y+4)$$
$$y^2=-2y$$
$$y^2+2y=0$$
$$y(y+2)=0$$
$$y=0 \text{ or } y+2=0$$
$$y=-2$$

Substitute these values into the equation $x = 2y + 4$.

$$y=0: x=2(0)+4=4$$
$$y=-2: x=2(-2)+4=0$$

The solutions are (4, 0) and (0, –2).

7. $\begin{cases} x^2+y^2=9 & (1) \\ 16x^2-4y^2=64 & (2) \end{cases}$

Multiply E1 by 4 and add to E2.

$$\begin{array}{r} 4x^2+4y^2=36 \\ 16x^2-4y^2=64 \\ \hline 20x^2=100 \end{array}$$

$$x^2=5$$
$$x=\pm\sqrt{5}$$

Substitute 5 for x^2 into E1.

$$5+y^2=9$$
$$y^2=4$$
$$y=\pm 2$$

The solutions are $\left(-\sqrt{5}, -2\right)$, $\left(-\sqrt{5}, 2\right)$, $\left(\sqrt{5}, -2\right)$, and $\left(\sqrt{5}, 2\right)$.

9. $\begin{cases} x^2+2y^2=2 & (1) \\ x-\ \ y=2 & (2) \end{cases}$

Solve E2 for *x*: $x=y+2$

Substitute into E1.

$$(y+2)^2+2y^2=2$$
$$y^2+4y+4+2y^2=2$$
$$3y^2+4y+2=0$$
$$y=\frac{-4\pm\sqrt{(4)^2-4(3)(2)}}{2(3)}=\frac{-4\pm\sqrt{-8}}{6}$$

There are no real solutions. The solution is ∅.

11. $\begin{cases} y=x^2-3 & (1) \\ 4x-y=6 & (2) \end{cases}$

Substitute x^2-3 for *y* in E2.

$$4x-(x^2-3)=6$$
$$4x-x^2+3=6$$
$$0=x^2-4x+3$$
$$0=(x-3)(x-1)$$
$$x-3=0 \text{ or } x-1=0$$
$$x=3 \text{ or } x=1$$

Substitute these values into E1.

$x=3: y=(3)^2-3=6$

$x=1: y=(1)^2-3=-2$

The solutions are (3, 6) and (1, –2).

13. $\begin{cases} y=x^2 & (1) \\ 3x+y=10 & (2) \end{cases}$

Substitute x^2 for *y* in E2.

$$3x+x^2=10$$
$$x^2+3x-10=0$$
$$(x+5)(x-2)=0$$
$$x+5=0 \text{ or } x-2=0$$
$$x=-5 \text{ or } x=2$$

Substitute these values into E1.

$x=-5: y=(-5)^2=25$

$x=2: y=(2)^2=4$

The solutions are (–5, 25) and (2, 4).

15. $\begin{cases} y=2x^2+1 & (1) \\ x+y=-1 & (2) \end{cases}$

Substitute $2x^2+1$ for *y* in E2.

$$x+2x^2+1=-1$$
$$2x^2+x+2=0$$
$$x=\frac{-1\pm\sqrt{(1)^2-4(2)(2)}}{2(2)}=\frac{-1\pm\sqrt{-15}}{4}$$

There are no real solutions. The solution is ∅.

17. $\begin{cases} y=x^2-4 & (1) \\ y=x^2-4x & (2) \end{cases}$

Substitute x^2-4 for *y* in E2.

$$x^2-4=x^2-4x$$
$$-4=-4x$$
$$1=x$$

Substitute this value into E1.

$y=(1)^2-4=-3$

The solution is (1, –3).

19. $\begin{cases} 2x^2+3y^2=14 & (1) \\ -x^2+\ \ y^2=3 & (2) \end{cases}$

Multiply E2 by 2 and add to E1.

$$\begin{array}{r} 2x^2+3y^2=14 \\ \underline{-2x^2+2y^2=6} \\ 5y^2=20 \\ y^2=4 \\ y=\pm 2 \end{array}$$

Substitute 4 for y^2 into E2.

$$-x^2+4=3$$
$$-x^2=-1$$
$$x^2=1$$
$$x=\pm 1$$

The solutions are (–1, –2), (–1, 2), (1, –2), and (1, 2).

21. $\begin{cases} x^2+y^2=1 & (1) \\ x^2+(y+3)^2=4 & (2) \end{cases}$

Multiply E1 by –1 and add to E2.

$$\begin{array}{r} -x^2-y^2=-1 \\ x^2+(y+3)^2=4 \\ \hline (y+3)^3-y^2=3 \end{array}$$

$$y^2+6y+9-y^2=3$$
$$6y=-6$$
$$y=-1$$

Replace y with –1 in E1.

$$x^2+(-1)^2=1$$
$$x^2=0$$
$$x=0$$

The solution is (0, –1).

23. $\begin{cases} y=x^2+2 & (1) \\ y=-x^2+4 & (2) \end{cases}$

Add E1 and E2.

$$\begin{array}{l} y=x^2+2 \\ y=-x^2+4 \\ \hline 2y=6 \end{array}$$

$$y=3$$

Substitute this value into E1.

$$3=x^2+2$$
$$1=x^2$$
$$\pm 1=x$$

The solutions are (–1, 3) and (1, 3).

25. $\begin{cases} 3x^2+y^2=9 & (1) \\ 3x^2-y^2=9 & (2) \end{cases}$

Add E1 and E2.

$$\begin{array}{l} 3x^2+y^2=9 \\ 3x^2-y^2=9 \\ \hline 6x^2 \quad =18 \end{array}$$

$$x^2=3$$
$$x=\pm\sqrt{3}$$

Substitute 3 for x^2 in E1.

$$3(3)+y^2=9$$
$$y^2=0$$
$$y=0$$

The solutions are $\left(-\sqrt{3}, 0\right)$, $\left(\sqrt{3}, 0\right)$.

27. $\begin{cases} x^2+3y^2=6 & (1) \\ x^2-3y^2=10 & (2) \end{cases}$

Solve E2 for x^2: $x^2=3y^2+10$.

Substitute into E1.

$$(3y^2+10)+3y^2=6$$
$$6y^2=-4$$
$$y^2=-\frac{2}{3}$$

There are no real solutions. The solution is ∅.

29. $\begin{cases} x^2+y^2=36 & (1) \\ y=\frac{1}{6}x^2-6 & (2) \end{cases}$

Solve E1 for x^2: $x^2=36-y^2$.

Substitute into E2.

$$y=\frac{1}{6}(36-y^2)-6$$
$$y=6-\frac{1}{6}y^2-6$$
$$6y=-y^2$$
$$y^2+6y=0$$
$$y(y+6)=0$$
$$y=0 \text{ or } y=-6$$

Substitute these values into the equation $x^2=36-y^2$.

$$y=0: x^2=36-(0)^2$$
$$x^2=36$$
$$x=\pm 6$$
$$y=-6: x^2=36-(6)^2$$
$$x^2=0$$
$$x=0$$

The solutions are (–6, 0), (6, 0) and (0, –6).

31. $x>-3$

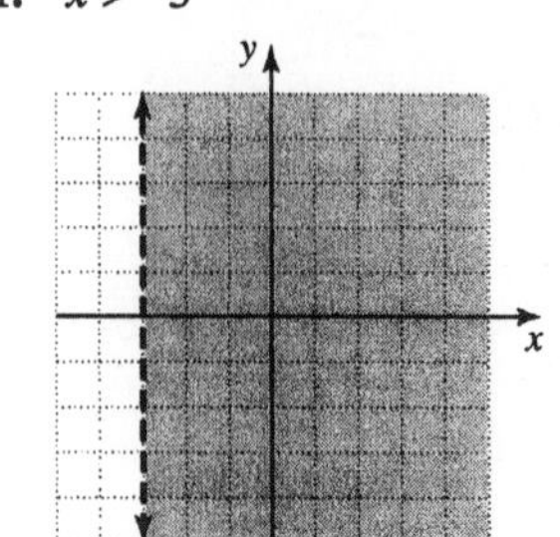

33. $y < 2x - 1$

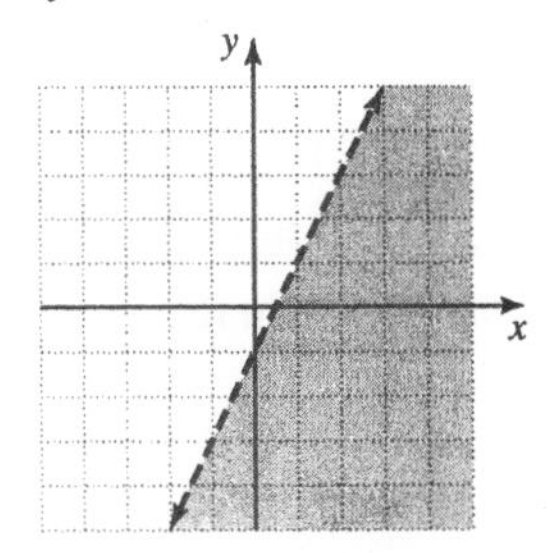

35. $P = x + (2x - 5) + (5x - 20) = (8x - 25)$ inches

37. $P = 2(x^2 + 3x + 1) + 2(x^2)$
$= 2x^2 + 6x + 2 + 2x^2$
$= (4x^2 + 6x + 2)$ meters

39. Answers may vary

41. There are 0, 1, 2, 3, or 4 possible real solutions. Answers may vary

43. Let x and y represent the numbers.

$$\begin{cases} x^2 + y^2 = 130 \\ x^2 - y^2 = 32 \end{cases}$$

Add the equations.

$$\begin{aligned} x^2 + y^2 &= 130 \\ x^2 - y^2 &= 32 \\ \hline 2x^2 \quad &= 162 \\ x^2 &= 81 \\ x &= \pm 9 \end{aligned}$$

Replace x^2 with 81 in the first equation.

$$\begin{aligned} 81 + y^2 &= 130 \\ y^2 &= 49 \\ y &= \pm 7 \end{aligned}$$

The numbers are –9 and –7, –9 and 7, 9 and –7, and 9 and 7.

45. Let x and y be the length and width.

$$\begin{cases} xy = 285 \\ 2x + 2y = 68 \end{cases}$$

Solve the first equation for y: $y = \frac{285}{x}$.

Substitute into the second equation.

$$\begin{aligned} 2x + 2\left(\frac{285}{x}\right) &= 68 \\ x + \frac{285}{x} &= 34 \\ x^2 + 285 &= 34x \\ x^2 - 34x + 285 &= 0 \\ (x - 19)(x - 15) &= 0 \\ x = 19 \text{ or } x &= 15 \end{aligned}$$

Using $x = 19$, $y = \frac{285}{x} = \frac{285}{19} = 15$.

Using $x = 15$, $y = \frac{285}{x} = \frac{285}{15} = 19$.

The dimensions are 19 cm by 15 cm.

47. $\begin{cases} p = -0.01x^2 - 0.2x + 9 \\ p = 0.01x^2 - 0.1x + 3 \end{cases}$

Substitute.

$$\begin{aligned} -0.01x^2 - 0.2x + 9 &= 0.01x^2 - 0.1x + 3 \\ 0 &= 0.02x^2 + 0.1x - 6 \\ 0 &= x^2 + 5x - 300 \\ 0 &= (x + 20)(x - 15) \end{aligned}$$

$x + 20 = 0$ or $x - 15 = 0$
$x = -20$ or $x = 15$

Disregard the negative.

$p = -0.01(15)^2 - 0.2(15) + 9$
$p = 3.75$

The equilibrium quantity is 15,000 compact discs, and the corresponding price is $3.75.

49. $\begin{cases} x^2 + 4y^2 = 10 \\ y = x \end{cases}$

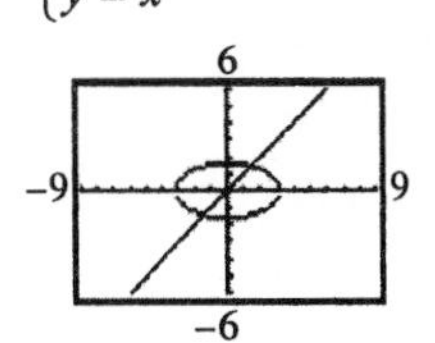

51. $\begin{cases} y = x^2 + 2 \\ y = -x^2 + 4 \end{cases}$

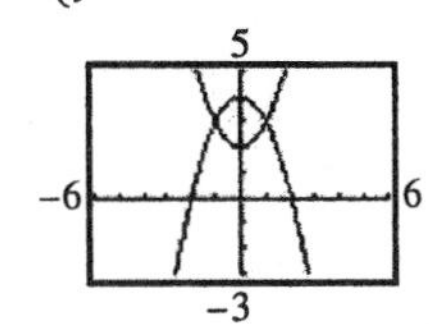

Section 13.4

Practice Exercises

1. $\frac{x^2}{36}+\frac{y^2}{16}\ge 1$

   First graph the ellipse $\frac{x^2}{36}+\frac{y^2}{16}=1$ as a solid curve. Choose (0, 0) as a test point.

   $\frac{x^2}{36}+\frac{y^2}{16}\ge 1$

   $\frac{0^2}{36}+\frac{0^2}{16}\ge 1$

   $0\ge 1$ False

   The solution set is the region that does not contain (0, 0).

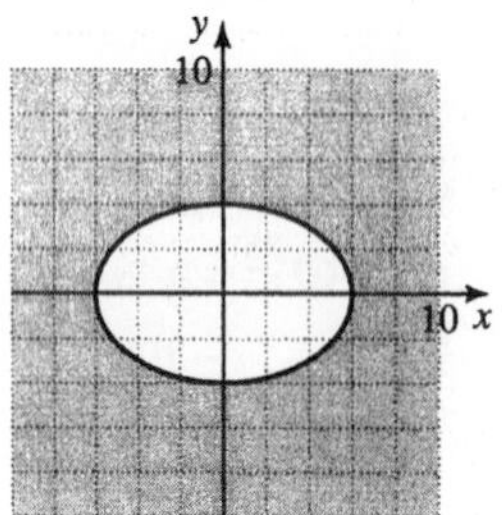

2. $16y^2 > 9x^2+144$

   The related equation is $16y^2 = 9x^2+144$.

   $16y^2-9x^2=144$

   $\frac{y^2}{9}-\frac{x^2}{16}=1$

   Graph the hyperbola as a dashed curve. Choose (0, 0), (0, 4), and (0, –4) as test points.

   (0, 0): $16(0)^2 > 9(0)^2+144$

   $0>144$ False

   (0, 4): $16(4)^2 > 9(0)^2+144$

   $256>144$ True

   (0, –4): $16(-4)^2 > 9(0)^2+144$

   $256>144$ True

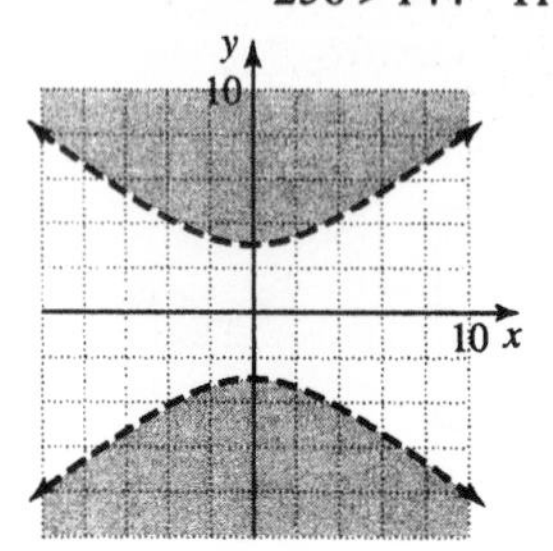

3. $\begin{cases} y\ge x^2 \\ y\le -3x+2 \end{cases}$

   Solve the related system $\begin{cases} y=x^2 \\ y=-3x+2 \end{cases}$.

   Substitute $-3x + 2$ for y in the first equation.

   $x^2=-3x+2$

   $x^2+3x-2=0$

   $x=\frac{-3\pm\sqrt{3^2-4(1)(-2)}}{2(1)}$

   $=\frac{-3\pm\sqrt{17}}{2}$

   ≈ 0.56 or -3.56

   $y=-3x+2\approx -3(0.56)+2=0.32$

   $y\approx -3(-3.56)+2=12.68$

   The points of intersection are approximately (0.56, 0.32) and (–3.56, 12.68).

   Graph $y=x^2$ and $y=-3x+2$ as solid curves.

   The region of the solution set is above the parabola but below the line.

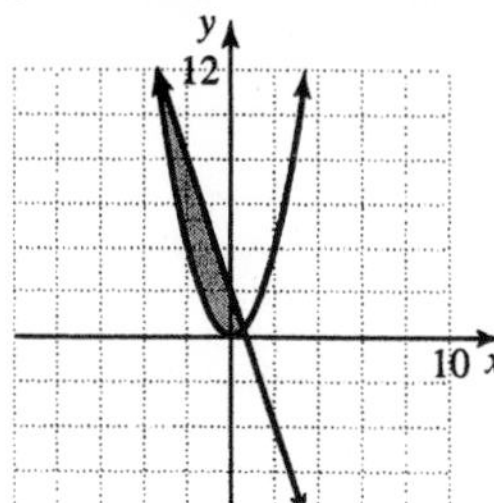

4. $\begin{cases} x^2+y^2<16 \\ \frac{x^2}{4}-\frac{y^2}{9}<1 \\ y<x+3 \end{cases}$

   Graph $x^2+y^2=16$, $\frac{x^2}{4}-\frac{y^2}{9}=1$, and $y=x+3$.

   The test point (0, 0) gives true statements for all three inequalities; thus, the innermost region is the solution set.

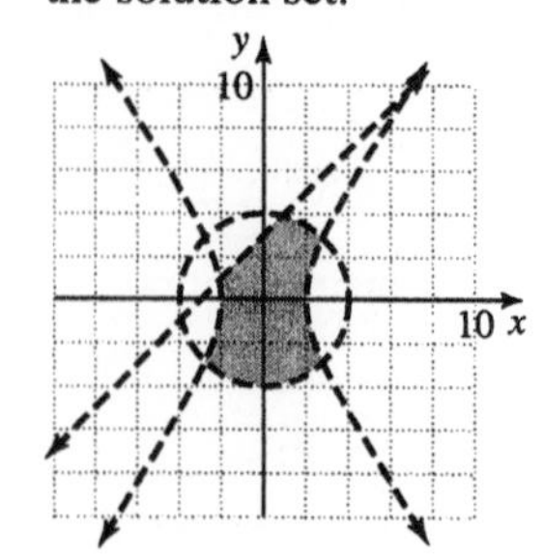

Exercise Set 13.4

1. $y < x^2$

First graph the parabola as a dashed curve.

Test Point	$y < x^2$; Result
(0, 1)	$1 < 0^2$; False

Shade the region which does not contain (0, 1).

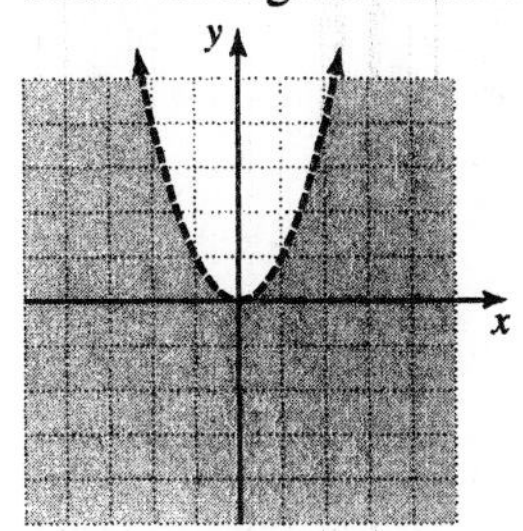

3. $x^2 + y^2 \geq 16$

First graph the circle as a solid curve.

Test Point	$x^2 + y^2 \geq 16$; Result
(0, 0)	$0^2 + 0^2 \geq 16$; False

Shade the region which does not contain (0, 0).

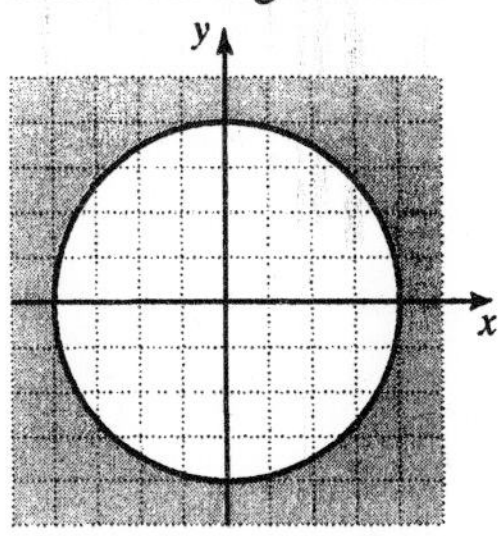

5. $\frac{x^2}{4} - y^2 < 1$

First graph the hyperbola as a dashed curve.

Test Point	$\frac{x^2}{4} - y^2 < 1$; Result
(−4, 0)	$\frac{(-4)^2}{4} - 0^2 < 1$; False
(0, 0)	$\frac{(0)^2}{4} - 0^2 < 1$; True
(4, 0)	$\frac{(4)^2}{4} - 0^2 < 1$; False

Shade the region containing (0, 0).

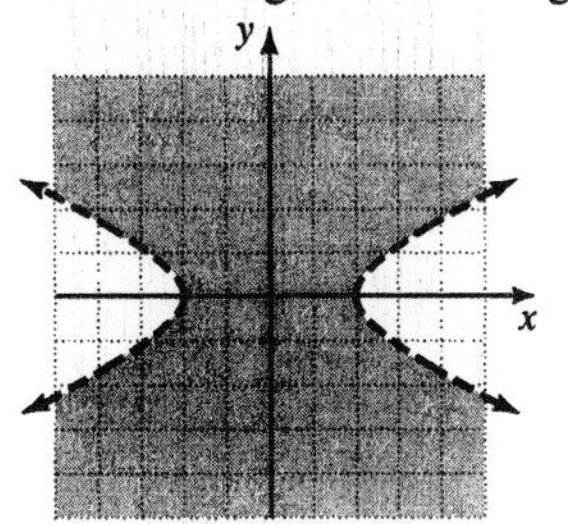

7. $y > (x-1)^2 - 3$

First graph the parabola as a dashed curve.

Test Point	$y > (x-1)^2 - 3$; Result
(0, 0)	$0 > (0-1)^2 - 3$; True

Shade the region containing (0, 0).

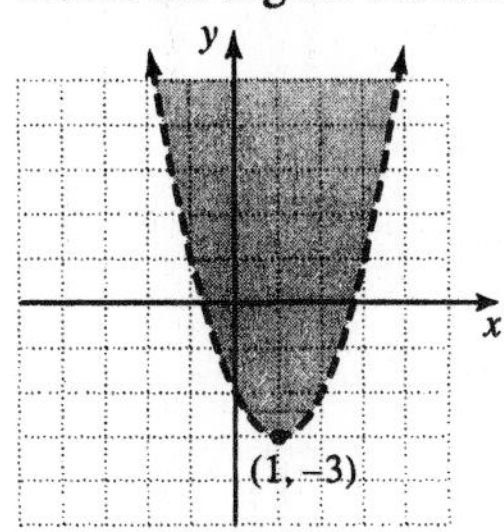

9. $x^2 + y^2 \leq 9$

First graph the circle as a solid curve.

Test Point	$x^2 + y^2 \leq 9$; Result
(0, 0)	$0^2 + 0^2 \leq 9$; True

Shade the region containing (0, 0).

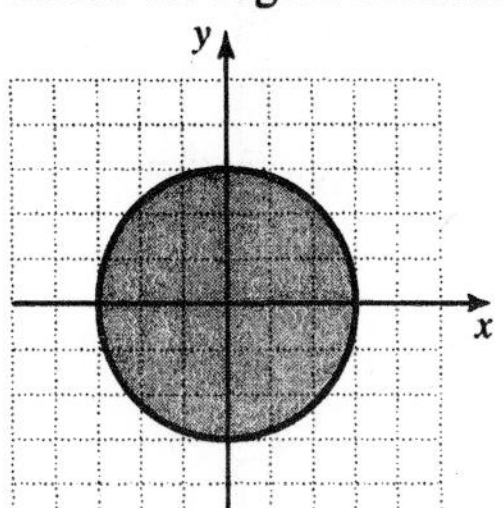

11. $y > -x^2 + 5$

First graph the parabola as a dashed curve.

Test Point	$y > -x^2 + 5$; Result
(0, 0)	$0 > -(0)^2 + 5$; False

Shade the region which does not contain (0, 0).

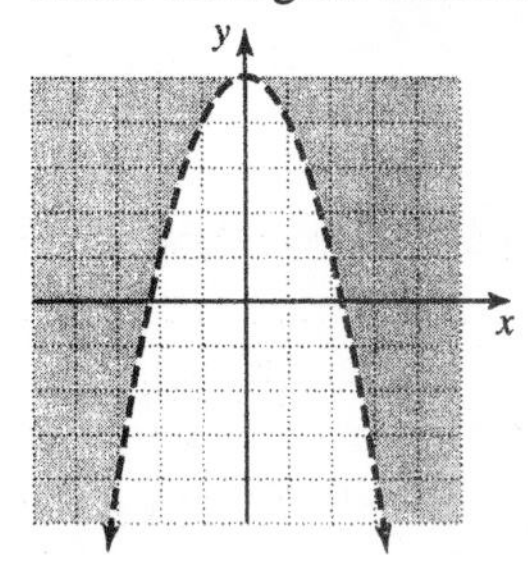

13. $\frac{x^2}{4} + \frac{y^2}{9} \le 1$

First graph the ellipse as a solid curve.

Test Point	$\frac{x^2}{4} + \frac{y^2}{9} \le 1$; Result
(0, 0)	$\frac{(0)^2}{4} + \frac{(0)^2}{9} \le 1$; True

Shade the region containing (0, 0).

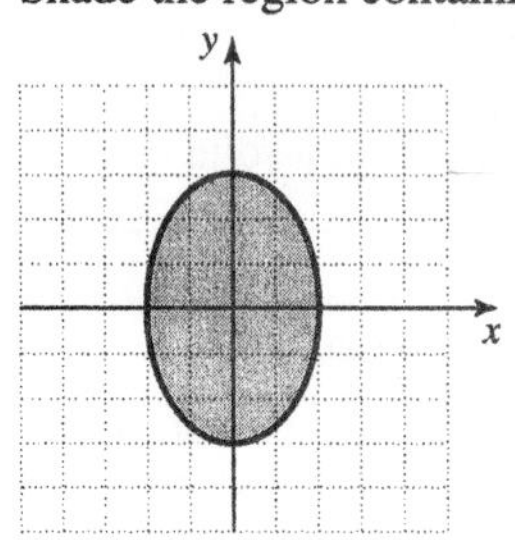

15. $\frac{y^2}{4} - x^2 \le 1$

First graph the hyperbola as solid curves.

Test Point	$\frac{y^2}{4} - x^2 \le 1$; Result
(0, −4)	$\frac{(-4)^2}{4} - 0^2 \le 1$; False
(0, 0)	$\frac{(0)^2}{4} - 0^2 \le 1$; True
(0, 4)	$\frac{(4)^2}{4} - 0^2 \le 1$; False

Shade the region containing (0, 0).

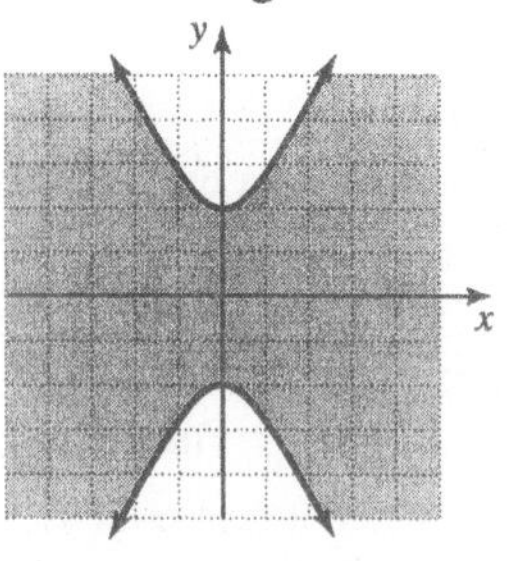

17. $y < (x-2)^2 + 1$

First graph the parabola as a dashed curve.

Test Point	$y < (x-2)^2 + 1$; Result
(0, 0)	$0 < (0-2)^2 + 1$; True

Shade the region containing (0, 0).

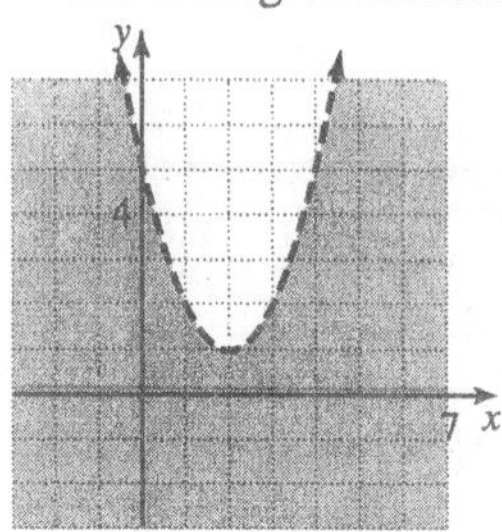

19. $y \le x^2 + x - 2$

First graph the parabola as a solid curve.

Test Point	$y \le x^2 + x - 2$; Result
(0, 0)	$0 \le (0)^2 + (0) - 2$; False

Shade the region which does not contain (0, 0).

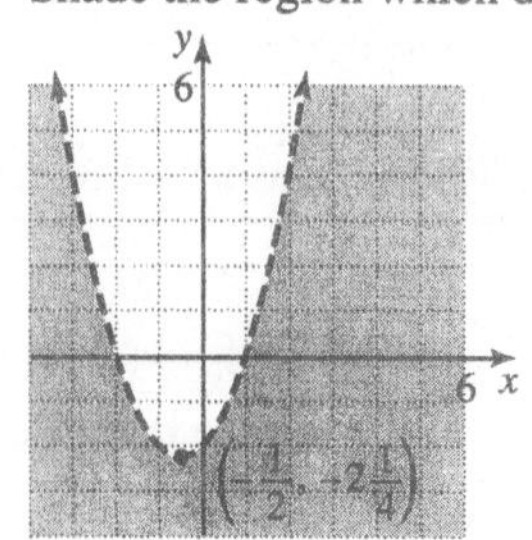

21. $\begin{cases} 4x+3y \geq 12 \\ x^2+y^2 < 16 \end{cases}$

First graph $4x+3y=12$ as a solid line.

Test Point	$4x + 3y \geq 12$; Result
(0, 0)	$4(0) + 3(0) \geq 12$; False

Shade the region which does not contain (0, 0). Next, graph the circle $x^2+y^2=16$ as a dashed curve.

Test Point	$x^2+y^2<16$; Result
(0, 0)	$0^2+0^2<16$; True

Shade the region containing (0, 0). The solution to the system is the intersection.

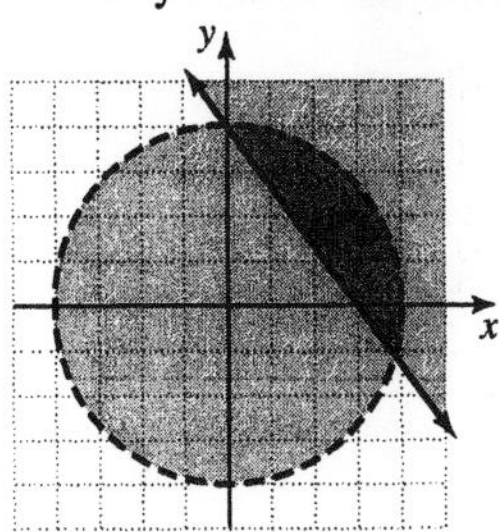

23. $\begin{cases} x^2+y^2 \leq 9 \\ x^2+y^2 \geq 1 \end{cases}$

First graph the circle with radius 3 as a solid curve.

Test Point	$x^2+y^2 \leq 9$; Result
(0, 0)	$0^2+0^2 \leq 9$; True

Shade the region containing (0, 0). Next, graph the circle with 1 as a dashed curve.

Test Point	$x^2+y^2 \geq 1$; Result
(0, 0)	$0^2+0^2 \geq 1$; False

Shade the region which does not contain (0, 0). The solution to the system is the intersection.

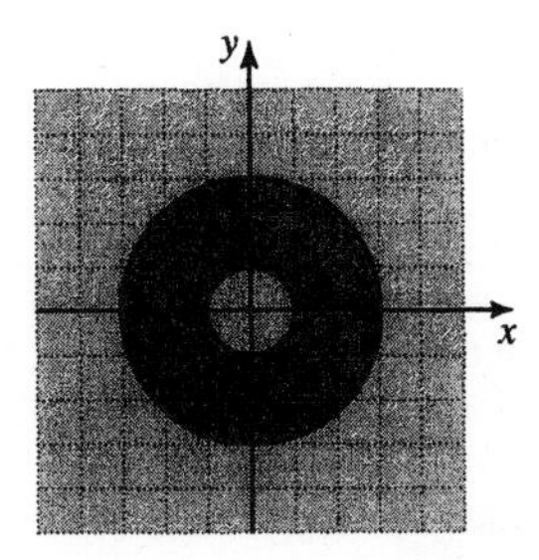

25. $\begin{cases} y > x^2 \\ y \geq 2x+1 \end{cases}$

First graph the parabola as a dashed curve.

Test Point	$y > x^2$; Result
(0, 1)	$1 > 0^2$; True

Shade the region containing (0, 1). Next, graph $y=2x+1$ as a solid line.

Test Point	$y \geq 2x + 1$; Result
(0, 0)	$0 \geq 2(0) + 1$; False

Shade the region which does not contain (0, 0). The solution to the system is the intersection.

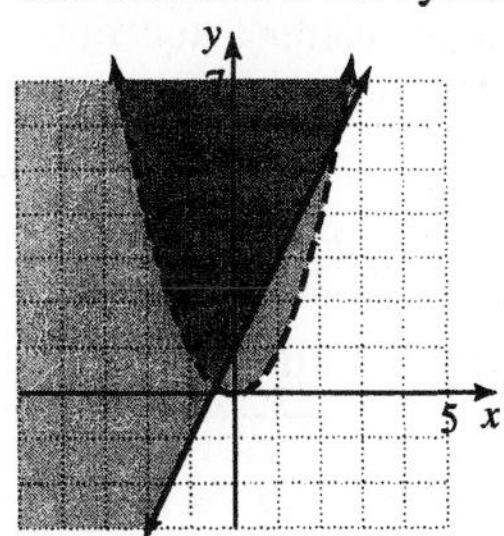

27. $\begin{cases} x^2+y^2 > 9 \\ y > x^2 \end{cases}$

First graph the circle as a dashed curve.

Test Point	$x^2+y^2>9$; Result
(0, 0)	$0^2+0^2>9$; False

Shade the region which does not contain (0, 0). Next, graph the parabola as a dashed curve.

Test Point	$y > x^2$; Result
(0, 1)	$1 > 0^2$; True

Shade the region containing (0, 1). The solution to the system is the intersection.

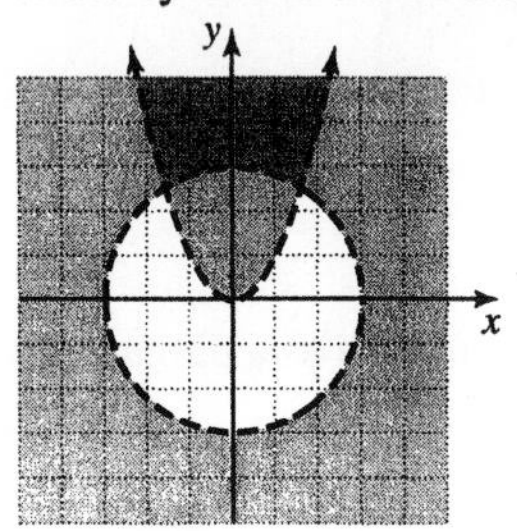

29. $\begin{cases} \frac{x^2}{4}+\frac{y^2}{9} \ge 1 \\ x^2+y^2 \ge 4 \end{cases}$

First graph the ellipse as a solid curve.

Test Point	$\frac{x^2}{4}+\frac{y^2}{9} \ge 1$; Result
(0, 0)	$\frac{0^2}{4}+\frac{0^2}{9} \ge 1$; False

Shade the region which does not contain (0, 0). Next, graph the circle as a solid curve.

Test Point	$x^2+y^2 \ge 4$; Result
(0, 0)	$0^2+0^2 \ge 4$; False

Shade the region which does not contain (0, 0). The solution to the system is the intersection.

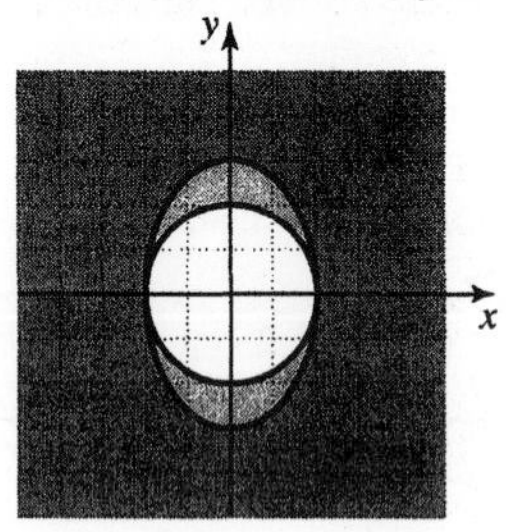

31. $\begin{cases} x^2-y^2 \ge 1 \\ y \ge 0 \end{cases}$

First graph the hyperbola as solid curves.

Test Point	$x^2-y^2 \ge 1$; Result
(−2, 0)	$(-2)^2-0^2 \ge 1$; True
(0, 0)	$0^2-0^2 \ge 1$; False
(2, 0)	$2^2-0^2 \ge 1$; True

Shade the region which does not contain (0, 0). Next, graph $y = 0$ as a solid line.

Test Point	$y > 0$; Result
(0, 1)	$1 \ge 0$; True

Shade the region containing (0, 1). The solution to the system is the intersection.

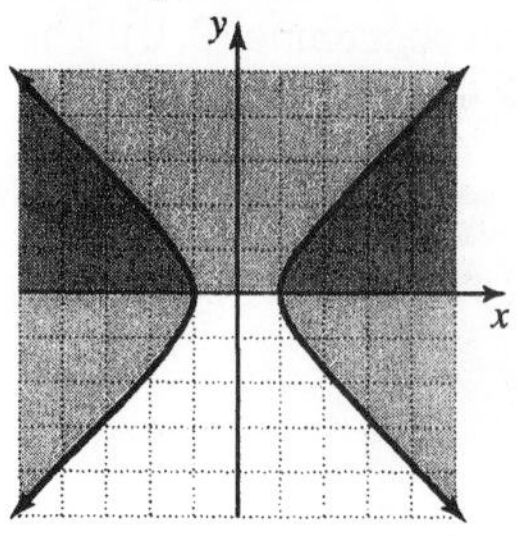

33. $\begin{cases} x+y \ge 1 \\ 2x+3y < 1 \\ x > -3 \end{cases}$

First graph $x+y=1$ as a solid line.

Test Point	$x+y \ge 1$; Result
(0, 0)	$0+0 \ge 1$; False

Shade the region which does not contain (0, 0). Next, graph $2x+3y=1$ as a dashed line.

Test Point	$2x+3y<1$; Result
(0, 0)	$2(0)$ 1+ $3(0) < 1$; True

Shade the region containing (0, 0). Now graph the line $x=-3$ as a dashed line.

Test Point	$x > -3$; Result
(0, 0)	$0 > -3$; True

Shade the region containing (0, 0). The solution to the system is the intersection.

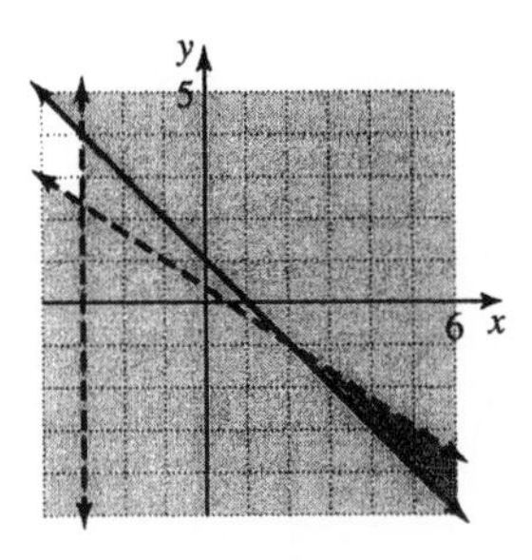

35. $\begin{cases} x^2 - y^2 < 1 \\ \frac{x^2}{16} + y^2 \le 1 \\ x \ge -2 \end{cases}$

First graph the hyperbola as dashed curves.

Test Point	$x^2 - y^2 < 1$; Result
(–2, 0)	$(-2)^2 - 0^2 < 1$; False
(0, 0)	$0^2 - 0^2 < 1$; True
(2, 0)	$2^2 - 0^2 < 1$; False

Shade the region containing (0, 0). Next, graph the ellipse as a solid curve.

Test Point	$\frac{x^2}{16} + y^2 \le 1$; Result
(0, 0)	$\frac{0^2}{16} + 0^2 \le 1$; True

Shade the region containing (0, 0). Now graph the line $x = -2$ as a solid line.

Test Point	$x \ge -2$; Result
(0, 0)	$0 \ge -2$; True

Shade the region containing (0, 0). The solution to the system is the intersection.

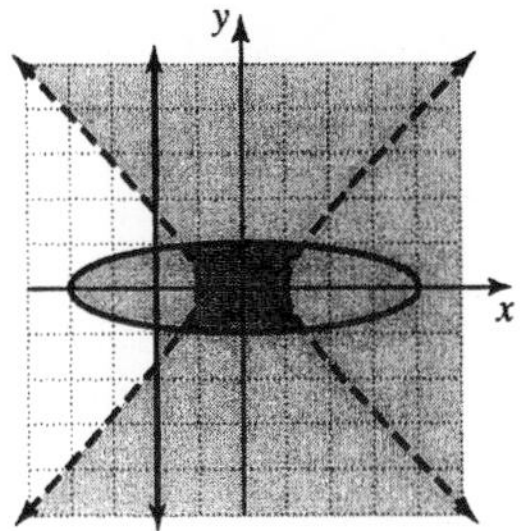

37. This is not a function because a vertical line can cross the graph in more than one place.

39. This is a function because a vertical line can cross the graph in no more than one place.

41. $f(x) = 3x^2 - 2$
$f(-1) = 3(-1)^2 - 2 = 3 - 2 = 1$

43. $f(x) = 3x^2 - 2$
$f(a) = 3(a)^2 - 2 = 3a^2 - 2$

45. Answers may vary

47. $\begin{cases} y \le x^2 \\ y \ge x + 2 \\ x \ge 0 \\ y \ge 0 \end{cases}$

First graph $y = x^2$ as a solid curve.

Test Point	$y \le x^2$; Result
(0, 1)	$1 \le 0^2$; False

Shade the region which does not contain (0, 1). Next, graph $y = x + 2$ as a solid line.

Test Point	$y \ge x + 2$; Result
(0, 0)	$0 \ge 0 + 2$; False

Shade the region which does not contain (0, 0). Next graph the line $x = 0$ as a solid line, and shade to the right. Now graph the line $y = 0$ as a solid line, and shade above. The solution to the system is the intersection.

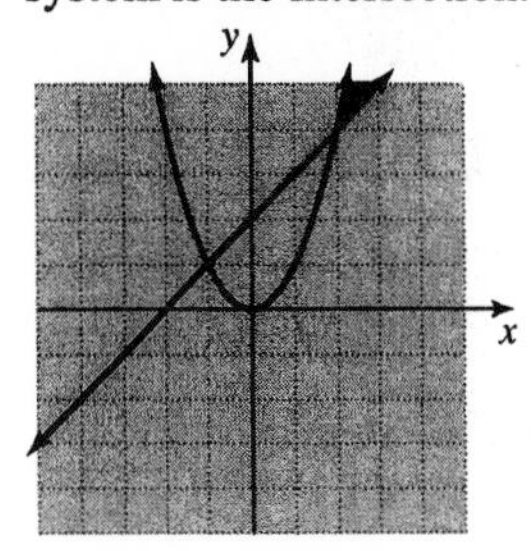

Chapter 13 Vocabulary Check

1. A circle is the set of all points in a plane that are the same distance from a fixed point, called the center.

2. A nonlinear system of equations is a system of equations at least one of which is not linear.

3. An ellipse is the set of points on a plane such that the sum of the distances of those points from two fixed points is a constant.

4. In a circle, the distance from the center to a point of the circle is called its radius.

5. A hyperbola is the set of points in a plane such that the absolute value of the difference of the distance from two fixed points is constant.

Chapter 13 Review

1. center (–4, 4), radius 3

$$[x-(-4)]^2+(y-4)^2=3^2$$
$$(x+4)^2+(y-4)^2=9$$

2. center (5, 0), radius 5

$$(x-5)^2+(y-0)^2=5^2$$
$$(x-5)^2+y^2=25$$

3. center (–7, –9), radius $\sqrt{11}$

$$[x-(-7)]^2+[y-(-9)]^2=\left(\sqrt{11}\right)^2$$
$$(x+7)^2+(y+9)^2=11$$

4. center (0, 0), radius $\frac{7}{2}$

$$(x-0)^2+(y-0)^2=\left(\frac{7}{2}\right)^2$$
$$x^2+y^2=\frac{49}{4}$$

5. $x^2+y^2=7$

Circle; center (0, 0), radius $r=\sqrt{7}$

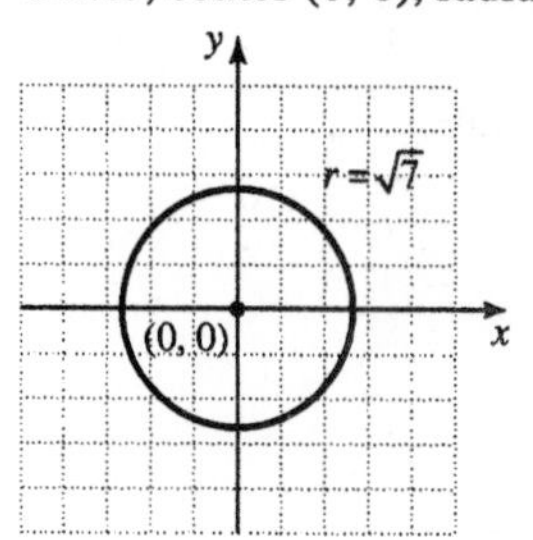

6. $x=2(y-5)^2+4$

Parabola; vertex: (4, 5)

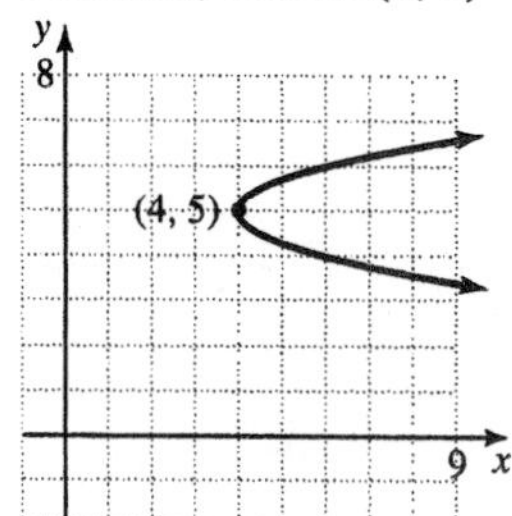

7. $x=-(y+2)^2+3$

Parabola; vertex: (3, –2)

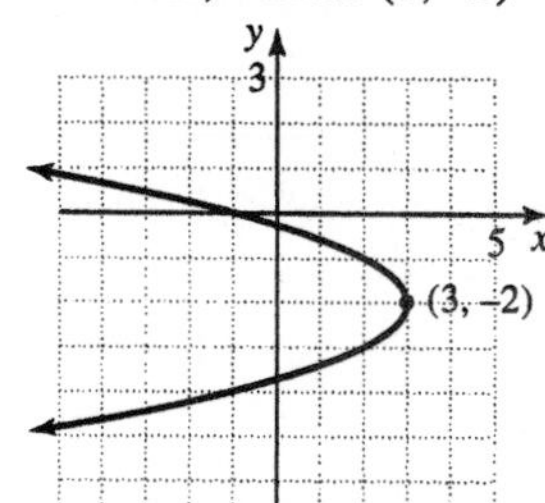

8. $(x-1)^2+(y-2)^2=4$

Circle; center (1, 2), radius $r=\sqrt{4}=2$

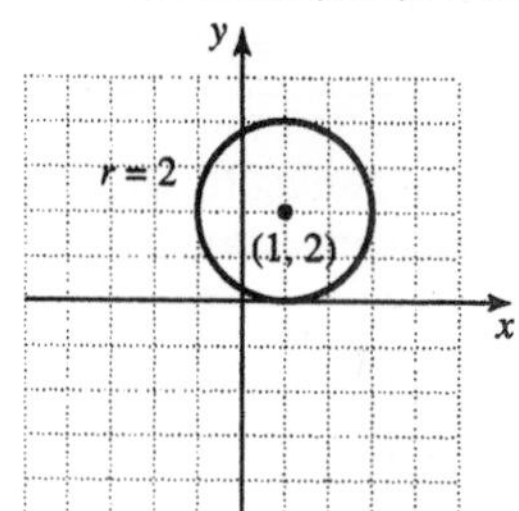

9. $y = -x^2 + 4x + 10$

Parabola; $x = \frac{-b}{2a} = \frac{-4}{2(-1)} = 2$

$y = -(2)^2 + 4(2) + 10 = 14$

Vertex: (2, 14)

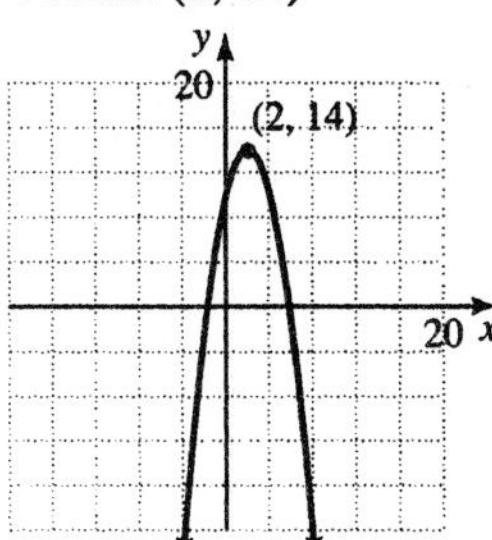

10. $x = -y^2 - 4y + 6$

Parabola; $y = \frac{-b}{2a} = \frac{-(-4)}{2(-1)} = -2$

$x = -(-2)^2 - 4(-2) + 6 = 10$

Vertex: (10, –2)

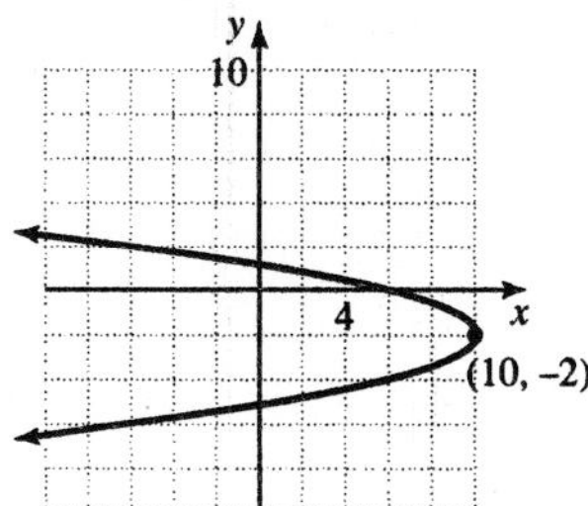

11. $x = \frac{1}{2}y^2 + 2y + 1$

Parabola; $y = \frac{-b}{2a} = \frac{-2}{2\left(\frac{1}{2}\right)} = -2$

$x = \frac{1}{2}(-2)^2 + 2(-2) + 1 = -1$

Vertex: (–1, –2)

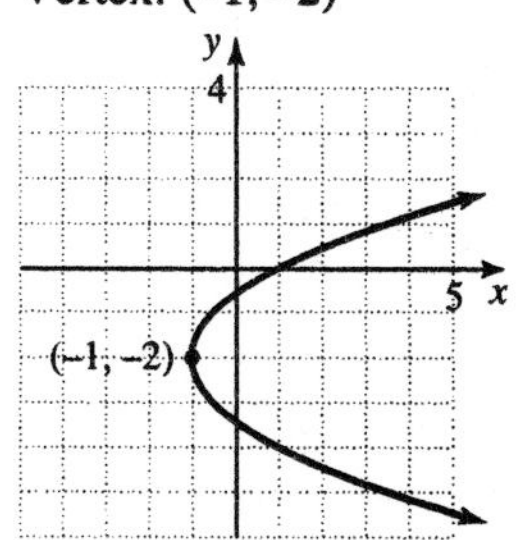

12. $y = -3x^2 + \frac{1}{2}x + 4$

Parabola; $x = \frac{-b}{2a} = \frac{-\frac{1}{2}}{2(-3)} = \frac{1}{12}$

$y = -3\left(\frac{1}{12}\right)^2 + \frac{1}{2}\left(\frac{1}{12}\right) + 4 = \frac{193}{48}$

Vertex: $\left(\frac{1}{12}, \frac{193}{48}\right)$

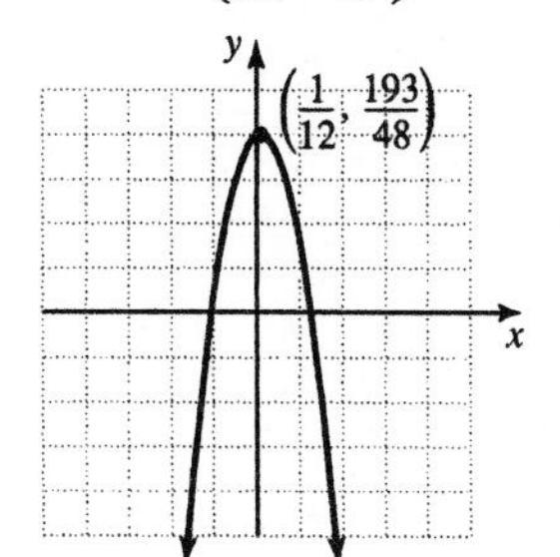

13.

$$x^2 + y^2 + 2x + y = \frac{3}{4}$$
$$(x^2 + 2x) + (y^2 + y) = \frac{3}{4}$$
$$(x^2 + 2x + 1) + \left(y^2 + y + \frac{1}{4}\right) = \frac{3}{4} + 1 + \frac{1}{4}$$
$$(x+1)^2 + \left(y + \frac{1}{2}\right)^2 = 2$$

Circle; center $\left(-1, -\frac{1}{2}\right)$, radius $r = \sqrt{2}$

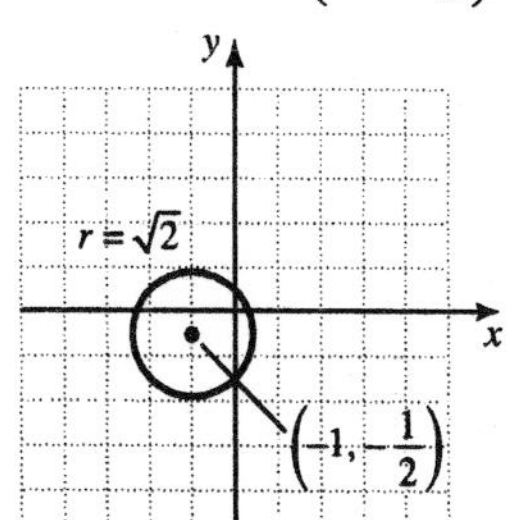

14.

$$x^2 + y^2 - 3y = \frac{7}{4}$$
$$x^2 + \left(y^2 - 3y + \frac{9}{4}\right) = \frac{7}{4} + \frac{9}{4}$$
$$x^2 + \left(y - \frac{3}{2}\right)^2 = 4$$

Circle; center $\left(0, \frac{3}{2}\right)$, radius $r = \sqrt{4} = 2$

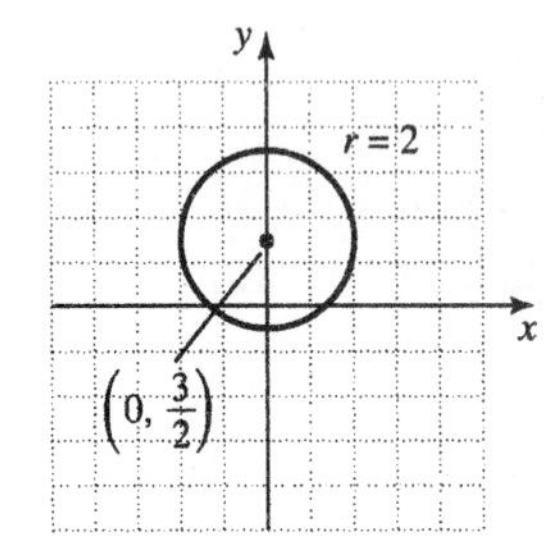

15.
$$4x^2+4y^2+16x+8y=1$$
$$(x^2+4x)+(y^2+2y)=\frac{1}{4}$$
$$(x^2+4x+4)+(y^2+2y+1)=\frac{1}{4}+4+1$$
$$(x+2)^2+(y+1)^2=\frac{21}{4}$$

Circle; center (–2, –1), radius $r=\sqrt{\frac{21}{4}}=\frac{\sqrt{21}}{2}$

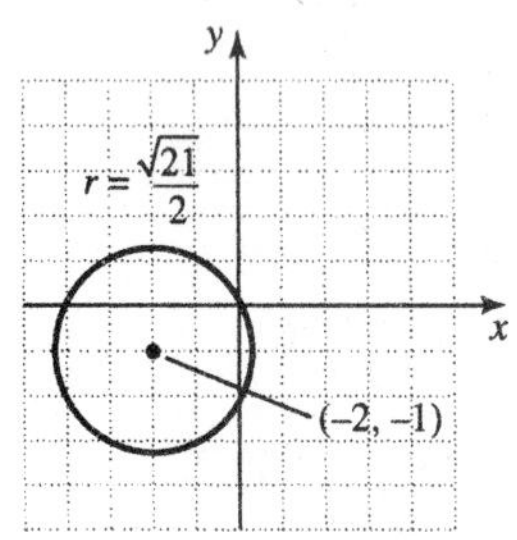

16. $x^2+\frac{y^2}{4}=1$

Center: (0, 0); $a = 1$, $b = 2$

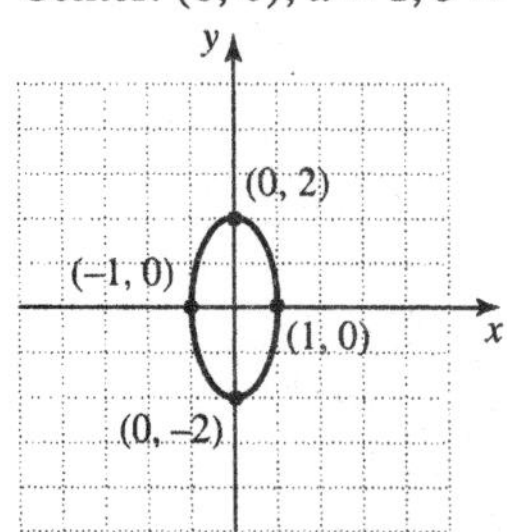

17. $x^2-\frac{y^2}{4}=1$

Center: (0, 0); $a = 1$, $b = 2$

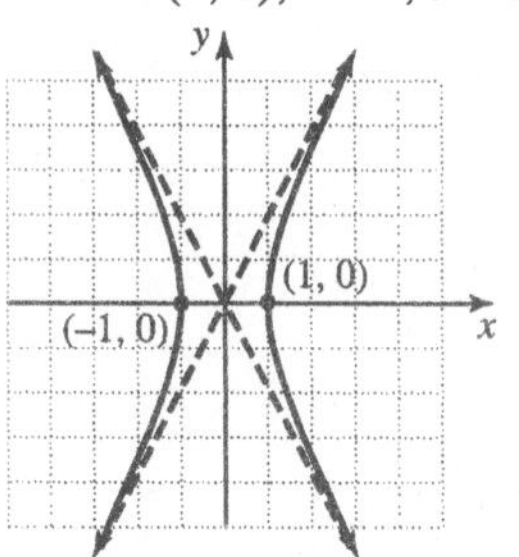

18. $\frac{x^2}{5}+\frac{y^2}{5}=1$

$x^2+y^2=5$

Center: (0, 0); radius $r=\sqrt{5}$

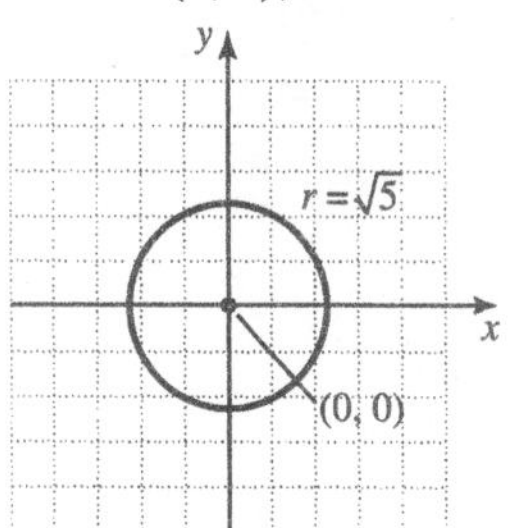

19. $\frac{x^2}{5}-\frac{y^2}{5}=1$

Center: (0, 0); $a=\sqrt{5}$, $b=\sqrt{5}$

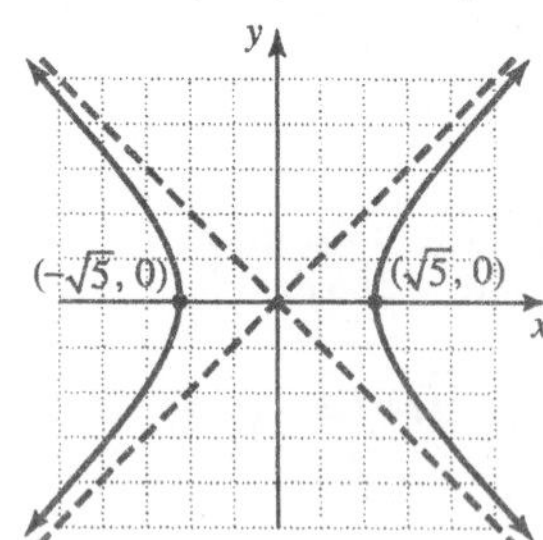

20. $-5x^2 + 25y^2 = 125$

$$\frac{y^2}{5} - \frac{x^2}{25} = 1$$

Center: (0, 0); $a = 5$, $b = \sqrt{5}$

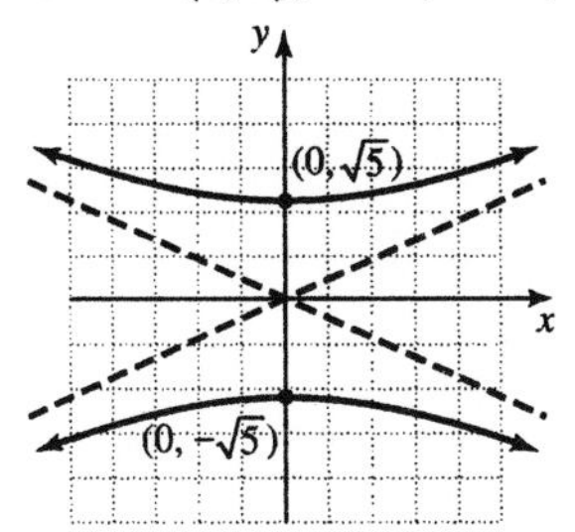

21. $4y^2 + 9x^2 = 36$

$$\frac{y^2}{9} + \frac{x^2}{4} = 1$$

Center: (0, 0); $a = 2$, $b = 3$

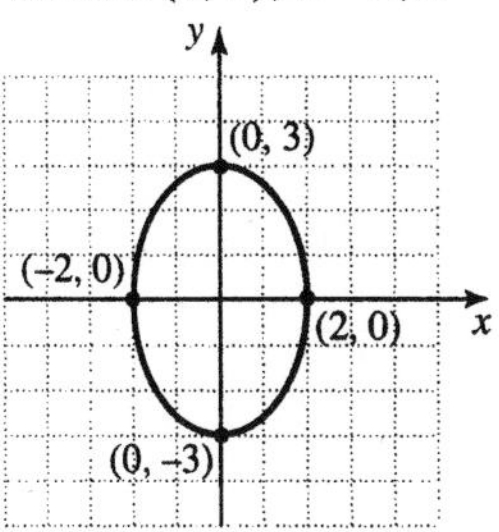

22. $x^2 - y^2 = 1$

Center: (0, 0); $a = 1$, $b = 1$

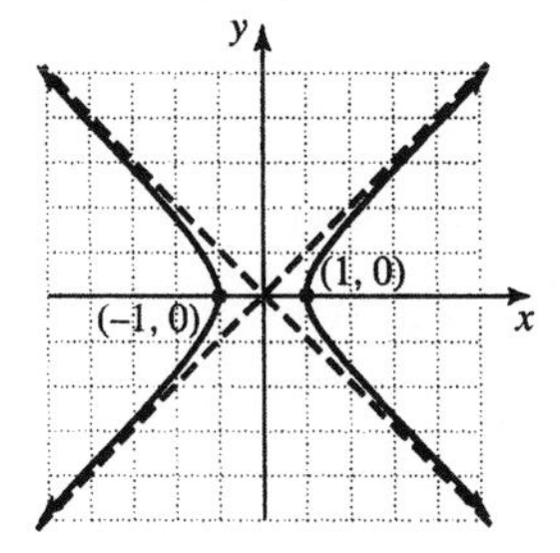

23. $\frac{(x+3)^2}{9} + \frac{(y-4)^2}{25} = 1$

Center: (–3, 4); $a = 3$, $b = 5$

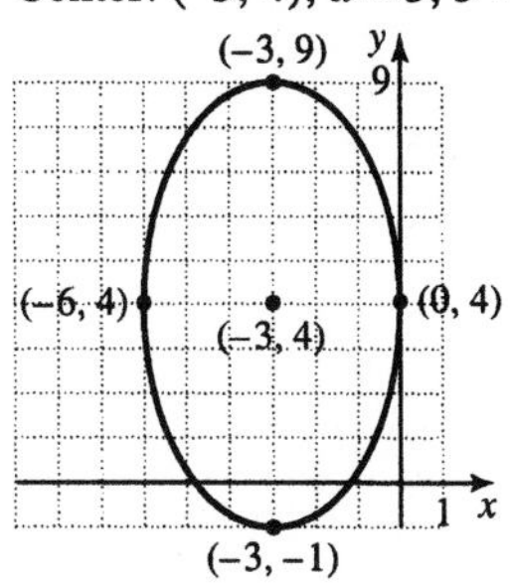

24. $y^2 = x^2 + 9$

$$y^2 - x^2 = 9$$

$$\frac{y^2}{9} - \frac{x^2}{9} = 1$$

Center: (0, 0); $a = 3$, $b = 3$

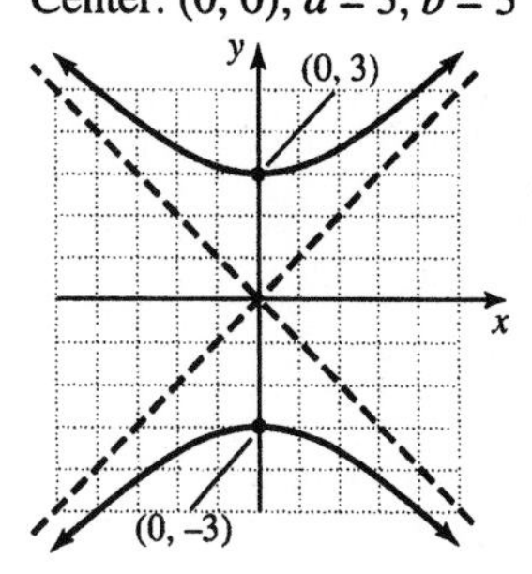

25. $x^2 = 4y^2 - 16$

$$16 = 4y^2 - x^2$$

$$1 = \frac{y^2}{4} - \frac{x^2}{16}$$

Center: (0, 0); $a = 4$, $b = 2$

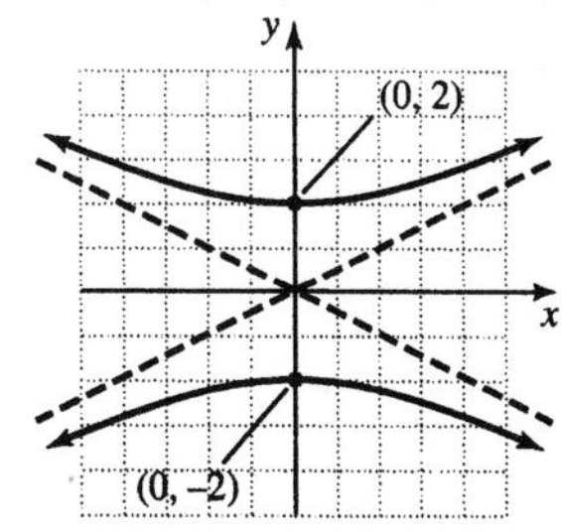

26. $100-25x^2=4y^2$

$100=25x^2+4y^2$

$1=\frac{x^2}{4}+\frac{y^2}{25}$

Center: (0, 0); $a=2$, $b=5$

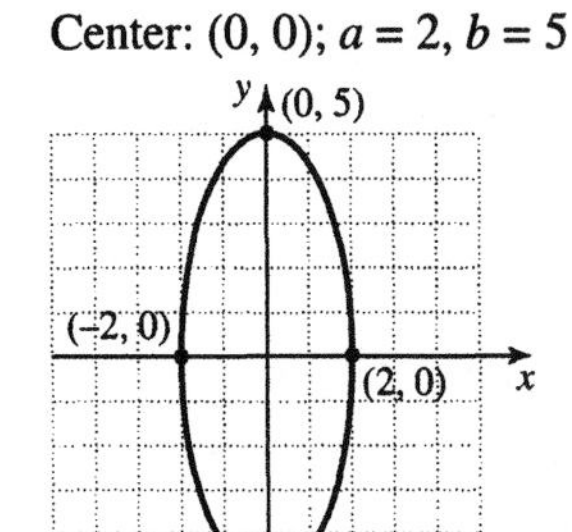

27. $\begin{cases} y=2x-4 & (1) \\ y^2=4x & (2) \end{cases}$

Substitute $2x-4$ for y in E2.

$(2x-4)^2=4x$

$4x^2-16x+16=4x$

$4x^2-20x+16=0$

$x^2-5x+4=0$

$(x-4)(x-1)=0$

$x=4$ or $x=1$

Use these values in E1.

$x=4: y=2(4)-4=4$

$x=1: y=2(1)-4=-2$

The solutions are (4, 4) and (1, −2).

28. $\begin{cases} x^2+y^2=4 & (1) \\ x-y=4 & (2) \end{cases}$

Solve E2 for x: $x=y+4$.

Substitute into E1.

$(y+4)^2+y^2=4$

$(y^2+8y+16)+y^2=4$

$2y^2+8y+12=0$

$y^2+4y+6=0$

$y=\frac{-4\pm\sqrt{(4)^2-4(1)(6)}}{2(1)}=\frac{-4\pm\sqrt{-8}}{2}$

There are no real solutions. The solution is ∅.

29. $\begin{cases} y=x+2 & (1) \\ y=x^2 & (2) \end{cases}$

Substitute $x+2$ for y in E2.

$x+2=x^2$

$0=x^2-x-2$

$0=(x-2)(x+1)$

$x=2$ or $x=-1$

Use these values in E1.

$x=2: y=2+2=4$

$x=-1: y=-1+2=1$

The solutions are (2, 4) and (−1, 1).

30. $\begin{cases} x^2+4y^2=16 & (1) \\ x^2+y^2=4 & (2) \end{cases}$

Multiply E2 by −1 and add to E1.

$x^2+4y^2=16$

$\underline{-x^2-y^2=-4}$

$3y^2=12$

$y^2=4$

$y=\pm 2$

Replace y^2 with 4 in E2.

$x^2+4=4$

$x^2=0$

$x=0$

The solutions are (0, 2) and (0, −2).

31. $\begin{cases} 4x-y^2=0 & (1) \\ 2x^2+y^2=16 & (2) \end{cases}$

Solve E1 for y^2: $y^2=4x$.

Substitute into E2.

$2x^2+4x=16$

$2x^2+4x-16=0$

$x^2+2x-8=0$

$(x+4)(x-2)=0$

$x=-4$ or $x=2$

Use these values in the equation $y^2=4x$.

$x=-4: y^2=4(-4)$

$y^2=-16$ (no real solutions)

$x=2: y^2=4(2)$

$y^2=8$

$y=\pm\sqrt{8}=\pm 2\sqrt{2}$

The solutions are $\left(2, -2\sqrt{2}\right)$ and $\left(2, 2\sqrt{2}\right)$.

32. $\begin{cases} x^2 + 2y = 9 & (1) \\ 5x - 2y = 5 & (2) \end{cases}$

Add E1 and E2.

$$\begin{array}{r} x^2 + 2y = 9 \\ 5x - 2y = 5 \\ \hline x^2 + 5x = 14 \end{array}$$

$x^2 + 5x - 14 = 0$

$(x+7)(x-2) = 0$

$x = -7$ or $x = 2$

Use these values in E1.

$x = -7: (-7)^2 + 2y = 9$

$49 + 2y = 9$

$2y = -40$

$y = -20$

$x = 2: (2)^2 + 2y = 9$

$4 + 2y = 9$

$2y = 5$

$y = \frac{5}{2}$

The solutions are (–7, –20) and $\left(2, \frac{5}{2}\right)$.

33. $\begin{cases} y = 3x^2 + 5x - 4 & (1) \\ y = 3x^2 - x + 2 & (2) \end{cases}$

Substitute.

$3x^2 + 5x - 4 = 3x^2 - x + 2$

$6x = 6$

$x = 1$

Use this value in E1.

$y = 3(1)^2 + 5(1) - 4 = 4$

The solution is (1, 4).

34. $\begin{cases} x^2 - 3y^2 = 1 & (1) \\ 4x^2 + 5y^2 = 21 & (2) \end{cases}$

Multiply E1 by –4 and add to E2.

$$\begin{array}{r} -4x^2 + 12y^2 = -4 \\ 4x^2 + 5y^2 = 21 \\ \hline 17y^2 = 17 \end{array}$$

$y^2 = 1$

$y = \pm 1$

Replace y^2 with 1 in E1.

$x^2 - 3(1) = 1$

$x^2 = 4$

$x = \pm 2$

The solutions are (–2, –1), (–2, 1), (2, –1) and (2, 1).

35. Let x and y be the length and width.

$\begin{cases} xy = 150 \\ 2x + 2y = 50 \end{cases}$

Solve the first equation for y: $y = \frac{150}{x}$.

Substitute into E2.

$2x + 2\left(\frac{150}{x}\right) = 50$

$x + \frac{150}{x} = 25$

$x^2 + 150 = 25x$

$x^2 - 25x + 150 = 0$

$(x-15)(x-10) = 0$

$x = 15$ or $x = 10$

Substitute these values into E1.

$15y = 150$	$10y = 150$
$y = 10$	$y = 15$

The room is 15 feet by 10 feet.

36. Four real solutions are possible.

37. $y \le -x^2 + 3$

Graph $y = -x^2 + 3$ as a solid curve.

Test Point	$y \le -x^2 + 3$; Result
(0, 0)	$0 \le -(0)^2 + 3$; True

Shade the region containing (0, 0).

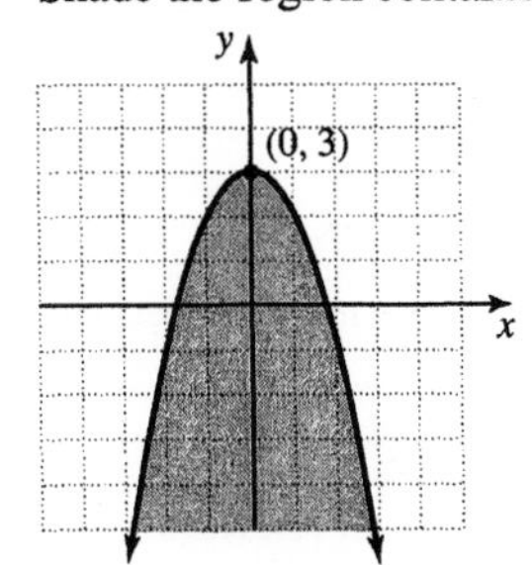

38. $x^2 + y^2 < 9$

First graph the circle as a dashed curve.

Test Point	$x^2 + y^2 < 9$; Result
(0, 0)	$0^2 + 0^2 < 9$; True

Shade the region containing (0, 0).

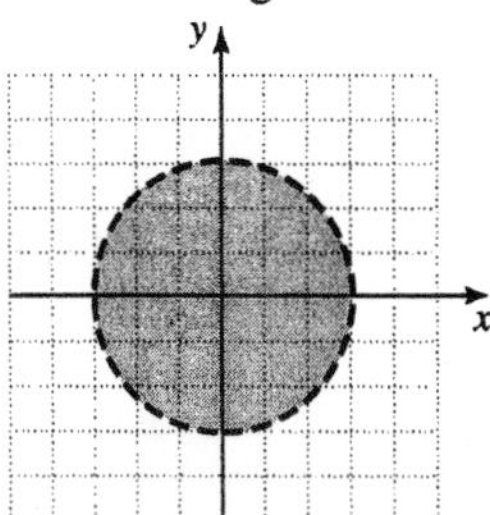

39. $\begin{cases} 2x \le 4 \\ x + y \ge 1 \end{cases}$

First graph $2x = 4$, or $x = 2$, as a solid line, and shade to the left of the line. Next, graph $x + y = 1$ as a solid line.

Test Point	$x + y \ge 1$; Result
(0, 0)	$0 + 0 \ge 1$; False

Shade the region which does not contain (0, 0). The solution to the system is the intersection.

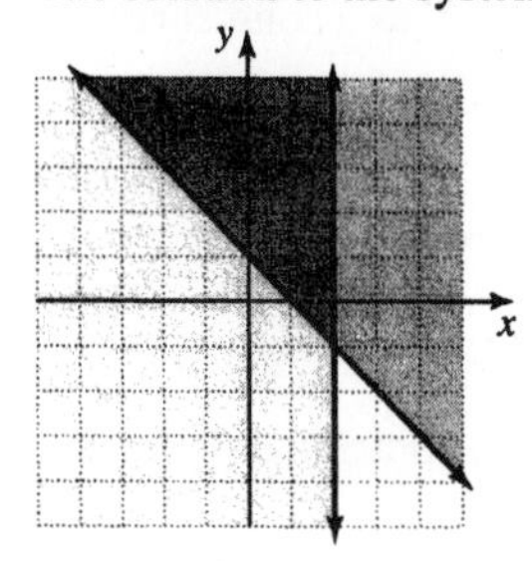

40. $\frac{x^2}{4} + \frac{y^2}{9} \ge 1$

First graph the ellipse as a solid curve.

Test Point	$\frac{x^2}{4} + \frac{y^2}{9} \ge 1$; Result
(0, 0)	$\frac{(0)^2}{4} + \frac{(0)^2}{9} \ge 1$; False

Shade the region that does not contain (0, 0).

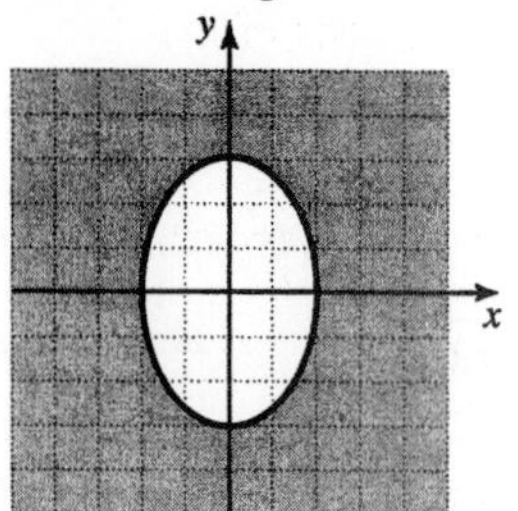

41. $\begin{cases} x^2 + y^2 < 4 \\ x^2 - y^2 \le 1 \end{cases}$

First graph the first circle as a dashed curve.

Test Point	$x^2 + y^2 < 4$; Result
(0, 0)	$0^2 + 0^2 < 4$; True

Shade the region containing (0, 0). Next, graph the hyperbola as a solid curve.

Test Point	$x^2 - y^2 \le 1$; Result
(−2, 0)	$(-2)^2 - 0^2 \le 1$; False
(0, 0)	$0^2 - 0^2 \le 1$; True
(2, 0)	$2^2 - 0^2 \le 1$; False

Shade the region containing (0, 0). The solution to the system is the intersection.

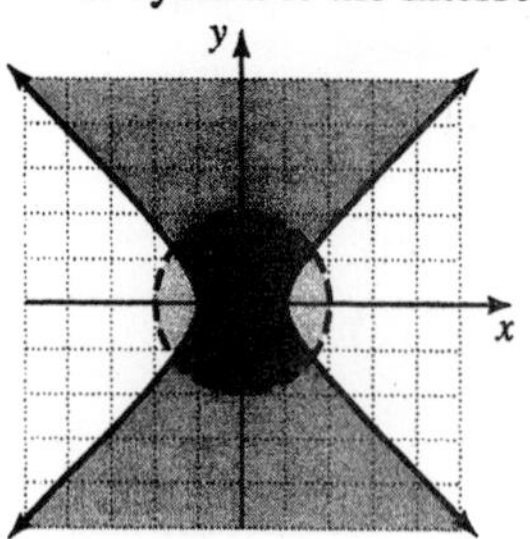

42. $\begin{cases} x^2 + y^2 \le 16 \\ x^2 + y^2 \ge 4 \end{cases}$

First graph the first circle as a solid curve.

Test Point	$x^2 + y^2 \le 16$; Result
(0, 0)	$0^2 + 0^2 \le 16$; True

Shade the region containing (0, 0). Next, graph the second circle as a solid curve.

Test Point	$x^2 + y^2 \geq 4$; Result
(0, 0)	$0^2 + 0^2 \geq 4$; False

Shade the region which does not contain (0, 0). The solution to the system is the intersection.

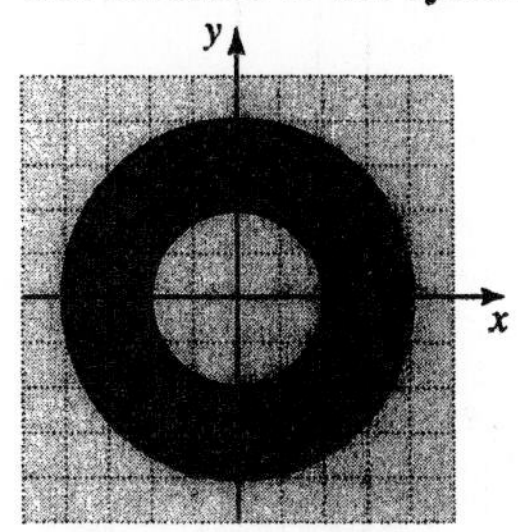

43. center: (−7, 8); radius = 5

$(x-h)^2 + (y-k)^2 = r^2$

$(x+7)^2 + (y-8)^2 = 25$

44. $3x^2 + 6x + 3y^2 = 9$

$x^2 + 2x + y^2 = 3$

$x^2 + 2x + 1 + y^2 = 3 + 1$

$(x+1)^2 + y^2 = 4$

This is a circle with center (−1, 0) and radius 2.

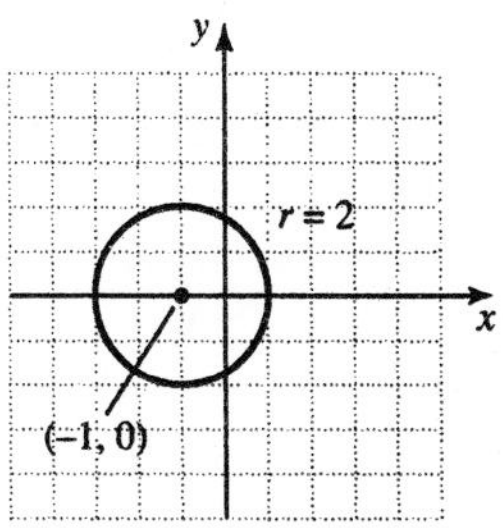

45. $y = x^2 + 6x + 9$

$y = (x+3)^2$

This is a parabola that opens upward with vertex (−3, 0).

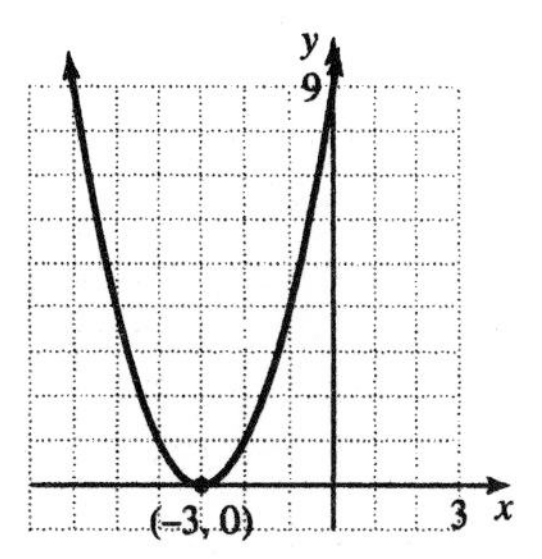

46. $x = y^2 + 6y + 9$

$x = (y+3)^2$

This is a parabola that opens to the right with vertex (0, −3).

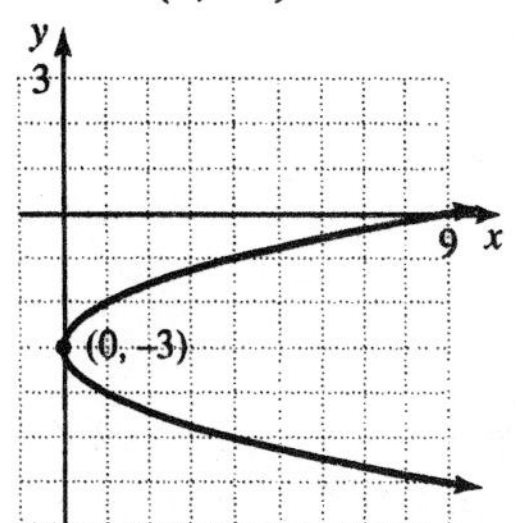

47. $\dfrac{y^2}{4} - \dfrac{x^2}{16} = 1$

This is a hyperbola with center (0, 0), $a = 4$ and $b = 2$.

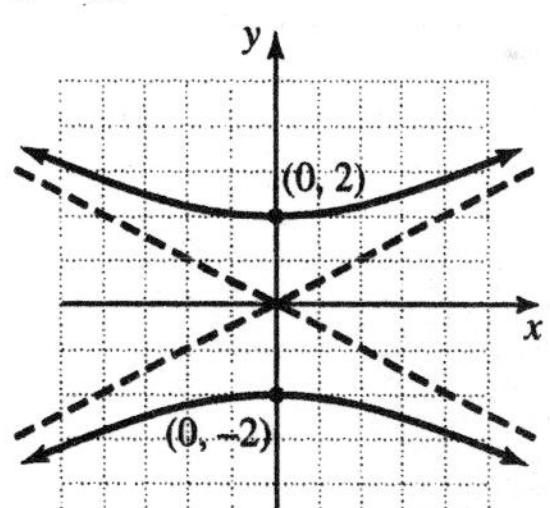

48. $\dfrac{y^2}{4} + \dfrac{x^2}{16} = 1$

This is an ellipse with center (0, 0), $a = 4$ and $b = 2$. The intercepts are (4, 0), (−4, 0), (0, 2), and (0, −2).

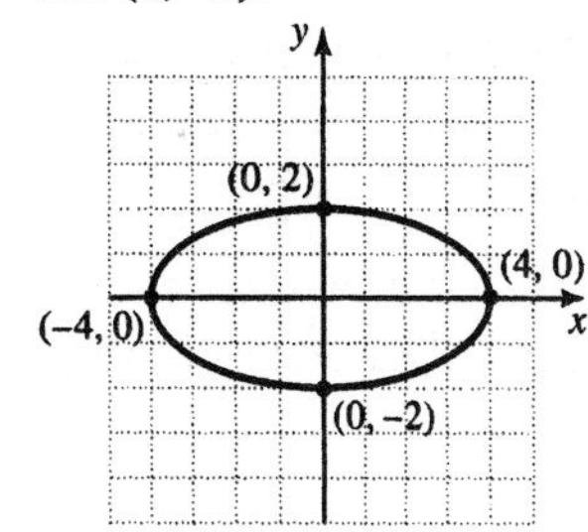

49. $\dfrac{(x-2)^2}{4}+(y-1)^2=1$

This is an ellipse with center (2, 1), $a = 2$ and $b = 1$.

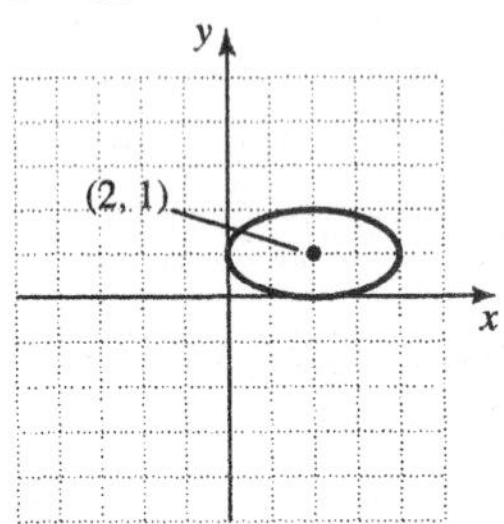

50.
$$y^2 = x^2 + 6$$
$$y^2 - x^2 = 6$$
$$\frac{y^2}{6} - \frac{x^2}{6} = 1$$

This is a hyperbola with center (0, 0), $a = \sqrt{6}$ and $b = \sqrt{6}$.

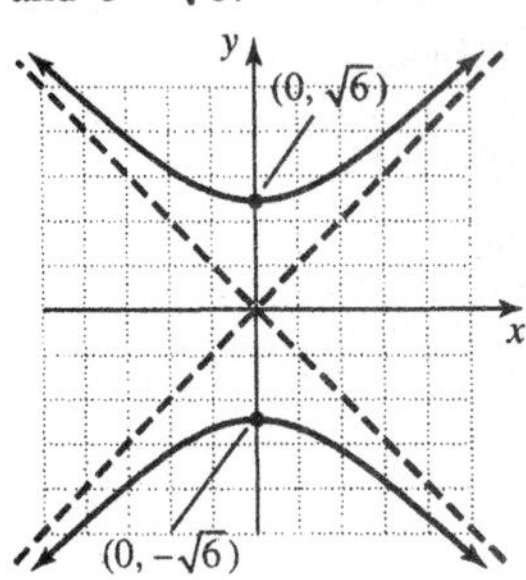

51.
$$y^2 + x^2 = 4x + 6$$
$$y^2 + (x^2 - 4x) = 6$$
$$y^2 + (x^2 - 4x + 4) = 6 + 4$$
$$y^2 + (x-2)^2 = 10$$

This is a circle with center (2, 0) and radius $\sqrt{10}$.

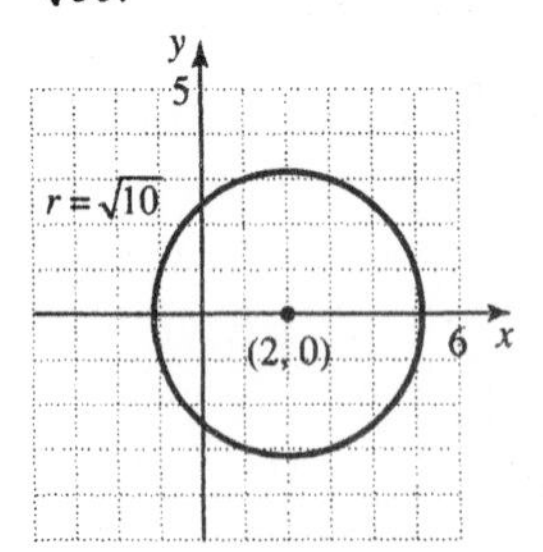

52.
$$x^2 + y^2 - 8y = 0$$
$$x^2 + y^2 - 8y + 16 = 16$$
$$x^2 + (y-4)^2 = 16$$

This is a circle with center (0, 4) and radius $\sqrt{16} = 4$.

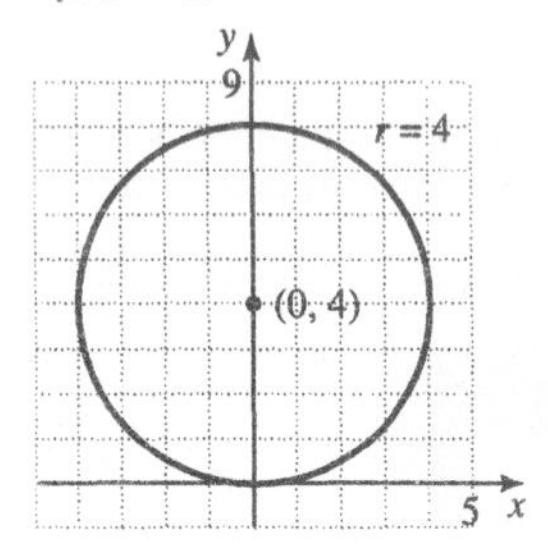

53. $6(x-2)^2 + 9(y+5)^2 = 36$
$$\frac{(x-2)^2}{6} + \frac{(y+5)^2}{4} = 1$$

This is an ellipse with center (2, –5), $a = \sqrt{6}$, and $b = 2$.

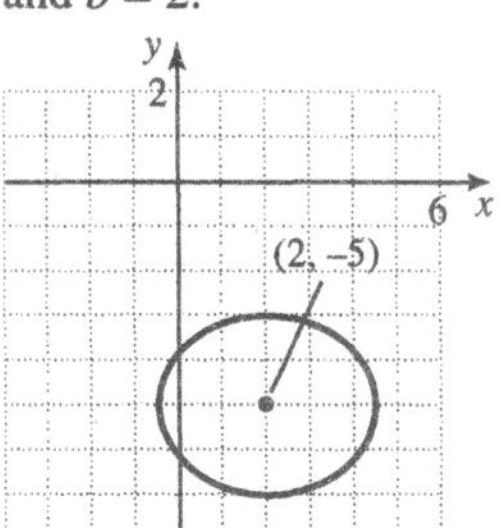

54. $\dfrac{x^2}{16} - \dfrac{y^2}{25} = 1$

This is a hyperbola with center (0, 0), $a = 4$, $b = 5$, and x-intercepts (4, 0) and (–4, 0).

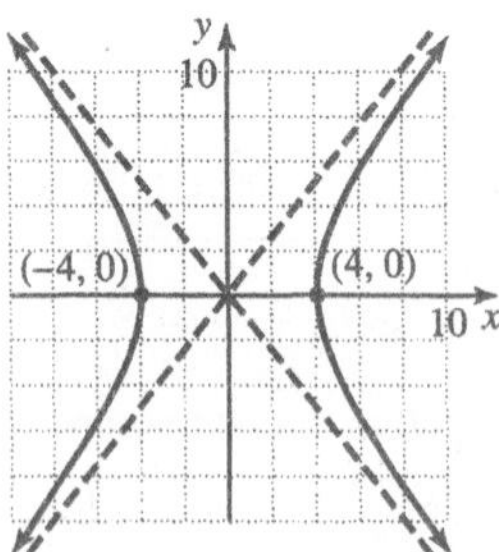

55. $\begin{cases} y = x^2 - 5x + 1 & (1) \\ y = -x + 6 & (2) \end{cases}$

Substitute $-x + 6$ for y in E2.

$$-x+6=x^2-5x+1$$
$$0=x^2-4x-5$$
$$0=(x-5)(x+1)$$
$x=5$ or $x=-1$

Use these values in E2.

$x=5: y=-(5)+6=1$

$x=-1: y=-(-1)+6=7$

The solutions are (5, 1) and (–1, 7).

56. $\begin{cases} x^2+y^2=10 & (1) \\ 9x^2+y^2=18 & (2) \end{cases}$

Multiply E1 by –1 and add to E2.

$$-x^2-y^2=-10$$
$$\underline{9x^2+y^2=18}$$
$$8x^2 \quad = 8$$
$$x^2=1$$
$$x=\pm 1$$

Replace x^2 with 1 in E1.

$$1+y^2=10$$
$$y^2=9$$
$$y=\pm 3$$

The solutions are (–1, –3), (–1, 3), (1, –3) and (1, 3).

57. $x^2-y^2<1$

First graph the hyperbola as dashed curves.

Test Point	$x^2-y^2<1$; Result
(–2, 0)	$(-2)^2-0^2<1$; False
(0, 0)	$0^2-0^2<1$; True
(2, 0)	$2^2-0^2<1$; False

Shade the region containing (0, 0).

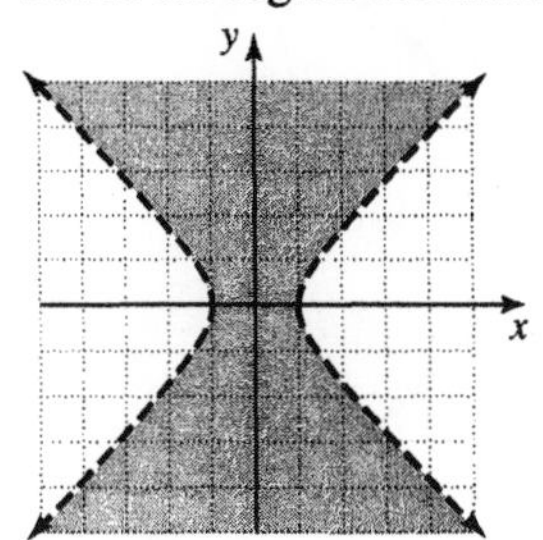

58. $\begin{cases} y>x^2 \\ x+y\geq 3 \end{cases}$

First graph the parabola as a dashed curve.

Test Point	$y>x^2$; Result
(0, 1)	$1>0^2$; True

Shade the region containing (0, 1). Next, graph $x+y=3$ as a solid line.

Test Point	$x+y\geq 3$; Result
(0, 0)	$0+0\geq 3$; False

Shade the region which does not contain (0, 0). The solution to the system is the overlapping region.

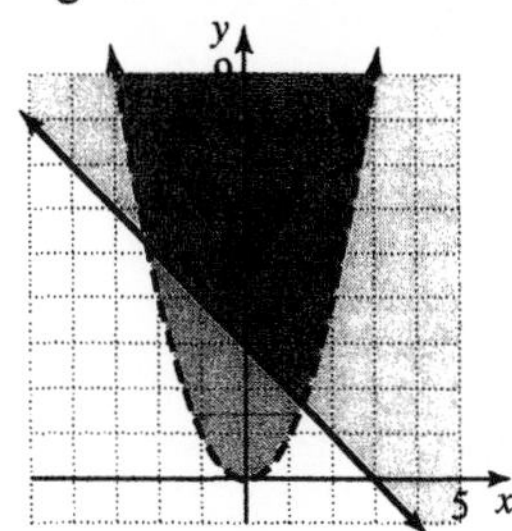

Chapter 13 Test

1. $x^2+y^2=36$

Circle; center: (0, 0), radius $r=\sqrt{36}=6$

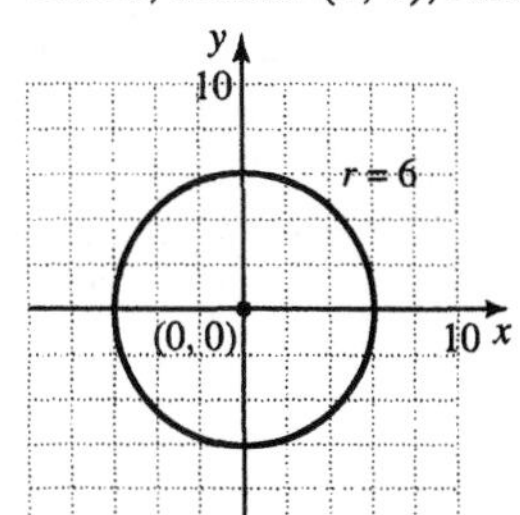

2. $x^2 - y^2 = 36$

$\frac{x^2}{36} - \frac{y^2}{36} = 1$

Hyperbola; center: (0, 0), $a = 6$, $b = 6$

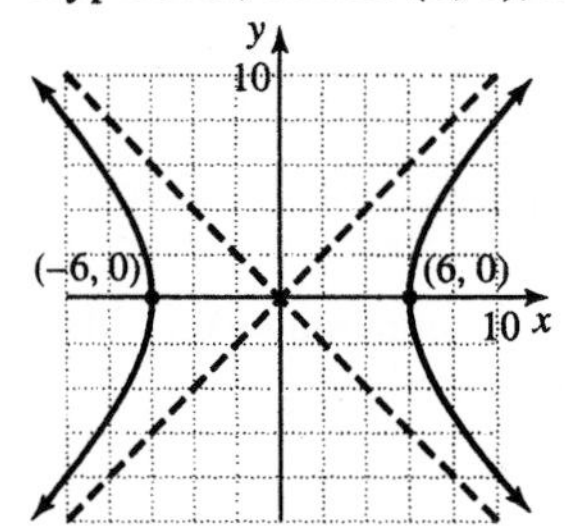

3. $16x^2 + 9y^2 = 144$

$\frac{x^2}{9} + \frac{y^2}{16} = 1$

Ellipse; center: (0, 0), $a = 3$, $b = 4$

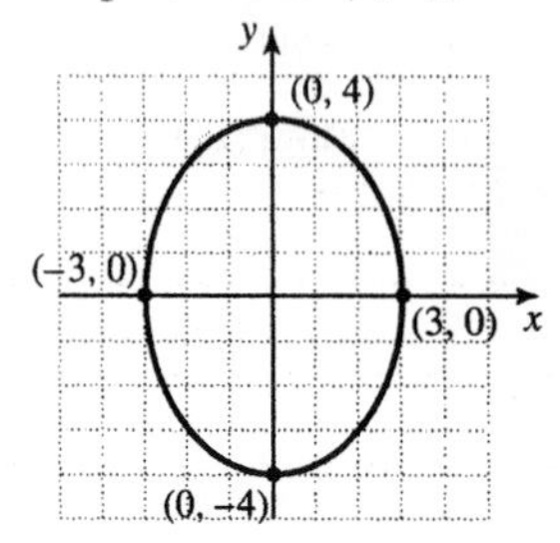

4. $y = x^2 - 8x + 16$

$y = (x-4)^2$

Parabola; vertex: (4, 0)

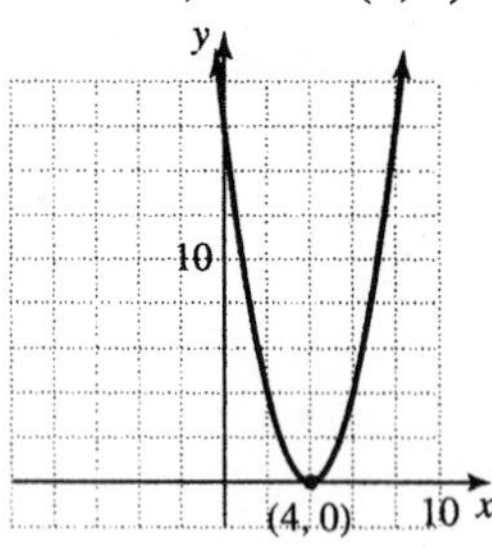

5. $x^2 + y^2 + 6x = 16$

$(x^2 + 6x) + y^2 = 16$

$(x^2 + 6x + 9) + y^2 = 16 + 9$

$(x+3)^2 + y^2 = 25$

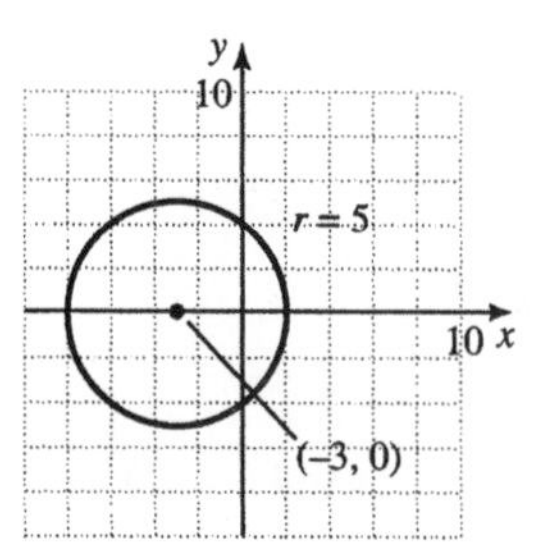

6. $x = y^2 + 8y - 3$

$x + 16 = (y^2 + 8y + 16) - 3$

$x = (y+4)^2 - 19$

Parabola; vertex: (−4, −19)

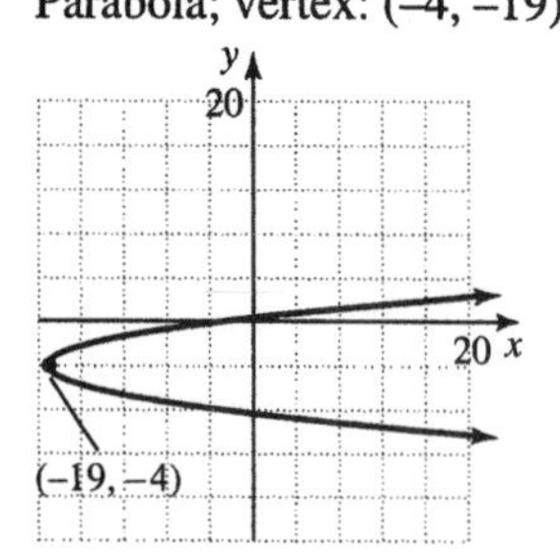

7. $\frac{(x-4)^2}{16} + \frac{(y-3)^2}{9} = 1$

Ellipse: center: (4, 3), $a = 4$, $b = 3$

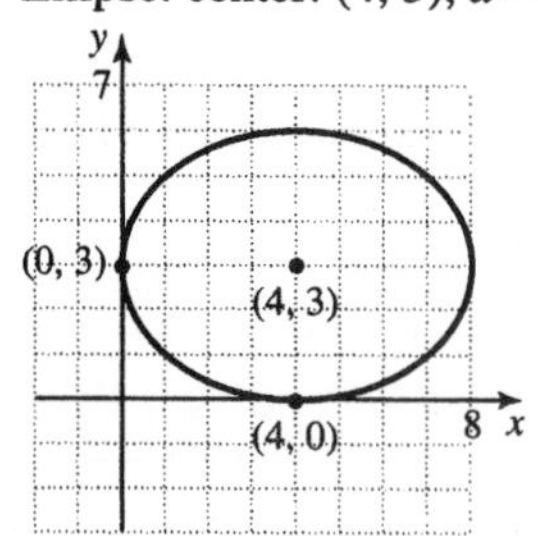

8. $y^2 - x^2 = 1$

Hyperbola: center: (0, 0), $a = 1$, $b = 1$

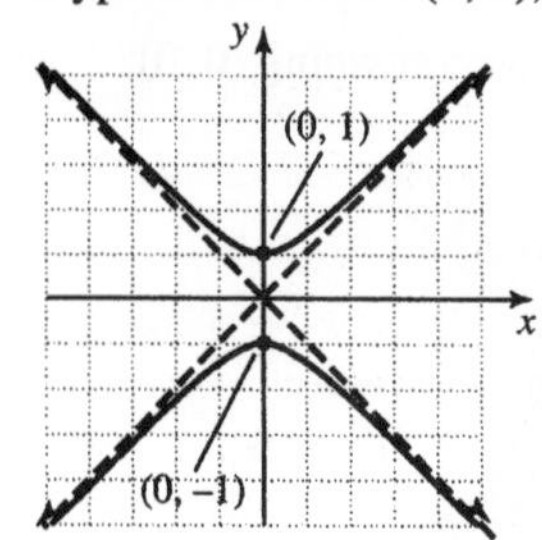

9. $\begin{cases} x^2 + y^2 = 26 & (1) \\ x^2 - 2y^2 = 23 & (2) \end{cases}$

Solve E1 for x^2: $x^2 = 26 - y^2$.
Substitute into E2.

$$(26 - y^2) - 2y^2 = 23$$
$$-3y^2 = -3$$
$$y^2 = 1$$
$$y = \pm 1$$

Replace y^2 with 1 in E1.

$$x^2 + 1 = 26$$
$$x^2 = 25$$
$$x = \pm 5$$

The solutions are (–5, –1), (–5, 1), (5, –1), and (5, 1).

10. $\begin{cases} y = x^2 - 5x + 6 & (1) \\ y = 2x & (2) \end{cases}$

Substitute $2x$ for y in E1.

$$2x = x^2 - 5x + 6$$
$$0 = x^2 - 7x + 6$$
$$0 = (x-6)(x-1)$$

$x = 6$ or $x = 1$
Use these values in E2.
$x = 6: y = 2(6) = 12$
$x = 1: y = 2(1) = 2$
The solutions are (1, 2) and (6, 12).

11. $\begin{cases} 2x + 5y \geq 10 \\ y \geq x^2 + 1 \end{cases}$

First graph $2x + 5y = 10$ as a solid line.

Test Point	$2x + 5y \geq 10$; Result
(0, 0)	$2(0) + 5(0) \geq 10$; False

Shade the region which does not contain (0, 0). Next, graph $y = x^2 + 1$ as a solid curve.

Test Point	$y \geq x^2 + 1$; Result
(0, 0)	$0 \geq 0^2 + 1$; False

Shade the region which does not contain (0, 0). The solution to the system is the intersection.

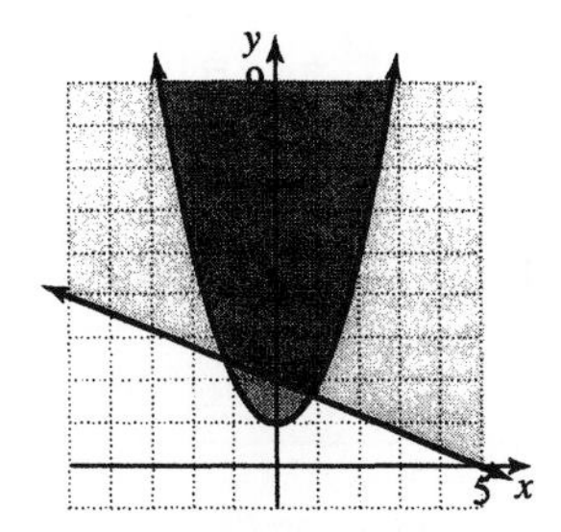

12. $\begin{cases} \frac{x^2}{4} + y^2 \leq 1 \\ x + y > 1 \end{cases}$

First graph the ellipse as a solid curve.

Test Point	$\frac{x^2}{4} + y^2 \leq 1$; Result
(0, 0)	$\frac{0^2}{4} + 0^2 \leq 1$; True

Shade the region containing (0, 0). Next, graph $x + y = 1$ as a dashed line.

Test Point	$x + y > 1$; Result
(0, 0)	$0 + 0 > 1$; False

Shade the region which does not contain (0, 0). The solution to the system is the intersection.

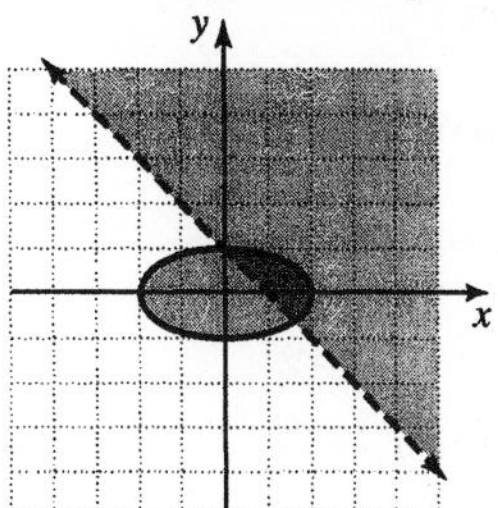

13. $\begin{cases} x^2 + y^2 \geq 4 \\ x^2 + y^2 < 16 \\ y \geq 0 \end{cases}$

First graph the circle $x^2 + y^2 = 4$ as a solid curve.

Test Point	$x^2 + y^2 \geq 4$; Result
(0, 0)	$0^2 + 0^2 \geq 4$; False

Shade the region which does not contain (0, 0).

Next graph the circle $x^2 + y^2 = 16$ as a dashed curve.

Test Point	$x^2 + y^2 < 16$; Result
(0, 0)	$0^2 + 0^2 < 16$; True

Shade the region containing (0, 0). Now graph the inequality $y \geq 0$ by shading the region above the x-axis. The solution to the system is the intersection.

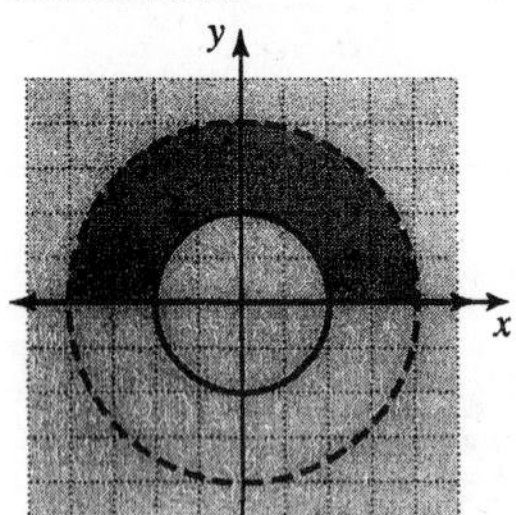

14. $100x^2 + 225y^2 = 22,500$

$$\frac{x^2}{225} + \frac{y^2}{100} = 1$$
$$a = \sqrt{225} = 15$$
$$b = \sqrt{100} = 10$$

Width = 15 + 15 = 30 feet
Height = 10 feet

Chapter 13 Cumulative Review

1. $2x \geq 0$ and $4x - 1 \leq -9$
$x \geq 0$ and $4x \leq -8$
$x \geq 0$ and $x \leq -2$
There is no solution, or ∅.

2. $3x + 4 > 1$ and $2x - 5 \leq 9$
$3x > -3$ and $2x \leq 14$
$x > -1$ and $x \leq 7$
$-1 < x \leq 7$
$(-1, 7]$

3. $5x - 3 \leq 10$ or $x + 1 \geq 5$
$5x \leq 13$ or $x \geq 4$
$x \leq \frac{13}{5}$ or $x \geq 4$

The solution set is $\left(-\infty, \frac{13}{5}\right] \cup [4, \infty)$.

4. (3, 2), (1, –4)

$$m = \frac{-4-2}{1-3} = \frac{-6}{-2} = 3$$

5. $|5w + 3| = 7$
$5w + 3 = 7$ or $5w + 3 = -7$
$5w = 4$ or $5w = -10$
$w = \frac{4}{5}$ or $w = -2$

The solutions are –2 and $\frac{4}{5}$.

6. Let x = speed of one plane. Then $x + 25$ = speed of the other plane.

$$d_{\text{plane 1}} + d_{\text{plane 2}} = 650 \text{ miles}$$
$$2x + 2(x + 25) = 650$$
$$2x + 2x + 50 = 650$$
$$4x = 600$$
$$x = 150$$
$$x + 25 = 150 + 25 = 175$$

The planes are traveling at 150 mph and 175 mph.

7. $\left|\frac{x}{2} - 1\right| = 11$

$\frac{x}{2} - 1 = 11$ or $\frac{x}{2} - 1 = -11$
$\frac{x}{2} = 12$ or $\frac{x}{2} = -10$
$x = 24$ or $x = -20$

The solutions are –20 and 24.

8. a. $\frac{4^8}{4^3} = 4^{8-3} = 4^5$

b. $\frac{y^{11}}{y^5} = y^{11-5} = y^6$

c. $\frac{32x^7}{4x^6} = \frac{32}{4}x^{7-6} = 8x$

d. $\frac{18a^{12}b^6}{12a^8b^6} = \frac{18}{12}a^{12-8}b^{6-6} = \frac{3}{2}a^4b^0 = \frac{3a^4}{2}$

9. $|3x+2| = |5x-8|$

$$\begin{aligned} 3x+2 &= 5x-8 &\text{ or }&& 3x+2 &= -(5x-8) \\ 2 &= 2x-8 &\text{ or }&& 3x+2 &= -5x+8 \\ 10 &= 2x &\text{ or }&& 8x+2 &= 8 \\ 5 &= x &\text{ or }&& 8x &= 6 \\ &&&& x &= \frac{3}{4} \end{aligned}$$

The solutions are $\frac{3}{4}$ and 5.

10. a. $3y^2+14y+15 = (3y+5)(y+3)$

b. $20a^5+54a^4+10a^3$
$= 2a^3(10a^2+27a+5)$
$= 2a^3(2a+5)(5a+1)$

c. $(y-3)^2-2(y-3)-8$

Let $u = y-3$. Then $u^2 = (y-3)^2$ and

$$\begin{aligned} u^2-2u-8 &= (u-4)(u+2) \\ &= [(y-3)-4][(y-3)+2] \\ &= (y-7)(y-1) \end{aligned}$$

11. $|m-6| < 2$
$-2 < m-6 < 2$
$4 < m < 8$
The solution set is (4, 8).

12.
$$\begin{aligned} &\frac{2}{3a-15}-\frac{a}{25-a^2} \\ &= \frac{2}{3(a-5)}+\frac{a}{a^2-25} \\ &= \frac{2}{3(a-5)}+\frac{a}{(a+5)(a-5)} \\ &= \frac{2(a+5)+3a}{3(a+5)(a-5)} \\ &= \frac{2a+10+3a}{3(a+5)(a-5)} \\ &= \frac{5a+10}{3(a+5)(a-5)} \end{aligned}$$

13.
$$\begin{aligned} \frac{x^{-1}+2xy^{-1}}{x^{-2}-x^{-2}y^{-1}} &= \frac{\frac{1}{x}+\frac{2x}{y}}{\frac{1}{x^2}-\frac{1}{x^2y}} \\ &= \frac{x^2y\left(\frac{1}{x}+\frac{2x}{y}\right)}{x^2y\left(\frac{1}{x^2}-\frac{1}{x^2y}\right)} \\ &= \frac{xy+2x^3}{y-1} \end{aligned}$$

14. a.
$$\begin{aligned} (a^{-1}-b^{-1})^{-1} &= \left(\frac{1}{a}-\frac{1}{b}\right)^{-1} \\ &= \left(\frac{b-a}{ab}\right)^{-1} \\ &= \frac{ab}{b-a} \end{aligned}$$

b.
$$\begin{aligned} \frac{2-\frac{1}{x}}{4x-\frac{1}{x}} &= \frac{\left(2-\frac{1}{x}\right)x}{\left(4x-\frac{1}{x}\right)x} \\ &= \frac{2x-1}{4x^2-1} \\ &= \frac{2x-1}{(2x+1)(2x-1)} \\ &= \frac{1}{2x+1} \end{aligned}$$

15. $|2x+9|+5 > 3$
$|2x+9| > -2$

The absolute value is never negative, so all real numbers are solutions. The solution set is $(-\infty, \infty)$.

16.
$$\begin{aligned} \frac{2}{x+3} &= \frac{1}{x^2-9}-\frac{1}{x-3} \\ \frac{2}{x+3} &= \frac{1}{(x+3)(x-3)}-\frac{1}{x-3} \\ 2(x-3) &= 1-1(x+3) \\ 2x-6 &= 1-x-3 \\ 2x-6 &= -x-2 \\ 3x &= 4 \\ x &= \frac{4}{3} \end{aligned}$$

17.
$$\begin{array}{r|rrrrrrr} 4 & 4 & -25 & 35 & 0 & 17 & 0 & 0 \\ & & 16 & -36 & -4 & -16 & 4 & 16 \\ \hline & 4 & -9 & -1 & -4 & 1 & 4 & 16 \end{array}$$

Thus, $P(4) = 16$.

18. $y = \frac{k}{x}$

$3 = \frac{k}{\frac{2}{3}}$

$k = 3\left(\frac{2}{3}\right) = 2$

Thus, the equation is $y = \frac{2}{x}$.

19. a. $\sqrt[3]{1} = 1$

b. $\sqrt[3]{-64} = -4$

c. $\sqrt[3]{\frac{8}{125}} = \frac{\sqrt[3]{8}}{\sqrt[3]{125}} = \frac{2}{5}$

d. $\sqrt[3]{x^6} = x^2$

e. $\sqrt[3]{-27x^9} = -3x^3$

20. a. $\sqrt{5}\left(2+\sqrt{15}\right) = 2\sqrt{5} + \sqrt{5}\cdot\sqrt{15}$
$= 2\sqrt{5} + \sqrt{75}$
$= 2\sqrt{5} + 5\sqrt{3}$

b. $\left(\sqrt{3}-\sqrt{5}\right)\left(\sqrt{7}-1\right)$
$= \sqrt{3}\cdot\sqrt{7} - \sqrt{3}\cdot 1 - \sqrt{5}\cdot\sqrt{7} + \sqrt{5}\cdot 1$
$= \sqrt{21} - \sqrt{3} - \sqrt{35} + \sqrt{5}$

c. $\left(2\sqrt{5}-1\right)^2 = \left(2\sqrt{5}\right)^2 - 2\cdot 2\sqrt{5}\cdot 1 + 1^2$
$= 4(5) - 4\sqrt{5} + 1$
$= 21 - 4\sqrt{5}$

d. $\left(3\sqrt{2}+5\right)\left(3\sqrt{2}-5\right) = \left(3\sqrt{2}\right)^2 - 5^2$
$= 9(2) - 25$
$= 18 - 25$
$= -7$

21. a. $z^{2/3}(z^{1/3} - z^5) = z^{2/3+1/3} - z^{2/3+5}$
$= z^{3/3} - z^{2/3+15/3}$
$= z - z^{17/3}$

b. $(x^{1/3} - 5)(x^{1/3} + 2)$
$= x^{1/3}\cdot x^{1/3} + 2x^{1/3} - 5x^{1/3} - 5(2)$
$= x^{2/3} - 3x^{1/3} - 10$

22. $\frac{-2}{\sqrt{3}+3} = \frac{-2\left(\sqrt{3}-3\right)}{\left(\sqrt{3}+3\right)\left(\sqrt{3}-3\right)}$
$= \frac{-2\left(\sqrt{3}-3\right)}{\left(\sqrt{3}\right)^2 - 3^2}$
$= \frac{-2\left(\sqrt{3}-3\right)}{3-9}$
$= \frac{-2\left(\sqrt{3}-3\right)}{-6}$
$= \frac{\sqrt{3}-3}{3}$

23. a. $\frac{\sqrt{20}}{\sqrt{5}} = \sqrt{\frac{20}{5}} = \sqrt{4} = 2$

b. $\frac{\sqrt{50x}}{2\sqrt{2}} = \frac{1}{2}\sqrt{\frac{50x}{2}} = \frac{1}{2}\sqrt{25x} = \frac{5\sqrt{x}}{2}$

c. $\frac{7\sqrt[3]{48x^4y^8}}{\sqrt[3]{6y^2}} = 7\sqrt[3]{\frac{48x^4y^8}{6y^2}}$
$= 7\sqrt[3]{8x^4y^6}$
$= 7\sqrt[3]{8x^3y^6\cdot x}$
$= 7\cdot 2xy^2\sqrt[3]{x}$
$= 14xy^2\sqrt[3]{x}$

d. $\frac{2\sqrt[4]{32a^8b^6}}{\sqrt[4]{a^{-1}b^2}} = 2\sqrt[4]{\frac{32a^8b^6}{a^{-1}b^2}}$
$= 2\sqrt[4]{32a^9b^4}$
$= 2\sqrt[4]{16a^8b^4\cdot 2a}$
$= 2\cdot 2a^2b\sqrt[4]{2a}$
$= 4a^2b\sqrt[4]{2a}$

24. $\sqrt{2x-3} = x-3$

$$\left(\sqrt{2x-3}\right)^2 = (x-3)^2$$
$$2x-3 = x^2-6x+9$$
$$0 = x^2-8x+12$$
$$0 = (x-6)(x-2)$$
$$x-6=0 \text{ or } x-2=0$$
$$x=6 \text{ or } x=2$$

Discard 2 as an extraneous solution. The solution is 6.

25. a. $\frac{\sqrt{45}}{4} - \frac{\sqrt{5}}{3} = \frac{3\sqrt{5}}{4} - \frac{\sqrt{5}}{3}$

$$= \frac{9\sqrt{5}-4\sqrt{5}}{12}$$
$$= \frac{5\sqrt{5}}{12}$$

b. $\sqrt[3]{\frac{7x}{8}} + 2\sqrt[3]{7x} = \frac{\sqrt[3]{7x}}{2} + 2\sqrt[3]{7x}$

$$= \frac{\sqrt[3]{7x}}{2} + \frac{4\sqrt[3]{7x}}{2}$$
$$= \frac{5\sqrt[3]{7x}}{2}$$

26. $9x^2-6x=-4$

$$9x^2-6x+4=0$$
$$a=9, b=-6, c=4$$
$$b^2-4ac = (-6)^2-4(9)(4)$$
$$= 36-144$$
$$= -108$$

Two complex but not real solutions

27. $\sqrt{\frac{7x}{3y}} = \frac{\sqrt{7x}}{\sqrt{3y}} = \frac{\sqrt{7x}\cdot\sqrt{3y}}{\sqrt{3y}\cdot\sqrt{3y}} = \frac{\sqrt{21xy}}{3y}$

28. $\frac{4}{x-2} - \frac{x}{x+2} = \frac{16}{x^2-4}$

$$\frac{4}{x-2} - \frac{x}{x+2} = \frac{16}{(x+2)(x-2)}$$
$$4(x+2)-x(x-2) = 16$$
$$4x+8-x^2+2x = 16$$
$$0 = x^2-6x+8$$
$$0 = (x-4)(x-2)$$
$$x-4=0 \text{ or } x-2=0$$
$$x=4 \text{ or } x=2$$

Discard the solution 2 as extraneous. The solution is 4.

29. $\sqrt{2x-3} = 9$

$$\left(\sqrt{2x-3}\right)^2 = 9^2$$
$$2x-3 = 81$$
$$2x = 84$$
$$x = 42$$

The solution is 42.

30. $x^3+2x^2-4x \ge 8$

$$x^3+2x^2-4x-8 \ge 0$$
$$x^2(x+2)-4(x+2) \ge 0$$
$$(x+2)(x^2-4) \ge 0$$
$$(x+2)(x+2)(x-2) \ge 0$$
$$(x+2)^2(x-2) \ge 0$$
$$(x+2)^2 = 0 \text{ or } x-2=0$$
$$x+2=0 \text{ or } x=2$$
$$x=-2$$

Region	Test Point	$(x+2)^2(x-2) \ge 0$ Result
A: $(-\infty, -2)$	-3	$(-1)^2(-5) \ge 0$ False
B: $(-2, 2)$	0	$(2)^2(-2) \ge 0$ False
C: $(2, \infty)$	3	$(5)^2(1) \ge 0$ True

Solution: $[2, \infty)$

31. a. $i^7 = i^4 \cdot i^3 = 1\cdot(-i) = -i$

b. $i^{20} = (i^4)^5 = 1^5 = 1$

c. $i^{46} = i^{44}\cdot i^2 = (i^4)^{11}\cdot(-1) = 1^{11}(-1) = -1$

d. $i^{-12} = \frac{1}{i^{12}} = \frac{1}{(i^4)^3} = \frac{1}{1^3} = 1$

32. 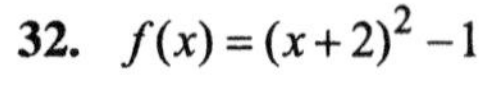
$f(x)=(x+2)^2-1$

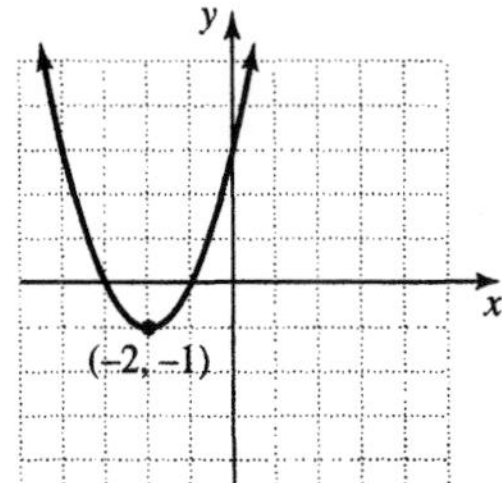

33.
$$p^2+2p=4$$
$$p^2+2p+\left(\frac{2}{2}\right)^2=4+1$$
$$p^2+2p+1=5$$
$$(p+1)^2=5$$
$$p+1=\pm\sqrt{5}$$
$$p=-1\pm\sqrt{5}$$

The solutions are $-1+\sqrt{5}$ and $-1-\sqrt{5}$.

34. $f(x)=-x^2-6x+4$

The maximum will occur at the vertex.
$$x=\frac{-b}{2a}=\frac{-(-6)}{2(-1)}=-3$$
$$f(-3)=-(-3)^2-6(-3)+4=13$$
The maximum value is 13.

35.
$$\frac{1}{4}m^2-m+\frac{1}{2}=0$$
$$4\left(\frac{1}{4}m^2-m+\frac{1}{2}\right)=4(0)$$
$$m^2-4m+2=0$$
$a=1, b=-4, c=2$
$$m=\frac{-(-4)\pm\sqrt{(-4)^2-4(1)(2)}}{2(1)}$$
$$=\frac{4\pm\sqrt{16-8}}{2}$$
$$=\frac{4\pm\sqrt{8}}{2}$$
$$=\frac{4\pm2\sqrt{2}}{2}$$
$$=2\pm\sqrt{2}$$

The solutions are $2+\sqrt{2}$ and $2-\sqrt{2}$.

36.
$$f(x)=\frac{x+1}{2}$$
$$y=\frac{x+1}{2}$$
$$x=\frac{y+1}{2}$$
$$2x=y+1$$
$$2x-1=y$$
$$f^{-1}(x)=2x-1$$

37.
$$p^4-3p^2-4=0$$
$$(p^2-4)(p^2+1)=0$$
$$(p+2)(p-2)(p^2+1)=0$$
$p+2=0$ or $p-2=0$ or $p^2+1=0$

$p=-2$ or $p=2$ or $p^2=-1$

$p=\pm i$

The solutions are −2, 2, −*i*, and *i*.

38. a.
$$\frac{\sqrt{32}}{\sqrt{4}}=\sqrt{\frac{32}{4}}=\sqrt{8}=\sqrt{4\cdot 2}=2\sqrt{2}$$

b.
$$\frac{\sqrt[3]{240y^2}}{5\sqrt[3]{3y^{-4}}}=\frac{1}{5}\sqrt[3]{\frac{240y^2}{3y^{-4}}}$$
$$=\frac{1}{5}\sqrt[3]{80y^6}$$
$$=\frac{1}{5}\sqrt[3]{8y^6\cdot 10}$$
$$=\frac{2y^3\sqrt[3]{10}}{5}$$

c.
$$\frac{\sqrt[5]{64x^9y^2}}{\sqrt[5]{2x^2y^{-8}}}=\sqrt[5]{\frac{64x^9y^2}{2x^2y^{-8}}}$$
$$=\sqrt[5]{32x^7y^{10}}$$
$$=\sqrt[5]{32x^5y^{10}\cdot x^2}$$
$$=2xy^2\sqrt[5]{x^2}$$

39.
$$\frac{x+2}{x-3}\le 0$$
$x+2=0$ or $x-3=0$

$x=-2$ or $x=3$

Region	Test Point	$\frac{x+2}{x-3} \le 0$ Result
A: $(-\infty, -2)$	-3	$\frac{-1}{-6} \le 0$; False
B: $(-2, 3)$	0	$\frac{2}{-3} \le 0$; True
C: $(3, \infty)$	4	$\frac{6}{1} \le 0$; False

Solution: $[-2, 3)$

40. $4x^2 + 9y^2 = 36$

$\frac{x^2}{9} + \frac{y^2}{4} = 1$

Ellipse: center (0, 0), $a = 3$, $b = 2$

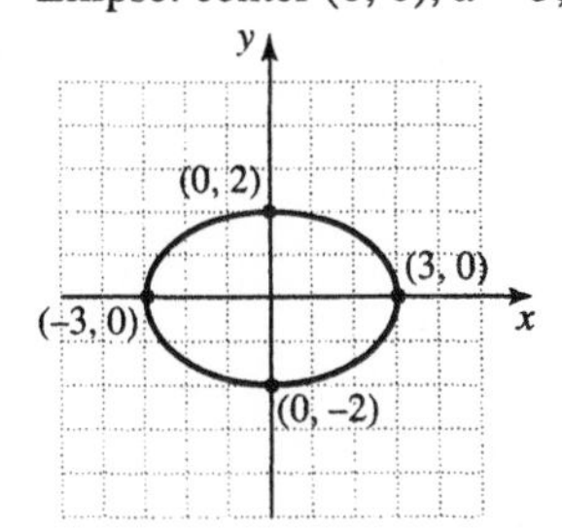

41. $g(x) = \frac{1}{2}(x+2)^2 + 5$

Vertex: $(-2, 5)$, axis: $x = -2$

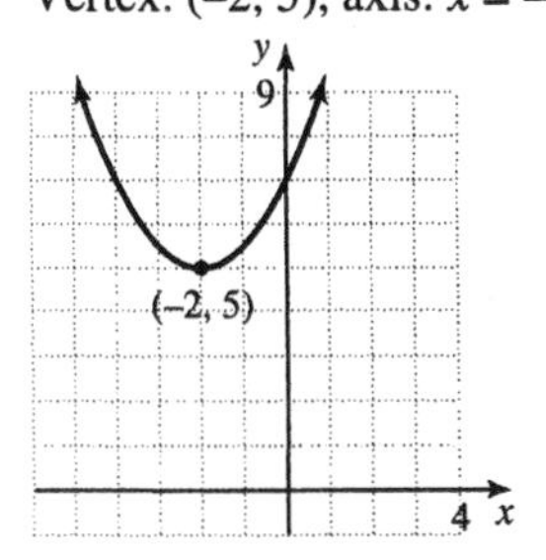

42. a.
$$64^x = 4$$
$$(4^2)^x = 4$$
$$4^{2x} = 4$$
$$2x = 1$$
$$x = \frac{1}{2}$$

b.
$$125^{x-3} = 25$$
$$(5^3)^{x-3} = 5^2$$
$$5^{3x-9} = 5^2$$
$$3x - 9 = 2$$
$$3x = 11$$
$$x = \frac{11}{3}$$

c.
$$\frac{1}{81} = 3^{2x}$$
$$3^{-4} = 3^{2x}$$
$$-4 = 2x$$
$$-\frac{4}{2} = x$$
$$-2 = x$$

43. $f(x) = x^2 - 4x - 12$

$x = \frac{-b}{2a} = \frac{-(-4)}{2(1)} = 2$

$f(2) = (2)^2 - 4(2) - 12 = -16$

Vertex: $(2, -16)$

44. $\begin{cases} x + 2y < 8 \\ y \ge x^2 \end{cases}$

First, graph $x + 2y = 8$ as a dashed line.

Test Point	$x + 2y < 8$; Result
(0, 0)	$0 + 2(0) < 8$; True

Shade the region containing (0, 0). Next, graph the parabola $y = x^2$ as a solid curve.

Test Point	$y \ge x^2$; Result
(0, 1)	$1 \ge 0^2$; True

Shade the region containing (0, 1). The solution to the system is the intersection.

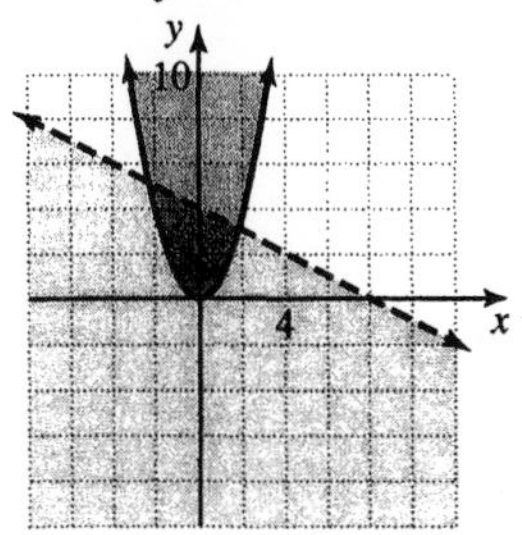

45. (2, –5), (1, –4)

$$d = \sqrt{[-4-(-5)]^2 + (1-2)^2}$$
$$= \sqrt{1^2 + (-1)^2}$$
$$= \sqrt{2} \approx 1.414$$

46. $\begin{cases} x^2 + y^2 = 36 & (1) \\ y = x + 6 & (2) \end{cases}$

Substitute $x+6$ for y in E1.

$$x^2 + (x+6)^2 = 36$$
$$x^2 + (x^2 + 12x + 36) = 36$$
$$2x^2 + 12x = 0$$
$$2x(x+6) = 0$$

$2x = 0$ or $x + 6 = 0$

$x = 0$ or $x = -6$

Use these values in E2 to find y.

$x = 0: y = 0 + 6 = 6$

$x = -6: y = -6 + 6 = 0$

The solutions are (0, 6) and (–6, 0).

Chapter 14

Section 14.1

Practice Exercises

1. $a_n = 5 + n^2$

   $a_1 = 5 + 1^2 = 5 + 1 = 6$

   $a_2 = 5 + 2^2 = 5 + 4 = 9$

   $a_3 = 5 + 3^2 = 5 + 9 = 14$

   $a_4 = 5 + 4^2 = 5 + 16 = 21$

   $a_5 = 5 + 5^2 = 5 + 25 = 30$

   Thus, the first five terms of the sequence are 6, 9, 14, 21, and 30.

2. $a_n = \dfrac{(-1)^n}{5n}$

   a. $a_1 = \dfrac{(-1)^1}{5(1)} = -\dfrac{1}{5}$

   b. $a_4 = \dfrac{(-1)^4}{5(4)} = \dfrac{1}{20}$

   c. $a_{30} = \dfrac{(-1)^{30}}{5(30)} = \dfrac{1}{150}$

   d. $a_{19} = \dfrac{(-1)^{19}}{5(19)} = -\dfrac{1}{95}$

3. a. 1, 3, 5, 7, ...
   These numbers are the first four odd natural numbers, so a general term might be $a_n = (2n-1)$.

   b. 3, 9, 27, 81, ...
   These numbers are all powers of 3 ($3 = 3^1$, $9 = 3^2$, $27 = 3^3$, and $81 = 3^4$), so a general term might be $a_n = 3^n$.

   c. $\dfrac{1}{2}, \dfrac{2}{3}, \dfrac{3}{4}, \dfrac{4}{5}, \ldots$
   The numerators are the first four natural numbers and each denominator is one greater than the numerator, so a general term might be $a_n = \dfrac{n}{n+1}$.

   d. $-\dfrac{1}{2}, -\dfrac{1}{3}, -\dfrac{1}{4}, -\dfrac{1}{5}, \ldots$
   The denominators are consecutive natural numbers beginning with 2 and each term is negative, so a general term might be

   $a_n = -\dfrac{1}{n+1}$.

4. $v_n = 3950(0.8)^n$

   $v_3 = 3950(0.8)^3$
   $= 3950(0.512)$
   $= 2022.4$

   The value of the copier after three years is \$2022.40.

Vocabulary and Readiness Check

1. The nth term of the sequence a_n is called the general term.

2. A finite sequence is a function whose domain is $\{1, 2, 3, 4, \ldots, n\}$ where n is some natural number.

3. An infinite sequence is a function whose domain is $\{1, 2, 3, 4, \ldots\}$.

4. $a_n = 7^n$

   $a_1 = 7^1 = 7$

5. $a_n = \dfrac{(-1)^n}{n}$

   $a_1 = \dfrac{(-1)^1}{1} = -1$

6. $a_n = (-1)^n \cdot n^4$

   $a_1 = (-1)^1 \cdot 1^4 = -1$

Exercise Set 14.1

1. $a_n = n + 4$

   $a_1 = 1 + 4 = 5$
   $a_2 = 2 + 4 = 6$
   $a_3 = 3 + 4 = 7$
   $a_4 = 4 + 4 = 8$
   $a_5 = 5 + 4 = 9$

   Thus, the first five terms of the sequence $a_n = n + 4$ are 5, 6, 7, 8, 9.

3. $a_n = (-1)^n$

$a_1 = (-1)^1 = -1$
$a_2 = (-1)^2 = 1$
$a_3 = (-1)^3 = -1$
$a_4 = (-1)^4 = 1$
$a_5 = (-1)^5 = -1$

Thus, the first five terms of the sequence $a_n = (-1)^n$ are $-1, 1, -1, 1, -1$.

5. $a_n = \dfrac{1}{n+3}$

$a_1 = \dfrac{1}{1+3} = \dfrac{1}{4}$
$a_2 = \dfrac{1}{2+3} = \dfrac{1}{5}$
$a_3 = \dfrac{1}{3+3} = \dfrac{1}{6}$
$a_4 = \dfrac{1}{4+3} = \dfrac{1}{7}$
$a_5 = \dfrac{1}{5+3} = \dfrac{1}{8}$

Thus, the first five terms of the sequence $a_n = \dfrac{1}{n+3}$ are $\dfrac{1}{4}, \dfrac{1}{5}, \dfrac{1}{6}, \dfrac{1}{7}, \dfrac{1}{8}$.

7. $a_n = 2n$

$a_1 = 2(1) = 2$
$a_2 = 2(2) = 4$
$a_3 = 2(3) = 6$
$a_4 = 2(4) = 8$
$a_5 = 2(5) = 10$

Thus, the first five terms of the sequence $a_n = 2n$ are 2, 4, 6, 8, 10.

9. $a_n = -n^2$

$a_1 = -1^2 = -1$
$a_2 = -2^2 = -4$
$a_3 = -3^2 = -9$
$a_4 = -4^2 = -16$
$a_5 = -5^2 = -25$

Thus, the first five terms of the sequence $a_n = n^2$ are $-1, -4, -8, -16, -25$.

11. $a_n = 2^n$

$a_1 = 2^1 = 2$
$a_2 = 2^2 = 4$
$a_3 = 2^3 = 8$
$a_4 = 2^4 = 16$
$a_5 = 2^5 = 32$

Thus, the first five terms of the sequence $a_n = 2^n$ are 2, 4, 8, 16, 32.

13. $a_n = 2n+5$

$a_1 = 2(1)+5 = 2+5 = 7$
$a_2 = 2(2)+5 = 4+5 = 9$
$a_3 = 2(3)+5 = 6+5 = 11$
$a_4 = 2(4)+5 = 8+5 = 13$
$a_5 = 2(5)+5 = 10+5 = 15$

Thus, the first five terms of the sequence $a_n = 2n+5$ are 7, 9, 11, 13, 15.

15. $a_n = (-1)^n n^2$

$a_1 = (-1)^1(1)^2 = -1(1) = -1$
$a_2 = (-1)^2(2)^2 = 1(4) = 4$
$a_3 = (-1)^3(3)^2 = -1(9) = -9$
$a_4 = (-1)^4(4)^2 = 1(16) = 16$
$a_5 = (-1)^5(5)^2 = -1(25) = -25$

Thus, the first five terms of the sequence $a_n = (-1)^n n^2$ are $-1, 4, -9, 16, -25$.

17. $a_n = 3n^2$

$a_5 = 3(5)^2 = 3(25) = 75$

19. $a_n = 6n-2$

$a_{20} = 6(20)-2 = 120-2 = 118$

21. $a_n = \dfrac{n+3}{n}$

$a_{15} = \dfrac{15+3}{15} = \dfrac{18}{15} = \dfrac{6}{5}$

23. $a_n = (-3)^n$

$a_6 = (-3)^6 = 729$

25. $a_n = \dfrac{n-2}{n+1}$

$a_6 = \dfrac{6-2}{6+1} = \dfrac{4}{7}$

27. $a_n = \dfrac{(-1)^n}{n}$

$a_8 = \dfrac{(-1)^8}{8} = \dfrac{1}{8}$

29. $a_n = -n^2 + 5$

$a_{10} = -10^2 + 5 = -100 + 5 = -95$

31. $a_n = \dfrac{(-1)^n}{n+6}$

$a_{19} = \dfrac{(-1)^{19}}{19+6} = -\dfrac{1}{25}$

33. 3, 7, 11, 15, or 4(1) – 1, 4(2) – 1, 4(3) – 1, 4(4) – 1. In general, $a_n = 4n - 1$.

35. –2, –4, –8, –16, or –2, $-2^2, -2^3, -2^4$

In general, $a_n = -2^n$.

37. $\dfrac{1}{3}, \dfrac{1}{9}, \dfrac{1}{27}, \dfrac{1}{81}$, or $\dfrac{1}{3}, \dfrac{1}{3^2}, \dfrac{1}{3^3}, \dfrac{1}{3^4}$

In general, $a_n = \dfrac{1}{3^n}$.

39. $a_n = 32n - 16$

$a_2 = 32(2) - 16 = 64 - 16 = 48$ ft

$a_3 = 32(3) - 16 = 96 - 16 = 80$ ft

$a_4 = 32(4) - 16 = 128 - 16 = 112$ ft

41. 0.10, 0.20, 0.40, or 0.10, 0.10(2), $0.10(2)^2$

In general, $a_n = 0.10(2)^{n-1}$

$a_{14} = 0.10(2)^{13} = \$819.20$

43. $a_n = 75(2)^{n-1}$

$a_6 = 75(2)^5 = 75(32) = 2400$ cases

$a_1 = 75(2)^0 = 75(1) = 75$ cases

45. $a_n = \dfrac{1}{2}a_{n-1}$ for $n > 1, a_1 = 800$

In 2000, $n = 1$ and $a_1 = 800$.

In 2001, $n = 2$ and $a_2 = \dfrac{1}{2}(800) = 400$.

In 2002, $n = 3$ and $a_3 = \dfrac{1}{2}(400) = 200$.

In 2003, $n = 4$ and $a_4 = \dfrac{1}{2}(200) = 100$.

In 2004, $n = 5$ and $a_5 = \dfrac{1}{2}(100) = 50$.

The population estimate for 2004 is 50 sparrows. Continuing the sequence:

In 2005, $n = 6$ and $a_6 = \dfrac{1}{2}(50) = 25$.

In 2006, $n = 7$ and $a_7 = \dfrac{1}{2}(25) \approx 12$.

In 2007, $n = 8$ and $a_8 = \dfrac{1}{2}(12) = 6$.

In 2008, $n = 9$ and $a_9 = \dfrac{1}{2}(6) = 3$.

In 2009, $n = 10$ and $a_{10} = \dfrac{1}{2}(3) \approx 1$.

In 2010, $n = 11$ and $a_{11} = \dfrac{1}{2}(1) \approx 0$.

The population is estimated to become extinct in 2010.

47. $f(x) = (x-1)^2 + 3$

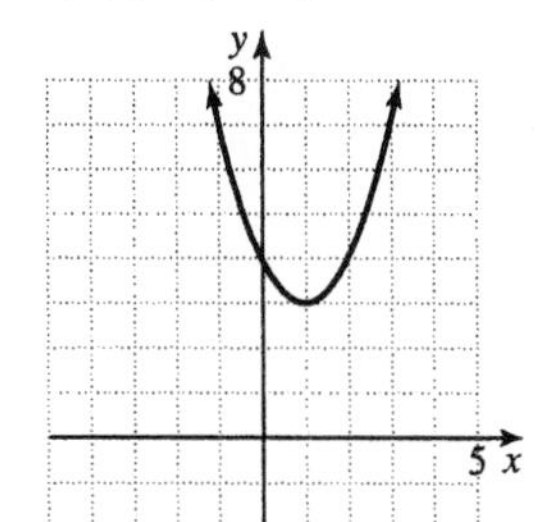

49. $f(x) = 2(x+4)^2 + 2$

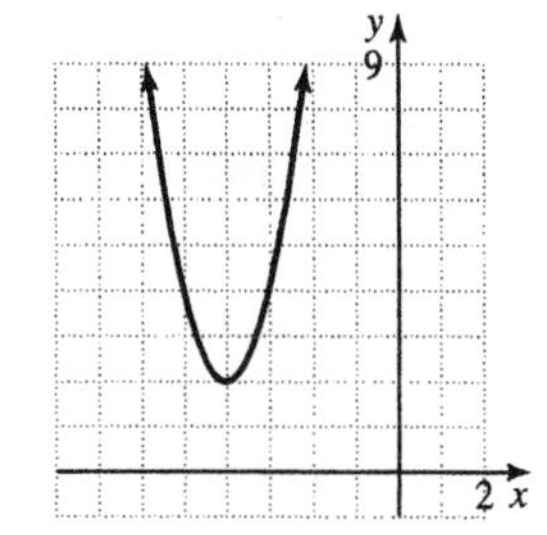

51. $(-4,-1)$ and $(-7,-3)$

$$\begin{aligned} d &= \sqrt{[-7-(-4)]^2+[-3-(-1)]^2} \\ &= \sqrt{(-7+4)^2+(-3+1)^2} \\ &= \sqrt{(-3)^2+(-2)^2} \\ &= \sqrt{9+4} \\ &= \sqrt{13} \text{ units} \end{aligned}$$

53. $(2,-7)$ and $(-3,-3)$

$$\begin{aligned} d &= \sqrt{(-3-2)^2+[-3-(-7)]^2} \\ &= \sqrt{(-5)^2+(-3+7)^2} \\ &= \sqrt{(-5)^2+(4)^2} \\ &= \sqrt{25+16} \\ &= \sqrt{41} \text{ units} \end{aligned}$$

55. $a_n = \dfrac{1}{\sqrt{n}}$

$$\begin{aligned} a_1 &= \frac{1}{\sqrt{1}} = \frac{1}{1} = 1 \\ a_2 &= \frac{1}{\sqrt{2}} \approx 0.7071 \\ a_3 &= \frac{1}{\sqrt{3}} \approx 0.5774 \\ a_4 &= \frac{1}{\sqrt{4}} = \frac{1}{2} = 0.5 \\ a_5 &= \frac{1}{\sqrt{5}} \approx 0.4472 \end{aligned}$$

Thus, the first five terms of the sequence $a_n = \dfrac{1}{\sqrt{n}}$ are 1, 0.7071, 0.5774, 0.5, 0.4472.

57. $a_n = \left(1+\dfrac{1}{n}\right)^n$

$$\begin{aligned} a_1 &= \left(1+\frac{1}{1}\right)^1 = (2)^1 = 2 \\ a_2 &= \left(1+\frac{1}{2}\right)^2 = \left(\frac{3}{2}\right)^2 = 2.25 \\ a_3 &= \left(1+\frac{1}{3}\right)^3 = \left(\frac{4}{3}\right)^3 \approx 2.3704 \\ a_4 &= \left(1+\frac{1}{4}\right)^4 = \left(\frac{5}{4}\right)^4 \approx 2.4414 \\ a_5 &= \left(1+\frac{1}{5}\right)^5 = \left(\frac{6}{5}\right)^5 \approx 2.4883 \end{aligned}$$

Thus, the first five terms of the sequence $a_n = \left(1+\dfrac{1}{n}\right)^n$ are 2, 2.25, 2.3704, 2.4414, 2.4883.

Section 14.2

Practice Exercises

1. $a_1 = 4$

$$\begin{aligned} a_2 &= 4+5 = 9 \\ a_3 &= 9+5 = 14 \\ a_4 &= 14+5 = 19 \\ a_5 &= 19+5 = 24 \end{aligned}$$

The first five terms are 4, 9, 14, 19, 24.

2. a. $a_n = a_1 + (n-1)d$

Here, $a_1 = 2$ and $d = -3$.

$a_n = 2+(n-1)(-3) = 2-3n+3 = 5-3n$

b. $a_n = 5-3n$

$a_{12} = 5-3\cdot 12 = 5-36 = -31$

3. Since the sequence is arithmetic, the ninth term is $a_9 = a_1 + (9-1)d = a_1 + 8d$.

a_1 is the first term of the sequence, so $a_1 = 3$. d is the constant difference, so $d = a_2 - a_1 = 9-3 = 6$. Thus, $a_9 = a_1 + 8d = 3+8\cdot 6 = 51$.

4. We need to find a_1 and d. The given facts, $a_3 = 23$ and $a_8 = 63$, lead to a system of linear equations.

$$\begin{cases} a_3 = a_1 + (3-1)d \\ a_8 = a_1 + (8-1)d \end{cases} \text{ or } \begin{cases} 23 = a_1 + 2d \\ 63 = a_1 + 7d \end{cases}$$

We solve the system by elimination. Multiply both sides of the second equation by -1.

$$\begin{cases} 23 = a_1 + 2d \\ -1(63) = -1(a_1 + 7d) \end{cases} \text{ or } \begin{cases} 23 = a_1 + 2d \\ -63 = -a_1 - 7d \end{cases}$$

$$\begin{aligned} -40 &= -5d \\ 8 &= d \end{aligned}$$

To find a_1, let $d = 8$ in $23 = a_1 + 2d$.

$$\begin{aligned} 23 &= a_1 + 2(8) \\ 23 &= a_1 + 16 \\ 7 &= a_1 \end{aligned}$$

Thus, $a_1 = 7$ and $d = 8$, so $a_n = 7+(n-1)(8) = 7+8n-8 = -1+8n$ and $a_6 = -1+8\cdot 6 = 47$.

5. The first term, a_1, is 57,000, and d is 2200.

$$a_n = 57,000 + (n-1)(2200)$$
$$= 54,800 + 2200n$$
$$a_3 = 54,800 + 2200 \cdot 3 = 61,400$$

The salary for the third year is $61,400.

6. $a_1 = 8$

$$a_2 = 8(-3) = -24$$
$$a_3 = -24(-3) = 72$$
$$a_4 = 72(-3) = -216$$

The first four terms are 8, −24, 72, and −216.

7. $a_n = a_1 r^{n-1}$

Here, $a_1 = 64$ and $r = \frac{1}{4}$.

Evaluate a_n for $n = 7$.

$$a_7 = 64\left(\frac{1}{4}\right)^{7-1}$$
$$= 64\left(\frac{1}{4}\right)^6$$
$$= 64\left(\frac{1}{4096}\right)$$
$$= \frac{1}{64}$$

8. Since the sequence is geometric and $a_1 = -3$, the seventh term must be $a_1 r^{7-1}$, or $-3r^6$. r is the common ratio of terms, so r must be $\frac{6}{-3}$, or −2.

$$a_7 = -3r^6$$
$$a_7 = -3(-2)^6 = -192$$

9. Notice that $\frac{27}{4} \div \frac{9}{2} = \frac{3}{2}$, so $r = \frac{3}{2}$.

$$a_2 = a_1\left(\frac{3}{2}\right)^{2-1}$$
$$\frac{9}{2} = a_1\left(\frac{3}{2}\right)^1, \text{ or } a_1 = 3$$

The first term is 3, and the common ration is $\frac{3}{2}$.

10. Since the culture is reduced by one-half each day, the population sizes are modeled by a geometric sequence. Here, $a_1 = 4800$ and $r = \frac{1}{2}$.

$$a_n = a_1 r^{n-1} = 4800\left(\frac{1}{2}\right)^{n-1}$$
$$a_7 = 4800\left(\frac{1}{2}\right)^{7-1} = 75$$

The bacterial culture should measure 75 units at the beginning of day 7.

Vocabulary and Readiness Check

1. A geometric sequence is one in which each term (after the first) is obtained by multiplying the preceding term by a constant r. The constant r is called the common ratio.

2. An arithmetic sequence is one in which each term (after the first) differs from the preceding term by a constant amount d. The constant d is called the common difference.

3. The general term of an arithmetic sequence is $a_n = a_1 + (n-1)d$ where a_1 is the first term and d is the common difference.

4. The general term of a geometric sequence is $a_n = a_1 r^{n-1}$ where a_1 is the first term and r is the common ratio.

Exercise Set 14.2

1. $a_n = a_1 + (n-1)d$

$$a_1 = 4;\ d = 2$$
$$a_1 = 4$$
$$a_2 = 4 + (2-1)2 = 6$$
$$a_3 = 4 + (3-1)2 = 8$$
$$a_4 = 4 + (4-1)2 = 10$$
$$a_5 = 4 + (5-1)2 = 12$$

The first five terms are 4, 6, 8, 10, 12.

3. $a_n = a_1 + (n-1)d$

$$a_1 = 6,\ d = -2$$
$$a_1 = 6$$
$$a_2 = 6 + (2-1)(-2) = 4$$
$$a_3 = 6 + (3-1)(-2) = 2$$
$$a_4 = 6 + (4-1)(-2) = 0$$
$$a_5 = 6 + (5-1)(-2) = -2$$

The first five terms are 6, 4, 2, 0, −2.

5. $a_n = a_1r^{n-1}$
$a_1 = 1, r = 3$
$a_1 = 1(3)^{1-1} = 1$
$a_2 = 1(3)^{2-1} = 3$
$a_3 = 1(3)^{3-1} = 9$
$a_4 = 1(3)^{4-1} = 27$
$a_5 = 1(3)^{5-1} = 81$
The first five terms are 1, 3, 9, 27, 81.

7. $a_n = a_1r^{n-1}$
$a_1 = 48, r = \frac{1}{2}$
$a_1 = 48\left(\frac{1}{2}\right)^{1-1} = 48$
$a_2 = 48\left(\frac{1}{2}\right)^{2-1} = 24$
$a_3 = 48\left(\frac{1}{2}\right)^{3-1} = 12$
$a_4 = 48\left(\frac{1}{2}\right)^{4-1} = 6$
$a_5 = 48\left(\frac{1}{2}\right)^{5-1} = 3$
The first five terms are 48, 24, 12, 6, 3.

9. $a_n = a_1 + (n-1)d$
$a_1 = 12, d = 3$
$a_n = 12 + (n-1)3$
$a_8 = 12 + 7(3) = 12 + 21 = 33$

11. $a_n = a_1r^{n-1}$
$a_1 = 7, d = -5$
$a_n = a_1r^{n-1}$
$a_4 = 7(-5)^3 = 7(-125) = -875$

13. $a_n = a_1 + (n-1)d$
$a_1 = -4, d = -4$
$a_n = -4 + (n-1)(-4)$
$a_{15} = -4 + 14(-4) = -4 - 56 = -60$

15. 0, 12, 24
$a_1 = 0$ and $d = 12$
$a_n = 0 + (n-1)12$
$a_9 = 8(12) = 96$

17. 20, 18, 16
$a_1 = 20$ and $d = -2$
$a_n = 20 + (n-1)(-2)$
$a_{25} = 20 + 24(-2) = 20 - 48 = -28$

19. 2, −10, 50
$a_1 = 2$ and $r = -5$
$a_n = 2(-5)^{n-1}$
$a_5 = 2(-5)^4 = 2(625) = 1250$

21. $a_4 = 19, a_{15} = 52$
$\begin{cases} a_4 = a_1 + (4-1)d \\ a_{15} = a_1 + (15-1)d \end{cases}$ or
$\begin{cases} 19 = a_1 + 3d \\ 52 = a_1 + 14d \end{cases}$
$\begin{cases} -19 = -a_1 - 3d \\ 52 = a_1 + 14d \end{cases}$
Adding yields 33 = 11*d* or *d* = 3. Then
$a_1 = 19 - 3(3) = 10.$
$$\begin{aligned} a_n &= 10 + (n-1)3 \\ &= 10 + 3n - 3 \\ &= 7 + 3n \end{aligned}$$
and $a_8 = 7 + 3(8) = 7 + 24 = 31$

23. $a_2 = -1, a_4 = 5$
$\begin{cases} a_2 = a_1 + (2-1)d \\ a_4 = a_1 + (4-1)d \end{cases}$ or
$\begin{cases} -1 = a_1 + d \\ 5 = a_1 + 3d \end{cases}$
$\begin{cases} 1 = -a_1 - d \\ 5 = a_1 + 3d \end{cases}$
Adding yields 6 = 2*d* or *d* = 3. Then
$a_1 = -1 - 3 = -4.$
$$\begin{aligned} a_n &= -4 + (n-1)3 \\ &= -4 + 3n - 3 \\ &= -7 + 3n \end{aligned}$$
and $a_9 = -7 + 3(9) = -7 + 27 = 20$

25. $a_2 = -\frac{4}{3}$ and $a_3 = \frac{8}{3}$

Notice that $\frac{8}{3} \div \frac{-4}{3} = \frac{8}{3} \cdot -\frac{3}{4} = -2$, so $r = -2$.

Then

$a_2 = a_1(-2)^{2-1}$

$-\frac{4}{3} = a_1(-2)$

$\frac{2}{3} = a_1$.

The first term is $\frac{2}{3}$ and the common ratio is -2.

27. Answers may vary

29. $2, 4, 6$ is an arithmetic sequence.

$a_1 = 2$ and $d = 2$

31. $5, 10, 20$ is a geometric sequence.

$a_1 = 5$ and $r = 2$

33. $\frac{1}{2}, \frac{1}{10}, \frac{1}{50}$ is a geometric sequence.

$a_1 = \frac{1}{2}$ and $r = \frac{1}{5}$

35. $x, 5x, 25x$ is a geometric sequence.

$a_1 = x$ and $r = 5$

37. $p, p+4, p+8$ is an arithmetic sequence.

$a_1 = p$ and $d = 4$

39. $a_1 = 14$ and $d = \frac{1}{4}$

$a_n = 14 + (n-1)\frac{1}{4}$

$a_{21} = 14 + 20\left(\frac{1}{4}\right) = 14 + 5 = 19$

41. $a_1 = 3$ and $r = -\frac{2}{3}$

$a_n = 3\left(-\frac{2}{3}\right)^{n-1}$

$a_4 = 3\left(-\frac{2}{3}\right)^3 = 3\left(-\frac{8}{27}\right) = -\frac{8}{9}$

43. $\frac{3}{2}, 2, \frac{5}{2}, \ldots$

$a_1 = \frac{3}{2}$ and $d = \frac{1}{2}$

$a_n = \frac{3}{2} + (n-1)\frac{1}{2}$

$a_{15} = \frac{3}{2} + 14\left(\frac{1}{2}\right) = \frac{17}{2}$

45. $24, 8, \frac{8}{3}, \ldots$

$a_1 = 24$ and $r = \frac{1}{3}$

$a_n = 24\left(\frac{1}{3}\right)^{n-1}$

$a_6 = 24\left(\frac{1}{3}\right)^5 = 24\left(\frac{1}{243}\right) = \frac{8}{81}$

47. $a_3 = 2$, $a_{17} = -40$

$\begin{cases} a_3 = a_1 + (3-1)d \\ a_{17} = a_1 + (17-1)d \end{cases}$ or

$\begin{cases} 2 = a_1 + 2d \\ -40 = a_1 + 16d \end{cases}$

$\begin{cases} -2 = -a_1 - 2d \\ -40 = a_1 + 16d \end{cases}$

Adding yields $-42 = 14d$ or $d = -3$. Then

$a_1 = 2 - 2(-3) = 8$.

$a_n = 8 + (n-1)(-3) = 8 - 3n + 3 = 11 - 3n$

and

$a_{10} = 11 - 3(10) = 11 - 30 = -19$

49. $54, 58, 62$

$a_1 = 54$ and $d = 4$

$a_n = 54 + (n-1)4$

$a_{20} = 54 + 19(4) = 54 + 76 = 130$

The general term of the sequence is $a_n = 4n + 50$. There are 130 seats in the twentieth row.

51. $a_1 = 6$ and $r = 3$

$a_n = 6(3)^{n-1} = 2 \cdot 3 \cdot (3)^{n-1} = 2(3)^n$

The general term of the sequence is $a_n = 6(3)^{n-1}$ or $a_n = 2(3)^n$.

53. $a_1 = 486$ and $r = \frac{1}{3}$

Initial Height $= a_1 = 486\left(\frac{1}{3}\right)^{1-1} = 486$

Rebound $1 = a_2 = 486\left(\frac{1}{3}\right)^{2-1} = 162$

Rebound $2 = a_3 = 486\left(\frac{1}{3}\right)^{3-1} = 54$

Rebound $3 = a_4 = 486\left(\frac{1}{3}\right)^{4-1} = 18$

Rebound $4 = a_5 = 486\left(\frac{1}{3}\right)^{5-1} = 6$

The first five terms of the sequence are 486, 162, 54, 18, 6.

The general term is $a_n = 486\left(\frac{1}{3}\right)^{n-1}$ or

$a_n = \frac{486}{3^{n-1}}$. Since $a_6 = 2$ and $a_7 = \frac{2}{3}$, a_7 is the first term less than 1. Since a_7 corresponds to the 6th bounce, it takes 6 bounces for the ball to rebound less than 1 foot.

55. $a_1 = 4000$ and $d = 125$

$a_n = 4000 + (n-1)125$ or

$a_n = 3875 + 125n$

$a_{12} = 4000 + 11(125) = 5375$

His salary for his last month of training is \$5375.

57. $a_1 = 400$ and $r = \frac{1}{2}$

12 hours = 4(3 hours), so we seek the fourth term after a_1, namely a_5.

$a_n = a_1 r^{n-1}$

$a_5 = 400\left(\frac{1}{2}\right)^4 = \frac{400}{16} = 25$

25 grams of the radioactive material remain after 12 hours.

59. $$\frac{1}{3(1)} + \frac{1}{3(2)} + \frac{1}{3(3)} = \frac{1}{3} + \frac{1}{6} + \frac{1}{9} = \frac{6}{18} + \frac{3}{18} + \frac{2}{18} = \frac{11}{18}$$

61. $3^0 + 3^1 + 3^2 + 3^3 = 1 + 3 + 9 + 27 = 40$

63. $$\frac{8-1}{8+1} + \frac{8-2}{8+2} + \frac{8-3}{8+3} = \frac{7}{9} + \frac{6}{10} + \frac{5}{11} = \frac{770}{990} + \frac{594}{990} + \frac{450}{990} = \frac{1814}{990} = \frac{907}{495}$$

65. $a_1 = \$11,782.40$

$r = 0.5$

$a_2 = (11,782.40)(0.5) = \$5891.20$

$a_3 = (5891.20)(0.5) = \$2945.60$

$a_4 = (2945.60)(0.5) = \$1472.80$

The first four terms of the sequence are \$11,782.40, \$5891.20, \$2945.60, \$1472.80.

67. $a_1 = 19.652$ and $d = -0.034$

$a_2 = 19.652 - 0.034 = 19.618$

$a_3 = 19.618 - 0.034 = 19.584$

$a_4 = 19.584 - 0.034 = 19.550$

69. Answers may vary

Section 14.3

Practice Exercises

1. a. $$\sum_{i=0}^{4} \frac{i-3}{4} = \frac{0-3}{4} + \frac{1-3}{4} + \frac{2-3}{4} + \frac{3-3}{4} + \frac{4-3}{4} = \left(-\frac{3}{4}\right) + \left(-\frac{2}{4}\right) + \left(-\frac{1}{4}\right) + 0 + \frac{1}{4} = -\frac{5}{4}, \text{ or } -1\frac{1}{4}$$

b. $$\sum_{i=2}^{5} 3^i = 3^2 + 3^3 + 3^4 + 3^5 = 9 + 27 + 81 + 243 = 360$$

2. a. Since the difference of each term and the preceding term is 5, the terms correspond to the first six terms of the arithmetic sequence $a_n = 5 + (n-1)5 = 5n$. Thus, in summation notation,

$$5 + 10 + 15 + 20 + 25 + 30 = \sum_{i=1}^{6} 5i.$$

b. Since each term is the product of the preceding term and $\frac{1}{5}$, these terms correspond to the first four terms of the geometric sequence $a_n = \frac{1}{5}\left(\frac{1}{5}\right)^{n-1} = \left(\frac{1}{5}\right)^n$.

In summation notation,

$$\frac{1}{5}+\frac{1}{25}+\frac{1}{125}+\frac{1}{625}=\sum_{i=1}^{4}\left(\frac{1}{5}\right)^i.$$

3. $S_4 = \sum_{i=1}^{4}\frac{2+3i}{i^2}$

$$= \frac{2+3\cdot1}{1^2}+\frac{2+3\cdot2}{2^2}+\frac{2+3\cdot3}{3^2}+\frac{2+3\cdot4}{4^2}$$
$$= \frac{5}{1}+\frac{8}{4}+\frac{11}{9}+\frac{14}{16}$$
$$= 5+2+\frac{11}{9}+\frac{7}{8}$$
$$= \frac{655}{72}, \text{ or } 9\frac{7}{72}$$

4. $S_5 = \sum_{i=1}^{5} i(2i-1)$

$$= 1(2\cdot1-1)+2(2\cdot2-1)+3(2\cdot3-1)+4(2\cdot4-1)+5(2\cdot5-1)$$
$$= 1+6+15+28+45$$
$$= 95$$

There are 95 plants after 5 years.

Vocabulary and Readiness Check

1. A series is an <u>infinite</u> series if it is the sum of all the terms of the sequence.

2. A series is a <u>finite</u> series if it is the sum of a finite number of terms.

3. A shorthand notation for denoting a series when the general term of the sequence is known is called <u>summation</u> notation.

4. In the notation $\sum_{i=1}^{7}(5i-2)$, the Σ is the Greek uppercase letter <u>sigma</u> and the i is called the <u>index of summation</u>.

5. The sum of the first n terms of a sequence is a finite series known as a <u>partial sum</u>.

6. For the notation in Exercise 4 above, the beginning value of i is <u>1</u> and the ending value of i is <u>7</u>.

Exercise Set 14.3

1. $\sum_{i=1}^{4}(i-3) = (1-3)+(2-3)+(3-3)+(4-3)$

$$= -2+(-1)+0+1$$
$$= -2$$

3. $\sum_{i=4}^{7}(2i+4) = [2(4)+4]+[2(5)+4]+[2(6)+4]+[2(7)+4]$

$$= 12+14+16+18$$
$$= 60$$

5. $\sum_{i=2}^{4}(i^2-3) = (2^2-3)+(3^2-3)+(4^2-3)$

$$= 1+6+13$$
$$= 20$$

7. $\sum_{i=1}^{3}\left(\frac{1}{i+5}\right) = \frac{1}{1+5}+\frac{1}{2+5}+\frac{1}{3+5}$

$$= \frac{1}{6}+\frac{1}{7}+\frac{1}{8}$$
$$= \frac{28}{168}+\frac{24}{168}+\frac{21}{168}$$
$$= \frac{73}{168}$$

9. $\sum_{i=1}^{3}\frac{1}{6i} = \frac{1}{6(1)}+\frac{1}{6(2)}+\frac{1}{6(3)}$

$$= \frac{1}{6}+\frac{1}{12}+\frac{1}{18}$$
$$= \frac{6+3+2}{36}$$
$$= \frac{11}{36}$$

11. $\sum_{i=2}^{6}3i = 3(2)+3(3)+3(4)+3(5)+3(6)$

$$= 6+9+12+15+18$$
$$= 60$$

13. $\sum_{i=3}^{5} i(i+2) = 3(3+2)+4(4+2)+5(5+2)$
$= 15+24+35$
$= 74$

15. $\sum_{i=1}^{5} 2^i = 2^1+2^2+2^3+2^4+2^5$
$= 2+4+8+16+32$
$= 62$

17. $\sum_{i=1}^{4} \frac{4i}{i+3} = \frac{4(1)}{1+3}+\frac{4(2)}{2+3}+\frac{4(3)}{3+3}+\frac{4(4)}{4+3}$
$= 1+\frac{8}{5}+2+\frac{16}{7}$
$= \frac{105}{35}+\frac{56}{35}+\frac{80}{35}$
$= \frac{241}{35}$

19. $1+3+5+7+9$
$a_1 = 1,\ d = 2$
$a_n = 1+(n-1)2 = 2n-1$
$\sum_{i=1}^{5}(2i-1)$

21. $4+12+36+108 = 4+4(3)+4(3)^2+4(3)^3$
$= \sum_{i=1}^{4} 4(3)^{i-1}$

23. $12+9+6+3+0+(-3)$
$a_1 = 12,\ d = -3$
$a_n = 12+(n-1)(-3) = -3n+15$
$\sum_{i=1}^{6}(-3i+15)$

25. $12+4+\frac{4}{3}+\frac{4}{9} = \frac{4}{3^{-1}}+\frac{4}{3^0}+\frac{4}{3}+\frac{4}{3^2}$
$= \sum_{i=1}^{4}\frac{4}{3^{i-2}}$

27. $1+4+9+16+25+36+49$
$= 1^2+2^2+3^2+4^2+5^2+6^2+7^2$
$= \sum_{i=1}^{7} i^2$

29. $a_n = (n+2)(n-5)$
$S_2 = \sum_{i=1}^{2}(i+2)(i-5)$
$= (1+2)(1-5)+(2+2)(2-5)$
$= 3(-4)+4(-3)$
$= -12-12$
$= -24$

31. $a_n = (-1)^n$
$S_6 = \sum_{i=1}^{6}(-1)^i$
$= (-1)^1+(-1)^2+(-1)^3+(-1)^4+(-1)^5 +(-1)^6$
$= -1+1+(-1)+1+(-1)+1$
$= 0$

33. $a_n = (n+3)(n+1)$
$S_4 = \sum_{i=1}^{4}(i+3)(i+1)$
$= (1+3)(1+1)+(2+3)(2+1)+(3+3)(3+1) +(4+3)(4+1)$
$= 4(2)+5(3)+6(4)+7(5)$
$= 8+15+24+35$
$= 82$

35. $a_n = -2n$
$S_4 = \sum_{i=1}^{4}(-2i)$
$= -2(1)+(-2)(2)+(-2)(3)+(-2)(4)$
$= -2-4-6-8$
$= -20$

37. $a_n = -\frac{n}{3}$
$S_3 = \sum_{i=1}^{3} -\frac{i}{3} = -\frac{1}{3}-\frac{2}{3}-\frac{3}{3} = -2$

39. 1, 2, 3,...,10

$a_n = n$

$S_{10} = \sum_{i=1}^{10} i = 1+2+3+...+10 = 55$

A total of 55 trees were planted.

41. $a_1 = 6$ and $r = 2$

$a_n = 6 \cdot 2^{n-1}$

$a_5 = 6 \cdot 2^4 = 6 \cdot 16 = 96$

There will be 96 fungus units at the beginning of the 5th day.

43. The general term of the sequence is $a_n = 50(2)^n$, where n represents the number of 12-hr periods.

$a_4 = 50(2)^4 = 50(16) = 800$

There are 800 bacteria after 48 hours.

45. $a_n = (n+1)(n+2)$

$a_4 = (4+1)(4+2) = 5(6) = 30$

30 opossums were killed in the fourth month.

$$S_4 = \sum_{i=1}^{4}(i+1)(i+2)$$
$$= 2(3)+(3)(4)+(4)(5)+(5)(6)$$
$$= 6+12+20+30$$
$$= 68$$

68 opossums were killed in the four months.

47. $a_n = 100(0.5)^n$

$a_4 = 100(0.5)^4 = 6.25$

The decay in the fourth year is 6.25 pounds.

$$S_4 = \sum_{i=1}^{4} 100(0.5)^i$$
$$= 100(0.5)^1 + 100(0.5)^2 + 100(0.5)^3 + 100(0.5)^4$$
$$= 100(0.5) + 100(0.25) + 100(0.125) + 100(0.0625)$$
$$= 50 + 25 + 12.5 + 6.25$$
$$= 93.75$$

The decay over the four years is 93.75 pounds.

49. $a_1 = 40$ and $r = \frac{4}{5}$

$a_5 = 40\left(\frac{4}{5}\right)^4 = 16.384$

The length of the fifth swing is approximately 16.4 inches.

$$S_5 = \sum_{i=1}^{5} 40\left(\frac{4}{5}\right)^{i-1}$$
$$= 40\left(\frac{4}{5}\right)^0 + 40\left(\frac{4}{5}\right)^1 + 40\left(\frac{4}{5}\right)^2 + 40\left(\frac{4}{5}\right)^3 + 40\left(\frac{4}{5}\right)^4$$
$$= 40 + 32 + 25.6 + 20.48 + 16.384$$
$$= 134.464$$

The pendulum swings about 134.5 inches in five swings.

51. $\frac{5}{1-\frac{1}{2}} = \frac{5}{\frac{1}{2}} = 5 \cdot \frac{2}{1} = 10$

53. $\frac{\frac{1}{3}}{1-\frac{1}{10}} = \frac{\frac{1}{3}}{\frac{9}{10}} = \frac{1}{3} \cdot \frac{10}{9} = \frac{10}{27}$

55. $\frac{3(1-2^4)}{1-2} = \frac{3(1-16)}{-1} = \frac{3(-15)}{-1} = \frac{-45}{-1} = 45$

57. $\frac{10}{2}(3+15) = \frac{10}{2}(18) = \frac{180}{2} = 90$

59. a. $\sum_{i=1}^{7}(i+i^2)$

$$= (1+1^2)+(2+2^2)+(3+3^2)+(4+4^2) + (5+5^2)+(6+6^2)+(7+7^2)$$
$$= 2+6+12+20+30+42+56$$

b. $\sum_{i=1}^{7} i + \sum_{i=1}^{7} i^2$

$$= (1+2+3+4+5+6+7) + (1+4+9+16+25+36+49)$$

c. Answers may vary

d. True; answers may vary.

Integrated Review

1. $a_n = n-3$

$a_1 = 1-3 = -2$
$a_2 = 2-3 = -1$
$a_3 = 3-3 = 0$
$a_4 = 4-3 = 1$
$a_5 = 5-3 = 2$

Therefore, the first five terms are −2, −1, 0, 1, 2.

2. $a_n = \frac{7}{1+n}$

$a_1 = \frac{7}{1+1} = \frac{7}{2}$
$a_2 = \frac{7}{1+2} = \frac{7}{3}$
$a_3 = \frac{7}{1+3} = \frac{7}{4}$
$a_4 = \frac{7}{1+4} = \frac{7}{5}$
$a_5 = \frac{7}{1+5} = \frac{7}{6}$

The first five terms are $\frac{7}{2}, \frac{7}{3}, \frac{7}{4}, \frac{7}{5}$, and $\frac{7}{6}$.

3. $a_n = 3^{n-1}$

$a_1 = 3^{1-1} = 3^0 = 1$
$a_2 = 3^{2-1} = 3^1 = 3$
$a_3 = 3^{3-1} = 3^2 = 9$
$a_4 = 3^{4-1} = 3^3 = 27$
$a_5 = 3^{5-1} = 3^4 = 81$

The first five terms are 1, 3, 9, 27, and 81.

4. $a_n = n^2 - 5$

$a_1 = 1^2 - 5 = 1-5 = -4$
$a_2 = 2^2 - 5 = 4-5 = -1$
$a_3 = 3^2 - 5 = 9-5 = 4$
$a_4 = 4^2 - 5 = 16-5 = 11$
$a_5 = 5^2 - 5 = 25-5 = 20$

The first five terms are −4, −1, 4, 11, and 20.

5. $(-2)^n;\ a_6$

$a_6 = (-2)^6 = 64$

6. $-n^2 + 2;\ a_4$

$a_4 = -(4)^2 + 2 = -16+2 = -14$

7. $\frac{(-1)^n}{n};\ a_{40}$

$a_{40} = \frac{(-1)^{40}}{40} = \frac{1}{40}$

8. $\frac{(-1)^n}{2n};\ a_{41}$

$a_{41} = \frac{(-1)^{41}}{2(41)} = \frac{-1}{82} = -\frac{1}{82}$

9. $a_1 = 7;\ d = -3$

$a_1 = 7$
$a_2 = 7-3 = 4$
$a_3 = 4-3 = 1$
$a_4 = 1-3 = -2$
$a_5 = -2-3 = -5$

The first five terms are 7, 4, 1, −2, −5.

10. $a_1 = -3;\ r = 5$

$a_1 = -3$
$a_2 = -3(5) = -15$
$a_3 = -15(5) = -75$
$a_4 = -75(5) = -375$
$a_5 = -375(5) = -1875$

The first five terms are −3, −15, −75, −375, −1875.

11. $a_1 = 45;\ r = \frac{1}{3}$

$a_1 = 45$
$a_2 = 45\left(\frac{1}{3}\right) = 15$
$a_3 = 15\left(\frac{1}{3}\right) = 5$
$a_4 = 5\left(\frac{1}{3}\right) = \frac{5}{3}$
$a_5 = \frac{5}{3}\left(\frac{1}{3}\right) = \frac{5}{9}$

The first five terms are 45, 15, 5, $\frac{5}{3}, \frac{5}{9}$.

12. $a_1 = -12;\ d = 10$

$a_1 = -12$
$a_2 = -12 + 10 = -2$
$a_3 = -2 + 10 = 8$
$a_4 = 8 + 10 = 18$
$a_5 = 18 + 10 = 28$

The first five terms are −12, −2, 8, 18, 28.

13. $a_1 = 20;\ d = 9$

$a_n = a_1 + (n-1)d$

$a_{10} = 20 + (10-1)9$
$= 20 + 81$
$= 101$

14. $a_1 = 64;\ r = \frac{3}{4}$

$a_n = a_1 r^{n-1}$

$a_6 = 64\left(\frac{3}{4}\right)^{6-1}$
$= 64\left(\frac{3}{4}\right)^5$
$= 64\left(\frac{243}{1024}\right)$
$= \frac{243}{16}$

15. $a_1 = 6;\ r = \frac{-12}{6} = -2$

$a_n = a_1 r^{n-1}$

$a_7 = 6(-2)^{7-1} = 6(-2)^6 = 6(64) = 384$

16. $a_1 = -100;\ d = -85 - (-100) = 15$

$a_n = a_1 + (n-1)d$

$a_{20} = -100 + (20-1)(15)$
$= -100 + (19)(15)$
$= -100 + 285$
$= 185$

17. $a_4 = -5,\ a_{10} = -35$

$a_n = a_1 + (n-1)d$

$$\begin{cases} a_4 = a_1 + (4-1)d \\ a_{10} = a_1 + (10-1)d \end{cases}$$

$$\begin{cases} -5 = a_1 + 3d \\ -35 = a_1 + 9d \end{cases}$$

Multiply eq. 2 by −1, then add the equations.

$$\begin{cases} -5 = a_1 + 3d \\ (-1)(-35) = -1(a_1 + 9d) \end{cases}$$

$$\begin{cases} -5 = a_1 + 3d \\ 35 = -a_1 - 9d \end{cases}$$

$30 = -6d$
$-5 = d$

To find a_1, let $d = -5$ in

$-5 = a_1 + 3d$
$-5 = a_1 + 3(-5)$
$10 = a_1$

Thus, $a_1 = 10$ and $d = -5$, so

$a_n = 10 + (n-1)(-5) = -5n + 15$

$a_5 = -5(5) + 15 = -10$

18. $a_4 = 1;\ a_7 = \frac{1}{125}$

$a_n = a_1 r^{n-1}$

$a_4 = a_1 r^{4-1}$ so $1 = a_1 r^3$

$a_7 = a_1 r^{71}$ so $\frac{1}{125} = a_1 r^6$

Since $a_1 r^6 = (a_1 r^3) r^3$, $\frac{1}{125} = 1 \cdot r^3$ and $r = \frac{1}{5}$.

$a_5 = a_4 \cdot r$ so $a_5 = 1 \cdot \frac{1}{5} = \frac{1}{5}$

19. $\sum_{i=1}^{4} 5i = 5(1) + 5(2) + 5(3) + 5(4)$
$= 5 + 10 + 15 + 20$
$= 50$

20. $\sum_{i=1}^{7} (3i + 2)$
$= (3(1) + 2) + (3(2) + 2) + (3(3) + 2)$
$+ (3(4) + 2) + (3(5) + 2) + (3(6) + 2)$
$+ (3(7) + 2)$
$= 5 + 8 + 11 + 14 + 17 + 20 + 23$
$= 98$

21. $\sum_{i=3}^{7} 2^{i-4}$

$= 2^{3-4} + 2^{4-4} + 2^{5-4} + 2^{6-4} + 2^{7-4}$

$= 2^{-1} + 2^0 + 2^1 + 2^2 + 2^3$

$= \frac{1}{2} + 1 + 2 + 4 + 8$

$= 15\frac{1}{2}$

$= \frac{31}{2}$

22. $\sum_{i=2}^{5} \frac{i}{i+1} = \frac{2}{2+1} + \frac{3}{3+1} + \frac{4}{4+1} + \frac{5}{5+1}$

$= \frac{2}{3} + \frac{3}{4} + \frac{4}{5} + \frac{5}{6}$

$= \frac{61}{20}$

23. $S_3 = \sum_{i=1}^{3} i(i-4)$

$= 1(1-4) + 2(2-4) + 3(3-4)$

$= -3 - 4 - 3$

$= -10$

24. $S_{10} = \sum_{i=1}^{10} (-1)^i (i+1)$

$= (-1)^1(1+1) + (-1)^2(2+1)$

$+ (-1)^3(3+1) + (-1)^4(4+1)$

$+ (-1)^5(5+1) + (-1)^6(6+1)$

$+ (-1)^7(7+1) + (-1)^8(8+1)$

$+ (-1)^9(9+1) + (-1)^{10}(10+1)$

$= -2 + 3 - 4 + 5 - 6 + 7 - 8 + 9 - 10 + 11$

$= 5$

Section 14.4

Practice Exercises

1. 2, 9, 16, 23, 30
Use the formula for S_n of an arithmetic sequence, replacing n with 5, a_1 with 2, and a_n with 30.

$S_n = \frac{n}{2}(a_1 + a_n)$

$S_5 = \frac{5}{2}(2 + 30) = \frac{5}{2}(32) = 80$

2. Because 1, 2, 3, ..., 50 is an arithmetic sequence, use the formula for S_n with $n = 50$, $a_1 = 1$, and $a_n = 50$.

$S_n = \frac{n}{2}(a_1 + a_n)$

$S_5 = \frac{50}{2}(1 + 50) = 25(51) = 1275$

3. The list 6, 7, ..., 15 is the first 10 terms of an arithmetic sequence. Use the formula for S_n with $n = 10$, $a_1 = 6$, and $a_n = 15$.

$S_{10} = \frac{10}{2}(6 + 15) = 5(21) = 105$

There are a total of 105 blocks of ice.

4. 32, 8, 2, $\frac{1}{2}$, $\frac{1}{8}$
Use the formula for the partial sum S_n of the terms of a geometric sequence. Here, $n = 5$, the first term $a_1 = 32$, and the common ratio $r = \frac{1}{4}$.

$S_n = \frac{a_1(1 - r^n)}{1 - r}$

$S_5 = \frac{32\left[1 - \left(\frac{1}{4}\right)^5\right]}{1 - \frac{1}{4}}$

$= \frac{32\left(1 - \frac{1}{1024}\right)}{\frac{3}{4}}$

$= \frac{32 - \frac{1}{32}}{\frac{3}{4}}$

$= \frac{\frac{1023}{32}}{\frac{3}{4}}$

$= \frac{1023}{32} \cdot \frac{4}{3}$

$= \frac{341}{8}$

$= 42\frac{5}{8}$

5. The donations are modeled by the first seven terms of a geometric sequence. Evaluate S_n when $n = 7$, $a_1 = 250{,}000$, and $r = 0.8$.

$S_7 = \frac{250{,}000[1 - (0.8)^7]}{1 - 0.8} = 987{,}856$

The total amount donated during the seven years is $987,856.

6. $7, \frac{7}{4}, \frac{7}{16}, \frac{7}{64}, \ldots$

For this geometric sequence $r = \frac{1}{4}$. Since $|r| < 1$, use the formula for S_∞ of a geometric sequence with $a_1 = 7$ and $r = \frac{1}{4}$.

$$S_\infty = \frac{a_1}{1-r} = \frac{7}{1-\frac{1}{4}} = \frac{7}{\frac{3}{4}} = \frac{28}{3} = 9\frac{1}{3}$$

7. We must find the sum of the terms of an infinite geometric sequence whose first term, a_1, is 36 and whose common ratio, r, is 0.96. Since $|r| < 1$, we may use the formula for S_∞.

$$S_\infty = \frac{a_1}{1-r} = \frac{36}{1-0.96} = \frac{36}{0.04} = 900$$

The ball travels a total distance of 900 inches before it comes to a rest.

Vocabulary and Readiness Check

1. Each term after the first is 5 more than the preceding term; the sequence is arithmetic.
2. Each term after the first is 2 times the preceding term; the sequence is geometric.
3. Each term after the first is –3 times the preceding term; the sequence is geometric.
4. Each term after the first is 2 more than the preceding term; the sequence is arithmetic.
5. Each term after the first is 7 more than the preceding term; the sequence is arithmetic.
6. Each term after the first is –1 times the preceding term; the sequence is geometric.

Exercise Set 14.4

1. $1, 3, 5, 7, \ldots$

$d = 2;\ a_6 = 1 + (6-1)(2) = 11$

$$S_6 = \frac{6}{2}(1+11) = 3(12) = 36$$

3. $4, 12, 36, \ldots$

$a_1 = 4,\ r = 3,\ n = 5$

$$S_5 = \frac{4(1-3^5)}{1-3} = 484$$

5. $3, 6, 9, \ldots$

$d = 3;\ a_6 = 3 + (6-1)(3) = 18$

$$S_6 = \frac{6}{2}(3+18) = 3(21) = 63$$

7. $2, \frac{2}{5}, \frac{2}{25}, \ldots$

$a_1 = 2,\ r = \frac{1}{5},\ n = 4$

$$S_4 = \frac{2\left[1-\left(\frac{1}{5}\right)^4\right]}{1-\frac{1}{5}} = 2.496$$

9. $1, 2, 3, \ldots, 10$

The first term is 1 and the tenth term is 10.

$$S_{10} = \frac{10}{2}(1+10) = 5(11) = 55$$

11. $1, 2, 3, 7$

The first term is 1 and the fourth term is 7.

$$S_4 = \frac{4}{2}(1+7) = 2(8) = 16$$

13. $12, 6, 3, \ldots$

$a_1 = 12,\ r = \frac{1}{2}$

$$S_\infty = \frac{12}{1-\frac{1}{2}} = \frac{12}{\frac{1}{2}} = 12 \cdot \frac{2}{1} = 24$$

15. $\frac{1}{10}, \frac{1}{100}, \frac{1}{1000}, \ldots$

$a_1 = \frac{1}{10},\ r = \frac{1}{10}$

$$S_\infty = \frac{\frac{1}{10}}{1-\frac{1}{10}} = \frac{\frac{1}{10}}{\frac{9}{10}} = \frac{1}{10} \cdot \frac{10}{9} = \frac{1}{9}$$

17. $-10, -5, -\frac{5}{2}, \ldots$

$a_1 = -10,\ r = \frac{1}{2}$

$$S_\infty = \frac{-10}{1-\frac{1}{2}} = \frac{-10}{\frac{1}{2}} = -10 \cdot \frac{2}{1} = -20$$

19. $2, -\frac{1}{4}, \frac{1}{32}, \ldots$

$a_1 = 2,\ r = -\frac{1}{8}$

$$S_\infty = \frac{2}{1-\left(-\frac{1}{8}\right)} = \frac{2}{\frac{9}{8}} = 2 \cdot \frac{8}{9} = \frac{16}{9}$$

21. $\frac{2}{3}, -\frac{1}{3}, \frac{1}{6}, \ldots$

$a_1 = \frac{2}{3},\ r = -\frac{1}{2}$

$$S_\infty = \frac{\frac{2}{3}}{1-\left(-\frac{1}{2}\right)} = \frac{\frac{2}{3}}{\frac{3}{2}} = \frac{2}{3} \cdot \frac{2}{3} = \frac{4}{9}$$

23. $-4, 1, 6, \ldots, 41$

The first term is -4 and the tenth term is 41.

$$S_{10} = \frac{10}{2}(-4+41) = 5(37) = 185$$

25. $3, \frac{3}{2}, \frac{3}{4}, \ldots$

$a_1 = 3,\ r = \frac{1}{2},\ n = 7$

$$S_7 = \frac{3\left[1-\left(\frac{1}{2}\right)^7\right]}{1-\frac{1}{2}} = \frac{381}{64}$$

27. $-12, 6, -3, \ldots$

$a_1 = -12,\ r = -\frac{1}{2},\ n = 5$

$$S_5 = \frac{-12\left[1-\left(-\frac{1}{2}\right)^5\right]}{1-\left(-\frac{1}{2}\right)} = -\frac{33}{4} = -8.25$$

29. $\frac{1}{2}, \frac{1}{4}, 0, \ldots, -\frac{17}{4}$

The first term is $\frac{1}{2}$ and the twentieth term is $-\frac{17}{4}$.

$$S_{20} = \frac{20}{2}\left(\frac{1}{2} - \frac{17}{4}\right) = 10\left(\frac{-15}{4}\right) = -\frac{75}{2}$$

31. $a_1 = 8,\ r = -\frac{2}{3},\ n = 3$

$$S_3 = \frac{8\left[1-\left(-\frac{2}{3}\right)^3\right]}{1-\left(-\frac{2}{3}\right)} = \frac{56}{9}$$

33. The first five terms are 4000, 3950, 3900, 3850, 3800.

$a_1 = 4000,\ d = -50,\ n = 12$

$a_{12} = 4000 + 11(-50) = 3450$

3450 cars will be sold in month 12.

$$S_{12} = \frac{12}{2}(4000 + 3450) = 44{,}700$$

44,700 cars will be sold in the first year.

35. Firm *A*:

The first term is 22,000 and the tenth term is 31,000.

$$S_{10} = \frac{10}{2}(22{,}000 + 31{,}000) = \$265{,}000$$

Firm *B*:

The first term is 20,000 and the tenth term is 30,800.

$$S_{10} = \frac{10}{2}(20{,}000 + 30{,}800) = \$254{,}000$$

Thus, Firm *A* is making the more profitable offer.

37. $a_1 = 30{,}000,\ r = 1.10,\ n = 4$

$a_4 = 30{,}000(1.10)^{4-1} = 39{,}930$

She made \$39,930 during her fourth year of business.

$$S_4 = \frac{30{,}000(1-1.10^4)}{1-1.10} = 139{,}230$$

She made \$139,230 during the first four years of business.

39. $a_1 = 30,\ r = 0.9,\ n = 5$

$a_5 = 30(0.9)^{5-1} = 19.683$

Approximately 20 minutes to assemble the first computer.

$$S_5 = \frac{30(1-0.9^5)}{1-0.9} = 122.853$$

Approximately 123 minutes to assemble the first 5 computers.

41. $a_1 = 20,\ r = \frac{4}{5}$

$$S_\infty = \frac{20}{1-\frac{4}{5}} = 100$$

We double the number (to account for the flight up as well as down) and subtract 20 (since the first bounce was preceded by only a downward flight). Thus, the ball travels 2(100) – 20 = 180 feet.

43. Player *A*:
The first term is 1 and the ninth term is 9.

$$S_9 = \frac{9}{2}(1+9) = 45 \text{ points}$$

Player *B*:
The first term is 10 and the sixth term is 15.

$$S_6 = \frac{6}{2}(10+15) = 75 \text{ points}$$

45. The first term is 200 and the twentieth is 200 – 19(5) = 105.

$$S_{20} = \frac{20}{2}(200+105) = 3050$$

Thus, $3050 rent is paid for 20 days during the holiday rush.

47. $a_1 = 0.01,\ r = 2,\ n = 30$

$$S_3 = \frac{0.01\left[1-2^{30}\right]}{1-2} = 10,737,418.23$$

He would pay $10,737,418.23 in room and board for the 30 days.

49. $6 \cdot 5 \cdot 4 \cdot 3 \cdot 2 \cdot 1 = 720$

51. $\frac{3\cdot 2\cdot 1}{2\cdot 1} = \frac{3\cdot \not{2}\cdot \not{1}}{\not{2}\cdot \not{1}} = 3$

53. $(x+5)^2 = x^2 + 2\cdot x\cdot 5 + 5^2 = x^2 + 10x + 25$

55.
$$\begin{aligned}(2x-1)^3 &= (2x-1)^2(2x-1)\\ &= (4x^2-4x+1)(2x-1)\\ &= 8x^3-4x^2+2x-8x^2+4x-1\\ &= 8x^3-12x^2+6x-1\end{aligned}$$

57.
$$\begin{aligned}0.88\overline{8} &= 0.8+0.08+0.008+\cdots\\ &= \frac{8}{10}+\frac{8}{100}+\frac{8}{1000}+\cdots\end{aligned}$$

This is a geometric series with $a_1 = \frac{8}{10},\ r = \frac{1}{10}$.

$$S_\infty = \frac{\frac{8}{10}}{1-\frac{1}{10}} = \frac{\frac{8}{10}}{\frac{9}{10}} = \frac{8}{10}\cdot\frac{10}{9} = \frac{8}{9}$$

59. Answers may vary.

Section 14.5

Practice Exercises

1. $(p+r)^7$

The $n = 7$ row of Pascal's triangle is

1 7 21 35 35 21 7 1

Using the $n = 7$ row of Pascal's triangle as the coefficients, $(p + r)^7$ can be expanded as

$p^7 + 7p^6r + 21p^5r^2 + 35p^4r^3 + 35p^3r^4 + 21p^2r^5 + 7pr^6 + r^7$

2. a. $\frac{6!}{7!} = \frac{6\cdot5\cdot4\cdot3\cdot2\cdot1}{7\cdot6\cdot5\cdot4\cdot3\cdot2\cdot1} = \frac{1}{7}$

b.
$$\begin{aligned}\frac{8!}{4!2!} &= \frac{8\cdot7\cdot6\cdot5\cdot4!}{4!\cdot2\cdot1}\\ &= \frac{8\cdot7\cdot6\cdot5}{2\cdot1}\\ &= 4\cdot7\cdot6\cdot5\\ &= 840\end{aligned}$$

c. $\frac{5!}{4!1!} = \frac{5\cdot4\cdot3\cdot2\cdot1}{4\cdot3\cdot2\cdot1\cdot1} = 5$

d. $\frac{9!}{9!0!} = \frac{9!}{9!\cdot1} = 1$

3. $(a+b)^9$

Let $n = 9$ in the binomial formula.

$$\begin{aligned}(a+b)^9 &= a^9 + \frac{9}{1!}a^8b + \frac{9\cdot8}{2!}a^7b^2 + \frac{9\cdot8\cdot7}{3!}a^6b^3 + \frac{9\cdot8\cdot7\cdot6}{4!}a^5b^4 + \frac{9\cdot8\cdot7\cdot6\cdot5}{5!}a^4b^5 + \frac{9\cdot8\cdot7\cdot6\cdot5\cdot4}{6!}a^3b^6\\ &\quad + \frac{9\cdot8\cdot7\cdot6\cdot5\cdot4\cdot3}{7!}a^2b^7 + \frac{9\cdot8\cdot7\cdot6\cdot5\cdot4\cdot3\cdot2}{8!}ab^8 + b^9\\ &= a^9 + 9a^8b + 36a^7b^2 + 84a^6b^3 + 126a^5b^4 + 126a^4b^5 + 84a^3b^6 + 36a^2b^7 + 9ab^8 + b^9\end{aligned}$$

4. $(a+5b)^3$

Replace b with $5b$ in the binomial formula.

$$\begin{aligned}(a+5b)^3 &= a^3 + \frac{3}{1!}a^2(5b) + \frac{3\cdot2}{2!}a(5b)^2 + (5b)^3\\ &= a^3 + 3a^2(5b) + 3a(25b^2) + 125b^3\\ &= a^3 + 15a^2b + 75ab^2 + 125b^3\end{aligned}$$

5. $(3x-2y)^3$
Let $a = 3x$ and $b = -2y$ in the binomial formula.

$$\begin{aligned}(3x-2y)^3 &= (3x)^3 + \frac{3}{1!}(3x)^2(-2y) + \frac{3\cdot 2}{2!}(3x)(-2y)^2 + (-2y)^3 \\ &= 27x^3 + 3(9x^2)(-2y) + 3(3x)(4y^2) - 8y^3 \\ &= 27x^3 - 54x^2y + 36xy^2 - 8y^3\end{aligned}$$

6. $(x-4y)^{11}$
Use the formula with $n = 11$, $a = x$, $b = -4y$, and $r + 1 = 7$. Notice that, since $r + 1 = 7$, $r = 6$.

$$\begin{aligned}\frac{n!}{r!(n-r)!}a^{n-r}b^r &= \frac{11!}{6!5!}x^5(-4y)^6 \\ &= 462x^5(4096y^6) \\ &= 1,892,352x^5y^6\end{aligned}$$

Vocabulary and Readiness Check

1. $0! = \underline{1}$

2. $1! = \underline{1}$

3. $4! = 4\cdot 3\cdot 2\cdot 1 = \underline{24}$

4. $2! = 2\cdot 1 = \underline{2}$

5. $3!0! = 3\cdot 2\cdot 1\cdot 1 = \underline{6}$

6. $0!2! = 1\cdot 2\cdot 1 = \underline{2}$

Exercise Set 14.5

1. $(m+n)^3 = m^3 + 3m^2n + 3mn^2 + n^3$

3. $(c+d)^5 = c^5 + 5c^4d + 10c^3d^2 + 10c^2d^3 + 5cd^4 + d^5$

5. $$\begin{aligned}(y-x)^5 &= [y+(-x)]^5 \\ &= y^5 - 5y^4x + 10y^3x^2 - 10y^2x^3 + 5yx^4 - x^5\end{aligned}$$

7. Answers may vary

9. $\frac{8!}{7!} = \frac{8\cdot 7!}{7!} = 8$

11. $\frac{7!}{5!} = \frac{7\cdot 6\cdot 5!}{5!} = 7\cdot 6 = 42$

13. $\frac{10!}{7!2!} = \frac{10\cdot 9\cdot 8\cdot 7!}{7!2!} = \frac{10\cdot 9\cdot 8}{2\cdot 1} = 360$

15. $\frac{8!}{6!0!} = \frac{8\cdot7\cdot6!}{6!1} = 8\cdot7 = 56$

17. Let $n = 7$ in the binomial theorem.

$$(a+b)^7 = a^7 + \frac{7}{1!}a^6b + \frac{7\cdot6}{2!}a^5b^2 + \frac{7\cdot6\cdot5}{3!}a^4b^3 + \frac{7\cdot6\cdot5\cdot4}{4!}a^3b^4 + \frac{7\cdot6\cdot5\cdot4\cdot3}{5!}a^2b^5 + \frac{7\cdot6\cdot5\cdot4\cdot3\cdot2}{6!}ab^6 + b^7$$
$$= a^7 + 7a^6b + 21a^5b^2 + 35a^4b^3 + 35a^3b^4 + 21a^2b^5 + 7ab^6 + b^7$$

19. Let $b = 2b$ and $n = 5$ in the binomial theorem.

$$(a+2b)^5 = a^5 + \frac{5}{1!}a^4(2b) + \frac{5\cdot4}{2!}a^3(2b)^2 + \frac{5\cdot4\cdot3}{3!}a^2(2b)^3 + \frac{5\cdot4\cdot3\cdot2}{4!}a(2b)^4 + (2b)^5$$
$$= a^5 + 10a^4b + 40a^3b^2 + 80a^2b^3 + 80ab^4 + 32b^5$$

21. Let $a = q$, $b = r$, and $n = 9$ in the binomial theorem.

$$(q+r)^2 = q^9 + \frac{9}{1!}q^8r + \frac{9\cdot8}{2!}q^7r^2 + \frac{9\cdot8\cdot7}{3!}q^6r^3 + \frac{9\cdot8\cdot7\cdot6}{4!}q^5r^4 + \frac{9\cdot8\cdot7\cdot6\cdot5}{5!}q^4r^5 + \frac{9\cdot8\cdot7\cdot6\cdot5\cdot4}{6!}q^3r^6$$
$$+ \frac{9\cdot8\cdot7\cdot6\cdot5\cdot4\cdot3}{7!}q^2r^7 + \frac{9\cdot8\cdot7\cdot6\cdot5\cdot4\cdot3\cdot2}{8!}qr^8 + r^9$$
$$= q^9 + 9q^8r + 36q^7r^2 + 84q^6r^3 + 126q^5r^4 + 126q^4r^5 + 84q^3r^6 + 36q^2r^7 + 9qr^8 + r^9$$

23. Let $a = 4a$ and $n = 5$ in the binomial theorem.

$$(4a+b)^5 = (4a)^5 + \frac{5}{1!}(4a)^4b + \frac{5\cdot4}{2!}(4a)^3b^2 + \frac{5\cdot4\cdot3}{3!}(4a)^2b^3 + \frac{5\cdot4\cdot3\cdot2}{4!}(4a)b^4 + b^5$$
$$= 1024a^5 + 1280a^4b + 640a^3b^2 + 160a^2b^3 + 20ab^4 + b^5$$

25. Let $a = 5a$, $b = -2b$, and $n = 4$ in the binomial theorem.

$$(5a-2b)^4 = (5a)^4 + \frac{4}{1!}(5a)^3(-2b) + \frac{4\cdot3}{2!}(5a)^2(-2b)^2 + \frac{4\cdot3\cdot2}{3!}(5a)(-2b)^3 + (-2b)^4$$
$$= 625a^4 - 1000a^3b + 600a^2b^2 - 160ab^3 + 16b^4$$

27. Let $a = 2a$, $b = 3b$, and $n = 3$ in the binomial theorem.

$$(2a+3b)^3 = (2a)^3 + \frac{3}{1!}(2a)^2(3b) + \frac{3\cdot2}{2!}(2a)(3b)^2 + (3b)^3$$
$$= 8a^3 + 36a^2b + 54ab^2 + 27b^3$$

29. Let $a = x$, $b = 2$, and $n = 5$ in the binomial theorem.

$$(x+2)^5 = x^5 + \frac{5}{1!}x^4(2) + \frac{5\cdot4}{2!}x^3(2)^2 + \frac{5\cdot4\cdot3}{3!}x^2(2)^3 + \frac{5\cdot4\cdot3\cdot2}{4!}x(2)^4 + (2)^5$$
$$= x^5 + 10x^4 + 40x^3 + 80x^2 + 80x + 32$$

31. 5th term of $(c-d)^5$ corresponds to $r = 4$:

$$\frac{5!}{4!(5-4)!}c^{5-4}(-d)^4 = 5cd^4$$

33. 8th term of $(2c+d)^7$ corresponds to $r = 7$:

$$\frac{7!}{7!(7-7)!}(2c)^{7-7}(d)^7 = d^7$$

35. 4th term of $(2r-s)^5$ corresponds to $r=3$:

$$\frac{5!}{3!(5-3)!}(2r)^{5-3}(-s)^3 = -40r^2s^3$$

37. 3rd term of $(x+y)^4$ corresponds to $r=2$: $\frac{4!}{2!(4-2)!}(x)^{4-2}(y)^2 = 6x^2y^2$

39. 2nd term of $(a+3b)^{10}$ corresponds to $r=1$: $\frac{10!}{1!(10-1)!}(a)^{10-1}(3b)^1 = 30a^9b$

41. $f(x)=|x|$

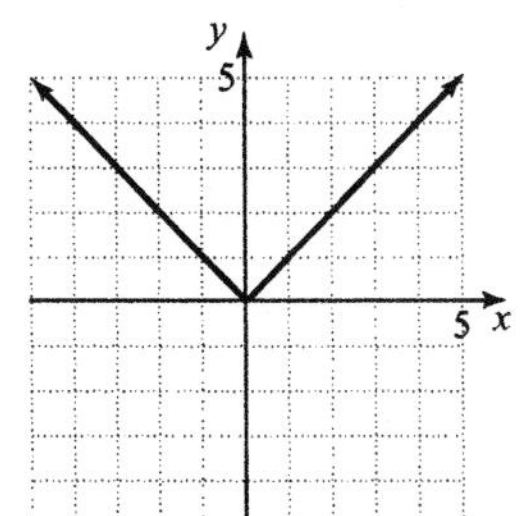

Not one-to-one

43. $H(x)=2x+3$

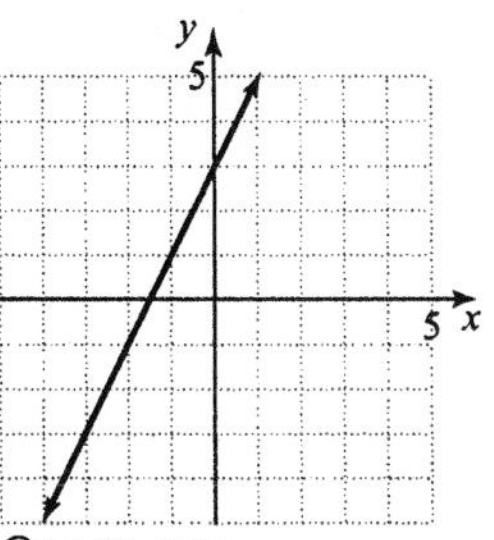

One-to-one

45. $f(x)=x^2+3$

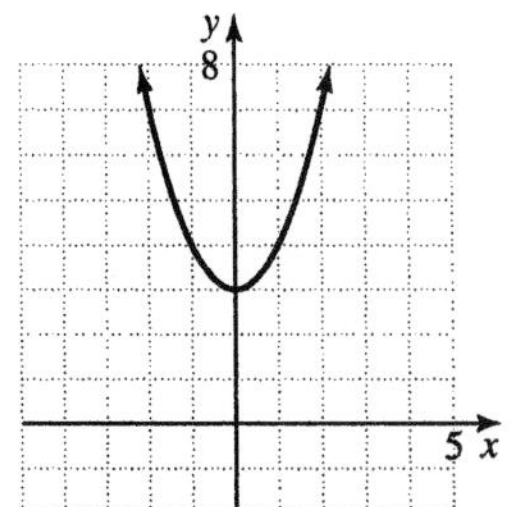

Not one-to-one

47. $(\sqrt{x}+\sqrt{3})^5$

Use the binomial theorem with $n=5$, $a=\sqrt{x}$, , and $b=\sqrt{3}$.

$$\left(\sqrt{x}+\sqrt{3}\right)^5 = \left(\sqrt{x}\right)^5 + 5\left(\sqrt{x}\right)^4\left(\sqrt{3}\right) + 10\left(\sqrt{x}\right)^3\left(\sqrt{3}\right)^2 + 10\left(\sqrt{x}\right)^2\left(\sqrt{3}\right)^3 + 5\left(\sqrt{x}\right)\left(\sqrt{3}\right)^4 + \left(\sqrt{3}\right)^5$$

$$= x^2\sqrt{x} + 5\sqrt{3}x^2 + 30x\sqrt{x} + 30\sqrt{3}x + 45\sqrt{x} + 9\sqrt{3}$$

49. $\binom{9}{5} = \frac{9!}{5!(9-5)!}$
$= \frac{9!}{5!4!}$
$= \frac{9\cdot8\cdot7\cdot6\cdot5\cdot4\cdot3\cdot2\cdot1}{(5\cdot4\cdot3\cdot2\cdot1)\cdot(4\cdot3\cdot2\cdot1)}$
$= 126$

51. $\binom{8}{2} = \frac{8!}{2!(8-2)!}$
$= \frac{8!}{2!6!}$
$= \frac{8\cdot7\cdot6\cdot5\cdot4\cdot3\cdot2\cdot1}{(2\cdot1)\cdot(6\cdot5\cdot4\cdot3\cdot2\cdot1)}$
$= 28$

53. Answers may vary.

Chapter 14 Vocabulary Check

1. A finite sequence is a function whose domain is the set of natural numbers $\{1, 2, 3, ..., n\}$, where n is some natural number.

2. The factorial of n, written $n!$, is the product of the first n consecutive natural numbers.

3. An infinite sequence is a function whose domain is the set of natural numbers.

4. A geometric sequence is a sequence in which each term (after the first) is obtained by multiplying the preceding term by a constant amount r. The constant r is called the common ratio of the sequence.

5. A sum of the terms of a sequence is called a series.

6. The nth term of the sequence a_n is called the general term.

7. An arithmetic sequence is a sequence in which each term (after the first) differs from the preceding term by a constant amount d. The constant d is called the common difference of the sequence.

8. A triangle array of the coefficients of the terms of the expansions of $(a+b)^n$ is called Pascal's triangle.

Chapter 14 Review

1. $a_n = -3n^2$
$a_1 = -3(1)^2 = -3$
$a_2 = -3(2)^2 = -12$
$a_3 = -3(3)^2 = -27$
$a_4 = -3(4)^2 = -48$
$a_5 = -3(5)^2 = -75$

2. $a_n = n^2 + 2n$
$a_1 = 1^2 + 2(1) = 3$
$a_2 = 2^2 + 2(2) = 8$
$a_3 = 3^2 + 2(3) = 15$
$a_4 = 4^2 + 2(4) = 24$
$a_5 = 5^2 + 2(5) = 35$

3. $a_n = \frac{(-1)^n}{100}$
$a_{100} = \frac{(-1)^{100}}{100} = \frac{1}{100}$

4. $a_n = \frac{2n}{(-1)^2}$
$a_{50} = \frac{2(50)}{(-1)^2} = 100$

5. $\frac{1}{6\cdot1}, \frac{1}{6\cdot2}, \frac{1}{6\cdot3}, ...$
In general, $a_n = \frac{1}{6n}$.

6. $-1, 4, -9, 16, ...$
$a_n = (-1)^n n^2$

7. $a_n = 32n - 16$
$a_5 = 32(5) - 16 = 144$ feet
$a_6 = 32(6) - 16 = 176$ feet
$a_7 = 32(7) - 16 = 208$ feet

8. $a_n = 100(2)^{n-1}$
$10,000 = 100(2)^{n-1}$
$100 = 2^{n-1}$
$\log 100 = (n-1)\log 2$
$n = \frac{\log 100}{\log 2} + 1 \approx 7.6$

Eighth day culture will be at least 10,000.
Since $n = 1$ corresponds to the end of the first day, the original amount corresponds to $n = 0$.

$$a_0 = 100(2)^{-1} = 100\left(\frac{1}{2}\right) = 50$$

The original measure of the culture was 50.

9. 2006: $a_1 = 660,000$

2007: $a_2 = 660,000(2) = 1,320,000$

2008: $a_3 = 1,320,000(2) = 2,640,000$

2009: $a_4 = 2,640,000(2) = 5,280,000$

2010: $a_5 = 5,280,000(2) = 10,560,000$

There will be 10,560,000 acres of infested trees in 2010.

10. $a_n = 50 + (n-1)8$

$a_1 = 50$
$a_2 = 50 + 8 = 58$
$a_3 = 50 + 2(8) = 66$
$a_4 = 50 + 3(8) = 74$
$a_5 = 50 + 4(8) = 82$
$a_6 = 50 + 5(8) = 90$
$a_7 = 50 + 6(8) = 98$
$a_8 = 50 + 7(8) = 106$
$a_9 = 50 + 8(8) = 114$
$a_{10} = 50 + 9(8) = 122$

There are 122 seats in the tenth row.

11. $a_1 = -2, r = \frac{2}{3}$

$a_1 = -2$

$$a_2 = -2\left(\frac{2}{3}\right) = -\frac{4}{3}$$

$$a_3 = \left(-\frac{4}{3}\right)\left(\frac{2}{3}\right) = -\frac{8}{9}$$

$$a_4 = \left(-\frac{8}{9}\right)\left(\frac{2}{3}\right) = -\frac{16}{27}$$

$$a_5 = \left(-\frac{16}{27}\right)\left(\frac{2}{3}\right) = -\frac{32}{81}$$

12. $a_n = 12 + (n-1)(-1.5)$

$a_1 = 12$
$a_2 = 12 + (1)(-1.5) = 10.5$
$a_3 = 12 + 2(-1.5) = 9$
$a_4 = 12 + 3(-1.5) = 7.5$
$a_5 = 12 + 4(-1.5) = 6$

13. $a_n = -5 + (n-1)^4$

$a_{30} = 5 + (30-1)4 = 111$

14. $a_n = 2 + (n-1)\frac{3}{4}$

$$a_{11} = 2 + 10\left(\frac{3}{4}\right) = \frac{19}{2}$$

15. 12, 7, 2,...

$a_1 = 12,\ d = -5,\ n = 20$
$a_{20} = 12 + (20-1)(-5) = -83$

16. $a_n = a_1 r^{n-1},\ a_1 = 4,\ r = \frac{3}{2}$

$$a_6 = 4\left(\frac{3}{2}\right)^{6-1} = \frac{243}{8}$$

17. $a_4 = 18,\ a_{20} = 98$

$$\begin{cases} a_4 = a_1 + (4-1)d \\ a_{20} = a_1 + (20-1)d \end{cases}$$

$$\begin{cases} 18 = a_1 + 3d \\ 98 = a_1 + 19d \end{cases}$$

$$\begin{cases} -18 = -a_1 - 3d \\ 98 = a_1 + 19d \end{cases}$$

Adding yields $80 = 16d$ or $d = 5$.
Then $a_1 = 18 - 3(5) = 3$.

18. $a_3 = -48,\ a_4 = 192$

$$r = \frac{a_4}{a_3} = \frac{192}{-48} = -4$$

$a_3 = a_1 r^{3-1}$
$-48 = a_1(-4)^2$
$-48 = 16a_1$
$-3 = a_1$
$r = -4,\ a_1 = -3$

19. $\frac{3}{10}, \frac{3}{10^2}, \frac{3}{10^3}, \ldots$

In general, $a_n = \frac{3}{10^n}$

20. 50, 58, 66, ...

$a_n = 50 + (n-1)8$ or $a_n = 42 + 8n$

21. $\frac{8}{3}, 4, 6, \ldots$

Geometric; $a_1 = \frac{8}{3}$,

$r = \frac{4}{\frac{8}{3}} = 4 \cdot \frac{3}{8} = \frac{12}{8} = \frac{3}{2}$

22. $-10.5, -6.1, -1.7$
Arithmetic; $a_1 = -10.5$,
$d = -6.1 - (-10.5) = 4.4$

23. $7x, -14x, 28x$

Geometric; $a_1 = 7x$, $r = -2$

24. neither

25. $a_1 = 8$, $r = 0.75$
$a_1 = 8$
$a_2 = 8(0.75) = 6$
$a_3 = 8(0.75)^2 = 4.5$
$a_4 = 8(0.75)^3 \approx 3.4$
$a_5 = 8(0.75)^4 \approx 2.5$
$a_6 = 8(0.75)^5 \approx 1.9$
Yes, a ball that rebounds to a height of 2.5 feet after the fifth bounce is good, since $2.5 \geq 1.9$.

26. $a_1 = 25$, $d = -4$
$a_n = a_1 + (n-1)d$
$a_n = 25 + (n-1)(-4) = 29 - 4n$
$a_7 = 25 + 6(-4) = 1$
Continuing the progression as far as possible leaves 1 can in the top row.

27. $a_1 = 1$, $r = 2$
$a_n = 2^{n-1}$
$a_{10} = 2^9 = 512$
$a_{30} = 2^{29} = 536{,}870{,}912$
You save $512 on the tenth day and $536,870,912 on the thirtieth day.

28. $a_n = a_1 r^{n-1}$, $a_1 = 30$, $r = 0.7$
$a_5 = 30(0.7)^4 = 7.203$
The length is 7.203 inches on the fifth swing.

29. $a_1 = 900$, $d = 150$
$a_n = 900 + (n-1)150 = 150_n + 750$
$a_6 = 900 + (6-1)150 = 1650$
Her salary is $1650 per month at the end of training.

30. $\frac{1}{512}, \frac{1}{256}, \frac{1}{128}, \ldots$

first fold: $a_1 = \frac{1}{256}$, $r = 2$

$a_{15} = \frac{1}{256}(2)^{15-1} = 64$
After 15 folds, the thickness is 64 inches.

31.
$$\begin{aligned}\sum_{i=1}^{5}(2i-1) &= [2(1)-1]+[2(2)-1]+[2(3)-1] \\ &\quad +[2(4)-1]+[2(5)-1] \\ &= 1+3+5+7+9 \\ &= 25\end{aligned}$$

32.
$$\begin{aligned}\sum_{i=1}^{5} i(i+2) &= 1(1+2)+2(2+2)+3(3+2) \\ &\quad +4(4+2)+5(5+2) \\ &= 3+8+15+24+35 \\ &= 85\end{aligned}$$

33.
$$\begin{aligned}\sum_{i=2}^{4}\frac{(-1)^i}{2i} &= \frac{(-1)^2}{2(2)}+\frac{(-1)^3}{2(3)}+\frac{(-1)^4}{2(4)} \\ &= \frac{1}{4}-\frac{1}{6}+\frac{1}{8} \\ &= \frac{5}{24}\end{aligned}$$

34.
$$\begin{aligned}\sum_{i=3}^{5} 5(-1)^{i-1} &= 5(-1)^{3-1}+5(-1)^{4-1}+5(-1)^{5-1} \\ &= 5(1)+5(-1)+5(1) \\ &= 5-5+5 \\ &= 5\end{aligned}$$

35. $a_n = (n-3)(n+2)$
$$\begin{aligned}S_4 &= \sum_{i=1}^{4}(i-3)(i+2) \\ &= (1-3)(1+2)+(2-3)(2+2) \\ &\quad +(3-3)(3+2)+(4-3)(4+2) \\ &= -6-4+0+6 \\ &= -4\end{aligned}$$

36. $a_n = n^2$

$$S_6 = \sum_{i=1}^{6} i^2$$
$$= (1)^2 + (2)^2 + (3)^2 + (4)^2 + (5)^2 + (6)^2$$
$$= 91$$

37. $a_n = -8 + (n-1)3 = 3n - 11$

$$S_5 = \sum_{i=1}^{5} (3i - 11)$$
$$= [3(1) - 11] + [3(2) - 11] + [3(3) - 11] + [3(4) - 11] + [3(5) - 11]$$
$$= -8 - 5 - 2 + 1 + 4$$
$$= -10$$

38. $a_n = 5(4)^{n-1}$

$$S_3 = \sum_{i=1}^{3} 5(4)^{i-1} = 5(4)^0 + 5(4)^1 + 5(4)^2 = 105$$

39. $1 + 3 + 9 + 27 + 81 + 243$

$$= 3^0 + 3^1 + 3^2 + 3^3 + 3^4 + 3^5$$
$$= \sum_{i=1}^{6} 3^{i-1}$$

40. $6 + 2 + (-2) + (-6) + (-10) + (-14) + (-18)$

$a_1 = 6,\ d = -4$

$a_n = 6 + (n-1)(-4)$

$$\sum_{i=1}^{7} [6 + (i-1)(-4)]$$

41. $$\frac{1}{4} + \frac{1}{16} + \frac{1}{64} + \frac{1}{256} = \frac{1}{4^1} + \frac{1}{4^2} + \frac{1}{4^3} + \frac{1}{4^4} = \sum_{i=1}^{4} \frac{1}{4^i}$$

42. $$1 + \left(-\frac{3}{2}\right) + \frac{9}{4} = \left(-\frac{3}{2}\right)^0 + \left(-\frac{3}{2}\right)^1 + \left(-\frac{3}{2}\right)^2 = \sum_{i=1}^{3} \left(-\frac{3}{2}\right)^{i-1}$$

43. $a_1 = 20,\ r = 2$

$a_n = 20(2)^n$ represents the number of yeast, where n represents the number of 8-hour periods. Since $48 = 6(8)$ here, $n = 6$.

$a_6 = 20(2)^6 = 1280$

There are 1280 yeast after 48 hours.

44. $a_n = n^2 + 2n - 1$

$a_4 = (4)^2 + 2(4) - 1 = 23$

$$S_4 = \sum_{i=1}^{4} (i^2 + 2i - 1)$$
$$= (1 + 2 - 1) + (4 + 4 - 1) + (9 + 6 - 1) + (16 + 8 - 1)$$
$$= 46$$

23 cranes are born in the fourth year and 46 cranes are born in the first four years.

45. For Job *A*: $a_1 = 39{,}500,\ d = 2200;$

$a_5 = 39{,}500 + (5-1)2200 = \$48{,}330$

For Job *B*: $a_1 = 41{,}000,\ d = 1400$

$a_5 = 41{,}000 + (5-1)1400 = \$46{,}600$

For the fifth year, Job *A* has a higher salary.

46. $a_n = 200(0.5)^n$

$a_3 = 200(0.5)^3 = 25$

$$S_3 = \sum_{i=1}^{3} 200(0.5)^i$$
$$= 200(0.5) + 200(0.5)^2 + 200(0.5)^3$$
$$= 175$$

25 kilograms decay in the third year and 175 kilograms decay in the first three years.

47. 15, 19, 23, ...

$a_1 = 15,\ d = 4,\ a_6 = 15 + (6-1)4 = 35$

$$S_6 = \frac{6}{2}[15 + 35] = 150$$

48. 5, −10, 20,...

$a_1 = 5,\ r = -2$

$$S_n = \frac{a_1(1 - r^n)}{1 - r}$$

$$S_9 = \frac{5(1 - (-2)^9)}{1 - (-2)} = 855$$

49. $a_1 = 1,\ d = 2,\ n = 30,\ a_{30} = 1 + (30-1)2 = 59$

$S_{30} = \frac{30}{2}[1+59] = 900$

50. 7, 14, 21, 28, ...

$a_n = 7 + (n-1)7$

$a_{20} = 7 + (20-1)7 = 140$

$S_{20} = \frac{20}{2}(7+140) = 1470$

51. 8, 5, 2, ...

$a_1 = 8,\ d = -3,\ n = 20$

$a_{20} = 8 + (20-1)(-3) = -49$

$S_{20} = \frac{20}{2}[8 + (-49)]$

$= -410$

52. $\frac{3}{4}, \frac{9}{4}, \frac{27}{4}, \ldots$

$a_1 = \frac{3}{4},\ r = 3$

$S_8 = \frac{\frac{3}{4}(1-3^8)}{1-3} = 2460$

53. $a_1 = 6,\ r = 5$

$S_4 = \frac{6(1-5^4)}{1-5} = 936$

54. $a_1 = -3,\ d = -6$

$a_n = -3 + (n-1)(-6)$

$a_{100} = -3 + (100-1)(-6) = -597$

$S_{100} = \frac{100}{2}(-3 + (-597)) = -30{,}000$

55. $5, \frac{5}{2}, \frac{5}{4}, \ldots$

$a_1 = 5,\ r = \frac{1}{2}$

$S_\infty = \frac{5}{1-\frac{1}{2}} = 10$

56. $18, -2, \frac{2}{9}, \ldots$

$a_1 = 18,\ r = -\frac{1}{9}$

$S_\infty = \frac{18}{1+\frac{1}{9}} = \frac{81}{5}$

57. $-20, -4, -\frac{4}{5}, \ldots$

$a_1 = -20,\ r = \frac{1}{5}$

$S_\infty = \frac{-20}{1-\frac{1}{5}} = -25$

58. 0.2, 0.02, 0.002,...

$a_1 = 0.2 = \frac{1}{5},\ r = \frac{1}{10}$

$S_\infty = \frac{\frac{1}{5}}{1-\frac{1}{10}} = \frac{2}{9}$

59. $a_1 = 20{,}000,\ r = 1.15,\ n = 4$

$a_4 = 20{,}000(1.15)^{4-1} = 30{,}418$

$S_4 = \frac{20{,}000(1-1.15^4)}{1-1.15} = 99{,}868$

He earned $30,418 during the fourth year and $99,868 over the four years.

60. $a_n = 40(0.8)^{n-1}$

$a_4 = 40(0.8)^{4-1} = 20.48$

$S_4 = \frac{40(1-0.8^4)}{1-0.8} = 118.08$

He takes 20 minutes to assemble the fourth television and 118 minutes to assemble the first four televisions.

61. $a_1 = 100, d = -7, n = 7$

$a_7 = 100 + (7-1)(-7) = 58$

$S_7 = \frac{7}{2}(100+58) = 553$

The rent for the seventh day is $58 and the rent for 7 days is $553.

62. $a_1 = 15, r = 0.8$

$S_\infty = \dfrac{15}{1-0.8} = 75$ feet downward

$a_1 = 15(0.8) = 12, r = 0.8$

$S_\infty = \dfrac{12}{1-0.8} = 60$ feet upward

The total distance is 135 feet.

63. 1800, 600, 200,...

$a_1 = 1800, r = \dfrac{1}{3}, n = 6$

$$S_6 = 1800\frac{\left(1-\left(\frac{1}{3}\right)^6\right)}{1-\frac{1}{3}} \approx 2696$$

Approximately 2696 mosquitoes were killed during the first six days after the spraying.

64. 1800, 600, 200, ...

For which n is $a_n < 1$?

$$a_n = 1800\left(\frac{1}{3}\right)^{n-1} < 1$$

$$\left(\frac{1}{3}\right)^{n-1} < \frac{1}{1800}$$

$$(n-1)\log\frac{1}{3} < \log\frac{1}{1800}$$

$$(n-1)\log 3^{-1} < \log 1800^{-1}$$

$$(n-1)(-\log 3) < -\log 1800$$

$$n-1 > \frac{-\log 1800}{-\log 3}$$

$$n > 1 + \frac{\log 1800}{\log 3}$$

$$n > 7.8$$

No longer effective on the 8th day

$$S_8 = \frac{1800\left(1-\left(\frac{1}{3}\right)^8\right)}{1-\frac{1}{3}} \approx 2700$$

About 2700 mosquitoes were killed.

65. $0.55\overline{5} = 0.5 + 0.05 + 0.005 + \cdots$

$a_1 = 0.5, r = 0.1$

$$S_\infty = \frac{0.5}{1-0.1} = \frac{0.5}{0.9} = \frac{5}{9}$$

66. 27, 30, 33, ...

$a_n = 27 + (n-1)(3)$

$a_{20} = 27 + (20-1)(3) = 84$

$S_{20} = \frac{20}{2}(27+84) = 1110$

There are 1110 seats in the theater.

67. $(x+z)^5 = x^5 + 5x^4z + 10x^3z^2 + 10x^2z^3 + 5xz^4 + z^5$

68.
$$\begin{aligned}(y-r)^6 &= y^6 + 6y^5(-r) + 15y^4(-r)^2 + 20y^3(-r)^3 + 15y^2(-r)^4 + 6y(-r)^5 + (-r)^6 \\ &= y^6 - 6y^5r + 15y^4r^2 - 20y^3r^3 + 15y^2r^4 - 6yr^5 + r^6\end{aligned}$$

69.
$$\begin{aligned}(2x+y)^4 &= (2x)^4 + 4(2x)^3y + 6(2x)^2y^2 + 4(2x)y^3 + y^4 \\ &= 16x^4 + 32x^3y + 24x^2y^2 + 8xy^3 + y^4\end{aligned}$$

70.
$$\begin{aligned}(3y-z)^4 &= (3y)^4 + 4(3y)^3(-z) + 6(3y)^2(-z)^2 + 4(3y)(-z)^3 + (-z)^4 \\ &= 81y^4 - 108y^3z + 54y^2z^2 - 12yz^3 + z^4\end{aligned}$$

71.
$$\begin{aligned}(b+c)^8 &= b^8 + \frac{8}{1!}b^7c + \frac{8\cdot7}{2!}b^6c^2 + \frac{8\cdot7\cdot6}{3!}b^5c^3 + \frac{8\cdot7\cdot6\cdot5}{4!}b^4c^4 + \frac{8\cdot7\cdot6\cdot5\cdot4}{5!}b^3c^5 \\ &\quad + \frac{8\cdot7\cdot6\cdot5\cdot4\cdot3}{6!}b^2c^6 + \frac{8\cdot7\cdot6\cdot5\cdot4\cdot3\cdot2}{7!}bc^7 + c^8 \\ &= b^8 + 8b^7c + 28b^6c^2 + 56b^5c^3 + 70b^4c^4 + 56b^3c^5 + 28b^2c^6 + 8bc^7 + c^8\end{aligned}$$

72.
$$\begin{aligned}(x-w)^7 &= x^7 + \frac{7}{1!}x^6(-w) + \frac{7\cdot6}{2!}x^5(-w)^2 + \frac{7\cdot6\cdot5}{3!}x^4(-w)^3 + \frac{7\cdot6\cdot5\cdot4}{4!}x^3(-w)^4 + \frac{7\cdot6\cdot5\cdot4\cdot3}{5!}x^2(-w)^5 \\ &\quad + \frac{7\cdot6\cdot5\cdot4\cdot3\cdot2}{6!}x(-w)^6 + (-w)^7 \\ &= x^7 - 7x^6w + 21x^5w^2 - 35x^4w^3 + 35x^3w^4 - 21x^2w^5 + 7xw^6 - w^7\end{aligned}$$

73.
$$\begin{aligned}(4m-n)^4 &= (4m)^4 + \frac{4}{1!}(4m)^3(-n) + \frac{4\cdot3}{2!}(4m)^2(-n)^2 + \frac{4\cdot3\cdot2}{3!}(4m)(-n)^3 + (-n)^4 \\ &= 256m^4 - 256m^3n + 96m^2n^2 - 16mn^3 + n^4\end{aligned}$$

74.
$$\begin{aligned}(p-2r)^5 &= p^5 + \frac{5}{1!}p^4(-2r) + \frac{5\cdot4}{2!}p^3(-2r)^2 + \frac{5\cdot4\cdot3}{3!}p^2(-2r)^3 + \frac{5\cdot4\cdot3\cdot2}{4!}p(-2r)^4 + (-2r)^5 \\ &= p^5 - 10p^4r + 40p^3r^2 - 80p^2r^3 + 80pr^4 - 32r^5\end{aligned}$$

75. The 4th term corresponds to $r = 3$.

$$\frac{7!}{3!(7-3)!}a^{7-3}b^3 = 35a^4b^3$$

76. The 11th term corresponds to $r = 10$.

$$\frac{10!}{10!0!}y^{10-10}(2z)^{10} = 1024z^{10}$$

77. $\sum_{i=1}^{4} i^2(i+1) = 1^2(1+1)+2^2(2+1)+3^2(3+1)+4^2(4+1)$

$$= 1\cdot 2+4\cdot 3+9\cdot 4+16\cdot 5$$
$$= 2+12+36+80$$
$$= 130$$

78. $a_1 = 14$

$d = 8 - 14 = -6$

$a_n = a_1 + (n-1)d$

$a_{15} = 14+(15-1)(-6) = 14+14(-6) = 14+(-84) = -70$

79. $a_1 = 27$

$r = \frac{9}{27} = \frac{1}{3}$

$S_\infty = \frac{27}{1-\frac{1}{3}} = \frac{27}{\frac{2}{3}} = 27\cdot\frac{3}{2} = \frac{81}{2} = 40.5$

80. $(2x-3)^4 = (2x)^4 + \frac{4}{1!}(2x)^3(-3) + \frac{4\cdot 3}{2!}(2x)^2(-3)^2 + \frac{4\cdot 3\cdot 2}{3!}(2x)(-3)^3 + (-3)^4$

$$= 16x^4 - 96x^3 + 216x^2 - 216x + 81$$

Chapter 14 Test

1. $a_n = \frac{(-1)^n}{n+4}$

$a_1 = \frac{(-1)^1}{1+4} = -\frac{1}{5}$

$a_2 = \frac{(-1)^2}{2+4} = \frac{1}{6}$

$a_3 = \frac{(-1)^3}{3+4} = -\frac{1}{7}$

$a_4 = \frac{(-1)^4}{4+4} = \frac{1}{8}$

$a_5 = \frac{(-1)^5}{5+4} = -\frac{1}{9}$

2. $a_n = 10+3(n-1)$

$a_{80} = 10+3(80-1) = 247$

3. $\frac{2}{5}, \frac{2}{25}, \frac{2}{125}, \ldots$

In general, $a_n = \frac{2}{5}\left(\frac{1}{5}\right)^{n-1}$ or $a_n = \frac{2}{5^n}$.

4. $(-1)^1 9\cdot 1, (-1)^2 9\cdot 2, \ldots, a_n = (-1)^n 9n$

5. $a_n = 5(2)^{n-1}, S_5 = \frac{5(1-2^5)}{1-2} = 155$

6. $a_n = 18 + (n-1)(-2)$

$a_1 = 18,\ a_{30} = 18 + (30-1)(-2) = -40$

$S_{30} = \frac{30}{2}[18 - 40] = -330$

7. $a_1 = 24,\ r = \frac{1}{6}$

$S_\infty = \frac{24}{1-\frac{1}{6}} = \frac{144}{5}$

8. $\frac{3}{2}, -\frac{3}{4}, \frac{3}{8}, \ldots$

$a_1 = \frac{3}{2},\ r = -\frac{1}{2}$

$S_\infty = \frac{\frac{3}{2}}{1-\left(-\frac{1}{2}\right)} = 1$

9. $\sum_{i=1}^{4} i(i-2) = 1(1-2) + 2(2-2) + 3(3-2) + 4(4-2)$

$= -1 + 0 + 3 + 8 - 20 + 40 - 80$

$= 10$

10. $\sum_{i=2}^{4} 5(2)^i(-1)^{i-1} = 5(2)^2(-1)^{2-1} + 5(2)^3(-1)^{3-1} + 5(2)^4(-1)^{4-1} = -20 + 40 - 80 = -60$

11. $(a-b)^6 = a^6 - 6a^5b + 15a^4b^2 - 20a^3b^3 + 15a^2b^4 - 6ab^5 + b^6$

12. $(2x+y)^5 = (2x)^5 + \frac{5}{1!}(2x)^4y + \frac{5\cdot4}{2!}(2x)^3y^2 + \frac{5\cdot4\cdot3}{3!}(2x)^2y^3 + \frac{5\cdot4\cdot3\cdot2}{4!}(2x)y^4 + y^5$

$= 32x^5 + 80x^4y + 80x^3y^2 + 40x^2y^3 + 10xy^4 + y^5$

13. $a_n = 250 + 75(n-1)$

$a_{10} = 250 + 75(10-1) = 925$

There were 925 people in the town at the beginning of the tenth year.

$a_1 = 250 + 75(1-1) = 250$

There were 250 people in the town at the beginning of the first year.

14. 1, 3, 5, ...

$a_1 = 1,\ d = 2,\ n = 8$

$a_8 = 1 + (8-1)2 = 15$

$1+3+5+7+9+11+13+15$

$S_8 = \frac{8}{2}[1 + 15] = 64$

There were 64 shrubs planted in the 8 rows.

15. $a_1 = 80$, $r = \frac{3}{4}$, $n = 4$

$$a_4 = 80\left(\frac{3}{4}\right)^{4-1} = 33.75$$

The arc length is 33.75 cm on the 4th swing.

$$S_4 = \frac{80\left(1-\left(\frac{3}{4}\right)^4\right)}{1-\frac{3}{4}} = 218.75$$

The total of the arc lengths is 218.75 cm for the first 4 swings.

16. $a_1 = 80$, $r = \frac{3}{4}$

$$S_\infty = \frac{80}{1-\frac{3}{4}} = 320$$

The total of the arc lengths is 320 cm before the pendulum comes to rest.

17. 16, 48, 80,...

$$a_{10} = 16 + (10-1)32 = 304$$

He falls 304 feet during the 10th second.

$$S_{10} = \frac{10}{2}[16+304] = 1600$$

He falls 1600 feet during the first 10 seconds.

18. $0.42\overline{42} = 0.42 + 0.0042 + 0.000042$

$$a_1 = 0.42 = \frac{42}{100},\ r = 0.01 = \frac{1}{100}$$

$$S_\infty = \frac{\frac{42}{100}}{1-\frac{1}{100}} = \frac{42}{100}\cdot\frac{100}{99} = \frac{14}{33}$$

Thus, $0.42\overline{42} = \frac{14}{33}$.

Chapter 14 Cumulative Review

1. a. $(-2)^3 = (-2)(-2)(-2) = -8$

b. $-2^3 = -(2)(2)(2) = -8$

c. $(-3)^2 = (-3)(-3) = 9$

d. $-3^2 = -(3)(3) = -9$

2. a. $3a - (4a+3) = 3a - 4a - 3 = -a - 3$

b. $(5x-3) + (2x+6) = 7x + 3$

c.
$$\begin{aligned} 4(2x-5) - 3(5x+1) &= 8x - 20 - 15x - 3 \\ &= -7x - 23 \end{aligned}$$

3. $(2x-3) - (4x-2) = 2x - 3 - 4x + 2 = -2x - 1$

4. Let x = the price before taxes, then

$$\begin{aligned} x + 0.06x &= 344.50 \\ 1.06x &= 344.50 \\ x &= 325 \end{aligned}$$

The price before taxes was $325.

5. $y = mx + b$; $m = \frac{1}{4}$, $b = -3$

$$y = \frac{1}{4}x - 3$$

6. If the line is to be parallel, then the slope has to be the same as the slope of the given line.

Therefore, $m = \frac{3}{2}$.

$$\begin{aligned} (y-(-2)) &= \frac{3}{2}(x-3) \\ y+2 &= \frac{3}{2}(x-3) \\ y &= \frac{3}{2}x - \frac{13}{2} \\ f(x) &= \frac{3}{2}x - \frac{13}{2} \end{aligned}$$

7. (2, 5), (−3, 4)

$$m = \frac{y_2 - y_1}{x_2 - x_1} = \frac{4-5}{-3-2} = \frac{-1}{-5} = \frac{1}{5}$$

$$\begin{aligned} y - y_1 &= m(x - x_1) \\ y - 5 &= \frac{1}{5}(x-2) \\ 5y - 25 &= x - 2 \\ -23 &= x - 5y \text{ or } x - 5y = -23 \end{aligned}$$

8.
$$\begin{aligned} y^3 + 5y^2 - y &= 5 \\ y^3 + 5y^2 - y - 5 &= 0 \\ (y^3 + 5y^2) + (-y - 5) &= 0 \\ y^2(y+5) - 1(y+5) &= 0 \\ (y^2 - 1)(y+5) &= 0 \\ (y+1)(y-1)(y+5) &= 0 \\ y &= -5, -1, 1 \end{aligned}$$

9.
$$\begin{array}{r|rrrrr} -2 & 1 & -2 & -11 & 5 & 34 \\ & & -2 & 8 & 6 & -22 \\ \hline & 1 & -4 & -3 & 11 & 12 \end{array}$$

Answer: $x^3-4x^2-3x+11+\frac{12}{x+2}$

10.
$$\frac{5}{3a-6}-\frac{a}{a-2}+\frac{3+2a}{5a-10}$$
$$=\frac{5}{3(a-2)}-\frac{a}{a-2}+\frac{3+2a}{5(a-2)}$$
$$=\frac{5\cdot 5-3\cdot 5a+3(3+2a)}{3\cdot 5(a-2)}$$
$$=\frac{25-15a+9+6a}{15(a-2)}$$
$$=\frac{34-9a}{15(a-2)}$$

11. a. $\sqrt{50}=\sqrt{2}\sqrt{25}=5\sqrt{2}$

b. $\sqrt[3]{24}=\sqrt[3]{8}\sqrt[3]{3}=2\sqrt[3]{3}$

c. $\sqrt{26}=\sqrt{26}$

d. $\sqrt[4]{32}=\sqrt[4]{16}\sqrt[4]{2}=2\sqrt[4]{2}$

12.
$$\sqrt{3x+6}-\sqrt{7x-6}=0$$
$$\sqrt{3x+6}=\sqrt{7x-6}$$
$$\left(\sqrt{3x+6}\right)^2=\left(\sqrt{7x-6}\right)^2$$
$$3x+6=7x-6$$
$$-4x=-12$$
$$x=3$$

13.
$$2420=2000(1+r)^2$$
$$\frac{2420}{2000}=(1+r)^2$$
$$\frac{121}{100}=(1+r)^2$$
$$\pm\sqrt{\frac{121}{100}}=1+r$$
$$\pm\frac{11}{10}=1+r$$
$$-1\pm\frac{11}{10}=r$$

Discard the negative value.

$$r=-1+\frac{11}{10}=\frac{1}{10}=0.10$$

The interest rate is 10%.

14. a.
$$\sqrt[3]{\frac{4}{3x}}=\frac{\sqrt[3]{4}}{\sqrt[3]{3x}}=\left(\frac{\sqrt[3]{9x^2}}{\sqrt[3]{9x^2}}\right)=\frac{\sqrt[3]{36x^2}}{3x}$$

b.
$$\frac{\sqrt{2}+1}{\sqrt{2}-1}=\frac{\sqrt{2}+1}{\sqrt{2}-1}\cdot\left(\frac{\sqrt{2}+1}{\sqrt{2}+1}\right)$$
$$=\frac{2+2\sqrt{2}+1}{2-1}$$
$$=3+2\sqrt{2}$$

15.
$$(x-3)^2-3(x-3)-4=0$$
$$x^2-6x+9-3x+9-4=0$$
$$x^2-9x+14=0$$
$$(x-2)(x-7)=0$$
$$x=2,7$$

16.
$$\frac{10}{(2x+4)^2}-\frac{1}{2x+4}=3$$
$$10-(2x+4)=3(2x+4)^2$$
$$10-2x-4=3(4x^2+16x+16)$$
$$-2x+6=12x^2+48x+48$$
$$12x^2+50x+42=0$$
$$6x^2+25x+21=0$$
$$(6x+7)(x+3)=0$$
$$x=-\frac{7}{6},-3$$

17.
$$\frac{5}{x+1}<-2$$
$$x+1=0$$
$$x=-1$$

Solve $\frac{5}{x+1}=-2.$

$$(x+1)\frac{5}{x+1}=(x+1)(-2)$$
$$5=-2x-2$$
$$7=-2x$$
$$-\frac{7}{2}=x$$

Region	Test Point	$\frac{5}{x+1} < -2$; Result
$\left(-\infty, -\frac{7}{2}\right)$	$x = -6$	$\frac{5}{-5} < -2$; False
$\left(-\frac{7}{2}, -1\right)$	$x = -2$	$\frac{5}{-1} < -2$; True
$(-1, \infty)$	$x = 4$	$\frac{5}{5} < -2$; False

The solution set is $\left(-\frac{7}{2}, -1\right)$.

18. $f(x) = (x+2)^2 - 6$
Axis of symmetry: $x = -2$
vertex: $(-2, -6)$

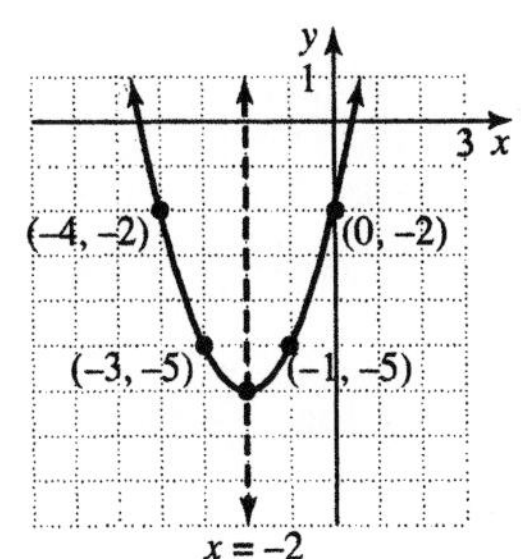

19. $f(t) = -16t^2 + 20t$
The maximum height occurs at the vertex.
$$t = \frac{-20}{2(-16)} = \frac{5}{8}$$
$$f\left(\frac{5}{8}\right) = -16\left(\frac{5}{8}\right)^2 + 20\left(\frac{5}{8}\right) = \frac{25}{4}$$
The maximum height of $\frac{25}{4}$ feet occurs at $\frac{5}{8}$ second.

20. $f(x) = x^2 + 3x - 18$
$a = 1, b = 3, c = -18$
$$x = \frac{-3}{2(1)} = -\frac{3}{2}$$
$$f\left(-\frac{3}{2}\right) = \left(-\frac{3}{2}\right)^2 + 3\left(\frac{3}{2}\right) - 18 = -\frac{81}{4}$$
The vertex is $\left(-\frac{3}{2}, -\frac{81}{4}\right)$.

21. a. $(f \circ g)(2) = f(g(2)) = f(5) = 5^2 = 25$
$(g \circ f)(2) = g(f(2)) = g(4) = 4 + 3 = 7$

b. $(f \circ g)(x) = f(x+3)$
$= (x+3)^2$
$= x^2 + 6x + 9$
$(g \circ f)(x) = g(x^2) = x^2 + 3$

22. $f(x) = -2x + 3$
$y = -2x + 3$
$x = -2y + 3$
$x - 3 = -2y$
$\frac{x-3}{-2} = y$
$f^{-1}(x) = -\frac{x-3}{2}$ or $f^{-1}(x) = \frac{3-x}{2}$

23. $f^{-1} = \{(1,0), (7,-2), (-6,3), (4,4)\}$

24. a. $(f \circ g)(2) = f(g(2)) = f(3) = 3^2 - 2 = 7$
$(g \circ f)(2) = g(f(2)) = g(2) = 2 + 1 = 3$

b. $(f \circ g)(x) = f(x+1)$
$= (x+1)^2 - 2$
$= x^2 + 2x - 1$
$(g \circ f)(x) = g(x^2 - 2) = x^2 - 2 + 1 = x^2 - 1$

25. a. $2^x = 16$
$2^x = 2^4$
$x = 4$

b. $9^x = 27$
$(3^2)^x = 3^3$
$2x = 3$
$x = \frac{3}{2}$

c. $4^{x+3} = 8^x$
$(2^2)^{x+3} = (2^3)^x$
$2^{2x+6} = 2^{3x}$
$2x + 6 = 3x$
$x = 6$

26. a. $\log_2 32 = x$
$$2^x = 32$$
$$2^x = 2^5$$
$$x = 5$$

b. $\log_4 \frac{1}{64} = x$
$$4^x = \frac{1}{64}$$
$$4^x = 4^{-3}$$
$$x = -3$$

c. $\log_{\frac{1}{2}} x = 5$
$$\left(\frac{1}{2}\right)^5 = x$$
$$x = \frac{1}{32}$$

27. a. $\log_3 3^2 = 2$

b. $\log_7 7^{-1} = -1$

c. $5^{\log_5 3} = 3$

d. $2^{\log_2 6} = 6$

28. a.
$$4^x = 64$$
$$\left(2^2\right)^x = 2^6$$
$$2x = 6$$
$$x = 3$$

b.
$$8^x = 32$$
$$\left(2^3\right)^x = 2^5$$
$$3x = 5$$
$$x = \frac{5}{3}$$

c.
$$9^{x+4} = 243^x$$
$$(3^2)^{x+4} = (3^5)^x$$
$$3^{2x+8} = 3^{5x}$$
$$2x + 8 = 5x$$
$$8 = 3x$$
$$x = \frac{8}{3}$$

29. a. $\log_{11} 10 + \log_{11} 3 = \log_{11}(10 \cdot 3) = \log_{11} 30$

b. $\log_3 \frac{1}{2} + \log_3 12 = \log_3\left(\frac{1}{2} \cdot 12\right) = \log_3 6$

c. $\log_2(x+2) + \log_2 x = \log_2[(x+2)x]$
$$= \log_2(x^2 + 2x)$$

30. a. $\log 100{,}000 = \log_{10} 10^5 = 5$

b. $\log 10^{-3} = \log_{10} 10^{-3} = -3$

c. $\ln \sqrt[5]{e} = \ln e^{1/5} = \frac{1}{5}$

d. $\ln e^4 = 4$

31. $A = Pe^{rt}$
$$A = 1600e^{0.09(5)} \approx 2509.30$$
\$2509.30 is owed after 5 years.

32. a. $\log_6 5 + \log_6 4 = \log_6(5 \cdot 4) = \log_6 20$

b. $\log_8 12 - \log_8 4 = \log_8 \frac{12}{4} = \log_8 3$

c. $2\log_2 x + 3\log_2 x - 2\log_2(x-1)$
$$= 5\log_2 x - \log_2(x-1)^2$$
$$= \log_2 x^5 - \log_2(x-1)^2$$
$$= \log_2 \frac{x^5}{(x-1)^2}$$

33.
$$3^x = 7$$
$$\log 3^x = \log 7$$
$$x \log 3 = \log 7$$
$$x = \frac{\log 7}{\log 3} \approx 1.7712$$

34. $10{,}000 = 5000\left(1 + \frac{0.02}{4}\right)^{4t}$
$$2 = (1.005)^{4t}$$
$$\ln 2 = \ln 1.005^{4t}$$
$$\ln 2 = 4t \ln(1.005)$$
$$t = \frac{\ln 2}{4 \ln 1.005} \approx 34.7$$
It takes 34.7 years.

35. $\log_4(x-2)=2$

$$\begin{aligned} 4^2 &= x-2 \\ x-2 &= 16 \\ x &= 18 \end{aligned}$$

36. $\log_4 10 - \log_4 x = 2$

$$\begin{aligned} \log_4 \frac{10}{x} &= 2 \\ 4^2 &= \frac{10}{x} \\ 16 &= \frac{10}{x} \\ 16x &= 10 \\ x &= \frac{5}{8} \end{aligned}$$

37. $\frac{x^2}{16} - \frac{y^2}{25} = 1$

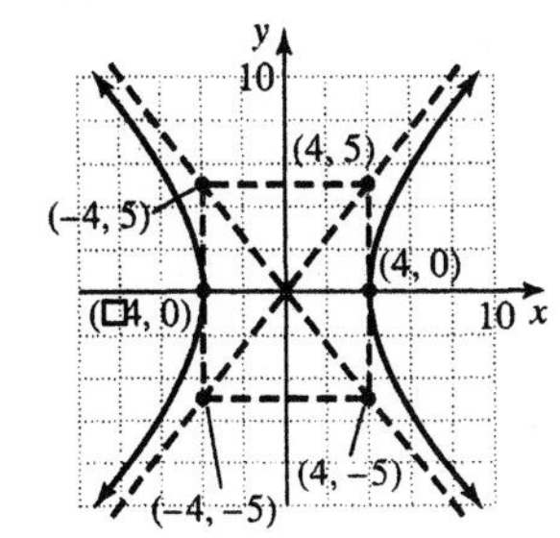

38. (8, 5), (−2, 4)

$d=\sqrt{(-2-8)^2+(4-5)^2}=\sqrt{101}$ units

39. $\begin{cases} y=\sqrt{x} \\ x^2+y^2=6 \end{cases}$

Replace y with $\sqrt{x}$ in the first equation.

$$\begin{aligned} (x)^2+\left(\sqrt{x}\right)^2 &= 6 \\ x^2+x-6 &= 0 \\ (x+3)(x-2) &= 0 \end{aligned}$$

$x=-3$(discard) or $x=2$

$x=2$: $y=\sqrt{x}=\sqrt{2}$

$\left(2, \sqrt{2}\right)$

40. $\begin{cases} x^2+y^2=36 \\ x-y=6 \Rightarrow x=y+6 \end{cases}$

Replace x with $y+6$ in the first equation.

$$\begin{aligned} (y+6)^2+y^2 &= 36 \\ 2y^2+12y &= 0 \\ 2y(y+6) &= 0 \end{aligned}$$

$y=0$ or $y=-6$

$x=0+6=6$ $\quad$ $x=-6+6=0$

$(0,-6);(6,0)$

41. $\frac{x^2}{9}+\frac{y^2}{16}\le 1$

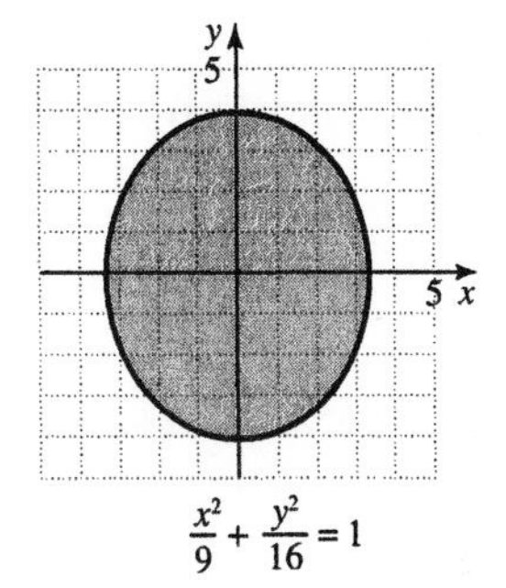

42.

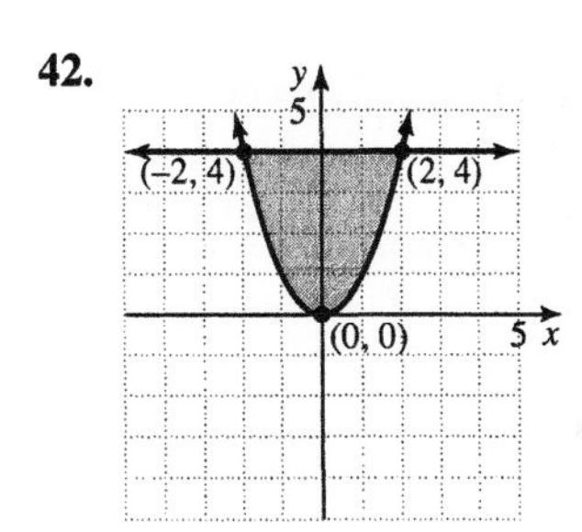

43. $a_n=n^2-1$

$$\begin{aligned} a_1 &= 1^2-1=0 \\ a_2 &= 2^2-1=3 \\ a_3 &= 3^2-1=8 \\ a_4 &= 4^2-1=15 \\ a_5 &= 5^2-1=24 \end{aligned}$$

44. $a_n=\frac{n}{n+4}$

$a_8=\frac{8}{8+4}=\frac{8}{12}=\frac{2}{3}$

45. $a_1=2,\ d=9-2=7$

$a_{11}=2+(11-1)(7)=72$

46. $a_1=2,\ r=\frac{10}{2}=5$

$a_6=2(5)^{6-1}=2(5)^5=6250$

47. a. $\sum_{i=0}^{6} \frac{i-2}{2} = \frac{0-2}{2} + \frac{1-2}{2} + \frac{2-2}{2} + \frac{3-2}{2} + \frac{4-2}{2} + \frac{5-2}{2} + \frac{6-2}{2}$

$$= -1 - \frac{1}{2} + 0 + \frac{1}{2} + 1 + \frac{3}{2} + 2$$

$$= \frac{7}{2}$$

b. $\sum_{i=3}^{5} 2^i = 2^3 + 2^4 + 2^5 = 8 + 16 + 32 = 56$

48. a. $\sum_{i=0}^{4} i(i+1) = 0(0+1) + 1(1+1) + 2(2+1) + 3(3+1) + 4(4+1)$

$$= 0 + 2 + 6 + 12 + 20$$

$$= 40$$

b. $\sum_{i=0}^{3} 2^i = 2^0 + 2^1 + 2^2 + 2^3 = 1 + 2 + 4 + 8 = 15$

49. $a_1 = 1$, $a_{30} = 30$

$$S_n = \frac{n}{2}(a_1 + a_n) = \frac{30}{2}(1 + 30) = 465$$

50. $(x-y)^6$ where $a = x$, $b = -y$, $n = 6$, and $r = 2$.

$$\frac{6!}{2!(6-2)!} x^{6-2} y^2 = 15x^4y^2$$

The third term in the expansion of $(x-y)^6$ is $15x^4y^2$.

Appendix A

A.2 Practice Final Exam

1. $6[5+2(3-8)-3] = 6[5+2(-5)-3]$
$= 6[5+(-10)-3]$
$= 6[-5-3]$
$= 6[-8]$
$= -48$

2. $-3^4 = -(3^4) = -81$

3. $4^{-3} = \frac{1}{4^3} = \frac{1}{64}$

4. $\frac{1}{2} - \frac{5}{6} = \frac{1}{2} \cdot \frac{3}{3} - \frac{5}{6} = \frac{3}{6} - \frac{5}{6} = \frac{-2}{6} = -\frac{1}{3}$

5. $(5x^3 + x^2 + 5x - 2) - (8x^3 - 4x^2 + x - 7)$
$= 5x^3 + x^2 + 5x - 2 - 8x^3 + 4x^2 - x + 7$
$= 5x^3 - 8x^3 + x^2 + 4x^2 + 5x - x - 2 + 7$
$= -3x^3 + 5x^2 + 4x + 5$

6. $(4x-2)^2 = (4x)^2 - 2(4x)(2) + 2^2$
$= 16x^2 - 16x + 4$

7. $(3x+7)(x^2+5x+2)$
$= 3x(x^2+5x+2) + 7(x^2+5x+2)$
$= 3x^3 + 15x^2 + 6x + 7x^2 + 35x + 14$
$= 3x^3 + 22x^2 + 41x + 14$

8. $y^2 - 8y - 48 = (y-12)(y+4)$

9. $9x^3 + 39x^2 + 12x = 3x(3x^2 + 13x + 4)$
$= 3x(3x+1)(x+4)$

10. $180 - 5x^2 = 5(36 - x^2)$
$= 5(6^2 - x^2)$
$= 5(6+x)(6-x)$

11. $3a^2 + 3ab - 7a - 7b = 3a(a+b) - 7(a+b)$
$= (a+b)(3a-7)$

12. $8y^3 - 64 = 8(y^3 - 8)$
$= 8(y^3 - 2^3)$
$= 8(y-2)(y^2 + y \cdot 2 + 2^2)$
$= 8(y-2)(y^2 + 2y + 4)$

13. $\left(\frac{x^2y^3}{x^3y^{-4}}\right)^2 = \left(\frac{y^{3-(-4)}}{x^{3-2}}\right)^2 = \left(\frac{y^7}{x^1}\right)^2 = \frac{y^{7 \cdot 2}}{x^{1 \cdot 2}} = \frac{y^{14}}{x^2}$

14. $-4(a+1) - 3a = -7(2a-3)$
$-4a - 4 - 3a = -14a + 21$
$-7a - 4 = -14a + 21$
$7a - 4 = 21$
$7a = 25$
$a = \frac{25}{7}$

15. $3x - 5 \geq 7x + 3$
$-5 \geq 4x + 3$
$-8 \geq 4x$
$-2 \geq x$
$x \leq -2$
$(-\infty, -2]$

16. $x(x+6) = 7$
$x^2 + 6x = 7$
$x^2 + 6x - 7 = 0$
$(x+7)(x-1) = 0$
$x + 7 = 0$ or $x - 1 = 0$
$x = -7$ $\quad x = 1$

17. $5x - 7y = 10$

x	y
2	0
0	$-\frac{10}{7}$

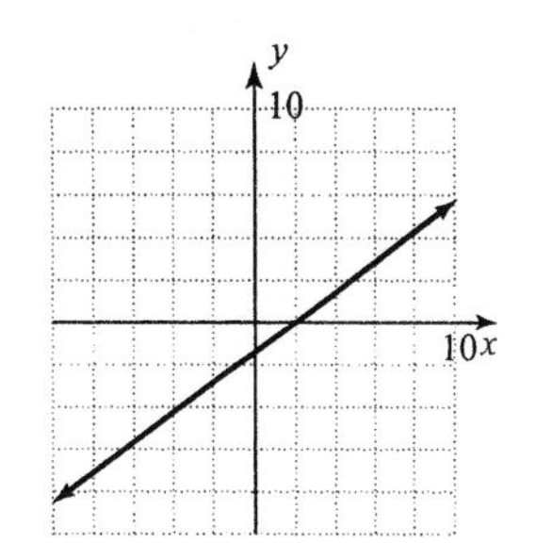

18. $x-3=0$

$x=3$

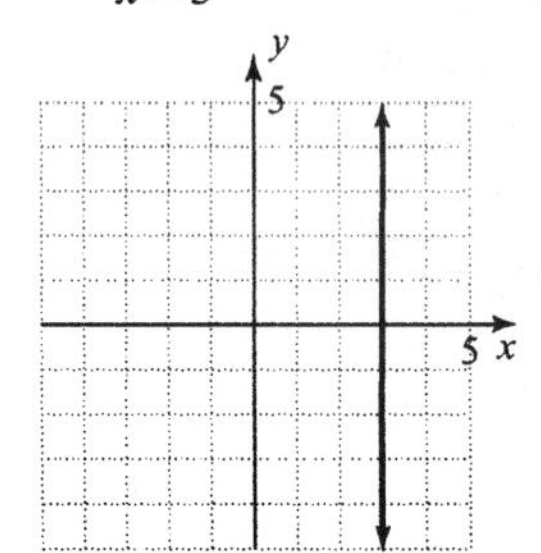

19. $m=\dfrac{y_2-y_1}{x_2-x_1}$

$m=\dfrac{2-(-5)}{-1-6}=\dfrac{2+5}{-7}=\dfrac{7}{-7}=-1$

20. $-3x+y=5$

$y=3x+5$

$m=3$

21. $(x_1, y_1)=(2, -5), (x_2, y_2)=(1, 3)$

$m=\dfrac{y_2-y_1}{x_2-x_1}=\dfrac{3-(-5)}{1-2}=\dfrac{3+5}{-1}=\dfrac{8}{-1}=-8$

$y-y_1=m(x-x_1)$

$y-(-5)=-8(x-2)$

$y+5=-8x+16$

$8x+y=11$

22. A line parallel to $x=7$ is a vertical line. The vertical line through $(-5, -1)$ has equation $x=-5$.

23. $\begin{cases}\dfrac{1}{2}x+2y=-\dfrac{15}{4}\\ 4x=-y\end{cases}$

Solve the second equation for y.

$y=-4x$

Substitute $-4x$ for y in the first equation.

$\dfrac{1}{2}x+2(-4x)=-\dfrac{15}{4}$

$\dfrac{1}{2}x-8x=-\dfrac{15}{4}$

$\dfrac{1}{2}x-\dfrac{16x}{2}=-\dfrac{15}{4}$

$-\dfrac{15}{2}x=-\dfrac{15}{4}$

$-\dfrac{2}{15}\left(-\dfrac{15}{2}x\right)=-\dfrac{2}{15}\left(-\dfrac{15}{4}\right)$

$x=\dfrac{1}{2}$

Let $x=\dfrac{1}{2}$ in $y=-4x$.

$y=-4x=-4\left(\dfrac{1}{2}\right)=-2$

The solution is $\left(\dfrac{1}{2}, -2\right)$.

24. $\begin{cases}4x-6y=7\\ -2x+3y=0\end{cases}$

Multiply the second equation by 2 and add the result to the first equation.

$$\begin{array}{r}4x-6y=7\\ \underline{-4x+6y=0}\\ 0=7\end{array}$$

The statement $0=7$ is false, so the system has no solution.

25.

$$\begin{array}{r}9x^2-6x+4\\ 3x+2\overline{)27x^3+0x^2+0x-8}\\ \underline{27x^3+18x^2}\\ -18x^2+0x\\ \underline{-18x^2-12x}\\ 12x-8\\ \underline{12x+8}\\ -16\end{array}$$

$\dfrac{27x^3-8}{3x+2}=9x^2-6x+4-\dfrac{16}{3x+2}$

26. $h(x)=x^3-x$

a. $h(-1)=(-1)^3-(-1)=-1+1=0$

b. $h(0)=0^3-0=0-0=0$

c. $h(4)=4^3-4=64-4=60$

27. Domain: $(-\infty, \infty)$
Range: $(-\infty, 4]$

28. Let x be the smaller area code. Then the other area code is $2x$.
$x + 2x = 1203$
$3x = 1203$
$x = 401$
$2x = 2(401) = 802$
The area codes are 401 (Rhode Island) and 802 (Vermont).

29. Let x be the number of hours since the trains left. One train will have traveled $50x$ miles and the other will have traveled $64x$ miles.
$50x + 64x = 285$
$114x = 285$
$x = \frac{285}{114} = \frac{5}{2}$ or $2\frac{1}{2}$
The trains are 285 miles apart after $2\frac{1}{2}$ hours.

30. Let x be the amount of 12% solution.

Amount	Percent	Total Saline
x	$12\% = 0.12$	$0.12x$
80	$22\% = 0.22$	$0.22 \cdot 80 = 17.6$
$80 + x$	$16\% = 0.16$	$0.16(80 + x)$ $= 12.8 + 0.16x$

$0.12x + 17.6 = 12.8 + 0.16x$
$4.8 = 0.04x$
$120 = x$
120 cc of 12% saline solution should be added.

31. $x^2 + 4x + 3 = 0$
$(x+1)(x+3) = 0$
$x + 1 = 0$ or $x + 3 = 0$
$x = -1$ $\quad x = -3$
The domain of $g(x)$ is
$\{x|x \text{ is a real number}, x \neq -1, x \neq -3\}$.

32. $\frac{15x}{2x+5} - \frac{6-4x}{2x+5} = \frac{15x-(6-4x)}{2x+5}$
$= \frac{15x-6+4x}{2x+5}$
$= \frac{19x-6}{2x+5}$

33. $\frac{x^2-9}{x^2-3x} \div \frac{xy+5x+3y+15}{2x+10}$
$= \frac{x^2-9}{x^2-3x} \cdot \frac{2x+10}{xy+5x+3y+15}$
$= \frac{(x+3)(x-3) \cdot 2(x+5)}{x(x-3) \cdot (y+5)(x+3)}$
$= \frac{2(x+5)}{x(y+5)}$

34. $a^2 - a - 6 = (a-3)(a+2)$
LCD = $(a - 3)(a + 2)$
$\frac{5a}{a^2-a-6} - \frac{2}{a-3}$
$= \frac{5a}{(a-3)(a+2)} - \frac{2 \cdot (a+2)}{(a-3) \cdot (a+2)}$
$= \frac{5a-2(a+2)}{(a-3)(a+2)}$
$= \frac{5a-2a-4}{(a-3)(a+2)}$
$= \frac{3a-4}{(a-3)(a+2)}$

35. $\frac{5-\frac{1}{y^2}}{\frac{1}{y}+\frac{2}{y^2}} = \frac{y^2\left(5-\frac{1}{y^2}\right)}{y^2\left(\frac{1}{y}+\frac{2}{y^2}\right)} = \frac{5y^2-1}{y+2}$

36. $\frac{4}{y} - \frac{5}{3} = -\frac{1}{5}$
$15y\left(\frac{4}{y}\right) - 15y\left(\frac{5}{3}\right) = 15y\left(-\frac{1}{5}\right)$
$60 - 25y = -3y$
$60 = 22y$
$\frac{60}{22} = y$
$\frac{30}{11} = y$

37. $\frac{5}{y+1} = \frac{4}{y+2}$
$5(y+2) = 4(y+1)$
$5y + 10 = 4y + 4$
$y + 10 = 4$
$y = -6$

38.
$$\frac{a}{a-3} = \frac{3}{a-3} - \frac{3}{2}$$
$$2(a-3)\left(\frac{a}{a-3}\right) = 2(a-3)\left(\frac{3}{a-3}\right) - 2(a-3)\left(\frac{3}{2}\right)$$
$$2a = 6 - 3(a-3)$$
$$2a = 6 - 3a + 9$$
$$2a = 15 - 3a$$
$$5a = 15$$
$$a = 3$$
In the original equation, $a = 3$ makes a denominator 0, so it is extraneous. The equation has no solution.

39. Let x be the number.
$$x + 5 \cdot \frac{1}{x} = 6$$
$$x\left(x + \frac{5}{x}\right) = x \cdot 6$$
$$x^2 + 5 = 6x$$
$$x^2 - 6x + 5 = 0$$
$$(x-1)(x-5) = 0$$
$$x - 1 = 0 \quad \text{or} \quad x - 5 = 0$$
$$x = 1 \qquad\qquad x = 5$$
The number is 1 or 5.

40. $\sqrt{216} = \sqrt{36 \cdot 6} = \sqrt{36} \cdot \sqrt{6} = 6\sqrt{6}$

41. $\left(\frac{1}{125}\right)^{-1/3} = 125^{1/3} = \sqrt[3]{125} = \sqrt[3]{5^3} = 5$

42.
$$\left(\frac{64c^{4/3}}{a^{-2/3}b^{5/6}}\right)^{1/2} = \frac{64^{1/2}c^{\frac{4}{3}\cdot\frac{1}{2}}}{a^{-\frac{2}{3}\cdot\frac{1}{2}}b^{\frac{5}{6}\cdot\frac{1}{2}}}$$
$$= \frac{8c^{2/3}}{a^{-1/3}b^{5/12}}$$
$$= \frac{8a^{1/3}c^{2/3}}{b^{5/12}}$$

43.
$$\sqrt{125x^3} - 3\sqrt{20x^3} = \sqrt{25x^2 \cdot 5x} - 3\sqrt{4x^2 \cdot 5x}$$
$$= 5x\sqrt{5x} - 3 \cdot 2x\sqrt{5x}$$
$$= 5x\sqrt{5x} - 6x\sqrt{5x}$$
$$= (5x - 6x)\sqrt{5x}$$
$$= -x\sqrt{5x}$$

44. $\left(\sqrt{5}+5\right)\left(\sqrt{5}-5\right) = \left(\sqrt{5}\right)^2 - 5^2 = 5 - 25 = -20$

45. $|6x-5| - 3 = -2$
$$|6x-5| = 1$$
$$6x - 5 = -1 \quad \text{or} \quad 6x - 5 = 1$$
$$6x = 4 \quad \text{or} \quad 6x = 6$$
$$x = \frac{4}{6} \quad \text{or} \quad x = 1$$
$$x = \frac{2}{3}$$
Both solutions check.

46.
$$-3 < 2(x-3) \le 4$$
$$-3 < 2x - 6 \le 4$$
$$-3 + 6 < 2x - 6 + 6 \le 4 + 6$$
$$3 < 2x \le 10$$
$$\frac{3}{2} < \frac{2x}{2} \le \frac{10}{2}$$
$$\frac{3}{2} < x \le 5$$
$$\left(\frac{3}{2}, 5\right]$$

47. $|3x + 1| > 5$
$$3x + 1 < -5 \quad \text{or} \quad 3x + 1 > 5$$
$$3x < -6 \qquad\qquad 3x > 4$$
$$x < -2 \qquad\qquad x > \frac{4}{3}$$
$$(-\infty, -2) \cup \left(\frac{4}{3}, \infty\right)$$

48.
$$y^2 - 3y = 5$$
$$y^2 - 3y - 5 = 0$$
$$y = \frac{-(-3) \pm \sqrt{(-3)^2 - 4(1)(-5)}}{2(1)}$$
$$y = \frac{3 \pm \sqrt{9 + 20}}{2}$$
$$y = \frac{3 \pm \sqrt{29}}{2}$$

49.
$$x = \sqrt{x-2} + 2$$
$$x - 2 = \sqrt{x-2}$$
$$(x-2)^2 = \left(\sqrt{x-2}\right)^2$$
$$x^2 - 4x + 4 = x - 2$$
$$x^2 - 5x + 6 = 0$$
$$(x-2)(x-3) = 0$$
$$x - 2 = 0 \quad \text{or} \quad x - 3 = 0$$
$$x = 2 \quad \text{or} \quad x = 3$$

50. $2x^2 - 7x > 15$

$2x^2 - 7x - 15 > 0$

$(2x+3)(x-5) > 0$

$2x+3=0 \quad \text{or} \quad x-5=0$

$x = -\dfrac{3}{2} \quad \text{or} \quad x = 5$

51. $y > -4x$

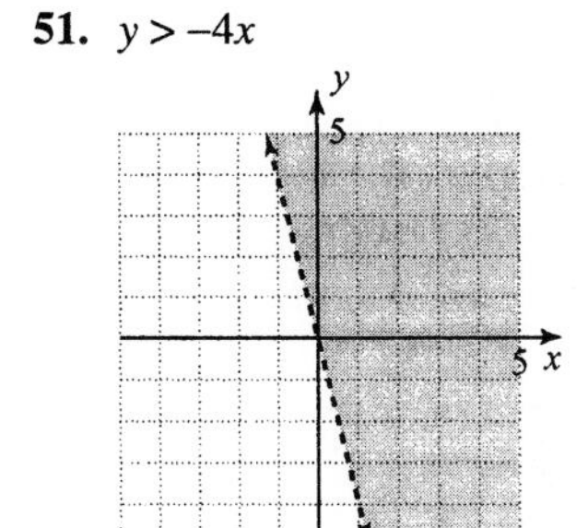

52. $g(x) = -|x+2| - 1$

x	$g(x) = -\|x+2\| - 1$	$g(x)$
−5	$-\|-5+2\| - 1 = -4$	−4
−4	$-\|-4+2\| - 1 = -3$	−3
−3	$-\|-3+2\| - 1 = -2$	−2
−2	$-\|-2+2\| - 1 = -1$	−1
−1	$-\|-1+2\| - 1 = -2$	−2
0	$-\|0+2\| - 1 = -3$	−3
1	$-\|1+2\| - 1 = -4$	−4

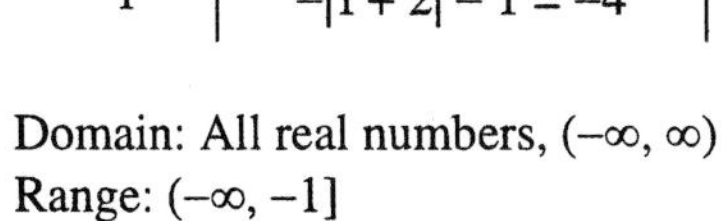

Domain: All real numbers, $(-\infty, \infty)$

Range: $(-\infty, -1]$

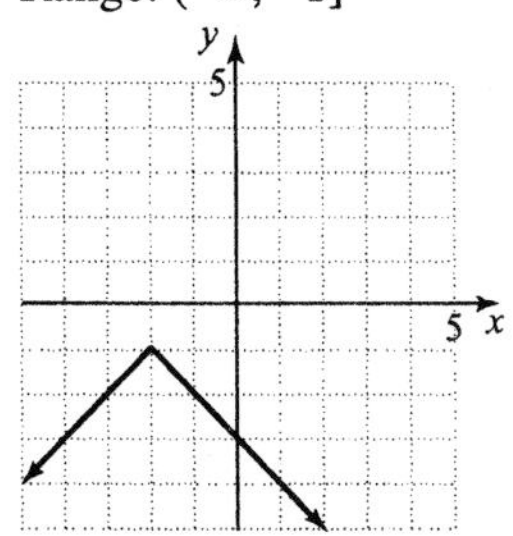

53. $h(x) = x^2 - 4x + 4$

x-intercept: Let $h(x) = 0$ and solve for x.

$0 = x^2 - 4x + 4$

$0 = (x-2)^2$

$x - 2 = 0$

$x = 2$

x-intercept: (2, 0)

y-intercept: Let $x = 0$.

$h(0) = 0^2 - 4(0) + 4 = 4$

y-intercept: (0, 4)

x-coordinate of vertex:

$-\dfrac{b}{2a} = -\dfrac{-4}{2(1)} = 2$

vertex: (2, 0)

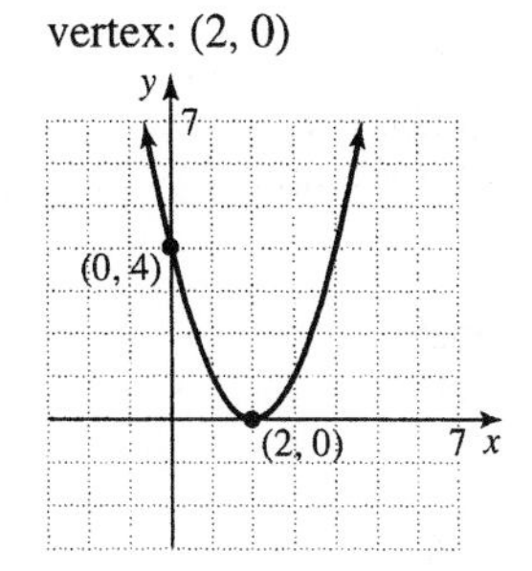

54. $f(x) = \begin{cases} -\dfrac{1}{2}x & \text{if } x \le 0 \\ 2x - 3 & \text{if } x > 0 \end{cases}$

If $x \le 0$

x	$-\frac{1}{2}x$	$f(x)$
0	$-\frac{1}{2}(0)$	0
−2	$-\frac{1}{2}(-2)$	1
−4	$-\frac{1}{2}(-4)$	2

If $x > 0$

x	$2x - 3$	$f(x)$
1	$2(1) - 3$	−1
2	$2(2) - 3$	1
3	$2(3) - 3$	3

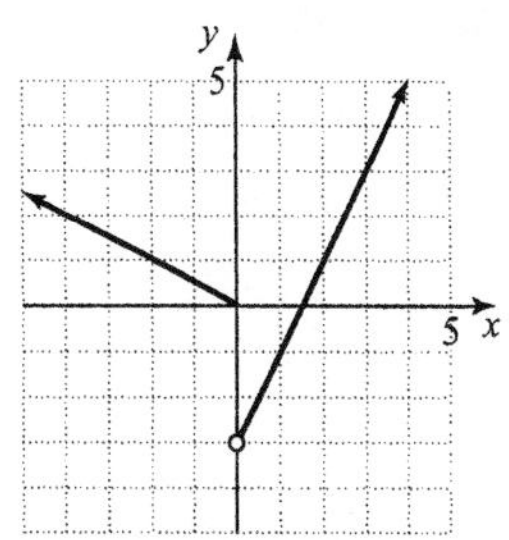

Domain: $(-\infty, \infty)$

Range: $(-3, \infty)$

55. through (4, –2) and (6, –3)

$$\text{slope} = m = \frac{y_2 - y_1}{x_2 - x_1} = \frac{-3-(-2)}{6-4} = \frac{-1}{2}$$

$$y - y_1 = m(x - x_1)$$
$$y-(-2) = -\frac{1}{2}(x-4)$$
$$y+2 = -\frac{1}{2}x+2$$
$$y = -\frac{1}{2}x$$
$$f(x) = -\frac{1}{2}x$$

56. through (–1, 2) and perpendicular to $3x - y = 4$
Find the slope of $3x - y = 4$ by writing the equation in slope-intercept form.

$$3x - y = 4$$
$$-y = -3x+4$$
$$y = 3x-4$$

The slope is 3. The slope of a line perpendicular to this line is $-\frac{1}{3}$.

Substitute $m = -\frac{1}{3}$ and $(x_1, y_1) = (-1, 2)$ in the equation:

$$y - y_1 = m(x - x_1)$$
$$y-2 = -\frac{1}{3}[x-(-1)]$$
$$y-2 = -\frac{1}{3}(x+1)$$
$$y-2 = -\frac{1}{3}x-\frac{1}{3}$$
$$y = -\frac{1}{3}x+\frac{5}{3}$$
$$f(x) = -\frac{1}{3}x+\frac{5}{3}$$

57. $(x_1, y_1) = (-6, 3)$; $(x_2, y_2) = (-8, -7)$

$$d = \sqrt{(x_2-x_1)^2+(y_2-y_1)^2}$$
$$= \sqrt{[-8-(-6)]^2+(-7-3)^2}$$
$$= \sqrt{(-2)^2+(-10)^2}$$
$$= \sqrt{4+100}$$
$$= \sqrt{104}$$
$$= 2\sqrt{26} \text{ units}$$

58. $(x_1, y_1) = (-2, -5)$; $(x_2, y_2) = (-6, 12)$

$$\text{midpoint} = \left(\frac{x_1+x_2}{2}, \frac{y_1+y_2}{2}\right)$$
$$= \left(\frac{-2+(-6)}{2}, \frac{-5+12}{2}\right)$$
$$= \left(\frac{-8}{2}, \frac{7}{2}\right)$$
$$= \left(-4, \frac{7}{2}\right)$$

59. $\sqrt{\frac{9}{y}} = \frac{\sqrt{9}}{\sqrt{y}} = \frac{\sqrt{9}}{\sqrt{y}}\cdot\frac{\sqrt{y}}{\sqrt{y}} = \frac{\sqrt{9}\cdot\sqrt{y}}{\sqrt{y}\cdot\sqrt{y}} = \frac{3\sqrt{y}}{y}$

60.

$$\frac{4-\sqrt{x}}{4+2\sqrt{x}} = \frac{4-\sqrt{x}}{4+2\sqrt{x}}\cdot\frac{4-2\sqrt{x}}{4-2\sqrt{x}}$$
$$= \frac{\left(4-\sqrt{x}\right)\left(4-2\sqrt{x}\right)}{\left(4+2\sqrt{x}\right)\left(4-2\sqrt{x}\right)}$$
$$= \frac{16-12\sqrt{x}+2x}{16-4x}$$
$$= \frac{2\left(8-6\sqrt{x}+x\right)}{2(8-2x)}$$
$$= \frac{8-6\sqrt{x}+x}{8-2x}$$

61. $W = \frac{k}{V}$

Find k by substituting $W = 20$ and $V = 12$.

$$20 = \frac{k}{12}$$
$$240 = k$$

Write the inverse relation equation.

$$W = \frac{240}{V}$$

Let $V = 15$ and find W.

$$W = \frac{240}{15}$$
$$W = 16$$

62. Use the Pythagorean Theorem.

$$c^2 = a^2 + b^2$$
$$20^2 = x^2 + (x+8)^2$$
$$400 = x^2 + x^2 + 16x + 64$$
$$0 = 2x^2 + 16x - 336$$
$$0 = 2(x^2 + 8x - 168)$$

$$x = \frac{-8 \pm \sqrt{8^2 - 4(1)(-168)}}{2(1)}$$
$$x = \frac{-8 \pm \sqrt{736}}{2}$$
$x \approx -17.6$ or $x \approx 9.6$
Discard a negative distance.
$x + 8 + x = 9.6 + 8 + 9.6 = 27.2$
$27.2 - 20 = 7.2$ or about 7
A person saves about 7 feet.

63. a. Find the vertex.
$s(t) = -16t^2 + 32t + 256$
t-value: $\frac{-b}{2a} = \frac{-32}{2(-16)} = 1$
$s(t)$-value:
$s(1) = -16(1)^2 + 32(1) + 256 = 272$
The maximum height is 272 feet.

b. Let $s(t) = 0$ and solve for t.
$0 = -16t^2 + 32t + 256$
$0 = -16(t^2 - 2t - 16)$
$$t = \frac{-(-2) \pm \sqrt{(-2)^2 - 4(1)(-16)}}{2(1)}$$
$$t = \frac{2 \pm \sqrt{68}}{2}$$
$$t = \frac{2 \pm 2\sqrt{17}}{2}$$
$t = 1 \pm \sqrt{17}$
$t \approx -3.12$ or $t \approx 5.12$
Discard a negative time.
The stone will hit the water in approximately 5.12 seconds.

64. $-\sqrt{-8} = -\sqrt{4 \cdot (-1) \cdot 2} = -\sqrt{4} \cdot \sqrt{-1} \cdot \sqrt{2} = -2i\sqrt{2}$

65.
$$\begin{aligned}(12 - 6i) - (12 - 3i) &= 12 - 6i - 12 + 3i \\ &= 12 - 12 - 6i + 3i \\ &= 0 - 3i \\ &= -3i\end{aligned}$$

66.
$$\begin{aligned}(4 + 3i)^2 &= (4 + 3i)(4 + 3i) \\ &= 16 + 12i + 12i + 9i^2 \\ &= 16 + 24i - 9 \\ &= 7 + 24i\end{aligned}$$

67.
$$\begin{aligned}\frac{1+4i}{1-i} &= \frac{1+4i}{1-i} \cdot \frac{1+i}{1+i} \\ &= \frac{(1+4i)(1+i)}{(1-i)(1+i)} \\ &= \frac{1+5i+4i^2}{1-i^2} \\ &= \frac{1+5i-4}{1-(-1)} \\ &= \frac{-3+5i}{2} \\ &= -\frac{3}{2} + \frac{5}{2}i\end{aligned}$$

68. $g(x) = x - 7$ and $h(x) = x^2 - 6x + 5$
$(g \circ h)(x) = (x^2 - 6x + 5) - 7 = x^2 - 6x - 2$

69. $f(x) = 6 - 2x$ is a one-to-one function since there is only one $f(x)$ value for each x-value.

Inverse:
$y = 6 - 2x \quad \Rightarrow \quad x = 6 - 2y$
$x + 2y = 6$
$2y = -x + 6$
$$y = \frac{-x+6}{2}$$
$$f^{-1}(x) = \frac{-x+6}{2}$$

70.
$$\begin{aligned}&\log_5 x + 3\log_5 x - \log_5(x+1) \\ &= \log_5 x + \log_5 x^3 - \log_5(x+1) \\ &= \log_5 x \cdot x^3 - \log_5(x+1) \\ &= \log_5 x^4 - \log_5(x+1) \\ &= \log_5 \frac{x^4}{x+1}\end{aligned}$$

71.
$$\begin{aligned}8^{x-1} &= \frac{1}{64} \\ (2^3)^{x-1} &= \frac{1}{2^6} \\ 2^{3(x-1)} &= 2^{-6} \\ 3(x-1) &= -6 \\ 3x - 3 &= -6 \\ 3x &= -3 \\ x &= -1\end{aligned}$$

72.
$$3^{2x+5}=4$$
$$\log 3^{2x+5}=\log 4$$
$$(2x+5)\log 3=\log 4$$
$$2x+5=\frac{\log 4}{\log 3}$$
$$2x=\frac{\log 4}{\log 3}-5$$
$$x=\frac{1}{2}\left(\frac{\log 4}{\log 3}-5\right)$$
$$x\approx -1.8691$$

73. $\log_8(3x-2)=2$
$$8^2=3x-2$$
$$64=3x-2$$
$$66=3x$$
$$22=x$$

74. $\log_4(x+1)-\log_4(x-2)=3$
$$\log_4\frac{x+1}{x-2}=3$$
$$4^3=\frac{x+1}{x-2}$$
$$64=\frac{x+1}{x-2}$$
$$64(x-2)=x+1$$
$$64x-128=x+1$$
$$63x=129$$
$$x=\frac{129}{63}=\frac{43}{21}$$

75. $\ln\sqrt{e}=x$
$$\ln e^{1/2}=x$$
$$\frac{1}{2}\ln e=x$$
$$\frac{1}{2}=x$$

76. $y=\left(\frac{1}{2}\right)^x+1$

x	$\left(\frac{1}{2}\right)^x+1$	y
-3	$\left(\frac{1}{2}\right)^{-3}+1=9$	9
-2	$\left(\frac{1}{2}\right)^{-2}+1=5$	5
-1	$\left(\frac{1}{2}\right)^{-1}+1=3$	3
0	$\left(\frac{1}{2}\right)^{0}+1=2$	2
1	$\left(\frac{1}{2}\right)^{1}+1=1\frac{1}{2}$	$1\frac{1}{2}$
2	$\left(\frac{1}{2}\right)^{2}+1=1\frac{1}{4}$	$1\frac{1}{4}$
3	$\left(\frac{1}{2}\right)^{3}+1=1\frac{1}{8}$	$1\frac{1}{8}$

77. Let $y_0=57{,}000$, $k=0.026$, $t=5$.
$$y=y_0e^{kt}$$
$$y=57{,}000e^{0.026(5)}$$
$$y\approx 64{,}913$$
In 5 years, there will be 64,913 prairie dogs.

78. $x^2-y^2=36$
$$x^2-y^2=6^2$$
hyperbola, with x-intercepts $(-6, 0)$, $(6, 0)$

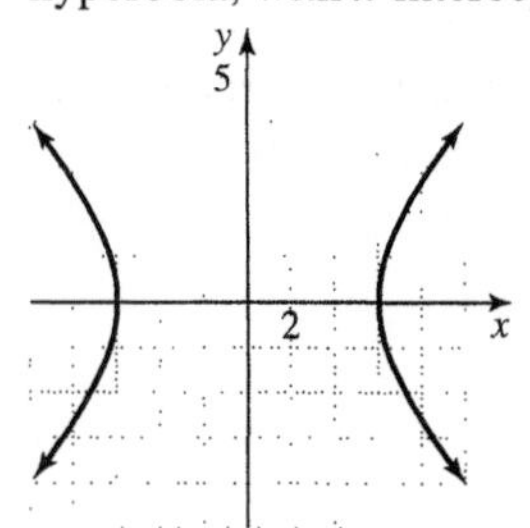

79. $16x^2 + 9y^2 = 144$

$$\frac{16x^2}{144} + \frac{9y^2}{144} = \frac{144}{144}$$

$$\frac{x^2}{9} + \frac{y^2}{16} = 1$$

Ellipse, x-intercepts $(-3, 0)$, $(3, 0)$
y-intercepts $(0, -4)$, $(0, 4)$

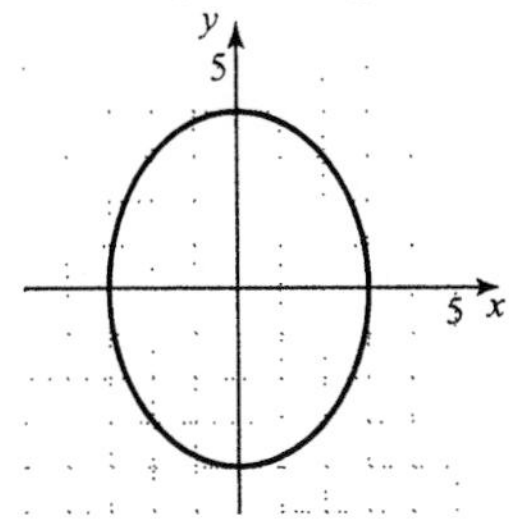

80.

$$x^2 + y^2 + 6x = 16$$
$$(x^2 + 6x + 9) + y^2 = 16 + 9$$
$$(x+3)^2 + y^2 = 25$$
$$[x-(-3)]^2 + (y-0)^2 = 5^2$$

circle with center $(-3, 0)$ and radius 5

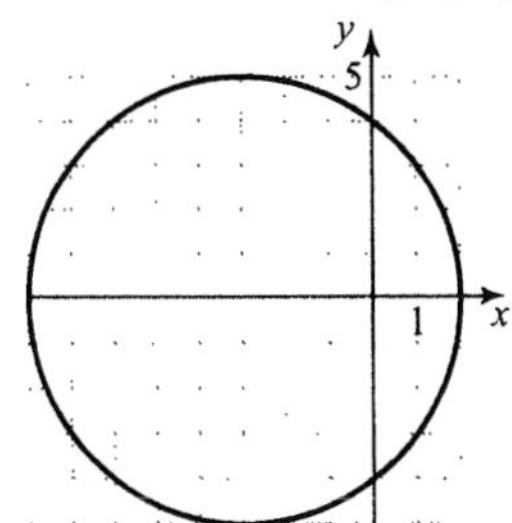

81. $\begin{cases} x^2 + y^2 = 26 \\ x^2 - 2y^2 = 23 \end{cases}$

Multiply equation (2) by -1 and add the equations.

$$x^2 + y^2 = 26$$
$$\underline{-x^2 + 2y^2 = -23}$$
$$3y^2 = 3$$
$$y^2 = 1$$
$$y = \pm 1$$

Substitute $y = -1$ and $y = 1$ into equation (1).

$$x^2 + (-1)^2 = 26$$
$$x^2 = 25$$
$$x = \pm 5$$

$$x^2 + 1^2 = 26$$
$$x^2 = 25$$
$$x = \pm 5$$

The solutions are $(-5, -1)$, $(-5, 1)$, $(5, -1)$, $(5, 1)$.

82. $a_n = \dfrac{(-1)^n}{n+4}$

$$a_1 = \frac{(-1)^1}{1+4} = -\frac{1}{5}$$
$$a_2 = \frac{(-1)^2}{2+4} = \frac{1}{6}$$
$$a_3 = \frac{(-1)^3}{3+4} = -\frac{1}{7}$$
$$a_4 = \frac{(-1)^4}{4+4} = \frac{1}{8}$$
$$a_5 = \frac{(-1)^5}{5+4} = -\frac{1}{9}$$

The first five terms are $-\frac{1}{5}, \frac{1}{6}, -\frac{1}{7}, \frac{1}{8}, -\frac{1}{9}$.

83. $a_n = 5(2)^{n-1}$

$$a_1 = 5(2)^{1-1} = 5(2)^0 = 5$$
$$r = 2$$
$$n = 5$$
$$S_n = \frac{a_1(1-r^n)}{1-r}$$
$$S_5 = \frac{5(1-2^5)}{1-2} = \frac{5(1-32)}{-1} = 155$$

84. Sequence $\frac{3}{2}, -\frac{3}{4}, \frac{3}{8}, \ldots$

$$a_1 = \frac{3}{2},\ r = -\frac{1}{2}$$
$$S_\infty = \frac{a_1}{1-r} = \frac{\frac{3}{2}}{1-\left(-\frac{1}{2}\right)} = \frac{\frac{3}{2}}{\frac{3}{2}} = 1$$

85.

$$\sum_{i=1}^{4} i(i-2) = 1(1-2) + 2(2-2) + 3(3-2) + 4(4-2)$$
$$= 1(-1) + 2(0) + 3(1) + 4(2)$$
$$= -1 + 0 + 3 + 8$$
$$= 10$$

86. $(2x+y)^5 = \binom{5}{0}(2x)^5 + \binom{5}{1}(2x)^4(y) + \binom{5}{2}(2x)^3(y)^2 + \binom{5}{3}(2x)^2(y)^3 + \binom{5}{4}(2x)^1(y)^4 + \binom{5}{5}y^5$

$= 2^5x^5 + 5 \cdot 2^4x^4y + 10 \cdot 2^3x^3y^2 + 10 \cdot 2^2x^2y^3 + 5 \cdot 2xy^4 + y^5$

$= 32x^5 + 80x^4y + 80x^3y^2 + 40x^2y^3 + 10xy^4 + y^5$

Appendix B

1.
$$\begin{array}{r} 9.076 \\ +\ 8.004 \\ \hline 17.080 \end{array}$$

3.
$$\begin{array}{r} 27.004 \\ -\ 14.200 \\ \hline 12.804 \end{array}$$

5.
$$\begin{array}{r} 107.92 \\ +\quad 3.04 \\ \hline 110.96 \end{array}$$

7.
$$\begin{array}{r} 10.0 \\ -\ 7.6 \\ \hline 2.4 \end{array}$$

9.
$$\begin{array}{r} 126.32 \\ -\ 97.89 \\ \hline 28.43 \end{array}$$

11.
$$\begin{array}{r} 3.25 \\ \times\quad 70 \\ \hline 227.50 \end{array}$$

13.
$$\begin{array}{r} 2.7 \\ 3\overline{)8.1} \\ \underline{6} \\ 2\,1 \\ \underline{2\,1} \\ 0 \end{array}$$

15.
$$\begin{array}{r} 55.4050 \\ -\quad 6.1711 \\ \hline 49.2339 \end{array}$$

17. $0.75\overline{)60}$ becomes
$$\begin{array}{r} 80 \\ 75\overline{)6000} \\ \underline{600} \\ 00 \\ \underline{00} \end{array}$$

19. $7.612 \div 100 = 0.07612$

21. $2.7\overline{)12.312}$ becomes
$$\begin{array}{r} 4.56 \\ 27\overline{)123.12} \\ \underline{108} \\ 15\,1 \\ \underline{13\,5} \\ 1\,62 \\ \underline{1\,62} \\ 0 \end{array}$$

23.
$$\begin{array}{r} 569.20 \\ 71.25 \\ +\quad 8.01 \\ \hline 648.46 \end{array}$$

25.
$$\begin{array}{r} 768.00 \\ -\quad 0.17 \\ \hline 767.83 \end{array}$$

27.
$$\begin{array}{r} 12.000 \\ +\ 0.062 \\ \hline 12.062 \end{array}$$

29.
$$\begin{array}{r} 76.00 \\ -\ 14.52 \\ \hline 61.48 \end{array}$$

31. $0.43\overline{)3.311}$ becomes
$$\begin{array}{r} 7.7 \\ 43\overline{)331.1} \\ \underline{301} \\ 30\,1 \\ \underline{30\,1} \\ 0 \end{array}$$

33.
$$\begin{array}{r} 762.12 \\ 89.70 \\ +\ 11.55 \\ \hline 863.37 \end{array}$$

35.
$$\begin{array}{r} 23.400 \\ -\ 0.821 \\ \hline 22.579 \end{array}$$

37.
$$\begin{array}{r} 476.12 \\ -\ 112.97 \\ \hline 363.15 \end{array}$$

39.
$$\begin{array}{r} 0.007 \\ +\ 7.000 \\ \hline 7.007 \end{array}$$

Appendix C

Appendix C.1

Practice Exercises

1.
$$\begin{aligned}3(x-5)&=6x-3\\3x-15&=6x-3\\3x-15-6x&=6x-3-6x\\-3x-15&=-3\\-3x-15+15&=-3+15\\-3x&=12\\\frac{-3x}{-3}&=\frac{12}{-3}\\x&=-4\end{aligned}$$

2.
$$\begin{aligned}\frac{y}{2}-\frac{y}{5}&=\frac{1}{4}\\20\left(\frac{y}{2}-\frac{y}{5}\right)&=20\left(\frac{1}{4}\right)\\20\left(\frac{y}{2}\right)-20\left(\frac{y}{5}\right)&=5\\10y-4y&=5\\6y&=5\\\frac{6y}{6}&=\frac{5}{6}\\y&=\frac{5}{6}\end{aligned}$$

3.
$$8(x^2+3)+4=-8x(x+3)+19$$
$$8x^2+24+4=-8x^2-24x+19$$
$$16x^2+24x+9=0$$
$$(4x+3)(4x+3)=0$$
$$4x+3=0 \quad\text{or}\quad 4x+3=0$$
$$4x=-3 \quad\text{or}\quad 4x=-3$$
$$x=-\frac{3}{4} \quad\text{or}\quad x=-\frac{3}{4}$$

The solution is $-\frac{3}{4}$.

4.
$$\begin{aligned}x-\frac{x-2}{12}&=\frac{x+3}{4}+\frac{1}{4}\\12\left(x-\frac{x-2}{12}\right)&=12\left(\frac{x+3}{4}+\frac{1}{4}\right)\\12\cdot x-12\left(\frac{x-2}{12}\right)&=12\left(\frac{x+3}{4}\right)+12\cdot\frac{1}{4}\\12x-(x-2)&=3(x+3)+3\\12x-x+2&=3x+9+3\\11x+2&=3x+12\\11x+2-3x&=3x+12-3x\\8x+2&=12\\8x+2-2&=12-2\\8x&=10\\\frac{8x}{8}&=\frac{10}{8}\\x&=\frac{5}{4}\end{aligned}$$

5.
$$4x^2=\frac{15}{2}x+1$$
$$2(4x^2)=2\left(\frac{15}{2}x+1\right)$$
$$8x^2=15x+2$$
$$8x^2-15x-2=0$$
$$(8x+1)(x-2)=0$$
$$8x+1=0 \quad\text{or}\quad x-2=0$$
$$8x=-1$$
$$x=-\frac{1}{8} \quad\text{or}\quad x=2$$

The solutions are $-\frac{1}{8}$ and 2.

Appendix C.1 Exercise Set

1.
$$x^2+11x+24=0$$
$$(x+3)(x+8)=0$$
$$x+3=0 \quad\text{or}\quad x+8=0$$
$$x=-3 \qquad x=-8$$

The solutions are –3 and –8.

3.
$$\begin{aligned}3x-4-5x&=x+4+x\\-2x-4&=2x+4\\-4&=4x+4\\-8&=4x\\-2&=x\end{aligned}$$

The solution is –2.

5. $12x^2+5x-2=0$
$(4x-1)(3x+2)=0$
$4x-1=0$ or $3x+2=0$
$4x=1$ $\quad 3x=-2$
$x=\frac{1}{4}$ $\quad x=-\frac{2}{3}$

The solutions are $-\frac{2}{3}$ and $\frac{1}{4}$.

7. $z^2+9=10z$
$z^2-10z+9=0$
$(z-1)(z-9)=0$
$z-1=0$ or $z-9=0$
$z=1$ $\quad z=9$
The solutions are 1 and 9.

9. $5(y+4)=4(y+5)$
$5y+20=4y+20$
$y+20=20$
$y=0$
The solution is 0.

11. $0.6x-10=1.4x-14$
$-10=0.8x-14$
$4=0.8x$
$40=8x$
$5=x$
The solution is 5.

13. $x(5x+2)=3$
$5x^2+2x=3$
$5x^2+2x-3=0$
$(5x-3)(x+1)=0$
$5x-3=0$ or $x+1=0$
$5x=3$ $\quad x=-1$
$x=\frac{3}{5}$

The solutions are -1 and $\frac{3}{5}$.

15. $6x-2(x-3)=4(x+1)+4$
$6x-2x+6=4x+4+4$
$4x+6=4x+8$
$6=8$
This is a false statement, so the equation has no solution.

17. $\frac{3}{8}+\frac{b}{3}=\frac{5}{12}$
$24\left(\frac{3}{8}+\frac{b}{3}\right)=24\left(\frac{5}{12}\right)$
$9+8b=10$
$8b=1$
$b=\frac{1}{8}$

The solution is $\frac{1}{8}$.

19. $x^2-6x=x(8+x)$
$x^2-6x=8x+x^2$
$-6x=8x$
$0=14x$
$0=x$
The solution is 0.

21. $\frac{z^2}{6}-\frac{z}{2}-3=0$
$z^2-3z-18=0$
$(z+3)(z-6)=0$
$z+3=0$ or $z-6=0$
$z=-3$ $\quad z=6$
The solutions are -3 and 6.

23. $z+3(2+4z)=6(z+1)+5z$
$z+6+12z=6z+6+5z$
$6+13z=11z+6$
$6+2z=6$
$2z=0$
$z=0$
The solution is 0.

25. $\frac{x^2}{2}+\frac{x}{20}=\frac{1}{10}$
$10x^2+x=2$
$10x^2+x-2=0$
$(2x+1)(5x-2)=0$
$2x+1=0$ or $5x-2=0$
$2x=-1$ $\quad 5x=2$
$x=-\frac{1}{2}$ $\quad x=\frac{2}{5}$

The solutions are $-\frac{1}{2}$ and $\frac{2}{5}$.

27. $\frac{4t^2}{5} = \frac{t}{5} + \frac{3}{10}$

$8t^2 = 2t + 3$

$8t^2 - 2t - 3 = 0$

$(2t+1)(4t-3) = 0$

$2t+1 = 0$ or $4t-3 = 0$

$2t = -1$ $\quad$ $4t = 3$

$t = -\frac{1}{2}$ $\quad$ $t = \frac{3}{4}$

The solutions are $-\frac{1}{2}$ and $\frac{3}{4}$.

29. $\frac{3t+1}{8} = \frac{5+2t}{7} + 2$

$56\left(\frac{3t+1}{8}\right) = 56\left(\frac{5+2t}{7} + 2\right)$

$7(3t+1) = 8(5+2t) + 112$

$21t + 7 = 40 + 16t + 112$

$21t + 7 = 16t + 152$

$5t + 7 = 152$

$5t = 145$

$t = 29$

The solution is 29.

31. $\frac{m-4}{3} - \frac{3m-1}{5} = 1$

$15\left(\frac{m-4}{3} - \frac{3m-1}{5}\right) = 15 \cdot 1$

$5(m-4) - 3(3m-1) = 15$

$5m - 20 - 9m + 3 = 15$

$-4m - 17 = 15$

$-4m = 32$

$m = -8$

The solution is –8.

33. $3x^2 = -x$

$3x^2 + x = 0$

$x(3x+1) = 0$

$x = 0$ or $3x+1 = 0$

$3x = -1$

$x = -\frac{1}{3}$

The solutions are $-\frac{1}{3}$ and 0.

35. $x(x-3) = x^2 + 5x + 7$

$x^2 - 3x = x^2 + 5x + 7$

$-3x = 5x + 7$

$-8x = 7$

$x = -\frac{7}{8}$

The solution is $-\frac{7}{8}$.

37. $3(t-8) + 2t = 7 + t$

$3t - 24 + 2t = 7 + t$

$5t - 24 = 7 + t$

$4t - 24 = 7$

$4t = 31$

$t = \frac{31}{4}$

The solution is $\frac{31}{4}$.

39. $-3(x-4) + x = 5(3-x)$

$-3x + 12 + x = 15 - 5x$

$-2x + 12 = 15 - 5x$

$3x + 12 = 15$

$3x = 3$

$x = 1$

The solution is 1.

41. $(x-1)(x+4) = 24$

$x^2 + 3x - 4 = 24$

$x^2 + 3x - 28 = 0$

$(x+7)(x-4) = 0$

$x+7 = 0$ or $x-4 = 0$

$x = -7$ $\quad$ $x = 4$

The solutions are –7 and 4.

43. $\frac{x^2}{4} - \frac{5}{2}x + 6 = 0$

$x^2 - 10x + 24 = 0$

$(x-6)(x-4) = 0$

$x-6 = 0$ or $x-4 = 0$

$x = 6$ $\quad$ $x = 4$

The solutions are 4 and 6.

45.
$$y^2 + \frac{1}{4} = -y$$
$$4y^2 + 1 = -4y$$
$$4y^2 + 4y + 1 = 0$$
$$(2y+1)^2 = 0$$
$$2y + 1 = 0$$
$$2y = -1$$
$$y = -\frac{1}{2}$$
The solution is $-\frac{1}{2}$.

47. **a.** Incorrect; answers may vary

b. Correct; answers may vary

c. Correct; answers may vary

d. Incorrect; answers may vary

49.
$$3.2x + 4 = 5.4x - 7$$
$$3.2x + 4 - 4 = 5.4x - 7 - 4$$
$$3.2x = 5.4x - 11$$
$K = -11$

51.
$$\frac{x}{6} + 4 = \frac{x}{3}$$
$$6\left(\frac{x}{6} + 4\right) = 6\left(\frac{x}{3}\right)$$
$$x + 24 = 2x$$
$K = 24$

53.
$$2.569x = -12.48534$$
$$\frac{2.569x}{2.569} = \frac{-12.48534}{2.569}$$
$$x = -4.86$$
Check:
$$2.569x = -12.48534$$
$$2.569(-4.86) \stackrel{?}{=} -12.48534$$
$$-12.48534 = -12.48534$$
The solution is −4.86.

55.
$$2.86z - 8.1258 = -3.75$$
$$2.86z - 8.1258 + 8.1258 = -3.75 + 8.1258$$
$$2.86z = 4.3758$$
$$\frac{2.86z}{2.86} = \frac{4.3758}{2.86}$$
$$z = 1.53$$

Check:
$$2.86z - 8.1258 = -3.75$$
$$2.86 \cdot 1.53 - 8.1258 \stackrel{?}{=} -3.75$$
$$4.3758 - 8.1258 \stackrel{?}{=} -3.75$$
$$-3.75 = -3.75$$

The solution is 1.53.

57. The quotient of 8 and a number is $\frac{8}{x}$.

59. The product of 8 and a number is $8x$.

61. 2 more than three times a number is $3x + 2$.

Appendix C.2

Practice Exercises

1. a.

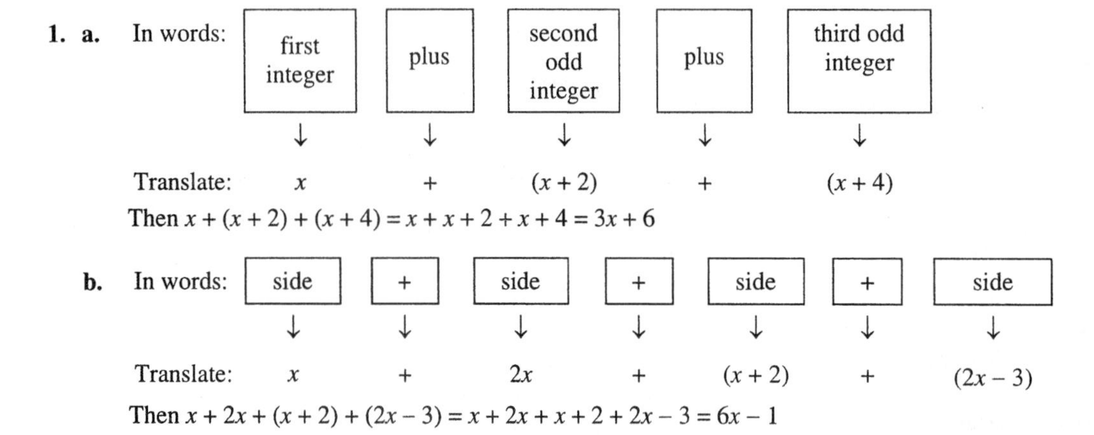

Then $x + (x + 2) + (x + 4) = x + x + 2 + x + 4 = 3x + 6$

Then $x + 2x + (x + 2) + (2x - 3) = x + 2x + x + 2 + 2x - 3 = 6x - 1$

2. If x = number of arrivals and departures at Frankfurt airport, then $x + 15.7$ = number at London, and $x + 1.6$ = number at Paris.

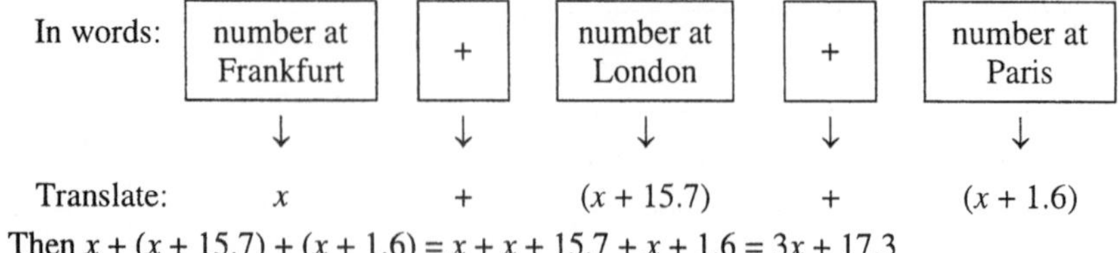

Then $x + (x + 15.7) + (x + 1.6) = x + x + 15.7 + x + 1.6 = 3x + 17.3$

3. Let x = the first number, then $3x - 8$ = the second number, and $5x$ = the third number.
   The sum of the three numbers is 118.
   $$\begin{aligned} x + (3x - 8) + 5x &= 118 \\ x + 3x + 5x - 8 &= 118 \\ 9x - 8 &= 118 \\ 9x &= 126 \\ x &= 14 \end{aligned}$$
   The numbers are 14, $3x - 8 = 3(14) - 8 = 34$, and $5x = 5(14) = 70$.

4. Let x = the original price. Then $0.4x$ = the discount. The original price, minus the discount, is equal to \$270.
   $$\begin{aligned} x - 0.4x &= 270 \\ 0.6x &= 270 \\ x &= \frac{270}{0.6} = 450 \end{aligned}$$
   The original price was \$450.

Vocabulary and Readiness Check

1. 130% of a number $\underline{>}$ the number.
2. 70% of a number $\underline{<}$ the number.
3. 100% of a number $\underline{=}$ the number.
4. 200% of a number $\underline{>}$ the number.

	First Integer	All Described Integers
5. Four consecutive integers	31	31, 32, 33, 34
6. Three consecutive odd integers	31	31, 33, 35
7. Three consecutive even integers	18	18, 20, 22
8. Four consecutive even integers	92	92, 94, 96, 98
9. Three consecutive integers	y	$y, y + 1, y + 2$
10. Three consecutive even integers	z(z is even)	$z, z + 2, z + 4$
11. Four consecutive integers	p	$p, p + 1, p + 2, p + 3$
12. Three consecutive odd integers	s (s is odd)	$s, s + 2, s + 4$

Appendix C.2 Exercise Set

1. The perimeter is the sum of the lengths of the four sides.
   $y + y + y + y = 4y$

3. Let z = first integer, then $z + 1$ = second integer, and $z + 2$ = third integer.
   $z + (z + 1) + (z + 2) = z + z + z + 1 + 2 = 3z + 3$

5. Find the sum of x nickels worth 5¢ each, and $(x + 3)$ dimes worth 10¢ each, and $2x$ quarters worth 25¢ each.
$$5x+10(x+3)+25(2x) = 5x+10x+30+50x$$
$$= 65x+30$$
The total amount is $(65x + 30)$ cents.

7. $4x + 3(2x + 1) = 4x + 6x + 3 = 10x + 3$

9. The length of the side denoted by ? is $10 - 2 = 8$. Similarly, the length of the unmarked side is $(x - 3) - (x - 10) = x - 3 - x + 10 = 7$. Thus the perimeter of the floor plan is given by $(x - 10) + 2 + 7 + 8 + (x - 3) + 10 = 2x + 14$.

11. Let x = the number.
$$4(x-2) = 2+4x+2x$$
$$4x-8 = 2+6x$$
$$-2x = 10$$
$$x = -5$$
The number is –5.

13. Let x = the first number, then
$5x$ = the second number, and
$x + 100$ = the third number.
$$x+5x+(x+100) = 415$$
$$7x+100 = 415$$
$$7x = 315$$
$$x = 45$$
$5x = 225$
$x + 100 = 145$
The numbers are 45, 225, and 145.

15. 29% of $2271 = 0.29 \cdot 2271 = 658.59$;
$2271 - 658.59 = 1612.41$.
Approximately 1612.41 million acres are not federally owned.

17. 85.3% of $2748 = 0.853 \cdot 2748 \approx 2344$
Approximately 2344 minor earthquakes occurred in 2006.

19. 15% of $1500 = 0.15 \cdot 1500 = 225$
$1500 - 225 = 1275$
1275 are willing to do business with any size retailer.

21. $100\% - (55\% + 7\% + 9\% + 7\%) = 100\% - 78\%$
$= 22\%$

23. 7% of $5957 = 0.07 \cdot 5957 = 416.99$
About 417 employees spend more than 3 hours per day using e-mail.

25. Let x = population in 2001.
$$x+0.054x = 31.6$$
$$1.054x = 31.6$$
$$x \approx 29.98$$
The population of Canada in 2001 was about 29.98 million.

27.
$$x+4x+(x+6) = 180$$
$$6x+6 = 180$$
$$6x = 174$$
$$x = 29$$
$4x = 4(29) = 116$
$x + 6 = 29 + 6 = 35$
The angles measure 29°, 35°, and 116°.

29.
$$(4x)+(5x+1)+(5x+3) = 102$$
$$14x+4 = 102$$
$$14x = 98$$
$$x = 7$$
$4x = 4(7) = 28$
$5x + 1 = 5(7) + 1 = 36$
$5x + 3 = 5(7) + 3 = 38$
The sides measure 28 meters, 36 meters, and 38 meters.

31.
$$x+(2.5x-9)+x+1.5x = 99$$
$$6x-9 = 99$$
$$6x = 108$$
$$x = 18$$
$1.5x = 1.5(18) = 27$
$2.5x - 9 = 2.5(18) - 9 = 36$
The sides measure 18 inches, 18 inches, 27 inches, and 36 inches.

33. Let x = first integer; then
$x + 1$ = next integer and
$x + 2$ = third integer.
$$x+(x+1)+(x+2) = 228$$
$$3x+3 = 228$$
$$3x = 225$$
$$x = 75$$
$x + 1 = 75 + 1 = 76$
$x + 2 = 75 + 2 = 77$
The integers are 75, 76, and 77.

35. Let x = first even integer, then
$x + 2$ = second even integer, and
$x + 4$ = third even integer.
$$2x+(x+4) = 268,222$$
$$3x+4 = 268,222$$
$$3x = 268,218$$
$$x = 89,406$$
$x + 2 = 89,408$
$x + 4 = 89,410$

Fallon's zip code is 89406, Fernley's zip code is 89408, and Gardnerville Ranchos's zip code is 89410.

37. $(2x-51)+\left(\frac{3}{2}x+3\right)+x=780$

$$\frac{9}{2}x-48=780$$
$$\frac{9}{2}x=828$$
$$x=828\cdot\frac{2}{9}$$
$$x=184$$

$2x-51=2(184)-51=317$

$\frac{3}{2}x+3=\frac{3}{2}(184)+3=279$

Occupation	Increase in Number of Jobs (in thousands) from 2000 to 2012
Security guards	317 thousand
Home health aides	279 thousand
Computer system analysts	184 thousand
Total	780 thousand

39. Let x = number of medical assistant jobs (in thousands), then $2x + 173$ = number of postsecondary teacher jobs (in thousands), and $3x - 22$ = number of registered nurse jobs (in thousands).

$$x+(2x+173)+(3x-22)=1441$$
$$6x+151=1441$$
$$6x=1290$$
$$x=215$$

$2x + 173 = 2(215) + 173 = 603$

$3x - 22 = 3(215) - 22 = 623$

The predicted job growth:
medical assistant: 215 thousand;
postsecondary teacher jobs: 603 thousand;
registered nurses: 623 thousand

41. Let x = no. of seats in the 737-200; then $x + 21$ = no. in the 737-300 and $2x - 36$ = no. in the 757-200.

$$x+(x+21)+(2x-36)=437$$
$$4x-15=437$$
$$4x=452$$
$$x=113$$

$x + 21 = 113 + 21 = 134$

$2x - 33 = 2(113) - 36 = 190$

The 737-200 has 113 seats. The 737-300 has 134 seats. The 757-200 has 190 seats.

43. Let x = price before taxes.

$$x+0.08x=464.40$$
$$1.08x=464.40$$
$$x=430$$

The price was $430 before taxes.

45. Let x = expected population.

$x = 44.2 - 0.056(44.2)$

$x = 44.2 - 2.4752$

$x \approx 41.7$

The expected population of South Africa in 2050 is 41.7 million.

47. Let x = measure of the angle; then $180 - x$ = measure of its supplement.

$$x=3(180-x)+20$$
$$x=540-3x+20$$
$$4x=560$$
$$x=140$$

$180 - x = 180 - 140 = 40$

The angles measure 140° and 40°.

49. Let x = measure of second angle; then $2x$ = measure of first angle and $3x - 12$ = measure of third angle.

$$x+2x+(3x-12)=180$$
$$6x-12=180$$
$$6x=192$$
$$x=32$$

$2x = 2(32) = 64$

$3x - 12 = 3(32) - 12 = 84$

The angles measure 64°, 32°, and 84°.

51. Let x = the length of a side of the square. Then $x + 6$ = the length of a side of the triangle.

$$4x=3(x+6)$$
$$4x=3x+18$$
$$x=18$$

The sides of the square are 18 cm and the sides of the triangle are 24 cm.

53. Let x = first even integer, then $x + 2$ = second even integer, and $x + 4$ = third even integer.

$$x+(x+4)=156$$
$$2x+4=156$$
$$2x=152$$
$$x=76$$

$x + 2 = 78$

$x + 4 = 80$

The integers are 76, 78, and 80.

55. $x+5x+(6x-3)=483$
$12x-3=483$
$12x=486$
$x=40.5$
$5x = 5(40.5) = 202.5$
$6x - 3 = 6(40.5) - 3 = 240$
The sides measure 40.5 feet, 202.5 feet, and 240 feet.

57. $3x+14.6=197.6$
$3x=183.0$
$x=61.0$
The arrivals and departures are as follows:
Los Angeles: $x = 61.0$ million
Atlanta: $x + 13.3 = 61.0 + 13.3 = 74.3$ million
Chicago: $x + 1.3 = 61.0 + 1.3 = 62.3$ million

59. Let x = hours for halogen; then
$25x$ = hours for fluorescent and
$x - 2500$ = hours for incandescent.
$x+25x+(x-2500)=105,500$
$27x-2500=105,500$
$27x=108,000$
$x=4000$
$25x = 100,000$; $x - 2500 = 1500$
The halogen has 4000 bulb hours.
The fluorescent has 100,000 bulb hours.
The incandescent has 1500 bulb hours.

61. Let x = height, then $2x + 12$ = length.
$2(x)+2(2x+12)=312$
$2x+4x+24=312$
$6x+24=312$
$6x=288$
$x=48$
$2x + 12 = 2(48) + 12 = 108$
The height is 48 inches and the length is 108 inches.

Appendix C.3

Practice Exercises

1. The six points are graphed as shown.

(−2, 4) (2.5, 3.5) (4, 0) $\left(-1\frac{1}{2}, -2\right)$ (0, −2) (3, −4)

a. (3, −4) lies in quadrant IV.

b. (0, −2) is on the y-axis.

c. (−2, 4) lies in quadrant II.

d. (4, 0) is on the x-axis.

e. $\left(-1\frac{1}{2}, -2\right)$ is in quadrant III.

f. (2.5, 3.5) is in quadrant I.

2. $y=-3x-2$
This is a linear equation. (In standard form, it is $3x + y = -2$.) Since the equation is solved for y, we choose three x-values.
Let $x = 0$.
$y=-3x-2$
$y=-3\cdot 0-2$
$y=-2$
Let $x=-1$.
$y=-3x-2$
$y=-3(-1)-2$
$y=1$
Let $x=-2$.
$y=-3x-2$
$y=-3(-2)-2$
$y=4$
The three ordered pairs (0, −2), (−1, 1), and (−2, 4) are listed in the table.

x	y
0	−2
−1	1
−2	4

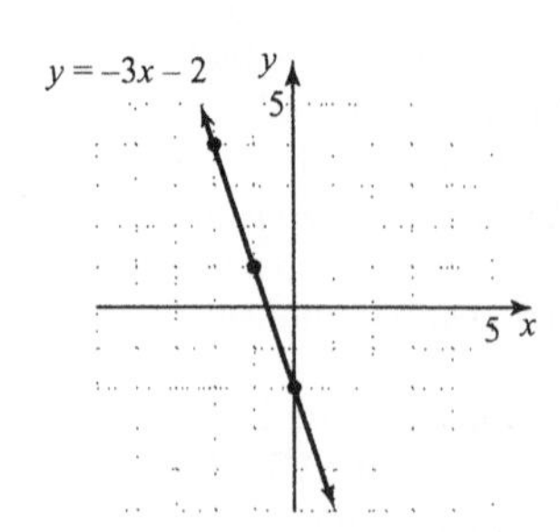

3. $y = -\frac{1}{2}x$

To avoid fractions, we choose x-values that are multiples of 2. To find the y-intercept, we let $x = 0$.

If $x = 0$, then $y = -\frac{1}{2}(0)$, or 0.

If $x = 2$, then $y = -\frac{1}{2}(2)$, or -1.

If $x = -2$, then $y = -\frac{1}{2}(-2)$, or 1.

x	y
0	0
2	−1
−2	1

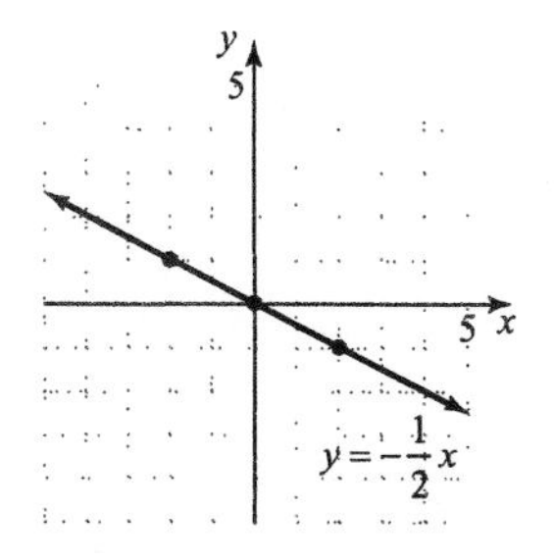

Appendix C.3 Exercise Set

1. Point A is (5, 2).

3. Point C is (3, 0).

5. Point E is (−5, −2).

7. Point G is (−1, 0).

9. (2, 3); QI

11. (−2, 7); QII

13. (−1, −4); QIII

15. (0, −100); y-axis

17. (−10, −30); QIII

19. (−87, 0); x-axis

21. $(x, -y)$ lies in quadrant IV.

23. $(x, 0)$ lies on the x-axis.

25. $(-x, -y)$ lies in quadrant III.

27. $y = -x - 2$
Let $x = 0$.
$y = -0 - 2 = -2$
Let $x = -1$.
$y = -(-1) - 2 = 1 - 2 = -1$
Let $x = 1$.
$y = -1 - 2 = -3$

x	y
0	−2
−1	−1
1	−3

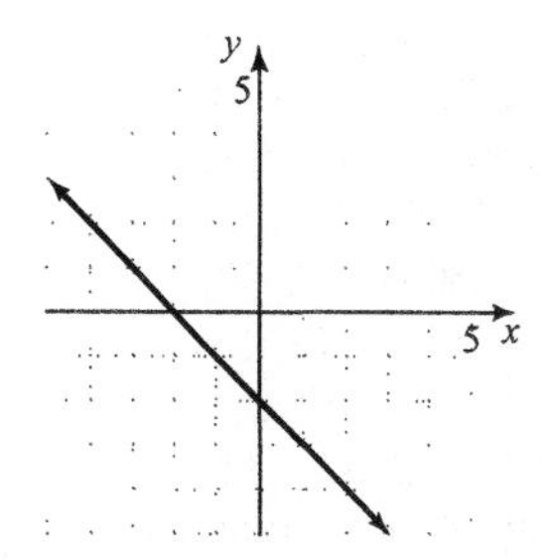

29. $3x - 4y = 8$
Let $x = 0$.
$$3(0) - 4y = 8$$
$$-4y = 8$$
$$y = -2$$
Let $y = 0$.
$$3x - 4(0) = 8$$
$$3x = 8$$
$$x = \frac{8}{3}$$
Let $x = 4$.
$$3(4) - 4y = 8$$
$$12 - 4y = 8$$
$$-4y = -4$$
$$y = 1$$

x	y
0	−2
$\frac{8}{3}$	0
4	1

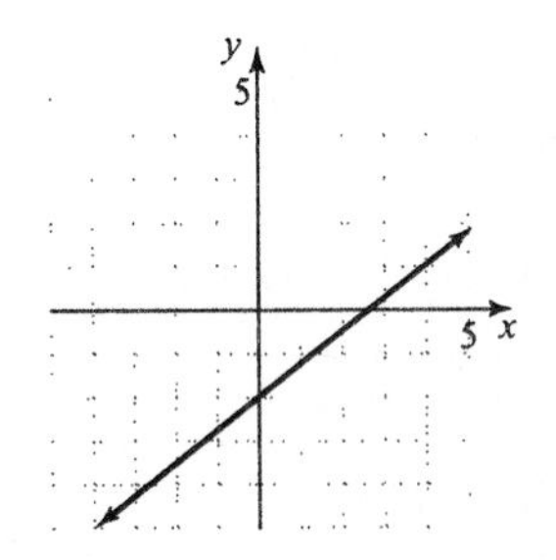

31. $y = \frac{1}{3}x$

Let $x = 0$.

$y = \frac{1}{3}(0) = 0$

Let $x = 3$.

$y = \frac{1}{3}(3) = 1$

Let $x = -3$.

$y = \frac{1}{3}(-3) = -1$

x	y
0	0
3	1
–3	–1

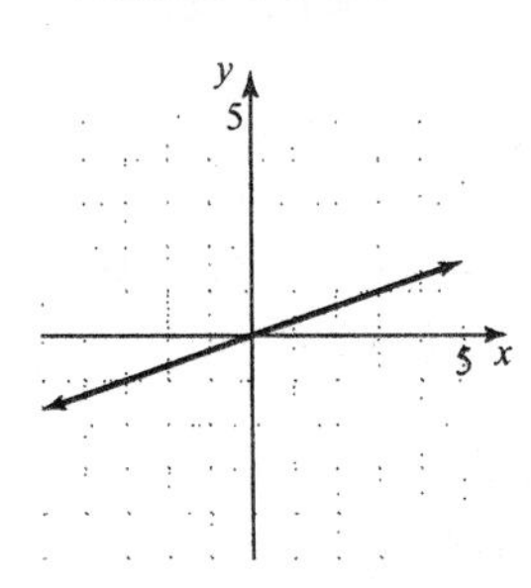

33. $y + 4 = 0$

$y = -4$

This is a horizontal line.

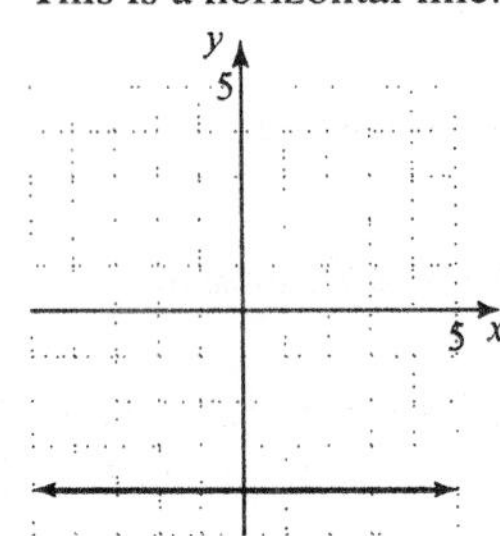

35. The point (4, 1) is on the graph of f. Thus, $f(4) = 1$.

37. The point (0, –4) is on the graph of g. Thus, $g(0) = -4$.

39. The points on the graph of f with y-value of 0 are (1, 0) and (3, 0). Thus, $f(x) = 0$ when $x = 1$ and $x = 3$.

41. $g(-1) = -2$

Appendix C.4

Practice Exercises

1. a. $12x^2y - 3xy = 3xy(4x) + 3xy(-1) = 3xy(4x - 1)$

b. $49x^2 - 4 = (7x)^2 - 2^2 = (7x + 2)(7x - 2)$

c. $5x^2 + 2x - 3 = (5x - 3)(x + 1)$

d. $3x^2 + 6 + x^3 + 2x = 3(x^2 + 2) + x(x^2 + 2) = (x^2 + 2)(3 + x)$

e. $4x^2 + 20x + 25 = (2x)^2 + 2 \cdot 2x \cdot 5 + 5^2 = (2x + 5)^2$

f. $b^2 + 100$ cannot be factored.

2. a. $64x^3 + y^3 = (4x)^3 + y^3 = (4x + y)[(4x)^2 - 4x \cdot y + y^2] = (4x + y)(16x^2 - 4xy + y^2)$

b. $7x^2y^2 - 63y^4 = 7y^2(x^2 - 9y^2) = 7y^2[x^2 - (3y)^2] = 7y^2(x - 3y)(x + 3y)$

c. $3x^2 + 12x + 12 - 3b^2$
$= 3(x^2 + 4x + 4 - b^2)$
$= 3[(x + 2)^2 - b^2]$
$= 3(x + 2 + b)(x + 2 - b)$

d. $x^5y^4+27x^2y$
$=x^2y(x^3y^3+27)$
$=x^2y[(xy)^3+3^3]$
$=x^2y(xy+3)(x^2y^2-3xy+9)$

e. $(x+7)^2-81y^2=(x+7)^2-(9y)^2$
$=(x+7+9y)(x+7-9y)$

Appendix C.4 Exercise Set

1. $(-y^2+6y-1)+(3y^2-4y-10)$
$=-y^2+6y-1+3y^2-4y-10$
$=2y^2+2y-11$

3. $(x^2-6x+2)-(x-5)=x^2-6x+2-x+5$
$=x^2-7x+7$

5. $(5x-3)^2=(5x)^2-2(5x)(3)+3^2$
$=25x^2-30x+9$

7.
$$\begin{array}{r} 2x^3-4x^2+5x-3 \\ x+2\overline{)\,2x^4+0x^3-3x^2\;\;+5x\;\;-2} \\ \underline{2x^4+4x^3} \qquad\qquad\qquad\quad \\ -4x^3-3x^2 \qquad\qquad \\ \underline{-4x^3-8x^2} \qquad\qquad \\ 5x^2\;\;+5x \qquad \\ \underline{5x^2+10x} \qquad \\ -5x\;\;-2 \\ \underline{-5x-10} \\ 8 \end{array}$$

$(2x^4-3x^2+5x-2)\div(x+2)$
$=2x^3-4x^2+5x-5+\dfrac{8}{x+2}$

9. $x^2-8x+16-y^2=(x-4)^2-y^2$
$=(x-4+y)(x-4-y)$

11. $x^4-x=x(x^3-1)=x(x-1)(x^2+x+1)$

13. $14x^2y-2xy=2xy(7x-1)$

15. $4x^2-16=4(x^2-4)=4(x+2)(x-2)$

17. $3x^2-8x-11=(3x-11)(x+1)$

19. $4x^2+8x-12=4(x^2+2x-3)$
$=4(x+3)(x-1)$

21. $4x^2+36x+81=(2x)^2+2\cdot 2x\cdot 9+9^2$
$=(2x+9)^2$

23. $8x^3+125y^3=(2x)^3+(5y)^3$
$=(2x+5y)(4x^2-10xy+25y^2)$

25. $64x^2y^3-8x^2=8x^2(8y^3-1)$
$=8x^2[(2y)^3-1^3]$
$=8x^2(2y-1)(4y^2+2y+1)$

27. $(x+5)^3+y^3$
$=[(x+5)+y][(x+5)^2-(x+5)y+y^2]$
$=(x+y+5)(x^2+10x+25-xy-5y+y^2)$
$=(x+y+5)(x^2+10x-xy-5y+y^2+25)$

29. Let $y=5a-3$. Then
$(5a-3)^2-6(5a-3)+9=y^2-6y+9$
$=(y-3)(y-3)$
$=(y-3)^2$
$=[(5a-3)-3]^2$
$=(5a-6)^2$

31. $7x^2-63x=7x(x-9)$

33. $ab-6a+7b-42=a(b-6)+7(b-6)$
$=(a+7)(b-6)$

35. $x^4-1=(x^2)^2-1^2$
$=(x^2+1)(x^2-1)$
$=(x^2+1)(x+1)(x-1)$

37. $10x^2-7x-33=(5x-11)(2x+3)$

39. $5a^3b^3-50a^3b=5a^3b(b^2-10)$

41. $16x^2+25$ is a prime polynomial.

43. $10x^3-210x^2+1100x=10x(x^2-21x+110)$
$=10x(x-11)(x-10)$

45. $64a^3b^4 - 27a^3b$
$= a^3b(64b^3 - 27)$
$= a^3b[(4b)^3 - 3^3]$
$= a^3b(4b-3)(16b^2 + 12b + 9)$

47. $2x^3 - 54 = 2(x^3 - 27)$
$= 2(x^3 - 3^3)$
$= 2(x-3)(x^2 + 3x + 9)$

49. $3y^5 - 5y^4 + 6y - 10 = y^4(3y-5) + 2(3y-5)$
$= (y^4 + 2)(3y-5)$

51. $100z^3 + 100 = 100(z^3 + 1)$
$= 100(z+1)(z^2 - z + 1)$

53. $4b^2 - 36b + 81 = (2b)^2 - 2 \cdot 2b \cdot 9 + 9^2$
$= (2b-9)^2$

55. Let $x = y - 6$. Then
$(y-6)^2 + 3(y-6) + 2 = x^2 + 3x + 2$
$= (x+2)(x+1)$
$= [(y-6)+2][(y-6)+1]$
$= (y-4)(y-5)$

57. Area $= 3^2 - 4x^2 = 3^2 - (2x)^2 = (3+2x)(3-2x)$

Appendix C.5

Practice Exercises

1. a. $\frac{2+5n}{3n} \cdot \frac{6n+3}{5n^2 - 3n - 2}$
$= \frac{2+5n}{3n} \cdot \frac{3(2n+1)}{(5n+2)(n-1)}$
$= \frac{2n+1}{n(n-1)}$

b. $\frac{x^3 - 8}{-6x + 12} \cdot \frac{6x^2}{x^2 + 2x + 4}$
$= \frac{(x-2)(x^2 + 2x + 4)}{-6(x-2)} \cdot \frac{6x^2}{x^2 + 2x + 4}$
$= \frac{(x-2)(x^2 + 2x + 4) \cdot 6 \cdot x^2}{-1 \cdot 6(x-2)(x^2 + 2x + 4)}$
$= \frac{x^2}{-1}$
$= -x^2$

2. a. $\frac{6y^3}{3y^2 - 27} \div \frac{42}{3-y} = \frac{6y^3}{3y^2 - 27} \cdot \frac{3-y}{42}$
$= \frac{6y^3(3-y)}{3(y+3)(y-3) \cdot 42}$
$= \frac{6y^3 \cdot (-1)(y-3)}{3(y+3)(y-3) \cdot 6 \cdot 7}$
$= -\frac{y^3}{21(y+3)}$

b. $\frac{10x^2 + 23x - 5}{5x^2 - 51x + 10} \div \frac{2x^2 + 9x + 10}{7x^2 - 68x - 20}$
$= \frac{10x^2 + 23x - 5}{5x^2 - 51x + 10} \cdot \frac{7x^2 - 68x - 20}{2x^2 + 9x + 10}$
$= \frac{(5x-1)(2x+5)}{(5x-1)(x-10)} \cdot \frac{(7x+2)(x-10)}{(2x+5)(x+2)}$
$= \frac{7x+2}{x+2}$

3. a. The LCD is $5p^4q$.
$\frac{4}{p^3q} + \frac{3}{5p^4q} = \frac{4 \cdot 5p}{p^3q \cdot 5p} + \frac{3}{5p^4q}$
$= \frac{20p}{5p^4q} + \frac{3}{5p^4q}$
$= \frac{20p + 3}{5p^4q}$

b. The LCD is the product of the two denominators: $(y + 3)(y - 3)$.

$$\frac{4}{y+3}+\frac{5y}{y-3}$$
$$=\frac{4\cdot(y-3)}{(y+3)\cdot(y-3)}+\frac{5y\cdot(y+3)}{(y-3)\cdot(y+3)}$$
$$=\frac{4y-12}{(y+3)(y-3)}+\frac{5y^2+15y}{(y+3)(y-3)}$$
$$=\frac{4y-12+5y^2+15y}{(y+3)(y-3)}$$
$$=\frac{5y^2+19y-12}{(y+3)(y-3)}$$

c. The LCD is either $z - 5$ or $5 - z$.

$$\frac{3z-18}{z-5}-\frac{3}{5-z}=\frac{3z-18}{z-5}-\frac{3}{-1(z-5)}$$
$$=\frac{3z-18}{z-5}-\frac{-1\cdot 3}{z-5}$$
$$=\frac{3z-18-(-3)}{z-5}$$
$$=\frac{3z-18+3}{z-5}$$
$$=\frac{3z-15}{z-5}$$
$$=\frac{3(z-5)}{z-5}$$
$$=3$$

4. $x^2-4=(x+2)(x-2)$

The LCD is $(x + 2)(x - 2)$.

$$\frac{2}{x-2}-\frac{5+2x}{x^2-4}=\frac{x}{x+2}$$
$$(x+2)(x-2)\cdot\frac{2}{x-2}-(x+2)(x-2)\cdot\frac{5+2x}{(x+2)(x-2)}$$
$$=(x+2)(x-2)\cdot\frac{x}{x+2}$$
$$2(x+2)-(5+2x)=x(x-2)$$
$$2x+4-5-2x=x^2-2x$$
$$x^2-2x+1=0$$
$$(x-1)(x-1)=0$$
$$x-1=0$$
$$x=1$$

Since 1 does not make any denominator 0, the solution is 1.

Appendix C.5 Exercise Set

1. $\frac{x}{2}=\frac{1}{8}+\frac{x}{4}$

The LCD is 8.

$$8\cdot\frac{x}{2}=8\cdot\frac{1}{8}+8\cdot\frac{x}{4}$$
$$4x=1+2x$$
$$2x=1$$
$$x=\frac{1}{2}$$

The solution is $\frac{1}{2}$.

3. $\frac{1}{8}+\frac{x}{4}=\frac{1}{8}+\frac{x}{4}\cdot\frac{2}{2}=\frac{1}{8}+\frac{2x}{8}=\frac{1+2x}{8}$

5. $\frac{4}{x+2}-\frac{2}{x-1}=\frac{4}{x+2}\cdot\frac{x-1}{x-1}-\frac{2}{x-1}\cdot\frac{x+2}{x+2}$

$$=\frac{4(x-1)}{(x+2)(x-1)}-\frac{2(x+2)}{(x+2)(x+1)}$$
$$=\frac{4x-4-2x-4}{(x+2)(x-1)}$$
$$=\frac{2x-8}{(x+2)(x-1)}$$
$$=\frac{2(x-4)}{(x+2)(x-1)}$$

7. $\frac{4}{x+2}=\frac{2}{x-1}$

The LCD is $(x + 2)(x - 1)$.

$$(x+2)(x-1)\cdot\frac{4}{x+2}=(x+2)(x-1)\cdot\frac{2}{x-1}$$
$$4(x-1)=2(x+2)$$
$$4x-4=2x+4$$
$$4x=2x+8$$
$$2x=8$$
$$x=4$$

The solution is 4.

9. $x^2-4=(x+2)(x-2)$

The LCD is $(x+2)(x-2)$.

$$\begin{aligned}
\frac{2}{x^2-4} &= \frac{1}{x+2}-\frac{3}{x-2}\\
(x+2)(x-2)\cdot\frac{2}{x^2-4} &= (x+2)(x-2)\cdot\frac{1}{x+2}-(x+2)(x-2)\cdot\frac{3}{x-2}\\
2 &= (x-2)-3(x+2)\\
2 &= x-2-3x-6\\
2 &= -2x-8\\
2x &= -10\\
x &= -5
\end{aligned}$$

The solution is -5.

11.
$$\begin{aligned}
\frac{5}{x^2-3x}+\frac{4}{2x-6} &= \frac{5}{x(x-3)}+\frac{4}{2(x-3)}\\
&= \frac{5}{x(x-3)}\cdot\frac{2}{2}+\frac{4}{2(x-3)}\cdot\frac{x}{x}\\
&= \frac{10}{2x(x-3)}+\frac{4x}{2x(x-3)}\\
&= \frac{4x+10}{2x(x-3)}\\
&= \frac{2(2x+5)}{2x(x-3)}\\
&= \frac{2x+5}{x(x-3)}
\end{aligned}$$

13. $x^2-1=(x-1)(x+1)$

The LCD is $(x-1)(x+1)$.

$$\begin{aligned}
\frac{x-1}{x+1}+\frac{x+7}{x-1} &= \frac{4}{x^2-1}\\
(x-1)(x+1)\cdot\frac{x-1}{x+1}+(x-1)(x+1)\cdot\frac{x+7}{x-1} &= (x-1)(x+1)\cdot\frac{4}{(x-1)(x+1)}\\
(x-1)(x-1)+(x+1)(x+7) &= 4\\
x^2-2x+1+x^2+8x+7 &= 4\\
2x^2+6x+8 &= 4\\
2x^2+6x+4 &= 0\\
2(x^2+3x+2) &= 0\\
2(x+1)(x+2) &= 0
\end{aligned}$$

$x+1=0$ or $x+2=0$

$x=-1$ or $x=-2$

The number -1 makes the denominator $x+1$ equal to 0, so it is not a solution. The solution is -2.

15.
$$\begin{aligned}
\frac{a^2-9}{a-6}\cdot\frac{a^2-5a-6}{a^2-a-6} &= \frac{(a+3)(a-3)}{a-6}\cdot\frac{(a-6)(a+1)}{(a-3)(a+2)}\\
&= \frac{(a+3)(a+1)}{a+2}
\end{aligned}$$

17. $\frac{2x+3}{3x-2}=\frac{4x+1}{6x+1}$

The LCD is $(3x - 2)(6x + 1)$.

$$(3x-2)(6x+1)\cdot\frac{2x+3}{3x-2}=(3x-2)(6x+1)\cdot\frac{4x+1}{6x+1}$$
$$(6x+1)(2x+3)=(3x-2)(4x+1)$$
$$12x^2+18x+2x+3=12x^2+3x-8x-2$$
$$12x^2+20x+3=12x^2-5x-2$$
$$20x+3=-5x-2$$
$$25x+3=-2$$
$$25x=-5$$
$$x=-\frac{5}{25}$$
$$x=-\frac{1}{5}$$

The solution is $-\frac{1}{5}$.

19. $\frac{a}{9a^2-1}+\frac{2}{6a-2}=\frac{a}{(3a-1)(3a+1)}+\frac{2}{2(3a-1)}$

$$=\frac{a}{(3a-1)(3a+1)}\cdot\frac{2}{2}+\frac{2}{2(3a-1)}\cdot\frac{(3a+1)}{(3a+1)}$$
$$=\frac{2a}{2(3a-1)(3a+1)}+\frac{6a+2}{2(3a-1)(3a+1)}$$
$$=\frac{8a+2}{2(3a-1)(3a+1)}$$
$$=\frac{2(4a+1)}{2(3a-1)(3a+1)}$$
$$=\frac{4a+1}{(3a-1)(3a+1)}$$

21. The LCD is x^2.

$$-\frac{3}{x^2}-\frac{1}{x}+2=0$$
$$x^2\cdot-\frac{3}{x^2}-x^2\cdot\frac{1}{x}+x^2\cdot 2=0$$
$$-3-x+2x^2=0$$
$$2x^2-x-3=0$$
$$(2x-3)(x+1)=0$$

$2x-3=0$ or $x+1=0$

$2x=3$ or $x=-1$

$x=\frac{3}{2}$ or $x=-1$

The solutions are -1 and $\frac{3}{2}$.

23. $\frac{x-8}{x^2-x-2}+\frac{2}{x-2}=\frac{x-8}{(x-2)(x+1)}+\frac{2}{x-2}$
$=\frac{x-8}{(x-2)(x+1)}+\frac{2}{x-2}\cdot\frac{x+1}{x+1}$
$=\frac{x-8}{(x-2)(x+1)}+\frac{2x+2}{(x-2)(x+1)}$
$=\frac{x-8+2x+2}{(x-2)(x+1)}$
$=\frac{3x-6}{(x-2)(x+1)}$
$=\frac{3(x-2)}{(x-2)(x+1)}$
$=\frac{3}{x+1}$

25. The LCD is a.
$\frac{3}{a}-5=\frac{7}{a}-1$
$a\cdot\frac{3}{a}-a\cdot 5=a\cdot\frac{7}{a}-a\cdot 1$
$3-5a=7-a$
$3=7+4a$
$-4=4a$
$-1=a$
The solution is -1.

27. a. $\frac{x}{5}-\frac{x}{4}+\frac{1}{10}$ is an expression.

b. The first step to simplify this expression is to write each rational expression term so that the denominator is the LCD, 20.

c. $\frac{x}{5}-\frac{x}{4}+\frac{1}{10}=\frac{x}{5}\cdot\frac{4}{4}-\frac{x}{4}\cdot\frac{5}{5}+\frac{1}{10}\cdot\frac{2}{2}$
$=\frac{4x}{20}-\frac{5x}{20}+\frac{2}{20}$
$=\frac{4x-5x+2}{20}$
$=\frac{-x+2}{20}$

29. $\frac{\triangle+\square}{\triangle}=\frac{\triangle}{\triangle}+\frac{\square}{\triangle}=1+\frac{\square}{\triangle}$
b is the correct answer.

31. $\frac{\triangle}{\square}\cdot\frac{\bigcirc}{\square}=\frac{\triangle\bigcirc}{\square\square}$
d is the correct answer.

33. $\frac{\frac{\triangle+\square}{\bigcirc}}{\frac{\triangle}{\bigcirc}}=\frac{\triangle+\square}{\bigcirc}\div\frac{\triangle}{\bigcirc}=\frac{\triangle+\square}{\bigcirc}\cdot\frac{\bigcirc}{\triangle}=\frac{\triangle+\square}{\triangle}$
d is the correct answer.

Appendix D

Viewing Window and Interpreting Window Settings Exercise Set

1. Yes, since every coordinate is between −10 and 10.

3. No, since −11 is less than −10.

5. Answers may vary. Any values such that Xmin < −90, Ymin < −80, Xmax > 55, and Ymax > 80.

7. Answers may vary. Any values such that Xmin < −11, Ymin < −5, Xmax > 7, and Ymax > 2.

9. Answers may vary. Any values such that Xmin < 50, Ymin < −50, Xmax > 200, and Ymax > 200.

11. Xmin = −12 Ymin = −12
Xmax = 12 Ymax = 12
Xscl = 3 Yscl = 3

13. Xmin = −9 Ymin = −12
Xmax = 9 Ymax = 12
Xscl = 1 Yscl = 2

15. Xmin = −10 Ymin = −25
Xmax = 10 Ymax = 25
Xscl = 2 Yscl = 5

17. Xmin = −10 Ymin = −30
Xmax = 10 Ymax = 30
Xscl = 1 Yscl = 3

19. Xmin = −20 Ymin = −30
Xmax = 30 Ymax = 50
Xscl = 5 Yscl = 10

Graphing Equations and Square Viewing Window Exercise Set

1. Setting A:

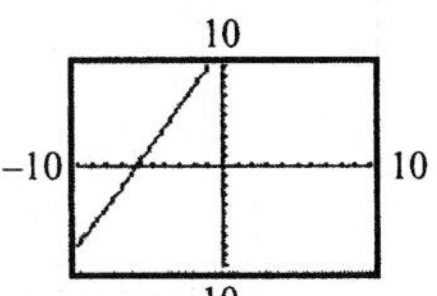

Setting B:

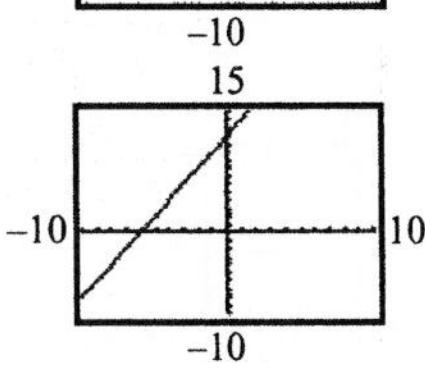

Setting B shows all intercepts.

3. Setting A:

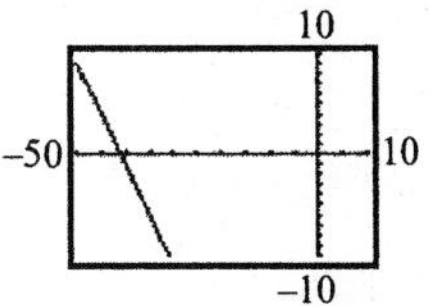

Setting B:

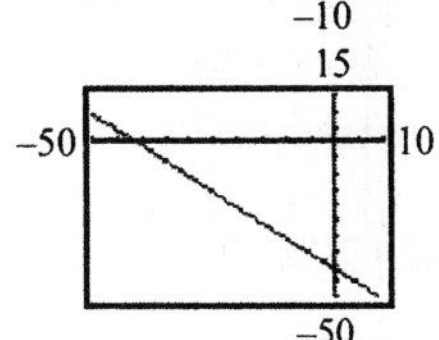

Setting B shows all intercepts.

5. Setting A:

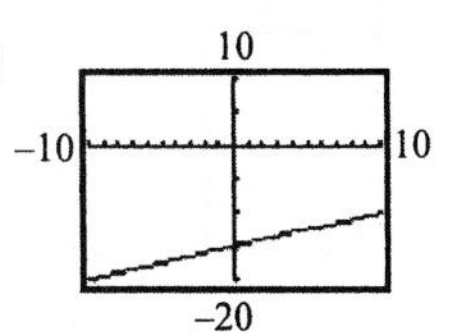

Setting B:

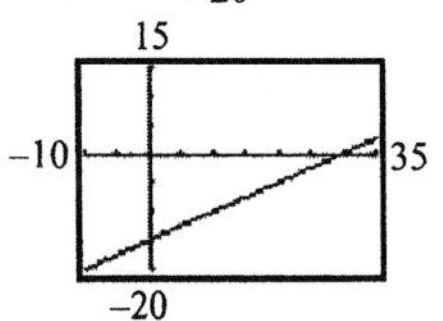

Setting B shows all intercepts.

7. $3x = 5y$

$y = \frac{3}{5}x$

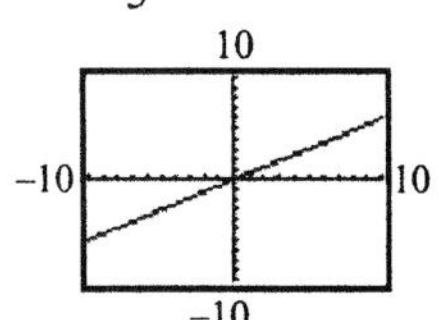

9. $9x - 5y = 30$

$$-5y = -9x + 30$$

$$y = \frac{9}{5}x - 6$$

10

−10 10

−10

11. $y = -7$

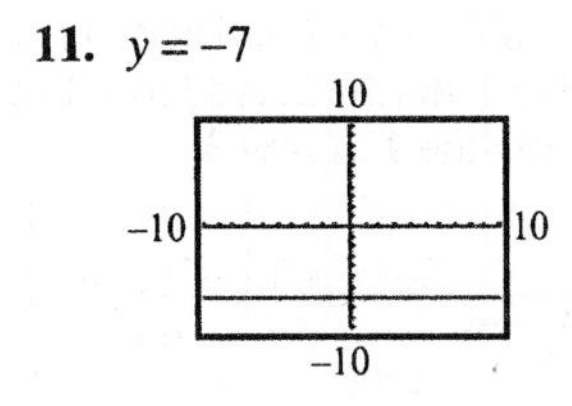

13. $x + 10y = -5$

$$10y = -x - 5$$

$$y = -\frac{1}{10}x - \frac{1}{2}$$

10

−10 10

−10

15. $y = \sqrt{x}$

8

−12 12

−8

17. $y = x^2 + 2x + 1$

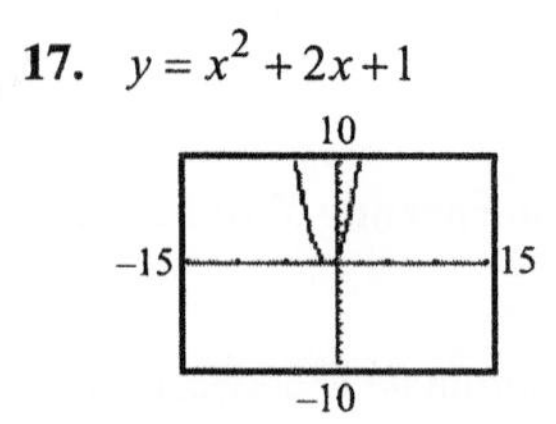

19. $y = |x|$

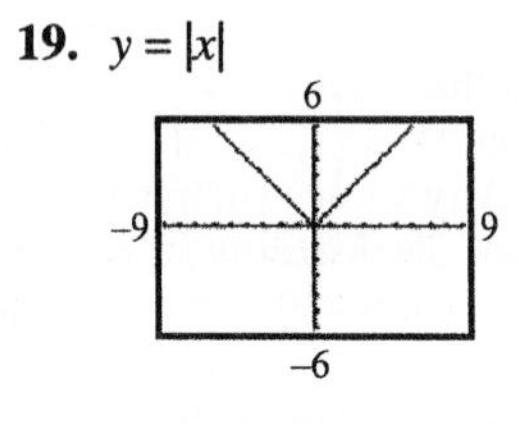

21. $x + 2y = 30$

$$2y = -x + 30$$

$$y = -\frac{1}{2}x + 15$$

Standard window:

Adjusted window:

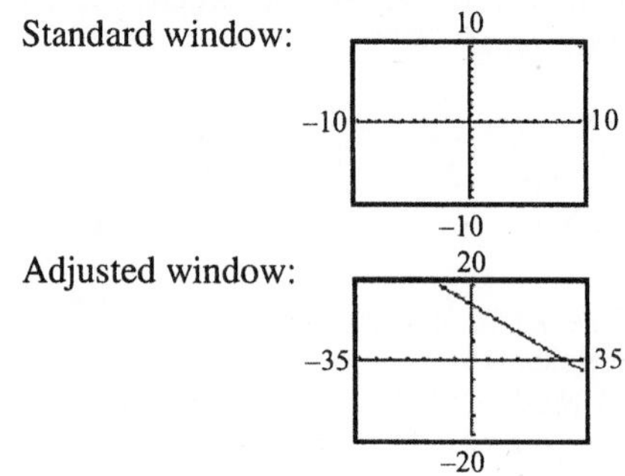

Appendix E

Appendix E

Practice Exercises

1. $\begin{cases} x+4y=-2 \\ 3x-y=7 \end{cases}$

The corresponding matrix is $\left[\begin{array}{cc|c} 1 & 4 & -2 \\ 3 & -1 & 7 \end{array}\right]$. The element in the first row, first column is already 1. Multiply row 1 by −3 and add to row 2 to get a 0 below the 1.

$$\left[\begin{array}{cc|c} 1 & 4 & -2 \\ -3(1)+3 & -3(4)+(-1) & -3(-2)+7 \end{array}\right]$$

$$\left[\begin{array}{cc|c} 1 & 4 & -2 \\ 0 & -13 & 13 \end{array}\right]$$

We change −13 to a 1 by dividing row 2 by −13.

$$\left[\begin{array}{cc|c} 1 & 4 & -2 \\ 0 & \frac{-13}{-13} & \frac{13}{-13} \end{array}\right]$$

$$\left[\begin{array}{cc|c} 1 & 4 & -2 \\ 0 & 1 & -1 \end{array}\right]$$

The last matrix corresponds to $\begin{cases} x+4y=-2 \\ y=-1 \end{cases}$

To find x, we let $y=-1$ in the first equation.

$$\begin{aligned} x+4y&=-2 \\ x+4(-1)&=-2 \\ x-4&=-2 \\ x&=2 \end{aligned}$$

The solution is (2, −1).

2. $\begin{cases} x-3y=3 \\ -2x+6y=4 \end{cases}$

The corresponding matrix is $\left[\begin{array}{cc|c} 1 & -3 & 3 \\ -2 & 6 & 4 \end{array}\right]$. The element in the first row, first column is already 1. Multiply row 1 by 2 and add to row 2 to get a 0 below the 1.

$$\left[\begin{array}{cc|c} 1 & -3 & 3 \\ 2(1)+(-2) & 2(-3)+6 & 2(3)+4 \end{array}\right]$$

$$\left[\begin{array}{cc|c} 1 & -3 & 3 \\ 0 & 0 & 10 \end{array}\right]$$

The corresponding system is $\begin{cases} x-3y=3 \\ 0=10 \end{cases}$

The equation 0 = 10 is false. Hence, the system is inconsistent and has no solution. The solution set is ∅.

3. $\begin{cases} x+3y-z=0 \\ 2x+y+3z=5 \\ -x-2y+4z=7 \end{cases}$

The corresponding matrix is $\left[\begin{array}{ccc|c} 1 & 3 & -1 & 0 \\ 2 & 1 & 3 & 5 \\ -1 & -2 & 4 & 7 \end{array}\right]$.

The element in the first row, first column is already 1. Multiply row 1 by −2 and add to row 2 to get a 0 below the 1 in row 2. Add row 1 to row 3 to get a 0 below the 1 in row 3.

$$\left[\begin{array}{ccc|c} 1 & 3 & -1 & 0 \\ -2(1)+2 & -2(3)+1 & -2(-1)+3 & -2(0)+5 \\ 1+(-1) & 3+(-2) & -1+4 & 0+7 \end{array}\right]$$

$$\left[\begin{array}{ccc|c} 1 & 3 & -1 & 0 \\ 0 & -5 & 5 & 5 \\ 0 & 1 & 3 & 7 \end{array}\right]$$

Now we want a 1 where the −5 is now. Interchange rows 2 and 3.

$$\left[\begin{array}{ccc|c} 1 & 3 & -1 & 0 \\ 0 & 1 & 3 & 7 \\ 0 & -5 & 5 & 5 \end{array}\right]$$

Now we want a 0 below the 1. Multiply row 2 by 5 and add to row 3.

$$\left[\begin{array}{ccc|c} 1 & 3 & -1 & 0 \\ 0 & 1 & 3 & 7 \\ 5(0)+0 & 5(1)+(-5) & 5(3)+5 & 5(7)+5 \end{array}\right]$$

$$\left[\begin{array}{ccc|c} 1 & 3 & -1 & 0 \\ 0 & 1 & 3 & 7 \\ 0 & 0 & 20 & 40 \end{array}\right]$$

Finally, divide row 3 by 20.

$$\left[\begin{array}{ccc|c} 1 & 3 & -1 & 0 \\ 0 & 1 & 3 & 7 \\ 0 & 0 & \frac{20}{20} & \frac{40}{20} \end{array}\right]$$

$$\left[\begin{array}{ccc|c} 1 & 3 & -1 & 0 \\ 0 & 1 & 3 & 7 \\ 0 & 0 & 1 & 2 \end{array}\right]$$

This matrix corresponds to the system

$$\begin{cases} x+3y-z=0 \\ y+3z=7 \\ z=2 \end{cases}$$

The z-coordinate is 2. Replace z with 2 in the second equation and solve for y.

$$\begin{aligned} y+3z&=7\\ y+3(2)&=7\\ y+6&=7\\ y&=1 \end{aligned}$$

To find x, we let $z = 2$ and $y = 1$ in the first equation.

$$\begin{aligned} x+3y-z&=0\\ x+3(1)-2&=0\\ x+1&=0\\ x&=-1 \end{aligned}$$

The solution is (−1, 1, 2).

Appendix E Exercise Set

1. $\begin{cases} x+\ \ y=1 \\ x-2y=4 \end{cases}$

$\left[\begin{array}{cc|c} 1 & 1 & 1 \\ 1 & -2 & 4 \end{array}\right]$

Multiply R1 by −1 and add to R2.

$\left[\begin{array}{cc|c} 1 & 1 & 1 \\ 0 & -3 & 3 \end{array}\right]$

Divide R2 by −3.

$\left[\begin{array}{cc|c} 1 & 1 & 1 \\ 0 & 1 & -1 \end{array}\right]$

This corresponds to $\begin{cases} x+y=1 \\ \quad\ \ y=-1 \end{cases}$.

$$\begin{aligned} x+(-1)&=1\\ x-1&=1\\ x&=2 \end{aligned}$$

The solution is (2, −1).

3. $\begin{cases} x+3y=2 \\ x+2y=0 \end{cases}$

$\left[\begin{array}{cc|c} 1 & 3 & 2 \\ 1 & 2 & 0 \end{array}\right]$

Multiply R1 by −1 and add to R2.

$\left[\begin{array}{cc|c} 1 & 3 & 2 \\ 0 & -1 & -2 \end{array}\right]$

Multiply R2 by −1.

$\left[\begin{array}{cc|c} 1 & 3 & 2 \\ 0 & 1 & 2 \end{array}\right]$

This corresponds to $\begin{cases} x+3y=2 \\ \qquad y=2 \end{cases}$.

$$\begin{aligned} x+3(2)&=2\\ x+6&=2\\ x&=-4 \end{aligned}$$

The solution is (−4, 2).

5. $\begin{cases} \ \ x-2y=4 \\ 2x-4y=4 \end{cases}$

$\left[\begin{array}{cc|c} 1 & -2 & 4 \\ 2 & -4 & 4 \end{array}\right]$

Multiply R1 by −2 and add to R2.

$\left[\begin{array}{cc|c} 1 & -2 & 4 \\ 0 & 0 & -4 \end{array}\right]$

This corresponds to $\begin{cases} x-2y=4 \\ \qquad 0=-4 \end{cases}$.

This is an inconsistent system. The solution is ∅.

7. $\begin{cases} 3x-3y=9 \\ 2x-2y=6 \end{cases}$

$\left[\begin{array}{cc|c} 3 & -3 & 9 \\ 2 & -2 & 6 \end{array}\right]$

Divide R1 by 3.

$\left[\begin{array}{cc|c} 1 & -1 & 3 \\ 2 & -2 & 6 \end{array}\right]$

Multiply R1 by −2 and add to R2.

$\left[\begin{array}{cc|c} 1 & -1 & 3 \\ 0 & 0 & 0 \end{array}\right]$

This corresponds to $\begin{cases} x-y=3 \\ \quad\ \ 0=0 \end{cases}$.

This is a dependent system. The solution is $\{(x, y)|x-y=3\}$.

9. $\begin{cases} x+\ \ y \qquad\ =3 \\ \quad\ 2y \qquad\ =10 \\ 3x+2y-4z=12 \end{cases}$

$\left[\begin{array}{ccc|c} 1 & 1 & 0 & 3 \\ 0 & 2 & 0 & 10 \\ 3 & 2 & -4 & 12 \end{array}\right]$

Multiply R1 by −3 and add to R3.

$\left[\begin{array}{ccc|c} 1 & 1 & 0 & 3 \\ 0 & 2 & 0 & 10 \\ 0 & -1 & -4 & 3 \end{array}\right]$

Divide R2 by 2.

$\left[\begin{array}{ccc|c} 1 & 1 & 0 & 3 \\ 0 & 1 & 0 & 5 \\ 0 & -1 & -4 & 3 \end{array}\right]$

Add R2 to R3.

$\left[\begin{array}{ccc|c} 1 & 1 & 0 & 3 \\ 0 & 1 & 0 & 5 \\ 0 & 0 & -4 & 8 \end{array}\right]$

Divide R3 by −4.

$$\left[\begin{array}{ccc|c}1 & 1 & 0 & 3\\0 & 1 & 0 & 5\\0 & 0 & 1 & -2\end{array}\right]$$

This corresponds to $\begin{cases}x+y=3\\ y=5\\ z=-2\end{cases}$.

$x+5=3$
$x=-2$

The solution is (–2, 5, –2).

11. $\begin{cases}2y-z=-7\\ x+4y+z=-4\\ 5x-y+2z=13\end{cases}$

$$\left[\begin{array}{ccc|c}0 & 2 & -1 & -7\\1 & 4 & 1 & -4\\5 & -1 & 2 & 13\end{array}\right]$$

Interchange R1 and R2.

$$\left[\begin{array}{ccc|c}1 & 4 & 1 & -4\\0 & 2 & -1 & -7\\5 & -1 & 2 & 13\end{array}\right]$$

Multiply R1 by –5 and add to R3.

$$\left[\begin{array}{ccc|c}1 & 4 & 1 & -4\\0 & 2 & -1 & -7\\0 & -21 & -3 & 33\end{array}\right]$$

Divide R2 by 2.

$$\left[\begin{array}{ccc|c}1 & 4 & 1 & -4\\0 & 1 & -\frac{1}{2} & -\frac{7}{2}\\0 & -21 & -3 & 33\end{array}\right]$$

Multiply R2 by 21 and add to R3.

$$\left[\begin{array}{ccc|c}1 & 4 & 1 & -4\\0 & 1 & -\frac{1}{2} & -\frac{7}{2}\\0 & 0 & -\frac{27}{2} & -\frac{81}{2}\end{array}\right]$$

Multiply R2 by $-\frac{2}{27}$.

$$\left[\begin{array}{ccc|c}1 & 4 & 1 & -4\\0 & 1 & -\frac{1}{2} & -\frac{7}{2}\\0 & 0 & 1 & 3\end{array}\right]$$

This corresponds to $\begin{cases}x+4y+z=-4\\ y-\frac{1}{2}z=-\frac{7}{2}\\ z=3\end{cases}$.

$y-\frac{1}{2}(3)=-\frac{7}{2}$
$y-\frac{3}{2}=-\frac{7}{2}$
$y=-2$
$x+4(-2)+3=-4$
$x-8+3=-4$
$x=1$

The solution is (1, –2, 3).

13. $\begin{cases}x-4=0\\ x+y=1\end{cases}$ or $\begin{cases}x=4\\ x+y=1\end{cases}$

$$\left[\begin{array}{cc|c}1 & 0 & 4\\1 & 1 & 1\end{array}\right]$$

Multiply R1 by –1 and add to R2.

$$\left[\begin{array}{cc|c}1 & 0 & 4\\0 & 1 & -3\end{array}\right]$$

This corresponds to $\begin{cases}x=4\\ y=-3\end{cases}$

The solution is (4, –3).

15. $\begin{cases}x+y+z=2\\ 2x-z=5\\ 3y+z=2\end{cases}$

$$\left[\begin{array}{ccc|c}1 & 1 & 1 & 2\\2 & 0 & -1 & 5\\0 & 3 & 1 & 2\end{array}\right]$$

Multiply R1 by –2 and add to R2.

$$\left[\begin{array}{ccc|c}1 & 1 & 1 & 2\\0 & -2 & -3 & 1\\0 & 3 & 1 & 2\end{array}\right]$$

Divide R2 by –2.

$$\left[\begin{array}{ccc|c}1 & 1 & 1 & 2\\0 & 1 & \frac{3}{2} & -\frac{1}{2}\\0 & 3 & 1 & 2\end{array}\right]$$

Multiply R2 by –3 and add to R3.

$$\left[\begin{array}{ccc|c}1 & 1 & 1 & 2\\0 & 1 & \frac{3}{2} & -\frac{1}{2}\\0 & 0 & -\frac{7}{2} & \frac{7}{2}\end{array}\right]$$

Multiply R3 by $-\frac{2}{7}$.

$$\left[\begin{array}{ccc|c}1 & 1 & 1 & 2\\0 & 1 & \frac{3}{2} & -\frac{1}{2}\\0 & 0 & 1 & -1\end{array}\right]$$

This corresponds to $\begin{cases} x+y+z=2 \\ y+\frac{3}{2}z=-\frac{1}{2} \\ z=-1 \end{cases}$.

$$y+\frac{3}{2}(-1)=-\frac{1}{2}$$
$$y-\frac{3}{2}=-\frac{1}{2}$$
$$y=1$$
$$x+1+(-1)=2$$
$$x=2$$

The solution is (2, 1, –1).

17. $\begin{cases} 5x-2y=27 \\ -3x+5y=18 \end{cases}$

$$\left[\begin{array}{cc|c} 5 & -2 & 27 \\ -3 & 5 & 18 \end{array}\right]$$

Divide R1 by 5.

$$\left[\begin{array}{cc|c} 1 & -\frac{2}{5} & \frac{27}{5} \\ -3 & 5 & 18 \end{array}\right]$$

Multiply R1 by 3 and add to R2.

$$\left[\begin{array}{cc|c} 1 & -\frac{2}{5} & \frac{27}{5} \\ 0 & \frac{19}{5} & \frac{171}{5} \end{array}\right]$$

Multiply R2 by $\frac{5}{19}$.

$$\left[\begin{array}{cc|c} 1 & -\frac{2}{5} & \frac{27}{5} \\ 0 & 1 & 9 \end{array}\right]$$

This corresponds to $\begin{cases} x-\frac{2}{5}y=\frac{27}{5} \\ y=9 \end{cases}$.

$$x-\frac{2}{5}(9)=\frac{27}{5}$$
$$x-\frac{18}{5}=\frac{27}{5}$$
$$x=9$$

The solution is (9, 9).

19. $\begin{cases} 4x-7y=7 \\ 12x-21y=24 \end{cases}$

$$\left[\begin{array}{cc|c} 4 & -7 & 7 \\ 12 & -21 & 24 \end{array}\right]$$

Divide R1 by 4.

$$\left[\begin{array}{cc|c} 1 & -\frac{7}{4} & \frac{7}{4} \\ 12 & -21 & 24 \end{array}\right]$$

Multiply R1 by –12 and add to R2.

$$\left[\begin{array}{cc|c} 1 & -\frac{7}{4} & \frac{7}{4} \\ 0 & 0 & 3 \end{array}\right]$$

This corresponds to $\begin{cases} x-\frac{7}{4}y=\frac{7}{4} \\ 0=3 \end{cases}$.

This is an inconsistent system. The solution set is ∅.

21. $\begin{cases} 4x-y+2z=5 \\ 2y+z=4 \\ 4x+y+3z=10 \end{cases}$

$$\left[\begin{array}{ccc|c} 4 & -1 & 2 & 5 \\ 0 & 2 & 1 & 4 \\ 4 & 1 & 3 & 10 \end{array}\right]$$

Divide R1 by 4.

$$\left[\begin{array}{ccc|c} 1 & -\frac{1}{4} & \frac{1}{2} & \frac{5}{4} \\ 0 & 2 & 1 & 4 \\ 4 & 1 & 3 & 10 \end{array}\right]$$

Multiply R1 by –4 and add to R3.

$$\left[\begin{array}{ccc|c} 1 & -\frac{1}{4} & \frac{1}{2} & \frac{5}{4} \\ 0 & 2 & 1 & 4 \\ 0 & 2 & 1 & 5 \end{array}\right]$$

Divide R2 by 2.

$$\left[\begin{array}{ccc|c} 1 & -\frac{1}{4} & \frac{1}{2} & \frac{5}{4} \\ 0 & 1 & \frac{1}{2} & 2 \\ 0 & 2 & 1 & 5 \end{array}\right]$$

Multiply R2 by –2 and add to R3.

$$\left[\begin{array}{ccc|c} 1 & -\frac{1}{4} & \frac{1}{2} & \frac{5}{4} \\ 0 & 1 & \frac{1}{2} & 2 \\ 0 & 0 & 0 & 1 \end{array}\right]$$

This corresponds to $\begin{cases} x-\frac{1}{4}y+\frac{1}{2}z=\frac{5}{4} \\ y+\frac{1}{2}z=2 \\ 0=1 \end{cases}$.

This is an inconsistent system. The solution set is ∅.

23. $\begin{cases} 4x+y+\ \ z=3 \\ -x+y-2z=-11 \\ x+2y+2z=-1 \end{cases}$

$$\left[\begin{array}{ccc|c} 4 & 1 & 1 & 3 \\ -1 & 1 & -2 & -11 \\ 1 & 2 & 2 & -1 \end{array}\right]$$

Interchange R1 and R3.

$$\left[\begin{array}{ccc|c} 1 & 2 & 2 & -1 \\ -1 & 1 & -2 & -11 \\ 4 & 1 & 1 & 3 \end{array}\right]$$

Add R1 to R2. Multiply R1 by –4 and add to R3.

$$\left[\begin{array}{ccc|c} 1 & 2 & 2 & -1 \\ 0 & 3 & 0 & -12 \\ 0 & -7 & -7 & 7 \end{array}\right]$$

Divide R2 by 3.

$$\left[\begin{array}{ccc|c} 1 & 2 & 2 & -1 \\ 0 & 1 & 0 & -4 \\ 0 & -7 & -7 & 7 \end{array}\right]$$

Multiply R2 by 7 and add to R3.

$$\left[\begin{array}{ccc|c} 1 & 2 & 2 & -1 \\ 0 & 1 & 0 & -4 \\ 0 & 0 & -7 & -21 \end{array}\right]$$

Divide R3 by –7.

$$\left[\begin{array}{ccc|c} 1 & 2 & 2 & -1 \\ 0 & 1 & 0 & -4 \\ 0 & 0 & 1 & 3 \end{array}\right]$$

This corresponds to $\begin{cases} x+2y+2z=-1 \\ \qquad y \qquad\ \ =-4. \\ \qquad\qquad\ \ z=3 \end{cases}$

$$\begin{aligned} x+2(-4)+2(3) &= -1 \\ x-8+6 &= -1 \\ x &= 1 \end{aligned}$$

The solution is (1, –4, 3).

25. The matrix should have four columns, so (a) is not the correct matrix. The matrix should have a 0 in the first column, second row since the coefficient of x in the second equation is 0, so (b) is not the correct matrix. The correct matrix is (c).

Appendix F

Appendix F Exercise Set

1. $\begin{vmatrix} 3 & 5 \\ -1 & 7 \end{vmatrix} = ad - bc = 3(7) - 5(-1) = 21 + 5 = 26$

3. $\begin{vmatrix} 9 & -2 \\ 4 & -3 \end{vmatrix} = ad - bc$

$$= 9(-3) - (-2)(4)$$
$$= -27 + 8$$
$$= -19$$

5. $\begin{vmatrix} -2 & 9 \\ 4 & -18 \end{vmatrix} = ad - bc$

$$= -2(-18) - 9(4)$$
$$= 36 - 36$$
$$= 0$$

7. $\begin{cases} 2y - 4 = 0 \\ x + 2y = 5 \end{cases}$ or $\begin{cases} 0x + 2y = 4 \\ 1x + 2y = 5 \end{cases}$

$$D = \begin{vmatrix} 0 & 2 \\ 1 & 2 \end{vmatrix} = 0 - 2 = -2$$

$$D_x = \begin{vmatrix} 4 & 2 \\ 5 & 2 \end{vmatrix} = 8 - 10 = -2$$

$$D_y = \begin{vmatrix} 0 & 4 \\ 1 & 5 \end{vmatrix} = 0 - 4 = -4$$

$$x = \frac{D_x}{D} = \frac{-2}{-2} = 1$$

$$y = \frac{D_y}{D} = \frac{-4}{-2} = 2$$

The solution is (1, 2).

9. $\begin{cases} 3x + y = 1 \\ 2y = 2 - 6x \end{cases}$ or $\begin{cases} 3x + 1y = 1 \\ 6x + 2y = 2 \end{cases}$

$$D = \begin{vmatrix} 3 & 1 \\ 6 & 2 \end{vmatrix} = 6 - 6 = 0$$

Since $D = 0$, Cramer's Rule cannot be used. Notice that equation (2) is equation (1) multiplied by 2.
The solution is $\{(x, y) | 3x + y = 1\}$.

11. $\begin{cases} 5x - 2y = 27 \\ -3x + 5y = 18 \end{cases}$

$$D = \begin{vmatrix} 5 & -2 \\ -3 & 5 \end{vmatrix} = 25 - 6 = 19$$

$$D_x = \begin{vmatrix} 27 & -2 \\ 18 & 5 \end{vmatrix} = 135 + 36 = 171$$

$$D_y = \begin{vmatrix} 5 & 27 \\ -3 & 18 \end{vmatrix} = 90 + 81 = 171$$

$$x = \frac{D_x}{D} = \frac{171}{19} = 9 \text{ and } y = \frac{D_y}{D} = \frac{171}{19} = 9$$

The solution is (9, 9).

13. Expand by first row.

$$\begin{vmatrix} 2 & 1 & 0 \\ 0 & 5 & -3 \\ 4 & 0 & 2 \end{vmatrix} = 2\begin{vmatrix} 5 & -3 \\ 0 & 2 \end{vmatrix} - 1\begin{vmatrix} 0 & -3 \\ 4 & 2 \end{vmatrix} + 0\begin{vmatrix} 0 & 5 \\ 4 & 0 \end{vmatrix}$$
$$= 2(10 - 0) - 1(0 + 12) + 0$$
$$= 20 - 12$$
$$= 8$$

15. Expand by third column.

$$\begin{vmatrix} 4 & -6 & 0 \\ -2 & 3 & 0 \\ 4 & -6 & 1 \end{vmatrix} = 0\begin{vmatrix} -2 & 3 \\ 4 & -6 \end{vmatrix} - 0\begin{vmatrix} 4 & -6 \\ 4 & -6 \end{vmatrix} + 1\begin{vmatrix} 4 & -6 \\ -2 & 3 \end{vmatrix}$$
$$= 0 - 0 + 1(12 - 12)$$
$$= 0$$

17. Expand by first row.

$$\begin{vmatrix} 3 & 6 & -3 \\ -1 & -2 & 3 \\ 4 & -1 & 6 \end{vmatrix} = 3\begin{vmatrix} -2 & 3 \\ -1 & 6 \end{vmatrix} - 6\begin{vmatrix} -1 & 3 \\ 4 & 6 \end{vmatrix} - 3\begin{vmatrix} -1 & -2 \\ 4 & -1 \end{vmatrix}$$
$$= 3(-12 + 3) - 6(-6 - 12) - 3(1 + 8)$$
$$= -27 + 108 - 27$$
$$= 54$$

19. $\begin{cases} 3x \quad + z = -1 \\ -x - 3y + z = 7 \\ \quad 3y + z = 5 \end{cases}$

$$D = \begin{vmatrix} 3 & 0 & 1 \\ -1 & -3 & 1 \\ 0 & 3 & 1 \end{vmatrix} = 3\begin{vmatrix} -3 & 1 \\ 3 & 1 \end{vmatrix} - 0\begin{vmatrix} -1 & 1 \\ 0 & 1 \end{vmatrix} + 1\begin{vmatrix} -1 & -3 \\ 0 & 3 \end{vmatrix}$$
$$= 3(-3 - 3) - 0 + 1(-3 - 0)$$
$$= -18 - 3$$
$$= -21$$

$$D_x = \begin{vmatrix} -1 & 0 & 1 \\ 7 & -3 & 1 \\ 5 & 3 & 1 \end{vmatrix}$$
$$= -1\begin{vmatrix} -3 & 1 \\ 3 & 1 \end{vmatrix} - 0\begin{vmatrix} 7 & 1 \\ 5 & 1 \end{vmatrix} + 1\begin{vmatrix} 7 & -3 \\ 5 & 3 \end{vmatrix}$$
$$= -1(-3-3) - 0 + 1|21+15|$$
$$= 6 + 36$$
$$= 42$$

$$D_y = \begin{vmatrix} 3 & -1 & 1 \\ -1 & 7 & 1 \\ 0 & 5 & 1 \end{vmatrix} = 3\begin{vmatrix} 7 & 1 \\ 5 & 1 \end{vmatrix} + 1\begin{vmatrix} -1 & 1 \\ 5 & 1 \end{vmatrix} + 0\begin{vmatrix} -1 & 1 \\ 7 & 1 \end{vmatrix}$$
$$= 3(7-5) + 1(-1-5) + 0$$
$$= 6 - 6$$
$$= 0$$

$$D_z = \begin{vmatrix} 3 & 0 & -1 \\ -1 & -3 & 7 \\ 0 & 3 & 5 \end{vmatrix}$$
$$= 3\begin{vmatrix} -3 & 7 \\ 3 & 5 \end{vmatrix} - 0\begin{vmatrix} -1 & 7 \\ 0 & 5 \end{vmatrix} - 1\begin{vmatrix} -1 & -3 \\ 0 & 3 \end{vmatrix}$$
$$= 3(-15-21) - 0 - 1(-3-0)$$
$$= -108 + 3$$
$$= -105$$

$$x = \frac{D_x}{D} = \frac{42}{-21} = -2,\quad y = \frac{D_y}{D} = \frac{0}{-21} = 0,$$
$$z = \frac{D_z}{D} = \frac{-105}{-21} = 5$$

The solution is (−2, 0, 5).

21. $\begin{cases} x + y + z = 8 \\ 2x - y - z = 10 \\ x - 2y + 3z = 22 \end{cases}$

$$D = \begin{vmatrix} 1 & 1 & 1 \\ 2 & -1 & -1 \\ 1 & -2 & 3 \end{vmatrix}$$
$$= 1\begin{vmatrix} -1 & -1 \\ -2 & 3 \end{vmatrix} - 1\begin{vmatrix} 2 & -1 \\ 1 & 3 \end{vmatrix} + 1\begin{vmatrix} 2 & -1 \\ 1 & -2 \end{vmatrix}$$
$$= 1(-3-2) - 1(6+1) + 1(-4+1)$$
$$= -5 - 7 - 3$$
$$= -15$$

$$D_x = \begin{vmatrix} 8 & 1 & 1 \\ 10 & -1 & -1 \\ 22 & -2 & 3 \end{vmatrix}$$
$$= 8\begin{vmatrix} -1 & -1 \\ -2 & 3 \end{vmatrix} - 10\begin{vmatrix} 1 & 1 \\ -2 & 3 \end{vmatrix} + 22\begin{vmatrix} 1 & 1 \\ -1 & -1 \end{vmatrix}$$
$$= 8(-3-2) - 10(3+2) + 22(-1+1)$$
$$= -40 - 50 + 0$$
$$= -90$$

$$D_y = \begin{vmatrix} 1 & 8 & 1 \\ 2 & 10 & -1 \\ 1 & 22 & 3 \end{vmatrix}$$
$$= 1\begin{vmatrix} 10 & -1 \\ 22 & 3 \end{vmatrix} - 8\begin{vmatrix} 2 & -1 \\ 1 & 3 \end{vmatrix} + 1\begin{vmatrix} 2 & 10 \\ 1 & 22 \end{vmatrix}$$
$$= 1(30+22) - 8(6+1) + 1(44-10)$$
$$= 52 - 56 + 34$$
$$= 30$$

$$D_z = \begin{vmatrix} 1 & 1 & 8 \\ 2 & -1 & 10 \\ 1 & -2 & 22 \end{vmatrix}$$
$$= 1\begin{vmatrix} -1 & 10 \\ -2 & 22 \end{vmatrix} - 1\begin{vmatrix} 2 & 10 \\ 1 & 22 \end{vmatrix} + 8\begin{vmatrix} 2 & -1 \\ 1 & -2 \end{vmatrix}$$
$$= 1(-22+20) - 1(44-10) + 8(-4+1)$$
$$= -2 - 34 - 24$$
$$= -60$$

$$x = \frac{D_x}{D} = \frac{-90}{-15} = 6,\quad y = \frac{D_y}{D} = \frac{30}{-15} = -2,$$
$$z = \frac{D_z}{D} = \frac{-60}{-15} = 4$$

The solution is (6, −2, 4).

23. $\begin{vmatrix} 10 & -1 \\ -4 & 2 \end{vmatrix} = 10(2) - (-1)(-4) = 20 - 4 = 16$

25. Expand by first row.

$$\begin{vmatrix} 1 & 0 & 4 \\ 1 & -1 & 2 \\ 3 & 2 & 1 \end{vmatrix} = 1\begin{vmatrix} -1 & 2 \\ 2 & 1 \end{vmatrix} - 0\begin{vmatrix} 1 & 2 \\ 3 & 1 \end{vmatrix} + 4\begin{vmatrix} 1 & -1 \\ 3 & 2 \end{vmatrix}$$
$$= 1(-1-4) - 0 + 4(2+3)$$
$$= -5 + 20$$
$$= 15$$

27. $\begin{vmatrix} \frac{3}{4} & \frac{5}{2} \\ -\frac{1}{6} & \frac{7}{3} \end{vmatrix} = \left(\frac{3}{4}\right)\left(\frac{7}{3}\right) - \left(\frac{5}{2}\right)\left(-\frac{1}{6}\right)$

$$= \frac{7}{4} + \frac{5}{12}$$
$$= \frac{21}{12} + \frac{5}{12}$$
$$= \frac{26}{12}$$
$$= \frac{13}{6}$$

29. Expand by first row.

$$\begin{vmatrix} 4 & -2 & 2 \\ 6 & -1 & 3 \\ 2 & 1 & 1 \end{vmatrix} = 4\begin{vmatrix} -1 & 3 \\ 1 & 1 \end{vmatrix} - (-2)\begin{vmatrix} 6 & 3 \\ 2 & 1 \end{vmatrix} + 2\begin{vmatrix} 6 & -1 \\ 2 & 1 \end{vmatrix}$$
$$= 4(-1-3)+2(6-6)+2(6+2)$$
$$= -16+0+16$$
$$= 0$$

31. Expand by first row.

$$\begin{vmatrix} -2 & 5 & 4 \\ 5 & -1 & 3 \\ 4 & 1 & 2 \end{vmatrix} = -2\begin{vmatrix} -1 & 3 \\ 1 & 2 \end{vmatrix} - 5\begin{vmatrix} 5 & 3 \\ 4 & 2 \end{vmatrix} + 4\begin{vmatrix} 5 & -1 \\ 4 & 1 \end{vmatrix}$$
$$= -2(-2-3)-5(10-12)+4(5+4)$$
$$= 10+10+36$$
$$= 56$$

33. $\begin{cases} 2x-5y=4 \\ x+2y=-7 \end{cases}$

$$D = \begin{vmatrix} 2 & -5 \\ 1 & 2 \end{vmatrix} = 4+5=9$$
$$D_x = \begin{vmatrix} 4 & -5 \\ -7 & 2 \end{vmatrix} = 8-35=-27$$
$$D_y = \begin{vmatrix} 2 & 4 \\ 1 & -7 \end{vmatrix} = -14-4=-18$$
$$x = \frac{D_x}{D} = \frac{-27}{9} = -3 \text{ and } y = \frac{D_y}{D} = \frac{-18}{9} = -2$$

The solution is (−3, −2).

35. $\begin{cases} 4x+2y=5 \\ 2x+y=-1 \end{cases}$

$$D = \begin{vmatrix} 4 & 2 \\ 2 & 1 \end{vmatrix} = 4-4=0$$

Since $D = 0$, Cramer's rule cannot be used. Multiply the second equation by −2, then add.

$$\begin{array}{r} 4x+2y=5 \\ \underline{-4x-2y=2} \\ 0=7 \end{array}$$

This is a false statement, so the system has no solution, or ∅.

37. $\begin{cases} 2x+2y\ \ +z=1 \\ -x\ \ +y+2z=3 \\ x+2y+4z=0 \end{cases}$

$$D = \begin{vmatrix} 2 & 2 & 1 \\ -1 & 1 & 2 \\ 1 & 2 & 4 \end{vmatrix} = 2\begin{vmatrix} 1 & 2 \\ 2 & 4 \end{vmatrix} - 2\begin{vmatrix} -1 & 2 \\ 1 & 4 \end{vmatrix} + 1\begin{vmatrix} -1 & 1 \\ 1 & 2 \end{vmatrix}$$
$$= 2(4-4)-2(-4-2)+1(-2-1)$$
$$= 0+12-3$$
$$= 9$$

$$D_x = \begin{vmatrix} 1 & 2 & 1 \\ 3 & 1 & 2 \\ 0 & 2 & 4 \end{vmatrix} = 1\begin{vmatrix} 1 & 2 \\ 2 & 4 \end{vmatrix} - 3\begin{vmatrix} 2 & 1 \\ 2 & 4 \end{vmatrix} + 0$$
$$= 1(4-4)-3(8-2)$$
$$= -18$$

$$D_y = \begin{vmatrix} 2 & 1 & 1 \\ -1 & 3 & 2 \\ 1 & 0 & 4 \end{vmatrix} = 2\begin{vmatrix} 3 & 2 \\ 0 & 4 \end{vmatrix} - 1\begin{vmatrix} -1 & 2 \\ 1 & 4 \end{vmatrix} + 1\begin{vmatrix} -1 & 3 \\ 1 & 0 \end{vmatrix}$$
$$= 2(12-0)-1(-4-2)+1(0-3)$$
$$= 24+6-3$$
$$= 27$$

$$D_z = \begin{vmatrix} 2 & 2 & 1 \\ -1 & 1 & 3 \\ 1 & 2 & 0 \end{vmatrix} = 2\begin{vmatrix} 1 & 3 \\ 2 & 0 \end{vmatrix} - 2\begin{vmatrix} -1 & 3 \\ 1 & 0 \end{vmatrix} + 1\begin{vmatrix} -1 & 1 \\ 1 & 2 \end{vmatrix}$$
$$= 2(0-6)-2(0-3)+1(-2-1)$$
$$= -12+6-3$$
$$= -9$$

$$x = \frac{D_x}{D} = \frac{-18}{9} = -2,\ y = \frac{D_y}{D} = \frac{27}{9} = 3,$$
$$z = \frac{D_z}{D} = \frac{-9}{9} = -1$$

The solution is (−2, 3, −1).

39. $\begin{cases} \frac{2}{3}x - \frac{3}{4}y = -1 \\ -\frac{1}{6}x + \frac{3}{4}y = \frac{5}{2} \end{cases}$

$$D = \begin{vmatrix} \frac{2}{3} & -\frac{3}{4} \\ -\frac{1}{6} & \frac{3}{4} \end{vmatrix} = \frac{1}{2} - \frac{1}{8} = \frac{3}{8}$$
$$D_x = \begin{vmatrix} -1 & -\frac{3}{4} \\ \frac{5}{2} & \frac{3}{4} \end{vmatrix} = -\frac{3}{4} + \frac{15}{8} = \frac{9}{8}$$
$$D_y = \begin{vmatrix} \frac{2}{3} & -1 \\ -\frac{1}{6} & \frac{5}{2} \end{vmatrix} = \frac{10}{6} - \frac{1}{6} = \frac{9}{6}$$

$$x = \frac{D_x}{D} = \frac{\frac{9}{8}}{\frac{3}{8}} = 3 \text{ and } y = \frac{D_y}{D} = \frac{\frac{9}{6}}{\frac{3}{8}} = 4$$

The solution is (3, 4).

41. $\begin{cases} 0.7x - 0.2y = -1.6 \\ 0.2x \quad - y = -1.4 \end{cases}$

$$D = \begin{vmatrix} 0.7 & -0.2 \\ 0.2 & -1 \end{vmatrix} = -0.7 + 0.04 = -0.66$$

$$D_x = \begin{vmatrix} -1.6 & -0.2 \\ -1.4 & -1 \end{vmatrix} = 1.6 - 0.28 = 1.32$$

$$D_y = \begin{vmatrix} 0.7 & -1.6 \\ 0.2 & -1.4 \end{vmatrix} = -0.98 + 0.32 = -0.66$$

$$x = \frac{D_x}{D} = \frac{1.32}{-0.66} = -2$$

$$y = \frac{D_y}{D} = \frac{-0.66}{-0.66} = 1$$

The solution is (−2, 1).

43. $\begin{cases} -2x + 4y - 2z = 6 \\ x - 2y + z = -3 \\ 3x - 6y + 3z = -9 \end{cases}$

$$\begin{aligned} D &= \begin{vmatrix} -2 & 4 & -2 \\ 1 & -2 & 1 \\ 3 & -6 & 3 \end{vmatrix} \\ &= -2\begin{vmatrix} -2 & 1 \\ -6 & 3 \end{vmatrix} - 4\begin{vmatrix} 1 & 1 \\ 3 & 3 \end{vmatrix} - 2\begin{vmatrix} 1 & -2 \\ 3 & -6 \end{vmatrix} \\ &= -2(-6+6) - 4(3-3) - 2(-6+6) \\ &= 0 - 0 - 0 \\ &= 0 \end{aligned}$$

Since $D = 0$, Cramer's rule cannot be used. Note that the first equation is −2 times the second, and the third equation is 3 times the second. Thus, the system is dependent and the solution set is $\{(x, y, z) | x - 2y + z = -3\}$.

45. $\begin{cases} x - 2y + z = -5 \\ 3y + 2z = 4 \\ 3x - y \quad = -2 \end{cases}$

$$\begin{aligned} D &= \begin{vmatrix} 1 & -2 & 1 \\ 0 & 3 & 2 \\ 3 & -1 & 0 \end{vmatrix} = 1\begin{vmatrix} 3 & 2 \\ -1 & 0 \end{vmatrix} + 2\begin{vmatrix} 0 & 2 \\ 3 & 0 \end{vmatrix} + 1\begin{vmatrix} 0 & 3 \\ 3 & -1 \end{vmatrix} \\ &= 1(0+2) + 2(0-6) + 1(0-9) \\ &= 2 - 12 - 9 \\ &= -19 \end{aligned}$$

$$\begin{aligned} D_x &= \begin{vmatrix} -5 & -2 & 1 \\ 4 & 3 & 2 \\ -2 & -1 & 0 \end{vmatrix} \\ &= -5\begin{vmatrix} 3 & 2 \\ -1 & 0 \end{vmatrix} + 2\begin{vmatrix} 4 & 2 \\ -2 & 0 \end{vmatrix} + 1\begin{vmatrix} 4 & 3 \\ -2 & -1 \end{vmatrix} \\ &= -5(0+2) + 2(0+4) + 1(-4+6) \\ &= -10 + 8 + 2 \\ &= 0 \end{aligned}$$

$$\begin{aligned} D_y &= \begin{vmatrix} 1 & -5 & 1 \\ 0 & 4 & 2 \\ 3 & -2 & 0 \end{vmatrix} = 1\begin{vmatrix} 4 & 2 \\ -2 & 0 \end{vmatrix} + 5\begin{vmatrix} 0 & 2 \\ 3 & 0 \end{vmatrix} + 1\begin{vmatrix} 0 & 4 \\ 3 & -2 \end{vmatrix} \\ &= 1(0+4) + 5(0-6) + 1(0-12) \\ &= 4 - 30 - 12 \\ &= -38 \end{aligned}$$

$$\begin{aligned} D_z &= \begin{vmatrix} 1 & -2 & -5 \\ 0 & 3 & 4 \\ 3 & -1 & -2 \end{vmatrix} \\ &= 1\begin{vmatrix} 3 & 4 \\ -1 & -2 \end{vmatrix} + 2\begin{vmatrix} 0 & 4 \\ 3 & -2 \end{vmatrix} - 5\begin{vmatrix} 0 & 3 \\ 3 & -1 \end{vmatrix} \\ &= 1(-6+4) + 2(0-12) - 5(0-9) \\ &= -2 - 24 + 45 \\ &= 19 \end{aligned}$$

$$x = \frac{D_x}{D} = \frac{0}{-19} = 0, \quad y = \frac{D_y}{D} = \frac{-38}{-19} = 2,$$

$$z = \frac{D_z}{D} = \frac{19}{-19} = -1$$

The solution is (0, 2, −1).

47. $\begin{vmatrix} 1 & x \\ 2 & 7 \end{vmatrix} = -3$

$$\begin{aligned} (1)(7) - x \cdot 2 &= -3 \\ 7 - 2x &= -3 \\ -2x &= -10 \\ x &= 5 \end{aligned}$$

49. 0; answers may vary

51.
$$\begin{matrix} + & - & + & - \\ - & + & - & + \\ + & - & + & - \\ - & + & - & + \end{matrix}$$

53. Expand by first row.

$$\begin{vmatrix} 5 & 0 & 0 & 0 \\ 0 & 4 & 2 & -1 \\ 1 & 3 & -2 & 0 \\ 0 & -3 & 1 & 2 \end{vmatrix} = 5\begin{vmatrix} 4 & 2 & -1 \\ 3 & -2 & 0 \\ -3 & 1 & 2 \end{vmatrix} - 0\begin{vmatrix} 0 & 2 & -1 \\ 1 & -2 & 0 \\ 0 & 1 & 2 \end{vmatrix} + 0\begin{vmatrix} 0 & 4 & -1 \\ 1 & 3 & 0 \\ 0 & -3 & 2 \end{vmatrix} - 0\begin{vmatrix} 0 & 4 & 2 \\ 1 & 3 & -2 \\ 0 & -3 & 1 \end{vmatrix}$$

$$= 5\left[4\begin{vmatrix} -2 & 0 \\ 1 & 2 \end{vmatrix} - 2\begin{vmatrix} 3 & 0 \\ -3 & 2 \end{vmatrix} + (-1)\begin{vmatrix} 3 & -2 \\ -3 & 1 \end{vmatrix}\right] - 0 + 0 - 0$$

$$= 5[4(-4-0) - 2(6+0) - 1(3-6)]$$

$$= 5(-16-12+3)$$

$$= 5(-25)$$

$$= -125$$

55. Expand by first column.

$$\begin{vmatrix} 4 & 0 & 2 & 5 \\ 0 & 3 & -1 & 1 \\ 0 & 0 & 2 & 0 \\ 0 & 0 & 0 & 1 \end{vmatrix} = 4\begin{vmatrix} 3 & -1 & 1 \\ 0 & 2 & 0 \\ 0 & 0 & 1 \end{vmatrix} - 0\begin{vmatrix} 0 & 2 & 5 \\ 0 & 2 & 0 \\ 0 & 0 & 1 \end{vmatrix} + 0\begin{vmatrix} 0 & 2 & 5 \\ 3 & -1 & 1 \\ 0 & 0 & 1 \end{vmatrix} - 0\begin{vmatrix} 0 & 2 & 5 \\ 3 & -1 & 1 \\ 0 & 2 & 0 \end{vmatrix}$$

$$= 4\left[3\begin{vmatrix} 2 & 0 \\ 0 & 1 \end{vmatrix} - 0\begin{vmatrix} -1 & 1 \\ 0 & 1 \end{vmatrix} + 0\begin{vmatrix} -1 & 1 \\ 2 & 0 \end{vmatrix}\right] - 0 + 0 - 0$$

$$= 4[3(2-0) - 0 + 0]$$

$$= 4(6)$$

$$= 24$$

Appendix G

1. 21, 28, 16, 42, 38

$$\bar{x} = \frac{21+28+16+42+38}{5} = \frac{145}{5} = 29$$

16, 21, 28, 38, 42
median = 28
no mode

3. 7.6, 8.2, 8.2, 9.6, 5.7, 9.1

$$\bar{x} = \frac{7.6+8.2+8.2+9.6+5.7+9.1}{6} = \frac{48.4}{6} = 8.1$$

5.7, 7.6, 8.2, 8.2, 9.1, 9.6

$$\text{median} = \frac{8.2+8.2}{2} = 8.2$$

mode = 8.2

5. 0.2, 0.3, 0.5, 0.6, 0.6, 0.9, 0.2, 0.7, 1.1

$$\bar{x} = \frac{0.2+0.3+0.5+0.6+0.6+0.9+0.2+0.7+1.1}{9}$$
$$= \frac{5.1}{9}$$
$$= 0.6$$

median = 0.6
mode = 0.2 and 0.6

7. 231, 543, 601, 293, 588, 109, 334, 268

$$\bar{x} = \frac{231+543+601+293+588+109+334+268}{8}$$
$$= \frac{2967}{8}$$
$$= 370.9$$

109, 231, 268, 293, 334, 543, 588, 601

$$\text{median} = \frac{293+334}{2} = 313.5$$

no mode

9. 1454, 1250, 1136, 1127, 1107

$$\bar{x} = \frac{1454+1250+1136+1127+1107}{5}$$
$$= \frac{6074}{5}$$
$$= 1214.8 \text{ feet}$$

11. 1454, 1250, 1136, 1127, 1107, 1046, 1023, 1002

$$\text{median} = \frac{1127+1107}{2} = 1117 \text{ feet}$$

13. $\bar{x} = \frac{7.8+6.9+7.5+4.7+6.9+7.0}{6}$

$$= \frac{40.8}{6}$$
$$= 6.8 \text{ seconds}$$

15. 4.7, 6.9, 6.9, 7.0, 7.5, 7.8
mode = 6.9

17. 74, 77, 85, 86, 91, 95

$$\text{median} = \frac{85+86}{2} = 85.5$$

19. Sum = 78 + 80 + 66 + 68 + 71 + 64 + 82 + 71 + 70 + 65 + 70 + 75 + 77 + 86 + 72
= 1095

$$\bar{x} = \frac{1095}{15} = 73$$

21. 64, 65, 66, 68, 70, 70, 71, 71, 72, 75, 77, 78, 80, 82, 86
mode = 70 and 71

23. 64, 65, 66, 68, 70, 70, 71, 71, 72, 75, 77, 78, 80, 82, 86
↑ mean = 73
9 rates were lower than the mean.

25. __, __, 16, 18, __;

Since the mode is 21, at least two of the missing numbers must be 21. The mean is 20. Let the one unknown number be x.

$$\bar{x} = \frac{21+21+16+18+x}{5} = 20$$
$$\frac{76+x}{5} = 20$$
$$76 + x = 100$$
$$x = 24$$

The missing numbers are 21, 21, 24.

Appendix H

Appendix H Exercise Set

1. $90° - 19° = 71°$

3. $90° - 70.8° = 19.2°$

5. $90° - 11\frac{1}{4}° = 78\frac{3}{4}°$

7. $180° - 150° = 30°$

9. $180° - 30.2° = 149.8°$

11. $180° - 79\frac{1}{2}° = 100\frac{1}{2}°$

13.
$$m\angle 1 = 110°$$
$$m\angle 2 = 180° - 110° = 70°$$
$$m\angle 3 = m\angle 2 = 70°$$
$$m\angle 4 = m\angle 2 = 70°$$
$$m\angle 5 = m\angle 1 = 110°$$
$$m\angle 6 = m\angle 4 = 70°$$
$$m\angle 7 = m\angle 5 = 110°$$

15. $180° - 11° - 79° = 90°$

17. $180° - 25° - 65° = 90°$

19. $180° - 30° - 60° = 90°$

21. $90° - 45° = 45°$
$45°, 90°$

23. $90° - 17° = 73°$
$73°, 90°$

25. $90° - 39\frac{3}{4}° = 50\frac{1}{4}°$

$50\frac{1}{4}°,\ 90°$

27.
$$\frac{12}{4} = \frac{18}{x}$$
$$4x\left(\frac{12}{4}\right) = 4x\left(\frac{18}{x}\right)$$
$$12x = 72$$
$$x = 6$$

29.
$$\frac{6}{9} = \frac{3}{x}$$
$$9x\left(\frac{6}{9}\right) = 9x\left(\frac{3}{x}\right)$$
$$6x = 27$$
$$x = 4.5$$

31.
$$a^2 + b^2 = c^2$$
$$6^2 + 8^2 = c^2$$
$$36 + 64 = c^2$$
$$100 = c^2$$
$$10 = c$$

33.
$$a^2 + b^2 = c^2$$
$$5^2 + b^2 = 13^2$$
$$25 + b^2 = 169$$
$$b^2 = 144$$
$$b = 12$$